Handbook of
Components for Electronics

OTHER McGRAW-HILL HANDBOOKS OF INTEREST

American Institute of Physics • American Institute of Physics Handbook
Baumeister and Marks • Standard Handbook for Mechanical Engineers
Beeman • Industrial Power Systems Handbook
Blatz • Radiation Hygiene Handbook
Brady • Materials Handbook
Burington • Handbook of Mathematical Tables and Formulas
Burington and May • Handbook of Probability and Statistics with Tables
Condon and Odishaw • Handbook of Physics
Coombs • Basic Electronic Instrument Handbook
Coombs • Printed Circuits Handbook
Croft, Carr, and Watt • American Electricians' Handbook
The Diebold Group, Inc. • Automatic Data Processing Handbook
Etherington • Nuclear Engineering Handbook
Fink • Electronics Engineers' Handbook
Fink and Carroll • Standard Handbook for Electrical Engineers
Gruenberg • Handbook of Telemetry and Remote Control
Hamsher • Communication System Engineering Handbook
Harper • Handbook of Electronic Packaging
Harper • Handbook of Materials and Processes for Electronics
Harper • Handbook of Thick Film Hybrid Microelectronics
Harper • Handbook of Wiring, Cabling, and Interconnecting for Electronics
Henney • Radio Engineering Handbook
Hicks • Standard Handbook of Engineering Calculations
Hunter • Handbook of Semiconductor Electronics
Huskey and Korn • Computer Handbook
Ireson • Reliability Handbook
Jasik • Antenna Engineering Handbook
Juran • Quality Control Handbook
Kaufman and Seidman • Handbook for Electronics Engineering Technicians
Klerer and Korn • Digital Computer User's Handbook
Koelle • Handbook of Astronautical Engineering
Korn and Korn • Mathematical Handbook for Scientists and Engineers
Kurtz • The Lineman's and Cableman's Handbook
Landee, Davis, and Albrecht • Electronic Designer's Handbook
Machol • System Engineering Handbook
Maissel and Glang • Handbook of Thin Film Technology
Markus • Electronics and Nucleonics Dictionary
Markus • Handbook of Electronic Control Circuits
Markus and Zeluff • Handbook of Industrial and Electronic Circuits
Perry • Engineering Manual
Skolnik • Radar Handbook
Smeaton • Motor Application and Maintenance Handbook
Stout and Kaufman • Handbook of Operational Amplifier Circuit Design
Terman • Radio Engineers' Handbook
Truxal • Control Engineers' Handbook
Tuma • Engineering Mathematics Handbook
Tuma • Handbook of Physical Calculations
Tuma • Technology Mathematics Handbook
Watt and Summers • NFPA Handbook of the National Electrical Code

Handbook of Components for Electronics

CHARLES A. HARPER *editor-in-chief*

Westinghouse Electric Corporation
Baltimore, Maryland

McGRAW-HILL BOOK COMPANY

New York   St. Louis   San Francisco   Auckland   Bogotá
Düsseldorf   Johannesburg   London   Madrid
Mexico   Montreal   New Delhi   Panama
Paris   São Paulo   Singapore
Sydney   Tokyo   Toronto

Library of Congress Cataloging in Publication Data
Main entry under title:

Handbook of components for electronics.

Includes index.
1. Electronic apparatus and appliances—Handbooks,
manuals, etc. I. Harper, Charles A.
TK7870.H23   621.381'028   76-26117
ISBN 0-07-026682-4

234567890   KPKP   78654321098

*The editors for this book were Harold B. Crawford and Ruth Weine,
the designer was Naomi Auerbach, and the production supervisor
was Teresa F. Leaden.   It was set in Caledonia by Bi-Comp, Incorporated.*

Printed and bound by The Kingsport Press.

Contents

Contributors

BUCHLEITNER, JOHN R. *Systems Development Division, Westinghouse Electric Corporation:* CHAPTER 6, COMPONENT PARTS FOR MICROWAVE SYSTEMS

BYLANDER, E. G. *Semiconductor Group, Texas Instruments, Incorporated:* CHAPTER 5, COMPONENTS FOR ELECTROOPTICS

DREXLER, H. BENNETT *Aerospace Division, Martin Marietta Corporation:* CHAPTER 7, RESISTORS AND PASSIVE-PARTS STANDARDIZATION AND CHAPTER 8, CAPACITORS

HENRY, E. N. *Systems Development Division, Westinghouse Electric Corporation:* CHAPTER 9, TRANSFORMERS AND INDUCTIVE DEVICES

HIERHOLZER, EDWARD L. *Aerospace Division, Martin Marietta Corporation:* CHAPTER 7, RESISTORS AND PASSIVE-PARTS STANDARDIZATION (CONTRIBUTING AUTHOR: **John H. Powers** *IBM Corporation*) and CHAPTER 8, CAPACITORS (CONTRIBUTING AUTHORS: **Harold T. Cates, Stephen D. Das,** and **Joseph F. Rhodes** *Aerospace Division, Martin Marietta Corporation*)

JAMES, VIRGIL E. *Consultant:* CHAPTER 10, RELAYS AND SWITCHES

SCHMID, ERWIN R. *Bell Laboratories:* CHAPTER 4, DISCRETE SEMICONDUCTOR DEVICES

SCHWARTZ, BERNARD R. [*deceased*] *Government and Commercial Systems, RCA Corporation:* CHAPTER 11, CONNECTORS AND CONNECTIVE DEVICES

WILSON, LEO E. *Systems Development Division, Westinghouse Electric Corporation:* CHAPTER 9, TRANSFORMERS AND INDUCTIVE DEVICES

ZATZ, SAUL *Aerospace Division, Martin Marietta Corporation:* CHAPTER 1, PRINCIPLES OF INTEGRATED-CIRCUIT-COMPONENT TECHNOLOGY; CHAPTER 2, DIGITAL INTEGRATED CIRCUITS; and CHAPTER 3, LINEAR AND SPECIAL-PURPOSE INTEGRATED CIRCUITS

Preface

Ultimate success for an electronic or electrical product design can only be achieved through judicious selection of component parts for use in that design. The increasingly competitive economic environment no longer allows designer complacency if corporate survival is to be assured. As a minimum, the designer must develop an in-depth understanding of the component parts which must be considered for a given set of cost and performance objectives, and the trade-offs and comparisons which are vital for design optimization. While there exists a myriad of segmented data sheets for given component parts, nowhere does there exist a thorough, comprehensive, broad-based single reference and guideline source—so sorely needed by electronic and electrical designers. It is this need which prompted the publication of this *Handbook of Components for Electronics*. This work is the result of years of painstaking effort by leading experts in the components field, with care being taken to update information as required during the progress of this publication. The result is the most complete and up-to-date single reference and text to be found—a must for the desk of anyone involved in any aspect of electronic and electrical design or any aspect of component specification or utilization, and for every reference library.

This *Handbook of Components for Electronics* was prepared as a thorough and comprehensive sourcebook of practical data, guidelines, and information for all ranges of interests. It contains an extensive array of property and performance data for all the important component groups; these are presented as a function of the most important design and performance variables. Further, it presents comparison data and guidelines for best trade-off design decisions, extensive test and reliability data, detailed listings of important specifications and standards, a wealth of data and information on dimensions, configuration, and mechanical and environmental performance. The Handbook's other major features include a completely cross-referenced and easy-to-use index, a comprehensive glossary, and end-of-chapter reference lists which have been exhaustively gathered and which cover years of important work not equally presented in any other source.

In addition to its thorough contents, the chapter organization and coverage of the *Handbook of Components for Electronics* are equally well suited to user convenience, and include the broad spectrum ranging from components for advanced semiconductor, microwave, and electrooptical systems to all of the types of conventional components. The first four chapters deal with semiconductor components, covering principles of integrated circuits, digital integrated circuits, linear and special-purpose integrated circuits, and discrete semiconductor devices such as transistors and diodes. The next two chapters deal with components for electrooptical systems and components for microwave systems. The following four chapters cover the more conventional components such as resistors and related passive components, capacitors, transformers and inductive devices, and relays and switches. Modern technology for these conventional components is fully covered. The final chapter deals with the all-important component category of connectors and connective devices. Each chapter of the *Handbook of Components for Electronics* is indeed a book in itself.

Length and coverage of a Handbook of this magnitude are necessarily measured compromises. Inevitably, varying degrees of shortages and excesses will exist, depending on the needs of the individual user. Then too, the time required to complete a major work such as this necessarily demands that some of the most recent data may not be fully covered. Further, in spite of the tremendous effort that has been made to minimize such shortcomings, it is my greatest desire to improve each successive edition. Toward this end, any and all comments will be welcomed and appreciated.

Charles A. Harper

Chapter 1

Principles of Integrated-Circuit-Component Technology

SAUL ZATZ
Martin Marietta Aerospace, Orlando, Florida

INTRODUCTION AND HISTORY

The integrated circuit is a unique component part, since it is both a subsystem and a part. Component parts are normally thought of as something fabricated in batch process which must be combined with other components to perform a function or subfunction. They build up to a hierarchy called a system. However, the integrated circuit is an exception; it in itself is a subsystem, a functional block, a unit performing a complete function. The range of functions varies from simple logic gates to amplifiers to complex central processing units and complex calculators with all functions of memory and arithmetic. The complexities of integrated circuits range from the equivalent of a few transistors and diodes to the equivalent of thousands of component parts. These equivalent component parts are not made separately but share with each other the same processing, being fabricated on the same piece of silicon crystal. This is monolithic processing, processing that is "from one stone." On this single crystal, through the processes of diffusion, ion implantation, metallization, and epitaxial growth, a complex structure is fabricated in one step with areas that can be functionally identified as diodes, transistors, resistors, field-effect devices, capacitors, and interconnection wires.

The technical basis for fabricating integrated circuits dates to the 1948 invention of the transistor by William Shockley et al. at Bell Telephone Laboratories. Practical fabrication, however, required a repeatable and dependable process, which was the Planar° process invented by Dr. J. Hoerni at Fairchild Instrument approximately 15 years later. The Planar process allowed fabrication of semiconductor devices with controlled geometries, and allowed their fabrication from a single side of crystal. This in turn permitted batch processes, photolithography, high yields, and repeatable devices, all of which are needed for production of integrated circuits. The first integrated circuits were logic devices, simple resistor transistor gates that met the power-consumption problems of military space systems. A chip photograph of an early integrated circuit is shown in Fig. 1. They were initially very expensive; a single logic gate cost about $100.

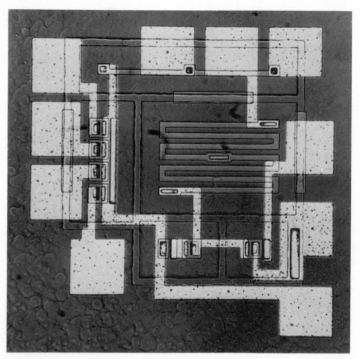

Fig. 1 Early integrated-circuit chip. Microphotograph of the SE 101 four input gate. (*Signetics Corporation.*)

If the aerospace industry gave birth to the integrated circuit, then the computer industry brought it to complexity, maturity, and domination of the electronic industry. Modern electronic equipment uses integrated circuits to perform the standard analog and digital functions, to regulate power-supply voltage, to sense fault, to be memory, to be logic, and to be decision. In short, the integrated circuit is the heart of modern electronics.

Deciding whether or not to use integrated circuits for an application is to determine the proper technology for the application. One thing should be made very clear: there is no best integrated circuit, just the best one for a particular application. As in any other form of high technology, trade-offs can be made. Where one circuit can give better speed and have higher power, another may be smaller and another may be cheaper; therefore, the questions of performance, cost, interchangeability, and compatibility must be considered before completing the process of selection. The use of integrated circuits requires special design rules, reliability criteria, handling precautions, and design expertise to capitalize on all the features designed into the

° Trademark of Fairchild Camera and Instruments.

circuits and to use them without degrading their characteristics. To some degree, the design engineer must be cognizant of fabrication to realize the limitations of integrated circuits.

FABRICATION

The fabrication of integrated circuits requires extensive use of complex, tightly controlled, fully understood, scientific processes. These processes require a great degree of sophistication in such fields as physical chemistry, solid-state physics, optics, photography, and metallurgy. As a result, and because of the relatively small size of a completed integrated-circuit chip, batch processes can yield an extremely large number of devices from a single production run. From a typical processing lot of 50 slices, each of which can contain 2,000 different integrated-circuit chips, a single production lot (a single diffusion run) can include 100,000 different integrated circuits.

To many people unfamiliar with integrated circuits and semiconductors, the industry seems to involve a strange craft similar to modern-day black magic, and it seems to center around a handful of wizards. These wizards are thought to be uniquely talented technical experts who gained their major knowledge through some strange combination of insight and technology. This is not the case today; there are limits and there are controls, as integrated circuits are fabricated by mature, structured processes. These processes are monitored and assessed by the techniques of various sciences as well as numerous statistical and physical control techniques. There is no wizardry here, for the technology is firmly established and the processes cannot be easily and practically modified.

The difference between integrated-circuit devices is created by a particular family of processing and application of specific tooling. This tooling is not tooling in the conventional sense of lays, jigs, and dies but a set of photolithographic artwork, used to place, size, and shape the regions to which processing will be applied. A set of photolithographic artwork is the result of a deliberate design cycle at a semiconductor manufacturing facility. This design cycle starts with the creation of a circuit design and circuit schematic. This schematic is then transformed into a large map defining the locations of various devices to be formed into the crystal. Each subdevice takes its own design and shape and multiple layers form transistors, two layers form diodes, and other structures form field-effect devices. Interconnection pads are also defined on these maps in the form of a layer of positive metal or polycrystalline silicon. This overall map is an extremely accurate drawing, through either a hand-drafting process or a computer-aided design process where an automatic plotter generates the final map. Either automatic or manual processes transform the map onto a series of sheets of a material called rubylith. This is a very rigid, perfectly clear plastic film which carries a surface of ruby-red film which can be removed by a cutting process. Layers of the integrated circuit are defined as individual layers on the finished rubylith, and through a series of photoreductions this rubylith map is reduced from several square feet to a small photographic pattern approximately ten or twenty times the size of the actual silicon crystal. A final photographic-reduction process reduces the pattern to actual size.

Through a step-and-repeat photographic process, the map pattern is reproduced many times on a slide equal in size to a finished slice of semiconductor material. The end result, a set of photomasks of chrome or conventional emulsion patterns on glass, produces the layers that then make the integrated circuits on the silicon wafer. This set of photomasks, as shown in Fig 2, can again be photographically reproduced for use as a production tooling in the semiconductor processes.

Each of the basic semiconductor processes uses one of two approaches to define the active areas, or areas of metallization. In the first approach, the basic process is applied uniformly over the entire area of the wafer, and a coat of photosensitive liquid is spread over the entire surface by a process called spinning. This coated material is conditioned onto the surface by baking or aging, and it is then exposed photographically by a single member of the set of masks. The pattern is developed, and chemicals remove all the photosensitive material, except where it has been developed. Where the material is developed, a chemical etchant is used to etch

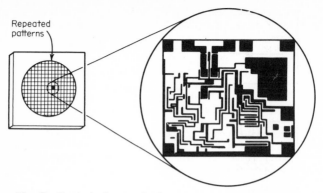

Repeated patterns

Fig. 2 Typical photomask showing one pattern in detail.

through the small areas of the material to be defined. These areas are left on the crystal, and the photographic material, which is the photoresist, is removed by a chemical-etching process. These two chemical processes must be compatible with each other. This type of process is used for etching away metal conductors, beam-lead attachments, and polycrystalline silicon material.

The second approach involves a secondary material and a mask. In most cases, this material is an oxide of the basic silicon. The initial step takes the silicon and oxidizes the entire surface, forming a layer of silicon dioxide, which is glass. A photolithographic process is then applied over the surface where masks may be used to define particular areas. An etchant is brought in to eliminate the oxide in areas not defined by the total lithographic process. A chemical cleanup removes the photolithographic material, leaving the basic silicon with a silicon dioxide pattern on its surface. Any number of high-temperature processes, diffusion, etc., may then be applied, with the silicon dioxide serving as the mask to define the areas into which the diffusion goes. Figure 3 illustrates a typical diffusion furnace. Major advances

Fig. 3 Integrated-circuit diffusion furnace. (*Thermco Corporation.*)

in chemical and photographic techniques such as these have allowed the definition of smaller devices on the surface of crystals and have increased the yield by reducing the number of defects. Figure 4 illustrates the integrated-circuit fabrication process.

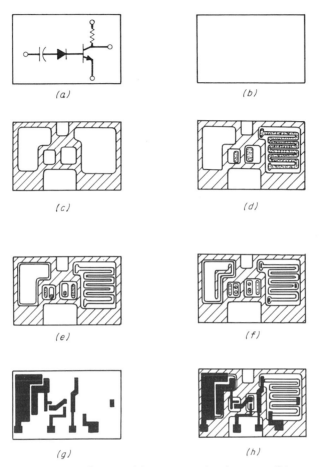

Fig. 4 Simplified integrated-circuit fabrication. (*a*) Schematic. (*b*) Starting wafer. (*c*) Top view after isolation diffusion. (*d*) Top view after base diffusion. (*e*) Top view after emitter diffusion. (*f*) Top view after preohmic etch. (*g*) Metallization. (*h*) Top view of completed circuit.

ASSEMBLY

Integrated circuits produce functional blocks that attain a degree of miniaturization not previously attainable in the electronic industry. However, these circuits are so small that interconnection and system packaging is difficult. An integrated-circuit chip is less than 0.1 in. in its largest dimension. To make it practical for inclusion in hand-assembled systems and to protect it from the adverse effects of the macroenvironment about it, specific packaging must be used. The finished package provides practical, usable interconnection pathways for leads and protects the integrated-circuit chip. Also, the package serves as a means to heat-sink the thermal energy produced during operation. The integrated-circuit chip is too small to govern the size of the package enclosure; so the only trade-off to be made in selecting package size is that of volume-density efficiency against ease of assembly.

Flatpack packaging has found its most common application in high-density airborne and spacecraft systems, and in ultraminiaturized portable equipment. Most large systems fabricated from integrated circuits use either dual-in-line plug-in or grid-pattern plug-in packages. Use of round-pin location packages has been primarily limited to specific circuit functions which are available only in these packages.   These specialized circuits are for the most part linear circuits.

Integrated-circuit packaging has been a major area of research in the industry. This research has been directed primarily at the perfection of more cost-effective, superior materials and material-fabrication techniques.   Much of this effort has been directed primarily at perfecting a reliable high-quality-plastic integrated-circuit package.   Unfortunately, this effort to perfect a high-quality-plastic package has not yet attained its goals, limiting plastic packages to situations of either low-cost, easily repairable electronic systems or systems where a high degree of environmental control can be maintained.

Now visible to the user of integrated circuits is the methodology by which the chip-device terminal has been interconnected to the external-device terminal.   This interconnection is usually accomplished by placing a microwire approximately 0.001 in. in diameter from the chip terminal to the inside edge of the external terminal. This microwire, or as it is properly called, bonding wire, is made of either aluminum or gold.   Gold wire permits ball bonding, a technique producing a highly reliable connection that requires a somewhat large bond pad on the chip.   However, it can be the source of gold/aluminum intermetallic crystals on a chip.   These intermetallics form between the gold bonding wire and the aluminum metallization on the chip, commonly as purple or white plague.   The plague can be minimized through specialized process controls or can be totally eliminated by restricting the choice of materials to a single metal for both the bond wire and the chip metallization.

Aluminum wires are used in both ultrasonic and thermal-compression wire bonding. Ultrasonic wire bonding has gained greater acceptance because of its high reliability and ease of production.

Research in the integrated-circuit industry is currently directed at methods other than bond wires for connecting the chip and the internal edge package pins.   Three techniques have become fruitful sources of investigation.   They use beam leads, deformable bumps, and spider bonds.   All three, along with conventional wire bonds, are illustrated in Fig. 5.   In the beam-lead concept, a deposited malleable-gold beam protrudes from the edges of the semiconductor chip after it has been chemically separated from adjacent chips on the wafer.   The beams can then be attached in a single process where all the beams on the chip are attached to metallized substrate regions which have been previously connected to the device pins.   Deformable bumps are also used for face-down bonding.   The bumps are electrochemically formed on the interconnection regions of the chip using a low-temperature eutectic or solder containing additional metallic material to make them semirigid.   The bumps are attached to metallized regions on a substrate by inverting the chip, going to the substrates, placing it in exact position, and then applying heat to flow the eutectic material.

The spider-bond technique is most directly suited to individual integrated circuits, while the two face-down bonding techniques are primarily used for integrated circuits in larger hybrid multichip arrays.   A spider bond is produced by simultaneously welding all connections on the chip and the package to a chemically machined microlead frame.   The shape of the stamping which results from the interconnection of the chip to the pins of the package is the source of the name spider.

All the developing methods of intrapackage interconnection serve their function and add direction to the integrated-circuit industry.   However, a more reliable and economical system must yet be developed to replace the present methodology of placing single bond wires from the package interconnection to external pin termination within the package.

Plastic and hermetic packages   *Plastic Packages.*   A variety of plastic materials have been used to encapsulate integrated circuits.   The most commonly used materials have been silicones, epoxies, and phenolics.   However, use of phenolic material has been discarded with the advent of superior-quality silicones and epoxies, and a great number of specific material types fall within this grouping.   The number of plastic types is further expanded by modifications with compounds called mold-release agents and with variations in terminal and mechanical stimuli for package formation.

Ideally, plastic should be the perfect package for integrated circuits, since plastic

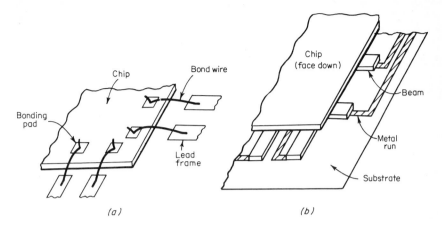

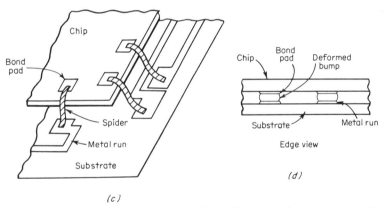

Fig. 5 Comparison of internal lead-attachment techniques. (*a*) Wire bond. (*b*) Beam lead. (*c*) Spider bond. (*d*) Deformable bump.

packages can be batch-fabricated along with the integrated-circuit chip. In addition, plastic is simple to fabricate and low in cost. Unfortunately, plastics have four primary deficiencies. First, injection molding of plastic tends to disrupt the placement of the bond wires by sweeping them along in the flow of the plastic into the mold cavity. Second, the thermal-expansion characteristics of plastic are much different from those of the lead frame, the lead wires, and the integrated-circuit chip itself. In fabricating a molded integrated-circuit package, the resultant assembly is voidless and all the materials are in intimate physical contact. This contact provides leverage by which differential expansion can become destructive.

Third, plastic packaging is only a partial barrier against moisture and contamination of the surface of the integrated-circuit chip. Plastics are not dense or impervious as compared with glass, ceramic, or metal. Actually, plastic encapsulation material but, given sufficient time and vapor pressure, will allow the complete transit of moisture from the external surface of the package to the surface of the integrated-circuit chip. Another problem which introduces moisture into the package is a lack of adherence to the lead frame. To assure that the plastic molding will release from the mold in which it is formed, materials called mold-release compounds are added.

Unfortunately, it is practically impossible to induce enough mold release to guarantee clean release of the completed part from the mold without causing incomplete adherence of the encapsulating material to the lead frame. This lack of adherence produces a fine via that parallels the surface of the lead frame from outside the package to the interior of the mold close to the integrated-circuit chip. Much work has been performed to design lead frames that minimize the effect of building in a continuous via by including a step or bump in the lead frame near the chip sight. This serves to disrupt the placement of the via.

Fourth, plastic packaging material and, in particular, the catalyst used to form the polarization of this material are sources of ionic contamination. This ionic contamination, along with moisture introduced into the package, can cause electrolytic corrosion.

These four deficiencies have severely restricted use of plastic integrated circuits. On the other hand, these properties can be reduced to levels that are consistent with the requirements of particular applications in many instances. Process-control techniques have been developed to minimize initial damage to the integrated circuit by bond wires being disturbed by the injection of molding compound. Furthermore, electrical screens separate out the percentage of the production population which has torn or shorted bond wires. Newer formulations of plastic materials have reduced thermal-expansion strains in plastic encapulation. Epoxies have been most promising in this area and, as such, are now gaining nearly universal acceptance as the preferred material for encapsulation of integrated circuits. Effects of any remaining thermal-expansion strain are further minimized by increasing the strength of mechanical bonds in the package prior to encapsulation. Most work in this area has been directed at improving the strength of the wire bonds to make them less likely to be pulled off the chip or lead frame by the encapsulation process or any resultant thermal mechanical strain.

The contamination problems, however, are less easily controlled. There is no way to modify an injectable plastic to make it form a true hermetic moisture seal. And there is no absolute technique to guarantee that a plastic will be produced without traces of free ionic material. These are simply two generic properties of plastic. Though the plastics can be modified and altered to some degree, they cannot be eliminated as sources of degradation. Work has been done to seal the chip hermetically using high-density voidless passivations, such as sputtered quartz and silicon nitride. These materials can serve as effective barriers to electromigration and the spread of contamination. Yet the processes themselves remain quite expensive, and the additional cost may make the choice of a plastic package less price-competitive.

Controlled-atmosphere applications and applications that do not require moderate or high levels of reliability are acceptable for plastic packaging. The choice to use plastics depends upon full definition of what appears to be a nonabusive atmosphere.

This definition requires full consideration of all details of the parts application and the environment. Parts that function under constant power dissipation are excellent choices for plastic packaging, since heat dissipated during steady-state operation keeps the package dry and prevents absorption of moisture. On the other hand, applications where power and temperature are cycled cause the integrated circuit to breathe in moisture on each cooling cycle and then further drive the moisture inward during the next heating cycle.

The real reliability and serviceability requirements for an integrated circuit must also be quantitatively considered before deciding whether to package it in plastic. Plastic integrated circuits, although less reliable than hermetic-sealed ones, can be orders of magnitude more reliable than the simple electromechanical assemblies they replace. This does not mean that plastic packaging should be an automatic decision because the using program is commercial. There are basically three applications for integrated circuits: military usage, commercial usage, and junk usage. Past history has dictated that most commercial computers be fabricated from hermetic-sealed integrated circuits. Some military applications have found the need for plastic integrated circuits. Plastic packages, being voidless, are more capable of withstanding severe environmental abuse, such as the mechanical shock of being fired from a cannon or the shock of being dropped from an airplane.

Hermetic-sealed Packages. Hermetic-sealed integrated-circuit packages are defined as packages made exclusively from a combination of glass, ceramic, or metal. The materials are fabricated into a structure which separates the outside environment from the closed, sealed microenvironment of the integrated-circuit package. There is a common misunderstanding that all hermetic packages are equally good in protecting an integrated-circuit chip. Packages vary as to the quality of hermetic seal, tendency to entrap internal contamination, and mechanical stress.

Early work in hermetic-sealed packages extended from the packaging technology evolved from both the vacuum-tube industry and the semiconductor industry. Therefore, early integrated-circuit packages were extensions of transistor packages. Metal cylindrical packages with round leads entering the lower face through glass hermetic seals were used. These glass hermetic seals were formed with borosilicate glasses, such as Corning 4052 or 4053, and special alloys which matched their thermal-expansion coefficients, such as Kovar* or Dumet.† The same specialized metal was used to fabricate the bottom of the enclosure, called a header. The chip was mounted on the header, and the package was then sealed by brazing or welding the remaining cylindrical-shape piece part to the header. The assembly of this package is illustrated in Fig 6.

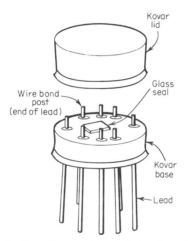

Fig. 6 Round-can integrated-circuit package.

Flatpacks and dual in-line packages answered the requirement for other form factors in packaging, being forms more suited to the flat structural shape of the integrated circuit and forms making a larger number of leads available. For these packages, most ceramic integrated-circuit packages are made by sealing a lead frame stamped from Kovar into borosilicate glass which seals the upper and lower ceramic parts of the package. This is illustrated in Fig. 7. Although this packaging technique is the most common technique for hermetic packaging, it is the least desirable. The package suffers from a basic fragility, since the lead frame goes through the glass that forms the package sidewalls. This glass is not fully devitrified; so it does not have great strength. Therefore, the lead frame becomes a lever by which external forces are applied to the sidewall of minimal strength. The package develops microcracking in use. The package also suffers moisture generated by devitrifying the glass. This moisture becomes entrapped in the package cavity and can develop into a source of contamination and degradation of the part.

This hermetic package is incapable of withstanding extended temperature cycles or thermal shocks owing to the inherent thermal mismatch of the piece parts and

* Trademark of Westinghouse Electric Corporation.
† Trademark of General Electric Corporation.

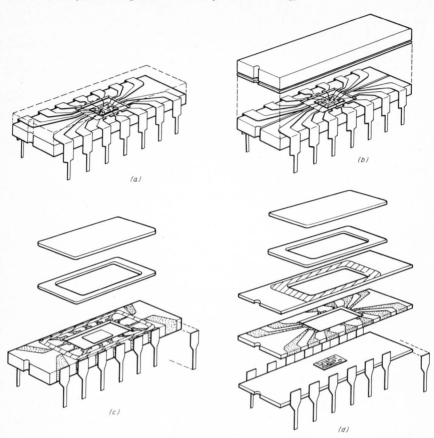

Fig. 7 Comparison of dual-in-line packaging technologies. (a) Plastic. (b) Cerdip. (c) Single-layer. (d) Multilayer.

the lack of package strength. There is also a great variation in package quality from vendor to vendor and from lot to lot. This variation traces to the different processes used to apply glass to ceramic prior to seal. The glass can be applied by taping, spraying, or silk screening. Further variation is the result of using different glasses with the range of thermal properties now being offered. The exact choice of sealing temperature and environment used to remelt the glass and partially devitrify it for a final seal is most significant. Unfortunately, the user normally does not have access to manufacturing information about the lots he buys. Without detailed acceptance test of individual lots, the user takes significant risk in specifying integrated circuits in this type of package. Cerdip and Cerpack are commercial examples of this packaging technology.

More successful techniques which use glass-to-metal seals for flat and dual-in-line packages are represented by the Kilburn package. The Kilburn package uses a lead frame that goes through glass, but the glass is fully devitrified and strengthened before final assembly and lidding of the package. Modern packaging has evolved two techniques which use multilayer ceramics to form packages. The basic construction is that of Fig. 7. In this construction, multilayers with interconnects printed onto them are placed in intimate contact. These layers of green ceramic are then fired together to form a single structure which has a printed lead pattern embedded. Final seal on these packages is made by emplacement of a braze lid. The other technique results from recent developments by the SCS Corporation. SCS has perfected methods of multilayer interconnections on the surface of a single-piece part-ceramic package. Thin layers of glass and conductors are alternately built up to serve as the lead frame. Then a weld ring or braze ring is secured to the package, which ultimately allows a final braze sealing of the lid. This package style and its construction are also illustrated in Fig. 7. Although multilayer packages are more expensive than glass-seal packages, they can be sealed at lower temperatures with less contamination in the package than glass-sealed packages. Therefore, they are used extensively in products sensitive to contamination and/or high-temperature extremes.

Some work in the industry has been directed at fabrication of ceramic packages using an epoxy seal. The nondesirable characteristics of plastic in intimate contact with the integrated-circuit chip are eliminated, but this style of package does not give a true hermetic seal. This makes the package a poor choice for applications in which a totally benign atmosphere cannot be guaranteed.

The technological impact of material technology in packaging integrated circuits is most important. To understand fully the reliability and performance consequences of a package choice, the design engineer must fully study the particular package and its characteristics. The package should be specified in detail in integrated-circuit procurement to guarantee the planned level of reliability and operation. Just as there is no best integrated circuit, only one best suited for a particular application, there is also no best technology to fabricate an integrated-circuit package. There is only a technology best suited to the requirements and economics of a particular integrated-circuit application.

Form factors The design engineer must properly consider package form factor of an integrated circuits if he is going to take full advantage of the miniaturization of today's integrated circuit. The design engineer must also weigh the appropriateness of an integrated-circuit package against planned cost of the system under construction, thermal-management questions, and desired overall packaging density.

In reality, the apparent high density of packaging with integrated circuits in flatpack is hard to attain without extremely expensive and sophisticated system-packaging techniques. The requirements for this sophistication are generated by the fact that miniature flatpacks have electrical terminals spaced on 50-mil centers. This tight placement of terminals normally requires expensive multilayer printed-circuit boards, which are usually employed to gain the next level of system integration. All components on printed-wiring boards consume a volume equal to the area on the printed-wiring board multiplied by the height between printed-wiring boards in the overall system mechanical configuration. Therefore, while flatpacks occupy approximately one-twentieth of the volume of an equivalent duel-in-line package, the flatpack configuration yields only a packaging-density improvement of about 3:1, if both are mounted on similar printed-wiring boards where interboard spacing is dictated by taller nonintegrated circuit components. Greater improvements in packaging density with flatpacks can be generated by using cordwood or, comb modules as an intermediate level of system integration between the integrated circuit and the printed-wiring board. Misuse of flatpacks has resulted in attempts to improve system packaging on small systems through their use. In reality, smaller system size is realized by conventional dual-in-line packaging and miniaturization of power supplies and discrete circuitry. In modern electronic systems, flatpack packaging is used only for miniature portable equipment, such as transceivers and pocket pagers, and airborne electronic equipment which require low system weight.

The round-can package is commonly called the multilead TO-5 package, and it is primarily used for specialized linear integrated circuits. This package is quite inconvenient to use, since its round lead pattern is not well suited to printed-wiring-board layout, and it is most cumbersome to apply in those cases where printed-wiring-board holes are laid out on a rectilinear grid. Therefore, the round-can integrated-circuit package is usually avoided, its use restricted to cases where the circuit is not available in a more practical package.

Most of the integrated circuits used in modern electronic equipment are packaged in either dual-in-line or rectangular-grid placed-pin packages. These packages yield a reasonably high level of system packaging density along with the advantages of simple and economical system assembly. These packages have pins placed on 0.1-in rectilinear centers and are readily adaptable to automated systems of circuit placement, printed-wiring-board artwork generation, and automated insertion into the printed-wiring board. For the most part, these packages are designed with lead frames that have a shoulder to raise the device off the printed-wiring board to allow proper cleaning of fluxes after assembly. The dual-in-line·package has also been modified to increase thermal dissipation for certain commercial applications through inclusion of an integral heat sink. The dual-in-line package is available with a variety of lead counts from 6 to as many as 64. One additional practical advantage of dual-in-line packaging is that the pins of the package protrude from the side of the package before bending toward the printed-wiring boards. This allows ready access to these terminals during troubleshooting or acceptance test of completed assemblies.

RELIABILITY

Integrated circuits are quite often selected for application because of greater reliability than discrete semiconductor circuits, vacuum-tube circuits, or electromechanical configuration. However, to establish a meaningful quantitative measure of integrated-circuit reliability in a given system is a formidable task. It has been established that integrated-circuit reliability is affected by process controls imposed during manufacture, by high-reliability processing which eliminates units prone to infant mortality, by assembly-level screening, by thermal and atmospheric conditions of the system, by the type and degree of derating applied to the application, and by the type and sophistication of electrical test used to specify an acceptable unit. Predicting failure-rate sensitivities to each of these factors with a reasonable level of confidence would require a controlled test program in the form of an eight-way factorial experiment where each test cell contains over one billion hours of life data. The experiment is not only financially but physically impossible. However, papers have been presented which proposed various schemes for calculating the reliability of integrated circuits in particular applications. Reliability prediction techniques are quite effective when applied to system reliability analysis.

Failure rates of 0.1 to 100 failures per trillion hours have been successfully associated with the use of integrated circuits. However, quick collection of accurate failure data is difficult, since integrated circuits, like all other solid-state devices, have no short-term wear-out mechanism. The activation energy associated with failure of silicon solid-state devices is approximately 1.6 eV. Other mechanisms, such as migration of metallic conductors, mechanical fracture of the crystal chip, chemical contamination of the chip, and mechanical failure of the wire-bond attachment system, cause premature failures of integrated circuits. These have the potential to be screened out. Individuals in the electronic industry have successfully presented the viewpoint that it is better to find the causes of failure and eliminate them than to try to establish a quantitative measure of the resultant reliability. Integrated circuits therefore represent a technological sophistication directed at reducing the fabrication anomalies which cause failure.

Process control Properly designed integrated circuits which are used within their rating have no significant wear-out mechanism. This feature means that the integrated circuit in its ideal form would be a component without a failure rate. In its practical form, the integrated circuit as it is fabricated today does have a failure rate which is quite low. Laboratory analysis of failed integrated circuits has consistently found the three significant sources of these relatively few failures to be operating overstress, improper circuit design, and variant-uncontrolled or at least not properly controlled processing steps. The key element in producing integrated circuits of high reliability is maintaining a system of exacting process control. And, because integrated circuits are batch-fabricated, they are responsive to the reliability-improvement methodologies which center upon process control. Proper processing control is a function of both classical quality control and measurement techniques and advancement in the science of integrated-circuit processing.

In the earliest phases of the semiconductor industry, many processes were highly nonrepeatable, which led to the fabrication of many devices at very low yields. This is no longer true today. The current trends in the art and science of integrated-circuit processing have led to refinement and control of these processes. Yields are now both repeatable and high, and the required performance characteristics of integrated circuits are obtained consistently as a result. Maintaining process control of this type calls for the various techniques of in-process monitoring. Quality of the chemicals, even the water and air available to the process, must be stabilized at control limits more severe than ever used before in an industrial environment. A collection of unique process-control methods had to be developed to satisfy the singular requirements of the integrated-circuit industry.

A good example of unique process control is formation of test devices on a wafer of fabricated integrated-circuit chips. These test devices are formed in a pattern separate from the patterns that form functional integrated circuits (see Fig. 8). The

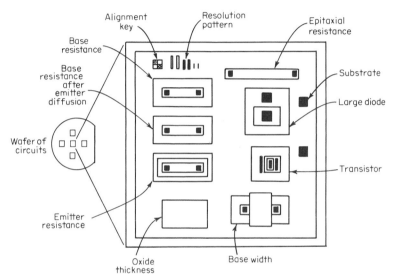

Fig. 8 Process control with a test pattern. Note the separate devices on the chip.

patterns of test devices are placed as a group in four, five, or sometimes nine different regions of the slice of silicon wafer. Each pattern is composed of typical component structures that would normally form an integrated circuit. However, by forming the components separately and independently, and by metallizing them to allow their separate tests, the process-control engineer allows a unique opportunity to monitor the effects of each process step on the performance of each component in the integrated circuit. These test patterns permit early detection of process variances, allowing the removal of a product circuit from further processing and cost when it will not be functional after total fabrication. Test patterns also present the mechanism for the timely implementation of corrective action.

An important additional element in the process-control scheme is automated measurement of device parameters by computer-controlled test equipment at the final stages of the device fabrication. Through techniques of computer-aided data collection, it becomes possible to use the qualitative acceptance and rejection data and the qualitative parametric data generated by automated electrical test to monitor overall effects of process variance. A typical process-control scheme is illustrated in Fig. 9.

Derating Derating is a process that improves in-use reliability of a component by either reducing the life stress on the component or making numerical allowances for minor degradation in the performance of the component. This technique is applied to integrated circuits in two separate and distinct ways.

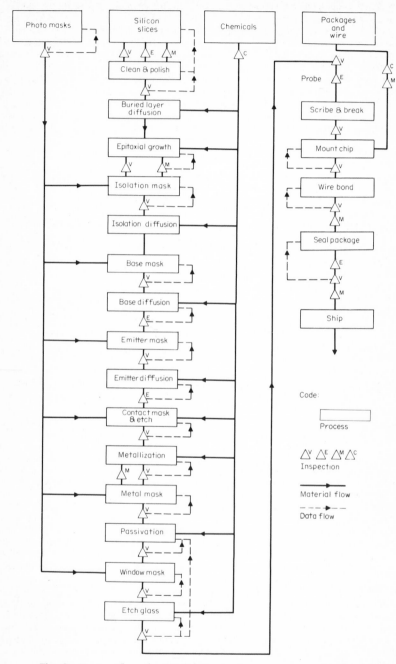

Fig. 9 Process flow chart indicating major process-control points.

The first is to specify a safety factor in the application of the component. Safety factors, or derating factors as they are more appropriately known, are applied to the voltage, current, and power stresses to which the integrated circuit is subjected during operation. Derating factors must be applied knowledgeably and singly; that is, they must be applied only to a degree which improves reliability, and they must be applied only once throughout the entire cycle which stretches from the design of the integrated circuit to its application in a system. From the outset, integrated circuits are designed to a set of reasonably conservative design-rating criteria. The currents that flow through the conductors and through the wire bonds on a chip, the voltages applied to the semiconductor junctions, and the overall power stress on the entire chip are limited during the design of an integrated circuit. It is therefore not appropriate to derate the integrated circuit further in its application. Derating power consumption of a digital integrated circuit is usually not possible in the design cycle, since the circuit must operate at a specified level of power-supply voltage for maximum performance. In application, some linear circuits, which have been designed to operate over an extended range of power-supply voltages and power dissipations, may accept some degree of derating when appropriately applied.

Therefore, the main area of derating in integrated circuits is not in derating the stresses applied to the circuit, but in derating the expected and required performance. The engineer must realize that performance of an integrated circuit is not guaranteed by circuit design but guaranteed only for a statistical subdistribution of the entire production lot of the integrated circuits made. This subdistribution is separated from the entire production lot by electrical parametric testing. Since the actual performance of an individual integrated circuit depends upon the parametric values of such on-chip subcomponents as diodes, transistors, and resistors, any shift in the performance of these subcomponents changes the overall performance of the integrated circuit. It is a recognized fact that the performance of solid-state components drifts measurably over their lives. In the subcomponents of an integrated circuit, this means that the overall distribution of components meeting the black-box-component specification is shifted.

In applying integrated circuits, the engineer must fully recognize potential performance degradation of integrated circuits over their life. He therefore applies parametric derating to the application of integrated circuits. This parametric degradation can mean using a digital circuit at less than its full fanout. It can mean designing for an extra noise margin and sacrificing some of it to the degradation of the integrated circuit. It can also mean applying the integated circuit at performance levels below those guaranteed by the circuit's characterization-specification sheet. Any establishment of derating factors for integrated-circuit parameters must be made after careful analysis of each source of a particular parameter within the circuit. Parameters depending directly on transistor beta, resistor value, or junction leakage are most prone to shift during life. Parameters depending directly on the saturation voltages of junctions and on the ratios of resistors are most likely to remain stable. Typically, device fanout should be derated by a factor of 20 percent, and logic noise-margin levels should be derated by a factor of 10 percent. The severity of the application further establishes the degree of proper derating. It is customary not to derate ac parameters, such as delay times or rates, as these parameters do not vary greatly over the life of an integrated circuit. Allowances should, however, be made for unit-to-unit variation within a given integrated-circuit chip. In many cases, the delay times of separate gates within one integrated-circuit package can vary greatly. These parameters are not normally measured on 100 percent of units, and as such, this variation should be allowed for in the design of equipment fabricated from integrated circuits.

Even though a design engineer may derate an integrated circuit for reliability in specified special cases, the reverse is not true. At no time may the design engineer take advantage of the derating designed into the integrated circuit and use it beyond its rating or specified capability. It is impossible to reconstruct, after the fact, the analysis used to derate the integrated circuit adequately during its fabrication. In applying the ratings of an integrated circuit, or for that matter any component, the engineer must simultaneously apply the worst case of all criteria contained within the device ratings. A more optimistic application of the device ratings will result in the misapplication of the part and potential degradation of both its reliability and its performance.

Military specifications Although 90 or more percent of the integrated circuits produced are used in nonmilitary applications, there is a broad-based interest in the military specifications and standards for integrated circuits. These documents represent a common yardstick by which integrated circuits and the various techniques for improving their reliability can be measured. However, much confusion exists in clearly identifying the relationship between MIL-M-38510 and MIL-STD-883. These two documents, as a functional pair, provide the framework for procurement of a broad range of standard integrated-circuit types. They also provide a set of standardized test and acceptance conditions by which other integrated-circuit types can be specified. The two documents should be examined separately, because they fulfill two completely distinct but complementary functions.

MIL-M-38510 is the general specification for integrated circuits. It, like many other general specifications in the military-specification system, is structured to have slash sheets, each of which gives detailed specifications for a limited number of similar circuits. As a general document, MIL-M-38510 requires a level of process and test control and documentation significantly superior to the levels customarily used in the integrated-circuit industry. MIL-M-38510 requires that a fully documented product-assurance plan be established and be approved by the cognizant military agency. It further requires that all facilities used for the fabrication and assembly of the circuits be surveyed and approved by military agencies. The main effect of this requirement is that integrated circuits manufactured in compliance with MIL-M-38510 are fabricated and tested at facilities within the United States, while most integrated circuits are assembled and/or tested at foreign plants. The requirement for documentation of procedures has required integrated-circuit manufacturing companies further to tighten and control process variables associated with their integrated circuits.

The slash sheets to MIL-M-38510 specify in detail the requirements for electrical performance, as represented by electrical acceptance test for the appropriate integrated-circuit types. The slash sheets define all particular attributes of the integrated circuits they concern. Besides the electrical, these attributes include performance, pin-out, available package choices, particular gradings, and suggested operating conditions.

Table 1 gives a complete listing of the device types available under the slash sheets of MIL-M-38510. This table relates the military type number to the common catalog type number for the same functional circuits. The table also gives some description of the functional nature of the part. MIL-M-38510 is not a complete listing of parts; the part types as specified in MIL-M-38510 represent prime selections from the basic commercial part type. In almost all cases, the parameters tested have been tested to tighter limits, and even the number of parameters tested on a 100 percent basis has been significantly increased. It is anticipated that there will be a continuing process of adding additional part types to the group specified by the MIL-M-38510 slash sheets. These additions will be made on the basis of requirements for military system use and the availability and suitability of the part for such use.

There are important distinctions among military-specification integrated circuits, military temperature-range integrated circuits, MIL-M-38510 integrated circuits, MIL-STD-883 integrated circuits, and vendor internal specification high-reliability integrated circuits. While MIL-M-38510 controls the full range of important device characteristics, including performance, plant quality procedures, preconditioning techniques and interchangeability, all other descriptions represent a less complete set of controls. Military temperature-range parts are just that, parts which are specified to operate over the temperature range of -55 to $125°C$. MIL-STD-883 parts are normally parts preconditioned to a particular preconditioning scheme specified within MIL-STD-883. Vendor high-reliability process parts, such as Unique-38510* or MACH-IV† or other such vendor trade names, are in reality cost-effective or cost-

---

* Trademark of Fairchild Semiconductor.
† Trademark of Motorola Semiconductor Products Division.

TABLE 1   MIL-M-38510 Cross Reference

Family	Function	Commercial-device type No.	Military specification type No.
Linear	Operational amplifier	LM101A	M38510/10103*
	Voltage-follower amplifier	LM102	M38510/10601
	Voltage comparator	LM106	M38510/10303
	Operational amplifier	LM108A	M38510/10104
	Voltage regulator	LM109	M38510/10701
	Voltage-follower amplifier	LM110	M38510/10602
	Voltage comparator	LM111	M38510/10304
	Voltage comparator	UA710	M38510/10301
	Dual-voltage comparator	UA711	M38510/10302
	Voltage regulator	UA723	M38510/10201
	Video amplifier	UA733	M38510/10501
	Compensated operational amplifier	UA741	M38510/10101
	Dual compensated operational amplifier	UA747	M38510/10102
CMOS	NOR gate, inverter	4000A	M38510/05201
	NOR gate	4001A	M38510/05202
	NOR gate	4002A	M38510/05203
	Static shift register	4006A	M38510/05701
	Dual complementary pair	4007A	M38510/05301
	4-bit full adder	4008A	M38510/05401
	Hex inverter-buffer	4009A	M38510/05501
	Hex buffer	4010A	M38510/05502
	NAND gate	4011A	M38510/05001
	NAND gate	4012A	M38510/05002
	Dual D flip-flop	4013A	M38510/05101
	Static shift register	4014A	M38510/05702
	Dual static shift register	4015A	M38510/05703
	Quad bilateral switch	4016A	M38510/05801
	Decade counter	4017A	M38510/05601
	Divide by N counter	4018A	M38510/05602
	Quad and/or select gate	4019A	M38510/05302
	Binary/ripple counter	4020A	M38510/05603
	Static shift register	4021A	M38510/05704
	Divide by 8 counter	4022A	M38510/05604
	NAND gate	4023A	M38510/05003
	Binary counter	4024A	M38510/05605
	NOR gate	4025A	M38510/05204
	Dual J-K flip-flop	4027A	M38510/05102
	Static shift register	4031A	M38510/05705
	Hex inverter-buffer	4049A	M38510/05503
	Hex buffer	4050A	M38510/05504
DTL	Expandable NAND gate	930	M38510/03001
	Hex inverter	935	M38510/03002
	U hex inverter	936	M38510/03003
	NAND gate	946	M38510/03004
	NAND gate	962	M38510/03005
Interface	Dual line receiver	55107	M38510/10401
	Dual line receiver	55108	M38510/10402
	Line driver	55113	M38510/10405
	Dual differential line driver	9614	M38510/10403
	Dual differential line receiver	9615	M38510/10404
TTL	NAND gate	5400	M38510/00104
	NAND gate	54H00	M38510/02304

* Three additional letters (numerals and/or letter) are added to complete the specification number.

TABLE 1   MIL-M-38510 Cross Reference (Continued)

Family	Function	Commercial-device type No.	Military specification type No.
TTL	NAND gate	54L00	M38510/02004
	NAND gate	54S00	M38510/07001
	NAND gate—open col.	5401	M38510/00107
	NAND gate—open col.	54H01	M38510/02306
	NOR gate	5402	M38510/00401
	NOR gate	54L02	M38510/02701
	NOR gate	54S02	M38510/07301
	NAND gate	5403	M38510/00109
	NAND gate	54L03	M38510/02006
	NAND gate	54S03	M38510/07002
	Hex inverter	5404	M38510/00105
	Hex inverter	54H04	M38510/02305
	Hex inverter	54L04	M38510/02005
	Hex inverter	54S04	M38510/07003
	Hex inverter—open col.	5405	M38510/00108
	Hex inverter—open col.	54S05	M38510/07004
	Hex inverter	5406	M38510/00801
	Hex buffer	5407	M38510/00803
	AND gate	5408	M38510/01601
	AND gate—open col.	5409	M38510/01602
	NAND gate	5410	M38510/00103
	NAND gate	54H10	M38510/02303
	NAND gate	54L10	M38510/02003
	NAND gate	54S10	M38510/07005
	AND gate	54S11	M38510/08001
	NAND gate—open col.	5412	M38510/00106
	AND gate	54S15	M38510/08002
	Hex inverter	5416	M38510/00802
	Hex buffer	5417	M38510/00804
	NAND gate	5420	M38510/00102
	NAND gate	54H20	M38510/02302
	NAND gate	54L20	M38510/02002
	NAND gate	54S20	M38510/07006
	NAND gate—open col.	54H22	M38510/02307
	NAND gate—open col.	54S22	M38510/07007
	NOR gate with strobe	5423	M38510/00402
	NOR gate with strobe	5425	M38510/00403
	NOR gate	5427	M38510/00404
	NAND gate	5430	M38510/00101
	NAND gate	54H30	M38510/02301
	NAND gate	54L30	M38510/02001
	NAND gate	54S30	M38510/07008
	NAND gate	5437	M38510/00302
	NAND gate—open col.	5438	M38510/00303
	NAND gate	5440	M38510/00301
	NAND gate	54H40	M38510/02401
	NAND gate	54S40	M38510/07201
	BCD to decimal decoder	5442	M38510/01001
	BCD to decimal decoder	54L42	M38510/02901
	Excess 3 to decimal decoder	5443	M38510/01002
	Excess 3 to decimal decoder	54L43	M38510/02902
	Excess 3 gray to decimal decoder	5444	M38510/01003
	Excess 3 gray to decimal decoder	54L44	M38510/02903
	1 of 10 decoder	5445	M38510/01004
	BCD to seven-segment decoder	5446	M38510/01006

TABLE 1   MIL-M-38510 Cross Reference (Continued)

Family	Function	Commercial-device type No.	Military specification type No.
TTL	BCD to seven-segment decoder	54L46	M38510/02904
	BCD to seven-segment decoder	5447	M38510/01007
	BCD to seven-segment decoder	54L47	M38510/02905
	BCD to seven-segment decoder	5448	M38510/01008
	BCD to seven-segment decoder	5449	M38510/01009
	AND/OR gate	5450	M38510/00501
	AND/OR gate	54H50	M38510/04001
	AND/OR invert gate	5451	M38510/00502
	AND/OR invert gate	54H51	M38510/04002
	AND/OR invert gate	54L51	M38510/04101
	AND/OR invert gate	54S51	M38510/07401
	Expandable AND/OR invert gate	5453	M38510/00503
	Expandable AND/OR invert gate	54H53	M38510/04003
	AND/OR invert gate	5454	M38510/00504
	AND/OR invert gate	54H54	M38510/04004
	AND/OR invert gate	54L54	M38510/04102
	AND/OR invert gate	54H55	M38510/04005
	AND/OR invert gate	54L55	M38510/04103
	AND/OR invert gate	54S64	M38510/07401
	AND/OR invert gate	54S65	M38510/07402
	JK flip-flop	5470	M38510/00206
	JK-M/S flip-flop	54L71	M38510/02101
	JK-M/S flip-flop	5472	M38510/00201
	JK-M/S flip-flop	54H72	M38510/02201
	JK-M/S flip-flop	54L72	M38510/02102
	Dual JK M/S flip-flop	5473	M38510/00202
	Dual JK M/S flip-flop	54H73	M38510/02202
	Dual JK M/S flip-flop	54L73	M38510/02103
	Dual D flip-flop	5474	M38510/00205
	Dual D flip-flop	54H74	M38510/02203
	Dual D flip-flop	54L74	M38510/02105
	Dual D flip-flop	54S74	M38510/07101
	4-bit latch	5475	M38510/01501
	Dual JK M/S flip-flop	5476	M38510/00204
	4-bit latch	5477	M38510/01502
	Dual JK M/S flip-flop	54L78	M38510/02104
	Dual D flip-flop	5479	M38510/00207
	2-bit full adder	5482	M38510/00601
	4-bit full adder	5483	M38510/00602
	4-bit magnitude comparator	54S85	M38510/08201
	Exclusive OR gate	5486	M38510/00701
	Exclusive OR gate	54L86	M38510/02601
	Exclusive OR gate	54S86	M38510/07501
	Decade counter	5490	M38510/01307
	Decade counter	54L90	M38510/02501
	Divide by 12 counter	5492	M38510/01301
	Binary counter	5493	M38510/01302
	Binary counter	54L93	M38510/02502
	4-bit bidirectional shift register	5495	M38510/00901
	4-bit bidirectional shift register	54L95	M38510/02801
	5-bit shift register	5496	M38510/00902
	Dual JK M/S flip-flop	54107	M38510/00203
	Dual JK flip-flop	54S112	M38510/07102
	Dual JK flip-flop	54S113	M38510/07103
	Dual JK flip-flop	54S114	M38510/07104

TABLE 1 MIL-M-38510 Cross Reference (Continued)

Family	Function	Commercial-device type No.	Military specification type No.
TTL	One-shot multivibrator	54121	M38510/01201
	One-shot multivibrator	54122	M38510/01202
	Dual one-shot multivibrator	54123	M38510/01203
	NAND gate	54S133	M38510/07009
	NAND gate	54S134	M38510/07010
	Exclusive OR gate	54S135	M38510/07502
	Decoder	54S138	M38510/07701
	Decoder	54S139	M38510/07702
	Line driver	54S140	M38510/08101
	1 of 10 decoder/driver	54145	M38510/01005
	16-input multiplexer	54150	M38510/01401
	Multiplexer	54S151	M38510/07901
	Dual multiplexer	54153	M38510/01403
	Dual multiplexer	54S153	M38510/07902
	Multiplexer	54S157	M38510/07903
	Multiplexer	54S158	M38510/07904
	Decade counter	54160	M38510/01303
	Binary counter	54161	M38510/01306
	Synchronous 4-bit counter	54162	M38510/01305
	Synchronous 4-bit counter	54163	M38510/01304
	8-bit converter	54164	M38510/00903
	8-bit converter	54L164	M38510/02802
	8-bit converter	54165	M38510/00904
	Dual JK flip-flop	54S174	M38510/07105
	Dual JK flip-flop	54S175	M38510/07106
	4-bit arithmetic logic unit	54181	M38510/01101
	4-bit arithmetic logic unit	54S181	M38510/07801
	Carry look ahead block	54S182	M38510/07802
	Up-down decade counter	54192	M38510/01308
	Up-down binary counter	54193	M38510/01309
	4-bit shift register	54194	M38510/00905
	4-bit shift register	54S194	M38510/07601
	4-bit shift register	54195	M38510/00906
	4-bit shift register	54S195	M38510/07602
	Multiplexer	54S251	M38510/07905
	Multiplexer	54S257	M38510/07906
	Multiplexer	54S258	M38510/07907
	Decoder	54S280	M38510/07703
	Dual 4-bit latch	9308	M38510/01503
	Dual 4-input multiplexer	9309	M38510/01404
	8-input multiplexer	9312	M38510/01402
	4-bit latch	9314	M38510/01504
	2-input multiplexer	9322	M38510/01405
	8-bit shift register	93L28	M38510/02803
Memory circuits	512-bit PROM	MCM5303	M38510/20101
	512-bit PROM	MCM5304	M38510/20102
	1024-bit PROM		M38510/20201
ECL	OR/NOR gate	10501	M38510/06001
	NOR gate	10502	M38510/06002
	OR/NOR gate	10505	M38510/06003
	NOR gate	10506	M38510/06004
	Exclusive OR/NOR gate	10507	M38510/06005
	OR/NOR gate	10509	M38510/06006

reductive compromises between the requirements of the military specification and the vendor's standard method of processing and fabricating parts. Figure 10 describes the formulation or interpretation of a part number, as used in MIL-M-38510. Significantly, all possible combinations represented by combinations of digits in the

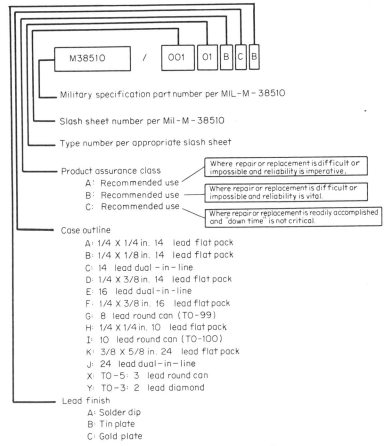

Fig. 10   MIL-M-38510 part number.

part-number format are not necessarily specified or available. An appropriate qualified-product list, QPL-38510, will list which type numbers of those specified in the slash sheet to MIL-M-38510 are actually available. The QPL will also indicate the name or names of a qualified source or sources.

MIL-STD-883 is a military standard of test methods for integrated circuits. Basically, it is a shopping list of separate documented test procedures which can be applied to integrated circuits. Table 2 lists the test methods available in this MIL standard. The most common reference to MIL-STD-883 is a reference to processing in accordance with one of its particular product-assurance levels. These references are to sections 5004 and 5005. These two sections specify in detail the appropriate screening and preconditioning sequences, respectively. The two sequences are fully presented in Tables 3 to 6. The application of conditions in these four tables must be modified to allow for any unique attributes, requirements, or sensitivities of a particular part, or for any special requirements of a system in which the part is used. Like any other general standard, MIL-STD-883 is designed to cover a large middle ground of commonality between military parts and applications. The preconditioning and screening methodologies of MIL-STD-883 are used in MIL-M-38510. In reality, the reference to product-assurance class in the MIL-M-38510 part number is a direct reference to the levels contained in Methods 5004 and 5005.

Since MIL-STD-883 is a full listing of test methods to be applied to integrated circuits, it is used most extensively within the slash sheets and body of MIL-M-

TABLE 2   Index of MIL-STD-883 Test Methods

Method No.	Title
Environmental Tests	
1001	Barometric pressure, reduced (altitude operation)
1002	Immersion
1003	Insulation resistance
1004	Moisture resistance
1005	Steady-state life
1006	Intermittent life
1007	Agree life
1008	High-temperature storage
1009	Salt atmosphere (corrosion)
1010	Temperature cycling
1011	Thermal shock
1012	Thermal characteristics
1013	Dew point
1014	Seal
1015	Burn-in screen
Mechanical Tests	
2001	Constant acceleration
2002	Shock
2003	Solderability
2004	Lead integrity
2005	Vibration fatigue
2006	Vibration noise
2007	Vibration, variable-frequency
2008	Visual and mechanical
2009	External visual
2010	Internal visual (precap)
2011	Bond strength
2012	Radiography
2013	Internal visual
Electrical Tests (Digital)	
3001	Drive source, dynamic
3002	Load conditions
3003	Propagation delay
3004	Delay and transition-time measurements
3005	Power-supply current
3006	High-level output voltage
3007	Low-level output voltage
3008	Breakdown voltage, input or output
3009	Input current, low-level
3010	Input current, high-level
3011	Output short-circuit current
3012	Terminal capacitance
3013	Noise-margin measurements for microelectronic logic gating circuits

TABLE 2   Index of MIL-STD-883 Test Methods (Continued)

Method No.	Title
	Electrical Tests (Linear)
4001	Input offset voltage and current and bias current
4002	Phase-margin and slew-rate measurements
4003	Common-mode input-voltage range
	Common-mode rejection ratio
	Supply-voltage rejection ratio
4004	Open-loop performance
4005	Output performance
4006	Power gain and noise figure
4007	Automatic-gain-control range
	Special Tests
5001	Parameter mean-value control
5002	Parameter distribution control
5003	Failure-analysis procedures for microcircuits
5004.1	Screening procedures
5005.1	Qualification and quality-conformance procedures
5006	Limit testing

38510. These two documents can be used in the preparation of user specifications for specific integrated circuits. This use can be taken from one of two forms, the first of which is to use MIL-M-38510 as the general specification and to supply information of similar content and format to the MIL-M-38510 slash sheets. Otherwise, where the user-prepared specification is not designed to procure integrated circuits for a military system, it is still useful to use MIL-STD-883 as a source for test methods. In this way, much of the repetition of commonly accepted test procedures and techniques can be eliminated. By using these two military documents, the chances of misinterpretation of the intended requirements of the particular specification are reduced. In addition, the use of standard conditions specified in the military documents allows the vendor to implement common procedures and equipment, and hence provides the tools to make effort effective.

Generally, the use of integrated circuits specified by MIL-M-38510 reduces the overall cost of systems in which they are being used. ·They eliminate the cost of parts documentation and impose standardization in the form of a limited number of part types available to the MIL specification. They further reduce cost over the period of system use because of the higher availability and lower maintenance costs associated with higher-reliability components.

The design engineer must be careful in specifying parts to MIL-M-38510, since the document takes into account that different military programs have different requirements for reliability. As such, the document has three product-assurance classes. Class A is directed at space systems, where maintenance is impossible and reliability is foremost. Class B represents a middle level, where preconditioning is still applied but the special procedures of line certification and the intended controls are not required. Class B represents high-reliability parts suitable for use in systems where maintenance is limited and difficult, and where there is a major requirement for high reliability. Class C represents a class of parts where reliability is above that of commercial integrated circuits but below the normal levels of higher-reliability components. Class C integrated circuits do not undergo a postassembly-stress screening program. However, class C components can be upgraded to class B merely by application of the screening sequence.

High-reliability specifications Many practical techniques can be incorporated into a specification to improve the reliability of an integrated circuit. These techniques are similar to those used in generating high-reliability components of any type. Most

TABLE 3 MIL-STD-883 Method 5004 Screening Sequence

Screen	Class A Method	Class A Requirement	Class B Method	Class B Requirement	Class C Method	Class C Requirement
Internal visual (precap)	2010, test condition A	100%	2010, test condition B	100%	2010, test condition B	100%
Stabilization bake	1008 24 h, min, test condition C min	100%	1008 24 h, min, test condition C min	100%	1008 24 h, min, test condition C min	100%
Thermal shock	1011, test condition A min	100%				
Temperature cycling	1010, test condition C min	100%	1010, test condition C min	100%	1010, test condition C min	100%
Mechanical shock	2002, test condition F one shock pulse in Y_1 plane only or five shock pulses at condition B in Y_1 plane	100%				
Constant acceleration	2001, test condition E (min), Y_2 plane, then Y_1 plane	100%	2001, test condition E (min), Y_1 plane	100%	2001, test condition E (min), Y_1 plane	100%
Seal: (a) Fine (b) Gross	1014	100%	1014	100%	1014	100%
Interim electrical parameters	Per applicable procurement document	100%	Per applicable procurement document			
Burn-in test	1015 240 h at 125°C min	100%	1015 168 h at 125°C min	100%		

Interim electrical parameters	Per applicable procurement document	100%		
Final electrical test:	Per applicable procurement document	Per applicable procurement document	Per applicable procurement document	Per applicable procurement document
(a) Static tests				
(1) 25°C		100%	100%	100%
(2) Max and min rated operating temp		100%	100%	100%
(b) Dynamic tests and switching tests 25°C		100%	100%	100%
(c) Functional test 25°C		100%	100%	100%
Radiographic	2012	100%		
Qualification or quality-conformance inspection	5005	Per applicable document	5005	Per applicable document
External visual	2009	100%	2009	100%

TABLE 4   MIL-STD-883 Method 5004 Group A Electrical Test

Subgroups	Class A LTPD	Class B LTPD	Class C LTPD
Subgroup 1: Static tests at 25°C	5	5	5
Subgroup 2: Static tests at maximum rated operating temperature	5	7	10
Subgroup 3: Static tests at minimum rated operating temperature	5	7	10
Subgroup 4: Dynamic tests at 25°C	5	5	5
Subgroup 5: Dynamic tests at maximum rated operating temperature	5	7	10
Subgroup 6: Dynamic tests at minimum rated operating temperature	5	7	10
Subgroup 7: Functional tests at 25°C	3	5	5
Subgroup 8: Functional tests at maximum and minimum rated operating temperatures	5	10	15
Subgroup 9: Switching tests at 25°C	5	7	10
Subgroup 10: Switching tests at maximum rated operating temperature	5	10	15
Subgroup 11: Switching tests at minimum rated operating temperature	5	10	15

The specific parameters to be included for tests in each subgroup shall be as specified in the applicable procurement document. Where no parameters have been identified in a particular subgroup or test within a subgroup, no group A testing is required for that subgroup or test to satisfy group A requirements.

of the techniques commonly used to improve integrated-circuit reliability are offshoots from methods generated to improve the reliability of discrete semiconductors.

The three main areas covered by high-reliability specifications are (1) requirements for adequate and documented process control, (2) specified plans of device screening or preconditioning, and (3) specified criteria for lot-acceptance test, and lot-acceptance criteria applied to the results of lot screening. These three techniques are normally specified in combination, as is done in MIL-M-38510, and they distinguish themselves by their purposes. While process control seeks a direct and immediate effect on the consistency and regularity of fabrication and assembly processes, screening is directed at culling out inferior units in the overall population of finished parts which appear to be acceptable from a reliability standpoint. On the other hand, lot acceptance does not seek either to improve the control of the product or even to separate out the inferior units; it merely removes as unacceptable those lots which have reached the customer populated with a disproportionate percentage of inferior products. Lot screening removes the infant-mortality subpopulation of the integrated-circuit lot in question. During screening, a series of stresses which are normally within the design ratings of the part are applied. Screening elements are included to simulate stresses which occur in the intended application of the part, or they apply stresses which separate proper construction from improper construction by failure, although the stresses are not part of the intended part application.

Stabilization bake is commonly part of a preconditioning sequence. High-temperature stabilization bake is performed to accelerate the chemical activity of any contaminants or active ionic material present within the device case. Both thermal shock and temperature cycling stress the mismatch of thermal-expansion coefficients within the package and chip materials. Mechanical shock and constant acceleration verify physical attachment of the chip to the substrate and the bond wires to the chip. Centrifuge is an excellent test for gold-ball wire bonds, since a large mass within each wire bond is accelerated during the centrifuge or constant acceleration. In cases where aluminum wire bonds are used, greatly reduced force is applied to the bond in this test owing to its reduced wire mass and the lack of a ball-shaped structure. The test may be omitted, since the force applied to an aluminum wire

TABLE 5 MIL-STD-883 Method 5005 Group B Environmental Test

Test	MIL-STD-883 Method	MIL-STD-883 Condition	Class A LTPD	Class B LTPD	Class C LTPD
Subgroup 1: Physical dimensions	2016		10	15	20
Subgroup 2: (a) Resistance to solvents	2015		3 devices (no failures) 1 device (no failures)	3 devices (no failures) 1 device (no failures)	3 devices (no failures) 1 device (no failures)
(b) Internal visual and mechanical	2014	Failure criteria from design and construction requirements of applicable procurement document			
(c) Bond strength* (1) Thermocompression (2) Ultrasonic or wedge (3) Flip-chip (4) Beam lead	2011	(1) Test condition C or D (2) Test condition C or D (3) Test condition F (4) Test condition H	5	15	20
Subgroup 3: Solderability†	2003	Soldering temperature of 260 ± 10°C	10	15	15
Subgroup 4: Lead integrity‡ Seal: (a) Fine (b) Gross	2004 1014	Test condition B₂, lead fatigue As applicable	10	15	15

Electrical reject devices from the same inspection lot may be used for all subgroups when end-point measurements are not required.
* Unless otherwise specified, at the manufacturer's option, test samples for bond strength may be selected randomly immediately following internal visual (method 5004) prior to sealing. The LTPD applies to the number of leads inspected, except in no case shall less than three devices be used to provide the number of leads required.
† All devices must have been through the temperature/time exposure in burnin.
‡ When fluorocarbon gross leak testing is utilized, test condition C₂ shall apply as a minimum.

TABLE 6 MIL-STD-883 Method 5005 Group C Environmental and Life Test

Test	MIL-STD-883 Method	MIL-STD-883 Condition	Class A LTPD	Class B LTPD	Class C LTPD
Subgroup 1:[a]			10	15	15
Thermal shock	1011	Test condition B as a minimum			
Temperature cycling	1010	Test condition C			
Moisture resistance	1004				
Seal	1014	As applicable			
(a)					
(b)[g]					
Visual examination[b]					
End-point electrical parameters		As specified in the applicable procurement document			
Subgroup 2:[a]			10	15	15
Mechanical shock	2002	Test condition B			
Vibration, variable frequency	2007	Test condition A			
Constant acceleration	2001	Test condition E			
Seal	1014	As applicable			
(a)					
(b)[g]					
Visual examination[c]					
End-point electrical parameters		As specified in the applicable procurement document			
Subgroup 3:			10	15	15
Salt atmosphere[d]	1009	Test condition A			
Visual examination[e]					
Subgroup 4:			7	7	7
High-temperature storage[f]	1008	Test condition C 1,000 h			
End-point electrical parameters		As specified in the applicable procurement document			

	Method	Conditions			
Subgroup 5:					
Operating-life test[f]	1005	Test condition to be specified in the applicable procurement document (1,000 h)	5	5	5
End-point electrical parameters		As specified in the applicable procurement document			
Subgroup 6:					
Steady-state reverse bias	1005	Test condition A, 72 h at 150°C	7		
End-point electrical parameters		As specified in the applicable procurement document			

[a] Devices used for environmental tests in subgroup 1 may be used for mechanical tests in subgroup 2.

[b] Visual examination shall be in accordance with method 1010 or 1011 at a magnification of 5X to 10X.

[c] Visual examination shall be performed at a magnification of 5X to 10X for evidence of defects or damage to case, leads, or seals resulting from testing (not fixturing) such damage shall constitute a failure.

[d] Electrical reject devices from the same inspection lot may be used for samples.

[e] Visual examination shall be performed in accordance with 3.3.1 of method 1009.

[f] See 40.4 of appendix B of MIL-M-38510.

[g] When fluorocarbon gross leak testing is utilized, test condition C_2 shall apply as minimum.

bond at 30,000 g's acceleration is actually less than the drag of the bonding capillary used in the wire-bonding operation.

Fine and gross hermetic-seal tests are added for two reasons in most screens. The first is to remove from a lot all improperly sealed products. The second is to remove from the lot those parts whose seal has been ruptured by the preceding steps in the screen. Therefore, it is most important to place seal leak tests after all mechanical and thermal tests. A number of techniques are used to test the hermetic seal of integrated circuits. Fine leak tests can be done with a mass spectrometer to measure the amount of helium given off by package leaks during package evacuation after a helium pressure soak. Krypton radioactive gas can also be used as a pressure vehicle. Leaks are then detected by geiger counter to measure the radiation given off by the gas. The radiation is proportional to the amount of radioactive gas entrapped within the package. Although the equipment for krypton leak detection is more expensive and requires an Atomic Energy Commission license, the technique is more reliable and consistent. It does not generate the false readings generated in the helium method through surface absorption of the tracer gas.

Measurement of gross leak rates involves one of three techniques, all of which center upon liquid flow through gross leaks. In one technique, the part is placed in heated fluoroinert or mineral oil, and expansion of air in the package causes a series of bubbles to form in the liquid, which is then observed by microscope. The second gross leak technique is a slight modification of the first, in that the part is first pressure-bombed with a liquid that has relatively low boiling point. The part is then removed from its first liquid and placed in a second that has a boiling point above the boiling point of the first liquid. The temperature of the second bath causes any of the first liquid entrapped in the package to boil, generating a stream of bubbles which can again be observed through a microscope. The third technique removes the subjective nature of microscope examination by weighing the package prior to its being placed in a liquid pressure bomb and then weighing it again afterward. The difference in weight is calculated, and this difference represents the amount of leak into the package. The exact choice of leak-seal methods and limits is dictated by the sensitivity of the integrated-circuit chip to contamination and the ultimate capability of the package used.

Most schemes for preconditioning include as their central step a simulation of operating life, that is, the simultaneous application of electrical potential and temperature. This step is commonly called burnin, and a limited period of this simulated life can be quite effective in improving the reliability of the screened product. Burnin can be performed in a circuit that simulates the actual system environment, for example, by placing a logic circuit at its operating potential and then applying digital pulses to the inputs and a typical load to the outputs. Such active burnin can be done using external generators. This method is called parallel excitation. In many cases, similar results can be achieved by configuring a group of circuits into a ring-counter configuration and then allowing one circuit to drive another. A typical burnin system is illustrated in Fig. 11. Burnin is sometimes done using nonoperating circuits which try to either forward-bias or reverse-bias the majority of the junctions in the device. This technique is quite often applied to metal oxide semiconductors (MOS) and linear circuits, but it is rarely applied to digital circuits. In some reliability screening systems, multiple burnins are used that involve burnins designed to generate particular failure modes. For example, the standard dynamic functional burnin could be followed by a reverse-bias burnin designed to generate channeling failure modes. It is crucial to integrated circuits that a test such as burnin be followed by a complete and adequate 100 percent electrical test so that all failed devices generated in the process can be removed from the lot.

Radiographic inspection of the internal package cavity is sometimes included in the highest level of high-reliability specifications. This inspection identifies poor workmanship and loose particles sealed within the package. Some effort has been made to use radiographic inspection to determine the quality of die attachment by examination of the shadow of the die-attach eutectic. However, whether any of these processes produce a cost-effective yield is questionable. Workmanship can best be inspected prior to the package seal, and it should be. Loose particles can be detected through acoustic-vibration techniques. The quality of the die attach-

Fig. 11 Integrated-circuit burnin system.

ment appears in the results of operating burnin in cases where degeneration is significant.   To date, the industry has found it impossible to describe objective and repeatable standards for radiographic inspection of integrated circuits.

Lot acceptance is an entire sequence removed from lot production and screening.
Lot acceptance is a follow-up measure using criteria that can be applied to the
percent defective allowed (PDA) at a particular acceptance gate within the screening
process. In this way, a normal range of allowed fallout within a lot may be established
before the entire lot is called nonacceptable.   Most lot-acceptance criteria are separate in themselves, and they are statistically applied to a sample drawn from each
lot or from a lot within a fixed interval of continuous production.   The most commonly applied lot-acceptance criteron is that of group A, or electrical performance.
Electrical acceptance tests are typically divided into separate subgroups by type,
such as ac, dc, or functional, and by temperature range.   Separate statistical criteria
are applied to each.   These separate criteria allow cost-effective testing, and they
simplify the methodologies necessary to rescreen lots in order to accept them after
initial failure.

Lot-acceptance tests have, as another major class, life and environmental tests.
These tests are normally called group B and group C tests and included extensive
exposure to a variety of environments within the capability of the part.   Although
these tests do nothing in themselves to improve the quality of the parts produced,
they do serve a major quality function.   This function is the establishment of objective criteria for the reliability and physical strength of the parts produced.   Including
a fixed set of such criteria obligates the manufacturer of these parts to define and
enforce the proper design rules, construction techniques, and process controls to build
a high-reliability part. Environmental acceptance tests are expensive in that they
destroy a large number of samples and that the laboratory effort required is extensive.   There is no method of guaranteeing the acceptability and quality of any product, including integrated circuits, other than to establish fixed and proper lot-acceptance criteria. The exact selection of criteria is dictated by the capabilities of the
parts selected and the requirements of the system in which they are to be used.

Assembly-level screening The screening techniques specified and explained in
previous sections are designed to simulate the application environment on the integrated circuit.   This removes the units which would normally fail during the early
life of the system.   However, it is sometimes appropriate to perform screening during
a specified predelivery operating mode of the overall ·system.   The overall system
is in one sense the best emulator of the actual system environment because it is

the actual system environment. The major shortcoming of using assembly-level screening is that it is sometimes quite difficult to operate an assembly at worst-case operating conditions. Also, extensive costs can be incurred in operating a complete and assembled system.

One solid advantage of assembly-level screening is that defects which are part of the assembly process or which can only be induced in components through assembly are now weeded out. A good example of this consideration involves the subdistribution of integrated circuits that crack or mechanically self-destruct during soldering or assembly. These are screened out through assembly-level screening. Assembly-level screening is normally conducted for intervals between 24 and 168 h and uses the full operating diagnostic hardware which is part of the equipment fabricated. Assembly-level screening is most appropriate to large systems, but the application of it in this case pays the penalty of tying up expensive hardware during the test period and complicating the debugging operation by introducing additional failures which must be removed.

The unique advantage of assembly-level screening is that it becomes an especially effective tool in improving the reliability of misapplied and marginally misapplied components. These components will have high failure rates only in the particular system under test, in a particular socket within the particular system. Assembly-level screening provides a vehicle to select the subpopulation of the overall available components which function in this particular socket without failure. If accurate data are collected during assembly-level screening, future misapplication of the integrated circuit is eliminated by establishing the pathway for their correction. The design engineer learns through analysis of assembly-level screening data which parts are reliability risks, either because of inherent defect in the part or because of part misapplication.

The final benefit is effectiveness in reducing short-term warranty costs associated with electronic systems. Assembly-level screening is most effective in weeding out early system failures.

Handling precautions   The integrated circuit is a sturdy, useful semiconductor component which must not be exposed to extremes of mechanical abuse, static discharge, improper currents or voltages, improper use of active chemicals, or operation in any mode beyond which it was originally designed. Like any creation of the design engineer, it must be used within the limits prescribed during the design-fabrication cycle.

The integrated circuit must be properly protected during test-handling insertion and all steps of manufacturing if it is to meet the reliability and performance capabilities designed and fabricated into it. Not that the integrated circuit is a frail component; it can be shot out of cannons, go into orbit, be irradiated, be stressed with voltage and current, and be operated at high-megahertz speeds and never suffer any damage. Yet it can fail when abused. That means that several simple precautions must be taken in handling it.

Many integrated circuits use high-impedance input stages, particularly metal oxide semiconductor devices. This class of parts is quite sensitive to the discharge of static electricity, which is easily created in the normal motion of operators and technicians. Fortunately, the precautions are very simple. First, parts should be stored in some fixture that shorts all the leads together. Common approaches use a piece of conductive plastic foam, aluminum foil pads, or shorting bars. Second, the circuits must be handled with care, preferably with technicians and assembly operators grounded by a ground strap during work operations. To reduce static electricity at the source, assembly and test areas should not be carpeted, since carpet generates static electricity. The moisture in the air should be regulated at a 40 or 50 percent humidity level. Soldering irons should be designed to prevent electric potential to be collected or generated at the tip during use. Grounded soldering tips may be required. Equipment containing these sensitive components should not be handled while wires are still hanging or loose. Partially completed assemblies should be safeguarded from open connections and devices that could be exposed to adjacent wiring. Bipolar, digital, and linear devices do not require precautions against static electricity that are this severe and do not require most of the special precautions previously mentioned.

Many types of hermetic-sealed integrated-circuit packages use glass-to-metal seals as the final package closure. Since the leads and lead frame go through this glass closure and since the glass which is used is not maximized for strength, great care must be taken in handling leads. Excessive bending of the lead or the transmittal of forces into the lead during a clipping operation can cause the glass to develop microcracks which expose the internal cavity of the device to the contamination of the outside environment.

A quite common practice is to preprocess the leads of integrated circuits before soldering. Two such processes which lead to problems are hot-tin dipping of the leads and chemical lead brightening. Hot-tin dipping must be precisely controlled to have a short dwell time, and the solder dip must not touch the body of the part or the lead frame above the shoulder of the lead. Lack of control exposes the part, in particular the glass-to-metal seal, to an excessive temperature change, causing degradation of the device internally or microcracking of the glass. The use of solder brights, materials that act as an acid etch, removes the oxidation which has occurred with storage on the lead surface. These materials do an excellent job in brightening the leads and improving the solderability, but it is most important that extensive, complete, and active cleanings follow these steps. Any residue of this material releases the plating or the plating materials that are used on the lead. This, in the presence of the ionic material from the solder flux, causes the growth of conductive salt and metal icicles between the leads. The end result is nonfunctionality of the device after system-life exposure.

In many cases, it is necessary to form the leads of integrated circuits into a pattern different from the placement of the leads on the integrated-circuit-pattern package. This lead forming is done to improve package density or to allow for proper placement of wire runs or printed circuits. The forming of integrated-circuit leads requires great care in the design of proper tools that minimize the strain during attachment of leads to the package. In cases where it is questionable whether any process step has damaged the hermetic-seal capability of the package, a good practice is to run sample parts through the process in question. The sample parts can then be submitted to a fine leak test, using either a krypton or helium test.

An additional precaution in using integrated circuits is to handle carefully all surfaces which are electrically connected to one of the terminals to make sure they are not inadvertently energized. It is quite common for metallized regions of the top and bottom ceramic dual-in-line packages to be connected to one terminal of a device. These surfaces may require protection with Mylar tape or other insulating material to remove the danger of such connection. In using integated circuits, it is essential that they not be inserted into a socket or equipment that is energized. It is essential that the ground pin be connected firmly before any other leads have voltage current applied to them. In many integrated-circuit designs, a ground pin is connected to a semiconductor substrate which must be reverse-biased from all other pins to prevent a large current flow across the substrate diode, which destroys the internal bonding. It is essential that integrated circuits be operated totally within the voltage-current ratings established by their manufacturer. For the most part, these ratings are derived with a safety factor for the capability of the part. However, a design engineer should not conclude that just because in a single case the device rating was exceeded without destroying it, this could be done again in future cases. The device capabilities have a normal distribution, and there is variance from unit to unit. For this reason, the safety factor between actual capability and the specification limit varies from device to device.

It is particularly important that logic circuits, such as the many families of transistor-transistor logic, be operated only at their nominal voltage rating. Although some manufacturer's specification sheets state that operation can occur at 7 and 8 V, the limitations of input voltage are now such that they cannot be run close to the higher values of the collect power-supply voltage, or V_{cc}. A circuit schematic for a typical transistor-transistor logic gate shows that the inputs are all emitters of transistors, and voltages at these inputs are directly applied as base-emitter reverse voltages. It is a documented fact that operation of transistors in the base-emitter breakdown mode causes degradation of the transistor beta and eventually impairs the functionality of the logic gate. Another misapplication of transistor-transistor logic gates is to use these circuits with a negative voltage supply, that is, connecting the V_{cc} terminal of the device to system ground and then connecting the device ground to a −5 V supply. This process, although at first glance consistent with the rules of good engineering, algebra, and logic, leads to a number of modes of operation that could be quite destructive to the device. The problem is created by depending on the power supply to reverse-bias the substrate diodes; secondary problems can also occur from ground-loop potentials being generated and causing an input of one of the devices to be biased at a potential higher than V_{cc}.

This is not the only common and destructive application of an integrated circuit. The complementary metal oxide semiconductor (CMOS) family is built on a structure of using n-channel and p-channel field-effect transistor switches which are set in series pairs between the source power-supply voltage V_{ss} and the drain power-supply voltage V_{DD}. If both these semiconductor switches are turned on at the same time, a short occurs between the power supplies, and large amounts of current can be consumed by the device. A destructive mode ensues. The design of such devices guarantees that this type of state cannot occur during normal operation. But this destructive mode can occur if a CMOS gate or circuit is powered up and one or more of the inputs is allowed to float. Therefore, all inputs must be returned to a fixed-potential level, either V_{ss} or V_{DD}, as appropriate. It is equally important when a package containing more than one functional circuit is used that the extra circuits in the package be properly biased to prevent this destructive mode.

A different but equally important set of handling procedures has been established for all plastic-packaged integrated circuits. Plastic has reduced the cost of packaging many integrated circuits by a significant amount, since properly applied and properly controlled plastic packaging can be both efficient and economical. But plastic has limited temperature capability and for this reason must be protected from the heat of soldering baths and soldering irons. A plastic package cannot give 100 percent protection to the integrated circuit against various contaminating liquids; so these parts must be sealed and stored in a relatively dry and uncontaminated atmosphere. When active cleaning baths are used as part of the assembly process, additional less active cleaning baths must follow to remove all traces of the active cleaning baths, which may become a source of destruction to the plastic integrated circuit. Most plastic packages have a small area on the end of the package through which a small metal zone is exposed. The metal zone is a continuation of a lead frame to which the integrated-circuit chip has been mounted. It is electrically connected to the circuit, and proper precautions must prevent bringing it into contact with any other potential.

A final precaution must be taken in troubleshooting. Integrated circuits, like most small semiconductor devices, are not capable of withstanding excessive flows of current or voltage when randomly applied to the device. The engineer, in troubleshooting a completed piece of hardware or in trying to verify whether or not a particular integrated circuit is functional, should avoid the use of simple but destructive test equipment, such as volt-ohmmeters and curve tracers. Without proper regulation, these instruments can generate currents or voltages beyond the breakdown voltages of many integrated circuits, forcing the integrated circuit into a destructive operating mode.

Parametric-test equipment Parametric test of integrated circuits requires a great deal of electronic sophistication. Integrated circuits have the complexity traditionally associated with electronic subsystems, but they do not have the advantage of being fabricated from components whose individual performance has been individually tested. As such, an integrated circuit is a complex network whose performance must be measured over a range of electrical stimuli representing the full range of practical part application. In recent years, a major evolution of specialized, sophisticated test equipment has been brought forward. This equipment varies from simple functional test sets programmed by slide switches, to complex computer-controlled automatic systems with test rates and sophistication suited to the most complex of circuits and the most sophisticated of users. Figure 12 shows typical application of a computer-controlled test system in the facility of a systems manufacturer. The machine shown is used for go no-go receiving inspection, characterization and test

Fig. 12 Integrated-circuit test system. The Teradyne J258 circuit test system. (*Teradyne, Inc.*)

of custom circuits, and engineering studies on circuits. Through the use of the wafer probe station and the electrical probe, integrated-circuit chips may be tested.

Computer-controlled systems have risen to the forefront of integrated-circuit test technology because of their flexibility. In a computer-controlled system, an artificial language instructs a series of programmable stimulus sources and a smaller group of measurement instruments. These instruments are coupled together, and to the pins of the device, by a matrix of relays and solid-state switches. Instructions contained in the computer language configure the switching matrix, apply the stimuli, direct the measuring equipment, and finally set the limits, if any, for the measurement. Use of a computer-controlled test system requires an individual programming effort for each type of integrated circuit used. This programming effort is, in reality, the reduction of parametric measurements to the artificial language of the automatic test set. Automatic test sets operate in a pulse mode, thereby reducing dissipation of the circuits, and for the most part, are capable of measuring only dc parameters. Some systems, however, have incorporated special test heads which use either sampling-scope techniques or single-shot measuring techniques to measure ac parameters of interest.

Many integrated circuits are commonly tested in an exerciser mode, that is, with the test set operating the circuit through its various logic states at a speed corresponding to the operating speed in the circuit application. Testing in this mode simulates both ac and dc testing and combines them in one high-speed operation. This technique is most appropriate for large-scale memories and random logic, and it bears a hard correlation to direct application of large-scale circuits in electronic systems. Table 7 lists test-system manufacturers.

There is no such thing as a universal electronic test set, since each set is optimized

TABLE 7   Integrated-Circuit Test-Equipment Manufacturers

Manufacturer	Address	Capability
AAI, Inc.	Box 6767, Baltimore, Md. 21204	AC/DC test
Adar Associates	85 Bolton Avenue, Cambridge, Mass. 02140	Exerciser
Daymarc Corp.	40 Bear Hill Road, Waltham, Mass. 02154	Handler
Fairchild Systems Technology	974 East Arques Avenue, Sunnyvale, Calif. 94086	Complete
General Radio Co.	300 Baker Avenue, Concord, Mass. 07142	DC test
Hewlett Packard	1501 Page Mill Road, Palo Alto, Calif. 94304	AC/DC test
Tektronix, Inc.	Box 500, Beaverton, Ore. 97005	AC/DC test
Teledyne TAC	10 Forbes Road, Woburn, Mass. 01801	Handlers
Temptronics	40 Glenn Avenue, Newton, Mass. 02159	Temperature probes
Teradyne, Inc.	183 Essex Street, Boston, Mass. 02111	Complete
Texas Instruments	13500 N. Central Expressway, Dallas, Tex. 75231	DC test

for a particular group of applications, circuits, and test rates. For example, to make full use of the capability of an automated test set also requires automated component handlers. If an automated test set can test a part in 100 ms, the majority of its capability is wasted if it has to wait for an operator to insert an integrated circuit manually into a socket, an operation taking 2 s. However, automatic test sets which are computer-controlled have the further advantage of computer use for both reducing the data generated during the test program and running the tester through specified subroutines that measure such step-duration parameters as logic thresholds. Less complex automatic test equipment uses sequential memories, such as disk files and drum memories or even punched paper tape, to store the test program for a given part. Machines without computer control are unable to respond to complex instructions, such as branch and calculate deltas.

Three primary types of non-dc test heads are available on automatic test sets. The first of these uses a sampling technique, such as that available in a sampling scope, to make precision measurements of small time intervals. Such techniques require a train of 400 to 1,000 pulses for each measurement. These measurements require long test intervals, and they can result in overheating of the components. The long test interval makes this technique not well suited to complex devices. A second class of techniques has been developed to use single-shot pulse measurements. These techniques involve gathering full-time data associated with a pulse during a single event. The overall accuracies must be evaluated in such systems. The third major class of non-dc test sets are designed to measure analog networks. These sets can be minor modifications of dc automatic test sets which can be appropriately applied to standard circuits, such as operational amplifiers and communication circuits. Other more complex systems incorporate rf signal generators, vector impedance meters, and other extremely complex equipment into a large, fully controlled test array. This kind of set in use to test complex custom linear networks is shown in Fig. 13.

The measurement of integrated-circuit parameters is quite often related to specific temperatures. Most testing is performed at a room-temperature ambient of 25°C. Testing integrated circuits at temperatures other than room temperature again requires special techniques. The two most common techniques are to use a temperature-controlled handler and a temperature-controlled probe. Either technique allows a controlled ambient temperature to be brought directly to the surface of the test head of an automatic set. The temperature-controlled probe is illustrated in Fig. 14.

Fig. 13 Computer-controlled test system capable of complex ac measurements.

Fig. 14 Temperature-controlled probe system (*Temptronics, Inc.*)

All test intervals for integrated circuits must be minimized to reduce dissipation and not significantly change the chip temperature from the impressed ambient. It is essential not to apply extraneous conditions during the test of integrated circuits. The use of ohmmeters and improperly designed custom testers has resulted in large numbers of good integrated circuits being destroyed through test. The integrated circuit is simply incapable of surviving misapplication during the test interval. Full consideration of applied biases and loads on integrated circuits must be made before beginning any test series.

Failure analysis   The integrated-circuit industry has spawned requirements that have advanced other major areas of technology.   One of them is the field of failure-analysis technology.   This scientific discipline was required to gather together elements of the sciences of metallurgy, material science, electron microscopy, photography, chemistry, and circuit theory.   These disciplines, applied together, were used to generate definitive data regarding the cause of failure of integrated circuits, either during reliability testing or during application in the system.

In the earlier days, much failure analysis was simply related to cutting open the package and looking inside.   In the case of integrated circuits, the inside is now microscopic.   To look inside requires the ability to examine carefully the submicroscopic physical attributes of the circuits.   And modern analysis of integrated circuits

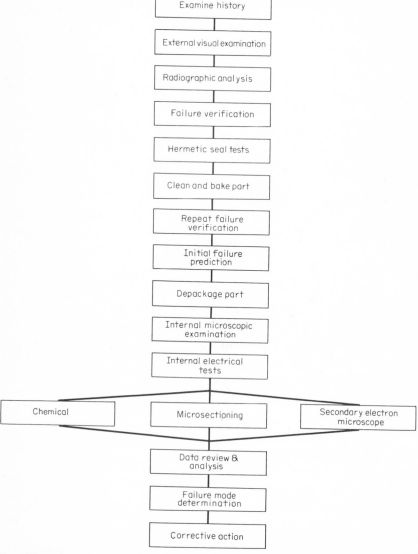

Fig. 15   Failure-analysis flow chart.   (*Chart adapted from procedure developed by E. Holtgrefe of the Martin Marietta Physics of Failure Laboratory.*)

must go into sufficient detail to separate obvious surface failures from the more esoteric hidden failures which caused them.   Figure 15 describes a typical failure-analysis sequence.   The sequence is structured to maintain all captured evidence, while pursuing the necessary steps.   Figure 16 shows the failure-analysis facility of a typical sophisticated user of circuits.

Fig. 16 Photograph of a complete failure-analysis laboratory. (*Martin Marietta Aerospace, Orlando, Fla.*)

It is essential in the application of failure-analysis techniques that great care be taken to isolate the cause of failure, relating it to both the mode of failure in the system and the history of the integrated circuit.   Too often, failure analysis is directed at finding all anomalies within the integrated-circuit structure without relating these to the failure.   By concentrating on the single anomaly which is coupled to the mode of failure, it can be detemined whether the failure was designed into the circuit, latent in the individual chip, or caused by some definable external stimulus.

The techniques of failure analysis have been successfully applied at stages earlier than failure of the first sample.   This technique, known as attributes analysis, is a detailed physical and electrical characterization of a controlled product that establishes a baseline to which all pending process and material changes are referenced. By performing this kind of analysis upon receipt of integrated circuits, the user gains absolute knowledge of any changes and/or modifications to the product that have occurred since he last procured and qualified it.   Attributes analysis can be further supplemented by using its information as the strengths and weaknesses of the construction of the device.   The information gained from attributes analysis, coupled with stress analysis, can be directed at environmental test methods to clarify the projected weaknesses of the part and to impose the necessary corrective action.

Techniques such as failure analysis, attributes analysis, and stress analysis are useful only when coupled with intensive communication between the user of the circuit and the manufacturer.   It is of only academic interest to isolate the cause of failure if the cause is not to be eliminated by changes in manufacture, test, or inspection.

Chapter **2**

Digital Integrated Circuits

SAUL ZATZ
Martin Marietta Aerospace, Orlando, Florida

DIGITAL INTEGRATED CIRCUITS

The first integrated circuits to be fabricated were digital integrated circuits. Today, they dominate the integrated-circuit field and are most frequently used. This is a natural consequence of the relative ease with which digital circuits are fabricated. Much of the large requirement for digital integrated circuits stems from the present trend to convert large amounts of analog electrical equipment to digital equipment. In this way, the greater availability and lower expense of the digital integrated circuit have spawned the requirement for larger production. The inputs and outputs of digital integrated circuits are designed to relate particular parametric values of voltage and current to particular logic states. Most digital integrated circuits are two-state devices; inputs and outputs correspond to the presence or absence of a binary digit.

Digital integrated circuits can be grouped together to fabricate a hierarchy of progressively more complex logic circuits which culminate in an overall logic network, the logic system. The most common example of this kind of completely digital system is the electronic computer.   All logic circuits used in a system must interface to each other to permit the transfer of logic state data from one circuit to another without the loss of intelligence.

Digital Families—A Comparison

Digital-logic families existed prior to the invention and production of digital integrated circuits. These families were composed of standard circuit arrangements which were repeated in rubber-stamp fashion throughout the design of early electronic computers. The designers of these early computers first partitioned their system into a block diagram of commonly repeated functional blocks, and then they set out to design each of these blocks and to optimize its performance with respect to such system parameters as speed, power, and noise immunity. As technology for fabrication of digital integrated circuits became available, the technique of designing functional circuit blocks and repeating their application was extended into a concept of designing a digital-integrated-circuit functional block which would then be batch-fabricated in large numbers and reapplied to fill the multiple requirements.

The first digital family produced was resistor-transistor logic (RTL). RTL was chosen for fabrication as the first family of digital integrated circuits because of its simplicity. This simplicity was well suited to the early limitations of integrated-circuit-fabrication capability.

Diode-transistor logic (DTL) was the second family of digital integrated circuits to be fabricated. Although more complex to fabricate, diode-transistor logic had improved speed, power, and stability characteristics. Complementary transistor logic was the first of the high-speed families to gain commercial acceptance, and it was used in the Burroughs B-3500 computer family. Further evolution of digital circuits initially extended in the direction of modifications to basic cell types to produce the optimum logic cell. This evolution and extended engineering development produced a broad range of logic based on the transistor-transistor logic (TTL) cell. In the transistor-transistor logic, the diode inputs of the DTL logic cell are replaced by multiple emitters on an input transistor. The multiple-emitter transistor structure is unique to integrated-circuit technology and has no analogous function in the family of discrete semiconductor devices. The evolution and development of transistor-transistor logic has seen variations in the basic cell, which have included the use of differing output drive structures, the use of internal Darlington transistors, the use of smaller geometries for higher speeds, and even the use of a variety of resistor values to optimize speed or power for a particular range of applications. The final extension of technologies applied to transistor-transistor logic has been the inclusion of Schottky-diode clamps which serve to keep the saturated transistors in the logic cells from going into deep saturation. These Schottky clamps are connected in the manner of Baker clamps, and they significantly reduce the amount of charge stored in the transistor junction. This reduction in stored charge reduces the switching time of the transistor-transistor logic cell without increasing its power consumption.

For higher logic speeds, emitter-coupled logic (ECL), which is also called current-mode logic, has been used as an alternative to the previously mentioned logic families. However, ECL has gained only limited acceptance. Its use has been restricted, properly, to applications that require its high operating speed, and it is typically more difficult to use than any of the families of simple saturated logic. But its use can yield higher system performance.

Complementary metal-oxide semiconductor (CMOS) devices present an alternative to the saturated bipolar logic for applications requiring less speed or greatly reduced power consumption. Various other types of MOS circuitry are available for use as digital integrated circuits, but their use has been sufficiently restricted that a complete and usable family of MOS integrated circuits has not been provided.

Throughout the remainder of this chapter, a detailed analysis of available digital integrated circuits is presented. However, it is recommended that system usage be restricted to three particular families. These families are the 4000 CMOS family, the 5400 transistor-transistor logic family, and the 10,000 emitter-coupled logic family. These three families have gained broad acceptance in commercial, industrial, and military systems because they represent logical compromises and trade-offs in design performance. No one circuit or circuit family may meet all the requirements of any possible system, but a particular circuit family should represent a viable set of practical trade-offs for the system under consideration. These three families do indeed represent three particular sets of trade-offs that are appropriate for three distinct types of system application.

The 5400 TTL family is the most commonly used family of integrated circuits today. The exceptionally broad range of available logic functions in the 5400 family make it most useful for building large, relatively high-speed digital systems. These circuits are economical, and they are complemented by high- and low-speed versions and by the inclusion of Schotty-clamp circuits which make available, at higher cost, high-speed circuits that meet subsystem requirements for high speed yet maintain the same logic levels as the rest of the family. 5400 TTL logic is simply used because it requires only one power supply. It does not require exotic techniques for circuit interconnection, such as matched-impedance transmission lines. It includes circuit subelements which provide sufficient current drive to operate external devices, such as clamps and relays. In addition, 5400 TTL logic is the most economical family of logic offered on the market today. Its use is also favored in military systems through the availability of circuits qualified to MIL-M-38510.

The ECL 10,000 family represents the evolution of emitter-coupled logic to a form which can give logic delays of approximately 2 ns without requiring highly controlled transmission lines. ECL 10,000 is an expanding family of logic circuits, but its application is properly limited to those systems requiring operating frequency in excess of 10 MHz. The ECL 10,000 family can be readily interfaced with other logic types.

The 4000 family complementary metal-oxide semiconductor devices may be operated at speeds four or five times slower than the transistor-transistor logic family, but its power consumption is approximately 10,000 times less than TTL. Furthermore, CMOS logic circuits may operate over a range of power-supply voltages from 3 to 15 V, and they tolerate in operation a great degree of nonregulation and ripple in the power supply used. These characteristics of complementary metal-oxide semiconductor circuits make them especially appropriate for portable battery-powered equipment and other noncomplex systems. These three preferred families are described among other logic families in Table 1.

Direct-coupled transistor logic and resistor-transistor logic Historically, the first form of digital logic offered as an integrated circuit was direct-coupled transistor logic (DCTL). A basic DCTL NOR-gate schematic is illustrated in Fig. 1. Although this structure is fairly simple to fabricate, it is not fully suitable for use in logical systems. Yield requires that all the transistors used in the system have similar characteristics, and this is not readily attained. In a practical system, fanout problems occur because of current hogging at the output node when one gate is used to drive several others. The nonsuitability of this logic form has been sufficient to prevent its popular acceptance.

A slight modification of this scheme is called a resistor-transistor logic. A basic RTL gate structure is illustrated in Fig. 2. The base resistors on the input serve to reduce current hogging, making the RTL NOR gate usable in logic systems. Although RTL was the first commonly accepted form of digital logic, it is still being used today in modern systems. It has the advantage of being able to function with only a 3-V power supply, and it is relatively immune to high-frequency noise in logic systems. The full range of available RTL circuits is presented in Table 2. The RTL gate relies on having resistors track each other with a very tight tolerance of 1 or 2 percent, although the absolute value of these resistors can vary 10 to 20 percent. These tolerances make them suitable for monolithic fabrication. However, RTL circuits are not recommended for use in any newly designed system, as they offer neither operating nor economic advantages.

Diode-transistor logic—the 930 family The next evolution in digital integrated circuits was the perfection of the diode-transistor logic family. A schematic of the

TABLE 1 Functional Comparison of Digital Integrated-Circuit Families

Family	Basis gate	Gate delay, nS	Flip-flop toggle rate, MHz	Power-dissipation per gate, mW (dc only)	No. of small-scale integration circuits	No. of medium-scale integration circuits	Table	Comment
Direct-coupled logic	AND							Not available
Resistor-transistor logic (700)	NOR	12	15	18	34	9	2	
Milliwatt RTL	NOR	27	3	6.5	31	2	3	
Diode-transistor logic (930)	NAND	30	30	12	44	1	4	
Complementary transistor logic (956)	AND-OR	4.5	30	40	12	14	5	
CTL II (9856)	AND-OR	4.0	50	40	12	4	6	
High-threshold logic	NAND	110	3	28	27		7	Industrial
Suhl I TTL	NAND	10	30	15	23		8	
Suhl II TTL	NAND	6	70	22	20	12 Compatible		
9000 TTL	NAND	8	50	10	18			
9L TTL	NAND	20	10	2	5			
5400 TTL	NAND	10	35	10	81	183	10, 11	Preferred
54H high-speed TTL	NAND	6	50	22	32	2	12, 13	
54L low-power TTL	NAND	33	3	1	30	42	14, 15	
54S Schottky TTL	NAND	3	125	19	26	50	16, 17	
54LS low-power Schottky	NAND	10	45	2	40	65	18, 19	
4000 CMOS	NOR	25	5	10	50	60	26, 27, 28	Preferred
54C CMOS	NOR	25	5	10	15	30	30	
MECL I	OR/NOR	8	30	31	21		20	
MECL II	OR/NOR	4	70	22	41	12	21	
MECL III	OR/NOR	1	500	60	20	9	22	
ECL 10K	OR/NOR	2	125	25	45	60	25	Preferred
ECL 95K	OR/NOR	2	250	25	19	1	24	
ECL 2500	OR/NOR	2.5	150	36	25	3	23	

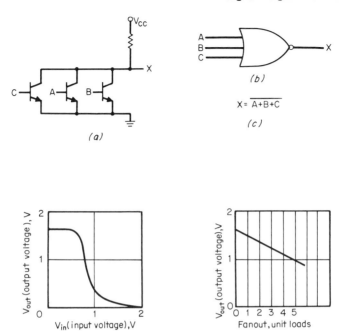

$$X = \overline{A+B+C}$$

(c)

Fig. 1 Characterization of the direct-coupled logic family. (*a*) Circuit schematic. (*b*) Logic diagram. (*c*) Logic equation. (*d*) Transfer curve, a plot of output voltage vs. input voltage. (*e*) Fanout curve, a plot of output voltage vs. fanout in unit loads.

basic NAND gate is shown in Fig. 3. In this circuit, the inputs are a group of diodes connected with their anodes at a common node. High-level voltage applied to all anode inputs denies a path for the current that would otherwise flow through the resistor to the power-supply bus. Therefore, the current is available to the output transistor, and it generates an inverter signal at the output node. As a result, the basic function is a NAND gate having moderately high speed and capable of operating from a 5- or 6-V power supply. The transfer function for DTL is illustrated in Fig. 3, and the full range of available DTL families is presented in Table 3. The most commonly accepted and broad-based family of DTL circuits is the 930 family introduced by Fairchild. This family consists of a range of basic NAND gates, OR gates, AND gates, NOR gates, latches, RS and JK flip-flops, and a small amount of MSI (medium-scale integration), which includes a divide-by-16 counter, a decade counter, and a monostable multivibrator. This range of circuits allows for complete fabrication of computer subsystems.

A typical DTL gate can drive a maximum fanout of 10, which applies some limitation to its use in complex logic trees. However, DTL has the advantage of wire OR-ing; that is, multiple outputs can be tied together to perform the OR function at the output node. Only gates which do not have a collector resistor may be used for wire OR-ing. The table of the family indicates which circuits may be used in this manner. This flip-flop is actually composed of two DTL NAND gates bowtied together. One of the most limiting factors to DTL has been the lack of a reliable one-shot multivibrator, but the 9601 monostable multivibrator is commonly used to perform a one-shot function with DTL. DTL may be intermixed with TTL, provided careful consideration is made to fanout and all requirements are met before OR tying.

Complementary transistor logic CTL was the first family of integrated digital logic to be used on a production commercial computer. It was designed to take

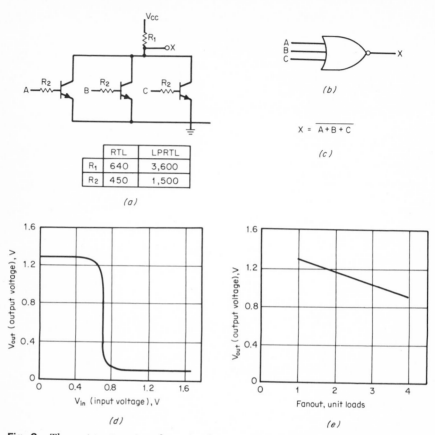

	RTL	LPRTL
R_1	640	3,600
R_2	450	1,500

(a)

(b)

$$X = \overline{A + B + C}$$

(c)

(d)

(e)

Fig. 2   The resistor-transistor logic family. NOR-gate characterization.   (a) Circuit schematic.   (b) Logic diagram.   (c) Logic equation.   (d) Transfer curve, a plot of output voltage vs. input voltage.   (e) Fanout curve, a plot of output voltage vs. fanout in unit loads.

advantage of easily fabricated substrate pnp transistors.   By using these transistors in the input stage, a very high-speed gate function was realized, and the typical CTL gate had a delay time of about 4 ns.   The simplicity and speed of the circuit are indicated by the schematic diagram of the basic gate in Fig. 4.   The CTL gate was an AND gate structure with an output that could be OR tied to other outputs without disrupting the structure.   One of the major limitations of this gate structure was that it did not have an integral logic-level reference.   In reality, it was a simple nearly-unity-gain amplifier with multiple inputs.   When a series of gates was connected to form a logic tree, an offset would occur in the logic levels, which would degrade the usable noise margin.   In order to allow structuring of large logic trees, two restoring elements, a 952 inverter and a 956 AND buffer, were added to the family.   These two circuits were not actually CTL circuits, but saturated logic elements that completed the functionality of the family.   These two additional circuits, however, were much slower than the high-speed gates in the family, and they tended to increase the average gate delay associated with a logic trace.   The flip-flops which were also added to the family, a 967 and a 957, were also quite slow compared with the basic gate.

The combination of the high-speed gates, slow restorers, and even slower flip-flops gave an average system logic-level delay of 10 or 12 ns.   Although this delay is slow compared with present standards of emitter-coupled logic, it was in 1968 a

TABLE 2   The 700 Family of Resistor-Transistor Logic, Complete Selection Guide of Circuit Types

Function	Circuit form	Type No. 0 to 70°C	Type No. −55 to +125°C	Total delay, ns	Power dissipation, mW	Fanout (unit loads)	Package*
NOR gate	3-input	703	903	12	18	5	C-8, F-10
NOR gate	4-input	707	907	12	19	5	C-8, F-10
NOR gate	Dual 2-input	714	914	12	33	5	C-8, F-10
NOR gate	Dual 3-input	715	915	12	35	5	C-10, F-10, P-14
NOR gate	Quad 2-input	724	924	12	65	5	F-14, P-14
NOR gate	Dual 4-input	725	925	12	37	5	F-14, P-14
NOR gate	5-input	729	929	12	20	5	C-8, F-10
NOR gate	Triple 3-input	792	992	12	53	5	F-14, P-14
OR gate	Quad 2-input	9715		14	64	5	P-14
Exclusive OR gate	Quad 2-input	771	971	12	28	5	F-14, P-14
AND	Quad 2-input	9713		28	100	5	P-14
NAND	Quad 2-input	9714		14	145	5	P-14
Buffer		700	900	20	37	25	C-8, F-14
Buffer	Dual 3-input	788	988	24	100	25	F-14, P-14
Buffer	Dual	799		15	70	25	C-10, F-10, P-14
Flip-flop inverter buffer	JK-1-2	779			130		P-14
Inventer	Quad	727	927	12	59	5	C-10, F-10
Inventer	Hex	789	989	12	72	5	F-14, P-14
Flip-flop	RS	702	902	14	32	4	C-8
Flip-flop	JK	723	916	30	85	2	C-8, F-10, P-14
Flip-flop	JK	726	926	35	93	5	C-10, F-10, P-14
Flip-flop	JK	774	974	35	93	5	C-8
Flip-flop	Dual JK	790	990	35	116	3	F-14, P-14
Flip-flop	Dual JK	791	991	40	175	5	F-14, P-14
Flip-flop	Dual JK	9702		35	170	3	P-14
Expander	Quad 2-input	785	985			12	F-14, P-14
Expander	Dual 4-input	786	986			12	F-14, P-14
	Hex	9719	9919	12			F-14, P-14
Schmitt trigger	Quad	9709		30	95	5	P-14
Counter adapter		700	900	22	80	5	C-8, F-10
Counter	Decade up	780			250	3	P-14
Adder	Half	704	904	14	65	5	C-8, F-10
Adder	Dual half	775	975	20	120	5	F-14, P-14
Adder	Dual full	796	996	60	225	5	F-14, P-14
Adder	4-bit	9704		125	265	2	P-16
Subtractor	Dual full	797	997	60	225	5	F-14, P-14
Shift register	Half	705	905	22	75	4	C-8, F-10
Shift register	Half without inverter	706	906	22	52	4	C-8, F-10
Shift register	Dual half	783	983	22	140	4	F-14, P-14
Shift register	Dual half	784	984	22	100	4	F-14, P-14
Shift register	Serial-parallel	794		55	225	5	P-14
Data selector	Dual 4-channel	9701		25	100	5	P-16
Data multiplexer	Dual 4-channel	9707		25	150	5	P-16

* Package entry letter is form factor: F = flatpack, D = hermetic dual-in-line, P = plastic dual-in-line, C = round can, S = studded flatpack.   Package entry number is terminal count.

high speed for a practical digital integrated-circuit family. This full CTL family is described in Table 4. CTL II followed the initial family by about 4 years and included the same basic gate structure with a modification in the internal clamp to reduce the tendency toward self-oscillation. Two primary additions were included in CTL II, the inclusion of high-speed flip-flops and restoration elements, and an expansion of the family to include some MSI functions. The inclusion of the high-speed elements was a major improvement to the family, since they could reduce the average delay associated with the level of gating in a fabricated system to approximately 5 ns. This delay was gained while maintaining a circuit, like CTL I, with limited requirements for transmission lines and other techniques normally required with systems of high-speed logic.

The inclusion of MSI elements in the CTL II family was quite limited, since the family was not well suited for fabrication of MSI elements. The basic CTL gates consumed too much power for use as on-chip gates in large arrays. The use of low-power CTL and TTL gates on chip was required to make most of the MSI

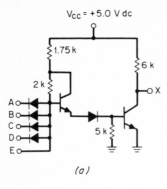

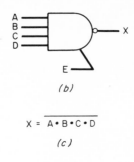

$$X = \overline{A \cdot B \cdot C \cdot D}$$

(c)

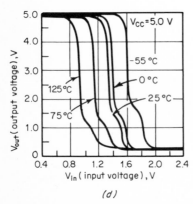

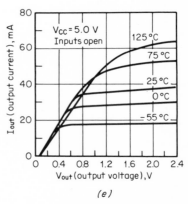

Fig. 3 Characterization of the 930 family of diode transistor logic. NAND-gate characterization. (*a*) Circuit schematic. (*b*) Logic diagram. (*c*) Logic equation. (*d*) Transfer curve, a plot of output voltage vs. input voltage. (*e*) Fanout curve, a plot of output voltage vs. output current.

practical. Using the mixture of gate structures on a chip required provisions within the chip for the conversion of logic levels to make the circuit compatible with the family of CTL circuits. It is strange that the CTL family did not gather more popular acceptance than it did. It represented a uniquely clever approach to implementing, at reasonably low cost, a high-speed digital system. Analysis indicates that the highest potential speed obtainable in digital logic will come from structures that use nonsaturated complementary logic, and further development in this area will be produced by the research now going on.

High-threshold logic All integrated-circuit logic families are directed at applications where there are low levels of electrical noise. As such, these circuits operate with 3- or 5-V power supplies and have noise margins in the range of 1 V. Such logic structures are well suited to applications in electronic computers and other tightly controlled electronic systems. There is, however, a need for logic circuits to function in high-electrical-noise environments, such as those present in numerically controlled machinery and other industrial controls. The DTL and TTL families have been modified for use in these types of environments by the inclusion of a Zener diode in the input network, which adds the Zener voltage of 5 or 6 V to the logic threshold. Further modification of these circuits has been made to minimize their response to short-duration noise spikes. This modification has increased the logic delay, but extended delays are rarely a consideration in industrial equipment.

TABLE 3   The 930 Family of Diode-Transistor logic, Complete Selection Guide of Circuit Types

Function	Circuit form	Type No. 0 to 70°C	Type No. -55 to +125°C	Total delay, ns	Power dissipation, mW	Fanout (unit loads)	Package‡
NAND gate*	Dual 4-input	830	930	30	22	8	D-14, P-14, F-14
NAND gate*	Dual 4-input	832	932	35	25	25	D-14, P-14, F-14
NAND gate	Dual 4-input	844	944	30	65	27	D-14, P-14, F-14
NAND gate	Quad 2-input	846	946	30	44	8	D-14, P-14, F-14
NAND gate†	Quad 2-input	849	949	25	66	7	D-14, P-14, F-14
NAND gate	Quad 2-input	857	957	35	170	25	D-14, P-14, F-14
NAND gate	Quad 2-input	858	958	30	130	27	D-14, P-14, F-14
NAND gate*,†	Dual 4-input	861	961	25	33	7	D-14, P-14, F-14
NAND gate	Triple 3-input	862	962	30	33	8	D-14, P-14, F-14
NAND gate†	Triple 3-input	863	963	25	50	7	D-14, P-14, F-14
NAND gate	Dual 6-input	1800	1900	30	22	8	D-14, P-14, F-14
NAND gate†	Dual 5-input	1801	1901	25	33	7	D-14, P-14, F-14
NAND gate*,†	8-input	1803	1903	25	16.5	7	D-14, P-14, F-14
NAND gate	10-input	1804	1904	30	11	8	D-14, P-14, F-14
NAND gate†	10-input	1805	1905	25	16.5	7	D-14, P-14, F-14
NAND gate	Quad 2-input	1818	1918	30	32	8	D-14, P-14, F-14
AND gate†	Quad 2-input	1807	1907	30	85	7	D-14, P-14, F-14
AND gate	Quad 2-input	1806	1906	35	72	8	D-14, P-14, F-14
OR gate	Quad 2-input	1808	1908	35	97	8	D-14, P-14, F-14
OR gate†	Quad 2-input	1809	1909	30	115	7	D-14, P-14, F-14
NOR gate	Quad 2-input	1810	1910	30	60	8	D-14, P-14, F-14
NOR gate†	Quad 2-input	1811	1911	25	72	7	D-14, P-14, F-14
Exclusive OR gate	Quad 2-input	1812	1912	40	120	8	D-14, P-14, F-14
Expander	Dual 4-input	833	933				D-14, P-14, F-14
Inverter	Hex	834	934	30	66	8	D-14, P-14, F-14
Inverter	Hex without R	835	935	30	42	8	D-14, P-14, F-14
Inverter	Hex	836	936	30	66	8	D-14, P-14, F-14
Inverter	Hex	837	937	25	90	7	D-14, P-14, F-14
Inverter	Hex without diodes	840	940	30	66	8	D-14, P-14, F-14
Inverter	Hex without diode, R	841	941	30	42	8	D-14, P-14, F-14
Flip-flop	Clocked	831	931	40	55	7	D-14, P-14, F-14
Flip-flop	Type D + AOI	842	942	40	110	10	D-14, P-14, F-14
Flip-flop	Clocked	845	945	40	60	11	D-14, P-14, F-14
Flip-flop	Clocked	848	948	40	70	10	D-14, P-14, F-14
Flip-flop	Dual JK	852	952	40	120	11	D-14, P-14, F-14
Flip-flop	Dual JK	853	953	40	120	11	D-14, P-14, F-14
Flip-flop†	Dual JK	855	955	40	140	10	D-14, P-14, F-14
Flip-flop†	Dual JK	856	956	40	140	10	D-14, P-14, F-14
Flip-flop	Para gated	1815	1915	40	65	11	D-14, P-14, F-14
Flip-flop	Para gated	1816	1916	40	75	10	D-14, P-14, F-14
Binary	Pulse trig	850	950	15	50	9	D-14, P-14, F-14
Multivibrator	Monstable	851	951	40	30	10	D-14, P-14, F-14
Quad latch		1813	1913	35	220	7	D-16, P-16
Quad latch		1814	1914	35	220	7	D-16, P-16
Counter	Decade	838	938	30 MHz	150	8	D-14, P-14, F-14
Counter	4-bit binary	839	939	30 MHz	150	8	D-14, P-14, F-14

* Expandable.
† 2 kΩ pullup.
‡ Package entry letter is form factor: F = flatpack, D = hermetic dual-in-line, P = plastic dual-in-line, C = round can, S = studded flatpack. Package entry number is terminal count.

Table 5 presents the available high-threshold logic circuits, and Fig. 5 illustrates the schematics for a high-threshold DTL NAND gate. Use of high-threshold logic should be limited only to specific cases where standard logic cannot be used. Typical gate delays of 50 or 70 ns are associated with high-threshold logic, making it unsuitable for most computer applications.

Transistor-transistor logic Transistor-transistor logic has become the most widely used form of digital integrated circuitry. TTL includes a great number of separately identifiable families, all of which have certain characteristics in common. Most of the logic circuits use a similar gate structure. Schematics, with the exception of the 54LS, have in common that all inputs of the logic gate are multiple emitters of the input transistor. One circuit type uses Schottky-diode inputs in place of the

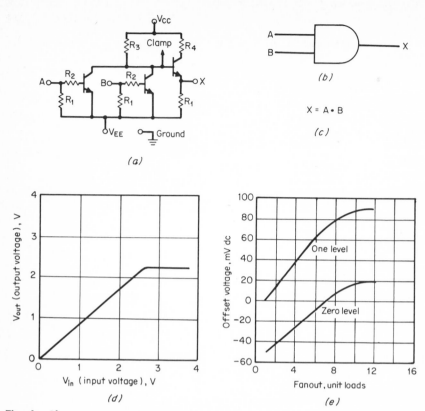

Fig. 4 Characterization of the 956 family of complementary transistor logic. AND-gate characterization. (*a*) Circuit schematic. (*b*) Logic diagram. (*c*) Logic equation. (*d*) Transfer curve, a plot of output voltage vs. input voltage. (*e*) Fanout curve, a plot of offset voltage vs. fanout in unit loads.

TABLE 4 The 956 Family of Complementary Transistor Logic, Complete Selection Guide of Circuit Types

Function	Circuit form	Type No.	Total delay, ns	Power dissipation, mW	Fanout (unit loads)	Package‡
AND-OR gate	Triple 2-2-3 input	953	4.5	195	11	D-14
AND-OR gate	Dual 4-input	954	4.5	162	11	D-14
AND-OR gate	8-input	955	4.5	135	11	D-14
AND-OR gate	Triple 3-3-1 input	964	4.5	195	11	D-14
AND-OR gate	Quad 1-input	965	4.5	216	11	D-14
AND-OR gate*	Quad 2-input	966	4.5	235	11	D-14
AND-OR gate*	Quad 2-input	971	4.5	220	11	D-14
AND-OR gate*	Quad 2-input	972	4.5	95	11	D-14
NOR†	Dual 2-input	952	7.0	166	12	D-14
AND†	Dual 2-input	956	12.0	280	25	D-14
Flip-flop	JK	967	16	365	12	D-14
Flip-flop	RS	957	20		9	D-14
Memory	8-bit	9030	25		3	D-14

  * No-load input.
  † Restored output.
  ‡ Package entry letter is form factor: F = flatpack, D = hermetic dual-in-line, P = plastic dual-in-line, C = round can, S = studded flatpack. Package entry number is terminal count.

TABLE 5   The 660 Family of High-Threshold Logic, Complete Selection Guide of Circuit Types

Function	Circuit form	Type No.	Fanout (unit loads)	Total delay, ns	Power dissipa-tion, mW	Package‡
NAND gate	Dual 4-input	660	10	110	57	D-14, P-14
NAND gate*	Dual 4-input	661	10	125	57	D-14, P-14
NAND gate*	Quad 2-input	668	10	125	114	D-14, P-14
NAND gate*	Triple 3-input	670	10	125	86	D-14, P-14
NAND gate	Triple 3-input	671	10	110	86	D-14, P-14
NAND gate	Quad 2-input	672	10	110	114	D-14, P-14
Expander	Dual 4-input	669				D-14, P-14
AND-OR invert gate	Dual 2-input	673	10	110	105	D-14, P-14
AND-OR invert gate	Dual 2-input	674	10	125	105	D-14, P-14
Inverter	Hex with strobe	677	10	110	170	D-16, P-16
Inverter†	Hex with strobe	678	10	125	144	D-16, P-16
Inverter	Hex	680	10	110	170	D-14, P-14
Inverter†	Hex	681	10	125	144	D-14, P-14
Inverter†	Hex high voltage	689	10	150	114	D-14, P-14
Inverter	Hex	690	10	150	114	D-14, P-14
Exclusive OR gate	Quad 2-input	683	10		380	D-14, P-14
Flip-flop	Dual JK	663	9	3.0 MHz	200	D-14, P-14
Flip-flop	Master-slave RS	664	8	3.0 MHz	160	D-14, P-14
Flip-flop	Dual JK	688	10	2.5 MHz	375	D-16, P-16
Multivibrator	Dual monostable	667	10	140	240	D-14, P-14
Pulse stretcher	Dual	675	10	150	180	D-14, P-14
Level translator	Triple HTL to TTL	665	5 TTL	40	85	D-14, P-14
Level translator	Triple TTL to HTL	666	10	75	105	D-14, P-14
Latch	Quad	682	10	250	375	D-16, P-16
Transceiver	Dual line	696	10	750	142	D-16, P-16
Lamp driver	Dual	679	125	500	140	D-14, P-14
Decoder-driver	BCD to decimal	676			380	D-16, P-16
Counter	Decade	684	10	0.5 MHz	480	D-16, P-16
Counter	Binary	685	10	0.5 MHz	480	D-16, P-16
Shift register		686	10	0.5 MHz	375	D-16, P-16

* Passive pullup.
† Open collector.
‡ Package entry letter is form factor: F = flatpack, D = hermetic dual-in-line, P = plastic dual-in-line, C = round can, S = studded flatpack. Package entry number is terminal count.

multiple-emitter transistor structure with the logic-gate inputs. This structure is more properly called diode transistor-transistor logic. The basic gate structures are all NAND gates, and operation of the gate requires that all emitters be pulled down for logic transition to occur. The output structure is one of three types, the most common being the totem-pole structure. In the totem-pole structure, the device functions with both active pullup and pulldown, which improves the switching speed of the circuit. Active pullup and pulldown suffers from the adverse effect of causing a momentary short across the power-supply bus when one member of the totem pole turns on before the other half turns off. Therefore, it is essential in using TTL circuits to provide a great degree of power-supply decoupling to reduce noise margin by the current spikes generated during the output totem-pole switch. The totem-pole output has limited current drive and is primarily suited to driving other logic gates. Totem-pole outputs cannot be tied together in parallel in any case.

A modification of the totem-pole output is the tristate output. This output structure is, in reality, a standard TTL output which has been coupled internally to the output pin through a transistorized switch. This switch, which is controlled by an enable circuit, allows complete isolation of unused outputs from a data bus. This technique permits OR tying the outputs, which is otherwise forbidden within the TTL gate family. Tristate outputs are most suited to memory circuits which would be configured in large arrays to interface circuits that drive data buses in a data-bus computer organization.

The third type of output structure is the uncommitted output. This output operates at a much slower switching rate but is capable of both OR tying and current sinking to high current loads. As such, the primary use of uncommitted collector outputs is in driving displays and relays. The variations between the different fami-

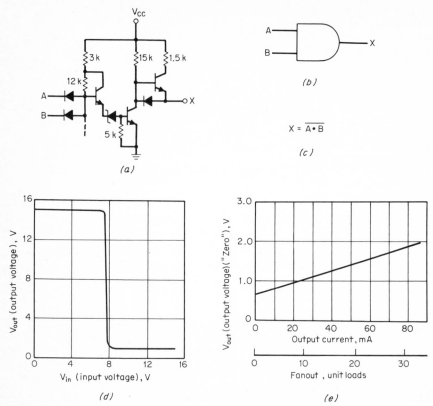

Fig. 5 Characterization of the high-threshold family of logic. NAND-gate characterization. (*a*) Circuit schematic (*b*) Logic diagram. (*c*) Logic equation. (*d*)
Transfer curve, a plot of output voltage vs. input voltage. (*e*) Fanout curve, a plot
of output voltage in the zero state vs. output current in unit loads.

lies of TTL primarily center upon different choices of drive capability, power-supply
consumption, and operating speed.

The families which use Schottky-diode clamps are able to operate at an improved
speed-power product. The Schottky clamps are used as Baker clamps in the switching transistors to minimize stored charge, thereby reducing switching-time transitions
within the logic circuits. Conventional TTL circuits use gold doping to reduce stored
charge, but this technique is not as effective as including Schottky-diode clamps.
The term Schottky transistor is improperly used to describe a conventional transistor
which uses a Schottky barrier diode as a Baker clamp.

The various families within the overall transistor-transistor logic family have found
specific areas of application. For example, low-power circuits have been used on
miniaturized and portable equipment and in military avionics systems. High-speed
and Schottky TTL circuits have gained some degree of acceptance in computer main-
frame applications. The evolution of a low-power Schottky family has been required
to make high-complexity MSI practical with TTL. Conventional transistor-transistor
logic is not suited for extremely complex MSI because of the excessive power dissipation
that standard transistor-transistor logic consumes.

Full operating characteristics and attributes of each family of transistor-transistor
logic are detailed in separate sections, including full listings of available circuits.
Certain commercial families have been combined on the basis of functional equality.

The use of a mixture of different families of transistor-transistor logic is appropriate
only in certain specified cases, because mixing usually requires complex allocation

of fanout and loading rules. Therefore, it is most appropriate that an entire system, or a subsystem that operates at a particular logic-data rate, be fabricated from a single family. Including a second family to meet the higher-speed requirements of a particular subsection demands special care when interfacing it so that the fanout rules of either family of circuits are not violated. The use of standard 5400 and 9300 circuits is most appropriate from the standpoint of availability and price. 5400 logic is to a great degree covered by slash sheets to MIL-M-38510, making its application to military systems most appropriate. Intermixing 5400 with 9300, 8200, and 7800 MSI represents a good choice in areas where the appropriate MSI element has not been induced in the 5400 family. It is not recommended, however, to include 9000 SSI in any system. The 9000 family has not gained popular acceptance, and the limited availability of the family should preclude its use, even though it fills a gap between the power and speed of 5400 and 5400H families.

The application of transistor-transistor logic must take into account factors to minimize the ac noise margin of the circuits. The three main factors are the tendency of inputs to oscillate when driven from long lines, the transmission of both current and voltage spikes along the V_{CC} power bus, and the tendency of crosstalk to develop in lines owing to the extremely fast front edge of the transistor-transistor logic. Oscillation of the inputs caused by long lines can be minimized by a number of techniques. The simplest and most straightforward is to reduce the length of uncontrolled lines between circuits. When this is not possible, higher-quality transmission lines, such as coax and twisted-pair, should be used for extremely long transmission lines. An alternate technique is to use line drivers and receivers to transmit information over long lines within a digital system. A real problem in transistor-transistor logic is transmission of voltage spikes from the V_{CC} power bus. The spikes are produced in the circuit by the output totem-pole structure, and they are caused in turning off one of the elements and turning on the other to switch the output-stage differences in the storage and delay times of the two transistors. This causes an interval where both transistors are momentarily turned on, which shorts the V_{CC} bus through a limiting resistor to the ground bus. Practically, it is impossible to improve this situation in the circuit design by using a higher-value resistor; this only increases the delay times associated with switching the totem pole. The prime method for improving and usually eliminating this noise source is to include adequate power-supply decoupling. Placing individual 10- to 50-μF capacitors on the printed-circuit board, with smaller-value capacitors placed between groups of circuits, is most effective. Decoupling requirements can be reduced by using multilayer printed-circuit boards for V_{CC} distribution, as these boards become a source of parasitic decoupling capacitance.

Problems associated with the crosstalk generated by the high DV/DT transition of typically 10 or 20 V μs^{-1} on the front edge of a TTL signal is best attacked by proper printed-wiring-board layout, since the problem must be faced in any circuit design. Most modern TTL circuits include a diode clamp across each input terminal, which is designed to minimize oscillation from all sources, especially that caused by long lines. Placing diode clamps in the circuit has limited effectiveness, since these clamps function slowly and become effective only when the amplitude of the oscillation exceeds 1 V below the ground level. Oscillations of smaller magnitude are not to be clamped, although they reduce the zero-level noise immunity.

Addition caution must be taken with circuits having the input-diode clamp. This input diode is of very small geometry to reduce parasitic capacitance on the input emitter nodes. This small geometry reduces the power-handling capability of these diodes; so they must be protected against excessive currents flowing in their forward-bias direction. In the case of transistor-transistor logic circuits which were initially designed without the input-diode clamps, later design alterations have in many cases included them. At the same time, vendors have been updating their specification sheets to indicate the input-diode clamps, while a significant quantity of stock which does not have the input-diode clamps remains available from distributors. Design engineers should confirm that all circuits being used have the input-diode clamp if it is required for operation.

5400 transistor-transistor logic is one of the three families of integrated circuits which are recommended for the great majority of digital-logic applications. While the 10,000 emitter logic provides the most important contribution in high-speed systems and the 4000 family of complementary metal-oxide semiconductors fulfills the requirements for small portable low-power systems, it is the 5400 family of transistor-transistor logic that meets the majority of application requirements. TTL can be combined with complementary oxide semiconductors operating from 5-V power supply with a minimum of difficulty, as discussed in the section on complementary metal-oxide semiconductors. Use of TTL with either MOS and/or emitter-coupled logic requires special interface circuits to make transition between logic levels without the loss of noise margin. TTL can be used directly with diode-transistor logic, except that the ability of DTL outputs to be or-tied is limited by the inability of TTL output to perform the same functions.

Transistor-transistor logic requires the use of highly regulated 5 or 10 percent power supplies to gain the maximum noise margin from the circuits. The requirement for tightly regulated power supplies is most deeply shown in MSI circuits, which have a lesser degree of noise immunity in their internal logic gates. Requirements for power-supply regulation are specified at the pin of the device, and not at the output terminal of the power supply. Therefore, all sources of modification of the power-supply voltages must be considered in a large system. It is essential that distribution losses and ground loops be accounted for, since these two problems can reduce the functionality of a system using TTL.

Transistor-transistor logic is used in a broad range of functions, from mass application in many computers and subsections of large main frames to single applications as power-supply control elements. All these applications must be made with full consideration for the functionality and limitations of the operation of the particular circuit family. A great degree of care must be taken in applying TTL circuits to drive discrete circuits. Full consideration of the load these discrete circuits present to the output must be made, since current beyond the design limit reduces the noise margin available to drive other outputs at the same time. Operation beyond the drive capability of the circuit reduces circuit life. Therefore the inputs of transistor-transistor logic must be tied to the true level in a particular manner and certain limitations must be placed on the possible ways of tying an input to this level. Many design engineers have noted that an emitter input which is left open acts as if it has been returned to the true level. This technique is not recommended; it reduces the noise margin and makes the circuit prone to false operation. Also discouraged is directly tying the inputs to the V_{CC} terminal. Transients on the V_{CC} terminal are then applied to the emitter diodes, which could result in a destructive emitter-base breakdown. This destructive mode progresses in stages which reduce the fanout and functionality of the circuit.

Care must also be taken in applying TTL to one-shot multivibrators. These circuits have found wide application, but in using a one-shot multivibrator in the retriggering mode, an interval must be allowed after the initial triggering before retriggering is attempted. This interval allows the timing capacitor to discharge fully before commencing an additional cycle. The interval for recharging the capacitor should approximately equal a time in nanoseconds that is numerically equivalent to one-third of the value of the timing capacitor in microfarads. This time varies from unit to unit and is a function of the switch resistance through which the capacitor is discharged. If the retriggering interval cannot be directly controlled, provision must be made to generate a double pulse for retriggering such that the interval between the two pulses in the double pulse is greater than the recharge interval.

The Suhl Family of Transistor-Transistor Logic. The Suhl I° family was the first widely accepted family in the transistor-transistor logic-circuit trend. It was introduced by the Sylvania Electric Company in the middle 1960s, and it gained a moderate level of success through its adoption in both military and commercial digital electronic systems. Like all other transistor-transistor logic families, the basic gate of the Suhl I family is a NAND gate. This gate is illustrated in the schematic in Fig. 6.

The Suhl I gate has many operating advantages, including a reasonable short gate delay of approximately 12 ns, which was complemented by the availability of 35-MHz

° Trademark of Sylvania Electric Company.

flip-flops. The basic circuit is simple, requiring only four transistors and four resistors to perform a saturated logic function. The high operating speed of this circuit results from a totem-pole output structure which provides both active pullup and active pulldown. The Suhl I family was a family completely composed of small-scale integration, although it did have a broad range of available circuits, such as the NAND gates and NOR gates, OR expander gates, exclusive-OR gates, line and bus drivers, and an array of AND input expanders. Completing the range of available circuits were noninverting AND gates, line and lamp drivers, and both D and JK flip-flops. The full range of available Suhl I circuits is presented in Table 6.

TABLE 6   The Suhl I Family of Transistor-Transistor Logic, Complete Selection Guide of Circuit Types

Function	Circuit form	Type No. −55 to +125°C	Type No. 0 to 75°C	Fanout (unit loads)	Total delay, ns	Power dissipation, mW	Package*
NAND gate	Dual 4-input	40	42	15/12	10	30	D-14, F-14
NAND gate	Dual 4-input	41	43	7/6	10	30	D-14, F-14
NAND gate	Single 8-input	60	62	15/12	10	15	D-14, F-14
NAND gate	Single 8-input	61	63	7/6	10	15	D-14, F-14
NAND gate	Quad 2-input	140	142	15/12	10	60	D-14, F-14
NAND gate	Quad 2-input	141	143	7/6	10	60	D-14, F-14
NAND gate, expandable	Single 8-input	120	122	15/12	10	15	D-14, F-14
NAND gate, expandable	Single 8-input	121	123	7/6	10	15	D-14, F-14
NAND gate	Triple 3-input	190	192	15/12	10	45	D-14, F-14
NAND gate	Triple 3-input	191	193	7/6	10	45	D-14, F-14
Inverter	Hex	370	372	15/12	10	90	D-14, F-14
Inverter	Hex	371	373	7/6	10	90	D-14, F-14
AND/OR invert gate	2 × 4	50	52	15/12	12	39	D-14, F-14
AND/OR invert gate	2 × 4	51	53	7/6	12	39	D-14, F-14
AND/OR invert gate	Dual 2 × 2	70	72	15/12	12	52	D-14, F-14
AND/OR invert gate	Dual 2 × 2	71	73	7/6	12	52	D-14, F-14
AND/OR invert gate	Expandable 3 × 3	100	102	15/12	12	35	D-14, F-14
AND/OR invert gate	Expandable 3 × 3	101	103	7/6	12	35	D-14, F-14
AND/OR invert gate	Expandable 4 × 2	110	112	15/12	12	23	D-14, F-14
AND/OR invert gate	Expandable 4 × 2	11	113	7/6	12	23	D-14, F-14
NOR gate	Quad 2-input	330	332	15/12	12	110	D-14, F-14
NOR gate	Quad 2-input	331	333	7/6	12	110	D-14, F-14
Gate expander	AND/OR 2/3 × 4	150	152				D-14, F-14
Gate expander	AND/OR 2/3 × 4	151	153				D-14, F-14
Gate expander	Dual 4-input AND	170	172				D-14, F-14
Gate expander	Dual 4-input AND	171	173				D-14, F-14
Exclusive-OR gate		90	92	15/12	11	35	D-14, F-14
Exclusive-OR gate		91	93	7/6	11	35	D-14, F-14
Bus driver	Dual 4-input NAND	130	132	30/15	25	30	D-14, F-14
Bus driver	Dual 4-input NAND	131	133	24/12	25	30	D-14, F-14
Open collector driver	Quad 2-input NAND	160	162		30	80	D-14, F-14
Open collector driver	Quad 2-input NAND	161	163		30	80	D-14, F-14
AND gate	Dual 4-input	280	282	15/12	10	76	D-14, F-14
AND gate	Dual 4-input	281	283	7/6	10	76	D-14, F-14
AND gate	Quad 2/3-input	290	292	15/12	10	76	D-14, F-14
AND gate	Quad 2/3-input	291	293	7/6	10	76	D-14, F-14
Flip-flop	AND JK	50	52	15/12	20	40	D-14, F-14
Flip-flop	AND JK	51	53	7/6	20	40	D-14, F-14
Flip-flop	OR JK	60	62	15/12	20	40	D-14, F-14
Flip-flop	OR JK	61	63	7/6	20	40	D-14, F-14
Flip-flop	Dual D	80	82	10	16	96	D-14, F-14
Flip-flop	Dual D	81	83	10	16	96	D-14, F-14
Flip-flop	Dual JK	100	102	16/13	10	110	D-14, F-14
Flip-flop	Dual JK	101	103	8/7	10	110	D-14, F-14
Flip-flop	Dual JK	110	112	16/13	10	110	D-14, F-14
Flip-flop	Dual JK	111	113	8/7	10	110	D-14, F-14

* Package entry letter is form factor: F = flatpack, D = hermetic dual-in-line, P = plastic dual-in-line, C = round can, S = studded flatpack. Package entry number is terminal count.

The Suhl family was available in four grades, according to output, fanout, and temperature range. In reality, these four grades represented postfabrication selections of devices. The four grades were military prime with a full temperature range of −55 to +125°C and a gate fanout of 15 in the 1 state. Military-standard grades had the same temperature range with a reduced fanout of 7.   The industrial-tempera-ture grade of 0 to 75°C was similarly divided into a prime grade with a fanout of 12 and an industrial grade having a fanout of 6.   The part numbers assigned in the Suhl I family were in groups of four consecutive numbers for each circuit function.   The four consecutive numbers were assigned in the following order: mili-tary prime, military standard, industrial prime, and industrial standard.   Hence, if the number TG-40 is assigned to a particular function in the military prime grade, the same function in the military standard is TG-41.   In the industrial prime grade, it is TG-42 and in the industrial standard, it is TG-43.

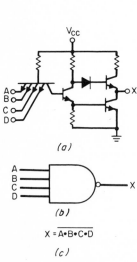

(a)

(b)

$$X = \overline{A \cdot B \cdot C \cdot D}$$

(c)

Fig. 6   Characterization of the Suhl I family of transis-tor-transistor logic.   NAND-gate characterization.   (a) Circuit schematic.   (b) Logic diagram.   (c) Logic equation.

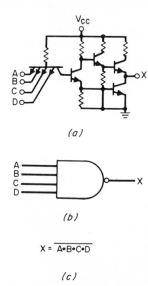

(a)

(b)

$$X = \overline{A \cdot B \cdot C \cdot D}$$

(c)

Fig. 7   Characterization of the Suhl II family of transis-tor-transistor logic.   NAND-gate characterization.   (a) Circuit schematic.   (b) Logic diagram.   (c) Logic equation.

Although complex logic arrays are not available within the Suhl I family, it is possible to assemble large logic arrays in multiple packages by using the expansion nodes, which are featured in particular gate types and which enable direct connection of the OR expanders to the AND/OR gates and the NAND gates.   These expanders allow for both logical expansion and expanding the direct fanin.   The use of multiple expansion gates was somewhat limited, since the additional capacitance applied to the expansion node caused the overall circuit to operate with increased delay.   Al-though the Suhl I family was later supplemented with the Suhl II family, which had a typical gate delay of 6 ns and a group of flip-flops which could effectively toggle at frequencies exceeding 50 MHz, the combined family never gained broad acceptance.   The lack of acceptance stems primarily from two sources, a relatively high power dissipation of typically 20 mW per gate, which made it impossible to fabricate higher-complexity devices practically, and the initial inability of the circuit suppliers to meet the quantity requirements of the electronics industry.

The typical Suhl II NAND gate schematic is illustrated in Fig. 7.   An additional transistor is added into the circuit to provide additional gain for the output totem pole.   Suhl II made use of smaller transistor geometries to reduce the parasitic capacitance, and gold doping was employed to reduce stored charge to lower gate-delay times.   The range of available Suhl II circuits is presented in Table 7.

The 9000 Family of Transistor-Transistor Logic.   The next major development in

TABLE 7   The Suhl II Family of Transistor-Transistor Logic, Complete Selection Guide of Circuit Types

Function	Circuit form	Type No. −55 to +125°C	Type No. 0 to 75°C	Fanout (unit loads)	Total delay, ns	Power dissipation, mW	Package*
NAND gate	Dual 4-input	240	242	11/9	6	44	D-14, F-14
NAND gate	Dual 4-input	241	243	6/5	6	44	D-14, F-14
NAND gate	Single 8-input	260	262	11/9	6	22	D-14, F-14
NAND gate	Single 8-input	261	263	6/5	6	22	D-14, F-14
NAND gate	Quad 2-input	220	222	11/9	6	88	D-14, F-14
NAND gate	Quad 2-input	221	223	6/5	6	88	D-14, F-14
NAND gate expandable	Single 8-input	200	202	11/9	6	22	D-14, F-14
NAND gate expandable	Single 8-input	201	203	6/5	6	22	D-14, F-14
NAND gate	Triple 3-input	320	322	11/9	6	66	D-14, F-14
NAND gate	Triple 3-input	321	323	6/5	6	66	D-14, F-14
Inverter	Hex	380	382	11/9	6	198	D-14, F-14
Inverter	Hex	381	383	6/5	6	198	D-14, F-14
AND/OR invert gate	2 × 4	250	252	11/9	7	55	D-14, F-14
AND/OR invert gate	2 × 4	251	253	6/5	7	55	D-14, F-14
AND/OR invert gate	Dual 2 × 2	310	312	11/9	7	75	D-14, F-14
AND/OR invert gate	Dual 2 × 2	311	313	6/5	7	75	D-14, F-14
AND/OR invert gate	3 × 3	300	302	11/9	7	47	D-14, F-14
AND/OR invert gate	3 × 3	301	303	6/5	7	47	D-14, F-14
AND/OR invert gate	4 × 2	210	212	11/9	7	38	D-14, F-14
AND/OR invert gate	4 × 2	211	213	6/5	7	38	D-14, F-14
NOR gate	Quad 2-input	340	342	11/9	7	150	D-14, F-14
NOR gate	Quad 2-input	341	343	6/5	7	150	D-14, F-14
Gate expander	AND/OR 2/3 × 4	230	232				D-14, F-14
Gate expander	AND/OR 2/3 × 4	231	233				D-14, F-14
Gate expander	AND dual 4-input	270	272				D-14, F-14
Gate expander	AND dual 4-input	271	273				D-14, F-14
Flip-flop	AND JK	250	252	10/8	20	40	D-14, F-14
Flip-flop	AND JK	251	253	5/4	20	40	D-14, F-14
Flip-flop	OR JK	260	262	10/8	20	40	D-14, F-14
Flip-flop	OR JK	261	263	5/4	20	40	D-14, F-14
Flip-flop	Dual D	90	92	10	10	150	D-14, F-14
Flip-flop	Dual D	91	93	10	10	150	D-14, F-14
Flip-flop	Dual JK	120	122	11/9	9	110	D-14, F-14
Flip-flop	Dual JK	121	123	6/5	9	110	D-14, F-14
Flip-flop	Dual JK	130	132	11/9	9	110	D-14, F-14
Flip-flop	Dual JK	131	133	6/5	9	110	D-14, F-14
Flip-flop	AND JK	200	202	11/9	9	55	D-14, F-14
Flip-flop	AND JK	201	203	6/5	9	55	D-14, F-14
Flip-flop	OR JK	210	212	11/9	9	65	D-14, F-14
Flip-flop	OR JK	211	213	6/5	9	65	D-14, F-14

* Package entry letter is form factor: F = flatpack, D = hermetic dual-in-line, P = plastic dual-in-line, C = round can, S = studded flatpack.   Package entry number is terminal count.

transistor-transistor logic was the introduction of the 9000 family of small-scales gates and flip-flops by Fairchild Semiconductor. This family represented an improvement over the basic Suhl gate by offering an 8-ns gate delay with a typical 10-mW gate power dissipation. The schematic diagram typical of a NAND gate in this family is shown in Fig. 8. The figure shows the addition of both input clamp diodes and a Darlington stage in the output. The clamp diodes were introduced to minimize or eliminate oscillations caused by termination effects, while the Darlington output stage both improved switching speed and increased the ability of the circuit to handle capacitive loading on an output. This family included NAND gates and AND/OR invert gates, and it introduced a new function called the buffer. The function of the buffer was to provide higher fanout, while sacrificing delay time and power dissipation for critical applications requiring exceptionally high fanout. The family was completed with exclusive-OR gates, NOR gates, JK flip-flops, and dual JK flip-flops. The complete range of available circuits in the family is presented in Table 8.

Again, this family gained only limited acceptance, and its use is now very restricted. The 9000 family was supplemented by the inclusion of a low-power 9L family, whose schematic is illustrated in Fig. 9. This schematic is identical to that of the 9000 gate, except for the higher-value resistors in the 9000 L

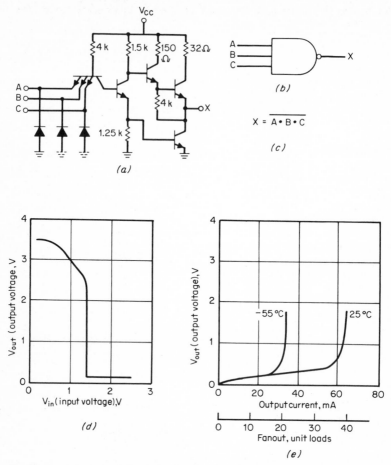

Fig. 8 Characterization of the 9000 family of transistor-transistor logic. NAND-gate characterization. (*a*) Circuit schematic. (*b*) Logic diagrams. (*c*) Logic equation. (*d*) Transfer curve, a plot of output voltage vs. input voltage. (*e*) Fanout curve, a plot of output voltage vs. fanout in unit loads.

family. In reality, the 9 L family was fabricated using the mask sets for the 9000 family, with only a minor modification during the base diffusion to yield a set of resistors having values four times larger.

The 9000 family included three circuits which gained acceptance much broader than the basic logic elements within the family. These three circuits were the 9600, 9601, and 9602 monostable multivibrators. These three circuits are widely used in combination with other TTL and DTL components to provide a one-shot function.

The 5400 Family of Transistor-Transistor Logic and Compatible Circuits. Frequently, 5400 and transistor-transistor logic are assumed to be synonyms. This family of transistor-transistor logic was developed by Texas Instrument Corporation and has gained broad acceptance in the United States, Europe, and Asia for commercial, industrial, and military electronic systems. The family combines the attributes of a large multitude of simple and complex functions and the parallel availability of these differing function in a range of speed and power dissipations. There are five subfamilies within the overall 5400 family. These are the basic 5400 family with a gate delay of 10 ns and a power dissipation of 10 mW, the 5400 high-speed family with a gate delay of 6 ns and a dissipation of 22 mW, the low-speed and

TABLE 8   The 9000 Family of Transistor-Transistor Logic, Complete Selection Guide of Circuit Types

Function	Circuit form	Type No.	Fanout (unit loads)	Total delay, ns	Power dissipation, mW	Package†
NAND gate	Quad 2-input	9002	10	8	40	D-14, P-14, F-14
NAND gate*	Quad 2-input	9012	10	8/25		D-14, P-14, F-14
NAND gate	Triple 3-input	9003	10	8	30	D-14, P-14, F-14
NAND gate	Dual 4-input	9004	10	8	20	D-14, P-14, F-14
NAND gate	Single 8-input	9007	10	8	10	D-14, P-14, F-14
NAND gate		9009	30	10	53	D-14, P-14, F-14
NOR gate	Quad (2-, 2-, 2-, 4-input)	9015	10	8	74	D-16, P-16, F-16
Exclusive-OR/NOR gate	Quad 2-input	9014	10/9	8/16	110	D-16, P-16, F-16
Inverter	Hex	9016	10	8	60	D-14, P-14, F-14
Inverter*	Hex	9017	10	8/25	60	D-14, P-14, F-14
AND-OR invert gate	Dual 2 × 2	9005	10	8	43	D-14, P-14, F-14
AND-OR invert gate	4 × 2 expandable	9008	10	8	40	D-14, P-14, F-14
Expander	Dual 4-input	9006				D-14, P-14, F-14
Flip-flop	JK-MS	9000	10	$f = 20$ MHz	140	D-14, P-14, F-14
Flip-flop	JK-MS	9001	10	$f = 50$ MHz	165	D-14, P-14, F-14
Flip-flop	Dual JK-MS	9020	10	$f = 50$ MHz	300	D-16, P-16, F-16
Flip-flop	Dual JK-MS	9022	10	$f = 50$ MHz	300	D-16, P-16, F-16
Flip-flop	Dual JK	9024	10	$f = 25$ MHz		D-16, P-16, F-16

* Uncommitted collector. Temperature range indicated by part number suffix, $C = 0$ to $70°C$, $M = -55$ to $+125°C$.
† Package entry letter is form factor: F = flatpack, D = hermetic dual-in-line, P = plastic dual-in-line, C = round can, S = studded flatpack. Package entry number is terminal count.

low-power family with a gate delay of 33 ns and a 1-mW power dissipation, the Schottky family with a 3-ns delay and a 19-mW power dissipation, and the low-power Schottky with a 9.5-ns gate delay and a 2-mW power dissipation. These subfamilies can be successfully integrated together, provided the different loads of the subfamilies are accounted for. The characteristics of the families within the 5400 grouping are compared in Table 9.

TABLE 9   Comparison of Transistor-Transistor Logic Families.   Compatibility of Inputs, Outputs, and Thresholds

Value	5400	54H	54L	54LS	54S	9000
Input low-level current, mA	−1.6	−2.0	−0.18/−0.8	−0.36	−2.0	−1.6
Input high-level current, μA	40	50	10/20	20	50	40
Input pullup resistor, kΩ	4	2.8	40/8	25	2.8	4
Output high-level voltage, V	2.4	2.4	2.4	2.5	2.5	2.4
Input high-level voltage, V	2.0	2.0	2.0	2.0	2.0	2.0
Output low-level voltage, V	0.4	0.4	0.3	0.5	0.5	0.4
Input low-level voltage, V	0.8	0.8	0.7	0.8	0.8	0.8

A broad range of functions are available within the 5400 families. The basic gate structure in this family, like all other transistor-transistor logic families, is a NAND gate. The basic NAND-gate schematic for the five subfamilies is illustrated in Figs. 10 and 13 to 16. The totem-pole output structure is supplemented by a group of devices which have outputs that are either uncommitted or open-collector outputs. Such devices are further illustrated in Fig. 11. The speed advantages of the totem-pole output sacrificed by the open-collector structure, but the open-collector structure is capable of driving large voltages and current at the load output. In addition, the open-collector output can be used in wired OR applications in a manner similar to the 930 DTL output. The 5400 family includes NOR gates, Schmitt-trigger input NAND gates, totem-pole output buffers, line drivers, OR gates, AND/OR invert gates, expandable gates, gate expanders, and an extremely broad range of flip-flops, including D types and JK types, both edge-triggered and pulse-triggered.

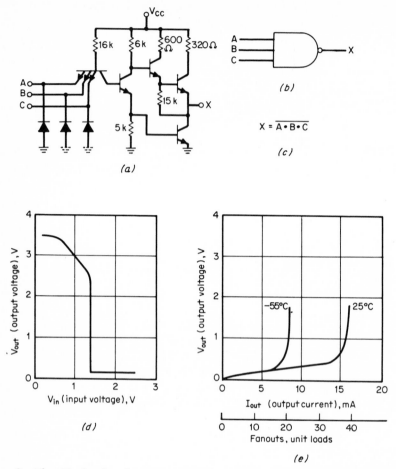

Fig. 9 The 9L family of transistor-transistor logic. NAND-gate characterization. (*a*) Circuit schematic. (*b*) Logic diagrams. (*c*) Logic equation. (*d*) Transfer curve, a plot of output voltage vs. input voltage. (*e*) Fanout curve, a plot of output voltage vs. output current.

The family has recently been expanded to include three-state output structures on a limited range of gates. A typical three-state output structure is illustrated in Fig. 12. The three-state outputs function through separate control nodes and operate in either a high, low, or disconnected mode. The use of three-state outputs simplifies fabrication of computer data buses. The range of small-scale integrated circuits available within the 5400 families is quite large, but it represents a small part of the total spectrum of available circuit functions contained within the 5400 families. Available small-scale integration devices are listed in Tables 10, 12, 14, 16, and 18.

The most complete range of medium-scale integration functions available within any logic family is available within the 5400 families. These functions range from the lower edge of MSI complexity through all ranges of MSI up to and including the higher levels of complexity which are properly called large-scale integration (LSI). The use of MSI and LSI functions serves to simplify the fabrication and test of any large digital system. They reduce the number of interconnects and the power consumption required to operate signals across them. MSI and LSI functions have required the use of low-dissipation internal gates within the circuit to minimize

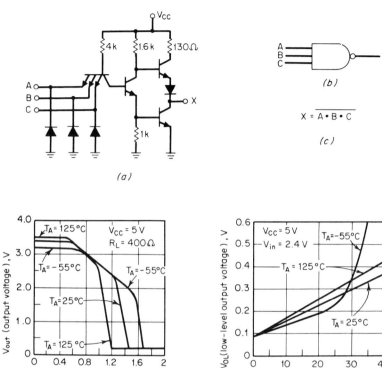

Fig. 10   Characterization of the 5400 family of transistor-transistor logic. NAND-gate characterization. (a) Circuit schematic. (b) Logic diagrams. (c) Logic equation. (d) Transfer curve, a plot of output voltage vs. input voltage. (e) Fanout curve, a plot of low-level output voltage vs. low-level output current.

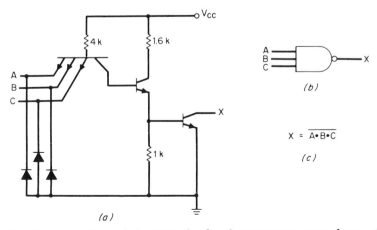

Fig. 11   Characterization of the 5400 family of transistor-transistor logic. Open collector. NAND-gate characterization. (a) Circuit schematic. (b) Logic diagrams. (c) Logic equation.

TABLE 10   The 5400 Family of Standard-Speed Transistor-Transistor Logic, Complete Selection Guide of Small-Scale Integration Circuit Types

Function	Circuit form	Type No. 0 to 70°C	Type No. −55 to +125°C	Fan-out (unit loads)	Total delay ns	Power dissipation, mW	Package §
NAND gate	Quad 2-input	7400	5400	10	10	40	D-14, F-14, P-14
NAND gate*	Quad 2-input	7401	5401	10	10/35	40	D-14, F-14, P-14
NAND gate*	Quad 2-input	7403	5403	10	10/35	40	D-14, F-14, P-14
NAND gate	Triple 3-input	7410	5410	10	10	30	D-14, F-14, P-14
NAND gate*	Triple 3-input	7412	5412	10	10/35	30	D-14, F-14, P-14
NAND gate†	Dual 4-input	7413	5413	10	18/15	85	D-14, F-14, P-14
NAND gate†	Quad 2-input	74132	54132	10	15	101	D-14, F-14, P-14
NAND gate	Dual 4-input	7420	5420	10	10	20	D-14, F-14, P-14
NAND gate*	Dual 4-input	7422	5422	10	10/35	20	D-14, F-14, P-14
NAND gate*	Quad 2-input high-voltage	7426	5426	10	16/11	40	D-14, F-14, P-14
NAND gate	Single 8-input	7430	5430	10	10	10	D-14, F-14, P-14
NAND gate	Quad 2-input	7437	5437	30	13/8	108	D-14, F-14, P-14
NAND gate*	Quad 2-input	7438	5438	30	14/11	98	D-14, F-14, P-14
NAND gate	Dual 4-input	7440	5440	30	13/8	52	D-14, F-14, P-14 D-14, P-14, F-14
NAND gate	Dual 2-input + inverter	8090	7090	10	10	60	D-16, P-16
NAND gate	Quad 2-input	8091	7091	30	13/8	61	D-14, P-14, F-14
NAND gate	Dual 5-input	8092	7092	10	10	20	D-14, P-14, F-14
AND gate	Quad 2-input	7408	5408	10	18/12	78	D-14, P-14, F-14
AND gate‡	Quad 2-input	7409	5409	10	12	78	D-14, P-14, F-14
NOR gate	Quad 2-input	7402	5402	10	21/16	55	D-14, P-14, F-14
NOR gate	Dual 4-input, expandable	7423	5423	10	13/8	45	D-14, P-14, F-14
NOR gate	Dual 4-input	7425	5425	10	13/8	23	D-14, P-14, F-14
NOR gate	Triple 3-input	7427	5427	10	7/10	65	D-14, P-14, F-14
NOR gate	Quad 2-input	7428	5428	30	7	112	D-14, P-14, F-14
NOR gate*	Quad 2-input	7433	5433	30	10	113	D-14, P-14, F-14
OR gate	Quad 2-input	7432	5432	10	10/14	95	D-14, P-14, F-14
Exclusive-OR gate	Quad 2-input	7486	5486	10	14	150	D-14, P-14, F-14
Buffer*	Hex high-voltage	7407	5407	25	6/10	125	D-14, P-14, F-14
Buffer*	Hex high-voltage	7417	5417	25	6/10	125	D-14, P-14, F-14
Buffer‡	Quad	74125	54125	10	8	160	D-14, P-14, F-14
Buffer‡	Quad	74126	54126	10	8	180	D-14, P-14, F-14
Buffer‡	Hex	8095	7095	20	15	325	D-16, P-16, F-16
Buffer‡	Hex	8097	7097	20	15	325	D-16, P-16, F-16
Inverter*	Hex high-voltage	7416	5416	25	10/15	155	D-14, P-14, F-14
Inverter†	Hex	7414	5414	10	15	152	D-14, P-14, F-14
Inverter	Hex	7404	5404	10	10	60	D-14, P-14, F-14
Inverter*	Hex	7405	5405	10	10/45	60	D-14, P-14, F-14
Inverter*	Hex high-voltage	7406	5406	25	10/15	155	D-14, P-14, F-14
Inverter‡	Hex	8096	7096	20	15	295	D-16, F-16, P-16
Inverter‡	Hex	8098	7098	20	15	295	D-16, F-16, P-16
AND-OR invert gate	Dual 2 × 2 expandable	7450	5450	10	13/8	28	D-14, F-14, P-14
AND-OR invert gate	Dual 2 × 2	7451	5451	10	13/8	28	D-14, F-14, P-14
AND-OR invert gate	Single 4 × 2 expandable	7453	5453	10	13/8	23	D-14, F-14, P-14
AND-OR invert gate	Single 4 × 2	7454	5454	10	13/8	23	D-14, F-14, P-14
Expander	Dual 4-input	7460	5460				D-14, F-14, P-14
Flip-flop	JK edge trigger	7470	5470	10	$f = 35$ MHz	65	D-14, F-14, P-14
Flip-flop	JK master slave	7472	5472	10	$f = 20$ MHz	50	D-14, F-14, P-14
Flip-flop	Dual JK	7473	5473	10	$f = 20$ MHz	100	D-14, F-14, P-14
Flip-flop	Dual D edge trigger	7474	5474	10	$f = 25$ MHz	85	D-14, F-14, P-14
Flip-flop	Dual JK	7476	5476	10	$f = 20$ MHz	100	D-14, F-14, P-14
Flip-flop	Dual JK master slave	74107	54107	10	$f = 20$ MHz	100	D-14, F-14, P-14
Flip-flop	Dual JK edge trigger	74109	54109	10	$f = 33$ MHz	90	D-14, F-14, P-14
Flip-flop	JK master slave	74110	54110	10	$f = 25$ MHz	100	D-14, F-14, P-14
Flip-flop	Dual JK master slave	74111	54111	10	$f = 25$ MHz	140	D-14, F-14, P-14
Flip-flop‡	Quad	8551	7551	10	$f = 30$ MHz	250	D-16, F-16, P-16
Latch	Quad	7475	5475	10	7/16	64	D-14, F-14, P-14
Latch	Quad	7477	5477	10	7/16	64	D-14, F-14, P-14
Latch	Quad set-reset	74279	54279	10	12	90	D-14, F-14, P-14
Monostable multi		74121	54121	10		115	D-14, F-14, P-14
Monostable multi	Retriggerable	74122	54122	10		115	D-14, F-14, P-14
Monostable multi	Dual retriggerable	74123	54123	10		230	D-16, F-16, P-16
Level translator	Dual TTL to MOS	8800	7800		70		C-10

TABLE 10 The 5400 Family of Standard-Speed Transistor-Transistor Logic, Complete Selection Guide of Small-Scale Integration Circuit Types (Continued)

Function	Circuit form	Type No. 0 to 70°C	Type No. −55 to +125°C	Fan-out (unit loads)	Total delay ns	Power dissipa-tion, mW	Package §
Level translator NAND	Quad TTL to MOS	8810	7810		12/29	40	F-14, P-14, D-14
Level translator NAND	Quad TTL to MOS	8811	7811		12/29	40	F-14, P-14, D-14
Level translator	Hex TTL to MOS	8812	7812		12/29	60	F-14, P-14, D-14
Level translator NAND	Quad TTL to MOS	8819	7819		16	62	F-14, P-14, D-14
Level translator	Dual MOS to TTL	8806	7806	10	10	220	F-14, P-16, D-14
Line driver	Quad NOR 75 Ω	74128	54128	30	7	113	
Line driver‡	Quad	8831	7831	25	17	32	D-16, P-16, F-16
Line driver‡	Quad	8832	7832	25	17	32	D-16, P-16, F-16
Line driver	Dual differential	8830	7830		12	22	D-14, F-14, P-14
Line receiver	Dual differential	8820	7820	2	100	50	D-14, P-14, F-14
Line receiver	Dual	8822	7822	2	65	120	D-14, P-14, F-14
Line transceiver‡	Quad	8833	7833	20			D-16, P-16
Line transceiver‡	Quad	8834	7834	20			D-16, P-16
Line transceiver‡	Quad	8835	7835	20			D-16, P-16
Line transceiver	Quad, bus, 120 Ω	8838	7838	10	20	250	D-16, P-16, F-16
Line transceiver‡	Quad	8839	7839	20			D-16, P-16
Bus receiver	Quad NOR 120 Ω	8836	7836	10	20	125	D-14, P-14
Bus receiver	Hex 120 Ω	8837	7837	10	20	225	F-16, D-16, P-16

* Uncommitted collector output.
† Internal Schmitt trigger.
‡ Three-stage output.
§ Package entry letter is form factor: F = flatpack, D = hermetic dual-in-line, P = plastic dual-in-line, C = round can, S = studded flatpack. Package entry number is terminal count.

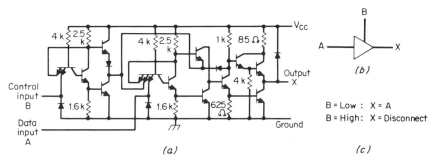

Fig. 12 Characterization of the 5400 family of transistor-transistor logic. Three-state output characterization. (*a*) Circuit schematic. (*b*) Logic diagrams. (*c*) Logic equation.

power dissipation and stage delay for logic routes which do not have to drive external interfaces directly. The more complex MSI and LSI elements have been primarily directed toward the lower-power members of the 5400 family, such as the straight 5400, the 54L, and the 54LS. The 54LS is most promising for future developments in regions of highest gate complexity. Tables 11, 13, 15, 17, and 19 summarize the available functions and their basic operating characteristics. These functions have been selected to meet the functional requirements inherent in the design of a digital computer or digital subsystem. It is most effective to modify, wherever possible, the requirements of system functions for MSI and LSI blocks, instead of using large numbers of small-scale integrated circuits.

It is essential that total power consumption of the circuit be considered whenever MSI or LSI blocks are used. Many of these circuits consume and dissipate power in excess of ½ W. Therefore, adequate cooling must be provided to the package to ensure reliable operation. Further packaging densities of high-dissipation circuits must not be so tight that circuits are subjected to the adverse effects of mutual heating. High-power dissipating MSI circuits must not be placed in close proximity

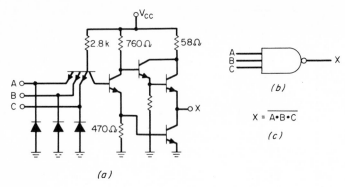

Fig. 13 Characterization of the 54H family of transistor-transistor logic. NAND-gate characterization. (*a*) Circuit schematic. (*b*) Logic diagrams. (*c*) Logic equation.

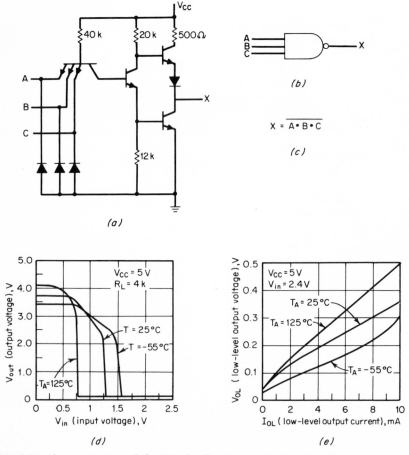

Fig. 14 Characterization of the 54L family of transistor-transistor logic. NAND-gate characterization. (*a*) Circuit schematic. (*b*) Logic diagrams. (*c*) Logic equation. (*d*) Transfer curve, a plot of output voltage vs. input voltage. (*e*) Fanout curve, a plot of low-level output vs. low-level output current.

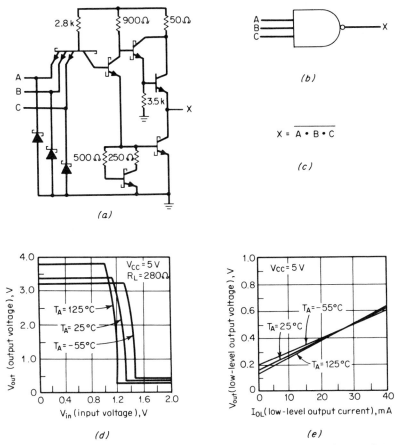

Fig. 15 Characterization of the 54S family of transistor-transistor logic. (*a*) Circuit schematic. (*b*) Logic diagram. (*c*) Logic equation. (*d*) Transfer curve, a plot of output voltage vs. input voltage. (*e*) Fanout curve, a plot of low-level output voltage vs. low-level output current.

to other circuit components which dissipate large amounts of power. Many of the higher dissipation circuits are not available in full temperature range or in all possible package styles owing to the adverse or marginal effects of high power dissipation. Adequate power-supply decoupling should be provided all MSI and LSI circuits to maintain the noise margin available to the internal gates. In using MSI blocks, the timing of all input and output signals must be considered. Quite often, different data paths within the circuit can have markedly different delay times. Consideration of signal timing is also quite important in using any of the sequential MSI blocks to avoid any internal rate problems. The range of available functions includes multiplexers and demultiplexers, adders, multipliers, digital comparators, parity generators and checkers, shift registers, register files, data files, latches, decoders, display drivers, counters, and priority encoders. Tables 10 through 19 should be initially consulted for selection of circuit types, but final selection and application of a given circuit type should await full analysis of the circuit specification. Tables 10 through 19 also include circuit types which are similar in gate structure and level to the 5400 family but which were not assigned 5400 numbers because of their initial design at National, Fairchild, or Signetics. The Fairchild parts are listed as 9300, Signetics as 8000, and National as 7000.

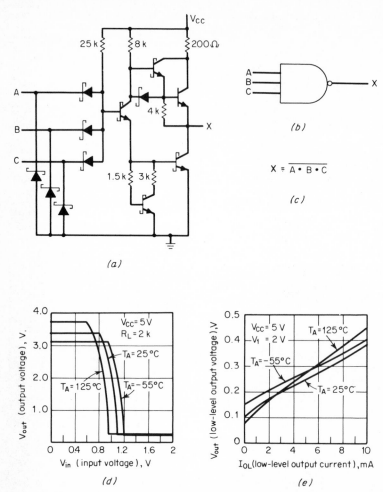

(a)

(b)

$$X = \overline{A \cdot B \cdot C}$$

(c)

(d)

(e)

Fig. 16 Characterization of the 54LS family of transistor-transistor logic. NAND-gate characterization. (*a*) Circuit schematic. (*b*) Logic diagram. (*c*) Logic equation. (*d*) Transfer curve, a plot of output voltage vs. input voltage. (*e*) Fanout curve, a plot of low-level output voltage vs. low-level output current.

Emitter-coupled Logic

In the initial phases of the evolution of integrated-circuit families, each of the major semiconductor houses focused its research in a different direction. The result was that each of these companies pioneered different forms of digital logic. The contribution of the Motorola Semiconductor Products Corporation was the early introduction of emitter-coupled logic (ECL) in 1962. The use of emitter-coupled logic had been widely accepted as a methodology for high speed in discrete logic arrays. The goal of emitter-coupled logic was to design a nonsaturated logic cell so that the input section is a classic differential amplifier with the emitters coupled onto a common bus. The output of an ECL cell is normally configured to use an emitter-follower output, such that the common internal logic nodes are then coupled emitters and the outputs are emitter followers, i.e., emitter-coupled logic.

In operation, the basic ECL gate functions by having the differential-amplifier section make the actual logic decision and provide a degree of voltage gain necessary for a narrow logic threshold. A voltage reference, internally or externally generated,

TABLE 11. The 5400 Family of Standard-Speed Transistor-Transistor Logic, Complete Selection Guide of Medium-Scale Integration Circuit Types

Function	Circuit form	Type No. 0 to 70°C	Type No. −55 to +125°C	Total delay ns	Power dissipation, mW	Package*
Counter	Asynchronous, decade	7490	5490	32 MHz	160	F-14, P-14, D-14
Counter	Base 12, asynchronous	7492	5492	32 MHz	160	F-14, P-14, D-14
Counter	4-bit binary, asynchronous	7493	5493	32 MHz	160	F-14, P-14, D-14
Counter	6-bit binary, synchronous	7497	5497	25 MHz	345	F-16, P-16, D-16
Counter	Decade, synchronous	74160	54160	25 MHz	305	F-16, P-16, D-16
Counter	4-bit binary, synchronous	74161	54161	25 MHz	305	F-16, P-16, D-16
Counter	Decade, synchronous	74162	54162	25 MHz	305	F-16, P-16, D-16
Counter	4-bit binary, synchronous	74163	54163	25 MHz	305	F-16, P-16, D-16
Counter	Decade, synchronous	74167	54167	25 MHz	270	F-16, P-16, D-16
Counter	Decade, asynchronous	74176	54176	35 MHz	150	F-14, P-14, D-14
Counter	4-bit, synchronous	74177	54177	35 MHz	150	F-14, P-14, D-14
Counter	Decade, up/down, synchronous	74190	54190	20 MHz	325	F-16, P-16, D-16
Counter	4-bit binary, synchronous	74191	54191	20 MHz	325	F-16, P-16, D-16
Counter	Decade, up/down, synchronous	74192	54192	25 MHz	325	F-16, P-16, D-16
Counter	4-bit binary, synchronous	74193	54193	25 MHz	325	F-16, P-16, D-16
Counter	Decade, asynchronous	74196	54196	50 MHz	240	F-14, P-14, D-14
Counter	4-bit binary, asynchronous	74197	54197	50 MHz	240	F-14, P-14, D-14
Counter	4-bit binary, asynchronous	74293	54293	32 MHz	160	F-14, P-14, D-14
Counter	Decade, presettable	8280	7280	45 MHz	130	F-14, P-14, D-14
Counter	Binary, presettable	8281	7281	45 MHz	130	F-14, P-14, D-14
Counter	Divide by 12, presettable	8288	7288	45 MHz	130	F-14, P-14, D-14
Counter	Decade, latch, three-state	8552	7552	23 MHz	330	F-16, P-16, D-16
Counter	Binary, latch, three-state	8554	7554	23 MHz	330	F-16, P-16, D-16
Counter	Decade, three-state	8555	7555	50 MHz	375	F-16, P-16, D-16
Counter	Binary, three-state	8556	7556	50 MHz	375	F-16, P-16, D-16
Counter	4-bit binary up/down	8284	8284	30 MHz	315	F-14, P-14, D-14
Counter	Decade BCD up/down	8285	8285	30 MHz	315	F-14, P-14, D-14
Counter	Decade presettable	8290	8290	60 MHz	190	F-14, P-14, D-14
Counter	4-bit binary presettable	8291	8291	60 MHz	190	F-14, P-14, D-14
Counter	Decade presettable	8292	8292	10 MHz	53	F-14, P-14, D-14
Counter	4-bit binary presettable	8293	8293	10 MHz	53	F-14, P-14, D-14
Counter	Variable modulo to 16	9305	9305	26 MHz	210	F-14, D-14
Counter	BCD decade	9310	9310	45 MHz	325	F-16, D-16
Counter	4-bit binary	9316	9316	45 MHz	325	F-16, D-16
Counter	Decade	9350	9350	18 MHz	160	D-14
Adder	1-bit full	7480	5480	52	105	F-14, D-14, P-14
Adder	2-bit full	7482	5482	25	174	F-14, D-14, P-14
Adder	4-bit full	7483A	5483A	16	304	F-16, D-16, P-16
Adder	4-bit full	74283	54283	16	304	F-16, D-16, P-16
Adder	1-bit full gated	8268	8268	30	150	F-14, D-14, P-14
Adder	Dual 1-bit	9304	9304	25	150	D-16, F-16
Multiplier	6-bit binary rate	7497	5497	32 MHz	345	D-16, F-16, P-16
Multiplier	Decade rate	74167	54167	32 MHz	270	D-16, F-16, P-16
Multiplier	4-bit parallel	74284	54284	40	305	D-16, F-16, P-16
Multiplier	4-bit parallel	74285	54285	40	305	D-16, F-16, P-16
Multiplier	4-bit by 2-bit	9344	9344	40	550	D-24, F-24
Comparator	4-bit magnitude	7485	5485	21	275	D-16, P-16, F-16
Comparator	10-bit magnitude	8130	7130	20	240	D-24, P-24, F-24
Comparator	6-bit magnitude	8160	7160	20	205	D-16, P-16, F-16
Comparator	4-bit magnitude	8200	7200	20	175	D-14, P-14, F-14
Comparator	5-bit magnitude	9324	9324	40	200	D-16, F-16
Arithmetic logic unit	4-bit	74181	54181	24	445	D-24, P-24, F-24
Arithmetic logic unit	4-bit	8260	8260	24	455	D-24, P-24, F-24
Arithmetic logic unit	4-bit, with carry look ahead	9340	9340	23	425	D-24, F-24
Carry look ahead	4 bit for 54181	74182	54182	13	180	D-16, P-16, F-16
Carry look ahead	4 bit for 8260	8261	8261	13	115	D-14, P-14, F-14
Priority encoder	Full BCD	74147	54147	10	225	D-16, P-16, F-16
Priority encoder	Cascadable octal	74148	54148	10	190	D-16, P-16, F-16
Priority encoder	4-bit cascadable	74278	54278	35	275	D-14, P-14, F-14
Priority encoder	8-input	9318	9318	35	250	D-16, F-16
Parity generator	8-bit odd/even	74180	54180	35	170	D-14, F-14, P-14
Parity generator	9-bit odd/even	8220	7220	30	130	D-14, F-14, P-14
Parity generator	12-bit	9348	9348	40	235	D-16, F-16
Programmable logic	96 term	8575	7575	90	550	D-24, F-24, P-24
Programmable logic	96 term	8575	7576	90	550	D-24, F-24, P-24

TABLE 11   The 5400 Family of Standard-Speed Transistor-Transistor Logic, Complete Selection Guide of Medium-Scale Integration-Circuit Types (continued)

Function	Circuit form	Type No. 0 to 70°C	Type No. −55 to +125°C	Total delay, ns	Power dissipation, mW	Package*
Shift register	8-bit, serial in serial out	7491A	5491A	10 MHz	175	D-14, F-14, P-14
Shift register	4-bit, parallel in parallel out	7495A	5495A	25 MHz	195	D-14, F-14, P-14
Shift register	5-bit, parallel in parallel out	7496	5496	10 MHz	240	D-16, F-16, P-16
Shift register	8-bit, serial in parallel out	74164	54164	25 MHz	167	D-14, F-14, P-14
Shift register	8-bit, parallel in serial out	74165	54165	25 MHz	210	D-16, F-16, P-16
Shift register	8-bit, parallel in serial out	74166	54166	20 MHz	360	D-16, F-16, P-16
Shift register	4-bit, parallel in parallel out	74178	54178	25 MHz	230	D-14, F-14, P-14
Shift register	4-bit, parallel in parallel out	74179	54179	25 MHz	230	D-16, F-16, P-16
Shift register	4-bit, bidirectional, parallel in-out	74194	54194	25 MHz	195	D-16, F-16, P-16
Shift register	4-bit, parallel in-parallel out	74195	54195	30 MHz	195	D-16, F-16, P-16
Shift register	8-bit, bidirectional, parallel in-out	74198	54198	25 MHz	360	D-24, F-24, P-24
Shift register	8-bit, parallel in parallel out	74199	54199	25 MHz	360	D-24, F-24, P-24
Shift register	Dual 3-bit	8200	8200	35 MHz	409	D-24, F-24, P-24
Shift register	Dual 5-bit	8201	8201	35 MHz	409	D-24, F-24, P-24
Shift register	10-bit	8202	8202	35 MHz	409	D-24, F-24, P-24
Shift register	10-bit	8203	8203	35 MHz	409	D-24, F-24, P-24
Shift register	8-bit scaler	8243	8243	25 MHz	315	D-24, F-24, P-24
Shift register	4-bit bidirectional	8270	8270	22 MHz	168	D-14, F-14, P-14
Shift register	4-bit bidirectional	8271	8271	22 MHz	271	D-16, F-16, P-16
Shift register	10-bit, serial in parallel out	8273	8273	35 MHz	341	D-16, F-16, P-16
Shift register	10-bit, parallel in serial out	8274	8274	30 MHz	380	D-16, F-16, P-16
Shift register	8-bit serial in serial out	8276	8276	20 MHz	205	D-14, F-14, P-14
Shift register	Dual 8-bit serial in serial out	8277	8277	20 MHz	540	D-16, F-16, P-16
Shift register	4-bit universal	9300	9300	38 MHz	300	D-16, F-16
Shift register	Dual 8-bit, serial in-out	9328	9328	20 MHz	300	D-16, F-16
Shift register	8-bit multiport	9338	9338	25 MHz	425	D-16, F-16
Register file	Quad bus register	74173	54173	25 MHz	250	D-16, F-16, P-16
Register file	Hex D register	74174	54174	25 MHz	225	D-16, F-16, P-16
Register file	Quad D register	74175	54175	25 MHz	150	D-16, F-16, P-16
Register file	Quad storage MUX	74298	54298	25 MHz	195	D-16, F-16, P-16
Memory–RAM	16-bit 16 × 1	7481A	5481A	15	224	D-14, F-14, P-14
Memory–RAM	16-bit 16 × 1	7484A	5484A	15	224	D-16, F-16, P-16
Memory–RAM	64-bit 16 × 4	7489		35	380	D-16, F-16, P-16
Memory–RAM	16-bit 4 × 4	74170	54170	30	640	D-16, F-16, P-16
Memory–RAM	16-bit 8 × 2	74172		30	560	D-24, F-24, P-24
Memory–RAM	256-bit 256 × 1	74200		45	460	D-16, F-16, P-16
Memory–RAM	8-bit	8553	7553	25	330	D-16, F-16, P-16
Memory–RAM	64-bit 16 × 4	8599	7599	28	200	D-16, P-16
Memory–RAM	4-bit	8275	8275	25	205	D-16, P-16, F-16
Memory–RAM	4-bit	9314	9314	25	175	D-16, F-16
Memory–ROM	256-bit 32 × 8	7488A	5488A	25	282	D-16, F-16, P-16
Memory–ROM	512-bit 64 × 4 prog.	74186	54186	50	307	D-24, F-24, P-24
Memory–ROM	1,024-bit 256 × 4	74187	54187	40	470	D-16, F-16, P-16
Memory–ROM	256-bit 32 × 8 prog.	74188A		30	334	D-16, F-16, P-16
Memory–ROM	1,024-bit prog.	8573	7573	60	400	D-16, P-16
Memory–ROM	1,024-bit prog., three-state	8574	7574	60	400	D-16, P-16
Memory–ROM	4,096-bit	8595	7595	90	515	D-24, P-24
Memory–ROM	4,096-bit	8596	7596	90	515	D-24, P-24
Memory–ROM	4,096-bit, three-state	8597	7597	90	530	D-24, P-24
Memory–ROM	4,096-bit, three-state	8598	7598	90	530	D-24, P-24
Multiplexer	16-line to 1-line	74150	54150	11	200	D-24, P-24, F-24
Multiplexer	8-line to 1-line	74151A	54151A	8/16	145	D-16, P-16, F-16
Multiplexer	8-line to 1-line	74152	54152	8	130	D-14, P-14, F-14
Multiplexer	Dual 4-line to 1-line	74153	54153	15	180	D-16, P-16, F-16
Multiplexer	Quad 2-line to 1-line	74157	54157	9	150	D-16, P-16, F-16
Multiplexer	8-line to 1-line (three-state)	74251	54251	20	250	D-16, P-16, F-16
Multiplexer	Quad 2-line to 1-line (storage)	54298	54298	20	195	D-16, P-16, F-16
Multiplexer	8-line to 1-line (three-state)	8121	7121	15	150	D-16, P-16, F-16
Multiplexer	Dual 4-line to 1-line (three-state)	8214	7214	20	170	D-16, P-16, F-16
Multiplexer	16-line to 1-line (three-state)	8219	7219	10	225	D-24, P-24, F-24
Multiplexer	8-line to 1-line	8230	8230	20	185	D-16, P-16, F-16
Multiplexer	8-line to 1-line	8231	8231	20	185	D-16, P-16, F-16
Multiplexer	8-line to 1-line	8232	8232	20	175	D-16, P-16, F-16
Multiplexer	Quad 2-line to 1-line	8233	8233	25	200	D-16
Multiplexer	Quad 2-line to 1-line	8234	8234	25	160	D-16, P-16, F-16

TABLE 11   The 5400 Family of Standard-Speed Transistor-Transistor Logic, Complete Selection Guide of Medium-Scale Integration-Circuit Types (Continued)

Function	Circuit form	Type No. 0 to 70°C	Type No. −55 to +125°C	Total delay, ns	Power dissipation, mW	Package*
Multiplexer	Quad 2-line to 1-line	8235	8235	25	235	D-16, P-16, F-16
Multiplexer	Quad 3-line to 1-line	8263	8263	25	378	D-24, P-24, F-24
Multiplexer	Quad 3-line to 1-line	8264	8264	25	400	D-24, P-24, F-24
Multiplexer	Quad 2-line to 1-line	8266	8266	30	200	D-14, P-14, F-14
Multiplexer	Quad 2-line to 1-line	8267	8267	30	200	D-14, P-14, F-14
Multiplexer	Dual 4-line to 1-line	9309	9309	25	150	D-16, F-16
Multiplexer	8-line to 1-line	9312	9312	25	300	D-16, F-16
Multiplexer	Quad 2-line to 1-line	9322	9322	20	150	D-16, F-16
Decoder	BCD to decimal	7442A	5442A	17	140	D-16, P-16, F-16
Decoder	Excess 3 to decimal	7443A	5443A	17	140	D-16, P-16, F-16
Decoder	Excess 3 gray to decimal	7444A	5444A	17	140	D-16, P-16, F-16
Decoder	4-line to 16-line	74154	54154	20	170	D-24, P-24, F-24
Decoder	Dual 2-line to 4-line	74155	54155	20	250	D-16, P-16, F-16
Decoder	Dual 2-line to 4-line	74156	54156	20	250	D-16, P-16, F-16
Decoder	4-line to 16-line	74159	54159	20	170	D-24, P-24, F-24
Decoder	1-line to 8-line	8210	7210	20	100	D-16, P-16, F-16
Decoder	1-line to 8-line	8211	7211	20	100	D-14, P-14, F-14
Decoder	1-line to 8-line	8223	7223	25	140	D-16, P-16
Decoder	2-line to 4-line	8230	7230	20	240	D-16, P-16, F-16
Decoder	1-line to 10-line	9301	9301	20	145	D-16, F-16
Decoder	1-line to 10-line	9302	9302	25	155	D-16, F-16
Decoder	Dual 1-line to 4-line	9321	9321	20	150	D-16, F-16
Decoder	1-line to 16-line	9311	9311	20	175	D-24, F-24
Code converter	BCD to binary (6-line)	74184	54184	25	280	D-16, P-16, F-16
Code converter	Binary to BCD (6-bit)	74185A	54185A	25	280	D-16, P-16, F-16
Display driver	BCD to decimal (30-V)	7445	5445		215	D-16, P-16, F-16
Display driver	BCD to 7-segment	7446A	5446A		320	D-16, P-16, F-16
Display driver	BCD to 7-segment	7447A	5447		320	D-16, P-16, F-16
Display driver	BCD to 7-segment	7448	5448		265	D-16, P-16, F-16
Display driver	BCD to 7-segment	7449	5449		165	D-14, P-14, F-14
Display driver	BCD to decimal (60-V)	74141			80	D-16, P-16, F-16
Display driver	BCD to decimal (latch) (55-V)	74142			340	D-16, P-16, F-16
Display driver	BCD to 7-segment (latch)	74143	54143		280	D-24, P-24, F-24
Display driver	BCD to 7-segment (latch)	74144	54144		280	D-24, P-24, F-24
Display driver	BCD to decimal	74145	54145		215	D-16, P-16, F-16
Display driver	BCD to 7-segment (16 × 7)	8880	7880		135	D-16, P-16
Display driver	BCD to 7-segment	8884	7884		200	D-18, P-18
Display driver	BCD to 7-segment	9307	9307		300	D-16, F-16
Display driver	BCD to 7-segment	9317	9317			D-16, F-16
Pulse synchronizer	Dual 30 MHz	74120	54120	16	255	D-16, F-16, P-16

* Package entry letter is form factor: F = flatpack, D = hermetic dual-in-line, P = plastic dual-in-line, C = round can, S = studded flatpack. Package entry number is terminal count.

provides a level with which to compare the input signal. This comparison then sets the levels which drive the emitter-follower output. These emitter followers provide the final degree of level shifting and provide sufficient current drive to operate transmission lines. These transmission lines are well suited to the fast rise times associated with high-speed circuitry. The emitter-follower circuits are designed both with and without pulldown resistors. The use of pulldown resistors is intended to drive high fanout loads, while those designed without pulldown resistors can be used for the wired OR function, which is possible with ECL. Emitter-coupled logic tends to have a smaller voltage difference between the 1 and 0 levels than the various forms of saturated logic that it is normally compared with. This smaller voltage margin between logic levels is not necessarily a disadvantage, since most noise in logic systems is generated by the voltage transition between levels.

Emitter-coupled logic has primarily gained acceptance in high-speed applications as the most advanced form of high-speed logic. It has always been faster than the newest forms of saturated logic at any point along the development of logic families. ECL allows for great logic flexibility; so gates can be internally tied into OR and AND strings without introducing additional levels of gating. Many emitter-coupled-logic gate structures allow for simultaneous generation of logic data and

TABLE 12   The 54H Family of High-Speed Transistor-Transistor Logic, Complete Selection Guide of Small-Scale Integration-Circuit Types

| Function | Circuit form | Type No. | | Fanout (unit loads) | Total delay, ns | Power dissipation, mW | Package† |
		0 to 70°C	−55 to +125°C				
NAND gate	Quad 2-input	74H00	54H00	10	6	90	D-14, F-14, P-14
NAND gate	Quad 2-input*	74H01	54H01	10	10	82	D-14, F-14, P-14
NAND gate	Triple 3-input	74H10	54H10	10	6	68	D-14, F-14, P-14
NAND gate	Dual 4-input	74H20	54H20	10	6	45	D-14, F-14, P-14
NAND gate	Dual 4-input*	74H22	54H22	10	10	41	D-14, F-14, P-14
NAND gate	8-input	74H30	54H30	10	6	23	D-14, F-14, P-14
NAND gate	Dual 4-input	74H40	54H40	20	8	80	D-14, F-14, P-14
AND gate	Triple 3-input	74H11	54H11	10	8	120	D-14, F-14, P-14
AND gate	Triple 3-input*	74H15	54H15	10	12	83	D-14, F-14, P-14
AND gate	Dual 4-input	74H21	54H21	10	8	80	D-14, F-14, P-14
Inverter	Hex	74H04	54H04	10	6	135	D-14, F-14, P-14
AND-OR invert gate	Dual 2 × 2*	74H50	54H50	10	8	58	D-14, F-14, P-14
AND-OR invert gate	Dual 2 × 2	74H51	54H51	10	7	58	D-14, F-14, P-14
AND-OR invert gate	4 × 2*	74H53	54H53	10	9	41	D-14, F-14, P-14
AND-OR invert gate	4 × 1	74H54	54H54	10	7	41	D-14, F-14, P-14
AND-OR invert gate	2 × 4*	74H55	54H55	10	9	30	D-14, F-14, P-14
AND-OR gate	2 × 4	74H52	54H52	10	10	88	D-14, F-14, P-14
Expander	Dual 4-input	74H60	54H60				D-14, F-14, P-14
Expander	Triple 3-input	74H61	54H61				D-14, F-14, P-14
Expander	4-wide AND/OR	74H62	54H62				D-14, F-14, P-14
Flip-flop	JK	74H71	54H71	10	30 MHz	95	D-14, F-14, P-14
Flip-flop	JK	74H72	54H72	10	30 MHz	80	D-14, F-14, P-14
Flip-flop	Dual JK	74H73	54H73	10	30 MHz	160	D-14, F-14, P-14
Flip-flop	Dual D	74H74	54H74	10	43 MHz	150	D-14, F-14, P-14
Flip-flop	Dual JK	74H76	54H76	10	30 MHz	160	D-14, F-14, P-14
Flip-flop	Dual JK	74H78	54H78	10	30 MHz	160	D-14, F-14, P-14
Flip-flop	JK-ET	74H101	54H101	10	50 MHz	100	D-14, F-14, P-14
Flip-flop	JK-ET	74H102	54H102	10	50 MHz	100	D-14, F-14, P-14
Flip-flop	Dual JK-ET	74H103	54H103	10	50 MHz	200	D-14, F-14, P-14
Flip-flop	Dual JK-ET	74H106	54H106	10	50 MHz	200	D-14, F-14, P-14
Flip-flop	Dual JK-ET	74H108	54H108	10	50 MHz	200	D-14, F-14, P-14

* Uncommitted collector output.
† Package entry letter is form factor: F = flatpack, D = hermetic dual-in-line, P = plastic dual-in-line, C = round can, S = studded flatpack. Package entry number is terminal count.

TABLE 13   The 54H Family of High-Speed Transistor-Transistor Logic, Complete Selection Guide of Medium-Scale Integration-Circuit Types

| Function | Circuit form | Type No. | | Total delay, ns | Power dissipation, mW | Package* |
		0 to 70°C	−55 to +125°C			
Adder	Dual 1-bit carry save	74H183	54H183	11	220	D-14, F-14, P-14
Shift register	4-bit universal	93H00	93H00	55 MHz	350	D-16, F-16

* Package entry letter is form factor: F = flatpack, D = hermetic dual-in-line, P = plastic dual-in-line, C = round can, S = studded flatpack. Package entry number is terminal count.

its complement.   These functions, which are sometime called series gates, make ECL an excellent choice for high-speed MSI arrays.   ECL can function as internal gates at very low power levels, and this feature has furthered its broad acceptance.
    The 10,000 family introduced by Motorola in 1970 is the preferred ECL family for high-speed digital-computer applications.   The 10,000 family is available in both limited-temperature-range and military-temperature-range versions.   The family is specified by a set of slash sheets to MIL-M-33510, allowing it to gain broad acceptance both in high-speed commercial data-processing equipment and in high-speed military digital equipment.   This family is being expanded to contain a great number

TABLE 14   The 54L Family of Low-Power Transistor-Transistor Logic, Complete Selection Guide of Small-Scale Integration-Circuit Types

Function	Circuit form	Type No. 0 to 70°C	Type No. −55 to +125°C	Fanout (unit loads)	Total delay, ns	Power dissipation, mW	Package†
NAND gate	Quad 2-input	74L00	54L00	10	35	4	D-14, F-14, P-14
NAND gate*	Quad 2-input	74L01	54L01	10	60	4	D-14, F-14, P-14
NAND gate*	Quad 2-input	74L03	54L03	10	60	4	D-14, F-14, P-14
NAND gate	Triple 3-input	74L10	54L10	10	35	3	D-14, F-14, P-14
NAND gate	Dual 4-input	74L20	54L20	10	35	2	D-14, F-14, P-14
NAND gate	8-input	74L30	54L30	10	35	1	D-14, F-14, P-14
AND gate	Triple 3-input	74L11	54L11				D-14, F-14, P-14
NOR gate	Quad 2-input	74L02	54L02	10	32	5	D-14, F-14, P-14
Exclusive-OR gate	Quad 2-input	74L86	54L86	10	55	15	D-14, F-14, P-14
Inverter	Hex	74L04	54L04	10	35	6	D-14, F-14, P-14
Inverter	Hex three-state	70L96	80L96	10	30	17	D-16, F-16, P-16
Inverter	Hex three-state	70L98	80L98	10	30	17	D-16, F-16, P-16
AND-OR invert gate	Dual 2 × 2	74L51	54L51	10	50	3	D-14, F-14, P-14
AND-OR invert gate	Dual 2 × 4	74L54	54L54	10	50	2.5	D-14, F-14, P-14
AND-OR invert gate	Dual 4 × 2	74L55	54L55	10	50	1.5	D-14, F-14, P-14
Flip-flop	R-S	74L71	54L71	10	3 MHz	3.8	D-14, F-14, P-14
Flip-flop	JK	74L72	54L72	10	3 MHz	3.8	D-14, F-14, P-14
Flip-flop	Dual JK	74L73	54L73	10	3 MHz	7.6	D-14, F-14, P-14
Flip-flop	Dual D	74L74	54L74	10	3 MHz	8.0	D-14, F-14, P-14
Flip-flop	Dual JK	74L78	54L78	10	3 MHz	7.6	D-14, F-14, P-14
Flip-flop	Dual D	85L11	75L11	10	6 MHz	18	D-16, F-16, P-16
Flip-flop	Dual JK	85L12	75L12	10	6 MHz	22	D-16, F-16, P-16
Flip-flop	Quad D three-state	85L51	75L51	10	6 MHz	30	D-16, F-16, P-16
Flip-flop	Quad D gated	86L13	76L13	10	6 MHz	25	D-16, F-16, P-16
Monostable multi		74L121	54L121	10	150	45	D-16, F-16, P-16
Monostable multi	With clear	74L122	54L122	10	70	55	D-14, F-14, P-14
Monostable multi	Dual with clear	74L123	54L123	10	70	55	D-16, F-16, P-16
Level translator	TTL to MOS hex	88L12	78L12			60	7	D-14, F-14, P-14

* Uncommitted collector output.
† Package entry letter is form factor: F = flatpack, D = hermetic dual-in-line, P = plastic dual-in-line, C = round can, S = studded flatpack. Package entry number is terminal count.

of logic gates, line drivers, and level converters, and a full spectrum of MSI functions. It is currently available from four major integrated-circuit manufacturers.

The primary advantage of the ECL 10,000 family is the sensible combination of propagation delay times and gate-edge speeds. Crosstalk, reflected signals, and oscillation all worsen with decreasing rise time during the logic transition, and the ideal logic-gate family would have the longest rise time consistent with the shortest propagation delay time. This optimization goal has been nearly obtained in the 2-ns delay time and the 3.5-ns gate-edge rise time of the ECL 10,000 family.

The prime features of ECL are high logic speeds, low output impedance, capability to drive a great number of loads, power-supply current independence from data frequency of logic state, low noise generation of the slow front-edge rise time, availability of collector dotting, emitter dotting, and direct complementary outputs, the large size of the family, and the growing availability of MSI circuits. These advantages are counterbalanced by the disadvantages of higher circuit cost and required use of controlled-impedance transmission lines. ECL circuits have found successful military avionic systems, large real-time computers, digital communication systems, multiplex data-transmission systems, and electronic instrumentation such as counters.

MECL I   MECL I° is the most rudimentary of the ECL families, and the family is now obsolete. The family had a basic gate delay of 8 ns per logic decision. It was offered in two temperature ranges, the MC 300 series of full MIL temperature-range devices (−55 to +125°C) and the MC 350 series, which functioned over the limited temperature range of 0 to 75°C. The basic logic cell with the OR/NOR gate is illustrated in Fig. 17. This basic gate contained only the differential-amplifier input and the emitter-follower outputs with fixed pulldown resistors to V_{EE}. The basic gate circuit did not include a voltage reference point. The circuit required

° Trademark of Motorola Semiconductor Products.

TABLE 15   The 54L Family of Low-Power Transistor-Transistor Logic, Complete Selection Guide of Medium-Scale Integration-Circuit Types

Function	Circuit form	Type No. 0 to 70°C	Type No. −55 to +125°C	Total delay, ns	Power dissipation, mW	Package*
Counter	Decade asynchronous	74L90	54L90	3 MHz	20	D-14, P-14, F-14
Counter	4-bit binary asynchronous	74L93	54L93	3 MHz	20	D-14, P-14, F-14
Counter	Decade, up-down synchronous	74L192	54L192	3 MHz	42	D-16, P-16, F-16
Counter	4-bit binary synchronous	74L193	54L193	3 MHz	42	D-16, P-16, F-16
Counter	Decade latch	85L52	75L52	6 MHz	38	D-16, P-16, F-16
Counter	4-bit binary latch	85L54	75L54	6 MHz	38	D-16, P-16, F-16
Counter	Decade, presettable	86L75	76L75	13 MHz	32	D-16, P-16, F-16
Counter	4-bit binary, presettable	86L76	76L76	13 MHz	32	D-16, P-16, F-16
Counter	Decade	93L10	93L10	20 MHz	85	D-16, F-16
Counter	4-bit binary	93L16	93L16	20 MHz	85	D-16, F-16
Comparator	4-bit magnitude	74L85	54L85	82	20	D-16, P-16, F-16
Comparator	5-bit magnitude	93L24	93L24	55	52	D-16, F-16
Priority encoder	8-bit	93L18	93L18	55	75	D-16, F-16
Shift register	8-bit serial in serial out	74L91	54L91	3 MHz	17.5	D-14, P-14, F-14
Shift register	4-bit parallel in parallel out	74L95	54L95	3 MHz	19	D-14, P-14, F-14
Shift register	5-bit parallel in parallel out	74L96	54L96	5 MHz	120	D-16, P-16, F-16
Shift register	4-bit parallel in parallel out	74L99	54L99	3 MHz	19	D-16, P-16, F-16
Shift register	8-bit serial in parallel out	74L164	54L164	12 MHz	84	D-14, P-14, F-14
Shift register	4-bit universal	93L00	93L00	15 MHz	75	D-16, F-16
Shift register	Dual 8-bit	93L28	93L28	10 MHz	80	D-16, F-16
Shift register	4-bit parallel in parallel out	86L70	76L70	3 MHz	19	D-14, P-14, F-14
Register file	Quad MUX with storage	74L98	54L98	3 MHz	25	D-16, P-16, F-16
Memory—RAM	64-bit, 16 × 4	74L89	54L89	50	55	D-16, P-16, F-16
Memory—RAM	64-bit, 16 × 4 three-state	86L99	76L99	50	80	D-16, P-16, F-16
Memory—RAM	4-bit	93L14	76L14	68	50	D-16, F-16
Memory—ROM	1,024-bit	86L97	76L97	70	60	D-16, P-16, F-16
Multiplexer	Dual 4-line to 1-line	74L153	54L153	27	90	D-16, P-16, F-16
Multiplexer	Quad 2-line to 1-line	74L157	54L157	18	75	D-16, P-16, F-16
Multiplexer	Dual 4-line to 1-line	93L09	93L09	48	40	D-16, F-16
Multiplexer	8-line to 1-line	93L12	93L12	80	45	D-16, F-16
Multiplexer	Quad 2-line to 1-line	93L22	93L22	44	45	D-16, F-16
Decoder	BCD to decimal	74L42	54L42	34	70	D-16, F-16, P-16
Decoder	Excess 3 to decimal	74L43	54L43	34	70	D-16, F-16, P-16
Decoder	Excess 3 gray to decimal	74L44	54L44	34	70	D-16, F-16, P-16
Decoder	4-line to 16-line	74L154	54L154	40	85	D-24, F-24, P-24
Decoder	1-line to 10-line	93L01	93L01	63	45	D-16, F-16
Decoder	Dual 1-line to 4-line	93L21	93L21	50	45	D-16, F-16
Decoder	1-line to 16-line	93L11	93L11	70	74	D-24, F-24
Display driver	BCD to 7-segment (30-V)	74L46	54L46		133	D-16, P-16, F-16
Display driver	BCD to 7-segment (15-V)	74L47	54L47		133	D-16, P-16, F-16

* Package entry letter is form factor: F = flatpack, D = hermetic dual-in-line, P = plastic dual-in-line, C = round can, S = studded flatpack. Package entry number is terminal count.

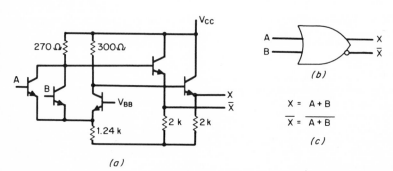

Fig. 17   Characterization of the MECL I family of emitter-coupled logic. OR/NOR-gate characterization. (*a*) Circuit schematic. (*b*) Logic diagram. (*c*) Logic equation.

TABLE 16   The 54S Family of Schottky Clamped Transistor-Transistor Logic, Complete Selection Guide of Small-Scale Integration-Circuit Types

Function	Circuit form	Type No. 0 to 70°C	Type No. −55 to +125°C	Fanout (unit loads)	Total delay, ns	Power dissipation, mW	Package §
NAND gate	Quad 2-input	74S00	54S00	10	4	75	D-14, P-14, F-14
NAND gate*	Quad 2-input	74S03	54S03	10	5	65	D-14, P-14, F-14
NAND gate	Triple 3-input	74S10	54S10	10	4	57	D-14, P-14, F-14
NAND gate	Dual 4-input	74S20	54S20	10	4	38	D-14, P-14, F-14
NAND gate*	Dual 4-input	74S22	54S22	10	5	33	D-14, P-14, F-14
NAND gate	8-input	74S30	54S30	10	4	21	D-14, P-14, F-14
NAND gate	Dual 4-input	74S40	54S40	20	6	44	D-14, P-14, F-14
NAND gate	13-input	74S133	54S133	10	4	21	D-14, P-14, F-14
NAND gate†	12-input	74S134	54S134	10			D-14, P-14, F-14
NAND gate‡	Quad 2-input	74S132	54S132	10	8	180	D-14, P-14, F-14
AND	Triple 3-input	74S11	54S11	10	5	94	D-14, P-14, F-14
AND*	Triple 3-input	74S15	54S15	10	6	86	D-14, P-14, F-14
NOR	Quad 2-input	74S02	54S02	10	5	104	D-14, P-14, F-14
NOR	Dual 5-input	74S260	54S260	10	3.5	93	D-14, P-14, F-14
Inverter	Hex	74S04	54S04	10	4	112	D-14, P-14, F-14
Inverter*	Hex	74S05	54S05	10	5	98	D-14, P-14, F-14
AND-OR invert	Dual 2 × 2	74S51	54S51	10	4	54	D-14, P-14, F-14
AND-OR invert	Dual 2 × 4	74S64	54S64	10	4	38	D-14, P-14, F-14
AND-OR invert	Dual 2 × 4	74S65	54S65				D-14, P-14, F-14
Flip-flop	Dual D	74S74	54S74	10	110 MHz	150	D-14, P-14, F-14
Flip-flop	Dual JK-ET	74S112	54S112	10	125 MHz	150	D-14, P-14, F-14
Flip-flop	Dual JK-ET	74S113	54S113	10	125 MHz	150	D-14, P-14, F-14
Flip-flop	Dual JK-ET	74S114	54S114	10	125 MHz	150	D-14, P-14, F-14
Latch	Hex D	74S174	54S174	10	110 MHz	450	D-14, P-14, F-14
Latch	Quad D	74S175	54S175	10	110 MHz	300	D-14, P-14, F-14
Line driver	Dual 4-input NAND	74S140	54S140	6		88	D-14, F-14, P-14

* Uncommitted collector output.
† Three-state output.
‡ Schmitt-trigger input.
§ Package entry letter is form factor: F = flatpack, D = hermetic dual-in-line, P = plastic dual-in-line, C = round can, S = studded flatpack. Package entry number is terminal count.

an external bias driver, which was part of the family. The family consisted of a small range of small-scale integration products, which were packaged in 10-lead enclosures. Functions offered included OR/NOR gates, gate expanders, the bias driver, a half adder, JK flip-flops, line drivers, lamp drivers, and a pair of level translators to make the transition between MECL I and DTL and transition between DTL and MECL I. The family also included two highspeed gates, or clock drivers, which were capable of driving a fanout of 100, a factor of 4 greater than the standard 25 fanout of a MECL I gate. The MECL I clock drivers were in reality the foreshadowing of the MECL II family with its decreased propagation delay time. The clock drivers, the MC 369F and the MC 369G, had a propagation delay time of 3 ns. The three power-supply nodes must operate with a fixed relationship between them. It does not matter which of the nodes is returned to circuit ground, although it is usually most convenient to ground the V_{CC} node. In this case, V_{CC} equals 0 V, V_{BB} equals −1.15 V, and V_{EE} equals −5.2 V. In this configuration, the output logic swing then makes the transition from a high state of approximately −0.7 V to a low state of approxmiately −1.6 V.

In using the ac-coupled functions, the dynamic 0 is then defined as the negative-going voltage excursion and the positive as the positive-going excursion. The bias voltage applied to the bias input should be obtained from a temperature-compensated bias driver, type MC 304 or MC 354. The voltage-temperature characteristics of this bias are directly suited to compensate the threshold points of the logic gate. A bias driver may be used to drive up to 25 logic elements. The available circuits are listed in Table 20.

MECL II The MECL II family was introduced approximately 4 years later than the MECL I family, and it represented a basic improvement and expansion in the capability of the initial offering, MECL I. MECL II included an internal bias driver

TABLE 17   The 54S Family of Schottky Clamped Transistor-Transistor Logic, Complete Selection Guide of Medium-Scale Integration-Circuit Types

Function	Circuit form	Type No. 0 to 70°C	Type No. −55 to +125°C	Total delay, ns	Power dissipation, mW	Package*
Counter	Decade	82S90	82S90	100 MHz	308	D-14, P-14
Counter	4-bit binary	82S91	82S91	100 MHz	308	D-14, P-14
Counter	Settable to 16	93S05	93S05	100 MHz	400	D-14, P-14
Counter	BCD decade	93S10	93S10	100 MHz	400	D-16, F-16
Counter	4-bit binary	93S16	93S16	100 MHz	400	D-16, F-16
Counter	4-bit binary	74S163	54S163	40 MHz	475	D-16, F-16, P-16
Counter	4-bit up/down binary	74S169	54S169	40 MHz	500	D-16, F-16, P-16
Counter	Decade	74S162	54S162	40 MHz	475	D-16, F-16, P-16
Counter	Decade up/down	74S168	54S168	40 MHz	500	D-16, F-16, P-16
Adder	BCD decimal	82S83	82S83	25		D-16, P-16
Comparator	4-bit magnitude	74S85	54S85	11	365	D-16, P-16, F-16
Arithmetic logic	4-bit unit	74S181	54S181	10	600	D-24, P-24, F-24
Arithmetic logic	4-bit with carry	82S82	82S82	30		D-24, P-24
Accumulator	4-bit binary	74S281	54S281	20	720	D-24, P-24, F-16
Carry look ahead	Use with 54S181	74S182	54S182	7	260	D-16, P-16, F-16
Multiplier	4-bit by 4-bit	74S274		45		P-20
Wallace tree	7-bit slice	74S275	54S275	45		D-16, P-16
Parity generator	9-bit	74S280	54S280	13	335	D-14, P-14, F-14
Parity generator	9-bit	93S62	93S62	16	300	D-14
Shift register	4-bit parallel in-out	74S194	54S194	70 MHz	450	D-16, P-16, F-16
Shift register	4-bit parallel in-out	74S195	54S195	70 MHz	375	D-16, P-16, F-16
Shift register	4-bit, serial-parallel in	82S70	82S70	60 MHz	400	D-14, P-14, F-14
Shift register	4-bit, serial-parallel in	82S71	82S71	60 MHz	400	D-16, P-16, F-16
Shift register	4-bit universal	93S00	93S00	100 MHz		D-16, F-16
Shift register	8-bit multiport	93S39	93S39			D-16, F-16
Register file	Hex D register	74S174	54S174	75 MHz	450	D-16, P-16, F-16
Register file	Quad D register	74S175	54S175	75 MHz	300	D-16, P-16, F-16
Memory—RAM	256-bit 256 × 1	74S200	54S200	30	435	D-16, F-16, P-16
Memory—RAM	256-bit 256 × 1 open col.	74S206	54S206	32	435	D-16, F-16, P-16
Multiplexer	8-line to 1-line	74S151	54S151	7	225	D-16, F-16, P-16
Multiplexer	Dual 4-line to 1-line	74S153	54S153	6	225	D-16, F-16, P-16
Multiplexer	Quad 2-line to 1-line	74S157	54S157	5	250	D-16, F-16, P-16
Multiplexer	Quad 2-line to 1-line	74S158	54S158	4	195	D-16, F-16, P-16
Multiplexer	8-line to 1-line	74S251	54S251	6	275	D-16, F-16, P-16
Multiplexer	Quad 2-line to 1-line	74S257	54S257	5	320	D-16, F-16, P-16
Multiplexer	Quad 2-line to 1-line	74S258	54S258	4	280	D-16, F-16, P-16
Multiplexer	8-line to 1-line	82S30	82S30	12	325	D-16, P-16
Multiplexer	8-line to 1-line	82S31	82S31	12	325	D-16, P-16
Multiplexer	8-line to 1-line	82S32	82S32	14	325	D-16, P-16
Multiplexer	Dual 4-line to 1-line	82S33	82S33	10	340	D-16, P-16
Multiplexer	Dual 4-line to 1-line	82S34	82S34	10	340	D-16, P-16
Multiplexer	Quad 2-line to 1-line	82S66	82S66	6	350	D-16, P-16
Multiplexer	Quad 2-line to 1-line	82S67	82S67	6	350	D-16, P-16
Decoder	3-line to 8-line	74S138	54S138	8	225	D-16, P-16, F-16
Decoder	Dual 2-line to 4-line	74S139	54S139	7	300	D-16, P-16, F-16

* Package entry letter is form factor: F = flatpack, D = hermetic dual-in-line, P = plastic dual-in-line, C = round can, S = studded flatpack. Package entry number is terminal count.

in each circuit to simplify the use of the circuit, and the circuits included gates with and without output pulldown resistors to provide additional options. Like MECL I, it could drive 25 loads, and it had similar input characteristics. The improvement in the circuit reduced the propagation delay time from 8 to 4 ns, although the edge speed was similarly reduced from 10 to 4 ns.

A group of higher-speed flip-flops introduced with this family allowed frequencies approaching 200 MHz. The basic MECL II gate had a dissipation of 22 mW, which is approximately two-thirds the dissipation of the MECL I gate. The use of 14- and 16-pin packages and small geometries allowed introduction of more complex logic functions. MECL II included OR/NOR gates with and without pulldown resistors, JK and RS flip-flops, adders, line receivers, subtractors, D flip-flops, clock drivers, data selectors, exclusive-OR gates, memories, level translators, AND gates, NAND gates, and latches. The full range of available circuits and their basic characteristics is presented in Table 21.

TABLE 18 The 54LS Family of Low-Power Schottky Clamped Transistor-Transistor Logic, Complete Selection Guide of Small-Scale Integration-Circuit Types

Function	Circuit form	Type No.		Fanout (unit loads)	Total delay, ns	Power dissipation, mW	Package†
		0 to 70°C	−55 to +125°C				
NAND gate	Quad 2-input	74LS00	54LS00	10	10	8	D-14, P-14, F-14
NAND gate*	Quad 2-input	74LS01	54LS01	10	16	8	D-14, P-14, F-14
NAND gate*	Quad 2-input	74LS03	54LS03	10	16	8	D-14, P-14, F-14
NAND gate	Triple 3-input	74LS10	54LS10	10	10	6	D-14, P-14, F-14
NAND gate	Dual 4-input	74LS20	54LS20	10	16	4	D-14, P-14, F-14
NAND gate*	Dual 4-input	74LS22	54LS22	10	16	8	D-14, P-14, F-14
NAND gate	8-input	74LS30	54LS30	10	10	2.4	D-14, P-14, F-14
NAND gate	Quad 2-input	74LS37	54LS37	20	12	17	D-14, P-14, F-14
NAND gate*	Quad 2-input	74LS38	54LS38	20	18	17	D-14, P-14, F-14
NAND gate	Dual 4-input	74LS40	54LS40	20	12	8	D-14, P-14, F-14
AND gate	Quad 2-input	74LS08	54LS08	10	12	17	D-14, P-14, F-14
AND gate*	Quad 2-input	74LS09	54LS09	10	20	17	D-14, P-14, F-14
AND gate	Triple 3-input	74LS11	54LS11	10	12	12	D-14, P-14, F-14
AND gate*	Triple 3-input	74LS15	54LS15	10	20	13	D-14, P-14, F-14
AND gate	Dual 4-input	74LS21	54LS21	10	12	8	D-14, P-14, F-14
NOR gate	Quad 2-input	74LS02	54LS02	10	10	11	D-14, P-14, F-14
NOR gate	Triple 3-input	74LS27	54LS27	10	10	14	D-14, P-14, F-14
NOR gate	Quad 2-input	74LS28	54LS28	10	12	21	D-14, P-14, F-14
NOR gate*	Quad 2-input	74LS33	54LS33	10	20	22	D-14, P-14, F-14
OR gate	Quad 2-input	74LS32	54LS32	10	14	20	D-14, P-14, F-14
Inverter	Hex	74LS04	54LS04	10	10	12	D-14, P-14, F-14
Inverter*	Hex	74LS05	54LS05	10	16	12	D-14, P-14, F-14
AND-OR invert gate	Dual 2 × 2	74LS51	54LS51	10	12	5	D-14, P-14, F-14
AND-OR invert gate	Dual 2 × 4	74LS54	54LS54	10	16	4.5	D-14, P-14, F-14
AND-OR invert gate	Dual 4 × 2	74LS55	54LS55	10	12	2.5	D-14, P-14, F-14
Flip-flop	Dual JK	74LS73	54LS73	10	45 MHz	20	D-14, P-14, F-14
Flip-flop	Dual D	74LS74	54LS74	10	33 MHz	20	D-14, P-14, F-14
Flip-flop	Dual JK	74LS76	54LS76	10	45 MHz	20	D-14, P-14, F-14
Flip-flop	Dual JK	74LS78	54LS78	10	45 MHz	20	D-14, P-14, F-14
Flip-flop	Dual JK-ET	74LS109	54LS109	10	33 MHz	20	D-14, P-14, F-14
Flip-flop	Dual JK-ET	74LS112	54LS112	10	45 MHz	20	D-14, P-14, F-14
Flip-flop	Dual JK-ET	74LS113	54LS113	10	45 MHz	20	D-14, P-14, F-14
Flip-flop	Dual JK-ET	74LS114	54LS114	10	45 MHz	20	D-14, P-14, F-14
Latch	Hex D	74LS174	54LS174	10	40 MHz	66	D-14, F-14, P-14
Register	4-bit	25LS08	25LS08	24	13	55	D-16, P-16
Register	6-bit	25LS07	25LS07	24	13	80	D-16, P-16

* Uncommitted collector.
† Package entry letter is form factor: F = flatpack, D = hermetic dual-in-line, P = plastic dual-in-line, C = round can, S = studded flatpack. Package entry number is terminal count.

The broad range of small-scale integration available in this family simplified the systems application of the circuits. Inclusion of a limited number of more complex circuits indicated the direction of further expansion within the MECL families. The prime features of MECL II were 4-ns delay times, the high fanout, and the availability of data and data complement simultaneously.

The basic MECL II gate is illustrated in Fig. 18. This gate provides for multiple inputs and contains the basic differential-amplifier stage to compare the input contents with the reference, which is not internally generated by a bias circuit. Emitter-follower outputs eliminate saturated delays in making the transition between logic states. The MECL II circuits, which include the on-chip pulldown resistors, can be used without requiring external components, but they are not capable or wire OR-ing. The availability of both types of circuits increases the flexibility of application of this family. The reduced rise time of the logic transition in the MECL II requires that open-wire lengths be restricted to 12 in to minimize undershoot and reflection. This restriction requires controlled-impedance transmission lines for all extended-length actions.

In applying the wired-OR technique to MECL II circuits, a maximum of two output load resistors are recommended per wire or connection to limit the output current. It is possible, when using the wire OR connection, to get noise spikes generated on the output when all gates are at one output and all of the gates except one can

TABLE 19   The 54LS Family of Low-Power Schottky Clamped Transistor-Transistor Logic, Complete Selection Guide of Medium-Scale Integration-Circuit Types

Function	Circuit form	Type No. 0 to 70°C	Type No. −55 to +125°C	Total delay, ns	Power dissipation, mW	Package*
Counter	Decade, set to 9, asynchronous	74LS90	54LS90	32 MHz	40	D-14, P-14, F-14
Counter	Divide by 12 asynchronous	74LS92	54LS92	32 MHz	39	D-14, P-14, F-14
Counter	4-bit binary asynchronous	74LS93	54LS93	32 MHz	39	D-14, P-14, F-14
Counter	Decade synchronous	74LS160	54LS160	25 MHz	93	D-16, P-16, F-16
Counter	4-bit binary synchronous	74LS161	54LS161	25 MHz	93	D-16, P-16, F-16
Counter	Decade synchronous	74LS162	54LS162	25 MHz	93	D-16, P-16, F-16
Counter	4-bit binary	74LS163	54LS163	25 MHz	93	D-16, P-16, F-16
Counter	4-bit binary up/down synchronous	74LS169	54LS169	25 MHz	100	D-16, P-16, F-16
Counter	Decade up/down synchronous	74LS190	54LS190	20 MHz	90	D-16, P-16, F-16
Counter	4-bit binary synchronous	74LS191	54LS191	20 MHz	90	D-16, P-16, F-16
Counter	Decade up/down synchronous	74LS192	54LS192	25 MHz	85	D-16, P-16, F-16
Counter	4-bit binary synchronous	74LS193	54LS193	25 MHz	85	D-16, P-16, F-16
Counter	Decade asynchronous	74LS196	54LS196	50 MHz	240	D-14, P-14, F-14
Counter	4-bit binary asynchronous	74LS197	54LS197	50 MHz	240	D-14, P-14, F-14
Counter	Decade asynchronous	74LS290	54LS290	32 MHz	40	D-14, P-14, F-14
Counter	4-bit binary asynchronous	74LS293	54LS293	32 MHz	39	D-14, P-14, F-14
Adder	4-bit full	74LS83A	54LS83A	15	24	D-16, P-16, F-16
Adder	4-bit full	74LS283	54LS283	15	24	D-16, P-16, F-16
Arithmetic logic unit	4-bit	74LS181	54LS181	24	410	D-24, P-24, F-24
Multiplier	2-bit by 4-bit	74LS261	54LS261	26	110	D-16, P-16, F-16
Comparator	4-bit magnitude	74LS85	54LS85	23	52	D-16, P-16, F-16
Register	Hex D	74LS174	54LS174	30 MHz	65	D-16, P-16, F-16
Register	Quad D	74LS175	54LS175	30 MHz	45	D-16, P-16, F-16
Register file	16-bit (4 × 4)	74LS170	54LS170	30	125	D-16, P-16, F-16
Register file	16-bit (4 × 4) tristate	74LS670	54LS670	24	135	D-16, P-16, F-16
Shift register	8-bit serial	74LS91	54LS91	10 MHz	60	D-14, P-14, F-14
Shift register	4-bit parallel in, parallel out	74LS95B	54LS95B	20 MHz	50	D-14, P-14, F-14
Shift register	5-bit parallel in, parallel out	74LS96	54LS96	10 MHz	60	D-16, P-16, F-16
Shift register	8-bit serial in, serial out	74LS164	54LS164	25 MHz	80	D-14, P-14, F-14
Shift register	4-bit parallel in, parallel out	74LS194	54LS194	20 MHz	60	D-16, P-16, F-16
Shift register	4-bit parallel in, parallel out	74LS195A	54LS195A	20 MHz	50	D-16, P-16, F-16
Shift register	4-bit parallel in, parallel out	74LS295A	54LS295A	20 MHz	65	D-14, P-14, F-14
Shift register	4-bit parallel in, parallel out	74LS395	54LS395	25 MHz	75	D-16, P-16, F-16
Multiplexer	8-line to 1-line	74LS151	54LS151	11	30	D-16, P-16, F-16
Multiplexer	8-line to 1-line	74LS152	54LS152	11	28	D-14, P-14, F-14
Multiplexer	Dual 4-line to 1-line	74LS153	54LS153	14	31	D-16, P-16, F-16
Multiplexer	Quad 2-line to 1-line	74LS157	54LS157	14	49	D-16, P-16, F-16
Multiplexer	Quad 2-line to 1-line	74LS158	54LS158	12	24	D-16, P-16, F-16
Multiplexer	8-line to 1-line	74LS251	54LS251	17	35	D-16, P-16, F-16
Multiplexer	Quad 2-line to 1-line	74LS257	54LS257	20	50	D-16, P-16, F-16
Multiplexer	Quad 2-line to 1-line	74LS258	54LS258	20	35	D-16, P-16, F-16
Multiplexer	Quad 2-line to 1-line	74LS298	54LS298	20	65	D-16, P-16, F-16
Decoder	4-line to 10-line	74LS42	54LS42	17	35	D-16, P-16, F-16
Decoder	3-line to 8-line	74LS138	54LS138	22	31	D-16, P-16, F-16
Decoder	Dual 2-line to 4-line	74LS139	54LS139	22	34	D-16, P-16, F-16
Decoder	Dual 2-line to 4-line	74LS155	54LS155	18	30	D-16, P-16, F-16
Decoder	Dual 2-line to 4-line	74LS156	54LS156	33	31	D-16, P-16, F-16
Display driver	BCD to 7-segment	74LS47	54LS47		35	D-16, P-16, F-16
Display driver	BCD to 7-segment	74LS48	54LS48		125	D-16, P-16, F-16
Display driver	BCD to 7-segment	74LS49	54LS49		40	D-14, P-14, F-14
Display driver	BCD to decimal	74LS145	54LS145		35	D-16, P-16, F-16
Display driver	BCD to 7-segment	74LS247	54LS247		35	D-16, P-16, F-16
Display driver	BCD to 7-segment	74LS248	54LS248		125	D-16, P-16, F-16
Display driver	BCD to 7-segment	74LS249	54LS249		40	D-16, P-16, F-16

* Package entry letter is form factor: F = flatpack, D = hermetic dual-in-line, P = plastic dual-in-line, C = round can, S = studded flatpack.   Package entry number is terminal count.

make the transition to the 0 state. The noise spike is generated by the odd gate being required to switch to full-current source to provide the output current previously supplied by other circuits. This current spiking results in pulse widths which are narrow, compared with the circuit propagation delay times, but should not result in logic errors being transmitted beyond the next flip-flop.

When using MECL II, full consideration must be given the effects of capacitor

TABLE 20 The MECL I Family of Emitte-Coupled Logic, Complete Selection Guide of Circuit Types

Function	Circuit form	Type No. 0 to 70°C	Type No. −55 to +125°C	Fanout (unit loads)	Total delay, ns	Power dissipation, mW	Package*
OR/NOR gate	5-input	351	301	25	7.5	37	F-10, C-10
OR/NOR gate	3-input	356	306	25	7.5	37	F-10, C-10
OR/NOR gate	3-input*	357	307	25	7.5	15	F-10, C-10
NOR gate	Dual 2-input	359	309	25	7.0	54	F-10, C-10
NOR gate	Dual 2-input*	360	310	25	7.0	54	F-10, C-10
NOR gate	Dual 2-input*	361	311	25	7.0	41	F-10, C-10
NOR gate (internal bias)	Dual 3-input	362A	312A	25	7.5	70	F-10, C-10
NOR gate	Dual 3-input	362	312	25	7.5	54	F-14, C-12
NOR gate	Quad 2-input	363	313	25	7.0	125	F-14
Clock driver	Dual 4-input	369		100	3.0	250	F-14
Clock driver	Dual 2-input	369					C-10
Flip-flop	RS	352	302	25	11	42	F-10, C-10
Flip-flop	JK†	358A	308	25	8.5	87	F-10, C-10
Flip-flop	JK†	364	314	25	12	118	F-10, C-10
Half adder		353	303	25	7.5	63	F-10, C-10
Gate expander		355	305		4.5		F-10, C-10
Line driver	50-Ω	365	315		14	180	F-10, C-10
Lamp driver	100 mA	366	316			135	F-10, C-10
Bias driver		354	304	25		18	F-10, C-10
Level translator	MECL to TTL	367	317	7 (TTL)	27.5	63	F-10, C-10
Level translator	TTL to MECL	368	318	25	17	105	F-10, C-10

* Pulldown resistors omitted.
† AC-coupled.
‡ Package entry letter is form factor: F = flatpack, D = hermetic dual-in-line, P = plastic dual-in-line, C = round can, S = studded flatpack. Package entry number is terminal count.

loading on delay times and output rise times. Placing load resistors between the output and the V_{EE} supply improves the fall and negative delay times at the cost of increased power dissipation. High-speed clock drivers should be used to drive large numbers of parallel flip-flops to reduce the clock skew generated by using a logic tree to drive them. Care should be taken to allow for differences in delay times in parallel logic paths to prevent false outputs at the points at which they are tied together. The use of good system grounding, large system ground planes, controlled transmission lines, twisted-pair lines, and other high-speed wiring techniques becomes necessary with MECL II. The delay time in extended lengths of such transmission lines must be considered in all clock-skew and race-condition analyses of the logic system.

Applications that require high fanout should be restricted to one card to reduce accumulation of large parasitic capacitance. All power supplies should be adequately decoupled for every four or five circuits. When breadboarding circuits where transmission lines cannot be optimized, decoupling should be applied at every power-supply pin. For on-board decoupling, 0.01-μF capacitors are adequate, but a larger capacitance should be provided at the entry of power supply to the board. Capacitors of approximately 1 μF are normally used in this application.

In MECL II applications, careful layout must minimize undershoot ringing at the logic 1 level, since this directly subtracts from noise immunity. Proper application should ideally limit this undershoot to 100 mV, and no more than 150 mV in any case. Greater care must be taken with the ac-coupled flip-flops, the MC 1027 and the MC 1032, since they are more prone to false triggering by either undershoot or overshoot. System interconnection and ground planes should further be optimized for high-speed transmission lines. The use of ground screen is preferred to reduce parasitics and to control the transmission-line impedances. Ferrite beads in back planes can be used to reduce overshoot, but they tend to increase delay times, resulting in reduced capability of the MECL II. For all long transmission lines on back planes, twisted pairs are recommended. Signal lines on back planes should be independently strung, and the use of laced harnesses should be minimized or totally avoided. Line drivers and line receivers should be used in all cases where short-

TABLE 21   The MECL II Family of Emitter-Coupled Logic, Complete Selection Guide of Circuit Types

Function	Circuit form	Type No. 0 to 70°C	Type No. −55 to +125°C	Fanout (unit loads)	Total delay, ns	Power dissipation, mW	Package‡
OR/NOR gate	6-input, 3T, 3C	1001	1201	25	4.0	115	D-14, F-14, P-14
OR/NOR gate	6-input, 3T, 3C†	1002	1202	25	4.0	80	D-14, P-14
OR/NOR gate	6-input, 3T†, 3C†	1003	1203	25	4.0	40	D-14; P-14
OR/NOR gate	Dual 4-input, 1T, 1C	1004	1204	25	4.0	95	D-14, F-14, P-14
OR/NOR gate	Dual 4-input, 1T, 1C†	1005	1205	25	4.0	65	D-14, P-14
OR/NOR gate	Dual 4-input, 1T†, 1C†	1006	1206	25	4.0	45	D-14, F-14, P-14
Exclusive-OR gate	Quad 2-input	1030	1230	25	5.0	130	D-14, F-14, P-14
Exclusive-NOR gate	Quad 2-input	1031	1231	25	5.0	130	D-14, F-14, P-14
NOR gate	Triple 3-input	1007	1207	25	4.0	110	D-14, F-14, P-14
NOR gate	Triple 3-input (1) (2†)	1008	1208	25	4.0	75	D-14, P-14
NOR gate	Triple 3-input†	1009	1209	25	4.0	60	D-14, P-14
NOR gate	Quad 2-input	1010	1210	25	4.5	115	D-14, F-14, P-14
NOR gate*	Quad 2-input	1062	1262	25	2.0	320	D-16, F-16, P-16
NOR gate	Quad 2-input (2) (2†)	1011	1211	25	4.5	95	D-14, F-14, P-14
NOR gate	Quad 2-input†	1012	1212	25	4.5	65	D-14, F-14, P-14
NOR gate*	Quad 2-input	1063	1263	25	2.0	320	D-14, F-14
AND gate	Quad 2-input	1047	1247	25	5.0	130	D-14, F-14, P-14
NAND gate	Quad 2-input	1048	1248	25	5.0	13	D-14, F-14, P-14
Expandable OR/NOR gate	Dual 2-input	1024	1224	25	4.0	95	D-14, P-14
Gate expander	4-input, 5-input	1025	1225				D-14, F-14, P-14
Flip-flop	JK ET	1013	1213	25	6.0	125	D-14, F-14, P-14
Flip-flop	Dual RS (positive clock)	1014	1214	25	6.0	140	D-14, F-14, P-14
Flip-flop	Dual RS (negative clock)	1015	1215	25	6.0	140	D-14, F-14, P-14
Flip-flop	Dual RS (single rail)	1016	1216	25	6.0	140	D-14, F-14, P-14
Flip-flop	Type D	1022	1222	25	8.0	110	D-14, F-14, P-14
Flip-flop	JK ET (120 MHz)	1027	1227	25	4.0	250	D-14, F-14, P-14
Flip-flop*	Dual JK ET (100 MHz)	1032	1232	25	4.5	180	D-16, F-16, P-16
Flip-flop	Dual RS	1033	1233	25	6.0	140	D-14, F-14, P-14
Flip-flop*	Type D	1034	1234	25	4.0	185	D-14, F-14, P-14
Latch	Quad	1040	1240	25	8.0	250	D-14, F-14, P-14
Latch	Quad	1070	1270	25	8.0	250	D-14, F-14, P-14
Level translator	DTL to MECL	1017	1217	25	15	105	D-14, F-14, P-14
Level translator	MECL to DTL	1018	1218	25	19	55	D-14, F-14, P-14
Level translator	Quad MECL to DTL	1039	1239	7 (DTL)	12	200	D-16, F-16, P-16
Level translator	Quad TTL to MECL	1067	1267	1	5.0	300	D-16, F-16, P-16
Level translator	Quad MECL to TTL	1068	1268	10 (TTL)	5.0	340	D-16, F-16, P-16
Adder	Full, single-bit	1019	1219	25	3.0/8.0	145	D-14, F-14, P-14
Subtractor	Full, single-bit	1021	1221	25	4.0/11.0	145	D-14, F-14, P-14
Adder*	Dual full	1059	1259	25	9.0	375	D-16, F-16, P-16
Data selector*	Dual 4-channel	1028	1228	25	5.0	170	D-16, F-16, P-16
Data selector*	8-channel	1038	1238	25	7.0/18.0	150	D-14, P-14
Data distributor	2 × 3 OR net	1029	1229	25	4.0	160	D-14, F-14, P-14
Memory*	16-bit X, Y decode	1036	1236	5	17	250	D-14, F-14, P-14
Memory*,†	16-bit X, Y decode	1037	1237	5	17	250	D-14, F-14, P-14
Decoder*	Dual binary/1 of 4	1042	1242	25	6.5	245	D-16, P-16
Decoder*	Binary/1 of 8	1043	1243	25	6.0/11.0	210	D-14, P-14
Decoder*	Binary/1 of 10	1044	1244	25	6.0	245	D-16, P-16
Decoder-driver*	NIXIE	1045	1245			178	D-16, F-16, P-16
Clock driver*	3-input, 4-input	1026	1226	25	2.0	140	D-14, F-14, P-14
Clock driver*	Dual 4-input OR/NOR	1023	1223	25	2.0	250	D-14, F-14, P-14
Line receiver	Quad	1020	1220	25	4.0	115	D-14, F-14, P-14
Line receiver	Triple (TC)	1035	1235	25	5.0	140	D-14, F-14, P-14
Line receiver*	Triple 2 (TC) 1 (C)	1066	1266	25	2.0	350	D-14, F-14, P-14

T = true output, C = complementary output, ET = edge-triggered.
* Noise margin 150 mV.
† Pulldown resistor omitted.
‡ Package entry letter is form factor: F = flatpack, D = hermetic dual-in-line, P = plastic dual-in-line, C = round can, S = studded flatpack. Package entry number is terminal count.

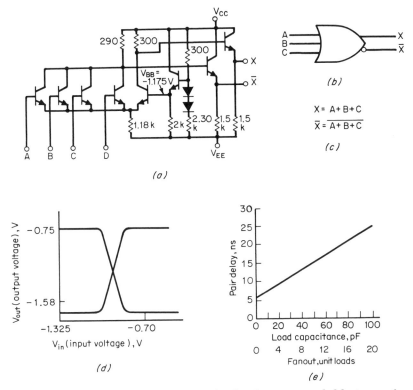

$$X = A + B + C$$

$$\overline{X} = \overline{A + B + C}$$

(c)

Fig. 18 Characterization of the MECL II family of emitter-coupled logic. OR/NOR-gate characterization. (*a*) Circuit schematic. (*b*) Logic diagrams. (*c*) Logic equation. (*d*) Transfer curve, a plot of output voltage vs. input voltage. (*e*) Fanout curve, a plot of pair delay vs. load capacitance and fanout.

length lines cannot be used. In all cases where cabinets are operated at different temperatures, the twisted-pair line drivers must again be used for connections between cabinets to eliminate the thermal common-mode effect.

MECL III   MECL III is the fastest form of digital logic available as a standard production family. MECL III has an average gate delay of 1 ns and is thus suited to processing data at very high rates, up to and including those above 200,000,000 bits per second. MECL III requires state-of-the-art manufacturing techniques to produce very small internal chip geometries, very low junction capacitance, and fast-rising current waveforms on chip. The use of this family requires that the MECL III circuits be used in an environment of tightly controlled transmission lines. This limitation results in more expensive system-fabrication cost. The basic MECL III schematic for a gate is illustrated in Fig. 19. This schematic differs slightly from the MECL II schematic, but the basic circuit concepts are the same. The lower values of bias resistors are consistent with the drive modes of the MECL III circuits, and like MECL II, MECL III includes an internal bias network to remove the need for a separate bias driver. MECL III gates are supplied with base pulldown resistors. These resistors provide a path for base leakage current to the unused input bases, causing them to be fully turned off. Therefore, unused inputs must be directly tied to the V_{EE} power supply. All MECL II outputs are open-transistor outputs without pulldown resistors. By removing the pulldown resistors, the circuits are suitable for external OR tying.

In MECL III, internal circuit gates operate with reduced logic swing. Because of this reduction, the relatively small margin of logic swing in external gates, and

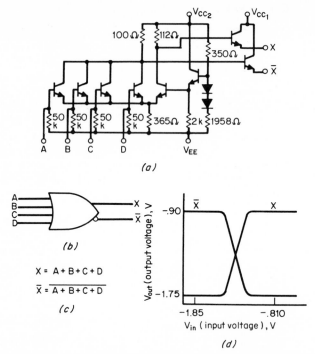

(a)

(b)

$$X = A + B + C + D$$

$$\overline{X} = \overline{A + B + C + D}$$

(c)

(d)

Fig. 19 Characterization of the MECL III family of emitter-coupled logic. OR/NOR-gate characterization. (*a*) Circuit schematic. (*b*) Logic diagram. (*c*) Logic equation. (*d*) Transfer curve, a plot of output voltage vs. input voltage.

the relationship of the margin to the temperature dependence of the logic levels, a great degree of thermal management must be provided throughout the system. MECL III gates dissipate reasonably high levels of power per gate. This high level of power is then multiplied by the large number of gates in the larger arrays to produce an exceptionally high level of power per package compared with the requirements for tight thermal management. To make the problem of thermal management a workable one, MECL III is packaged in a special flatpack which incorporates a heat-sink stud that doubles as the V_{EE} connection. Although MECL III is available in packages other than the stud flatpack, its applications are best controlled in this package. In many circuit types, the use of dual-in-line packages requires individual circuit heat sinks for proper thermal management.

The family of MECL III circuits is presented in Table 22. This table shows that the range of circuits includes basic OR/NOR gates, NOR gates, flip-flops, latches, exclusive-OR gates, counters, memories, and high-frequency prescaling flip-flops. These circuits are primarily selected for use in high-speed communication equipment and in digital instrumentation.

MECL III is a very specialized form of logic. It is suited only to limited applications where the requirements for exceptionally high operating speeds outweigh the costs of the circuits themselves, their thermal management, and a fully controlled transmission-line interconnection system.

The 2500 ECL family The 2500 ECL family was introduced by Texas Instruments as an improvement over the basic MECL I family. It is a high-speed logic family with a basic gate delay of 2 to 3 ns per gate function. As shown in the schematic of Fig. 20, the basic 2500 ECL gate uses input resistors and open outputs. The 2500 gates do not internally generate the V_{BB} reference level for the differential-amplifier stage. This requires system distribution of this reference level. This family is designed for operation with controlled transmission lines of 50-Ω impe-

TABLE 22   The MECL III Family of Emitter-Coupled Logic, Complete Selection Guide of Circuit Types

Function	Circuit form	Type No. 0 to 75°C	Fanout (unit loads)	Total delay, ns	Power dissipation, mW	Package†
OR/NOR gate	Dual 4-input (high z)	1660	7/70	1.1	120	D-16, S-15, F-16
OR/NOR gate	Dual 4-input (low z)	1661	7/70	1.1	120	D-16, S-15
OR gate	Quad 2-input (high z)	1664	7/70	1.1	240	D-16, S-15, F-16
OR gate	Quad 2-input (low z)	1665	7/70	1.1	240	D-16, S-15
NOR gate	Quad 2-input (high z)	1662	7/70	1.1	240	D-16, S-15, F-16
NOR gate	Quad 2-input (low z)	1663	7/70	1.1	240	S-15, D-16
Exclusive-OR gate	Triple 2-input (high z)	1672	7/70	1.3	220	D-16, S-15, F-16
Exclusive-OR gate	Triple 2-input (low z)	1673	7/70	1.3	250	S-15, D-16
Exclusive-NOR gate	Triple 2-input (high z)	1674	7/70	1.3	220	D-16, S-15, F-16
Exclusive-NOR gate	Triple 2-input (low z)	1675	7/70	1.3	250	S-15, D-16
Flip-flop	Dual RS (high z)	1666	7/70	1.8	220	D-16, S-15, F-16
Flip-flop	Dual RS (low z)	1667	7/70	1.8	230	D-16, S-15
Flip-flop, master/slave	Type D (high z)	1670	7/70	350 MHz	220	D-16, S-15, F-16
Flip-flop, master/slave	Type D (low z)	1671	7/70	350 MHz	220	D-16, S-15
Flip-flop	Type D	1690	7/70	500 MHz	200	D-16, S-15, F-16
Latch clocked	Dual (high z)	1668	7/70	1.8	220	D-16, S-15, F-16
Latch clocked	Dual (low z)	1669	7/70	1.8	220	D-16, S-15
Voltage-controlled oscillator	225 MHz	1648			150	D-14, F-14, P-14
Voltage-controlled multivibrator	150 MHz	1638	7/70		125	D-16, F-16, P-16
Comparator	Dual A/D	1650	7/70	3.5	275	S-15, D-16
Comparator	Dual A/D	1651	7/70	2.5	275	S-15, D-16
Counter, binary	325 MHz (high z)	1654	7/70		750	D-16*
Counter, bi-quinary	350 MHz (high z)	1678	7/70		750	D-16*
Counter, bi-quinary	350 MHz (high z)	1679	7/70		750	D-16*
Memory (high z)	Random access	1680	7/70	2.5/3.5	270	D-16
Memory (high z)	Content addressable	1682	7/70	2.8/4.0	270	D-16
Memory (high z)	Content addressable	1684	7/70	2.8/4.0	270	D-16
Register (high z)	4-bit shift	1694	7/70	325 MHz	750	D-16*
Line receiver	Quad	1692	7/70	1.1	220	D-16, F-16, S-15

* Requires heat sink.
† Package entry letter is form factor: F = flatpack, D = hermetic dual-in-line, P = plastic dual-in-line, C = round can, S = studded flatpack. Package entry number is terminal count.

dance.   All lines must be terminated to operate the circuit with minimum oscillation and maximum ac fanout.   Collector dotting (internal AND) and emitter dotting (internal OR) reduce the number of logic gates required to perform a complete logic function.

In operation, the circuit produces approximately 200 mV of noise margin.   This noise margin is reduced in applications using emitter dotting.   To minimize reflection on transmission lines from the sharp-rising wavefronts, all lines must be terminated. Both lumped and distributed loads may be employed separately or in combination, but in any case, the lines must be terminated at the end of the transmission line.   All unused inputs must be returned to V_{EE} supply.   As is the case in any high-speed logic, all transmission lines should also be minimized in length to prevent excess delays in the interconnection system.   Delay times in 2500 logic are degraded approximately 75 ps for each additional load.   This degradation, and the additional degradation that comes from capacitance loading, is a limiting factor in achieving high system speeds.   The full range of available circuits in the 2500 family is presented in Table 23.   This table shows that the basic family includes a wide range of OR/NOR gates, (AND/OR, and more complex functions, including arithmetic and other modules, complex multilevel gating circuits, line drivers and receivers, level converters, and a small number of simple memory elements.   Drive is improved in the 2500 family by including multiple emitter-follower outputs.   In these units, three outputs are available, each of which has the same logic information but separate drive capability.   These circuits are useful in clock-drive networks and where large parallel computations must be made.   The 2536 and 2537 logic-level converters are designed for directly driving 2500 logic elements into TTL and driving TTL into 2500 ECL.

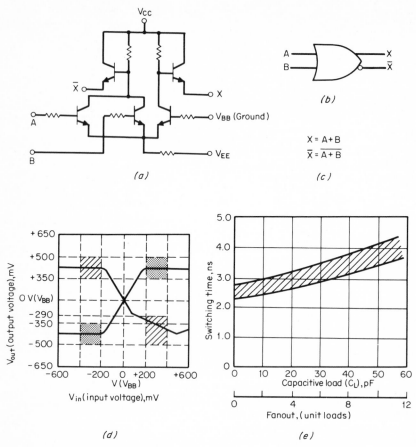

Fig. 20 Characterization of the 2500 family of emitter-coupled logic. OR/NOR-gate characterization. (*a*) Circuit schematic. (*b*) Logic diagrams. (*c*) Logic equation. (*d*) Transfer curve, a plot of output voltage vs. input voltage. (*e*) Fanout curve, a plot of switching time vs. fanout and load capacitance.

The 95K family Fairchild has introduced, as an improved variant of the 10,000 family, its own 95K and the F-10K family. This family of circuits represents a modification of the ECL 10,000 family by modifying the V_{BB} reference level source and generating a second reference level with which to drive the active current source in the differential amplifier. In doing this, Fairchild has minimized the effects of both temperature and supply-voltage variation on the logic levels. Comparison of the Fairchild circuit shown in Fig. 21 with the ECL 10,00 basic OR/NOR gate circuit shows the application of the voltage and temperature control V_{BB} source to be most appropriate. However, the absolute gains in using the Fairchild temperature- and voltage-compensated sources are not to any great extent realizable in a real system.

It is preferable to provide a high degree of voltage control in the distributor power-supply system of the overall logic system. It is also appropriate to provide proper levels of thermal management to minimize temperature variations within given cabinets of a large computer system. Since ECL 10,000 can be made to operate as differential line drivers between cabinets, the effects of temperature variations can be removed from these applications. It is not recommended that temperature-compensated and non-temperature compensated circuits be intermixed in a working system, since the temperature-compensated-system circuits do not provide the uniform

TABLE 23   The 2500 Family of Emitter-Coupled Logic, Complete Selected Guide of Circuit Types

Function	Circuit form	Type No.	Total delay, ns	Power dissipa- tion, mW	Fanout (unit loads)	Package*
OR/NOR gate	9-input, T, C	2501	2.5	36	10	P-16
OR/NOR gate	Dual 4-input, T, C	2500	2.5	77	10	P-16
OR/NOR gate	Triple 2-input, T, C	2502	2.5	95	10	P-16
OR/NOR gate	Dual 2-input, 3T, C	2520	2.5/3.5	122	10	P-16
OR/NOR	Quad, 1-input, T, C	2504		126		P-16
NOR gate	Triple, 3-input	2505	2.5	95	10	P-16
NOR gate	Quad, 2-input	2503	2.5	126	10	P-16
NOR gate	Dual 3-input, 3-output	2523	2.5/3.5	122	10	P-16
NOR gate	Dual 4-input, 2-output	2522	2.5/3.5	77	10	P-16
OR gate	Quad, 2-input, 1 common	2511	2.5	126		P-16
OR gate	Dual 2-input, 3 output	2521	2.5/3.5	122	10	P-16
Multilevel gate	OR-AND/NOR-OR 2 × 5	2509	3.0	126	10	P-16
Multilevel gate	NOR-OR 3 × 4	2506	3.0	126	10	P-16
Multilevel gate	OR-AND/NOR-OR 3 × 4	2510	3.0	126	10	P-16
Multilevel gate	NOR-OR 2 × 5	2507	3.0	158	10	P-16
Multilevel gate	NOR-OR 2 × 6	2508	3.0	189	10	P-16
Multilevel gate	Dual OR-AND/NOR-OR 2 × 2	2513	3.0	126	10	P-16
Multilevel gate	Dual NOR-OR 2 × 3	2512	3.0	189	10	P-16
Full adder	2-bit	2516	3.0	202	4	P-16
Carry block	5-bit	2515	3.0	162	4	P-16
Decoder	3-bit/8-line	2517	5.5	598	10	P-16
Latch		2540	3.0	153	10	P-16
Latch		2541	3.0/8.0	555	10	P-16
Latch		2542	2.5/7.0	742	10	P-16
Line receiver	Dual OR/NOR	2530	2.5	77	10	P-16
Line driver	Dual OR/NOR	2531	2.5	99	10	P-16
Level translator	Dual TTL to ECL T, C	2536	4.5	182	10	P-16
Level translator	Dual ECL to TTL T, C	2537	3.5	136	5 TTL	P-16

T = true output, C = complementary output.

* Package entry letter is form factor: F = flatpack, D = hermetic dual-in-line, P = plastic dual-in-line, C = round can, S = studded flatpack.   Package entry number is terminal count.

temperature drift of the standard ECL 10,000 circuits. This standard drift is used to maintain a constant noise margin over temperature in a working large logical system.   The circuits available in the 95K family and in the F-10K family are listed together in Table 24.

The ECL 10,000 family   ECL 10,000 is becoming the most widely accepted and widely used family of emitter-coupled logic.   ECL 10,000 represents a good set of technological trade-offs among speed, power, complexity, rise time, and ease of application.   Through its availability in both limited-temperature-range and full military-temperature-range products, it has gained application in test equipment, communication equipment, computer main frames, and military digital systems.   Comparison of the ECL 10,000 family with competing forms of nonsaturated logic shows that it, like all the families except MECL I and 2500 ECL, includes an internal reference source for the differential-amplifier stange.   ECL 10,000 does not include output pulldown resistors, and it functions with open emitter-follower transistors.   It is designed to operate into transistor lines, but it is tolerant of short-length uncontrolled lines.   It has a basic gate delay time of 2 ns, although the gate-edge speed is approximately 3.5 ns.   The family includes flip-flops that function to 125 MHz, which compares favorably with the other families, except for MECL III, which is able to function at frequencies to 500 MHz.   The basic high-speed ECL gate has a dissipation of 25 mW, which compares most favorably with the slower logic gates. V_{cc} operation which ECL 10,000 offers.

Application of ECL 10,000 requires reasonable controls to be maintained on the transmission-line system.   Doubled-sided plated through-hole printed-circuit boards may be used with ECL 10,000, although multilayer boards are better suited to the high functional density of the ECL 10,000 circuit elements.   The high density of multilayer boards also reduces interconnection length to improve system speeds further.   ECL 10,000's low edge speed of 3.5 ns permits wire wrapping in back-plant interconnection.   When using double-sided printed-circuit boards, a ground plane

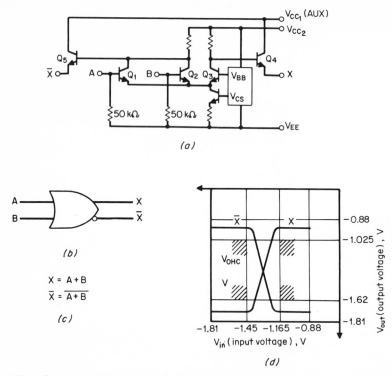

Fig. 21   Characteristics of the 95K family of emitter-coupled logic. OR/NOR-gate characterization. (*a*) Circuit schematic. (*b*) Logic diagrams. (*c*) Logic equation. (*d*) Transfer curve, a plot of output voltage vs. input voltage.

should be maintained on one side of the boards to provide a reasonably well-controlled transmission line. If a ground plane cannot be designed into the board, ground buses should be used to minimize the effects of noncontrolled transmission. Interconnections on one side of the printed-circuit board should be run perpendicular to those on the other side to minimize the amount of crosstalk between leading edges. Connections to the two V_{cc} pins on each package should be made separately and should be separately decoupled to gain the full advantage of the separated V_{cc} operation which ECL 10,000 offers.

The power-supply distribution should be tightly controlled to minimize the difference in power-supply voltage from device to device throughout the entire system. This minimizes degradation of noise margin. Back-plane interconnection should be as short as possible, and a ground screen should be included to improve the transmission-line characteristics of the back-plane wiring. Coaxial cables or twisted pairs may be used in back planes to improve the quality of the transmission lines further. Series termination resistors can be used to terminate interconnection lines. All lines up to 20 ft can be driven directly by ECL 10,000 gates if these lines are composed of twisted pairs and a line receiver is used at the end of the line. The OR/NOR outputs of the gates allow the direct drive of twisted-pair lines.

The basic MECL 10,000 gate is illustrated in Fig. 22. From the schematic, the basic circuit functions with two separate positive power supplies, V_{EE_1} and V_{cc_2}. The separation of the two power supplies minimizes the effect of the output-drive currents on the input switching thresholds. This separation improves the functional noise immunity of the device and is a major basic improvement to the ECL gate. The two power-supply connections must be separated in the system layout to avoid an external coupling which has been eliminated internally. The reference level for

TABLE 24 The 95K Family of Emitter-Coupled Logic, Complete Selection Guide of Circuit Types

Function	Circuit form	Type No. 0 to 75°C	Total delay, ns	Power dissipation, mW	Package*
OR/NOR gate	Dual 4-input	95002	2.0	60	D-16
OR/NOR gate	Triple 2-input	95003	2.0	90	D-16
OR/NOR gate	Triple 2/3/2-input	95105	2.0	90	D-16
OR/NOR gate	Dual 4/5-input	95109	2.0	60	D-16
OR/NOR gate	Quad 2-input (1-bus)	95101	2.0	100	D-16
NOR gate	Quad 2-input	95004	2.0	100	D-16
NOR gate	Quad 2-input	95102	2.0	100	D-16
NOR gate	Dual 3-input, 3-output	95111	2.4	160	D-16
OR gate	Triple 4/3/3-input	95106	2.0	90	D-16
OR	Dual 3-input, 3-output	95110	2.4	160	D-16
Exclusive-OR/NOR gate	Triple	95107	2.5	110	D-16
OR/AND gate	Dual 3 × 2	95118	2.3	100	D-16
OR/AND gate	Single 3 × 4	95119	2.3	100	D-16
OR/AND invert gate	Dual 2/3 × 2	95117	2.3	100	D-16
OR/AND invert gate	Single 3 × 4	95121	2.5	100	D-16
Line receiver	Quad	95115	2.0	110	D-16
Line receiver	Triple w/invert	95116	2.0	85	D-16
Flip-flop	JK edge trigger	95029	250 MHz	208	D-16
Flip-flop	Dual D	95231	200 MHz	260	D-16
Shift register	4-bit universal	95000	190 MHz	345	D-16

* Package entry letter is form factor: F = flatpack, D = hermetic dual-in-line, P = plastic dual-in-line, C = round can, S = studded flatpack. Package entry number is terminal count.

the differential amplifier is provided by a resistor-resistor divider network, which is upbiased by two diodes which overcompensate the voltage-temperature characteristic of the base-emitter drop. The nominal reference level of the bias network is -1.29 V in the normal ECL configuration. The use of small-geometry transistors in the ECL 10,000 gate minimizes junction capacitance, providing for the high operating speed of the circuit.

Over 70 basic circuits are available in the MECL 10,000 family. The full range of the family is described by Table 25. The family includes OR/NOR gates, NOR gates, AND gates, line drivers and receivers, (AND/OR invert gates, level converters between TTL and ECL 10,000, bus drivers, latches and flip-flops, counters, shift registers, decoders, multiplexers, priority encoders, read-only memories, voltage-controlled oscillators, high-speed flip-flops, random-access memories, and content addressable random-access memories. In the numbering applied to ECL 10,000, the part number for parts available in a full temperature range and military temperature range of -55 to $+125°C$ have the basic part number increased by the addition of 400. Thus, the ECL 10101 in limited temperature range becomes 10501 in full military temperature range.

ECL 10,000 is the preferred high-speed logic family.

Complementary Metal-Oxide Semiconductor Devices

The complementary metal-oxide semiconductor (CMOS) family is a family characterized by operation over a wide power-supply voltage range of 3 to 18 V. This family has moderate speed and is able to operate with very low power consumption; so it is widely used in portable, battery-powered, and automotive equipment. The basic gate structure, illustrated in Fig. 23, is a NOR gate. This gate structure is the basic building block for all CMOS circuits.

The second most important gate structure in CMOS is the transmission gate. The transmission gate is illustrated in Fig. 24. The transmission gate is a structure avail-

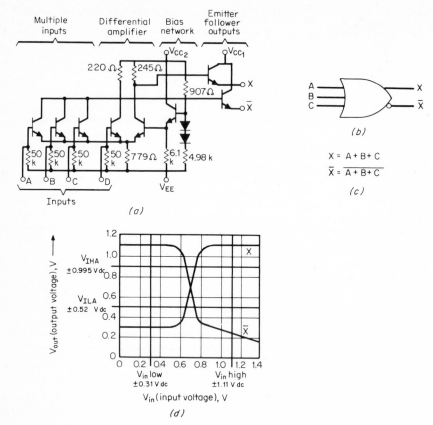

$$X = A + B + C$$

$$\overline{X} = \overline{A + B + C}$$

(c)

Fig. 22 Characterization of the 10,000 family of emitter-coupled logic. OR/NOR-gate characterization. (*a*) Circuit schematic. (*b*) Logic diagrams. (*c*) Logic equation. (*d*) Transfer curve, a plot of output voltage vs. input voltage.

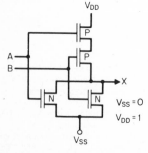

Fig. 23 Complementary metal-oxide semiconductor, NOR gates; schematic diagram.

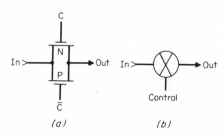

Fig. 24. Complementary metal-oxide semiconductor, transmission gate. (*a*) Schematic diagram. (*b*) Logic symbol.

TABLE 25   The 10,000 Family of Emitter-Coupled Logic, Complete Selection Guide of Circuit Types

Function	Circuit form	Type No. –30 to +85°C	Type No. –55 to +125°C	Total delay, ns	Power dissipa- tion, mW	Package†
OR/NOR gate	Quad 2-input (1-bus)	10101	10501	2.0	100	D-16, F-16, P-16
OR/NOR gate	Triple 2-3-2-input	10105	10505	2.0	90	D-16, F-16, P-16
OR/NOR gate	Dual 4-5-input	10109	10509	2.0	60	D-16, F-16, P-16
OR gate	Dual 3-input, 3-output	10110		2.4	160	D-16, P-16
OR gate	Dual 3-input, 3-output	10210		1.5	160	D-16
NOR gate	Quad 2-input	10102	10502	2.0	100	D-16, F-16, P-16
NOR gate	Triple 4-3-3-input	10106	10506	2.0	90	D-16
NOR gate	Dual 3-input, 3-output	10111		2.4	160	D-16
NOR gate	Dual 3-input, 3-output	10211		1.5	160	D-16
AND gate	Quad 2-input	10104		2.7	140	D-16
Exclusive-OR/NOR gate	Triple 2-input	10107	10507	2.5	110	D-16, F-16
OR-AND gate, T, C	Dual 2-wide, 2,3-input	10117	10517	2.3	100	D-16, P-16
OR-AND gate	Dual 2-wide, 3-input	10118	10518	2.3	100	D-16
OR-AND gate	4-wide, 4,3,3,3-input	10119	10519	2.3	100	D-16
OR-AND gate, T, C	4-wide, 3-input	10121	10521	2.5	100	D-16
Parity generator	12-bit	10160	10560	5.0	320	D-16
Flip-flop, type D	Dual, master-slave	10131	10531	160 MHz*	235	D-16, F-16
Flip-flop, JK	Dual, master-slave	10135		140 MHz*	280	D-16
Flip-flop, type D	Hex, master-slave	10176		150 MHz*	460	D-16
Flip-flop, type D	Dual, master-slave	10231	10631	225 MHz*	270	D-16
Latch	Dual	10130		2.5	155	D-16
Latch	Quad	10133		4.0	310	D-16
Latch	Quint	10175		2.5	400	D-16
Counter, universal	Hexadecimal	10136		150 MHz*	625	D-16
Counter, universal	Decade	10137		150 MHz*	625	D-16
Multiplexer	Dual with latch and reset	10132		3.0	225	D-16
Multiplexer	With latch	10134		3.0	225	D-16
Multiplexer	8-line	10164	10564	3.0	310	D-16
Multiplexer	Quad, 2-line, latch	10173		2.5	275	D-16
Multiplexer	Dual 4-line	10174		3.5	350	D-16
Register file	64-bit	10145		10.0	625	D-16
Decoder (low)	Binary/1 of 8	10161		4.0	315	D-16
Decoder (high)	Binary/1 of 8	10162		4.0	315	D-16
Decoder (low)	Dual, binary/1 of 4	10171		4.0	325	D-16
Decoder (high)	Dual, binary/1 of 4	10172		4.0	325	D-16
Look-ahead block	4-bit	10179		3.0/4.0	300	D-16
Adder/subtractor	Dual	10180		4.5	360	D-16
Arithmetic logic unit	4-bit	10181	10581		600	D-24, P-24
Memory, random-access	64-bit, 90-Ω	10140		10.0	420	D-16
Memory, random-access	64-bit, 50-Ω	10148		10.0	420	D-16
Shift register	4-bit, universal	10141		200 MHz*	425	D-16
Level translator	Quad TTL to ECL	10124		3.5	380	D-16
Level translator	Quad ECL to TTL	10125		5.0	380	D-16
Level translator	Dual ECL to MOS	10127				D-16
Bus driver NOR	Triple 4,3,3-input	10123		3.0	310	D-16
Bus driver flip-flop	Dual type D	10128		12.0	700	D-16
Bus receiver	Quad	10129		10.0	750	D-16
Line receiver	Triple (T, C)	10114		2.4	145	D-16
Line receiver	Quad (T)	10115	10515	2.0	110	D-16, P-16
Line receiver	Triple (T, C)	10116	10516	2.0	85	D-16
Line receiver	Triple (T, C)	10216		1.8	100	D-16
Encoder, priority	8-input	10165		7.0	545	D-16
NOR gate	Quad 2-input with strobe	10100		2.0	100	D-16
Exclusive-OR gate	Quad	10113		2.5	175	D-16
Counter	Biquinary	10138		150 MHz*	370	D-16
Latch	Quad	10153		4.0	310	D-16
Detection circuit	Error correction	10163		5.0	510	D-16
Detection circuit	Error correction	10193		3.5	520	D-16
Inverter	Hex buffer	10195		2.0	200	D-16
AND gate	Hex	10197		2.8	200	D-16
OR/NOR gate	3-input, 3-output	10212		1.5	160	D-16
Multiplier	2-bit by 1-bit	10287			400	D-16

T = true output, C = complement output.

* Toggle frequency.

† Package entry letter is form factor: F = flatpack, D = hermetic dual-in-line, P = plastic dual-in-line, C = round can, S = studded flatpack. Package entry number is terminal count.

able in a CMOS but has no analog in any other digital family. The transmission gate is in many ways the solid-state equivalent of a relay. It has an input and output terminal, indistinguishable from each other. These two terminals are joined by either a high- or a low-impedance path, depending on the bias applied to the control terminal. The transmission gate is used in two places in the CMOS family. Primarily, it is used as an additional circuit element within complex digital arrays. This circuit element, as an example, allows cross coupling and switching of gates to form a flip-flop. The transmission gate is also used in CMOS circuits to control analog signals. The large and growing class of CMOS circuits that include a transmission gate in this manner are not true digital circuits, but analog circuits with digital control. The most basic circuit of this type is the 4016, which is a set of four independent single-pole, single-throw, analog switches. When a CMOS analog-switch circuit is used, the limitations of the transmission gate come into play. First and foremost, the signal voltages applied to either the input or output terminal or terminals must be limited to the range between the V_{DD} and V_{SS} supplies. Second, the ac frequency of the signal must be within the range of pass frequencies for these devices, which is typically in the low-megahertz region. Third, the speed of response of the CMOS transmission gate must be considered when signals are switched from one channel to another.

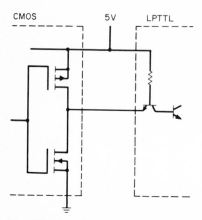

Fig. 25 Interface circuitry between CMOS and transistor-transistor logic when both are operating from a 5-V power supply.

Interface circuits between complementary metal-oxide semiconductors and any other logic form are reasonably simple. The easiest conversions are those between CMOS and TTL, in cases where both operate from the same 5-V power supply. Figures 25 and 26 illustrate recommended circuits for making the transition from CMOS to TTL and TTL to CMOS. As shown, the transition circuitry is inserted to allow the necessary currents for operating the output stages. The circuits in themselves allow the proper noise margin for operating one into the other. In cases where CMOS circuits must drive a higher fanout than is available, the transition

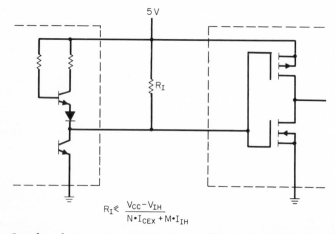

$$R_I \lessgtr \frac{V_{CC} - V_{IH}}{N \cdot I_{CEX} + M \cdot I_{IH}}$$

Fig. 26 Interface between transistor-transistor logic and CMOS when both are operating from a 5-V power supply.

should be made to a single TTL buffer, whose outputs can drive the additional fanout. The difference in fanout or drive capability within the CMOS family itself should be noted, and the design engineer should not and cannot expect a simple gate to perform the function of a CMOS buffer. Even in cases where sufficient drive can be obtained with a gate without the use of a buffer in the hope of saving delay time, the additional output load increases the *RC* switching time associated with a gate driving a large amount of current, and adds an unacceptable amount to the delay associated with this stage.

The 4000 family of complementary metal-oxide semiconductors The 4000 family is a large and developing family of logic and analog switch circuits fabricated according to the complementary metal-oxide semiconductor technique. This family is divided into two subfamilies. The first is composed of standard logic circuits. Within this subfamily are also included logic circuits of similar cell structure, impedances, and delays which have not been assigned a 4000 number but which could be mixed with 4000 circuits to build a proper system hierarchy.

Certain circuits introduced by the original CMOS manufacturer, Radio Corporation of America, and which were assigned 4000 numbers, are really not members of the 4000 family. These circuits are special-purpose circuits that were originally designed as custom products and do not have general application in system design. The second category of CMOS circuits are circuits which use transmission gates to switch analog signals. Both subfamilies are illustrated in Tables 26 to 28. These tables show the range of functions available, the delay times associated with each, and the power-supply current consumed. Delay times and power-supply current are a function of bias. In addition, the actual power consumption is dominated by the ac components; so quiescent power is not a major consideration in selecting CMOS circuits. Figure 27 illustrates the relationship of the delay time to applied bias of a typical circuit. Figure 28 shows the relationship between power dissipation and

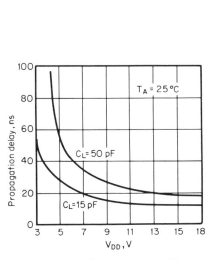

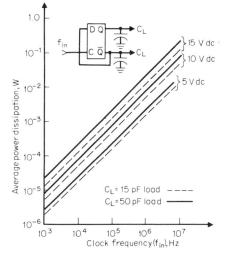

Fig. 27 Typical propagation delay vs. supply voltage for complementary metal-oxide semiconductor devices.

Fig. 28 Typical power-supply voltage vs frequency for complementary metal-oxide semiconductor devices.

clock frequency for a typical CMOS gate structure. CMOS consumes very low power levels in most cases, and exceptionally low power in cases where low power frequencies are applied. The standby power consumption of a CMOS circuit is negligible. These combinations together make CMOS ideal for battery-powered systems. The basic NOR-gate characteristics of this family are shown in Fig. 29. A chip photograph is presented in Fig. 30.

TABLE 26   The 4000 Family of Complementary Metal-Oxide Semiconductor Logic, Complete Selection Guide of Small-Scale Integration-Circuit Types

Function	Circuit form	Type No.	Fanout (unit loads)	Total delay, ns (at 5 V)	Power dissipation, μW	Package*
NOR gate	Dual 3-input	4000	10	35	0.01	D-14, P-14, F-14
NOR gate	Dual 4-input	4002	10	35	0.01	D-14, P-14, F-14
NOR gate	Triple 3-input	4025	10	35	0.01	D-14, P-14, F-14
NOR gate	Quad 2-input	4001	10	35	0.01	D-14, P-14, F-14
NOR gate	8-input	4078	10	35	0.01	D-14, P-14, F-14
NAND gate	Dual 4-input	4012	10	100	0.01	D-14, P-14, F-14
NAND gate	Triple 3-input	4023	10	50	0.01	D-14, P-14, F-14
NAND gate	Quad 2-input	4011	10	50	0.01	D-14, P-14, F-14
NAND gate	8-input	4068	10	100	0.01	D-14, P-14, F-14
OR gate	Quad 2-input	4071	10	100	0.01	D-14, P-14, F-14
OR gate	Triple 3-input	4075	10	100	0.01	D-14, P-14, F-14
OR gate	Dual 4-input	4072	10	100	0.01	D-14, P-14, F-14
AND gate	Quad 2-input	4081	10	100	0.01	D-14, P-14, F-14
AND gate	Triple 3-input	4073	10	100	0.01	D-14, P-14, F-14
AND gate	Dual 4-input	4082	10	100	0.01	D-14, P-14, F-14
Exclusive-OR gate	Quad	4030	10	100	0.1	D-14, P-14, F-14
Exclusive-OR gate	Quad	4070	10	100	0.1	D-14, P-14, F-14
Exclusive-OR gate	Quad	4507	10	35	0.01	D-14, P-14
Exclusive-NOR gate	Quad	4077	10	50	0.1	D-14, P-14, F-14
Multifunction gate	8-input	4048	9 mA	1,000	0.5	D-16, P-16, F-16
AND-OR gate	Quad 2-input	4019	10	100	0.5	D-16, P-16, F-16
AND-OR gate	Triple	4037	10	250	0.5	D-14, P-14, F-14
AND-OR invert gate	Dual	4506	10	55	0.1	D-16, P-16
Gate NAND/NOR	Dual plus 1	4501	10	25	0.1	D-16, P-16
Inverter	Hex split V_{CC}	4009	8 mA	15	0.1	D-16, P-16, F-16
Buffer	Hex split V_{CC}	4010	8 mA	15	0.1	D-16, P-16, F-16
Inverter	Hex	4049	3 mA	15	0.1	D-16, P-16, F-16
Buffer	Hex	4050	3 mA	55	0.1	D-16, P-16, F-16
Buffer inverter	Quad	4041	3 mA	65	0.05	D-14, P-14, F-14
Buffer inverter	Strobed	4502	3 mA	50	0.1	D-16, P-16
Inverter	Hex	4069	3 mA	50	0.1	D-14, P-14, F-14
Inverter	Plus pair	4007	10	35	0.01	D-14, P-14, F-14
Flip-flop	Dual D	4013	10	5 MHz	0.05	D-14, P-14, F-16
Flip-flop	Dual JK	4027	10	3 MHz	0.05	D-16, P-16, F-16
Flip-flop	Dual JK	4096	10	5 MHz	0.05	D-14, P-14, F-14
Flip-flop	Dual JK	4095	10	5 MHz	0.05	D-14, P-14, F-14
Latch	Quad D	4042	10	150	0.2	D-16, P-16, F-16
Latch	Quad RS NOR	4043	10	175	0.05	D-16, P-16, F-16
Latch	Quad RS NAND	4044	10	175	0.05	D-16, P-16, F-16
Multivibrator	Monostable	4047	10	125	0.5	D-14, P-14, F-14
Multivibrator	Dual monostable	4528	10	125	0.5	D-14, P-14, F-14
Schmitt trigger	Quad 2-input	4093				D-14, P-14, F-14
Buffer	Hex 3-state	340097	10	25	10	D-16, P-16
Inverter	Hex 3-state	340098	10	25	10	D-16, P-16
Flip-flop	Hex D	340174	10	16 MHz	100	D-16, P-16
Flip-flop	Quad D	340175	10	16 MHz	100	D-16, P-16

* Package entry letter is form factor: F = flatpack, D = hermetic dual-in-line, P = plastic dual-in-line, C = round can, S = studded flatpack. Package entry number is terminal count.

Speed-enhancement techniques for complementary metal-oxide semiconductors   The use of complementary metal-oxide semiconductors has been somewhat limited because of the rather slow operating speed of circuits in this logic family.   In a CMOS circuit, the delay times are not associated with the storage of charge in a semiconductor junction, as is the case in most types of saturated bipolar logic.   In CMOS, delay times are simply the time required to move the charge from one storage element to another.   As such, delay times are the RC time constant associated with the circuit nodes being switched.   In order to improve the speed of such a circuit, the appropriate RC time constant must be lowered.   The significant RC time constant for an internal gate is the product of the channel "on" resistance and the parasitic capacitances and functional capacitances associated with the appropriate source or drain.   Any scheme for reduction of the parasitic capacitances cannot merely be a methodology of scaling down the physical dimensions of the semiconductor device, as such a technique would generate both a reduction in the parasitic capacitance and an equivalent increase in the channel "on" resistance.

TABLE 27 The 4000 Family of Complementary Metal-Oxide Semiconductor Logic, Complete Selection Guide of Medium-Scale Integration-Circuit Types

Function	Circuit form	Type No.	Total delay, ns (at 5 V)	Power dissipation, μW	Package*
Shift register	Dual, 4-bit, serial in, parallel out	4015	2.5 MHz	10	D-16, P-16, F-16
Shift register	64-bit serial in-out	4031	2.0 MHz	10	D-16, P-16, F-16
Shift register	8-bit serial-parallel	4014	2.5 MHz	10	D-16, P-16, F-16
Shift register	8-bit parallel/serial in	4021	2.5 MHz	10	D-16, P-16, F-16
Shift register	18-bit tapped	4006	2.5 MHz		D-14, P-14, F-14
Shift register	4-bit parallel in-out	4035	2.5 MHz	5	D-16, P-16, F-16
Shift register	8-bit bidirectional bus	4034	2.5 MHz	5	D-24, F-24
Shift register	8-bit bidirectional	4058	1.5 MHz	5	D-24, F-24
Shift register	200-bit dynamic	4062			D-16, F-16
Bus register	8-bit	4094			D-16, F-16
Counter	7-bit binary	4024	2.5 MHz	5	D-14, P-14, F-14, C-12
Counter	12-bit binary	4040	2.5 MHz	10	D-16, P-16, F-16
Counter	14-bit binary	4020	2.5 MHz	20	D-16, P-16, F-16
Counter	21-bit binary	4045	5.0 MHz	15	D-16, P-16, F-16
Counter	Decade	4017	2.5 MHz	5	D-16, P-16, F-16
Counter	Divide by 8, decoded	4022	2.5 MHz	5	D-16, P-16, F-16
Counter	Divide by N (2-10)	4018	2.5 MHz	5	D-16, P-16, F-16
Counter presettable	Up/down, binary decade	4029	2.5 MHz	5	D-16, P-16, F-16
Counter presettable	Up/down, BCD	4510	2.5 MHz	1	D-16, P-16
Counter and oscillator	14-bit binary	4060	2.5 MHz	20	D-16, P-16, F-16
Counter	Dual BCD up	4518	6 MHz	4	D-16, P-16
Counter	Dual binary up	4520	6 MHz	4	D-16, P-16
Counter	Binary up/down	4516	12 MHz	1	D-16, P-16
Counter	Divide by N	4059	2 MHz		D-24, F-24
Counter display driver	Decoded decade 7-segment	4026	1 MHz	5	D-16, P-16
Counter, display driver	Decoded decade 7-segment	4033	1 MHz	5	D-16, P-16
Adder	4-bit full	4008	350	5	D-16, P-16, F-16
Adder	Triple serial positive	4022	400	5	D-16, P-16, F-16
Adder	Triple serial negative	4038	400	5	D-16, P-16, F-16
Parity tree	12-bit	4531	140	0.1	D-16, P-16, F-16
Arithmetic logic unit	4-bit	4057	1,000	500	D-28
Decoder	BCD to decimal	4028	250	10	D-16, P-16, F-16
Data selector	8-channel	4512	75	0.5	D-16, P-16
Decoder latch	4-line to 16-line	4514	300	0.2	D-16, P-16
Decoder latch	4-line to 16-line	4515	300	0.2	D-16, P-16
Decoder	Dual 1 of 4	4555	80	0.1	D-16, P-16
Decoder	Dual 1 of 4	4556	90	0.1	D-16, P-16
Magnitude comparator	4-bit	4063	250	5	D-16, P-16
Memory—RAM	32-bit 4 × 8 binary	4036	500	10	D-24, F-24
Memory—RAM	32-bit 4 × 8 direct	4039	500	10	D-24, F-24
Memory—RAM	64-bit 4 × 16	4505	200	0.3	D-14, P-14
Phase-lock loop	500 kHz	4046		600	D-16, P-16, F-16
Multiplexer	Quad 2-input	4019	50	50	D-16, P-16
Latch	8-bit addressable	4099	40	20	D-16, P-16
AND-OR invert	Dual 2-wide 2-input	4085	20	10	D-16, P-16
Rate bit generator	Programmable	4702	24 MHz	10,000	D-16, P-16
Memory—RAM	256-bit	4720	70	40	D-16, P-16
Latch	4-bit addressable	4723	45	40	D-16, P-16
Comparator	4-bit	40085	70	100	D-16, P-16

* Package entry letter is form factor: F = flatpack, D = hermetic dual-in-line, P = plastic dual-in-line, C = round can, S = studded flatpack. Package entry number is terminal count.

Table 29 describes some of the basic techniques that have been used to improve the speed of CMOS circuits. It should be noted that these techniques can also be applied to n-channel MOS and p-channel MOS. The techniques basically fall into three categories. The first three techniques are designed to limit the amount of bulk material available for junction spreading region and the associated parasitic capacitance, while the silicon-gate technique is a methodology for allowing the finer definition of the channel region. The third category is ion implant, which gives benefits similar to those of the silicon gate along with a precision control of the channel doping and the corresponding "on" resistance.

Silicon-on-sapphire and dielectric isolation are techniques for fabricating CMOS devices on small separated islands of silicon material such that reverse-biased junctions are not necessary to separate one device from the other. In the case of silicon

TABLE 28   The 4000 Family of Complementary Metal-Oxide Semiconductor Logic, Complete Selection Guide of Analog Switch-Circuit Types

Circuit form	Type No.	Isolation leakage current, pA	On resistance, Ω	Power dissipation, μW	Package*
Quad, SPST switch	4016A	100	300	0.01	D-14, P-14, F-14
Single 8-channel switch	4051A	10	50	0.1	D-16, P-16, F-16
Dual, coupled 4-channel switch	4052A	10	50	0.1	D-16, P-16, F-16
Triple, coupled 2-channel switch	4053A	10	50	0.1	D-16, P-16, F-16
Quad, SPST switch	4066A	100	50	0.01	D-14, P-14, F-14
Single 8-channel switch	L02	50	250	50,000	D-16
4 × 4 cross point	INS001	50	250		D-16
Single 8/dual 4 switch	L05	500	15	150,000	D-16
Single 8-channel switch	L021	50	250	75,000	D-16
Single 8-channel switch	7501	500	170	20	D-16
Dual coupled 4-channel switch	7502	500	170	20	D-16
Quad, SPST switch	7510	200	75	20	D-16

* Package entry letter is form factor: F = flatpack, D = hermetic dual-in-line, P = plastic dual-in-line, C = round can, S = studded flatpack.   Package entry number is terminal count.

TABLE 29   Comparison of Speed-Improvement Techniques for Complementary Metal-Oxide Semiconductor Logic

Technique	Methodology	Cost	Comment
Silicon on sapphire	Reduced back capacitance	High	Radiation-resistant
Dielectric isolation	Reduced back capacitance	High	
Isoplanar isolation	Reduced gate capacitance	Moderate	
Silicon gate	Reduced gate geometry	Low	Low voltage
Ion implant	Reduced gate geometry and controlled doping	Low	

on sapphire a thin film of single-crystal silicon is formed on the surface of a host wafer of single-crystal sapphire.   The thin film is then etched into separate distinct islands.   These islands are then separately diffused into separate p- and n-channel field-effect transistors.   Since the silicon material is extremely thin and is formed on a dielectric surface, a minimal parasitic capacitance is associated with such devices.

Dielectric isolation is described in detail in a later section on radiation-hardened devices.   The use of dielectric isolation to form CMOS devices has benefits similar to those when such devices are fabricated on a sapphire substrate.   The two techniques have produced approximate gains of speed in the order of a factor of 3 for on-chip gates where the capacitance is dominated by the device capacitance.   It must be emphasized that such gains are not realizable on small-scale integration devices where the delay times are dominated by the interconnection capacitance from integrated circuit to integrated circuit.   Hence, the main gain for speed improvement is associated with MSI and LSI devices.   Fabrication techniques of silicon on sapphire and dielectric isolation are not fully suited to large chips and high density because of the reduced yield associated with these less mature processes.   At the present state of the art neither technique is used for volume-production standard product.

Much work is being done to bring the techniques of dielectric isolation and silicon on sapphire for CMOS fabrication into maturity so they may be applied to the speci-

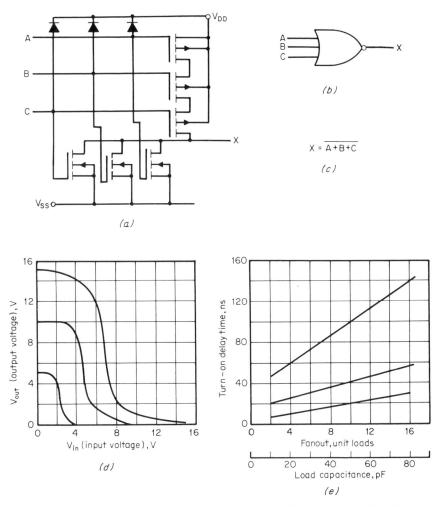

Fig. 29 Characterization of the 4000 family of complementary metal-oxide semi-conductors. NOR-gate characterization. (*a*) Circuit schematic. (*b*) Logic diagrams, (*c*) Logic equation. (*d*) Transfer curve, a plot of output voltage vs. input voltage. (*e*) Fanout curve, a plot of turn-on delay time vs. load capacitance.

fied radiation requirements of military systems. High-radiation environments along with specified limited high-speed applications will serve as the only avenue of use, leaving the mainstream of CMOS production restricted to standard technologies such as isoplanar, silicon-gate, and ion-implant. The latter three techniques are part of the main stream of semiconductor-production technology and are readily handled in volume batch processing. The isoplanar technique isolates the sidewalls of junctions by depositing a V-shaped groove of silicon dioxide insulator and polycrystalline silicon. This groove serves to limit the spread of the junction and the associated parasitic capacitance.

The silicon-gate technique allows finer definition of the gate region by using a gate contact of polycrystalline silicon. This gate can then be used as a self-masking tool during the source-drain diffusion. Hence the technique is sometimes called self-aligning gate. Ion implant is a technique for gaining precise control of the doping densities by use of high-energy accelerators for direct implant of the dopant ions

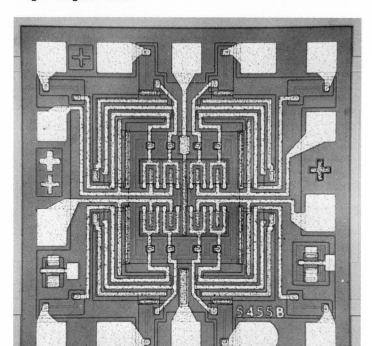

Fig. 30 A chip microphotograph of a complementary metal-oxide semiconductor device CD 4018. (*RCA, Solid State Division.*)

without the use of high-temperature diffusion. This technique can be used to apply precisely controlled metal-gate geometries and to allow for extreme precision on the control of carriers. All these techniques can be applied for the improvement of speed in CMOS circuits. One other point must be emphasized. The elements of circuit design can be applied for the improvement of speed possibly to the detriment of power dissipation. Speed-improvement techniques through circuit design are available, and much use has centered upon the technique of double buffering which allows a single inverter to drive two succeeding inverters each of larger geometries so as to minimize the overall delay and allow a symmetric input-output waveform. This technique will realize an improvement in the overall system speed.

The 54C family of complementary metal-oxide semiconductors The 54C family of complementary metal-oxide semiconductor devices was introduced by the National Semiconductor Corporation in 1973 to combine the best attributes of the 4000 family of CMOS devices with the 5400 family of transistor-transistor logic. The purpose of this family was to further expand the matrix of available speed and power combinations of the 5400 family by emulating its functions within CMOS technology. The concept was directed at providing interchangeable replacement of 5400 logic elements with an equivalent 54C element.

The concept, although quite appealing, has not been satisfactorily implemented owing to the basic differences between the CMOS logic cell and the transistor-transistor logic cell. The approaches to pinout and logic partition are necessarily different in the two families because of the different efforts required to implement differing logic functions. CMOS is basically a NOR family, while TTL uses the NAND gate as its basic logic. In addition, TTL is limited to operation with a single 5-V power supply, while CMOS has the advantage of being operative over an extended range. The difference in fanout drive capability between the CMOS elements and the TTL elements represents a major difficulty in intermixing the two elements.

The logic cell of this family is similar to the logic cell used in the 4000 family, with the major modification being inclusion of a larger output-drive transistor pair to approximate more closely the drive capability of the 5400 family.   The range of available circuits in the 54C family is presented in Table 30.   The use of this

TABLE 30   The 54C Family of Complementary Metal-Oxide Semiconductor Logic, Complete Selection Guide of Circuit Types

Function	Circuit form	Type No. 0 to 70°C	Type No. −55 to +125°C	Fanout (unit loads)	Total delay, ns (at 5 V)	Power dissipation, nW	Package*
NAND gate	Quad 2-input	74C00	54C00	10	50	10	D-14, P-14
NAND gate	Triple 3-input	74C10	54C10	10	60	10	D-14, P-14
NAND gate	Dual 4-input	74C20	54C20	10	75	10	D-14, P-14
Inverter	Hex	74C04	54C04	10	40	10	D-14, P-14
NOR gate	Quad 2-input	74C02	54C02	10	50	10	D-14, P-14
Flip-flop	Dual JK	74C73	54C73	10	5 MHz	50	D-14, P-14
Flip-flop	Dual D	74C74	54C74	10	5 MHz	50	D-14, P-14
Flip-flop	Dual JK	74C76	54C76	10	5 MHz	50	D-16, P-16
Flip-flop	Dual JK	74C107	54C107	10	5 MHz	50	D-14, P-14
Flip-flop	Quad D, 3-state	74C173	54C173	10	5 MHz	10	D-16, P-16
Decoder	BCD to decimal	74C42	54C42	10	200	50	D-16, P-16
Decoder	Quad 1-line to 4-line	74C154	54C154	10	275	100,000	D-24, P-24
Shift register	4-bit, bidirection	74C95	54C95	10	5 MHz	100	D-14, P-14
Shift register	8-bit, parallel out	74C164	54C164	10	5 MHz	500	D-14, P-14
Multiplexer	8-line to 1-line	74C151	54C151	10	240	50	D-16, P-16
Multiplexer	Quad 2-line to 1-line	74C157	54C157	10	180	50	D-16, P-16
Counter	Synchronous decade	74C160	54C160	10	5 MHz	50	D-16, P-16
Counter	Synchronous binary	74C161	54C161	10	5 MHz	50	D-16, P-16
Counter	Synchronous decade	74C162	54C162	10	5 MHz	50	D-16, P-16
Counter	Synchronous binary	74C163	54C163	10	5 MHz	50	D-16, P-16
Counter	4-bit up/down	74C192	54C192	10	4 MHz	200	D-16, P-16
Counter	4-bit up/down	74C193	54C193	10	4 MHz	200	D-16, P-16
Register file	4-bit	74C195	54C195	10	5 MHz	400	D-16, P-16
NAND gate	8-input	74C30	54C30	10	90	10	D-14, P-14
AND gate	Quad 2-input	74C08	54C08	10	90	10	D-14, P-14
OR gate	Quad 2-input	74C32	54C32	10	90	10	D-14, P-14
Exclusive-OR gate	Quad	74C86	54C86	10	90	50	D-14, P-14
Schmitt trigger	Dual	74C13	54C13	10	100	50	D-14, P-14
Multivibrator	Dual monostable	74C123	54C123	10	100	50	D-16, P-16
Flip-flop	Hex D	74C174	54C174	10	4 MHz	10	D-16, P-16
Flip-flop	Quad D	74C175	74C175	10	4 MHz	10	D-16, P-16
Buffer	Hex tristate	80C97	70C97	10	50	10	D-16, P-16
Inverter	Hex tristate	80C98	70C98	10	50	10	D-16, P-16
Decoder	BCD to 7-segment	74C48	54C48	10	200	50	D-16, P-16
Shift register	8-bit, parallel	74C165	54C165	10	5 MHz	500	D-16, P-16
Counter	Decade, presettable	85C55	75C55	10	5 MHz	50	D-16, P-16
Counter	Decade, presettable	85C56	75C56	10	5 MHz	50	D-16, P-16
Memory	RAM 64-bit	74C89	54C89	10	100	100	D-16, P-16
Adder	4-bit full	74C83	54C83	10	100	100	D-16, P-16
Comparator	4-bit	74C85	54C85	10	100	100	D-16, P-16

* Package entry letter is form factor: F = flatpack, D = hermetic dual-in-line, P = plastic dual-in-line, C = round can, S = studded flatpack. Package entry number is terminal count.

family is not recommended, since the 4000 family takes better advantage of the unique attributes of complementary metal-oxide semiconductor devices.

Metal-oxide Semiconductor Devices—A Non-family Approach to Digital Logic

Metal-oxide semiconductor (MOS) devices play a specialized but important role in the implementation of many specialized digital functions in modern electronics. MOS devices are integrated circuits which are fabricated from a family of field-effect transistors whose gate region is insulated by a thin oxide layer.   They are fabricated in one of two structures, p-channel or n-channel.   The p-channel devices were the first devices produced as commercial product.   A cross-sectional diagram of the p-channel structure is shown in Fig. 31.   The figure shows that fabrication of a p-channel device is inherently much simpler and smaller a bipolar integrated circuit. The structure of an n-channel integrated circuit is also shown in Fig. 31.   This structure is the exact opposite of the p-channel structure, but its practical high-yield fabrication requires greater controls on the surface states of the silicon material used to fabricate the integrated-circuit chip.

PMOS, as it is sometimes called, or p-channel MOS, is available in two basic types, high-threshold and low-threshold. The high-threshold product was the initial form introduced and requires high-voltage power supplies and signal levels to overcome the switching threshold of the MOS transistor structure. It was desirable to reduce the requirements of three power supplies approximately 24 V apart to more manageable levels which could really be interfaced with bipolar logic. Analysis of the solid-state physics equation that relates device threshold to the physical characteristics of the structure shows that the threshold can be modified by changing the dielectric constant of the gate insulating material, by changing the surface potentials of the gate material, by changing the crystal orientation of the silicon substrate, or by changing and precisely controlling the impurity concentrations within the gate region. The various processes for reducing the threshold of PMOS have used these technologies either individually or in some combinations.

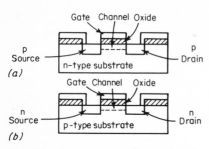

Fig. 31 Cross section of metal-oxide semiconductor devices. (a) p-channel device. (b) n-channel device.

The first methodology perfected for low-threshold MOS was the use of 1,0,0 silicon substrates, which reduced thresholds while sacrificing some device performance. The use of polycrystalline silicon material to form the gate structure served as a second method of fabricating low-threshold MOS, offering a reduced contact potential compared with the standard metal gate. Ion implantation has also been used as a methodology for lowering threshold by precisely controlling the dopant concentration in the channel region. The use of silicon-gate technology has gained the widest acceptance because it offers the advantages of both low threshold and increased speed through reduced gate parasitic capacitance. The reduction in gate parasitic capacitance is generated by the polycrystalline silicon serving as a direct self-aligning mask for the formation of the source and drain regions. This self-aligning feature allows fabrication of smaller-geometry devices having lower parasitic capacitance. The technology of the silicon gate is sometimes referred to as self-aligning gate technology. Silicon nitride has been applied as a dielectric material for the reduction in threshold. This technology is complicated by the high-temperature processes necessary for the formation of thin layers of silicon nitride.

NMOS's, or n-channel metal-oxide semiconductor devices, have the further advantage of being able to function with a single power supply and, as such, are directed and completely compatible with bipolar integrated circuits. NMOS devices should offer higher speeds, but at present, their production is somewhat limited by the technological problems in mass production. This processing can be combined with the technology of fabricating MOS devices on silicon, which is deposited on the surface of a single crystalline insulating substrate, such as sapphire or spinel. This technique reduces the parasitic capacitance and improves the operating speed of MOS devices.

MOS devices are most properly suited for fabrication of large arrays in either MSI or LSI. MOS devices which have on-chip interface only and which are not required to drive or receive signals from lines not located on the single chip can be fabricated using very small geometries. MOS devices require exceptionally small areas, since the transistors have small areas and MOS devices can be used directly as loads on chip. Structures having in excess of 10,000 MOS transistors on a single chip are possible with present technology.

MOS technology is suited primarily to two separate categories of devices. The first is for circuits that have repetitious nonrandom patterns that can readily be produced in high-density, high-complexity arrays. Good examples of this kind of regular circuit array are memories and shift registers. MOS devices are also suited to irregular arrays that perform system or subsystem functions, such as keyboard encoders, clock-divider circuits, microprocessor chips, calculator logic chips, and universal asynchronous receiver-transmission circuits. These irregular arrays cannot be laid out with the optimum density of the regular memory arrays. Since MOS devices can be made very small when designed only to function on a chip without an off-chip interface, it is essential to minimize the number of external connections. It is quite common for an output-driver buffer transistor and the associated protective clamp network to consume more area than 25 internal devices. It is also essential in MOS design to perform partitioning in a most expeditious and thoughtful manner.

Because MOS devices cannot gain high efficiency unless the number of internal devices is much larger than the number of inputs and outputs, MOS has not seen application as the standard family of digital circuits in the manner by which TTL or DTL gained acceptance. MOS has gained wide acceptance in custom arrays, memories, systems on a chip, shift registers, and similar functions.

The selection and application of MOS devices must be made with a great degree of care. MOS devices are exceptionally high impedance devices and are easily destroyed by current waveforms. Although most MOS devices are designed with input-protection clamps, these clamps are of limited effectivity when stressed by quickly rising current waveforms. MOS devices are sometimes categorized as TTL- or DTL-compatible. However, claims of compatibility must be examined in detail to determine whether all input-output and clock lines are truly compatible with the bipolar forms of logic. Many MOS circuits have inputs which are directly compatible with TTL and outputs which are compatible with TTL, but which function at a greatly reduced fanout of only one or two standard TTL loads. In many cases, the clock lines of MOS devices must be driven at MOS levels by a separate clock-driving circuit. MOS devices are more limited in speed than all bipolar technologies, except for the high-threshold logic. MOS devices normally function at frequencies below 1 MHz, and this range of operating frequencies covers a broad range of digital equipment. MOS systems do not require high degrees of power-supply decoupling because of the small currents drawn by the circuits in operation. However, MOS circuits are more prone to the damaging effects of contamination because of sensitivity of the surface states of the MOS devices. For long life, it is essential that MOS devices be procured in high-quality hermetic-sealed enclosures.

A number of manufacturers of MOS devices have documented a complete set of design rules that enable persons outside the company to become familiar with these design rules and then be capable of designing additional functional MOS circuits. These approaches have included the use of computer-aided design programs and computer-design centers to which a user may come and activate the equipment and software to design his own circuit. Published sets of standard test cells have also been used. These may be selected and interconnected by the user in a particular manner to make fabrication of a particular functional block possible. The availability of formalized procedures for the design of MOS circuits is evidence that most MOS design is a cooperative function between user and manufacturer.

Integrated-Injection Logic—A Bipolar Alternative to MOS Logic

Integrated-injection logic, or I^2L as it is commonly called, is an emerging technology allowing for great densities, low power, and high speeds in bipolar-circuit fabrication. It is said that the newest of techniques are sometimes the oldest of techniques, and this is true in integrated-injection logic. This technology is really a reintroduction of the simplest and original form of integrated-circuit logic, direct-coupled transistor logic. In I^2L all circuitry is reduced to basic transistors, for all coupling is transistor to transistor and all switching is within a single transistor. This reduces the total number of components needed to form an on-chip gate. The basic simplicity of this simple geometric system of transistors driving transistors achieves densities of eight to twelve times that which is customary with either CMOS or TTL. Further, integrated-injection logic offers the unique advantage of operation at a fixed-speed power product with the availability of varying speed by varying power over five or six decades of linearity. To change the operating speed of an I^2L structure, one must only change the current by which it is biased. This bias current is externally applied and can thus be varied from application to application. In some special applications the circuit could be reduced to a slower hold mode to conserve power and then brought back up to high speed by the corresponding increase in the injector current. I^2L gates are simple, basic, and static and require no specified recharge or polyphase clock signals.

Examination of Fig. 32 will show the basic simplicity of the I^2L logic block. The *pnp* transistor is used as a current source to drive current from the base to the emitter of each of the logic-cell *npn* transistors. The *npn* transistors are formed

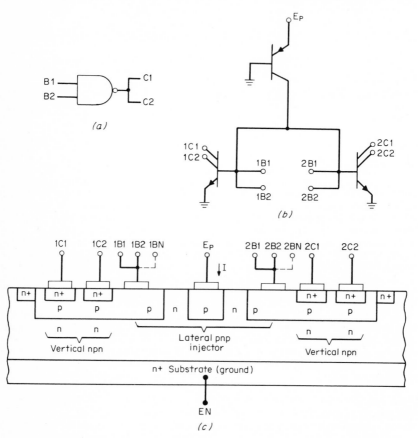

Fig. 32 Integrated-injection logic cell. (*a*) Logic diagram. (*b*) Schematic diagram. (*c*) Cross-section diagram.

as inverted structures with all emitters common to the substrate. The biasing of the expanded base lead on these transistors serves as input to the main gate function, which is now contained within the transistor. The multiple collector leads serve as logic-mode outputs. Basically, collectors drive adjacent bases to form complex logic cells. An increase in the injector current will increase the operating speed and decrease the appropriate switching times. By fabrication in standard bipolar processing, the *pnp* current injectors and the inverted *npn* switches are formed as part of the same physical structure, requiring no separate isolation. The overall simplicity allows for extreme density. It should be noted that a control in inverse betas is necessary so as to achieve reasonable fanin-fanout characteristics on a chip. This control of transistor betas has been recently achieved through the sophisticated bipolar-process technology of the mid 1970s. Hence they apply the current state-of-the-art process technology and the basic previous logic-cell design of 15 years before to achieve high-density, high-speed, low-power logic systems. It should be noted that this technique is best applied to LSI circuits, as the efficiency of these logic cells can be applied only on a chip within a given circuit. In typical operation a logic swing of approximately 700 mV is assured as the base-emitter mode switches from the VCE_{sat} condition of 50 mV to the VBE_{sat} condition of one silicon work function, or approximately 750 mV. It should also be noted that the base region of the npn transistor serves as the collection point for all the inputs of the corresponding gate. Any number of inputs can be connected at this base as all of them are common and each is driven by a separate output source. Hence, logic ambiguity of the drivers can be avoided only if each driver exists as an individual source. Hence in design each collector must be separated for each node.

A technique for the use of Schottky-barrier diodes has been applied by Motorola in a technology called constant-current complementary logic, or C^3L. This technique offers high speed and the potentiality for further use in radiation-hardened environments.

The techniques of integrated-injection logic will see wide application both in standard-product large-scale circuts such as Texas Instruments SBP0400 processor element and in custom circuits such as the watch-logic circuit designed by Texas Instruments.

SOLID-STATE MEMORIES

The earliest electronic computers used a combination of semiconductor diodes and vacuum tubes to perform logic functions, and magnetic-core circuits to perform memory functions. The initial transition to integrated circuits for computer logic functions still left main-frame computer memories in the form of core stacks. The last area within the computer to adopt solid-state integrated circuitry was bulk memory. Much of the reason for the delay was economic considerations. Core-memory systems could be built at cost below that of solid-state memories. However, the economic conditions that favored core memories were reversed during the early 1970s by a combination of the increasing labor cost for core-stack assembly and the availability of low-cost MOS memories. Solid-state memories also gained acceptance in high-performance applications for which core memories were not capable of fabrication.

A solid-state memory is a specially configured circuit capable of storing some particular pattern and array of digital information. This array can be either fixed or variable, and it can be stored in a number of different operating configurations. The smallest memory circuits store 16 bits, and the largest store 12,000 bits.

Memories must be considered on the basis of a number of criteria. The main criteria for specifying solid-state memories are memory size in bits, memory organization, provision for expanding organization, access and/or write speed, method of programming, data format, and simplicity of interface with other electronics. It is quite common that the memory circuits used in large arrays do not interface directly with the system logic but use an intermediate level of a memory-sense amplifier and a memory driver.

The computer-memory market has an annual volume of approximately one-half billion dollars. Progressively, solid-state memories have taken a growing share of this market. Having made their initial impact only in the late 1960s, they are projected to dominate the memory market by the end of the 1970s. Growth rates approaching 100 percent per year have been projected for the solid-state-memory market during the first half of the 1970s. This growth has been triggered by continuing improvement of solid-state memory capability and its continuous decrease in price per bit.

The hierarchy of computer memory must be understood before the potential of the impact of solid-state memory can be realized. Solid-state memory gained its initial acceptance for applications in what is called scratch-pad memory. Scratch-pad memory is small arrays of memory contained within the central processing unit, which is used to store information temporarily during the arithmetic or fetch operations within the central processor. The second major category is main-frame memory. This is also located within the central processing unit, and it is used for storing data and conditions associated with the computer program being processed to reduce processing time. Main-frame memory is in a continuing state of flux, with both core memory and solid-state memory being used. The third category of memory is on-line memory stack. The term core stacks is commonly used for this memory, and it is memory separate from the central processing unit which can be directly accessed by the central processing unit. The area of this type of memory is for the most part dominated by core, although there have been limited applications using slower MOS memory for this function. The final two areas of memory within a computer hierarchy are the auxiliary storage sites, both on-line and off-line. Auxiliary storage sites are used to store a large data base which can be retrieved manually or by the computer to be entered into a functioning program. Electromechanical systems, such as disk, drum, and magnetic tape, are primarily used for off-line storage and on-line auxiliary storage. The requirements for massive memory base in these areas and necessity of extremely low cost per bit have restricted auxiliary storage to electromagnetic and electromechanical mechanisms.

The computer is required to store permanently a set of instructions and subroutines in memory is a nonvolatile structure, and its earliest forms consisted of rope core and diode matrices. Solid-state read-only memories have come into use just in the past few years. The six categories of solid-state memories are (1) shift registers which are serial-entry, serial-read multibit storage elements; (2) random-access memories which are address-accessible read/write memories; (3) read-only memories, which can be either electrically programmable or mask programmable; (4) read-mostly memories, which are essentially read-only memories that can be altered by specified electrical programming; (5) first-in, first-out memories, which are a complex form of memory buffering; and (6) content-addressable random-access memories, which represent the forefront of very specialized memory technology.

Shift Registers

The shift register is the most basic form of solid-state memory. It is the most easily fabricated, and therefore was the first form of solid-state memory to be mass-produced. A shift register is in reality a sequential chain of flip-flops configured such that the output of each flip-flop drives the input of the next flip-flop in the chain. In operation, a common clock line is used to toggle each flip-flop simultaneously, causing the data bit to be shifted one step forward. Shift registers are of two basic types, dynamic shift registers and static shift.

A dynamic shift register is a shift register whose cell is configured to store and hold a data bit for only a specified, limited time interval. All cells in dynamic registers accomplish storage by means of charge storage on a capacitor, and there is an RC time constant associated with this storage. The RC time constant represents an interval after which the stored data bit is lost. In operation, a dynamic shift register must be clocked at some minimum frequency to prevent the loss of information stored in the register. Dynamic-shift-register cells are basically much simpler to fabricate than static-shift-register cells; so more complex registers having a larger number of bits can be fabricated at the same level of technological evolution and economics by using this technique.

Static shift registers are shift registers whose storage is accomplished using a complete flip-flop capable of permanent bit storage as long as the power-supply currents are maintained. Static shift registers typically consume more power in operation, since power is dissipated during the storage. Data are entered into a shift register sequentially and retrieved sequentially. The shift register must be clocked for an interval of clock pulses equal to the length of the message subtracted from the length of the shift register for the data to be available at the output. Data in dynamic shift registers can be retained for longer intervals if the data are recirculated from the output back to the input. This recirculating of data requires that a count be maintained of the data bit position within the shift register.

A few small-sized transistor-transistor logic shift registers have been introduced which have additional features that separate them from ordinary shift registers. These shift registers can be controlled by control pins to shift either left or right. They have capability for both parallel and serial entry and parallel or serial output. However, these specialized shift registers are not properly classified in the general category of shift registers.

A shift register is the solid-state analog of a tape recorder in that data are entered and played back sequentially without a direct (within circuit) measure of the data-bit position. On this basis, a recirculating flip-flop is analogous to an endless tape recorder. Shift registers are commonly used to store information for a limited interval which is repetitously used and reused. A good example of this use is storage of the display bits used in a cathode-ray-tube computer terminal. In this configuration, the message is decoded from alphanumeric binary code, such as ASCII, to a sequential 5 × 7 dot matrix with which the tube electronics are modulated. In operation, as the beam sweeps across the face of the tube, the information to pattern the display is sequentially produced by the shift register. Shift registers have also seen application as solid-state delay lines, since data can be entered into the shift register while it is clocked at a fixed frequency. The end result is that the data are available at a later interval whose delay time is the product of the shift-register length in bits times the clock period.

Many shift registers are designed to operate using multiphase clock systems. A multiphase clock system is a method of operating a shift-register array at an apparent frequency which is greater than the maximum operating frequency of a single shift register fabricated from the same technology. In multiphase clock-shift registers, the array contains the same number of shift registers as there are phases to the clock. Data are alternately entered to the inputs of succeeding shift registers with each phase of the clock. In a four-phase shift register where the initial clock is 2 MHz, the collection of four clocks then becomes 8 MHz, and an 8-MHz data stream can be entered and output through a shift register. In reality, the register is internally functioning as four 2-MHz shift registers, each 90° out of phase with the other. The range of available static shift registers is presented in Table 31,

TABLE 31   Static Shift Register Integrated Circuits, Selected Circuit Types and Descriptions

Type No.	Size, bits	Organization	TTL compatibility	Shift rate	Package*
3300	25	16 + 8 + 1	No	0–250 kHz	C-10
1003	64	Dual 32	Yes	0–1.5 MHz	D-14
2104	100	Quad 25	Yes	0–2.0 MHz	D-14
1012	100	Dual 50	Yes	0–3.0 MHz	C-10
2105	128	Dual 64	Yes	0–2.0 MHz	C-8
1008	160	Dual 80	Yes	0–3.0 MHz	C-10
3348	192	Hex 32	Yes	0–1.0 MHz	D-24
2010	200	Dual 100	No	0–3.0 MHz	C-10
3383	256	Single 256	No	0–2.0 MHz	C-10
2107	256	Dual 128	Yes	0–1.5 MHz	C-8
3347	320	Quad 80	Yes	0–2.0 MHz	D-16

* Package entry letter is form factor: F = flatpack, D = hermetic dual-in-line, P = plastic dual-in-line, C = round can, S = studded flatpack. Package entry number is terminal count.

where shift registers are listed in the ascending order of data bits. The comments in the table point out special features of the shift register. Table 32 similarly covers the available range of dynamic shift registers. They are again arranged in order of the number of data bits within the shift register, and the table completely describes whether or not the shift register is directly compatible with transistor-transistor logic. As pointed out in the table, many shift registers have TTL-type inputs and outputs but have clock lines that are highly capacitive and must be driven by special clock-driver circuits.

Random-Access Memories

Random-access memories are basic arrays of alterable read-write memory storage used in a digital system. A random-access memory is in reality a location-addressable digital storage memory. As such, each bit is directed to a particular location

TABLE 32   Dynamic-Shift-Register Integrated Circuits, Selected Circuit Types and Descriptions

Type No.	Size, bits	Organization	TTL compatibility	Shift rate	Package*
1200	128	Quad 32	Yes	0.01–3.0 MHz	D-14
1506	200	Dual 10	Yes	100 Hz–2.0 MHz	C-8
1205	256	Single 256	Yes	0.01–1.0 MHz	C-10, D-14
7780	320	Quad 80	Yes	0.01–2.5 MHz	D-16
1405	512	Single 512	Yes	200 Hz–2.0 MHz	C-10
1403	1,024	Single 1,024	Yes	10 kHz–5.0 MHz	C-10, P-8

* Package entry letter is form factor: F = flatpack, D = hermetic dual-in-line, P = plastic dual-in-line, C = round can, S = studded flatpack. Package entry number is terminal count.

in the memory during the write operation and is recallable by that location during the read operation. Random-access memories vary in size from 64 to 4,096 bits on a single monolithic chip. Further development will expand single-chip random-access memories to 16,000 bits before the 1980s arrive. A broad range of technologies have been applied to the fabrication of memory circuits, and all the basic technologies of semiconductor processing and circuit design have been applied to the fabrication of dynamic and static memory circuits. In the case of dynamic storage cells memory is actually an array of single-bit recirculating shift registers; the recirculation operation occurs every time a particular column is accessed. This recirculation causes the data which are stored on a temporary capacitor to be rewritten and the level properly established. Static random-access memories require only that power supply be maintained in order to maintain the storage of data. Random-access memories can also be fabricated with a variety of output structures such as the three-state or open collector in order to allow for the simple expansion of memory circuits into large memory arrays.

Table 33 presents the available random-access memory circuits. These circuits fall into three basic speed categories: the large MOS circuits that have access and

TABLE 33   Random-Access Memory Integrated Circuits, Selected Types and Descriptions

Type No.	Size, bits	Organization	TTL compatibility	Static/dynamic	Read time, ns	Write time, ns	Package*
3101	64	16 × 4	Yes	Static	30	25	D-16, P-16
31L01	64	16 × 4	Yes	Static	70	100	D-16, P-16
27S02	64	16 × 4	Yes	Static	22	25	D-16, P-16
2701	256	256 × 1	Yes	Static	70	60	D-16, P-16
27LS01	256	256 × 1	Yes	Static	35	30	D-16, P-16
1101A	256	256 × 1	Yes	Static	850	800	D-16, P-16
3532	512	512 × 1	Yes	Static	1,000	600	D-16, P-16
1103	1,024	1,024 × 1	No	Dynamic	300	580	D-18, P-18
2102	1,024	1,024 × 1	Yes	Static	50	100	D-16, P-16
93415	1,024	1,024 × 1	Yes	Static	60	45	D-16
2548	2,048	2,048 × 1	Partial	Dynamic	345	560	D-22
4402	4,096	4,096 × 1	Yes	Static	250	250	D-22
5101	1,024	256 × 4	Yes	Static	800	800	D-22
6605	4,096	4,096 × 1	Yes	Dynamic	370	490	D-22
4096	4,096	4,096 × 1	Yes	Dynamic	425	425	D-16
2107B	4,096	4,096 × 1	Yes	Dynamic	270	220	D-22

* Package entry letter is form factor: F = flatpack, D = hermetic dual-in-line, P = plastic dual-in-line, C = round can, S = studded flatpack. Package entry number is terminal count.

cycle times of up to $\frac{1}{2}$ μs, the TTL circuits that have access and cycle times of 30 to 50 ns, and the ECL arrays that have access times of only a few nanoseconds. The choice of random-access memory type is totally governed by the system constraints and requirements. However, four random-access memory types have gained great acceptance within the engineering community. The first of these is the 1103, which is a 1,024-bit dynamic memory. This memory is fabricated using low-threshold silicon-gate technology and a depletion-mode p-channel MOS device. Memory is fully decoded with a chip-select control. It requires a refresh cycle of every 2 ms and has a full rewrite-cycle time of 500 ns. The 1103 is well suited to building large-scale memory systems and has found application in computer memory extenders, main-frame memories, and peripheral-memory applications. The 1102 was the circuit that first channeled the economic viability of magnetic-core memory systems for computer applications.

The second memory-circuit type is the 2102 which is also a 1,024-bit random-access memory. This, however, is static memory. It is fabricated using n-channel technology and is designed for operation from a single power supply. All inputs and outputs are directly compatible with TTL loads and levels. Since it is a static memory, there is no requirement for clocks or refresh cycles. The 2102 is somewhat slower than the 1103, but simplicity of operation makes it an exceptionally good choice in most cases except for the fabrication of large memory stacks. Only in large memory stacks is the cost of the overhead circuits required for refresh recovered by the reduced price of the 1103-type circuits.

The third random-access memory is the 3101, 64-bit fully decoded random-access memory. This memory is built using TTL technology and is organized in the form of 16 words, each containing 4 bits. All outputs are open-collector transistor-transistor logic. This feature allows for expansion to build larger memory arrays. This circuit contains a chip-select feature to make memory expansion simpler. Although the 3101 has a much higher cost and consumes much more power than either the 1103 or the 2102 on a bit-by-bit basis, it is the preferred choice where high speed is required in main-frame applications. In applying the 3101, provision must be made to allow for the data inversion that occurs between the data inputs and the data outputs, i.e., the negation of the logic data is directly available as the output-data level.

The fourth category of random-access memory circuits is the 4,096-bit single-chip memory circuit as typified by the 2107B. Table 33 shows the basic types of 4,096-bit memories that are available and their interrelationship within a set of selected random-access memory circuits. The 2107B is a 250-ns dynamic memory organized as 4,096 words, each of a single bit. Memory is dynamic; it is packaged in a 22-pin package. Work has now progressed toward circuits from Mostek that provide similar circuitry in a 16-pin package using more complicated decoding techniques. The smaller packages allow for both more economical fabrication of the memory integrated circuit and denser memory-system fabrication, as smaller printed-circuit boards can be used for the memory circuits. The application of the 4,096-bit dynamic memory is purely restricted to computer system memories where the requirements for overhead timing-control circuits are insignificant when compared with the overall saving in system cost by the use of such circuits.

It must be emphasized that random-access memories represent a unique problem in integrated-circuit testing technologies, as many such circuits are sensitive to bit pattern. That is, certain combinations of data may generate errors because of adjacent lines or pickup within the circuit. Hence it is necessary to test the circuits for such sensitivity by exercising them with critical test patterns. Such test patterns must be designed on the basis of a knowledge of the circuit design, the circuit layout, and the sensitivity of the various circuit elements to adjacent circuit elements. Reliance on bit-pattern schemes that have been generated for core-memory testing is not appropriate.

Read-Only Memories

A read-only memory is a location-addressable fixed-content memory. It is used to store digital information within an electronic system when this information is fixed and unchanged. Read-only memories store instruction sets, look-up tables, subroutines, and microprogramming sets. They have gained wide acceptance in computer peripherals for the purpose of storing code-conversion tables and display-font tables.

Read-only memories are segregated into three functional categories, mask-programmable read-only memories, electrically programmable nonerasable read-only memories, and electrically programmable erasable read-only memories. Mask-programmable read-only memories represent manufacturing economy where large volumes of repeated memory are to be fabricated. In most designs of read-only memories, only a single mark is modified to enter the data input onto the memory during fabrication. In many cases, the making of this particular mask and its changes can be controlled directly by the computer program. If the mask is also selected for one of the later production steps, wafers can be partially fabricated and held until program receipt so that custom-programmed read-only memories can be produced with a short turnaround time. Run rates in excess of 1,000 are required to make mask-programmable read-only memories practical. Any alteration of the bit pattern requires generation of a complete new mask. Many standard read-only-memory bit patterns have been produced, and these are offered as catalog products by the semiconductor vendors. The range of standard mask-programmable read-only memory patterns includes ASCII to EBCDIC code conversion, ASCII to 35-dot font generation, both upper- and lowercase, both row and column entry, trigonometric-value tables, and logarithmic tables. Read-only memories have been fabricated on a single chip to sizes in excess of 16,000 bits, and the structure of a read-only memory is primarily a matrix of storage cells, which are maintained separately in fixed logic states and which are accessed by row, column-decoding networks.

Programmable read-only memories are memories that are fabricated with all bits in one state but are designed such that the state of each bit can be electrically altered prior to the application of the part. This alteration is made by either opening an electrical fuse on the circuit surface or causing a field inversion in a specialized MOS storage element. A separate control or lead is available that must be exercised to make the circuit available for programming. If an error is made during the programming of a programmable read-only memory, it becomes useless, and the circuit must be destroyed. Three materials have been used for fusable links in programmable read-only memories, aluminum-film fuses, Nichrome-film fuses, and polycrystalline-silicon-film fuses. The aluminum fuses have not gained engineering acceptance because they are unreliable. Both the Nichrome and silicon fuses function by passing a high current density through a thin defined strip, and the high-current pulse causes the circuit to open by metal migration. This mode of opening the circuit is preferable to vapor fusion, since no out products are generated. The field-inversion technique is suitable only to MOS devices, and it has the advantage of not causing any structural disturbance to the chip.

Erasable-programming read-only memories are fabricated using the field-inversion technique, but include a quartz window on the package so that ultraviolet radiation can be used to restore the state of all the memory cells to their original condition. Programmable read-only memory cells can be manually programmed, programmed by special memory programmers, or programmed by standard computer-controlled test equipment. They are properly used for small production runs and in the early phases of system design for systems built in large production runs. In the latter case, the programmable read-only can be replaced by mask-programmable memories after the design is complete and proved and the bit pattern is stabilized. Programmable read-only memories have the advantage of instant availability in any bit pattern. Some distributors have begun to offer the service of programming read-only memories on automatic equipment at their facilities to enable the small user to get programmed read-only memories without having to provide the specialized test equipment. The available range of product of read-only memories is indicated in Table 34. In cases where the same form of read-only memory is available as both programmable and mask-programmable, the two are placed adjacent to each other in the table, and their interchangeability has been noted. The cases where large numbers of standard patterns are available in the same memory structure are also listed together. Read-only memories are playing an increasingly important part in the design of modern digital systems.

TABLE 34   Read-Only Memories, Selected Circuit Types and Descriptions

Type No.	Size, bits	Organization	TTL compat- ibility	Read time, ns	Mask	Pro- gram- able	Package §
82S23	256	32 × 8	Yes	25		X	D-16, P-16
8224	256	32 × 8	Yes	50	X		D-16, P-16
82S26	1,024	256 × 4	Yes	35		X	D-16, P-16
3301	1,024	256 × 4	Yes	45	X		D-16, P-16
8204	2,048	256 × 8	Yes	35	X		D-24, P-24
1302	2,048	256 × 8	Yes	700	X		D-24
1602	2,048	256 × 8	Yes	700		X	D-24
1702*	2,048	256 × 8	Yes	700		X	D-24
2513	2,560	64 × 7 × 5	Yes	450	X		D-24, P-24
2513/ CM2140†	2,560	64 × 7 × 5	Yes	450	X		D-24, P-24
3304	4,096	512 × 8/1,024 × 4	Yes	65	X		D-24
S8772‡	4,096	512 × 8	Yes	250	X		D-28
2556†	5,184	64 × 9 × 9	Yes	625	X		P-24
2580	8,192	2,048 × 4	Yes	625	X		P-24
3800	12,288	1,024 × 12	Yes	2,000	X		P-28, D-28
4800	16,384	2,048 × 8	Yes	700			P-24, D-24
2708*	8,196	1,024 × 8	Yes	500		X	D-24
3601	1,024	256 × 4	Yes	50		X	D-16
3604	4,096	512 × 4	Yes	70		X	D-24

* Erasable and reprogrammable.
† ASCII font generator.
‡ Sine table.
§ Package entry letter is form factor: F = flatpack, D = hermetic dual-in-line, P = plastic dual-in-line, C = round can, S = studded flatpack. Package entry number is terminal count.

Read-Mostly Memories

The read-mostly memory is a structure unique to a single processing technology called ovonic technology. They serve the middle ground between read-only memories and random-access memories. They are memories with read and access times comparable with those of random-access memories but with much slower write times. And they have the unique advantage of nonvolatile storage during an interruption of power supply, since they hold data for extended intervals without loss of data content. The Ovonic technique is an application of amorphous memory cells. The amorphous cells are fabricated using amorphous silicon, and this is combined with decoding circuitry built from single-crystal silicon. Available read-mostly memory circuits are presented in Table 35.

TABLE 35   Read-Mostly Memories, Selected Circuit Types and Descriptions

Type No.	Size, bits	Organization	Write time, ms	Read time, ns	Dual-in-line package pins
RM256A	256	16 × 16	20	50	40
HRM2048	2,048	64 × 32	0.0005	400	42 (module)
RM32	32	8 × 4	0.08	50	16
RM15	15	1 × 15	0.08	50	16

The high cost of read-mostly memories severely limits their applicability, but even without that consideration, their use would be restricted to cases where variable data in a computer program must be maintained during power interruption. The requirement for excessively high voltages to operate the write lines of read-mostly memories complicates the circuitry to interface it with other types of electronic circuits.

First-In, First-Out Registers

First-in, first-out memory, or FIFO, is a unique class of large-scale integration circuits that combines the functions of a shift register and a random-access memory. In operation, the FIFO serves as a data buffer between two asynchronous data points. It smooths the flow of data between the two asynchronous data points, for which it has gained acceptance in computer peripheral applications. In operation, data are clocked into the input of a FIFO and then bubble through the circuit to make the data available at the output when the output is clocked in the order that it was entered. Data are handled in a FIFO in same way that sheets of paper are handled in a file folder, since data are entered from the back and removed from the front without placing blank or space data bits.

FIFOs are fabricated from two circuit-design schemes. In the first, a FIFO is directly fabricated from a series of registers and peripheral control logic such that the data are directly moved forward within the FIFO. A second, more efficient technique has evolved which enters the data sequentially into random-access memory locations and uses control logic to generate markers which signify the beginning and end of the stored data train. The control logic also serves to generate flags for empty-register and register-overflow situations. Available FIFO circuits are presented in Table 36.

TABLE 36   First-in, First-out Memories, Selected Circuit Types and Descriptions

Type No.	Size, bits	Organization	TTL compatibility	Data rate	Dual-in-line package pins
S1709	104	13 × 8	Yes—resister	0.1 MHz	24
2535	256	32 × 8	Yes	1.0 MHz	28
MS-618	96	24 × 4		2.5 MHz	24
3341	256	64 × 4	Yes	1.0 MHz	16

Content-Addressable Random-Access Memories

A content-addressable random-access memory, or CARAM, is a unique integrated-circuit function that has been only recently realized using MSI technology. The CARAM performs the natural logic function of comparing the content of a digital word presented to the input terminals with the content of a digital word or words stored within the memory. The organization for CARAM allows the comparison operation to be completed on all bits in parallel. The CARAM memory can be operated directly as a random-access memory by entering and reading bits. It has found application in data-to-memory comparison, pattern recognition, high-speed information retrieving, autocorrelation, virtual memory, and the newer techniques of self-learning memory. A limited range of content-addressable random-access mem-

TABLE 37   Content-addressable Random-Access Memories, Selected Circuit Types and Descriptions

Type No.	Size, bits	Organization	TTL compatibility	Delay time, ns	Dual-in-line package pins
3104	16	4 × 4	Yes	30	24
8220	8	4 × 2	Yes	20	16

ory is available today, and this range is presented in Table 37. Much of the evolution of memory organizations within integrated circuits is anticipated to be directed toward other methods of implementing the content-addressable random-access memory. The function of content-addressable random-access memory can be performed by conventional random-access memories and computer subroutines, either hardware or software.

Charge-coupled Device Memories

A new concept for the realization of memory systems was invented at the Bell Telephone Laboratories in 1970. This technique, called charge-coupled devices, is an offshoot of MOS technology and is fabricated from structures similar to MOS structures. In fact most CCD circuits include MOS circuits on the same chip. The basic substrate of n-minus silicon with separate gates being fabricated on top of a a cross section of the basic CCD memory cell. The material is fabricated on a basic substrate of n-minus silicon with separate gates being fabricated on top of a dielectric film. These gates are biased so as to form separate potential wells. If charge is introduced into a given potential well and then the external potentials applied to the gates are changed so as to modify the wells as illustrated in Fig. 33, then the charge trapped in a given well, like the one under gate G-2 can

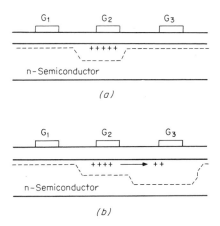

Fig. 33 Cross section of a charge-coupled device. (*a*) Charge stored under gate 2. (*b*) Change in potential transfer charge from gate 2 region to gate 3 region.

be caused to flow into the well under gate G-3. This flow of charge is linear and reversible by the application of the appropriate potential. The function of a linear string of CCD cells is similar to that of a shift register except that they can be made to function as a linear bucket-brigade shift register with the quantity charge transferred remaining intact. The CCD cells used in memories are not optimized for carrying analog data but rather are used to move a digital-data bit represented by a specified amount of charge. Arrays of shift registers formed as CCD structures can then be created with extreme density on a single chip of silicon.

Figure 34 is a chip photograph of the 2416 CCD serial memory circuit. This circuit was introduced and perfected by the Intel Corporation and is the first commercial introduction of a CCD serial digital memory circuit. This circuit is organized as an array of 64 independent recirculating shift registers. Each shift register is 256 bits long. Internal decoding circuit on the chip but external to the CCD array is used for the selection of a given shift register. All shift registers are clocked by a common clock line and up to 16 of them can be accessed during a given clock-pulse period. The data generated by such a circuit are similar in form to the data coming off a rotating memory system such as a disk or magnetic drum. The CCD memory allows for extreme density in a comparatively small chip, as illustrated in the photograph. The large dark areas on the chip are the memory array, and the other irregular circuits are the decoding, clock, and current-injection circuitry.

In operation any bit in the memory could be accessed in less than 100 μs. Serial data-transfer rates as high as 2 megabits per second are possible. In operation it is required that one of the shift registers be used as a timing track in order to

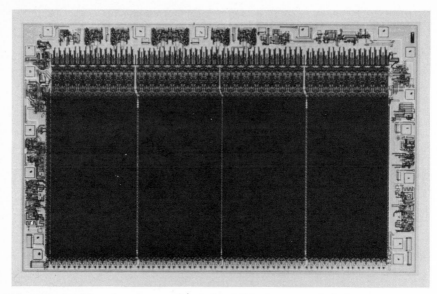

Fig. 34 A chip microphotograph of the 2416 charge-coupled device, 16,384-bit serial memory array. (*Intel Corporation.*)

synchronize the data available from the other shift registers.   The extreme density of these circuits is made more evident by the fact that 64 of them could be placed on a single moderately sized printed-circuit board to give an overall array of 1 million bits of digital data on this single card.   Other organizations of CCD memories are in the process of introduction by Fairchild, Signetics, and others in the industry.

Chapter **3**

Linear and Special-Purpose Integrated Circuits

SAUL ZATZ
Martin Marietta Aerospace, Orlando, Florida

LINEAR INTEGRATED CIRCUITS
The second major subclass of integrated circuits is commonly called linear integrated circuits. The term linear is actually a misnomer, since the class includes both linear and nonlinear analog integrated circuits, but the subclass includes all circuit types

whose inputs and/or outputs can vary meaningfully over a continuum of electrical voltages or currents.   On the other hand, the integrated-circuit industry defines linear integrated circuits in terms of the processing technology applied to their fabrication. By this definition, linear integrated circuits are circuits processed from technology that yields on-chip transistors with a higher breakdown voltage (VCBO of 40 vs. 15 V) than is customarily used for processing digital integrated circuits.   This processing difference includes the omission of gold doping and yields significantly slower storage time and increased transistor beta.   This processing difference is similar to that between processes for amplifier transistors and switching transistors.   Either definition covers a range of circuits that includes comparators, differential amplifiers, operational amplifiers, linear amplifiers, analog multipliers, analog phase-lock loops, digital-to-analog converters, voltage regulators, and an extensive variety of entertainment circuits.

The design, manufacture, and test of linear integrated circuits is quite different from the corresponding operations for digital integrated circuits.   Digital integrated circuits are configured in families which interfere easily to each other, while linear integrated circuits are usually used in independent nonfamily applications.   Linear integrated circuits are designed to meet separately a great variety of different system-circuit applications.   Therefore, a detailed understanding of the test parameters and test methods applied to the characterization of a linear integrated circuit is essential. Since linear integrated circuits operate over continuous ranges of input and output voltages, no characterization can be made over this full range; it must be made at carefully selected intervals.

Linear integrated circuits have historically had a greater number of reliability problems because of the higher voltages impressed during their operation and the greater variety of load and stimulus variables.   Their viability in a given application requires that full consideration be given to total power consumption and impressed voltages so that the recommended ratings are not exceeded.   Proper limitation of supply voltages and proper sequencing of their connection is a basic requirement.   As a rule, power supplies should be energized in ascending order of voltage; i.e., the most negative supply must be energized before the ground connection is completed, and the ground connection must be completed before the positive supply is energized. This requirement avoids forward biasing the substrate isolation diode, which would otherwise send a destructive current surge through the internal ground-bond wire. All supplies should be brought up to specified operating levels and stabilized before any signal voltages are impressed on the circuit.

Many linear circuits have extremely high-gain stages which are prone to oscillation if proper care is not taken in layout.   Since gain-bandwidth products in excess of 5 MHz are quite often available within linear circuits, power-supply bypassing and layout techniques must be adequate for these frequencies.

In applying linear circuits, full consideration should be given to the complete range of recommended voltage limits, including power-supply voltage, input voltage, differential input voltage, and common-mode input-voltage limits.   Exceeding the recommended ratings for the voltage limits on the device could initially lead to sacrifices in operating characteristics, and ultimately to total destruction.   The inputs of linear integrated circuits quite often have direct access to the emitter-base junctions of input transistors.   Reverse biasing of these inputs into the breakdown mode causes gradual and nonrepairable destruction of the linear integrated circuit.

Comparators

The voltage comparator is a circuit that has two analog inputs and a single digital output.   This output properly interfaces with a particular digital family with sufficient current-logic levels and drive capability to operate digital circuits.   In application, the voltage comparator makes high-speed comparisons between the levels at the two inputs.   Comparators are fabricated using both linear and digital processing in that high-resistivity material with high breakdown voltages is used, but gold doping is also used to minimize switching time.   Comparators can be made using Schottky clamped transistors to reduce the switching time further.   The direct application of the comparator makes it a single-bit analog-to-digital converter, and it can be used in the variety of applications illustrated in Fig. 1.   Simple connection of the

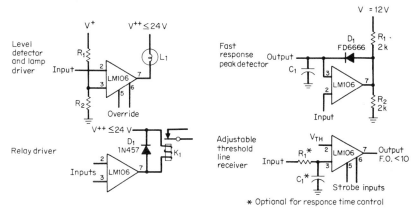

Fig. 1   Applications of comparator circuits.

comparator allows application as a Schmitt trigger, pulse-width modulator, line receiver, or level detector.   Comparators may also be used to detect pulse heights or dc levels, to operate high-speed analog-to-digital converters, and even as the basis for automatic electronic test equipment, such as that used to characterize integrated circuits.

Two types of comparators which have gained broad acceptance are the UA710 and the LM111   These circuits are readily available, are produced by a great number of suppliers, and meet a broad range of circuit applications.   The LM111 is a higher-performance circuit with lower input currents, and it operates over a wider range of supply voltages.   The output of the LM111 may be interfaced with RTL, DTL, and TTL directly and with minor modifications can be used to drive lamps, relays, and MOS circuits.   Specialized selections of comparators may sometimes be required to obtain the required precision or the required switching time.   The full range of available comparator circuits is detailed in Table 1.   For voltage comparator appli-

TABLE 1   Analog Comparators, Selected Circuit Types and Description

Type No.	Gain	Input offset voltage	Output voltage		Delay time, ns	Power-supply voltage	Package*
			False	True			
710	1,000	5 mV	−1.0/0.0	2.5/4.0	40	+12, −6	C-8, D-14, F-10
711 (dual)	1,000	5 mV	−1.0/0.0	2.5/	40	+12, −6	C-10, D-14, F-10
734	60,000	5 mV	TTL	TTL	200	±5	C-10, D-14
760		6 mV	TTL	TTL	25	±5	D-14
111	200	0.7 mV	TTL	TTL	200	5–30	C-8
1514	1,250	2.0	−1.0/0.0	2.5/4.0	40	+12, −6	D-14
160		1.0	TTL	TTL	20	±5	C-8, P-14, D-14
685		2.0	ECL	ECL	6.5	±5	C-10, D-16

* Package entry letter is form factor: F = flatpack, D = hermetic dual-in-line, P = plastic dual-in-line, C = round can, S = studded flatpack.   Package entry number is terminal count.

cation, the important characteristics to consider are input offset voltage, response time, drift, input-voltage range, and differential-input-voltage range.

Sense Amplifiers

Sense amplifiers are a special class of comparator used in digital systems primarily to read small voltages associated with core memory stacks and plated-wire memory stacks.   They accurately measure the very small voltage levels, typically only a few millivolts, associated with the output of a core stack and then compare that with an externally generated reference level.   After comparison, they generate in a minimal time a digital voltage level which corresponds to the logic state of the memory stack.

Sense amplifiers are specialized to interface properly with the range of available memory configurations.   Table 2 describes representative sense amplifiers which are available.

TABLE 2   Sense Amplifiers, Selected Circuit Types and Description

| Type No. | Application | Threshold at V_{ref} | | | Delay time | Package* |
		Min, mV	Max, mV	V_{ref}		
1440	Core memory	14	20	–6.0 V	30	C-10, F-10, D-14
1443	Core memory—ECL	17	23	540 mV	35	D-14
1444	Plated wire	1.0			25	D-16
7520	Core memory, dual	11	19	15 mV	55	D-16
7524	Core memory, dual	11	19	15 mV	40	D-16
3541	Dual input, core	10	35		20	

* Package entry letter is form factor: F = flatpack, D = hermetic dual-in-line, P = plastic dual-in-line, C = round can, S = studded flatpack.   Package entry number is terminal count.

Operational Amplifiers

The operational amplifier is the most widely used linear integrated circuit.   As an integrated circuit, it is an outgrowth from the modular solid-state operational amplifiers which were introduced in the early 1960s by Burr-Brown Research Corporation and G. A. Philbrick Researches, Inc., and they were initially used in analog computers.   The positive and negative feedback loops and nonlinear feedback loops around operational amplifiers brought the technology for analog computers to solve integral and differential equations.   After a characterized amplifier became available in operational-amplifier form, numerous other applications appeared for this circuit. The operational amplifier is simply a block of gain with both inverting and noninverting inputs, with some provision for compensating the gain characteristics of the amplifier and with internal or external balance for any offset generated within the circuit.

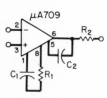

The first monolithic operational-amplifier circuit to gain acceptance was the UA709, the simplest basic operational amplifier in use.   It requires all compensation balance to be done externally with discrete components.   Figure 2 illustrates the set of external components required to give the amplifier unity gain. This set of compensation components makes the amplifier unconditionally stable in all feedback configurations.   The design of the UA709, like most other integrated operational amplifiers, has minimal drift, based on the emitter-base forward drops and current transfer ratios of the two input transistors remaining at the same level.   This is a reasonable assumption, since monolithic construction of the amplifier guarantees similarity in the two transistors.   However, there are minor variations in the base-emitter forward drops of the two transistors because of the differences in applied input bias to the device and minor variations in the constant-current sources

Fig. 2 709 operational-amplifier compensation components.   Use $R_2 = 50\Omega$ when the amplifier is operated with capacitive loading.

that bias the two transistors.   Frequency characteristics and stability criteria vary for each configuration the feedback and compensation components.   In Fig. 3, a variety of feedback loops and the voltage gain vs. frequency characteristics are shown. The basic schematic of the UA709 amplifier is illustrated in Fig. 4.

The simplicity of the UA709 leads to three limiting characteristics that discourage its use.   The first is the presence of "popcorn" noise, noise generated by thermal

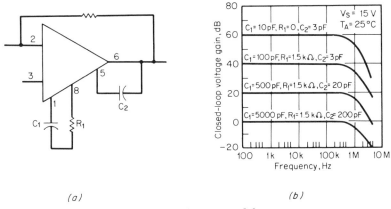

(a)   (b)

Fig. 3   709 feedback-loop schematic and frequency response.

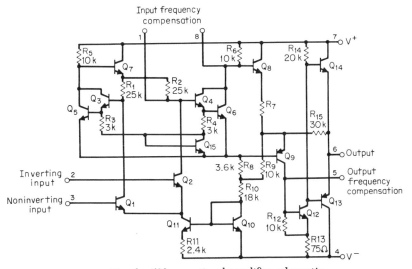

Fig. 4   709 operational-amplifier schematic.

feedback or interstate noise as a result of the particular operating characteristics of the amplifier. Popcorn noise is an attribute of circuit design and fabrication variables, and it can have seriously degrading effects on the operation of this class of integrated circuit. The second is the requirement that a complete set of compensating components be used. The third is that later-design operational amplifiers offer improvements in slew rates, voltage gain, and input offset characteristics and are protected against both output short circuit and output latch-up. Output latch-up is characteristic of many of the early operational amplifiers. When input common-mode voltage is exceeded on a UA709 type of circuit, the output is driven to the level of the corresponding power supply and is maintained at that potential, regardless of further changes in the input bias. The only way to remove the latch-up condition is to deenergize the integrated circuit completely.

The LM101A is illustrated in Fig. 5 and is a major improvement over the design of the UA709. Basically, the input-biasing circuitry was redesigned to minimize unfavorable circuit effects. The result was an offset voltage of less than 3 mV over the full operating-temperature range. The higher-impedance, higher-gain input stage

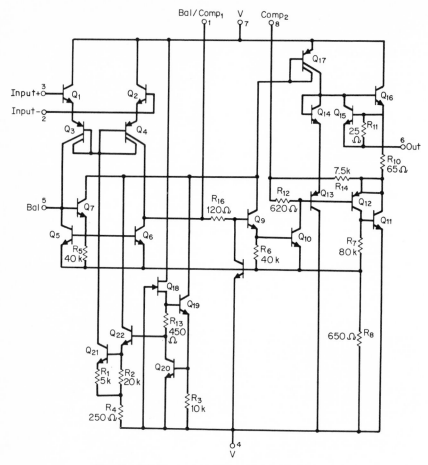

Fig. 5   101A operational-amplifier schematic.

demanded only an input current of 100 nA, which further improved application of the circuit. In many cases, it removed the requirement for nulling circuitry to be inserted. The compensation of the LM101A was simplified so that only a single 30-pF capacitor was required. The circuit was further modified to minimize popcorn noise, eliminate latch-up, and provide complete protection against output short circuit, as long as the case temperature stays within the ratings of the device. Figure 6 illustrates some typical applications of the LM101A type of operational amplifier.

The UA741 represents further development of the science of operational-amplifier production. The UA741 is similar to the LM101, except that it now includes the required 30-pF capacitor as an internal circuit element. This inclusion simplifies application of the operational amplifier, although it makes it less versatile than the LM101A or the UA709. The UA741 has all the basic characteristics of the LM101A; it is fully short-circuit-protected, has a low offset voltage, is protected against power-supply latch-up, and does not suffer from popcorn noise. The UA741 is also available in a dual version called the UA747. Applications of dual operational amplifiers should be limited to cases where it is desirable to have amplifiers with drift characteristics that track together. Dual operational amplifiers do not represent a good economic trade-off, since they reduce the circuit-fabrication yield without improving simplicity of operation.

A broad range of specialized operational amplifiers are available for high slew

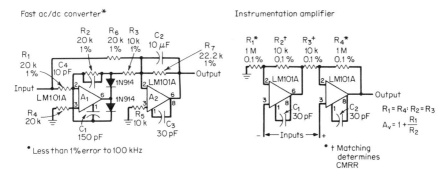

Fast ac/dc converter*   Instrumentation amplifier

* Less than 1% error to 100 kHz   * † Matching
                                           determines
                                           CMRR

Fig. 6 101A operational-amplifier applications.

rate and precision applications. Programmable operational amplifiers are now available, and dual and quad amplifiers are available for specialized cases. The full range of these available operational amplifiers is listed in Table 3. However, certain

TABLE 3   Operational Amplifiers, Selected Circuit Types and Description

Type no.	Feature	Offset voltage, mV	Voltage gain	Band-width	Package*
101A	No latch-up	3.0	160 K	5 MHz	C-8, F-10, D-14
741	Fully compensated	1.0	160 K	1.0 MHz	C-8, F-10, P-8, P-14, D-14
709	Economical	5.0	45 K	2 MHz	C-8
747	Dual 741	1.0	160 K	1.0 MHz	C-10, F-14, D-14, P-14
1556	Fully compensated	4.0	200 K	1.0 MHz	
1536	Fully compensated	2.0	500 K	1.0 MHz	C-8
LH001	Low power	0.2	60 K	5 MHz	C-10
108	Precision	3.0	300 K	2 MHz	C-8, F-10, D-14
118	High slew	6.0	200 K	15 MHz	C-8, F-10, D-14
725	Instrumentation	0.5	3,000 K	2 MHz	C-8, P-8
740	FET input	10	100 K	1 MHz	C-8
2900	Quad gain block		2.8 K	2.5 MHz	P-14
776	Programmable	6.0	50 K	1.0 MHz	C-8, D-14, P-8
3080	OTA	0.4		2 MHz	C-8

* Package entry letter is form factor: F = flatpack, D = hermetic dual-in-line, P = plastic dual-in-line, C = round can, S = studded flatpack. Package entry number is terminal count.

specialized characteristics cannot be met by simple monolithic operational amplifiers. These characteristics are ultratight drift specifications, high-voltage operation, high-current application, and extremely high slew rates. Modular operational amplifiers are best suited for these applications, but for economic and reliability considerations, the application of modular operational amplifiers should be severely limited.

A significant class of circuits, even though they are not true operational amplifiers, includes the 3401 and the LM3900 quad gain blocks. These circuits are designed to operate from a single power supply and are designed for automotive applications. The inverting input of these circuits is not a true operational-amplifier inverting input, but a current mirror input. This input structure severely limits the amount of input current and voltage that can be applied, but it simplifies the processing and fabrication of these circuits. The circuit schematic is presented in Fig. 7. The extreme economy of these circuits should make their use quite prevalent in automotive and commercial equipment.

Video Amplifiers

Differential video amplifiers are a small class of monolithic amplifiers characterized by differential inputs and outputs and extremely high bandwidth for integrated circuits. Video amplifiers typically have bandwidths in excess of 100 MHz. The most broadly applied video amplifier is the 733, first produced by Fairchild Semiconductor. The 733 has basic characteristics of a very high input impedance of 250,000 Ω, and it contains built-in adjustable-feedback components which allow variation of

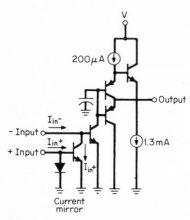

Fig. 7 Quad-gain-blocks schematic.

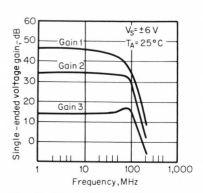

Fig. 8 733 video amplifier, plot of gain vs. frequency.

the device gain to be set at 10, 100, or 400. With an external resistor, the gain can be varied to any value between 10 and 400. All feedback components necessary for stability are included within the device. The exceptionally good gain stability, wide bandwidth, and low phase distortion of this circuit result from series shunt feedback from the output stage to the input of the second stage. The output stage, which is an emitter follower, provides the capability to drive capacitance loads while providing an extremely low output impedance.

Variations of circuit gain modify the frequency characteristics of the circuit. These characteristics are shown in Fig. 8 for three values of fixed gain. Gain set by external resistors can be interpreted from this graph. To vary the gain of the circuit, an external resistor is applied, as shown in Fig. 9. The selection of this resistor sets the gain of the circuit. Variation of gain with resistance value is shown in Fig. 10.

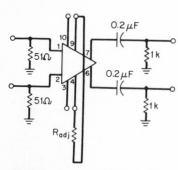

Fig. 9 Schematic of external gain-control-resistor circuitry for 733 video amplifier.

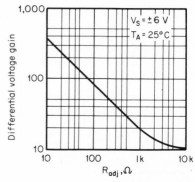

Fig. 10 Gain vs. external resistance value of 733 video amplifier.

The 733 and similar circuits are suitable for a limited range of application, which includes disk file read amplifiers, high-speed thin-film memory amplifiers, video and pulse amplifiers, and other amplifier applications where wide bandwidth, low phase shift, and repeatable gain stability are required. Table 4 lists the range of available video amplifiers and characterizes their basic operating parameters.

TABLE 4   Video Amplifiers, Selected Circuit Types and Description

Type no.	Bandwidth, MHz	Gain, dB	Input resistance	Propagation decay, ns	Package*
733	120	10–400	250 kΩ	10	C-10, F-10, D-14
1590	100	44	3 kΩ		C-8
501	150	24	700 Ω	15	D-14, P-10, C-10
592	120	0–400	30 kΩ	6.0	D-14, C-10

* Package entry letter is form factor: F = flatpack, D = hermetic dual-in-line, P = plastic dual-in-line, C = round can, S = studded flatpack. Package entry number is terminal count.

Function Modules

Analog-to-digital converters   Precision conversion of analog signals to digital signals, and digital signals to analog signals is most commonly done by modular assemblies fabricated from discrete semiconductor devices. The extremely tight tolerances of digital-to-analog (D-to-A) converters with 10 or more bits of content precludes their fabrication from monolithic technology. A limited range of small (6- or 8-bit) D-to-A converters is available. These converters are useful for simplified applications not requiring extreme accuracy. Characteristics of the available D-to-A converters are presented in Table 5. Analog-to-digital (A-to-D) conversion is normally per-

TABLE 5   Digital-to-Analog Converters and Analog-to-Digital Converters, Selected Circuit Types and Description

Type No.	Function	Features	Package*
1506	6-bit D-to-A	TTL compatible, current output	D-14
1508L8	8-bit D-to-A	TTL compatible, current output	D-16
1507	6-bit A-to-D	Requires 1506 D-to-A	D-16
SH8090	10-bit D-to-A	Hybrid subsystem	F-30
722	10-bit	Current source for D-to-A	F-24
7520	10-bit D-to-A	CMOS, low power, complete requires only reference	D-16

* Package entry letter is form factor: F = flatpack, D = hermetic dual-in-line, P = plastic dual-in-line, C = round can, S = studded flatpack. Package entry number is terminal count.

formed by comparator circuits. A fully fabricated A-to-D converter is not available as a single integrated circuit, but integrated-circuit D-to-A converters are sometimes configured into A-to-D converters using the successive-approximation technique and external logic for control.

Special integrated circuits are available for specific portions of the analog-to-digital or digital-to-analog conversion function, in addition to comparators. These include multibit switched current sources, such as the 722 10-bit current source. The block diagram in Fig. 11 shows how this current source is configured with an external data register, external precision resistor set, and a buffer amplifier to yield a complete D-to-A subsystem with precision limited only by the accuracy and precision of the resistor set. The 9650 represents a similar approach that includes internal logic and reference-signal compensation within the device. Although the 9650 is only a 4-bit current source, it can be placed in arrays of three to yield 12-bit accuracy. It is possible to assemble a combination of integrated-circuit chips into a complex hybrid

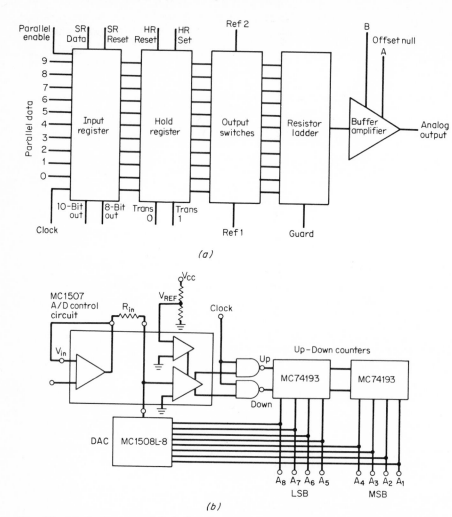

Fig. 11 Block diagram of (*a*) the SH 8090 digital-to-analog circuit and (*b*) the 1508/1509 analog-to-digital converter.

D-to-A converter. A standard product example of this is the SH8090 offered by Fairchild. In this case, MOS technology is used for decoding, linear current switches control the current, and the resistor array is thick-film. Figure 11 shows the block diagram of this circuit, which is typical of the approach to a hybrid D-to-A converter.

It is important in specifying a D-to-A converter to understand its operating parameters fully. The prime parameters of a D-to-A converter are resolution, relative accuracy, absolute accuracy, linearity, monotonicity, offset error, propagation delay time, slew rate, glitch, and overshoot. Resolution is the number of divisions into which the analog range can be discretely divided by separate digital states. Therefore, resolution represents the overall number of digital divisions that can be applied. Relative accuracy is the error at each output level when referenced to the full-scale analog output. Absolute accuracy is the overall combination of all sources of static error in the output. Linearity is the relative difference between the analog and digital level; ideally, it should be less than one-half the least significant bit, to assure monotonicity. Monotonicity is the characteristic of the output to change always in the same direction as the input; i.e., an increasing input always increases the output. Lack of monotonicity is the inverse. Offset error is the initial difference between the output with an all zero input and the zero level. Propagation-delay time is defined as the time between the change in the digital code and the time that the output makes 50 percent of the necessary change. Propagation delay is generally defined for the condition when the inputs change from all zeros to all ones. Slew rate is defined as the rate of change of the analog output voltage or current due to change in the reference-power-supply level. A glitch is a false spike or transient in the output which is generated internally by the switching network. Glitches are normally generated during certain transitions in which a large number of digital-code inputs make changes at the same time. They are caused by some current sources turning on faster than others turn off, leaving a momentary state with a large number of the wrong sources being on.

One additional type of circuitry which is now available for A-to-D conversion is the 1507 control block. With the 1507, the designer may directly fabricate A-to-D converters from a D-to-A converter such as the 1506. The circuit consists of a buffer amplifier, dual-threshold comparators, and the proper interface circuitry to transistor-transistor logic. When combined with a TTL up-down counter such as the 54193, the three circuits jointly function as a complete A-to-D converter, using the successive-approximation technique.

Phase-locked loops A phase-locked loop is an error-correcting, frequency-controllable oscillator source. They are used in the generation of frequencies and in such applications as signal generators and frequency synthesizers. They can be used for tracking filters, wideband phase detectors, and in modems, FM detection, and synchronization. Phase-locked loops fall into two basic categories. The first is the fully linear phase-locked loop represented by the Signetics 560 circuit. In this circuit, a voltage-controlled oscillator, in-phase comparator, amplifier, and low-pass filter are interconnected, as shown in the block diagram of Fig. 12. The capture charac-

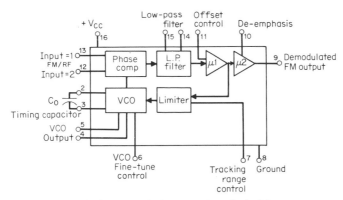

Fig. 12 Block diagram of the 560 phase-locked-loop circuit.

teristics of the phase-locked loop are totally determined by the low-pass filter, which is formed by connecting passive components to the phase-comparator output. In operation, error signals are generated by the difference between the phase of the input reference signal and the generated signal. This error then modifies the signal generated by the voltage-controlled oscillator. Characteristics of available linear phase-locked loops are given in Table 6.

The second category of phase-locked loop is the analog-to-digital phase-locked loop. In these, the frequency comparison is made on a digital basis using a set of cross-coupled NAND gates, as illustrated in Fig. 13. The outputs of these NAND gates are fed into an analog section which generates a dc voltage proportional to the frequency pump-up or pump-down logic outputs. The analog-section output is then applied to a voltage-controlled oscillator. Advantages of the digital-analog

TABLE 6   Phase-lock Loops, Selected Circuit Types and Description

Type No.	Function	Family	Frequency	Package*
4324	Dual voltage-control oscillator	TTL	30 MHz	F-14, D-14, P-14
4344	Phase detector	TTL	8 MHz	F-14, D-14, P-14
54416	Divide-by-N counter	TTL	N:0–9, 8 MHz	D-16, P-16
54417	Divide-by-N counter	TTL	N:0–4, 8 MHz	D-16, P-16
54418	Divide-by-N counter	TTL	N:0–15, 8 MHz	D-16, P-16
54419	Divide-by-N counter	TTL	N:0–3, 8 MHz	D-16, P-16
1648	Oscillator	ECL	225 MHz	F-14, D-14, P-14
1658	Voltage-controlled oscillator	ECL	250 MHz	D-16, P-16
12000	Mixer-translator	ECL	250 MHz	D-14
560	Phase-locked loop	Linear	0–30 MHz	D-16
565	Phase-locked loop	Linear	0–500 kHz	D-14, C-10
12060	Oscillator	TTL-ECL	100 kHz–2.0 MHz	D-16
12061	Oscillator	TTL-ECL	2.0–20 MHz	D-16
12040	Phase detector	ECL	80 MHz	D-14

* Package entry letter is form factor: F = flatpack, D = hermetic dual-in-line, P = plastic dual-in-line, C = round can, S = studded flatpack.   Package entry number is terminal count.

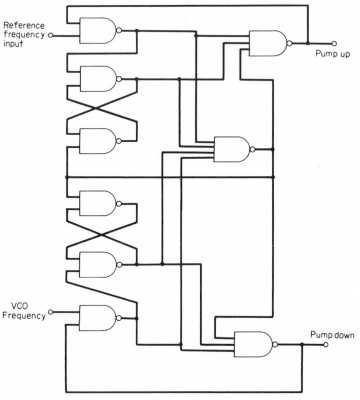

Fig. 13 Logic diagram of digital frequency comparator for use with phase-locked loop.

approach are that control can be placed on signals that can be counted down from higher frequencies, so that phase comparison of extremely high frequency signals can be made by simply scaling and comparing. Cascade flip-flops provide the scaling or countdown function. In transistor-transistor logic, the set of components required to perform the phase-locked function are the MC4344 and MC4324 offered by the Motorola Semiconductor Products Division. Application of these two circuits into a controllable frequency-synthesizer system with programmable divide function is illustrated in Fig. 14. The figure shows that a very simple array of components

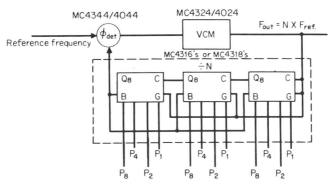

Fig. 14 Block diagram of the application of a phase-locked loop to form a frequency synthesizer.

can perform the complex function of frequency synthesization. Higher-frequency phase-locked loops using emitter-coupled logic are also available, such as the 1648 and the 1658.

Analog multipliers The analog multiplier has two inputs and one output that makes analog output proportional to the product of two analog inputs. This function is very useful, both in analog-computer functions and in modulation techniques. The technology for analog multiplication comes from a technique called variable transconductance. In the schematic of Fig. 15, the basic circuit for the analog multiplier is shown. In this circuit, a differential amplifier composed of two matched transistors is driven separately by one input into a transistor base and the other input as a

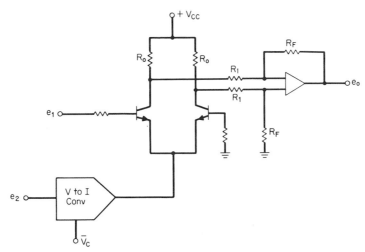

Fig. 15 Schematic diagram of variable-transconductance multiplier.

current into the summed emitters of the device. As shown in the literature,[*] the output voltage e_o is proportional to the product of e_1 and e_2 by

$$e_o = \frac{AR_oR_ce_1e_2}{R_1}$$

where A is an equivalency constant. The technology of analog multipliers has presented a variety of circuits. The most commonly used is the MC1595. This circuit can be directly applied to multiply, divide, and take square roots and squares. It can also be used as a frequency doubler, phase detector, electronic gain control, balanced modulator, or demodulator. The limited range of available multiplier circuits is illustrated in Table 7.

TABLE 7   Analog Multipliers, Selected Circuit Types and Description

Type No.	Function	Description	Package*
1594	4-quadrant	0.5% linearity	D-16
1595	4-quadrant	1.0% linearity	D-14
1596	Modulator	0.5 to 10 MHz, 85-dB CMR	D-14, C-10
AD530	4-quadrant	0.5%, 750 kHz	D-14
AD531	4-quadrant	1%, 750 kHz	D-14, C-10
AD532	4-quadrant	Fully trimmed, 0.8%	D-14, C-10

* Package entry letter is form factor: F = flatpack, D = hermetic dual-in-line, P = plastic dual-in-line, C = round can, S = studded flatpack. Package entry number is terminal count.

The MC1596 is particularly characterized as a balanced modulator-demodulator for use in frequencies up to 10 MHz. The 1596 can be used for amplitude modulation and suppressed carrier modulation, FM detection, phase detection, and even as an analog chopper. The prime characteristics of this circuit are a high common-mode rejection ratio of approximately 85 dB, adjustable gain, high impedance, and fully balanced inputs and outputs. This circuit applied as a typical modulator is illustrated in Fig. 16.

Analog multiplier circuits require a number of external precision balancing components to balance the internal offsets of all amplifiers. Some circuits have been offered by Analog Devices, Inc., that include internal laser-trimmed balancing resistors which reduce the number of external components necessary for application. However, they raise the cost significantly.

Voltage-regulator Integrated Circuits

The regulation of power-supply voltage is a most critical functional requirement for operation of most integrated circuits and solid-state circuit assemblies. It is quite natural that the function of voltage regulation would be developed as an integrated circuit. Voltage regulators basically require three elements: a voltage-reference source, an error amplifier, and a series-pass element.

In the case of discrete voltage regulators, a Zener diode or temperature-compensated Zener diode is used for the reference source. A constant-current source provides the appropriate bias for this reference, and the temperature-compensated Zener is then coupled using a voltage-reference amplifier. An operational amplifier or a simple two-transistor gain stage is used as an error amplifier, and a power transistor forms the series-pass element. More elaborate voltage regulators include elements for fault detection, overcurrent shutdown, and even overcurrent voltage-foldback circuits. All these elements can be built from a similar set of processes. Therefore, they can be fabricated as a single integrated circuit.

Implementing voltage regulators as integrated circuits brought basically three types of integrated-circuit voltage-regulator designs. The three types include all the basic

[*] J. Graeme, "Operational Amplifiers," McGraw-Hill, New York, 1971.

Typical Modulator Circuit

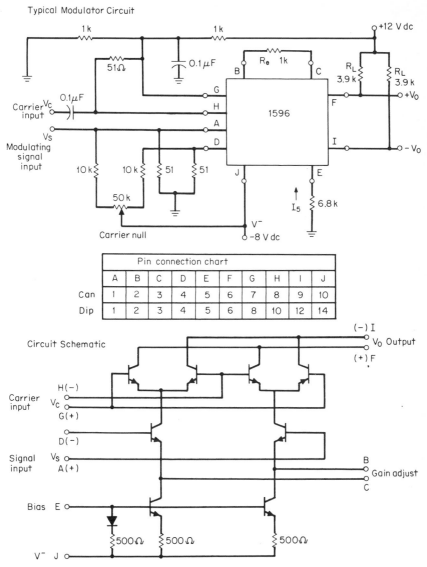

Fig. 16   Typical modulator applications of the 1596.

elements of the voltage regulators, but they vary the overall configuration of the available regulator. These three classes are (1) single precision voltage regulators, which are typified by the 723; (2) three-terminal regulators, which are typified by the 109; and (3) dual-tracking regulators, which are typified by the 1568. A description of the available regulators and their characteristics is given in Table 8.

The precision regulator is the most flexible. It is designed to separate the voltage reference from the amplifier, the pass elements, and the current-limiting circuitry. This separation allows the design engineer to configure the regulator as either a positive or negative regulator, allows it to be configured as a series or shunt regulator, allows operation as a switching regulator, and even allows regulator operation at a potential floating from the ground potential. The 723 offers approximately 0.01

TABLE 8   Voltage-Regulator Integrated Circuits, Selected Circuit Types and Description

Type No.	Function	Output voltage, V	Output current	Regu-lation	Package‡
723	Precision	2–37	150 mA, 10 A†	0.05%	P-14, D-14, C-10
105	Precision	+4.5–+40	12 mA, 10 A†	0–1%	C-8, F-10
104	Precision	0 to –40		0.05%	C-10, F-10
109	3-terminal	+5.05	1.5 A	0.5%	TO-5, TO-3
120	3-terminal	–5.05 to –15	1 A	1%	TO-5, TO-3
123	3-terminal	+5	3 A	0.6%	TO-3
78XX	3-terminal	5, 6, 8, 12, 15, 18, 24	1 A	2%	TO-3
79XX	3-terminal	–2, –5, –6, –12, –15, –18, –24	1 A	2%	TO-3
1568	Dual tracking	±15	100 mA	0.1%	C-10, D-14, diamond-10
125	Dual tracking	±15	100 mA, 5 A†		C-10
126	Dual tracking	±12	100 mA, 5 A†		C-10
127	Dual tracking	+5, –12	100 mA, 5 A†		C-10
4194	Dual tracking	±0.5–±42	200 mA	0. %	D-14, diamond-10
103	2-lead	+1.8–+5.6*	20 mA		TO-46
113	2-lead	+1.22	50 mA		TO-46

XX is voltage value.
* In discrete fixed values.
† With external-pass transistor.
‡ Package entry letter is form factor: F = flatpack, D = hermetic dual-in-line, P = plastic dual-in-line, C = round can, S = studded flatpack. Package entry number is terminal count.

percent regulation for variation of both line and load. Depending upon the available configuration, output voltage can be adjusted from 2 to 37 V. The internal-pass transistor of the 723 is capable of handling output loads of 150 mA, and larger current loads can be controlled with an external-pass transistor. Figure 17 shows the basic applications of the 723 regulator, each of which is suited to a particular requirement. Table 9 gives appropriate values of passive components to set output

TABLE 9   Value of Passive External Components for the 723 Voltage-Regulator Integrated Circuit

Positive output voltage, Vdc	Fig. 17	R_1, kΩ	R_2, kΩ	Negative output voltage, Vdc	Fig. 17	R_1, kΩ	R_2, kΩ
+ 3.0	a, e	4.12	3.01	– 6.0	c	3.57	2.43
+ 3.6	a, e	3.57	3.65	– 9.0	c, f	3.48	5.36
+ 5.0	a, e	2.15	4.99	–12	c, f	3.57	8.45
+ 6.0	a, e	1.15	6.04	–15	c, f	3.65	11.5
+ 9.0	b, d, e	1.87	7.15	–28	c, f	3.57	24.3
+12	b, d, e	4.87	7.15				
+15	b, d, e	7.87	7.15				
+28	b, d, e	21.0	7.15				

voltage to a particular value. Table 10 gives equations for finding output-voltage characteristics in terms of the passive component values. Rating limits for voltage, power, and current must be considered separately when applying voltage-regulator circuits. Operation beyond these limits is destructive.

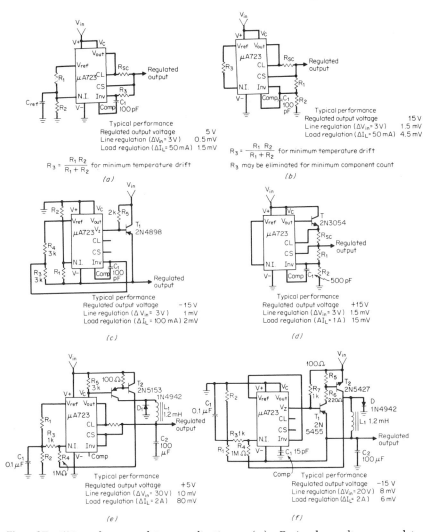

Fig. 17 723 voltage-regulator applications. (*a*) Basic low-voltage regulator (V_{out} = 2 to 7V). (*b*) Basic high-voltage regulator (V_{out} = 7 to 37V). (*c*) Negative-voltage regulator. (*d*) Positive-voltage regulator (external *npn* pass transistor). (*e*) Positive-switching regulator. (*f*) Negative-switching regulator.

The use of three-terminal regulator circuits, such as the LM109, is a most simple situation compared with the use of a precision regulator like the 723. Three-terminal regulators combine all the elements of the voltage-regulator system into a preconfigured operating mode. Figure 18 shows the block-diagram function of this voltage regulator. In a system, the three-terminal regulator has the unregulated power-supply input delivered to one terminal, a ground connected to the second terminal, and the regulated power-supply voltage available at the third. Three-terminal voltage regulators are preconfigured to a particular reference voltage; so they must be tailored to each particular class of application. They are most suitable for operating integrated circuits that require standard power-supply voltage levels. They are also well suited as point-of-load regulators; i.e., one regulator may be placed on each printed-circuit card that contains digital integrated circuits to provide the necessary

TABLE 10   Operating Equations for the 723 Voltage-Regulator Integrated Circuit

For output from +2 to +7 V	Fig. 17a	$V_{\text{out}} = \dfrac{V_{\text{ref}} \times R_2}{R_1 + R_2}$
Current-limiting		$I_{\text{limit}} = \dfrac{V_{\text{sense}}}{R_{sc}}$
For output from +7 to +37 V	Fig. 17b, d, e	$V_{\text{out}} = \dfrac{V_{\text{ref}}(R_1 + R_2)}{R_2}$
For outputs from −6 to −250 V	Fig. 17c, f	$V_{\text{out}} = \dfrac{V_{\text{ref}}(R_1 + R_2)}{2R_1}$
Foldback current-limiting		$I_{\text{knee}} = \dfrac{V_{\text{out}} R_3}{R_{sc} R_4} + \dfrac{V_{\text{sense}}(R_3 + R_4)}{R_{sc} R_4}$
		$I_{\text{sc}} = \dfrac{V_{\text{sense}}(R_3 + R_4)}{R_{sc} R_4}$

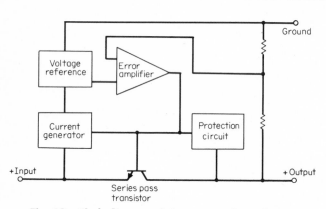

Fig. 18   Block diagram of three-terminal regulator.

regulated potential for operation. This scheme simplifies the control and distribution of power-supply voltage, since the effect of distributing and sensing power-supply voltages over a long distance is eliminated. Three-terminal voltage regulators are also appropriately applied to systems where the basic power supply is either unregulated or at the wrong potential for a small, limited number of circuits which require a limited power-supply voltage within the range of the three-terminal regulator. Three-terminal regulators are available as both positive and negative regulators, and they are available in a wide range of current and voltage. Available three-terminal voltage-regulator circuits are presented in Table 8.

The third class of voltage-regulator circuits is dual-tracking voltage regulators. The concept of the dual-tracking regulator is to provide in a single circuit two voltage supplies for simultaneous operation. These regulators are directed at dual power-supply-potential applications, which are required for the operation of such linear circuits as comparators and operational amplifiers. Dual-tracking regulators reduce the complexity of providing power-supply voltages and provide for symmetry of the two power-supply voltages, since they are both generated within the same source. As is illustrated in Fig. 19, tracking voltage regulators use a single voltage reference which separately drives two error amplifiers. The single reference guarantees the symmetry of the two output voltages. With a tracking regulator, simultaneous application of both power supplies is guaranteed. Tracking regulators are also summarized in Table 8.

The simplest integrated circuit fabricated is the LM103, a two-terminal voltage regulator. It functionally replaces a Zener diode, but it is merely a simple integrated circuit using an extremely low level internal-reference-source voltage divider and an error amplifier. The LM103 uses a reversed-biased emitter-base junction as its internal reference source. Normally, these junctions cannot be used as voltage-reference sources because they rapidly degrade and vary if a significant reverse current is applied to them. In the LM103, the emitter-base current is

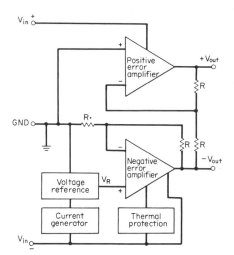

Fig. 19 Block diagram of dual-voltage regulator.

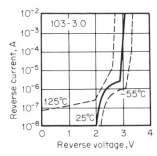

Fig. 20 Diode-like characteristic curve of the 103 integrated circuit.

kept at an extremely low level, typically in the picoampere range, and this low current is safe for operation. The LM103 appears as an extremely sharp-knee, low-dynamic-impedance voltage-regulator diode. Figure 20 shows the diodelike characteristics of the voltage-current plot of this device.

Line Drivers and Receivers

Line drivers and line receivers are specialized functions that allow data transmission over extended distances. These circuits normally operate with higher bias voltages; so they are fabricated using linear fabrication techniques, although appropriate input or output terminals are interchangeable with the levels of common digital families. Table 11 shows a broad range of available line drivers and receivers that have digital

TABLE 11   Line Drivers and Line Receivers, Selected Circuit Types and Description

Type No.	Function	Line form	Logic compat-ability	Description	Package*
9614	Driver	Pair	TTL	Dual, short circuit, 50 Ω	D-16, P-16
9621	Driver	Pair	TTL	Dual differential	D-14, P-14
55109	Driver	Pair	TTL	Dual, high-speed	D-14, P-14
1488	Driver	RS232C	TTL	Quad, slew rate adjustable	D-14, P-14
9615	Receiver	Pair	TTL	Dual, high CMR	D-16, P-16
9620	Receiver	Pair	TTL	Differential	D-14, P-14
55107	Receiver	Pair	TTL	High-speed, CMR	D-14, P-14
1489	Receiver	RS232C	TTL	Quad, hysteresis	D-14, P-14

* Package entry letter is form factor: F = flatpack, D = hermetic dual-in-line, P = plastic dual-in-line, C = round can, S = studded flatpack. Package entry number is terminal count.

terminals interchangeable with the levels of saturated logic. These line drivers and receivers are characterized for transmission on particular line lengths and line types, and these qualities are detailed in the table. Line drivers and receivers are required to maintain sufficient noise margin for transmission over extended lengths, or to allow transmission of digital signals in communication systems that have logic levels defined at voltages different from standard logic circuits. Important characteristics which must be considered in the selection of line drivers are (1) the logic levels at the logic side of the circuit, (2) drive mode and levels at the line-transmission side of the circuit, (3) the maximum data-throughput rate, and (4) the ability to withstand specified stresses on the data-input and data-output terminals.

Entertainment and Communication Circuits

The entertainment industry generated high-volume requirements for special-purpose integrated circuits. The circuits comprise almost the full range of system functions in entertainment systems, including audio amplifiers, power-frequency amplifiers, dc volume controls, and preamplifier circuits. Video is represented by video amplifiers, discriminator sections, and automatic-gain-control and automatic-fine-tuning circuitry. Specialized circuitry for color television includes chroma detection, chroma amplification, and chroma demodulation functions. Horizontal amplifiers and circuitry providing the incidental functions for TV, normally called the jungle circuits, are also offered. Low-noise, high-quality stereo and quadrasonic preamplifiers and power amplifiers are available for audio circuitry. Limited-availability IF amplifiers, limiters, detectors, and IF preamplifiers can also be obtained. Decoders are available to decode and detect both stereo multiplex an four-channel quadrasonic multiplex. Additional special-range MOS circuits provide the digital-control functions necessary for electronic organs and are fabricated primarily as special proprietary circuits. A full family of digital logic for rhythm generation and control is also available. Electronic attenuators for use in organ circuits are also entertainment circuits.

There has been very little standardization in circuits for use in consumer electronics, and the evolution of newer, more complex, higher-performance products is continuing at too fast a rate to make this standardization possible. Therefore, much of the growth of specialized consumer electronic entertainment circuits centers around proprietary circuits designed for and used by a single manufacturer of entertainment equipment. In working with this class of equipment, it is essential to select a group of circuits that are compatible for operation, in terms of both power-supply requirements and levels of inputs and outputs.

Special care should be taken when interfacing integrated-circuit components to discrete components to limit properly the impressed voltages and signals applied to the integrated-circuit element. This becomes extremely important in cases where tubes or high-voltage transistor circuits are used. Integrated-circuit preamplifiers and amplifiers are quite often sources of extremely high gain; so great care must be taken in laying these circuits out to prevent development of sources of oscillation or signal pickup or crosstalk. Many entertainment circuits that operate at relatively large power dissipations are provided in packages which either have integral heat sinks or require that heat sinks be properly attached to the power package. In these cases, the mechanical design of the equipment must allow for adequate airflow to lower the operating-circuit temperature.

SPECIAL-PURPOSE INTEGRATED CIRCUITS

The integrated circuit combines, in a single crystal produced by a batch process, a microcombination of active and passive circuit elements. Most integrated circuits are designed to perform specified digital or logical functions. Only a small percentage of the integrated circuits are specially designed to perform unique, special functions. These special-function integrated circuits can be used to unique advantage when properly applied by a design engineer. Special-purpose integrated circuits are available today in an ever-increasing range of types and configurations, as they have become offshoots of the expanding electronics industry and the expanding technological capability of the integrated-circuit industry. Special-purpose integrated circuits have found use in clocks, calculators, complete central-processing units, timers, automotive-seat-belt interlock systems, and a host of other special applications.

Many problems are associated with fabrication of custom integrated circuits. For the majority of circuit applications, the most appropriate circuits are standard-product circuits. It is not unlike the situation of trying to select the best nut or bolt to be used to secure a framework. Even though analysis of the dimensions indicates that the best bolt may be a specially sized bolt 0.997 in long, full consideration of available standard hardware suggests the use of a 1-in bolt and letting the 3 mils of excess bolt length hang over the end. Just like bolts, integrated circuits are fabricated in batch processes to fixed tooling, and they are best and most efficiently procured to designs that are bought in high volume. Custom designs which in themselves will generate high volume are appropriate. However, those which are limited in potential sales volume should be discarded. Also, standard circuits are frequently noticed to be capable of operation outside the scope of the initial specification. Questions of this kind should be directed to the vendor of the part, and the user should supplement that analysis by design analysis of the internal schematic of the circuit. Quite often, selections proved to be high-yield from one production run are not high-yield from the next, and the custom circuit that was selected from the general population of parts then becomes no longer selectable and no longer procurable.

Custom Integrated Circuits

A custom integrated circuit is an integrated circuit designed to perform the functions for a single special-purpose application. They can be digital or linear, and can fall into any number of technologies. Custom logic circuits can be part of custom families, i.e., families using logic cells that are different from any of the standard cell families, or they can be digital-logic circuits which are fabricated from standard logic cells but perform nonstandard logic functions. It is strongly recommended that design engineers do not, in most cases, pursue custom integrated circuits. A custom monolithic integrated circuit costs between \$20,000 and \$100,000 in initial tooling and design engineering. This cost must, in the end, be amortized over the application. Careful examination of standard integrated circuits offered on the market usually indicates the availability of a standard circuit that meets the full requirements of the vast majority of applications.

There are three particular areas where custom integrated circuits do make good engineering sense. The first of these is in the design of commercial digital computers. These computers are produced in large numbers and, as such, represent the production volumes necessary to amortize the cost of circuit design. Here, custom circuits allow the design team to add unique features to their system. A second area in which custom circuits play a large and proper role is the area of custom LSI arrays. These arrays are fabricated from standard MOS, or occasionally bipolar processes using standard cells, and they are most useful in designing systems that are produced in large annual volumes. The cost of designing circuits within the limitations and design rules of a given MOS family tends to optimize, so that production rates of 5,000 to 10,000 circuits per year become economical. This area is comprised of the many mass-produced small digital systems, such as point-of-sale machines, calculators, and numerically controlled factory machinery. This is also a proper size of marketplace to justify a number of custom designs.

Custom designs of this second type can be obtained through two methodologies. One is primarily manual design with subsidiary use of computer placement of cells and interconnects. This, at the current state of technological science, is the most efficient way to use and design such circuits. However, the cost of design is higher, and only a limited amount of specially trained technical people are available for such designs. A second and cheaper methodology has been under active development during the early 1970s. This is a technique of computer-aided design. Computer-aided design, in its total optimization, allows automatic transition from a logic diagram to a completed set of masks for fabrication of integrated circuits, along with the computer-aided test programs to exercise these circuits from automatic test equipment. Computer-aided design, at the current state of the art, is marginally efficient. Although initial engineering costs are reduced by computer-aided design, in many cases silicon surface area is used so inefficiently that the cost of individual-circuit manufacture is raised. Many integrated-circuit companies have combined computer-aided design with manual-design expertise to enable them to operate interactive design systems. Future work needs to be done in this area.

The third major class of important custom circuits are the many linear consumer circuits that have been developed in the past few years. Most of these had their origin as custom circuits for a particular manufacturer of clocks, televisions, or audio equipment. In many cases, such circuits have evolved into standard catalog products, offered by the various semiconductor vendors to the general market. The appropriateness of custom circuits for the consumer industry comes again from the large volume of circuits required by the industry.

Certain additional considerations must be made to justify the appropriateness of designing a custom integrated circuit. The design cycle of a custom integrated circuit requires from as few an 10 to as many as 52 weeks, depending on the availability of the proper technical talent and the complexity of the design. In addition, more time must be allowed for fabrications of a pilot run. This fabrication takes between 2 and 16 weeks, depending on the processes required for the circuit and scheduling within the semiconductor manufacturer's facility. Further time must be allowed for cases where the initial design may not be efficient or not even operative. To allow additional time for a second design stage and a second pilot fabrication run, the design of a custom integrated circuit may consume more than 16 months. Minor design changes are implemented with difficulty. Just as the great efficiency of an integrated circuit is that it is batch-processed from fixed tooling, the great inefficiency is that any changes in the circuit must be preceded by tooling changes. This tooling, being microphotographic artwork, is not easily altered. For example, in a discrete amplifier, to change a biasing resistor from 10 to 15 Ω would require only reaching into a box, selecting the resistor, and making the replacement. In the case of an integrated circuit, a complete and total relayout of the circuit may be required along with generation of new artwork and masks. Therefore, changing from a 10- to a 15-Ω resistor may occupy the entire effort of a total redesign.

Quite often, a design engineer conceives a circuit that he would like to see as an integrated circuit and presents his design to an integrated-circuit manufacturer so that the circuit can be produced for his use and the use of other engineers in similar situations. However, most design engineers are totally unaware of the constraints in the design of an integrated circuit. They vary greatly from process to process and even from vendor to vendor within the same process or process family. These constraints have been self-imposed by the vendor to minimize the design task for himself and to maximize the yield of finished product from the design and process. As a result, before undertaking any task of designing a custom integrated circuit, the engineer must become fully cognizant of design rules associated with the particular process within which he is trying to design. Careful analysis of the design rules and the processes on which these rules were based allows a design engineer to be most efficient in the design of custom circuits. He must train himself not to be limited by the fixed methodology of designing circuits from discrete components. He must learn to exploit the inherent design advantages in an integrated circuit, while realizing and being confined by the process limitations of integrated-circuit manufacture. For example, in most digital bipolar integrated-circuit processes, the design rules require resistor tolerances no tighter than 25 percent. But the same set of process rules allow resistors to be used in pairs, or sets with ratios as tightly controlled as 2 percent. These requirements result from the resistors being fabricated from the same sheet material during the same diffusion, having similar temperature coefficients and similar bulk resistance characteristics. The clever design engineer realizes that circuits with tight voltage tolerances internal to the integrated circuit are best fabricated using voltage dividers, rather than constant-current sources driving into fixed-value resistors. A major advantage of integrated-circuit processing is the inherent match of device characteristics within a chip. The design engineer must, however, realize that to gain this natural match to its tightest degree, he must restrict the geometries of matched components so that they are similar. To gain efficient use of the surface area on silicon chips for the integrated circuit, he must understand the areas consumed by active devices, resistors, metallization runs, and even bonding pads. He must further realize the restrictions that circuit packaging places upon him.

One cost of custom integrated circuits which is rarely considered is the cost of maintaining adequate testing on the circuit. This cost first appears at the facility of the circuit manufacturer in preparing custom programs and possibly even special test equipment to exercise the new custom circuit fully and properly to guarantee its functionality. The user of the circuit must also expend similar effort. The cost of this effort multiplies, since the user, of necessity, is unable to base his test procedures, test plans, test program, or even test equipment on previous experience. The custom circuit is a newborn, a child of engineering know-how without history. All testing must therefore be of a complete and preemptive nature if nonfunctional circuits are to be avoided.

A custom circuit is manufactured to order by a semiconductor vendor. It cannot be stocked by the vendor or a distributor to make it available on short-term need to the user. Also, a custom circuit quite often requires large minimum orders for its production; so the integrated-circuit vendor is not hampered in scheduling his manufacturing facility. It is not uncommon for a single production run of wafers of an integrated circuit to exceed 100,000 chips. This kind of small specialized lot is highly inefficient, and the cost of these inefficiencies must be passed on to the user of custom integrated circuits. A further restraint on the user is that a custom design jointly developed by the user and a manufacturer is not easily transferred to another manufacturer. The situation becomes one of strictly single-source procurement. Developing a second source almost always necessitates a separate design cycle and the reconciliation of any differences in the two designs to make them interchangeable in the equipment for which they are being fabricated.

Custom hybrid integrated circuits Custom multichip or hybrid custom circuits are erroneously thought of as an intermediate step between discrete solid-state systems and systems employing monolithic integrated circuits. Hybrid circuits compare in only one way with monolithic circuits, and that is that they are both methods of obtaining extremely high packaging density in electronic systems. Circuit configurations and system concepts suitable for one approach are not necessarily suitable for the other. The primary advantage of hybrid circuits is that they, like monolithic circuits, reduce the packing volume of a system. None of the cost or reliability advantages of monolithic circuits are available, since hybrid circuits are not batch-processed material but rather a microassemblage of separately fabricated devices, although many hybrid circuits contain monolithic integrated-circuit chips to obtain the highest levels of packaging density. Other methods of packaging should be considered before adopting the use of hybrid integrated circuits. Also, owing to the lack of full testing capability for chip devices, it is not possible to select electrically the chips used for hybrid circuits to the exacting criteria by which discrete semiconductor pieces can be selected.

System on a Chip

With the advent of increasing complexity in integrated circuits, many engineers projected this growth and predicted the eventual fabrication of a complete self-contained system on a single integrated-circuit chip. The early 1970s brought realization of this prediction in four specialized areas of standard-product integrated circuits. These four areas are clocks and watches, calculators, central processing units, and universal asynchronous receiver-transmitter circuits. Each represents a system that is fully obtainable in a purely digital form and whose complexity is consistent with the large-scale integration capabilities of the early 1970s. All four are standard products where system on a chip has been produced and sold.

Naturally, there are other cases where similar or even greater complexity has been used to generate system on a chip for specialized custom applications, such as pocket pagers. The relatively high volumes of production of the devices selected for system on a chip has reduced the cost of the single large-scale integration element to the point where electronic calculators and watches now retail for as little as $10 to $20. In the near future, system on a chip will probably be applied to the production of numerous other electronics.

Clock and watch circuits Timekeeping has always involved the selection of natural or scientific phenomena with a fixed and specified time interval. This interval was then used to measure the indeterminate interval of other events. Leonardo

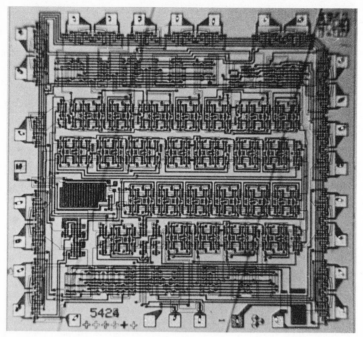

Fig. 21 Microphotograph of integrated circuit watch chip. (*Solid State Scientific, Inc.*)

da Vinci first became aware of the regularity of a swinging pendulum by comparing its time period with his own pulse rate. Manfacturers of clocks and watches proceeded through fabrication of constant time sources, such as pendulums and balance wheels. The mechanism of a watch or clock is simply a mechanical assemblage designed to count accumulated periods of an oscillatory source and then display the total as the elapsed time interval. The first electronic watches used mechanical time sources, such as tuning forks, and then used an electronic system to sense their motion and produce an amplified mechanical impulse to operate a mechanical counter. This is the scheme employed by the Accutron° watch.

More recent electronic watches use a solid-state source of oscillatory interval, which is a crystal. The electronic impulses from the crystal are sensed within a logic-countdown circuit contained on a chip circuit. This countdown element is crucial, since the crystal frequency has a period significantly shorter than any standard unit of time. Electronic watches have been fabricated using crystal frequencies as high as 6 MHz, although the lower frequency of 32,768 Hz is commonly used. The oscillatory frequency is then divided down to a useful frequency by a series of cascaded flip-flops. For example, in the case of a crystal at 32,768 Hz, 16 flip-flops are required to generate a frequency of ½ Hz. The resulting low-frequency signal may then be used to drive a stepping motor directly, and the clock is counted down through mechanical means. Some clock circuits use a more sophisticated technique of applying the low-frequency signal generated from a countdown from the crystal and then applying the signal to a solid-state counter assemblage which keeps track of time. The outputs of the solid-state counter must then be decoded and driven into electronic displays, such as liquid crystal or light-emitting diodes. An example of modern electronic watch circuits is shown in Fig. 21. The complexity of this integrated-circuit chip is evident by inspection.

Electronic clocks are similar to electronic watches, except that the packaging constraints of a small wristwatch case are not applied. In addition, electronic clocks

° Trademark of Bulova Watch Company.

are customarily connected to a 50- or 60-Hz ac power line. The power-line frequency then becomes the source of oscillatory signal; so an electronic clock need only generate and decode the time information by keeping count of the impulses of the controlled line frequency. Electronic clock circuits can include provisions for operating alarms, generating tones for use in alarms, and directly driving displays through internal decoding.

Designs of electronic watches must allow time setting, since the status of the internal counters is arbitrarily set when power is first applied to the circuit. Setting schemes for electronic watches involve either external generation of pulses which are directly applied to the counters, or the bridging of stages within the counter structure to allow the watch to run at greatly increased speeds while it is being set. Electronic clocks often allow temporary operation on battery power so that the clock setting is not disturbed during a power outage.

Electronic watches have been designed to interface with two basic types of digital displays, liquid-crystal displays and light-emitting-diode displays. The simplification of the fabrication of digital watches has required that most if not all of the circuitry needed for display decoding and driving be moved into the main watch chip. This has generated the need for high-voltage and high-current drive circuitry for either the liquid-crystal or the light-emitting-diode display. Newer high-density watch circuits have allowed watches to be fabricated from a single integrated-circuit chip which has the function of timekeeping, date keeping, switch interface, and display output-drive capability.

Calculator circuits and circuit sets Electronic calculator circuits have been developed to provide the standard functions of business and scientific numerical calculation. The calculators depend upon either a single circuit or a set of circuits which contains all the necessary logic to perform the arithmetic operations of the calculator and to interface the calculator to both its keyboard and its displays. Calculator circuits have ranged from a relatively simple four-function calculator, i.e., add, subtract, multiply, and divide, up to complex calculators which provide complete use of scientific notation, complex functions, conversion from units, and self-contained electronic memory. Many calculator designs depend on a complementary set of LSI circuits that work in combination to perform the required calculator function. A representative sample of the available calculator circuits is presented in Table 12, where the primary characteristics of these calculator circuits are included.

TABLE 12   Calculator Integrated Circuits and Calculator Integrated-Circuit Sets, Selected Types and Description

Type No.	Description	Package*
EAS-100	6-chip set for 8-digit calculator, 4-function	D-24
EAS-114	4-chip set for 16-digit (8-digit alternate display)	D-24
EAS-129	2-chip set for 12-digit, memory, constant, floating decimal point	D-28, D-40
ML-1003	Single-chip, 8-digit, 4-function, memory, internal led driver	D-28
ML-1007	Single-chip, 10-digit, 4-function, fixed and floating decimal point	D-40
MPS-2521	Single-chip, 8-digit, memory, 4-function,	D-28, P-28
MPS-2523	Single-chip, 8-digit, memory, 8-function,	D-28, P-28
3820-4	5-chip accounting calculator set	

* Package entry letter is form factor: F = flatpack, D = hermetic dual-in-line, P = plastic dual-in-line, C = round can, S = studded flatpack. Package entry number is terminal count.

Many calculator circuits have been developed as proprietary custom products by calculator manufacturers. Therefore, adequate understanding of the input and interface circuits of the calculator chips must precede their proper application. In particular, examination must be made to determine if provision for contact, bounce, and multiple key depression has been made in the input circuitry. The drive and voltage capabilities of the output stage must be selected for consistency with the display technology selected, and proper display-driver interface circuits must be designed and included in the overall calculator system. In the selection of calculator circuits, it is appropriate to examine the chip pinout for simplicity of printed-circuit-board layout for the end item calculator.

Two features have made a limited group of calculator designs much more valuable than others. The first feature is an accessible keyboard memory that allows constants to be stored and used repetitively in specified mathematical manipulations. The second is a special register or memory which carriers a running total of products and quotients. This carry function is most useful in business and accounting.

A new class of calculator chip has been developed to include a large bank of programmed read-only memory as a source of permanently stored numerical constants. One application of this technique makes a complete range of conversion factors from metric to English units available within the circuit. Other potential applications of this technique allow it to be a storage source for a set of commonly used constants for a particular industry or scientific discipline.

Some of the more complex electronic calculators use specialized sets of circuits to perform complex calculations in scientific notation. In the future, these more complex calculators will be attainable in a single-chip design as capabilities for more complex large-scale integration increase.

Central processor units No discussion of the concept of a system on a chip could be complete without a detailed examination of monolithic central processor units, or as they are now known, microprocessors. Microprocessors were developed in their early stages by both Intel and Rockwell International. These circuits were initially very simple 4-bit central processing units fabricated as a standard logic block for use in simplified calculating equipment such as point-of-sale machines and scientific calculators. Figure 22 is a photograph of the Intel 8008 p-channel MOS

Fig. 22 Microphotograph of a microprocessor chip, the 8080. (*Intel Corporation*)

processor. Examination of the photograph will clearly display the extreme complexity of this circuit.

The basic circuit includes output buffers, I/O buffers, two separate 8-bit registers, a complete 8-bit arithmetic-logic block, a computer-control logic, status-decoding elements, timing-generation circuitry, stack-address decoding and multiplexing circuits, an 8-word by 14-bit address stack, read and write multiplexers, complete memory control, memory-cycle control coding, an 8-bit instruction register, cycle-control flip-flops, scratch-pad memory, accumulators, and a complete set of interconnecting logic.

A microprocessor is a simplified digital central processing unit fabricated from a single or a limited number of special-purpose digital integrated circuits. A microcomputer is a computer fabricated from a microprocessor and the necessary peripheral, memory, and I/O circuits. It must be emphasized that a major difference exists between a custom calculator circuit and a microprocessor in that the microprocessor is not dedicated by design to a particular application but rather its application is a function of the program software developed for the particular application. Therefore, a microprocessor can be dedicated to any application by the proper set of software. This assumes that there is sufficient capability within the microprocessor for the application.

In using microprocessors one must first select one. In the first 4 years of microprocessor production an amazing range of over two dozen standard-product microprocessors have been introduced. These circuits are indicated in Table 13 and their

TABLE 13 Central-Processor-Unit Integrated Circuits, Microprocessor Selected Circuit Types and Description

Type No.	Word length	Technology	Clock frequency, MHz	Feature	Package*
F-8	8-bit	NMOS	2.0	Built-in clock	D-40
PPS-25	4-bit	PMOS	0.4		Multiple
PPS-4	4-bit	PMOS	0.2		F-42
PPS-8	8-bit	PMOS	0.250		F-42
2650	8-bit	NMOS	1.2		D-40
IMP-4	4-bit	PMOS	0.5		D-24, D-40
IMP-8	8-bit	PMOS	0.7		D-24
IMP-16	16-bit	PMOS	0.7		D-24
6800	8-bit	NMOS	1.0		D-24, D-40
6701	4-bit	TTL	6.5	Slice organization	D-40
6100	12-bit	CMOS	4.0	PDP-8 code	D-28, D-40
4004	4-bit	PMOS	0.7		D-16
4040	4-bit	PMOS	0.7		D-16, D-24
8008	8-bit	PMOS	0.8		D-18
8080	8-bit	NMOS	2.1		D-40
3001	2-bit	TTL	6.0	Slice organization	D-28, D-40
COSMAC	8-bit	CMOS	3.0		D-28, D-40
SPB0400	4-bit	I^2L	1.0	Slice organization	P-40

* Package entry letter is form factor: F = flatpack, D = hermetic dual-in-line, P = plastic dual-in-line, C = round can, S = studded flatpack. Package entry number is terminal count.

basic operating characteristics are described. Selection of a microprocessor for given application will require significantly greater data than are available in this table, as many performance, software availability, hardware support, and system considerations must be made in the final selection of a microprocessor.

The basic characteristics of a processor are its performance in terms of speed and power consumption and its architecture in terms of the instruction set, registers, I/O, and word length. Detailed consideration must be made of these characteristics

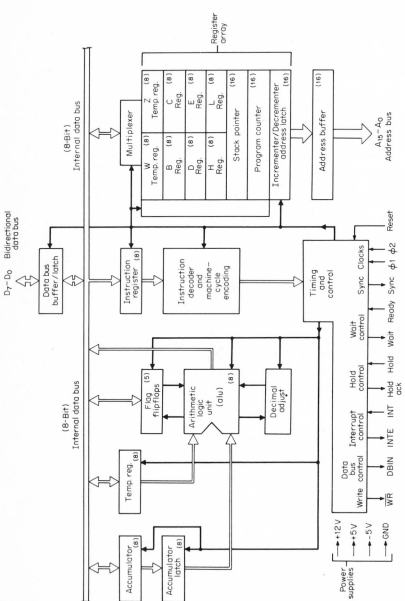

Fig. 23 Block diagram of the 8080 microprocessor chip.

in order to exploit the microprocessor technology properly. Examination of Fig. 23 will show you the basic architecture of the Intel 8080. The Intel 8080 is discussed not necessarily because it is the best microprocessor but rather because it is one of the prime choices for 8-bit high-speed microprocessors. Discussion of this microprocessor is made as an example for which comparison can be made with other microprocessors. In Fig. 23 we can see the basic structure of the 8080 microprocessor. It is an 8-bit processor which carries a 16-bit address bus. A complete arithmetic-logic unit is available for doing 8-bit arithmetic directly and double-precision arithmetic in multiple cycles. The machine has a single accumulator and has an internal register array for storing the immediately operated data. A 16-bit program counter and stack pointer are included as part of this register array and are directly accessible for operation. Indexed addressing is not available in this processor, but direct addressing is. The set of control signals available are reasonably functional, although some external logic decode must be made from these signals in order to generate all the timing signals necessary for final system operation. It should be noted that the processor has no internal stack but rather uses a stack pointer to address a user-selected range of random-access memory as the stack. It should further be noted that the control signals are not available on direct pins but are rather available on the data line at a specified place in the machine cycle time. External decoding and latches must be used to accumulate these control signals for other use. The Intel 8080 operates with a clock cycle of $\frac{1}{2}$ μs. This speed is consistent with a full range of applications. The 16-bit address line allows for the direct addressing of 64,000 words of memory. The memory bank can be a combination of read-only memory and random-access memory as is necessary for the application.

It is most important to examine the instruction set in order to understand fully the capability and limitations of a given microprocessor. Unlike many computers the instruction set of a microprocessor is basic, simple, and does not include many of the niceties such as hardware multiplication and direct instructions for indexed addressing. The Intel 8080 instruction set is described in Fig. 24. The instructions have been grouped in a manner to simplify their use and understanding. The initial mnemonic descriptor for each instruction heads the tabular description. Tabular description also includes the function of the instruction and the 8-bit code for the instruction. As can be seen from the figure, the ability to move data from register to memory and memory to register, to load the stack pointer, and to load direct data are all available. Further, input and output instructions, jump, call, and return instructions and a full set of arithmetic instructions are available. It should be emphasized that the "DAD" instruction allows for double-precision addition between register pairs or within a given register to accumulate a left shift. Instructions for enabling and disabling the interrupt, halt, and no operation are also available. Data from the register pairs and a processor status word can all be popped and pushed to the stack. Finally, a set of logical instructions, register rotates, increments and decrements for both registers, and register pairs are available within the machine. The overall instruction set, although somewhat limited, is most useful. One further point should be emphasized; the instruction set includes specified instructions for doing decimal arithmetic, in which case each 4-bit grouping of data is treated as a binary-coded decimal.

The application and selection of microprocessors must be based on the real availability of the required units. Few microprocessors are second-sourced, most being available only from a single supplier. Some microprocessors have been announced, data published, and yet the microprocessors remain only a paper design not fully implemented as a semiconductor device. Care should be taken in microprocessors introduced in a preliminary hardware version which is scheduled for final redesign and reimplementation at a later date. This situation could be most risky. One further point must be made in availability. Most microprocessors are not designed for operation at full military-temperature range. At present the Intel 8080, the RCA Cosmac, and the Texas Instruments 400 are all available for full military-temperature-range operation. It should be noted that the operating characteristics are somewhat compromised for full military-temperature-range operation.

In using microprocessors the application is software- rather than hardware-dominated. The necessary tools for the software design are most important to consider.

MICROPROCESSOR
8080

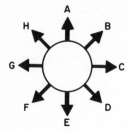

Symbol	Meaning
M	Memory eferenced by H&L registers
addr	16 – bit address
data	8 – bit data
data 16	16 – bit data
r, r₁, r₂	Any register
rp	Any register pair
DDD, SSS	Bit pattern for register
	(DDD = destination, SSS = source)
	B= 000
	C = 001
	D = 010
	E = 011
	H = 100
	L = 101
	Accumulator = 111
RP	Bit pattern for register pairs
	B – C = 00
	D – E = 01
	H – L = 10
	Stack pointer = 11
X	Bit pattern for register pairs
	B – C = 0
	D – E = 1

A

		Arithmetic Group	Code
ADD	r	(Add register to accumulator)	1 0 0 0 0 S S S
ADD	M	(Add memory to accumulator)	1 0 0 0 0 1 1 0
ADI	data	(Add immediate to accumulator)	1 1 0 0 0 1 1 0
ADC	r	(Add with carry register to accumulator)	1 0 0 0 1 S S S
ADC	M	(Add with carry memory to accumulator)	1 0 0 0 1 1 1 0
ACI	data	(Add with carry immediate to accumulator)	1 1 0 0 1 1 1 0
SUB	r	(Subtract register from accumulator)	1 0 0 1 0 S S S
SUB	M	(Subtract memory from accumulator)	1 0 0 1 0 1 1 0
SUI	data	(Subtract immediate from accumulator)	1 1 0 1 0 1 1 0
SBB	r	(Subtract with borrow register from accumulator)	1 0 0 1 1 S S S
SBB	M	(Subtract with borrow memory from accumulator)	1 0 0 1 1 1 1 0
SBI	data	(Subtract with borrow immediate from accumulator)	1 1 0 1 1 1 1 0
DAD	rp	(Double precision add register to HL)	0 0 R P 1 0 0 1
DAA		(Decimal adjust accumulator)	0 0 1 0 0 1 1 1

B

		Increment/Decrement Group	Code
INC	r	(Increment register)	0 0 D D D 1 0 0
INC	M	(Increment memory)	0 0 1 1 0 1 0 0
DCR	r	(Decrement register)	0 0 D D D 1 0 1
DCR	M	(Decrement memory)	0 0 1 1 0 1 0 1
INX	rp	(Increment register pair)	0 0 R P 0 0 1 1
DCX	rp	(Decrement register pair)	0 0 R P 1 0 1 1

C

		Logical	Code
ANA	r	(AND register with accumulator)	1 0 1 0 0 S S S
ANA	M	(AND memory with accumulator)	1 0 1 0 0 1 1 0
ANI	data	(AND immediate with accumulator)	1 1 1 0 0 1 1 0
XRA	r	(Exclusive or register with accumulator)	1 0 1 0 1 S S S
XRA	M	(Exclusive or memory with accumulator)	1 0 1 0 1 1 1 0
XRI	data	(Exclusive or immediate with accumulator)	1 1 1 0 1 1 1 0
ORA	r	(OR register with accumulator)	1 0 1 1 0 S S S
ORA	M	(OR memory with accumulator)	1 0 1 1 0 1 1 0
ORI	data	(OR immediate with accumulator)	1 1 1 1 0 1 1 0
CMP	r	(Compare register with accumulator)	1 0 1 1 1 S S S
CMP	M	(Compare memory with accumulator)	1 0 1 1 1 1 1 0
CPI	data	(Compare immediate with accumulator)	1 1 1 1 1 1 1 0
CMA		(Complement accumulator)	0 0 1 0 1 1 1 1

D

	Rotate	Code
RLC	(Rotate accumulator left)	0 0 0 0 0 1 1 1
RRC	(Rotate accumulator right)	0 0 0 0 1 1 1 1
RAL	(Rotate accumulator left through carry)	0 0 0 1 0 1 1 1
RAR	(Rotate accumulator right through carry)	0 0 0 1 1 1 1 1

E

Machine Control Group		Code								
EI	(Enable interrupts)	1	1	1	1	1	0	1	1
DI	(Disable interrupts)	1	1	1	1	0	0	1	1
HLT	(Halt)	0	1	1	1	0	1	1	0
NOP	(No operation)	0	0	0	0	0	0	0	0

Carry Control Group		Code								
CMC	(Complement carry)	0	0	1	1	1	1	1	1
STC	(Set carry bit = 1)	0	0	1	1	0	1	1	1

Stack Group			Code								
PUSH	rp	(Push register pair)	1	1	R	P	0	1	0	1
PUSH	PSW	(Push program status word)	1	1	1	1	0	1	0	1
POP	rp	(Pop register pair)	1	1	R	P	0	0	0	1
POP	PSW	(Pop program status word)	1	1	1	1	0	0	0	1
XTHL		(Exchange stack with H L)	1	1	1	0	0	0	1	1

F

Branch Group			Code								
JMP	addr	(Jump)	1	1	0	0	0	0	1	1
JZ	addr	(Jump on zero)	1	1	0	0	1	0	1	0
JNZ	addr	(Jump on no zero)	1	1	0	0	0	0	1	0
JC	addr	(Jump on carry)	1	1	0	1	1	0	1	0
JNC	addr	(Jump on no carry)	1	1	0	1	0	0	1	0
JPO	addr	(Jump on odd parity)	1	1	1	0	0	0	1	0
JPE	addr	(Jump on even parity)	1	1	1	0	1	0	1	0
JP	addr	(Jump on positive)	1	1	1	1	0	0	1	0
JM	addr	(Jump on minus)	1	1	1	1	1	0	1	0
CALL	addr	(Call)	1	1	0	0	1	1	0	1
CZ	addr	(Call on zero)	1	1	0	0	1	1	0	0
CNZ	addr	(Call on no zero)	1	1	0	0	0	1	0	0
CC	addr	(Call on carry)	1	1	0	1	1	1	0	0
CNC	addr	(Call on no carry)	1	1	0	1	0	1	0	0
CPO	addr	(Call on odd parity)	1	1	1	0	0	1	0	0
CPE	addr	(Call on even parity)	1	1	1	0	1	1	0	0
CP	addr	(Call on positive)	1	1	1	1	0	1	0	0
CM	addr	(Call on minus)	1	1	1	1	1	1	0	0
RET		(Return)	1	1	0	0	1	0	0	1
RZ		(Return on zero)	1	1	0	0	1	0	0	0
RNZ		(Return on no zero)	1	1	0	0	0	0	0	0
RC		(Return on carry)	1	1	0	1	1	0	0	0
RNC		(Return on no carry)	1	1	0	1	0	0	0	0
RPO		(Return on odd parity)	1	1	1	0	0	0	0	0
RPE		(Return on even parity)	1	1	1	0	1	0	0	0
RP		(Return on positive)	1	1	1	1	0	0	0	0
RM		(Return on minus)	1	1	1	1	1	0	0	0
RST	n	(Restart)	1	1	X	X	X	1	1	1
PCHL		(Load H&L to program counter)	1	1	1	0	1	0	0	1

G

Input/Output Group		Code								
IN port	Input	1	1	0	1	1	0	1	1
OUT port	Output	1	1	0	1	0	0	1	1

H

Data Transfer Group			Code								
MOV	r1, r2	(Move register to register)	0	1	D	D	D	S	S	S
MOV	r, M	(Move memory to register)	0	1	D	D	D	1	1	0
MOV	M, r	(Move register to memory)	0	1	1	1	0	S	S	S
SPHL		(Load stack pointer from HL)	1	1	1	1	1	0	0	1
MVI	r, data	(Move immediate to register)	0	0	D	D	D	1	1	0
MVI	M, data	(Move immediate to memory)	0	0	1	1	0	1	1	0
LXI	rp, data 16	(Load register pair immediate)	0	0	R	P	0	0	0	1
LDA	addr	(Load accumulator direct)	0	0	1	1	1	0	1	0
STA	addr	(Store accumulator direct)	0	0	1	1	0	0	1	0
LHLD	addr	(Load H & L direct)	0	0	1	0	1	0	1	0
SHLD	addr	(Store H&L direct)	0	0	1	0	0	0	1	0
LDAX	rpp	(Load accumulator)	0	0	0	X	1	0	1	0
STAX	rpp	(Store accumulator)	0	0	0	X	0	0	1	0
XCHG		(Exchange H-L with D-E)	1	1	1	0	1	0	1	1

Fig. 24 Instruction set for the 8080 microprocessor. Instructions are grouped by function. Mnemonics are recognized by either the Intellec Assembler or the Intel furnished cross-assembler.

Some microprocessors are available with only minimal software support, and the designer is forced to use the processor by hand coding all instructions in binary ones and zeros. Minimal requirements of software for successful and economically sensible application are a working assembler and the basic utility software such as a loader and editor. Further ease of program preparation can be had through the use of a higher-order language such as the PLM developed for the Intel 8080 or by the use of a simulator allowing the emulation of the microprocessor on a larger computer. Assemblers fall into two categories: assemblers which can be run on the processor itself and cross assemblers which require the use of a larger host computer for operation. Much of the cross software is available on time-sharing services such as G.E. and Tymeshare. The cost and availability of such software should be considered in detail before beginning the application.

Software support is a major element in microprocessor applications development but a second major need is the need for hardware support, that is, a readily available prefabricated development system. Such a system allows for the test and debugging of software and may even include special interface circuits and special timing circuits to get around the need for programmable read-only memories in early development stages of a program. A system should be made complete enough such that developmental software can be run directly on the system. System software would include an assembler, a loader, an editor, and software necessary for the programming of read-only memories. An important consideration in the developmental system is the ease of interfacing this system with standard interface circuits, standard memory circuits, and standard peripheral devices.

In considering microprocessors, an ultimate consideration should be system size. A microprocessor system will not just be a single microprocessor but could be anywhere between 4 and 60 individual integrated circuits. The ease of building the system out of a limited number of circuits will greatly depend on the availability of special-purpose interface circuits such as those described in Table 14. Examination of the

TABLE 14   Special-Purpose Integrated Circuits for Micro-processor Applications, Selected Circuit Types

Type No.	Function	Process
8102	1,024-bit RAM memory	NMOS
8308	8,192-bit RAM memory	NMOS
8205	1 of 8 decoder	TTL
8214	Priority interrupt control	TTL
8216	4-bit bidirectional bus driver	TTL
8224	Clock generator—power up circuit	NMOS
8228	System controller—bus driver	TTL
8212	8-bit input/output port	TTL
8255	Peripheral interface	NMOS
8251	Universal communications interface	NMOS
74S225	80-bit FIFO memory	TTL
74S373	8-bit data latch	TTL
74S124	Dual VCO clock driver	TTL
6820	Peripheral interface adapter	NMOS

table will show the variety of circuits available. It should be noted that circuits can be used from multiple vendors to be combined into the end user's microcomputer. The only thing that must be checked is the compatibility of signal levels and timing. In considering the microprocessor system application, one should consider the overall power consumption and size of the microprocessor, its memory circuits, peripheral circuits, and power-supply subsystem. Overall size and cost will be dominated in most applications by the memory and peripheral circuits. Therefore, it is most important that the cost, power consumption, size, and number of these circuits be fully considered during the design phase.

The design of a system using microprocessor technology can reap many advantages if microprocessors are properly and judiciously applied. A microprocessor, being a programmable central processing unit, allows the substitution of a microcomputer for a random network of logic. Microprocessors can be used in two classes of applications, those applications where a microcomputer is used to replace a minicomputer which has previously been underutilized and those cases where a microprocessor replaces a network of random logic used for a particular dedicated digital application. Most microprocessor applications are in the second category. This category of applications is possible because any network of logic is equivalent to any computer with the proper set of software called a program. In using a microprocessor to replace a network of random logic, one replaces the hardware of the design by the software of the program. This is not that strange a thing when one considers that the design was initially an outgrowth of a specified kind of software called system-design requirements. The unique thing of effecting a design in software rather than hardware is the fact that software is much more adaptable to change than is hardware. When an error or a new application or change in concept is made in a hardware application of random logic, a complete redesign must be done, new printed-circuit boards made, new integrated circuits procured, and possibly even the system power supplies reconfigured. In the case of a microprocessor system a change in design concept is quite often a change in software; hence the change now becomes a modification in the software. The implementation of this change will be the programming of new read-only memories. Further, in developing a microprocessor system, if a developmental system is used such as the Intel Intellec,* the original program is first tried out by temporarily storing it in random-access memory to simulate the program store ROM. If changes, modifications, or error corrections are made at this point, they can be implemented in the first set of read-only memories programmed for the application.

Many people are bewildered by the astonishingly quick success of the microprocessor endeavor. These people look at the limited capability of the microprocessor, compare it with the minicomputer, and wonder what all this commotion is all about. They realize that a microcomputer is an extremely limited computer. What they do not realize is that the microprocessor is the exploitation of LSI technology into a standard product. Like other LSI techniques, low cost and extreme density of circuit function are both realized. A single microprocessor will cost over $200,000 to design, but this cost can be shared by the multiple users. Further the added cost of the software development, the design of the developmental systems, and even the design of the various software aids can be shared by the expansive user base. The size of the user base for microprocessors is growing at such an extreme rate that by the end of 1974 the monthly production rate of microprocessors exceeded the total number of standard computers that had been produced up to that date. A survey of the modern world will show us microprocessors used in industrial control, traffic control, point-of-sale machines, scientific calculators, bowling-alley scoring computers, and even military missile-guidance systems.

Universal asynchronous receiver-transmitter circuits   The large requirements for transmission and reception of digital data have resulted in the need for a specialized circuit that could perform the repetitious interface requirement. This circuit is the universal asynchronous receiver transmitter. The circuit is fabricated using large-scale integration techniques and the MOS fabrication technology. It is a complex array of digital logic that stores, controls, and compares a strain of data in a peripheral communications unit. A typical block diagram of this circuit is shown in Fig. 25. The complexity of this circuit is shown by the interlocking complex elements within the device. The universal asynchronous receiver-transmitter circuits available today offer a variety of features and functions which make them suitable for computer peripherals, terminals, multiplexers, controllers, modems, and remote data-acquisition systems.

The 2536 is typical of second-generation universal asynchronous receiver transmitters, and its characteristics are similar to and compatible with the 6010, AY-5-1012, 1402A, 1757, and COM 2502. These circuits are all directly compatible with transistor-transistor logic levels; so no interface circuits are required. They have the ability

* Trademark of Intel Corporation.

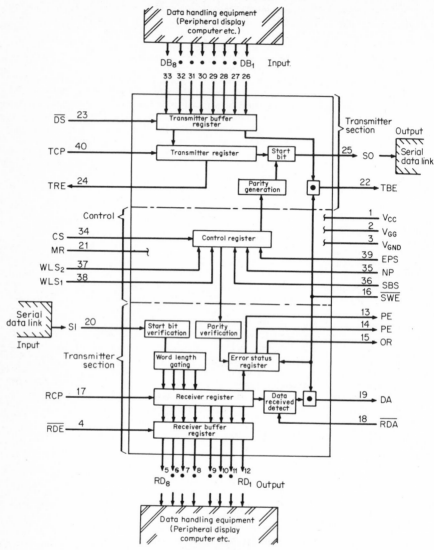

Fig. 25 Block diagram of universal asynchronous receiver-transmitter circuit.

to operate in a mode of alternately receiving and transmitting data, or they may operate full duplex in which they both receive and transmit simultaneously. They are fully buffered, eliminating the need for sychronization of remote systems. Universal asynchronous receiver transmitters have the ability to program externally and select specified operating features, such as word length, throughput-data rate, odd or even parity, presence or absence of parity check, and the option of either single- or double-bit generation. Other features of these circuits include data outputs and status flags to indicate internal operating conditions. The inclusion of a start-bit-verification scheme minimizes error rates in data transmission. The internal logic of the circuit is totally static, further minimizing internally generated errors. Universally asynchronous receiver-transmitter circuits are customary packaged in a 40-pin dual-in-line package. The effectiveness of this circuit in application has generated an extremely high usage rate, which in turn has appreciably reduced its cost. Use of this circuit will continue to grow because of its outstanding features.

Timers

The timer integrated circuit is an unusual phenomenon in the history of integrated circuits, because a single circuit, which is not part of any family and is not part of the evolution of circuit types within a circuit class, has in a very short interval gained broad-based acceptance. The 555 is a circuit capable of producing accurate and repeatable oscillations or delays. Basically, it is the control circuitry necessary to drive and monitor an external precision resistor-capacitor delay circuit. Internal organization of the 555 is shown in the block diagram of Fig. 26. This block diagram

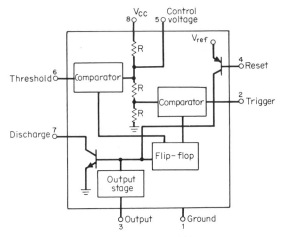

Fig. 26 Block diagram of the 555 timer circuit.

shows that the 555 is a comparator-actuated flip-flop with a buffer output stage and a built-in loop for discharging the external capacitor.

The 555 can be applied to a broad range of applications, including pulse generation, sequential timing, time-delay generation, pulse-position modulation, pulse-width modulation, and detection of missing pulses. Since either the oscillation period or the delay interval is totally determined by the external passive circuitry, there is complete freedom to establish any necessary time interval from microseconds to hours. Proper choice of external components can determine the necessary duty cycles for the requirements. The output stage of the 555 is capable of both sourcing and sinking extremely high currents, with the current limit being 200 mA. The 555 is designed to operate on any power-supply voltage between 5 and 15 V dc.

Figure 27 illustrates the configuration of the 555 for operation as a monostable delay source. In the configuration, the time delay is set by the value of the external timing resistor R_1 and the charge capacitor C. Figure 28 shows the relationship between these parametric values.

A common application of this circuit is to use it as the source of a continuous clock signal. This signal is generated by configuring the 555 in the astable or free-running mode. The circuit configuration for this mode is illustrated by Figure 29. In this mode, the frequency of oscillation is given by

$$F = \frac{1.4}{(R_1 + R_2)C}$$

and the duty cycle is given by

$$D = \frac{R_2}{R_1 + 2R_2}$$

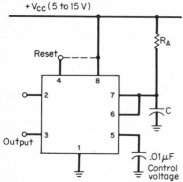

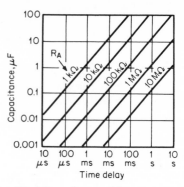

Fig 27 Schematic diagram of the 555 timer circuit configured for monostable delay.

Fig. 28 Plot of time delay vs. external components for the 555 timer.

Most applications of the 555 circuit have been in situations where integrated circuits would not normally have been considered for use. The 555 can be used to modulate audio tones, to control timing-sequence circuitry for appliances and vending machines, to provide low-frequency modulation of signals in control loops, and even to establish operating sequences in simplified test equipment.

The 555 is also available in a dual configuration, the 556. In this configuration, the two timers operate independently of each other, having only the V_{cc} and ground connections in common. The advantage of the 556 is twofold. First, it offers increased packaging density. Second, it has the advantage of similarity between the operating characteristics of the two timers. The dual timer is most advantageous in applications where timers are cascaded and interconnected to give sequential control and sequential delay functions.

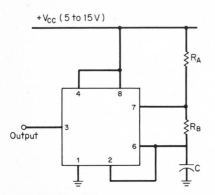

Fig 29 The 555 timer as a free-running multivibrator.

Automotive Circuits

Automotive use of integrated circuits has rapidly expanded because of complex electronic control systems for newly imposed safety and pollution-abatement requirements. This use is supplemented by new entertainment-equipment options, such as tape decks, AM/FM radios, and 4-channel sound. Almost all automotive circuits are proprietary circuits designed to the particular requirements of a single user. The development work has primary used two areas of integrated-circuit technology, linear integrated-circuit technology and complementary metal-oxide semiconductor technology.

Table 15 lists potential functions which may be implemented by automotive integrated-circuit subsystems. The most complex of these is skid-control circuitry. Skid control requires a microcomputer and multiple sensors to control electronically the braking and steering mechanisms of a car in order to eliminate modes of operation that cause skids. In operation, the circuits allow easy recovery from any skid induced by such improper road conditions as ice, water, or oil road films. Electronic tachometer circuits consist of one-shot multivibrators triggered by a direct motor output, such as the closure of a distributor contact. This switch closure triggers the one-shot multivibrator, and the current-pulse outputs are integrated on a capacitor. The established voltage level maintained on this capacitor is then translated into an equivalent engine-revolution rate.

TABLE 15   Automotive Functions as They Apply to the Integrated Circuit

Function	Technology	Availability
Seat-belt interlock	CMOS	Yes
Seat-belt interlock	Linear	Yes
Tachometer	Digital	Yes
Voltage regulator	Hybrid	Yes
Antiskid control	LSI	Planned
Digital meter	LSI	Planned
Pollution control	LSI	Planned
Driver-altertness test	LSI	Planned

Radiation-hardened Integrated Circuits

The integrated circuit has been readily adapted to the numerous environments through which electronic equipment has been forced to function. Now, integrated circuits are capable of successful operation over a range of environmental and thermal stresses that permit use under the sea, in the air, in vibration, in portable equipment, and in space. Recent applications for deep space and orbit missions and weapon systems designed to operate during- nuclear-blast events required integrated circuits to be designed for operation under radiation environments. Specific designs have met both the blast environment and the natural-space radiation environment. The blast environment is a combination of high-energy neutrons, x-ray radiation, gamma radiation with a high gamma rate, and an electromagnetic pulse. For the natural-space radiation environment, the prime constituent is a large total gamma dose, which can be collected over an extended interval. The effects of radiation can be either transient or permanent. The transient mode affects operative status of the circuits only temporarily, while permanent modes do damage that cannot be repaired during the mission.

Radiation-hardened integrated circuits are circuits specially designed and processed to minimize the effects of both permanent and transient radiation, which guarantees operation of the system during the mission. However, many of the details of radiation level and the nature of blast threats are classified. In designing a radiation-hardened integrated circuit, two basic concepts are used. One is to use structures which are inherently less sensitive to change and damage. Otherwise the circuit may be overdesigned so that degradation does not reduce the functionality of the circuit. Most radiation-hardened integrated circuits use dielectric isolation, a technique to separate electrically the various elements on the chip from each other by forming a silicon dioxide dielectric around the isolated areas of silicon, which are called tubs. Figure 30 shows a cross-sectional view of this structure and its fabri-

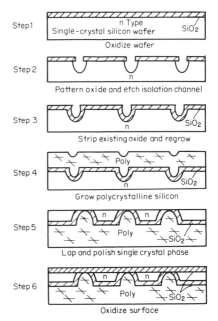

Fig. 30 Cross section of the dielectric-isolation structure used in radiation-hardened integrated circuits. Note: the wafer is turned over between step 4 and step 5. Standard processing then proceeds after step 6 to form active devices in the wells prior to defining the film resistors and conductors on the surface.

cation.   This dielectric isolation eliminates the possibility of a four-layer *pnpn* latch-up during a radiation event.

The transistors in the radiation-hardened circuit are normally fabricated with a very small geometry that has very thin base regions to minimize the volume of capture for gamma rays.   As a result, the generation of photocurrents is minimized.   Compensating diodes are quite often connected to the base nodes of the transistors to reduce the magnitude of the photocurrent flowing through the transistor.   The thin base regions further tend to limit degradation in transistor beta from permanent neutron damage.   Great care is taken in preparing the semiconductor surfaces to minimize total gamma-dose susceptibility.   To eliminate the total current associated with the normally diffused resistors used in conventional integrated circuits, a thin-film technique is used to form the necessary resistors on the surface of the chip.   Many materials have been used for the formation of these thin-film resistors, including nickel-chrome (Nichrome), chrome silicide, molybdenum silicide, and tantalum nitride.   The Nichrome process is the most commonly used, although recent literature indicates that a reliability problem may be associated with it, since entrapped package moisture causes a failure mode of electrolytic corrosion.

The metallurgical system used to attach the integrated-circuit chip to the package substrate, to connect the device pads to the internal terminals of the external pins, must be selected with great care in a radiation-hardened integrated circuit.   Thermal-mechanical stresses can be generated in high-atomic-number materials during intense x-ray radiation.   Therefore, materials such as gold are not used in the construction of these radiation-hardened integrated circuits.   In many cases where gold-bond wires are required, special precautions must be applied to give the wire bond sufficient strength to withstand thermal-mechanical x-ray abuse.

Four basic groups of radiation-hardened circuits are now available in the marketplace.   However, these circuits cannot be considered standard products, since radiation-hardened circuits are made in relatively small batches under tightly controlled conditions to meet the praticular requirements of systems for which they are manufactured.   The first group of circuits is analogous in function and schematic to the 930 family of diode transistor logic.   These circuits, along with the other commonly available radiation-hardened circuits, are presented in Table 16.   The second class

TABLE 16   The 930 Family of Radiation-hardened Diode Transistor Logic, Complete Selection Guide of Circuit Types

Function	Circuit form	Type No.*
NAND gate	Dual 4-input	930
Buffer NAND gate	Dual 4-input	932
NAND power gate	Dual 4-input	944
NAND gate	Quad 2-input	946
NAND gate	Triple 3-input	962
Inverter	Hex	936
Expander	Dual 4-input	933
Flip-flop		945
Flip-flop		948

* See Table 3 in Chap. 2.

of circuits are the 5400L and H families.   Members of this family available as radiation-hardened circuits are presented in Table 17.   More recently, a third family of more complex radiation-hardened circuits has been developed.   This family uses Schottky processes to give performance similar to the 54LS family, and they are presented in Table 18.   The circuits in this family are fabricated using the beam-lead technique for interconnection between chips.   This allows fabrication of hybrid circuits having the equivalent density LSI functions.   These classes of digital integrated circuits are for the most part not identical to their non-radiation-hardened counter-

TABLE 17 The 5400 Family of Radiation-hardened Transistor-Transistor Logic, Complete Selection Guide of Circuit Types

Function	Circuit form	Type No.*
NAND gate	Quad 2-input	54H00, 5400, 54L00
NAND gate	Quad 2-input	54H01
NAND gate	Triple 3-input	54H10, 5410, 54L00
NAND gate	Dual 4-input	54H20, 5420, 54L20
NAND gate	11-input	54H31, 5431
Buffer NAND gate	Dual 4-input	54H40, 5440
NAND gate	Dual 3-input	54L130
NAND gate	Expandable 3-input	54L131
Inverter	Hex	5404, 54H04
AND/OR invert gate	Dual 2 × 3, 2 × 2	5456, 54H56
AND/OR invert gate	3 × 2	5457, 54H57, 54L57
AND/OR invert gate	2 × 4	5458, 54H58
Flip-flop	RS master-slave	54L71
Flip-flop	JK master-slave	54L72˙
Flip-flop	Dual D	5474, 54H74, 54L74
Flip-flop	Dual JK	54H103

* See Tables 10, 12, and 14 in Chap. 2.

TABLE 18 The 54LS Family of Beam-Lead Radiation-hardened Transistor-Transistor Logic, Complete Selection Guide of Circuit Types

Function	Circuit form	Type No.*
NAND gate	Quad 2-input	54LS00
NAND gate	Dual 4-input	54LS20
Flip-flop	Dual JK	54LS112
Multiplexer	Dual 4-line to 1-line	54LS253
Decoder	Dual 2-line to 4-line	54LS155
Register file	4 × 4	54LS270
Arithmetic logic unit	4-bit	54LS181
Memory—ROM	1,024-bit	54LS187
Counter	4-bit up/down	54LS193
Shift register	4-bit universal	54LS194
Memory—RAM	64-bit	54LS189

* See Tables 18 and 19 in Chap. 2.

parts. Typically, in the preirradiated state these circuits are superior in performance in both speed and fanout to the non-radiation-hardened circuit. Radiation-hardened digital integrated circuits are complemented by a limited number of available linear functions that comprise the fourth class. These linear circuits are functionally limited to operational amplifiers and sense amplifiers. There is a complete absence of voltage-regulator circuits. The linear circuits are described by function in Table 19.

In addition to these circuits, other integrated circuits have found use in radiation environments, although they were not specially designed for this purpose. Circuits that will not reach the level of a blast environment can allow such circuit choices. In particular, circuits used in space missions can be fabricated from normal processes with only some additional special care. A good example of this is the widespread use of complementary metal oxide semiconductor devices in satellite electronic systems. Experience has shown that only circuits fabricated under the most extensive control of surface cleanliness are able to survive without degradation through long periods of accumulated gamma doses. Other applications at reduced overall radiation doses can be met by conservative application of standard integrated-circuit processes. Therefore, the design engineer must be aware of degrading phenomena and

TABLE 19   Radiation-hardened Linear Circuits, Selected Circuit Types and Description

Circuit function	Type No.
Uncompensated operational amplifier	709
Threshold detector	55900
Dual-channel preamplifier	55910
4-channel sense amplifier	55920
Compensated operational amplifier	7041
Compensated operational amplifier	7042

should experimentally gather some measure of the degradation to be expected at the specified radiation environmental level.

PART NUMBERS

While integrated circuits are generally referred to by the numeric section of the part number, a full part number must be used when ordering. This full part number includes the numeric root as well as prefixes and suffixes. These prefixes and suffixes indicate the vendor, the class of parts, the package, and sometimes even the temperature range of this circuit. Table 20 lists major vendors and their lettering system. Each vendor has a different system of appropriate prefixes and suffixes. Therefore, the same part ordered from different vendors has different part numbers. Although all vendors recognize and are able to work with the prefixes and suffixes used by their competition, it is far preferable to order the parts by appropriate vendor letters. Design engineers should also learn to recognize markings on products shipped by the various vendors. Not all vendors make a full range of package and temperature styles for the products that they manufacture; so it is up to the design engineer to minimize these mistakes. It is most appropriate to consult a full listing of type numbers, package styles, and temperature ranges sold by each manufacturer. The most authoritative source for this is the vendor's published price list.

THE INTEGRATED-CIRCUIT INDUSTRY

The second half of the twentieth century will doubtless be identified as the age of the electronic industrial revolution. Three important industries have played key roles in creating this revolution, the computer industry, the electronic consumer-goods industry, and the semiconductor industry. The semiconductor industry is the source of the broad range of new and original active electronic components that are the basis for the products of the other two industries.

The integrated-circuit industry is dominated by manufacturers independent of large-electronic-system manufacturing corporations. Although a few integrated-circuit manufacturers are divisions of such corporations, they do not dominate the industry. Unfortunately, it is not possible for a small manufacturer of integrated circuits to operate efficiently owing to requirements for a broad range of varied technical talent and a moderately large financial base on which to support expensive and sophisticated capital equipment.

As a result, the integrated-circuit industry is composed of two primary types of suppliers. The first are broad-based integrated-circuit manufacturers who produce a great variety of different circuit types, while the second are smaller companies that specialize in one or two areas of technical endeavor. The smaller companies may be limited by their lack of a large financial and technical base, but in many cases they may be able to respond better to specialized requirements. Therefore, it is not possible to state a clear preference for large or small companies as a rule of thumb for the design engineer. The particular capability and strength of any integrated-circuit vendor and the applicability of this capability to the particular circuit requirements must be separately considered for each integrated-circuit application.

Table 21 lists the primary noncaptive integrated-circuit sources in the free world.

TABLE 20 Integrated-Circuit Part Numbers, Cross Reference and Interpretation

Vendor	Prefix	Meaning	Suffix	Meaning
Advance Micro Devices	AM	Advanced micro devices	P—	Plastic dual-in-line
			H—	Metal can
			D—	Hermetic dual-in-line
			F—	Hermetic flatpack
			—C	Commercial temperature range
			—M	Military-temperature range
Fairchild	F	Fairchild	P—	Plastic dual-in-line
	SH	Hybrid	D—	Hermetic dual-in-line
			F—	Hermetic flatpack
			H—	Metal can
			—C	Commercial-temperature range
			—M	Military-temperature range
			J—	Power pack (TO-66)
			K—	Power pack (TO-3)
			T—	Mini dip
Motorola	MC	Motorola	L	Hermetic dual-in-line
	MC1	McMOS	P	Plastic dual-in-line
	MCM	Memory	F	Hermetic flatpack
	MLM	Motorola version of nat.	C	Chip
	MFC	Entertainment		
	MCE	Radiation-hardened		
	MCB	Beam-lead		
National	LM	Linear monolithic	D	Hermetic dual-in-line
	DM	Digital monolithic	F	Hermetic flatpack
	LH	Linear hybrid	G	TO-8 metal can
	DH	Digital hybrid	H	Metal can
	MM	MOS monolithic	J	Cerdip dual-in-line
	AH	Analog hybrid	N	Plastic dual-in-line
	AM	Analog monolithic	W	Flatpack
RCA	CA	Linear circuit	E	Plastic dual-in-line
	CD	Digital circuit	T	Metal can
			S	Metal can—formed
			D	Ceramic dual-in-line
			F	Hermetic dual-in-line
			L	Beam-lead chip
Signetics	SE	Military-temp. range		
	NE	Commercial-temp. range		
Texas Instruments	SN	Texas Instruments	F	Flatpack
	TM	MOS circuit	J	Ceramic dual-in-line
	RSN	Radiation-hardened	N	Plastic dual-in-line
			L	Metal can
			W	Flatpack

TABLE 21   The Integrated-Circuit Industry

Company name	Mailing address
Advanced Memory Systems, Inc.	1276 Hammerwood Ave., Sunnyvale, Calif. 94086
Advanced Micro Devices, Inc.	901 Thompson Place, Sunnyvale, Calif. 94086
American Micro-Systems, Inc.	3800 Homestead Rd., Santa Clara, Calif. 95051
Analog Devices	P.O. Box 280, Norwood, Mass. 02062
Antex Industries	1059 E. Meadow Circle, Palo Alto, Calif. 94303
Bell and Howell Control Products	706 Bostwick Ave., Bridgeport, Conn. 06605
Burr-Brown Research Corp.	P.O. Box 11400, Tuscon, Ariz. 85706
Cal-Tex Semiconductor, Inc.	P.O. Box 2808, Santa Clara, Calif. 94040
Circuit Technology, Inc.	160 Smith St., Farmingdale, N.Y. 11735
Collins Radio Co.	4311 Jamboree Rd., Newport Beach, Calif. 92663
Custom Technology Corp.	3100 Coronado Drive, Santa Clara, Calif. 95051
CTS Micro Electronics	1201 Cumberland Ave., West Lafayette, Ind.
Electronic Arrays, Inc.	501 Ellis St., Mountain View, Calif. 94040
Energy Conversion Devices, Inc.	1675 West Maple Rd., Troy, Mich. 48084
Epitek Electronics, Inc.	19 Grenfell Crescent, Ottawa 12, Ontario, Canada
Exar Integrated Systems, Inc.	750 Palomar Ave., Sunnyvale, Calif. 94086
Fairchild Semiconductor	464 Ellis St., Mountain View, Calif. 94040
Frontier Manufacturing Co.	2955 North Airway Ave., Costa Mesa, Calif. 92626
General Instruments Corp.	600 West John St., Hicksville, N.Y. 11802
Halex, Inc.	3500 West Torrance Blvd., Torrance, Calif. 90509
Harris Semiconductor	P.O. Box 883, Melbourne, Fla. 32901
Hughes Aircraft	500 Superior Ave., Newport Beach, Calif. 92663
ILC Data Devices	100 Tec St., Hicksville, N.Y. 11801
Integrated Circuit Engineering	6710 E. Camelback Rd., Scottsdale, Ariz. 85251
Integrated Circuits International	1008 Stewart Drive, Sunnyvale, Calif. 94086
Integrated Microsystems, Inc.	1215 Terra Bella Ave., Mountain View, Calif. 94043
Intel Corporation	3065 Bowers Ave., Santa Clara, Calif. 95051
Interdesign	1255 Reamwood, Sunnyvale, Calif. 94086
Intersil	10900 Tantau Ave., Cupertino, Calif. 95014
Intronics, Inc.	57 Chapel St., Newton, Mass. 02158
ITT Semiconductor	Electronics Way, West Palm Beach, Fla. 33407
Ledex, Inc.	123 Weber St., Dayton, Ohio 45402
Lithic Systems, Inc.	P.O. Box 478, Saratoga, Calif. 95070
Litronix, Inc.	4900 Homestead Rd., Cupertino, Calif. 95014
LSI Computer Systems, Inc.	22 Cain Drive, Plainview, N.Y. 11735
Martin Marietta Aerospace	P.O. Box 5837, Orlando, Fla. 32805
Matsushita Electronics Corp.	1 Kotari, Yakimachi, Nagaokakyo, Kyoto, Japan
Micro Networks Corp.	5 Barbara Lane, Worchester, Mass. 01604
Micropac Industries, Inc.	905 East Walnut, Garland, Tex. 75040
Micro Power Systems, Inc.	3100 Alfred, Santa Clara, Calif. 95050
Mini Systems, Inc.	20 David Rd., North Attleboro, Mass. 02761
Mitsubishi Electronic Corp.	4-1 Zuhara, Itami, Hyogo, Japan, 644

Diode transistor logic	Resistor-transistor logic	Transistor-transistor logic	Emitter-coupled logic	Other bipolar logic	Custom monolithic	Custom hybrid	Standard hybrid	Linear	Entertainment	MOS	CMOS	Military	M38510 qualified	Memories	Radiation-hardened	Automotive	System on a chip	Microprocessor	CCD	Technical strength
										X				X						Memory
		X			X			X	X	X		X	X	X			X	X		Linear, TTL
					X				X	X	X	X	X	X			X	X		MOS
					X			X		X	X									Linear, CMOS
					X					X							X			
						X	X													
						X	X	X						X						
										X	X			X			X			
						X								X						Hybrids
					X					X	X	X								
					X					X	X	X					X			Custom
						X	X	X	X											MOS memory
										X					X		X	X		
											X				X					Oshinsky devices
						X	X	X												
				X	X				X	X		X								Custom
X	X	X	X	X	X	X	X	X	X	X	X	X	X	X	X	X	X	X	X	Industry leader
										X							X			
						X	X	X		X	X				X		X	X		MOS, hybrids
						X	X													
X		X			X			X		X	X	X	X	X	X		X	X		Linear, radiation-hardened
						X	X			X	X	X					X			
						X						X								
						X														Consulting
						X				X	X									
										X										
	X		X							X	X	X			X		X	X	X	MOS, Memory, Microprocessor
						X		X												Custom chip
						X		X		X	X	X			X		X	X		
					X															
		X	X	X	X			X	X	X			X	X		X				
					X	X	X	X				X								
						X		X	X								X			
										X	X						X			
						X		X	X	X	X						X	X		
						X	X			X		X								
X	X							X	X			X								
						X	X	X	X			X					X			
												X					X			
X	X					X	X	X	X	X		X	X							

TABLE 21   The Integrated-Circuit Industry (Continued)

Company name	Mailing address
Monolithic Memories, Inc.	1165 East Arques Ave., Sunnyvale, Calif. 94086
MOSFET Micro Labs, Inc.	Penn Center Plaza, Quakertown, Pa. 18951
MOS Technology	Valley Forge, Pa. 19481
Mostek Corporation	1215 West Crosby Rd., Carrolton, Tex. 75006
Motorola Semiconductor Products	5005 E. McDowell Rd., Phoenix, Ariz. 85008
Mullard, Limited	Torrington Place, London, WCIE YHD, U.K.
National Semiconductor	2900 Semiconductor Drive, Santa Clara, Calif. 95051
Nortec Electronics Corp.	3697 Tahoe Way, Santa Clara, Calif. 95051
Optical Electronics, Inc.	Box 11140, Tucson, Ariz. 85734
Precision Monolithics, Inc.	1500 Space Park Drive, Santa Clara, Calif. 95050
RCA Solid State Division	Route 202, Somerville, N.J. 08876
Radio Technique Compelec	130 Ave. Ledru-Rolin, Paris, France
Ragen Semiconductor	53 S. Jefferson Rd., Whippany, N.J. 07981
Raytheon Semiconductor	350 Ellis St., Mountain View, Calif. 94040
Reticon Corporation	910 Benicia Avenue, Sunnyvale, Calif. 94086
Rockwell Micro Electronics	Box 3669, Anaheim, Calif. 93803
Sanken Electric Company	1-22-8 Nishi-Ikebukuro, Toshima-Ku, Tokyo, Japan
Signetics	811 Arques Ave., Sunnyvale, Calif. 94086
Siliconix, Inc.	2201 Laurelwood Rd., Santa Clara, Calif. 95054
Silicon General, Inc.	7382 Bolsa Ave., Westminster, Calif. 92683
Solid State Scientific	Commerce Drive, Montgomeryville, Pa. 18936
Sprague Electric Co.	115 N.E. Cutoff, Worcester, Mass. 01606
Stewart Warner Microcircuits	730 E. Evelyn Ave., Sunnyvale, Calif. 94086
Synertek	3050 Coronado Drive, Santa Clara, Calif. 95051
Teledyne Crystalonics	147 Sherman St., Cambridge, Mass. 02140
Teledyne Philbrick	Allied Drive, Dedham, Mass. 02026
Teledyne Semiconductor	1300 Terra Bella Ave., Mountain View, Calif. 94040
Texas Instruments	19500 North Central Expressway, Dallas, Tex. 75222
Transitron Electronic Corp.	168 Albion Place, Wakefield, Mass. 01880
TRW Microelectronics	14520 Aviation Blvd., Lawndale, Calif. 90260
Valvo GmhB Hauptniederlassung	2000 Hamburg 1, West Germany
Western Digital Corp.	19242 Red Hill Ave., Newport Beach, Calif. 92663

Each company is listed by its official name and its mailing address for information; a matrix presents the primary capabilities of each company. This table sums the data gathered during a survey of the industry by Zatz during the latter part of 1975.

The integrated-circuit industry, in general, has suffered from the reputation of not being fully responsive to business ethics and morality. This reputation is not fully unearned in many cases. Like many other high-technology industries, it has been dominated by enthusiastic and progressive but immature management personnel. These personnel have tended to overestimate the capability of the product being designed and produced. This overestimation has caused the industry to produce

Product range (columns 1–20)

Diode transistor logic	Resistor-transistor logic	Transistor-transistor logic	Emitter-coupled logic	Other bipolar logic	Custom monolithic	Custom hybrid	Standard hybrid	Linear	Entertainment	MOS	CMOS	Military	M38510 qualified	Memories	Radiation-hardened	Automotive	System on a chip	Microprocessor	CCD	Technical strength
	X									X		X		X				X		Memory
										X	X									Custom processing
										X							X	X		Calculator circuits
				X						X							X	X		MOS
X	X	X	X	X	X			X	X	X	X	X	X	X	X	X	X	X		Industry leader
X	X	X								X	X	X			X					
X		X	X	X	X	X	X	X	X	X	X	X	X	X	X	X		X		Industry leader
												X						X		
						X	X	X												
								X												Linear
	X			X	X			X	X	X	X	X	X	X	X	X	X	X	X	CMOS linear
X	X									X	X	X								
												X								
X		X	X	X	X			X				X					X			CCD
										X	X								X	SOS MOS
					X												X	X		
						X	X	X												
X		X	X	X				X	X	X		X		X	X		X	X	X	
										X										
										X	X									
												X			X	X	X	X		CMOS
	X				X	X	X									X				
X	X	X										X				X				
					X					X	X	X			X	X	X	X		
						X	X	X												
								X												
				X								X	X							
X		X	X	X	X	X	X	X	X	X	X	X	X	X	X	X	X	X		
X		X	X							X	X							X		
		X	X									X						X		
X		X	X	X	X	X	X	X	X	X						X		X		EFL
					X						X							X		

products that have not been fully characterized, measured, or qualified to levels of reasonable functionality. The result of this lack of control has been a great series of product introductions and manufacturing ventures that terminated in recall, redesign, and finally reissue of the product.

The early 1970s have seen the beginning of a maturation in the integrated-circuit industry, evidenced by more conservative marketing and greater attention to quality on the part of the principal integrated-circuit vendors. Many of the smaller firms that have recently entered the industry have cleverly learned from the past mistakes of the older firms and have applied more mature business criteria to their initial operation. The industry, however, is still plagued with a small number of integrated-circuit firms which are unwilling or unable to produce quality products.   Like any growth situation which can yield high profits to contributors, the integrated-circuit industry has attracted its share of profiteers and opportunists.   To be safe, the user of integrated circuits is well advised to select vendors carefully on the basis of direct vendor surveys and laboratory product evaluation.   The user must maintain firm control on the product produced by his suppliers through continuous monitoring of the product received.

The user of integrated circuits has himself sometimes been responsible for poor quality in the circuits he procures.   Many times, the user forces a supplier to offer integrated circuits at prices below the level at which they can be properly made while still yielding adequate profit to the manufacturer.   This abuse is foolish, since lack of proper specifications for tests and design control invites the supplier to compromise the quality and suitability of the product being manufactured.

ACKNOWLEDGMENTS

The author acknowledges the fine cooperation that he has received from the integrated-circuit industry in gathering the data for these three handbook chapters. Most of the information in this handbook is based upon published noncopyright catalog specification sheets.   No direct reference to this information has been given in the handbook because of its public nature.   The fine cooperation in all direct inquiries for information and photographs is also greatly appreciated.   I further thank the management of the Components Engineering Department of Martin Marietta Aerospace for their support, encouragement, and assistance during the many months that I have spent in compiling and writing these handbook chapters Particular thanks is given to Mrs. Estelle Terry for typing the original manuscript and to my wife, Lea Zatz, for her encouragement and editorial assistance.

Chapter **4**

Discrete Semiconductor Devices

ERWIN R. SCHMID
Bell Laboratories, Allentown, Pa.

INTRODUCTION

When integrated circuits became a practical reality and economically competitive, it was predicted that they would shortly replace discrete semiconductor devices. However, consumption of discrete semiconductors in U.S.-manufactured equipment reached a record high of over 3 billion parts in 1972, equivalent to sales of integrated circuits in dollar volume, with even greater numbers predicted for 1973.[1]

Some discrete semiconductors whose functions have been largely taken over by integrated circuits are indeed dying out: low-current, low-gain, low-power devices such as computer diodes, general-purpose transistors and diodes, and *npn* switches. Others have reached a demand peak and can be expected to start declining: rectifier and Zener diodes, 1- to 5-W silicon-controlled rectifiers, thyristors, and other power devices. But there is also a group of devices still in the ascendancy: high-frequency, low-noise MOSFET's* and JFET'st, high-speed diode arrays, high-voltage devices, all power functions above 5 W, and high-frequency and microwave FET's and diodes.

In this chapter, greatest emphasis is placed on those devices which are likely to be in use for a long time, and particularly on devices which are used in conjunction with integrated circuits. Little will be said about the techniques by which these devices are manufactured, other than to identify them where appropriate. These techniques have reached a high degree of sophistication in the manufacture of integrated circuits. They are discussed in some detail elsewhere.

PHYSICS OF SEMICONDUCTOR JUNCTIONS

Junction formation   Practical semiconductor devices make use of single-crystal materials (germanium or silicon) whose conductivity has been modified by the addition of impurities during the crystal-forming stage and subsequent processing of the material. The introduction of atoms of elements such as phosphorus, arsenic, and bismuth, each of which has five electrons in its valence band, into a crystal of a semiconductor, which has four electrons in its valence band, creates a supply of excess electrons which are free to move about. The crystal now has a net negative charge and is called an n-type semiconductor. On the other hand, if a substance with three valence electrons, such as boron, gallium, indium, and aluminum, is introduced into the semiconductor crystal, a deficiency of one electron exists between each impurity atom and its neighbors. This deficiency or vacancy, which can capture an electron from a semiconductor atom, is called a hole and is equivalent to a positive charge; the material is termed a p-type semiconductor.

In a practical semiconductor, imperfections and impurities cause both electrons and holes to exist simultaneously in n- and p-type material. However, in n-type material electrons predominate and are the majority carriers, while holes are the minority carriers. The reverse is true in p-type material, where holes are now the majority carriers.

If p- and n-type materials were to be joined, the existence of net positive and negative charges would cause an exchange of charges across the boundary until a balance is achieved. When this occurs, a region containing no free charges separates the p and n regions. Balance between the two parts is maintained by offsetting majority and minority currents across the charge-free layer (variously called a depletion layer, junction barrier, potential barrier, or simply a pn junction). Practical junctions are classified according to the method by which they are formed: grown (during crystal growth), alloy (by donor or acceptor material alloyed into the basic semiconductor), or diffused (by diffusion of impurities into a semiconductor wafer, at high temperatures).

Junction operation   The important characteristics of a pn junction and how they are affected by the application of external voltages (bias) are depicted graphically in Fig. 1. The electrical field which exists across the barrier can be visualized as an equivalent battery. In the quiescent state, the barrier voltage is approximately 0.3 V in germanium and 0.6 V in silicon. If we apply an external voltage with the negative terminal connected to the p side of the junction, as in Fig. 1b, the barrier widens and current flow across the depletion layer is effectively cut off. On the other hand, a forward-bias condition, shown in Fig. 1c, causes the barrier to become narrower, the external and internal batteries now aid each other, and current flows readily across the barrier. Depending on the polarity of the applied voltage, the junction either blocks or passes current; i.e., it rectifies.

Transistor action   A transistor can be thought of as two pn junctions placed back to back, with either the p or n side common to both junctions, as shown in Fig. 2. Depending on which polarity material (p or n) is common, the transistor is said to be an npn or pnp type, respectively. If the emitter side of the transistor is forward-biased, majority carriers (in the emitter) are injected into the base region. Some are lost because of recombination in the base region, but most diffuse through the base region toward the collector-base junction. Applying a relatively high reverse-bias potential to the collector terminal causes these charges to be attracted

* Metal-oxide system field-effect transistors.
† Junction field-effect transistors.

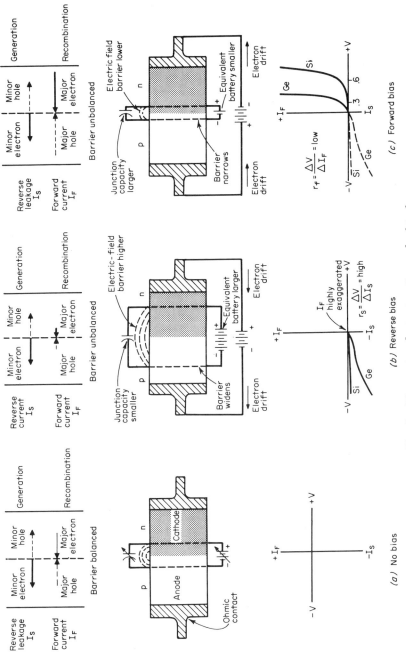

Fig. 1 The semiconductor junction under bias.[2]

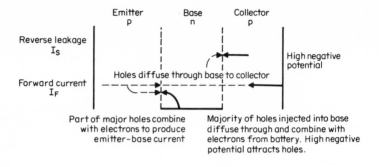

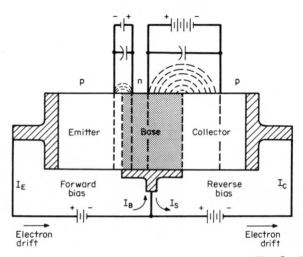

Fig. 2 Basic

into the collector region. Although the current arriving at the collector terminal is actually smaller than the current at the emitter (because some charges are lost to recombination), power gain has been achieved because collection takes place across the high impedance of the reverse-biased collector junction.

SEMICONDUCTOR DEVICES

Device Data Sheets

The most important sources of device information available to the designer are the data books published by semiconductor manufacturers. They are variously called data books, catalogs, data handbooks, etc. Although they contain a wealth of other information, such as selection guides, interchangeability directories, and mechanical data, they consist mainly of a collection of data sheets for all the devices currently made by a particular manufacturer. Data sheets contain two classes of information: ratings and characteristics.

Ratings A rating is a limiting condition or capability, either a maximum or minimum, established for a device, within which the device will perform in accordance with its design objectives. It is important to realize that such conditions cannot usually be achieved simultaneously. For example, it is usually not possible to operate a device at both maximum current and maximum voltage without exceeding some other limiting condition, such as maximum power or maximum junction temperature.

Characteristics A characteristic is a measurable device parameter which is used to describe the performance of the device, such as gain vs. current or temperature.

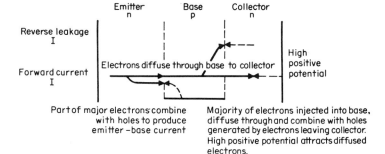

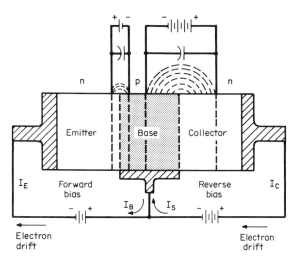

transistor action.[2]

Device characteristics are determined by the design of the device and are therefore not under the control of the circuit designer. The format and content of data sheets vary from manufacturer to manufacturer, but most include some graphical data. The terminology and symbols used in data sheets are listed in Tables 12 through 16. The overall Glossary for this Handbook includes a general glossary of semiconductor terms, with Tables 12 through 16 being specific glossaries for classes of devices, such as diodes and transistors.

Diodes

Classification Diodes, and semiconductor devices in general, may be classified in many ways, depending on which characteristics one wishes to emphasize. There are the obvious categories of type of semiconductor material (germanium, silicon), construction (point-contact, diffused, alloy, etc.), and function (rectifier, regulator, reference, control). Diodes may be further categorized by power level, voltage rating, and operating frequency. The approach taken here is to give the reader entry from a user's viewpoint, beginning with the general function to be performed (rectification, reference, or control), and then dividing further into current or voltage ranges, structures, encapsulations, etc. Table 1 lists the major groupings of diodes and related devices.

Rectifiers
Selenium. Historically, the rectifying properties of semiconductors have been known for nearly a century. The first practical semiconductor rectifiers, using copper

TABLE 1   Major Types of Diodes

Function	Type of material	Approx. ranges	Usage
Rectification:			
Low power	Si, Ge	< 1 A	Signal diodes, high-speed switching
	Si, Se	To 12 A, 1,000 V	Rectifiers, blocking diodes
	Se	20–600 V	Transient suppressors
Medium power	Si, Se	To 40 A, 1,000 V	Rectifiers
High power	Si	To 1,500 A, 1,000 V	Rectifiers
Reference and regulation (Zener diodes):			
Reference	Si	750 mW, 20 V	Precision voltage reference
Regulation	Si	250 mW to 50 W, to 200 V	Output-voltage control

oxide and selenium, made their appearances in the late 1920s and early 1940s, respectively. These polycrystalline materials have been largely replaced by germanium and silicon, especially the latter, which is today the basic material for the bulk of all semiconductor devices. However, selenium rectifiers are still in use, notably in two areas: low-cost entertainment applications, especially circuitry operating directly off the ac power line, and transient or overvoltage protection applications. Their relatively low cost and ability to recover from voltage transients make them preferable to silicon diodes in such applications. Figure 3 lists characteristics for a typical small selenium rectifier, of the type generally used in off-the-line entertainment cir-

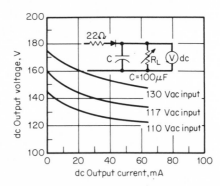

Fig. 3   Typical selenium-rectifier (clip-type) characteristics.[5]

Characteristics for capacitive load:

Nominal rms input voltage	117
Maximum rms input voltage	130
Maximum peak reverse voltage	380
Maximum peak current, mA	650
Maximum rms current, mA	175
Maximum dc output current, mA	65
Minimum recommended series resistance, Ω	22
Maximum cell operating temperature, °C	85
Maximum forward voltage drop, dc	10
Dielectric strength, V for 1 min	900
Output voltage with 117 V input, 100-μF capacitor and 65 mA dc, V	130

Environmental characteristics:

Humidity—withstands 120 h in 95% + RH at 65°C
Vibration—withstands 10 to 55 Hz, 0.060-in displacement for 2 h in each plane.

cuits. Note the large forward voltage drop of 10 V, as compared with a typical value of 1 V for silicon.

Germanium diodes find application mostly as general-purpose or signal diodes. They have the lowest forward voltage drop of all semiconductors (hence the lowest loss), typically 0.25 to 0.65 V for low-current diodes, and have short recovery times (as low as a few nanoseconds). These characteristics make them especially useful in high-frequency circuits, such as video detectors, demodulators, and signal-steering applications into the uhf range. Manufacturers list some 300 different types, ranging in peak inverse voltage (PIV) from 5 to 250 V and forward currents up to 0.5 A. They are generally hermetically sealed in glass (DO-7 package; see Fig. 44).

Silicon rectifiers, along with thyristors, are the most widely used power semiconductors. Typical applications include dc motors, generator and magnet excitation, electrochemical processes, battery charging, welding, dust precipitators, electronic power supplies for systems ranging from computers to tabletop radios, and automotive electrical systems.[4] Figure 4 shows a cutaway view of a typical 20-A rectifier, and

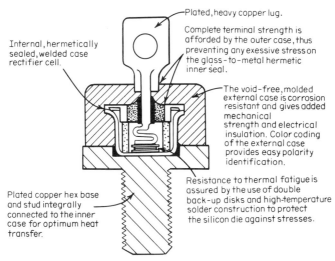

Plated, heavy copper lug.

Complete terminal strength is afforded by the outer case, thus preventing any exessive stress on the glass-to-metal hermetic inner seal.

Internal, hermetically sealed, welded case rectifier cell.

The void-free, molded external case is corrosion resistant and gives added mechanical strength and electrical insulation. Color coding of the external case provides easy polarity identification.

Resistance to thermal fatigue is assured by the use of double back-up disks and high-temperature solder construction to protect the silicon die against stresses.

Plated copper hex base and stud integrally connected to the inner case for optimum heat transfer.

Fig. 4 Cutaway view of typical silicon rectifier.[6]

Fig. 5 gives performance curves for the same type of rectifier. Nearly all the important rectifier operating curves may be generated from the typical forward characteristics (graph *I* of Fig. 5). The curves represent maximum conditions, so that all devices with this specification will have less power dissipation than that estimated from the values of V_F and I_F at the chosen operating point on the curve. Note the relatively high surge current permissible—as high as 350 A for a 1/60-s duration. The choice of a rectifier for a specific application is made on the basis of the maximum peak repetitive reverse voltage and average forward current. These values are calculated for the specific application, i.e., half-wave or full-wave rectification, bridge rectifiers, single or multiphase, type of load, etc. Design guides for this purpose may be found in data handbooks published by the various semiconductor manufacturers. Several of these are listed among the references at the end of this chapter.

Major manufacturers of silicon rectifiers list hundreds of different devices in their catalogs. An industry-wide compilation of diodes and silicon-controlled rectifiers[8] lists no fewer than 13,000 silicon rectifiers. The designer who is searching for a device for a particular application will find the task greatly simplified if he begins with a listing of *preferred* devices which all major manufacturers make available. One such listing is shown in Table 2. As can be seen, the entire range of available voltage and current combinations can be covered with approximately 150 devices.

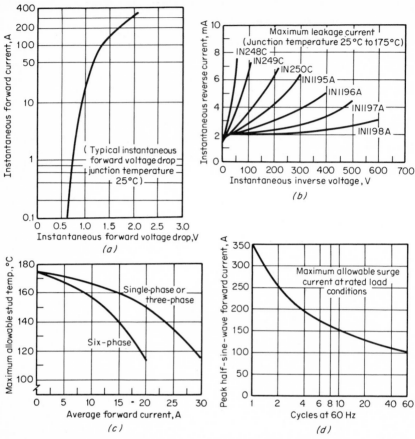

Fig. 5 Silicon-rectifier performance curves.[7] (*a*) Typical forward characteristics. (*b*) Reverse characteristics. (*c*) Maximum allowable stud temperature. (*d*) Surge rating.

Zener diodes are unique among semiconductor devices in that they are intended to operate in the reverse-breakdown region. Strictly speaking, the term Zener breakdown should be applied to a breakdown mechanism characterized by a negative temperature coefficient (the breakdown voltage decreases as the *pn*-junction temperature increases). Such characteristics are obtained by designing a junction with an extremely narrow depletion region. Junctions with wide depletion regions break down by an *avalanche breakdown* mechanism, characterized by a positive temperature coefficient, before Zener conditions are reached. Zener breakdown predominates in diodes with breakdown voltages up to 5 V; between 5 and 8 V both Zener and avalanche mechanisms are involved, and above that the avalanche mechanism alone takes over. However, in practice, all diodes operated in the reverse-breakdown region are called Zener diodes, regardless of the actual breakdown voltage.

Figure 6 illustrates the volt-ampere characteristics of a 30-V Zener diode. The breakdown voltage V_Z is called the zener voltage, V_Z, I_{ZT}, and Z_{TE} represent the voltage, current, and impedance values at the test condition, 420 mA in this case, I_{ZK} and Z_{ZK} represent a set of minimum conditions for reasonable regulation, and I_{ZM} is the maximum reverse current recommended for this device, 1.4 A in this instance.

Zener diodes are used principally in one of two applications: as *regulator* or *reference* devices. Regulators are used to minimize output variations with respect to

TABLE 2 Silicon-Rectifier Selection Guide

$V_{RM(REP)}$ Max peak repetitive reverse voltage	1A Surmetic (Case 59)	1A Fast recovery (Case 52)	1.5A (Case 55)	3A (Case 60)	3A (Case 70)	6A Fast recovery (Case 56A)	6A (Case 56A)	12A (Case 56A)	12A Fast recovery (Case 56A)	15A (Case 42)	20A (Case 42)	25A (Case 43)	30A (Case 43)	35A (Case 42)	50A (Case 100)	80A (Case 101)	160A (Case 102)	240A (Case 102)	400A (Case 103)	650A (Case 104)	1000A (water cooled) (Case 105)
50 V	1N4001	MR1337-1		1N4719 (MR1030A)	1N4997 (MR1030B)	1N3879		MR1120	1N3889	1N3208	1N248B	1N3491 (MR322)	1N3659	1N1183	MR1200	MR1210	MR1220	MR1230	MR1240	MR1260	MR1290
100 V	1N4002	MR1337-2	1N1563	1N4720 (MR1031A)	1N4998 (MR1031B)	1N3880		MR1121	1N3890	1N3209	1N249B	1N3492 (MR323)	1N3660	1N1184	MR1201	MR1211	MR1221	MR1231	MR1241	MR1261	MR1291
150 V				MR1033A	MR1033B			MR1122			1N193			1N1185	MR1202	MR1212	MR1222	MR1232	MR1242	MR1262	MR1292
200 V	1N4003 / 1N3611	MR1337-3	1N1564	1N4721 (MR1032A)	1N4999 (MR1032B)	1N3881		MR1123	1N3891	1N3210	1N250B	1N3493 (MR324)	1N3661	1N1186	MR1203	MR1213	MR1223	MR1233	MR1243	MR1263	MR1293
250 V															MR1204	MR1214	MR1224	MR1234	MR1244	MR1264	MR1294
300 V		MR1337-4	1N1565			1N3882		MR1124	1N3892	1N3211	1N195	1N3494 (MR325)	1N3662	1N1187	MR1205	MR1215	MR1225	MR1235	MR1245	MR1265	MR1295
350 V															MR1206	MR1216	MR1226	MR1236	MR1246	MR1266	MR1296
400 V	1N4004 / 1N3612	MR1337-5	1N1566	1N4722 (MR1034A)	1N5000 (MR1034B)	1N3883		MR1125	1N3893	1N3212	1N196	1N3495 (MR326)	1N3663	1N1188	MR1207	MR1217	MR1227	MR1237	MR1247	MR1267	MR1297
500 V		MR1337-6	1N1567					MR1126		1N3213	1N197	MR327		1N1189	MR1208	MR1218	MR1228	MR1238	MR1248	MR1268	MR1298
600 V	1N4005 / 1N3613	MR1337-7	1N1568	1N4723 (MR1036A)	1N5001 (MR1036B)			MR1128		1N3214	1N198	MR328		1N1190	MR1209	MR1219	MR1229	MR1239	MR1249	MR1269	MR1299
800 V	1N4006			1N4724 (MR1038A)	1N5002 (MR1038B)			MR1130				MR330									
1000 V	1N4007			1N4725 (MR1040A)	1N5003 (MR1040B)							MR331									

This table covers only the most popular rectifiers. The manufacturer should be consulted for special requirements.

Courtesy Motorola Semiconductor Products, Inc.

variations in input, temperature, and load requirements. Zener diodes used for this purpose are required to cover wide ranges of currents at many different voltages. Reference diodes, on the other hand, are required to provide a precise voltage and are not generally expected to dissipate large amounts of power. Temperature-compensated Zener diodes are available which provide voltage variations as low as 0.0005 percent per °C. This is accomplished by combining the Zener diode with one or more forward-biased diodes whose negative temperature coefficient is used to offset the positive coefficient of the Zener diode.

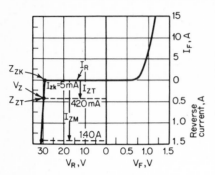

Fig. 6 Zener-diode characteristics.[9]

Miscellaneous diodes Solid-state *pn* junctions possess a number of useful properties other than those associated with rectification. Many of these nonrectifying features find application at uhf and microwave frequencies. Table 3 lists the various special-purpose diodes, with a summary of their nature, useful properties, and applications. Photodiodes and microwave diodes are described more fully in Chaps. 5 and 6, which deal specifically with these subjects. Special mention should be made of Schottky (hot carrier)

TABLE 3   Summary of Special-Purpose Diodes

Type	Nature	Useful properties	Applications
Tunnel	Extremely narrow depletion layer permits minority carriers to flow easily, or "tunnel" through barrier	Negative resistance, operation of 50 GHz	uhf and microwave oscillators, amplifiers, and switches
Varactor	Specially constructed, controllable depletion layer	Capacity varies with voltage, operation to 400 GHz	Tuning circuits, frequency multipliers, switches, pulse circuits, parametric amplifiers
PIN	Contains thin layer of intrinsic material. Acts as a variable conductance	Does not rectify at high frequencies	Modulators and switches at uhf
Schottky	Metal-to-semiconductor junction	High-speed recovery	uhf switches, detectors
Backward	Tunnel diode with "knee" at zero voltage. Supports heavy reverse current	Low rf impedance, current-sensitive	uhf low-level detector
Noise	Operated in avalanche region	Self-generating white noise	Measurements and testing
Photo	Light-sensitive conductive or voltaic diodes	Conductive or generative modulation	Light sensors and counters

diodes. Their ability to recover very rapidly (i.e., dissipate stored charges) upon reversal of applied voltage is used extensively in input circuits. They have also found their way into many integrated circuits.

Thyristors

The term thyristor applies to a family of semiconductor devices known as *pnpn* devices. Figure 7 lists the major members of this family, grouped according to

Type	Number of leads	IEC Official name	Common name	Schematic symbol Usage	Schematic symbol USASI	Equivalent cross section	Main trigger means	Maximum ratings available	Major applications
Unidirectional (reverse blocking)	2 (diode)	Reverse blocking diode thyristor	Four-layer (Shockley) diode			Anode / p n p n / Cathode	Exceeding anode breakover voltage	1,200 V / 300 A peak pulse	Triggers for SCRs, overvoltage protection, timing circuits, pulse generators
		Reverse blocking diode thyristor	Light-activated switch (LAS)	*		Anode / p n p n / Cathode	Infrared and visible radiation	200 V / 0.5 A	Static switches, triggers for high-voltage SCR applications, photoelectric controls
	3 (triode)	Reverse blocking triode thyristor	Silicon controlled rectifier (SCR)	†	†	Anode / p n p / p n / Gate Cathode	Gate signal	1,800 V / 550 A avg	Phase controls, inverters choppers, pulse modulators, static switches
		Reverse blocking triode thyristor	Light-activated SCR (LASCR)	*		Anode / p n p / p n / Gate Cathode	Gate signal or radiation	200 V / 1 A avg	Position monitors static switches, limit switches, trigger circuits, photoelectric controls
		Turnoff thyristor	Gate controlled switch (GCS,GTO)			Anode / p n p / p n / Gate Cathode	Gate signal turns GCS off as well as on	500 V / 10 A	Inverters, pulse generators, choppers dc switches
		—	Silicon unilateral switch (SUS)		—	Anode Cathode / p n n+ p p p+ / Gate	Exceeding breakover voltage or gate signal	10 V / 0.2 A	Timer circuits, trigger circuits threshold detector
		—	Complementary unijunction transistor (CUJT)		—	Base1 / p n / n p / n / Emitter Base2	When B-1 emitter vol. reaches predetermined fraction of B1–B2 voltage	30 V / 2 A peak pulse	Interval timing trigger circuits level detector, oscillator
	4 (tetrode)	Reverse blocking tetrode thyristor	Silicon controlled switch (SCS)	‡	‡	Cathode gate / Anode Cathode / p p / n / Anode gate	Gate signal on either gate lead	200 V / 1 A avg	Lamp drivers, logic circuits, counters, alarm and control circuits
Bidirectional	2 (diode)	Bidirectional diode thyristor	Biswitch, diac, SSS		—	n p / p n	Exceeding breakover voltage in either direction	400 V / 60 A rms	Overvoltage protection, ac phase control, triac trigger
	3 (triode)	Bidirectional triode thyristor	Triac		—	Anode 2 / n p n / n / n n p / Gate Anode1	Gate signal or exceeding breakover voltage	500 V / 20 A rms	Switching and phase control of ac power
		—	Silicon bilateral switch (SBS)		—	Two SUS structures inverse-parallel on same chip	Exceeding breakover voltage in either direction or gate signal	10 V / 0.2 A	Threshold detector, trigger circuits, overvoltage protection

Fig. 7 Characteristics of major types of thyristors.[4]

unidirectional or bidirectional characteristics and number of leads. The *silicon-controlled rectifier* (*SCR*) is by far the most widely used type of thyristor, largely because of its early introduction into commercial use and its versatility. Figure 8 is a diagrammatic representation of an SCR and of its analogy to two interconnected *pnp* and *npn* transistors.

SCR operation Figure 9 shows the voltage and current relationships in an SCR. At low voltages, the three junctions of the *pnpn* assembly block current flow

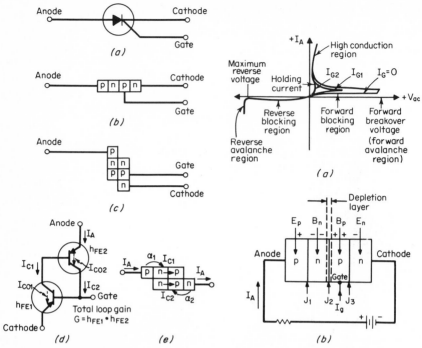

Fig. 8 SCR construction and transistor analogy.[10] (a) Symbol. (b) Basic junction arrangement. (c) Transistor-junction analogies. (d) Transistor-analogy schematic. (e) Alpha relationships.

Fig. 9 SCR voltage and current characteristics.[10] (a) General current-voltage characteristics. (b) Junction conducting relationships (gate open).

because the middle junction is reverse-biased. Forward blocking extends up to the forward breakover voltage, at which point avalanche breakdown takes place and the entire assembly goes into heavy conduction; the SCR is said to be "turned on" at this point. Turn-on can be made to occur at lower voltages by injecting carriers through the gate terminal into the p region next to the cathode. As gate current is increased, the forward blocking region all but disappears and the SCR approaches an ordinary diode in its behavior—except for a small negative-resistance region in the forward current characteristic just before normal diode current begins. At this point, there is a minimum value of current, called "holding current," below which the current must fall before the SCR can be turned off. It is important to remember that the SCR cannot be turned off by gating: the current in the circuit must be reduced below the holding-current level. There are, however, other thyristors which can be gated off (silicon-controlled switches and gate-controlled switches).

Figure 10 gives ratings and characteristics for a family of low-current SCR's. Manufacturers' data sheets also include graphs of changes in voltages, currents, and other characteristics vs. temperature. Gating characteristics are shown separately, as in Fig. 11, which gives the triggering requirements for a 35-A SCR. Figure 11a shows an area of preferred gate drive, as well as a shaded area denoting combinations of gate-triggering voltage and current which will not work. Figure 11b and c shows how to construct a gate-triggering load line and determine if the chosen operating point lies within allowable limits.

SCR applications The major applications of SCR's are shown in Fig. 7. Of these, ac phase control is the most important. Five basic controls are illustrated in Fig. 12. Control is achieved by governing the phase angle on the ac wave at which the SCR is triggered and turns on for the remainder of the half-cycle. The

Type	Minimum forward breakover vltage $V_{(BR)FR}$ $T_J = -65\,°C$ to $+100\,°C$ $R_{GK} = 40,000\,\Omega$	Repetitive peak reverse voltage V_{ROM} (rep) $T_J = -65\,°C$ to $+100\,°C$	Nonrepetitive peak reverse voltage V_{ROM} (non-rep) (<5 ms) $T_J = -65\,°C$ to $+100\,°C$
2N2344	25 V	25 V	40 V
2N2345	50 V	50 V	75 V
2N2346	100 V	100 V	150 V
2N2347	150 V	150 V	225 V
2N2348	200 V	200 V	300 V

Maximum allowable ratings (all types)

Repetitive peak forward blocking voltage (PFV)_____300 V
RMS forward current_____1.6 A
Average forward current, on-state, I_F(av)_____Depends on conduction angle
Peak one cycle forward surge current, (nonrepetitive) I_{FM}(surge)____15 A
Peak gate power, P_{GM}_____0.1 W
Average gate power, $P_{G(av)}$_____0.01 W
Peak gate current, I_{GFM}_____0.1 A
Peak gate voltage, forward and reverse, V_{GFM} and V_{GRM}_____6 V
Storage temperature, T_{stg}_____$-65\,°C$ to $+125\,°C$
Operating temperature, T_J_____$-65\,°C$ to $+100\,°C$

Characteristics

Test	Symbol	Min	Typ	Max	Units	Test conditions
Reverse blocking current	I_{RX}	—	40	100	μA	V_{RX} = rated V_{ROM} (rep) $T_J = 100\,°C$ $R_{GK} = 40,000\,\Omega$
Forward blocking current	I_{FX}	—	40	100	μA	V_{FX} = rated V_{FOM} $T_J = 100\,°C$ $R_{GK} = 40,000\,\Omega$
Gate current to trigger	I_{GT}	––	5	20	μA dc	$V_{FX} = +6$ V D-C $T_J = 25\,°C$ $R_L = 100\,\Omega$ max
Gate supply current to trigger	I_{GS}	—	10	40	μA dc	$V_{FR} = +6$ V D-C, $T_J = 25\,°C$ $R_L = 100\,\Omega$ max $R_{GK} = 40,000\,\Omega$
Holding current	I_{HX}	—	0.2	1.0	mA	$R_{GK} = 40,000\,\Omega$, $T_J = 25\,°C$
Turn-on time	$t_d + t_r$	—	1.4	—	μs	$I_F = 1$ amp, $T_J = 25\,°C$
Turn-off time	t_0	—	20	—	μs	$i_F = 1$ A, $i_R = 1$ A $dv/dt = 20$ V/μs $R_{GK} = 100\,\Omega$, $T_J = 100\,°C$ (See application notes)

Fig. 10 Typical low-current SCR ratings and characteristics.[7]

SCR is particularly flexible and efficient in regulated power supplies, motor-speed control, and polyphase circuits. The reader is referred to the list of references for further application information on SCR's and related devices.

Transistors

Types and structures During the early years of semiconductor development, it was customary to classify transistors according to the process by which the junctions were formed, and the early transistors were named in this manner (point-contact, surface-barrier, alloy-junction, etc.). Many of these types have become obsolete. Transistors today are made almost exclusively by a diffusion process and are typically

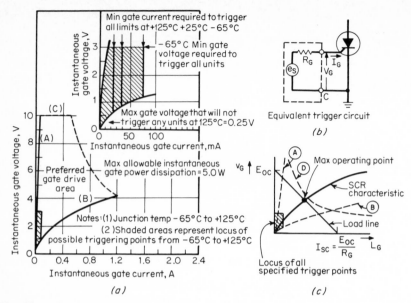

Fig. 11 SCR gate triggering.[10] (*a*) Characteristic for typical 35-A unit. (*b*) Gate trigger unit. (*c*) Load lines on characteristic in part *a*.

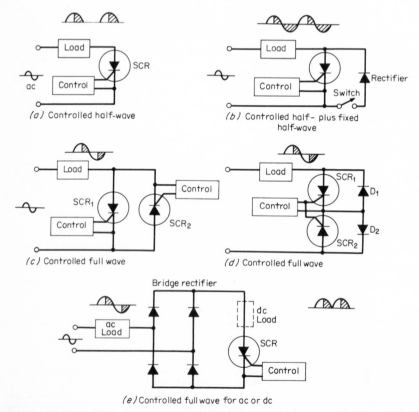

Fig. 12 SCR ac phase control.[10] (*a*) Controlled half-wave. (*b*) Controlled half-plus fixed half-wave. (*c*), (*d*) Controlled full wave. (*e*) Controlled full wave for ac or dc.

classified according to the geometry of their structure (mesa, planar, annular, inter-digitated, etc.). A brief description of the most common types now in use follows.

Mesa transistors (Fig. 13) get their name from the elevated area of the base and emitter region which remains when the semiconductor material is etched away. The chief virtue of this process is improved high-frequency performance.

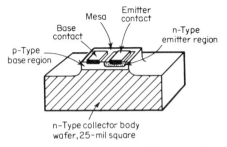

Fig. 13 Mesa-transistor fabrication.[3]

Planar transistors (Fig. 14) are produced by repeated sequences of silicon dioxide (SiO_2) deposition, masking, "cutting" of windows through the oxide, and diffusion of impurity materials into the exposed silicon regions. Unlike the mesa transistor, the edges of whose junctions are exposed at the surface, the planar transistor has junctions protected by a layer of SiO_2, resulting in lower leakage currents, and there-fore improved performance, as well as greater immunity to contamination, and hence improved reliability. This is the basic process used in most integrated circuits as well.

Epitaxial transistors (Fig. 15), usually also planar types, feature a very thin high-resistivity collector region grown directly on low-resistivity material which forms a backing for the wafer. It has the advantages of high resistivity in the junction areas, allowing for high breakdown voltages, and low resistivity in the collector body, resulting in low saturation voltages. This is particularly useful in switching and power transistors.

Geometries used in the fabrication of transistors vary widely, depending on the use for which a particular design is intended; in addition, each manufacturer has his own guidelines and processing methods which influence the designs. Most geometries, however, are variations on two designs. One of these is the annular design, shown in Fig. 16. The rings which surround the base area are low-resistance barriers which prevent base-current leakage under the silicon dioxide layer (so-called channels). Depending on the design objectives, it may be desirable to have a large emitter perimeter, or a large base area, or a large perimeter-to-area ratio. These concerns lead to an evolution of design illustrated in Fig. 17, and to the most com-monly used design: the interdigitated design, so named because emitter and base areas are fingerlike shapes intertwined in each other. Examples of both designs are shown in Fig. 18, which depicts some of the basic shapes found in one manufac-turer's line of transistors.

Data sheets Transistor data sheets generally contain varying amounts of informa-tion, presented in a variety of ways. They always contain a section on maximum ratings and a table of electrical characteristics. Table 4 lists typical ratings and characteristics for a number of transistors, grouped by functional families, and Fig. 19 is an example of the type of information contained in an individual data sheet. This data sheet is representative of a group of transistors, all using the same chip, which is shown in outline form on the data sheet. Also shown are the transistor type numbers and package outline numbers in which this particular ship is available, and some information on structure, processing, and application. Table 15 defines the symbols encountered in transistor terminology. In addition to a listing of char-acteristics, data sheets usually contain sets of curves showing the behavior of certain characteristics for various conditions. Perhaps the most common set of curves is the common-emitter family (see Transistor circuits, configurations). A set of such

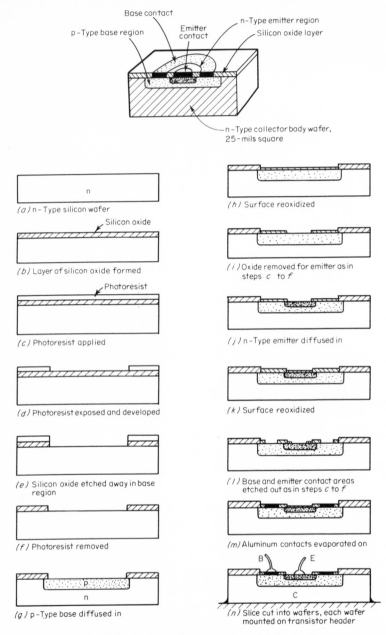

Fig. 14   Steps in planar-transistor fabrication.[3]

curves is shown in Fig. 20. The three regions of transistor operation are defined in the graph:

    1. The clear area is the active region, in which the transistor amplifies. The line *ABC* is a load line, representing a load resistance of approximately 3,000 Ω. (See Applications, Amplifier design guide.)

    2. The cutoff region, at the bottom of the graph, where collector current is

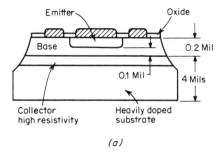

Fig. 15 Epitaxial transistor structures.[23] (*a*) Double-diffused epitaxial transistor structure. (*b*) Double-diffused epitaxial planar transistor structure. (*c*) Epitaxial-base transistor structure.

essentially shut off, only leakage currents flow, and the collector voltage is high, approaching the supply voltage. The transistor is said to be off.

3. The saturation region, to the left of the graph, where conduction is heavy and collector voltage is low. The transistor is said to be turned on or in the on state.

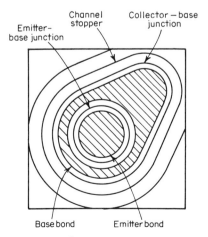

Fig. 16 Annular power-transistor geometry.

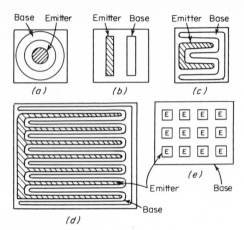

Fig. 17 Basic transistor geometries.[23] (*a*) Ring-dot. (*b*) Line. (*c*) Early inter-digitated. (*d*) Interdigitated. (*e*) Overlay.

The active region defines the operating area for amplification or linear operation, where the object is to reproduce at the output an amplified version of the signal at the input. The other two regions define the switching capabilities or nonlinear applications of a transistor. (See Applications, below.)

Figure 21 is a complete set of characteristic curves, such as are available for modern transistors. (The curves are for the 2N3564, a general-purpose small-signal rf amplifier.)

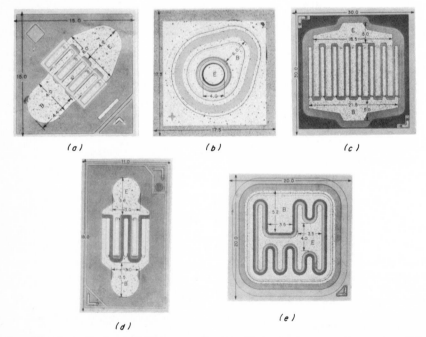

Fig. 18 Typical transistor geometries.[12] (*a*) High-speed saturated switch. (*b*) Low-level, low-noise, high-gain amplifier. (*c*) High-current core driver. (*d*) General-purpose amplifier. (*e*) General-purpose amplifier and switch.

Process 22 NPN Small Signal

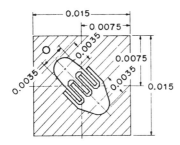

DESCRIPTION

Process 22 is an overlay, double-diffused, gold doped-silicon epitaxial device. Complement to Process 64.

APPLICATION

This device was designed for high-speed logic and core driver applications to 300 mA.

PRINCIPAL DEVICE TYPES:

TO-52	2N3013
TO-92	MPS3646
TO-106	2N3646

PARAMETER	TEST CONDITIONS	MIN	TYP	MAX	UNITS	NOTES
t_s	$I_C = 10$ mA, $I_{B1} = I_{B2} = 10$ mA		12	18	ns	Fig. 1
t_{on}	$I_C = 300$ mA, $I_{B1} = I_{B2} = 30$ mA		10	18	ns	Fig. 2
t_{off}	$I_C = 300$ mA, $I_{B1} = I_{B2} = 30$ mA		18	30	ns	
C_{ob}	$V_{CB} = 5$V		3.2	5.0	pF	TO-18
C_{ob}	$V_{EB} = 0.5$V		6.2	8.0	pF	TO-18
h_{fe}	$I_C = 30$ mA, $V_{CE} = 10$V, $f = 100$ MHz	3.5	7.0	10		
h_{FE}	$V_{CE} = 1$V, 10 mA	20	50	150		
h_{FE}	$V_{CE} = 1$V, $I_C = 30$ mA	20	50	150		
h_{FE}	$V_{CE} = 1$V, $I_C = 100$ mA	20	48	150		
h_{FE}	$V_{CE} = 1$V, $I_C = 300$ mA	15	30	120		
h_{FE}	$V_{CE} = 0.4$V, $I_C = 30$ mA	20	50	150		
h_{FE}	$V_{CE} = 0.5$V, $I_C = 100$ mA	20	50	150		
$V_{CE(SAT)}$	$I_C = 30$ mA, $I_B = 3$ mA		0.14	0.20	V	
$V_{CE(SAT)}$	$I_C = 100$ mA, $I_B = 10$ mA		0.20	0.28	V	
$V_{CE(SAT)}$	$I_C = 300$ mA, $I_B = 30$ mA		0.40	0.50	V	
$V_{BE(SAT)}$	$I_C = 30$ mA, $I_B = 3$ mA		0.80	0.95	V	
$V_{BE(SAT)}$	$I_C = 100$ mA, $I_B = 10$ mA		0.92	1.2	V	
$V_{BE(SAT)}$	$I_C = 300$ mA, $I_B = 30$ mA		1.1	1.7	V	
BV_{CBO}	$I_C = 100$ μA	40	50		V	
BV_{CEO}	$I_C = 10$ mA	15	18		V	
BV_{EBO}	$I_E = 100$ μA	5.0	5.7		V	
I_{CBO}	$V_{CB} = 20$V			50	nA	
I_{EBO}	$V_{EB} = 3$V			50	nA	

Fig. 19 Typical transistor data sheet.[13]

Transistor circuits

Configurations. The transistor is a three-element, three-lead device, and it is therefore possible to use it in any one of three circuit configurations in which two of the leads serve as input and output connections, and the third lead is common to both input and output. They are called common-base, common-emitter, and common-collector circuits. The three configurations are shown in Fig. 22. Although the common-emitter form is the most widely employed of the three forms, each

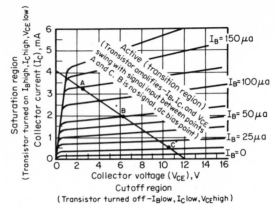

Fig. 20 Typical common-emitter collector curves.[2]

has useful circuit properties which may be uniquely suited to a particular design. Table 5 summarizes the important characteristics of the three basic configurations. The common-emitter configuration is also used almost exclusively in integrated circuits.

Equivalent Circuits. Several approaches have been taken in developing equivalent circuits to represent the transistor as a circuit element, and although there are many variations, there are three basic forms of equivalent circuits:

1. The *generic equivalent circuit.* It looks at the device from the device de-

TABLE 4   Transistor Ratings and Characteristics

General-Purpose, Small-Signal

	Max ratings		h_{FE}		$V_{CE(\text{sat})}$		f_T min, MHz
Package	P_D, mW	V_{CEO}, V	min/max	$I_{C,\text{mA}}$	V_{max}	$I_{C,\text{mA}}$	
TO-18	300	25	20/160	10	0.5	10	70
TO-92	310	50	200/600	0.1	...	...	30
TO-5	800	30	30/120	150	0.4	150	250
TO-46	600	50	70/225	150	0.4	150	200
TO-39	1,000	40	50/250	150	1.4	150	100

Low-Frequency Power

	Max ratings		h_{FE}		$V_{CE(\text{sat})}$		f_T min, MHz
Package	P_D, W	BV_{CEO}, V	min/max	I_C, A	V_{max}	I_C	
TO-5	5	60	30/150	0.25	0.6	1	3
TO-5	10	100	40/150	0.25	0.6	0.25	30
TO-66	40	120	25/200	0.75	1.0	0.75	
TO-3	100	400	20/100	1.0	2.5	2.5	
TO-114	300	80	20/100	20	1.0	20	20

Table 4   Transistor Ratings and Characteristics (Continued)

RF, Small-Signal

| Package | Max ratings | | | Characteristics | | |
| | P_D, mW | V_{CB}, V | f_T min, MHz | Power gain, dB | Noise figure | |
					Typical/max dB	f, MHz
TO-72	200	25	250	16	–/8.0	200
TO-72	200	30	600	6.0	–/6.0	60
TO-72	200	30	1,000	12.5	3.7/4.5	450
TO-72	200	30	1,200	17	–/2.5	450
TO-39	3,500	40	700	11.4	2.7/	200

RF, Power

| Package | Max ratings | | f_T, MHz typical | Characteristics | |
| | P_D, W | V_{CB}, V | | P_{out} | |
				W_{min}	f, MHz
TO-18	1.0	56	350	0.2	100
TO-102	6.0	60	175	3.5	50
TO-60	11.6	65	. . .	10	100
TO-60	20	60	500	30	175
Plastic	80	65	. . .	75	150

signer's viewpoint; i.e., it attempts to describe the transistor in terms of its physical mechanisms and uses circuit elements identifiable with physical attributes of the device.

2. The *T-equivalent circuit*. This circuit still maintains close touch with the physical transistor mechanisms but uses a single current—or voltage generator—to represent the amplifying property of the transistor. It has the advantage of being easy to analyze. The T-equivalent circuits for the three basic transistor circuit configurations are shown in Fig. 23.

3. The *hybrid-parameter equivalent circuit* treats the transistor as a four-terminal linear network (a so-called black box). Thus the transistor is viewed entirely from the circuit point of view and is characterized by what is measured at its terminals. In a generalized circuit (Fig. 24), the parameters used to describe the transistor use numerical subscripts, such as h_{12}, where the first indicates that the element lies in the input mesh and the second signifies that it is caused by current or voltage in

TABLE 5   Characteristics of Basic Transistor Circuits

Characteristic	Common base	Common emitter	Common collector
Power gain	Yes	Highest	Yes
Voltage gain	Yes	Highest	No (less than 1)
Current gain	No (less than 1)	Yes	Yes
Input impedance	Lowest (50 Ω)	Medium (10 kΩ)	Highest (300 kΩ)
Output impedance	Highest (1 MΩ)	Medium (50 kΩ)	Low (300 Ω)
Phase	No inversion	Inversion	No inversion

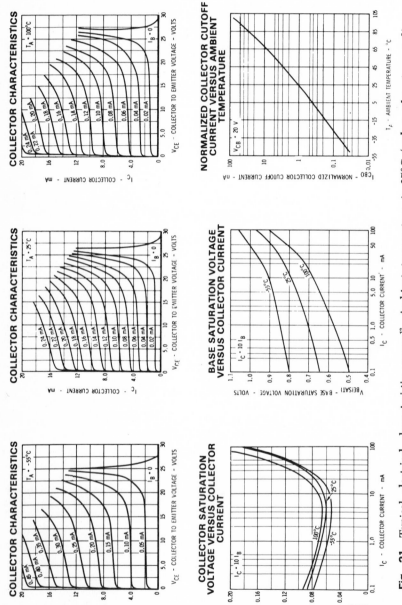

Fig. 21 Typical electrical characteristic curves.[12] Ambient temperature is 25°C (unless otherwise noted).

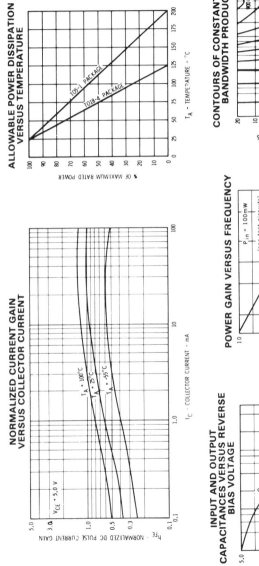

ALLOWABLE POWER DISSIPATION VERSUS TEMPERATURE

NORMALIZED CURRENT GAIN VERSUS COLLECTOR CURRENT

CONTOURS OF CONSTANT GAIN BANDWIDTH PRODUCT

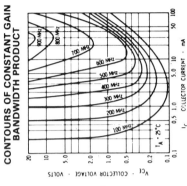

POWER GAIN VERSUS FREQUENCY

INPUT AND OUTPUT CAPACITANCES VERSUS REVERSE BIAS VOLTAGE

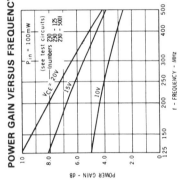

For legend see p. 4-22.

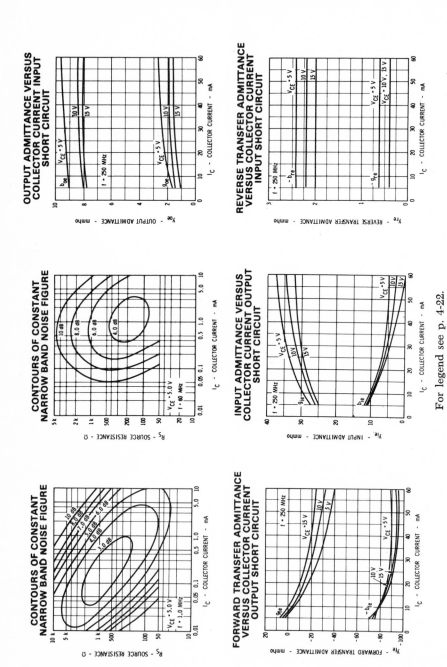

For legend see p. 4-22.

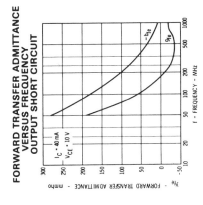

FORWARD TRANSFER ADMITTANCE VERSUS FREQUENCY OUTPUT SHORT CIRCUIT

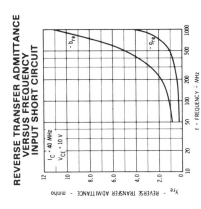

REVERSE TRANSFER ADMITTANCE VERSUS FREQUENCY INPUT SHORT CIRCUIT

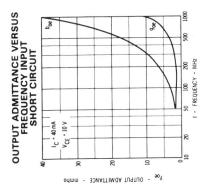

OUTPUT ADMITTANCE VERSUS FREQUENCY INPUT SHORT CIRCUIT

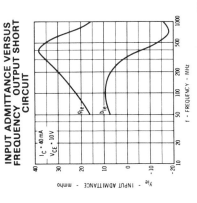

INPUT ADMITTANCE VERSUS FREQUENCY OUTPUT SHORT CIRCUIT

For legend see p. 4-22.

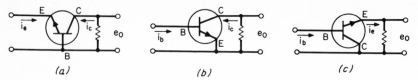

Fig. 22 Transistor circuit configurations. (*a*) Common base. (*b*) Common emitter. (*c*) Common collector.

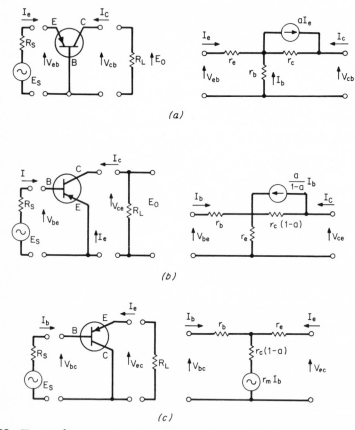

Fig. 23 T-equivalent circuits. (*a*) Common base. (*b*) Common emitter. (*c*) Common collector.

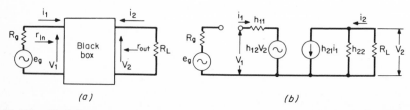

Fig. 24 Black-box representation of transistor circuit. (*a*) Black-box representation. (*b*) Equivalent circuit.

the output mesh. Since these parameters are a mixture of impedances, admittances, and ratios, they are termed *hybrid parameters.* When they are applied to a particular transistor circuit configuration (common-base, etc.), the numerical subscripts are replaced by appropriate letters, as shown in Fig. 25. The complete list, with defini-

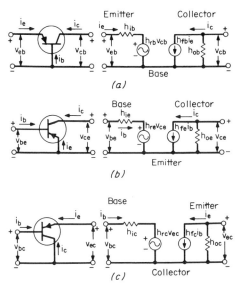

(a)

(b)

(c)

Fig. 25 Hybrid-parameter equivalent circuits. (*a*) Common base. (*b*) Common emitter. (*c*) Common collector.

tions and equivalent names, appears in Table 6. Some of these parameters are widely used in transistor data sheets; for example, h_{FE} and h_{fe} are commonly used to designate dc and small-signal ac gain, respectively, for the common-emitter circuit.

TABLE 6   Hybrid Parameters

Parameter	Definition	Common-base, common-emitter common-collector symbols
h_{11}	Input impedance	h_{ib}, h_{ie}, h_{ic}
h_{21}	Forward-current transfer ratio (forward-current gain)	$h_{fb}(\alpha), h_{fe}(\beta), h_{fc}$
h_{12}	Reverse-voltage transfer ratio (reverse-voltage feedback)	h_{rb}, h_{re}, h_{rc}
h_{22}	Output admittance	h_{ob}, h_{oe}, h_{oc}

Table 7 is a tabulation of conversion formulas to go from one transistor configuration to another or to the T-equivalent circuit, other designations for some of the parameters, and approximate numerical values for a typical small-signal transistor.

Applications Transistor circuits can be broadly classified in three categories:

1. *Linear amplifiers.* The term linear means that the output is a distortionless reproduction of the input signal. Although the transistor is basically a nonlinear device, as evidenced, for example, by the fact that the I_B curves in Fig. 20 are neither parallel nor uniformly spaced, near-linear operation can be achieved by operating over a relatively small region of the curves. Amplifiers are further described as small-signal or large-signal, low- or high-frequency.

TABLE 7   Conversion Formulas: h Parameters and T-equivalent Circuits[2]

IEEE	Other	Common emitter	Common base	Common collector	T-equivalent circuit (approximate)
h_{ie}	h_{11e}, $\dfrac{1}{Y_{11e}}$	1,400 Ω	$\dfrac{h_{ib}}{1+h_{fb}}$	h_{ic}	$r_b + \dfrac{r_e}{1-\alpha}$
h_{re}	h_{12e}, μ_{bc}, μ_{re}	3.37×10^{-4}	$\dfrac{h_{ib}h_{ob}}{1+h_{fb}} - h_{rb}$	$1 - h_{rc}$	$\dfrac{r_e}{(1-\alpha)r_c}$
h_{fe}	h_{21e}, β	44	$-\dfrac{h_{fb}}{1+h_{fb}}$	$-(1+h_{fc})$	$\dfrac{\alpha}{1-\alpha}$
h_{oe}	h_{22e}, $\dfrac{1}{Z_{22e}}$	27×10^{-6} mhos	$\dfrac{h_{ob}}{1+h_{fb}}$	h_{oe}	$\dfrac{1}{(1-\alpha)r_c}$
h_{ib}	h_{11}, $\dfrac{1}{Y_{11}}$	$\dfrac{h_{ie}}{1+h_{ie}}$	31 Ω	$-\dfrac{h_{ic}}{h_{fc}}$	$r_e + (1-\alpha)r_b$
h_{rb}	h_{12}, μ_{ec}, μ_{rb}	$\dfrac{h_{ie}h_{oe}}{1+h_{fe}} - h_{re}$	5×10^{-4}	$h_{rc} - 1 - \dfrac{h_{ic}h_{oc}}{h_{fc}}$	$\dfrac{r_b}{r_c}$
h_{fb}	h_{21}, α	$-\dfrac{h_{fe}}{1+h_{fe}}$	-0.978	$-\dfrac{1+h_{fc}}{h_{fc}}$	$-\alpha$
h_{ob}	h_{22}, $\dfrac{1}{Z_{22}}$	$\dfrac{h_{oe}}{1+h_{fe}}$	0.60×10^{-6} mhos	$-\dfrac{h_{oc}}{h_{fc}}$	$\dfrac{1}{r_c}$
h_{ic}	h_{11c}, $\dfrac{1}{Y_{11c}}$	h_{ie}	$\dfrac{h_{ib}}{1+h_{fb}}$	1,400 Ω	$r_b + \dfrac{r_e}{1-\alpha}$
h_{rc}	h_{12c}, μ_{bc}, μ_{rc}	$1 - h_{re}$	1	1.00	$1 - \dfrac{r_e}{(1-\alpha)r_c}$
h_{fc}	h_{21c}, α_{eb}	$-(1+h_{fe})$	$-\dfrac{1}{1+h_{fb}}$	-45	$-\dfrac{1}{1-\alpha}$
h_{oc}	h_{22c}, $\dfrac{1}{Z_{22c}}$	h_{oe}	$\dfrac{h_{ob}}{1+h_{fb}}$	27×10^{-6} mhos	$\dfrac{1}{(1-\alpha)r_c}$
α		$\dfrac{h_{fe}}{1+h_{fe}}$	$-h_{fb}$	$\dfrac{1+h_{ic}}{h_{fc}}$	0.978
r_c		$\dfrac{1+h_{fe}}{h_{oe}}$	$\dfrac{1-h_{rb}}{h_{ob}}$	$-\dfrac{h_{fc}}{h_{oc}}$	1.67 MΩ
r_e		$\dfrac{h_{re}}{h_{oe}}$	$h_{ib} - \dfrac{h_{rb}}{h_{ob}}(1+h_{fb})$	$\dfrac{1-h_{rc}}{h_{oc}}$	12.5 Ω
r_b		$h_{ie} - \dfrac{h_{re}}{h_{oe}}(1+h_{fe})$	$\dfrac{h_{rb}}{h_{ob}}$	$h_{ic} + \dfrac{h_{fc}}{h_{oc}}(1-h_{rc})$	840 Ω

Numerical values are typical for the 2N525 at 1 mA, 5 V.

2. *Oscillators*, which are self-generating, free-running circuits in which a feedback network is used with a high-gain amplifier to generate ac voltages.

3. *Switching circuits* in which the transistor is operated between the cutoff region (off state) and saturation region (on state).

Small-Signal Amplifiers. A general amplifier design is best illustrated graphically (Fig. 26). A load resistor R_L is chosen, on the basis of circuit requirements and from an approximate location of the operating point (for example, we may wish to operate at a collector-current level at or near maximum gain). The resistor value and a chosen collector supply voltage V_{cc} determine the load line AC (point A is determined by dividing V_{cc} by R_L). The operating point B is then chosen such that the excursions of the input current I_B about it are as nearly symmetric as possible. The performance of the amplifier can now be approximately determined. For the case illustrated, we note that a total swing of 40 μA in base current produces a 9-mA change in output current, so that the current gain is 225. Similarly, a voltage swing of $I_B \times R_i$ produces a voltage change of 4.5 V at the collector, so that the voltage gain is 4.5 V divided by 40 μA $\times$ 1,000 Ω, or 112.5. The power gain may be similarly estimated from the graph. More exact calculations of the performance characteristics of an amplifier may be made by considering the internal transistor parameters. Table 8 summarizes the appropriate relations for the three transistor

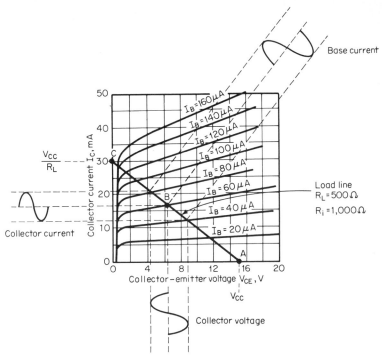

Fig. 26 Graphical amplifier design.

TABLE 8 Approximate Amplifier Relations

	Common base	Common emitter	Common collector
R_i	$r_e + r_b(1-a)$	$r_b + r_e/(1-a)$	$R_L/(1-a)$
R_i range	30–1,000	200–1,500	$10^3 - 5 \times 10^5$
R_o	$r_e - \dfrac{r_b r_m}{r_e + r_b + R_s}$	$r_e(1-a) + \dfrac{r_e r_e}{r_b + r_e + R_s}$	$r_e + (1-a)(r_b + R_s)$
R_o range	10^5–10^6	5,000–100,000	10^2–10^4
Av	$\dfrac{a R_L}{r_s + R_s + r_b(1-a)}$	$\dfrac{-a R_L}{r_e + (1-a)(r_b + R_s)}$	1
A_i	$-a$	$a/(1-a) = b$	$-1/(1-a)$
Power gain	$\dfrac{a^2 R_L}{r_e + R_s + r_b(1-a)}$	$\dfrac{a^2 R_L}{(1-a)r_e + (1-a)^2(r_b + R_s)}$	$1/(1-a)$
Power gain, range, dB	15–30	30–40	10–16

Assumed: $(r_e + r_b) \ll R_L \ll r_e(1-a)$.

configurations.   Note that a simplifying assumption·has been made.   The terminology used here is that of the T-equivalent circuits of Fig. 23.

Biasing Circuits.   One of the basic problems of amplifier design is that of establishing stable operating conditions, or bias conditions (also referred to as the quiescent operating point).   These are the dc-emitter current and collector-voltage conditions which exist in the absence of an input signal.   Factors which enter into the design of bias circuits are:

1. The range of current gain h_{FE} at the operating point.
2. The variation of h_{FE} with temperature.
3. The variation of collector leakage current I_{co} with temperature—this current roughly doubles with a temperature rise of 9 to 11°C.

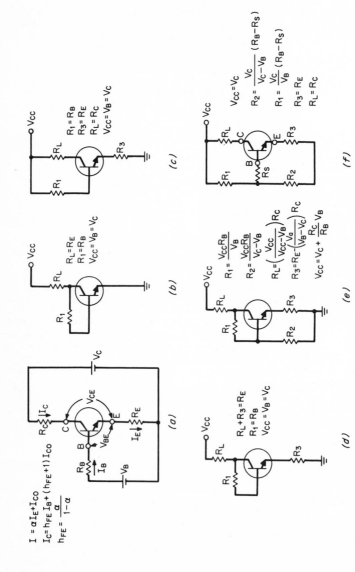

Fig. 27 Development of transistor bias circuits.[2]

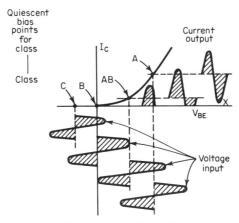

Fig. 28 Transistor bias classes.

Class A: Bias point in linear region 360° (full) conduction
            Low distortion
Class B: Bias point at cutoff 180° conduction
            Used in push-pull operation
Class C: Bias point beyond cutoff
            Less than 180° conduction
            High distortion
            Used primarily in rf work
Class AB: Bias point just beyond cutoff
            Small-signal class A operation and large-signal class B operation

4. The variation of base-emitter voltage drop V_{BE} with temperature. V_{BE} is normally about 0.2 V for germanium transistors and 0.7 V for silicon transistors and has a temperature coefficient of about -2.5 mV per °C.

5. Tolerances of resistors and supply voltages.

Many circuits have been used to provide stable operation. A few of them, and how they may be derived from a general bias circuit, are shown in Fig. 27. Figure 27a is a generalized bias circuit in which the parameters are chosen to satisfy the requirements of the circuit. The supply voltages and resistors for the other bias circuits can be obtained from this circuit, as shown in Fig. 27b through f. The circuits of b, c, and d are progressive forms of self-bias circuits, while e and f are examples of voltage-divider bias networks. Figure 27f is perhaps the most commonly used bias network for small-signal linear amplifiers.

Large-Signal and Power Amplifiers. Large-signal amplifiers are generally power amplifiers. They may be operated single-ended (i.e., a single transistor, or several in parallel are used) or double-ended push-pull (in which two transistors operate back to back). Biasing for these circuits is generally class A (linear operation) for single-ended and some push-pull circuits, or it may be class AB, B, or C for push-pull circuits. The various classes of operation are summarized in Fig. 28. In the design of power amplifiers, it is also necessary to observe limits on allowable power dissipation. The limit can be plotted as a curve on the collector characteristics, as in Fig. 29,

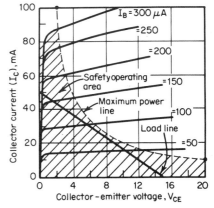

Fig. 29 Safe power-dissipation area.

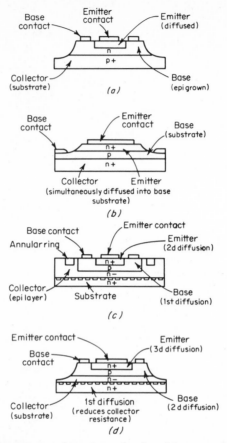

Fig. 30 Four paths to power have been developed by way of four different pathways. These are briefly described here.

The oldest technique puts a simple epitaxially grown base area on a standard collector substrate. This epitaxial base process (*a*) is used for general-purpose power devices ranging in output up to 25 W and intended for a host of amplifier applications, with both *npn* and *pnp* polarities available for complementary design. Most audio amplifier circuits are still served by this process.

Taking the place of the epitaxial-base method for newer high-powered applications are the single- and multidiffused processes.

The single-diffused process (*b*), which is best suited for rugged, low-frequency applications requiring powers up to 50 W, is the mainstay of power designs. Here, collector and emitter areas are diffused simultaneously into the base substrate so that low junction temperatures are maintained at high powers.

For higher-frequency operation at moderate power outputs, a double-diffused device (*c*) is generally fabricated on the basis of techniques developed for integrated circuits. The base is first diffused on a collector epitaxial layer, and then a second emitter diffusion is made. Because the base and emitter require independent process steps, they can be separately doped to make them yield the high currents at high frequencies necessary in uhf and vhf applications.

The newest power-transistor process is a triple-diffused (mesa) system (*d*) that was developed principally for high-voltage applications in TV deflection circuits, automobile ignition, and power-supply switching. The key here is the ability of the process to reduce collector resistance so as to obtain breakdown voltages of up to 1,600 V. Here, a first diffusion lowers the collector substrate resistance, after which the conventional base and emitter diffusions are made.

and is simply the locus of all points for which the product of V_c and I_c equals the maximum power allowable. For many commercial power transistors, the constant-power curve appears on the data sheet. As long as the load line lies below this curve, the maximum power will not be exceeded.

In spite of the rapid growth of the integrated-circuit market, perhaps even because of it, the discrete power transistor is truly a long-term growth area, since IC designers are unlikely to be able to integrate power circuits above 10 W in the audio range and 1 W in the rf range for some time. Major applications are in audio and TV equipment and mobile radio and automotive equipment. The automobile industry alone uses over 100 million power devices per year. Some of the power transistors being developed for these applications are:

2-A, 40-V plastic-encapsulated to 5-A, 300-V metal can for autoignition systems and control devices

1- to 2-A plastic-encapsulated transistors to operate buzzers and interlock systems in cars

3- to 4-A, 100-V complementary pairs for the vertical drive in TV sets

Up to 2,000 V for TV horizontal-deflection circuits.

These demands are being met by new techniques and processes, some of them developed for integrated circuits. Figure 30 illustrates the four major processes used in manufacturing power transistors.

Oscillators are widely used in electronic equipment, as signal generators and clocks. They are generally classified as sine-wave or relaxation oscillators. The basic circuits have been in existence for a long time, having been designed around vacuum tubes, and transistors have simply been adapted to them. A few of the most familiar circuits are shown in Figs. 31 and 32. Any transistor-design handbook may be consulted for more details.

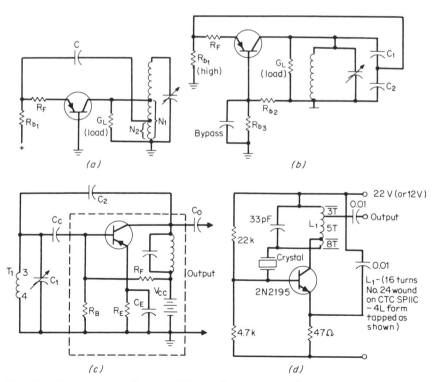

Fig. 31 Sine-wave oscillators. (*a*) Hartley circuit. (*b*) Colpitts circuit. (*c*) Tuned-base collector circuit. (*d*) Crystal oscillator.

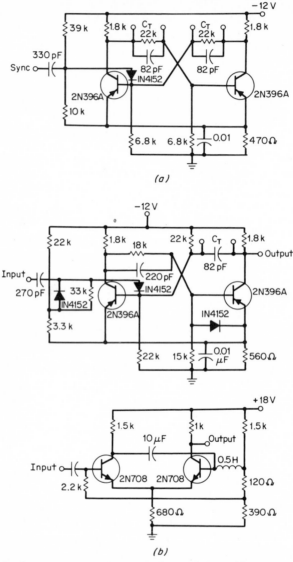

Fig. 32 Multivibrator circuits.   (*a*) Cathode-coupled astable circuit.   (*b*) Monostable circuits.

High-Frequency Limitations.   So far, the transistor has been characterized as a network of linear circuit elements.   As soon as it is operated outside the low-frequency range (audio), the presence of reactive components within the transistor becomes apparent.   Each of the barrier layers separating the areas of the transistor has capacity associated with it, and these capacities must be taken into account.   In Fig. 33 they are shown added to the T-equivalent circuit.   Figure 34 illustrates the hybrid-π high-frequency equivalent circuit, a very popular circuit, since it preserves close ties to the physical structure of the transistor.   It is interesting to note that, as the frequency becomes higher, the capacitive elements become so dominant as to simplify the equivalent circuit, as in Fig. 34*b*.   A useful concept for expressing

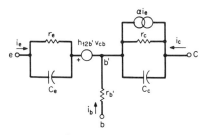

Fig. 33 Low-to medium-frequency T-equivalent circuits.

a transistor's frequency limitation is f_T (or f_t), the frequency at which the common-emitter gain h_{FE} becomes unity (or the power gain becomes zero). f_T is also called the gain-bandwidth product. The quantity is often given in transistor data sheets, or a gain at a specified frequency is given. Knowing this, and the fact that the gain vs. frequency curve falls off at 6 dB per octave, allows us to construct a gain vs. frequency curve, as in Fig. 35.

Switching Circuits. Transistor switching circuits are most often operated in the common-emitter configuration. The signals driving them are usually large enough to cause the transistor to operate as an overdriven large-signal amplifier. Depending on their mode of operation, transistor switching circuits fall into three broad cate-

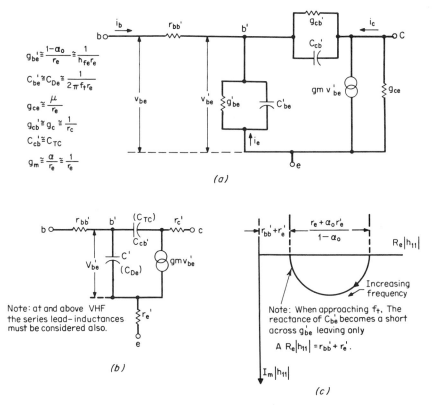

Fig. 34 Hybrid-π equivalent circuit at high frequencies.[2] (*a*) Medium frequencies. (*b*) Simplification at high frequencies. (*c*) Characteristics of the short-circuit input impedance vs. frequency.

gories: *saturated mode, current mode,* and *avalanche mode.* These are determined by the portion of the transistor output-characteristic curve utilized. Curve A in Fig. 36 represents operation in the saturated mode. Here the transistor is driven from cutoff into the saturation region. It is the most commonly used mode, in spite of severe speed limitations because of long storage delay times (see Table 17).

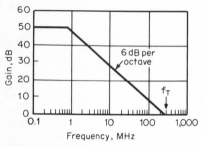

Fig. 35 Gain-frequency curve.

However, it permits switching of large currents with low power dissipation and is therefore most attractive for power-switching applications (see Applications, large-signal and power amplifiers).

Current-mode switching (curve B in Fig. 36) overcomes the problem of storage times by not allowing the transistor to become saturated. If V_{EE} and R_E in the circuit in Fig. 36c are properly chosen, a constant current is produced by V_{EE}, flowing through the diode when Q_1 is off and through the transistor when it is on. If V_{EE}/R_E is chosen to be less than V_{CC}/R_L (the limit of collector current in a saturated circuit), the collector current cannot enter the saturation region. Of course, this mode of operation has the disadvantage of greater power dissipation than saturated switching.

Avalanche-mode switching (curve C in Fig. 36) utilizes the negative-resistance region of the transistor characteristic. The transistor in Fig. 36d is normally held off by a small current I_{BR}. The transition from off to on is initiated by a positive pulse, and a negative pulse is required to return the transistor to the off state. The circuit has the advantage of very high speed operation, but dissipation is very high, particularly if operated in a bistable mode where the transistor can remain indefinitely in the on state. It also suffers from instability problems associated with operation in the negative-resistance region. Primary applications for this mode are pulse generators. Aside from *power* switching circuits already mentioned, the major application of transistor switching is in *digital circuitry.* Broadly classified, digital circuits fall into these categories:

1. Triggered circuits (multivibrators, flip-flops, Schmitt triggers)
2. Gate circuits
3. Separators and isolators (buffers, restorers, impedance changers)

Of these, gate circuits are the largest family, comprising the basic gates (AND, OR, NOR) and the many combinations thereof. Gates are also classified by the type of logic (actually the type of device and/or coupling used), such as RTL (for resistor-transistor logic), DTL (diode-transistor logic), and TTL (transistor-transistor logic). Although all these circuits were originally developed for discrete devices, they have been almost completely taken over into integrated circuits. They are discussed in detail in Chap. 1 and will not be further treated here.

Field-effect transistors Like the power transistor, field-effect transistors seem destined to grow alongside the integrated-circuit field, and in fact they are also growing in usage in integrated form. They are unipolar devices (meaning that charge carriers of only one polarity are used), as contrasted with conventional transistors, which are bipolar (both polarities of charge carriers are used), and they take their name from the fact that current flow in the device is controlled by varying an electric field. There are two types of field-effect transistors (FET's): the *junction field-effect transistor (JFET)*, and the *metal-oxide semiconductor field-effect transistor (MOSFET)*, also called an *insulated-gate field effect transistor (IGFET)*. (We shall henceforth use abbreviations for all of them, and use IGFET in preference to MOSFET.)

Junction Field-Effect Transistor (JFET). In its simplest form the JFET starts out as a bar of doped silicon with contacts (Fig. 37a), behaving like a resistor. The terminal into which current is injected is called the *source;* the other terminal is called the *drain.* In Fig. 37b, *p*-type regions are diffused into the *n*-type substrate, leaving a channel between source and drain. These *p* regions, called *gates,* are

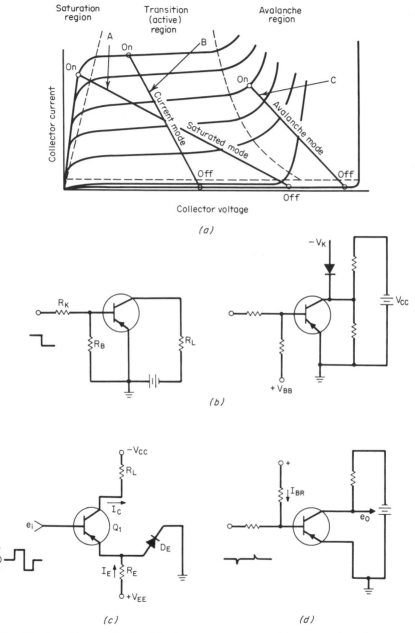

Fig. 36 Switching modes.[15] (a) Operating regions. (b) Saturated-mode circuits. (c) Current-mode circuit. (d) Avalanche-mode circuit.

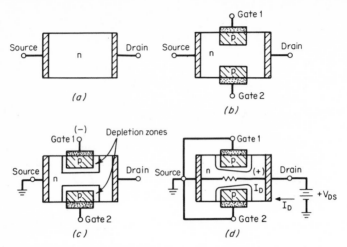

Fig. 37   Development of JFET.[16]

used to control the current between source and drain by changing the width of the depletion regions which surround the junctions (Fig. 37c). The device in Fig. 37 is called an *n*-channel *JFET*. A *p*-channel device is produce by reversing the polarity of the materials. Figure 38a shows the conventional symbols for *p*-channel

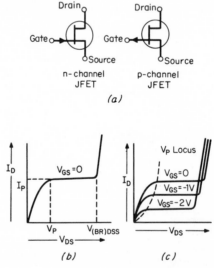

Fig. 38   JFET symbols and characteristics.[16]   (a) JFET symbols.   (b), (c) Drain-current characteristics.

and *n*-channel JFET's. If an external voltage is applied between source and drain (Fig. 37d), with the gate at zero volts, current at first increases nearly linearly (Fig. 38b), until the depletion regions meet at the *pinch-off voltage* V. The current remains constant at this value because the channel resistance increases and causes the channel to saturate. Finally, avalanche breakdown occurs, at $V_{BR(DSS)}$. By making the gate-source voltage negative, channel pinch-off is made to occur at lower

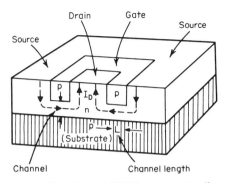

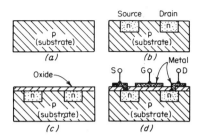

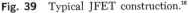

Fig. 39 Typical JFET construction.[16]

Fig. 40 Typical IGFET construction.[16]

values of I_D (Fig. 38c). Normally, a single-ended structure (diffused from one side only) is used, as shown in Fig. 39.

Insulated-gate field-effect transistors (*IGFET*) operate on a slightly different principle. Figure 40 shows the development of an *n*-channel IGFET. It begins with a *p*-type substrate (Fig. 40a), into which two separate *n*-type regions, the source and drain, are diffused (Fig. 40b). An insulating layer of oxide covers the structure, and a protective layer of silicon nitride is placed over that. The gate is a metal area covering the entire channel area (Fig. 40c), and contacts are made. The metal insulating layers and silicon form a capacitor. If a positive voltage is applied to the gate (Fig. 41a), negative charges are "induced" in the semiconductor, increasing until current flows between drain and source through this *induced channel*. This is termed an *enhancement* mode of operation, since the current is enhanced by application of voltage. Alternately, an *n*-type channel can be diffused into the device

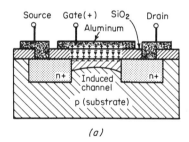

(a)

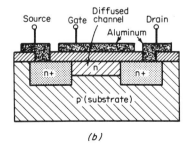

(b)

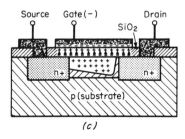

(c)

Fig. 41 IGFET operation.[16] (*a*) Operation of typical IGFET (channel enhancement). (*b*) Typical depletion-mode IGFET construction. (*c*) Operation of typical IGFET (channel depletion).

(Fig. 41*b*), and the current in the channel can be controlled by depleting the charges in the channel with a negative gate potential, as in Fig. 41*c*. This is called a *depletion-mode IGFET*. The symbols for *n*-channel and *p*-channel IGFET's are shown in Fig. 42 (*p*-channel devices are made by reversing all the material polarities above).

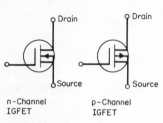

n-Channel
IGFET

p-Channel
IGFET

Fig. 42   IGFET symbols.[16]

Modes of Operation. FET's can be operated in three modes whose characteristics are illustrated in Fig. 43:

1. Depletion only, called type A in Fig. 43, usually IGFET devices
2. Depletion/enhancement, type B, mostly IGFET devices
3. Enhancement only, type C, always IGFET's

Applications. FET's have many advantages over conventional bipolar transistors: low noise, radiation and burnout resistance, high input and output impedances.

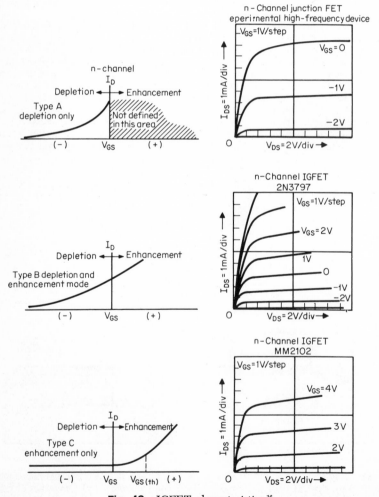

Fig. 43   IGFET characteristics.[16]

These and other characteristics are compared with those of bipolar transistors and tubes in Table 9. Their chief shortcoming is a relatively small gain-bandwidth product, which limits their use at high frequencies. Typical applications include switches and choppers, voltage-controlled oscillators, rf stages in radios, vhf amplifiers in TV circuits, and Zener current sources.

TABLE 9   JFET and IGFET Comparisons

Characteristics	Vacuum tube	JFET	IGFET	Bipolar
Input impedance	High	High	Very high	Low
Noise	Low	Low	Unpredictable	Low
Warm-up time	Long	Short	Short	Short
Size	Large	Small	Small	Small
Power consumption	Large	Small	Small	Small
Aging	Noticeable	Not notice-able	Noticeable	Not notice-able
Bias voltage temp coefficient	Low, not predictable	Low, predictable	High, not predictable	Low, predictable
Typical gate/grid current	1 nA	0.1 nA	10 pA	
Gate/grid current change with temp	High, un-predictable	Medium, predictable	Low, un-predictable	
Reliability	Low	High	High	High
Sensitivity to overload	Very good	Good	Poor	Good

Miscellaneous devices

Multiple Devices. There is a general trend in the discrete-component industry toward multiple devices—dual and quad transistors with various matching of characteristics, diode arrays, Darlington pairs, complementary *npn-pnp* devices. More and more of them are appearing in integrated-circuit-type packages, such as the various dual-in-line packages, making them mechanically compatible with IC's while at the same time taking advantage of cost and space savings.

Diode arrays with up to 16 diodes in a single package have replaced the discrete computer diode, offering a transition step toward complete integration.

Darlington pairs, essentially consisting of two transistors direct-coupled to create a high-gain device, are available in gains of up to 150,000. Power Darlingtons are now also available, with 5 to 10 A available current for such applications as motor drives, converters, servos, and relay drives. Although many *dual* and *quad transistors* are not matched and exist mainly for the sake of cost and space savings, the real attraction of multiple transistors lies in that they can be matched for gain, input current, speed, or almost anything one uses to characterize transistors. The following are examples of matched devices available:

Differential amplifiers, gain matched to 5 percent, V_{BE} matched to 1 mv, and a maximum change in the V_{BE} difference of as low as 0.5 mV over a 100°C temperature range

Complementary pairs offering near-identical characteristics in *npn* and *pnp* transistors plus common packaging, reducing temperature variations between the chips and thus retaining better match over a temperature range

Four transistors on what is essentially a single chip (a continuous area on the silicon slice, not fully separated), usually with some common elements, closely matched in V_{BE}, gain, speed, temperature effects, or combinations of these

Dual JFET's with matched impedances, eliminating the need for external balancing resistors

Chips. Diodes and transistors may be purchased from some manufacturers in chip form. (Terminology varies: chips, or unencapsulated individual devices, are also called dice, and the silicon slice of which they were a part is called a wafer, or simply a slice.) These devices are all of silicon planar construction, with aluminum contact areas and a gold backing. They are usually shipped in one of three ways:

In special containers with cavities to accommodate individual chips
Loosely poured into Freon*-filled vials
Entire wafers, each chip in its original position

Beam-lead sealed-junction devices are a recent development (1966). This structure seals the silicon surface against atmospheric contaminants with a layer of silicon nitride over the planar silicon oxide, thus eliminating the need for a housing or "package" to protect the chip from the external environment. Contacts consist of a multimetal layer onto which gold leads are plated, extending beyond the device periphery (hence the name "beam-lead"), in order to allow direct bonding of these leads or beams to film circuit connectors. Although this technology is used mainly on integrated circuits, some beam-leaded discrete devices are available. They are useful in hybrid-circuit assemblies (mixtures of discrete and integrated devices) and are compatible with thin-film circuits.

Mechanical and Thermal Properties

Encapsulations Semiconductors are encapsulated in a bewildering number of packages—usually referred to as "cases" in manufacturers' catalogs. References 8 and 18 list no fewer than 800 diode and 400 transistor outlines. A number of them are standardized throughout the industry through the process of registering the devices with the Joint Electron Device Engineering Council (JEDEC). A JEDEC-registered device bears a number beginning with 1N, 2N, 3N, and its outline, likewise registered, will be identified as DO-1, DO-2, etc. (for diode outline), or TO-1, TO-2, etc. (for transistor outline). Each manufacturer has numerous additional shapes and sizes, some of which are eventually registered, but many will always be identified by so-called in-house numbers and encapsulated in packages bearing identifications like Case 58 or Outline 23. Figures 44 and 45 reproduce a representative sampling of diode and SCR, and transistor outlines. The outlines shown are the more familiar ones, mainly JEDEC types, but include some new shapes which are likely to become standards.

There are basically two types of packages: *hermetically sealed packages*, which include the familiar metal-canned devices as well as so-called glass diodes, and *plastic packages*, more recent arrivals on the scene.

Hermetically sealed devices are the original form of encapsulation, consisting of metal cans with leads sealed in glass. The environment inside the can is carefully controlled, ranging from a vacuum to "backfilling" with a gas (mostly dry nitrogen in modern devices). They are still widely used, especially where reliability requirements are high and where devices are exposed to high temperature or high humidity.

Plastic packages were made possible with the development of planar technology and surface passivation. Plastic devices are used by the millions in the entertainment-circuit market and have been instrumental in bringing about almost complete "transistorization" of first the portable radio, then music systems, and now TV receivers. The plastic most often used is a form of epoxy, although silicone compounds are also in use. Epoxies place severe temperature limitations on the devices—manufacturers usually place a 150°C limit on such devices—while silicones can withstand temperatures of 300°C. All plastics, however, permit moisture to pass through them, thus exposing devices to the risk of corrosion. (See Reliability, below.) Nevertheless, their advantages of low cost, ease of handling, electrical isolation, and high packing density make them extremely attractive. Plastic packages are also widely used in the integrated-circuit industry.

Thermal considerations The temperature dependence of some properties of semiconductors imposes limitations on them which can be broadly classified as temperature limitations and thermal instabilities.

Temperature and Power Limitations. Manufacturers generally specify a storage-temperature range and a maximum operating temperature. A lower storage-temperature limit, usually in the range of −50 to −75°C, is imposed to prevent mechanical damage which could be caused by strains as a result of different thermal-expansion coefficients of the various materials used in semiconductors. An upper temperature

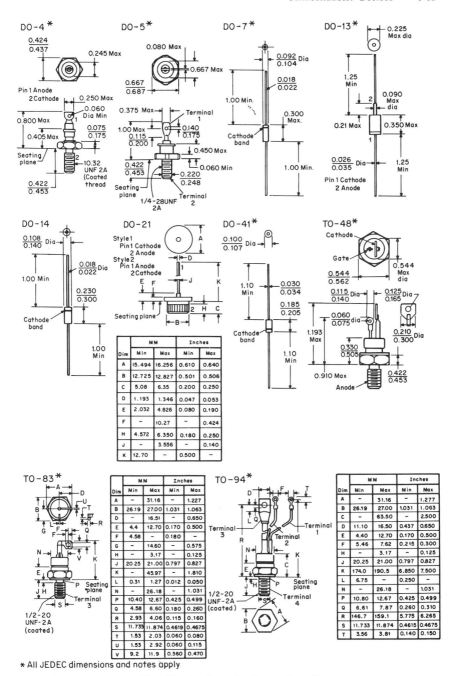

Fig. 44 Diode and SCR outlines.[21]

* All JEDEC dimensions and notes apply

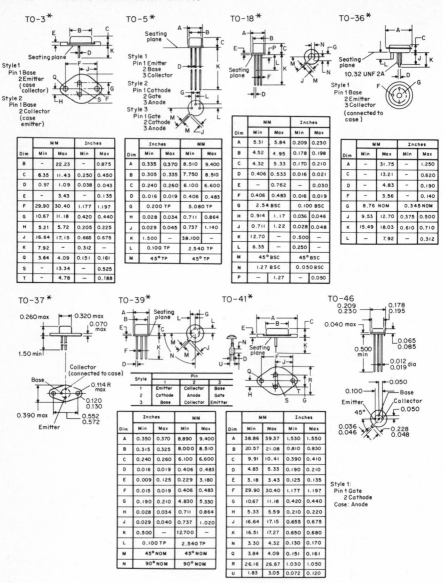

Fig. 45   Transistor outlines.[21]

limit involves two considerations: the softening temperature of solders or metal alloys, and the increased chemical activity of contaminants at higher temperatures. Limits range from as low as 85 to 120°C for germanium transistors to as high as 300°C for planar silicon transistors. The operating-temperature range for a given semiconductor device generally lies within the storage range, as might be expected. Typical ranges may be 0 to 70°C for low-cost devices and −55 to 125°C for military types, and as high as 200°C in some silicon devices. These limits are often less a physical limitation on the operation of the devices than they are a limitation of the temperature range over which the manufacturer guarantees the electrical characteristics in the data sheet. It is important to realize that the maximum operating temperature refers

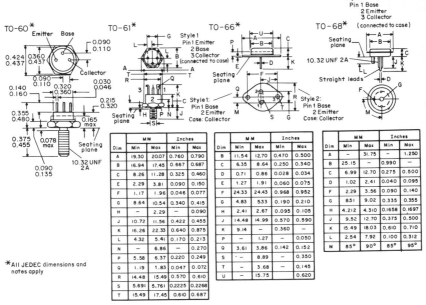

to the temperature in the active region of the transistor, the *junction temperature,* as it is usually called. Manufacturers usually specify a thermal derating factor or present a derating curve such as Fig. 46. Derating factors in terms of thermal impedance, the junction-temperature rise per power dissipated, may be given in one of two forms:

θ_{J-A}: the junction-to-air thermal impedance, to be used when the device is used in free air (no heat removal by an auxiliary means, such as a heat sink). For example, a small transistor may have a θ_{J-A} value of 100°C W^{-1} and, if it has a 125°C operating limit, can dissipate 1 W of power in a 25°C ambient temperature. A power transistor with a θ_{J-A} of 0.5°C W^{-1} could dissipate 200 W under the same circumstances.

θ_{C-J}: the junction-to-case thermal impedance, intended for use with a heat sink which conducts heat away from the transistor case. The curves of Fig. 46 show the differences between a small transistor case (TO-18) and a somewhat larger one (TO-5), as well as between no heat sink (free air) and an infinite heat sink. The term infinite heat sink refers to the limiting situation where the heat sink is so large that the case temperature and ambient temperature are the same. In practice, a few square inches of copper or aluminum sheet may approach the effect of an infinite heat sink for a milliwatt device, whereas for a large power transistor this condition may not be attainable.

Thermal Instabilities. Under certain conditions, the junction temperature and the power dissipation in a transistor both increase without apparent limit until the transistor is destroyed. This condition, referred to as thermal runaway, has several contributing causes:

1. I_{co}, which is a part of I_C, increases exponentially with temperature, at a rate of 10 to 16 percent per °C.

2. If the base current is held constant, I_C increases with temperature because the gain increases with temperature.

3. If V_{BE} is held constant, I_E, and thus I_C, increases with temperature, at a rate of about 8 percent per °C.

In germanium devices, I_{co} is the principal factor involved, because of the large I_{co} in these devices. Silicon transistors have very small values of I_{co}. Thermal

TO-91* TO-92* TO-102* TO-114*

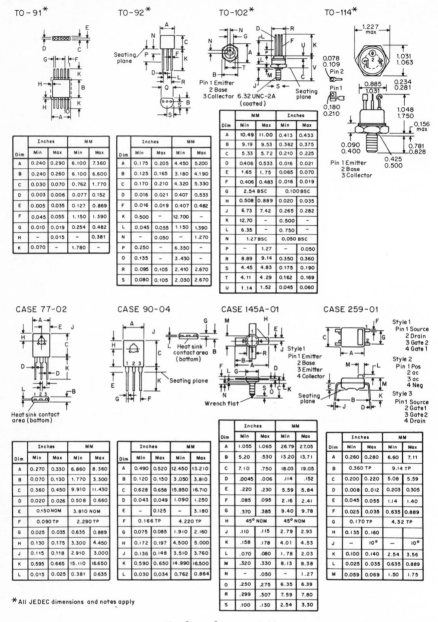

Pin 1 Emitter
2 Base
3 Collector 6.32 UNC-2A (coated)

Pin 1 Emitter
2 Base
3 Collector

TO-91

Dim	Inches Min	Inches Max	MM Min	MM Max
A	0.240	0.290	6.100	7.360
B	0.240	0.260	6.100	6.600
C	0.030	0.070	0.762	1.770
D	0.003	0.006	0.077	0.152
E	0.005	0.035	0.127	0.869
F	0.045	0.055	1.150	1.390
G	0.010	0.019	0.254	0.482
H	–	0.015	–	0.381
K	0.070	–	1.780	–

TO-92

Dim	Inches Min	Inches Max	MM Min	MM Max
A	0.175	0.205	4.450	5.200
B	0.125	0.165	3.180	4.190
C	0.170	0.210	4.320	5.330
D	0.016	0.021	0.407	0.533
E	0.016	0.019	0.407	0.482
K	0.500	–	12.700	–
L	0.045	0.055	1.150	1.390
N	–	0.050	–	1.270
P	0.250	–	6.350	–
O	0.135	–	3.430	–
R	0.095	0.105	2.410	2.670
S	0.080	0.105	2.030	2.670

TO-102

Dim	MM Min	MM Max	Inches Min	Inches Max
A	10.49	11.00	0.413	0.433
B	9.19	9.53	0.362	0.375
C	5.33	5.72	0.210	0.225
D	0.406	0.533	0.016	0.021
E	1.65	1.75	0.065	0.070
F	0.406	0.483	0.016	0.019
G	2.54 BSC		0.100 BSC	
H	0.508	0.889	0.020	0.035
J	6.73	7.42	0.265	0.282
K	12.70	–	0.500	–
L	6.35	–	0.750	–
N	1.27 BSC		0.050 BSC	
P	–	1.27	–	0.050
R	8.89	9.14	0.350	0.360
S	4.45	4.83	0.175	0.190
T	4.11	4.29	0.162	0.169
U	1.14	1.52	0.045	0.060

CASE 77-02 CASE 90-04 CASE 145A-01 CASE 259-01

Heat sink contact area (bottom)

Heat sink contact area (bottom)

Seating plane

Wrench flat

Style 1
Pin 1 Emitter
2 Base
3 Emitter
4 Collector

Seating plane

Style 1
Pin 1 Source
2 Drain
3 Gate 2
4 Gate 1

Style 2
Pin 1 Pos
2 ac
3 ac
4 Neg

Style 3
Pin 1 Source
2 Gate 1
3 Gate 2
4 Drain

CASE 77-02

Dim	Inches Min	Inches Max	MM Min	MM Max
A	0.270	0.330	6.860	8.360
B	0.070	0.130	1.770	3.300
C	0.360	0.450	9.910	11.430
D	0.020	0.026	0.508	0.660
E	0.150 NOM		3.810 NOM	
F	0.090 TP		2.290 TP	
G	0.025	0.035	0.635	0.889
H	0.130	0.175	3.300	4.450
J	0.115	0.118	2.910	3.000
K	0.595	0.665	15.110	16.650
L	0.015	0.025	0.381	0.635

CASE 90-04

Dim	Inches Min	Inches Max	MM Min	MM Max
A	0.490	0.520	12.450	13.210
B	0.120	0.150	3.050	3.810
C	0.628	0.658	15.950	16.710
D	0.043	0.049	1.090	1.250
E	–	0.125	–	3.180
F	0.166 TP		4.220 TP	
G	0.075	0.085	1.910	2.160
H	0.172	0.197	4.500	5.000
J	0.136	0.148	3.510	3.760
K	0.590	0.650	14.990	16.500
L	0.030	0.034	0.762	0.864

CASE 145A-01

Dim	Inches Min	Inches Max	MM Min	MM Max
A	1.055	1.065	26.79	27.05
B	.520	.530	13.20	13.71
C	.710	.750	18.03	19.05
D	.0045	.006	.114	.152
E	.220	.230	5.59	5.84
F	.085	.095	2.16	2.41
G	.370	.385	9.40	9.78
H	45° NOM		45° NOM	
J	.110	.115	2.79	2.93
K	.158	.178	4.01	4.53
L	.070	.080	1.78	2.03
M	.320	.330	8.13	8.38
N	–	.050	–	1.27
O	.250	.275	6.35	6.39
R	.299	.307	7.59	7.80
S	.100	.130	2.54	3.30

CASE 259-01

Dim	Inches Min	Inches Max	MM Min	MM Max
A	0.260	0.280	6.60	7.11
B	0.360 TP		9.14 TP	
C	0.200	0.220	5.08	5.59
D	0.008	0.012	0.203	0.305
E	0.045	0.055	1.14	1.40
F	0.025	0.035	0.635	0.889
G	0.170 TP		4.32 TP	
H	0.135	0.160		
J	–	10°	–	10°
K	0.100	0.140	2.54	3.56
L	0.025	0.035	0.635	0.889
M	0.059	0.069	1.50	1.75

*All JEDEC dimensions and notes apply

For legend see p. 4-44.

runaway is less common in silicon transistors, and when it occurs, it is generally caused by emitter-current runaway in constant V_{BE} situations.

Transistors can also fail because of *internal thermal instabilities* which cause the base current to be crowded into a small area, developing a hot spot. In severe cases, local melting occurs, causing the conductivity of the molten area to increase abruptly, and the collector voltage drops suddenly, as shown in Fig. 47. This mechanism is believed to be one of the causes of a phenomenon called *second breakdown* or secondary breakdown, although there may also be other, nonthermal mechanisms besides hot-spot formation that can cause it. A transistor which undergoes second breakdown usually fails catastrophically (collector-emitter shorts), but even if it does not, its electrical characteristics frequently undergo permanent changes.

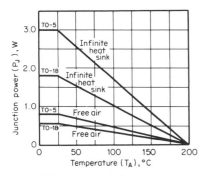

Fig. 46 Transistor derating curve.

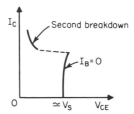

Fig. 47 Second breakdown.[19]

Handling, storage, and installation precautions Semiconductor devices are inherently long-lived, once they are safely installed and have survived an initial period of higher failure rate (see Reliability, p. 4-48). It is safe to say that more semiconductors die before they are installed and during installation and equipment testing than during operating life. Many such failures are caused by improper handling. Following is a listing of the more common abuses encountered.

Mechanical. Dropping of the device and cutting of leads send shock waves through the devices. Some examples of the severity of common shocks[2] (rated in g's, g being the acceleration of gravity):

4½-in drop on hard bench: 500 g
30-in drop onto concrete: 7,000 to 20,000 g
Snapping device into clip: 600 g
Cutting lead with pliers: several thousand g

Such shocks can cause fracture of the semiconductor material and should of course be avoided. Any transistor that has been dropped from working-bench height to a hard floor should be assumed to have been damaged unless proved otherwise by test. Leads should be trimmed with a *shearing tool,* rather than wire cutters.

Lead bends, if repeated, cause work hardening of the lead metal and fracture or breakage of the leads. If leads must be bent, they should be bent once and then left in that shape. Bends should never be made within ⅟₁₆ in of the metal-glass seals on a "header" (the bottom of a hermetically sealed device) to prevent cracks in the glass which would allow moisture and contaminants to enter the encapsulation. This is especially important because failures induced by this process would not be immediate but would result in gradual degradation of the device.

Soldering creates hazards if the maximum junction temperature of the device is exceeded during heating. The danger of doing this is greater for germanium than for silicon devices. Problems can be avoided by following these precautions:

Use a small-wattage iron (20 to 50 W).
Solder quickly and cleanly.

If soldering to a socket, remove the device first.   Heat travels along wires; therefore, also protect neighboring devices.

If soldering directly to a device lead, use a heat shunt (pliers, clips) between heat source and device.

Electrical.   Failures can be caused by excessive voltage, either by exceeding recommended absolute maximum voltages or by transients or other unintentional surges.   Most common among these are:

Inductive or capacitive kicks when removing or inserting a device into a live circuit.

Accidental shorts while working on a live circuit (shorting a device lead to ground or to a high voltage with a screwdriver has caused many "unexplained" failures).

Testing with an ohmmeter.   Many ohmmeters use 22½- and even 45-V batteries.   For example, applying such voltages without current limiting to the emitter-base junction of a transistor in the reverse direction will cause breakdown and will probably destroy the transistor.   A good rule is that ohmmeters are not safe for transistor testing (with the exception of some newer solid-state meters which have very low voltage ohmmeter ranges, as low as 75 mV).

Static-electricity discharges while handling the devices.   This can be particularly damaging to high-frequency transistors, FET's, and rf diodes. It is a good idea to wear a grounding bracelet when handling and installing such devices.

Making or opening connections on a live circuit.   Opening the base of a common-emitter transistor with collector-emitter voltage still applied can cause thermal runaway (second breakdown, Fig. 47).

Design Precautions.   Some dos and don'ts for proper design:

Do not exceed ratings.

Limit maximum junction temperature to stay within recommended limit.

Apply a derating factor for operation above 25°C ambient.

Stabilize emitter current (large resistor or constant-current supply).

Keep base-emitter resistance as low as possible.

Use low source impedance to drive base.

Consider transient voltages, and design to limit or eliminate them.

Reliability

Causes of Semiconductor-Device Failure.   Failure mechanisms in semiconductors can be broadly classified into three categories:

1. Mechanical or workmanship defects
2. Surface defects
3. Bulk defects

Mechanical defects are perhaps the easiest to detect—certainly they are the easiest to analyze.   Among them are:

1. Poor bonds.   Thermocompression bonding is used to attach fine lead wires to contact areas.   This is a critical operation requiring good control, inspection, and testing.

2. Faulty die attachment.   The die or chip is usually attached to the header by soldering.   A poor contact can result in increased thermal impedance and overheating.

3. Use of dissimilar metals in lead and contact area (i.e., gold and aluminum) can result in embrittlement and unwanted compounds.   An example of this is the formation of a gold-aluminum compound when devices using gold wires bonded to aluminum contact areas are heated to 200 to 300°C.   This phenomenon has been called the "purple plague."

4. Nonhermetic seals, allowing moisture and contaminants to enter and cause surface problems or metallization corrosion.

Surface defects are probably the most prevalent cause of poor device reliability. Surface defects may be caused by imperfections of the transistor surface itself, or by external contaminants trapped within the housing or having entered because of an encapsulation failure, or by a combination of both.   Certain stresses to which a device may be subjected can initiate failure mechanisms of this type:

1. Release of gases from internal structures or the can, particularly at high operating temperatures

2. Trapped moisture

3. Leaks in the encapsulation, at time of manufacture or occurring later

Bulk defects are usually defects in the crystal structure of the semiconductor, undesired impurities, and diffusion faults. They are generally detectable in the final electrical testing of devices. However, to the degree that they are not detectable, and also because diffusion goes on, however, slowly, at ordinary operating temperatures, these are thought to contribute to the eventual "wear-out" failure of the device (see Failure Rates below).

Failure Modes. The external manifestations of device failures are referred to as failure modes. In the analysis of semiconductor failures to determine the cause of failure, and in the design of the tests and device-screening methods, a knowledge of how these modes relate to device-design and fabrication methods is essential. Some of the common electrical-failure modes, and their probable origin, are listed below:

1. I_{CBO} is the most sensitive indicator of surface defects. A continous increase in this parameter, especially when also accompanied by decreasing gain h_{FE}, is a good indication of surface contamination.

2. Catastrophically occurring shorts, particularly a collector-emitter short, may indicate hot-spot melting caused by either a bulk problem or faulty circuit use.

3. Open circuits could be an indication of poor bonds or of a lead melted by excessive current.

4. Combinations of an open circuit in one lead and short circuits in the others could be the result of a melted lead and splattered metal from the lead covering the device.

5. Erratic or intermittent open circuits or abrupt shifts in device characteristics, especially at higher temperatures, are usually caused by bonding problems.

Failure Rates. Failures in semiconductor devices are measured as a failure rate, which is the proportion of devices in operation failing in a given unit of time. It is usually expressed as percent per 1,000 h. Thus a failure rate of 1 percent per 1,000 h means that we can expect 1 device out of 100 to fail in a 1,000-h period. A unit called FIT (short for failure unit) has recently come into use. It is defined as 1 failure per 10^9 device-hours, where a device-hour is 1 unit operating for 1 h. A comparison shows that 1 percent per 1,000 h equals 10^4 FIT's. (So-called FIT rates are more manageable for expressing failure rates in highly reliable devices, such as recent integrated circuits: 10 FIT's vs. 0.001 percent per 1,000 h. FIT's are also useful to calculate failure rates for systems using many devices, since they can be directly added.)

Failure rates are not constant with time but are found generally to fit a curve such as that shown in Fig. 48. This curve, sometimes called a bathtub curve, has

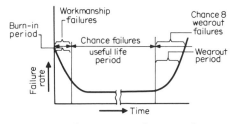

Fig. 48   Failure rate as a function of time.

three distinct portions. The first represents a period of sharply decreasing failure rate. It is variously referred to as burn-in period or early-failure period. Failures during this period are usually caused by manufacturing defects. They are called workmanship failures in Fig. 48. The middle portion of the curve represents the

TABLE 10   Screening Procedures

Test	Typical conditions	Purpose
Preseal visual inspection	100X examination	Workmanship defects
High-temperature bake	200 to 300°C, 16 h	Surface and bond faults, stabilization
Power life defect	T_J to 315°C, 16 h	Bias-related surface defects
Reverse-bias life defect	To 250°C, V dc > 5V, 16 h	Bias-related surface defects
Temperature cycling	−65 to 200°C, 5 cycles	Thermal-expansion strains
Centrifuge	20,000 g, 1 cycle	Internal weak bonds
Shock	2,000 g, 0.5 ms	Structural resonances
Fine leak	Radiflo or He, 10^{-8} atm,cm^3/s	Hermetic-seal leaks
Gross leak	Dye test, 10^{-6} atm,cm^3/s	Hermetic-seal leaks

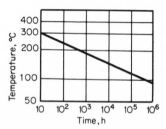

Fig. 49 Temperature acceleration of failures.[20] (*Copyright 1971, Bell Telephone Laboratories, Inc.   Reprinted by permission.*)

TABLE 11   Environmental Tests

Test	Typical conditions	Purpose
Physical dimensions		Assure mechanical fit
Solderability	230°C	Control lead solderability
Thermal shock	0–100°C, 5 cycles	Induce seal strains
Fine-leak test	Radiflo 10^{-8} atm,cm^3/s	Detect leaks induced by previous tests
Shock	2,000 g, 0.5 ms	Control structural strength
Vibration	50 g, 20–2,000 Hz, 1 cycle	Detect intermittent connections
Centrifuge	20,000 g, 1 cycle	Control structural strength
Terminal strength	Tension, bending, fatigue	Control terminal strength
Fine leak	Radiflo or He, 10^{-8} atm,cm^3/s	Hermetic-seal leaks
Gross leak	Dye test, 10^{-6} atm,cm^3/s	Hermetic-seal leaks
Operating life		Control life distribution
Reverse-bias life	To 250°C, V_{dc} > 5 V, 100 h	Control life distribution

TABLE 12   Signal-Diode and Rectifier Symbols[22]

Symbol	Term	Definition
$I_{F(rms)}$, I_f, I_F, $I_{F(av)}$, i_F, I_{FM}	Forward current	The respective value of current that flows through a semiconductor diode or rectifier diode in the forward direction
I_{FRM}	Forward current, repetitive peak	The peak value of the forward current including all repetitive transient currents
I_{FSM}	Forward current, surge peak	The maximum (peak) surge forward current having a specified waveform and a short specified time interval
I_O	Average rectified forward current	The value of the forward current averaged over a full cycle of half-sine-wave operation at 60 Hz with a conduction angle of $180°$
$I_{R(rms)}$, I_r, I_R, $I_{R(av)}$, i_R, I_{RM}	Reverse current	The respective value of current that flows through a semiconductor diode or rectifier diode in the reverse direction
$i_{R(rec)}$, $I_{RM(rec)}$	Reverse recovery current	The transient component of reverse current associated with a change from forward conduction to reverse voltage
I_{RRM}	Reverse current, repetitive peak	The maximum (peak) repetitive instantaneous reverse current
I_{RSM}	Reverse current, surge peak	The maximum (peak) surge reverse current having a specified waveform and a short specified time interval
P_F, $P_{F(av)}$, p_F, P_{FM}	Forward power dissipation	The power dissipation resulting from the flow of the respective forward current
P_R, $P_{R(av)}$, p_R, P_{RM}	Reverse power dissipation	The power dissipation resulting from the flow of the respective reverse current
Q_S	Stored charge	The total amount of charge recovered from a diode minus the capacitive component of that charge when the diode is switched from a specified conductive condition to a specified nonconductive condition with other circuit conditions (as described in EIA-JEDEC Suggested Standard No. 1) optimized to recover the largest possible amount of charge
R_θ	Thermal resistance	See General Glossary*
T_J	Junction temperature	See General Glossary
t_{fr}	Forward recovery time	The time required for the current or voltage to recover to a specified value after instantaneous switching from a stated reverse-voltage condition to a stated forward-current or -voltage condition in a given circuit
t_p	Pulse time	See General Glossary
t_r	Rise time	See General Glossary

* This refers to the glossary at the end of this volume.

TABLE 12   Signal-Diode and Rectifier Symbols[22] (Continued)

Symbol	Term	Definition
t_{rr}	Reverse recovery time	The time required for the current or voltage to recover to a specified value after instantaneous switching from a stated forward-current condition to a stated reverse-voltage or -current condition in a given circuit
t_w	Pulse average time	See General Glossary
$V_{(BR)}, v_{(BR)}$	Breakdown voltage (dc, instantaneous total value)	The value of voltage at which breakdown occurs
$V_{F(\text{rms})}, V_f, V_F,$ $V_{F(\text{av})}, v_F, V_{FM}$	Forward voltage	The voltage drop in a semiconductor diode resulting from the respective forward current
$V_{R(\text{rms})}, V_r, V_R,$ $V_{R(\text{av})}, v_R, V_{RM}$	Reverse voltage	The voltage applied to a semiconductor diode which causes the respective current to flow in the reverse direction
V_{RWM}	Working peak reverse voltage	The maximum instantaneous value of the reverse voltage, excluding all transient voltages, which occurs across a semiconductor rectifier diode
V_{RRM}	Repetitive peak reverse voltage	The maximum instantaneous value of the reverse voltage, including all repetitive transient voltages but excluding all nonrepetitive transient voltages, which occurs across a semiconductor rectifier diode
V_{RSM}	Nonrepetitive peak reverse voltage	The maximum instantaneous value of the reverse voltage including all nonrepetitive transient voltages but excluding all repetitive transient voltages, which occurs across a semiconductor rectifier diode

main-life period of a device and features a lower failure rate, usually assumed to be constant. Failures during this time tend to be more or less random, both in nature and in time of occurrence—chance failures, as they are called in Fig. 48. The final period, a so-called wear-out period, again has an increasing failure rate, presumably continuing until all devices have failed. Failures in this period are thought to be caused by basic limitations of semiconductor materials and design. Under ordinary design and environment conditions, failures of this type should not be experienced in the lifetime of a practical well-designed system.

Screening Techniques. Since the high failure rate during the early life of devices is generally attributable to workmanship defects, it is possible to remove some potential early failures by screening procedures. While it is true that reliability cannot be "tested into a product," it is possible to improve the reliability of a group of devices by removing those potential early failures which escape even the most careful process control. Every reputable manufacturer performs a number of such screening operations as a part of the device-fabrication process. The number of steps, their severity, and whether they are performed on 100 percent of the product, or only on a sample, depend on the intended end use of the device. Naturally, they also bear a direct relationship to device costs.

TABLE 13   Regulator and Reference-Diode Symbols[22]

Terms and Definitions

Term	Definition
Anode	The electrode to which the reverse current flows within the device when it is biased to operate in its breakdown region
Cathode	The electrode from which the reverse current flows within the device when it is biased to operate in its breakdown region
Voltage-reference diode	A diode which is normally biased to operate in the breakdown region of its voltage-current characteristic and which develops across its terminals a reference voltage of specified accuracy, when biased to operate throughout a specified current and temperature range (IEC 147-0, Par. 0-2.3) Graphic symbol for voltage-reference diode (ANSI Y32.2) Reverse current → Cathode — Anode Envelope optional
Voltage-regulator diode	A diode which is normally biased to operate in the breakdown region of its voltage-current characteristic and which develops across its terminals an essentially constant voltage throughout a specified current range (IEC 147-0, Par. 0-2.4) Graphic symbol for voltage-regulator diode (ANSI Y32.2) Reverse current → Cathode — Anode Envelope optional

Each screening test is designed to accelerate some specific failure mechanism, so that a device which would have failed in too short a time in its real-use situation will do so in minutes or hours in the test. The test must also ensure that the useful life of the passing devices is not materially impaired. A brief description of some common screening tests, typical conditions, and the defect to be detected is contained in Table 10. Major manufacturers usually offer a number of processing options on a selected list of components. Among these are devices qualified under military specifications (MIL-S-19500, typically, for discrete semiconductors). Such devices bear the designation JAN. JAN devices are screened, using some form of the steps in Table 10 (except burn-in and preseal visual inspection). JAN TX devices receive a 100 percent burn-in test, and JAN TXV devices are also given visual inspection. In addition, each manufacturer has a high-reliability program for nonmilitary applications, designated by a proprietary name but generally following the steps in Table 10 to varying degrees.

High-Reliability Specifications. The ultimate reliability of a device is of course determined by actual use. Data of this nature are acquired very slowly, and only after more than 10 years of experience is enough information becoming available to support the extrapolations and predictions made earlier. Underlying any reliability test is the assumption that the test accelerates a normal condition and that no additional types of failures are induced. *Acceleration factors* have been developed by several methods, notably matrix testing and step-stress testing. In *matrix testing,* a number of devices are tested under a number of combinations of stress factors such as current, voltage, junction temperature, and ambient temperature. From the analysis of failure data from such tests, acceleration factors are developed which permit the prediction of failure rates at other stress conditions, i.e., the use conditions. *Step-stress testing* subjects devices to successively increasing levels of stress until most of the devices have failed. Methods of analyzing the results have been

TABLE 13   Regulation and Reference-Diode Symbols (Continued)

Letter Symbols, Terms, and Definitions

Symbol	Term	Definition
I_F	Forward current, dc	The value of direct current that flows through the diode in the forward direction
I_R	Reverse current, dc	The value of direct current that flows through the diode in the reverse direction
I_Z, I_{ZK}, I_{ZM}	Regulator current, reference current (dc, dc near breakdown knee, dc maximum-rated current)	The value of dc reverse current that flows through the diode when it is biased to operate in its breakdown region and at a point on its voltage-current characteristic as follows: I_Z: a specified operating point between I_{ZK} and I_{ZM} I_{ZK}: a specified point near the breakdown knee I_{ZM}: a specified point based on the maximum-rated power
T_J	Junction temperature	See General Glossary*
V_F	Forward voltage, dc	The voltage drop in the diode, resulting from the dc forward current
V_R	Reverse voltage, dc	The voltage applied to the diode which causes the direct current to flow in the reverse direction
V_Z, V_{ZM}	Regulator voltage, reference voltage (dc, dc at maximum-rated current)	The value of dc voltage across the diode when it is biased to operate in its breakdown region and at a specified point in its voltage-current characteristic as follows: V_Z: at I_Z V_{ZM}: at I_{ZM}
z_z, z_{zk}, z_{zm}	Regulator impedance, reference impedance, (small-signal, at I_Z, at I_{ZK}, at I_{ZM})	The small-signal impedance of the diode wheniit is biased to operate in its breakdown region and at a specified point in its voltage-current characteristic as follows: z_z: at I_Z z_{zk}: at I_{ZK} z_{zm}: at I_{ZM}

* This refers to the glossary at the end of this volume.

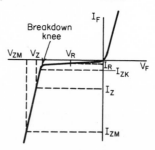

developed to take account of the cumulative effect of the steps.   Step-stress testing has the advantage of fewer devices being required, and relatively short test times.

Acceleration factors are either used directly as such or, in the case of temperature-accelerated failure mechanisms, may take the form of a time-to-failure vs. temperature plot, such as that shown in Fig. 49.   Actually the regression curve of Fig. 49 is a plot of the natural logarithm of time vs. the reciprocal of absolute temperature, based on the so-called Arrhenius relationship, which says that time to failure is a function of

$$e^{-E_a/kT}$$

where E_a = activation energy of failure mechanism being accelerated
k = Boltzmann's constant
T = absolute temperature

TABLE 14   Thyristor Symbols[11]

Symbol	Term	Definition
$I_{(BO)}$ $i_{(BO)}$	Static breakover current Instantaneous breakover current	The principal current at the breakover point
$I_{(BR)R}$ $i_{(BR)R}$	Static reverse breakdown current Instantaneous reverse breakdown current	The principal current at the reverse breakdown voltage
$I_{D(rms)}$ I_D $I_{D(av)}$ i_D I_{DM}	rms off-state current Static off-state current Average off-state current Instantaneous off-state current Peak off-state current	The principal current when the thyristor is in the off state
I_{DRM}	Repetitive peak off-state current	The maximum instantaneous value of the off-state current that results from the application of repetitive peak off-state voltage
I_G $I_{G(av)}$ i_G I_{GM}	Static gate current Average gate current Instantaneous gate current Peak gate current	The current that results from the gate voltage 1. Positive gate current refers to conventional current entering the gate terminal 2. Negative gate current refers to conventional current leaving the gate terminal
I_{GD} i_{GD} I_{GDM}	Static gate nontrigger current Instantaneous gate nontrigger current Peak gate nontrigger current	The maximum gate current which will not cause the thyristor to switch from the off state to the on state
I_{GQ} i_{GQ} I_{GQM}	Static gate turn-off current Instantaneous gate turn-off current Peak gate turn-off current	The minimum gate current required to switch a thyristor from the on state to the off state
I_{GT} i_{GT} I_{GTM}	Static gate trigger current Instantaneous gate trigger current Peak gate trigger current	The minimum gate current required to switch a thyristor from the off state to the on state
I_H i_H	Static holding current Instantaneous holding current	The minimum principal current required to maintain the thyristor in the on state
I_L i_L	Static latching current Instantaneous latching current	The minimum principal current required to maintain the thyristor in the on state immediately after switching from the off state to the on state has occurred and the triggering signal has been removed
$I_{R(rms)}$ I_R $I_{R(av)}$ i_R I_{RM}	rms reverse current Static reverse current Average reverse current Instantaneous reverse current Peak reverse current	The current for negative anode-to-cathode voltage
I_{RRM}	Repetitive peak reverse current	The maximum instantaneous value of the reverse current that results from the application of repetitive peak reverse voltage

TABLE 14   Thyristor Symbols[11] (Continued)

Symbol	Term	Definition
$I_{T(rms)}$ I_T $I_{T(av)}$ i_T I_{TM}	rms on-state current Static on-state current Average on-state current Instantaneous on-state current Peak on-state current	The principal current when the thyristor is in the on state
$I_{T(OV)}$	Overload peak on-state current	An on-state current of substantially the same waveshape as the normal on-state current and having a greater value than the normal on-state current
I_{TRM}	Repetitive peak on-state current	The peak value of the on-state current including all repetitive transient currents
I_{TSM}	Surge (nonrepetitive) peak on-state current	An on-state current of short-time duration and specified waveshape
P_G $P_{G(av)}$ p_G P_{GM}	Static gate power dissipation Average gate power dissipation Instantaneous gate power dissipation Peak gate power dissipation	
T_A	Free-air temperature (ambient temperature)	The air temperature measured below a device, in an environment of substantially uniform temperature, cooled only by natural air convection and not materially affected by reflective and radiant surfaces (MIL-S-19500D, par. 20.20.1)
T_C	Case temperature	The temperature measured at a specified location on the case of a device (MIL-S-19500D, par. 20.20.2)
T_J	Virtual junction temperature (junction temperature)	A theoretical temperature based on a simplified representation of the thermal and electrical behavior of the semiconductor device. This term (and its definition) is taken from IEC standards. It is particularly applicable to multijunction semiconductors and is used to denote the temperature of the active semiconductor element when required in specifications and test methods. The term junction temperature is used interchangeably with the term virtual junction temperature
T_{stg}	Storage temperature	The temperature at which the device, without any power applied, is stored (MIL-S-19500D, par. 20.20.3)
t_{gt}	Gate-controlled turn-off time	The time interval between a specified point at the beginning of the gate pulse and the instant when the principal voltage (current) has dropped (risen) to a specified low (high) value during switching of a thyristor from the off state to the on state by a gate pulse

TABLE 14   Thyristor Symbols[11] (Continued)

Symbol	Term	Definition
t_{gq}	Gate-controlled turn-off time	The time interval between a specified point at the beginning of the gate pulse and the instant when the principal current has decreased to a specified value during switching from the on state to the off state by a gate pulse
t_q	Circuit-commutated turn-off time	The time interval between the instant when the principal current has decreased to zero after external switching of the principal voltage circuit, and the instant when the thyristor is capable of supporting a specified principal voltage without turning on
R_θ $R_{\theta\,JA}$ $R_{\theta JC}$ $R_{\theta\,CA}$	Thermal resistance Thermal resistance, junction-to-ambient Thermal resistance, junction-to-case Thermal resistance, case-to-ambient	The temperature difference between two specified points or regions divided by the power dissipation under conditions of thermal equilibrium
$V_{(BO)}$ $v_{(BO)}$	Static breakover voltage Instantaneous breakover voltage	The principal voltage at the breakover point
$V_{(BR)R}$ $v_{(BR)R}$	Static reverse breakdown voltage Instantaneous reverse breakdown voltage	The value of negative anode-to-cathode voltage at which the differential resistance between the anode and cathode terminals changes from a high value to a substantially lower value
$V_{D(rms)}$ V_D $V_{D(av)}$ v_D V_{DM}	rms off-state voltage Static off-state voltage Average off-state voltage Instantaneous off-state voltage Peak off-state voltage	The principal voltage when the thyristor is in the off state
V_{DRM}	Repetitive peak off-state voltage	The maximum instantaneous value of the off-state voltage which occurs across a thyristor, including all repetitive transient voltages, but excluding all nonrepetitive transient voltages
V_{DSM}	Nonrepetitive peak off-state voltage	The maximum instantaneous value of any nonrepetitive transient off-state voltage which occurs across the thyristor
V_{DWM}	Working peak off-state voltage	The maximum instantaneous value of the off-state voltage which occurs across a thyristor, excluding all repetitive and nonrepetitive transient voltages
V_G $V_{G(av)}$ v_G V_{GM}	Static gate voltage Average gate voltage Instantaneous gate voltage Peak gate voltage	The voltage between a gate terminal and a specified main terminal. Gate voltage polarity is referenced to the specified main terminal

TABLE 14   Thyristor Symbols[11] (Continued)

Symbol	Term	Definition
V_{GD} v_{GD} V_{GDM}	Static gate nontrigger voltage Instantaneous gate nontrigger voltage Peak gate nontrigger voltage	The maximum gate voltage which will not cause the thyristor to switch from the off state to the on state
V_{GQ} v_{GQ} V_{GQM}	Static gate turn-off voltage Instantaneous gate turn-off voltage Peak gate turn-off voltage	The gate voltage required to produce the gate turn-off current
V_{GT} v_{GT} V_{GTM}	Static gate trigger voltage Instantaneous gate trigger voltage Peak gate trigger voltage	The gate voltage required to produce the gate trigger current
$V_{R(rms)}$ V_R $V_{R(av)}$ v_R V_{RM}	rms reverse voltage Static reverse voltage Average reverse voltage Instantaneous reverse voltage Peak reverse voltage	A negative anode-to-cathode voltage
V_{RRM}	Repetitive peak reverse voltage	The maximum instantaneous value of the reverse voltage which occurs across the thyristor, including all repetitive transient voltages, but excluding all nonrepetitive transient voltages
V_{RSM}	Nonrepetitive peak reverse voltage	The maximum instantaneous value of any nonrepetitive transient reverse voltage which occurs across a thyristor
V_{RWM}	Working peak reverse voltage	The maximum instantaneous value of the reverse voltage which occurs across the thyristor, excluding all repetitive and nonrepetitive transient voltages
$V_{T(rms)}$ V_T $V_{T(av)}$ v_T V_{TM}	rms on-state voltage Static on-state voltage Average on-state voltage Instantaneous on-state voltage Peak on-state voltage	The principal voltage when the thyristor is in the on state
$V_{T(min)}$	Static minimum on-state voltage	The minimum positive principal voltage for which the differential resistance is zero with the gate open-circuited
$Z_{\theta(t)}$ $Z_{\theta JA(t)}$ $Z_{\theta JC(t)}$ $Z_{\theta JC(t)}$	Transient thermal impedance Transient thermal impedance, junction-to-ambient Transient thermal impedance, junction-to-case	The change of temperature difference between two specified points or regions at the end of a time interval divided by the step-function change in power dissipation at the beginning of the same time interval causing the change of temperature difference

TABLE 15   Transistor Symbols[22]

Symbol	Term	Definition
C_{cb}, C_{ce}, C_{eb}	Interterminal capacitance (collector-to-base, collector-to-emitter, emitter-to-base)	The direct interterminal capacitance between the terminal indicated by the first subscript and the reference terminal indicated by the second subscript, with the respective junction (collector-base, collector-emitter, emitter-base) reverse-biased and with the remaining terminal (emitter, base, collector) open-circuited to dc, but ac-connected to the guard terminal of a three-terminal bridge. This capacitance includes the interelement capacitances plus capacitance to the shield where the shield is connected to one of the terminals under measurement
C_{ibo}, C_{ieo}	Open-circuit input capacitance (common-base, common-emitter)	The capacitance measured across the input terminals (emitter and base, base and emitter) with the collector open-circuited for ac (IEEE 255)
C_{ibs}, C_{ies}	Short-circuit input capacitance (common-base, common-emitter)	The capacitance measured across the input terminals (emitter and base, base and emitter) with the collector short-circuited to the reference terminal for ac (IEEE 255)
C_{obo}, C_{oeo}	Open-circuit output capacitance (common-base, common-emitter)	The capacitance measured across the output terminals (collector and base, collector and emitter) with the input open-circuited to ac (IEEE 255)
C_{obs}, C_{oes}	Short-circuit output capacitance (common-base) common-emitter)	The capacitance measured across the output terminals (collector and base, collector and emitter) with the third terminal short-circuited to the reference terminal for ac (IEEE 255)
C_{rbs}, C_{res}	Short-circuit reverse transfer capacitance (common-base, common-emitter)	The capacitance measured from the output terminal to the input terminal with the respective reference terminal (base or emitter) and the case (unless connected internally to another terminal) connected to the guard terminal of a three-terminal bridge and with the device biased into the active region
C_{tc}, C_{te}	Depletion-layer capacitance (collector, emitter)	The part of the capacitance across the (collector-base, emitter-base) junction that is associated with its depletion layer. This capacitance is a function of the total potential difference across the depletion layer (IEC 147-0, Par. 11-4.8, 4.9)
$\overline{F}$ or F	Noise figure, average or spot	See General Glossary*
f_{hfb}, f_{hfe}	Small-signal short-circuit forward current transfer ratio cutoff frequency (common-base, common-emitter)	The lowest frequency at which the modulus (magnitude) of the small-signal short-circuit forward current transfer ratio is 0.707 of its value at a specified low frequency (usually 1 kHz or less) (IEEE 255)
f_{max}	Maximum frequency of oscillation	The maximum frequency at which a transistor can be made to oscillate under specified conditions. This approximates the frequency at which the maximum available power gain has decreased to unity (IEC 147-0, Par. 11-4.17)
f_T	Transition frequency or frequency at which small-signal forward current transfer ratio (common-emitter) extrapolates to unity	The product of the modulus (magnitude) of the common-emitter small-signal short-circuit forward current transfer ratio $|h_{fe}|$ and the frequency of measurement when this frequency is sufficiently high so that $|h_{fe}|$ is decreasing with a slope of approximately 6 dB/octave (IEEE 255)

* This refers to the glossary at the end of this volume.

TABLE 15   Transistor Symbols[22] (Continued)

Symbol	Term	Definition
f_1	Frequency of unity current transfer ratio	The frequency at which the modulus (magnitude) of the common-emitter small-signal short-circuit forward current transfer ratio $|h_{fe}|$ has decreased to unity (IEC 147-0, Par. 11-4.19)
G_{PB}, G_{PE}	Large-signal insertion power gain (common-base, common-emitter)	The ratio, usually expressed in dB, of the signal power delivered to the load to the large-signal power delivered to the input
G_{pb}, G_{pe}	Small-signal insertion power gain (common-base, common-emitter)	The ratio, usually expressed in dB, of the signal power delivered to the load to the small-signal power delivered to the input
G_{TB}, G_{TE}	Large-signal transducer power gain (common-base, common-emitter)	The ratio, usually expressed in dB, of the signal power delivered to the load to the maximum large-signal power available from the source
G_{tb}, G_{te}	Small-signal transducer power gain (common-base, common-emitter)	The ratio, usually expressed in dB, of the signal power delivered to the load to the maximum small-signal power available from the source
h_{FB}, h_{FE}	Static forward current transfer ratio (common-base, common-emitter)	The ratio of the dc output current to the dc input current (MIL-S-19500D, Par. 30.28)
h_{fb}, h_{fe}	Small-signal short-circuit forward current transfer ratio (common-base, common-emitter)	The ratio of the ac output current to the small-signal ac input current with the output short-circuited to ac (MIL-S-19500D, Par. 30.20)
h_{ib}, h_{ie}	Small-signal short-circuit input impedance (common-base, common-emitter)	The ratio of the small-signal ac input voltage to the ac input current with the output short-circuited to ac (MIL-S-19500D, Par. 30.24)
$h_{ie(imag)}$ or $Im(h_{ie})$	Imaginary part of the small-signal short-circuit input impedance (common-emitter)	The ratio of the out-of-phase (imaginary) component of the small-signal ac base-emitter voltage to the ac base current with the collector terminal short-circuited to the emitter terminal for ac
$h_{ie(real)}$ or $Re(h_{ie})$	Real part of the small-signal short-circuit input impedance (common-emitter)	The ratio of the in-phase (real) component of the small-signal ac base-emitter voltage to the ac base current with the collector terminal short-circuited to the emitter terminal for ac
h_{ob}, h_{oe}	Small-signal open-circuit output admittance (common-base, common-emitter)	The ratio of the ac output current to the small-signal ac output voltage applied to the output terminal, with the input open-circuited to ac (MIL-S-19500D, Par. 30.15)
$h_{oe(imag)}$ or $Im(h_{oe})$	Imaginary part of the small-signal open-circuit output admittance (common-emitter)	The ratio of the ac collector current to the out-of-phase (imaginary) component of the small-signal collector-emitter voltage with the base terminal open-circuited to ac
$h_{oe(real)}$ or $Re(h_{oe})$	Real part of the small-signal open-circuit output admittance (common-emitter)	The ratio of the ac collector current to the in-phase (real) component of the small-signal collector-emitter voltage with the base terminal open-circuited to ac
h_{rb}, h_{re}	Small-signal open-circuit reverse voltage transfer ratio (common-base, common-emitter)	The ratio of the ac input voltage to the small-signal ac output voltage with the input open-circuited to ac (MIL-S-19500D, Par. 30.18)
I_B, I_C, I_E	Current, dc (base-terminal, collector-terminal, emitter-terminal)	The value of the direct current into the terminal indicated by the subscript
I_b, I_c, I_e	Current, rms value of alternating component (base-terminal, collector-terminal, emitter-terminal)	The root-mean-square value of alternating current into the terminal indicated by the subscript

TABLE 15   Transistor Symbols[22] (Continued)

Symbol	Term	Definition
i_B, i_C, i_E	Current, instantaneous total value (base-terminal, collector-terminal, emitter-terminal)	The instantaneous total value of current into the terminal indicated by the subscript

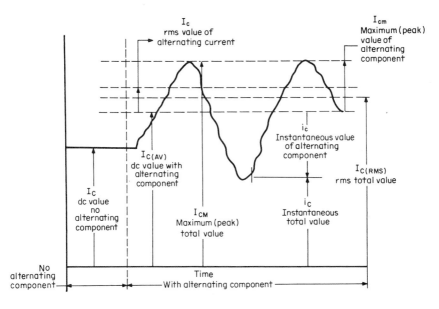

Symbol	Term	Definition
I_{BEV}	Base cutoff current, dc	The direct current into the base terminal when it is biased in the reverse direction with respect to the emitter terminal and there is a specified voltage between the collector and emitter terminals
I_{CBO}	Collector cutoff current, dc, emitter open	The direct current into the collector terminal when it is biased in the reverse direction with respect to the base terminal and the emitter terminal is open-circuited (IEEE 255)
I_{CEO}	Collector cutoff current, dc, with base open	The direct current into the collector terminal when it is biased in the reverse direction* with respect to the emitter terminal and the base terminal is indicated by the last subscript letter as follows (IEEE 255):
I_{CER}	With resistance between base and emitter	
I_{CES}	With base short-circuited to emitter	O = open-circuited
I_{CEV}	With voltage between base and emitter	R = returned to the emitter terminal through a specified resistance
I_{CEX}	With circuit between base and emitter	S = short-circuited to the emitter terminal V = returned to the emitter terminal through a specified voltage X = returned to the emitter terminal through a specified circuit
$I_{E1E2(off)}$	Emitter cutoff current	The current into the emitter-1 terminal of a double-emitter transistor when the emitter-1 terminal is biased with respect to the emitter-2 terminal and the transistor is in the off state (the collector-base diode is not forward-biased) with specified termination of the collector and base terminals

TABLE 15   Transistor Symbols[22] (Continued)

Symbol	Term	Definition
I_{EBO}	Emitter cutoff current, dc, collector open	The direct current into the emitter terminal when it is biased in the reverse direction with respect to the base terminal and the collector terminal is open-circuited (IEEE 255)
$I_{EC(ofs)}$	Emitter-collector offset current	The external short-circuit current between the emitter and collector when the base-collector diode is reverse-biased
I_{ECS}	Emitter cutoff current, dc, base short-circuited to collector	The direct current into the emitter terminal when it is biased in the reverse direction† with respect to the collector terminal and the base terminal is short-circuited to the collector terminal (IEEE 255)
$\text{Im}(y_{ie})$		See preferred symbol $y_{ie(\text{imag})}$
$\text{Im}(y_{oe})$		See preferred symbol $y_{oe(\text{imag})}$
I_n	Noise current, equivalent input	See General Glossary
$\overline{NF}$ or NF‡	Noise figure, average or spot	See General Glossary
P_{IB}, P_{IE}	Large-signal input power (common-base, common-emitter)	The product of the large-signal ac input current and voltage with the common reference terminal circuit configuration
P_{ib}, P_{ie}	Small-signal input power (common-base, common-emitter)	The product of the small-signal ac input current and voltage with the common reference terminal circuit configuration
P_{OB}, P_{OE}	Large-signal output power (common-base, common-emitter)	The product of the large-signal ac output current and voltage with the common reference terminal circuit configuration
P_{ob}, P_{oe}	Small-signal output power (common-base, common-emitter)	The product of the small-signal ac output current and voltage with the common reference terminal circuit configuration
P_T	Total nonreactive power input to all terminals	The sum of the products of the dc input currents and voltages, i.e., $V_{BE}I_B + V_{CE}I_C$ or $V_{BE}I_E + V_{CB}I_C$
$r_b{}'C_c$	Collector-base time constant	The product of the intrinsic base resistance and collector capacitance under specified small-signal conditions
$r_{CE(\text{sat})}$	Saturation resistance, collector-to-emitter	The resistance between the collector and emitter terminals for the saturation conditions specified (IEEE 255)
$\text{Re}(y_{ie})$		See preferred symbol $y_{ie(\text{real})}$
$\text{Re}(y_{oe})$		See preferred symbol $y_{oe(\text{real})}$
$r_{e1e2(\text{on})}$	Small-signal emitter-emitter on-state resistance	The small-signal resistance between the emitter terminals of a double-emitter transistor when the base-collector diode is forward biased
R_θ	Thermal resistance	See General Glossary
s_{fb} or s_{21b}, s_{fe} or s_{21e}	Forward transmission coefficient (common-base, common-emitter)	The respective forward or reverse transmission coefficient with the transistor in the indicated configuration. See General Glossary
s_{rb} or s_{12b}, s_{re} or s_{12e}	Reverse transmission coefficient (common-base, common-emitter)	
s_{ib} or s_{11b}, s_{ie} or s_{11e}	Input reflection coefficient (common-base, common-emitter)	The respective input or output reflection coefficient with the transistor in the indicated configuration. See General Glossary
s_{ob} or s_{22b}, s_{oe} or s_{22e}	Output reflection coefficient (common-base, common-emitter)	
T_J	Junction temperature	See General Glossary
t_d	Delay time	See General Glossary
t_f	Fall time	See General Glossary

TABLE 15   Transistor Symbols[22] (Continued)

Symbol	Term	Definition
t_{off}	Turn-off time	The sum of $t_s + t_f$. See General Glossary
t_{on}	Turn-on time	The sum of $t_d + t_r$. See General Glossary
t_p	Pulse time	See General Glossary
t_r	Rise time	See General Glossary
t_s	Storage time	See General Glossary
t_w	Pulse average time	See General Glossary
$V_{BB}, V_{CC},$ V_{EE}	Supply voltage, dc (base, collector, emitter)	The dc supply voltage applied to a circuit connected to the reference terminal
$V_{BC}, V_{BE},$ $V_{CB}, V_{CE},$ V_{EB}, V_{EC}	Voltage, dc or average (base-to-collector, base-to-emitter, collector-to-base, collector-to-emitter, emitter-to-base, emitter-to-collector)	The dc voltage between the terminal indicated by the first subscript and the reference terminal (stated in terms of the polarity at the terminal indicated by the first subscript)
$v_{bc}, v_{be},$ $v_{cb}, v_{ce},$ v_{eb}, v_{ec}	Voltage, instantaneous value of alternating component (base-to-collector, base-to-emitter, collector-to-base, collector-to-emitter, emitter-to-base, emitter-to-collector)	The instantaneous value of ac voltage between the terminal indicated by the first subscript and the reference terminal
$V_{(BR)CBO}$ (formerly BV_{CBO})	Breakdown voltage, collector-to-base, emitter-open	The breakdown voltage between the collector terminal and the base terminal when the collector terminal is biased in the reverse direction with respect to the base terminal and the emitter terminal is open-circuited (IEEE 255)
$V_{(BR)CEO}$ (formerly BV_{CEO})	Breakdown voltage, collector-to-emitter, with base open	The breakdown voltage between the collector terminal and the emitter terminal when the collector terminal is biased in the reverse direction§ with respect to the emitter terminal and the base terminal is indicated by the last subscript letter as follows (IEEE 255):
$V_{(BR)CER}$ (formerly BV_{CER})	With resistance between base and emitter	
$V_{(BR)CES}$ (formerly BV_{CES})	With base short-circuited to emitter	O = open circuited R = returned to the emitter terminal through a specified resistance
$V_{(BR)CEV}$ (formerly BV_{CEV})	With voltage between base and emitter	S = short-circuited to the emitter terminal V = returned to the emitter terminal through a specified voltage
$V_{(BR)CEX}$ (formerly BV_{CEX})	With circuit between base and emitter	X = returned to the emitter terminal through a specified circuit
$V_{(BR)E1E2}$	Emitter-emitter breakdown voltage	The breakdown voltage between the emitter terminals, of a double-emitter transistor, with specified termination between collector and base
$V_{(BR)EBO}$ (formerly BV_{EBO})	Breakdown voltage, emitter-to-base, collector open	The breakdown voltage between the emitter and base terminals when the emitter terminal is biased in the reverse direction with respect to the base terminal and the collector terminal is open-circuited (IEEE 255)
$V_{(BR)ECO}$ (formerly BV_{ECO})	Breakdown voltage, emitter-to-collector, base open	The breakdown voltage between the emitter and collector terminals when the emitter terminal is biased in the reverse direction¶ with respect to the collector terminal and the base terminal is open-circuited
$V_{CB(fl)},$ $V_{CE(fl)},$ $V_{EB(fl)},$ $V_{EC(fl)}$	dc open-circuit voltage (floating potential) (collector-to-base, collector-to-emitter, emitter-to-base, emitter-to-collector)	The dc open-circuit voltage (floating potential) between the terminal indicated by the first subscript and the reference terminal when the remaining terminal is biased in the reverse direction with respect to the reference terminal (IEEE 255)

TABLE 15   Transistor Symbols[22] (Continued)

Symbol	Term	Definition
V_{CBO}	Collector-to-base voltage, dc, emitter open	The dc voltage between the collector terminal and the base terminal when the emitter terminal is open-circuited
$V_{CE(ofs)}$	Collector-emitter offset voltage	The open-circuit voltage between the collector and emitter terminals when the base-emitter diode is forward-biased
$V_{CE(sat)}$	Saturation voltage, collector-collector-to-emitter	The dc voltage between the collector and the emitter terminals for specified saturation conditions (IEEE 255)
V_{CEO}	Collector-to-emitter voltage, dc, with base open	The dc voltage between the collector terminal and the emitter terminal when the base terminal is indicated by the last subscript letter as follows:
V_{CER}	With resistance between base and emitter	
V_{CES}	With base short-circuited to emitter	O = open-circuited R = returned to the emitter terminal through a specified resistance
V_{CEV}	With voltage between base and emitter	S = short-circuited to the emitter terminal
V_{CEX}	With circuit between base and emitter	V = returned to the emitter terminal through a specified voltage X = returned to the emitter terminal through a specified circuit
V_{EBO}	Emitter-to-base voltage, dc, collector open	The dc voltage between the emitter terminal and the base terminal with the collector terminal open-circuited
$V_{EC(ofs)}$	Emitter-collector offset voltage	The open-circuit voltage between the emitter and collector when the base-collector diode is forward-biased
$|V_{E1E2^-}$ $(ofs)|$	Magnitude of the emitter-emitter offset voltage	The absolute value of the open-circuit voltage between the two emitters of a double-emitter transistor when the base-collector diode is forward-biased
$|\Delta V_{E1E2^-}$ $(ofs)|\Delta|_B$	Magnitude of the change in offset voltage with base current	The absolute value of the algebraic difference between the emitter-emitter offset voltages of a double-emitter transistor at two specified base currents
$|\Delta V_{E1E2^-}$ $(ofs)|\Delta T_A$	Magnitude of the change in offset voltage with temperature	The absolute value of the algebraic difference between the emitter-emitter offset voltages of a double-emitter transistor at two specified ambient temperatures
V_n	Noise voltage, equivalent input	General Glossary
V_{RT}	Reach-through (punch-through) voltage	That value of reverse collector-to-base voltage at which the space-charge region of the collector-base junction extends to the space-charge region of the emitter-base junction (IEEE 255)
y_{fb}, y_{fe}	Small-signal short-circuit forward-transfer admittance (common-base, common-emitter)	The ratio of rms output current to rms input voltage with the output short-circuited to ac
y_{ib}, y_{ie}	Small-signal short-circuit input admittance (common-base, common-emitter)	The ratio of rms input current to rms input voltage with the output short-circuited to ac
$y_{ie(imag)}$ or $Im(y_{ie})$	Imaginary part of the small-signal short-circuit input admittance (common-emitter)	The ratio of rms input current to the rms out-of-phase (imaginary) component of the input voltage with the output short-circuited to ac
$y_{ie(real)}$ or $Re(y_{ie})$	Real part of the small-signal short-circuit input admittance (common-emitter)	The ratio of rms input current to the rms in-phase (real) component of the input voltage with the output short-circuited to ac

TABLE 15 Transistor Symbols[22] (Continued)

Symbol	Term	Definition
y_{ob}, y_{oe}	Small-signal short-circuit output admittance (common-base, common-emitter)	The ratio of rms output current to rms output voltage with the input short-circuited to ac
y_{oe}(imag) or $\mathrm{Im}(y_{oe})$	Imaginary part of the small-signal short-circuit output admittance (common-emitter)	The ratio of rms output current to the out-of-phase (imaginary) component of the rms output voltage with the input short-circuited to ac
y_{oe}(real) or $\mathrm{Re}(y_{oe})$	Real part of the small-signal short-circuit output admittance (common-emitter)	The ratio of rms output current to the in-phase (real) component of the rms output voltage with the input short-circuited to ac
y_{rb}, y_{re}	Small-signal short-circuit reverse transfer admittance (common-base, common-emitter)	The ratio of rms input current ot rms output voltage with the input short-circuited to ac

* For these parameters, the collector terminal is considered to be biased in the reverse direction when it is made positive for *npn* transistors or negative for *pnp* transistors with respect to the emitter terminal.

† For this parameter the emitter terminal is considered to be biased in the reverse direction when it is made positive for *npn* transistors or negative for *pnp* transistors with respect to the collector terminal.

‡ $\overline{NF}$ and NF abbreviations are often used for symbols $\overline{F}$ and F; however, the symbols F and F are preferred.

§ For these parameters, the collector terminal is considered to be biased in the reverse direction when it is made positive for *npn* transistors or negative for *pnp* transistors with respect to the emitter terminal.

¶ For this parameter the emitter terminal is considered to be biased in the reverse direction when it is made positive for *npn* transistors or negative for *pnp* transistors with respect to the collector terminal.

The slope of the curve is a measure of the activation energy, and for the types of defects in silicon devices which are usually accelerated by temperature, it has been found to be constant for a great variety of devices. Other slopes of course apply to such devices as germanium transistors and the newer sealed-junction devices.

A device specification for high reliability, in addition to the screening procedures designed to eliminate weak product, also contains provisions for assuring that the life of the device meets the prediction based on accelerated testing. These provisions take the form of environmental testing, and unlike the screening tests, which are performed on 100 percent of the product, they are generally performed on a sample from each production lot. This is necessary not only because of cost, but because many of these tests are destructive. Table 11 lists the more important types of such tests. In military specifications they are called *group B tests*, a term which is widely used in the industry. (Group A tests are the usual electrical tests performed to guarantee conformance to the data sheet.)

Using these techniques, specifications can be written to control the failure rate of semiconductor devices to a specified maximum for a particular usage temperature and over the required operating life. Experience has shown that failure rates as low as 10 FIT's (0.001 percent per 1,000 h) or less can be attained for silicon transistors, and 1 FIT (0.0001 percent per 1,000 h) for silicon diodes. The reader is referred to the proceedings of two national symposia, the Annual Symposium on Reliability and the Annual Reliability Physics Symposium, which trace the history of the development of device reliability.

TABLE 16 Field-Effect Transistor Symbols[22]

Term	Definition
Channel	A region of semiconductor material in which current flow is influenced by a transverse electrical field. A channel may physically be an inversion layer, a diffused layer, or bulk material. The type of channel is determined by the type of majority carriers during conduction; i.e., p-channel or n-channel
Depletion-mode operation	The operation of a field-effect transistor such that changing the gate-source voltage from zero to a finite value decreases the magnitude of the drain current
Depletion-type field-effect transistor	A field-effect transistor having appreciable channel conductivity for zero gate-source voltage; the channel conductivity may be increased or decreased according to the polarity of the applied gate-source voltage
Drain (D, d)	A region into which majority carriers flow from the channel
Dual-gate field-effect transistor	Alternate term for tetrode field-effect transistor
Enhancement-mode operation	The operation of a field-effect transistor such that changing the gate-source voltage from zero to a finite value increases the magnitude of the drain current
Enhancement-type field-effect transistor	A field-effect transistor having substantially zero channel conductivity for zero gate-source voltage; the channel conductivity may be increased by the application of a gate-source voltage of appropriate polarity
Field-effect transistor	A transistor in which the conduction is due entirely to the flow of majority carriers through a conduction channel controlled by an electric field arising from a voltage applied between the gate and source terminals
Gate (G, g)	The electrode associated with the region in which the electric field due to the control voltage is effective
Insulated-gate field-effect transistor	A field-effect transistor having one or more gate electrodes which are electrically insulated from the channel
Junction (junction-gate) field-effect transistor	A field-effect transistor that uses one or more gate regions that form pn junction(s) with the channel
Metal-oxide-semiconductor (MOS) field-effect transistor	An insulated-gate field-effect transistor in which the insulating layer between each gate electrode and the channel is oxide material
n-channel field-effect transistor	A field-effect transistor that has a n-type conduction channel
p-channel field-effect transistor	A field-effect transistor that has a p-type conduction channel
Source (S, s)	A region from which majority carriers flow into the channel
Substrate (U, u) (of a junction field-effect transistor or an insulated-gate field-effect transistor)	A semiconductor material that contains a channel, a source, and a drain and which may be connected to a terminal
Substrate (of a thin-film field-effect transistor)	An insulating material that supports the thin semiconductor layer, the insulating layer, and the source, gate, and drain electrodes
Tetrode field-effect transistor	A field-effect transistor having two independent gates, a source, and a drain. An active substrate terminated externally and independently of other elements is considered a gate for the purpose of this definition
Triode field-effect transistor	A field-effect transistor having a gate, a source, and a drain

TABLE 16   Field-Effect Transistor Symbols[22] (Continued)

		Junction-gate	Insulated–gate	
		Depletion–type		Enhancement–type
n–channel	Triode	(symbol) G—S D	(symbol) G—S D	(symbol) G—S D
n–channel	Tetrode	(symbol) G1—G2/S D	(symbol) G—S D G2—G1/S D	(symbol) G—S D G2—G1/S D
p–channel	Triode	(symbol) G—S D	(symbol) G—S D	(symbol) G—S D
p–channel	Tetrode	(symbol) G1—G2/S D	(symbol) G—S D G2—G1/S D	(symbol) G—S D G2—G1/S D

Passive substrate

Symbol	Term	Definition
b_{fs}, b_{is}, b_{os}, b_{rs}	Common-source small-signal (forward transfer, input, output, reverse transfer) susceptance	The imaginary part of the corresponding admittance. See y_{fs}, y_{is}, y_{os}, and y_{rs}. Symbols in the forms b_{xx} and $y_{xx(\text{imag})}$ are equivalent
C_{ds}	Drain-source capacitance	The capacitance between the drain and source terminals with the gate terminal connected to the guard terminal of a three-terminal bridge
C_{du}	Drain-substrate capacitance	The capacitance between the drain and substrate terminals with the gate and source terminals connected to the guard terminal of a three-terminal bridge
C_{iss}	Short-circuit input capacitance, common-source	The capacitance between the input terminals (gate and source) with the drain short-circuited to the source for alternating current (IEEE 255)
C_{oss}	Short-circuit output capacitance, common-source	The capacitance between the output terminals (drain and source) with the gate short-circuited to the source for alternating current (IEEE 255)
C_{rss}	Short-circuit reverse transfer capacitance, common-source	The capacitance between the drain and gate terminals with the source connected to the guard terminal of a three-terminal bridge
$\overline{F}$ or F	Noise figure, average or spot	See General Glossary*
g_{fs}, g_{is}, g_{os}, g_{rs}	Common-source small-signal (forward transfer, input, output, reverse transfer) conductance	The real part of the corresponding admittance. See y_{fs}, y_{is}, y_{os}, and y_{rs}. Symbols in the forms g_{xx} and $y_{xx(\text{real})}$ are equivalent

*This refers to the glossary at the end of this volume.

TABLE 16   Field-Effect Transistor Symbols[22] (Continued)

Symbol	Term	Definition
G_{pg}, G_{ps}	Small-signal insertion power gain (common-gate, common-source)	The ratio, usually expressed in dB, of the signal power delivered to the load to the signal power delivered to the input
G_{tg}, G_{ts}	Small-signal transducer power gain (common-gate, common-source)	The ratio, usually expressed in dB, of the signal power delivered to the load to the maximum signal power available from the source
I_D	Drain current, dc	The direct current into the drain terminal
$I_{D(off)}$	Drain cutoff current	The direct current into the drain terminal of a depletion-type transistor with a specified reverse gate-source voltage applied to bias the device to the off state
$I_{D(on)}$	On-state drain current	The direct current into the drain terminal with a specified forward gate-source voltage applied to bias the device to the on state
I_{DSS}	Zero-gate-voltage drain current	The direct current into the drain terminal when the gate-source voltage is zero. This is an on-state current in a depletion-type device, an off-state current in an enhancement-type device
I_G	Gate current, dc	The direct current into the gate terminal
I_{GF}	Forward gate current	The direct current into the gate terminal with a forward gate-source voltage applied. See V_{GSF}
I_{GR}	Reverse gate current	The direct current into the gate terminal with a reverse gate-source voltage applied. See V_{GSR}
I_{GSS}	Reverse gate current, drain short-circuited to source	The direct current into the gate terminal of a junction-gate field-effect transistor when the gate terminal is reverse-biased with respect to the source terminal and the drain terminal is short-circuited to the source terminal
I_{GSSF}	Forward gate current, drain short-circuited to source	The direct current into the gate terminal of an insulated-gate field-effect transistor with a forward gate-source voltage applied and the drain terminal short-circuited to the source terminal. See V_{GSF}
I_{GSSR}	Reverse gate current, drain short-circuited to source	The direct current into the gate terminal of an insulated-gate field-effect transistor with a reverse gate-source voltage applied and the drain terminal short-circuited to the source terminal. See V_{GSR}
I_n	Noise current, equivalent input	See General Glossary
$\text{Im}(y_{fs}), \text{Im}(y_{is}), \text{Im}(y_{os}), \text{Im}(y_{rs})$		See preferred symbols: b_{fs} or $y_{fs(imag)}$, b_{is} or $y_{is(imag)}$, b_{os} or $y_{os(imag)}$, b_{rs} or $y_{rs(imag)}$
I_S	Source current, dc	The direct current into the source terminal
$I_{S(off)}$	Source cutoff current	The direct current into the source terminal of a depletion-type transistor with a specified gate-drain voltage applied to bias the device to the off state
I_{SDS}	Zero-gate-voltage source current	The direct current into the source terminal when the gate-drain voltage is zero. This is an on-state current in a depletion-type device, an off-state current in an enhancement-type device
$\overline{NF}$ or NF†	Noise figure, average or spot	See General Glossary

TABLE 16   Field-Effect Transistor Symbols[22] (Continued)

Symbol	Term	Definition
$r_{ds(on)}$	Small-signal drain-source on-state resistance	The small-signal resistance between the drain and source terminals with a specified gate-source voltage applied to bias the device to the on state. For a depletion-type device, this gate-source voltage may be zero
$r_{DS(on)}$	Static drain-source on-state resistance	The dc resistance between the drain and source terminals with a specified gate-source voltage applied to bias the device to the on state. For a depletion-type device, this gate-source voltage may be zero
$Re(y_{fs})$, $Re(y_{is})$, $Re(y_{os})$, $Re(y_{rs})$		See preferred symbols: g_{fs} or $y_{fs(real)}$, g_{is} or $y_{is(real)}$, g_{os} or $y_{os(real)}$, g_{rs} or $y_{rs(real)}$
R_θ	Thermal resistance	See General Glossary
s_{fg} or s_{21g}, s_{fs} or s_{21s}	Forward transmission coefficient (common-gate, common-source)	The respective forward or reverse transmission coefficient with the transistor in the indicated configuration. See General Glossary
s_{rg} or s_{12g}, s_{rs} or s_{12s}	Reverse transmission coefficient (common-gate, common-source)	
s_{ig} or s_{11g}, s_{is} or s_{11s}	Input reflection coefficient (common-gate, common-source)	The respective input or output reflection coefficient with the transistor in the indicated configuration. See General Glossary
s_{og} or s_{22g}, s_{os} or s_{22s}	Output reflection coefficient (common-gate, common-source)	
T_J	Junction temperature	See General Glossary
$t_{d(off)}$	Turn-off delay time	The time interval from a point 90% of the maximum amplitude on the trailing edge of the input pulse to a point 90% of the maximum amplitude on the trailing edge of the output pulse. This corresponds to storage time for a multijunction transistor. This definition assumes a device initially in the off state with an input pulse applied of proper polarity to switch the device to the on state
$t_{d(on)}$	Turn-on delay time	The time interval from a point 10% of the maximum amplitude on the leading edge of the input pulse to a point 10% of the maximum amplitude on the leading edge of the output pulse. This corresponds to delay time for a multijunction transistor. This definition assumes a device initially in the off state with an input pulse applied of proper polarity to switch the device to the on state
t_f	Fall time	See General Glossary
t_{off}	Turn-off time	The sum of $t_{d(off)} + t_f$
t_{on}	Turn-on time	The sum of $t_{d(on)} + t_r$
t_p	Pulse time	See General Glossary
t_r	Rise time	See General Glossary
t_w	Pulse average time	See General Glossary

TABLE 16   Field-Effect Transistor Symbols[22] (Continued)

Symbol	Term	Definition
$V_{(BR)GSS}$	Gate-source breakdown voltage	The breakdown voltage between the gate and source terminals with the drain terminal short-circuited to the source terminal. The symbol $V_{(BR)GSS}$ is primarily used with junction-gate field-effect transistors. The symbols $V_{(BR)GSSR}$ or $V_{(BR)GSSF}$ should be used with insulated-gate transistors having shunting diodes or similar voltage-limiting devices
$V_{(BR)GSSF}$	Forward gate-source breakdown voltage	The breakdown voltage between the gate and source terminals with a forward gate-source voltage applied and the drain terminal short-circuited to the source terminal. See V_{GSF}
$V_{(BR)GSSR}$	Reverse gate-source breakdown voltage	The breakdown voltage between the gate and source terminals with a reverse gate-source voltage applied and the drain terminal short-circuited to the source terminal. See V_{GSR}
V_{DD}, V_{GG}, V_{SS}	Supply voltage, dc (drain, gate, source)	The dc supply voltage applied to a circuit connected to the reference terminal
V_{DG}	Drain-gate voltage	The dc voltage between the drain and gate terminals
V_{DS}	Drain-source voltage	The dc voltage between the drain and source terminals
$V_{DS(on)}$	Drain-source on-state voltage	The dc voltage between the drain and source terminals with a specified forward gate-source voltage applied to bias the device to the on state
V_{DU}	Drain-substrate voltage	The dc voltage between the drain and substrate terminals
V_{GS}	Gate-source voltage	The dc voltage between the gate and source terminals
V_{GSF}	Forward gate-source voltage	The dc voltage between the gate and source terminals of such polarity that an increase in its magnitude causes the channel resistance to decrease
V_{GSR}	Reverse gate-source voltage	The dc voltage between the gate and source terminals of such polarity that an increase in its magnitude causes the channel resistance to increase
$V_{GS(off)}$	Gate-source cutoff voltage	The reverse gate-source voltage at which the magnitude of the drain current of a depletion-type field-effect transistor has been reduced to a specified low value
$V_{GS(th)}$	Gate-source threshold voltage	The forward gate-source voltage at which the magnitude of the drain current of an enhancement-type field-effect transistor has been increased to a specific low value
V_{GU}	Gate-substrate voltage	The dc voltage between the gate and substrate terminals
V_n	Noise voltage, equivalent input	See General Glossary
V_{SU}	Source-substrate voltage	The dc voltage between the source and substrate terminals

TABLE 16   Field-Effect Transistor Symbols[22] (Continued)

Symbol	Term	Definition
y_{fs}	Common-source small-signal short-circuit forward transfer admittance	The ratio of rms drain current to rms gate-source voltage with the drain terminal ac short-circuited to the source terminal
y_{is}	Common-source small-signal short-circuit input admittance	The ratio of rms gate current to rms gate-source voltage with the drain terminal ac short-circuited to the source terminal
y_{os}	Common-source small-signal short-circuit output admittance	The ratio of rms drain current to rms drain-source voltage with the gate terminal ac short-circuited to the source terminal
y_{rs}	Common-source small-signal short-circuit reverse transfer admittance	The ratio of rms gate current to rms drain-source voltage with the gate terminal ac short-circuited to the source terminal
$y_{fs(imag)}$, $y_{is(imag)}$, $y_{os(imag)}$, $y_{rs(imag)}$	Common-source small-signal (forward transfer, input, output, reverse transfer) susceptance	The imaginary part of the corresponding admittance. See y_{fs}, y_{is}, y_{os}, and y_{rs}. Symbols in the forms $y_{xx(imag)}$ and b_{xx} are equivalent
$y_{fs(real)}$, $y_{is(real)}$, $y_{os(real)}$, $y_{rs(real)}$	Common-source small-signal (forward transfer, input, output, reverse transfer) conductance	The real part of the corresponding admittance. See y_{fs}, y_{is}, y_{os}, and y_{rs}. Symbols in the forms $y_{xx(real)}$ and g_{xx} are equivalent

† $\overline{NF}$ and NF abbreviations are often used for symbols $\overline{F}$ and F; however, the symbols $\overline{F}$ and F are preferred.

REFERENCES

1. Altman, L.: Discrete Semiconductor Devices Proliferate and Prosper, *Electronics,* Apr. 26, 1973.
2. "Transistor Manual," 7th ed., General Electric Co.
3. Hibberd, R. G.: "Solid State Electronics," McGraw-Hill, New York, 1968.
4. Gutzwiller, F. W.: Thyristors and Rectifier Diodes—the Semiconductor Work-horses, *IEEE Spectrum,* August 1967.
5. ITT Product Catalog, 1972–1973.
6. "Silicon Rectifier Handbook," Motorola, Inc., 1966.
7. "Semiconductor Data Handbook," General Electric Co., 1971.
8. "D.A.T.A. Book—Diode and SCR," D.A.T.A., Inc., A Cordura Company, Orange, N.J.
9. "Zener Diode Handbook," Motorola, Inc., 1967.
10. "SCR Handbook," General Electric Co., 1967.
11. "The Power Semiconductor Data Book for Engineers," Texas Instruments, Inc., 1973.
12. "Discrete Products Databook," Fairchild Semiconductors, July 1973.
13. "Transistors," National Semiconductor Corp., March 1973.
14. "Handbook of Basic Transistor Circuits and Measurements," S.E.E.C. Series, vol. 7, Wiley, New York.
15. "Switching Transistor Handbook," Motorola, Inc., 1963.
16. Application Note 211-A, Motorola, Inc.
17. Application Note AN-455, Motorola, Inc.
18. "D.A.T.A. Book—Transistors," D.A.T.A., Inc., A Cordura Company, Orange, N.J.
19. "Characteristics and Limitations of Transistors," S.E.E.C. Series, vol. 4, Wiley, New York.
20. Peck, D. S., and C. H. Zierdt, Jr.: Testing Techniques That Assure Reliable Semiconductor Devices, *Bell Lab. Rec.,* November 1971.
21. "The Semiconductor Data Library—Reference Volume," 3d ed., Motorola Semiconductor Products, Inc.
22. "The Transistor and Diode Data Book for Design Engineers," Texas Instruments, Inc., 1973.
23. "Solid-State Power Circuits," RCA Designer's Handbook, 1971.

Chapter **5**

Components for Electrooptics

E. G. BYLANDER
Texas Instruments, Incorporated, Semiconductor Group,
Dallas, Tex.

SPECIAL DESIGN CONSIDERATIONS

The components described in this chapter differ from the others in this book in several respects. One difference is that the economics of the electrooptical manufacturing industry can affect the designer's choices. Also demand is low, which affects volume, price, and competition. Probably the production rate for a popular EO component will not exceed, in a year, a month's production of a popular transistor type. Important classes of EO devices are built at a rate of ones to tens per year.

EO devices are high-technology types. Frequently from efficiency or noise considerations, designers will demand that an EO device operate near theoretical limits. Most EO devices incorporate a semiconductor, which is probably not as well understood or as economical as silicon or germanium.

As in all devices which have small volume and high technical cost, EO device specifications tend to be inflexible.

Because EO devices have limited application, designers and manufacturers may be relatively unfamiliar with detailed interface requirements.

A few examples are:

1. Infrared-system designs are sensitive to detector cutoff wavelengths by as little as ± 0.1 μm. Cutoff values for some types such as Ge-Hg detectors may vary in an unknown fashion by as much as $\pm\frac{1}{2}$ μm from data-sheet values.

2. An infrared-detector cold-filtering design may not be reliable because of Johnson noise, dead areas, shunts, or housing radiation leaks. Few design rules are available to the components designer to aid design realization.

3. Thick, fiber-optic vidicon faceplates may be used for coupling to other fiber-optics devices. They must be lapped after vidicon assembly because of vacuum distortion. Common optical polishing equipment would lap the faceplate with the vidicon vertical. Vertical positioning of a vidicon is not recommended, since loose particles may blemish the photocathode. Common solutions are either horizontal hand lapping or vertical polishing and acceptance of some blemishes.

4. Image converters must be potted, since unpotted tubes arc in use. There appear to be few design rules for potting.

5. Many types of infrared detectors cannot be used in a real-time system which has a low-frequency limit below 3 Hz because spike transients, of unknown origin, cause large level shifts. Common testing techniques do not reveal these pulses.

6. Many camera and display devices have large off-axis astigmatism and shading effects, which are not noticeable to the eye but reduce dynamic range and resolution. These effects complicate the interfacing with other systems. Many suppliers are not routinely prepared to accept control specifications for each parameter.

Such factors as those described above influence an EO systems design in this way. Usually, by the time a design is complete, several worker-years of effort are committed. However, the definition and solution of a problem area such as those illustrated above, will usually require expenditure of between $\frac{1}{2}$ and 6 worker-years of additional engineering. For the reasons of component inflexibility, changes may involve extensive conventional-circuit redesign.

Problems of designing with EO components may be minimized in the usual way through comprehensive specifications and a vendor acceptance-test plan. Some dos and don'ts are:

1. Do not overspecify.

2. Do not leave out second-order requirements such as overload, overload recovery, and crosstalk.

3. Do leave space for interface changes.

4. Do include off-axis, blemish, shading, and uniformity requirements.

5. Do plan to be creative in defining components measurements as related to systems requirements.

6. Do plan to make the components supplier a member of the working team.

7. Do expect a more restricted environmental range than usual for other components.

8. Do plan to go beyond normal design practice to accommodate production variations of EO devices.

The last point is particularly important when one parameter is specified to be near its optimum or theoretical value; it may be possible to relax other specifications at cost savings which more than offset the costs of increased circuit complexity.

While the considerations of this chapter can be a start toward the definition of a particular component, many aspects of the application must be determined by modeling or simulation testing.

PHYSICAL PROPERTIES OF RADIATION

Electromagnetic Spectrum and Light

Optics is the study of light. Traditionally light is considered to comprise the visible, ultraviolet, and infrared portions of the electromagnetic spectrum. Figure 1 shows the electromagnetic spectrum with the optical portion expanded. Photon wavelength λ and frequency ν have the relation

$$\nu = \frac{c}{\lambda}$$

where c is the speed of light (Fig. 2 and Table 1). Wavenumber $\sigma = 1/\lambda$. Pho-

TABLE 1   Some Physical Constants[20]

Speed of light in free space c	$2.997924462 \times 10^8 \text{ m s}^{-1}$
	($= 0.9835712 \text{ ft ns}^{-1}$)
Elementary charge e	$1.60210 \times 10^{-19} \text{ C}$
Planck's constant h	$6.6256 \times 10^{-34} \text{ J s}$
(energy per hertz of a photon)	
Boltzmann's constant k	$1.38054 \times 10^{-23} \text{ J K}^{-1}$
Stefan-Boltzmann constant σ	$5.6697 \times 10^{-8} \text{ W m}^{-2} \text{ K}^{-4}$

Some Derived Constants

$$2\pi c^2 h = 3.7413 \times 10^8 \text{ W } \mu\text{m}^4 \text{ m}^{-2}$$
$$2c^2 h = 1.1909 \times 10^8 \text{ W } \mu\text{m}^4 \text{ m}^{-2} \text{ sr}^{-1}$$
$$2c = 5.9958 \times 10^{26} \text{ photons s}^{-1} \mu\text{m}^3 \text{ m}^{-2} \text{ sr}^{-1}$$
$$2h/c^2 = 1.4744 \times 10^{-50} \text{ W Hz}^{-4} \text{ m}^{-2} \text{ sr}^{-1}$$
$$ch/k = 1.4388 \times 10^4 \text{ } \mu\text{m K}$$
$$\lambda_m T = 2,898 \text{ } \mu\text{m K}$$
$$\lambda'_m T = 3,670 \text{ } \mu\text{m K}$$
$$\sigma = \frac{2\pi^5 \text{ K}^4}{15 \text{ } c^2 h^3} = 1.4338 \times 10^4 \text{ } \mu\text{m K}$$

ton energy E is

$$E = \frac{hc}{\lambda}$$

where h is Planck's constant. The value of hc is very nearly 1.24 eV/μm (1.2398). Electrooptical components are those optical components which have electrical inputs or outputs.

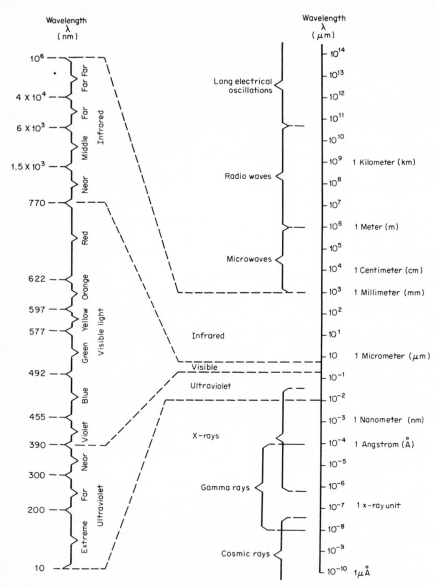

Fig. 1 The electromagnetic spectrum with the electrooptical portion expanded.[59]

Radiometry and Photometry

Radiometry is the science of radiation-source measurement. It is based on the radiometer, whose design is based in turn on Kirchhoff's law. Photometry is concerned with the apparent brightness to a specified receiver, usually the eye, of a source. Present practice is to carry out all calculations, where possible, with the use of radiometric units and convert the final form to physiological estimates where required. For sensors other than the eye which are specified in photometric units (based on a 2870°C or 2845K tungsten source) such as amperes per lumen or nanoamperes per footcandle, it is preferable to convert the response to radiometric units initially.

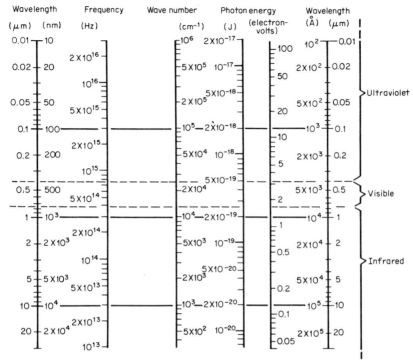

Fig. 2   Radiation conversion chart.[60]

Table 2 lists the symbols, names, and units of commonly used radiometric units. This table is based on the recommendations of the Working Group on Infrared Backgrounds.[1]   The units are consistent with ANSI Y10.19-1969.   Other names and symbols are in use,[2] but their correspondence with those in Table 2 is usually apparent. While mks units are preferred for consistency, many results are commonly reported in mixed cgs-mks units.

Some radiometric properties of materials are:

$$\text{Absorptance } \alpha = \frac{\text{radiation absorbed (at a single reflectance)}}{\text{radiation incident}}$$

$$\text{Reflectance } \rho = \frac{\text{radiation reflected}}{\text{radiation incident}}$$

$$\text{Transmittance } \tau = \frac{\text{radiation transmitted (unscattered)}}{\text{radiation incident}}$$

$$\text{Emittance } \epsilon = \frac{\text{radiation emitted}}{\text{radiation emitted by a blackbody}}$$

These constants typically vary with wavelength, and their local values, such as α_λ, are known as the spectral values.

By conservation of energy, a body with only radiation-energy gains or losses (radiative equilibrium) and whose temperature is constant with time has the property that

$$\alpha + \tau + \rho = 1$$

For an opaque material, $\tau = 0$ and

$$\alpha + \rho = 1$$

TABLE 2   Radiometric Symbols and Units[20]

Symbol	Name	Description	Units
S, S_p	Area	Area, projected area	m²
Ω	Solid angle	. . . .	sr
ν	Frequency	. . . .	Hz
λ	Wavelength	. . . .	μm
E	Photon energy	. . . .	eV
U	Radiant energy	. . . .	J
P	Radiant power	Rate of transfer of radiant energy $\dfrac{\partial U}{\partial t}$	W
W	Radiant emittance	Radiant power per unit area emitted from a surface $\dfrac{\partial P}{\partial S}$	W m⁻²
H	Irradiance	Radiant power per unit area incident upon a surface $\dfrac{\partial P}{\partial S}$	W m⁻²
J	Radiant intensity	Radiant power per unit solid angle from a point source $\dfrac{\partial P}{\partial \Omega}$	W sr⁻¹
N	Radiance*	Radiant power per unit solid angle per unit projected area $\dfrac{\partial^2 P}{\partial \Omega\, \partial S_p}$	W m⁻² sr⁻¹
P_λ	Spectral radiant power	Radiant power per unit wavelength interval $\dfrac{\partial P}{\partial \lambda}$	W μm⁻¹
P_ν	Spectral radiant power	Radiant power per unit frequency interval $\dfrac{\partial P}{\partial \nu}$	W s or W Hz⁻¹
W_λ	Spectral radiant emittance	Radiant emittance per unit wavelength interval $\dfrac{\partial W}{\partial \lambda}$	W m⁻² μm⁻¹
H_λ	Spectral irradiance	Irradiance per unit wavelength interval $\dfrac{\partial H}{\partial \lambda}$	W m⁻² μm⁻¹
J_λ	Spectral radiant intensity	Radiant intensity per unit wavelength interval $\dfrac{\partial J}{\partial \lambda}$	W sr⁻¹ μm⁻¹
N_λ	Spectral radiance	Radiance per unit wavelength interval $\dfrac{\partial N}{\partial \lambda}$	W m⁻² sr⁻¹ μm⁻¹
n_ν	Quantum flux	Photons per unit area emitted from a surface $\dfrac{\nu}{n_\nu}$	Photon m⁻² s⁻¹

* The above definition of radiance is used throughout this handbook. Some authors define radiance in terms of actual surface area rather than projected area; hence they include a factor cos θ in their formulas, θ being the angle from normal incidence.

The Nomenclature Committee of the Optical Society of America has recommended that the flux density radiated from a source be called radiant exitance; the ratio of such flux from a sample to that of a blackbody be called emittance; and that of an opaque sample with perfect surfaces be called emissivity.

A diffuse or Lambertian surface has constant radiance in all directions about normals to the surface and

$$N = \frac{W}{\pi}$$

Opposed to a diffuse surface is a specular surface (mirror) for which the angle of incidence equals the angle of exitance. Of course, if $\rho = 1$, $\alpha = \tau = 0$.

A blackbody is defined to be a surface in radiative equilibrium, and where $\epsilon = 1$. One form of Kirchhoff's law states that the radiation from a surface in radiative equilibrium w_λ as compared with that from a blackbody is proportional to its absorption α_λ.

$$w_\lambda = \alpha_\lambda w_{BB\lambda}$$

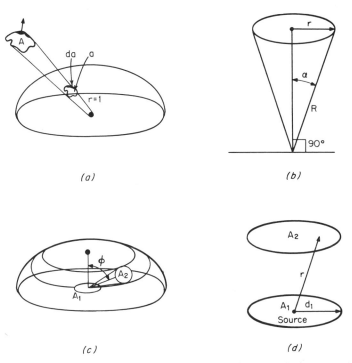

(a)

(b)

(c)

(d)

Fig. 3 Definition of solid angle and representative radiation calculations. (a) The solid angle Ω is the projected area a on unit sphere and is measured in steradians. The solid angle may be calculated by integrating a differential solid angle which is related to that differential area dA by $d\Omega = dA\,(\vec{r}\cdot\vec{n}/r^3)$ where $\vec{n}$ is the normal to dA. (b) In terms of the half-angle α, $\Omega = 2\pi\,(1 - \cos 2\alpha) = 4\pi \sin 2\alpha$. (c) A small areal or point source radiates J W sr⁻¹ through area A_2 at an angle θ. The flux density H through (or at) A_2 will be $H = (J/r_1^2)\cos\theta$ W m⁻², and the power P at H_2 is $P = HA_2$ W. Other quantities may be derived in a like manner. Note that N is defined so that $W = \pi N$ for an areal source, not $w = 2\pi N$. Thus if a power is calculated for an area A, where $r_1 \gg A_1$, the $J = NA_1 = (W/H)A_1$ radiated into 2π steradians (sr). (d) If, for an area source A_1 of diameter d_1 where $d_1 = 2/\pi\,\sqrt{A_1}$ is the order of r, then

$$H = \frac{JA_1}{r^2 + (d_1/2)^2} \approx \frac{JA_1}{(d_1/2)^2}$$

Substituting A_1, H may be written

$$H = \pi J = W$$

Therefore, the incident radiation H at A_2 close to source A_1 is equal to the emitted radiation W.

where $w_{BB\lambda}$ is the blackbody radiant emittance. Another form of Kirchhoff's law is that $\alpha = \epsilon$. Then a blackbody may be defined as a body in equilibrium for which $\alpha = \epsilon = 1$ for all wavelengths. A blackbody source may be constructed with additional energy inputs if it is arranged so that $\alpha = 1$. A blackbody receiver or radiometer may have energy losses if $\alpha = 1$ (and is calibrated).

Radiation geometry To calculate the power or irradiance at a distance from a source, the geometry of a test receiver, relative to the source, is an important concept. The solid angle Ω is defined in Fig. 3.

Radiation measurements Measurements of radiant power or energy are based upon a standard source (standard radiation field) or standard detector if one exists. A standard source should be (but may not be) characterized by spectral intensity as a function of viewing angle. A standard detector should be characterized by spectral detectivity as a function of angle. One of the most convenient sources and a recognized standard at certain wavelengths is the blackbody. Blackbodies are thermodynamically well understood and practically readily constructed.

Useful transfer standards are quantum or near quantum detectors such as a solar cell (0.4 to 1.1 μm) or indium antimonide photovoltaic detector (3 to 5 μm).

Spectral measurements are based upon the use of spectrometers: grating, prism, interferometric.

Laser measurements are, for the most part, based on techniques adapted from older areas of optics. Tables 3 through 7, summarized from Ref. 3, are suggested

---

TABLE 3   Principal Laser Parameters

External Beam Parameters

1. Power
   a. Power density in the near and far field
   b. Power distribution in the beam
2. Energy
3. Angular divergence and beam spot size
4. Output wavelength and time dependence
5. Coherence
   a. Spatial
   b. Temporal
   c. Mutual
6. Polarization

Internal Beam Parameters

1. Mode spectrum
2. Gain
3. Noise
4. Modulation

TABLE 4   Beam-Sampling Techniques*

Method	Technique	Components
Amplitude splitting	Beam splitter	Fresnel reflector, transparent flat plate, or film grating
Beam scattering	Smoke	Dust, dielectric slab
Induced fluorescence	Phosphor coating	Stopcock grease, phosphors, acrylic resins
Resonant transition	Spectroscopy	Grating or prism (observe line-intensity-ratio changes) in cavity
Harmonic conversion	Frequency doubling	Nonlinear crystalline material
Photochemical decomposition	Fluorescence	Ammonia cell

* Must not allow light reflected back to cavity to perturb measurement.

TABLE 5   Measurement of Beam Parameters

Parameter	Method
Transverse-mode structure	Photography
	Scanning detector
	Image converter
	Evaporograph
Beam divergence	Photography
	Scanning detector
	Image converter
	Evaporograph
Polarization	Polariscope

TABLE 6   Measurement of Energy and Power

Measurement of Energy (Time-integrated)

Method:
  Calorimetry
  Detector (integrated output)
  Radiation thermopile (long-time constant)
  Photographic

Measurement of Power (Time-resolved)

Detector
Thermopile

Attenuation (to Bring Energy or Power into Range)

Beam splitter
Scattering from white diffuse (Lambertian) surface
Integrating sphere
Fine-wire scatterer
Polarizer

TABLE 7   Measurement of Other Parameters

Measurement of Wavelength

Spectrometers and spectrographs
  Grating
  Prism
  Interferometer

Measurement of Bandwidth and Temporal Coherence

Interferometer

Measurement of Frequency Stability

Multibeam interferometer
Photoelectric mixing
Homodyne (self-mixing) beating
Spectrometer

Measurement of Noise

Low-noise detector

methods for various laser measurements. Further details may be found in Ref. 3 and references therein.

Distribution functions A statistically distributed variable in a small range of the independent variable (iv) will have no expected value at any given value of the iv. For example, a source which emits a (probable) number of photons with wavelengths between 1 and 2 μm will (probably) emit no photons having a wavelength exactly 1.5 μ. It will emit a countable number between 1.5 and 1.6 μm in any finite time interval. The number of photons per unit wavelength interval $d\lambda$ between λ and $\lambda + d\lambda$ is written n_λ, $n(\lambda)$, or $dn/d\lambda$. Such distribution functions are known as the photon spectrum. Other distribution functions apply to other radiation parameters derived from photon distribution functions.

The wavelength response of a receiver to radiation is known as its spectral response and is the response to some incident-radiation spectrum:

$$R_\lambda = \frac{v_0}{f_\lambda}$$

where R = detector responsivity
$f(\lambda)$ = quantity of radiation between λ and $\lambda + d\lambda$
v_0 = detector output

The output of a detector to an arbitrary source with spectrum $F(\lambda)$ will be

$$v_0 = \int_0^\infty R_\lambda F_\lambda \, d\lambda$$

The integration may be done to the accuracy of most measurements by point-by-point multiplication of their $R_\lambda F_\lambda$ graphs (data sheet or measured) and numerical integration. Commonly source and detector spectra are reported as relative response with their peak values normalized to 1. In such cases it is necessary to renormalize source strength and detector responsivity to their measured or reported values as follows:

Source spectrum reported as $F_n(\lambda)$, where

$$F_n(\lambda) = \frac{F_\lambda}{F_{\lambda p}}$$

where $F_{\lambda p}$ is the peak value for λ_p to $\lambda_p + d\lambda$. Source strength is reported as F_0, where

$$F_0 = A \int_0^\infty F_\lambda \, d\lambda$$

where A takes into account the operating conditions.

The spectrum for the source of interest is obtained, then, by calculating A:

$$A = \frac{F_0}{\int_0^\infty F_n \, d\lambda}$$

and multiplying all points of F_n by A. A similar procedure is followed to renormalize the detector spectral response, with sometimes one exception. Sometimes R_0 is obtained by measurement from a blackbody spectrum. R_0 is then known as R_{BB} if a cover window is used and corrected for. Since

$$R_0 = \int_0^\infty R_\lambda \tau_\lambda \, d\lambda$$

where τ_λ is the window transmission and

$$R_{BB} = \int_0^\infty R_{BB\lambda} \, d\lambda$$

R_{BB} is found as

$$R_{BB} = \int_0^\infty \frac{R_\lambda \tau_\lambda}{\tau_\lambda} \, d\lambda$$

Another term has been proposed by IRIS for the responsivity measured with a blackbody source and where no window correction is applied:

$$R_{\beta\beta} = \frac{\int_0^\infty R_{BB\lambda} H_{BB\lambda} S_{\tau\lambda} \, d\lambda}{\int H_{BB\lambda} S \, d\lambda}$$

where $H_{\beta\beta\lambda}$ is the blackbody spectrum of measurement, S is the detector area, and the responsivity is normalized to unity.

The deconvolution of R_λ from a normalized R_λ and measured $R_{\beta\beta}$ is then carried out as follows: v_0 is measured and normalized to unit power for the source as

$$R_{\beta\beta} = \frac{v_0}{\int H_{BB\lambda} S \, d\lambda}$$

by numerical integration of $H_{BB\lambda}$ where $v_0 = \int R_\lambda H_\lambda S_{\tau\lambda} \, d\lambda$.

The $R_{n\lambda}$ is measured (reported) as

$$R_{n\lambda} = \frac{v_\lambda'}{v_\lambda''}$$

and further normalized to 1 "at" λ_p, where v_λ' is the response to the spectrometer and v_λ'' is the response of a blackbody detector used to normalize the response of the spectrometer to constant irradiance. Then

$$R_{\beta\beta} = \frac{A \int R_{n\lambda} H_{BB\lambda} S_{\tau\lambda} \, d\lambda}{\int H_{BB\lambda} S_{\tau\lambda} \, d\lambda}$$

and the renormalization proceeds as before. (τ_λ may be assumed constant in many cases.)

Radiometry   Classically, radiometers have consisted of blackened surfaces so that $\alpha \cong 1$ for the incident spectrum. One example is the familiar vane radiometer sold in novelty shops. Other instruments involve measuring the temperature reached when the receiver achieves equilibrium with the incident radiation. Originally dc-bridge techniques were used, and the reference receiver was compared with the radiometer. At present, ac-coupled, chopped detectors (referenced to the chopper temperature) are preferred. Chopped blackbody sources are prefered as standards for source temperatures up to about 1000 K. (Peak wavelength about 2.5 μm.) Graybody tungsten sources are available for 1400 and 2200 K standards and are the basis for NBS standards in the visible.[4] Above 1000 K, blackbody thermopile detectors are also useful but are insensitive and have nonuniform sensitivity over their surface. Their calibration is valid, provided that they are uniformly illuminated over their total aperture. As a result of their insensitivity, they are most conveniently used to calibrate solar-cell transfer standards (1.1- to 0.3-μm range). Commercial solar cells are uniform within 5 percent across their surface and may be purchased with 1 percent uniformity. In addition their short-circuit current is linear as a function of intensity from near their noise limit up to where conductivity modulation occurs (provided their temperature is controlled). Also they are quite rugged, stable, and convenient to use.

Blackbody sources are available from:
    Barnes Engineering Company
    Perkin-Elmer Corporation
    International Telephone and Telegraph Corporation
    Infrared Industries
    Radiation Electronics Company
    Electronics Communications, Inc.
    Eppley Laboratories, Inc.
    Williamson Development Company
These sources are designed so their emissivity exceeds 0.99 over their stated wavelength range.

Since radiometry depends ultimately on the temperature of a blackbody, temperature-standard errors place an upper limit on accuracy. A concise summary of temperature-standard status is given by Biberman.[5] The status and availability of NBS standards is discussed in Ref. 6.

Display brightness and photometry Originally photometry was performed by comparing the brightness of thermal sources (graybodies). The brightness of the un-known was compared with that of a standard such as a sperm candle. Since it is difficult to compare sources directly, their images may be compared on a diffuse surface such as a grease spot on paper or by using a Lummer-Brodhun or flicker photometer.[7] More recently the response of a standard observer's eye has been established (Fig. 4). Recent photometers consist of a detector filtered so that its spectral response approximates the CIE curve of Fig. 4. Illumination symbols and nomenclature are listed in Table 8 and explained in Fig. 5.

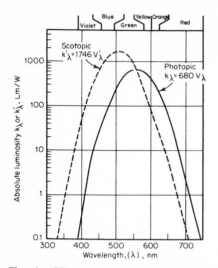

Fig. 4 CIE eye-response curves; absolute luminosity dependence on wavelength.[61]

Some engineering applications of photometric measurements are:

1. Comparison of production light sources to ensure uniformity

2. Illumination engineering

3. Direct and projection display engineering

4. Determination of illumination requirements for sensors other than the eye, such as for film or television cameras

These applications will have different requirements. Application 4, for example, attempts to compare one detector (the eye) with another with an entirely different spectral response. Such a procedure is not considered good practice.[5]

Display engineering (application 3) usually takes as a starting point the adjudged illumination (in a "well-lighted" room, direct sunlight, or darkened cockpit; Table 8). This point is then multiplied by the desired contrast ratio, dynamic range, or "gray-shades" range to obtain the peak-highlight brightness. Or a design may start with the peak-highlight brightness in a comparable application (Table 9). The goal is to obtain a design brightness equal to the design requirement. For thermal sources, the multiplication of the renormalized-source spectral irradiance by the CIE curve is usually satisfactory

$$H = \int \frac{dp}{d\lambda} v_\lambda \, d\lambda \qquad \text{(lumens)}$$

where $dp/d\lambda$ is the source spectral irradiance and v_λ is the spectral CIE value. A similar procedure for colored sources such as a CRT or fluorescent-tube phosphor will usually yield satisfactory estimates.

Photometric measurements and units apply to the bright adapted eye (photopic or cone vision, intensity greater than 3cdm^{-3}; see Fig. 4). For lesser illuminations, corrections may be estimated from the scotopic curve, of Fig. 4 or may be based on the apparent brightness of the 1 cm² area blackbody operated at 2042 K.

The response of detectors other than the eye such as film or a detector may be estimated from their spectral sensitivity to the illumination-source spectrum, as described earlier.

Other units Many lamps are rated by their color temperature, which is the temperature of a blackbody whose radiation has the same visible color as that lamp. Color temperature may conveniently be obtained from an optical pyrometer which superimposes the image of the unknown onto the image of calibrated tungsten filament. A rough visual estimate of color temperature is given in Table 10.

Fig. 5 Photometric-unit chart.[62] This chart, developed by Francis Clark, Lighting Services, Waterbury, Conn., and L. E. Barbrow, National Bureau of Standards, Washington, D.C., illustrates the interrelationship among the six most frequently used photometric units.

Units of Luminous Intensity, Luminous Flux, Luminous Energy, Illumination, Luminance, and Luminous Existance

The °candela (cd) is the luminous intensity of $\frac{1}{60}$ of 1 cm² of projected area of a blackbody radiator operating at the temperature of solidification of platinum, 2045 K. The °lumen (lm) is the luminous flux from 1 condela in all directions through a unit solid angle of 1 steradian.

The °lumen-second (which is 1 talbot) is the luminous energy delivered by 1 lumen in 1 second.

Point A to any B is 1 cm	Point A to any C is 1 ft	Point A to any D is 1 m
BBBB is 1 cm²	CCCC is 1 ft²	DDDD is 1 m²

Units of illumination

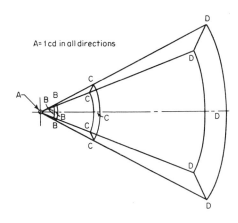

A= 1 cd in all directions

The phot is the unit of illumination resulting from the flux of 1 lumen falling on BBBB from A = 1 candela.

The footcandle is the unit of illumination resulting from the flux of 1 lumen falling on CCCC from A = 1 candela.

The °lux (lx) is the unit of illumination resulting from the flux of 1 lumen falling on DDDD from A = 1 candela.

Units of luminance

Assume 100% of he luminous flux from A (1 cd in all directions) is reflected in a perfectly diffuse manner by BBBB — CCCC — or — DDDD — then

BBBB will have a directionally uniform luminance of 1 lm/cm² which is 1 lambert or $1/\pi$ stilb

CCCC will have a directionally uniform luminance of 1 lm/ft² which is 1 foot-lambert or $1/\pi$ cd/ft²

DDDD will have a directionally uniform luminance of 1 lm/m² which is 1 apostilb or $1/\pi$ °cd/m² which is $1/\pi$ nit

Units of luminous exitance

Assume 100% of the luminous flux from A (1 cd in all directions) is reflected by BBBB —— CCCC —— or —— DDDD —— then

BBBB will have an exitance of 1 lm/cm²

CCCC will have an exitance of 1 lm/ft²

DDDD will have an exitance of 1 °lm/m²

° SI Unit

TABLE 8 Illumination Standards[21]

Quantity[a]	Symbol[c]	Defining equation[b]	Commonly used units[i]	Unit
Radiometric:				
Radiant energy	$Q(Q_e)$	— —	erg † joule kilowatt hour	J kWh
Radiant density	$w(w_e)$	$w = dQ/dV$	† joule per cubic meter erg per cubic centimeter	J m^{-3} erg cm^{-3}
Radiant flux	$\Phi(\Phi_e)$	$\Phi = dQ/dt$	erg per second † watt	erg s^{-1} W
Radiant flux density at a surface:				
Radiant exitance[c]	$M(M_e)$	$M = d\Phi/dA$	watt per square centimeter	W cm^{-2}
Irradiance	$E(E_e)$	$E = d\Phi/dA$	† watt per square meter, etc.	W m^{-2}
Radiant intensity	$I(I_e)$	$I = d\Phi/d\omega^d$	† watt per steradian	W sr^{-1}
Radiance	$L(L_e)$	$L = d^2\Phi/d\omega(dA\cos\theta)^d$ $= dI/(dA\cos\theta)^d$	watt per steradian and square centimeter † watt per steradian and square meter	W sr^{-1} cm^{-2} W sr^{-1} m^{-2}
Emissivity	ϵ	$\epsilon = M/M_{\text{blackbody}}^e$	one (numeric)	
Photometric:				
Absorptance	$\alpha(\alpha_v, \alpha_e)$	$\alpha = \Phi_\alpha/\Phi_i^f$	one (numeric)	
Reflectance	$\rho(\rho_v, \rho_e)$	$\rho = \Phi_r/\Phi_i^f$	one (numeric)	
Transmittance	$\tau(\tau_v, \tau_e)$	$\tau = \Phi_l/\Phi_i^f$	one (numeric)	
Luminous energy (quantity of light)	$Q(Q_v)$	$Q_v = \int_{380}^{700} K(\lambda)Q_{e\lambda}\, d\lambda$	lumen-hour † lumen-second (talbot)	lm h lm s

Quantity	Symbol	Equation	Unit	
Luminous density	$w(w_v)$	$w = dQ/dV$	† lumen-second per cubic meter	lm s m⁻³
Luminous flux	$\Phi(\Phi_v)$	$\Phi = dQ/dt$	† lumen	lm
Luminous flux density at a surface:				
Luminous exitance[c]	$M(M_v)$	$M = d\Phi/dA$	lumen per square foot	lm ft⁻²
Illumination (illuminance)	$E(E_v)$	$E = d\Phi/dA$	footcandle (lumen per square foot)	fc
			† lux (lm m⁻²)	lx
			phot (lm cm⁻²)	ph
Luminous intensity (candlepower)	$I(I_v)$	$I = d\Phi/d\omega$[g]	† candela (lumen per steradian)	cd
Luminance (photometric brightness)	$L(L_v)$	$L = d^2\Phi/d\omega(dA \cos\theta)$[d] $= dI/(dA \cos\theta)$[d]	† candela per unit area	cd in⁻², etc.
			stilb (cd cm⁻²)	sb
			nit (cd m⁻²)	nt
			footlambert (cd/πft²)	fL
			lambert (cd/πcm²)	L
			apostilb (cd/πm²)	asb
Luminous efficacy	K	$K = \Phi_v/\Phi_e$	† lumen per watt	lm W⁻¹
Luminous efficiency	V	$V = K/K_{maximum}$[h]	one (numeric)	

a The symbols for photometric quantities are the same as those for the corresponding radiometric quantities. When it is necessary to differentiate them, the subscripts v and e, respectively, should be used, e.g., Q_v and Q_e. Quantities may be restricted to a narrow-wavelength band by adding the word spectral and indicating the wavelength. The corresponding symbols are changed by adding a subscript λ, e.g., Q_λ for a spectral concentration or a λ in parentheses, e.g., $K(\lambda)$, for a function of wavelength.

b The equations in this column are given merely for identification.

c The Nomenclature Committee of the Optical Society of America deprecates the use of the alternative terms radiant emittance and luminous emittance.

d θ is the angle between line of sight and normal to surface considered.

e M and $M_{blackbody}$ are, respectively, radiant exitance of measured specimen and of a blackbody at the same temperature as the specimen.

f Φ_i is incident flux, Φ_a is absorbed flux, Φ_r is reflected flux, Φ_t is transmitted flux.

g ω is the solid angle through which flux from point source is radiated.

h $K_{maximum}$ is the maximum value of the $K(\lambda)$ function.

i International Systems (SI) unit indicated by dagger (†).

TABLE 9   Approximate Brightness Values[22]

Illumination Earths at Surface

Sun at zenith = 10,000 fc
Full moon   = 0.03 fc

Highlights, 35-mm movie	0.004 L
Page brightness for reading fine print	0.011 L
November football field	0.054 L
Surface of moon seen from Earth	1.6 L
Summer baseball field	3 L
Surface of 40-W vacuum bulb, frosted	8 L
Crater of carbon arc	45,000 L
Sun seen from Earth	520,000 L

TABLE 10   Color Temperatures

Color	Temp, °C
Incipient red heat	500–550
Dark red heat	650–750
Bright red heat	850–950
Yellowish-red heat	1,050–1,150
Incipient white heat	1,250–1,350
White heat	1,450–1,550

The mired is the reciprocal color temperature times one million. Mired stands for microreciprocal degrees:

$$\text{mired value} = \frac{10^6}{\text{color temperature (K)}}$$

Contrast   The contrast between an object and its background is important in detecting and recognizing the object with a sensor. Since sensors have a limited dynamic range, the usable contrast is a combined source-sensor function. There are a number of definitions of contrast (Table 11). Definition (6) is important in the

TABLE 11   Definitions of Contrast

Definition	Application
(1) $\dfrac{\|A-B\|}{A}$	TV
(2) $\dfrac{\|A-B\|}{B}$	Physiology
(3) $\|A-B\|$	
(4) A/B	
(5) $0.5\|A-B\|/\bar{B}$	
(6) $\|A-B\|/(A+B)$	TV; commonly called modulation or depth of modulation, (5) reduces to (6) when $\bar{B} = 1/2(A+B)$, i.e., checkerboard or stripe pattern
(7) $\dfrac{\|A-B\|}{1/2(A+B)}$	Infrared target

A = target or highlight brightness or irradiance
B = background brightness or irradiance
$\bar{B}$ = average scene brightness or irradiance

design of many test situations, since a uniformly balanced scene will test the sensor's internal target resistance and its power-supply regulation where common test chart, knife-edge, and slit tests may give only ideal results.

Scene contrast referenced to the background is

$$\frac{C_0 - C_b}{C_b}$$

where the C's are in suitable units and preferably linearized i.e., (temperature)^4 for source power. The use of one or the other of the definitions of Table 11 depends in part on whether the sensor (system) operates on the total scene brightness or uses the background as a reference.

These definitions are an upper limit to scene contrast, since sensor noise or limited sensor dynamic range may introduce a reduction.

Contrast-enhancement methods usually involve (1) reducing the background (B or $\bar{B}$ of the table) or (2) increasing the source intensity A relative to the background by increasing source directionality.

A relatively unstudied field is the dependence of an image-uniformity criterion on the peak-highlight to background contrast. For example, the segment of a display may be specified to be uniform to the eye, to within a factor of 2 in brightness; yet the apparent segment-brightness contrast is clearly dependent on the background illumination relative to the segment brightness.

LIGHT-SOURCE PROPERTIES

Introduction

Only a few years ago few light sources were readily available for engineering purposes, but today a multiplicity of sources are available. Thermal sources include chopped blackbody sources with operating temperature from 230 to 1000 K, tungsten lamps at 1400 and 2870 K, a variety of cored carbon arcs at about 5000 K, and a xenon arc at about 6000 K (Table 12). Molecular hydrogen-, deuterium-, and helium-band sources operate into the UV (Table 13). Line sources encompass a

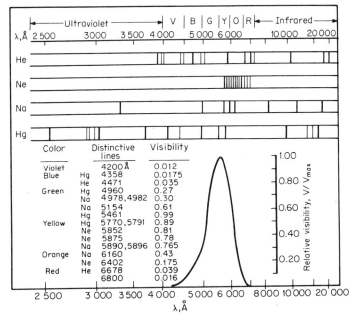

Fig. 6 Characteristics of line spectra.[63]

TABLE 12   Brightness of Representative Sources[23]

Type	Output, W	Efficiency, %	λ, μm	Brightness $(k = 1)$ MW cm^{-2} sr	Pulsed energy, J
CW Lasers					
CO_2	1,000	20	10.6	900	
YAG	100	3	1.06	9,000	
GaAs*	1	20	0.84	140	
HeNe	0.05	0.05	0.6328	12.5	
A II	5	0.05	0.5	2,000	
Ne II	0.05	....	0.3324	45	
Pulsed: ms					
Nd-glass	10^6	1	1.06	9×10^7	1,000
Ruby	4×10^5	1	0.6943	3×10^7	400
Pulsed: 20 ns					
CO_2	2×10^7	....	10.6	1.8×10^7	0.4
Nd-YAG†	5×10^3	0.2	1.06	4.5×10^5	5×10^{-4}
Nd-glass	3×10^9	....	1.06	2.7×10^{11}	60
Ruby	10^9	....	0.6943	2.1×10^{11}	20
Dye	2×10^6	....	0.5	8×10^8	0.04
N_2	2×10^5	....	0.3371	1.9×10^8	0.004
GaAs	100†	10	0.9	1.2×10^4	2×10^{-6}
Pulsed: 20 ps					
Nd-glass	10^{11}	....	1.06	9×10^{12}	2
Nd-YAG§	10^2	....	1.06	9,000	2×10^{-9}
High-pressure mercury arc				5×10^{-4}	
The sun				2.3×10^{-3}	

* Low-temperature operation.   Output from both ends.
† Continuous train of 100-ns pulses at 1,000 pps.
‡ Much higher power is available from arrays of these lasers.
§ Continuous train of pulses.

variety of hollow-cathode discharges and arcs of the most common metals and gases (Table 14).   Other sources include tunable dye lasers, solid-state, semiconductor lasers, and others (Tables 15 to 19 and Figs. 6 to 8).

Efficiency

Conduction and convection losses in many light sources will be small, and nearly all the energy will be radiated.   "Black" thermal-source efficiency will then be nearly 100 percent.   Envelope enclosures will absorb some of this energy, but as they come to an equilibrium temperature, they will reradiate that portion not lost by conduction or convection.   Practical efficiencies are about 90 percent.

The efficiency of uncooled spectral sources is imprecisely known at best.   Only recently has it been possible to measure line-intensity ratios for gas-discharge sources with any degree of accuracy, and there have been few reported measurements. Table 12 lists some source efficiencies.   Estimates may be made from the manufac-

TABLE 13   Vacuum Ultraviolet Sources[24]

Source	Emission and range	Excitation
High-pressure Xe arc	Continuum with superimposed lines 10,000–2,000 Å	dc 50 V, 10 A
Low-pressure D_2 arc	Continuum: 6,000–2,000 Å	dc 200 V, 250 mA
Low-pressure H_2 discharge (0.1–4 mm)	Continuum: 6,000–1,650 Å Line spectrum 1,650–900 Å	dc 10 kV, 50 mA dc
High-pressure He discharge (50–500 mm)	Continuum: 4,000–1,050 Å 1,000–600 Å	ac disruptive discharge in kilocycle range
Low-pressure He discharge	Emission line at 584 Å	dc 10 kV, 50 mA
High-pressure Xe discharge	Continuum 2,200–1,500 Å	ac disruptive discharge or
High-pressure Kr discharge	Continuum 1,800–1,250 Å	microwave excitation
High-pressure Ar discharge	Continuum 1,600–1,090 Å	200 W, 2,450 MHz
Electron synchrotron	Continuum, visible–60 Å (polarized radiation)	Peak emission depends on beam energy, e.g., 300 MeV gives max at 85 Å

All the sources shown here are operable under a wide range of both pressure and excitation conditions. The numbers given are intended only as a guide to typical conditions.

TABLE 14   Commercially Available Hollow-Cathode Sources Designed to Emit Characteristic Atomic Spectra of the Listed Atoms*

Ag	Fe	Na
Al, Sb	K	Ni
Bi	Li	Pb
Cd	Mg	Sn
Co	Mn	Sr
Cu	Mo	Zn, Ca

* Available from most scientific supply houses.

TABLE 15   Glow-Discharge Lamp Color

Gas	Color
Neon	Red
CO_2	White
Mercury	Green
Xenon	Bluish-white
Helium	Purplish-white

TABLE 16   Commercially Available Diode Sources

Material	Wavelength, μm	Typical vendors
SiC	0.45	General Electric
GaP	0.56	
$Ga_xAs_{1-x}P$	0.65, 0.7	Most semiconductor vendors
$In_xGa_{1-x}As$	0.85–3.15	
InAs	3.15	Texas Instruments
$Pb_xSn_{1-x}Te$	6, 10	Texas Instruments special order

TABLE 17   Wavelengths, in Angstrom Units, of Some Useful Spectral Lines[25]

Sodium	Mercury	Helium	Cadmium	Hydrogen
5,889.95 s 5,895.92 m	4,046.56 m 4,077.81 m 4,358.35 s 4,916.04 w 5,460.74 s 5,769.59 s 5,790.65 s	4,387.93 w 4,437.55 w 4,471.48 s 4,713.14 m 4,921.93 m 5,015.67 s 5,047.74 w 5,875.62 s 6,678.15 m	4,678.16 m 4,799.92 s 5,085.82 s 6,438.47 s	6,562.82 s 4,861.33 m 4,340.46 w 4,101.74 w

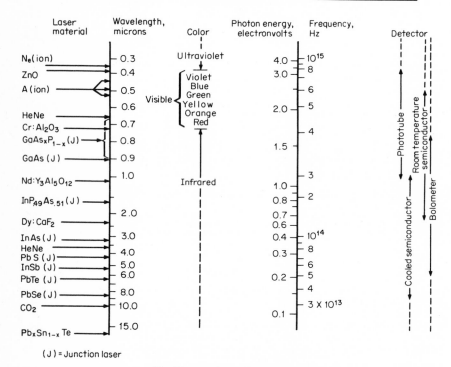

Fig. 7   User's guide to lasers.[64]

turer's data-sheet spectra and by using the input power and electrode and envelope blackbody spectra as reference factors in the spectra.

Spectral efficiencies of mercury lamps can be about 60 percent; xenon and sodium lamps can be about 30 percent efficient. Note that the lamp output energy is an upper limit to the system radiated energy; i.e., the lamp energy must be collected and focused.

Thermal Sources: Black- and Graybodies

Blackbody distribution functions   The spectral radiant emittance of a blackbody is given by Planck's equation as W_λ:

$$W_\lambda = \frac{2\pi c^2 h}{\lambda^5 (e^{hc/\lambda kT} - 1)}$$

TABLE 18   Variable and Tunable Lasers

Laser	Wavelength, μm	Notes
He-Cd	0.6328	Tube life 1,000 h
	0.4416	
	0.3250	
CO	5.1–6.5	
CO_2	9.2–9.8	
	10.4–10.1	
H_2O, D_2O, H_2S, SO_2, and OCS	33–220	By changing gases
HCN	337	
DCN	311	
Dye	0.35–0.74	Pumps: Nd-YAG, N_2 (superradiant) with KDP doubling
YAG	0.47–1.36	With $LiIO_3$ doubling

where λ = wavelength
h = Planck's constant
c = velocity of light in free space
k = Boltzmann's constant
T = absolute temperature, K

Since spectral radiance N_λ is equal to W_λ/π and since blackbodies are Lambertian sources,

$$N_\lambda = \frac{2c^2h}{\lambda^5(e^{hc/\lambda kT} - 1)}$$

The number of photons per second is

$$n_\lambda = \frac{2c}{\lambda^4(e^{hc/\lambda kT} - 1)}$$

where n_λ is $N_\lambda/h\nu$ and ν is the frequency of the photon of wavelength λ.

The total integrated power from a blackbody of unit area is

$$W_{BB} = \sigma T^4$$

and is known as the Stefan-Boltzmann law.

The radiant-power transfer from one surface to another, where their temperatures differ, is the difference between their radiated powers. For blackbodies this difference is proportional to $T_2^4 - T_1^4$.

Figure 9 shows N_λ as a function of wavelength. Figure 10 shows n_λ as a function of wavelength.

References to tabulated blackbody equations are given in Ref. 2.

Slide rules are available from:
General Electric Company
Block Associates
A. G. Thornton Company
Jarrell Ash Company
International Scientific and Precision Instrument Company

A more convenient procedure may be to incorporate the necessary calculations from Planck's equation in the computer program being used to aid other calculations.

Comparison of practical sources with blackbody spectra The solar-continuum radiation is shown in Fig. 11a along with a comparative blackbody spectrum. Departures from an ideal blackbody are explained by negative hydrogen-ion absorption (Fig. 11b). A carbon-arc spectrum is compared with a blackbody spectrum in Fig. 12. Departures from the blackbody distribution are attributable to radiation from

TABLE 19 Some Common Gas-Laser Sources[26]

Wavelength, μm	Medium	Operation	Design parameters Length	Design parameters Power in	Design parameters Power out	Comments
0.2358	NeIV	Pulsed	1 m	15 MW peak	· · · · · · ·	Shortest wavelength known in laser transition. Requires large currents
0.3324	NeII	Pulsed	1 m	300 W avg	10 mW	Other NeII and ArIII lines obtainable on same basis in this wavelength region
0.3371	N₂	Pulsed	2 m	10 W/pulse	200 kW peak	Requires very high voltages
0.4880	ArII	CW	60 cm	5 kW	1 W	High-power visible—blue
0.5145	ArII	CW	60 cm	5 kW	1 W	High-power visible—green
0.5682	KrII	CW	60 cm	3 kW	0.5 W	Strong line in yellow part of spectrum
0.6150	HgII-He	Pulsed	60 cm	50 W avg	100 mW	Narrow Doppler width—good interferometer source
0.6328	He-Ne	CW	15 cm	10 W	0.1 mW	Single-frequency—wavelength standard
0.6328	He-Ne	CW	2 m	100 W	100 mW	General-purpose—high power red
1.1523	He-Ne	CW	2 m	100 W	40 mW	General-purpose
2.0261	Xe	CW	2 m	100 W	10 mW	High-gain
3.3912	He-Ne	CW	2 m	100 W	20 mW	Very high gain per pass—competition with 0.6328 useful in plasma interferometry
3.507	Xe	CW	1 m	50 W	1 mW	Very high gain
10.6	CO₂-N₂-He	CW or Q-sw	2 m	1,000 W	100 W	High efficiency and high power
27.9	H₂O	Pulsed	1 m	100 W	10 W peak	
118.6	H₂O	Pulsed	1 m	100 W	1 mW	

This list (not necessarily complete or unbiased) of gas-laser wavelengths may be useful for purposes other than laser research. The design parameters indicate the approximate scaling between size, power in, and power out, for lasers that might be practical in an average laboratory.

S = solid-state laser
N = neutral gas laser
I = ion gas laser
M = molecular gas laser
L = liquid laser
MM = microwave laser

Period*	IA	IIA Alkali and alkaline earth metals	IIIB	IVB	VB	VIB	VIIB	VIIIB	VIIIB	VIIIB	IB	IIB	IIIA Boron family	IVA Carbon family	VA Nitrogen and oxygen family	VIA Nitrogen and oxygen family	VIIA Halogens	O Inert noble gases	
1	1 H																	2 He N	1
2	3 Li	4 Be											5 B I	6 C N,I	7 N N,I,M	8 O N,I	9 F N,I	10 Ne N,I	2
3	11 Na N	12 Mg											13 Al	14 Si N,I	15 P I	16 S N,I	17 Cl N,I	18 Ar N,I	3
4	19 K N	20 Ca N,I	21 Sc	22 Ti	23 V S^{2+} MM^{3+}	24 Cr S^{3+} MM^{3+}	25 Mn N	26 Fe MM^{3+}	27 Co S^{2+}	28 Ni S^{2+} MM^{2+}	29 Cu N	30 Zn N,I	31 Ga	32 Ge I	33 As I	34 Se I	35 Br N,I	36 Kr N,I	4
5	37 Rb MM	38 Sr N,I	39 Y	40 Zr	41 Nb	42 Mo	43 Tc	44 Ru	45 Rh	46 Pd	47 Ag	48 Cd N,I	49 In I	50 Sn N,I	51 Sb I	52 Te I	53 I N,I	54 Xe N,I	5
6	55 Cs N	56 Ba	Rare earth metals 57–71	72 Hf	73 Ta	74 W	75 Re	7 Os	77 Ir	78 Pt	79 Au N	80 Hg N,I^{2+}	81 Tl N	82 Pb N,I	83 Bi I	84 Po	85 At	86 Rn	6
7	87 Fr	88 Ra	Actinide metals 89–103	104 Ku															

Transition metals: IIIB to IIB.

*Number of electron shells surrounding the atom nucleus.

Rare earth metals 6	57 La	58 Ce	59 Pr S^{3+}	60 Nd S^{3+} L^{3+}	61 Pm	62 Sm S^{3+}	63 Eu S^{3+} L^{3+}	64 Gd MM^{3+}	65 Tb L^{3+}	66 Dy S^{2+}	67 Ho S^{3+}	68 Er S^{3+}	69 Tm S^{3+} S^{2+}	70 Yb S^{3+}	71 Lu	6
Actinide metals 7	89 Ac	90 Th	91 Pa	92 U S^{3+}	93 Np	94 Pu	95 Am	96 Cm	97 Bk	98 Cf	99 Es	100 Fm	101 Md	102 No	103 Lr	7

Fig. 8 Elements in which laser action has been observed. Atomic number is in the upper left-hand corner of the element box.[65]

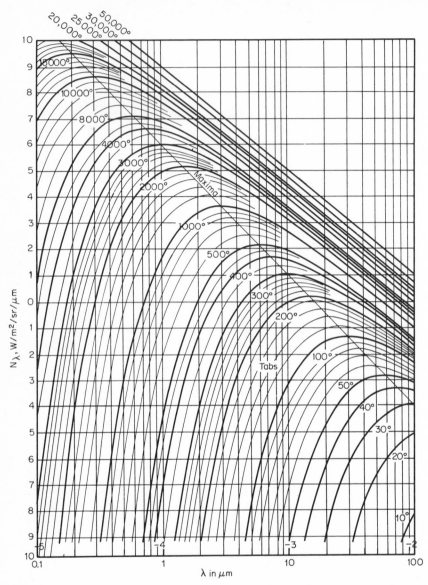

Fig. 9 N_λ as a function of wavelength for blackbodies at the temperature shown.[66]

cyanogen formed from the air. Tungsten-filament and xenon short-arc spectra are shown in Figs. 13 and 14. Other blackbody sources are the Globar (bonded SiC) and the Nernst glower (Z_rO_2, Y_2O_3, and SrO_2 or ThO_3). Departures from an ideal blackbody spectrum for the Globar are illustrated in Fig. 15.

The National Bureau of Standards has designated a group of commercially available lamps as spectral and total-irradiance standards for wavelengths from 0.25 to 2.6 μm.[8] Calibrated lamps may be obtained from Optronics Laboratories, Silver Spring, Md., among other places. Spectral irradiances for representative quartz-iodine lamps are reproduced in Table 20.

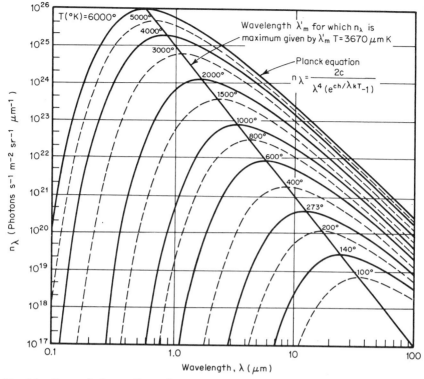

Fig. 10 Spectral photon flux radiated from blackbodies at the temperature shown.[67]

Properties of Gas Discharges and Gaseous Light Sources

A gas may be excited by light (fluorescence), by an electron beam, or by an electrical discharge. The latter technique is the traditional one; the first is important for self-excitation processes in lasers, and electron-beam excitation is relatively new. The discharge characteristics of a gas are shown in Fig. 16, where the need for a load ballast to stabilize the operating point may be seen. Characteristics of gas discharges as a function of pressure are shown in Fig. 17. Energy transitions in a gas, leading to light emission, may involve atomic, ionic, and molecular energy levels. The spectra from a source can consist of blackbody radiation from the electrodes; atomic and molecular transitions, and an apparent continuum from electronic molecular transitions and ion-electron recombination in the gas.

The line spectra are characteristic of the atomic species. The occurrence and strength of the lines for each species may be estimated from tabulated data.[9,10] Molecular processes are treated by Herzberg.[11]

Spectroscopy Transitions between stationary states of a gaseous atom are related by Bohr's frequency relation:

$$\nu_{mn} = \frac{E_n - E_m}{h}$$

where E_n and E_m are the energies of states n and $m(n > m)$ and h is Planck's constant. The calculation of E_n and E_m for hydrogen and sodium atoms is given in standard quantum-mechanical texts[12] and is the subject of spectroscopy.

In practice spectroscopic studies have yielded a great deal of information concerning atomic processes but little on spectroscopy itself. For example, only recently have line strengths been carefully measured for some gases, and most tabulated

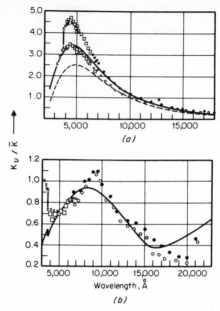

$K_\nu / \bar{K}$

(a)

(b)

Wavelength, Å

Fig. 11 (*a*) Intensity distribution in the continuous spectrum of the sun. Solid points are measured at the center. The full curve is calculated; open points are the measured emergent flux. The dashed curve is calculated based on the temperature of 5470 K. (*b*) Absorption coefficient of the solar atmosphere based on the experimental points of (*a*) above (points). The solid curve is the theoretical-absorption curve for H⁻ at 5740 K.[68]

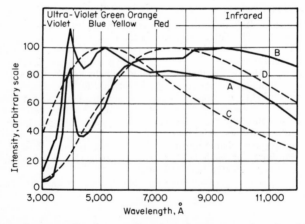

Fig. 12 Spectral-energy distribution of carbon arcs compared with a blackbody at the temperature given. The peaks are the radiation from cyanogen formed in the discharge. Curve *A* is a 13.6-mm dc high-intensity arc of 125 A, 6.3 V compared with (*c*) a 5600 K blackbody spectrum. Curve *B* is a 12-mm dc low-intensity arc of 30 A, 55 V compared with (*d*) 3810 K blackbody spectrum.[69]

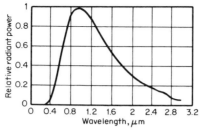

Fig. 13   Relative spectral radiant power of a 2854 K tungsten lamp.[70]

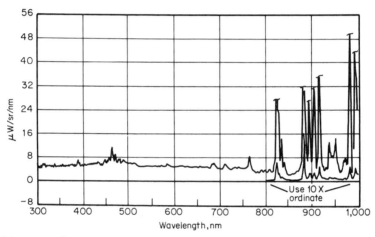

Fig. 14   Spectral output of a 5-W xenon short-arc lamp.   Note the 10X scale reduction in the near infrared.[71]

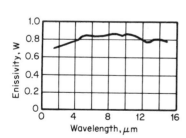

Fig. 15   Spectral   emissivity   of Globar.[72]

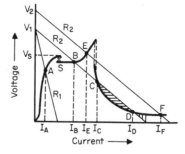

Fig. 16   Discharge characteristics of a gas.   Region *A* is the prebreakdown region; the plateau is the Townsend region; *S* is the spark; *B* is the normal glow region; *E* is the abnormal glow region.   The arc discharge occurs at the right.[73]

TABLE 20   NBS Spectral Irradiance Standard Lamp[27]

Wavelength, nm	Lamp QL-2	Lamp QL-5	Lamp QL-10
250	0.0051	0.0052	0.0051
260	0.0093	0.0093	0.0090'
270	0.0158	0.0159	0.0155
280	0.0253	0.0252	0.0244
290	0.0380	0.0380	0.0369
300	0.0545	0.0548	0.0532
320	0.104	0.105	0.102
350	0.237	0.242	0.234
370	0.366	0.374	0.363
400	0.643	0.647	0.630
450	1.26	1.26	1.23
500	2.04	2.04	2.02
550	2.93	2.96	2.91
600	3.88	3.94	3.88
650	4.79	4.91	4.80
700	5.54	5.72	5.58
750	6.11	6.32	6.14
800	6.51	6.69	6.49
900	6.72	6.94	6.71
1,000	6.51	6.73	6.53
1,100	6.07	6.25	6.11
1,200	5.53	5.67	5.55
1,300	4.97	5.00	4.98
1,400	4.44	4.52	4.44
1,500	3.93	4.00	3.93
1,600	3.46	3.51	3.45
1,700	3.03	3.06	3.01
1,800	2.63	2.65	2.61
1,900	2.29	2.28	2.26
2,000	1.98	1.97	1.95
2,100	1.73	1.71	1.70
2,200	1.52	1.51	1.50
2,300	1.36	1.34	1.33
2,400	1.22	1.21	1.21
2,500	1.12	1.10	1.11
2,600	1.04	1.03	1.04

In μW cm^{-2} nm^{-1} at a distance of 43 cm (measured from the axis of the lamp filament and normal to the plane of the lamp press) when operated at 6.50 A.

results are estimates only. As a result many cases of engineering interests must be established from vendor information or by direct measurement.

Systems of atoms, ions, or molecules are characterized by a ground state that can absorb energy and excited levels that can absorb or emit energy. There are two types of emission: stimulated and spontaneous. Spontaneous emission results from internal atomic processes and is random and incoherent. Stimulated emission results from external radiation and is coherent. The power per unit volume from spontaneous and stimulated processes is described by Einstein's relation

$$p(\nu_{nm}) = h\nu_{nm}[n_n A_{nm} + (n_n - n_m)B_{nm}\mu(\nu_{nm})]$$

where A_{nm} = coefficient for spontaneous emission
B_{nm} = coefficient for stimulated emission

Laser action will result when the cavity gain for stimulated emission exceeds 1.

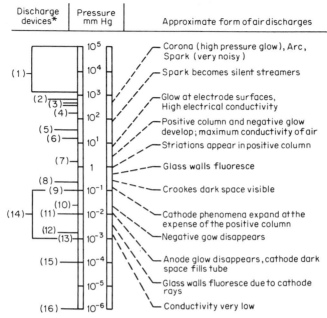

Fig. 17 Discharge characteristics as a function of pressure.[74] °(1) High-pressure mercury-capillary lamps. (2) High-intensity mercury-arc lamps (400-W type) and carbon arc. (3) Gas-filled incandescent lamps (630 mmHg). (4) Medium-pressure mercury-arc lamp (250-W type), (380 mmHg). (5) Tungar starting pressure (argon pressure = 50 mmHg). (6) Neon lamps (20 mmHg). (7) Starting pressure of sodium lamp (neon pressure = 1 to 3 mmHg). (8) Gas-filled tubes (argon pressure = 0.15 to 0.5 mmHg). (9) Gas-filled phototubes (0.1 mmHg). (10) Early carbon-filament lamps (0.023 mmHg). (11) Cold-cathode oscillograph discharge tube (0.01 mmHg). (12) Gas x-ray tubes (0.0015 mmHg). (13) Operating pressure of sodium vapor in sodium lamp (0.001 mmHg). (14) Mercury-arc tubes (0.1 to 0.001 mmHg). (15) Deflection chamber of cold-cathode oscillograph (0.0001 mmHg). (16) High-vacuum tubes, sealed cathode-ray tubes, Coolidge x-ray tube (0.000001 mmHg).

The energy levels of atomic electrons and their optical transitions, states, lifetimes, and line strengths are among the concerns of spectroscopy. The optical electrons are those in the outer shell of the atom. Spectroscopy of gaseous atoms is perhaps the best-understood area.

Spectroscopy is of interest to the optical-systems designer, since an understanding of the terminology will enable him to unravel some details of vendors' data sheets and other literature of interest.

One-Electron Atoms. The one-electron atoms are hydrogen, the alkalies, the like ions such as Be^+, Mg^+, Ca^+, and the rare-gas ions.

Spectroscopic notation for the electron state of alkali atoms is

$$n^{(2s+1)}L_J$$

where n = principal quantum number*
s = spin angular momentum
L = orbital angular momentum
J = total angular momentum (vector sum of $L + s$)

* The quantum numbers assigned to the electronic-energy levels or states are called momentum quantum numbers, since when appropriately multiplied by $\hbar\omega$ they yield the corresponding electronic momentum.

All are written as numbers except L.   Translation of L is

L number	0	1	2	3	4	5	6
Letter designation	S	P	D	F	G	H	I

Electron spin is $\pm \frac{1}{2}$; so for the alkalies, the s or ground state $L = 0$, $J = \frac{1}{2}$.   For example, the ground state of cesium is $6^2S_{1/2}$.   All other states will be doublets, since J is $L \pm \frac{1}{2}$.   The energy levels of sodium are shown in Fig. 18.

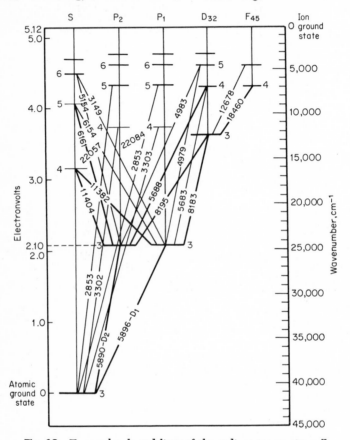

Fig. 18   Energy levels and lines of the sodium-arc spectrum.[75]

The potential of the ground state is taken to be 0 V, and the ionization potential is at the top line and is the zero of the wavenumber scale.   Allowed electronic transitions are shown by arrows on the figure.   The use of Bohr's relation and the energy difference of the initial and final state will yield the wavelength of the absorbed light ($\lambda = hc/\Delta E$) for raised orbital electron energies or wavelength of the emitted light for lowered energies.   The $n = 1$ and 2 shells are filled by core electrons, and the first levels are the $n = 3$ levels.   The doublet 5,890 and 5,896-A lines are the familiar sodium D lines.

Allowed transitions between states are given by the selection rules:

$$\Delta s = 0$$
$$\Delta L = 1$$
$$\Delta J = 0, \pm 1$$

There are no selection rules on n, but large changes usually lead to weak line strength.

Two-Electron Atoms. The two-electron atoms are the rare gases (He, Ne, Kr, Ar, Xe), Be, Mg, Ca, Sr, Ba, Ra, Zn, Cd, Hg, C, Ge, Sn, Pb, and ions such as Li^+. The momentum of the two electrons can couple. If the momentum quantum numbers couple individually, the coupling is called Russell-Saunders or L-S coupling. Generally L-S coupling occurs in the lighter elements. If certain momenta cross-couple, the resulting momentum is no longer a good quantum number and the coupling is referred to as J-J coupling. Combined quantum numbers for the optical electrons are represented by capital letters, S or L, etc. (and for single-electron atoms the lowercase momenta letters may be and often are shown capitalized). The angular momenta then are $L = l_1 + l_2$ and $S = s_1 + s_2$, and J is the vector sum of the resultants. S is integral and odd for even-number electron atoms and is 0 or 1 for two-electron atoms. When $S = 0$, $J = L$ and when $S = 1$, $J = L + 1$, L, and $L - 1$. Optical transitions for $S = 0$ are called singlet and for $S = 1$ are triplet states because three closely spaced lines occur. Triplet to singlet-state transitions are forbidden by the selection rule $\Delta s = 0$ and therefore are rare (and weak). Selection rules for atoms with Russell-Saunders coupling are:

1. Transitions are limited to changes in n and l of a single electron. Two (or more) electrons cannot simultaneously change subshells.
2. $\Delta l = \pm 1$
3. $\Delta s = 0$
   $\Delta L = 0, \pm 1$
   $\Delta J = 0, \pm 1$ (but not $J = 0 \rightarrow J = 0$)

Energy levels for mercury and some optical transitions are shown in Fig. 19. Here

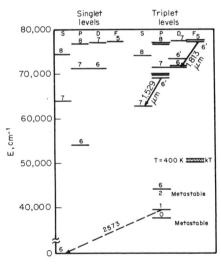

Fig. 19 Mercury energy levels. Two lasing transitions are shown by solid arrows.[76]

the wavenumber zero is taken as the ground state. A compilation of spectral lines is given in Table 21. Spectra at different gas pressures, shown in Table 21 and Fig. 20, reflect line broadening due to increased pressure. Calcium energy levels are shown in Fig. 21, and carbon in Fig. 22. Here the energy zero is taken at the ionization limit.

The rare-gas atoms (He, Ne, Kr, Xe) form a special case, since the two optical electrons form a closed shell. Promotion of one electron leaves an optically active hole with decreasing energy levels. The momenta coupling is referred to as j-l coupling, and one of several types of spectral notation may be used to designate the states.

TABLE 21   Relative Spectral Energy of a Typical Mercury-Arc Lamp[28]

λ, nm	Continuum	λ, nm	Principal lines above continuum
380	7.81		
90	5.33		
400	4.62	404.7	5.15
10	5.95		
20	4.62		
30	4.57		
40	7.19	435.8	10.4
450	4.09		
60	3.02		
70	2.66		
80	1.17		
90	2.66	491.6	0.27
500	2.22		
10	1.87		
20	1.95		
30	2.39		
40	5.15		
550	7.28	546.1	13.7
60	5.15		
70	4.53		
80	5.86	578.0	19.3
90	4.71		
600	2.98		
10	2.49		
20	2.39		
30	2.31		
40	2.32		
650	2.33		
60	2.31		
70	2.12		
80	1.77		
90	2.13	693.8	2.31
700	2.59		
10	2.84		

* From a private communication from C. N. Clark, General Electric Company, Nela Park, Cleveland, Ohio. The values given for the continuum as well as the lines refer to a constant bandwidth of $\Delta\lambda = 10$ nm.

Paschen notation based on the spectra is the most commonly used notation.[13] Spectral lines where both helium electrons are excited are rare, since the ionization energy for one electron is less than the two-electron excited-state energy.[14] Therefore, the helium spectrum is, in many respects, a shifted hydrogen spectrum, but with modified allowed transitions due to the presence of the second electron.

The helium energy-level diagram is shown in Fig. 23. The 2^1S_0 and 2^3S_1 states are metastable states, and 1^1S_0 is the ground state. There is no p state between the ground and metastable states. The metastable states are long-lived, since to deexcite them $\Delta L \neq \pm 1$. Also to deexcite the 2^3S, $\rightarrow 1^1S_0$, the $\Delta s \neq 0$.

The excited states of the other rare gases are mixed states, and the selection rules tend to breakdown.

Molecular Spectra. Molecular spectra may be termed immensely complex (Figs. 24 and 25). There are three excitation modes: rotational vibrational, and electronic. Diatomic atoms are the least complex. J is the rotational quantum number, the

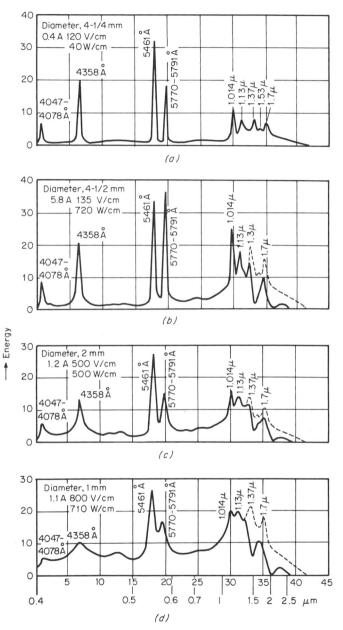

Fig. 20 Spectra of mercury lamps as a function of pressure. (*a*) 20 atm, air cooling. (*b*) 20 atm, water cooling. (*c*) 130 atm, water cooling. (*d*) 200 atm, water cooling. The dashed curve has been corrected for water absorption.[77]

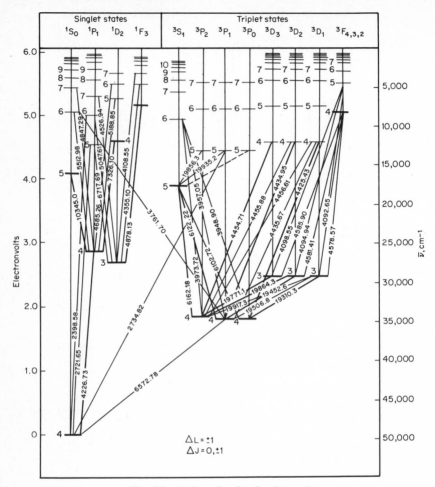

Fig. 21 Energy levels of calcium.[78]

energy levels are proportional to $J(J + 1)$, and the radiation will be in the far infra-red. The vibrational quantum number is ν; the energy levels are proportional to $(\nu + \frac{1}{2})$ and may be coupled to the vibrational levels.

The selection rule for pure rotation is $\Delta J = \pm 1$. The vibration-rotation selection rule is $\Delta \nu = \pm 1$, $\Delta J = 0, \pm 1$, If $\Delta \nu$ and ΔJ have the same sign, the spectrum is known as the R branch; if they have the opposite sign, their branch is known as the P branch. An example for CO_2 is shown in Fig. 24. In practice the vibrational levels are not equally spaced because of anharmonicity.

Polyatomic atoms may be arranged in order of the increasing complexity of their spectra.

1. Linear molecules: example, CO_2
2. Spherical top molecules: example, CH_4
3. Symmetrical top molecules: example, $CHCl_3$
4. Asymmetric top molecules: example, H_2O (described by three vibrational and three rotational quantum numbers)

The CO_2 molecule is a special case of the linear molecule. It is arranged as O—C—O. Four normal modes are allowed: symmetrical stretch, asymmetrical stretch, and two degenerate bending modes. In order, the quantum numbers are ν_1,

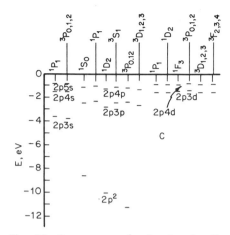

Fig. 22 Some energy levels of carbon.[79]

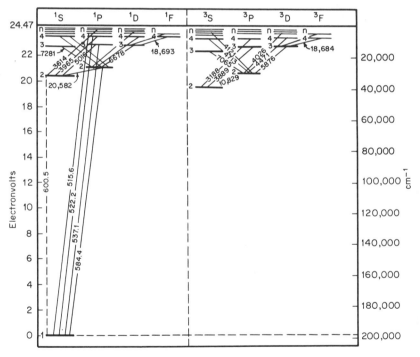

Fig. 23 Energy levels of helium.[80]

ν_2, and ν_3. The degeneracy of the last, ν_3, may be split. For example, the 010 mode is unsplit; 020, 030 are twofold split and 040 and 050 are threefold split. The splitting is labeled with a superscript attached to ν_2. The vibrational states are labeled with a symmetry symbol Σ_u^+ or Σ_g^+ (similar symbols are used for electronic excitations and should not be confused). For example, (ν_1 0° 0) states are all Σ_g^+ states like the ground state. For more details see Ref. 11.

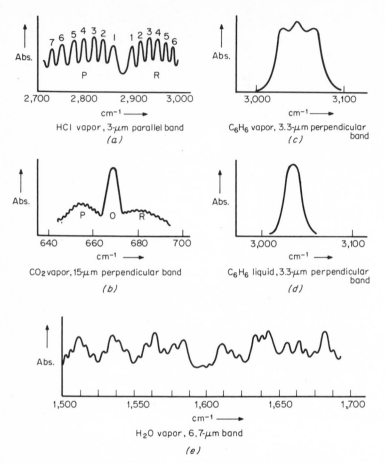

Fig. 24   Typical molecular vibration-rotation bands.[81]

Electronic Excitations of Molecules.   In addition to the discrete line spectra, there is a continuous electron spectrum associated with the electronic excitation of the molecule.   The spectrum, which is an important UV source, from H_2 and D_2 discharges (Fig. 26), is actually a closely spaced series of lines.   The good quantum numbers associated with the electronic excitation are the components of orbital and spin angular momentum along the molecule axis.

	Value			
Orbital component Λ	0	1	2	3
Letter associated	Σ	Π	Δ	Φ

The spin component S defines a level multiplicity $(2S + 1)$ which is written as a superscript, for example, $^3\Sigma$.   There is a total electronic angular momentum Ω analogous to the J for the atom which will vary from $\Lambda + S$ to $|\Lambda - S|$.   It is written as a subscript; an example is $^2\Pi_{3/2}$.   There are two other rules:

1. With homonuclear atoms such as H_2 or D_2 the electron wavefunction may be either even or odd.   A g designates the even and a u designates the odd state, for example, $^1\Sigma_g$ and $^1\Sigma_u$ states for hydrogen.

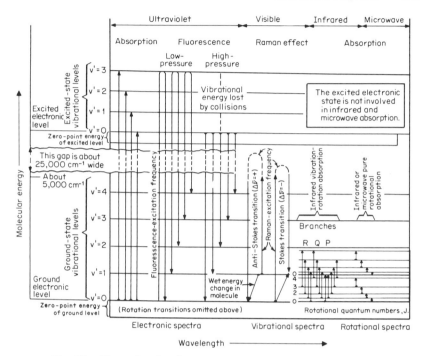

Fig. 25 Chart of molecular spectra showing source of radiation.[82]

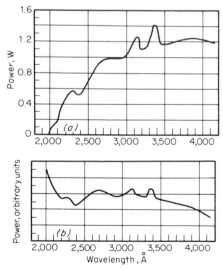

Fig. 26 Hydrogen-arc spectrum. (*a*) dc arc operated at 3.4 A, 70 V. (*b*) Corrected for envelope absorption.[83]

2. For Σ states of any molecule a $+$ or $-$ is used to distinguish molecular symmetry or antisymmetry relative to a plane containing the molecular axis. The rotational-selection rule is:

$$\Delta J = 0, \pm 1$$
$$\Delta \nu = \pm 1$$
$$g \rightarrow u$$

All three may not be required. Further rules depend on rank order of coupling. Most laser transitions involve $\Delta \Lambda = 0, \pm 1$ and $\Delta S = 0$. It usually holds that Σ^+ couples only with other Σ^+ states and Σ^- with Σ^-. Other complications are introduced when nuclear-spin effects are included.

Gas lasers   Any of the previously discussed spectral lines are candidates for stimulated emission (lasing) with a gain exceeding 1. (Nonlasing lines are useful and available from a gas-discharge lamp.)

The laser operates by increasing the number of excited electrons in an excited state (inverted population) until the stimulated emission generated exceeds the losses. For more details of operation see Garrett,[15] Bloom,[16] and Smith and Sorokin.[17]

Additional considerations for a transition to be a lasing transition involve the availability of convenient energy-transfer processes to the upper lasing transition level and the lifetimes of the associated states. To obtain gain, a high-Q cavity is required. Such tuned structures are obtained through the use of mirror structures, and the useful light is extracted by making one of the mirrors partially transmitting. Mirror life is extended if the mirrors are outside the discharge, and typically windows inclined at the Brewster angle are used to accomplish this purpose. The mirrors may also be shuttered by one of a number of means and the cavity Q can be switched to achieve greater outputs. Discharges are excited by using one of a variety of cathode and anode structures which are placed out of the positive-column region used for the laser or are excited by means of an rf discharge which may be either capacitively or inductively coupled. Chemical means may also be employed. Optical pumping has not been commercially important for gas lasers other then cesium. Water or other cooling may be provided, and a solenoidal magnetic field may be used to pinch the discharge and increase the efficiency.

Tunable gas lasers provide a means for selecting various discrete lasing lines. Lasing transitions have been observed in various gases ranging from the UV to the far infrared. Lasers are identified by the gas species used to obtain the lasing transition as atomic, ionic, or molecular (Table 22).

When the cavity has high enough gain for spontaneous radiation to be amplified enough to saturate the medium in less than one pass, the phenomenon is called superradiance. Superradiant-nitrogen lasers are commonly used as dye-laser pumps.

Modes and Cavities. The normal modes of a laser cavity are designated as TEM_{mnq}, where m and n are the transverse modes and q is the longitudinal or temporal mode. (Some authors reverse the order qnm.) The q is the number of half-wavelengths separating the mirrors and is usually dropped in reporting the mode number (especially since many lasers operate over a multitude of longitude modes).

Usually the lower-order modes 00 and 01 are used in applications because they lead to the cleanest spot and highest spot power and brightness. Higher-order modes are usually suppressed with an aperture (Fig. 27).

Other advantages of lowest-order mode are

1. Phase does not change across spot.
2. Easy to understand theoretically.
3. Unchanged near and far from exit aperture.
4. Smallest beam divergence at exit apertures.

TEM°_{01} or doughnut mode is difficult to suppress. The two modes TEM_{01} and TEM_{10} are 90° out of phase and rotate at the optical frequency. Their use leads to phase reversal in the wavefront at all times.

Originally plane-parallel mirrors, or Fabry-Perot resonators, were used to form the optical cavity. Severe planarity requirements for this structure have led to the adoption of nearly confocal structures.

Should the lifetime of the inverted population state greatly exceed the photon transit time of the cavity, Q switching may be used. The CO_2 molecular laser (about 20 percent efficient) may be adapted to this technique. The population is allowed

TABLE 22   Characteristics of Practical Gas Lasers and Gain in Lasers of Laboratory Size[29]

Laser Parameters

Laser	Length, cm	Power out	Gain % double pass	Bore diam, mm
He-Ne	15	1 mW	5	1
6,328 Å	180	80 mW	40	3
Ar⁺:				
4,880 Å		4,880–1 W	100	
	60			2–3
5,145 Å		5,145–2 W	$\doteq$ 40	
CO_2, 10.6 μm	200	100 W	≈100	>10

Doppler Widths

Laser	δ, MHz	Width at half-maximum points, MHz	Effective temp, K
He-Ne:			
6,328 Å	1,020	1,700	
1.15 μm	552	920	400
3.39 μm	186	310	
Ar⁺, visible	2,100	3,500	3,000
CO_2, 10.6 μm	36	60	300

to build up in excess of what would occur in an optical cavity, then is dumped by completing the cavity requirements such as by use of a rotating mirror at one end. The reverse Q switch uses a clear mirror that becomes more opaque as the level builds up.

A prism may be used as one mirror for line selection or tuning.

Coherence. An infinite plane wave has perfect spatial coherence. When this wave is incident on an aperture of diameter D (such as a laser-cavity window), it will diffract and the central zone will spread at an angle $\alpha \doteq \lambda/D$, where λ is the wavelength. The smallest system aperture also sets the limiting spot size to which the source may be focused.

The time coherence is measured by the frequency or wavelength spread of the exit beam and is primarily governed by the number of modes the cavity supports. Mode-locking techniques are commonly used to stabilize the cavity around a single mode.

A combined measure of the spatial and temporal coherence of a beam is the spectral radiance N_λ.

Polarization of the beam is well determined when a Brewster-angle exit window is used. In other cases the polarization must be specified by the manufacturer.

Noise.[15] Noise in laser sources is a complex subject, since major sources of noise are strongly influenced by construction techniques. In general, there are two sources of noise: extrinsic and intrinsic. The extrinsic noise is that introduced external to the laser by mechanical vibration, path variations, etc. The intrinsic noise is the noise of the laser itself and is in both the amplitude and the phase. For gas lasers, plasma noise can be as much as 20 percent of the output intensity of a He-Ne laser, and mode-interaction noise, with a long time constant, can be as high as 5 to 10 percent. The noise limit of a laser will be set by the statistics of the basic emission process. Most manufacturers specify the noise in their lasers.

Helium-Neon Laser. The widely used helium-neon laser is an example of the atomic laser. The energy-level interaction diagram is shown in Fig. 28. Helium

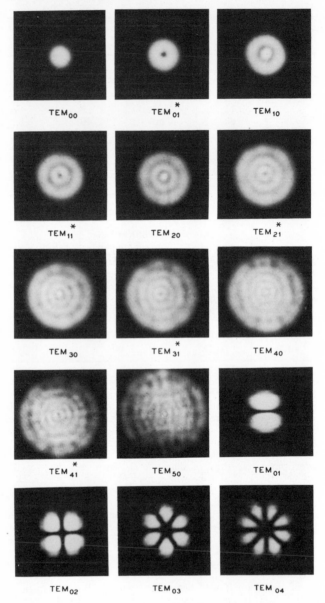

Fig. 27 Transverse-mode patterns obtained with an aperture in a He-Ne laser.[84]

is widely used in lasers and in particular is used in this laser because transitions from the 1S and 3S levels are long-lived metastable states. Transfer of energy to excite neon electrons to the $3S_2$ and 2S states readily occurs, whereas these states are relatively unpopulated thermally from the neon ground state.

There are also two sets of lower P states relatively unpopulated, and inversion between them and the higher S states is readily achieved through helium–metastable-neon collisions.

The three major laser transitions shown in the figure are from $3S_2$, high-gain

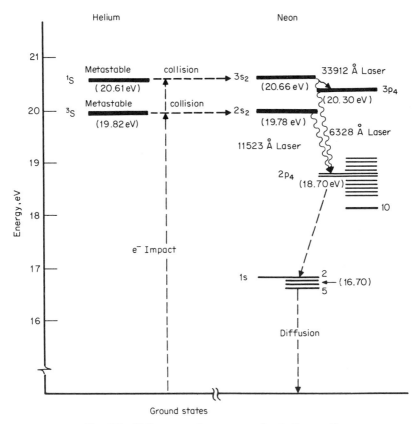

Fig. 28 Helium-neon-laser energy-level diagram.[85]

3.39-μm laser, from $2S$, 1-μm laser, and from $3S_2$, 0.06328-μm laser. A 90 percent helium–10 percent neon mixture will generate all three, whereas the 1-μm transition can be achieved in pure neon.

The 3-μm infrared transition leads to high gain, and in order to obtain efficient operation at 6,328 Å, it is necessary to suppress the 3-μm transition. Prismoidal end mirrors may be used, possibly in conjunction with small magnets (to secure Zeeman broadening of the 3-μm line) to reduce the cavity gain at 3 μm and lead to its suppression.

CW Argon-Ion Laser. A laser that has been popular in the blue and green spectral regions is the argon-ion laser. It may be operated in the pulsed or CW mode, where the strong CW lines lend themselves to CW operation. Figure 29 shows the Ar^+ levels and lasing transitions.

CO_2 Molecular Laser. The CO_2 molecular energy levels are shown in Fig. 30. Nitrogen or helium may be added to increase the power, as shown in the energy-level transfer sketch of Fig. 31.

Liquid, Solid-State, and Semiconductor Sources

The principal engineering applications of liquid and solid fluorescence are given in Table 23.

Dye laser Dye lasers use organic fluorescent materials as a lasing medium. Because of their low efficiency, another laser, such as a superradiant nitrogen or neodymium glass laser, is used as a pump. Their principal advantages are that they operate in the visible and are tunable. Details may be found in Ref. 132.

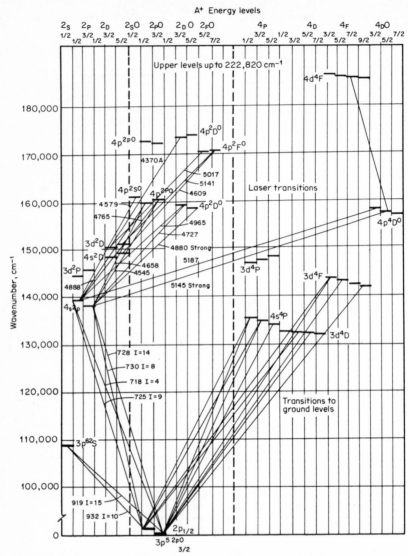

Fig. 29 Energy levels of Ar$^+$ showing laser transitions.[86]

TABLE 23 Engineering Applications of Fluorescence in Liquid and Solid Materials

Material	Application
Liquid	Dye laser
Insulating solid	Solid-state laser
Semiconductor	CRT and fluorescent, light phosphor, light sources, laser

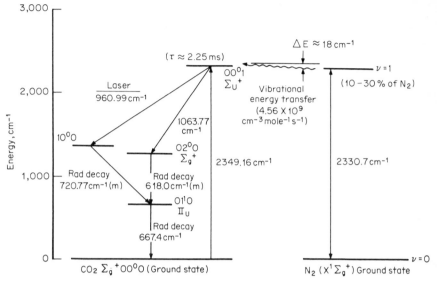

Fig. 30 Partial energy levels of CO_2 and N_2 showing relevant lasing transitions.[87]

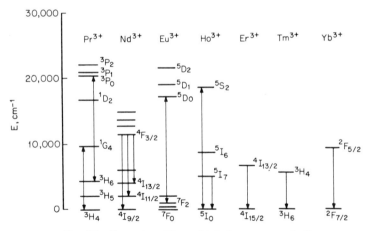

Fig. 31 Laser transitions of trivalent rare earths.[88]

Harmonic generation Harmonics may be generated by using a nonlinear electro-optic material such as a harmonic generator just as is done at rf frequencies. Frequency doublers are often used to expand a laser's capabilities and are now catalog items.

Solid-state lasers—crystalline and glass Solid-state lasers operate on the principle of dissolving a fluorescent ion, usually a rare-earth ion, in an insulating host. Simple spectroscopy of these ions is shown in Fig. 31. There are three-level and four-level varieties (Fig. 32). Since these lasers are optically pumped, the fluorescent ion should have broad absorption bands for the pump source. In addition, the host material should have good thermal and mechanical properties (Table 24). Table 25 summarizes these requirements and shows commonly employed wavelengths. The optical cavity is usually formed by polishing the ends of the laser rod itself. Q switching may also be employed.

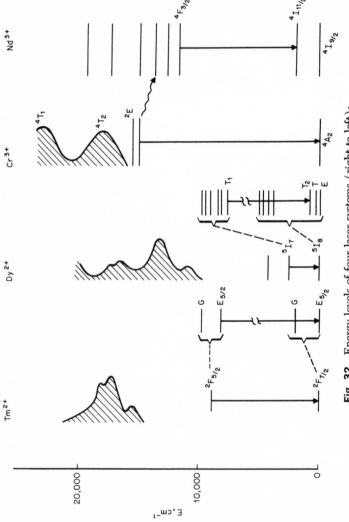

Fig. 32 Energy levels of four laser systems (right to left): CaF_2-Tm^{2+}, three level; CaF_2-Dy^{2+}, four level; Al_2O_3-Cr^{3+}, three level; and YAG-Nd^{3+}, four level.[89]

TABLE 24   Laser Host Materials[30]

Host	Symmetry	Lattice constants, A	Melting point, °C	Refractive index n	Hardness, Moh	Thermal conductivity at room temp, cal/cm^{-1} °C^{-1}	Thermal-expansion coefficients 10^{-6}
Al_2O_3	D_{3d}^5 $R3C$	5.12	2,040	1.765	9	0.11	5.8
CaF_2	O_h^5 $Fm3m$	5.451	1,360	1.4335	4		19.5
SrF_2	O_h^5 $Fm3m$	5.78	1,400	1.438			
BaF_2	O_h^5 $Fm3m$	6.19	1,280	1.475			
$SrCl_2$	O_h^5 $Fm3m$	7.00	873	1.6			
LaF_3	D_{6h}^3 $C6/mcm$	$a_0 = 4.148$ $c_0 = 7.354$	1,493				
CeF_3	D_{6h}^3 $C6/mcm$	$a_0 = 4.115$ $c_0 = 7.288$	1,324				
$CaWO_4$	C_{4h}^6 $I4/a$	5.24 11.38	1,570	1.918 1.934	4.5		
$SrWO_4$	C_{4h}^6 $I4/a$		1,566				
$CaMoO_4$	C_{4h}^6 $I4/a$	5.23 11.44	1,430	1.967 1.978	6	0.0095	25.5 c axis 19.4 a axis
$PbMoO_4$	C_{4h}^6 $I4/a$	5.41 12.08	1,070				
Y_2O_3	T_h^7 $Ia3$	10.6	2,450				
Gd_2O_3	T_h^7 $Ia3$	10.79	2,330				
Er_2O_3	T_h^7 $Ia3$	10.54					
$Y_3Al_5O_{12}$	O_h^{10} I_a3d	12.00	1,970	1.83	8.5	0.030	9.3
$Y_3Ga_5O_{12}$	O_h^{10} I_a3d	12.27		1.93	7.5		
$Gd_3Ga_5O_{12}$	O_h^{10} I_a3d		1,825				
MgF_2	D_{4h}^{14} $P4/mnm$	$a_0 = 4.6213$ $c_0 = 3.0529$	1,255	1.38			
ZnF_2	D_{4h}^{14} $P4/mnm$	$a_0 = 4.715$ $c_0 = 3.131$	872				
$Ca(NbO_3)_2$			1,560	2.07–2.20			

Glass lasers may use Nd^{3+}, Yb^{3+}, Er^{3+}, and Ho^{3+} as lasing ions (Table 26). Advantages of the use of glass as a host are:

1. Easily worked in a variety of shapes; easily polished; fibers are flexible.
2. Quite uniform index of refraction.
3. Index easily varied from 1.5 to 2.0.
4. Temperature coefficient of index and other important properties may be varied.

Disadvantages are low thermal conductivity and high thresholds. Crystalline lasers have complementary properties which allow design trade-offs to be made between the two systems.

Increased-efficiency sensitized lasers are made by adding a second impurity whose absorption bands are a better match to the pump-source spectrum (Table 27).

TABLE 25   Crystalline Laser System[31]

Host	Dopant	Frequency of laser, Å	Transition	Energy of terminal state, cm^{-1}	Mode and highest temp of operation, K
Al_2O_3	0.05% Cr^{3+}	6,934	$^2E(\bar{E}) \to {}^4A_2$	0	CW 350
		6,929	$^2E(A) \to {}^4A_2$	0	300
Al_2O_3	0.5% Cr^{3+}	7,009			77
		7,041	Pair lines	100	77
		7,670			300
MgF_2	1% Ni^{2+}	16,220	$^3T_2 \to$	340	77
MgF_2	1% Co^{2+}	17,500	$^4T_2\text{-}^4T_1$	1,087	77
		18,030		1,256	77
ZnF_2	1% Co^{2+}	26,113			77
$CaWO_4$	1% Nd^{3+}	10,580	$^4F_{3/2} \to {}^4I_{11/2}$	2,000	CW 300
		9,145	$^4F_{3/2} \to {}^4I_{9/2}$	471	77
		13,392	$^4F_{3/2} \to {}^4I_{13/2}$	4,004	300
CaF_2	1% Nd^{3+}	10,460	$^4F_{3/2} \to {}^4I_{11/2}$	~2,000	77
$CaMoO_4$	1.8% Nd^{3+}	10,610	$^4F_{3/2} \to {}^4I_{11/2}$	~2,000	CW 300
$Y_3Al_5O_{12}$	Nd^{3+}	10,648	$^4F_{3/2} \to {}^4I_{11/2}$	2,111	CW 360 / 440
LaF_3	1% Nd^{3+}	10,633	$^4F_{3/2} \to {}^4I_{11/2}$	2,187	300
LaF_3	1% Pr^{3+}	5,985	$^3P_0 \to {}^3H_6$	~4,200	77
$CaWO_4$	0.5% Pr^{3+}	10,468	$^1G_4 \to {}^3H_4$	377	77
Y_2O_3	5% Eu^{3+}	6,113	$^5D_0 \to {}^7F_2$	859	220
CaF_2	Ho^{3+}	5,512	$^5S_2 \to {}^5I_8$	~370	77
$CaWO_4$	0.5% Ho^{3+}	20,460	$^5I_7 \to {}^5I_8$	250	77
$Y_3Al_5O_{12}$	Ho^{3+}	20,975	$^5I_7 \to {}^5I_8$	518	CW 77 / 300
$CaWO_4$	1% Er^{3+}	16,120	$^4I_{13/2} \to {}^4I_{15/2}$	375	77
$Ca(NbO_3)_2$	Er^{3+}	16,100	$^4I_{13/2} \to {}^4I_{15/2}$		77
$Y_3Al_5O_{12}$	Er^{3+}	16,602	$^4I_{13/2} \to {}^4I_{15/2}$	525	77
$CaWO_4$	Tm^{3+}	19,110	$^3H_4 \to {}^3H_6$	325	77
$Y_3Al_5O_{12}$	Tm^{3+}	20,132	$^3H_4 \to {}^3H_6$	582	CW 77 / 300
Er_2O_3	Tm^{3+}	19,340	$^3H_4 \to {}^3H_6$		CW 77
$Y_3Al_5O_{12}$	Yb^{3+}	10,296	$^2F_{5/2} \to {}^2F_{7/2}$	623	77
CaF_2	0.05% U^{3+}	26,130	$^4I_{11/2} \to {}^4I_{9/2}$	609	300 / CW 77
SrF_2	U^{3+}	24,070	$^4I_{11/2} \to {}^4I_{9/2}$	334	90
CaF_2	0.01% Sm^{2+}	7,083	$^5D_0 \to {}^7F_1$	263	20
SrF_2	0.01% Sm^{2+}	6,969	$^5D_0 \to {}^7F_1$	270	4.2
CaF_2	0.01% Dy^{2+}	23,588	$^5I_7 \to {}^5I_8$	30	CW 77 / 145
CaF_2	0.01% Tm^{2+}	11,160	$^2F_{5/2} \to {}^2F_{7/2}$	0	27 / CW 4.2

TABLE 26   Laser Ions in Glass[33]

Ion	Glass	Transition	Wavelength, μm	$E_1 - E_0$, cm^{-1}	Inversion for 1% gain cm^{-1}, cm^{-3}
Nd^{3+}	K-Ba-Si	$^4F_{3/2}$-$^4I_{11/2}$	1.06	1,950	0.7×10^{18}
	La-Ba-Th-B	$^4F_{3/2}$-$^4I_{13/2}$	1.37	4,070	
	Na-Ca-Si	$^4F_{3/2}$-$^4I_{9/2}$	0.92	470	3.5×10^{18}
Yb^{3+}	Li-Mg-Al-Si	$^2F_{5/2}$-$^2F_{7/2}$	1.015	400	2.8×10^{18}
	K-Ba-Si	$^2F_{5/2}$-$^2F_{7/2}$	1.06	830	11.0×10^{18}
Ho^{3+}	Li-Mg-Al-Si	$^5I_7$-$^5I_8$	1.95	230	
Er^{3+}	Yb-Na-K-Ba-Si	$^4I_{13/2}$-$^4I_{15/2}$	1.54	0	1.8×10^{18}
	Li-Mg-Al-Si	$^4I_{13/2}$-$^4I_{15/2}$	1.55	110	

TABLE 27   Sensitized Emission of Laser Ions[34]

Laser ion	Sensitizer
Nd^{3+}	UO$_2$$^{2+}$
	Mn^{2+}
	Ag
	Ce^{3+}
	Tb^{3+}
	Eu^{3+}
	Cr^{3+}
Yb^{3+}	Nd^{3+}
	Ce^{3+}
	Cr^{3+}
Ho^{3+}	Yb^{3+}
Er^{3+}	Yb^{3+}
	Mo^{3+}

The fluorescent yield of the rare-earth ions is near 100 percent because the weak transitions are in an f shell which lies inside the fully filled outer electron shell. Spectroscopy of rare-earth atoms is similar to that of the simple atoms described earlier and is described, for example, in Smith and Sorokin.[18] Examples of the crystalline host laser are the three-level ruby laser at 6,900 Å and the 1.06-μm Nd^{3+}-YAG (yttrium aluminum garnet) laser (Table 28). The most familiar glass laser is the Nd^{3+} at 1.06 μm.

TABLE 28   Continuous Solid-State Laser Powers and Efficiencies[32]

Material active system	Sensitizer	Optical pump	λ, μm	Efficiency %	Power, W	Operating temp, K
Dy^{2+}CaF$_2$	....	W	2.36	0.06	1.2	77
Cr^{3+}Al$_2$O$_3$	....	Hg	0.69	0.1	1.0	300
Nd^{3+}Y$_3$Al$_5$O$_{12}$	....	W	1.06	0.2	2	300
			1.06	0.6	15	300
Nd^{3+}Y$_3$Al$_5$O$_{12}$	....	Plasma arc	1.06	0.2	200	300
Nd^{3+}Y$_3$Al$_5$O$_{12}$	....	Na-doped Hg	1.06	0.2	0.5	300
Nd^{3+}Y$_3$Al$_5$O$_{12}$	Cr^{3+}	Hg	1.06	0.4	10	300
Ho^{3+}Y$_3$Al$_5$O$_{12}$	[Er^{3+}, Yb^{3+}, Tm^{3+}]	W	2.12	5.0	15	77

Semiconductor sources   Semiconductor sources are listed in Table 16. Pumping may be by injection at a junction, by electron beam, or by optical excitation. The most efficient radiation will be at an energy near the band gap.

The spectroscopy of semiconductors is similar to that for solid-state lasers so far as deep-lying impurities are concerned (wavelengths corresponding to energies below

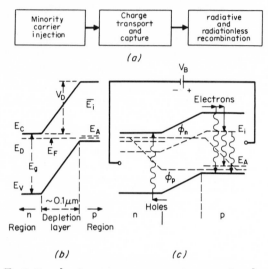

Fig. 33   (*a*) Excitation luminescence processes in a semiconductor. (*b*) Semiconductor-junction zero-bias energy levels (ordinate is distance). (*c*) Forward-biased junction where E_F is the Fermi level; φ_n and φ_p are the quasi-Fermi levels.[90]

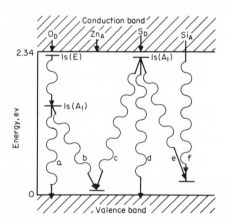

Fig. 34   Impurity-induced radiative recombination processes; O_D, oxygen donor, Zn_A, zinc acceptor; S_D, sulfur donor; Si_A, silicon acceptor.[91]

the band edge), but semiconductors have additional features near or above the edge that are exploited. Figures 33 to 36 show energy-level diagrams for semiconductor materials and devices and their associated radiation processes. Recombination at shallow impurities is the dominant commerically exploited process.

The spectral output as well as the efficiency of the alloy $GaAs_{1-x}P_x$ varies with x (Fig. 37*a*). Figure 37*b* shows how brightness may be optimized; this curve is obtained by multiplying the CIE visual-response curve by the efficiency curve.

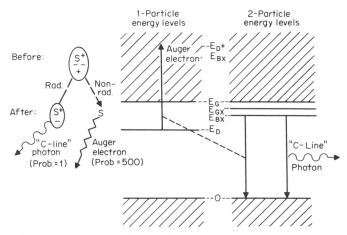

Fig. 35 Schematic comparison of the radiationless (Auger) and resonance-radiation radiative recombination processes for excitons (bound hole-electron pair). The excitons are bound to sulfur donors in gallium phosphide; E_{gx} is the lower energy limit of the free exciton and E_{bx} the energy of the bound exciton.[92]

Laser sources are listed in Table 29. Cavities are most conveniently formed by cleaving opposite crystal faces normal to the junction.

Electroluminescence   Electroluminescent sources are generally II-VI semiconductor materials (CdS, CdSe) whose band-edge energy lies in the visible. They are generally ac-excited, and their operation depends on high field emission from tiny copper-needle precipitates along grain boundaries. As a result, such devices have short life ($\sim$200 to 1,000) in average-brightness applications.

TABLE 29   Semiconductor Laser Materials[35]

Material	Photon energy, eV	Method of excitation
ZnS	3.82	Electron beam
ZnO	3.30	Electron beam
CdS	2.50	Electron beam, optical
GaSe	2.09	Electron beam
CdS_xSe_{1-x}	1.80–2.50	Electron beam
CdSe	1.82	Electron beam
CdTe	1.58	Electron beam
$Ga(As_xP_{1-x})$	1.41–1.95	pn junction
GaAs	1.47	pn junction, electron beam, optical, avalanche
InP	1.37	pn junction
$In_xGa_{1-x}As$	1.5	pn junction
GaSb	0.82	pn junction, electron beam
InP_xAs_{1-x}	1.40	pn junction
InAs	0.40	pn junction, electron beam, optical
InSb	0.23	pn junction, electron beam, optical
Te	0.34	Electron beam
PbS	0.29	pn junction, electron beam
PbTe	0.19	pn junction, electron beam, optical
PbSe	0.145	pn junction, electron beam
$Hg_xCd_{1-x}Te$	0.30–0.33	Optical
$Pb_xSn_{1-x}Te$	0.075–0.19	Optical

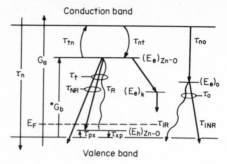

Fig. 36 Schematic recombination scheme for electrons in p-type GaP doped with zinc and oxygen. Ga and Gp are generation processes, Auger processes are τ_{NR}, τ_{INR}; thermal-trap ionization processes are τ_{tn} and τ_{xp}.[93]

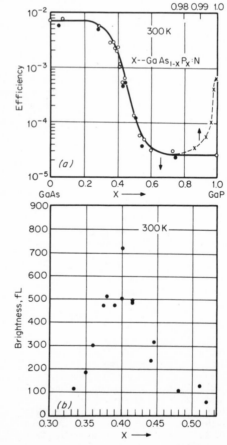

Fig. 37 (a) Room-temperature external quantum efficiency of $GaAs_{1-x}P_x$ as a function of phosphorus fraction. Dashed curve shows effect of degenerate nitrogen doping (scale at top). (b) Experimental brightness for diodes above.[94]

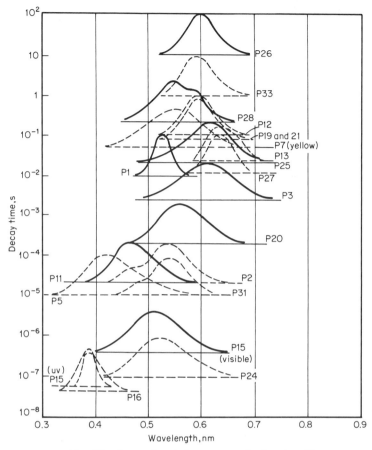

Fig. 38 Spectral-emission curves of phosphors.[95]

Phosphors   Phosphors are electron, ion, or optically excited semiconductors which are usually II–VI compounds. Figure 38 shows spectra and decay times of important phosphors.

SOURCE-SELECTION AND -APPLICATION GUIDE

Spectral Sources

Many applications require sources with particular spectral lines as bands without regard to efficiency or size. The usual procedure is to examine the mercury, sodium, or xenon spectra for the particular line and select a filter, where necessary. Appropriate commercial projection or arc lamps may then be selected in consultation with the manufacturer. A typical data sheet for one such lamp is given in Table 35. Where appropriate lines are not available from the above sources, hollow-cathode

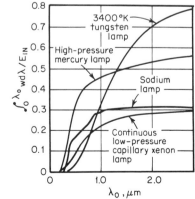

Fig. 39 The fractional lamp output above a given wavelength for four sources.[96]

lamps with appropriate element additions may be considered. Spectra for the sources may be obtained by reference to earlier figures and tables. Some efficiencies are shown in Fig. 39.

Lamps

Table 30 shows a comparison of some trade-off criterion for neon, incandescent, and solid-state sources. Because of long life compared with the product, solid-state lamps may be permanently wired into place. Where replacement costs of indicators are considered, solid-state lamps may be more economical than other choices.

TABLE 30   Lamp-Selection Guide[36]

Criterion	Neon	Incandescent	Solid-state
Operation	Gas ionization	Heating of a tungsten filament in vacuum to white heat (incandescence)	Light-emitting diode. Operates by junction electroluminescence
Light output	Relatively low. Amber-red color only	Low to high, depending on lamp. Generally white light	Relatively low; ir, red, orange, green
Operating voltage	65 Vac, 90 Vdc; for standard brightness types; 95 Vac, 135 Vdc and up for high brightness	1–28 V (operation in IDI pilot lights generally not recommended above 24 V)	1.5–3 V
Operating current	0.0005–0.003 A depending on lamp	From 0.015 A up, depending on lamp	0.005–0.025 A; normal operation at 0.015 A
Ballast required (current-limiting resistor)	Yes. Normally built into housing of pilot light or installed in one lead	No (voltage-dropping resistor sometimes used)	Yes; rectifier diode also required for ac service
Life	Min 5,000 h; up to 50,000+ depending on current	Depends on lamp-life rating and conditions of use; 750–10^5 h	Reduced at high current and temp; 20,000–100,000 h
Resistance to vibration, shock, voltage transients	Very high	Low to moderate depending on lamp, type of shock, etc.	Very high
Cost	Low	Low to high, depending on lamp	Relatively high
Applications	Line-voltage-operated applicances, instruments, etc.	Very broad range	Battery-operated devices with integrated low-current applications, etc.
Efficiency	. . . .	10–20 lm/W, 1.5–3%	0.2–45%
f_{co}	10–100 kHz	. . . .	1–100 MHz

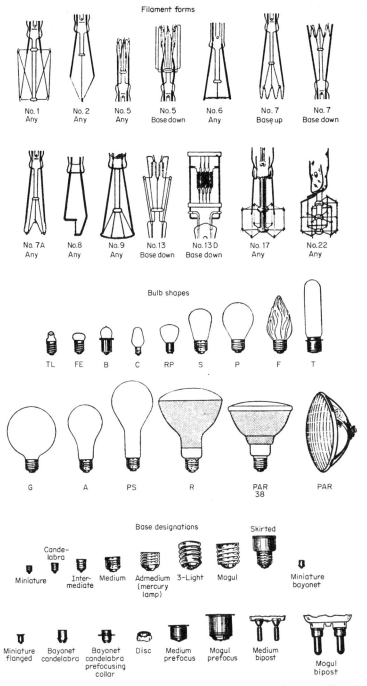

Fig. 40   Tungsten-lamp filament, bulb, and base configuration.[97]

Glow Lamps

Glow lamps, in general, will require ballasting to set their operating point; the operating point will be approximately 20 percent below their firing voltage. Reliable firing dictates, in turn, a 20 to 50 percent overvoltage; i.e., the lamp supply V_{KK} should be rated higher than the firing voltage. Lamp current is set by choosing a proper load line which when drawn from V_{KK} to $V_{KA(on)}$ (the tube drop at the rated current) holds the tube within the rated dissipation. An integral resistor is often used as the ballast. Other options include: (1) inclusion of resistors and capacitors to form a relaxation oscillator used as a flashing indicator and (2) the use of the tube as a combined indicator and voltage-reference source, particularly in the base of a series-pass transistor regulator.

Basings and packages are available in almost any configuration used for tungsten lamps.

Red and orange (as set by gas pressure) neon or 99.5 percent neon, 0.5 percent argon are the most common mixtures; the latter is preferred for lowest firing and sustaining voltages (in combination with a suitable cold cathode).

The lamps are stabilized as delivered, since a burn-in and aging procedure is part of the manufacturing process. Lamp life is determined by loss of gas pressure due to absorption and sputtering and electrode-wear effects. Life is typically 5,000 to 10,000 h in applications that require visibility with high ambient illumination.

Tungsten-Filament Lamps

Some typical tungsten-lamp designs are shown in Fig. 40. Schematic miniature basing is shown in Fig. 41. Brightness figures are given in Table 31 and Fig. 42. Rerating rules of thumb for filamentary lamps are given in Table 32. Note that

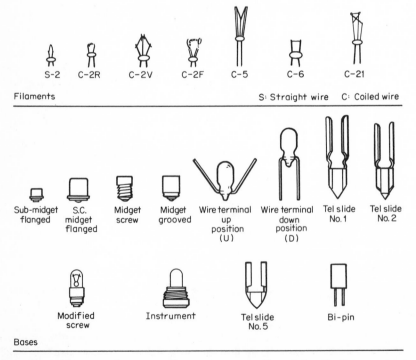

Fig. 41 Schematic subminiature-lamp construction.[98] While all three categories overlap to some degree, bead-sealed, butt-sealed, and stem-sealed lamps are each predominant in certain size classifications.

TABLE 31   Characteristics of Multiple Incandescent Lamps[37]

Watts	Bulb	Base	Volts	Initial A	Fila-ment con-struction	Approx initial lm	Rated av life, h	Rated initial lm W^{-1}	lm W^{-1} at 70% of rated life	Fila-ment temp, °F
6*	S-6	Cand.	120	0.050	C-7A	41	1,500	6.8		3,860
10*	S-14	Med.	120	0.083	C-9	79	1,500	7.9	7.2	3,900
25*	A-19	Med.	120	0.21	C-9	260	1,000	10.5	9.0	4,190
40	A-19	Med.	120	0.34	C-9	465	1,000	11.7	10.7	4,490
60	A-19	Med.	120	0.50	CC-6	835	1,000	13.9	13.2	4,530
75	A-19	Med.	120	0.63	CC-6	1,150	750	15.4	14.4	4,610
100	A-21	Med.	120	0.83	CC-6	1,630	750	16.3	15.3	4,670
100	A-23	Med.	240	0.42	C-7A	1,260	1,000	12.6	11.8	4,470
100	A-23	Med.	30	3.12	C-9	1,850	1,000	18.5	16.9	4,660
150	PS-25	Med.	120	1.25	C-9	2,600	750	17.2	15.7	4,710
200	PS-30	Med.	120	1.67	C-9	3,700	750	18.4	16.3	4,750
300	PS-30	Med.	120	2.50	C-9	5,900	750	19.6	17.3	4,825
500	PS-40	Mog.	120	4.17	C-7A	9,900	1,000	19.8	17.2	4,840
750	PS-52	Mog.	120	6.25	C-7A	15,600	1,000	20.8	17.2	
1,000	PS-52	Mog.	120	8.3	C-7A	21,500	1,000	21.5	17.3	4,930
1,000	PS-52	Mog.	240	4.2	C-7A	19,500	1,000	19.5	16.9	4,760
1,500	PS-52	Mog.	120	12.5	C-7A	33,000	1,000	22.0	15.6	5,010
3,000	T-32	Mog. Bip.	32	93.8	C-13B	88,500	1,000	29.5		5,390
5,000	T-64	Mog. Bip.	120	41.7	C-13	164,000	75	32.8		5,360
10,000	G-96	Mog. Bip.	120	83.4	C-13	328,000	75	32.8		5,540

	Electrical characteristics					Energy-emission characteristics		
Lamp no.	Volts	Amperes	Avg mscp†	Avg life, h	Filament temp (approx), K	UV, 0.3–0.4 μm, %	Visible, 0.4–0.7 μm, %	IR‡ radiation 0.7 μm and up, %
680	5.0	0.060	0.03	100,000+	1,850	0.0005	0.4095	99.59
683	5.0	0.060	0.05	100,000	1,950	0.001	0.649	99.35
713	5.0	0.075	0.088	25,000	2,100	0.003	1.157	98.84
715	5.0	0.15	0.115	40,000	2,125	0.004	1.246	98.75
328	5.0	0.18	0.34	3,000	2,275	0.010	1.94	98.05
327	28.0	0.040	0.34	7,000	2,200	0.006	1.554	98.44

* Vacuum.
† mscp (mean spherical candlepower) is approximately one foot-candle at one foot. mscp X 4π = lumens.
‡ Includes heat loss by conduction.

over a 20 percent or so range, these rules are useful for either extending life or increasing brightness.

Lamp life is typically rated for the point at which the brightness is reduced to 50 percent by filament-coil shorting, evaporation, and vacuum deterioration. It is important to mount lamps to minimize transmitted vibration. Low-voltage, high-current lamps will also be more immune to vibration because of their more rugged filaments. It is important to operate lamps below 100°C to prevent vacuum deterioration and increased aging. These lamps should not be connected across inductive loads.

Inrush current, important for transistor drive-circuitry design, may be estimated from Fig. 43.

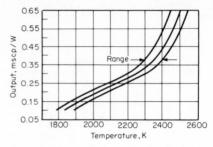

Fig. 42   Lamp output as a function of filament temperature.[99]

TABLE 32   Rerating Rules for Tungsten Lamps[38]

$$\text{Rerated mscp} = \left(\frac{V}{V_1}\right)^{3.5} \times \text{mscp at design volts}$$

$$\text{Rerated life} = \left(\frac{V_1}{V}\right)^{12} \times \text{life at design volts}$$

$$\text{Rerated current} = \left(\frac{V}{V_1}\right)^{0.55} \times \text{current at design volts}$$

V = application voltage
V_1 = design voltage

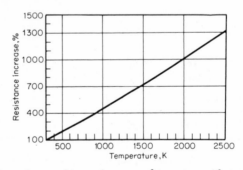

Fig. 43   Percentage resistance increase of tungsten with temperature.[100]

Solid-State Lamps

Table 33 and Fig. 44 compare various available semiconductor sources. With suitable packages and drive circuits GaAs and red GaAsP can be modulated at frequencies above 30 MHz. Rise times for optical pulses for GaAs and InAs (3-μm emission) can be in the nanosecond region. Alloy emitters of lead tin telluride, as "specials," are available in the 6- to 20-μm region, but these units must be cryogenically cooled. They have the advantage of being temperature- (current-) tunable.

The color of visible emitters is an important selection criterion. Red is often associated with danger. Human vision is most sensitive to green. (The green of visible lamps is described by some observers as yellowish-green or lime green.) About 8 percent of the male population sees red as gray. Older men may see double images when attempting to focus on red.

TABLE 33 Comparison of Semiconductor Sources[39]

Material	Voltage	Wavelength, nm	Efficiency, %	Response time, ns
GaAs	1.25	900	0.7 (A/W)	1
Si-GaAs	1.35	940	4.2 (A/W)	300
GaP-Zn,0		698		
GaP-N		565	0.01	
GaP-Zn		553	1	
$GaAs_{1-x}P_x$		670	0.1	
		660	0.03	
		610	0.001	
		500		
$Ga_{1-x}Al_xAs$		688		
SiC		590		

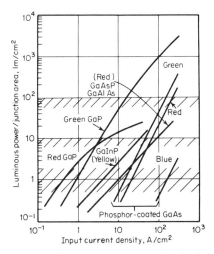

Fig. 44 LED brightness and efficiency comparison for shaded area. (I) Symbolic-display power requirement. (II) Single-indicator-light requirement. (III) Distant indicators.[101]

Manufacturers may use various coupling optics (Fig. 45) to increase light intensity with a possible included-viewing-angle trade-off.

Typical packages are shown in Fig. 46. Life figures for well-burned-in units (24 to 48 h) are in excess of 30,000 h and may well range to above 10^6. Units operated above the rated currents are subject to unpredictable life variations.

Spectral and power output of diodes will vary with temperature. Figure 47 shows the typical variation for GaAs.

Since these diodes have low reverse ratings, protection diodes may be added in series. Series diodes may also be added for ac operation.

Since light output is proportional to current, with about a 1-mA offset, diode sources are best operated in critical applications from a constant-current source such as a 1-kΩ resistor or a series regulator. $GaAs_{1-x}P_x$ sources designed for high visible brightness may not be the most efficient alloy for sensors other than the eye. For example, a vidicon or plumbicon may more efficiently use another alloy composition (Fig. 48). Recommended interface devices are given in Table 34.

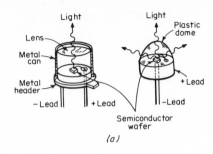

(a)

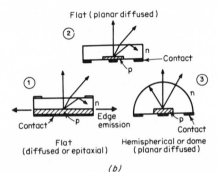

(b)

Fig. 45   (*a*) Indicator-light packages.   (*b*) Light-extraction geometries and their effectiveness.[102]   Various IR LED geometries enhance GaAs-junction radiation.   The configurations can be flat, as in (1) and (2).   The hemispherical, or dome, geometry (3) is another alternative.   The following table provides the trade-offs:

Emitter structure	Figure applicable	Power output, mW	Collection optics	Cost	Beam spread, °
Flat..............	1, 2	0.5–1.0	Lens	Low	80
Edge..............	1	0.5–3.0	Mini-parabola	Low	80
Flat/edge.........	1	3.0	Lens/mini-parabola	Low	
Dome.............	3	10.0	Parabola	High	18

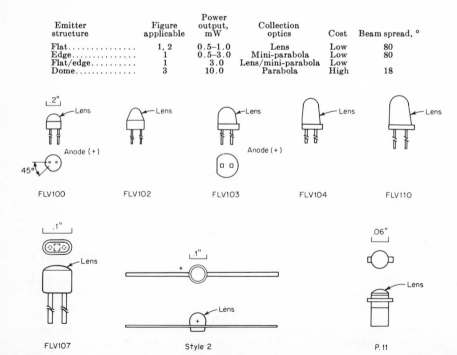

Fig. 46   Representative LED packages.[103]

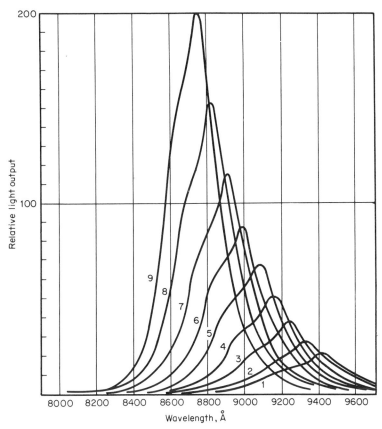

Fig. 47 Relative spectral output of a GaAs emitter with temperature. Curve 1 is 50°C; curve 2 is 25°C, etc.[131]

TABLE 34   Infrared Emitter and Visible-Light-Emitting Diode Interface Charts[131]

Device type	Description	Used with
2N4300	*npn* epitaxial planar silicon power transistor	Laser diodes
2N5385	*pnp* epitaxial planar silicon power transistor	Laser diodes
2N5450	*npn* epitaxial planar silicon transistor	Infrared emitters
TIP33	*npn* single-diffused silicon power transistor	Infrared emitters

Device type	Description
2N2905	*pnp* epitaxial planar silicon transistor
ECL2502	Emitter-coupled-logic triple two-input OR/NOR gate

High-Intensity Sources

Many applications such as microscopy, microfilm printing, glass and industrial inspection, projection, artificial aging, monochrometers, scanning, medical, lighting, and copy machines require high-intensity sources for which mercury, Xenon, and argon (Fig. 49 and Table 38) arc lamps are most commonly used. Table 35 shows the characteristics of one line of such lamps. Note that all arcs must be stabilized with a ballast.

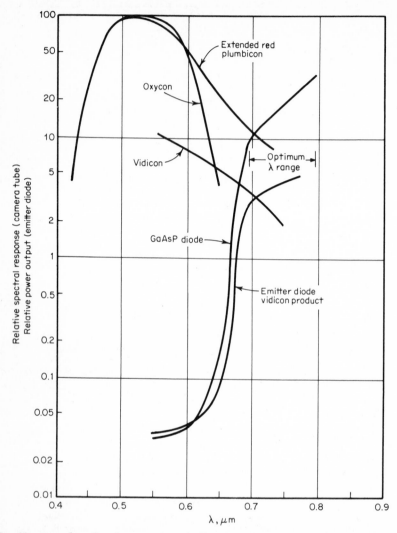

Fig. 48 Procedure for choosing GaAsP alloy (wavelength) to optimize its use with a vidicon sensor.

High-intensity glow modulator tubes are used for a variety of scanning applications, but they are characterized by useful lives as short as 50 h (Table 36).

High-intensity flash lamps find applications where nanosecond pulses are required. These are generally operated in a cold-cathode arc mode. A standard life for such sources is one million shots. Xenon is normally used as a fill gas because of its high conversion efficiency. However, krypton or argon may offer advantages in some spectral regions where they have stronger lines than xenon.

Design parameters are listed in Table 37, and typical packages are shown in Fig. 50.

Laser Sources

Laser selection Many laser selections are made on the basis of size; therefore, if an uncooled semiconductor laser of appropriate wavelength and power is available,

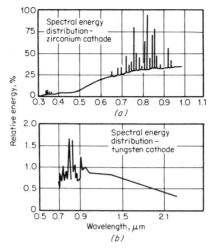

Fig. 49 Spectrum of argon-filled, high-intensity, enclosed arc lamps. (*a*) Spot sizes: ZrO cathode, 3200K, color temperature; (*b*) Tungsten cathode.[104]

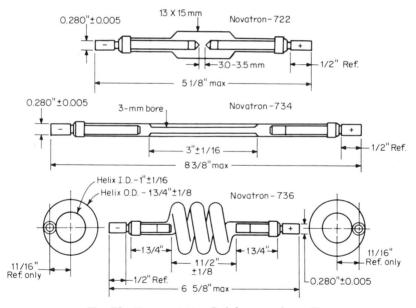

Fig. 50 Representative flash-lamp packages.[106]

it will usually suffice. Choices thereafter are usually the economical helium-neon and glass lasers. Should these not be suitable, a selection will begin with choice by wavelength and tunability and thereafter will be narrowed by power, size, life, and reliability considerations.

Semiconductor The various laser materials as characterized by operating mode, temperature, spectral output, and efficiency are shown in Table 29. Note that all lasers below threshold will operate as light-emitting diodes. At the threshold current the output is characterized by a narrow high-intensity spectral spike of stimulated

TABLE 35   Representative Xenon Short-Arc Lamps[40]

Performance specifications						Mechanical specifications		
Lamp identification	Rated power, W	Av luminance cd mm^{-2}	Luminous flux, lm	Operating voltage, V	Nominal operating current, A	Max length A, in	Diam B, in	Cold-arc gap C, in
SA—75TX	75	500	1,400	14 ± 2	5.3	1.95	0.250	0.018
SA—150TX	150	150	3,200	18 ± 2	8.3	2.28	0.312	0.075
SA—300TX	300	320	7,050	18 ± 2	16.7	2.28	0.375	0.085
SA—1000TX	1,000	450	33,000	22 ± 2	45.5	4.00	0.750	0.150

5-W Xenon Short-Arc Point-Source Lamp

Performance:
Intensity, cd	0.5 cd
Average brightness, cd cm^{-2}	1,500
Luminous flux, lm	5
Rated life, h at 5 W	1,000
Spectral output	Approximates solar spectrum (see curve of spectral distribution). Intense white light to an observer

Electrical:
Operating voltage, Vdc	10–14
Operating current, mA dc	350–500
Open-circuit voltage, Vdc min	30
Ignition voltage, kV	6–8
Input power, W:	
Min	2
Rated	5
Max	6
Operating modes	CW typically. Pulsed or modulated also

Mechanical:
Mounting orientation	Any
Arc length:	
Operating	0.25 mm (0.010 in)
Cold	0.30 mm (0.012 in)
Dimensions	Envelope, 40 mm long × 4 mm OD (1.58 × 0.16 in) plus 20-mm (0.8 in) leads at each end. Tip-off located near cathode
Cooling	Natural convection
Envelope material	Clear fused quartz

radiation superimposed on the spontaneous emission. Current densities can be above 10^4 A cm^{-2} for spontaneous emission and above 4×10^4 A cm^{-2} for stimulated emission. Threshold-current requirements are a sensitive function of temperature, and the device should be temperature-stabilized. Table 39 emphasizes this point.

Diodes may be stacked in series or arranged side by side in series to increase the output flux (Fig. 51). The combination of Al arsenide with GaAs allows for close confinement of radiation and lower thresholds. A selection guide is given in Table 40.

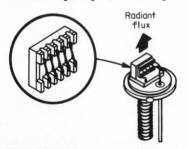
Radiant flux

Fig. 51 Series-connected injection-laser array.[107]

Gas-discharge, solid-state, and dye lasers As opposed to most electrooptical components,

TABLE 36 Representative Glow-Modulator Specifications[42]

Type no.	Max operating voltage	Current Avg	Current Peak	Min starting voltage	Crater diam, in	Approx L.C.L., in	Light output, cd	Brightness, cd/in²	Rated life, h	Base type	Bulb type	M.O.L., in	Max diam, in	Color of discharge
GM-514	160	5–25	55	240	0.056	1¾	0.1 at 25 mA	41 at 25 mA	100 at 15 mA	3-pin miniature*	T-4½	2⅝	19/32	Blue-red
GM-514C	160	5–15	35	240	0.093	1¾	0.1 at 15 mA	15 at 15 mA	25 at 10 mA	3-pin miniature*	T-4½	2⅝	19/32	White
1B59/R-1130B	150	5–35	75	225	0.056	2	0.13 at 30 mA	43 at 30 mA	250 at 20 mA	Intermediate shell, octal†	T-9	3 1/16	1 9/32	Blue-violet
R-1131C	150	3–25	55	225	0.093	2	0.2 at 25 mA	29 at 25 mA	150 at 15 mA	Intermediate shell, octal†	T-9	3 1/16	1 9/32	White
R-1166	150	3–25	55	225	0.093	2	0.2 at 25 mA	29 at 25 mA	150 at 15 mA	Intermediate shell, octal†	T-9	3 1/16	1 9/32	White
R-1168	150	5–15	30	225	0.015	2	0.023 at 15 mA	132 at 15 mA	150 at 15 mA	Intermediate shell, octal†	T-9	3 1/16	1 9/32	Blue-violet
R-1169	150	5–25	45	225	0.025	2	0.036 at 15 mA	72 at 15 mA	250 at 15 mA	Intermediate shell, octal†	T-9	3 1/16	1 9/32	Blue-violet

Type R-1166 is opaque-coated with the exception of a circle ⅜ in in diameter at the end of the lamp. All other types have a clear-finish bulb.
* Pin 1 and anode 3, Pin 2 cathode.
† Pin 7 anode, Pin 3 cathode.

TABLE 37   Design Factors for Flash Lamps[43]

l = arc length, in, measured from electrode face to electrode face

K_0 = lamp impedance parameter, Ω $(A)^{1/2}$

For the use of this parameter see Ref. 133

K_e = single-shot explosion energy constant

$E_x = K_e\, T^{1/2}$ where E_x is explosion energy in joules and $T = \sqrt{LC}$ = one-third of the pulse duration (10% current points) in seconds for a critically damped pulse. The loading factor $\Lambda = E_0/E_x$ is related exponentially to lamp life. The relationship can be usefully approximated by

$$\text{Life in flashes} = \Lambda^{-8.5}$$

$\Lambda = E_0/E_x$	1	0.76	0.58	0.44	0.33	0.25	0.19
Average life (flashes)	1	10	100	1,000	10^4	10^5	10^6

In long-life applications this must be derated for current densities above approx. 7,500/ A cm^{-2} because of the erosion of the quartz wall material. Conversely, at lower current densities the predicted life may be conservative. Lamp life is not normally guaranteed beyond 10^6 flashes ($\Lambda = 0.2$), but many customers report more than 10^7 flashes using properly designed driving circuits at low loading factors

Maximum Average Power

L-series lamps are made with high-temperature quartz-tungsten seals and so may be operated at considerably higher ratings than those given at some sacrifice of long-term life. At temperatures above 300°C the metal parts must be protected from oxidation by an inert-gas atmosphere. The seal can be operated for short periods at 600°C, and with proper cooling of the external metal parts and seal region the full temperature capability of the fused quartz can be utilized

$$V_{min}$$

Minimum recommended capacitor voltage for reliable external triggering

Recommended Trigger Pulse

Flash lamps are sensitive to the duration of the trigger pulse. The suggested trigger pulse assumes that the trigger voltage will be at or above the suggested value for the time suggested. The trigger-pulse durations are minimum in all cases; however, operation at capacitor voltages above V_{min} may substantially reduce the voltage required. L-series lamps are supplied with trigger wires for use with external trigger transformers or as the "reference plane" with series triggering. If other considerations require the use of a reference plane removed from the lamp surface, required trigger voltages will be substantially higher. Information on trigger units and trigger transformers (both external and series) may be obtained from the manufacturer

Operating Parameters

These have been calculated to give a critically damped pulse of 500 μs duration (10% current points) and a lamp life of approximately 1,000,000 flashes if no external derating factors are present, e.g., cavity operation, excessive average power. For other values of energy E_0 and pulse duration $3T$ use the circuit-design techniques described in Ref. 133. The use of discharge circuits where the damping factor a is less than 0.7 or greater than 1.1 voids any warranties regarding lamp life

$$a = \frac{K_0}{(V_0 Z_0)^{1/2}} = \frac{K_0 C^{1/4}}{V_0^{1/2} L^{1/4}}$$

TABLE 38 Characteristics of High-Intensity Arc Sources[44]

Watts	Lamp volts	Lamp amperes	Supply volts min	Starting volts min	Mean light-source diam, in	Avg brightness, cd mm²	Avg brightness, cd in⁻²	Avg axial candle power	Mean candle power/watt	Avg lumens in 90° angle	Bulb type	Base type	Max temp, °F Bulb	Max temp, °F Base	Avg life, h
2*	27	0.090	200	1,000	0.007	18	11,300	0.30	0.13	0.47	T5	Min 3-pin	140	100	150
2	38	0.055	200	1,000	0.005	25	15,500	0.30	0.13	0.47	T5	Min 3-pin	140	100	150
5	20	0.25	50	1,000	0.010	37	24,000	1.9	0.38	3.5	T9	Octal 8-pin	225	130	450
10	20	0.5	50	1,000	0.015	47	29,300	4.7	0.47	7.4	T9	Octal 8-pin	225	130	450
25	20	1.25	50	1,000	0.030	36	22,500	16.0	0.64	25.0	ST19	Octal 8-pin	355	145	350
100	16	6.25	50	2,000	0.072	39	24,500	100.0	1.0	157.0	ST19	Med. 4-pin	470	160	375‡
300	20	15.0	40	2,000	0.110	46	29,000	275.0	0.92	432.0	GT26	Med. 6-pin	520	180	250
300†	24	12.5	40 rms	2,000	0.110	46	29,000	275.0	0.92	432.0	GT26	Med. 6-pin	520	180	200

* Tungsten source.
† 60 Hz ac operating data.
‡ K100PA–200 h.

TABLE 39   Comparison of Single-Diode, 9-Mil-Width Laser Performance at Cryogenic and Room Temperatures[45]

	Room temp, 27°C	Cryogenic temp, 77 K
Peak forward current, A	40	6
Threshold current, A	10	0.6
Peak radiant flux (radiant power), W	12	3.5
Pulse duration (max), μs	0.2	2
Duty factor (max), %	0.1	4
Radiant efficiency, %	4	30
Forward voltage, V	8	2
Wavelength of peak radiant intensity, nm	904	852.5

TABLE 40   Injection-Laser and Luminescent-Diode Comparison[45]

Characteristic	Infrared-emitting diode	Injection laser
Peak radiant flux (radiant power)	40 mW	12 W
Avg radiant flux (radiant power), mW	3	12
		70*
Radiant efficiency, %	4	4
		30*
Max duty factor, %	100	0.1
		4*
Rise time of radiant flux, ns	300	<1
Half-angle beam spread, °	<15	<10
Spectral width, nm	40	3.5

Choose	When
An infrared-emitting diode	The pulse duration must exceed 200 ns at room temperature
	The application requires CW operation
A single-diode laser	The application is general
A stacked-diode laser	High radiant flux in a compact emitting area is required
An injection-laser array	Very high peak radiant flux is required
A cryogenic-laser array	High average radiant flux is required and cryogenic cooling is suitable

The generation of light by lasing takes place in the p-GaAs region. The lasing radiation propagates in the combined p-GaAs and n-GaAs regions. Because the n-GaAs region can be lightly doped, the light absorption is reduced. This reduction leads to improved laser efficiency. The GaAlAs-GaAs–GaAs heterojunctions prevent loss of laser radiation perpendicular to the plane of the mode guiding regions. The step in energy gap at the p-GaAs–p-GaAlAs heterojunction also tends to contain the injected electron density to the p-GaAs region.

* Values for cryogenic operation.

lasers (except semiconductor) are generally supplied as systems. Power supplies, cabinets, cooling provisions, and pump sources are typically packaged as a unit with the laser itself. As a result, the major design effort lies in choosing a suitable source for the desired spectral region. Spectral regions of lasers were summarized earlier. Reliability and life figures must be obtained from the manufacturer.

SYMBOLIC-DISPLAY SELECTION AND APPLICATION GUIDE

Formats and Types

Various formats for symbolic displays are shown in Fig. 52. Seven-segment displays produce digits 0 to 9; 5 × 7 dot matrix produces any letter of the alphabet

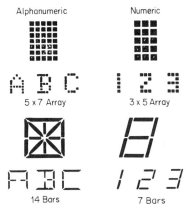

Fig. 52 Symbolic-array geometries.[105]

and digits 0 to 9; the 4 × 7 dot matrix is referred to as hexadecimal and provides letters A to F, digits 0 to 9, and a decimal. Display formats available with a typical interface chip are shown in Fig. 53.

Table 41 lists the advantages and disadvantages of various displays. Figure 54

TABLE 41   Comparison of Display Characteristics[46]

Electro-optical characteristics	Electro-luminescent panel	Incandescent lamp	Gas discharge	GaP diode	GaAlAs diode	Vacuum fluorescent
Brightness, fl	8	500	90*	200	1,000	200
Life, h	400	10,000	3,000–15,000	40,000	>40,000	10,000
Voltage, V	115 ac	4.5	180 ac	2 dc	2 dc	24 V
Current, mA	1	72	0.2	20	15	3
Speed	1 μs	1 ms	85 μs	100 ns	60 ns	40 μs
Color	White	White	Neon-red	Green-red	Red	Green

* Multiplexed at $^1/_{10}$ duty cycle.

shows some displays by range of feasible character heights. Figure 55 compares their spectral characteristics. Table 42 lists display interface IC's.

Most displays are driven by a multiplexer, which leads to the requirements that segments driven at a 1/N duty cycle be N times brighter than the apparent brightness. Avoidance of flicker requires bright displays to be pulsed at least 60 Hz and dim displays at least 120 Hz.

Symbolic displays have some parameters in common with image displays; some defects that are specified to be at an acceptable level are listed in Table 43.

LED Display

LED display parameters are much the same as for individual diodes of the same material. Figure 56 shows a typical driving circuit with resistors used to set the operating point. An LED equivalent circuit is shown in Fig. 58a.

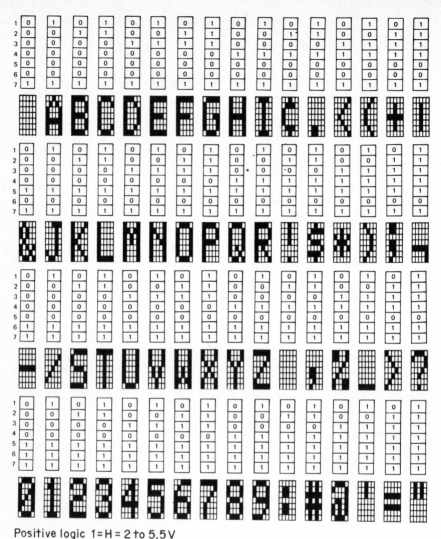

Positive logic 1 = H = 2 to 5.5 V

0 = L = 0 to 0.8 V

Fig. 53 5 × 7 alphanumeric display type TIL305. Resultant displays use TMS-4179JC or TMS4179NC chips with EBDIC coded inputs.[131]

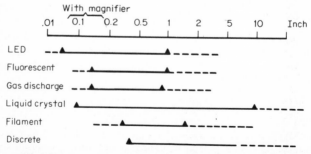

Fig. 54 Comparison of character heights by display.

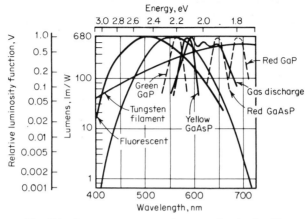

Fig. 55   Comparative spectral content by display.[108]

TABLE 42   Visible-Display Interface Chart[131]

Device type	Description	Used with
2N3980	pn planar unijunction silicon transistor	Visible displays
2N5449	npn epitaxial planar silicon transistor	Visible displays
A5T2907	pnp epitaxial planar silicon transistor	Visible displays
SN5404/SN7404	Hex inverter	Visible displays
SN54L04/SN74L04	Low-power hex inverter	Visible displays
SN5416/SN7416	Hex inverter buffer/driver with open-collector high-voltage outputs	Visible displays
SN5442/SN7442	4-line-to-10-line decoder (1 of 10)	Visible displays
SN5447A/SN7447A	BCD-to-seven segment decoder/driver	Seven-segment displays
SN5449/SN7449	BCD-to-seven-segment decoder/driver	Seven-segment displays
SN5470/SN7470	Edge-triggered J-K flip-flop	Visible displays
SN5490/SN7490	Decade counter	Visible displays
SN5492/SN7492	Divide-by-twelve counter (divide-by-two and divide-by-six)	Visible displays
SN5496/SN7496	5-bit shift register	Visible displays
SN54143/SN74143	4-bit counter/latch, seven-segment LED driver	Seven-segment displays
SN75491	VLED segment driver	Visible displays
SN75492	VLED digit driver	Visible displays
TMS 1802 NC	One-chip calculator circuit	Visible displays
TMS 2500 JC, NC	2,560-bit static read-only memory	Alphanumeric displays
TMS 2900 JC, NC	1,280-bit static read-only memory	Alphanumeric displays
TMS 4100 JC, NC	Character generator	Alphanumeric displays

Gas-Discharge Display

The gas-discharge display is a multiplexed segment or dot-matrix display.   Starting is aided by the addition of a dc keep-alive glow or by the addition of a small amount of radiokrypton.   Keep-alive circuits are operated like voltage-regulator tubes; a dc supply of about 180 V is dropped through a ballast resistor by this tube element. Operating segments of a 7-bar display are characterized by a starting voltage V_k, operating voltage $V_{KA(on)}$ reionization time, and new-entry time.   A cathode-off voltage $V_{k(off)}$ is provided to prevent extraneous glow from off cathodes in on anodes.   An anode off voltage $V_{A(off)}$ is provided to prevent on cathodes in off anodes from glowing.   These parameters are related and specified as given in Table 44.

TABLE 43   Display-Tube Uniformity Criterion

	Specification
Spatial uniformity:	
Bright spots	No more than twice as bright
Broken segments, faded segments, partial segments	Differential brightness less than twice. Not more than 3 minor or no major per display
Missing segments	None allowed
Temporal uniformity:	
Flickering segment	Nondistracting
Slow-to-enter	$< \frac{1}{3}$ s max

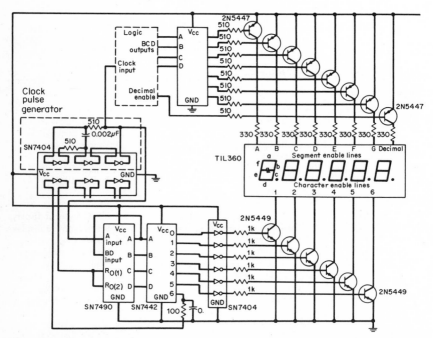

Fig. 56   Typical LED drive circuitry.   Decoder driver circuitry is shown that can interface the TIL360 six-digit display with TTL logic.   It also shows a multiplex circuit used to turn each digit on in sequence at a one-sixth duty cycle.   The BCD code, generated by the user's specific logic circuitry and applied as input to the SN7447, will be decoded into a seven-segment output.   This output drives *pnp* transistors which supply current to operate *pn*-junction segments of the display. The 330-Ω resistor in series with each segment limits the peak current to 9 mA. The display brightness may be controlled by selection of the resistor value.   Multiplexing or strobing the digits sequentially is accomplished by use of the SN7490 counter and SN7442 4-to-10-line decoder.   After counting to 6, the output from the SN7442 resets the SN7490 to zero, thus giving a duty cycle of one-sixth.[131]

The parameter E_{RI} in a tube without keep-alive depends on the MUX cycle time, the pulse rise time, direction of scan and entry, and the number of digits (distance between first and last digit).   Typical limits on cycle time are 2 to 4 ms.   If possible, cathode blanking between adjacent digits should be supplied to prevent crosstalk and arcing.   Typical values are 10 percent of the digit time.   Rise times can be as long as 20 μs, but values in the neighborhood of 2 to 5 μs yield more reliable

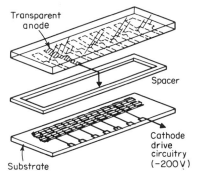

Fig. 57 Exploded view of the Panaplex II package.[109] (Trademark of Burroughs Corp.)

firing. The popular doubler drive circuit, which grounds a doubler capacitor during off cathode times and holds the cathode, through a diode, at one-half the firing voltage, leads to long rise times and varying doubler capacitor voltages because of high associated capacitances. A more satisfactory circuit is a constant-current series pass driver driven directly from the MOS switch.

A typical package is shown in Fig. 57. An equivalent circuit for the gas-discharge tube which shows associated discharge paths is given in Figs. 58 and 59.

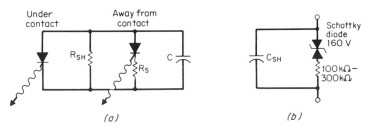

Fig. 58 (*a*) LED equivalent diode showing junction shunt effects and the effects of debiasing away from the contact. (*b*) Gas-discharge cathode-anode pair.

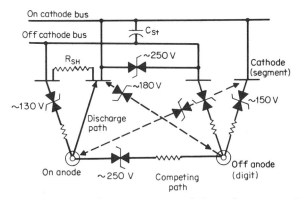

Fig. 59 Gas-discharge equivalent circuit expanded to show competing discharge paths. On and off states shown to aid in separating effects. Supply-voltage regulation and tight component tolerances are often used to maintain design performance and prevent crosstalk, as indicated here.

TABLE 44   Gas-Discharge 0.4-in-High Character Ten Digit Display-Tube Parameters[131]

Absolute maximum ratings at 25°C free-air temperature (unless otherwise noted):

Peak cathode voltage, V*	−250
Avg cathode current, mA†	−1.3
Operating free-air temp range, °C	0–55
Storage-temp range	−40 to 70

Recommended operating conditions	Min	Max
Cathode supply voltage, V_{KK}, V	−170	−210
Cathode current I_K, after firing, μA	−480	−750
Anode off-state voltage $V_{A(off)}$, V	$V_{KA}(on) + 125$	−125
Cathode off-state voltage $V_{K(off)}$, V	$V_{KK} + 125$	−125
Digit time period t_{digit}, μs	150	350
Segment blanking interval t_{blank}, μs	25	55

Operating characteristics over operating free-air temperature range:

Parameter	Test conditions	Min	Typical max
Luminance L, cd m^{-2}	$I_K = -550\ \mu$A	175	
Segment on-state voltage, $V_{KA(on)}$, V	$I_K = -450\ \mu$A	−129	−148
	$I_K = -1,300\ \mu$A	−140	−160
Initial ionization time, s	$V_{KK} = -180$ V, ambient		5
	illuminance = 50–500 L_x		
New-digit ionization time, s			0.3
Reionization time, μs		60	

Definition of Terms

Cathode supply voltage V_{KK}: The total voltage, with respect to an energized anode terminal, applied to the cathode terminal through the ballast resistor.

Cathode current I_K: The current into the cathode terminal. Negative values indicate current out of the terminal.

Anode off-state voltage $V_{A(off)}$: The voltage at the terminal of an unselected anode with respect to an energized anode terminal.

Cathode off-state voltage $V_{K(off)}$: The voltage at the terminal of an unselected cathode with respect to an energized anode terminal.

Digit time period t_{digit}: The time interval starting when one digit or character is selected and ending when the next digit is selected.

Segment blanking interval t_{blank}: The time interval between successive address pulses provided to ensure glow extinction.

Segment on-state voltage $V_{KA(on)}$: The voltage at the cathode terminal with respect to the anode terminal after the segment is fired.

Ionization time: The time interval between the initiation of conditions for and the establishment of conduction.

Initial ionization time: The ionization time following nonoperating storage.

New-digit ionization time: The ionization time for a segment in a digit adjacent to a digit that was operating during the previous cycle.

Reionization time: The ionization time for a segment that was operating during the previous cycle.

---

* All voltage values are with respect to an energized anode terminal. This value applies for a pulse width ≤350 μs, duty cycle ≤1/11.

† This value is determined by averaging over the time the current flows.

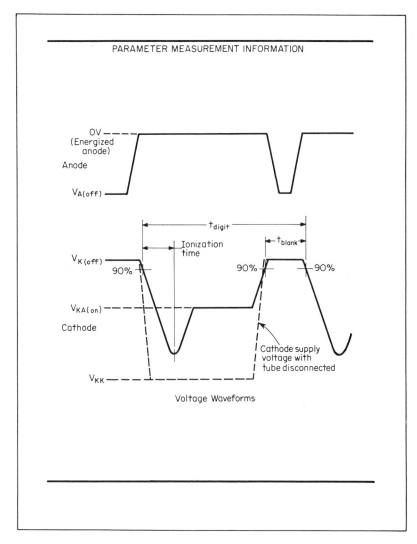

PARAMETER MEASUREMENT INFORMATION

Voltage Waveforms

Life expectancy and reliability decrease with increased cathode current or operation below 10°C. Typically failure is degradation of uniformity after 3,000 to 12,000 operational hours.

Dot-matrix tubes use a three-phase circuit to accomplish reliable dot shift and are recommended by the manufacturer for each tube.

Incandescent Displays

Incandescent displays achieve long life and low driving requirements by operating at a low temperature of 1450 K. These displays have the advantage of high brightnesses of up to 7,000 fL (Fig. 60) and white or filtered colors. Also they are operable over wide temperature ranges of at least −40 to 120°C.

Directly viewed filaments are compatible with 5-V logic; however, ballast resistors are generally required to reduce the inrush current. One drawback is the large power consumption. Packages vary from flat rectangular to NIXIE-tube-like to rear-projection type (Fig. 61).

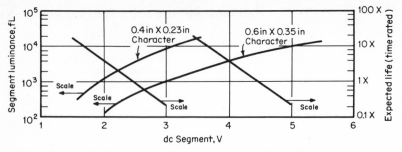

Fig. 60 Filament display segment luminescence and life trade-off factors. Normal life is about 10^5 h.[110]

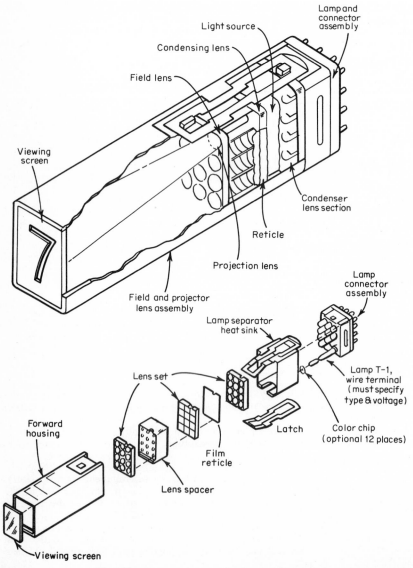

Fig. 61 Rear-projection display package.[111]

Liquid Crystals

Liquid-crystal displays use either reflected or transmitted light which may be provided by
- Ambient light
- Filament, LED, or gas-discharge source
- Radio-activated phosphor

Major advantages of the liquid-crystal display are virtually unlimited size, low power, and high contrast.

Major disadvantages may be limited temperature range, reliability questions, ac operation for long life, and possible objectionable appearance.

Such displays may be either field-effect or dynamic-scattering mode. The field effect is 5-V compatible, whereas the other may require two to six times the driving voltage. Life is enhanced by ac operation. Response times of these displays are in the millisecond range.

When a supplier is selected, the hermetic-package seal should be a major consideration, since the seal will have a major impact on the life.

Other Display Types

Other display types include vacuum-fluorescent, fluorescent-gas-discharge, and electroluminescent types. The vacuum fluorescent is an attractive option because the anode voltage of 24 V may allow direct MOS drive. The principal failure mode is phosphor brightness deterioration, apparently from bombardment by residual gas ions. CRT's are widely used for displays larger than a few inches. Scanned laser systems have been developed for displays larger than a few feet.

DETECTORS

Principles of Operation

Spectral regions and modes of operation The spectral regions for detector operation are much the same as for emitters but have some important differences. These are repeated in Table 45 for reference. The mode of operation changes from one

TABLE 45 Detector Spectral Ranges and Visual Detector Calibration Source by Range

Spectral region	Wavelengths	Calibration technique	Reason
Ultraviolet	Less than 0.3 μm	Calibrated visible source	Above visible
Visible and near	0.4–0.6	Calibrated visible source	Visible, silicon cutoff,
	0.4–1.1		glass-transmission cutoff,
	0.4–1.3		background crossover
	0.4–2.5		
Infrared	3, 5, 14 μm	Chopped blackbody	Atmospheric windows
Far infrared	Greater than 20 μm	Chopped blackbody	
	16–300 μm		Overlap with mm sources
			($10^3 \mu$m = 1 mm)

of high contrast with the background to one of low contrast at about 2.5 μm. For example, target detection at 14 μm is akin to observing a neon bulb against the sun, since a target a few degrees hotter than a 300 K background would contribute only 10^{12} photons per cm²-s more than the 10^{18} photons per cm²-s from the background. In order to remove the background, the usual procedure is to chop the target mechanically with a blade nearly the same temperature as the background [in practice, alternate the detector field of view (FOV) with the target and a background near the target]. These considerations are outlined in Tables 46 to 48.

For applications such as card reading, where the designer has control over the source strength, and therefore over the contrast, the principal considerations will be detector sensitivity and stability. He generally will not be concerned with noise and parameters based on noise.

TABLE 46   Detector Operational Modes

Mode	Factor
Noise-limited	Detector noise exceeds amplifer noise
Photon-noise-limited	Detector noise is less than photon-fluctuation noise
Background-limited (BLIP)	Internal detector noise equals background photon noise
Responsivity-limited	Detector noise not a factor; operation determined by source strength

TABLE 47   Characteristics of Natural Radiation by Region and Properties of Detectors

	Wavelength regions of interest	
	Near IR (0.7–2 μm)	Far IR (2–5.5 μm, 8–14 μm)
Characteristics	Few photons, target externally illuminated, high contrast, detectivity limited by device noise	Many photons, target a radiative source, low contrast, detectivity limited by fluctuations in background flux
Sensor:		
Image device	Vidicon	Mechanically scanned sensor arrays
Point sensors	Vacuum phototube, semiconductor, thermal	Thermal, semiconductor
Special considerations	Active source or reflected light detection. Black level set by low contrast point (detector noise or scene black level)	Passive source. Detects both above and below background ambient contrast. DC restoration to "ambient" difficult

TABLE 48   Factors Influencing Detector Choice

Parameter	Factor
Responsivity	Choose high responsivity or moderate and stable with appropriate spectral response
Photon-noise-limited	Choose:
	No longer spectral wavelength than necessary because noise generated out to cutoff (if detector unfiltered)
	Highest-temperature operation possible because of efficiency and size consideration
	Lowest cutoff frequency possible because of responsivity and noise considerations

Fluctuations in photon emission from a source lead to a photon noise current. If the photon number has a Poisson time distribution, the fluctuation is the square root of the number of photons (effective in producing a response).

Detectors may be generally classed by mode of operation as power detectors or photon detectors. A power detector responds to the power in the incident beam, i.e., $nh\nu/\Delta t$ where n is the photon flux, $h\nu$ is the mean photon energy, and Δt is the effective detector response time. Photon detectors essentially count the photons. Examples of power detectors are pyroelectric and bolometric detectors. Photoconductor, photovoltaic, and photoemitter detectors are photon detectors. The primary noise source in thermal detectors is Johnson noise. Shot noise dominates in junction photovoltaic detectors, and G-R noise dominates in photoconductors; low-frequency $1/f$ noise is contributed by all electronic devices including these. Usual practice is to cool the photon detector until the internal noise is less than or equal to the photon noise. This temperature will vary as the lowest photon energy to be detected. Therefore, long-wavelength detectors must be operated at low temperatures to achieve photon noise-limited operation. Table 49 provides typical device and temperature correlations for various spectral regions.

TABLE 49   Common Detector Candidates for Various Spectral Regions

Spectral region	Device	Operating temp
Ultraviolet	Vacuum phototube	Cooled or ambient
	Silicon PV, UV extended	
Visible to	CdS PC	Ambient
1.1 μm	Si PV	
1-3 μm	PbS PC	$-40\,^{\circ}$C
	InSb (filtered) PV	77 K
3-5 μm	HgCdTe (tailored) PC	-40 to $-100\,^{\circ}$C
	InSb PV	77 K
5-8 μm	PbTe PC	-40 to $-100\,^{\circ}$C
	HgCdTe (tailored) PC	-40 to $-100\,^{\circ}$C
8-14 μm	PbSnTe PV	77-100 K
	HgCdTe PC	77 K

Noise may also be reduced by restricting the number of photons, particularly background photons, falling on the detector. Two methods are commonly used to reduce the background contribution: reduction in the detector field of view, known as cold shielding (Fig. 3b), or restricting the viewed wavelengths with a cold filter.

Interface requirements   Interface requirements depend on detector type. Some requirements are listed in Table 50. Requirements for semiconductor detectors are further specified in Table 51.

TABLE 50   Detector Interface and Amplifier Considerations

Scene contrast	Detector	Interface consideration
High (visible)	Photovoltaic	Stable zero reference of zero current
	Photoconductive	Dark current; short-term cancellation by zeroing with detector covered. Amplifier referenced by blanking
Low (infrared)	General	Background or ambient reference levels are:
		Amplifier-detector
		Detector-enclosure/optics
		Detector-scene average
		Normalization or dc detector amplifier coupling:
		DC or direct couple (dynamic range problems)
		AC or capacitively couple (amplifier-detector reference problems)
		Ambient reference methods:
		Chop between interfaces and reference dc (ambient) to housing, scene average or chopper blade
		Supply references with IR source

Figure of merit and measurement techniques   Detectors are characterized by their responsivity, which is signal-to-power ratio (Table 52), and a normalized S/N parameter. NEI and NEP are measures of the signal parameter required to equal the rms detector noise. In the case of wideband system measurements the noise

TABLE 51   Amplifiers for Semiconductor Detectors

Detector	Bias choices	Amplifier choices	Remarks
PC	Constant current	Voltage amplifier	High input impedance susceptible to stray pickup
	Constant voltage	Current amplifier	Transistor may be biased to lowest noise point
PV	High impedance	FET	Logarithmic response with light intensity
	Low impedance	Operational amplifier for low capacitance and minimum low detector noise	Low offset current required to minimize amplifier noise
		Current amplifier	Transistor may be biased to lowest noise point
	Back bias	Charge amplifier in gated operation	Nonavalanche
		Self-amplifying	Avalanche—provides amplification
			Higher-noise transistors usable

bandwidth is the amplifier 3-dB point, which is set by the highest system signal frequency. In general, the parameter D is $1/NEP$ and D° is D normalized to unit area and bandwidth. Spectral D° or D_{λ} is required in order to compare system design predictions accurately with manufacturers' reported D°. D° measurements are best made by use of a calibrated source (Table 53), a system amplifier (calibrated), and a bias supply as shown in Fig. 62. The incident power is calculated

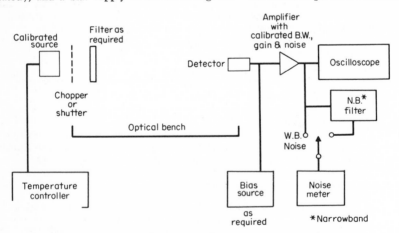

Fig. 62 Detector calibration and measurement arrangement.

from the blackbody temperature, aperture size, and distance. Two numbers are then obtained: signal and noise. The signal is just

$$\int_{\lambda_1}^{\lambda_2} R_{\lambda} H_{\lambda} S \tau_{\lambda} \, d\lambda$$

where the limits of integration are the half-power window transmission points or the R_{λ} long-wave cutoff point, S is the detector area, and τ_{λ} is the window spectral transmission.

A monochrometer chopped against a spectrally flat blackbody detector may now be used to obtain the relative response of the detector $AR_{\lambda} \tau_{\lambda}$. Point-by-point multiplication of this normalized curve with the blackbody spectrum used for measurement and subsequent integration yields $A\int R_\lambda\ H_\lambda S_{T\lambda}\ d\lambda$, a signal for $\int H_{BB\lambda}\tau_\lambda S\ d\lambda$ power. Normalizing constant A is then obtained as described under Blackbody Distribution Functions.

The noise is independent of the signal and may be measured at any time.

Detector types and applications

Power Detectors. Power detectors considered are generally the thermopile and pyroelectric types. These are basically blackbody or near-blackbody receivers which measure the small temperature rise due to the incident radiation as it comes into equilibrium with the source. Spectral response is shown in Fig. 63. Series-

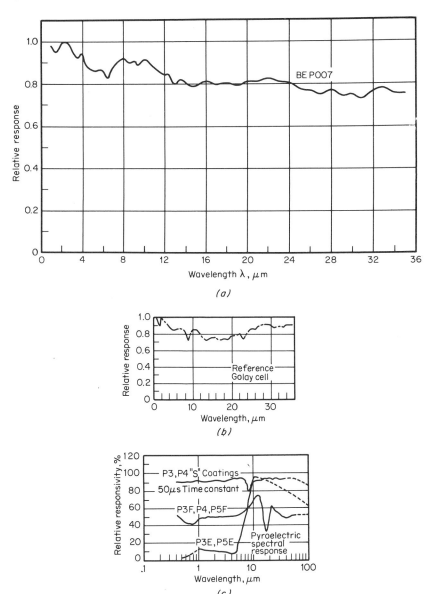

Fig. 63 Thermopile- and pyroelectric-detector spectral response.[112]

TABLE 52 Detector Parameters[47]

Parameter symbol and preferred units	Definition	Defining equation	Functional relationship
Responsive area A, cm²	For responsive elements made of thin films or single crystals used in the photoconductive and photoelectromagnetic mode, the responsive area is the region between the electrodes. For photovoltaic detectors and for detectors using integrating chambers, the responsive area is the effective area A_e	$A = \text{length} \times \text{width}$ $$A_e = \iint_s \frac{R(x,y)\,dx\,dy}{R_{max}}$$ where s = aperture area, R_{max} = maximum value of $R(x,y)$, R = responsivity	The aperture is in the (x,y) plane
Impedance Z, Ω	The slope of the voltage-current curve at bias voltage E_1	$Z = \dfrac{dE}{dI}\Big]_{E_1}$	Z is a function of the bias voltage, the interelectrode capacitance, and the level of background irradiance
Resistance $\bar{R}$, Ω	The ratio of the dc voltage across the detector to the direct current through it	$\bar{R} = E_{dc}/I_{dc}$	$\bar{R}$ is a function of the detector temperature and in some cases of Ω and T_B
Background temp T_B, K	The effective temperature of all radiation sources viewed by the detector exclusive of the signal source		where $R_{max}(0,0)$ is the maximum value of $R(x,y,0,0)$. ϕ and θ are spherical coordinates with ϕ being the azimuthal angle. The Z axis is normal to the plane of the responsive element. If the responsivity is not a function of ϕ, the element is said to have circular symmetry, and
Operating temp T, K	For uncooled detectors the operating temperature is simply the ambient temperature, and for cooled detectors it is the temperature of the coolant or the heat sink		
Detector solid angle Ω, sr	The solid angle (field of view) from which the detector receives radiation	$$\Omega = \iint_s \left[\int_0^{\pi/2} \int_0^{2\pi} \frac{\cos\theta\sin\theta\, R(x,y,\phi,\theta)}{A R_{max}(0,0)}\, d\phi\, d\theta \right] dx\, dy$$	$\Omega = \pi \sin^2 \Theta/2$ where Θ is the total cone angle

rms signal voltage (or current) ($V_{s,\text{rms}}$ (or $I_{s,\text{rms}}$) rms (or A rms)	That component of the electrical output voltage (or current) which is coherent with P_s, the input-signal radiation power. P_s can be monochromatic or have a blackbody character	If the incident radiation power is periodic in time, $$P_s(t) = P_0 + P_1 \cos(\omega_1 t + \phi_1) + \cdots$$ then $$V_s(t) = V_0 + V_1 \cos(\omega_1 t + \theta_1) + \cdots$$ and if the dc gain of the associated electronics is zero, $$V_{s,\text{rms}} = (2)^{-1/2}V_1$$ assuming that $\Delta f << f$ and that f lies in the region of Δf; i.e, $\Delta f = f_a - f_b; f_b < f < f_a$	The signal voltage is a function of electrical frequency f. For a single-time-constant detector $$V_{s,\text{rms}} = \frac{V_{s,\text{rms}}|_{f=0}}{(1 + \omega^2\tau^2)^{1/2}}$$
Spectral noise equivalent power NEP$_\lambda$, W	That value of monochromatic incident rms signal power of wavelength λ required to produce an rms signal to rms noise ratio of unity. The chopping frequency, the electrical bandwidth used in the measurement, and the detector area should be specified	$$\text{NEP}_\lambda = P_{s\lambda,\text{rms}}; \frac{V_{n,\text{rms}}}{V_{s,\text{rms}}} = \frac{V_{n,\text{rms}}}{R_\lambda}$$	Depends upon λ, A, f, Δf, and in some cases Ω and T_B

TABLE 52   Detector Parameters[47] (Continued)

Parameter symbol and preferred units	Definition	Defining equation	Functional relationship
Blackbody noise equivalent power NEP_{BB}, W	That value of incident rms signal power (with a blackbody spectral character) required to produce an rms signal to rms noise ratio of unity. The blackbody temperature must be specified along with the detector area, the electrical bandwidth used in the measurement, and the chopping frequency	$NEP_{BB} = P_{sBB \cdot rms}; \dfrac{V_{n,\,rms}}{V_{s,\,rms}} = \dfrac{V_{n,\,rms}}{R_{BB}}$	Depends upon blackbody temperature, A, f, Δf, and in some cases Ω and T_B
Spectral detectivity D_λ, W^{-1}	The reciprocal of spectral noise equivalent power. The chopping frequency, the electrical bandwidth used in the measurement, and the detector sensitive area should be specified	$D_\lambda = 1/NEP_\lambda$	Depends upon λ, A, f, Δf, and in some cases Ω and T_B
Blackbody detectivity D_{BB}, W^{-1}	The reciprocal of the blackbody noise equivalent power. The blackbody temperature should be specified, along with the electrical bandwidth used in the measurement, the detector area, and the chopping frequency	$D_{BB} = 1/NEP_{BB}$	Depends upon blackbody temperature, A, f, Δf, and in some cases Ω and T_B
Spectral D-star $D^*(\lambda,f_0)$, cm (Hz)$^{1/2}$ W^{-1}	A normalization of spectral detectivity to take into account the area and electrical bandwidth dependence. The chopping frequency f_0 used in the measurement is specified by inserting it in the parentheses as indicated in the last column. For de-	$D^*(\lambda,f_0) = \sqrt{A\,\Delta f}\,D_\lambda$	For background-noise-limited detectors, $D^*(\lambda,f_0)$ depends upon Ω and T_B

Quantity	Description	Equation	Remarks
Blackbody D-star $D^*(T_B,f_0)$, cm $(Hz)^{1/2}$ W^{-1}	tectors limited by the fluctuation in arrival rate of background photons, Ω and T_B must be specified. A normalization of blackbody detectivity to take into account the detector area and the electrical bandwidth. The chopping frequency f_0 and the blackbody temperature are specified in the parentheses as indicated. For detectors that are background-noise-limited, Ω and T_B must be specified	$D^*(T_{BB},f_0) = \sqrt{A\,\Delta f}\, D_{BB}$	For background-noise-limited detectors, $D^*(T_{BB},f_0)$ depends upon Ω and T_B
Maximized D-star $D^*_{\mu m}(\lambda_p,f_0)$, cm $(Hz)^{1/2}$ W^{-1}	A quantity obtained when the wavelength is λ_p, and the chopping frequency used yields a maximum rms signal to rms noise ratio		Same as for $D^*(\lambda,f)$
Spectral D-double star $D^{**}(\lambda,f_0)$, cm $(Hz)^{1/2}$ W^{-1} $sr^{1/2}$	A normalization of $D^*(\lambda,f_0)$ to account for the detector field of view Ω. It is used only when the detector is radiation-noise-limited. (*Note:* if $\Omega = \pi$, $D^{**} = D^*$)	$D^{**}(\lambda,f_0) = (\Omega/\pi)^{1/2} D^*(\lambda,f_0)$	
Peak wavelength (λ_p), μm	The wavelength at which detectivity is a maximum		
rms noise voltage (or current) $V_{n,\,rms}$ (or $I_{n,\,rms}$), V rms (or A rms)	That component of the electrical output voltage (or current) which is incoherent with the radiation signal power. This value is determined with the signal radiation power removed.	If the dc gain of the associated electronics is zero, $V_{n,\,rms} = (\langle V_n^2 \rangle_{av})^{1/2}$	Depends upon cell temperature and detector material used $V_{n,\,rms}$ is related to the detector area, Δf, f, and in some cases to Ω and T_B
Spectral responsivity R_λ, V rms, W rms (or A rms, W rms)	The ratio between the rms signal voltage (or current) and the rms value of the monochromatic incident signal power, referred to an infinite load impedance and to the terminals of the detector	$R_\lambda = \dfrac{V_{s,\,rms}}{P_{s\lambda,\,rms}}$	Responsivity is a function of λ, f, T, and bias voltage

TABLE 52   Detector Parameters[47] (Continued)

Parameter symbol and preferred units	Definition	Defining equation	Functional relationship
Blackbody responsivity R_{BB}, same units as above	Same as above except that the incident-signal radiation power has a blackbody spectrum	$$R_{BB} = \frac{V_{s,\,\text{rms}}}{P_{sBB,\,\text{rms}}}$$	Same as above
Time constant τ	A measure of the detector's speed of response. The alternative equations for τ (next column) become identical if the noise has a flat power spectrum and if the responsivity varies with frequency according to the relation $$R_\lambda = \frac{R_\lambda\|_{f=0}}{(1 + \Omega^2 \tau^2)^{1/2}}$$	$$\tau = 1/2\pi f_c$$ where f_c is that chopping frequency at which the responsivity has fallen to 0.707 of its maximum value τ_p is the time required for the signal voltage (or current) to rise to 0.63 times its asymptotic value. It is measured by the light-pulse method: exposing the detector to a "square-wave" pulse of radiation Responsive time constant $$\tau_r = \frac{R_{\max}{}^2}{4 \int_0^\infty [R(f)]^2 \, df}$$ Detective time constant $$\tau_d = \frac{R_{\max}{}^2}{4 \int_0^\infty [D^*(f)]^2 \, df}$$ Empirical responsive time constant $$\tau_{rs} = \frac{1}{2\pi} \left\{ \frac{[R(f_1)]^2 - [R(f_2)]^2}{[f_2 R(f_2)]^2 - [f_1 R(f_1)]^2} \right\}^{1/2}$$ f_1 and f_2 must be specified Empirical detective time constant	

		$$\tau_{ds} = \frac{1}{2\pi}\left\{\frac{[D^*(f_1)]^2 - [D^*(f_2)]^2}{[f_2 D^*(f_2)]^2 - [f_1 D^*(f_1)]^2}\right\}^{1/2}$$ f_1 and f_2 must be specified	
Cutoff wavelength λ_c, μm	The wavelength at which $D^*(\lambda,f_0)$ has degraded to one-half its peak value		Depends upon cell temperature and detector material used
Responsive quantum efficiency RQE	The ratio of the number of countable output events to the number of incident photons	$$RQE = N_0/N_p$$	Depends upon bias voltage, time constant, and cell geometry
Detective quantum efficiency DQE	The square of the ratio of measured detectivity to the theoretical limit of detectivity	$$DQE = \left[\frac{D(\lambda_0)_{\text{measured}}}{D(\lambda_0)_{\text{theoretical limit}}}\right]^2$$	

TABLE 53   Common Calibration Sources by Spectral Range

Spectral range	Detector	Source	Standard	Amplifier	Notes
Visible to Near IR	Si PV Ge PV CdS PC	Filtered tungsten or xenon	Standard solar cell or standard bolometer	None	Photovoltaic or PC current measured with ammeter. Shutter used to determine CdS background current
3 μm	InAs PV	Chopped 700–1,000 K blackbody	Calculated blackbody from aperture and distance	Operational amplifier	
5 μm	InSb	Chopped 500 K blackbody	Calculated blackbody from aperture and distance	Operational amplifier	
14 μm	PbSnTe PV	350 K blackbody	Calculated blackbody from aperture and distance	Operational amplifier	Low reverse impedance
	HgCdTe PC	350 K blackbody		Current amplifier	Low-impedance detector with low responsivity
	Ge-Hg PC	350 K blackbody		Current amplifier	High-impedance detector with high responsivity

connected thermocouples are used in the thermopile detector. Characteristics of evaporated thermopiles are given in Table 54. A low-mass electrostrictive material is used in the pyroelectric detector. It will have a large change in dielectric constant near its curie point; and if the detector operating temperature is chosen near this

TABLE 54   Characteristics of Evaporated Thermopiles[48]

Characteristic	1 X 1 mm	0.25 X 0.25 mm	2 mm diam	0.12 X 0.12 mm
Responsivity R (vacuum), V/W	50	220	160	280
Time constant τ (vacuum), μs	100	75	150	13
Impedance Z, kΩ	6.3	10	47	5
NEP, W	2.1×10^{-10}	5.9×10^{-11}	1.7×10^{-10}	3.3×10^{-11}
D^*, cm Hz$^{1/2}$/W	5.0×10^{8}	4.2×10^{8}	1.0×10^{9}	3.6×10^{9}

point, capacity changes can be used as a measure of temperature rise. This may be done by biasing and observing the change in voltage V with capacitance C for a fixed charge q:

$$\Delta V = \frac{q}{\Delta C}$$

The ideal behavior of the voltage and current response with frequency is shown in Fig. 64.

The noise from these detectors is primarily Johnson noise with little $1/f$ noise for the thermopile. D° is commonly used as a figure of merit, although noise may not depend on the square root of the area for these detectors.

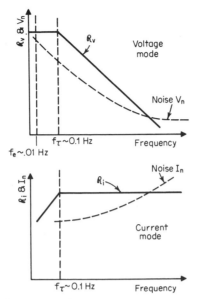

Fig. 64 Pyroelectric detector idealized responsivity and noise. (*a*) Pyroelectric. (*b*) Thermopile.

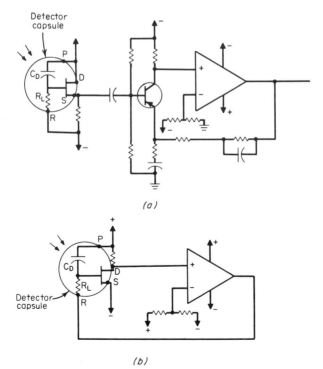

Fig. 65 Pyroelectric-detector amplifier configurations. (*a*) Voltage mode. (*b*) Current mode.[114]

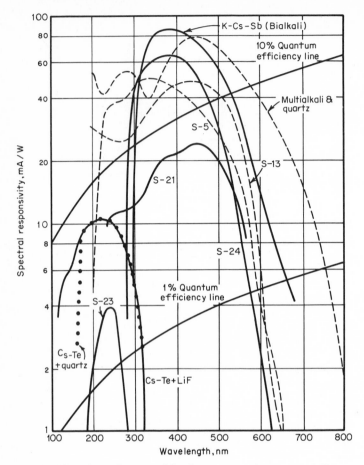

Fig. 66 Ultraviolet to visible photoemitter characteristics.[115]

The major advantage of these detectors is that they operate far into the infrared without cooling, although pyroelectric detectors are limited to a maximum temperature, which may be as low as 40°C.

The thermopile is a low-impedance voltage source and may use standard current or voltage amplifiers. The pyroelectric detector must maintain q constant and therefore requires a high dc input impedance. A field-effect transistor is often recommended as an input stage and is often incorporated in the detector package. It is also used in a source-follower configuration to obtain an impedance transformation. The use of an operational amplifier as a current amplifier is also possible, but a large feedback resistor is required for isolation. The noise will be the same in either case, but the detector impedance varies with signal strength and offsets the capacitive rolloff. Representative amplifiers are shown in Fig. 65.

Photoemitters. Photoemitters emit electrons into a vacuum when the photon energy exceeds the surface work function. Curves for typical photoemitters are given in Figs. 66 to 70 and Tables 55 and 56.

Detectors that use a photoemissive active surface are phototubes and photomultiplier tubes. A comparison of high-frequency photomultiplier characteristics is given in Table 57.

Photovoltaic Detector and Transistor. The photovoltaic detector uses the junction

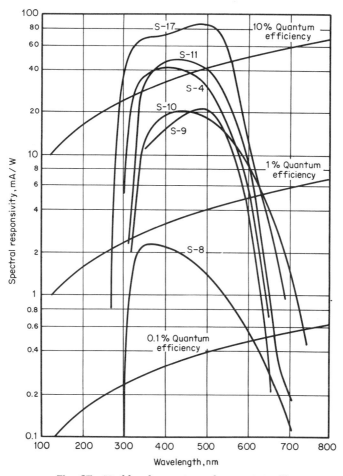

Fig. 67 Visible photoemitter characteristics.[115]

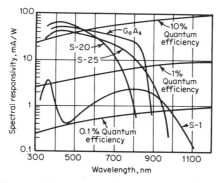

Fig. 68 Visible and near-infrared photoemitter characteristics.[115]

TABLE 55 Characteristics of Photocathodes[49]

Spectral response designation[a]	Photo-element	Type of sensor	Window	Wavelength of max response, nm	Conversion factor,[b] lm W^{-1}	Typical luminous responsivity,[c] μA lm^{-1}	Typical radiant responsivity[d] β_λ(max), mA W^{-1}	Typical quantum efficiency,[d] %	Typical photocathode dark emission at 25°C, A cm^{-2}
S-1	Ag-O-Cs	Photoemitter[e]	Lime glass	800	92.7	25	2.32	0.36	10^{-11}–10^{-12}
S-3	Ag-O-Rb	Photoemitter[e]	Lime glass	420	285	6.5	1.85	0.55	10^{-12}
S-4	Cs-Sb	Photoemitter[e]	Lime glass	400	1,044	40	41.8	12.5	10^{-14}–10^{-15}
S-5	Cs-Sb	Photoemitter[e]	9741 glass	340	1,262	40	50.5	18	10^{-14}–10^{-15}
S-8	Cs-B$_1$	Photoemitter[e]	Lime glass	365	757	3	2.27	0.78	10^{-16}
S-9	Cs-Sb	Photoemitter	Lime glass	480	683	30	20.5	5.3	10^{-14}–10^{-15}
S-10	Ag-Bi-O-Cs	Photoemitter	Lime glass	450	509	40	20.4	5.6	10^{-13}–10^{-14}
S-11	Cs-Sb	Photoemitter	Lime glass	440	808	60	48.5	14	10^{-14}–10^{-15}
S-12f	CdS	Crystal photoconductor	Epoxy and lime glass	502					
S-13	Cs-Sb	Photoemitter	Fused silica	440	799	60	48.0	13	10^{-14}–10^{-15}
S-14	Ge	pn alloy junction	Lime glass	1,500	41.8	12,400[h]	520.8	43.8	
S-15f	CdS	Polycrystalline photoconductor	Lime glass	580					
S-16	CdSe	Polycrystalline photoconductor	Lime glass	730	158.7				
S-17	Cs-Sb	Photoemitter[e] reflecting substrate	Lime glass	490	667	125	83.4	21	10^{-14}–10^{-15}

S-19	Cs-Sb	Photoemitter[e]	Fused silica	330	1,603	40	64	11	10^{-14}–10^{-15}
S-20	K-Na-Cs-Sb	Photoemitter	Lime glass	420	428	150	64.2	18	10^{-15}–10^{-16}
S-21	Cs-Sb	Photoemitter	9741 glass	440	783	30	23.5	6.6	10^{-14}–10^{-15}
S-23	Rb-Te	Photoemitter	Fused silica	240	.....	.....	4	2	10^{-17}
S-24	K-Na-Sb	Photoemitter	7056 glass	380	1,505	45	67	21.8	10^{-16}
g	K-Cs-Sb	Photoemitter	7740 pyrex	385	1,117	77	86	28	2×10^{-17}
g	CdS	Polycrystalline Photoconductor	Lime glass	510	643				
g	Cd(S-Se)	Polycrystalline Photoconductor	Lime glass	615	276				
g	Si	n-on-p photovoltaic	No window	860	75.9	7,650[i]	580[i]	83.6[i]	
S-25	K-Na-Cs-Sb	Photoemitter	Lime glass	420	215	200	43	12.7	3×10^{-16}
g	Si	PIN photoconductor	No window	1,060	.....	.....	600[i]	71[i]	$3 \times 10^{-16}(^{10})$

a The S number is the designation of the spectral-response characteristic of the device and includes the transmittance of the device envelope. It characterizes only the relative response. The absolute response of different devices with the same S number can be different. However, representative absolute values are given wherever possible.

b These conversion factors are the ratio of the radiant responsivity at the peak of the spectral-response characteristic in amperes per watt to the luminous responsivity in amperes per lumen from a tungsten source having a 2854 K color temperature.

c The luminous responsivity for the photocathode for 2854 K color temperature tungsten source. In the case of a multiplier phototube, output responsivity is obtained by multiplying the listed responsivity by the tube amplification factor.

d At the wavelength of maximum response.

e Opaque substrate, photoemission is from the side of incident radiation.

f The spectral responses indicated by S-12 and S-15 are essentially obsolete. S-12 represents the response of a crystal photoconductive cell which is no longer being manufactured. S-15 represents a polycrystalline photoconductor whose peak response at 580 nm is believed to have been the result of impurities; the CdS curve, which has a maximum at 510 nm is more typical of the state of the art.

g Spectral-response data for these photoelements have not been standardized to date by the EIA (Electronic Industries Association). The data presented are typical of results of a number of spectral-response measurements, but they have not been coordinated by the JEDEC Committee of the EIA. Therefore, they must be considered as tentative data only.

h With 45 V polarizing voltage.

i Photovoltaic short-circuit responsivity.

j 3.2-mm-thick depletion layer.

TABLE 56 Photocathode Responsivity* to Various Sources[50]

Source	Source luminosity K, lm W^{-1}	Photoconductors		Photoemitters				
		Sb$_2$S$_3$	ASOS	S-1	S-10	K-Cs-Sb Bi-alkali	S-20	S-25
P-11 phosphor	140	8.0×10^{-2}	1.4×10^{-1}	5.1×10^{-4}	1.9×10^{-2}	6.7×10^{-2}	5.7×10^{-2}	4.1×10^{-2}
P-20 phosphor	476	9.0×10^{-2}	2.5×10^{-1}	9.3×10^{-4}	1.2×10^{-2}	2.8×10^{-2}	3.8×10^{-2}	3.4×10^{-2}
2854 K tungsten	23	8.1×10^{-3}	1.8×10^{-2}	5.8×10^{-4}	9.1×10^{-4}	2.2×10^{-3}	3.5×10^{-3}	4.8×10^{-3}
5500 K blackbody	88	3.1×10^{-2}	6.1×10^{-2}	1.0×10^{-3}	5.5×10^{-3}	8.6×10^{-2}	1.9×10^{-2}	1.7×10^{-2}

† 5500 K blackbody radiation is approximately representative of sunlight.

* Responsivity computed for each source from: $\beta = \dfrac{\displaystyle\int_0^\infty \beta_\lambda P_\lambda\, d\lambda}{P}$

where β = responsivity in AW^{-1}
β_λ = spectral responsivity (AW^{-1})
$P = \displaystyle\int_0^\infty P_\lambda\, d\lambda$ = radiant power on photo cathods (W)
P_λ = spectral radiant power, (Wμm^{-1})

TABLE 57   Comparison of High-Frequency Photomultiplier Characteristics[51]

Device	Current multiplication	R_{eff}	$M^2 R_{eff}$	Comment
Dynamic cross-field electron multiplier (DCFEM)	10^7	50	5×10^{15}	Very sensitive and potentially capable of very high frequency operation, but requires a microwave pump source. Can be used as a gated detector with variable gating interval
Static crossed-field photomultiplier	2×10^5	50	2×10^{12}	Very sensitive and seems best choice for most base-band applications
Electrostatic, cancellation in pairs photomultiplier	10^3	50	5×10^7	Can be very sensitive. Suitable for large-area cathodes if desired, and can use multigap or helical output circuits
TSEM-type multiplier traveling-wave phototube	64	10^6	6×10^7	A very sensitive bandpass detector. Capable of very high (>10 GHz) frequency response. However (1) signal-to-noise- ratio is limited by low current-density capability of the TSEM films, and (2) very high voltages are required to achieve large multiplication factors
Reflection dynode traveling-wave phototube	10^3	10^1	10^{10}	A very sensitive, wide dynamic-range bandpass detector; probably a best choice for most bandpass applications

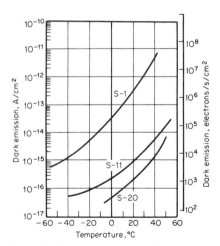

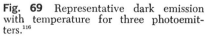

Fig. 69 Representative dark emission with temperature for three photoemitters.[116]

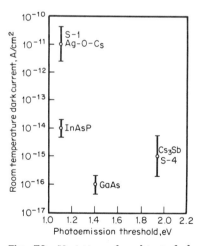

Fig. 70 Variation of ambient dark current with the photoemission threshold for some photoemitters.[117]

to separate the charges. The photon-induced current i_s is just

$$i_s = \eta q \, N A$$

where η = quantum efficiency
q = electronic charge
N = photon flux
A = detector area

The rms shot is

$$i_n^2 = 2qi_s \, \Delta f$$

The equivalent circuit is shown in Fig. 71.

The phototransistor amplifies the collector base current above by the transistor gain. Its equivalent circuit is shown in Fig. 72. To achieve high quantum efficiency, response times for these devices seem to be low; pulse-decay times tend to be particularly long. Avalanche detectors amplify the current i by the back-biased

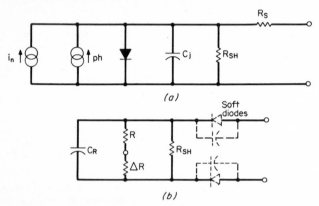

(a)

(b)

Fig. 71 Detector equivalent circuits. (*a*) Photovoltaic-detector equivalent circuit. (*b*) Photoconductive-detector equivalent circuit. Noise generation occurs in the diodes R_{SH}, and $R + \Delta R$. The R^{-1} is the photoconductivity.

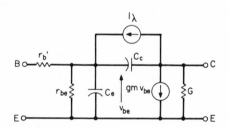

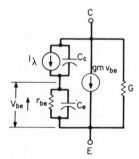

Fig. 72 Phototransistor equivalent circuits: g_m is the forward transconductance, i_λ is the collector photocurrent, and r_{be} is the effective base-emitter resistance. (*a*) Hybrid-pi model. (*b*) Floating-base approximate model.[118]

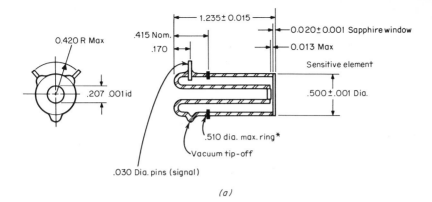

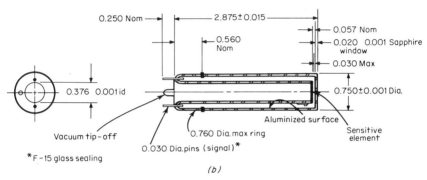

Fig. 73 Standard infrared detector Dewars.[131] (*a*) Sensitive element concentric with 0.207 bore within 0.015 TIR. (*b*) Sensitive element concentric with 0.376 bore within 0.015 TIR.

avalanche multiplication M. As long as M is below 10, little noise is added to the rms shot noise above (also multiplied by M).

Photoconductive Detector. Photoconductive detectors are of two types: intrinsic and extrinsic. The extrinsic detector has impurities added whose energies fall in the gap and allow absorption of long-wavelength radiation. These detectors will also act as intrinsic detectors above the band edge; therefore, care must be used in making measurements to prevent significant amounts of radiation shorter than the energy gap from falling on the detector. Intrinsic detectors detect photons with energies greater than the semiconductor energy gap. The photoconductivity effect is due to the increase in conductivity of a semiconductor slab when light falls on it. This occurs because of the creation of electron-hole pairs and an arrangement to collect one at the expense of the other: conductivity σ is $\sigma = nq\mu$, where n is the concentration of the collected carrier, q is the charge on the electron, and μ is the mobility. In a simplified fashion, the change in σ from a change in n is $\Delta\sigma = \Delta n q \mu$.

There will be a gain due to the lifetime of the carriers

$$G = \frac{\mu\tau V}{L^2}$$

where μ = an appropriate mobility
  τ = an appropriate charge-carrier lifetime
  V = bias voltage
  L = distance between contacts

TABLE 58 Performance of Detectors[52]

Material	Photon or thermal	Mode of operation	Film or single crystal	n-type, p-type, or intrinsic	Operating temp, K	Wavelength of peak response λ_p, μm	Cutoff wavelength (50% value) λ_o, μm	D^* (500 K, f, 1) cm Hz$^{1/2}$/W (measuring frequency indicated)	$D\lambda_p^*$ (λ_p, f, 1) cm Hz$^{1/2}$/W (measuring frequency indicated)	Response time, μs	Calculated optimum chopping frequency, Hz	Resistance per square, Ω	Noise mechanism
PbS[a]	P	PC	F	I	295	2.1	2.5	4.5×10^8 90 Hz	1.0×10^{11} 90 Hz	250	640	1.47 MΩ	Current[dd]
PbS[a]	P	PC	F	I	195	2.5	3.0	4.0×10^9 1,000 Hz	1.7×10^{11} 1,000 Hz	455	350	4 MΩ	Current
PbS[a]	P	PC	F	I	77	2.5	3.3	4.0×10^9 90 Hz	8.0×10^{10} 90 Hz	455	350	5 MΩ	Current
PbSe	P	PC	F	I	295	3.4	4.2	3.0×10^7 90 Hz	2.7×10^8 90 Hz	4	40 kHz	50 MΩ	Current
PbSe	P	PC	F	I	195	4.6	5.4	7.5×10^8 900 Hz	6×10^9 900 Hz	125	1,270	40 MΩ	Current below 6 kHz
PbSe[b]	P	PC	F	I	77	4.5	5.8	2.2×10^9 90 Hz	1.1×10^{10} 90 Hz	48	3,300	5 MΩ	Current
PbTe[c]	P	PC	F	I	77	4.0	5.1	3.8×10^8 90 Hz	2.7×10^9 90 Hz	25	6,500	32 MΩ	Current
Ge-Au[d]	P	PC	SC	p	77	5.0 (excluding intrinsic peak)	7.1	7.5×10^9 900 Hz	1.75×10^{10} 900 Hz	<1	Frequency-independent above 40 Hz	1.0 MΩ	Current below 40 Hz, gr above
Ge-Au[e]	P	PC	SC	p	65	4.7 (excluding intrinsic peak)	6.9	1.7×10^{10} 900 Hz	4×10^{10} 900 Hz	<1	Frequency-independent above 40 Hz		Current below 40 Hz, gr above[ee]
Ge-Au, Sb[f]	P	PC	SC	n	77	No clearly defined peak exists except for intrinsic excitation		2.9×10^9 90 Hz	2.5×10^{10} at 3μ 90 Hz	110	1,500	1.0 MΩ	Current
Ge-Zn (zip)	P	PC	SC	p	4.2	36	39.5	4.0×10^9 800 Hz	1.0×10^{10} 800 Hz	<0.01		300 kΩ	Current
Ge-Zn, Sb[g]	P	PC	SC	n	50	12	15	2×10^9 900 Hz	3×10^9 900 Hz				
Ge-Cu	P	PC	SC	p	<20	20	27	1×10^{10} (60° field of view) 900 Hz	2.5×10^{10} (60° field of view) 900 Hz			0.1 MΩ	Current below 1 kc; gr above 1 kc

Material				Type	T (K)	λ₁ (μm)	λ₂ (μm)	D* (a)	D* (b)	τ (μs)	Frequency response	Resistance	Noise
Ge-Cd	P	PC	SC	p	<25	16	21.5	7×10^{9} (60° field of view) 500 Hz; 3.1×10^{9} 90 Hz	1.8×10^{10} (60° field of view) 500 Hz; 7.0×10^{9} 90 Hz	0.1		10 MΩ	Current below 500 Hz, gr above 500 Hz
Ge-Si-Au	P	PC	SC	p	50	7.3	10.1				Frequency-independent to approximately 1 MHz	20 MΩ	gr
Ge-Si-Zn, Sb[i]	P	PC	SC	p	50	10	13.3	4.0×10^{9} 100 Hz	1.0×10^{10} 100 Hz	0.1	Frequency-independent below approximately 1 MHz		gr
InSb	P	PC	SC	I	295	6.5	7.3	1.4×10^{7} 800 Hz	4.3×10^{7} 800 Hz	0.2	Frequency-independent to 500 kHz	20	Thermal
InSb	P	PC	SC	I	195	5.0	6.1	5×10^{8} 900 Hz	2.5×10^{9} 900 Hz	<1	Frequency-independent above 500 Hz	60	Current below 400 Hz
InSb[j]	P	PC	SC	p	77	5.0	5.4	1.2×10^{10} (60° field of view) 900 Hz	6×10^{10} (60° field of view) 900 Hz	<2		10 kΩ	Current
InSb[k]	P	PV	SC	pn	77	5.3	5.6	8.6×10^{9} 900 Hz	4.3×10^{10} 900 Hz	<1	Frequency-independent above 500 Hz	1 kΩ	Current below 100 Hz, gr above
InSb[l]	P	PEM	SC	I	295	6.2	7.0	1.0×10^{8} 400 Hz	3.0×10^{8} 400 Hz	0.2	Frequency-independent below 100 kHz	20	Thermal
InAs[m]	P	PC	SC	n	295	3.6	3.8	1.4×10^{7} 90 Hz	1.4×10^{8} 90 Hz	0.2	Frequency-independent below 100 Hz		
InAs[n]	P	PV	SC	pn	295	3.4	3.7	2.5×10^{8} 90 Hz	2.5×10^{9} 750 Hz	<2		50	Assumed thermal
InAs[o]	P	PEM	SC	n	295	2.5	3.4	1.4×10^{7} 90 Hz	1.4×10^{8} 90 Hz	0.2	Frequency-independent below 100 Hz		Assumed thermal

TABLE 58 Performance of Detectors[52] (Continued)

Material	Photon or thermal	Mode of operation	Film or single crystal	n-type, p-type, or intrinsic	Operating temp, K	Wavelength of peak response λ_p, μm	Cutoff wavelength (50% value) λ_o, μm	D^* (500 K, f, 1) cm Hz$^{1/2}$/W (measuring frequency indicated)	$D_{\lambda p}^*$ (λ_p, f, 1) cm Hz$^{1/2}$/W (measuring frequency indicated)	Response time, μs	Calculated optimum chopping frequency, Hz	Resistance per square Ω	Noise mechanism
Te[p]	P	PC	SC	p	77	3.5	3.8	4.0×10^9 900 Hz	6.0×10^{10} 900 Hz	60	2,700	2 kΩ	Current
Tl$_2$S[q]	P	PC	F	I	295	0.9	1.1		2.2×10^{12} 90 Hz	530	300	5 MΩ	Current
86% HgTe-14% CdTe[r]	P	PC	SC	I	295	6	6.5	5×10^6 cc	1.5×10^7			≈1	
Thermistor bolometer[s]	T	Bolometer			295			1.95×10^8 (1.5 ms) 10 Hz	1.95×10^8 (1.5 ms) 10 Hz	1,500	Frequency-independent below 30 Hz	2.4 MΩ	Thermal
Radiation thermocouple[t]	T	Thermoelectric effect			295			1.4×10^9 5 Hz	1.4×10^9 5 Hz	3.6×10^1	<5	5	Thermal
Golay cell[u]	T	Expansion of air			295			1.67×10^9 10 Hz	1.67×10^9 10 Hz	2×10^4	<5		Temperature
NbSn bolometer[v]	T	Super conducting bolometer			15			4.8×10^9 360 Hz	4.8×10^9 360 Hz	500		0.2	Unknown
Carbon bolometer[w]	T	Bolometer			2.1			4.25×10^{10} 13 Hz	4.25×10^{10} 13 Hz	10^4	16	0.12 MΩ	Current
CdS[x]	P	PC	SC, F sintered	n	295	0.5	0.51	......	3.5×10^{14} 90 Hz	5.3×10^4	3	5×10^{11}	Current
CdSe[y]	P	PC	SC sintered		295	0.7	0.72		2.1×10^{11} 90 Hz	1.2×10^4	13	1.5×10^{11}	Current
Se-SeO[z]	P	PV	SC	pn	295	0.55	0.69		1.2×10^{11} 90 Hz	910	160	3 kΩ area-dependent	
GaAs[aa]	P	PV	SC	pn	295	0.8	0.89	4.5×10^{11} 400 Hz	4.5×10^{11} 400 Hz	<1	160	4.6 MΩ area-dependent	Current
IN 2175[bb] photoduodiode	P	PC	SC	pn	295	0.95	1.07	2.5×10^{10} 400 Hz	2.5×10^{10} 400 Hz	8	20 kHz	4×10^9	
1P21 photomultiplier	P	PE	F		295	0.40	0.53	5×10^{14} 1,000 Hz	5×10^{14}	<0.01	Frequency-independent to about 100 MHz		Shot

a Detectors with time constants ranging from about 1 to 10,000 μs are available. The detectivity will vary with time constant according to the McAllister relation. The cutoff wavelength may also be shifted to greater values with a sacrifice in detectivity. Detectors operating at 77 K may exhibit double time constants.

b May exhibit double time constant.

c Resistance may be reduced by grid type of electrodes. Detector is background-limited. Performance at 90 K same as at 77 K. May have second time constant for 1.5-μm radiation.

d Exhibits long time constant for intrinsic excitation (less than 2 μm). Detectivity improved by cooling to 65 K. See below.

e Exhibits long time constant for intrinsic excitation (less than 2 μ).

f Detectivity at 90 K equal that at 77 K. Exhibits wavelength-dependent time constant.

g Not readily available.

h Spectral response can be changed by varying alloy composition. Frequency response may be limited by *RC* time constant.

i Spectral response can be changed by varying alloy composition. Frequency response may be limited by time *RC* constant.

j Responsivity is superior to InSb PV, 77 K.

k May be either broad-area diffused junction or line type of grown junction. May be operated with or without bias voltage.

l Maximum dimensions approx. 2 × 10 mm. Should be transformer-coupled to amplifier. Sensitive to magnetic pickup from electrical mains.

m Not readily available.

n Detector is sapphire-immersed.

o Not readily available.

p Peak detectivity in solar and earth background minimum.

q Not readily available.

r Not readily available.

s Detectors with time constants ranging from about 1 to 50 ms are available.

t Widely used in infrared spectroscopy.

u Fragile, microphonic.

v Not readily available. Noise appears to arise from some unknown mechanism associated with superconductivity.

w Not readily available. Quartz and paraffin filters cut out response at wavelengths shorter than 40 μm.

x Highest responsivity of any photoconductor.

y Responds to longer wavelengths and is faster than CdS.

z Used in exposure meters.

aa Useful for star tracking.

bb Very small overall size.

cc Considerably higher values may be available.

dd Current noise is also known as 1/f noise

ee gr means generation-recombination noise resulting from photon or phonon fluctuations.

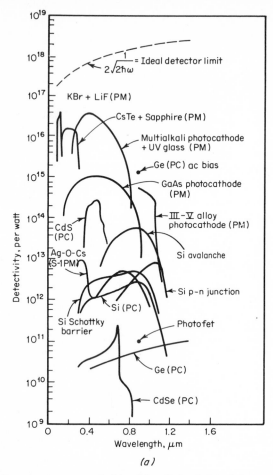

(a)

Fig. 74 (*a*) Comparative spectral detectivity *D* for 0.1 to 1.0-μm photoconductors, PC photoconductive; PM photomultiplier. (*b*) spectral D^*'s for some of the detectors above; photoconductive: (1) CdS (0 Hz); (2) CdSe (90 Hz); photovoltaic; (3) Se-SeO, (90 Hz); (4) GaAs (90 Hz); (5) IP21 photomultiplier; (6) IN2175 silicon *n*-π-*n* device (usually operated in a photoconductive mode).[119]

This gain will saturate at high biases so that the signal becomes a nonlinear function of the light flux. Detector biases must therefore be chosen to avoid this effect.

Detector Applications

Choices   The choice of a suitable detector will be determined by the available classes of detectors in a given spectral range (Table 58). First choice will usually be an uncooled semiconductor detector. Vacuum-tube devices may be necessary in the far UV, and cooled semiconductor devices or pyrometers are required in the far infrared.

Detector cooling is usually by one of three methods: cryogenic-liquid cooling, Joule-Kelvin cooler, or mechanical refrigeration. Common cryogenic materials are CO_2 at −78°C, liquid argon at 87 K, liquid nitrogen at 77 K, liquid neon at 27 K, liquid hydrogen at 20 K, and liquid helium at 4 K. The fluids are readily available from most gas suppliers in suitable Dewar storage containers. Special transfer

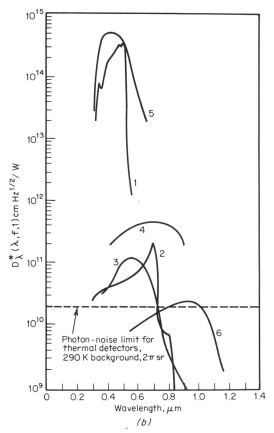

Fig. 74 (*b*) For legend see opposite page.

TABLE 59   Selection Guide for Silicon PV Detectors[53]

Choose	When
A single-element photodiode (standard configuration)	Extremes in temperature and humidity are expected
	Fast time response is required
A single-element photodiode (special wide-field-of-view configuration)	The detector is close to the emitting source
	Large f/number optics are used
	Fiber optics are employed
A quadrant-type photodetector (standard configuration)	A position-sensitive device is needed
	Fast time response is required
An avalanche photodiode	Small-signal detection is needed
	Very fast time response is required
	High responsivity is required
A hybrid photodetector-preamplifier assembly	Small-signal detection is needed
	High responsivity is required
	Very low noise is needed
	Small NEP levels independent of frequency up to 10 MHz are required.

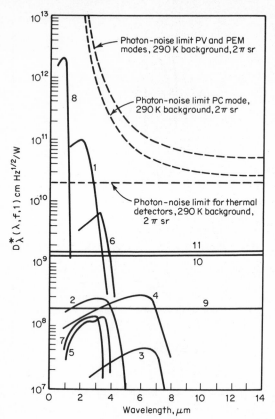

Fig. 75 Spectral D_λ^* of room-temperature detectors. (1) PbS, PC (250 μs, 90 Hz). (2) PbSe, PC (90 Hz). (3) InSB, PC (800 Hz). (4) InSb, PEM (400 Hz). (5) InAs, PC (90 Hz). (6) InAs, PV (frequency unknown, sapphire-immersed). (7) InAs, PEM (90 Hz). (8) Tl₂S, PC (90 Hz). (9) Thermistor bolometer (1,500 μs, 10 Hz). (10) Radiation thermocouple (36 ms, 1 Hz). (11) Golay cell (20 ms, 10 Hz).[120]

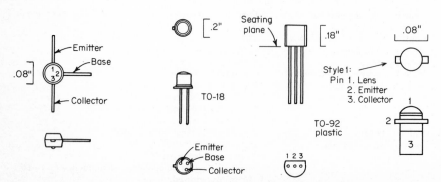

Fig. 76 Phototransistor and photodiode packages.[121]

TABLE 60   Light-Sensor Interface Chart[131]

Device type	Description	Used with
1N746 series	Silicon voltage-regulator diodes	Photodiodes, phototransistors
2N929	*npn* planar silicon transistor	Phototransistors
2N2222	*npn* epitaxial planar silicon transistor	Photodiodes, phototransistors
2N3707	*npn* planar silicon transistor	Phototransistors
2N5447	*pnp* epitaxial planar silicon transistor	Photodiodes, phototransistors
SN52107/SN72307	High-performance operational amplifier	Photodiodes, phototransistors
SN52710/SN72710	Differential comparator	Photodiodes, phototransistors
SN52741/SN72741	High-performance operational amplifier	Photodiodes, phototransistors
SN5404/SN7404	Hex inverter	Phototransistors
SN54L04/SN74L04	Low-power hex inverter	Phototransistors
SN5413/SN7413	Dual NAND Schmitt trigger	Photodiodes, phototransistors
SN5438/SN7438	Quadruple 2-input positive NAND buffer	Phototransistors
SN55107A/SN75107A	Dual line receiver	Photodiodes, phototransistors
SN75450/SN75450A	Dual peripheral driver	Photodiodes, phototransistors
TIS98	*npn* epitaxial planar silicon transistor	Photodiodes, phototransistors

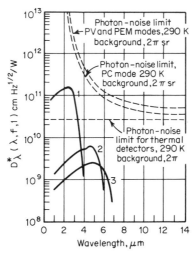

Fig. 77 Spectral D_λ^* of detectors operating at 195 K: (1) PbS, PC (1,000 Hz). (2) PbSe, PC (900 Hz) .   (3) InSb, PC (900 Hz).[122]

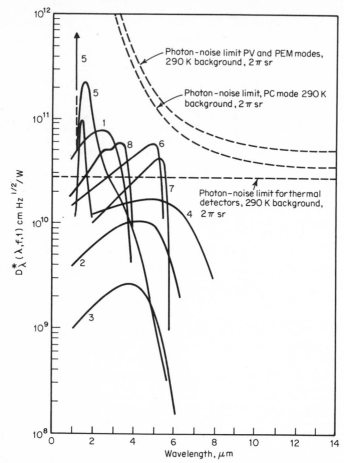

Fig. 78 Spectral D_λ^* of detectors operating at 77 K. (1) PbS, PC (90 Hz). (2) PbSe, (90 Hz). (3) PbTe, PC (90 Hz). (4) Ge-Au, PC (900 Hz). (5) Ge-Au, Sb, PC (90 Hz). (6) InSb, PC (900 Hz, 60° field of view). (7) InSb, PV (900 Hz). (8) Te, PC (900 Hz).[123]

tubes for neon and helium are also available from the same sources. Special detector Dewar packages for use with these materials are shown in Fig. 73.

The Joule-Kelvin cooler is a small expansion cooler operated from a pressurized-gas container. These typically use the same working fluids listed above and operate in the Dewar bore of the detector.

Mechanical refrigerators are available from a number of vendors: Hughes Aircraft, 500, Inc., and North American Philips Corp. among others. Transmitted vibration can be a problem, and care must be used to provide a stiff detector support.

Ultraviolet region Figure 74 shows representative spectral-sensitivity characteristics of UV detectors.

Visible and near infrared Figure 75 shows representative spectral-sensitivity curves. Silicon detectors are widely used with GaAs emitters because of the close spectral match. Phototransistor packages are shown in Fig. 76. A selection guide to silicon detectors is given in Table 59. Interface choices are given in Table 60.

Far infrared Curves for cooled IR detectors are given by temperature of operation in Figs. 77 to 79. Storage temperature of these devices is usually limited by construction materials to 125°C or less. Detector packages are shown in Fig. 73.

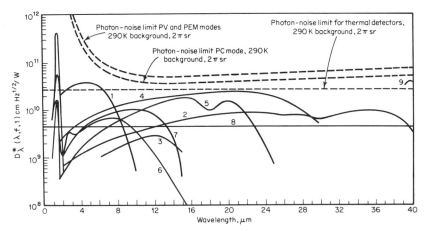

Fig. 79 Spectral D_λ^* of detectors operating at temperature below 77 K. (1) Ge-Au, 65 K, PC (900 Hz). (2) Ge-Zn, 4.2 K, PC (800 Hz). (3) Ge-Zn, Sb, 50 K, PC (900 Hz). (4) Ge-Cu, 4.2 K, PC (900 Hz, 60° field of view). (5) Ge-Cd, 4.2 K, PC (500 Hz, 60° field of view). (6) Ge–Si-Au, 50 K, PC (90 Hz). (7) Ge–Si-Zn, Sb, 50 K, PC (100 Hz). (8) NbSn superconducting bolometer, 15 K (360 Hz). (9) Carbon bolometer, 2.1 K (13 Hz).[124]

IMAGE DEVICES

Camera Devices

Classes, operation, and measurement   Various modes of camera-device operation are in use.   Figure 80 shows some of these.   The most widely used device is the electron-beam-scanned vidicon.

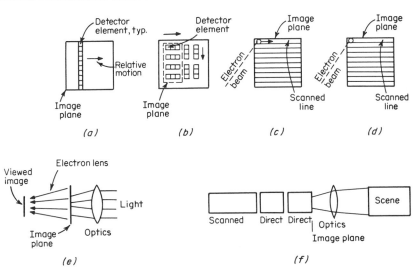

Fig. 80   Image-tube modes of operation. (*a*) Mechanically scanned linear array; example: silicon-diode array. (*b*) Staring array with storage; example: silicon charge-coupled array. (*c*) Electron-beam-scanned, no storage; example: image dissector. (*d*) Electron-beam-scanned with frame-to-frame storage; example: vidicon. (*e*) Direct conversion; example: image converter. (*f*) Combination; example: SEC vidicon coupled to image converter.

Except for the pyroelectric vidicon and the scanned thermal imager, camera tubes do not operate beyond 1.3 μm, and the choice is essentially one of operating mode and sensitivity.

Some image-device parameters under consideration by the IRIS specialty group on image devices are defined in the General Glossary.

The image device itself is a part of the larger camera system, which includes optics, power supply, and amplifiers and scanning circuits if used. The principal electrooptical performance functions for such systems are given in Fig. 81. These parameters must be specified both on and off the optical axis. A most important image criterion, which is rarely mentioned, is image uniformity. Uniformity is of two types: shading and cosmetic. Shading is usually specified to be less than a specified percentage variation over 80 percent of the imaged scene. Cosmetic defects are blemishes and are specified not to exceed certain numbers in zones of increasing radius from the center of the imaged scene. Tubes meeting various uniformity criteria may be classes by manufacturers in order of decreasing price as class (or type) I, Ia, Ib, II, and III.

Testing techniques are summarized in Table 61.

Mechanically scanned arrays   Linear multiplexed silicon arrays are available from Fairchild and Reticon which use silicon junction diodes to store the scene information. When mechanically scanned to form an image, they are read periodically at a frequency of at least twice the spatial-resolution frequency. Such multiplexing

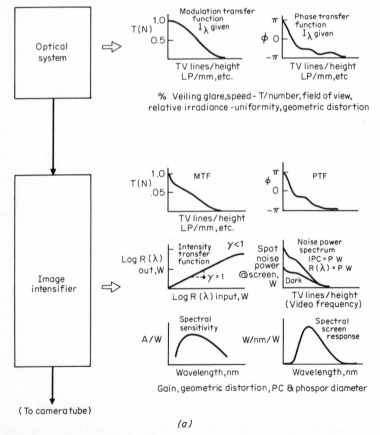

Fig. 81 Describing functions for electrooptical systems. (*a*) Image converter. (*b*) Camera tube.[125]

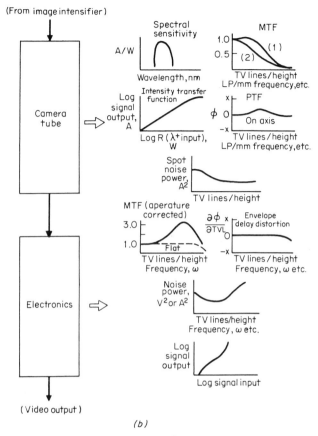

(From image intensifier)

Camera tube

Spectral sensitivity

A/W

Wavelength, nm

MTF

1.0
0.5 (2)   (1)

TV lines/height
LP/mm frequency, etc.

Log signal output, A

Intensity transfer function

Log R (λ⁺ input), W

x
φ 0   On axis
-x

PTF

TV lines/height
LP/mm frequency, etc.

Spot noise power, A²

TV lines/height

MTF (aperature corrected)

3.0
1.0   Flat

TV lines/height
Frequency, ω

$\frac{\partial \phi}{\partial TVL}$

x
0
-x

Envelope delay distortion

TV lines/height
Frequency, ω etc.

Electronics

Noise power, V² or A²

TV lines/height
Frequency, ω etc.

Log signal output

Log signal input

(Video output)

(b)

Fig. 81 (*b*) For legend see opposite page.

adds coherent noise in the form of an $n \times m$ array of picture elements, where m is the number of resolution elements in the scan direction and n is the number of detectors (normal to the scan direction). These devices are not well suited for photon-noise-limited operation because the FET readout causes switching spikes. However, these spikes may be filtered, provided that the amplifier recovery from overloading is sufficiently fast.

Dark leakage current in silicon charge-storage devices is an important factor; it is recommended that these units be operated within $\pm 10°C$ of their design value. A useful rule of thumb is that leakage current doubles every $10°C$. The useful signal for these devices may be estimated from the silicon vidicon curve of Fig. 86. The ratio of the vidicon diode area based on about ½ mil diameter to the storage transistor area and to the sample-to-storage-time ratio may be used as a correction.

Electron-scanned devices Some scanned-device categories are with and without storage, photoconductive target, photovoltaic target, or photoemissive target.

Nonstorage Devices. Nonstorage devices are insensitive and primarily used where the light-source intensity is under the designer's control. They may have an electron-beam-scanned target without storage or use a mechanically or electrically scanned photoemissive target. With the exception of sensitivity, the characteristics of these tubes and targets are similar to those of devices with storage.

Scanned Devices with Storage. Vidicon construction is shown in Fig. 83. The mesh target space is a diode structure which is the inverse of the cathode-G_1 structure.

TABLE 61   Camera-Device Testing

Test	Method	Notes
Uniformity	Uniform illumination from light box	Light box should be calibrated for uniformity with a solar cell
Square-wave response	Test chart and light box	Initially set up camera with black-to-white test chart High-contrast test-chart emulsions must be used Equal black-and-white areas must be provided to include accurately target resistance and PS regulation effects Camera amplitude response must be known and adjusted to eliminate high or low clipping Optics aperture response or MTF must be known to not be limiting Near-IR testing can be done with IR optics, diffused GaAs illuminated test chart MTF can be obtained from aperture response from square-wave Fourier expansion and deconvolving swr from highest spatial frequency to lowest
Lag	Chopped light source or "bounced" Strobotac and triggered oscilloscope	Camera must be linear for minimum lag
Absolute sensitivity	Measure dc target current using a calibrated light source	Light source may be calibrated by imaging uniform blackbody source on solar cell (known image size) with camera optics
Noise	Estimate from oscilloscope Determine optical intensity required to equal noise	No generally accepted technique

As a result, target potentials approximately 25 mV or more above the effective cathode potential will cause beam turnaround. Target current is a function of target potential, as indicated in Figs. 54d and e. Semi-insulating targets biased positively may be charged down to cathode potential by beam landing, and the output current may be used as a measure of the target potential during the scan. A photoconductive semi-insulating target will be discharged at illuminated portions between beam interrogations by an amount proportional to the illumination. Therefore, output current from a picture element will be related to the light intensity at that point. An equivalent circuit of a scanned picture element is shown in Fig. 84a. Output current is proportional to the illumination to about the three-halves power. This power is known as the tube gamma.

Photovoltaic targets, such as the multidiode silicon target or the polycrystalline diode lead oxide tube, discharge by means of the generated photocurrent. The equivalent circuit is shown in Fig. 84b and c. The output current will clearly be a linear function of illumination intensity.

Vidicons are ac devices, since the tube is cathode-blanked during retrace. Three black levels are associated with the tube-camera combination: amplifier black level, scanning-beam black level, and target black level. Dynamic range may be related to any one of these black levels, or the limiting noise may be divided into the peak-highlight target current. Peak-highlight current is determined by the amount of light required to discharge a target element during a frame time. For photoconduc-

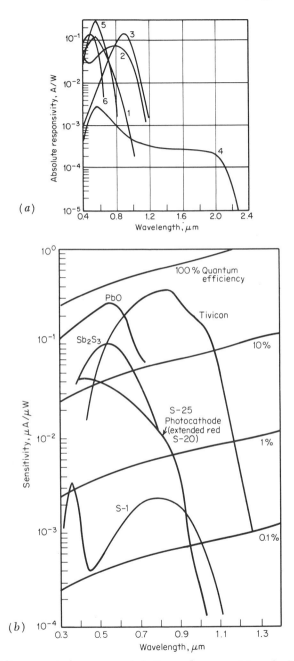

Fig. 82 Vidicon spectral response. (*a*) Scanned area = 1.2 cm², dark current = 20 nA (except as noted), signal current = 2 nA: (1) standard photoconductor commercial vidicon (Sb₂S₃); (2) S-1 photocathode intensifier vidicon; (3) silicon photoconductor = vidicon (estimated); (dark current = 50 nA); (4) lead sulfide photoconductor vidicon; (5) ASOS photoconductor vidicon; (6) lead monoxide photoconductor vidicon (approximate); dark current < 1 nA. (*b*) Spectral-sensitivity data on present TIVICON.[131]

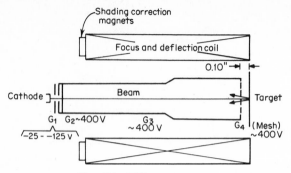

Fig. 83   Vidicon geometry.

tive targets, this current will depend on target voltage, and target voltage may be used for gain control.   Since the photocurrent in a diode will not depend on back bias,* target voltage is not a convenient control for photovoltaic target.   Gain control is best applied to the video preamplifier.

Vidicons may be electrostatic, magnetic, or combination focus and scan.   The all-magnetic tube with separate mesh is generally considered to have the highest limiting resolution.

Spectral responses for various types of targets are given in Fig. 82.   Vidicon noise is of three types: smoothed shot noise from the scanning beam, shot or G-R noise from the target, and amplifier input resistor noise.   The latter dominates practical tubes, and low-capacitance coupling to the amplifier is important.

Because a picture element may not be read "down" completely during beam sampling, partial image retention results.   Like noise, this retention is scene-dependent and as a result is measured for peak highlights.   Image retention shows up as image stickiness during camera or image motion and is referred to as lag.   Typical lag curves are shown in Fig. 85.

Opposed to lag is target burn, which occurs when an unchanging image is scanned for a length of time.   Minor burn may be removed by conventional operation for a time.   Severe burn such as "raster burn" (a burn the size of the electron imaged raster) may be permanent.   Raster burn eliminates the possibility of increasing the scanned raster size once burn has occurred, and repeated setup must ensure equal or smaller rasters.

Output may be taken from the target or from the mesh, which intercepts about half the scanning plus about half of the return beam.   Alternatively low noise amplification may be achieved by electron-multiplier amplication of the noninter-cepted portion of the return beam.

Operating temperatures of photoconductive-target vidicons are restricted to less than 71°C.   Storage temperatures are limited to less than 100°C.   Silicon-target vidicons are best operated within ±10°C of 25°C because their leakage and therefore dynamic range (equal to the peak-highlight current minus the leakage current divided by the scene average rms noise) will be limited otherwise (Fig. 86).   Where an indium front seal is used, storage temperatures are limited to less than 100°C.

Ruggedized vidicons are available with ruggedized gun structures, target mounting, and a coarse, stiff mesh.   Their resolution will generally suffer over their nonrug-gedized counterpart as a result of the coarser mesh.   Amplitude responses of typical camera tubes are given in Fig. 87.

Cascade or operational amplifiers have not been popular for use with vidicons; instead FET or low-noise transistor amplifiers are preferred.   Input impedance is generally reduced by bypassing the target resistor and compensating the amplifier frequency response at the postamplifier.   The noise will then be mostly what is termed peaked high-frequency noise and has a high spatial-frequency content

* There is a slight increase in the quantum efficiency as the depletion layer is widened.

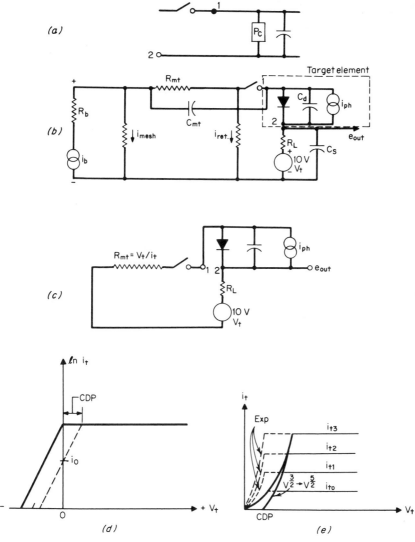

Fig. 84 Vidicon equivalent circuits. (*a*) Photoconductive target element. (*b*) Vidicon equivalent circuit with photovoltaic target element connected between points 1 and 2. (*c*) Simplified vidicon equivalent circuit. Note that V_t/I_t, the target acceptance, is not linear but follows an approximately $i_t \alpha V_t^{5/2}$ behavior. (*d*) Ideal target acceptance or I_t-V_t characteristics. (*e*) Depressed potential modification of I_t-V_t characteristics.

$\quad i_b$ = beam current
i_{mesh} = mesh interception current
$\quad i_{ret}$ = return beam
$\quad R_b$ = accelerated-beam resistance
$\quad R_{mt}$ = scanning-beam resistance
$\quad C_{mt}$ = mesh-target capacitance
$\quad C_s$ = target capacitance to ground
$\quad R_L$ = load resistance
$\quad V_t$ = target voltage
$\quad c_d$ = target diode-junction capacitance
$\quad i_{ph}$ = photogenerated target current

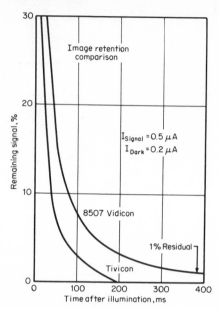

Fig. 85 Image-retention comparison of a silicon target vidicon and antimony trisulfide target vidicon.[131]

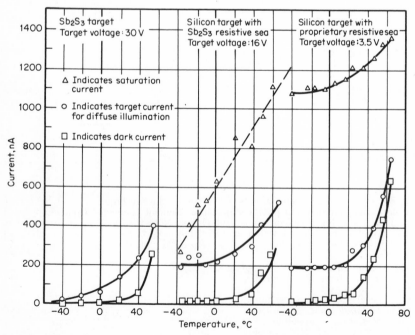

Fig. 86 Effect of operating temperature on dynamic range of three vidicons. (Here dynamic range will be peak highlight or saturation current minus the dark current divided by the dark current.) The resistive sea is a target coating designed to reduce lag.[131]

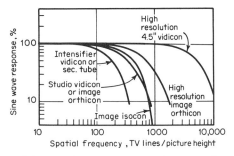

Fig. 87   Camera- and image-tube responses as a function of spatial frequency.[127]

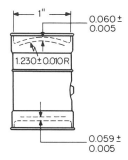

Fig. 88   A widely used image-converter/image-intensifier package.[128]

TABLE 62   Comparison of SEC and Silicon (S.V.) Vidicons

Characteristic	S.V. and camera	SEC and camera	Comments
AGC	External f/stop change required	10:1 change in AGC provides for 16:1 change in source power	SV target voltage is fixed
Temperature	No signal above 40°C marginal at 25°C owing to coherent noise. Recommended 0°C operation	71°C continuous (warranty)	SV coherent noise and dark leakage must be reduced from 25°C value by cooling
Modulation (resolution) at 400 lines	Est. 40%, depending on camera	30%+, depending on camera	No peaking. Multiply by 2 to 3 with peaking
Dynamic range	Between 30:1 and 10:1, depending on Δf and cooling	Between 20:1 and 30:1	Reduced if peaking used
Length L	L	$L + 3$ in	SEC will fit
Life	500 h (est.)	500 h warranty	SEC is unknown
Lag	$\approx 15\%$	Specified to be less than 7%	Lag source is camera nonlinearity
Lens	85% MTF (max)	90% MTF (max)	SEC has larger target
Target capacitance	≈ 36 pF	≈ 30 pF	SEC has increased noise and lower resolution

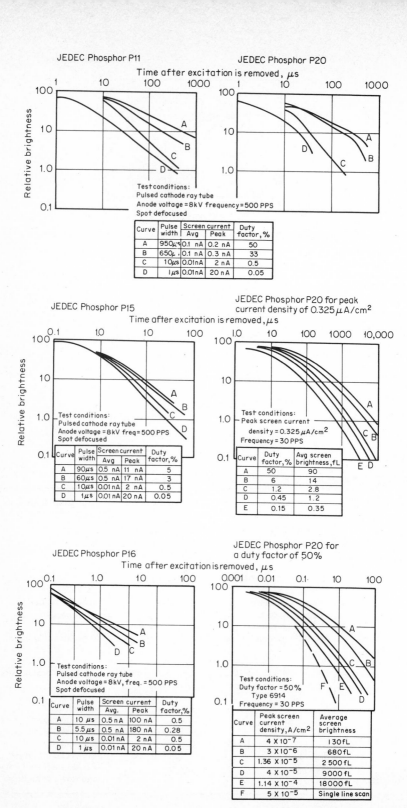

Fig. 89 Typical phosphor-persistence characteristics.[129]

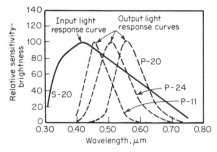

Fig. 90 Possible spectral-conversion characteristics of image intensifiers.[130]

TABLE 63 Summary of Characteristics[54]

	Length, in	Max diam, in	Eff. quantum, %	Lines/mm	Gain factor	Comments
One-stage intensifier	3.0	1.0–2.0	20	25–50	25–50	
Two-stage intensifier	4–10	3–8 (no coil)	20	20–18 18–25	500–1,500	With coils, diam 12 in
Three-stage intensifer	7–14	3–8 (no coil)	20	15–30	50,000– 100,000	With coils, diam 12 in
Image orthicon	10–25	2–4.0	20	4 horiz., 500 line 2 vertical, 100 line	1,000– 50,000	ANSI rating 10^{-3} fc 1.6×10^{-5} 10^{-6} fc 1.6×10^{-8}
SEC vidicon	17	3	20	1,500 lines/ in	200	
Film camera	8–15	8–15	2	60 fast, 150 slow	*NA*	
Eye dark-adapted	1	1	2	10 arc	*NA*	
Binoculars or telescope	8–36	2–4	*NA*	*NA*	*NA*	

TABLE 64 Image-Display Measurements

Measurement	Instrument/technique	Comments
Spot size	Calibrated microscope	
Spot intensity	Microphotometer	
Uniformity	Solar cell	
Resolution	High-resolution camera/test chart	Use with eye or measuring micro-photometer
	Electronic pattern	Alternatively use sine-wave input, measure bright/dark levels, and calculate MTF
Dynamic range or contrast ratio	Measure noise at input to display	Calculate according to system application
	Measure dark level and peak-high-light level with solar cell	
Gamma	Use linear drive voltage and solar cell	

TABLE 65 Phosphor Characteristics[55]

Screen type	Persistence	Decay time, μs	Fluorescence	Phosphorescence	Wavelength of peak radiant energy, Å	Applications	Remarks
P4	Medium or medium-short	112	White	White or blue-white	4,500	Television picture tubes	Some P4 fluorescent powders are available with a silica coating for burn resistance
P11	Short	30	Blue	Blue	4,550	Special oscilloscope for photographic recording	Produces a brilliant actinic spot. Widely used for photographic recording
P15	Extremely short	1.7	Blue-green and near UV	Blue-green and near UV	5,100	Flying-spot scanners, photographic recording	Helpful for high-resolution, high-frequency, continuous-motion recording. Not so actinic as P11
P16	Extremely short	0.1	Violet and near UV	Violet and near UV	3,850	Flying-spot scanners	Has stable exponential decay
P24	Extremely short	2.5	Light green	Light green	4,900	Flying-spot scanners	

Phosphor type	Screen material	Luminous equivalent (radiated lumens per radiated watt)	Absolute efficiency (radiated watts per watt excitation)	Quantum yield factor, photons/eV
P5	Calcium tungstate CaWO$_4$	90	0.025	0.009
P11	Zinc sulfide ZnS-Ag	140	0.10	0.038
P16	Calcium magnesium silicate	25	0.049	0.015

Decay time of a cathode-ray phosphor is usually measured as the time it takes for the light intensity to fall from its full value to $1/e$ (36.8 percent) of the full value. Sometimes, however, it is measured as the time required for the light intensity to fall to 10 percent of the full value. The decay times given in this table are to $1/e$ of the full value.

TABLE 66   Measurement Considerations for Storage CRT[56]

1. Saturation brightness is measured after removing erase pulses from the backing electrode and allowing the DVST to ion write to maximum brightness. Saturation brightnesses range from 100–7,500 fL for direct-view applications. Typical airborne cockpit displays run 1,000–2,000 fL at the tube faceplate.

2. Uniformity factor is measured as follows:
    Adjust erase duty cycle to give 3-s viewing time. Write tube to saturation and bias off writing gun. Measure time in seconds for first area to erase to visual cutoff. This time is noted as T_1. Repeat and measure time for complete display to erase. This time is designated T_2. Erase uniformity factor is defined as $(T_2 - T_1)/T_2$.

3. Resolution is measured using the shrinking-raster method, at a raster brightness of 50% saturated brightness.

4. Writing speed is measured at 50% saturated brightness with no line overlapping. Writing speeds can be tailored for any application up to 1 million in/s or more. It should be noted that as writing speed increases, viewing time decreases.

5. Erase time is equal to the maximum period required for a single pulse, of optimum amplitude, to erase stored information from 100% saturated brightness down to 10% saturated brightness. Erase time decreases as writing speed increases. As erase time decreases, viewing time decreases. Typical DVST's have erase times in the range of 1.0–200 ms. Some fast-erase tubes have erase times as low as 500 μs without dunking. With the use of dunking, the erase time of the typical DVST will be halved. Using dunking, erase times as low as 240 μs have been achieved.

6. Viewing time or persistence is the time required for written information to erase from 100% saturated brightness to visual extinction under specified erase pulse conditions, e.g., PRF, width.

7. Shades of gray are the number of clearly evident half-tones or brightness levels obtained by applying a suitable staircase video waveform to the write gun-control grid (black and white levels included). A typical DVST would offer 8 or more shades of gray.

DVST = direct-view storage tube.

TABLE 67   Useful Transmission (Exceeding 10%) Regions of IR-Transmitting Materials for 2-mm Thickness[57]

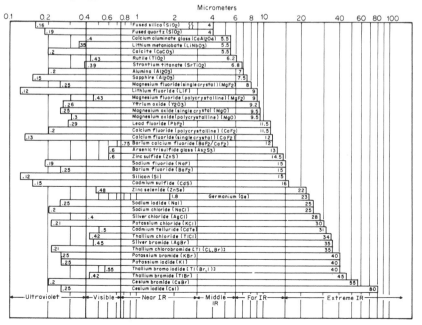

TABLE 68   Infrared-Transmitting Materials[58]

Material		Transmission — Uncoated - room temperature — Wavelength, μm		Temperature °C		Hardness		Solubility x 10⁻³ g/100g H₂O
		Thickness mm	2 4 6 8	Melting	Softening or deformation	Knoop	Mohs	Room temperature
As₂S₃	Arsenic trisulfide	4	12μm		195	100–109	2	.05
Al₂O₃	Sapphire	2		2030			9	.098
Si	Silicon	6.4		1420	1300		7	0
SiO₂	Quartz	10	1 2	1750	1585		7	0
	Corning #0160	2			610	25–460	2–3	0
SiO₂–B₂O₃ Na₂O–Al₂O₃	Pyrex	5			620		6	0
Mg O	Periclase	6.4		2800	2000	600–692	5	.012
CoF₂	Fluorite	33		1360			4	1.7
T/Br–1	KRS–5	2	70μm	414.5		200–40	2	50
CF₂–CFCl	KEL–F	6.3			260		~1	0
KCl	Sylvite	12	20μm →	776		100–93	0.5	.34700
TiO₂	Titania	6		1825			7	0
AgCl	Silver chloride	12.7	25μm →	475			1.3	.15
Ge	Germanium	6.4	~30μm →	958	500		6	0
Ca Al₂O₄	Calcium aluminate	4		1600			2	40

("snow" or "waterfall").   An 8-dB advantage is usually claimed for this arrangement over a flat head-amplifier response.

The pyroelectric vidicon uses a pyroelectric target to develop the scanning voltage. Because of the blacker-than-black nature of far-infrared radiant scenes, a synchronized mechanical scene chopper must be provided.

Nonscanned devices   Light amplifiers (matching input and output spectral ranges) and image converters (nonmatching spectral ranges) use a photoemissive target, electrostatic, channel-proximity focus, or magnetic lens, and phsophor-screen construction.   An outline of a typical tube is shown in Fig. 88.   Typical operational factors are given in Figs. 38, 66, 67, 68, 89, and 90.   The photoemissive surface operating and storage temperatures are restricted to ranges similar to those for vidicons.   The tube must be potted in a silicone rubber or a long leakage-path shield must be potted around the tube to prevent arc-over between cathode and anode.

Coupled devices

Scanned Device.   The image-converter device may be coupled to a vidicon via matching fiber-optics faceplates or may be internally coupled to a vidicon.   It is then known as an SEC vidicon or image orthicon, depending on the target.

The former device has fiber-optic faceplates lapped to a quarter-wavelength flatness (after tube assembly) which are clamped together.   The image-converter dc-to-dc power-supply oscillator is generally blanked during the active scan portion of the frame.   It is important to use a low-lag output phosphor for the image converter.

The SEC vidicon and image orthicon use photoemissive targets whose signals are imaged on a spongy KCl vidicon target in the former and a semi-insulating target in the latter.   Operational considerations are the same as for the coupled vidicon image converter.

Index of refraction Wavelength, μm			Density g/cc	Coefficient of linear thermal expansion Temperature $\times 10^6/^\circ C$ range $^\circ C$		Young's modulus dyn/cm^2	Remarks
2	6	10					
2.423	2.398	2.360	3.20	24	26	1.6×10^9	Servo Corp of America
1.750	1.746		3.98	50	50-67	344×10^{11}	Linde Air Products Co.
3.452	3.420	3.418	2.33	10-50	4.2	1.09×10^{11}	Single- and polycrystalline, 50 Ω cm
1.478			2.20	20-320	0.56	68.9×10^6	[1]Corning [2]General Electric
1.56			3.05				High lead content
1.709			2.23	0-300	3.2	6.76×10^{11}	Corning # 77YO
1.709	1.597		3.58	0-1000	13.9	25.1×10^{11}	Max diameter 2"
1.424	1.396	1.300	3.18	0-100	19.5	$7.6-14.5 \times 10^{11}$	
2.396	2.379	2.376	7.37	20-200	58	$16-2.5 \times 10^{11}$	Tonic
1.43			2.12	20-150	54.5		Rockwell hardness R-scale: III-115
1.475	1.468	1.457	1.98	(-50)-200	37	3.5×10^{11}	
2.4			4.25	40	7.1-9.2		
2.006	1.994	1.980	5.56	(-50)-100	31	$1.6-2.0 \times 10^{11}$	Affected by uv
4.116	4.102	4.005	5.3	20-100	7		
11.635			3.67	20-400	8.4	10.5×10^{11}	Bausch and Lomb

Comparison of camera tubes   Tables 62 and 63 show comparisons of some camera tubes and are representative of such comparisons.

Image Displays

Measurements   Some measurement techniques are summarized in Table 64.

Cathode-ray tubes   The cathode-ray tube is the most commonly used display Typical phosphors and characteristics are listed in Table 65, and phosphor spectral-emission curves have been shown in Fig. 38.   The P31 phosphor is capable of 500 fL operation under limited life conditions ($<$50 h).   While most CRT parameters are well known, the storage CRT requires additional attention to several parameters, as indicated in Table 66.

High-contrast operation is first obtained by reducing the black level, i.e., reducing reflected light.   This may be done with polarizing or scattering filters at some sacrifice in brightness or by the use of a hood.   Second, it may be provided by use of high-brightness phosphors.

Dynamic focusing and dynamic contrast control may be used to maintain off-center resolution and dynamic range.

AUXILIARY ELECTROOPTICAL COMPONENTS

A requirement for windows, filters, and specialized optics often arises in electrooptics-system designs.   Multilayer interference filters are most widely used, and the choice of a suitable substrate often involves factors similar to window choice.

Useful transmission ranges for common window materials are given in Tables 67 and 68.

REFERENCES

1. "Concepts and Units for the Presentation of Infrared Background Information," *Report of the Working Group on Infrared Backgrounds,* Part II, The University of Michigan, Institute of Science and Technology, Ann Arbor, Mich., *Rept.* 2389-3-S, 1956, AD 123 097.
2. Wolfe, W. L., and F. E. Nicodemus (eds.): "Handbook of Military Infrared Technology," GPO, Washington, D.C., 1965.
3. Heard, H. G. (ed.): "Laser Parameter Measurements Handbook," Wiley, New York, 1968.
4. Precision Measurement and Calibration, vol. 7, *Natl. Bur. Stand. (U.S.) Spec. Publ. 300,* 1971.
5. Biberman, L. M., and S. Nudelman (eds.): "Photoelectronic Imaging Devices," vol. 1, pp. 23, 28, Plenum, New York, 1971.
6. Precision Measurement and Calibration, vol. 7, *Natl. Bur. Stand. (U.S.) Spec. Publ. 300,* 1971.
7. Steeb, E. S., Jr., and W. E. Forsythe: "Handbook of Physics," [E. U. Condon and Hugh Odishaw (eds.)], p. 47, McGraw-Hill, New York, 1958. See also Kaufman, S. E., and J. F. Christensen: (eds.) "IES Lighting Handbook," 5th ed., Illuminating Engineering Society, New York, 1972.
8. Precision Measurement and Calibration, vol. 7, *Natl. Bur. Stand. (U.S.) Spec. Publ. 300,* 1971.
9. Weast, R. C. (ed.): "Handbook of Chemistry and Physics," Chemical Rubber Co., Cleveland, Ohio, 1973, E-207.
10. Crosswhite, H. M., and G. H. Dieke: "American Institute of Physics Handbook," 3d ed., Sec. 7, McGraw-Hill, New York, 1972.
11. Herzberg, G.: "Molecular Spectra and Molecular Structure," vol. I, "Diatomic Molecules, 1950; vol. II, "Infrared and Raman Spectra," 1945, Van Nostrand, Princeton, N.J.
12. Eisberg, R. M.: "Fundamentals of Modern Physics," pp. 293, 391, Wiley, New York, 1961.
13. Garrett, C. G. B.: "Gas Lasers," p. 10, McGraw-Hill, New York, 1967.
14. Bethe, H. A., and E. E. Salpeter: "Quantum Mechanics of One- and Two-Electron Atoms," p. 125, Academic, New York, 1957.
15. Garrett, C. G. B.: "Gas Lasers," McGraw-Hill, New York, 1967.
16. Bloom, A. L.: "Gas Lasers," Wiley, New York, 1968.
17. Smith, W. V., and P. P. Sorokin: "The Laser," McGraw-Hill, New York, 1966.
18. Smith, W. V., and P. P. Sorokin, "The Laser," p. 66, McGraw-Hill, New York, 1966.
19. White, E. L. C., and M. G. Harker: *Proc. IEEE,* vol. 97, III, p. 293, 1950. Schade, O. H., Jr.: *RCA Rev.,* vol. 26, p. 178, 1965. James, I. J. P.: *Proc. IEEE,* vol. 99, IIIA, p. 796, 1952. Sadashige, K.: *J. SMPTE,* vol. 173, p. 202, 1964.
20. Anonymous: "Electro-Optics Handbook," p. 3-1, RCA, Harrison, N.J., 1968.
21. ANSI Standard Z7.1—1967, UDC 653.014,8:621.32, sponsored by the Illumination Engineering Society. Adapted from Biberman.[5]
22. Adapted from "Reference Data for Radio Engineers," pp. 16–33, Sams, New York, 1969.
23. Smith, W. V.: "Laser Applications," p. 19, Artech House, Dedham, Mass., 1970.
24. Greenaway, D. L., and G. Harbeke: "Optical Properties and Band Structure of Semiconductors," p. 18, Pergamon, New York, 1968.
25. Jenkins, F. A., and H. E. White: "Fundamentals of Optics," p. 435, McGraw-Hill, New York, 1950.
26. Bloom, A. L.: *Proc. IEEE,* vol. 54, p. 1262, 1966.
27. Precision Measurement and Calibration, vol. 7, *Natl. Bur. Stand. (U.S.) Spec. Publ. 300,* pp. 276–1153 1971.
28. Wyszecki, Gunter, and W. S. Stiles: "Color Science," p. 35, Wiley, New York, 1967.
29. Bloom, A. L.: "Gas Lasers," p. 48, Wiley, New York, 1968.
30. Kiss, Z. J., and R. J. Pressley: *Proc. IEEE,* vol. 54, p. 1236, 1966.
31. Kiss, Z. J., and R. J. Pressley: *Proc. IEEE,* vol. 54, p. 1240, 1966.
32. Kiss, Z. J., and R. J. Pressley: *Proc. IEEE,* vol. 54, p. 1244, 1966.
33. Snitzer, E.: *Proc. IEEE,* vol. 54, p. 1249, 1966.
34. Snitzer, E.: *Proc. IEEE,* vol. 54, p. 1250, 1966.
35. Nathan, M. I.: *Proc. IEEE,* vol. 54, p. 1276, 1966.
36. Industrial Devices, Inc., Edgewater, N.J.

37. Chicago Miniature Lamp Works, Chicago, Ill. Boast, W. B.: "Illumination Engineering," p. 120, McGraw-Hill, New York, 1953.
38. Chicago Miniature Lamp Works, Chicago, Ill.
39. Adapted from Forrest Mims, *Electron. Des.*, vol. 19, p. 120, Sept. 14, 1972; and A. A. Bergh and P. J. Dean, *Proc. IEEE*, vol. 60, p. 156, 1972.
40. Xenon Corp., Medford, Mass.
41. ILC Technology, Sunnyvale, Calif.
42. GTE Sylvania, Inc., Salem, Mass.
43. ILC Technology, Sunnyvale, Calif.
44. GTE Sylvania, Inc., Salem, Mass.
45. RCA, Harrison, N.J.
46. Sher, Rober, and Rainer Zuleeg: "Semiconductors and Semimetals," vol. V, p. 520, in R. K. Willardson and A. K. Beer (eds.), "Infrared Detectors," Academic Press, New York, 1970. York, 1970.
47. Limperis, T.: in W. L. Wolfe (ed.), "Handbook of Military Infrared Technology," p. 462, GPO, Washington, D.C., 1965.
48. Stevens, N. B.: "Semiconductors and Semimetals," vol. V, p. 311, in R. K. Willardson and A. K. Beer, "Infrared Detectors," Academic Press, New York, 1970.
49. Anonymous: "Electro-Optics Handbook," p. 10-3, RCA, Harrison, N.J., 1968.
50. Anonymous: "Electro-Optics Handbook," p. 10-9, RCA, Harrison, N.J., 1968.
51. Seib, D. H., and L. W. Aukerman: "Advances in Electronics and Electron Physics," p. 148, vol. 34 in L. Marton (ed.), Academic, New York, 1973.
52. Heard, H. G. (ed.): "Laser Parameter Measurements Handbook," p. 126, Wiley, New York, 1968.
53. RCA, Harrison, N.J.
54. Soule, H. V.: "Electro-Optical Photography at Low Illumination Levels," p. 318, Wiley, New York, 1968.
55. Soule, H. V.: "Electro-Optical Photography at Low Illumination Levels," p. 154, Wiley, New York, 1968.
56. ITT Electron Tube Division, Roanoke, Va.
57. Barker, J. D.: *Electro-Optical Systems Design* p. 32, October 1970.
58. General Electric Co., Utica, N.Y.
59. Anonymous: "Electro-Optics Handbook," p. 3-3, RCA, Harrison, N.J., 1968.
60. Anonymous: "Electro-Optics Handbook," p. 2-2, RCA, Harrison, N.J., 1968. Blattner, D.: *Electron. Des.*, vol. 14, p. 94, Sept. 17, 1966.
61. Anonymous: "Electro-Optics Handbook," p. 5-4.
62. Clark, F., and L. E. Barlow: *Illum. Eng. (N.Y.)*, vol. 64, p. 603, September 1969.
63. Cobine, J. D.: "Gas Discharges," p. 514, Dover, New York, 1958.
64. Anonymous: "Electro-Optics Handbook," p. 9-2, RCA, Harrison, N. J., 1968. Blattner, D., and R. Wasserman: *Electron. Eng.* vol. 26, p. 91, August 1967.
65. Eleccion, M.: *IEEE Spectrum*, vol. 9, p. 32, March 1972.
66. Joos, G.: "Theoretical Physics, 2d ed.," p. 615, Haftner, vol. 9, p. 32, New York, 1950.
67. Adapted from S. L. Valley, "Handbook of Geophysics and Space Environments," McGraw-Hill, New York, 1965.
68. Chandrasekhar, S.: "Radiative Transfer," pp. 303, 308, Dover, New York, 1960.
69. Cobine, J. D.: "Radiative Transfer," p. 520, Dover, New York, 1960.
70. Anonymous: "Electro-Optics Handbook," p. 6-13, RCA, Harrison, N.J., 1968.
71. ILC Technology, Sunnyvale, Calif.
72. Silverman, S.: *J. Opt. Soc. Am.*, vol. 38, p. 989, 1948.
73. Cobine, J. D.: "Gas Discharges," p. 205, Dover, New York, 1958.
74. Cobine, J. D.: "Gas Discharges," p. 206, Dover, New York, 1958.
75. Cobine, J. D.: "Gas Discharges," p. 74, Dover, New York, 1958.
76. Smith, W. V., and P. P. Sorokin: *J. Opt. Soc. Am.*, vol. 38, p. 212, 1948.
77. Cobine, J. D.: "Gas Discharges," p. 538, Dover, New York, 1958.
78. Semat, H.: "Introduction to Atomic and Nuclear Physics," p. 271, Rinehart, New York, 1954.
79. Eisberg, R. M.: "Introduction to Atomic and Nuclear Physics," p. 439, Rinehart, New York, 1954.
80. Courtesy of Howard Huff.
81. Harrison, G. R., R. C. Lord, and J. R. Loofbourow: "Practical Spectroscopy," p. 271, Prentice-Hall, Englewood Cliffs, N.J., 1948.
82. Harrison, G. R., R. C. Lord, and J. R. Loofbourow, "Practical Spectroscopy," p. 298, Prentice-Hall, Englewood Cliffs, N.J., 1948.

83. Koller, L. R.: "Ultraviolet Radiation," p. 75, New York, 1952.
84. Rigrod, *Appl. Phys. Lett.* vol. 2, p. 51, 1963.
85. Bloom, A. L: "Gas Lasers," p. 53, Wiley, New York, 1968.
86. Bloom, A. L.: "Gas Lasers," p. 61, Wiley, New York, 1968.
87. Bloom, A. L.: "Gas Lasers," p. 66, Wiley, New York, 1968.
88. Kiss, Z. J., and R. J. Pressley: *Proc. IEEE*, vol. 54, p. 1239, 1966.
89. Kiss, Z. J., and R. J. Pressley: *Proc. IEEE*, vol. 54, p. 1237, 1966.
90. Bergh, A. A., and P. J. Dean: *Proc. IEEE*, vol. 60, p. 156, 1972.
91. Bergh, A. A., and P. J. Dean: *Proc. IEEE*, vol. 60, p. 162, 1972.
92. Bergh, A. A., and P. J. Dean: *Proc. IEEE*, vol. 60, p. 178, 1972.
93. Bergh, A. A., and P. J. Dean: *Proc. IEEE*, vol. 60, p. 182, 1972.
94. Bergh, A. A., and P. J. Dean: *Proc. IEEE*, vol. 60, p. 191, 1972.
95. Garlick, G. F. J.: *Sci. Prog. (London)*, vol. 52, p. 3, 1964.
96. Kiss, Z. J., and R. J. Pressley: *Proc. IEEE*, vol. 54, p. 1242, 1966.
97. Boast, W. B.: "Illumination Engineering," pp. 119, 121, McGraw-Hill, New York, 1953.
98. General Electric Company, Cleveland, Ohio.
99. Chicago Miniature Lamp Works, Chicago, Ill.
100. Chicago Miniature Lamp Works, Chicago, Ill.
101. Bergh, A. A., and P. J. Dean: *Proc. IEEE*, vol. 60, p. 213, 1972.
102. Bergh, A. A., and P. J. Dean: *Proc. IEEE*, vol. 60, p. 211, 1972.   Mims, F. M.: *Electron. Des.*, vol. 19, Sept. 14, 1972.
103. Fairchild Camera and Instrument Corp., Optoelectronics Div., Palo Alto, Calif. Motorola Semiconductor Products, Inc., Phoenix, Ariz.
104. GTE Sylvania, Inc., Salem, Mass.
105. Bergh, A. A., and P. J. Dean: *Proc. IEEE*, vol. 60, p. 215, 1972.
106. Xenon Corp., Medford, Mass.
107. RCA, Harrison, N.J.
108. Saxon, R.: *Electron. Eng.*, August 1972, p. 28.
109. Saxon, R.: *Electron. Eng.*, August 1972, p. 29.
110. Farine, P. L.: *Electron. Eng.*, vol. 34, p. 34, August 1972.
111. Shelley Associates, Inc., Santa Ana, Calif.
112. Barnes Engineering Co., Stamford, Conn., and Molectron, Inc., Sunnyvale, Calif.
113. Barnes Engineering Co., Stamford, Conn.
114. Barnes Engineering Co., Stamford, Conn.
115. Anonymous: "Electro-Optics Handbook," pp. 10-4, 10-5, 10-6, RCA, Harrison, N.J., 1968.
116. Heard, H. G. (ed.): "Laser Parameter Measurements Handbook," p. 127, Wiley, New York, 1968.
117. Spicer, W. E., and R. L. Bell: *Publ. Astron. Soc. Pac.* vol. 84, p. 110, 1972.
118. Bliss, J.: Theory and Characteristics of Phototransistors, *Appl. Note AN-440*, Motorola Semiconductor Products, Inc., Phoenix, Ariz.
119. Seib, D. H., and L. W. Aukerman: "Advances in Electronics and Electron Physics," p. 212, vol. 34 in L. Marton (ed.), Academic, New York, 1973.   Heard, H. G. (ed.): "Laser Parameter Measurements Handbook," p. 127, Wiley, New York, 1968.
120. Heard, H. G. (ed.): "Laser Parameter Measurements Handbook," p. 128, Wiley, New York, 1968.
121. Motorola Semiconductor Products, Inc., Phoenix, Ariz.
122. "Laser Parameter Measurements Handbook," p. 129, Wiley, New York, 1968.
123. Heard, H. G. (ed.): "Laser Parameter Measurements Handbook," p. 130, Wiley, New York, 1968.
124. Heard, H. G. (ed.): "Laser Parameter Measurements Handbook," p. 131, Wiley, New York, 1968.
125. Lavin, H. P.: in L. M. Bibermand and S. Nudleman (eds.), "Photoelectronic Imaging Systems," p. 373, Plenum, New York, 1972.
126. Anonymous: "Electro-Optics Handbook," p. 11-3, RCA, Harrison, N.J., 1968.
127. Anonymous: "Electro-Optics Handbook," p. 11-2, RCA, Harrison, N.J., 1968.
128. RCA, Harrison, N.J.
129. RCA, Harrison, N.J.
130. Soule, H. V.: "Electro-Optical Photography at Low Illumination Levels," p. 45, Wiley, New York, 1968.
131. Texas Instruments, Inc., Dallas, Tex.
132. Schafer, F. P. (ed.): "Dye Lasers," Springer-Verlag, New York, 1973.
133. Discharge Circuit Design Section, *ILC Tech. Bull.* 1; or Markiewicz and Emmett, Design of Flashlamp Driving Circuits, *IEEE J. Quantum Electron.*, vol. QE-2, no. 11, November, 1966.

Chapter **6**

Component Parts
for Microwave Systems

JOHN R. BUCHLEITNER

Westinghouse Electric Corp.,
Systems Development Division,
Defense and Space Center,
Baltimore, Maryland

INTRODUCTION

This chapter is concerned with the use of microwave parts by the designers and manufacturers of electronic and electrical systems and equipment. It is intended to give guidance to the user as to the capabilities and limitations of microwave parts and to provide the user with an aid in preparing specifications for circuit design parameters and part performance requirements. Theoretical treatments are presented at a very basic level to provide refresher orientation for those who need it and a basic language to the uninitiated for discourse with components manufacturers. Since many sources of information are available in handbook form on conventional microwave parts, the presentation in this area will be light. There will be a deeper discussion on the newer devices employing the use of semiconductors in the microwave field. The physics of the semiconductor device will be left for other chapters, and the discussion in this chapter relates only to the circuit equivalency of the device. It must be pointed out that the use of integrated techniques in microwave circuits is still in an advanced state. Discoveries on a daily basis make today's law tomorrow's history. Another point to be made is that the use of microwave semiconductors should be considered not for what it will replace but for what extras can be employed in the circuits. Increased versatility is the biggest advantage, especially when the increased cost must be justified.

WAVE-THEORY SUMMARY

The study of classical wave theory is based on the effects of electrical and magnetic interaction as derived from Maxwell's equations. These equations are treated in many theses and papers and are available in any technical library. Summarizing briefly, the behavior of time-varying electromagnetic fields is completely characterized by a set of four partial differential equations commonly referred to as Maxwell's equations. Solutions to these equations with the appropriate boundary conditions provide the dynamic relationships upon which microwave theory is based. These equations are as follows:

$$\nabla \times E = -j\omega\beta \tag{1}$$
$$\nabla \cdot \bar{D} = \rho \tag{2}$$
$$\nabla \cdot B = 0 \tag{3}$$
$$\nabla \times H = i + j\omega\bar{D} \tag{4}$$

Working from Maxwell's equations, the equations governing the motion of electromagnetic waves are derived. These equations are set forth as follows:

$$\nabla^2 \mathcal{E} = \mu\epsilon^1 \left(\frac{\sigma}{\epsilon^1} \frac{\partial \mathcal{E}}{\partial t} + \frac{\partial^2 \mathcal{E}}{\partial t^2} \right) \tag{5}$$

$$\nabla^2 H = \mu\epsilon^1 \left(\frac{\sigma}{\epsilon^1} \frac{\partial H}{\partial t} + \frac{\partial^2 H}{\partial t^2} \right) \tag{6}$$

(ϵ^1 is the relative dielectric constant of the medium.)

Solutions to the above equations for any particular set of boundary conditions must also be solutions for the Maxwell equations in order to have a satisfactory solution for describing the electromagnetic behavior. This is true because even though the wave equations are derived from Maxwell's equations, the converse does not deductively follow.

One of the more scholastic solutions to Maxwell's equations which typifies the manner in which the equations are solved is the uniform plane wave in a nonconducting medium. This is done by using a rationalization that at large distances from a point radiating source the surface of a sphere has no detectable spatial variation in the plane perpendicular to the direction of propagation. The study of the solution to the wave equations and the appropriate solutions to Maxwell's equations yields some interesting discoveries. It has been found that the locus of the electric vectors in the transverse plane is elliptical in nature, from which a description of the wave can be made. Such terms as linear, vertical, or horizontal describe a polarization depending upon which vector is zero, and circular describes a polarization when the vector components are equal. Further examinations of the solutions show the E and H vectors have an orthogonal relationship with each other. One of the fallouts from the solution for this problem is the term η, which has ohms for its dimension and is known as the intrinsic impedance. Typically, for free space the value for η is 377 Ω.

The next step in the academic treatments of the wave equations and Maxwell's equations eventually leads us to the rectangular waveguide. The assumptions that govern here are that the waveguide walls are perfect conductors filled with a nonconducting dielectric. This boundary condition demands that all normal components of the magnetic vector H and all tangential components of the electric vector E must be zero. Solutions to Maxwell's equations and the wave equations use the factor $e^{j\omega t - RZ}$ as propagation constant. The familiar results are shown in Table 1.

One very important characteristic of propagation through a waveguide is a phenomenon known as the skin effect. Through commonly available analysis the skin effect is shown to be related to the conduction properties of the material. It is accepted that this value is the depth at which the electric intensity has been reduced to 37 percent of its value at the surface of the conductor. The relationship for skin effect is given as follows:

$$\delta = \frac{1}{\sqrt{\pi f \mu_0}} \qquad \text{Meters} \tag{7}$$

The power has been dissipated by 13.6 percent of its initial value through 1δ. The depth of 5δ is usually regarded as the end of the penetration. Figure 1 shows the relationship of skin depth with frequency for some common conductors.

Another discovery in the solution of the waveguide equations is that the waveguide is a high-pass filter and the cutoff frequency is a value above which the propagation

TABLE 1   Summary of TE, TM, and Propagation Characteristics of Waveguide[3]

TM_{m,n} C<small>OMPONENT</small> S<small>UMMARY</small>*

$$E_z = E_1 \sin \frac{m\pi x}{a} \sin \frac{n\pi y}{b}$$

$$E_x = -\frac{y(m\pi/a)E_1}{(m\pi/a)^2 + (n\pi/b)^2} \cos \frac{m\pi x}{a} \sin \frac{n\pi y}{b}$$

$$= -\frac{(jm\lambda/2a)E_1 \sqrt{1 - (\lambda/\lambda_c)^2}}{(\lambda/\lambda_c)^2} \cos \frac{m\pi x}{a} \sin \frac{n\pi y}{b}$$

$$E_y = -\frac{(yn\pi/b)E_1}{(m\pi/a)^2 + (n\pi/b)^2} \sin \frac{m\pi x}{a} \cos \frac{n\pi y}{b}$$

$$= -\frac{(jn\lambda/2b)E_1 \sqrt{1 - (\lambda/\lambda_c)^2}}{(\lambda/\lambda_c)^2} \sin \frac{m\pi y}{a} \cos \frac{n\pi y}{b}$$

$$H_x = -\frac{-\omega\epsilon E_y}{\sqrt{\omega^2\mu\epsilon - (m\pi/a)^2 - (n\pi/b)^2}} = -\frac{\sqrt{\epsilon/\mu}}{\sqrt{1 - (\lambda/\lambda_c)^2}} E_y$$

$$H_y = \frac{\omega\epsilon E_x}{\sqrt{\omega^2\mu\epsilon - (m\pi/a)^2 - (n\pi/b)^2}} = \frac{\sqrt{\epsilon/\mu}}{\sqrt{1 - (\lambda/\lambda_c)^2}} E_x$$

$$H_z = H_1 \cos \frac{m\pi x}{a} \cos \frac{n\pi y}{b}$$

$$H_y = \frac{H_1\gamma(n\pi/b)}{(m\pi/a)^2 + (n\pi/b)^2} \cos \frac{m\pi x}{a} \sin \frac{n\pi y}{b} = \frac{j(n\lambda/2b)H_1 \sqrt{1 - (\lambda/\lambda_c)^2}}{(\lambda/\lambda_c)^2} \cos \frac{m\pi x}{a} \sin \frac{n\pi y}{b}$$

$$H_x = \frac{H_1\gamma(m\pi/a)}{(m\pi/a)^2 + (n\pi/b)^2} \sin \frac{m\pi x}{a} \cos \frac{n\pi y}{b}$$

$$= \frac{j(m\lambda/2a) \sqrt{1 - (\lambda/\lambda_c)^2}}{(\lambda/\lambda_c)^2} H_1 \sin \frac{m\pi x}{a} \cos \frac{n\pi y}{b}$$

$$E_y = -\frac{\omega\mu}{\sqrt{\omega^2\mu\epsilon - (m\pi/a)^2 - (n\pi/b)^2}} H_x = -\frac{\sqrt{\mu/\epsilon}}{\sqrt{1 - (\lambda/\lambda_c)^2}} H_x$$

P<small>ROPAGATIONAL</small> C<small>HARACTERISTICS OF</small> TE_{m,n} <small>AND</small> TM_{m,n} M<small>ODES</small>

$$\omega_c = \sqrt{\frac{1}{\mu\epsilon}\left[\left(\frac{m\pi}{a}\right)^2 + \left(\frac{n\pi}{b}\right)^2\right]}$$

$$\gamma = j\omega \sqrt{\mu\epsilon} \sqrt{1 - \frac{(m\pi/a)^2 + (n\pi/b)^2}{\omega^2\pi\epsilon}} = j\frac{2\pi}{\lambda}\sqrt{1 - \left(\frac{\lambda}{\lambda_c}\right)^2}$$

$$\lambda_c = \frac{2}{\sqrt{(m/a)^2 + (n/b)^2}}$$

$$\lambda = \frac{2\pi c}{\omega \sqrt{\mu_r \epsilon_r}}$$

---

* The quantity E, is merely a reference amplitude.   All amplitude values are crest values.

---

constant allows the particular mode to propagate without attenuation.   The equation generally expressed for a closed medium of rectangular pipe is

$$\lambda_c = \frac{2}{\sqrt{(m^2/a) + (n^2/b)}} \tag{8}$$

The observation of interest is that the cutoff frequency is dependent only upon the dimensions of the waveguide and the mode integers.   Table 2 gives the cutoff

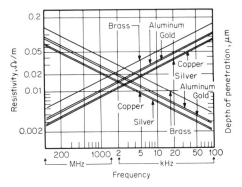

Fig. 1 Resistivity-penetration relationship.[1]

TABLE 2 Cutoff Wavelengths for Rectangular and Circular Waveguide[1]

Rectangular		Circular	
Mode	λ_c	Mode	λ_c
$TE_{0,1}$	$2b$	$TE_{1,1}$	$1.71d$*
$TE_{1,1}$ or $TM_{1,1}$	$\dfrac{2}{\sqrt{(1/a)^2 + (1/b)^2}}$	$TM_{0,1}$	$1.31d$
$TE_{0,2}$	b	$TE_{0,1}$	$0.82d$
$TE_{1,0}$	$2a$	$TE_{2,1}$	$1.03d$
$TE_{m,n}$ or $TM_{m,n}$	$\dfrac{2}{\sqrt{(m/a)^2 + (n/b)^2}}$	$TM_{1,1}$	$0.82d$

*The symbol d represents the diameter of the circular pipe.

frequencies for circular and rectangular pipe, Table 3 summarizes some useful relationships for waveguide in general, and Table 4 shows some typical field configurations for closed waveguide systems.

When the wavelength of signal passing through the waveguide is measured, the term is usually referred to as λ_g, the waveguide wavelength. This phenomenon occurs because of the manner in which the wave flow propagates in the closed system. The wavefronts actually reflect at the waveguide walls, and there are two wavefronts to consider: one that travels in the Z direction of the waveguide and one that is incident to and reflects from the waveguide walls. This ends up with two concepts of velocity and wavelength. The phase velocity is the wavefront propagated along the waveguide and is expressed as follows:

$$v_p = \frac{w}{2\pi} \lambda_g \qquad (9)$$

The group velocity relates the actual velocity of the wavefront which is folding up on itself between the walls of the waveguide. From these expressions comes the familiar formula for waveguide wavelength as follows:

$$\lambda_g = \frac{\lambda}{\sqrt{1 - (\lambda/\lambda_c)^2}} = \frac{2\pi}{\beta} \qquad (10)$$

TABLE 3 Basic Microwave Symbols and Equations[7]

Symbols		Equations

α Attenuation

α_0 Attenuation of air-filled copper transmission line = 0.35×10^{-9} nepers/meter = 0.3×10^{-5} dB/kilometer

Γ Coefficient of reflection = -1 for short circuit = $+1$ for open circuit = 0 for matched load skin depth

δ Skin depth

ϵ Dielectric constant

ϵ_0 Dielectric constant for air

λ Wavelength

λ_g Guide wavelength

λ_c Cutoff wavelength

μ Permeability

μ_0 Permeability of air = $4\pi \times 10^{-7}$ H/m

K Coupling coefficient

ρ Resistivity = 1.74×10^{-8} Ω-m for copper

σ Electrical conductivity

κ_0 Permittivity of air $- 8.854 \times 10^{-12}$ F/m

v Velocity of propagation

v_0 Velocity of propagation in air = 2.998×10^8 cm/s = 2.998×10^8 m/s = $186,280$ mi/s = 11.808×10^9 in/s

ϕ Phase angle

β Phase constant

Z_0 Characteristic impedance = 376.7 Ω for free space = 120π

D Directivity

P Power

V Voltage

I Current

R Resistance

C Capacitance

G Conductance

B Susceptance

X Reactance

Y Admittance

L Inductance

f Frequency

ω Angular frequency = $2\pi f$

Q Figure of merit of a resonator

$$2\pi = \frac{\text{energy stored}}{\text{energy dissipated per cycle}} = \frac{\Delta f}{f_0}$$

H Magnetic vector

E Electric vector

a Broad waveguide dimension

b Narrow waveguide dimension

z Direction of propagation

n Mode designation (for TE_{10} $m = 1$, $n = 0$)

m Mode designation (for TE_{10} $m = 1$, $n = 0$)

vswr Voltage standing-wave ratio

pswr Power standing-wave ratio

Wavelength in meters $\lambda_m =$
$$\frac{300,000}{f \text{ in kHz}} = \frac{300}{f \text{ in MHz}}$$

Wavelength in centimeters $\lambda_{cm} =$
$$\frac{30,000,000}{f \text{ in kHz}} = \frac{30,000}{f \text{ in MHz}}$$

Wavelength in inches $\lambda_{in} =$
$$\frac{11,808,000}{f \text{ in kHz}} = \frac{11,808}{f \text{ in MHz}}$$

$$\delta = \frac{1}{2\pi}\sqrt{\frac{4\pi\rho\lambda_{\text{meters}}}{\mu c_{\text{meters}}}}$$

$$Z_0 = (\mu_0/\kappa_0)^{1/2}$$

$$\text{dB} = 10 \log \frac{P_1}{P_2} = 20 \log \frac{V_1}{V_2} = 20 \log \frac{I_1}{I_2}$$

$$\lambda = \frac{v}{f} \qquad Z_0 = \sqrt{\frac{\mu}{\kappa}}$$

In a lossless medium of dielectric constant ϵ

$$v = \frac{v_0}{\sqrt{\epsilon}}$$

$$\lambda = \frac{v_0}{f\sqrt{\epsilon}} \qquad Z_0 = \sqrt{\frac{\mu_0}{\kappa_0\epsilon}} = \frac{120\pi}{\sqrt{\epsilon}}$$

$$|\Gamma| = \sqrt{\frac{P_{\text{reflected}}}{P_{\text{incident}}}} = \frac{\text{vswr} - 1}{\text{vswr} + 1}$$

$$\therefore \text{vswr} = \frac{1 + |\Gamma|}{1 - |\Gamma|}$$

% power reflected
$$= \left(\frac{\text{vswr} - 1}{\text{vswr} + 1}\right)^2 \times 100$$
$$\text{pswr} = (\text{vswr})^2$$

For waveguides

$$(\lambda_{mn})_c = \frac{2}{\sqrt{(m/a)^2 + (n/b)^2}}$$

$$\beta_{mn} = \frac{2\pi}{\lambda}\sqrt{\epsilon - \left[\frac{\lambda}{(\lambda_{mn})_c}\right]^2} = \frac{2\pi}{\lambda_q}$$

$$\lambda_g = \frac{\lambda}{\sqrt{\epsilon - [\lambda/(\lambda_{nm})_c]^2}}$$

For TE_{10} in air-filled waveguide

$$\lambda_g = \frac{\lambda}{\sqrt{1 - (\lambda/2a)^2}}$$

$$\lambda_c = 2a$$

$$\alpha = \frac{4\alpha_0 A}{a}\left(\frac{a}{2b} + \frac{\lambda^2}{\lambda_c^2}\right)$$

where $A = \dfrac{\sqrt{c/\lambda}}{\sqrt{1 - (\lambda/\lambda_c)^2}}$

TABLE 4*a* Summary of Wave Types for Rectangular Guides[3]

TABLE 4*b*   Summary of Wave Types for Circular Guides[a]

Wave type	TM_{01}	TM_{02}
Field distributions in cross-sectional plane, at plane of maximum transverse field		
Field distributions along guide		
Field components present	E_z, E_r, H_ϕ	E_z, E_r, H_ϕ
p_{nl} or p'_{nl}	2.405	5.52
$(k_c)_{nl}$	$\dfrac{2.405}{a}$	$\dfrac{5.52}{a}$
$(\lambda_c)_{nl}$	$2.61a$	$1.14a$
$(f_c)_{nl}$	$\dfrac{0.383}{a\sqrt{\mu\epsilon}}$	$\dfrac{0.877}{a\sqrt{\mu\epsilon}}$
Attenuation due to imperfect conductors	$\dfrac{R_e}{a\eta}\ \dfrac{1}{\sqrt{1-(f_c/f)^2}}$	$\dfrac{R_e}{a\eta}\ \dfrac{1}{\sqrt{1-(f_c/f)^2}}$

One way in which these velocities are sometimes visualized is to watch ocean waves striking on a beach at an oblique angle. The velocity of the point of contact of the wave with the beach represents the phase velocity. The velocity of the wave itself represents the group velocity.

TRANSMISSION LINES

The heart of transmission-line theory rests in the concept of distributed-element treatment as an engineering type of solution to the problem. Generally, the transmission lines discussed in the academic solutions are the coaxial line, the parallel pair, shielded pairs, and the stripline.

Ideal Transmission Theory

The representation of the transmission line by lumped elements has not been possible because the energy propagation in the line has time delays that are orders of

TM_{11}	TE_{01}	TE_{11}
$E_z, E_r, E_\phi, H_r, H_\phi$	H_z, H_r, E_ϕ	$H_z, H_r, H_\phi, E_r, E_\phi$
3.83	3.83	1.84
$\dfrac{3.83}{a}$	$\dfrac{3.83}{a}$	$\dfrac{1.84}{a}$
$1.64\,a$	$1.64\,a$	$3.41\,a$
$\dfrac{0.609}{a\sqrt{\mu\epsilon}}$	$\dfrac{0.609}{a\sqrt{\mu\epsilon}}$	$\dfrac{0.293}{a\sqrt{\mu\epsilon}}$
$\dfrac{R_e}{a\eta}\dfrac{1}{\sqrt{1-(f_c/f)^2}}$	$\dfrac{R_e}{a\eta}\dfrac{(f_c/f)^2}{\sqrt{1-(f_c/f)^2}}$	$\dfrac{R_e}{a\eta}\dfrac{1}{\sqrt{1-(f_c/f)^2}}\left[\left(\dfrac{f_c}{f}\right)^2+0.420\right]$

magnitude greater than those encountered in ordinary lumped-element theory. The distributed-element rationalization is a fiction that enables voltage-current relationships to be predicted. The equivalent circuits shown in Fig. 2 show how the distribution of the elements may be made in terms of π and T networks. One important note to be made at this time is that the development of the parameters shows that they are sensitive to frequency. For instance, the skin effect can cause the resistance to change with frequency because the effect decreases as very low frequencies are approached. The following characterization is used for the transmission-line parameters:

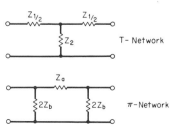

Fig. 2 Distributed parameters for transmission lines.

R = series resistance
L = series inductance
G = shunt conductance
C = shunt capacitance

The parameters are expressed as a function of length and usually in terms of meters. Table 5 lists some common parameter approximations for the coaxial and parallel-wire transmission line.

The steady-state solution for the impedance at any point in the line may be ex-

TABLE 5   Parameter Approximations for

	Coaxial

Capacitance C, farads/meter	$\dfrac{2\pi\epsilon}{\ln\left(\frac{r}{r}\right)}$
External inductance L, henrys/meter	$\dfrac{\mu}{2\pi}\ln\left(\frac{r_o}{r_i}\right)$
Conductance G, mhos/meter	$\dfrac{2\pi\sigma}{\ln\left(\frac{r_o}{r_i}\right)} = \dfrac{2\pi\omega\epsilon_0\epsilon''}{\ln\left(\frac{r_o}{r_i}\right)}$
Resistance R, ohms/meter	$\dfrac{R_3}{2\pi}\left(\frac{1}{r_o}+\frac{1}{r_i}\right)$
Internal inductance L_i, henrys/meter (for high frequency)	←
Characteristic impedance at high frequency Z_0, ohms	$\dfrac{\eta}{2\pi}\ln\left(\frac{r_o}{r_i}\right)$
Z_0 for air dielectric	$60\ln\left(\frac{r_o}{r_i}\right)$
Attenuation due to conductor α_a	←
Attenuation due to dielectric α_b	←
Total attenuation dB/meter	←
Phase constant for low-loss lines β	←

All units above are mks.
For the dielectric:

$$\epsilon = \epsilon'\epsilon_0 = \text{dielectric constant, F/m}$$

$$\mu = \mu'\mu_0 = \text{permeability, H/m}$$

$$\eta = \sqrt{\mu/\epsilon}\ \ \Omega$$

pressed as follows:

$$Z_x = \frac{V_x}{I_x} = Z_0 \left[\frac{Z_R \cosh v(l-x) + Z_0 \sinh v(l-x)}{Z_0 \cosh v(l-x) + Z_R \sinh v(l-x)} \right]^2 \tag{11}$$

In the discussions of the Maxwell equations it is revealed that the concept of the traveling wave is required for the solutions in order to have a time-varying term in the equation. With the aid of any convenient textbook on the subject solutions to these traveling-wave equations may be shown. One of the terms of the equations

Transmission Lines[3]

Turn lead	Shielded pair	Parallel plate
	$p = \frac{s}{d}$ $q = \frac{s}{D}$	Formula for $a \ll b$
$\dfrac{\pi \epsilon}{\cosh^{-1}\left(\frac{s}{d}\right)}$	. . . .	$\dfrac{\epsilon b}{a}$
$\dfrac{\mu}{\pi} \cosh^{-1}\left(\frac{s}{d}\right)$	. . . .	$\mu \dfrac{a}{b}$
$\dfrac{\pi \sigma}{\cosh^{-1}\left(\frac{s}{d}\right)} = \dfrac{\pi \omega \epsilon_0 \epsilon''}{\cosh^{-}\left(\frac{s}{d}\right)}$	. . . .	$\dfrac{\sigma b}{a} = \dfrac{\omega \epsilon_0 \epsilon' b}{a}$
$\dfrac{2R}{\pi d}\left[\dfrac{s/d}{\sqrt{(s/d)^2-1}}\right]$	$\dfrac{2R_{s2}}{\pi d}\left[1 + \dfrac{1+2p^2}{4p}(1-4q^2)\right]$ $\dfrac{8R_{s3}}{D}q^2\left[1+q^2 - \dfrac{1+4p^2}{8p^4}\right]$	$\dfrac{2R}{b}$
$\xleftarrow{\hspace{3cm}} \dfrac{R}{\omega} \xrightarrow{\hspace{3cm}}$		
$\dfrac{\eta}{\pi} \cosh^{-1}\left(\frac{s}{d}\right)$	$\dfrac{\eta_1}{\pi}\left\{\text{in}\left[2p\left(\dfrac{1-q^2}{1+q^2}\right)\right]\right.$ $\left. - \dfrac{1+4p^2}{10p^4}(1-4q)\right\}$	$\eta \dfrac{a}{b}$
$120\cosh^{-1}\left(\frac{s}{d}\right) = 120\,\text{in}\left(\frac{2s}{d}\right)$ $/d \gg$	$120\,\text{in}\left[2p\dfrac{1-q^2}{1+q^2}\right]$ $- \dfrac{1+4p^2}{16p^4}(1-4q^2)\right\}$	$120\,\pi\,\dfrac{a}{b}$
$\xleftarrow{\hspace{3cm}} \dfrac{2R}{Z_0} \xrightarrow{\hspace{3cm}}$		
$\xleftarrow{\hspace{2cm}} \dfrac{GZ_0}{2} = \dfrac{\sigma \eta}{2} = \dfrac{\pi\sqrt{\epsilon\mu}}{\lambda_0}\left(\dfrac{\epsilon''}{\epsilon'}\right) \xrightarrow{\hspace{2cm}}$		
$\xleftarrow{\hspace{3cm}} 8.686\,(\alpha_a + \alpha_d) \xrightarrow{\hspace{3cm}}$		
$\xleftarrow{\hspace{3cm}} \omega\sqrt{\mu\epsilon} = \dfrac{2\pi}{\lambda} \xrightarrow{\hspace{3cm}}$		

ϵ'' = loss factor of dielectric = $\sigma \epsilon / \omega \epsilon_0$

R_3 = skin-effect surface resistivity of conductor, Ω

λ = wavelength in dielectric = $\lambda_0 / \sqrt{\epsilon' \mu'}$

shows that a constant wavefront in the traveling wave is moving in the direction of propagation with a velocity of ω/β.   Because this wave is also attenuated as it moves in the forward direction of propagation, it is known as the incident wave or the forward wave. A similar solution for a reverse wavefront is known as the reflected wave.   The parameter that relates the forward and reflected waves is called the reflection coefficient and is symbolized by Γ.   The reflection coefficient is a complex term showing that the reflected wave is different in magnitude and phase from the forward wave.   The constant α is the real part of the propagation term and is usually expressed as attenuation per unit of length.   The β term is the phase constant and is expressed as degrees per unit length.

Several important relationhips must be presented at this point to show how the transmission line should be described:

$$Z_9 = \sqrt{\frac{L}{C}} \tag{12}$$

$$\lambda = \frac{2\pi}{\beta} \tag{13}$$

$$v = \sqrt{\frac{1}{LC}} \text{ for a lossless line} \tag{14}$$

The significance of the characteristic impedance is that it indicates the terminating impedance for the line when minimum reflection and maximum energy transfer are to be accomplished.   It also gives the rate at which the voltage change is moving in the line.

The concept of reflections in a transmission line is probably one of the most important to the system user because it is the tool by which all components are evaluated for use in the line.

Study of the transmission-line reflections shows that they usually assume a sinusoidal varying voltage.   Treatment of the voltage and current relationship reveals that in a dissipationless open-circuited line the impedances along the line are purely reactive and vary every quarter wavelength from capacitance to inductive reactances.   This is demonstrated in Fig. 3.

An analysis of the impedance characteristics shows that the voltage across the line varies from a minimum to a maximum in a periodic fashion.   This is the result of the vector summation of the voltage of the forward and reflected waves in their phasal relationship.

A similar approach for the short-circuited line shows the behavior is the same

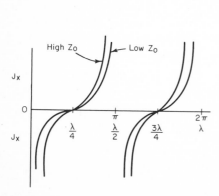

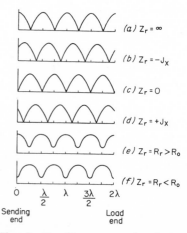

Fig. 3 Input impedance vs. distance from open-circuited end of a lossless line.

Fig. 4 Standing waves of voltage on a dissipationless line two wavelengths long.

except that it appears to be shifted ½λ. Figure 4 shows how the various line characteristics affect the standing-wave patterns.

Observation of the voltage patterns shows that termination in the same position of each voltage curve yields the same voltage standing wave regardless of the absolute value of the line lengths.

The importance of the standing-wave ratio is that it is a quantitative term that can be very easily measured and gives other parameters to describe the transmission line. One such illustration of this is that the voltage standing-wave ratio can be shown to be a function of the reflection coefficient. This yields the following result:

$$|\Gamma| = \frac{\text{vswr} - 1}{\text{vswr} + 1} \tag{15}$$

In the real world there is no such thing as a lossless line. Therefore, the transmission-line equations are handled by making some practical assumptions as to the conditions of the line, and solutions are presented as approximations. In working with the expansion for the characteristic impedance, an assumption is usually made in the case of a low-loss dielectric that the conductance term is negligible and may be dropped. This results in keeping only the first-order term, which ends up in the approximation that the characteristic impedance is as follows:

$$Z \cong \sqrt{\frac{L}{C}} \tag{16}$$

This familiar equation is the same result as in the lossless line. It may be shown that the attenuation factor and phase shift are similar to the lossless line.

Smith and Blanchard Charts

The Smith chart is the basic microwave graphic aid. Discovery of this tool is related to the solutions for the normalized impedance of the misterminated line. Solving the equations for the normalized impedance ends in the following:

$$\frac{Z}{Z_0} = \frac{1 + [E_R(l)/E_i(l)]}{1 - [I_R(l)/I_i(l)]} \tag{17}$$

Working further and assuming a lossless line, the following equation results:

$$\frac{Z}{Z_0} = \frac{R}{Z_0} + \frac{X}{Z_0} = \frac{1 + r\epsilon^{j\theta}}{1 - r\epsilon^{j\theta}} \tag{18}$$

This is known as the equation of the Smith chart.

The Smith chart is a plot of normalized resistance and reactance as a function of magnitude and phase of the reflection coefficient. The chart itself is made up of lines of constant resistance·and normalized reactance. Figure 5 shows how the lines look by themselves and how the combination is created.

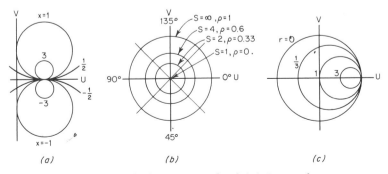

Fig. 5 Development of Smith chart constant loci,[2] (a) Locus of constant reactance. (b) Locus of constant standing-wave ratio S and line angle βd. (c) Locus of constant resistance.

With the Smith chart the transmission-line equation can be solved conveniently with a ruler and a compass instead of hyperbolic functions and tables.

The key to using the Smith chart is to remember that in a lossless line the changing line length describes a circle about the center of the Smith chart. This relationship is expressed as follows:

$$\Delta\Theta = 2\beta\,\Delta l = \frac{\Delta\pi\,\Delta l}{\lambda} \tag{19}$$

Note that a change of $\lambda/2$ is a complete trip around the circle. The outer circle of the Smith chart is marked off in fractions of wavelengths.

Some examples of the use of the Smith chart demonstrate its practical usefulness. One typical problem would be to find the sending and receiving impedance of a line. The characteristic impedance of the line is known because of its physical characteristics. A typical value is 50 Ω. Assume the line is terminated in an impedance of 100 Ω resistive and 40 Ω capacitive reactance. The first step is to normalize the impedance, which yields the following:

$$\frac{Z_1}{Z_0} = 2 - j0.8$$

Then entering the Smith chart (see Fig. 6) and plotting the point on a constant-resistance circle of 2.0 and a negative-reactance component of 0.8, the position of the load impedance is found. Now a circle is drawn using the $1.0 \pm 0j$ as the center and the distance from this center to the impedance point found above as the radius. This circle gives the loci of impedances through the transmission line. Direct reading from the circle gives the reflection coefficient from the maximum and minimum values of R that the circle crosses on the real impedance load line. Now traveling around the circumference of the circle, we discover the impedance point we selected in wavelengths toward the generator. If necessary one could find the value of impedance at any point in the line by knowing only the number of wavelengths toward the generator or away from the impedance. This can easily be found from the frequency and the physical dimensions of the transmission line.

Another typical problem is to find the value of the load impedance when the voltage standing-wave ratio is found by measurement as well as the position of the voltage minimum. In a manner that will be discussed later this is found easily and illustrates the usefulness of the Smith chart in making transmission-line problems simple.

Using the instrumentation, find the location of the voltage minimum with the unknown load attached. Then replace the load with a short circuit. Measure the shift and convert it to wavelengths toward the load or generator. Plot the voltage standing-wave ratio circle on the Smith chart, moving along the chart on the vswr circle the distance in wavelengths computed above, toward or away from the generator. The intersection of the radial line with the vswr circle is the load impedance. Figure 7 illustrates this on the transmission line.

Several other scales that are usually on the Smith chart deserve some explanation.

Voltage and current limits   The relationships that give maximum/minimum values of the voltage and current can be expressed as

$$E_{max} = \sqrt{Pz_0(\text{swr})} \quad I_{max} = \sqrt{\frac{P(\text{swr})I}{Z_0}} \tag{20}$$

$$E_{min} = \sqrt{Pz_0(\text{swr})} \quad I_{min} = \sqrt{\frac{P}{Z_0(\text{swr})}} \tag{21}$$

The scales on the chart are thus $(\text{swr})^{1/2}$ and $1/(\text{swr})^{1/2}$. Once the load impedance and the power are known, the voltage and current limits are easily found.

Voltage and current ratios are given as scales of swr and 20 logs. The reflections loss in dB is given as the scale of $-10\log(1 - r^2)$. The standing-wave loss coefficient is given in the scale of $(1 + r^2)/(1 - r^2)$. The significance of this term is that it

Impedance or Admittance Coordinates

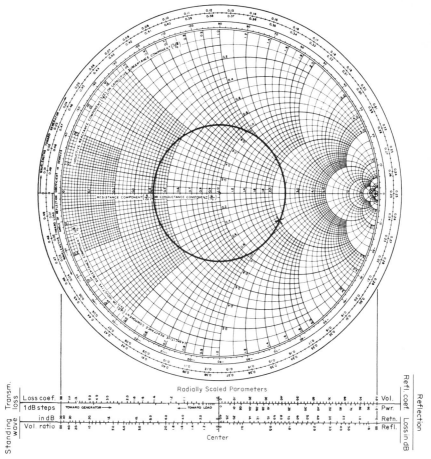

Fig. 6 The Smith chart.

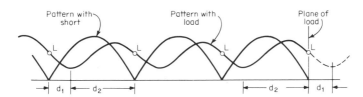

Fig. 7 Voltage standing-wave patterns as a function of load impedance. Input impedance = load impedance at points marked L.

is the factor by which the actual power dissipation in the line is increased by the swr. The "1-dB steps" scale is used to develop the line loss.

The use of the Smith chart up to this point has assumed a lossless transmission line. However, since this is not always a practical assumption, it must be dealt with to make the chart a useful instrument. Referring back to the bilinear transformation, it can easily be shown that the locus of input impedance as a function of line length is not a circle but a logarithmic spiral. The scale identified as the 1-dB

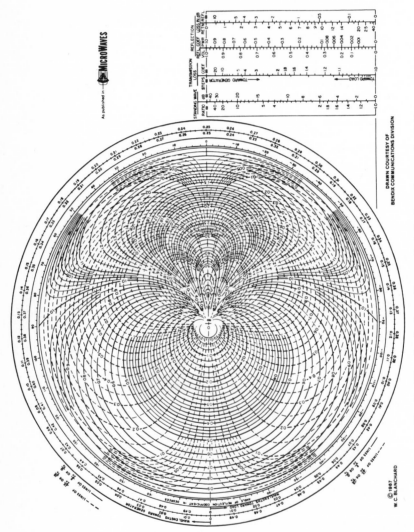

Fig. 8 Blanchard chart (inverted-circle impedance chart).[35]

scale is used to approximate the effect of these losses. All that need be known is the loss per wavelength. Using these factors and the distance moved toward the generator, the correction may be made for the loss term in the impedance by moving radially toward the center the scaled amount from the plotted lossless impedance point.

The following expression may be used to measure the loss of a line:

$$1 = 4.343 \log \frac{\text{swr} + 1}{\text{swr} - 1} \tag{22}$$

This expression is valid for low loss when used in the 1-dB steps scale on the transmission-line chart. For higher values of attenuation the actual measurement of loss is preferable.

Recently another graphic instrument has gained some usefulness in the microwave field. Known as the Blanchard chart, this instrument was first introduced in *Microwaves* (November 1967). Figures 8 and 9 demonstrate this chart and how it is used. The advantage of the Blanchard chart is that it increases the plotting area for small values of vswr and allows the display of the phase length of the transmission line. Slide rules and charts may be obtained from Greencastle Electronics, P.O. Box 9613, Rosedale, MD 21237.

Many useful aids are supplied to the microwave engineer by the use of graphic displays. One is computer-aided design using Smith-chart techniques. Such a program was demonstrated by Joseph W. Verzino.[44]

Another expanded use for the Smith chart was shown by Hansjuergen C. Blume.[45] Figure 10 demonstrates the manner in which this was done using the expanded Smith chart to allow the input-impedance display of negative-resistance amplifiers. It can be seen that negative resistance falls outside the realm of normal Smith-chart displays.

PARAMETERS

A variety of terms have been presented that are useful to the microwave engineer in describing hardware. Although often a mystery to an engineer who deals with conventional voltage, current, and impedance relationships, they turn out to be perhaps simpler in their usage than one might imagine.

Voltage Standing-Wave Ratio

This widely used term, commonly shortened to vswr, describes the quality of a component that affects its ability to match the transmission line it was designed to operate on. This particular parameter is usually very easy to measure and with some imagination in the particular measurement situation can be determined quite accurately. The parameter itself is a dimensionless quantity that is expressed as a ratio. One particular caution that must be noted here is that the vswr is often lumped with the insertion-loss parameter. The reflected signal is lost in the sense that it does not arrive at the load, but it is not a measure of the dissipation of the particular component in the line. However, it is sometimes convenient to express the vswr as a loss parameter in that it indicates the amount of power reflected. The percentage of power reflected in terms of the vswr is given by the following relationship:

$$P_r = \left| \frac{\text{vswr} - 1}{\text{vswr} - 1} \right|^2 \times 100 \tag{23}$$

Figure 16 shows a curve from which it is easy to pick the values desired in a particular problem.

Insertion Loss

The parameter described as insertion loss is usually referred to as an undesirable quality of the component. That is, it is a measure of the energy lost in the device, and one wants to keep it as low as possible. Another parameter very much like insertion loss but usually intended to be a descriptive requirement of a wanted characteristic is attenuation. Attenuation can mean both the reflected energy and the dissi-

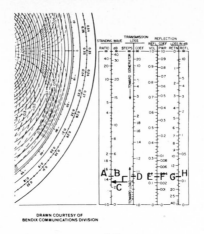

DRAWN COURTESY OF
BENDIX COMMUNICATIONS DIVISION

(a)

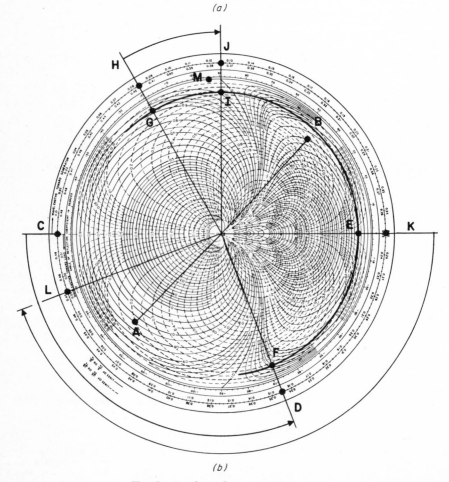

(b)

Fig. 9 For legend see opposite page.

pated energy, and it is necessary to include in a description just what it means. Many devices become reflective at higher attenuation values, and the reflection can include a large part of the value. Both parameters are expressed in dB and are simply a ratio of the incident energy to the energy leaving the output port.

Decibels

The use of decibels in microwave engineering is so commonplace that its importance is not always appreciated. Quite simply the decibel is the ratio of two powers expressed as the multiplicand of 10 and their log, as follows:

$$dB = 10 \log_{10} \frac{P_1}{P_2} \quad \left(also \ dB = 20 \log_{10} \frac{V_1}{V_2} \right) \tag{24}$$

Without too much difficulty it may be shown that a ratio of 3 dB results from a ratio of 2:1. Also, it is convenient to note that any ratio in a power of 10 results in a dB of a multiple of 10; that is, a ratio of 100 gives a dB value of 10. Recalling that exponents add in the logarithm, a useful technique allows the development of any whole ratio. For instance, knowing that 13 dB is the sum of 10 and 3 dB, one can determine the absolute value of the power from the known reference value. Actually knowing that 2 dB is approximately 16 percent, one can determine the absolute value of any whole value of dB with some quick arithmetic calculation.

dBm

The expression dBm is often used. It is a reference to one milliwatt and can be expressed as follows:

$$1 \ dBm = 10 \log \frac{P_1}{0.001 \ W} \tag{25}$$

Fig. 9 Blanchard chart instructions.[35] (a) Using the scales to the right of the Blanchard chart. If the load produces a 1.45:1 standing-wave voltage ratio, A, the measured probe coupled power at the maximum is 3.2 dB, B, above the power at the minimum. If 1 dB of loss, C, were inserted in the line, the generator would see a 1.35:1 vswr instead of the load vswr of 1.45:1. The copper and dielectric loss due to the increased current of the standing wave has caused the transmission-line loss to be increased by 1.07, D, over the loss that was expected for a matched line. The absolute value of the voltage reflection coefficient is 0.18, E, and the absolute value of the power reflection is 0.03 dB, F. The power being returned to the generator is 15 dB, G, below the incident power. The power being lost to the load due to the mismatch is 0.11 dB, H.
(b) Using the Blanchard chart to solve transmission-line problems. How to determine the representation of an impedance on the chart: If the load impedance is $Z_L = (37.5 - j10)\Omega$ and the line impedance is $Z_0 = 50\Omega$, then the normalized impedance of the load is $Z_n = Z_L/Z_0 = 0.75 - j0.2$, and the impedance is represented at point A. The normalized admittance is $Y_n = 1/Z_n = 1.25 = j0.33$, point B, and is diametrically opposite A.
How to determine the Z_L when the vswr it produces on a line is 1.2:1 and the distance from the standing-wave voltage minimum to the load is 0.155λ: Move CCW from C, 0.155λ, to D. Draw a radial to D. Draw a 1.2:1 vswr circle E. Z_n is at the intersection F, $Z_n = 1.06 - j0.19$. When $Z_0 = 50\Omega$, $Z_L = Z_nZ_0 = (53 - j9.5)\Omega$.
How to match a load of $Z_L = (53 - j9.5)\Omega$ to a line of $Z_0 = 50\Omega$ using a shorted Z_0 stub: $Z_n = Z_L/Z_0 = 1.06 - j0.19$ is at F. Construct a vswr circle and draw a line from F through the center to the vswr circle G and on through to the outer perimeter H. The scale reading at H is 0.09λ toward the generator. The admittance at G is $Y_n = 0.92 + j0.16$. Draw a radial through the intersection of the vswr circle and the $R = 1$ line, point I, through to the outer perimeter, J. The scale reading at J is 0.13λ toward the generator. At I, 0.037λ from the load which is the distance from H to J, the admittance is $Y_n = 1.0 + j0.17$ and is where the stub will be connected. The length of the stub must transform its short $Y_n = \infty$, point K, to a $Y_n \doteq -j0.17$. Draw a radial, tangential to $-j0.17$, to the outer perimeter L. The length of the $Z_0 = 50\Omega$ stub is 0.223λ, the distance from K to L. With the stub attached at I, $Y_n = 1 + j0.0$, point M.

This is a convenient way of expressing the power of an amplifier or an oscillator and of measuring the minimum sensitivity of a detector in terms of power. A traveling-wave tube, for instance, has a maximum power output of 53 dBm, which is 200 W.

One further use of the decibel is in expressing the dynamic range of a device. One of the problems in microwave measurement, for instance, is the definite limitation to the amount of power available for making measurements. Measurements made over a large attenuation range may exceed the dynamic range of the device. This means that the attenuation range reduces the power level to the detectors so that only noise is detectable. Suppose a piece of instrumentation has a maximum power output of 10 dBm, and the detectors lose their sensitivity at −50 dBm. The equipment is said to have a dynamic range of 60 dB. The significance of this is that attenuation measurements could not be made over a range of 60 dB. Table 6 and Figs. 11 through 17 are some useful tools to make calculations simpler.

THE S PARAMETERS

Most of the discussion in this chapter has attempted to relate the microwave functions in terms of the more easily recognizable voltage-current relationships. Another type of treatment, however, might be much simpler and a more suitable analytical tool for the microwave engineer. This particular technique involves the use of the S

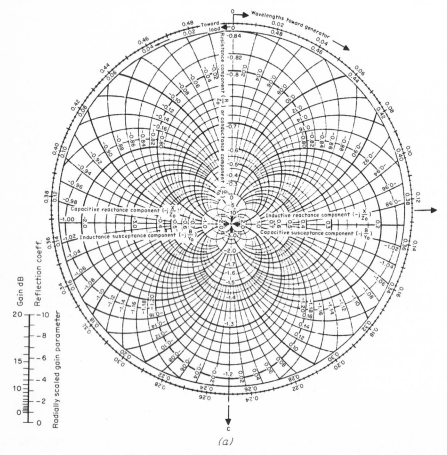

(a)

Fig. 10   For legend see opposite page.

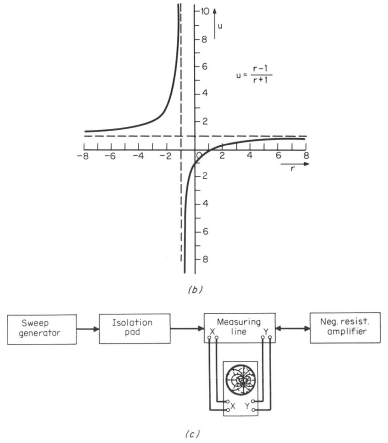

$$u = \frac{r-1}{r+1}$$

(b)

| Sweep generator | Isolation pad | Measuring line X   Y | Neg. resist. amplifier |

(c)

Fig. 10 Expanded Smith chart showing negative resistances.[45] (*a*) Impedance or admittance coordinates. (*b*) Impedance displays. (*c*) Block diagram of a direct-indicating measuring device for the projection of negative impedances or admittances in the extended Smith chart.

parameters, or scattering parameters.   The discussion will show the decided practical advantages of the S parameters over other network analysis.

For most network analysis an assumption has to be made that the functions are dealing with linear responses or with signal situations that can approximate linear responses.   The traditional network treatment of a two-port network assumes a four-variable situation, defined as v_1, i_1, v_2, i_2.   In the analysis any two of the variables can be made dependent and the other two independent.   Figure 18 describes their physical relationship.

Depending on which of the variables are selected to be dependent and which are independent, a set of parameters can be determined which are familiarly known as z, h, and y.   For illustration with reference to Fig. 18, the following relationships will be presented:

$$I_1 = y_{11}V_1 + y_{12}V_2 \tag{26}$$
$$I_2 = y_{21}V_1 + y_{22}V_2 \tag{27}$$

Selecting the port voltages as the independent variables and the currents as the dependent variables, a series of four measurements would have to be made to deter-

TABLE 6 Conversion of Voltage and Power Ratios to Decibels

Voltage ratio	Power ratio	-db+	Voltage ratio	Power ratio
1.000	1.000	0.0	1.000	1.000
0.989	0.977	0.1	1.012	1.023
0.977	0.955	0.2	1.023	1.047
0.966	0.933	0.3	1.035	1.072
0.955	0.912	0.4	1.047	1.096
0.944	0.891	0.5	1.059	1.122
0.933	0.871	0.6	1.072	1.148
0.923	0.851	0.7	1.084	1.175
0.912	0.832	0.8	1.096	1.202
0.902	0.813	0.9	1.109	1.230
0.891	0.794	1.0	1.122	1.259
0.881	0.776	1.1	1.135	1.288
0.871	0.759	1.2	1.148	1.318
0.861	0.741	1.3	1.161	1.349
0.851	0.724	1.4	1.175	1.380
0.841	0.708	1.5	1.189	1.413
0.832	0.692	1.6	1.202	1.445
0.822	0.676	1.7	1.216	1.479
0.813	0.661	1.8	1.230	1.514
0.804	0.646	1.9	1.245	1.549
0.794	0.631	2.0	1.259	1.585
0.785	0.617	2.1	1.274	1.622
0.776	0.603	2.2	1.288	1.660
0.767	0.589	2.3	1.303	1.698
0.759	0.575	2.4	1.318	1.738
0.750	0.562	2.5	1.334	1.778
0.741	0.550	2.6	1.349	1.820
0.733	0.537	2.7	1.365	1.862
0.724	0.525	2.8	1.380	1.905
0.716	0.513	2.9	1.396	1.950
0.708	0.501	3.0	1.413	1.995
0.700	0.490	3.1	1.429	2.042
0.692	0.479	3.2	1.445	2.089
0.684	0.468	3.3	1.462	2.138
0.676	0.457	3.4	1.479	2.188
0.668	0.447	3.5	1.496	2.239
0.661	0.437	3.6	1.514	2.291
0.653	0.427	3.7	1.531	2.344

Power ratio	Voltage ratio	-db+	Power ratio	Voltage ratio
5.012	2.239	7.0	0.200	0.447
5.129	2.265	7.1	0.195	0.442
5.248	2.291	7.2	0.191	0.437
5.370	2.317	7.3	0.186	0.432
5.495	2.344	7.4	0.182	0.427
5.623	2.371	7.5	0.178	0.422
5.754	2.399	7.6	0.174	0.417
5.888	2.427	7.7	0.170	0.412
6.026	2.455	7.8	0.166	0.407
6.166	2.483	7.9	0.162	0.403
6.310	2.512	8.0	0.159	0.398
6.457	2.541	8.1	0.155	0.394
6.607	2.570	8.2	0.151	0.389
6.761	2.600	8.3	0.148	0.385
6.918	2.630	8.4	0.145	0.380
7.079	2.661	8.5	0.141	0.376
7.244	2.692	8.6	0.138	0.372
7.413	2.723	8.7	0.135	0.367
7.586	2.754	8.8	0.132	0.363
7.762	2.786	8.9	0.129	0.359
7.943	2.818	9.0	0.126	0.355
8.128	2.851	9.1	0.123	0.351
8.318	2.884	9.2	0.120	0.347
8.511	2.917	9.3	0.118	0.343
8.710	2.951	9.4	0.115	0.339
8.913	2.985	9.5	0.112	0.335
9.120	3.020	9.6	0.110	0.331
9.333	3.055	9.7	0.107	0.327
9.550	3.090	9.8	0.105	0.324
9.772	3.126	9.9	0.102	0.320
10.000	3.162	10.0	0.100	0.316
10.23	3.199	10.1	0.0977	0.313
10.47	3.236	10.2	0.0955	0.309
10.72	3.273	10.3	0.0933	0.306
10.96	3.311	10.4	0.0912	0.302
11.22	3.350	10.5	0.0891	0.299
11.48	3.388	10.6	0.0871	0.295
11.75	3.428	10.7	0.0851	0.292

Power ratio	Voltage ratio	-db+	Power ratio	Voltage ratio
25.12	5.012	14.0	0.0398	0.200
25.70	5.070	14.1	0.0389	0.197
26.30	5.129	14.2	0.0380	0.195
26.92	5.188	14.3	0.0372	0.193
27.54	5.248	14.4	0.0363	0.191
28.18	5.309	14.5	0.0355	0.188
28.84	5.370	14.6	0.0347	0.186
29.51	5.433	14.7	0.0339	0.184
30.20	5.495	14.8	0.0331	0.182
30.90	5.559	14.9	0.0324	0.180
31.62	5.623	15.0	0.0316	0.178
32.36	5.689	15.1	0.0309	0.176
33.11	5.754	15.2	0.0302	0.174
33.88	5.821	15.3	0.0295	0.172
34.67	5.888	15.4	0.0288	0.170
35.48	5.957	15.5	0.0282	0.168
36.31	6.026	15.6	0.0275	0.166
37.15	6.095	15.7	0.0269	0.164
38.02	6.166	15.8	0.0263	0.162
38.90	6.237	15.9	0.0257	0.160
39.81	6.310	16.0	0.0251	0.159
40.74	6.383	16.1	0.0246	0.157
41.69	6.457	16.2	0.0240	0.155
42.66	6.531	16.3	0.0234	0.153
43.65	6.607	16.4	0.0229	0.151
44.67	6.683	16.5	0.0224	0.150
45.71	6.761	16.6	0.0219	0.148
46.77	6.839	16.7	0.0214	0.146
47.86	6.918	16.8	0.0209	0.145
48.98	6.998	16.9	0.0204	0.143
50.12	7.079	17.0	0.0200	0.141
51.29	7.161	17.1	0.0195	0.140
52.48	7.244	17.2	0.0191	0.138
53.70	7.328	17.3	0.0186	0.137
54.95	7.413	17.4	0.0182	0.135
56.23	7.499	17.5	0.0178	0.133
57.54	7.586	17.6	0.0174	0.132
58.88	7.674	17.7	0.0170	0.130

Voltage Ratio	Power Ratio	dB	Voltage Ratio	Power Ratio
0.646	0.417	3.8	1.549	2.399
0.638	0.407	3.9	1.567	2.455
0.631	0.398	4.0	1.585	2.512
0.624	0.389	4.1	1.603	2.570
0.617	0.380	4.2	1.622	2.630
0.610	0.372	4.3	1.641	2.692
0.603	0.363	4.4	1.660	2.754
0.596	0.355	4.5	1.679	2.818
0.589	0.347	4.6	1.698	2.884
0.582	0.339	4.7	1.718	2.951
0.575	0.331	4.8	1.738	3.020
0.569	0.324	4.9	1.758	3.090
0.562	0.316	5.0	1.778	3.162
0.556	0.309	5.1	1.799	3.236
0.550	0.302	5.2	1.820	3.311
0.543	0.295	5.3	1.841	3.388
0.537	0.288	5.4	1.862	3.467
0.531	0.282	5.5	1.884	3.548
0.525	0.275	5.6	1.905	3.631
0.519	0.269	5.7	1.928	3.715
0.513	0.263	5.8	1.950	3.802
0.507	0.257	5.9	1.972	3.890
0.501	0.251	6.0	1.995	3.981
0.496	0.246	6.1	2.018	4.074
0.490	0.240	6.2	2.042	4.169
0.484	0.234	6.3	2.065	4.266
0.479	0.229	6.4	2.089	4.365
0.473	0.224	6.5	2.113	4.467
0.468	0.219	6.6	2.138	4.571
0.462	0.214	6.7	2.163	4.677
0.457	0.209	6.8	2.188	4.786
0.452	0.204	6.9	2.213	4.898

Voltage Ratio	Power Ratio	dB	Voltage Ratio	Power Ratio
0.288	0.0832	10.8	3.467	12.02
0.285	0.0813	10.9	3.508	12.30
0.282	0.0794	11.0	3.548	12.59
0.279	0.0776	11.1	3.589	12.88
0.275	0.0759	11.2	3.631	13.18
0.272	0.0741	11.3	3.673	13.49
0.269	0.0724	11.4	3.715	13.80
0.266	0.0708	11.5	3.758	14.13
0.263	0.0692	11.6	3.802	14.45
0.260	0.0678	11.7	3.846	14.79
0.257	0.0661	11.8	3.890	15.14
0.254	0.0648	11.9	3.938	15.49
0.251	0.0631	12.0	3.981	15.85
0.248	0.0617	12.1	4.027	16.22
0.246	0.0603	12.2	4.074	16.60
0.243	0.0589	12.3	4.121	16.98
0.240	0.0575	12.4	4.169	17.38
0.237	0.0562	12.5	4.217	17.78
0.234	0.0550	12.6	4.266	18.20
0.232	0.0537	12.7	4.315	18.62
0.229	0.0525	12.8	4.365	19.05
0.227	0.0513	12.9	4.416	19.50
0.224	0.0501	13.0	4.467	19.95
0.221	0.0490	13.1	4.519	20.42
0.219	0.0479	13.2	4.571	20.89
0.216	0.0468	13.3	4.624	21.38
0.214	0.0457	13.4	4.677	21.88
0.211	0.0447	13.5	4.732	22.39
0.209	0.0437	13.6	4.786	22.91
0.207	0.0427	13.7	4.842	23.44
0.204	0.0417	13.8	4.898	23.99
0.202	0.0407	13.9	4.955	24.55

Voltage Ratio	Power Ratio	dB	Voltage Ratio	Power Ratio
0.129	0.0166	17.8	7.762	60.26
0.127	0.0162	17.9	7.852	61.66
0.126	0.0159	18.0	7.943	63.10
0.125	0.0155	18.1	8.035	64.57
0.123	0.0151	18.2	8.128	66.07
0.122	0.0148	18.3	8.222	67.61
0.120	0.0145	18.4	8.318	69.18
0.119	0.0141	18.5	8.414	70.79
0.118	0.0138	18.6	8.511	72.44
0.116	0.0135	18.7	8.610	74.13
0.115	0.0132	18.8	8.710	75.86
0.114	0.0129	18.9	8.811	77.62
0.112	0.0126	19.0	8.913	79.43
0.111	0.0123	19.1	9.016	81.28
0.110	0.0120	19.2	9.120	83.18
0.108	0.0118	19.3	9.226	85.11
0.107	0.0115	19.4	9.333	87.10
0.106	0.0112	19.5	9.441	89.13
0.105	0.0110	19.6	9.550	91.20
0.104	0.0107	19.7	9.661	93.33
0.102	0.0105	19.8	9.772	95.50
0.101	0.0102	19.9	9.886	97.72
0.100	0.0100	20.0	10.000	100.00
10^{-3}	30		10^{3}
10^{-2}	10^{-4}	40	10^{2}	10^{4}
10^{-5}	50		10^{5}
10^{-3}	10^{-6}	60	10^{3}	10^{6}
10^{-7}	70		10^{7}
10^{-4}	10^{-8}	80	10^{4}	10^{8}
10^{-9}	90		10^{9}
10^{-5}	10^{-10}	100	10^{5}	10^{10}
10^{-11}	110		10^{11}
10^{-6}	10^{-12}	120	10^{6}	10^{12}

This basic table indicates the number of decibels (dB) corresponding to the listed ratios of voltage or power over the range of -20 to $+20$ dB. For voltage or power ratios greater than those included in the chart, the ratio can be broken down into a product of two numbers, the value in dB for each is found separately, and the two results added. Example: To convert a power ratio of 2,000:1 to dB, express 2,000 as a product of 2×10^{3}; the number of dB corresponding to a power ratio of 2 is very nearly 3, and the number of dB for a power ratio of 10^{3} is 30. Therefore, the power ratio of 2,000:1 is approximately 30 dB + 3 dB = 33 dB. In the lower right-hand corner of the Table dB values for voltage and power ratios of integral powers of 10 are given.

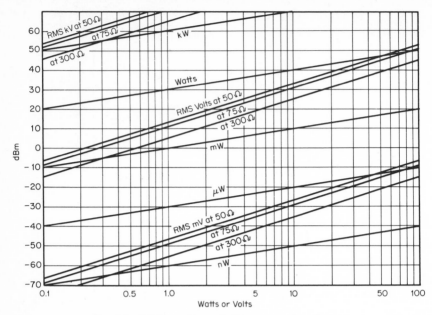

Fig. 11   Watts, volts, and dBm conversion chart.

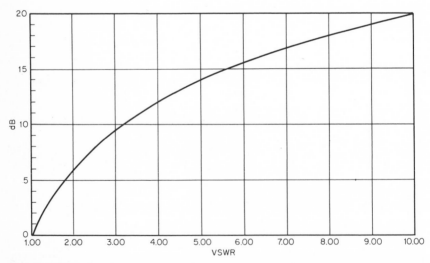

Fig. 12   Decibels (dB) vs. vswr. This graph can be used to find vswr when employing a voltmeter with a dB scale but not a vswr scale; however, if a square-law detector has been used, the dB meter reading must be divided by 2.   To find vswr for values >20 dB, subtract units of 20 until remainder is less than 20.   Multiply the vswr found by 10 for each subtraction of 20 dB.   Example: vswr (dB) = 35 dB; $43 - 20 - 20 = 3$.   From graph: $3.0 \rightarrow 1.41$; $1.41 \times 10 \times 10 = 141$.   To find dB for vswr >10.0, divide the ratio by tens until the remainder is less than 10.0.   To the number of dB found add 20 dB for each division of 10.   Example: Voltage ratio = 300; $300 \times 1/10 \times 1/10 = 3$.   From graph: $3.0 \rightarrow 9.54$; 9.54 dB $+ 20$ dB $+ 20$ dB $= 49.54$

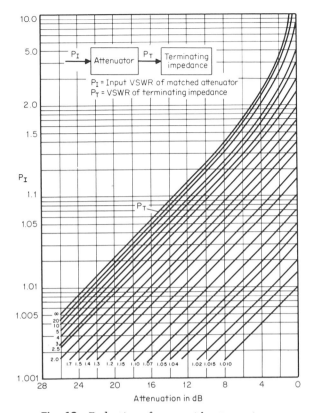

Fig. 13 Reduction of vswr with attenuation.

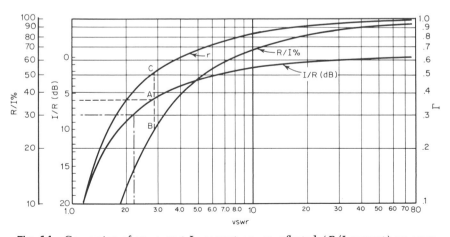

Fig. 14 Conversion of vswr, part I; percent power reflected (R/I percent) vs. vswr reflection coefficient (Γ) vs. vswr return loss (I/R dB) vs vswr. Example: Known: 6-dB difference between incident and reflected power (return loss), Find: vswr of unknown, percent power reflected (R/I), reflection coefficient (Γ). Locate 6 dB on I/R(dB) scale, find point A on I/R(dB) graph; vswr is 3.01 where the 3.01 ordinate intersects R/I percent. At point B read 25.1 percent power reflected. Also point C equals a reflection coefficient of 0.501.

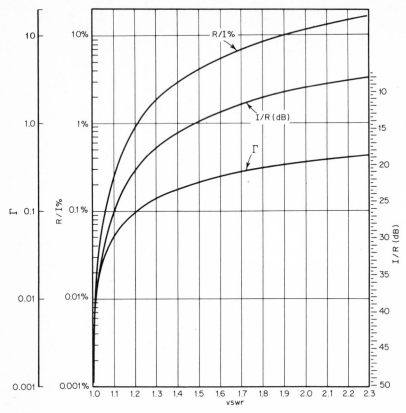

Fig. 15 Conversion of vswr, part II; R/I percent vs. vswr; Γ vs. vswr, I/R dB vs. vswr.

mine the short-circuit admittance parameters. Such a parameter is defined as follows:

$$y_{21} = \frac{1_2}{V_1} \quad V_2 = 0 \quad \text{(output short-circuited)} \tag{28}$$

Similarly other sets of short-circuit parameters may be established and defined for measurement and related to each other. However, the use of these parameters present problems at microwave frequencies. The circuit elements become very difficult to handle, and the linearity assumptions get awkward. Also, in amplifiers unstable situations can develop that can cause undesired oscillations. The use of the scattering parameters resolves these practical problems in that they are established for a situation when the devices are terminated in their characteristic impedances. Another advantage in the use of the scattering parameter is that, using low-loss lines and approximating the lossless transmission lines, one can assume that the magnitude of the traveling wave does not vary with the actual measuring line.

The basic paper in the generalized scattering parameter was presented by K. Kurokawa (cited in Ref. 46). The parameters discussed in his paper describe a new set of variables, as shown in Fig. 19.

The variables a_1 and b_1 are the complete voltage waves incident on and reflected from the ith port of the network. These terms are defined in terms of the terminal

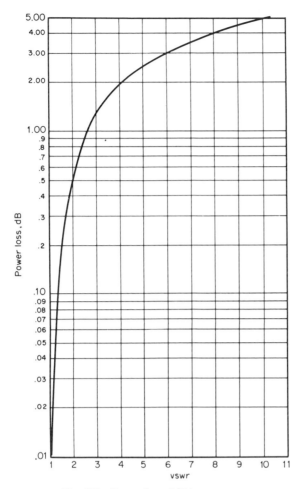

Fig. 16 Power loss (dB) vs. vswr.

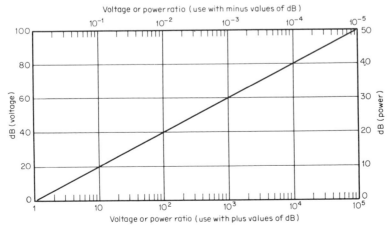

Fig. 17 Conversion of voltage or power ratio to decibels.

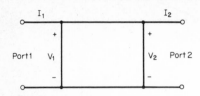
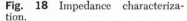
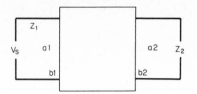

Fig. 18 Impedance characterization.   **Fig. 19** Power characterization.

voltage and the terminal current and the arbitrary reference impedance Z_1, as follows:

$$a_i = \frac{V_i + Z_i I_i}{2\sqrt{1 R_e Z_i}} \tag{29}$$

$$b_1 = \frac{V_i - Z^* I_i}{2\sqrt{|R_e Z_i|}} \tag{30}$$

Further textual development will show the S-parameters can be defined as follows:

$$a_1 = \frac{V_1 + I_1 Z_0}{2\sqrt{Z_0}} = \frac{\text{voltage wave incident on port 1}}{Z_0} \tag{31}$$

$$= \frac{V_{i1}}{\sqrt{Z_0}}$$

$$a_2 = \frac{V_2 + I_2 Z_0}{2\sqrt{Z_0}} = \frac{\text{voltage wave incident to port 2}}{Z_0} \tag{32}$$

$$= \frac{V_{i2}}{\sqrt{Z_0}}$$

$$b_1 = \frac{V_1 - I_1 Z_0}{\sqrt{Z_0}} = \frac{\text{voltage wave reflected from port 1}}{\sqrt{Z_0}} = \frac{V_{r1}}{\sqrt{Z_0}} \tag{33}$$

$$b_2 = \frac{V_2 - I_2 Z_0}{2\sqrt{Z_0}} = \frac{\text{voltage reflected from port 2}}{\sqrt{Z_0}} = \frac{V_{r2}}{\sqrt{Z_0}} \tag{34}$$

The linear equations for the variables are thus

$$b_1 = s_{11}a_1 + s_{12}a_2 \tag{35}$$
$$b_2 = s_{22}a_1 + s_{22}a_2 \tag{36}$$

The observation made from this definition of the wave variables is that $s_{11} = b_1/a_1 = (Z_1 - Z_0)/Z_1 + Z_0$. This appears in the familiar form of the relationship

TABLE 7 S Parameters

$S_{21} = \dfrac{b_2}{a_1}$	$a_2 = 0 =$	forward transmission (insertion) gain with the output port terminated in a matched load
$S_{12} = \dfrac{b_1}{a_2}$	$a_1 = 0 =$	reverse transmission (insertion) gain with the input port terminated in a matched load
$S_{11} = \dfrac{b_1}{a_1}$	$a_2 = 0 =$	input reflection coefficient with the output port terminated by a matched load ($Z_1 = Z_0$ sets $a_2 = 0$)
$S_{22} = \dfrac{b_2}{a_2}$	$a_1 = 0 =$	output reflection coefficient with the input terminated by a matched load ($Z_s = Z_0$ and $V_s = 0$)

* The actual is the complex conjugate.

between the reflection coefficient and the impedance which is the basis for the Smith chart transmission-line aid. Therefore, s_{11} and s_{22} can be plotted on the Smith chart to determine impedances and the required matching networks for optimizing circuit design. Table 7 summarizes the relationships.

Summarizing in words then, the S parameter and the a and b variables can be described in less than abstract terms to give a simple picture of what is happening in the network. Notice how this characterization is given in terms of power.

$|a_1|^2$ = power incident on the input of the network
       = power available from a source of impedance Z_0

$|a_2|^2$ = power incident on the output of the network
       = power reflected from the load

$|b_1|^2$ = power reflected from the input port of the network
       = power available from a Z_0 source minus the power delivered to the input of the network

$|b_2|^2$ = power reflected or emanating from the output of the network
       = power incident on the load
       = power that would be delivered to a Z_0 load

$$|S_{11}|^2 = \frac{\text{power reflected from the network input}}{\text{power incident on the network input}}$$

$$|S_{22}|^2 = \frac{\text{power reflected from the network output}}{\text{power incident on the network output}}$$

$$|S_{21}|^2 = \frac{\text{power delivered to a } Z_0 \text{ load}}{\text{power available from } Z_0 \text{ source}}$$

         = transducer power gain with Z_0 load and source

$|S_{12}|^2$ = reverse transducer power gain with Z_0 load and source

The use of S parameters has been most commonly associated with amplifier design at frequencies above which parasitics become significant. Generally, this frequency is around 100 MHz. Difficulties are encountered which make problems not only in characterization but also in the measurement of the transistor parameters. The design of a transistor amplifier stage involves some straightforward steps that are well documented in the literature. The following summary only outlines some of the steps involved to demonstrate the use of the S parameters.

From the previous discussion, s_{11} and s_{22} are the reflection coefficients and can be measured using conventional measurement techniques familiar to most microwave engineers. Two common methods are the reflectometer and the slotted line. These parameters are usually measured in well-matched lines so that they can be taken as voltage-reflection coefficients. Similarly the transmission coefficients can be measured into matched lines by using voltage relationships. Once the measurements are made, they are plotted on the Smith chart, since they are dimensionless quantities that are complex. On the Smith chart a series of constant-gain circles are plotted. Since these gain circles are measured at a function of frequency, the relationship of the gain response to frequency can be seen. Now a set of impedances is selected from the Smith chart to give a broadband constant-gain response. Recalling that the values on the Smith chart are normalized, it is now a matter of selecting the proper component value to try in the circuit.

Other parameters used in the description of the transistor provide additional design information to the amplifier designer. These parameters are necessary because the S parameter characterizations do not of themselves give the level of performance of the transistor amplifier nor do they take into account the possibility of parasitic contributions by header elements and other factors. One particular problem is the stability problem associated with the amplifier. This is usually handled by the use of constant K, defined in the literature as the stability factor. Generally, if $K > 1$, the amplifier is unconditionally stable with the presence of some external feedback path. For values of $K > 1$ the amplifier is potentially unstable and can be set into oscillation by some combinations of passive load and source impedances.

Table 8 lists some of the most important S parameter relationships used. Also listed are the conversion formulas for h, y, and z.

TABLE 8   Useful Scattering-Parameter Relationships[6]

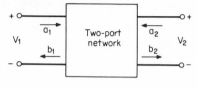

$$b_1 = s_{11}a_1 + s_{12}a_2$$
$$b_2 = s_{21}a_1 + s_{22}a_2$$

Input reflection coefficient with arbitrary Z_L

$$s'_{11} = s_{11} + \frac{s_{12}s_{21}\Gamma_L}{1 - s_{22}\Gamma_L}$$

Output reflection coefficient with arbitrary Z_S

$$s'_{22} = s_{22} + \frac{s_{12}s_{21}\Gamma_S}{1 - s_{11}\Gamma_S}$$

Voltage gain with arbitrary Z_L and Z_S

$$A_V = \frac{V_2}{V_1} = \frac{s_{21}(1 + \Gamma_L)}{(1 - s_{22}\Gamma_L)(1 + s'_{11})}$$

Power gain $= \dfrac{\text{power delivered to load}}{\text{power input to network}}$

$$G = \frac{|s_{21}|^2(1 - |\Gamma_L|^2)}{(1 - |s_{11}|^2) + |\Gamma_L|^2(|s_{22}|^2 - |D|^2) - 2\,\text{Re}\,(\Gamma_L N)}$$

Available power gain $= \dfrac{\text{power available from network}}{\text{power available from source}}$

$$G_A = \frac{|s_{21}|^2(1 - |\Gamma_S|^2)}{(1 - |s_{22}|^2) + |\Gamma_S|^2(|s_{11}|^2 - |D|^2) - 2\,\text{Re}\,(\Gamma_S M)}$$

Transducer power gain $= \dfrac{\text{power delivered to load}}{\text{power available from source}}$

$$G_T = \frac{|s_{21}|^2(1 - |\Gamma_S|^2)(1 - |\Gamma_L|^2)}{|(1 - s_{11}\Gamma_S)(1 - s_{22}\Gamma_L) - s_{12}s_{21}\Gamma_L\Gamma_S|^2}$$

Unilateral transducer power gain $(s_{12} = 0)$

$$G_{Tu} = \frac{|s_{21}|^2(1 - |\Gamma_S|^2)(1 - |\Gamma_L|^2)}{|1 - s_{11}\Gamma_S|^2|1 - s_{22}\Gamma_L|^2}$$
$$= G_0G_1G_2$$
$$G_0 = |s_{21}|^2$$
$$G_1 = \frac{1 - |\Gamma_S|^2}{|1 - s_{11}\Gamma_S|^2}$$
$$G_2 = \frac{1 - |\Gamma_L|^2}{|1 - s_{22}\Gamma_L|^2}$$

Maximum unilateral transducer power gain when $|s_{11}| < 1$ and $|s_{22}| < 1$

$$G_u = \frac{|s_{21}|^2}{|(1 - |s_{11}|^2)(1 - |s_{22}|)^2)|}$$
$$= G_0G_{1\,\text{max}}G_{2\,\text{max}}$$
$$G_{i\,\text{max}} = \frac{1}{1 - |s_{ii}|^2} \qquad i = 1, 2$$

This maximum attained for $\Gamma_S = s^*_{11}$ and $\Gamma_L = s^*_{22}$

TABLE 8   **Useful Scattering-Parameter Relationships[6] (Continued)**

Constant-gain circles (unilateral case: $s_{12} = 0$):
    Center of constant-gain circle is on line between center of Smith chart and point representing s^*_{ii}
    Distance of center of circle from center of Smith chart:

$$r_i = \frac{g_i|s_{ii}|}{1 - |s_{ii}|^2(1 - g_i)}$$

Radius of circle:

$$\rho_i = \frac{\sqrt{1 - g_i}\,(1 - |s_{ii}|^2)}{1 - |s_{ii}|^2(1 - g_i)}$$

where $i = 1,\ 2$ and $g_i = \dfrac{G_i}{G_{i\,\max}} = G_i(1 - |s_{ii}|^2)$

Unilateral figure of merit

$$u = \frac{|s_{11}s_{22}s_{12}s_{21}|}{|(1 - |s_{11}|^2)(1 - |s_{22}|^2)|}$$

Error limits on unilateral-gain calculation

$$\frac{1}{1 + u^2} < \frac{G_T}{G_{Tu}} < \frac{1}{1 - u^2}$$

Conditions for absolute stability
    No passive source or load will cause network to oscillate if a, b, and c are all satisfied:

a $|s_{11}| < 1,\ |s_{22}| < 1$

b $\left|\dfrac{|s_{12}s_{21}| - |M^*|}{|s_{11}|^2 - |D|^2}\right| > 1$

c $\left|\dfrac{|s_{12}s_{21}| - |N^*|}{|s_{22}|^2 - |D|^2}\right| > 1$

Condition that a two-port network can be simultaneously matched with a positive real source and load:

$$K > 1 \text{ or } C < 1$$
$$C = \text{Linvill } C \text{ factor}$$

Linvill C factor

$$C = K^{-1}$$
$$K = \frac{1 + |D|^2 - |s_{11}|^2 - |s_{22}|^2}{2|s_{12}s_{21}|}$$

Source and load for simultaneous match

$$\Gamma_{mS} = M^* \frac{B_1 \pm \sqrt{B_1{}^2 - 4|M|^2}}{2|M|^2}$$

$$\Gamma_{mL} = N^* \frac{B_2 \pm \sqrt{B_2{}^2 - 4|N|^2}}{2|N|^2}$$

where $B_1 = 1 + |s_{11}|^2 - |s_{22}|^2 - |D|^2$
          $B_2 = 1 + |s_{22}|^2 - |s_{11}|^2 - |D|^2$

Maximum available power gain

TABLE 8   Useful Scattering-Parameter Relationships[6] (Continued)

If $K > 1$,

$$G_{A\,max} = \left| \frac{s_{21}}{s_{12}} (K \pm \sqrt{K^2 - 1}) \right|$$

$$K = C^{-1}$$

$$C = \text{Linvill } C \text{ factor}$$

(Use plus sign when B_1 is positive, minus sign when B_1 is negative.   See definition of B_1 above)

$$D = s_{11}s_{22} - s_{12}s_{21}$$

$$M = s_{11} - Ds^*_{22}$$

$$N = s_{22} - Ds^*_{11}$$

s parameters in terms of h, y, and z parameters	h, y, and z parameters in terms of s parameters
$s_{11} = \dfrac{(z_{11} - 1)(z_{22} + 1) - z_{12}z_{21}}{(z_{11} + 1)(z_{22} + 1) - z_{12}z_{21}}$	$z_{11} = \dfrac{(1 + s_{11})(1 - s_{22}) + s_{12}s_{21}}{(1 - s_{11})(1 - s_{22}) - s_{12}s_{21}}$
$s_{12} = \dfrac{2z_{12}}{(z_{11} + 1)(z_{22} + 1) - z_{12}z_{21}}$	$z_{12} = \dfrac{2s_{12}}{(1 - s_{11})(1 - s_{22}) - s_{12}s_{21}}$
$s_{21} = \dfrac{2z_{21}}{(z_{11} + 1)(z_{22} + 1) - z_{12}z_{21}}$	$z_{21} = \dfrac{2s_{21}}{(1 - s_{11})(1 - s_{22}) - s_{12}s_{21}}$
$s_{22} = \dfrac{(z_{11} + 1)(z_{22} - 1) - z_{12}z_{21}}{(z_{11} + 1)(z_{22} + 1) - z_{12}z_{21}}$	$z_{22} = \dfrac{(1 + s_{22})(1 - s_{11}) + s_{12}s_{21}}{(1 - s_{11})(1 - s_{22}) - s_{12}s_{21}}$
$s_{11} = \dfrac{(1 - y_{11})(1 + y_{22}) + y_{12}y_{21}}{(1 + y_{11})(1 + y_{22}) - y_{12}y_{21}}$	$y_{11} = \dfrac{(1 + s_{22})(1 - s_{11}) + s_{12}s_{21}}{(1 + s_{11})(1 + s_{22}) - s_{12}s_{21}}$
$s_{12} = \dfrac{-2y_{12}}{(1 + y_{11})(1 + y_{22}) - y_{12}y_{21}}$	$y_{12} = \dfrac{-2s_{12}}{(1 + s_{11})(1 + s_{22}) - s_{12}s_{21}}$
$s_{21} = \dfrac{-2y_{21}}{(1 + y_{11})(1 + y_{22}) - y_{12}y_{21}}$	$y_{21} = \dfrac{-2s_{21}}{(1 + s_{11})(1 + s_{22}) - s_{12}s_{21}}$
$s_{22} = \dfrac{(1 + y_{11})(1 - y_{22}) + y_{12}y_{21}}{(1 + y_{11})(1 + y_{22}) - y_{12}y_{21}}$	$y_{22} = \dfrac{(1 + s_{11})(1 - s_{22}) + s_{12}s_{21}}{(1 + s_{22})(1 + s_{11}) - s_{12}s_{21}}$
$s_{11} = \dfrac{(h_{11} - 1)(h_{22} + 1) - h_{12}h_{21}}{(h_{11} + 1)(h_{22} + 1) - h_{12}h_{21}}$	$h_{11} = \dfrac{(1 + s_{11})(1 + s_{22}) - s_{12}s_{21}}{(1 - s_{11})(1 + s_{22}) + s_{12}s_{21}}$
$s_{12} = \dfrac{2h_{12}}{(h_{11} + 1)(h_{22} + 1) - h_{12}h_{21}}$	$h_{12} = \dfrac{2s_{12}}{(1 - s_{11})(1 + s_{22}) + s_{12}s_{21}}$
$s_{21} = \dfrac{-2h_{21}}{(h_{11} + 1)(h_{22} + 1) - h_{12}h_{21}}$	$h_{21} = \dfrac{-2s_{21}}{(1 - s_{11})(1 + s_{22}) + s_{12}s_{21}}$
$s_{22} = \dfrac{(1 + h_{11})(1 - h_{22}) + h_{12}h_{21}}{(h_{11} + 1)(h_{22} + 1) - h_{12}h_{21}}$	$h_{22} = \dfrac{(1 - s_{22})(1 - s_{11}) - s_{12}s_{21}}{(1 - s_{11})(1 + s_{22}) + s_{12}s_{21}}$

TABLE 8   Useful Scattering-Parameter Relationships[6] (Continued)

The h, y, and z parameters listed above are all normalized to Z_0.   If h', y', and z' are the actual parameters, then:

$$z_{11}' = z_{11}Z_0 \qquad y_{11}' = \frac{y_{11}}{Z_0} \qquad h_{11}' = h_{11}Z_0$$

$$z_{12}' = z_{12}Z_0 \qquad y_{12}' = \frac{y_{12}}{Z_0} \qquad h_{12}' = h_{12}$$

$$z_{21}' = z_{21}Z_0 \qquad y_{21}' = \frac{y_{21}}{Z_0} \qquad h_{21}' = h_{21}$$

$$z_{22}' = z_{22}Z_0 \qquad y_{22}' = \frac{y_{22}}{Z_0} \qquad h_{22}' = \frac{h_{22}}{Z_0}$$

Transistor frequency parameters:

$$f_t = \text{frequency at which } |h_{fe}|$$
$$= |h_{21} \text{ for common-emitter configuration}| = 1$$
$$f_{\max} = \text{frequency at which } G_{A\,\max} = 1$$

PASSIVE COMPONENTS

For the sake of defining categories, the components discussed will be divided into two classes, active and passive.   Passive components include any device that does not alter the character of the signal in terms of amplification, modulation, and frequency changes (mixing and converters).   Generally, devices in this class do not have an input such as bias voltage or trigger voltage, and their behavior is essentially linear.

Reactances in Transmission Lines

Some appreciation is first needed of what happens physically to make the transmission line behave in some electrically descriptive pattern.   Conventional circuit theory deals with resistances and reactances.   It has been shown that the transmission line can be treated as having these conventional parameters under certain conditions.   Also, the discussion of the Smith chart showed that the normalized impedances and susceptances may be found in a transmission line by some very simple measurements.   The problem is how these values of capacitance and inductance are physically introduced into the waveguide.   The development of the equations for derivation of the values is available in most basic texts on microwave circuits.   However, many charts and tables are available to provide the designer with a starting point.   Figures 20 through 35 give some values of iris commonly used in waveguide transmission lines.

Other popular transmission lines used in microwave work today are the stripline and the microstrip.   Their methods of propagation are discussed below under Transmission Lines.   Figures 36 through 40 show how the stripline equivalents are determined.

Couplers

The basic microwave component used in the transmission line is the coupler.   The coupler can be used for dividing and for combining power, sampling power, injecting power, and many other circuit functions useful to the microwave engineer.   Three parameters are usually used in defining the coupler: coupling, directivity, and insertion loss.   When insertion loss is described, it must be made clear whether it includes the coupled energy or not.   In couplers of high coupling value the amount of energy coupled out is usually negligible compared with the loss.   The determination of the parameters is shown in Fig. 41.   The coupler is a four-terminal network, and

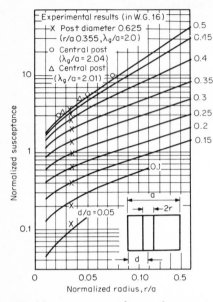

Fig. 20 Susceptance for single post in waveguide ($\lambda_g/a = 2.0$).[7]

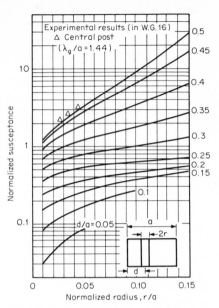

Fig. 21 Susceptance for single post in waveguide ($\lambda_g/a = 1.4$).[7]

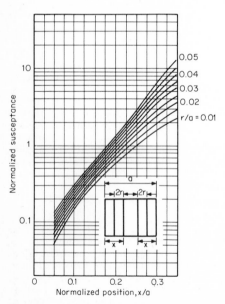

Fig. 22 Susceptance for post doublet in waveguide ($\lambda_g/a = 1.2$).[7]

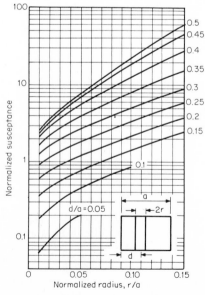

Fig. 23 Susceptance for single post in waveguide ($\lambda_g/a = 2.8$).[7]

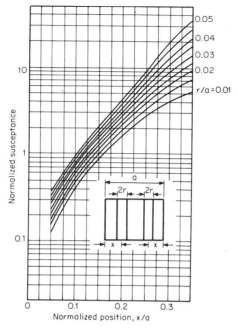

Fig. 24 Susceptance for post doublet in waveguide $(\lambda_g/a = 2.8)$.[7]

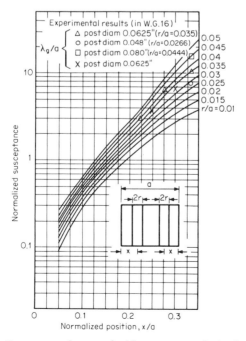

Fig. 25 Susceptance for post doublet in waveguide $(\lambda_g/a = 2.0)$.[7]

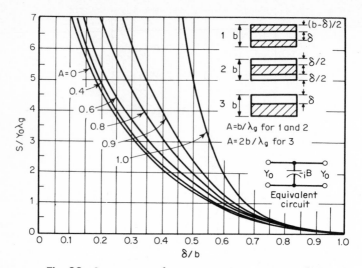

Fig. 26   Susceptance of capacitive irises in waveguide.[7]

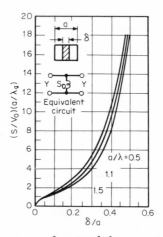

Fig. 27   Susceptance of centered thin vane in waveguide.[7]

from Fig. 42, the coupling is given as follows:

$$\text{Coupling factor} = 10 \log_{10} \left(\frac{P_a}{P_d} \right) \qquad \text{dB} \tag{37}$$

The directivity is defined as follows:

$$\text{Directivity} = 10 \log_{10} \left(\frac{P_d}{P_c} \right) \qquad \text{dB} \tag{38}$$

The insertion loss is defined as follows:

$$\text{Insertion loss} = 10 \log_{10} \left(\frac{P_a - P_b}{P_a} \right) \qquad \text{dB} \tag{39}$$

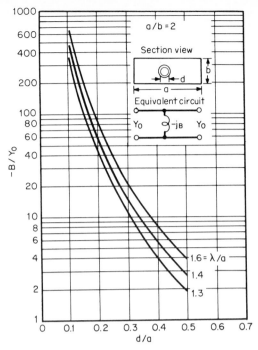

Fig. 28 Susceptance of hole in iris in waveguide.[7] (*Reprinted from Massachusetts Institute of Technology Radiation Laboratory Report 43, February 1944. Prepared under contract with the Office of Scientific Research and Development.*)

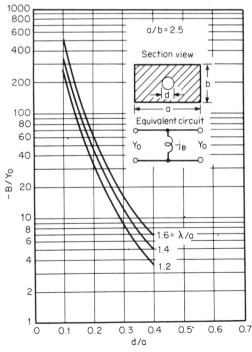

Fig. 29 Susceptance of hole in iris in waveguide.[7]

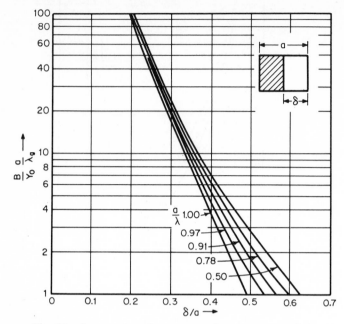

Fig. 30 Susceptance of asymmetrical iris in waveguide.[7]

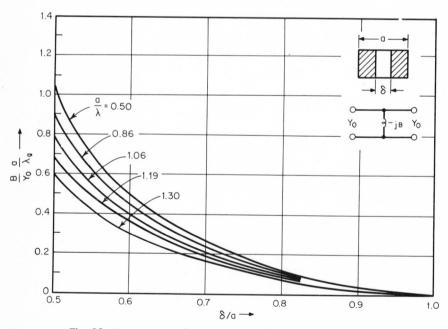

Fig. 31 Susceptance of two symmetrical irises in waveguide.[7]

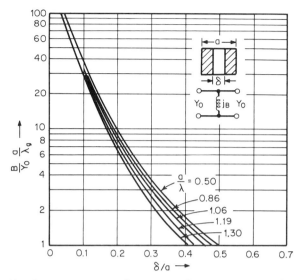

Fig. 32   Susceptance of symmetrical irises in waveguide.[7]

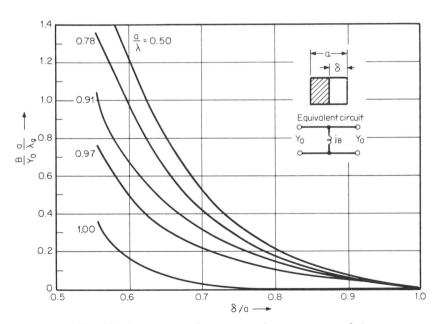

Fig. 33   Susceptance of asymmetrical iris in waveguide.[7]

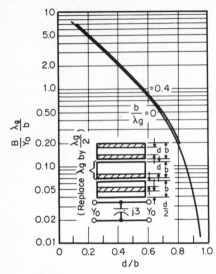

Fig. 34 The equivalent susceptance of thin capacitive windows in a rectangular waveguide. The curves are theoretical.[47]

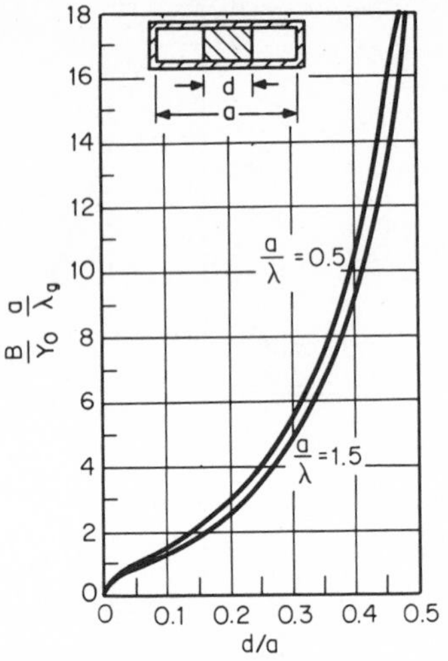

Fig. 35 The equivalent circuit of a thin inductive strip in a rectangular waveguide. The curves are theoretical.[47]

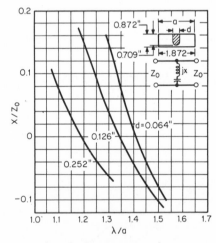

Fig. 36 The reactance-wavelength characteristics of a resonant post in a rectangular waveguide (experimental values).

Coupling can be physically accomplished in the waveguide in a variety of ways, and the following illustrates one simple device often used for study in the waveguide transmission line. With reference to Fig. 43, the ideal situation is to have all energy coupling at P_d combine in phase. Likewise, all energy reflected from the port at P_c should combine out of phase to cancel for maximum directivity. Therefore, ideally the value of s should be $(2N - 1)\lambda/4$, where N is any positive integer. It is immediately obvious that this coupler is frequency-sensitive. Other defects that are not so obvious are interaction between the holes, which affects spacing relationship, and also the fact that increasing hole diameter decreases directivity. These defects

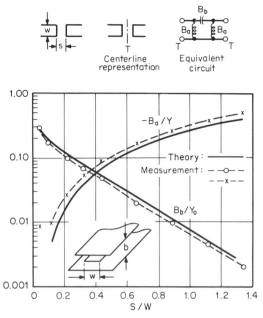

Fig. 37 Gap in stripline; 50-Ω line; centerline representation.[7]   Circuit parameters:
$B_a/Y_0 = 2b/\lambda \ln \cosh \pi s/2b$; $B_a/Y_0 = b/\lambda \ln \cot \pi s/2b$; $-B_a/Y_0$ and B_b/Y_0 vs. s/w.

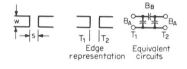

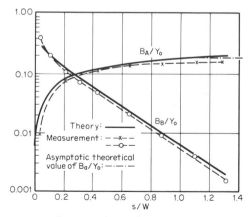

Fig. 38 Gap in stripline 50-Ω line; edge representation.[7]

Circuit parameters: $\dfrac{B_A}{Y_0} = \dfrac{1 + (B_a/Y_0)\cot(\pi s/\lambda)}{\cot(\pi s/\lambda) - (B_a/Y_0)}$

$\dfrac{B_B}{Y_0} = \dfrac{1}{2}\dfrac{1 + (2B_b/Y_0 + B_a/Y_0)\cot(\pi s/\lambda)}{\cot(\pi s/\lambda) - (2B_b/Y_0 + B_a/Y_0)} - \dfrac{1}{2}\dfrac{B_A}{Y_0}$

For $s \gg w$: $B_A/Y_0 \sim \tan[(2b/\lambda)\ln 2]$

B_A/Y_0 and B_B/Y_0 vs. s/w

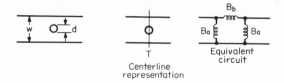

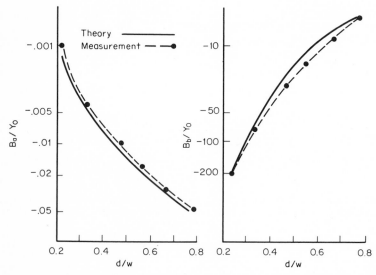

Fig. 39 Round hole in stripline; 50-Ω line; centerline representation.[7] $B_a/Y_0 =$ $1/[4(B_4/Y_0)]$.

are overcome with multihole couplers or the slot or cross-slot coupler. Figures 44 through 50 give some relationships used in waveguide-coupler design.

The Bethe-hole coupler illustrates a technique that employs both electric- and magnetic-vector coupling. Figure 51 shows how the coupling appears through the hole. In Fig. 51c the field is shown from the top of the wall. Waves traveling in the right direction reinforce each other and enhance the coupling, while waves in the opposite direction cause cancellation and thus accomplish the desired effect of directivity. The practical coupler optimizes the directivity by rotating the waveguide slightly, because the magnetic coupling is a function of the guide angles. With certain assumptions as to wall thickness and hole radii, the coupling varies approximately as a function of the sixth power of the hole radius.[8] Another system of couplers is the calibrated reflectometer coupler. Two couplers are calibrated back to back so that each measures the amounts of power in their respective directions. These couplers have a high value of directivity and precisely valued amounts of coupling.

Other types of transmission lines use couplers in the same manner as the waveguide, but of course they must use different schemes of coupling. Figure 52 shows a configuration for a coupler to use in stripline (or microstrip). Figure 52a and b demonstrates how the supported and canceling field exist in the coupler circuit. Because stripline does not exist in an ideal configuration, great reliance must be placed on charts and empirical curves available in the literature. Figures 53 through 55 give typical values from curves used by designers in the field. Figures 56–58 summarize a number of approaches used in couplers of special configuration. The

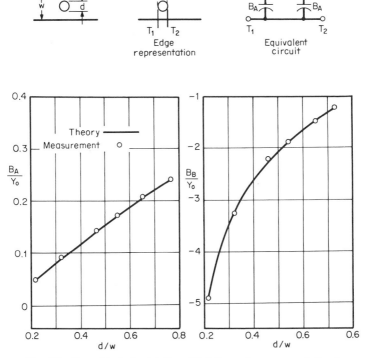

Fig. 40 Round hole in stripline; 50-Ω line; edge representation.[7]

Circuit parameters: $\dfrac{B_A}{Y_0} = \dfrac{1 + (B_a/Y_0)\cot(\pi d/\lambda)}{\cot(\pi d/\lambda) - (B_a/Y_0)}$

$\dfrac{B_B}{Y_0} = \dfrac{1}{2}\dfrac{1 + 2(B_b/Y_0)\cot(\pi d/\lambda)}{\cot(\pi d/\lambda) - 2(B_B/Y_0)} - \dfrac{1}{2}\dfrac{B_A}{Y_0}$

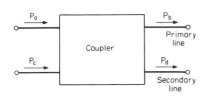

Fig. 41 Coupling network.

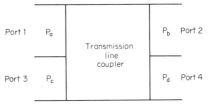

Fig. 42 The coupler as a four-terminal network.

notes under each of the approaches describe the situation when it is most desirable to seek this design technique. These designs are made readily available by the use of computer-aided design.

Attenuators

The next class of passive components is the attenuator. This device compares with the resistor in the conventional network in that it is a device usually intended to waste energy for an intended purpose. Attenuators are rated in values of dB

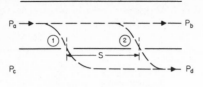

Fig. 43   The two-hole coupler.

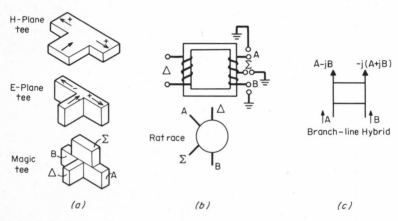

(a)   *(b)*   *(c)*

Fig. 44 Couplers, tees, rat race, and hybrids.   Coherent power division was first accomplished by means of simple tee junctions (*a*).   At microwave frequencies, waveguide tees have two possible forms—the *H*-plane or the *E*-plane tee.   These two junctions split power equally, but because of the different field configurations at the junction, the electric fields at the output arms are in phase for the *H*-plane tee and are antiphase for the *E*-plane tee.   The combination of these two tees to form a hybrid tee allowed the realization of a four-port component which could perform the vector sum Σ and difference Δ of two coherent microwave signals *A* and *B*.   This device is, of course, the magic tee.

Components which perform the same function as the magic tee have been realized in many different forms in balanced, coaxial, and strip transmission-line configurations.   Also, lumped-component devices which make use of a center-tapped transformer have been built at frequencies up to 1 GHz.   The frequency limitation in this device is principally due to the decline of the scalar permeability and the increase in loss of ferrite materials at microwave frequencies.   The distributed versions bear little or no physical resemblance to the waveguide hybrid but are still sometimes referred to as hybrids.   The rat race (*b*) is an example of a TEM version of the waveguide magic tee.

Another device which is also called a hybrid is a branch-line hybrid (*c*). This device, however, differs from the magic tee and the rat race in that the output signals are ±90° relative to each other instead of 0° and 180°. To differentiate between these two types of devices, one is called a 180° hybrid and the other a 90° hybrid.

and power.   The type of material and size are the limiting factors for the power rating.   Generally, there are three types of attenuators, the fixed, the continuously variable, and the step variable.

The attenuator is usually made with a resistor material of some type mounted in such a fashion as to minimize the impedance-mismatch effects in a transmission line.   Various types of base materials are used, and the resistance or energy-absorbing material may be deposited on a card of the base material by vacuum deposition or some similar technique.   The card is rated in ohms per square centimeter.   There

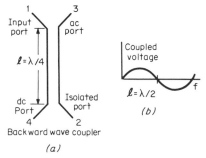

Fig. 45 Couplers; backward-wave coupler, (*Anaren Microwave, Inc.*) All the microwave devices already described are limited in bandwidth regarding input match and isolation. Another verson of the 90° hybrid which is superior in this regard is the backward-wave coupler (*a*). This device consists simply of two lengths of transmission line in close electrical proximity. Any two such lengths of line, whether or not of equal characteristic impedances, form a coupler. The input impedances to the device, however, and the amount of power coupled from one line to the other depend critically on the cross-sectional geometry in the coupled section. If this geometry is arranged so that the coupled section is matched to the impedance of the input lines, then power into port 1 divides between ports 3 and 4, and port 2 is isolated. This same pattern holds for power into any of the ports. The power at port 3 differs in phase by 90° relative to the power at port 4.

The input match, isolation, and relative phase are independent of frequency. The amplitude of the coupled voltage varies almost sinusoidally with frequency, being maximum when the electrical length of the coupled section is an odd number of quarter wavelengths (*a*). The slow variation of the amplitude of the coupled voltage at the maxima allows the device to be useful over an octave bandwidth or more centered at a quarter wavelength.

The frequency characteristics of a backward-wave coupler are most easily described by means of a coupling angle θ. This angle is a function of both the proximity of the coupled lines and the electrical length of the coupled section. The coupling angle θ varies almost sinusoidally with frequency as

$$\theta \approx \theta_{\max} \sin \frac{2\pi l}{\lambda}$$

The maximum coupling angle thus occurs when the coupler length is an odd multiple of quarter wavelengths. $\theta_{\max}$ depends only on the cross-sectional geometry.

If a signal of strength 1 V is applied to one port of the coupler, the signals appearing at the dc and the coupled ports are, respectively (ignoring a slight dispersion)

$$V_{\mathrm{dc}} = \cos \theta e^{-j\beta_1}$$
and
$$V_{\mathrm{coupled}} = j \sin \theta e^{-j\beta_1}$$

If the coupler geometry is arranged such that $\theta_{\max} = 45°$, then at the frequency where the coupler length is a quarter wavelength, the output voltages are

$$V_{\mathrm{dc}} = \frac{-j}{\sqrt{2}}$$
and
$$V_{\mathrm{coupled}} = \frac{1}{\sqrt{2}}$$

This is, of course, a 3-dB coupler.

The common couplers have coupling values of 3, 10, and 20 dB. The last two devices are normally called directional couplers. The first device is usually called a 3-dB quadrature coupler, or simply 3-dB coupler.

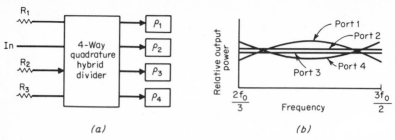

(a)   *(b)*

Fig. 46 Couplers; quadrature hybrid couplers. (*Anaren Microwave, Inc.*) Anaren Microwave Application Note (Combiner-Dividers for Power Amplifiers) discusses the match and transfer properties of quadrature hybrid couplers. The conclusions drawn are that poorly matched devices may be placed at the output ports of a hybrid divider without decreasing the input match to the divider. In a sense, then, the hybrid divider acts like an isolator provided that the devices have nearly identical reflection coefficients.

The case for a four-way hybrid power divider is illustrated (*a*). The divider is terminated by four devices having complex reflection coefficients ρ_1, ρ_2, ρ_3, and ρ_4. An expression for the input reflection coefficient of the divider is given by

$$\rho_{in} = \tfrac{1}{4}(\rho_1 - \rho_2 - \rho_3 + \rho_4)$$

It can be seen that if $\rho_1 = \rho_2 = \rho_3 = \rho_4$, then $\rho_{in} = 0$. The reflected power from the mismatch at the output goes to the terminations R_1, R_2, and R_3, as can be seen by the expression for the power to the loads:

$$P_{R1} = \frac{P_{in}}{8}(\rho_1 + \rho_2)^2$$

$$P_{R2} = \frac{P_{in}}{8}(\rho_3 + \rho_4)^2$$

$$P_{R3} = \frac{P_{in}}{16}(\rho_1 - \rho_2 + \rho_3 - \rho_4)^2$$

Another advantage of the quadrature hybrid divider is in its power-handling capability. The terminations may be brought out of the device so that high-power terminations may be used (i.e., relatively large heat sink or finned loads). The stripline circuitry has a power-handling capability up to 200 W CW. In any given application, the power handling is usually limited by the loads. The table following this note gives expressions for power dissipated in terminations in hybrid dividers and combiners.

The primary disadvantage of hybrid dividers is power imbalance between output ports. The imbalance which occurs for a four-way divider over an octave bandwidth is shown (*a*). As can be seen, two ports track closely while the other two diverge at the band edges and at the center of the band f_0. Specifications for a four-way divider are usually ±1 dB for an octave bandwidth device. Better results, of course, can be achieved over narrow bandwidths with optimization of the divider.

Another disadvantage of a quadrature hybrid divider (for certain applications) is that the output ports are not in phase. The ports have a 90° (of a multiple of 90°) phase relationship which remains constant over octave bands. Specifications for hybrid dividers are for the relative phase variation from 90° or multiple of 90° over the frequency range specified.

Quadrature dividers can be constructed only with a binary number (2^n) of output ports (i.e., 2, 4, 8, 16).

have been some successes in using higher-temperature materials for the card, but the more common approach is to use very good heat sinking, low thermal coefficients, and large amounts of material. The energy is distributed along large areas of the material. Another very popular material is chemically formed carbonyl iron spheres uniformly distributed on a plastic carrier. The material is uniform in the microwave spectrum, and its behavior is very predictable. The singular advantage of this material is that it is ideal for fabrication. Table 9 lists the physical constants of the

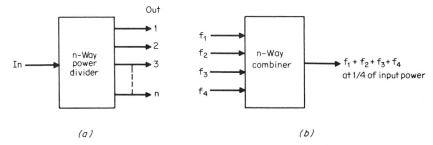

(a)   *(b)*

Fig. 47 Power dividers. (*Anaren Microwave, Inc.*) Although both in-phase (Wilkinson) and quadrature (90°) hybrid couplers may be used for coherent power divider/combiner applications, fundamental differences exist making each more suitable for specific applications.

An illustration of an n-way power divider is shown (*a*). The device has a single input port and n output ports. Ideally, input power would be divided equally between the output ports. The output phase relationship would depend upon the construction of the device. If the device were an in-phase divider, the output ports would be in phase. In a quadrature hybrid divider, the output ports would have a 90° (or multiple of 90°) phase relationship.

These devices, either quadrature hybrid or in-phase dividers, can be used as coherent combiners as well as dividers, provided that the reciprocity of the device is understood. For example, for lossless recombination, the same amplitude and phase relationships which exist at the output of the device when used as a divider must drive the n input ports of the device when used as a combiner. That is to say, a divider will losslessly combine n input signals provided that they are of the proper input phase and amplitude relationship. If this is not the case, power is lost in the combiner.

In many applications coherent addition of signals is not a requirement. An example of this is where n signals of different frequencies are applied to a device with a single output port (multiplexer). One wishes to see the sum of the signals at the output. This may be accomplished using hybrid or in-phase combiners provided losses can be tolerated. Using an n-way combiner in this manner, only $1/n$ of the input power will appear at the output. A four-way combiner used as a multiplexer is shown (*b*). Four signals of four different frequencies at unit power level each are applied at the inputs. The output is the sum of the four signals, but each has one-fourth unit of power. The signals have been combined, but at the expense of a loss in power. Lossless multiplexing can be done only with filter networks.

TABLE 9   Physical Constants for RF Absorbing Material

Characteristic	Value	Remarks
Attenuation, dB/in	6.25 at 1 GHz, 18.00 at 3 GHz, 80.00 at 10 GHz	TEM wave assumed
Dielectric dissipation factor	0.045	1–10 GHz
Magnetic dissipation factor	$0.05 \times F(\mathrm{GHz})$	1–10 GHz
Permeability	$4.5\text{-}\log_e F(\mathrm{GHz})$	1–10 GHz
Dielectric constant	10.0	1–10 GHz
Volume resistivity	2×10^{14}	Ω-cm³ at 25°C
Dielectric strength	375 kV/in	0.100-in sample
Water absorption	0.04% max	40 h immersion/25°C
Flexural strength	15,700 approx	lb in⁻²
Impact (Izod)	0.30	ft-lb/in notch
Softening temp	350°F min	

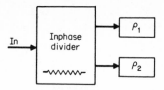

Fig. 48 In-phase power divider. (*Anaren Microwave, Inc.*) The in-phase (Wilkinson) power divider has the advantage of excellent output port amplitude balance over octave or wider bandwidths combined with in-phase power division. The power split and phase balance is theoretically perfect, and nearly ideal results can be achieved in practice.

The chief disadvantage (for power applications) is that the terminations cannot be brought out externally to the device. A constraint on the internal terminations is that the terminations must be much less than a wavelength in any dimension. This constraint limits power handling to approximately 100 mW at X-band and 20 to 50 W at uhf and vhf frequencies. These are "fail-safe" specifications. The final power-handling capability depends upon the external terminations of the divider. As an example, a two way divider with terminations is illustrated having complex reflection coefficients ρ_1 and ρ_2. The power lost in the internal termination is given by

$$P = \frac{P_{\text{in}}}{4} (\rho_1 - \rho_2)^2$$

It can be seen that if the reflection coefficients are nearly identical, no power will be lost in the internal termination. The worst case is for one reflection coefficient to be 180° out of phase with the other. In this case, all reflected power is dissipated in the internal termination. A good example is if both terminals are open ($\rho_1 = \rho_2 = 1.0$) or short ($\rho_1 = \rho_2 = -1$). In this case no power is dissipated in the external termination. If, however, one output port is open-circuited and the other short-circuited ($\rho_1 = 1.0$, $\rho_2 = -1.0$), then all the power input is dissipated in the internal termination.

Unlike the quadrature hybrid power divider, if the output ports are loaded with devices of nearly identical reflection coefficients, the input match degrades according to the magnitude of the reflection coefficients. This is seen by an expression for the input reflection coefficient.

$$\rho_{\text{in}} = \frac{1}{2}(\rho_1 + \rho_2)$$

Unlike the quadrature hybrid divider, if the outputs are open- or short-circuited ($\rho_1 = \rho_2 = 1.0$ or $\rho_1 = \rho_2 = -1$), the input reflection coefficient is unity.

In-phase power dividers, like the quadrature hybrid dividers, are most easily constructed with a binary (2^n) number of output ports. n-way dividers can be constructed but are most easily done in cylindrical geometries (more difficult from a construction point of view) than the binary dividers which lend themselves to planar geometries.

Both quadrature hybrid and in-phase power dividers have common properties of good match at all ports and high isolation between output ports. Devices may be constructed using combinations of both types of power dividers to gain the advantages of both. For example, an eight-way divider can be constructed from in-phase and quadrature hybrid dividers. Using this approach, input match under matched-load conditions can be guaranteed, and the divider would have better amplitude balance than a device made up of quadrature hybrid dividers alone.

In power applications (i.e., amplifiers), in-phase dividers can be kept in the low-power region of the device with hybrids in the higher-power region. Again better amplitude balance is obtained while maintaining the advantage of good input match.

Finally, where output phase is not important, n-way dividers for odd division or even division other than 2^n may be done using any combination of quadrature hybrid dividers, in phase power dividers, and backward-wave couplers. In this manner. 3-, 5-, 6-, . . . way dividers have been constructed for particular applications.

Two – way Quadrature Hybrid Combiner – Dividers

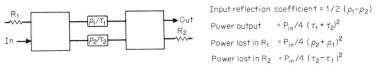

Input reflection coefficient $= 1/2\,(\rho_1 - \rho_2)$

Power output $= P_{in}/4\,(\tau_1 + \tau_2)^2$

Power lost in R_1 $= P_{in}/4\,(\rho_2 + \rho_1)^2$

Power lost in R_2 $= P_{in}/4\,(\tau_2 - \tau_1)^2$

Four-way Quadrature Hybrid Combiner – Dividers

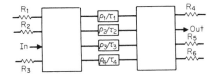

Input reflection coefficient $= 1/4\,(\rho_1 - \rho_2 - \rho_3 + \rho_4)$

Power output $= P_{in}/16\,(\tau_1 + \tau_2 + \tau_3 + \tau_4)^2$

Power lost in R_1 $= P_{in}/8\,(\rho_1 + \rho_2)^2$

Power lost in R_2 $= P_{in}/16\,(\rho_1 - \rho_2 + \rho_3 - \rho_4)^2$

Power lost in R_3 $= P_{in}/8\,(\rho_3 + \rho_4)^2$

Power lost in R_4 $= P_{in}/8\,(\tau_1 - \tau_2)^2$

Power lost in $R_5 = P_{in}/16\,(\tau_1 + \tau_2 - \tau_3 - \tau_4)^2$

Power lost in $R_6 = P_{in}/8\,(\tau_3 - \tau_4)^2$

Two -way in – phase Combiner – Dividers

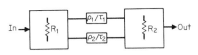

Input reflection coefficient $= 1/2\,(\rho_1 + \rho_2)$

Power output $= P_{in}/4\,(\tau_1 + \tau_2)^2$

Power lost in R_1 $= P_{in}/4\,(\rho_1 - \rho_2)^2$

Power lost in R_2 $= P_{in}/4\,(\tau_1 - \tau_2)^2$

Four-way in – phase Combiner – Dividers

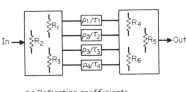

ρ = Reflection coefficients
τ = Transmission coefficient

Input reflection coefficient $= 1/4\,(\rho_1 + \rho_2 + \rho_3 + \rho_4)$

Power output $= P_{in}/16\,(\tau_1 + \tau_2 + \tau_3 + \tau_4)^2$

Power lost in R_1 $= P_{in}/8\,(\rho_1 - \rho_2)^2$

Power lost in R_2 $= P_{in}/16\,(\rho_1 + \rho_2 - \rho_3 - \rho_4)^2$

Power lost in R_3 $= P_{in}/8\,(\rho_3 - \rho_4)^2$

Power lost in R_4 $= P_{in}/8\,(\tau_3 - \tau_4)^2$

Power lost in $R_5 = P_{in}/16\,(\tau_1 + \tau_2 - \tau_3 - \tau_4)^2$

Power lost in $R_6 = P_{in}/8\,(\tau_3 - \tau_4)^2$

Fig. 49 Power dividers summary. (*Anaren Microwave, Inc.*)

material. Its energy-absorbing properties are usually taken as 1 W in⁻³. Figure 59 demonstrates various design approaches that may be used in attenuators with materials described above.

The various types of attenuators use as many innovative techniques as there are designers. The fixed attenuator uses a material such as those described above firmly attached to the transmission-like structure. The variable attenuator may be a card that is inserted into the transmission line, usually transformed into air-dielectric line. Variable attenuators are sometimes unstable in vibration environments because the designs are frequently used with cantilevered mechanisms. Also, a problem encountered with any variable design is rf leakage. One useful relationship that helps in heat calculations for energy-absorbing material used in attenuators is the concept of thermal resistivity. When the temperature change through an attenuator

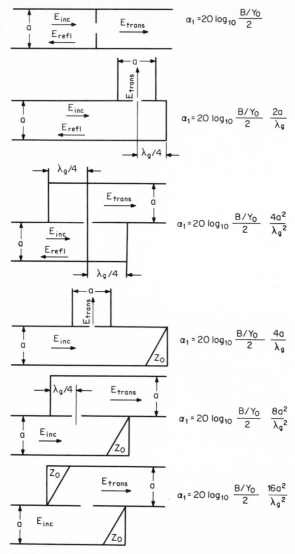

Fig. 50 Miscellaneous arrangements of a small coupling hole between two rectangular waveguides, and the corresponding formulas to be used in calculating the attenuation through the hole.[47]

to a heat sink is derived, the following relationship may be used:

$$\Delta T = \frac{1}{K}\frac{L}{A}Q \tag{40}$$

where $1/K$ = thermal resistivity (see Table 10)
L = path length
A = area of attenuation surface
Q = heat flow
ΔT = temperature drop

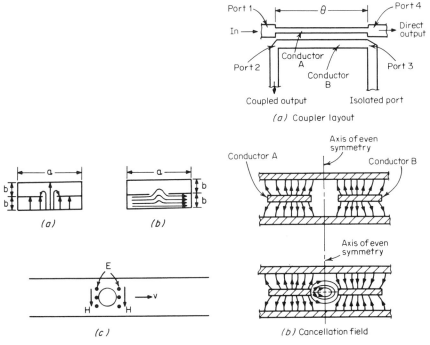

Fig. 51 Bethe-hole coupler. Single-hole directional complex. (*a*) *E*-field coupling. (*b*) *H*-field coupling. (*c*) Top view of Bethe-hole coupler, indicating direction of *E* field, *H* field, and energy flow.

Fig. 52 Stripline coupler.[1] (*a*) Coupler layout. (*b*) Cancellation fields.

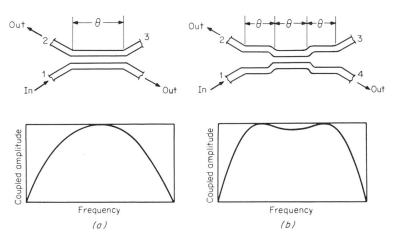

Fig. 53 Typical configurations and frequency responses for TEM-mode, coupled-transmission-line directional couplers of one and three sections.[7] (*a*) Quarter-wavelength coupler. (*b*) Three-quarter-wavelength coupler.

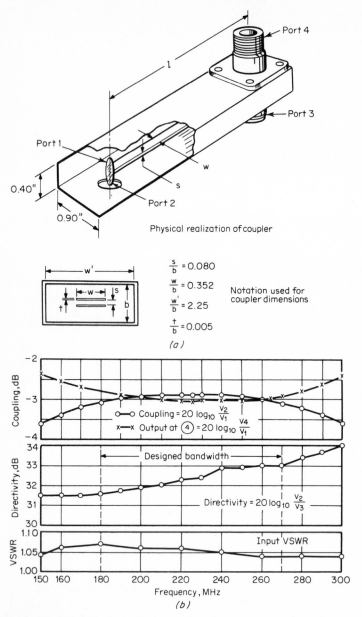

Fig. 54 Construction details and measured performance of a printed-circuit, 3-dB coupler.[7]   (*a*) Physical realization of coupler.   (*b*) measured performance.

Ferrite Components

In the earlier discussion of waves in transmission lines it was brought out that there is a system of forward and reverse waves.   In many applications the reflected waves cannot be tolerated because of the high vswr's that might be present.   Ferrite components used to get rid of these unwanted reflections are functionally named isolators and circulators.   Phase shifters will also be discussed.

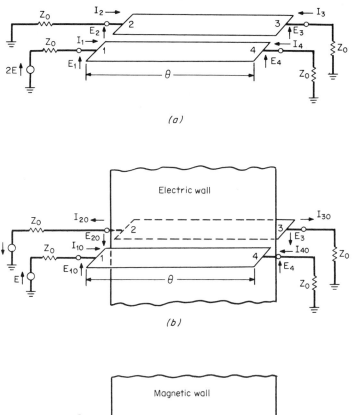

(a)

(b)

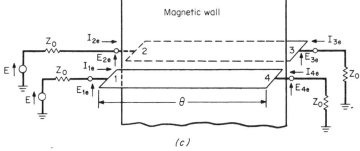

(c)

Fig. 55 Coupled-strip directional coupler.[7] (a) Equivalent circuit. (b) Odd excitation. (c) Even excitation.

Isolators The ferrite components all behave similarly in terms of the physics involved. Each electron spins about its own axis as it moves in its orbit around the nucleus. When a magnetic field is introduced in the path of the electron, its orbit becomes distorted and the axis seems to wobble around the nucleus. A natural frequency associated with this occurrence is referred to as the resonance frequency. When a field is passing through an isolator that is magnetically biased, the field created by the spinning electrons can either enhance or cancel the waves, dependent upon which way they are traveling. The electrons have to be polarized by the forward field, which excites the energy phenomenon at or near the resonance frequency. It is this nonreciprocal property that makes it possible to isolate the reflected wave. It is important to note that the energy of the reflected wave is absorbed by the ferrite material as well as the little bit of energy lost in the forward

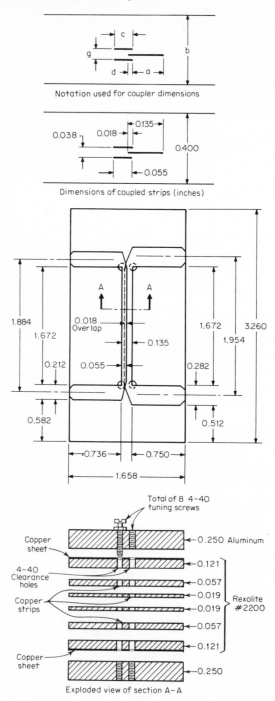

Notation used for coupler dimensions

Dimensions of coupled strips (inches)

Exploded view of section A–A

Fig. 56 Details of construction of 1,000-MHz, 3-dB coupled-transmission-line directional coupler using interleaved strips.[7]

Fig. 57 General formulas for the design of TEM-mode coupled-transmission-line directional couplers from the quarter-wave filter prototype.[7]

The midband vswr R of the quarter-wave filter is related to the overall coupling factor c_0 of the coupler by

$$R = \frac{1 + c_0}{1 - c_0} \quad \text{or} \quad c_0 = \frac{R - 1}{R + 1}$$

The midband vswr R of the quarter-wave filter is

$$R = \left(\frac{Z_1 Z_3 Z_5 \ldots}{Z_2 Z_4 Z_6 \ldots}\right)^{\pm 2} > 1$$

where the Z_i are the normalized impedances of the quarter-wave filter prototype sections.
  The coupling factors c_i of the several coupler sections are related to the normalized impedances Z_i of the quarter-wave filter prototype sections by

$$Z_i{}^2 = \frac{1 + c_i}{1 - c_i} \quad \text{or} \quad c_i = \frac{Z_i{}^2 - 1}{Z_i{}^2 + 1}$$

The even-mode and odd-mode impedances $(Z_{0e})_i$ and $(Z_{0o})_i$ of the coupler sections are given by

$$(Z_{0e})_i = Z_0 \sqrt{\frac{1 + c_i}{1 - c_i}} \quad \text{and} \quad (Z_{0o})_i = Z_0 \sqrt{\frac{1 - c_i}{1 + c_i}}$$

Certain simplifications result for couplers with end-to-end symmetry. In that case n is odd, and

$$c_i = c_{n+1-i} \qquad Z_i = Z_{n+1-i} \qquad V_i = V_{n+2-i} = \text{vswr of } i\text{th step in prototype}$$

and

$$R = (V_1 V_3 V_5 \ldots V_n)^2 = (V_2 V_4 V_6 \ldots V_{n+1})^2$$

TABLE 10   Typical Values of Thermal Resistivity R[14]

Material	R, °C-in./W	Material	R, °C-in./W	Material	R, °C-in./W
Diamond	0.06	Boron nitride	1.24	Epoxy—high	24
Silver	0.10	(isotropic)		conductivity	
Copper	0.11	Alumina	2.13	Quartz	27.6
Gold	0.13	ceramic		Glass (7740)	34.8
Aluminum	0.23	Kovar	2.34	Silicon thermal	46
Beryllia ceramic	0.24	Silicon carbide	2.3	grease	
Molybdenum	0.27	Steel (300	2.4	Water	63
Brass	0.34	series)		Mica (avg)	80
Silicon	0.47	Nichrome	3.00	Polyethylene	120
Platinum	0.54	Carbon	5.7	·Teflon	190
Tin	0.60	Ferrite	6.3	Nylon	190
Nickel	0.61	Pyroceramic	11.7	Silicone rubber	~190
Eutectic lead-	0.78	(9606)		Polyphenylene	205
tin solder				oxide	
Lead	1.14			Polystyrene	380
				Mylar	1,040
				Air	2,280

Conversion from thermal conductivity k to R:

$$R = \frac{22.8}{K} \text{ for } K \text{ in Btu ft}/(\text{h})(\text{ft}^2)(°\text{F})$$

$$R = \frac{0.094}{K} \text{ for } K \text{ in cal cm}/(\text{s})(\text{cm}^2)(°\text{C})$$

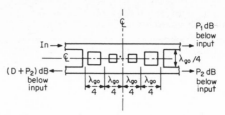

Fig. 58   Branch-line coupler schematic.[7]

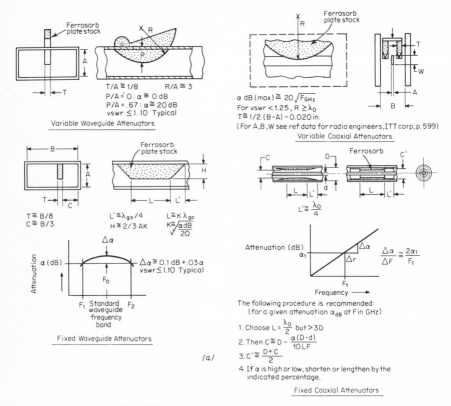

Fig. 59(a)   For legend see opposite page.

direction.   Figure 60 illustrates the manner of biasing a ferrite in a waveguide structure.   A similar scheme is used for other types of transmission lines.

The basic element of the isolator is the ferrite, the manufacture of which is an art all its own.   Ferrites are formed by a sintering process from oxides of iron, zinc, manganese, cobalt, aluminum, or nickel fired at temperatures of 2000 degrees or more.   The peculiar ratio of materials and firing temperatures requires such a special degree of skill that companies in the field seldom manufacture anything else.   Ferrites are made in a variety of shapes depending upon the type of transmission line and the impedance characteristics desired.   The shapes are achieved by sawing and grinding.   Because the ferrite material is characterized by low thermal conductivity and low ductility, these machining operations are very difficult.   The art of manufacturing ferrites is the basis of the microwave-ferrite-component industry.   When deal-

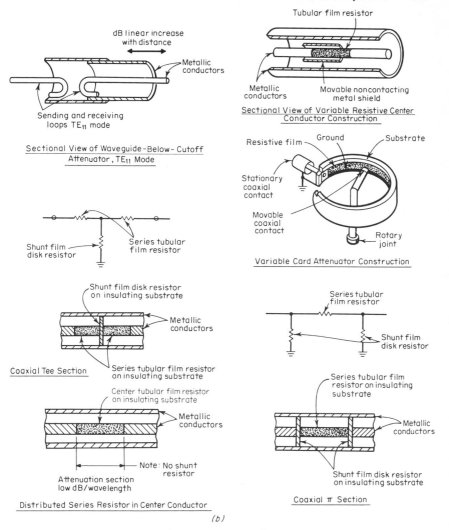

Fig. 59   (*a*) Some attenuator design routines.   (*b*) Some typical coaxial attenuator designs.   (*Wienschall Cap.*)

ing with microwave-component suppliers, it is important to find out if the supplier has good control of the material used in the device.

Circulators   Circulators operate from the same basic principles as the isolator except that the treatment in the circuit is different.   In the circulator the differences are involved in the manner in which the rf field is moved through the transmission line.   The field is bent into the adjoining guide by the polarization effects of the introduction of the ferrite.   The isolation effects are achieved by enhancement in one direction and cancellation in the other direction.   This bending of the field into the adjacent port gives the circulator one desirable effect when high-power situations are encountered.   The energy reflected down a transmission line can be diverted into a dummy load so that the only power consideration is the loss on the forward direction, which is usually on the order of a few tenths of a decibel.

Circulators are usually three- or four-port devices, depending on the construction, and allow energy to pass in one direction with nominal loss while providing isolation in the other. Reference to Fig. 61 shows that the energy passes from port 1 to port 2 and port 2 to port 3 and port 3 to port 1 with only nominal insertion loss. Energy in the opposite direction is isolated on the order of 20 to 25 dB. If an isolator were desired, it would only be necessary to terminate port 2 and all reflected

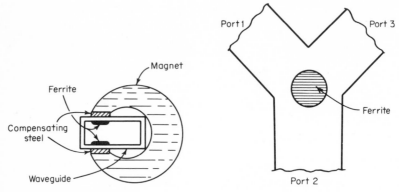

Fig. 60 Microwave-isolator cross section.

Fig. 61 Y-junction circulator cutaway showing ferrite location.

energy would be dumped into the termination. Another use would be in an amplifier circuit in which the amplifier would be attached to port 2. The incoming signal would pass into port 2 and be amplified and sent out through port 3.

Other ferrite components deserve some mention at this point. One such device is the differential phase-shift circulator commonly used in high-power applications. This device is usually used in duplexing functions when high power from a transmitter is sent to the antenna and low-level receiver power is sent to the mixers. Figure 62 shows some typical configurations of ferrite devices.

Fig. 62 Miscellaneous ferrite devices. (*Microwave Associates.*)

Phase shifter   Another class of ferrite device is the phase shifter. The magnetic field is introduced into the ferrites by means of pole pieces which are of soft iron. A special carefully wound coil is wrapped around the waveguide and provides the necessary magnetic field to bias the ferrite. See Fig. 63. The phase shift is introduced by changing the amount of permeability of the microwave path. Another class of phase shifter gaining prominence today is a version of the Fox phase shifter that is used in multielement phase-arrayed antennas. The Fox phase shifter generally involves the use of circular waveguide and some means of making a transition to the circular guide both in and out. The fields are rotated electrically in a dielectric medium by coils applied in quadrature.

Filters and Cavities

Filters Filters in microwave systems are used much like filters in other applications. There are generally the low-pass filter, the high-pass filter, and the bandpass filter. Another class of filters is the multiplexing filter, which is usually a combination of bandpass filters. These filters are used in a separator arrangement according to the bandpass characteristics. Figure 63 illustrates the filter-response curves for different classes of filter.

Two major parameters are of concern in deciding on filter type: the amount of rejection and the loss in the bandpass. Also, the slope of the skirt becomes important when a band of frequencies must be passed for the maximum rejection outside the band. The slope of the rejection curves is generally determined by the number of resonant elements in the filter. This usually causes the insertion loss to go up.

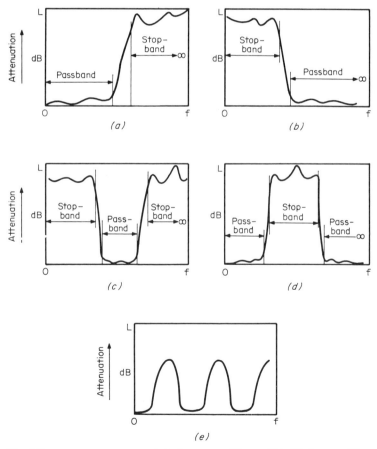

Fig. 63 Filter-response curves. (a) Low-pass filter. (b) High-pass filter. (c) Bandpass filter. (d) Bandstop filter. (e) Multiple-passband filter.

Thus the selection of filter parameters is generally a trade between needed performance and size.

The type of pass characteristic and the effect on signals through the filter are sometimes important considerations in the selection of a design. Figure 64 illustrates the different types of response commonly used. The Chebyshev and the Butterworth are the most common. The Butterworth is referred to as the maximally flat filter in that the signals are the least amplitude-distorted. The Chebyshev offers higher rejection but has a definite bandpass ripple. It must be remembered that, when specifying around the Chebyshev design, the amount of rippling in the passband should be controlled. The gaussian response is obviously the worst in terms of rejection and bandpass characteristics. However, it is less lossy for narrow-band application and is frequently used in these situations. The elliptic response filter offers

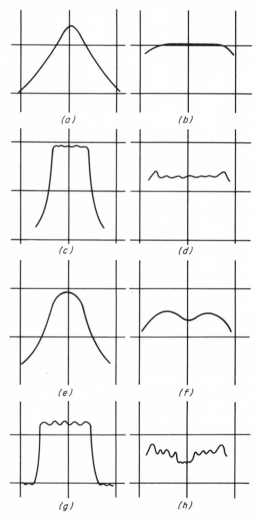

Fig. 64 Functional responses of filters. (*a*) Gaussian passband response. (*b*) Gaussian time-delay response. (*c*) Chebyshev passband response. (*d*) Chebyshev time-delay response. (*e*) Butterworth passband response. (*f*) Butterworth time-delay response. (*g*) Elliptic passband response. (*h*) Elliptic time-delay response.

the best slope behavior but poor ripple response.   A design known as the low-ripple Chebyshev is also available.

The waveguide filter has been treated in vast numbers of texts, most of which are readily available for examination.   However, a brief discussion is in order.

Earlier in this chapter a form of high-pass filter was discussed when the cutoff characteristics of waveguide were presented.   Below a certain frequency the waveguide would not propagate and thus acted like a high-pass filter.   Most waveguide filters consist of one or more sections of cavities.   These cavities represent equivalent resonant circuits, and their analysis can be treated like conventional filter design.   Figure 65 shows how such an equivalency is developed.   Each section of the filter is a

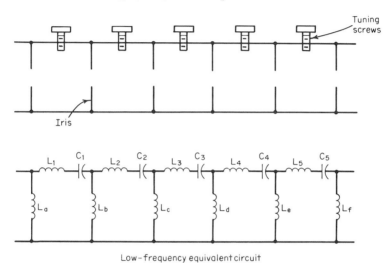

Low-frequency equivalent circuit

Fig. 65   Filter circuits.

cavity coupled by an iris to the next section.   The actual design of the filter, however, involves many empirical steps.   For instance, the coupling iris is usually accomplished by a cut-and-try method.   Also, the actual cavity dimensions are usually modified by the experienced designer since the cavity Q's are not exactly as calculated because losses in the walls are a function of manufacturing tolerances.   Figure 66 shows a typical design tool for the filter designer.   Accompanying this curve are some equations that may generally be used.   The cavities may be configured in coaxial, circular, or rectangular transmission line.   Figures 67 and 68 present some additional information for use of a low-pass design.

The development of the design of filters rests heavily with circuit treatment of impedances and admittances in S-plane geometry.   Analysis of this sort presents information as to pass-response behavior in simple form.   S-plane geometry deals with the pole-zero concept as representing impedances in a complex geometry plane.   The treatment of this analysis technique is available in many texts.   Figure 69 shows some pictorials and curves for common types of filters.   Filters known as constant-K filters are all pole filters.   Elliptic-function filters have zeros at finite frequencies. See Table 11.

Filters are available in other configurations than the waveguide cavity.   One very popular type is in stripline.   Stripline filters are constructed with half-wavelength resonators capacitively coupled.   They may be either end-coupled or parallel-coupled.   Figure 70 demonstrates how this is done.   The end-coupled filter is advantageous when a long, narrow geometry is desired.   Also the width of the strips is the same for each resonator.   The parallel coupler filter makes a shorter design and a symmetrical response curve.   There are other configurations such as that shown in Fig. 71.

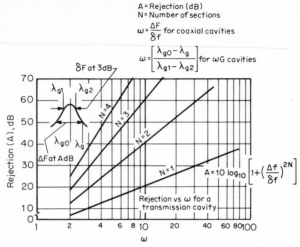

Fig. 66 Filter-response relationships.[21]

Maximally flat response:

$$\text{Rejection} = 10 \log_{10} (1 + \omega^{2N})$$

where N = number of filter sections

$$\omega = \frac{2(f - f_0)}{f_1 - f_2} \quad \text{for coaxial filters}$$

$$= \frac{2(\lambda_g - \lambda_{g0})}{\lambda_{g1} - \lambda_{g2}} \quad \text{for waveguide filters}$$

Chebyshev response:

$$\text{Rejection loss} = 10 \log_{10} \{1 + d[T_n(\omega_1)]^2\}$$

where

$$d = 10^{A_m/10} - 1$$

$$A_m = \text{peak-to-valley ratio, dB}$$
$$T_n(\omega_1) = \cosh [n \cosh (n \cosh^{-1} \omega_1)] \text{ for } \omega_1 - 1$$
$$\omega_1 = \frac{2(\lambda_g - \lambda_{g0})}{\lambda_{g1} - \lambda_{g2}}$$

$$\text{Insertion loss} = 20 \log_{10} \left[\frac{1}{1 - (QL/QU)} \right]$$

where $QL = \dfrac{f_0}{3\text{-dB bandwidth}}' = \text{loaded } Q$

Rejection loss vs. ω shown for maximally flat filters of one, two, three, and four sections.

Cavities   The microwave cavity has many uses, as discussed in the design of filters. One of its more useful advantages is in the use of frequency references such as one might use in a stabilized oscillator. The physical parameters are carefully selected, and the design is made such that the resonating frequency of the cavity can be used as reference and locked back to the oscillator.

Figure 72 illustrates how the fields are distributed in the basic resonance mode for the cavity. These field distributions give the information necessary for determining the manner in which the energy is coupled out of the cavity. Figure 73 shows how the cavity-resonance wavelengths may be calculated. The cavity description is made complete when the Q is specified. The Q of the cavity is the relationship

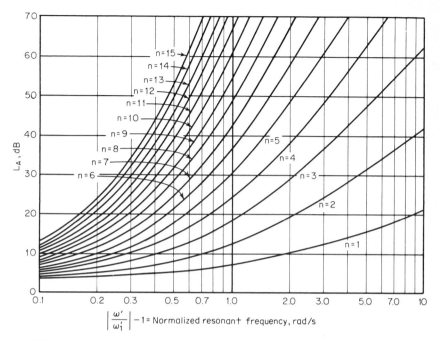

Fig. 67   Attenuation characteristics of maximally flat filters.   (The frequency ω_1' is the 3-dB band-edge point.)

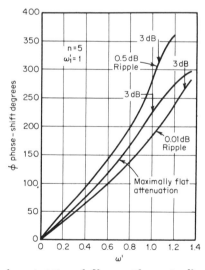

Fig. 68   Phase-shift characteristics of filters with maximally flat or Chebyshev attenuation responses and $n = 5$.[†]

of stored energy to lost energy.   However, since it is difficult to make the surfaces of the cavity walls perfect, more energy is usually lost than one wants and two values of Q are actually used.   The ideal or theoretical Q is called the unloaded Q and the actual Q is referred to as the loaded Q.   The latter value, as was indicated previously, is controlled by workmanship of the cavity—the surface finish and the

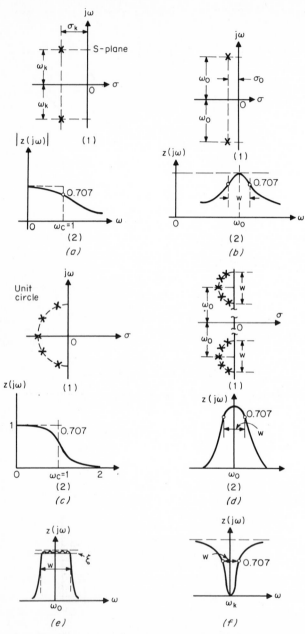

Fig. 69 S-plane geometry for typical filter.[17] (a) Pole-zero pattern (1) and impedance variation (2) for single pole-pair unit. (b) Pattern (1) impedance (2) for pole-pair unit with resonance curve. (c) Pattern for Butterworth low-pass characteristic. (d) Pattern for bandpass unit combination. (e) Equiripple passband behavior. (f) Notch unit.

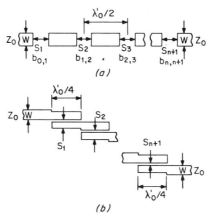

Fig. 70 Stripline filters.[19] (a) Half-wavelength end-coupled filter. (b) Half-wavelength side-coupled filter.

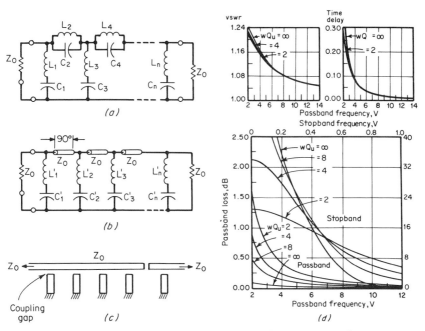

Fig. 71 Stripline filter and lumped-constant filter.[19] (a) Lumped-constant configuration of bandstop prototype filter contains both series and shunt branches. (b) Stripline design is simplified by absence of all series branches. (c) Pictorial layout of filter elements depicts simplicity in design of stripline bandstop filter. (d) Frequency characterization of bandstop filter for stripline design begins with three shunt branches.

TABLE 11a   Response Relationships for Filters[18]

Subclass (ladders only)	Amplitude-response relationships	Time-domain response
Cauer-Chebyshev (elliptic-function filter)	Has equal ripple, oscillating passband ripple with equiripple, peaked attenuation response in reject band. Very fast roll-off to reject band	Usually not a good candidate for critical time/phase response applications
Chebyshev	Has equal ripple, oscillating passband response with monotonic rejection responses, quite sharp in roll-off	Is very limited in its applications for critical time/phase applications
Butterworth (MFA)	Has flat (no ripple) passband (maximally flat amplitude) with monotonic attenuation response	A popular device if the signal components are restricted to the very flat amplitude and time-delay portion of the passband
TBT—transitional Butterworth-Thomson	Provides a compromise series of designs between the MFA and MFTD subclasses	Generally used only for interstaging active circuitry as a compromise of time/frequency-domain properties
Thomson (MFTD) (maximally flat time-delay response), also bessel filter	A monotonic (very slow) attenuation response	Has very linear phase response
Linear phase with equal ripple error (LPERE)	Somewhat sharper attenuation response than MFTD or gaussian filters	Has prespecified ripple in phase and time-delay response
Gaussian	Quite slow in monotonic attenuation roll-off	Has moderately good time and linear phase-response properties

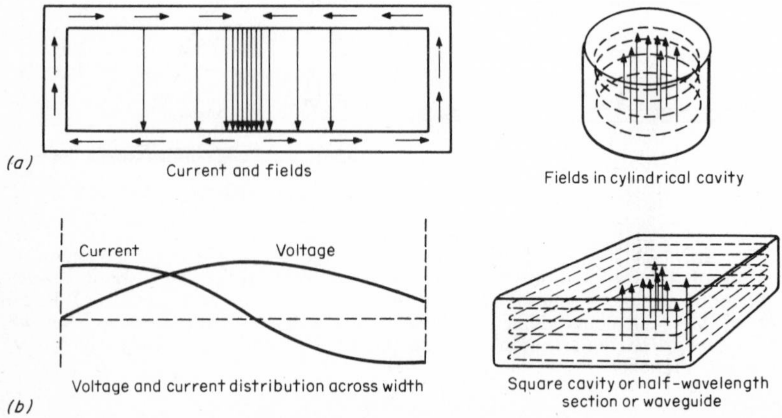

(a) Current and fields

Fields in cylindrical cavity

(b) Voltage and current distribution across width

Square cavity or half-wavelength section or waveguide

Fig. 72   Fields in cavities.

TABLE 11*b*   Some Basic Filter Network Forms[18]

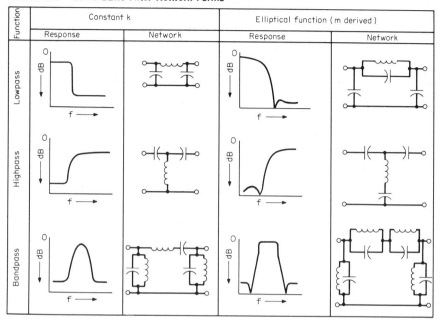

material used for plating the walls.   Figure 74 shows how the Q may be determined for some differently shaped cavities.

One consideration that must be given in selecting the cavity geometry is the mode that it is to operate in.   If the part to be used is to operate in a variety of frequency ranges which are not the one that is desired, care must be taken to use a mode that has no or few adjacent modes that can be excited.   Also, it might be necessary to design some mode suppressors into the cavity.   This is undesirable because the suppressors also reduce the Q of the cavity.   Figure 75 gives the mode chart for a circular cylinder cavity.   Figures 76 and 77 give some additional design information for cavities.

Waveguide Switches

The electromechanical switch used in the waveguide and coaxial transmission line almost always uses solenoids, which are linear-motion devices.   The switch must use some mechanical means of turning this linear motion into a circular motion.   The exception to this rule is in the relay-type devices frequently found in coaxial designs.

Figure 78 illustrates the concept used in waveguide switches.   This switch uses four ports as indicated and is referred to as a transfer switch.   The actuating mechanism is either a motor or a solenoid for this switch.   Because of difficult packaging the motor is generally less reliable and much slower-acting.   Some designs are available with a torque motor that employs an asymmetrical magnetic field in which the rotor is placed.

Several things must be considered by the user of a waveguide switch if acceptable performance and reliability are to be obtained.   First, the isolation between ports is important to the design because it affects the spacing between the rotor and stator.   The amount of isolation that is comfortable for the switch designer is usually 60 dB.   Many radar systems require 80 dB, especially if leakage levels are to be kept lower than that required for human safety and adjacent radar interference.

The waveguide-switch designer can achieve isolation either by using resonant chokes built into the rotor or by packing the rotor with an energy-absorbing material.   The latter approach is preferable for inexpensive construction and broadband

Shape	Mode	Resonant wavelength
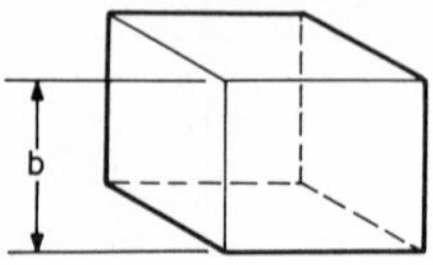 Cube	$TE_{m,n,p}$ or $TM_{m,n,p}$	$\lambda_k = \dfrac{2b}{\sqrt{p^2 + m^2 + n^2}}$
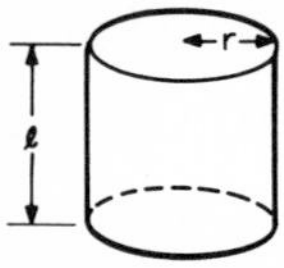 Cylinder	$TE_{0,1,1}$	$\lambda_k = \dfrac{2}{\sqrt{1/\ell^2 + 1.49/r^2}}$
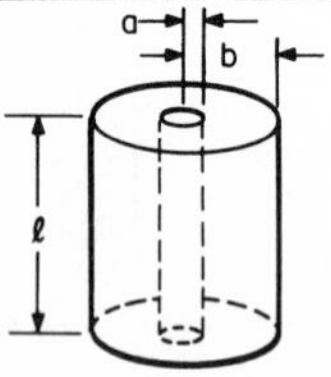 Coaxial cylinder	TEM_p	$\lambda_k = 2\dfrac{\ell}{p}$
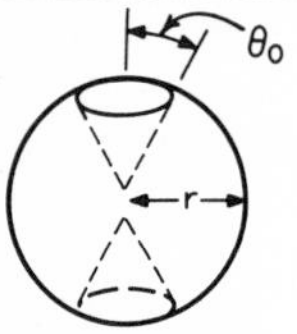 Sphere with reentrant cones	TEM_1	$\lambda_k = 4r$

Fig. 73 Cavity type and wavelength calculation.[1]

isolation. The resonant-choke design has advantages in that there is no concern over the instability of the isolation material and possible sticking problems caused by the expansion of the load material.

Waveguide switches can be provided with auxiliary contacts that can provide additional circuitry for such functions as indicating lights. Some radar systems interlock the transmitter with the auxiliary contacts, in which case it is important to tie down the amount of rotor rotation permissible before activation of the auxiliary switches. One technique is for the manufacturer to use an indirect drive from the solenoid such as a Geneva cam. Such mechanical linkage provides a lot of solenoid motion before the rotor moves. A further advantage of this system is that the solenoid air gap is reduced before the higher torque is required to move the rotor.

Further consideration in selecting a switch must be given to whether or not the solenoid and auxiliary circuits are to be sealed. Such a requirement demands the use of an active seal around the rotor shaft through the stator block.

In summary, it is important to make it very clear to the manufacturer what the intended use of the switch will be. For instance, use of the switch at high temperatures could be a problem because the coil resistance increases. Sufficient drive torque may not be available at the high temperatures, and it may be necessary to use another

Shape	Mode	Q	Maximum Q	Conditions for maximum Q
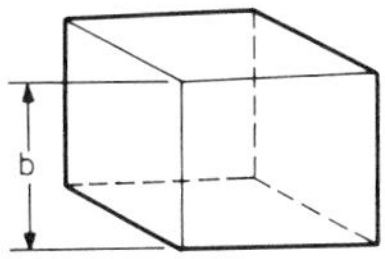 Cube	$TE_{m,n,p}$ $TM_{m,n,p}$	$Q_k = \frac{\lambda_k}{\Delta}\frac{\sqrt{p^2+m^2+n^2}}{8}$ provided p,m,n > 0		
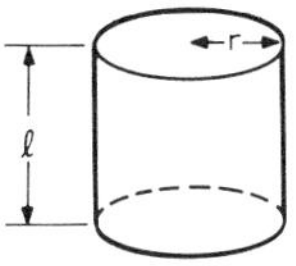 Cylinder	$TE_{0,1,1}$	$Q_k = \frac{\lambda_k}{\Delta}\frac{\sqrt{r^2/\ell^2+1.49}}{2}$ $\times \frac{r^2/\ell^2+1.46}{r^3/\ell^3+2.92}$	$Q_k = 0.67\frac{\lambda_k}{\Delta}$	$\frac{r}{\ell}=0.5$
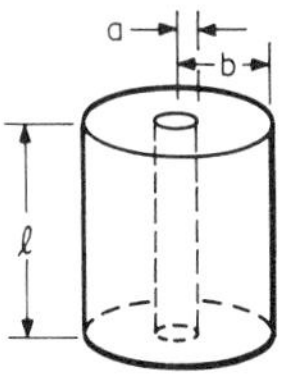 Coaxial Cylinder	TEM_p	$Q_k = \frac{\lambda_k}{\Delta}\frac{p}{4+\frac{\ell}{b}\frac{1+b/a}{\ln b/a}}$	$Q_k = \frac{\lambda_k}{\Delta}\frac{p}{4+3.6\frac{\ell}{b}}$	$\frac{b}{a}=3.6$
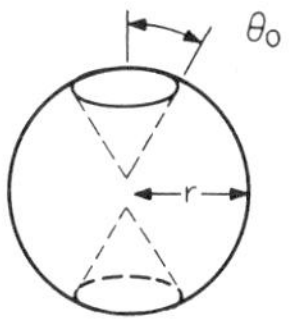 Sphere with reentrant cones	TEM_1	$Q_k = \frac{\lambda_k}{\Delta}\frac{1}{4+3.3\frac{\csc\theta_0}{\ln\cot\theta_0/2}}$	$Q_k =0.11\frac{\lambda_k}{\Delta}$	$\theta_0 = 33.5°$

Fig. 74 Cavity Q calculations.[1]

switch and a holding resistor, which may be internal or external. With such a scheme a high switching current may be used with a low value of holding current.

The switching functions may be described as fail-safe or latching. In the latter the switch stays in its switched position after activation. In the former it is spring-loaded so that after the voltage is removed it returns to its deenergized state. Figure 79 shows a schematic for a latching configuration. In some electronic systems one feature available in switch design is to have the solenoid assembly removable from the stator block. Figure 80 illustrates such a modular concept in a complex radar system.

Transmission Lines

The subject of transmission lines appropriately belongs in a handbook of its own. However, some mention should be made of the types of line available and their appropriate applications.

Rectangular waveguide This waveguide has been the subject of the microwave industry ever since the use of radar in the Second World War. There are many standards, papers, articles, and texts for any in-depth study. Rectangular guide, circular guide, and ridged guide, both single and double, are available. The rectangular guide is of course the most popular. As was discussed previously, the wave-

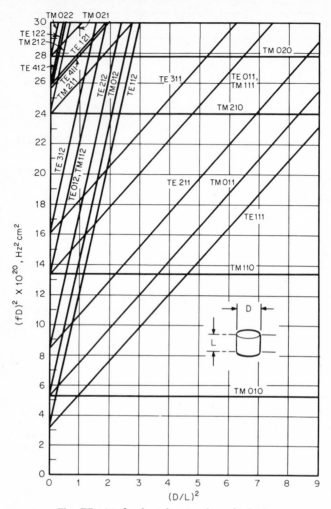

Fig. 75 Mode chart for circular cylinder.[7]

guide propagates in the TE$_{01}$ mode.   For any particular set of cross-section dimensions there is a cutoff frequency below which propagation cannot be sustained.   Propagation at higher than the rated frequency of the guide excites other modes, which create problems in matching and power transfer.   Tables 12 and 13 summarize the commercial and military versions of available waveguides and their ratings.

The use of waveguide involves an in-depth knowledge of fabrications and metallurgical techniques.   Common shapes are made by bending and casting, and fabrication.   When quantity is desired, the investment-casting process is preferred.   Usually in the rectangular configuration almost any type of shape, bend, and component is available as a standard part.   This is a very important economical factor to be considered by the systems user.   The expense of manufacturing waveguide parts can be controlled by selecting parts that have been produced by some cost-reduction scheme.   For a further discussion of manufacturing and designing waveguide assemblies, see Ref. 47.

Couplings Waveguides are joined by couplings which are of two types, the contact flange and the choke flange.   The flanges are attached to the end of the wave-

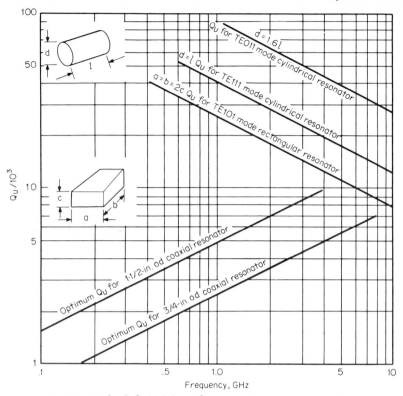

Fig. 76 Unloaded $Q(Q_u)$ vs. frequency for various resonators.[7]

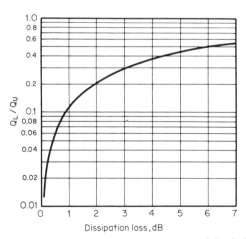

Fig. 77 Dissipation in a transmission cavity vs. ratio of loaded to unloaded Q.[7]

guide by either brazing or soldering or in some cases by bolting. Figure 81 illustrates these two types of couplings. Table 14 lists the flanges generally recognized as industry standards.

The contact flange is generally used in the larger-sized waveguide where it is desired to keep the size down. The disadvantage of the coupling is that gaps can

TABLE 12 Rigid Rectangular Waveguide (MIL-W-85)

Part number M85/1( )	Frequency range, GHz	Inside Width	Inside Height	Outside Width	Outside Height	Metallic alloy	Used with UG-( )/U flange	Type	Theoretical attenuation lowest to highest frequency, dB/100 ft	Theoretical CW power rating lowest to highest frequency, MW
001	0.32-0.49	23.000	11.500	23.376	11.876	AL	WR-2300	RG-290/U	0.040-0.026	528.3-753.8
003	0.35-0.53	21.000	10.500	21.376	10.876	AL	WR-2100	RG-291/U	0.046-0.031	439.3-625.4
005	0.41-0.62	18.000	9.000	18.250	9.250	AL	WR-1800	RG-201/U	0.057-0.038	325.1-461.4
007	0.49-0.75	15.000	7.500	15.250	7.750	AL	WR-1500	RG-202/U	0.076-0.051	224.1-320.4
009	0.64-0.96	11.500	5.750	11.750	6.000	AL	WR-1150	RG-203/U	0.113-0.076	132.0-186.9
011	0.75-1.12	9.750	4.875	10.000	5.125	AL	WR-975	RG-204/U	0.147-0.098	93.81-133.7
013	0.96-1.45	7.700	3.850	7.950	4.100	AL	WR-770	RG-205/U	0.205-0.139	59.67-84.18
017	1.12-1.70	6.500	3.250	6.660	3.410	CU	417B, 1362, 1714	RG-69/U	0.316-0.209	41.34-59.74
018	1.12-1.70	6.500	3.250	6.660	3.410	AL	418B, 1343, 1720	RG-103/U	0.273-0.180	41.34-59.74
020	1.12-1.70	6.500	3.250	6.660	3.410	MG		RG-206/U	0.484-0.320	41.34-59.74
023	1.45-2.20	5.100	2.550	5.260	2.710	CU	1715, 1718	RG-337/U	0.440-0.299	26.19-37.00
024	1.45-2.20	5.100	2.550	5.260	2.710	MG		RG-339/U	0.674-0.458	26.19-37.00
025	1.45-2.20	5.100	2.550	5.260	2.710	AL	1717, 1719	RG-338/U	0.380-0.258	26.19-37.00
029	1.70-2.60	4.300	2.150	4.460	2.310	AL	437B, 1345, 1711	RG-105/U	0.502-0.334	18.23-26.26
031	1.70-2.60	4.300	2.150	4.460	2.310	CU	435B, 1344, 1716	RG-104/U	0.583-0.387	18.23-26.26
032	1.70-2.60	4.300	2.150	4.460	2.310	MG		RG-207/U	0.892-0.592	18.23-26.26
035	2.20-3.30	3.400	1.700	3.560	1.860	AL	554A, 1347, 1713	RG-113/U	0.682-0.474	11.87-16.44
037	2.20-3.30	3.400	1.700	3.560	1.860	CU	553A, 1346, 1712	RG-112/U	0.791-0.550	11.87-16.44
038	2.20-3.30	3.400	1.700	3.560	1.860	MG		RG-208/U	1.211-0.842	11.87-16.44
041	2.60-3.95	2.840	1.340	3.000	1.500	AL	584, 585A, 1349, 1484, 1725	RG-75/U	0.950-0.651	7.645-10.85
043	2.60-3.95	2.840	1.340	3.000	1.500	CU	53, 54B, 1348, 1479, 1724	RG-48/U	1.102-0.754	7.645-10.85
044	2.60-3.95	2.840	1.340	3.000	1.500	MG	1196, 1197	RG-167/U	1.687-1.551	7.645-10.85

No.	Freq. Range					Mat'l	Cable Nos.	RG No.		
047	3.30–4.90	2.290	1.145	2.418	1.273	AL	1351, 1727	RG-341/U	1.211-0.858	5.475-7.549
049	3.30–4.90	2.290	1.145	2.418	1.273	CU	1350, 1726	RG-340/U	1.404-0.996	5.475-7.549
050	3.30–4.90	2.290	1.145	2.418	1.273	MG		RG-342/U	2.149-1.523	5.475-7.549
053	3.95–5.85	1.872	0.872	2.000	1.000	AL	406B, 407, 1353, 1480, 1729	RG-95/U	1.785-1.238	3.296-4.697
055	3.95–5.85	1.872	0.872	2.000	1.000	CU	148C, 149A, 1352, 1475, 1728	RG-49/U	2.071-1.436	3.296-4.697
056	3.95–5.85	1.872	0.872	2.000	1.000	MG	1198, 1199	RG-168/U	3.168-2.197	3.296-4.697
059	4.90–7.05	1.590	0.795	1.718	0.923	AL	1355, 1731	RG-344/U	1.988-1.485	2.792-3.719
061	4.90–7.05	1.590	0.795	1.718	0.923	CU	1354, 1730	RG-343/U	2.305-1.722	2.792-3.719
062	4.90–7.05	1.590	0.795	1.718	0.923	MG		RG-345/U	3.527-2.634	2.792-3.719
065	5.85–8.20	1.372	0.622	1.500	0.750	AL	440B, 441	RG-106/U	2.532-1.999	1.975-2.531
067	5.85–9.20	1.372	0.622	1.500	0.750	CU	1357, 1481, 1733; 343B, 344, 1356, 1476, 1732	RG-50/U	2.936-2.319	1.975-2.531
068	5.85–8.20	1.372	0.622	1.500	0.750	MG	1200, 1201	RG-169/U	4.492-3.548	1.975-2.531
071	7.05–10.0	1.122	0.497	1.250	0.625	AL	137B, 138, 1359, 1482, 1735	RG-68/U	3.548-2.756	1.284-.1702
073	7.05–10.0	1.122	0.497	1.250	0.625	CU	51, 52B, 1358, 1477, 1734	RG-51/U	4.114-3.197	1.284-1.702
074	7.05–10.0	1.122	0.497	1.250	0.625	MG	1202, 1203	RG-170/U	6.275-4.891	1.284-1.702
077	8.20–12.40	0.900	0.400	1.000	0.500	AL	135, 136B, 1361, 1483, 1737	RG-67/U	5.540-3.833	0.758-1.124
079	8.20–12.40	0.900	0.400	1.000	0.500	CU	39, 40B, 1360, 1476, 1736	RG-52/U	6.424-4.445	0.758-1.124
080	8.20–12.40	0.900	0.400	1.000	0.500	MG	1204, 1205	RG-171/U	9.830-6.801	0.758-1.124
083	10.00–15.00	0.750	0.375	.850	0.475	AL		RG-347/U	6.554-4.578	0.622-0.903
085	10.00–15.00	0.750	0.375	.850	0.475	CU		RG-346/U	7.601-5.309	0.622-0.903
086	10.00–15.00	0.750	0.350	.850	0.475	MG		RG-348/U	11.63-8.124	0.622-0.903
089	12.40–18.00	0.622	0.311	0.702	0.391	CU	419, 541A	RG-91/U	9.578-7.041	0.457-0.633
090	12.40–18.00	0.622	0.311	0.702	0.391	AL	1665, 1666	RG-349/U	8.259-6.071	0.457-0.633
092	12.40–18.00	0.622	0.311	0.702	0.391	MG	1206, 1207	RG-172/U	14.655-10.773	0.457-0.633
094	15.00–22.00	0.510	0.255	0.590	0.335	CU		RG-352/U	8.836-6.402	0.312-0.433
096	15.00–22.00	0.510	0.255	0.590	0.335	CU		RG-353/U	13.08-9.477	0.312-0.433

TABLE 12 Rigid Rectangular Waveguide (MIL-W-85) (Continued)

Part number M85/1( )	Frequency range, GHz	Inside Width	Inside Height	Outside Width	Outside Height	Metallic alloy	Used with UG-( )/U flange	Type	Theoretical attenuation lowest to highest frequency, dB/100 ft	Theoretical CW power rating lowest to highest frequency, MW
097	15.00-22.00	0.510	0.255	0.590	0.335	AL		RG-351/U	11.27-8.172	0.312-0.433
099	15.00-22.00	0.510	0.255	0.590	0.335	MG		RG-350/U	20.01-14.50	0.312-0.433
102	18.00-26.50	0.420	0.170	0.500	0.250	CU	595, 596A	RG-53/U	20.48-15.04	0.171-0.246
103	18.00-26.50	0.420	0.170	0.500	0.250	AL	597, 598A	RG-121/U	17.66-12.97	0.171-0.246
105	18.00-26.50	0.420	0.170	0.500	0.250	MG	1208, 1209	RG-173/U	31.34-23.02	0.171-0.246
109	22.00-33.00	0.340	0.170	0.420	0.250	CU		RG-354/U	25.03-17.41	0.139-0.209
110	22.00-33.00	0.340	0.170	0.420	0.250	AL		RG-355/U	21.58-15.01	0.139-0.209
112	22.00-33.00	0.340	0.170	0.420	0.250	MG		RG-356/U	38.29-26.63	0.139-0.209
113	22.00-33.00	0.340	0.170	0.420	0.250	CU-AG	1530	RG-357/U	16.18-11.25	0.139-0.209
114	26.50-40.00	0.280	0.140	0.360	0.220	AG	599, 600A	RG-96/U	24.55-16.80	0.096-0.146
117	26.50-40.00	0.280	0.140	0.360	0.220	CU-AG	599	RG-271/U	21.99-15.06	0.096-0.146
118	33.00-50.00	0.224	0.112	0.304	0.192	AG	383, 1521	RG-97/U	34.57-23.50	0.0644-0.097
121	33.00-50.00	0.224	0.112	0.304	0.192	CU-AG	383	RG-272/U	30.98-21.06	0.0644-0.097

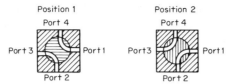

Fig. 78   Transfer-switch rf paths.   Positions as seen from the top of the switch.

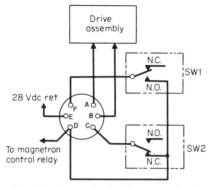

Fig. 79   Break-before-make latching-switch schematic.

Fig. 80   Modular-head waveguide switch

cause mismatch problems.   Also, high-voltage stresses can be developed in the gaps and cause arcing and sputtering.   The choke coupling uses a half-wave-series branching transmission line with one quarter wave being in the' form of a circular groove around the waveguide.   This is done to simplify manufacturing.   At a larger diameter around the choke groove is a groove intended to take an O-ring for additional rf sealing and pressure.   The purpose of the choke groove is to allow some misalignment of the waveguide in the E and H plane.

TABLE 13   Reference Table of Rigid Rectangular Waveguide Data and Fittings—Courtesy of Microwave Development Laboratories, Inc.

EIA WG WR ( )	MDL band	Recommended operating range for TE_{10} mode		Cutoff for TE_{10} mode		Range in $2\lambda/\lambda_c$	Range in λ_g/λ	Theoretical cw power rating lowest to highest frequency, mW
		Frequency, kHz	Wavelength, cm	Frequency, kHz	Wavelength, cm			
2,300		0.32–0.49	93.68–61.18	0.256	116.84	1.60–1.05	1.68–1.17	153.0–212.0
2,100		0.35–0.53	85.65–56.56	0.281	106.68	1.62–1.06	1.68–1.18	120.0–173.0
1,800		0.41–0.625	73.11–47.96	0.328	91.44	1.60–1.05	1.67–1.18	93.4–131.9
1,500		0.49–0.75	61.18–39.97	0.393	76.20	1.61–1.05	1.62–1.17	67.6–93.3
1,150		0.64–0.96	46.84–31.23	0.513	58.42	1.60–1.07	1.82–1.18	35.0–53.8
975		0.75–1.12	39.95–26.76	0.605	49.53	1.61–1.08	1.70–1.19	270.–38.5
770		0.96–1.45	31.23–20.67	0.766	39.12	1.60–1.06	1.66–1.18	17.2–24.1
650	L	1.12–1.70	26.76–17.63	0.908	33.02	1.62–1.07	1.70–1.18	11.9–17.2
510		1.45–2.20	20.67–13.62	1.157	25.91	1.60–1.05	1.67–1.18	7.5–10.7
430	W	1.70–2.60	17.63–11.53	1.372	21.84	1.61–1.06	1.70–1.18	5.2–7.5
340		2.20–3.30	13.63–9.08	1.736	17.27	1.58–1.05	1.78–1.22	3.1–4.5
284	S	2.60–3.95	11.53–7.59	2.078	14.43	1.60–1.05	1.67–1.17	2.2–3.2
229		3.30–4.90	9.08–6.12	2.577	11.63	1.56–1.05	1.62–1.17	1.6–2.2
187	C	3.95–5.85	7.59–5.12	3.152	9.510	1.60–1.08	1.67–1.19	1.4–2.0
159		4.90–7.05	6.12–4.25	3.711	8.078	1.51–1.05	1.52–1.19	0.79–1.0
137		5.85–8.20	5.12–3.66	4.301	6.970	1.47–1.05	1.48–1.17	0.56–0.71
112	X_L	7.05–10.0	4.25–2.99	5.259	5.700	1.49–1.05	1.51–1.17	0.35–0.46
102		7.05–10.0						
90	X	8.20–12.40	3.66–2.42	6.557	4.572	1.60–1.06	1.68–1.18	0.20–0.29
75		10.00–15.00	2.99–2.00	7.868	3.810	1.57–1.05	1.64–1.17	0.17–0.23
62	Ku	12.4–18.0	2.42–1.66	9.486	3.160	1.53–1.05	1.55–1.18	0.12–0.16
51		15.00–22.00	2.00–1.36	11.574	2.590	1.54–1.05	1.58–1.18	0.080–0.107
42	K	18.00–26.50	1.66–1.13	14.047	2.134	1.56–1.06	1.60–1.18	0.043–0.058
34		22.00–33.00	1.36–0.91	17.328	1.730	1.57–1.05	1.62–1.18	0.034–0.048
28	K_A	26.50–40.00	1.13–0.75	21.081	1.422	1.59–1.05	1.65–1.17	0.022–0.031
22	Q	33.00–50.00	0.91–0.60	26.342	1.138	1.60–1.05	1.67–1.17	0.014–0.020
19		40.00–60.00	0.75–0.50	31.357	0.956	1.57–1.05	1.63–1.16	0.011–0.015
15	V	50.00–75.00	0.60–0.40	39.863	0.752	1.60–1.06	1.67–1.17	0.0063–0.0090
12		60.00–90.00	0.50–0.33	48.350	0.620	1.61–1.06	1.68–1.18	0.0042–0.0060
10		75.00–110.0	0.40–0.27	59.010	0.508	1.57–1.06	1.61–1.18	0.0030–0.0041
8		90.00–140.00	0.333–0.214	73.840	0.406	1.64–1.05	1.75–1.17	0.0018–0.0026
7		110.00–170.00	0.272–0.176	90.840	0.330	1.64–1.06	1.77–1.18	0.0012–0.0017
5		140.00–220.00	0.214–0.136	115.750	0.259	1.65–1.05	1.78–1.17	0.00071–0.00107
4		170.00–260.00	0.176–0.115	137.520	0.218	1.61–1.05	1.69–1.17	0.00052–0.00075
3		220.00–325.00	0.136–0.092	173.280	0.173	1.57–1.06	1.62–1.18	0.00035–0.00047

SOURCE: Microwave Development Laboratories, Inc.
* Contact flange.

Theoretical attenuation lowest to highest frequency, dB/100 ft	JAN WG RG ( )	Material alloy	Jan flange Choke UG ( )/U	Cover UG ( )/U	Inside	Tolerance ±	Outside	Tolerance ±	Wall thickness (nom)
0.051-0.031		Aluminum			23.000-11.500	0.020	23.376-11.876	0.020	0.188
0.054-0.034		Aluminum			21.000-10.500	0.020	21.376-10.876	0.020	0.188
0.056-0.038	201	Aluminum			18.000-9.000	0.020	18.250-9.250	0.020	0.125
0.069-0.050	202	Aluminum			15.000-7.500	0.015	15.250-7.750	0.015	0.125
0.128-0.075	203	Aluminum			11.500-5.750	0.015	11.750-6.000	0.015	0.125
0.137-0.095	204	Aluminum			9.750-4.875	0.010	10.000-5.125	0.010	0.125
0.201-0.136	205	Aluminum			7.700-3.850	0.010	7.950-4.100	0.010	0.125
0.317-0.312	69	Brass		417A*	6.500-3.250	0.010	6.660-3.410	0.010	0.080
0.269-0.178	103	Aluminum		418A*					
					5.100-2.550	0.010	5.260-2.710	0.010	0.080
0.588-0.385	104	Brass		435A*	4.300-2.150	0.008	4.460-2.310	0.008	0.080
0.501-0.330	105	Aluminum		437A*					
0.877-0.572	112	Brass		553*	3.400-1.700	0.005	3.560-1.860	0.005	0.080
0.751-0.492	113	Aluminum		554*					
1.102-0.752	48	Brass	54B	53	2.840-1.340	0.005	3.000-1.500	0.005	0.080
0.940-0.641	75	Aluminum	585A	584					
					2.290-1.145	0.005	2.418-1.273	0.005	0.064
2.08-1.44	49	Brass	148C	149A	1.872-0.872	0.005	2.000-1.000	0.005	0.064
1.77-1.12	95	Aluminum	406B	407					
					1.590-0.795	0.004	1.718-0.923	0.004	0.064
2.87-2.30	50	Brass	343B	344	1.372-0.622	0.004	1.500-0.750	0.004	0.064
2.45-1.94	106	Aluminum	440B	441					
4.12-3.21	51	Brass	52B	51	1.122-0.497	0.004	1.250-0.625	0.004	0.064
3.50-2.74	68	Aluminum	137B	138					
	320	Brass	1494	1493	1.020-0.510	0.003	1.148-0.638	0.003	0.064
6.45-4.48	52	Brass	40B	39	0.900-0.400	0.003	1.000-0.500	0.003	0.050
5.49-3.83	67	Aluminum	136B	135					
					0.750-0.375	0.003	0.850-0.475	0.003	0.050
9.51-8.31	91	Brass	541A	419					
		Aluminum							
6.14-5.36	107	Silver			0.622-0.311	0.002	0.702-0.391	0.003	0.040
					0.510-0.255	0.0025	0.590-0.335	0.003	0.040
20.7-14.8	53	Brass	596A	595	0.420-0.170	0.0020	0.500-0.250	0.003	0.040
17.6-12.6	121	Aluminum	598A	597					
13.3-9.5	66	Silver							
		Brass		1530*	0.340-0.170	0.0020	0.420-0.250	0.003	0.040
		Brass	600A	599					
21.9-15.0	96	Aluminum			0.280-0.140	0.0015	0.360-0.220	0.002	0.040
		Silver							
31.0-20.9	97	Brass		383	0.224-0.112	0.0010	0.304-0.192	0.002	0.040
		Silver							
		Brass		1529*	0.188-0.094	0.0010	0.268-0.174	0.002	0.040
52.9-39.1	98	Brass		385	0.148-0.074	0.0010	0.228-0.154	0.002	0.040
		Silver							
93.3-52.2	99	Brass		387	0.122-0.061	0.0005	0.202-0.141	0.002	0.040
		Silver							
		Brass		1528*	0.100-0.050	0.0005	0.180-0.130	0.002	0.040
152-99	278	Silver laminate		1527*	0.0800-0.0400	0.0003	0.120-0.080	0.001	0.020
163-137	276	Silver laminate		1525*	0.0650-0.0325	0.00025	0.105-0.073	0.001	0.020
308-193	275	Silver laminate		1524*	0.0510-0.0255	0.00025	0.091-0.066	0.001	0.020
384-254	277	Silver laminate		1526*	0.0430-0.0215	0.00020	0.083-0.062	0.001	0.020
512-348		Silver			0.0340-0.0170	0.00020	0.156 diam	0.001	

TABLE 14   Preferred Waveguide Flanges

Type	Flange type	Material	For use with waveguide type	MIL-F-3922/	Type of coupling
UG-417A/U*	Contact (+gasket)	Copper alloy	RG-69/U	32	Contact
UG-418A/U*	Contact (+gasket)	Aluminum	RG-103/U	33	Contact
UG-435A/U*	Contact (+gasket)	Copper alloy	RG-104/U	34	Contact
UG-437A/U*	Contact (+gasket)	Aluminum	RG-105/U	35	Contact
UG-553/U*	Contact (+gasket)	Copper alloy	RG-112/U	14	Contact
UG-554/U*	Contact (+gasket)	Aluminum	RG-113/U	15	Contact
UG-53/U	Cover	Copper alloy	RG-48/U	3	Cover
UG-54B/U	Choke (+gasket)			26	Choke
UG-584/U	Cover	Aluminum	RG-75/U	16	Cover
UG-585A/U	Choke (+gasket)			17	Choke
UG-149A/U	Cover	Copper alloy	RG-49/U	29	Cover
UG-148C/U	Choke (+gasket)			37	Choke
UG-407/U	Cover	Aluminum	RG-95/U	12	Cover
UG-406B/U	Choke (+gasket)			31	Choke
UG-344/U	Cover	Copper alloy	RG-50/U	6	Cover
UG-343B/U	Choke (+gasket)			30	Choke
UG-441/U	Cover	Aluminum	RG-106/U	11	Cover
UG-440B/U	Choke (+gasket)			36	Choke
UG-51/U	Cover	Copper alloy	RG-51/U	2	Cover
UG-52B/U	Choke (+gasket)			25	Choke
UG-138/U	Cover	Aluminum	RG-68/U	5	Cover
UG-137B/U	Choke (+gasket)			28	Choke
UG-39/U	Cover	Copper alloy	RG-52/U	2	Cover
UG-40B/U	Choke (+gasket)			24	Choke
UG-135/U	Cover	Aluminum	RG-67/U	4	Cover
UG-136B/U	Choke (+gasket)			27	Choke
UG-419/U	Cover	Copper alloy†	RG-91,107/U	10	Cover
UG-541A/U	Choke (+gasket)			13	Choke
UG-595/U	Cover	Copper alloy†	RG-53,66/U	18	Cover
UG-596A/U	Choke (+gasket)			19	Choke
UG-597/U	Cover	Aluminum	RG-121/U	20	Cover
UG-598A/U	Choke (+gasket)			21	Choke
UG-599/U	Cover	Copper alloy†	RG/96/U	22	Cover
UG-600A/U	Choke (+gasket)			23	Choke
UG-383/U*	Contact (+gasket)		RG-97/U	7	Contact
UG-385/U*	Contact (+gasket)	Copper alloy†	RG-98/U	8	Contact
UG-387/U*	Contact (+gasket)		RG-99/U	9	Contact

* Two identical flanges must be used for a contact junction.   In a pressurized junction, one surplus gasket is used as a spare.

† Flanges are silver-plated after assembly when used with silver waveguides.

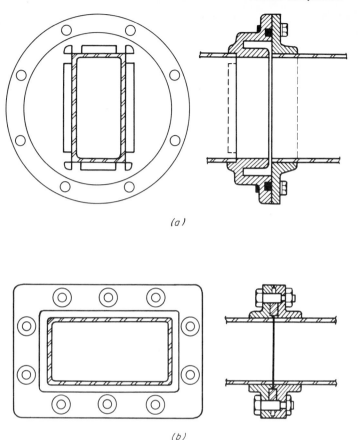

(a)

(b)

Fig. 81 Waveguide couplings. (*a*) Typical choke coupling. (*b*) Typical contact coupling.

Flexible waveguide Also available is a class of guide usually referred to as flexguide. Figure 82 shows some of the different types of flexible guide available. These configurations are of two classes, the resonant type and the nonresonant type. The latter is more common. Table 15 lists some of the important mechanical properties of the different types of flexible waveguides.

Circular waveguide Circular waveguides have not been in as great demand as their rectangular counterparts, mostly because it is difficult to maintain a fixed polarization for a long run of the guide. This is due to mechanical imperfections of the guide over which only so much control can be practically exercised. The circular waveguide has one distinct advantage over the rectangular guide in that it has six times as much power-handling capacity. It is thus used in such devices as rotary joints where size is an important factor. However, the transition from circular to rectangular might be the limiting factor.

Coaxial transmission line Coaxial transmission line consists of an outer conductor and an inner conductor, both of high strength and high conductivity. The dielectric may be gas or solid, depending on what is the desired feature of the line. For instance, if phase stability is important, the gas dielectric would be given serious consideration. Detailed treatment of this type of line may be found in Ref. 11, Chap. 4.

TABLE 15   Mechanical Properties of Flexible Waveguides

Type of flexible waveguide	Bend		Twist		Longitudinal	
	Relatively sharp	Moderately sharp	Appreciable	Negligible	Relatively large	Relatively small
			STATIC DEFORMATION			
Nonresonant types:						
Interlocked		X	X			X
Unsoldered convolute	X		X			X
Soldered convolute	X			X		X
Null-point seam	X			X		X
Seamless corrugated	X			X	X	
Resonant types:						
Vertebra		X	X		X	
Bellows		X			X	
			REPEATED DEFORMATION			
Nonresonant types:						
Interlocked		X	X			X
Unsoldered convolute	X		X			X
Soldered convolute		X		X		X
Null-point seam		X		X		X
Seamless corrugated		X		X		X
Resonant types:						
Vertebra	X		X		X	
Bellows	X			X	X	

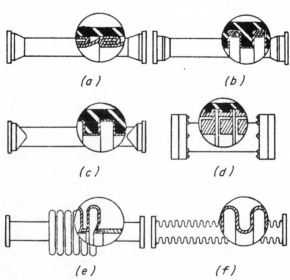

(a)   (b)

(c)   (d)

(e)   (f)

Fig. 82 Flexible waveguide construction. (*a*) Interlocked. (*b*) Soldered or unsoldered convolute. (*c*) Null-point seam. (*d*) Vertebra. (*e*) Spun bellows. (*f*) Seamless corrugated.

Ridged waveguide The last waveguide system is single-ridged and double-ridged guide. The ridge of the waveguide is placed symmetrically along the broad wall of the rectangular waveguide. The double-ridged guide is preferred for longer runs because it simplifies manufacturing tolerances. Single-ridged guide is more adaptable to the transition to coaxial lines.

The addition of the ridge to the waveguide has the effect of reducing the cutoff frequency without disturbing higher-order modes. This results in increased bandwidth of the guide.

Ideal gap spacings are associated with S/A (ridge width to waveguide width) ratios. The compromises are usually made between low attenuation and power-handling capabilities.

The use of ridged waveguide has some serious limitations when bandwidth is considered. The first of these is that the attenuation increases about 11.5 times. Second, the power rating is reduced to about 2 percent of the normal CW rating of the rectangular waveguide.

Figures 83 through 88 and Tables 16 to 18 give some curves that may be used for designing with ridged waveguide.

One important consideration in the use of ridged waveguide is that little or nothing is available in the component area except single transitions to coaxial transmission line. This can be an important limitation.

Dielectric gases The use of waveguide involves the use of gas as a dielectric, although some solid dielectric is used. Gas, like all dielectrics, has a breakdown phenomenon when used in high-power situations. The best single source for the prediction of waveguide behavior is Ref. 23. However, some simple tools are available to the microwave engineer for making rapid predictions in the more commonly used situations. One tool involves the use of three charts which allow the designer to pick factors that can be used to adjust the rated value of the waveguide. Table 19 lists the available breakdown power levels commonly found in handbooks. Table 20 is a chart that gives the scaling factors for different waveguide situations. Table 21 lists ways in which the breakdown conditions may be improved by the use of sulfur hexafluoride (SF_6) as a dielectric gas.

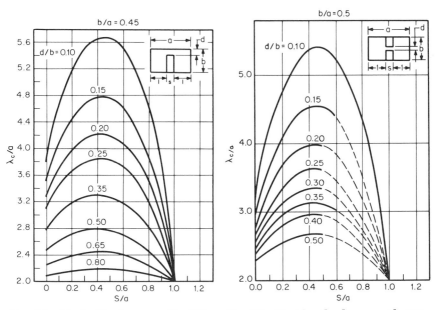

Fig. 83 Single-ridged waveguide TE_{10} mode cutoff wavelength.[7]

Fig. 84 Double-ridged waveguide TE_{10} mode cutoff wavelength.[7]

TABLE 16 Single-Ridged Waveguide, Bandwidth Ratio 3.6:1

Type*	Cross-sectional dimensions, in.				TE₁₀ mode frequency range, GHz	Metallic alloy*†	Used with UG-( )/U flange	$f = \sqrt{3}\,f c_{10}$	
	Inside		Outside					Attenuation, dB/ft‡	Power-handling capacity, kW§
	Width	Height	Width	Height					
WRS108U36A	31.218	14.048	31.468	14.298	0.108–0.390	Aluminum		0.0016	14,550
WRS108U36B						Brass			
WRS108U36C						Copper			
WRS108U36M						Magnesium			
WRS108U36S						Silver			
WRS270U36A	12.542	5.644	12.792	5.894	0.270–0.970	Aluminum		0.0065	2,348
WRS270U36B						Brass			
WRS270U36C						Copper			
WRS270U36M						Magnesium			
WRS270U36S						Silver			
WRS390U36A	8.677	3.905	8.927	4.155	0.390–1.400	Aluminum		0.0112	1,124
WRS390U36B						Brass			
WRS390U36C						Copper			
WRS390U36M						Magnesium			
WRS390U36S						Silver			
WRS970U36A	3.494	1.572	3.654	1.732	0.970–3.500	Aluminum	1604	0.0438	182.2
WRS970U36B						Brass	1605		
WRS970U36C						Copper	1606		
WRS970U36M						Magnesium			
WRS970U36S						Silver			
WRS140D36A	2.422	1.090	2.582	1.250	1.40–5.00	Aluminum	1607	0.0758	87.56
WRS140D36B						Brass	1608		
WRS140D36C						Copper	1609		
WRS140D36M						Magnesium			
WRS140D36S						Silver			

Type						Material	Type No.		
WRS350D36A	0.968	0.436	1.068	0.536	3.50–12.40	Aluminum	1610	0.300	13.99
WRS350D36B						Brass	1611		
WRS350D36C						Copper			
WRS350D36M						Magnesium	1612		
WRS350D36S						Silver			
WRS500D36A	0.678	0.305	0.778	0.405	5.00–18.00	Aluminum	1613	0.513	6.857
WRS500D36B						Brass	1614		
WRS500D36C						Copper			
WRS500D36M						Magnesium	1615		
WRS500D36S						Silver			
WRS124C36A	0.273	0.123	0.353	0.203	12.40–40.00	Aluminum	1616	2.008	1.115
WRS124C36B						Brass	1617		
WRS124C36C						Copper			
WRS124C36M						Magnesium	1618		
WRS124C36S						Silver			

NOTE: The electrical values are for information only.

* The last letter of the type number indicates the material.

† Caution: Magnesium is an incendiary agent and is highly susceptible to corrosion. Precautions must be taken to minimize the possibility of igniting the material and to prevent corrosion when assembling magnesium waveguides to the applicable flanges.

‡ The attenuation values given are for copper. Attenuation for other materials may be determined by the following formula:

$$\text{Attenuation for material } M = \sqrt{\frac{\text{resistivity of material } M}{\text{resistivity of copper}}} \times \text{attenuation for copper}$$

§ Based on breakdown of air — 15,000 V cm^{-1} (safety factor of approximately 2 at sea level), corner radii considered.

TABLE 17 Double-Ridged Waveguide, Bandwidth Ratio 3.6:1

| Type designation* | Cross-sectional dimensions, in. | | | | TE$_{10}$ mode frequency range, GHz | Metallic alloy*'† | Used with UG-( )/U flange | $f = \sqrt{3}\,f c_{10}$ | |
| | Inside | | Outside | | | | | Attenuation, dB/ft‡ | Power-handling capacity, kW§ |
	Width	Height	Width	Height					
WRD108U36A WRD108U36B WRD108U36C WRD108U36M WRD108U36S	34.638	14.894	34.888	15.144	0.108–0.390	Aluminum Brass Copper Magnesium Silver		0.0014	28,830
WRD270U36A WRD270U36B WRD270U36C WRD270U36M WRD270U36S	13.916	5.984	14.166	6.234	0.270–0.970	Aluminum Brass Copper Magnesium Silver		0.0055	4,653
WRD390U36A WRD390U36B WRD390U36C WRD390U36M WRD390U36S	9.628	4.140	9.878	4.390	0.390–1.400	Aluminum Brass Copper Magnesium Silver		0.0097	2,227
WRD970U36A WRD970U36B WRD970U36C WRD970U36M WRD970U36S	3.877	1.667	4.037	1.827	0.970–3.500	Aluminum Brass Copper Magnesium Silver	1589 1590 1591	0.0378	361.2
WRD140D36A WRD140D36B WRD140D36C WRD140D36M WRD140D36S	2.687	1.155	2.847	1.315	1.40–5.00	Aluminum Brass Copper Magnesium Silver	1592 1593 1594	0.0656	173.5

WRD350D36A	1.074	0.462	1.174	0.562	3.50-12.40	Aluminum	1595		27.74
WRD350D36B						Brass	1596	0.259	
WRD350D36C						Copper			
WRD350D36M						Magnesium	1597		
WRD350D36S						Silver			
WRD500D36A	0.752	0.323	0.852	0.423	5.00-18.00	Aluminum	1598		13.59
WRD500D36B						Brass	1599	0.443	
WRD500D36C						Copper			
WRD500D36M						Magnesium	1600		
WRD500D36S						Silver			
WRD124C36A	0.303	0.130	0.383	0.210	12.40-40.00	Aluminum	1601		2.210
WRD124C36B						Brass	1602	1.730	
WRD124C36C						Copper			
WRD124C36M						Magnesium	1603		
WRD124C36S						Silver			

NOTE: The electrical values are for information only.

* The last letter of the type number indicates the material.

† Caution: Magnesium is an incendiary agent and is highly susceptible to corrosion. Precautions must be taken to minimize the possibility of igniting the material and to prevent corrosion when assembling magnesium waveguides to the applicable flanges.

‡ The attenuation values given are for copper. Attenuation for other materials may be determined by the following formula:

$$\text{Attenuation for material } M = \sqrt{\frac{\text{resistivity of material } M}{\text{resistivity of copper}}} \times \text{attenuation for copper}$$

§ Based on breakdown of air, 15,000 V cm^{-1} (safety factor of approximately 2 at sea level), corner radii considered.

TABLE 17a Single-Ridged Waveguide, Bandwidth Ratio 2.4:1

Type designation*	Inside Width	Inside Height	Outside Width	Outside Height	TE$_{10}$ mode frequency range, GHz	Metallic alloy†	Used with UG-( )/U flange	Attenuation, dB/ft‡	Power-handling kW§
WRS175U24A	28.129	12.658	28.379	12.908	0.175–0.420	Aluminum		0.00024	32,870
WRS175U24B						Brass			
WRS175U24C						Copper			
WRS175U24M						Magnesium			
WRS175U24S						Silver			
WRS267U24A	18.421	8.289	18.671	8.539	0.267–0.640	Aluminum		0.00045	14,100
WRS276U24B						Brass			
WRS267U24C						Copper			
WRS267U24M						Magnesium			
WRS267U24S						Silver			
WRS420U24A	11.695	5.263	11.945	5.513	0.420–1.000	Aluminum		0.00087	5,682
WRS420U24B						Brass			
WRS420U24C						Copper			
WRS420U24M						Magnesium			
WRS420U24S						Silver			
WRS640U24A	7.682	3.457	7.932	3.707	0.640–1.530	Aluminum		0.00164	2,451
WRS640U24B						Brass			
WRS640U24C						Copper			
WRS640U24M						Magnesium			
WRS640U24S						Silver			
WRS840U24A	5.847	2.631	6.007	2.791	0.840–2.000	Aluminum	1541	0.00248	1,421
WRS840U24B						Brass	1542		
WRS840U24C						Copper	1543		
WRS840U24M						Magnesium			
WRS840U24S						Silver			
WRS150D24A	3.276	1.474	3.436	1.634	1.500–3.600	Aluminum	1544	0.00591	445.8
WRS150D24B						Brass	1545		
WRS150D24C						Copper	1546		
WRS150D24M						Magnesium			

$f = \sqrt{3}\, f_{c10}$

Type*					Frequency range (GHz)	Material	Ref. no.	Attenuation‡	§
WRS150D24S						Silver	1547		
WRS200D24A	2.456	1.105	2.616	1.265	2.000–4.800	Aluminum	1548	0.00908	250.6
WRS200D24B						Brass	1549		
WRS200D24C						Copper			
WRS200D24M						Magnesium			
WRS200D24S						Silver	1550		
WRS350D24A	1.404	0.632	1.532	0.760	3.500–8.200	Aluminum	1551	0.0212	81.87
WRS350D24B						Brass	1552		
WRS350D24C						Copper			
WRS350D24M						Magnesium			
WRS350D24S						Silver	1553		
WRS475D24A	1.034	0.465	1.134	0.565	4.750–11.000	Aluminum	1554	0.0333	44.43
WRS475D24B						Brass	1555		
WRS475D24C						Copper			
WRS475D24M						Magnesium			
WRS475D24S						Silver	1556		
WRS750D24A	0.655	0.295	0.755	0.395	7.500–18.000	Aluminum	1557	0.0661	17.82
WRS750D24B						Brass	1558		
WRS750D24C						Copper			
WRS750D24M						Magnesium			
WRS750D24S						Silver	1559		
WRS110C24A	0.446	0.2010	0.527	0.281	11.000–26.500	Aluminum	1560	0.117	8.285
WRS110C24B						Brass	1561		
WRS110C24C						Copper			
WRS110C24M						Magnesium			
WRS110C24S						Silver	1562		
WRS180C24A	0.2729	0.1228	0.353	0.203	18.000–40.000	Aluminum	1563	0.246	3.095
WRS180C24B						Brass	1564		
WRS180C24C						Copper			
WRS180C24M						Magnesium			
WRS180C24S						Silver			

NOTE: The electrical values are for information only.

* The last letter of the type number indicates the material.

† Caution: Magnesium is an incendiary agent and is highly susceptible to corrosion. Precautions must be taken to minimize the possibility of igniting the material and to prevent corrosion when assembling magnesium waveguides to the applicable flanges.

‡ The attenuation values given are for copper. Attenuation for other materials may be determined by the following formula:

$$\text{Attenuation for material } M = \sqrt{\frac{\text{resistivity of material } M}{\text{resistivity of copper}}} \times \text{attenuation for copper}$$

§ Based on breakdown of air, 15,000 V cm^{-1} (safety factor of approximately 2 at sea level), corner radii considered.

TABLE 18 Double-Ridged Waveguide, Bandwidth Ratio 2.4:1

Type designation*	Cross-sectional dimensions, in				TE$_{10}$ mode frequency range, GHz	Metallic alloy*,†	Used with UG-( )/U flange	$f = \sqrt{3}\,f_{c_{10}}$	
	Inside		Outside					Attenuation, dB/ft‡	Power-handling capacity, kW§
	Width	Height	Width	Height					
WRD175U24A	29.667	13.795	29.917	14.045	0.175–0.420	Aluminum		0.00023	61,960
WRD175U24B						Brass			
WRD175U24C						Copper			
WRD175U24M						Magnesium			
WRD175U24S						Silver			
WRD267U24A	19.428	9.034	19.678	9.284	0.267–0.640	Aluminum		0.00043	26,570
WRD267U24B						Brass			
WRD267U24C						Copper			
WRD267U24M						Magnesium			
WRD267U24S						Silver			
WRD420U24A	12.333	5.737	12.583	5.987	0.420–1.000	Aluminum		0.00085	10.710
WRD420U24B						Brass			
WRD420U24C						Copper			
WRD420U24M						Magnesium			
WRD420U24S						Silver			
WRD640U24A	8.100	3.767	8.350	4.017	0.640–1.530	Aluminum		0.0016	4,720
WRD640U24B						Brass			
WRD640U24C						Copper			
WRD640U24M						Magnesium			
WRD640U24S						Silver			
WRD840U24A	6.167	2.868	6.417	3.118	0.840–2.000	Aluminum	1565	0.0024	2,676
WRD840U24B						Brass	1566		
WRD840U24C						Copper	1567		
WRD840U24M						Magnesium			
WRD840U24S						Silver			
WRD150D24A	3.455	1.607	3.615	1.767	1.500–3.600	Aluminum	1568	0.0058	840.5
WRD150D24B						Brass	1569		
WRD150D24C						Copper	1570		
WRD150D24M						Magnesium			
WRD150D24S						Silver			

Waveguide					Freq. range (GHz)	Material	Type No.		
WRD200D24A	2.590	1.205	2.750	1.365	2.000–4.800	Aluminum	1571	0.0089	472.5
WRD200D24B						Brass	1572		
WRD200D24C						Copper	1573		
WRD200D24M						Magnesium			
WRD200D24S						Silver			
WRD350D24A	1.480	0.688	1.608	0.816	3.500–8.200	Aluminum	1574	0.0204	151.3
WRD350D24B						Brass	1575		
WRD350D24C						Copper			
WRD350D24M						Magnesium	1576		
WRD350D24S						Silver			
WRD475D24A	1.090	0.506	1.190	0.606	4.750–11.000	Aluminum	1577	0.0324	83.72
WRD475D24B						Brass	1578		
WRD475D24C						Copper			
WRD475D24M						Magnesium	1579		
WRD475D24S						Silver			
WRD750D24A	0.691	0.321	0.791	0.421	7.500–18.000	Aluminum	1580	0.0641	33.58
WRD750D24B						Brass	1581		
WRD750D24C						Copper	1582		
WRD750D24M						Magnesium			
WRD750D24S						Silver			
WRD110C24A	0.471	0.219	0.551	0.299	11.000–26.500	Aluminum	1583	0.114	15.63
WRD110C24B						Brass	1584		
WRD110C24C						Copper			
WRD110C24M						Magnesium	1585		
WRD110C24S						Silver			
WRD108C24A	0.288	0.134	0.368	0.214	18.000–40.000	Aluminum	1586	0.238	5.834
WRD108C24B						Brass	1587		
WRD108C24C						Copper			
WRD108C24M						Magnesium	1588		
WRD108C24S						Silver			

NOTE: The electrical values are for information only.

* The last letter of the type number indicates the material.

† Caution: Magnesium is an incendiary agent and is highly susceptible to corrosion. Precautions must be taken to minimize the possibility of igniting the material and to prevent corrosion when assembling magnesium waveguides to the applicable flanges.

‡ The attenuation values given are for copper.

$$\text{Attenuation for material } M = \sqrt{\frac{\text{resistivity of material } M}{\text{resistivity of copper}}} \times \text{attenuation for copper}$$

§ Based on breakdown of air — 15,000 V cm⁻¹ (safety factor of approximately 2 at sea level), corner radii considered.

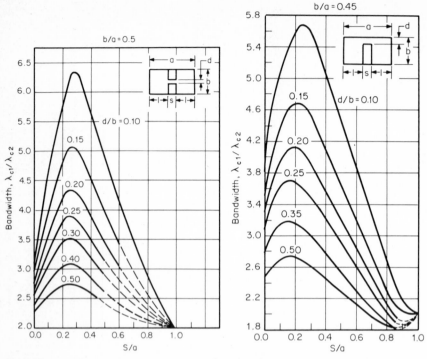

Fig. 85 Double-ridged waveguide band-width curves.[7]

Fig. 86 Single-ridged waveguide band-width curves.[7]

Sulfur hexafluoride is the most used of the insulating gases; more data are available, and its use is more familiar.   However, it is a member of a family of gases known as the electronegative gases.   In addition to SF_6, there are also the halogenated hydrocarbons.   Table 22 lists the dielectrics and some of their physical properties. Figure 89 demonstrates some of the relative dielectric strengths as a function of pressure.   The use of electronegative gases provides the definite advantage of improved high-power performance.   However, caution is needed in their use.   During corona discharge, should such a state be reached, corrosive by-products are given off.   If this situation is anticipated, some sort of alkaline getter should be used to minimize the effects of the highly electronegative halogens.

Microstrip and stripline   The last type of transmission line to be discussed is the stripline or microstrip line.   Though different in theory, it is similar for the system user.

Conventional stripline consists of two dielectric sheets with copper deposited on one side to form the circuits and conductors on the other to form a ground plane. Figure 90 demonstrates how the center conductor is sandwiched between the outer conducting planes.   The two outer conducting planes are usually clamped between two pressure plates to provide for rigidity and stability of the assembly.   Figure 91 shows fields for different modes in stripline.

Microstrip is like stripline except that it consists of one plane and conductor. Leakage with microstrip (Fig. 92) is probably the most difficult problem in instrumenting the design concept to hardware.

The circuits are usually prepared by using available design information on a positive drawing several times the size desired.   This positive is then reduced to a negative and becomes the master.   The laminates are prepared with a photosensitive material, and the circuits are etched by a conventional photographic process.

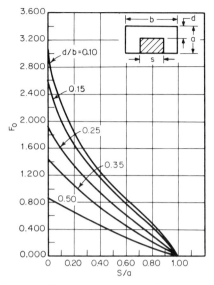

Fig. 87 Single-ridged waveguide correction curves for nonstandard b/a ratio.[7] The extension factor λ_c/a of any single-ridged waveguide of ratio b/a other than the standard ratio 0.45 is given by

$$\frac{\lambda_c}{a} = \left(\frac{\lambda_c}{a}\right)_0 + \left(\frac{b}{a} - 0.45\right) F_0$$

where λ_c/a_0 is the extension factor of the standard single-ridged guide and F_0 is obtained from the graph.

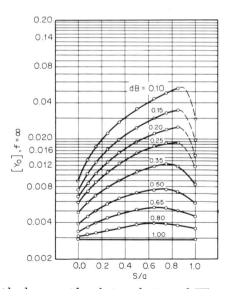

Fig. 88 Single-ridged waveguide relative admittance.[7] TE_{10} mode $b/a = 0.45$.

$$Y_{0\infty} = \frac{2P}{V_0{}^2} \qquad f = \infty$$

where P = average power carried, W
V_0 = instant peak voltage across center of guide

TABLE 19   Chart 1—Standard Rectangular Waveguide, TE_{01} Mode[23]
Dimensions, Recommended Frequencies, and Breakdown Power

JAN designation RG( )/u	RETMA designation	Frequency range, kHz		Inside dimensions, in		CW breakdown* power, 760 mm Hg at 30,000 V/cm^{-1}, MW	
		f_1	f_2	Width	Height	f_1	f_2
201	WR1800	0.41	0.625	18.00	9.00	313	435
203	WR1150	0.64	0.96	11.50	5.75	128	180
205	WR707	0.96	1.45	7.70	3.85	57.5	81.5
69, 103	WR650	1.12	1.70	6.50	3.25	40.2	58.0
	WR510	1.45	2.20	5.10	2.55	25.3	35.8
104, 105	WR430	1.70	2.60	4.30	2.15	17.5	25.2
112, 113	WR340	2.20	3.30	3.40	1.70	11.5	16.0
48, 75	WR284	2.60	3.95	2.84	1.34	7.30	10.4
	WR229	3.30	4.90	2.29	1.14	5.30	7.30
49, 95	WR187	3.95	5.85	1.87	0.872	3.20	4.50
	WR159	4.90	7.05	1.59	0.795	2.70	3.50
50, 106	WR137	5.85	8.20	1.37	0.622	1.90	2.50
51, 68	WR112	7.05	10.0	1.12	0.497	1.24	1.75
52, 67	WR90	8.20	12.4	0.900	0.400	0.730	1.10
	WR75	10.0	15.0	0.750	0.375	0.600	0.860
91, 107	WR62	12.4	18.0	0.622	0.311	0.440	0.600
	WR51	15.0	22.0	0.510	0.255	0.300	0.410
53, 121	WR42	18.0	26.5	0.420	0.170	0.160	0.240
	WR34	22.0	33.0	0.340	0.170	0.130	0.185
96	WR28	26.5	40.0	0.280	0.140	0.095	0.145
97	WR22	33.0	50.0	0.224	0.112	0.062	0.090
	WR19	40.0	60.0	0.188	0.094	0.047	0.064
98	WR15	50.0	75.0	0.148	0.074	0.029	0.042
99	WR12	60.0	90.0	0.122	0.061	0.020	0.029

* These values are for atmospheric air operating at the actual breakdown field strength of air, approximately 30,000 V cm^{-1}.

Until recently most integrated or stripline microwave circuits were of the hybrid type.   That is, the boards were prepared as described and the necessary components were added by soldering on epoxies or bonding.   Recent lumped-element designs tend toward the ideal of a monolithic circuit.   Development of thin-film techniques in microwave frequencies has made it possible for designers to use lumped-element design.

Such design consists of several layers of substrates stacked with the asscoiated artwork on each plate.   Active devices are usually compression-bonded to the substrates.   Figure 93 illustrates a series of such layers which combine to form an amplifier.   The biggest advantage of this technique compared with the distributed-element technique (stripline and microstrip) is that the package size can be reduced considerably.

The use of thick- or thin-filming techniques for integrated circuits depends upon the design.   Thick film is the use of the process discussed previously with the photographic etching.   The advantage of this method is usually cost and turnaround time. The thin-films system consists of printing the circuits by vacuum deposition of composite metals of gold and chrome.   Thin-film circuits are more uniform and better for high frequencies.   The use of photographics for making the artwork gives both systems high resolution.   Figure 94 demonstrates some of the methods used for these processes.

TABLE 20   Chart 2—System Parameters Affecting Waveguide Power-Handling Capacity[12]

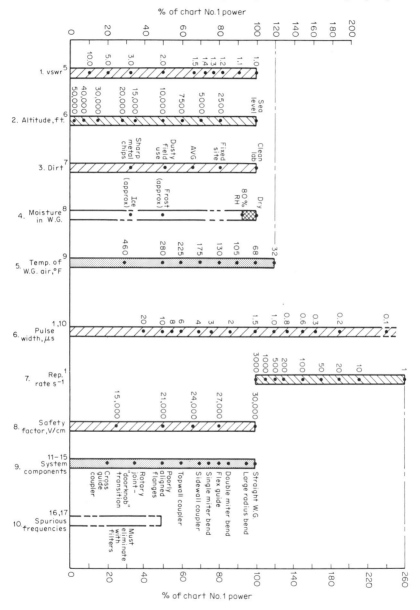

TABLE 21   Chart 3—Ways to Increase Power-Handling Capacity[12]

(a) Common methods

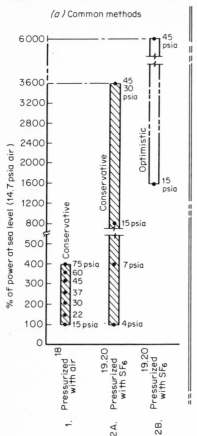

(b) Special methods

1. Create waveguide of special size (between standard sizes).

2. Cool waveguide and components by means of fins, heat transfer, etc.

3. Purge air through waveguide – large flow required.

4. Constant volume of air in waveguide, (rather than constant pressure), so pressure will increase with temperature.

TABLE 22a   Dielectric Strength of Electronegative Gases[34]

Average relative to air*

Air	1.00
Sulfur hexafluoride (SF_6)	2.35
Halogenated hydrocarbons:	
CCl_4	6.33
$CHCl_3$	4.24
CCl_3F	3.50
CCl_2F_2	2.42
$CClF_3$	1.43
$CHClF_2$	1.40
$CHCl_2F$	1.33
CH_2ClF	1.03
CF_4	1.01

* In uniform dc field.

TABLE 22b Typical Physical Properties of Sulfur Hexafluoride (SF$_6$)[34]

Molecular weight	146.06
Melting point, °F	−58 to −68
Sublimation temp, °F	−80
Specific heat (86°F), Btu/(lb)(°F)	0.143
Density:	
Solid (−58°F), g/ml^{-1}	2.51
Liquid (77° to −58°F), g/ml^{-1}	1.33–1.87
Gas (14.7 lb in^{-2}, 68°F), g/ml^{-1}	6.1–6.2
Relative (air = 1)	5.1
Critical points:	
Pressure, lb in^{-2} abs	540–605
Temp, °F	113–129
Density, g/ml	0.755
Surface tension (−58°F), dyn/cm^{-1}	11.63
Coefficient of thermal expansion (−4°F), per °F	0.015
Thermal conductivity, Btu/(h)(ft^2)(°F)(ft)	0.813
Viscosity (gats at 77°F), P	1.61×10^{-4}
Vapor pressure (68°F), lb in^{-2}	150
Refractive index (32°F), D	1.000783

TABLE 22c Typical Physical Properties of Some Proprietary Halogenated Hydrocarbon Gases[34]

Property	Freon* 12	Freon 11	Freon 21	Freon 113	Freon 114	Freon 22
Boiling point, °F	−7	73	48	118	39	41
Freezing point, °F	−252	−168	−197	−31	−137	−256
Critical points:						
Pressure, lb in^{-2}	582	635	750	495	550	716
Temp, °F	234	388	667	417	293	205
Density, g cm^{-3}	0.558	0.554	0.522	0.576	0.582	0.525
Specific heat of vapor (constant pressure), Btu/(lb)/(°F)	0.144	0.142	0.149	0.163	0.132	0.152
Toxicity by volume, %	>20	>10	>10	>13		
Thermal conductivity, Btu/(h)(ft^2)(°F)(ft):						
Liquid	0.0492	0.0609	0.0697	0.0521	0.0447	0.0595
Vapor (1 atm)	0.005	0.0048	0.0057	0.0045	0.0065	0.0068
Viscosity (liquid, 86°F), cP	0.251	0.405	0.330	0.619	0.356	0.229
Surface tension (77°F), dyn cm^{-1}	9	19	19	19	13	9
Solubility in water, g/100 g:						
86°F	0.012	0.013	0.160	0.013	0.011	0.15
32°F	0.002	0.003	0.055	0.003	0.002	0.060
Density (vapor, STP), 10^{-4} g/cm^{-3}	49.6	61.5		83.5	76.5	324.9

* Trademark of E. I. du Pont de Nemours & Co., Inc.

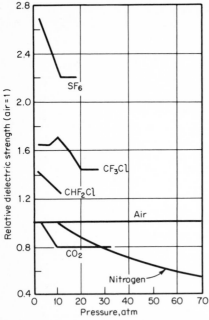

Fig. 89 Comparison of relative dielectric strengths.[34] Electronegative gases vs. air and nitrogen (air = 1).

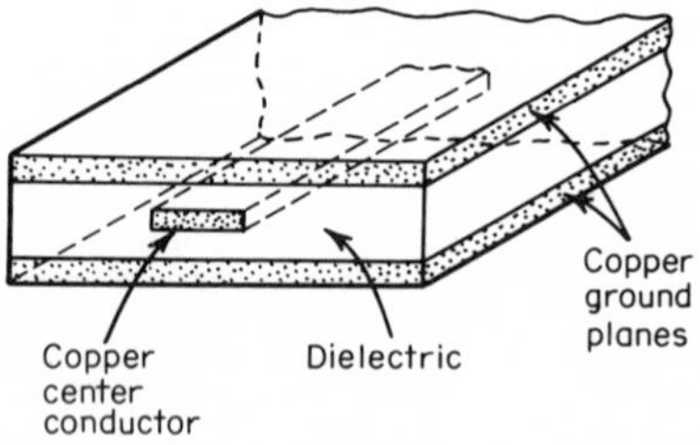

Fig. 90 Basic stripline construction.

At present microstrip is most popular for integrated-circuit design. It consists of a conductor on a laminate that has a high dielectric value to keep the stray fields at a minimum. The characteristic impedance of the microstrip is found from the following formula:

$$Z_0 = \frac{33}{\sqrt{K}} \frac{t}{0.178\ W + C_e t} \tag{41}$$

where K = dielectric constant
C_e = air-edge capacitance due to field lines beyond the strip
t = thickness of strip
W = width of strip

The wavelength in microstrip is found by the following formula:

$$\lambda = \frac{\lambda_0}{\sqrt{K}} \tag{42}$$

The loss in microstrip can be found by the following: $X = 27.3$ dB/λ. The relationship of loaded and unloaded Q is given as follows:

$$\frac{1}{Q} = \frac{1}{Q_c} + \tan \delta \tag{43}$$

where Q_c = unloaded Q due to copper loss alone (see Fig. 95)
$\tan \delta$ = loss tangent of the dielectric

Figure 95 shows some curves for characteristic impedance. It appears obvious from the dimensions of the copper strip that manufacturing tolerances must be closely

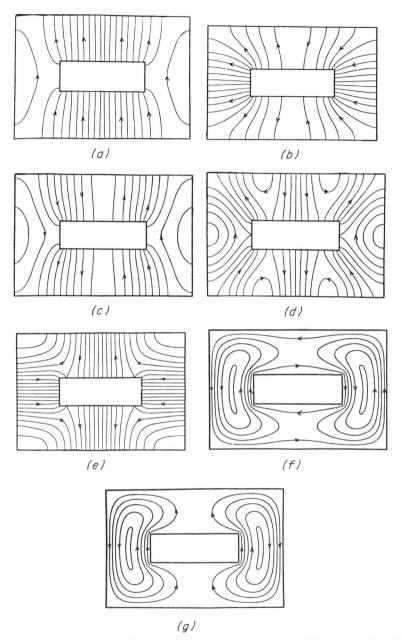

(a)

(b)

(c)

(d)

(e)

(f)

(g)

Fig. 91 Strip transmission-line mode charts.[7]   (a) Electric field lines of the TE_{10} mode.   (b) Electric-field lines of the TE_{01} mode.   (c) Electric field lines of the TE_{20} mode.   (d) Electric-field lines of the TE_{30} mode.   (e) Electric-field lines of the TE_{11} mode.   (f) Magnetic field lines of the TM_{11} mode.   (g) Magnetic-field lines of the TM_{21} mode.

TABLE 23 Characteristics of Stripline Laminates[26]

Laminate	Dielectric constant at 10 GHz	Dissipation factor at 10 GHz	Electrical variation with frequency	Useful temp range, °F	Dimensional stability	Chemical resistance	Physical qualities	Bond strength, lb in⁻¹	Thermal match	Blister resistance	Processing method	Relative cost
Cross-linked polystyrene	2.54	0.0005	Very small	-80 $+230$	Good	Good	Medium	5	Poor	Good	Modified Kodak	Medium
Cross-linked polystyrene/glass-reinforced	2.62	0.001	Small	-80 $+230$	Good	Good	Medium	5	Fair	Good	Modified Kodak	Low
Cross-linked polystyrene/quartz mat	2.60	0.0005	Small	-80 $+230$	Good	Good	Medium	5	Fair	Good	Modified Kodak	Medium to high
Cross-linked polystyrene/woven quartz	2.65	0.0005	Small	-80 $+230$	Good	Good	Medium high	5	Fair	Good	Modified Kodak	Medium to high
Teflon/unreinforced (unclad)	2.1	0.0004	Very small	-80 $+500$	Poor	Excellent	Low except flexure		Poor	Poor	Standard Kodak	Medium to high
Teflon/glass-reinforced	2.55	0.0015	Small	-80 $+500$	Good	Excellent	Medium high	8	Fair	Good	Standard Kodak	Medium
Teflon/quartz-reinforced	2.47	0.0006	Small	-80 $+500$	Good	Excellent	Medium high	8	Fair	Good	Standard Kodak	High
Teflon/ceramic-reinforced	2.30	0.001	Small	-70 $+500$	Fair to good	Excellent	Medium flexure	8	Poor	Medium	Standard Kodak	Medium to high
Cross-linked polystyrene/ceramic-powder-filled	3–15	0.0005–0.0015	Medium	-80 $+230$	Fair to medium	Fair	Low to medium	5	Fair	Fair	Modified Kodak	Medium
Silicone resin/ceramic-powder-filled	3–25	0.0005–0.004	Medium	-80 $+515$	Fair to medium	Good	Low to medium	5	Fair	Fair to good	Modified Kodak	High
Polyphenylene oxide (PPO)	2.55	0.0016	Medium	-80 $+380$	Good	Poor	Medium	3	Fair	Fair	Shipley	Medium
Irradiated polyolefin	2.32	0.0005	Small	-80 $+212$	Poor	Excellent	Low	3	Poor	Poor	Standard Kodak	Low
Irradiated polyolefin/glass-reinforced	2.42	0.001	Small	-80 $+212$	Fair	Excellent	Low to medium	3	Fair	Fair	Standard Kodak	Medium
Glass-bonded mica	7.5	0.002	Medium	-80 to $+1,100$ (unclad)	Excellent	Excellent	High	1–2	Good	Good	Standard Kodak	Medium to high
Ceramics (typical)	Typ 6.5	0.0006	Small	To $+3,000$	Excellent	Excellent	High except. impact		Good		Standard Kodak	Medium to high
Polyolefin/ceramic-powder-filled	3–10	0.001	Medium	-80 $+212$	Poor	Excellent	Low	5	Poor	Fair	Standard Kodak	High
Polyester/ceramic-powder-filled, glass-reinforced	6	0.017	Medium	-80 $+400$	Excellent	Excellent	High	5	Good	Good	Standard Kodak	Medium

Layer 1 – chromium, 200 Ω/□

Layer 2 – gold, 0.1 Ω/□

Layer 3 – SiO, 0.03 pF/mil²

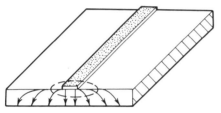

Layer 4 – Aluminum, standard

Fig. 92 Illustration of microstrip principle. Propagation of electromagnetic energy assumes distinct field patterns. Transverse electric mode is shown for waveguide structure, and transverse electromagnetic modes for coaxial configuration and IC microstrip.

Fig. 93 Lumped-circuit design.[10] Circuit-board layout. Note multiple-layer construction. Conductor pads are kept wide to avoid stray inductance.

TABLE 24   Microwave IC Materials

Dielectrics	Coefficient of expansion, 10^{-6} cm/(cm)(°C)	Thermal conductivity, cal/(cm²)(s)(°C)	Resistivity, Ω-cm	Dielectric constant ϵ	Conductors	Coefficient of expansion, 10^{-6} cm/(cm)(°C)	Thermal conductivity, cal/(cm²)(s)(°C)
Alumina, 96%	6–9	0.04	10^{14}	9	Copper	16	1.0
Beryllia	~6	0.50	10^{17}	9	Kovar	4.5	0.04
Silicon, high-resistivity	~2.5	0.37	10^3	11.8	Aluminum	25	0.5
					Steel	11	0.5
					Gold	14	0.7
Gallium arsenide, semi-insulating	~6	0.10	10^6	12	Silver	19	0.1

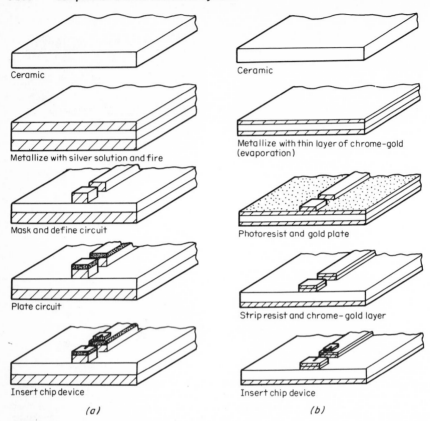

Ceramic

Ceramic

Metallize with silver solution and fire

Metallize with thin layer of chrome-gold (evaporation)

Mask and define circuit

Photoresist and gold plate

Plate circuit

Strip resist and chrome-gold layer

Insert chip device

Insert chip device

(a)

(b)

Fig. 94 Deposition of thick and thin films. (a) Film making: Starting material can be either a 96 percent alumina with 4 percent glass binder or 100 percent single-crystal sapphire; thickness of the substrate is usually 20 to 60 mils. In the thick-film etched process (left), the substrate is polished to a 5 to 10 μin finish and coated with silver or gold metal about 0.5 mil thick. The composite is fired at 850°C to bond the metal to the substrate. Photoresist is then applied, and the circuit is defined by the mask and exposed. Unprotected metal is etched off. Gold plating via a liquid-plating bath completes the passive formation. (b) In the thin-film process (right) a layer of chrome about 500 Å thick bonds conductive lines to the dielectric substrate. A layer of chrome and gold can be evaporated or sputtered on. The circuit is photographically defined. Gold is then plated in the conductive-circuit path to a thickness of 0.3 to 1 mil. Then resist and chrome-gold layer are stripped, and active devices are bonded to conductive lines.

controlled. Figures 96 through 100 give some convenient design information. It should be noted that, since there are many stray effects in the microstrip and strip-lines, many different types of correction factors and approximations are necessary for designing the components. Through computer-aided design the use of these transmission lines at higher frequencies is becoming more and more realistic every day. Tables 23 and 24 give some of the characteristics of more common stripline laminates and microstrip dielectrics.

Packaging Boards may be packaged for transition to the system circuitry in a variety of ways. Figure 101 illustrates some of these methods. Epoxy bonding is also a useful technique, but it should be pointed out that process control of epoxy is very important. Cleaning, proper curing, and the type of sealing used in the final package are extremely important in the use of epoxy. A moistureproof seal

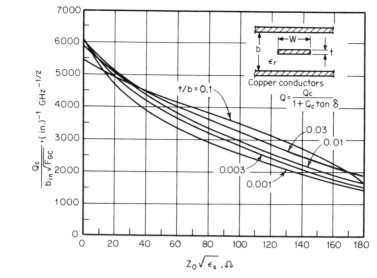

Fig. 95   Theoretical Q of copper-shielded stripline in a dielectric medium.[7]

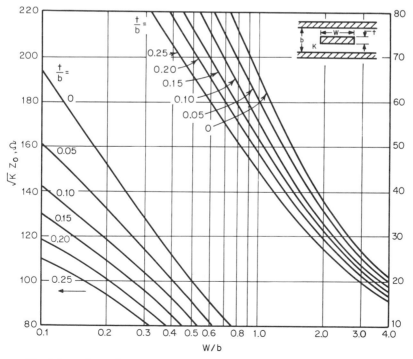

Fig. 96   Relationship of dielectric thickness to characteristic impedance.[35]

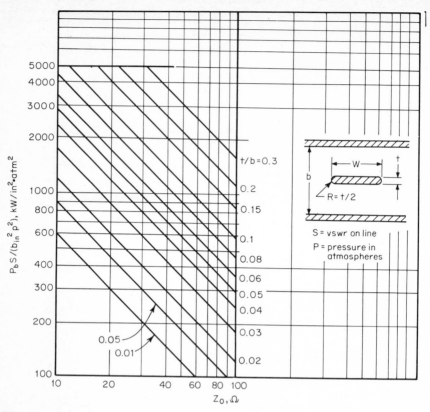

Fig. 97 Theoretical breakdown power of air-dielectric rounded-strip transmission line.[7]

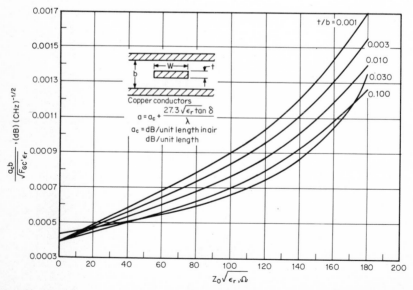

Fig. 98 Theoretical attenuation of copper-shielded stripline in a dielectric medium.[7]

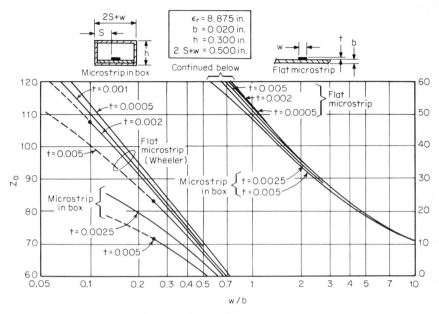

Fig. 99   Impedance of microstrip, ohms.[7]

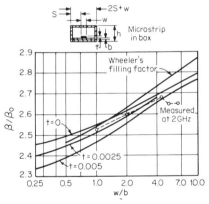

Fig. 100   Relative velocity β/β_0 vs. w/b (propagation constant).[7] $b = 0.020$ in, $h = 0.300$ in, $2S + w = 0.500$ in, $\epsilon_r = 8.875$ in.

or the use of critical sealing techniques might be required. Also, the amount of EMI sealing is important in a final design.

ACTIVE COMPONENTS

Active components involve the use of semiconductors. The scope of this discussion does not include any of the physics, which is covered in another chapter. The circuit equivalency of the semiconductors is the main feature. However, some discussion of the types of semiconductors and their peculiarities in microwave circuits is in order here.

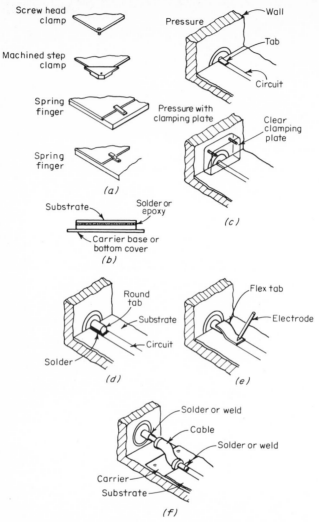

Fig. 101 Some construction techniques for microstrip.[36]   (*a*) Clamping.   (*b*) Soldering.   (*c*) Pressure transitions.   (*d*) Solder transitions.   (*e*) Compression or welding.   (*f*) Cable transition.

Fig. 102   Point-contact mixer diodes, ceramic package.   (*Microwave Assoc.*)

Point Contacts

Diodes The point-contact diode is the most common and probably the most basic diode. Figure 102 shows this diode. Figure 103 shows how it is assembled. The chip is made of silicon and the whisker is tungsten. The electrical properties of the diode are determined by the point shape and pressure of the whisker. The junction capacitances are small with this device because the point is both the contact and the junction. Although this device is generally looked down upon by those interested in high reliability, its useful properties should not be overlooked considering its price. It is also available in a coaxial configuration as shown in Figure 104.

Schottky Barrier

The Schottky-barrier diode has in recent years become the most-used device in across-the-board applications. It is made with a silicon-base material and has an ohmic contact that may be formed in several ways. Figure 105 illustrates how this is done, Figure 106 shows the details of the actual chip, and Figure 107 gives a typical set of characteristic curves for the Schottky diode. A common misconception about Schottky diodes is that they are more reliable than point-contact diodes in microwave receivers because they are less sensitive to leakage levels from the receiver protectors. As it turns out, this is not always the case. Table 25 gives some of the burnout ratings for a series of diodes. Figure 108 compares the pulse

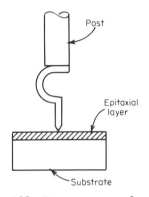

Fig. 103 Point-contact configuration.

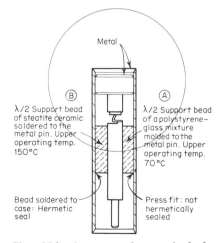

Fig. 104 Conventional coaxial diode.

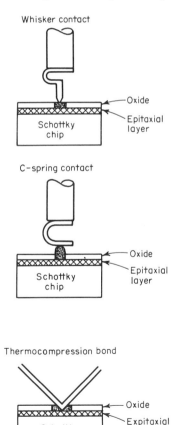

Fig. 105 Schottky contacting techniques.[33]

TABLE 25   CW Burnout of Point-Contact and Schottky-Barrier Diodes[35]

	Frequency	
	X band (cartridge mixer), W	*Ku* band (coaxial mixer), W
Regular point-contact, silicon (*p*-type)	0.10–0.50	0.10–0.25
High-burnout point-contact, silicon (*p*-type)	0.20–1.0	0.20–0.50
Nickel-silicon Schottky (*n*-type)	0.10–0.20	0.10–0.20
Ti-Mo-Au silicon Schottky (*n*-type)	0.15–0.50	0.10–0.40
Nickel-GaAs Schottky (*n*-type)	0.10–0.30	0.10–0.25

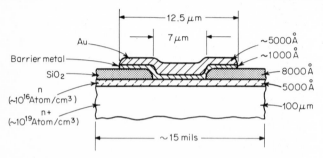

Fig. 106   Cross section of silicon Schottky-barrier diode.[33]

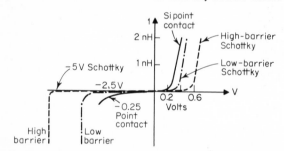

Fig. 107   Typical *I-V* characteristics and barrier heights.[33]

Experimental Values of Metal-
Semiconductor Barrier Heights

Metal	*Silicon* n-Type (III), eV
Cr	0.55
Mo	0.60
Ni	0.45
Pd	0.72
Ti	0.30
W	0.69

burnout ratings.   For silicon, the thermal-relaxation time is around 10^{-8}, which means that for pulse widths less than 10^{-8} the temperature is a function of the energy peak.   The ratings for diodes are then given terms of watts peak.   The advantage of the Schottky diode over the point contact is its peak inverse voltage (PIV).   In making a reliability comparison of the point-contact with the Schottky diode, it is important to evaluate the performance in the circuits and make the comparisons on that basis.   One such determination is the load resistance.   Table 26 shows a

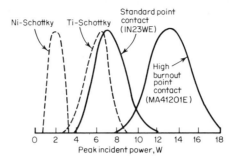

Fig. 108 *X*-band rf pulse burnout of silicon point-contact and Schottky diodes.[33] Pulse width = 10 ns. Pulse duty = 1,000 pulses/s. Frequency = 9.375 GHz. Noise-figure degradation for burnout = 1 dB.

TABLE 26   Comparison Table, Schottky vs. Point-Contact Diodes[13]

	Point contact	Schottky
Construction	Pressure contact offers possible nonuniformity	Deposited contact offers good fabrication control and therefore uniformity
Mixer applications: Noise figure	Better at low LO levels (e.g., 12 dBm)	Better at higher LO levels (e.g., +6 dBm). Less susceptible to LO change (−9 to +6 dBm). Higher NF at low LO levels (−12 dBm due to nonconduction)
With dc bias	Diodes are comparable depending on dc-bias levels used	
if impedance	Lower for PC at low LO levels (−12 dBm). Otherwise, comparable for both diodes	
rf impedance (vswr)		Offers better match to 50 Ω system. Better suited for broadband applications
Detector applications: Tangential signal sensitivity (TSS) (without dc bias)	Better performance	Worst by ≈30 dB
With dc bias	Diodes are comparable	
Voltage output vs. power input	Lower saturation level. Higher-voltage output at low levels	Higher saturation level. Lower-voltage output at low levels
Noise voltage at low if frequencies: With dc bias	Noise voltage curves show diodes comparable as they approach 1,000 kHz	Superior at 1 kc by a factor of 3 (~10 dB).
Microphonics		Offers vastly improved performance
DC characteristics	Higher R_s, lower conducting voltage, low peak inverse voltage (PIV 2–3 V, 100 μA)	Lower R_s, higher conducting voltage, high PIV (12–13 V, 100 μA)

comparison between the two diodes. One important factor in considering the use of these diodes is that their impedances in the transmission line are different.

Tunnel Diodes

The tunnel diode is very useful to the microwave engineer. It is a *pn* junction with a very narrow junction, which allows a tunneling effect that results in the current-voltage curve shown in Figure 109. These diodes are made from germanium,

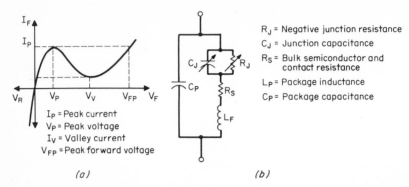

R_J = Negative junction resistance
C_J = Junction capacitance
R_S = Bulk semiconductor and contact resistance
L_P = Package inductance
C_P = Package capacitance

I_P = Peak current
V_P = Peak voltage
I_V = Valley current
V_{FP} = Peak forward voltage

(a)   (b)

Fig. 109 Typical current-voltage curve and circuit of tunnel diodes.[37] (*a*) Typical current-voltage characteristics of tunnel diodes. (*b*) Tunnel-diode equivalent circuit.

gallium arsenide, or gallium antimonide. Examination of the reverse-bias curve shows that the diode conducts at an exponentially increasing rate without any voltage offset. This means the diode can conduct without the need of external biasing for very low levels of incident power. This device is also known as a backward diode.

Because of the negative-resistance characteristic in the forward-bias curve the tunnel diode has other uses besides detecting for low-noise amplifiers. It does not have the performance of the parametric amplifier but is less complex to construct and manufacture. As with most negative-resistance amplifiers the design problem is in stablizing the amplifier. The greatest single limitation of the tunnel-diode amplifier is its power output. Tables 27 and 28 summarize some of the performance characteristics of tunnel diodes.

TABLE 27   Tunnel-Diode Features and Benefits[38]

Device features	User benefits
Can be used for high-frequency, low-noise, wideband amplifiers	Replaces various complex microwave tube and parametric amplifiers in many applications
Can be used for low-noise, high-frequency local oscillator source	Provides less noise than Gun or avalanche oscillators and much less complex than varactor sources
Capable of very high frequency operation	Reported applications to 40 GHz
Low flicker (1/f) and shot noise	Improved Doppler radar detectors which use audio-frequency IFs
Low local oscillator-drive requirements for mixers	Reduces LO complexity and power-output requirements
Good detector sensitivity without biasing	Simpler detector circuitry and no dc biasing required
Capable of very fast transitions from ON to OFF states	Requires very low trigger drive, is bistable, and suitable for TDR pulsers

TABLE 28   Summary of Important Tunnel-Diode Parameters[38]

Parameter	Importance
F_{ro}	Resistive cutoff frequency determines upper frequency limit at which diode no longer exhibits a negative resistance
F_{zo}	Self-resonant frequency which limits frequency range due to package inductance and junction capacitance; also affected by package capacitance if not properly designed into matching structure
V_v, V_p	Determines maximum voltage swing and power output in amplifier and oscillator circuits. Also affects switching speed
I_p/I_v	Affects switching speed and pulse output swing
C_j	Limits both high-frequency performance and switching speed
$-R_m$	Minimum negative resistance determines maximum available gain in amplifier circuits and load resistance in switching applications
K	Shot-noise constant affects amplifier noise figure
R_s	Measure of the diode internal losses
V_r	Measured at some reverse current (typically 1 to 5 mA); indicates suitability for backward-diode applications
β	Current sensitivity gives a measure at nonlinearity for backward diode detectors and mixers

Avalanche Diodes

The avalanche diode has recently become popular as a source of microwave oscillators. The theory of this device is based upon a phase shift between the rf field and the avalanche current resulting from the avalanche multiplication at the carriers. The significance of this device is in its conversion of dc power to rf power. This occurs in the negative-conduction region of the diode IV curve when the diode is reverse-biased. Figure 110 illustrates the typical construction of the avalanche.

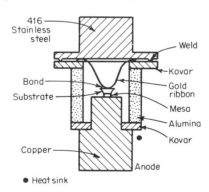

Fig. 110   Avalanche-diode construction.[33]

Note that the mesa is mounted to the heat sink rather than the substrate as is done in more conventional diodes. This is because most of the heat dissipation occurs at the junction.

Gunn Diodes

The bulk-effect device, or the Gunn diode, is different in construction from the avalanche device in that it is an n-type material with two ohmic contacts. Figure 111 illustrates the manner in which this is done. The Gunn diode is based on discoveries that when properly excited by a dc field the gallium arsenide material generates microwave frequencies without the use of ohmic contacts. One singular advantage of the diode over the avalanche is that it takes less voltage to operate. To make

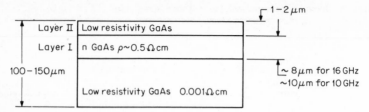

Fig. 111 Gunn-diode construction.[36] Cross section of typical slice of GaAs. Layers I and II are grown epitaxially in special reactors, providing a single crystal with three impurity concentrations.

an oscillator it is only necessary to surround the diode with a cavity of the proper frequency

PIN Diode

The PIN diode is a class of diode used for control devices such as attenuators and switches. The diode consists of a very heavily doped p region and n region separated by an intrinsic region, the base material without any doping. When reverse-biased, there is no current in the I region and the device acts like a capacitor. When there is conduction in the forward region, the device acts like a very low series resistance. The bias in a microwave application may be provided by the incident microwave energy or by an external dc source. Figure 112 illustrates the equivalent circuit of the PIN for different biasing conditions.

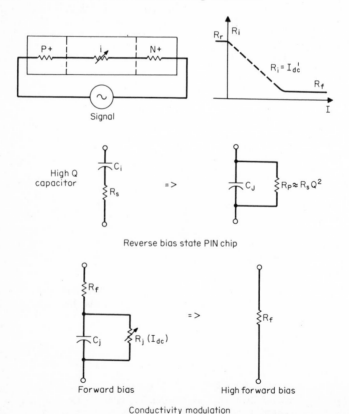

Fig. 112 Principal mechanism of PIN junction.[33]

Tuning Varactor

The tuning varactor is a useful device for the oscillator diodes described above because it can change the equivalent capacitance of a microwave cavity. The diode is a *pn*-junction device biased in a reverse condition and never allowed to be forward-biased. It acts like a capacitor that is a function of voltages. The response function is basically logarithmic. Figure 113 illustrates a typical curve. The diode is characterized by a figure of merit referred to as Q. Figure 114 illustrates how this is determined and gives a typical curve showing the capacitance relationship. It is important to keep the series resistance as low as possible.

Step-Recovery Diode

The step-recovery diode is used in pulse shaping and harmonic generation. By constructing the diode so that the high built-in field confines the stored charge close to the junction, a shutoff characteristic can be obtained. See Figure 115, I—V curve and equivalent circuits for step-recovery functions.

Attenuators and Switches

The microwave attenuator or switch is made from a PIN diode or varactor diode in a microwave transmission line. The circuit may be arranged in various ways depending upon the desired performance of the switch. The performance of the PIN diode demonstrates the operation of the microwave components.

Figure 116 represents the equivalent circuits of the two diodes used for control devices. The significant difference between the two is that the PIN device is represented by a constant capacitance. The PIN does not rectify at its operating point, and the varactor does. This characteristic makes the PIN ideal for a control device because it does not introduce unwanted harmonics.

The PIN performance is limited at the lower frequency by the minority-carrier lifetime, below which point the device acts as a typical rectifying diode. This frequency is expressed as $f_t = \frac{1}{2}\pi\tau$, where τ is the minority-carrier lifetime. The *I* layer in the device causes a large resistance ($10,000\Omega$) in the reverse direction.

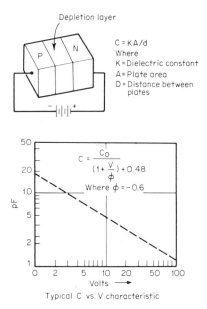

Depletion layer

$C = KA/d$
Where
K = Dielectric constant
A = Plate area
D = Distance between plates

$$C = \frac{C_0}{(1+\frac{V}{\phi})+0.48}$$

Where $\phi = -0.6$

pF

Volts

Typical C vs. V characteristic

Fig. 113 Reverse-bias PIN junction.[33] Forward bias causes reduction in depletion layer, increasing capacitance. Reverse bias widens depletion layer, decreasing capacitance. Applications: (1) Remote electronic tuning of resonant cavities and rf circuits. (2) Oscillator AFC tracking systems. (3) Oscillator main tuning systems. (4) Receiver preselector front ends. (5) Gunn-diode oscillator tuning.

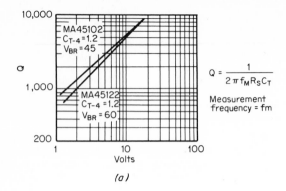

$$Q = \frac{1}{2\pi f_M R_S C_T}$$

Measurement
frequency = fm

(a)

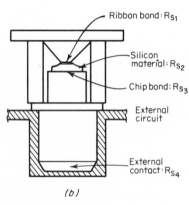

Ribbon bond: R_{S_1}

Silicon material: R_{S_2}

Chip bond: R_{S_3}

External circuit

External contact: R_{S_4}

(b)

Fig. 114 Tuning varactor.[33] (*a*) Diode losses and Q, Q vs. bias at 50 MHz. (*b*) Contact-resistance contributions.

Increase of this bias causes the resistance to vary because it expands the depletion layer. This layer of intrinsic material also allows the device to have a high PIV parameter, making it useful in high-power applications. Another transition frequency associated with the constant capacitance of the device is expressed as follows:

$$f_p = \tfrac{1}{2} e\rho$$

where ρ = resistivity of the material
e = permittivity of the material

The design of the control device must keep these two frequencies below the operating range desired. The speed of the switch is affected by the minority-carrier lifetime, and thus a trade-off results in speed v. power when a particular junction is selected. Figure 117 shows an equivalent circuit for the PIN package. The upper frequency limitation in terms of bandwidth is determined by the parasitic capacitances and inductances.

The control device may be designed as a shunt device or a series device. The advantage of the shunt design is that it can hold off high power for short periods of time. In the shunt state the diode is in forward conduction in the off state and short-circuits all incoming signal current. If the incoming pulse is short in pulse width to prevent the device from heating up, a change in charge current can be made equivalent to significant amounts of transmission-line peak power.

The switches are classed as either broadband or resonant. The broadband switch is as generally described above. The resonant type is a very narrow-band device

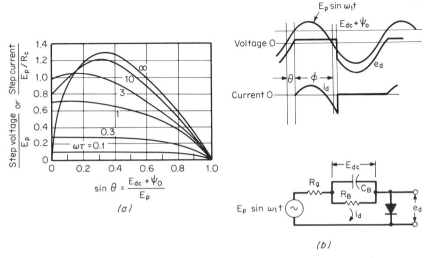

Fig. 115 Step-recovery diode.[33] (a) Step amplitude as a function of operating conditions. (b) Waveforms for biased step-recovery diode. The dependence of step amplitude upon circuit parameters is shown in (a).

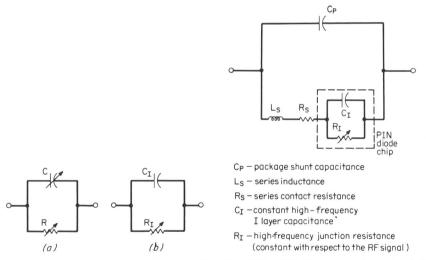

C_P — package shunt capacitance

L_S — series inductance

R_S — series contact resistance

C_I — constant high-frequency
I layer capacitance

R_I — high-frequency junction resistance
(constant with respect to the RF signal)

Fig. 116 Equivalent circuits for control diodes. (a) Varactor diode. (b) PIN diode.

Fig. 117 PIN-diode equivalent circuit.[33]

that is designed to present a large reflection in the open state. This is accomplished by varying the Q of the circuit, usually with a very small swing of current. This type of design is very sensitive to frequency, and the bandwidth must be sacrificed for isolation. Figures 118 and 119 summarize important design equations and the equivalent circuits for the different approaches.

The diodes in the switch have to be baised for the switching operation. This problem can be resolved in many ways. There are several important considerations. First, whatever technique is used, the rf circuit must be isolated from the biasing circuits. This can be done by clever transmission-line coupling through quarter-

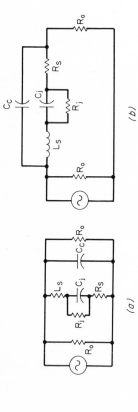

(a)

(b)

Fig. 118 Resonant-switch circuits. (a) Shunt switch. (b) Series switch.

Parameter	Insertion loss	Isolation	Max peak power P_0	Max average power P_0		Bias operation	
				Forward bias	Reverse bias	Forward	Reverse
Shunt	$\left[1 + \dfrac{R_0}{2R_S(1+Q^2)}\right]^2$	$\left(1 + \dfrac{R_0}{2R_S}\right)^2$	$\dfrac{R_0}{32}\left[\dfrac{V_b(1+2R_S/R_0)}{X_{c_j}}\right]^2$	$P_{D_F}\dfrac{R_S(1+Q^2)[1+R_0/2R_S(1+Q^2)]^2}{R_0}$	$P_{D_R}\dfrac{R_S[1+R_0/2R_S]^2}{R_0}$	On	Off
Series	$\left(1 + \dfrac{R_S}{2R_0}\right)^2$	$\left[1 + \dfrac{R_S(1+Q^2)}{2R_0}\right]^2$	$\dfrac{R_0}{8}\left[\dfrac{V_b(1+R_S/2R_0)}{X_{c_j}}\right]^2$	$P_{D_F}\dfrac{R_0[1+R_S(1+Q^2)/2R_0]^2}{R_S(1+Q^2)}$	$P_{D_R}\dfrac{R_0[1+R_S/2R_0]^2}{R_S}$	Off	On

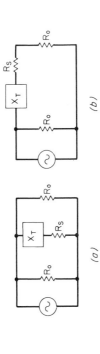

Fig. 119 Nonresonant-switch circuits. (a) Shunt switch. (b) Series switch.

(a)   (b)

Parameter	Transmission loss	Approx insertion loss	Approx isolation	Max peak power P_0	Max average power P_0	Bias operation Forward	Bias operation Reverse
Shunt	$1 + \dfrac{R_0(R_s + R_0/4)}{Rs^2 + X_T^2}$	$1 + \left(\dfrac{R_0}{2X_T}\right)^2$ Limit$\left[1 + \dfrac{R_0}{2(Rs + R_i)}\right]^2$	$\left(1 + \dfrac{R_0}{2Rs}\right)^2$	$\dfrac{V_b^2}{8R_0}$	$P_{dF}\dfrac{R_0}{4Rs}\left(1 + \dfrac{2Rs}{R_0}\right)^2$	Off	On
Series	$1 + \dfrac{Rs}{R_0} + \left(\dfrac{Rs}{2R_0}\right)^2 + \left(\dfrac{X_T}{2R_0}\right)^2$	$\left(1 + \dfrac{Rs}{2R_0}\right)^2$	$\left(\dfrac{X_T}{2R_0}\right)^2$ Limit $[1 + (R_s + R_i)/2R_0]^2$	$\dfrac{V_b^2}{32R_0}$	$P_{dF}\dfrac{R_T}{Rs}\left(1 + \dfrac{R_0}{2R_0}\right)^2$	On	Off

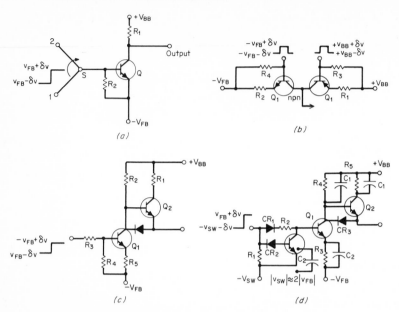

Fig. 120 Driving circuits for control devices.[43] (*a*) Simple PIN driver simulates a switch (*S*). (*b*) Complementary output stages have reduced charging resistance at the price of complex preamplifying stages. (*c*) Power-saving "totem pole" has improved switching time with addition of parallel capacitors. (*d*) Additional circuit improves speed but adds components.

and half-wavelength matching stubs.   Another way is to use conventional lumped-element filters.   The use of the PIN switch involves additional circuitry to drive the diodes.   Figure 120 illustrates versions of such a switch driver.   Recently the trend has been to TTL circuitry, and a highly specialized design approach has evolved.   The ideal is to minimize the effects of parasitics and match the drivers to the diodes.

Attenuators operate in much the same manner as the switches do except for the drive circuitry.   If simple current-attenuation relationships are desired, the attenuator can be supplied with just the microwave circuitry and a bias input with some sort of decoupling.   Figure 121 illustrates a packaged attenuator.   However, since the attenuator is more likely to be driven by voltage output signals and it is desired that the device be a voltage transfer function.   levelers are usually designed into the circuitry external to the microwave circuits.

Fig. 121 Coaxial diode attenuator. (*Microwave Associates.*)

The maximum power that can be handled by the diodes is the major limitation in the usefulness of the switch.   As was mentioned before, the standoff power for the diode with short pulses can be quite high if the junction does not heat up.   However, where the duty cycle is sufficient to cause the diodes to heat, equations have been developed that make it possible to predict fairly accurately the maximum heat-dissipation ability of the diodes in the switch.   The theory developed rests on an analogous treatment of a relaxation oscillator. The maximum peak power that can be dissipated by the diode is given by

$$P_m = P_c \frac{1 - e^{-t_r \theta_k}}{1 - e^{-t_p \theta_k}} \qquad (44)$$

where P_m = maximum pulse power
   P_c = maximum CW power the diode can dissipate
   t_r = pulse time and interpulse time
   t_p = interpulse interval
   $\theta_k = \theta_{ja} C_h$
   C_h = heat capacity of junction
   θ = thermal resistance of junction

The maximum power that can be switched is

$$P_1 = P_m \frac{n^2 Z_0}{4R_s} \qquad (45)$$

where n = number of diodes
   R_s = series resistance
   Z_0 = characteristic impedance

Broadband units have recently been especially designed to make use of modules of diodes. These configurations considerably simplify the microwave engineer's task when flatness is a critical specification. Such devices use diodes arranged to act like capacitances in the back-biased state, and the gold lead interconnecting the devices provides the inductance.

Mixers and Converters

Mixers and converters both operate on the principle that two frequencies put into a nonlinear device generate harmonics. The beat frequencies of the two signals are the ones of interest to the receiver. Mixers are used for heterodyning and Doppler detection.

Figure 122 illustrates the signal components for the operation of a mixer. The

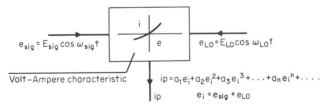

Fig. 122 Signal components of a mixer.[24]

$$
\begin{aligned}
i_p = {}& a_1 E_{LO} \cos \omega_{LO} t + \cdots && \text{Local oscillator} \\
& + a_1 E^2_{sig} \cos \omega^2_{sig} t + \cdots && \text{Signal} \\[4pt]
& + \frac{a_2}{2}(E_{LO}^2 + E^2_{sig}) + \cdots && \text{dc Component} \\[4pt]
& + a_2 E_{LO} E_{sig} \cos [\omega_{LO} - \omega_{sig}] t + \cdots && \text{Lower-side frequency} \\
& + a_2 E_{LO} E_{sig} \cos [\omega_{LO} + \omega_{sig}] t + \cdots && \text{Upper-side frequency} \\[4pt]
& + \frac{a_2}{2} E^2_{LO} + \cos 2\omega_{LO} t + \cdots && \text{Second harmonic of LO} \\[4pt]
& + \frac{a_2}{2} E^2_{sig} \cos 2\omega_{sig} t + \cdots && \text{Second harmonic of signal}
\end{aligned}
$$

diode is operating in a square-law region so that intermediate-frequency components can be calculated. In the past intermediate frequencies of 30 to 60 MHz have been used with point-contact diodes because the noise performance of the diodes has been best at those frequencies. Recently, backward and Schottky-barrier diodes have made the use of higher intermediate frequencies possible.

Good mixer performance obtains as much IF output power with as little noise introduction as possible. The important parameters of the mixer are noise figure, conversion loss, vswr, isolation, and spurious responses.

Noise figure   The introduction of noise to the system is of great concern to the receiver in the use of a mixer. As in all black-box treatment of noise, the measurement is usually made without regard to source. General treatments of noise deal with gaussian noise, which is stationary. Academic treatments of noise phenomena classify it as thermal noise and shot noise, the former having to do with the temperature of the device and the latter with electron phenomena.

The noise performance of a system is rated by its noise factor, which is a determination based upon the input impedance and the noise contribution of the system. The noise factor of a mixer at a specified input frequency is defined as the ratio of (1) the total noise power per unit bandwidth at a corresponding output frequency available at the output port when the noise temperature of the input termination is standard (290 K) to (2) that portion of (1) engendered at the input frequency by the input termination.[25] The standard noise temperature of 290 K approximates the actual noise temperature of most input terminations. Since frequency becomes important in measuring the noise, it is also part of the parameter. However, a more practical approach is to characterize the system by its average noise factor, designated as $\bar{F}$. For most mixer definitions the noise factor is simply defined as the ratio of signal to noise at the input to signal to noise at the output, expressed in decibels.

The noise factor is measured by introducing a known amount of noise at the input and measuring the noise with the source of noise on and off. The receiver has an adjustable attenuator which resets the level of a meter after the source is turned off. The difference in reading is the Y value. Figure 123 demonstrates how

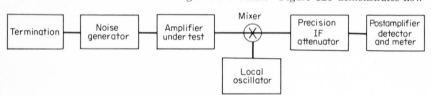

Fig. 123   Noise-figure measurements.[7]

$$F_{dB} = 10 \log \left(\frac{T_d}{T_0 - 1} \right) - 10 \log (Y - 1)$$

F_{dB} = noise figure
T_d = temperature of noise generator
T_0 = 290 K
Y^0 = IF attenuation, or ratio of signal power when noise
   generator is on to signal power when noise generator is off

For low-noise-figure measurements corrections can be made for heating of the terminating resistor as follows:

$$\text{Correction} = 10 \log \left(\frac{F}{F + t - 1} \right)$$

this is done. The noise figure of the mixer as measured above also includes the noise figure of the postamplifier. This number is generally 1.5 dB or less for most amplifiers used at the intermediate frequencies mentioned previously. The mixer used should indicate whether double or single sidebands are shown to the receiver because the final noise figure would be off by 3 dB, if equal sideband power were assumed. The noise figure should be given as a single-sideband figure. Figure 124 gives some additional information for determining receiver noise figures.

Conversion loss   The next most critical parameter in determining the mixer performance is conversion loss. Quite simply the conversion loss is the ratio of available signal power to IF output power. The measurements are usually done with conventional techniques. When specifying the parameter, it is important to distinguish whether the IF power is single- or double-sideband power.

Intermodulation products   Because the diode is a harmonic generator, many frequencies will be present in the mixer that could conceivably cause bad information

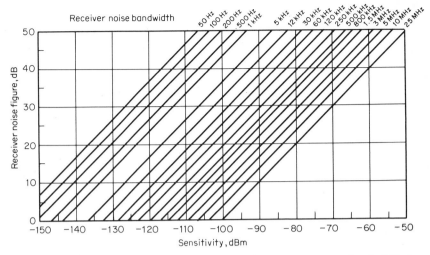

Fig. 124 Receiver noise figure vs. sensitivity for various receiver bandwidths.[7]

to the receiver. Thus it is important that these products be considered and controlled.

The usual manner of specifying the products is to define the multiple of the rf or LO base frequency. Once the desired sideband is selected, this is referred to as the 1×1 response. The other 1×1 is the image frequency, which will have to be rejected or filtered off. Increasing multiples of the LO frequency change the designation to 2×1, 3×1, and so on. Increasing multiples of radio frequency change the designation to 1×2, 1×2, and so on. The products have a definite power relationship with the rf input power, and it is essentially linear as long as the mixing device is below a saturation point. Thus, once the power levels are selected, the relative power levels of the intermodulation products may easily be determined. Figure 125 gives the performance for a typical mixer.

Coupling mechanism Four types of coupling mechanisms are used: capacitive, inductive, resistive, and directional. Figure 126 illustrates these devices. The capacitive coupling has been the most popular in recent designs.

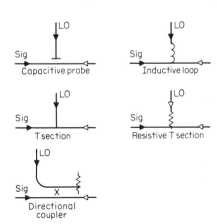

Fig. 125 Mixer intermodulation products.

Fig. 126 Coupling mechanisms.[24] Schematics of typical coupling mechanisms for single-diode mixers.

Balanced mixers   Balanced mixers are used to gain advantages in reducing LO-noise contributions and better overall efficiency due to less reflection of the LO signal. There are basically two ways to form the balanced configuration. One method is to use a 90° phase-shift device, which results in good broadband matching but poor isolation between the arms. The 180° scheme yields good isolation but poor match of the diodes, thus giving poor vswr's. The only way to resolve the difficulties on each of these designs is to match the holders to the transmission line as ideally as possible. Figures 127 and 128 show the double-balanced mixer configuration and

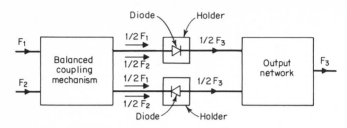

Fig. 127   Schematic of a balanced mixer.[24]

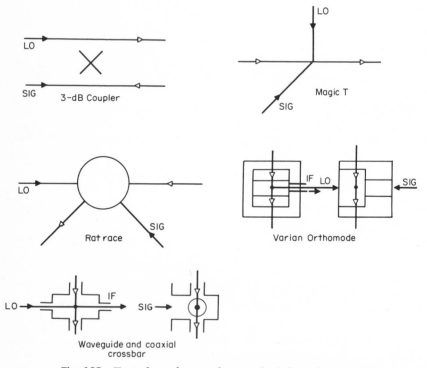

Fig. 128   Typical coupling mechanisms for balanced mixers.[24]

some typical coupling mechanisms. The LO-noise power contribution is achieved by reversing the polarity of the diodes in the respective areas of the mixer.

Image-rejection mixers   The image-rejection mixer has gained a great deal of recent attenton. In the past images were traditionally filtered out of the line, but recent developments with quadrature hybrids at intermediate frequencies have made it possible to reject the IF images by a technique that is not frequency-sensitive. Broadband mixers are thus easier to design. Figure 129 shows how this is done.

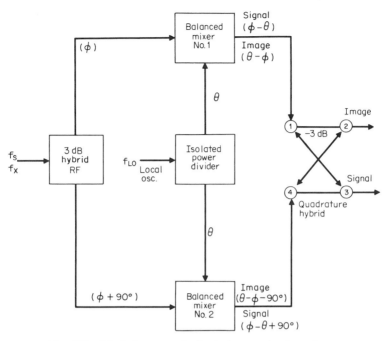

Fig. 129 Block diagram of a basic image-rejection mixer.

TABLE 29   Summary of Image Rejection, Mixer Characteristics

	Cancellation within IF quadrature hybrid	
	Output port 2	Output port 3
$f_s(f_s > f_{10})$ Desired signal	$(\varphi - \theta) + (\varphi - \theta + 180°) = 0$ $= (\varphi - \theta) - (\varphi - \theta) = 0$ Cancellation	$(\varphi - \theta + 90°) + (\varphi - \theta + 90°)$ $= \Sigma$ Combination
$f_x(f_x < f_{10})$ Image signal	$(\theta - \varphi) + (\theta - \varphi) = \Sigma$ Combination	$(\theta - \varphi + 90°) + (\theta - \varphi - 90°)$ $= 0$ Cancellation

Table 29 shows the phasing scheme for the cancellation, and Figure 130 shows how this rejection can be affected by error in the design.

Modulators Modulators and converters operate much like mixers. The only difference is usually in the way the frequencies are used. Modulators or converters are usually used to change a basic radio frequency slightly by the introduction of a frequency that would be the intermediate frequency in a mixer.

Solid-State Sources

There are three popular methods of developing oscillators at the microwave frequencies from solid-state devices: transistor oscillators through multiplier chains, Gunn oscillators, and avalanche oscillators.

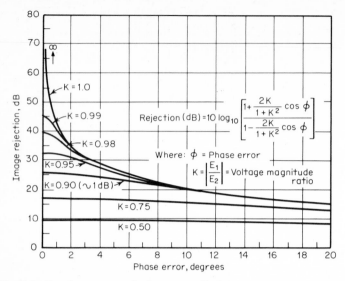

Fig. 130 Rejection by means of phase cancellation vs. errors in magnitude and phase.

Although very popular, the transistor oscillator will not be discussed because the Gunn and avalanche devices are at present the leaders in the field at the higher frequency. Also, the technology of the transistor devices requires considerations beyond the scope of this chapter.

Gunn oscillator The Gunn oscillator uses the Gunn diode discussed previously. The diode develops instabilities owing to the formation of traveling high-field domains. The behavior of the device is directly linked to the thickness of active region thickness. Figure 131 shows a relationship of this thickness for band center frequency. The efficiency of the oscillator is between 2 and 5 percent. The limiting factor on input dc power is operating temperature. Figure 132 shows the active-

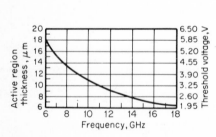

Fig. 131 Active-region thickness vs. frequency.[39] Active-region thickness usually used for Gunn diodes, as a function of the band center frequency. Wide variations from this relationship are possible.

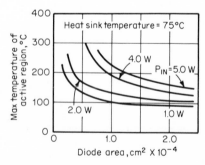

Fig. 132 Maximum diode temperature vs. area.[39] Maximum diode temperature vs. area for various dc power inputs with a Gunn-diode chip on a copper heat sink.

region temperature for the conditions described. The higher the operating temperature the more accelerated are impurity diffusion and ion migration. Data to date indicate that the best set of conditions for long life is a maximum active-region temperature of 200°C with a heat-sink temperature of 75°C. Packaging of the diode must be done to keep any resonant frequencies above the operating frequency.

The Gunn oscillator consists of the diode mounted in a microwave cavity and biased in some manner by a coupling mechanism. The bias voltage applied must be above the threshold to ensure oscillation. The threshold is defined as the point on the $I - V$ curve at which the conductivity becomes negative. Figure 133 shows a typical waveguide oscillator with a sapphire tuning rod. The tuning rod reduces the frequency as it enters the cavity so the length of the cavity is selected by the highest frequency the oscillator is supposed to operate at. The cavity operates in the TE_{110} mode and l is one-half the wavelength of the resonant frequency. The iris is the conventional waveguide susceptance device discussed previously. The resonating frequency of the rod must be higher than the operating frequency of the oscillator. Figure 134 shows the tuning effect of the rod for a typical design.

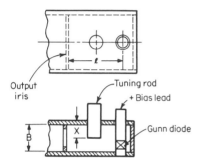

Fig. 133 Postcoupled waveguide cavity for Gunn diodes.[39]

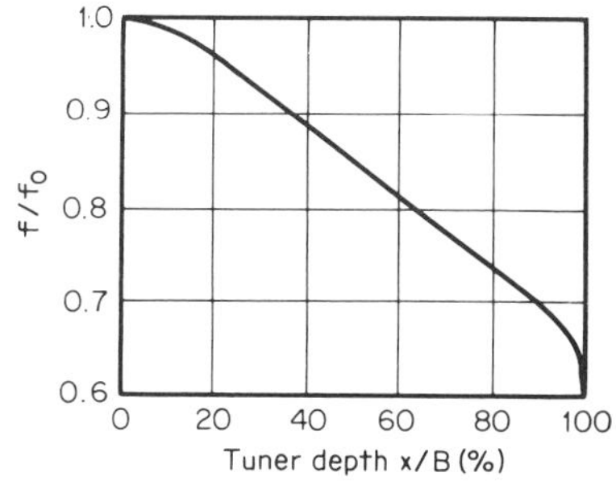

Fig. 134 Frequency vs. tuner depth for waveguide cavity of Fig. 133 with a dielectric-rod tuner.[39]

This means for feeding the bias to the oscillator has to be given consideration. Figure 135 illustrates such a device. The choke is designed to minimize the effects

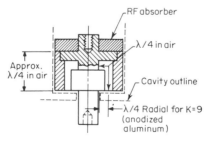

Fig. 135 Bias choke for feeding dc to a Gunn diode in a resonant cavity.[39]

of leakage. Just about everything affects the oscillating frequency and power output of the oscillator, and it is important that the operating conditions be properly described when the requirements are presented to the manufacturer. The cavity is coupled to the transmission line through the iris; so pulling is a problem with the device. It is recommended that isolators or pads be used before the oscillator in the transmission line.

The noise spectrum of the oscillator is always of concern to the system user. Figure 136a to d gives typical curves for devices in the AM and FM domains. Generally, steps can be taken to reduce noise. The quality of the cavity can be carefully controlled, thus improving the Q. Diodes can be carefully selected for minimum-noise characteristics, and ripples on supply voltages can be kept to an absolute mini-

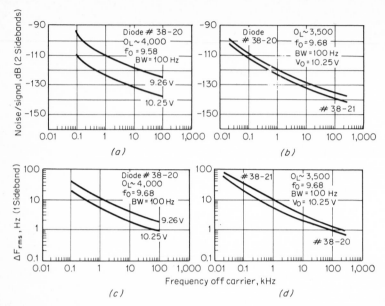

Fig. 136 Noise performance of the Gunn diode.[39]   (*a*) AM-noise spectrum of Gunn diode in high-*Q* cavity at two bias voltages.   (*b*) AM-noise spectrum of two different Gunn diodes in the same cavity and at the same bias voltage.   (*c*) FM-noise spectrum of Gunn diode in high-*Q* cavity at two bias voltages.   (*d*) FM-noise spectrum of two different Gunn diodes in the same cavity and at the same bias voltage.

mum.   Also, the pushing factor of the diode affects the FM noise, and it may be necessary to adjust the bias to keep this factor at the zero point (see Fig. 137).

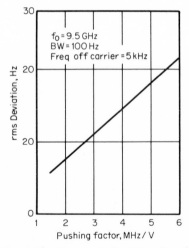

Fig. 137 Noise performance as a function of pushing factor.[39]

A varactor tuning diode has made it possible to tune the Gunn oscillator.   Figure 138 illustrates the manner in which the diode is placed in the waveguide.   Figure 139 describes the equivalent circuit of the oscillator with the varactor.   The fre-

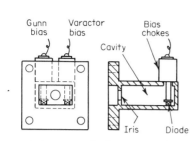

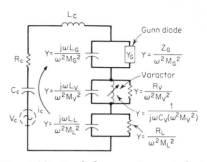

Fig. 138 Gunn-diode cavity with second tuning post for varactor.[39]

Fig. 139 Single-loop series-equivalent circuit for the varactor-tuned Gunn-diode oscillator.[36]

quency is changed because the varactor is a capacitive device whose capacitance varies with voltage. The tuning effect is given by the following relationship:

$$\Delta f \approx f_0 \tfrac{1}{2} \frac{C_c\,M^2_v}{C_v L^2_v} \tag{46}$$

where M = mutual inductance of the subscripted components

There are other ways to tune the oscillator. The two most discussed are the YIG and the ferrite. The YIG is a single-crystal sphere coupled to the diode and tuned by an applied field. The ferrite employs magnetic biasing, and tuning is accomplished by changing the magnetic permeability, thus perturbing the resonant frequency.

Avalanche oscillator The avalanche oscillator consists of a diode constructed by conventional techniques with a *pn* junction. As discussed in an earlier section, the diode is dc-biased beyond its breakdown into the avalanche region. Because of the nonlinear behavior of the diode in this region, under a certain set of conditions the rf current lags the rf voltage by 90°, thus giving negative-resistance characteristics required for oscillation. The load impedance of the oscillator determines whether the oscillation will occur. Figure 140 demonstrates these conditions. At higher load

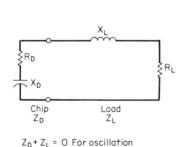

$Z_D + Z_L = 0$ For oscillation

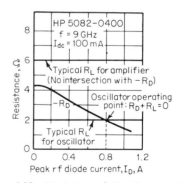

Fig. 140 Equivalent circuit for IMPATT diode chip.[40]

Fig. 141 Variation of R_D (a negative number) with current amplitude I_D. The operating point is determined by the condition $R_L = R_D$.[37]

impedances the device can be made to act as an amplifier. Typical operating voltages for the avalanche to occur are between 70 and 100 V. Figure 141 illustrates the load characteristic for the circuit conditions necessary for oscillation.

The avalanche oscillator illustrates one particular characteristic in its noise performance that is significantly different from the Gunn oscillator. The excess noise near the carrier, or more commonly the i/f noise, is nonexistent for the avalanche device. This is significant in evaluating the FM noise near the carrier in that it simplified calculations made from basic noise theory.

The basic expressions for FM and AM noise are as follows:

$$\Delta f_{rms} = \frac{f_0}{Q_{ext}} \frac{kT \, BM}{P_0} \tag{47}$$

$$(N/C)_{SSB}^{AM} = \frac{\frac{1}{2}(kT \, BM/P_0)}{(s/2)^2 + (Q_{ext} f_m/f_0)^2} \tag{48}$$

where Δf_{rms} = rms noise deviation that would be measured in a bandwidth B at the output of an FM discriminator with the oscillator at its input

$(N/C)_{SSB}^{AM}$ = single-sideband AM noise-to-carrier ratio in a bandwidth B measured at a distance f_m from the carrier

f_0 = oscillation frequency
P_0 = power output of the oscillator
B = measurement bandwidth
k = Boltzmann constant
T = ambient temperature
f_m = distance from carrier (commonly called the modulation frequency)
Q_{ext} = true external Q of the oscillator $\approx Q_{ext}'$
Q_{ext}' = external Q as derived from injection-phase locking measurements
s = a measure of diode nonlinearity; for maximum power output, $s = 2$
M = noise measure of the oscillator

Through the use of approximation, the equation for AM noise can be reduced to

$$(N/C)_{SSB}^{AM} = \frac{\frac{1}{2}(kT \, BM/P_0)}{1 + (Q_{ext} f_m/f_0)^2} \tag{49}$$

Figure 142 demonstrates the behavior of common microwave oscillators. Note that the FM and AM noise stays the same until a frequency of f_0/Q_{ext} is reached, at which point the AM noise begins to fall off at a rate of 6 dB per octave.

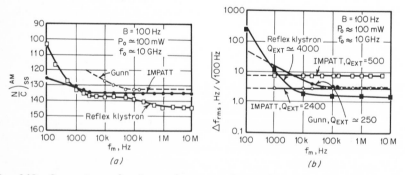

(a)　　　　　　　　　　　　　　　　　*(b)*

Fig. 142 Comparison of noise performance of typical Gunn, reflex klystron, and IMPATT oscillators.[37] (*a*) AM noise of typical X-band Gunn, reflex klystron, and IMPATT oscillators at 100-mW output. (*b*) FM noise performance of typical X-band Gunn, reflex klystron, and IMPATT oscillators with $P_0 \simeq 100$ mW.

The process of noise reduction is not very difficult with solid-state oscillators. In the case of the avalanche oscillator one step to be taken is to match the real impedance properly for maximum power transfer. Another is to build a low-Q oscillator and use rejection-phase locking by means of a low-power low-noise source as in Figure 143. In addition to low noise, frequency stability can be achieved by phase locking.

The extent of any locking circuit's effectiveness is found from the following equation:

$$\Delta f_{rms} = \frac{\Delta f_r{}^2 + \Delta f^2{}_{fro}(f_m/B_L)^2}{1 + (f_m/B_L)^2} \tag{50}$$

where Δf_{rms} = noise deviation of the injection-locked IMPATT oscillator
$\quad\quad \Delta f_{fro}$ = noise deviation of the unlocked (free-running) IMPATT oscillator
$\quad\quad \Delta f_r$ = noise deviation of the low-noise locking signal ("reference" signal)
$\quad\quad B_L$ = locking bandwidth defined in Fig. 144
$\quad\quad f_m$ = modulation frequency

It should be pointed out that phase locking does not affect noise far from the carrier or AM noise.

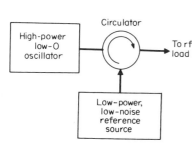

Fig. 143 Injection-phase locking.[37]

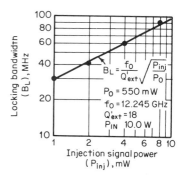

Fig. 144 Typical injection-locking behavior of circuit in Fig. 143.[37] An injected signal of only 10 mW is sufficient to lock the oscillator over a range of ±50 MHz in this case.

Microwave transistor amplifier Although other devices are available, the transistor amplifier is being used increasingly to replace traveling-wave tubes in broadband-amplifier applications with the smaller, more reliable transistor that consumes. The transistor used in the amplifier is of the planar type with one of four types of geometry, the interdigitated, overlay, mesh, and diamond. Different manufacturers use different geometries depending upon the advantages desired. The transistors consist of cells on one slice and usually have emitter resistors on each cell to equalize current distributions. These may be thin-film or diffused on the emitter itself. The better transistors are passivated with silicon nitride. In the selection of a particular ·geometry the tradeoffs are made by considering the following: current capacity is proportional to emitter periphery, base resistance is inversely proportional to emitter width and frequency, and f_T (where hfe goes to unity) is inversely proportional to emitter area. Emitter widths today are at 1 μm.

Four classes of parameters characterize the transistor for use in a microwave amplifier: gain, noise figure, characteristic frequencies, and parasitics. The manufacturer's test method influences the significance of the parameter.[27]

Gain. Several definitions of gain are used in describing the microwave transistor. The maximum available gain is obtained when input and output impedances are matched in both real and imaginary terms and the amplifier is unconditionally stable. The other gain parameter, called unilateral gain U, is obtained when the device is unilateralized with a lossless network and matched at both ports. The general form of gain equation is as follows:

$$G(f) \cong \left[\frac{G^0}{1 + G_0{}^2} \left(\frac{f}{f_m} \right)^4 \right]^{1/2} \tag{51}$$

Figure 145 gives the typical gain responses for different configurations.

Noise Figure. The noise figure has been defined. The contribution to the signal-to-noise ratio through the device is

$$\text{Noise figure } F = \frac{\text{signal-to-noise ratio at output}}{\text{signal-to-noise ratio at input}}$$

The noise-contributing factors are shot noise in the emitter, shot noise in the collector, and thermal noise in the base resistance. Other contributions to be considered are $1/f$ noise and noise developed from parasitics.

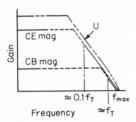

Fig. 145 Maximum available gain and U vs. frequency. Broken line represents regions of potential instability.[27]

Significant Frequencies. As the gain passes through the microwave regions, it passes through a number of phases. The first frequency is known as the lower critical frequency beyond which it becomes conditionally stable. After passing through this region, the gain becomes stable until it passes through $0.1\,f_T$ and falls off at 6 dB per octave. The term f_T is the point at which hfe goes to unity. The fT is determined by transit time from emitter to collector, emitter-base junction capacity charging time, base transit time, collector depletion-layer transit time, and collector capacitance-resistance charging time.

Matching Networks. The transistor is a low-impedance device and must be matched to the transmission line for maximum power transfer. Also, it is necessary to keep the Q's as low as possible to keep energy storage as low as possible. Two techniques are usually used, the networks of low-pass filters and, short-step Chebyshev transformers. The manufacturer of the transistor supplies some help by providing certain impedance and admittance values in the package of the transistor. These values must be carefully controlled for consistent amplifier performance.

Heat Sinking and Proper Mounting. Proper mounting of the transistor is critical to the ultimate performance of the amplifier. First, all leads must be short to reduce parasitic inductances and minimize stresses on the pack-

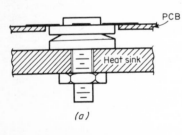

(a)

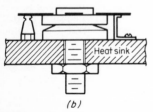

(b)

Fig. 146 Transistor mounting techniques.[22] (*a*) Proper mounting technique using printed-circuit board. (*b*) Proper mounting technique without printed-circuit board.

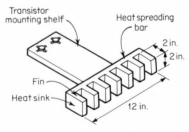

Fig. 147 Transistor heat sinking.[22] $q = hA\ \Delta T$ = heat flow, Btu/h; where h = coefficient of heat transfer, Btu/(h) (ft²) (°F); A = surface area, ft²; ΔT = temperature difference between heat sink and ambient, °F or °R; θ_{HS} = thermal resistance = $1/hA$, h/(°F) (Btu).

age. Second, thermal interfaces must be good to allow heat transfer to the heat sink. It is recommended that a silicone grease such as G.E. Insulgrease be used. Figure 146 illustrates a proper mounting technique.

A typical heat sink is shown in Fig. 147.   The common expression for temperature rise in a transistor is as follows:

Total thermal resistance times power dissipation = temperature rise

$$\Delta T(°C) = \theta(°C/W)P \text{ (watts)} \tag{52}$$

Amplifier devices should be given complete thermal consideration, and it is imperative that calculations be performed by competent heat-transfer techniques.

MICROWAVE TUBES

Throughout the industry two generally accepted classes of microwave tubes are crossed-field devices, including fall magnetrons and crossed-field amplifiers, and linear-beam devices, including klystrons and traveling-wave tubes.

Crossed-Field Tubes

The crossed-field oscillator or magnetron was the first source of transmitted power used in radar at microwave frequencies.   The magnetron, like all crossed-field devices, consists of a cylindrical cathode mounted on the axis of the tube surrounded by a series of cavities calculated for the frequency of the device and a magnetic field in the direction of the axis.   The device is basically a white-noise generator excited into a semicoherent mode by the dimensions of the surrounding cavity.   Once the emissions are started, the electrons set a space charge that bleeds off to the anodes.   Many electrons drift back to the cathode and cause further emission by bombardment.   Therefore, it is not necessary to keep heat on the cathode after it is warmed up to proper emission temperatures.

There are various configurations for the anode structures, the more common being the vane type with strap rings, hole-and-slot type with strap rings, and rising-sun type.   The literature and application notes more than adequately cover use of the magnetron.   However, a few important points are worth noting.

Magnetrons are an economical source of microwave energy and are generally reliable performers in the proper sockets.   However, there are limitations in their performance. · There is lack of uniformity in the rf pulses.   There are leading-edge jitter, amplitude jitter, and definite levels of FM which prevent the tube from being truly coherent.   Solution of the most persistent problem associated with magnetrons, arcing, is apparently as elusive as ever.

Coaxial magnetrons Some of the performance problems with magnetrons can be resolved with the use of a coaxial magnetron.   In the coaxial magnetron, alternate resonators are slot-coupled to a coaxial cavity used for stabilization which operates in the TE_{011} mode.   Figure 148 demonstrates a typical configuration of a coaxial magnetron.   The coaxial magnetron has less tendency to arc because there is a lower field gradient at the cathode.   Also, the efficiency is better without sacrificing pulling figure, as is done with conventional tubes, because tubes are overcoupled to the load.   The pulling figure is three to five times lower for the coaxial magnetron.   The pushing figure is an order of magnitude better.   However, one notable disadvantage of the coaxial tube is that FM spectrum during vibration is significantly worse.

Voltage-tunable magnetron With the use of a control anode, the magnetron can be made voltage-tunable in a fairly linear fashion.   The tube has its greatest usefulness at the lower portion of the microwave spectrum up to C band.   Noise can be a problem if it is not properly resolved between the manufacturer and the user.

Crossed-field amplifiers The crossed-field amplifier (CFA) is characterized by low or moderate gain, high perveance, small size, moderate bandwidth, and high efficiency.   Variations of the performance parameters are available if tradeoffs are made.   CFA's are categorized by two features, the direction of the interaction wave (forward or backward) and the source of emission.   The particular design approach selected for an application is best decided by the manufacturer and is partially a function of the current state of the art (see Fig. 149).

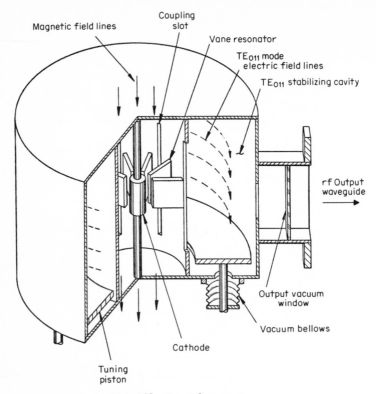

Magnetic field lines

Coupling slot

Vane resonator

TE_{011} mode electric field lines

TE_{011} stabilizing cavity

rf Output waveguide

Output vacuum window

Vacuum bellows

Cathode

Tuning piston

Fig. 148 Coaxial magnetron.

Users of the CFA are most suspicious of the noise performance of the tube. Because of the nature of the electron-velocity distribution the tube is essentially a white-noise generator. The spurious outputs of the tube can be controlled to a large extent by modulating pulse shapes. Fast rise times avoid the problems of low-voltage-mode oscillations. Noise during the interpulse period with no rf drive is commensurate with thermal levels. This is true for both the cold- and hot-cathode tubes. Table 30 lists some levels of actual measurements. When the tube is driven by an rf signal, the noise levels can be quite low. This phenomenon is much like the phase locking of an oscillator. Phase stability in this driver mode is a good feature of the CFA.

Linear-Beam Tubes

Linear-beam tubes include the klystron and the traveling-wave tube, which extract energy from the kinetic energy of a linear electron beam by a phenomenon known as velocity modulation.

Reflex klystron The first tube to be discussed is of course an exception. In this device the beam passes through a single resonator and is reflected back in bunched form by an electronic mirror. The self-oscillation is accomplished by the returning modulated beam, which excites the resonator. Figure 150 demonstrates the reflex oscillator.

The klystron is described as having modes that must satisfy requirements to deliver energy to the resonator. The node as described by the quantity

$$N_m = M - \frac{1}{4} \tag{53}$$

where N is the number of cycles through the center of the gap and return from the repeller. Increasing the repeller voltage increases the efficiency of power transfer

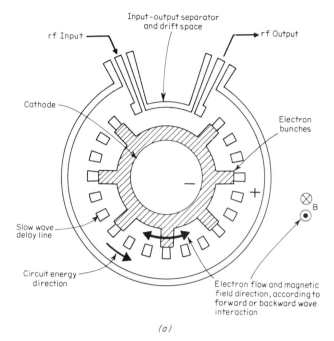

(a)

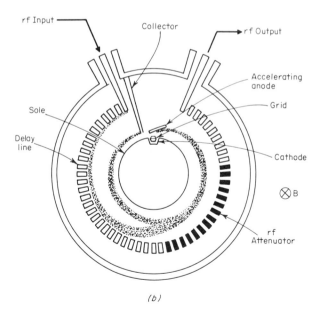

(b)

Fig. 149 Schematic diagrams of crossed-field amplifier.[29] (*a*) Continuous cathode emitting sole CFA, forward or backward wave. (*b*) Injected-beam CFA.

TABLE 30   CFA Intraspectrum PM Noise Measurements[29]

	P_0, kW	S_{ib}/N_{ib},* dB	BW,† Hz	S_0/N_0,‡ dB/ MHz	t_p, μs	PRF, kHz	D_u, dB	S/N,§ dB/ MHz
D band	100	71.8	10	21.8	11	1.3	18.38	40.2
D band	100	67.0	50	24.0	11	1.3	18.38	42.4
D band	100	74.4	150	37.2	30	1.0	15.22	52.4
D band	100	73.0	50	29.0	10	1.0	20.0	49.0
G band	500	69.0	200	32.0	1	5.0	23.0	55.0
G band	660	78.0	3	21.7	20	0.160	25.0	46.7
G band	600	78.9	3	22.6	20	0.150	25.2	47.8
F band	750	83.0	50	39.0	10	4.0	14.0	53.0
F band	750	88.0	50	44.0	10	4.0	14.0	58.0
F band	60	71.0	10	21.0	5.5	1.25	21.6	42.6
F band	60	76.0	10	26.0	5.5	1.25	21.0	47.6
F band	60	70.0	10	19.0	5.5	1.5	20.8	40.0
F band	60	77.0	10	26.0	5.5	1.5	20.8	47.0
F band	666	79.0	3	22.0	29.5	0.189	22.5	44.5
F band	60	76.7	10	25.7	10	0.585	23.3	49.0
F band	60	60.9	50	16.0	33.4	0.585	16.6	32.6

* Measured noise power relative to spectral line power.
† Filter bandwidth for measurement.
‡ Measurements normalized for 1-MHz bandwidth.
§ Equivalent CW signal-to-noise power-density ratio.

through the gap.   Usually the tube operates in three modes.   More detailed examination of klystron theory and admittance plots is necessary to see the reasons for this.

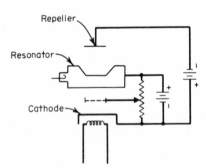

Fig. 150   Reflex-klystron schematic.

The more efficient the operating mode the greater the magnitude of the modulation sensitivity.   Power output and modulation sensitivity are frequently traded off.
   Klystron amplifier   The klystron amplifier has been the workhorse of the coherent radar industry for years.   It consists of several cavities, usually four or five, which accomplish the velocity modulation of the beam.   This type of device achieves an energy transformation from the kinetic energy of the beam in contrast to crossed-field devices, which use potential energy.   Figure 151 illustrates the cross section of a typical klystron.   Klystrons employ a very high-Q cavity in the modulation process and are narrow-bandwidth devices.   The collector catches the electron beam and must dissipate the unused energy.   Some tubes have a negative voltage on the collector to achieve a greater tube efficiency by decelerating the beam before impact.   Focusing of the klystron has to be achieved by a coaxial magnetic field.   This can

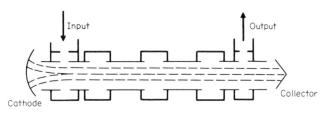

Fig. 151 Klystron-amplifier schematic.

be done by permanent-magnet focusing, periodic permanent-magnet focusing, and solenoid focusing. The trade-offs are usually made on a weight and economy basis. For the PM and PPM devices the long-term stability of the magnets is very important for reliability, since degradation of magnetic-field density causes defocusing.

Klystrons are a good source of high power if adequate cooling is available. The inherent efficiency of the device is normally 40 to 50 percent.

The weak link in the klystron is in the output windows. Two types are used most frequently: a circular ceramic disk with a transducer to rectangular guide, and a half-wavelength ceramic window that can be used directly in rectangular waveguides. The former has a better bandwidth response, usually 30 percent, and the latter is limited to 15 percent. Care must be taken to avoid mechanical stresses on the waveguide near the window. Arc protection in the waveguide line is mandatory because arcs travel to the source of power and will cause the window to crack. Noise contributed to amplified signals is related to the amplification of the shot noise caused by amplitude and velocity fluctuations of the beam current. The way to a low-noise klystron is tight control of the beam geometry. The noise requirements for the klystron are frequently specified in terms of power density per bandwidth. The power level is referred to in dB below the peak value of the carrier spectral line. It is important that the actual spectrum of interest be identified. Because klystrons are narrow-banded, harmonics of the amplified signals are seldom of any concern. The klystron user must be concerned with power-supply regulation. The manufacturer can supply voltage-pushing figures for each type of klystron.

One klystron parameter is perveance u, defined by the following equations:

$$\frac{I_0}{(V_0)^{3/2}} = u \tag{54}$$

The value of perveance determines the amount of voltage to be applied to the control electrode. Figure 152 illustrates some different types of control electrodes, and Table 31 shows some of the relative evaluations of each. The most advanced of these

TABLE 31 Typical Characteristics of Control Electrodes[28]

Type	μ	μ_c	Capacity, pF	Grid interception, %	Focusing at low voltage
Modulating anode	1–3		50	0	Good
Control-focus electrode	2–10	2–10	100	0	Poor
Intercepting grid	50	100	50	15	Fair
Shadow grid	30	300	50	0.1	Fair

control schemes is the shadow grid, which makes low-voltage pulsers realizable. Figure 153 illustrates a gun design with the shadow grid.

Advances in today's technology are being made in the effort to reduce weight, for instance, the use of PPM focusing and samarium cobalt. Figure 154 shows a

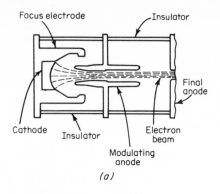

(a)

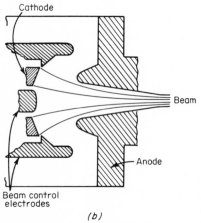

(b)

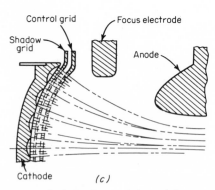

(c)

Fig. 152 Electron-gun control electrodes.   (a) Modulating anode gun.   (b) Control-focus electrode gun.   (c) Shadow-gridded gun.[28]

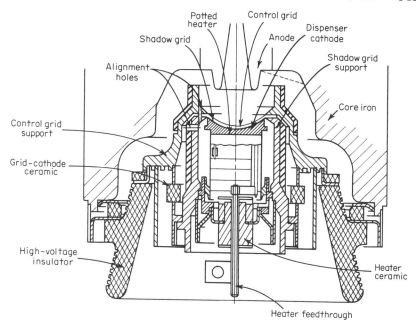

Fig. 153 Nonintercepting gridded gun. (*Varian, Inc.*)

Fig. 154 Klystrons with Alnico 5 and samarium cobalt magnets (VA880G).

Fig. 155 Klystrons with PM and PPM focusing.

tube with an alnico V magnet and a samarium cobalt design. Figure 155 illustrates a design with PPM focusing.

Traveling-wave tube The traveling-wave tube was invented by Kompfner at Birmingham and Oxford during the Second World War. Postwar developments by Pierce at the Bell Laboratories made the tube a useful concept.[30] The operation of the twt (see Fig. 156) involves the introduction of an electric beam with an rf field traveling on a concentric helix. As the electrons enter the helix structure,

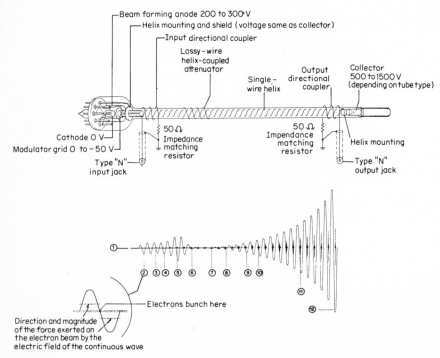

Fig. 156 Operation of traveling-wave tube (twt).

the varying polarity of the rf field at the entrance initiates the bunching of the electron beam. The relative speed of the beam with the field wave determines whether the beam will attenuate or amplify the field wave. If the beam travels slightly more slowly than the field wave, the field wave will be attenuated, usually by a negligible value somewhere between one-third and one-half of the way along the length of the helix.

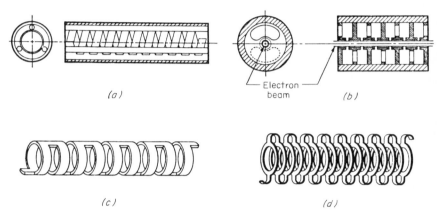

Fig. 157 Traveling-wave tube wave structures. (*a*) Helix. (*b*) Coupled cavity. (*c*) Ring bar. (*d*) Ring loop.

Solutions to the propagation equation show that, in order to satisfy the boundary conditions, there will be four waves, three forward and one backward.[30] Only one forward wave is amplified, which means the only waves of great concern are the amplified forward wave and the backward wave. In order to avoid the possibility of oscillation, an attenuator element is added to the tube at the point where the forward voltage wave is attenuated to a negligible value. This attentutor thus attentuates the backward wave. After energy is coupled out of the tube at the rf output

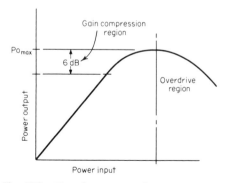

Fig. 158 Traveling-wave tube gain response.

connection, the kinetic energy of the beam is dissipated in the collector region. One limitation in the use of traveling-wave tubes has been in high-power applications. The helix structure is able to dissipate only so much energy within the confines of the envelope of the tube and can handle only a limited amount of high voltage. The ring bar (see Fig. 157) and the ring loop were used to handle the voltage problem owing to the high synchronous-beam voltages developed. However, the power limitation still existed with these devices.   The coupled-cavity scheme is radical but alleviates the power-dissipation problem.   Although these approaches sound familiar, there are differences in the klystron.   In the klystron there is significant coupling, while the twt has loosely coupled cavities.   When operated as an amplifier, the twt has some important features to be considered by the designer.   First, when the tube gain curve is examined, a nonlinear or saturated region is observed.   The gain-compression region is generally defined as the point 6 dB down from the maximum-gain point (see Fig. 158).

The twt in saturated gain is a harmonic generator, and it is important to control the waveguide at these harmonics in sensitive broadband applications.   Also, if the input signals are not linear, their harmonics will also be amplified, which may require the use of harmonic filters.   Examination of the gain curves shows that there is some notable variation in the response curve.   These variations are called gain-flatness and fine-gain variations.   The gain flatness of an octave-band tube can be typically 4 dB.   The fine-gain variation can be held to ±0.5 dB.   The problems of controlling these parameters have to do with matching the output circuits.

Noise in the traveling-wave tube can be treated in the same manner as in the klystron.   Starting with the thermal noise at the terminated input (-174 dB/Hz), the noise power is found by the following relationship:

$$NP_0 = -174 \text{ dB/Hz} + \log_{10}(\text{BW}) + G_{SS} + (\text{NF}) \tag{55}$$

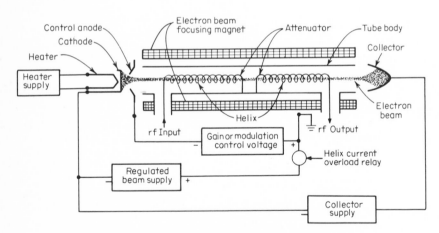

Fig. 159   Traveling-wave tube power-supply schematic.

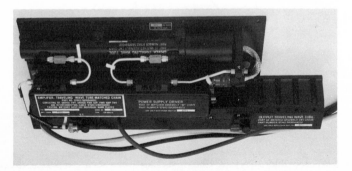

Fig. 160   Traveling-wave tube amplifier chain.   (*Varian, Inc.*)

Modulation of the traveling-wave tube can be accomplished in the same manner as in the klystron, by cathode, anode, and noninterfering gridded tubes. Shadow-gridded tubes for high-power applications are available. Figure 159 show a typical schematic for a typical schematic for a traveling-wave tube. Figure 160 illustrates how twt's may be supplied as a chain to control overall gain performance. In this chain equalizers are used to match the drive and output tube.

Gas Control Devices

In order to limit leakage energy on the transmitter incident on receiver diodes, gas-discharge devices are used. The gas is usually an argon–water vapor mixture. For devices requiring extremely rapid recovery times highly electronegative gases in quartz cylinders may be used followed by argon–water vapor cells and finally a diode limiter. The most common approach in today's radar systems is to use a ferrite duplexer as shown by Figure 161.

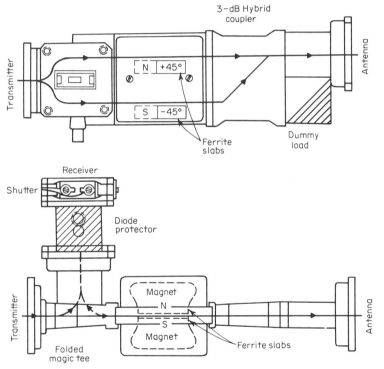

Fig. 161 Ferrite duplexer. Nonreciprocal differential phase-shift ferrite duplexer with diode protector and shutter, shown in the high-level (transmit) condition.

MICROWAVE MEASUREMENTS

Almost all microwave measurements consist of measuring either power or frequency. The decision will depend upon the speed of measurement, economics, and available accuracy.

Power

Power is usually measured by a transfer device such as a diode or thermally variable resistors. In the case of the diode it can be shown that the diode transfer function has a dc component that is proportional to the square of the barrier voltage in the diode. This square-law characteristic is the basis of most detector functions. The square-law response is limited in power up to a few milliwatts, when the diode becomes saturated. For measurements above this level it is necessary to use calibrated substitution methods.

Thermistors and barretters are devices that are frequently used with power bridges. The thermistor is a negative-coefficient device with long thermal time constant. The barretter has a short thermal time constant with a positive thermal coefficient. Both mount and element are calibrated with a correction factor that is related to the direct current. This is due to the vswr of the mount. For measurement at high-power levels calibrated attenuators and couplers are used. However, it is very important to make all matches with as low a vswr as possible to eliminate errors. Figure 162 presents the dynamic range for square-law response for some typical power-sensing transducers.

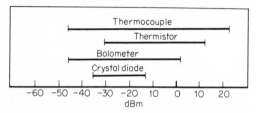

Fig. 162 Dynamic range of various power-measurement devices in the square-law region.

One other scheme that is often used for high-power measurements is to dump the energy into a water load and calculate the input energy by measuring the ΔT of the water. A variation is to measure the dc power needed to heat the water to an equivalent ΔT with a heating element. Often this is called a comparison method.

Measurement of power can also provide other pieces of information relative to the performance of the component under test. For instance, measurement of forward

TABLE 32   Summary of Maximum System-Measurement Errors[*, 31]

Attenuation range, dB	Single-channel audio substitution, dB	Dual-channel audio substitution, dB	Parallel IF substitution, dB	DC substitution, dB	RF substitution, dB
0–1	0.08	0.01	0.06	0.062	0.15
10	0.1	0.03	0.072	0.076	0.20
20	0.11	0.04	0.074	0.22	0.28
30	0.16	0.05	0.076		0.38
40			0.078		0.40
50			0.08		0.48
60			0.082		
70			0.084		
80			†		
90			†		
100			†		

* This table shows the sum of all errors in a particular measuring system as described in the text. In practice, the individual errors seldom all add in any single measurement. The probable system errors are therefore much smaller than the errors shown above. The table shows only the *system errors* and does not include the *microwave equipment errors* as defined in the text.

† Increases rapidly into noise.

power, reflected power, and power passed through the device can reveal the gain of the device (positive and negative) and the impedance matches at the input (vswr). Measurement can be made from a system-technique viewpoint in a variety of ways. The most direct way is to make measurements directly into the power system. Another method is to use direct substitution with calibrated parts. A third scheme is to use lower-frequency substitution schemes, often referred to as audio or IF substitution methods. Table 32 summarizes the accuracy available with these different methods.

Frequency

Frequency measurements are made by using cavity-type frequency meters in conjunction with a detector and oscilloscope. The input rf signal is pulsed at 1 kHz

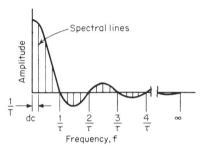

Fig. 163 Spectrum of perfectly rectangular pulse.[32] The envelope of this plot follows a function of the basic form $y = \sin x/x$. Amplitudes and phases of an infinite number of harmonics are plotted, resulting in a smooth envelope as shown.

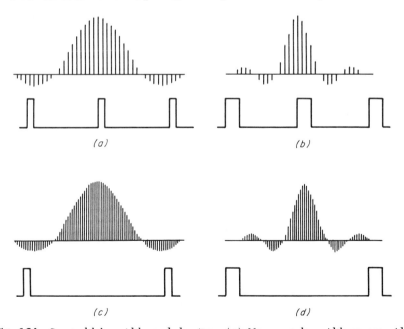

Fig. 164 Spectral-lobe widths and density. (a) Narrow pulse width causes wide-spectrum lobes; high prf results in low spectral-line density. (b) Wider pulse than (a) causes narrower lobes, but line density remains constant since prf is unchanged. (c) prf lower than (a) results in higher spectral density. Lobe width is same as (a), since pulse widths are identical. (d) Spectral density and prf unchanged from (c), but lobe widths are reduced by wider pulse.

and detected on the oscilloscope. When the frequency meter is tuned to operating frequency, a dip is observed on the oscilloscope. The accuracy of this scheme is about 0.02 percent.

Other measurement schemes include using phase comparators with stabilized frequency sources. Such a source may be generated from fixed crystals and multiplied up to the desired frequencies. The most useful frequency-measuring device is the spectrum analyzer. The analyzer is basically a receiver with oscilloscope indicator that displays Fourier transforms of the input frequency spectrum. The Fourier transform breaks periodic functions into their basic sinusoidal components. Appendix V gives a brief description and lists some properties of the transforms.

The envelope of the display follows the function $y = \text{sine } X/X$. The display consists of a series of lines separated in time by 1/prf. The intersection with the $y = 0$ axis is determined by the width of the pulse (see Figs. 163 and 164). For an excellent treatment on the use of the spectrum analyzer, see Ref. 39.

ELECTROMAGNETIC INTERFERENCE

Although properly a subject by itself, electromagnetic interference (EMI) has sufficient peripheral interaction with microwave components to merit a limited discussion.

Problems

EMI generally breaks down into three categories, electrical interference, and plane-wave interference. Electrical interference is the type associated with automotive ignitions or high-impedance sources, and magnetic interference is associated with such things as rotating electrical machinery or low-impedance sources, neither of which is of any concern here. Plane-wave interference is of interest in the microwave spectrum because it exists in the propagation region of air.

There are two concerns in EMI in its interaction with components. First there is radiated interference that can affect other parts. Normally this is caused by a hole or discontinuity that allows an electric field to be set up and causes radiation in a manner similar to a slot antenna. The source of energy is from within the part. The obvious cure is to make sure that there are perfectly conducting surfaces at all interfaces and to provide for rf chokes at points where it is difficult to obtain good interfaces. These chokes are usually quarter-wave shorts that present short circuits at the interface in the transmission line.

The second area of concern is in the susceptibility of the circuits to external fields. In complete waveguide systems the problem is usually encountered in biasing circuits or something similar where wires are exposed. The solution is to provide adequate filtering at the terminals of the component.

Shielding

EMI shield-excluding filtering in lines consists of using a material that absorbs and/or reflects the incident energy. The degree in dB to which this is done is known as shielding effectiveness. The reflection occurs from the incident surface and on the inside of the exit surface. However, when the attenuation is greater than 10 dB, secondary surface reflections are disregarded.

Absorption or attenuation is computed from the formula

$$A = 3.34t \sqrt{fG\mu} \tag{56}$$

where A = attenuation, dB
t = thickness of material, in. $\times 10^{-3}$
f = frequency, MHz
G = conductivity relative to copper
μ = relative magnetic permeability

The amount of reflection is determined by the following relationship:

$$R = 20 \lpg \frac{Z_w}{4Z_s} \tag{57}$$

TABLE 33 Properties of Electromagnetic Waves in Biological Media[42]

Muscle, skin, and tissues with high water content

Frequency, MHz	Wavelength in air, cm	Dielectric constant ϵ_H	Conductivity σ_H, mho/m	Wavelength λ_H, cm	Depth of penetration, cm	Reflection coefficient			
						Air–muscle interface		Muscle–fat interface	
						r	ϕ	r	ϕ
1	30,000	2,000	0.400	436	91.3	0.982	+179		
10	3,000	160	0.625	118	21.6	0.956	+178		
27.12	1,106	113	0.612	68.1	14.3	0.925	+177	0.651	−11.13
40.68	738	97.3	0.693	51.3	11.2	0.913	+176	0.652	−10.21
100	300	71.7	0.889	27	6.66	0.881	+175	0.650	−7.96
200	150	56.5	1.28	16.6	4.79	0.844	+175	0.612	−8.06
300	100	54	1.37	11.9	3.89	0.825	+175	0.592	−8.14
433	69.3	53	1.43	8.76	3.57	0.803	+175	0.562	−7.06
750	40	52	1.54	5.34	3.18	0.779	+176	0.532	−5.69
915	32.8	51	1.60	4.46	3.04	0.772	+177	0.519	−4.32
1,500	20	49	1.77	2.81	2.42	0.761	+177	0.506	−3.66
2,450	12.2	47	2.21	1.76	1.70	0.754	+177	0.500	−3.88
3,000	10	46	2.26	1.45	1.61	0.751	+178	0.495	−3.20
5,000	6	44	3.92	0.89	0.788	0.749	+177	0.502	−4.95
5,800	5.17	43.3	4.73	0.775	0.720	0.746	+177	0.502	−4.29
8,000	3.75	40	7.65	0.578	0.413	0.744	+176	0.513	−6.65
10,000	3	39.9	10.3	0.464	0.343	0.743	+176	0.518	−5.95

TABLE 34 Properties of Electromagnetic Waves in Biological Media[42]

Fat, bone, and tissues with low water content

Frequency, MHz	Wavelength in air, cm	Dielectric constant ϵL	Conductivity σL, mmho/m	Wavelength λL, cm	Depth of penetration, cm	Reflection coefficient Air-fat interface r	Air-fat interface ϕ	Fat-muscle interface r	Fat-muscle interface ϕ
1	30,000								
10	3,000								
27.12	1,106	20	10.9–43.2	241	159	0.660	+174	0.651	+169
40.68	738	14.6	12.6–52.8	187	118	0.617	+173	0.652	+170
100	300	7.45	19.1–75.9	106	60.4	0.511	+168	0.650	+172
200	150	5.95	25.8–94.2	59.7	39.2	0.458	+168	0.612	+172
300	100	5.7	31.6–107	41	32.1	0.438	+169	0.592	+172
433	69.3	5.6	37.9–118	28.8	26.2	0.427	+170	0.562	+173
750	40	5.6	49.8–138	16.8	23	0.415	+173	0.532	+174
915	32.8	5.6	55.6–147	13.7	17.7	0.417	+173	0.519	+176
1,500	20	5.6	70.8–171	8.41	13.9	0.412	+174	0.506	+176
2,450	12.2	5.5	96.4–213	5.21	11.2	0.406	+176	0.500	+176
3,000	10	5.5	110–234	4.25	9.74	0.406	+176	0.495	+177
5,000	6	5.5	162–309	2.63	6.67	0.393	+176	0.502	+175
5,900	5.17	5.05	186–338	2.29	5.24	0.388	+176	0.502	+176
8,000	3.75	4.7	255–431	1.73	4.61	0.371	+176	0.513	+173
10,000	3	4.5	324–549	1.41	3.39	0.363	+175	0.518	+174

where R = reflection, dB

Z_w = impedance of wave at shield

Z_s = impedance of the shield (usually CCZ_w)

Further analysis of the above reveals the following:
- Reflection waves with wave impedance.
- Wave impedance depends upon source impedance and distance from load.
- R depends upon source impedance and distance from source.

In order to select the proper sort of gasketing, the designer should have some information regarding the following:
1. Maximum joint unevenness
2. Available compression
3. Shielding required
4. Mating material

Two classes of gaskets are used in microwave structures: metal gaskets consisting of solid metal or metal-mesh screen interfaces and elastomer metal-filled gaskets. Manufacturers' catalogs provide pertinent data for consideration by a designer.

SAFETY HAZARDS

Most theoretical approaches to safety hazards break the harmful effects of electromagnetic radiation into two categories: effects from high-level radiation due to thermal heating and effects from low-level radiation not due to thermal heating. The treatment of these effects covers the frequency spectrum from 1 MHz to 100 GHz, with the greatest attention given to the highly populated range from 300 to 10,000 MHz.

Tremendous conflict exists today as to what are safe levels of electromagnetic radiation. The USSR advocates a level of 0.01 mWcm^{-2}, while the United States subscribes to a level of 10 mW cm^{-2}. Medical usage has prescribed levels up to 590 mWcm^{-2} in diathermy treatment.[42] If death occurs, it is usually the result of breakdown of the body-temperature-regulating system. VSWR's can cause localized heating, which causes hot-spot burns. Areas of poor circulation such as the lens of the eye, the gall bladder, and parts of the gastrointestinal tract are particularly susceptible. Because radiation can penetrate, burns can be quite deep. Some parts of the body can suffer more owing to greater destructive effects of small temperature rises. Russian work and recent efforts in the United States have shown effects on the central nervous system and possible effects on the brain that can cause behavior change.

Much of the work on these low-level effects is still undergoing analysis and conclusion. Work on animals is difficult to project onto effects on human beings because the propagation medium and paths change so radically. Regulations in industry today are controlled by the Radiation Safety Acts of 1968 and the Occupational Safety and Health Act of 1971. Tables 33 and 34 list some useful data on electromagnetic radiation in biological media.

REFERENCES

1. Thomas, Harry E.: "Handbook of Microwave Techniques and Equipment," © 1972. Reprinted by permission of Prentice-Hall, Inc., Englewood Cliffs, N.J.
2. Reich, Herbert J., et al.: "Microwave Theory and Techniques," Boston Technical Publishers, Inc., 1965.
3. Ramo, Simon, et al.: "Fields and Waves in Modern Radio," Wiley, New York, 1953.
4. Glasser, A.: Electric and Magnetic Fields, *Electro-Technology*, 1963.
5. Barney, O.: Waves on Transmission Lines, Hewlett-Packard Application Note 16.
6. S-Parameter Circuit Analysis and Design, Hewlett-Packard Application Note 95.
7. Microwave Engineers Technical and Buyers Guide Edition, *Microwave J. Int.*, February 1969.
8. Montgomery, C.: "Techniques of Microwave Measurements," MIT Radiation Laboratory Series, vol. 11, pp. 863–866, McGraw-Hill, New York, 1948.
9. Gullemin, E. P.: "Mathematical Circuits Analysis Synthesis of Passive Networks," Wiley, New York, 1949.

10. Gourse, S. J., et al.: How to Use Lumped Constants in Microwave Circuit Design, *Microwaves*, March 1968.
11. Harper, Charles A.: "Handbook of Wiring, Cabling, and Interconnecting for Electronics," McGraw-Hill, New York, 1972.
12. Ciavolella, J.: Take the Hassel out of High Power Design, *Microwaves*, June 1972.
13. Bayliss, R., et al.: Why a Schottky-Barrier, Why a Point Contact, *Microwaves*, March 1968.
14. Klein, G.: Thermal Resistancy Table, Simplified Temperature Calculations, *Microwaves*, February 1970.
15. Heins, H.: How Much Peak Power Can Switching Diodes Handle? *Microwaves*, January 1966.
16. Reid, M. J.: Microwave Switches and Attenuator Modules, *Microwave J.*, July 1973.
17. Gullemin, E. A.: Modern Filter Design Part 1—Impedance Behavior and S-Plane Geometry, *EEE*, February 1965.
18. Wainwright, R. A.: Filters in Communications, *Telecommunications*, November 1968.
19. Silver, R. L.: Save Space with Strip Line Design, *Microwaves*, September 1970.
20. Richardson, J. K.: Design for Stripline Band-Pass Filters, *Microwaves*, 1968.
21. Microwave Band Pass Filters, *Mil. Syst. Des.*, September–October 1968.
22. Johnson, J.: "Solid Circuits," Communications Transistor Corp., March 1973.
23. Gilden, M., et al.: "Handbook on High Power Capability of Waveguide Systems," Contact Nobson - 85190, June 1963.
24. Saad, T.: "The Microwave Mixer," Sage Laboratories, 1966.
25. IRE Standards on Electron Tubes, Methods of Testing, 1962.
26. Vossberg, W. A.: Stripping the Mystery from Strip-Line Laminates, *Microwaves*, January 1968.
27. Cooke, H. F.: Microwave Transistor: Theory and Design, *Proc. IEEE*, vol. 59, pp. 1163–1181, August 1971.
28. Strapans, A., et al.: High Power Linear-Beam Tubes, *Proc. IEEE*, vol. 61, March 1973.
29. Skowren, J. F.: The Continuous-Cathode (Emitting-Sole) Crossed Field Amplifier, *Proc. IEEE*, vol. 61, March 1973.
30. Beck, A. H. W.: "Thermionic Valves, Their Theory and Design," Cambridge, London, 1953.
31. Ebert, J. E., et al.: Survey of Precision Microwave Attenuation Measuring Techniques, *ISA Trans.*, vol. 3, July, 1964.
32. Oliver, B. M.: Square Wave and Pulse Testing of Linear Systems, Hewlett-Packard Application Note 17, 1954.
33. Products Seminar, Microwave Associates, Inc., 1972.
34. Clark, F. M.: The Newer Insulating Gases, *Mater. Des. Eng.*, February 1961, pp. 95–100.
35. Blanchard, W. C.: Transmission-Line Slide Rule, *Microwaves*, April 1969, 46.
36. "Gunn Diode Circuit Handbook," HB-9000, Microwave Associates, Inc., February 1971.
37. Microwave Power Generation and Amplification Using IMPATT Diodes, Hewlett-Packard Application Note 935.
38. Siegal, B.: Practical Guide to Microwave Semi-Conductors, *Microwave Syst. News*, August–September 1972.
39. Spectrum Analysis, Hewlett-Packard Application Note 150–21, November 1971.
40. Tehrfeld: Hints and Kinks for Your MIC Package Design, *Microwaves*, March 1973.
41. Cohn, S. B.: Problems in Strip Transmission Lines, *IRE Trans.*, vol. MTT-3(2), pp. 119–126, March 1955.
42. Johnson, C., et al.: Nonionizing Electromagnetic Effects in Biological Materials and Systems, *Proc. IEEE*, vol. 60, no. 6, June 1972.
43. Georgopoulos, Christos J.: Pin Driver Design, *Microwaves*, August 1972.
44. Verfino, J. W.: Computer Programs for Smith-Chart Solutions, *Microwaves*, September 1968.
45. Blume, H. C.: Input Impedance Display of One-Port Reflection Type Amplifiers, *Microwave J.*, November 1967.
46. Kurokawa, U.: Power Waves and the Scattering Matrix, *IEEE Trans.*, 5, vol. MTT-13, no. 2, March 1965.
47. Harvey, A. F.: Mechanical Design and Manufacture of Microwave Structures, *IRE Trans. Microwave Theory Tech.*, 1959.

APPENDIX I   LIST OF SYMBOLS[i]

NOTE: Boldface symbols indicate that the quantities are vectors.   Boldface symbols in italics indicate complex vectors.

$\mathbf{A}$ = magnetic-vector potential

$\mathbf{B}$ = magnetic induction

$\mathbf{D}$ = electric displacement

$\mathbf{E}$ = electric field vector

$\boldsymbol{\mathcal{E}}$ = electromotive force

f = frequency

$\mathbf{F}$ = force vector

$\mathbf{H}$ = magnetic-field-intensity vector

i = electric current

$\mathbf{i}$ = unit vector in rectangular-coordinate system

$\mathbf{j}$ = unit vector in rectangular-coordinate system

$\mathbf{J}$ = current-density vector

$\mathbf{J}_m$ = magnetization current density

k = dielectric constant

$\mathbf{k}$ = unit vector in rectangular-coordinate system

$\mathbf{m}$ = magnetic-dipole moment

$\mathbf{M}$ = macroscopic magnetic-dipole moment

$\mathbf{n}$ = unit vector normal to a surface

$\mathbf{p}$ = electric-dipole moment

$\mathbf{P}$ = macroscopic electric-dipole moment

$\mathbf{P}_m$ = electric-dipole moment of a molecule

$\mathbf{P}_n(\theta)$ = Legendre polynomial

q = electric charge

$\mathbf{r}$ = position vector

ds = incremental surface area

$d\mathbf{s}$ = incremental surface area in vector sense

S = Poynting's vector

T = transmission factor

w = energy density

W = energy

Z_0 = characteristic impedance of transmission line

Z_w = wave impedance

β_0 = propagation vector

γ = complex propagation vector

Γ = reflection factor

δ = skin depth

ϵ = permittivity

ϵ_0 = permittivity of free space

ϵ_c = complex dielectric constant

η = intrinsic impedance

λ = line-charge density

λ = wavelength

μ = permeability

μ_0 = permeability of free space

ρ = volume charge density

ρ_p = volume polarization charge density

σ = surface charge density

σ_p = surface polarization charge density

ϕ = electrostatic potential

ϕ_m = magnetic scalar potential

Φ = magnetic flux

χ = electric susceptibility

χ_m = magnetic susceptibility

ω = radian frequency

div = divergence, a vector operation

grad = gradient, a vector operation

∇ = del or nabla, a vector differential operator

∇^2 = the Laplacian

∇^2 = the vector Laplacian

APPENDIX II   USEFUL FORMULAS FROM VECTOR ANALYSIS[i]

$$\mathbf{A} \cdot \mathbf{B} = A_x B_x + A_y B_y + A_z B_z$$

$$\mathbf{A} \times \mathbf{B} = \begin{vmatrix} \mathbf{i} & \mathbf{j} & \mathbf{k} \\ A_x & A_y & A_z \\ B_x & B_y & B_z \end{vmatrix}$$

$$\mathbf{A} \cdot (\mathbf{B} \times \mathbf{C}) = \mathbf{B} \cdot (\mathbf{C} \times \mathbf{A}) = \mathbf{C} \cdot (\mathbf{A} \times \mathbf{B})$$

$$\mathbf{A} \times (\mathbf{B} \times \mathbf{C}) = (\mathbf{A} \cdot \mathbf{C})\mathbf{B} - (\mathbf{A} \cdot \mathbf{B})\mathbf{C}$$

curl grad $\phi = \nabla \times \nabla \phi = 0$

div curl $\mathbf{A} = \nabla \cdot (\nabla \times \mathbf{A}) = 0$

$\nabla(\phi + \psi) = \nabla \phi + \nabla \psi$

$\nabla \phi \psi = \phi \nabla \psi + \psi \nabla \phi$

div $(\mathbf{A} + \mathbf{B})$ = div $\mathbf{A}$ + div $\mathbf{B}$

curl $(\mathbf{A} + \mathbf{B})$ = curl $\mathbf{A}$ + curl $\mathbf{B}$

div $\phi \mathbf{A} = \phi$ div $\mathbf{A} + \mathbf{A} \cdot$ grad ϕ

div $(\mathbf{A} \times \mathbf{B}) = \mathbf{B} \cdot$ curl $\mathbf{A} - \mathbf{A} \cdot$ curl $\mathbf{B}$

curl $\phi \mathbf{A} = \phi$ curl $\mathbf{A} + \nabla \phi \times \mathbf{A}$

Definition of the vector Laplacian:

$$\text{curl curl } \mathbf{A} = \text{grad div } \mathbf{A} - \nabla^2 \mathbf{A}$$

Gauss' theorem:

$$\int_{\text{surface}} \mathbf{A} \cdot d\mathbf{s} = \int_{\text{volume}} \text{div } \mathbf{A} \, dv$$

Stokes' theorem:

$$\int_{\text{contour}} \mathbf{A} \cdot d\mathbf{l} = \int_{\text{surface}} \text{curl } \mathbf{A} \cdot d\mathbf{s}$$

Green's theorem:

$$\int_{\text{volume}} (\psi \nabla^2 \phi - \phi \nabla^2 \psi) dv = \int_{\text{surface}} \psi \text{ grad } \phi - \phi \text{ grad } \psi) \cdot d\mathbf{s}$$

APPENDIX III   VECTOR OPERATIONS IN ORTHOGONAL COORDINATES[1]

In rectangular coordinates:
x, y, and z are the orthogonal coordinates.
$\mathbf{i}$, $\mathbf{j}$, and $\mathbf{k}$ are the unit vectors in the x, y, and z directions.

$$\nabla \phi = \text{grad } \phi = \mathbf{i} \frac{\partial \phi}{\partial x} + \mathbf{j} \frac{\partial \phi}{\partial y} + \mathbf{k} \frac{\partial \phi}{\partial z}$$

$$\nabla \cdot \mathbf{A} = \text{div } \mathbf{A} = \frac{\partial \mathbf{A}}{\partial x} + \frac{\partial \mathbf{A}}{\partial y} + \frac{\partial \mathbf{A}}{\partial z}$$

$$\nabla \times \mathbf{A} = \text{curl } \mathbf{A} = \mathbf{i} \left(\frac{\partial A_z}{\partial y} - \frac{\partial A_y}{\partial z} \right) + \mathbf{j} \left(\frac{\partial A_x}{\partial z} - \frac{\partial A_z}{\partial x} \right) + \mathbf{k} \left(\frac{\partial A_y}{\partial x} - \frac{\partial A_x}{\partial y} \right)$$

$$\nabla^2 \phi = \text{grad div } \phi = \frac{\partial^2 \phi}{\partial x^2} + \frac{\partial^2 \phi}{\partial y^2} + \frac{\partial^2 \phi}{\partial z^2}$$

In circular cylindrical coordinates:
r, φ, and z are the circular cylindrical coordinates.
$\mathbf{i}_r$, $\mathbf{i}_\varphi$, and $\mathbf{i}_z$ are the corresponding unit vectors.

$$\text{grad } \phi = \nabla \phi = \mathbf{i}_r \frac{\partial \phi}{\partial r} + \mathbf{i}_\varphi \frac{\partial \phi}{\partial \varphi} + \mathbf{i}_z \frac{\partial \phi}{\partial z}$$

$$\text{div } \mathbf{A} = \nabla \cdot \mathbf{A} = \frac{1}{r} \frac{\partial}{\partial r} (r A_r) + \frac{1}{r} \frac{\partial A_\varphi}{\partial \varphi} + \frac{\partial A_z}{\partial z}$$

$$\text{curl } \mathbf{A} = \nabla \times \mathbf{A} = \mathbf{i}_r \left(\frac{1}{r} \frac{\partial A_z}{\partial \varphi} - \frac{\partial A_\varphi}{\partial z} \right) + \mathbf{i}_\varphi \left(\frac{\partial A_r}{\partial z} - \frac{\partial A_z}{\partial r} \right) + \mathbf{i}_z \left(\frac{1}{r} \frac{\partial}{\partial r} (r A_\varphi) - \frac{1}{r} \frac{\partial A_r}{\partial \varphi} \right)$$

$$\nabla^2 \phi = \text{grad div } \phi = \frac{1}{r} \frac{\partial r}{\partial} \left(r \frac{\partial \phi}{\partial r} \right) + \frac{1}{r^2} \frac{\partial^2 \phi}{\partial \varphi^2} + \frac{\partial^2 \phi}{\partial z^2}$$

In spherical coordinates:
r, θ, and φ are the spherical coordinates.
$\mathbf{i}_r$, $\mathbf{i}\theta$, and $\mathbf{i}\varphi$ are the corresponding unit vectors

$$\nabla \phi = \text{grad } \phi = \mathbf{i}_r \frac{\partial \phi}{\partial r} + \mathbf{i}_\theta \frac{1}{r} \frac{\partial \phi}{\partial \theta} + \mathbf{i}_\varphi \frac{1}{r \sin \theta} \frac{\partial \phi}{\partial \varphi}$$

$$\nabla \cdot \mathbf{A} = \text{div } \mathbf{A} = \frac{1}{r^2} \frac{\partial r}{\partial} (r^2 A_r) + \frac{1}{r \sin \theta} \frac{\partial}{\partial \theta} (A\theta \sin \theta) + \frac{1}{r \sin \theta} \frac{\partial A\varphi}{\partial \varphi}$$

$$\nabla \times \mathbf{A} = \text{curl } \mathbf{A} = i_r \left[\frac{1}{r \sin \theta} \frac{1}{\partial \theta} (A\varphi \sin \theta) - \frac{1}{r \sin \theta} \frac{\partial A_\theta}{\partial \varphi} \right]$$

$$+ i_\theta \left[\frac{1}{r \sin \theta} \frac{\partial A_r}{\partial \varphi} - \frac{1}{r} \frac{\partial}{\partial r} (rA\varphi) \right] + i_\varphi \left[\frac{1}{r} \frac{\partial}{\partial r} (rA_\theta) - \frac{1}{r} \frac{\partial A_r}{\partial \theta} \right]$$

$$\nabla^2 \phi = \text{div grad } \phi = \frac{1}{r^2} \frac{\partial}{\partial r} \left(r^2 \frac{\partial \phi}{\partial r} \right) + \frac{1}{r^2 \sin \theta} \frac{\partial}{\partial \theta} \left(\sin \theta \frac{\partial \phi}{\partial \theta} \right) + \frac{1}{r^2 \sin^2 \theta} \frac{\partial^2 \phi}{\partial \varphi^2}$$

APPENDIX IV   IMPORTANT RELATIONSHIPS FROM ELECTROMAGNETIC THEORY[4]

Coulomb's law:

$$\mathbf{F}_1 = \frac{q_1 q_2}{4 \pi \epsilon_0} \frac{\mathbf{r}_{21}}{|r_{21}|^3}$$

Definition of electric field vector:

$$\mathbf{E} = \lim_{q \to 0} \frac{\mathbf{F}}{q}$$

where q is a small test charge.

Gauss' law:
Integral:

$$\int_{\text{surface}} \epsilon \mathbf{E} \cdot \mathbf{n} \, ds = \int \mathbf{D} \cdot \mathbf{n} \, ds = q$$

Differential:

$$\text{div } \epsilon \mathbf{E} = \text{div } \mathbf{D} = \rho$$

Potential in an electrostatic field:

$$\mathbf{E} = -\text{grad } \phi$$

Dipole moment:

$$\mathbf{p} = \lim_{\substack{q \to \infty \\ l \to 0}} q\mathbf{l}$$

Poisson's equation:

$$\nabla^2 \phi = -\rho/\epsilon_0$$

Laplace's equation:

$$\nabla^2 \phi = 0$$

Polarization charge:

$$\int_{\text{volume}} (-\text{div } \mathbf{P}) \, dv + \int_{\text{actual surfaces}} \mathbf{P} \cdot \mathbf{n} \, ds = Q$$

Dielectric constant:

$$k = \epsilon/\epsilon_0 = 1 + \chi/\epsilon_0$$

Continuity equation:

$$\operatorname{div} \mathbf{J} + \frac{\partial \rho}{\partial t} = 0$$

Law of Biot and Savart:

$$\mathbf{F}_2 = \frac{\mu_0}{4\pi} I_1 I_2 \oint_1 \oint_2 \frac{d\mathbf{l}_2 \times (d\mathbf{l}_1 \times \mathbf{r}_{12})}{r_{12}}$$

$$= I_2 \oint_2 d\mathbf{l}_2 \times \mathbf{B}_2$$

Ampere's circuital law:

$$\oint_{\text{contour}} \mathbf{B} \cdot d\mathbf{l} = \mu_0 \oint_{\text{surface}} \mathbf{J} \cdot d\mathbf{s} = \mu_0 I$$

Maxwell's equations:

$$\operatorname{div} \mathbf{D} = \rho_{\text{free}}$$
$$\operatorname{div} \mathbf{B} = 0$$
$$\operatorname{curl} \mathbf{E} = -\frac{\partial \mathbf{B}}{\partial t}$$
$$\operatorname{curl} \mathbf{H} = \frac{\partial \mathbf{D}}{\partial t} + \mathbf{J}$$

Wave equation:

$$\nabla^2 \begin{bmatrix} \mathbf{H} \\ \mathbf{E} \end{bmatrix} - \mu\sigma \frac{\partial}{\partial t} \begin{bmatrix} \mathbf{E} \\ \mathbf{H} \end{bmatrix} - \mu\epsilon \frac{\partial^2}{\partial t^2} \begin{bmatrix} \mathbf{E} \\ \mathbf{H} \end{bmatrix} = 0$$

Boundary conditions:

$$E_{1T} = E_{2T}$$
$$D_{2N} - D_{1N} = \sigma$$
$$B_{1N} = B_{2N}$$
$$H_{1T} - H_{2T} = \sigma_j$$

Poynting's vector:

$$\mathbf{S} = \mathbf{E} \times \mathbf{H}$$

APPENDIX V   TABLE OF IMPORTANT TRANSFORMS[32]

Explanation of the Table

The time functions and corresponding frequency functions in this table are related by the following expressions:

$$F(\omega) = \int_{-\infty}^{\infty} f(t) e^{-i\omega t} \, dt \qquad \text{(direct transform)}$$

$$f(t) = \frac{1}{2\pi} \int_{-\infty}^{\infty} F(\omega) e^{i\alpha t} \, d\omega \qquad \text{(inverse transform)}$$

The $1/2\pi$ multiplier in the inverse transform arises merely because the integration is written with respect to ω, rather than cyclic frequency. Otherwise the expressions are identical except for the difference of sign in the exponent. As a result, functions and their transforms can be interchanged with only slight modification. Thus, if $F(\omega)$ is the direct transform of $f(t)$, it is also true that $2\pi f(-\omega)$ is the direct transform of $F(t)$. For example, the spectrum of a $\sin x/x$ pulse is rectangular (pair 6) while the spectrum of a rectangular pulse is of the form $\sin x/x$ (pair 7). Likewise pair 1S is the counterpart of the well-known fact that the spectrum of a constant (dc) is a spike at zero frequency.

The frequency functions in the table are in many cases listed as functions of both ω and p. This is done merely for convenience. $F(p)$ in all cases is found by substituting p for $i\omega$ in $F(\omega)$. [Not simply p for ω as the notation would ordinarily indicate. That is, in the usual mathematical convention one would write $F(\omega) = F(p/i) = G(p)$ where the change in letter indicates the resulting change in functional form. The notation used above has grown through usage and causes no confusion, once understood.] Thus, in the p notation

$$F(p) = \int_{-\infty}^{\infty} f(t)\, e^{-pt}\, dt \qquad f(t) = \frac{1}{2\pi i} \int_{-i\infty}^{i\infty} F(p)\, e^{pt}\, dp$$

The latter integral is conveniently evaluated as a contour integral in the p plane, letting p assume complex values.

The frequency functions have been plotted on linear amplitude and frequency scales and, where convenient, also on logarithmic scales. The latter scales often bring out characteristics not evident in the linear plot. Thus, many of the spectra are asymptotic to first- or second-degree hyperbolas on a linear plot. On a log plot these asymptotes become straight lines of slope -1 or -2 (i.e., -6 or -12 dB per octave).

The time functions in the table have all been normalized to convenient peak amplitudes, areas, or slopes. For any other amplitude, multiply both sides by the appropriate factor. Thus, the spectrum of a rectangular pulse 10 V in amplitude and 2 s long is (from pair 7) 20 sin ω/ω, V-s.

Again, upon multiplication by a constant having appropriate dimensions, the frequency functions become filter transmissions. Thus, if pair 1 is multiplied by α, the frequency function represents a simple RC cutoff. A 1-C impulse (pair 1S) applied to this filter would produce an output (impulse response) with the spectrum $[\alpha/(p + \alpha)] \times 1$ C, representing the time function $\alpha e^{-\alpha t}$ coulombs (which has the dimensions of amperes). Or a 1-V step function (pair 2S) would produce the output spectrum $[\alpha/(p + \alpha)] \times 1/p$ V, which represents the time function $(1 - e^{-\alpha t})$ volts (pair 4S).

The entries 1S through 6S in the table are singular functions for which the transforms as defined above exist only as a limit. For example, 1S may be thought of as the limit of pair 7 (multiplied by $1/\tau$) as $\tau \to 0$.

Properties of Transforms

There are a number of important relations which describe what happens to the transforms of functions when the functions themselves are added, multiplied, convolved, etc. These relations state mathematically many of the operations encountered in communications systems: operations such as linear amplification, mixing, modulation, filtering, and sampling. These relations are all readily deducible from the defining equations above; but for ready reference some of the more important ones are listed in the Table of Properties.

Again, because of the similarity of the direct and inverse transforms, a symmetry exists in these properties. Thus, delaying a function multiplies its spectrum by a complex exponential, while multiplying the function by a complex exponential delays its spectrum. Multiplying any two functions is equivalent to convolving their spectra; multiplying their spectra is equivalent to convolving the functions; etc.

Many of the pairs listed under Important Transforms can be obtained from others by using one or more of the rules of manipulation listed in the Table of Properties. For example, the time function in pair 8 is $1/\tau$ times the *convolution* of that in pair 7 with itself. The spectrum should therefore be $1/\tau$ times the *product* of that in pair 7 with itself, as it indeed is. Further, by using these properties, many pairs *not* in the table can be obtained from those given. For example, the spectrum of $f(t) = (1 - \alpha t)\, e^{-\alpha t}$ is (by the addition property)

$$F(p) = \frac{1}{p + \alpha} - \frac{\alpha}{(p + \alpha)^2} = \frac{p}{(p + \alpha)^2}$$

TABLE V-1 Important Transforms

Time functions	No.	Function	Frequency functions (Linear scales)	Log ampl.– log freq.
$f(t) = \begin{cases} 0, & t<0 \\ e^{-\alpha t}, & t>0 \end{cases}$	1	$F(\rho) = \dfrac{1}{\rho+\alpha}$ $F(\omega) = \dfrac{1}{\alpha+i\omega}$		
$f(t) = \begin{cases} 0, & t<0 \\ \alpha t\,e^{-\alpha t}, & t>0 \end{cases}$	2	$F(\rho) = \dfrac{\alpha}{(\rho+\alpha)^2}$ $F(\omega) = \dfrac{\alpha}{(\alpha+\omega)^2}$		
$f(t) = e^{-\alpha|t|}$	3	$F(\rho) = \dfrac{2\alpha}{\alpha^2-\rho^2}$ $F(\omega) = \dfrac{2\alpha}{\alpha^2+\omega^2}$		
$f(t) = \begin{cases} 0, & t<0 \\ e^{-\alpha t}\sin\beta t, & t>0 \end{cases}$	4	$F(\rho) = \dfrac{\beta}{(\rho+\alpha)^2+\beta^2}$ $F(\omega) = \dfrac{\beta}{(\alpha^2+\beta^2)-\omega^2+i2\alpha\omega}$		

		No.	$F(\omega)$
$f(t) = \begin{cases} 0, & t<0 \\ e^{-\alpha t}(\cos\beta t + \dfrac{\alpha}{\beta}\sin\beta t), & t>0 \end{cases}$		5	$F(\rho) = \dfrac{\rho}{(\rho+\alpha)^2+\beta^2}$ $F(\omega) = \dfrac{i\omega}{(\alpha^2+\beta^2)-\omega^2+i2\alpha\omega}$
$f(t) = \dfrac{\sin\left(\pi\frac{t}{\tau}\right)}{\left(\pi\frac{t}{\tau}\right)}$		6	$F(\omega) = \begin{cases} \tau, & |\omega| < \dfrac{\pi}{\tau} \\ 0, & |\omega| > \dfrac{\pi}{\tau} \end{cases}$
$f(t) = \begin{cases} 1, & |t| < \dfrac{\tau}{2} \\ 0, & |t| > \dfrac{\tau}{2} \end{cases}$		7	$F(\omega) = \tau\,\dfrac{\sin\left(\frac{\omega\tau}{2}\right)}{\left(\frac{\omega\tau}{2}\right)}$
$f(t) = \begin{cases} 1-\dfrac{|t|}{\tau}, & |t| < \tau \\ 0, & |t| > \tau \end{cases}$		8	$F(\omega) = \tau\,\dfrac{\sin^2\left(\frac{\omega\tau}{2}\right)}{\left(\frac{\omega\tau}{2}\right)^2}$
$f(t) = \begin{cases} \sqrt{1-\left(\dfrac{t}{\tau}\right)^2}, & |t| < \tau \\ 0, & |t| > \tau \end{cases}$		9	$F(\omega) = \dfrac{\pi}{2}\,\tau\,\dfrac{2J_1(\omega\tau)}{(\omega\tau)}$

TABLE V-1 Important Transforms (Continued)

Time functions		No.	Function	Frequency functions	
				Linear scales	Log ampl.-log freq.
	$f(t) = e^{-\frac{1}{2}\left(\frac{t}{\tau}\right)^2}$	10	$F(\omega) = \tau\sqrt{2\pi}$		
	$f(t) = \begin{cases} 0 & ,\ \|t\| < 0 \\ \dfrac{\alpha e^{-\alpha t} - \beta e^{-\beta t}}{\alpha - \beta} & ,\ \|t\| > 0 \end{cases}$	11	$F(\rho) = \dfrac{\rho}{(\rho+\alpha)(\rho+\beta)}$		
	$f(t) = \begin{cases} \cos \omega_0 t & ,\ \|t\| < \frac{\tau}{2} \\ 0 & ,\ \|t\| > \frac{\tau}{2} \end{cases}$	12	$F(\omega) = \dfrac{\tau}{2}\left[\dfrac{\sin\left(\frac{\omega-\omega_0}{2}\right)\tau}{\left(\frac{\omega-\omega_0}{2}\right)\tau} + \dfrac{\sin\left(\frac{\omega+\omega_0}{2}\right)\tau}{\left(\frac{\omega+\omega_0}{2}\right)\tau}\right]$		
	$f(t) = \lim_{\tau \to 0} \begin{cases} \frac{1}{\tau}, & \|t\| < \frac{\tau}{2} \\ 0, & \|t\| > \frac{\tau}{2} \end{cases}$ $= \delta(t)$ (delta function)	1S	$F(\rho) = F(\omega) = 1$		
	$f(t) = \int_{-\infty}^{t} \delta(\lambda)d\lambda = \begin{cases} 0, & t<0 \\ 1, & t>0 \end{cases}$ $= u(t)$ (unit step)	2S	$F(\rho) = \dfrac{1}{\rho}$		

	$f(t)$		$F(\rho)$ / $F(\omega)$	
	$f(t)=\displaystyle\int_{-\infty}^{t} u(\lambda)\,d\lambda = \begin{cases}0, & t<0\\ t, & t>0\end{cases}$ $= s(t)$ (unit slope)	3S	$F(\rho)=\dfrac{1}{\rho^{2}}$	
	$f(t)=\begin{cases}0, & t<0\\ 1-e^{-\alpha t}, & t>0\end{cases}$	4S	$F(\rho)=\dfrac{\alpha}{\rho(\rho+\alpha)}$	
	$f(t)=\cos\omega_0 t$	5S	$F(\omega)=\dfrac{\delta(\omega+\omega_0)-\delta(\omega-\omega_0)}{2}$	
	$f(t)=\displaystyle\sum_{-\infty}^{\infty}\delta(t-n\tau)$	6S	$F(\omega)=\displaystyle\sum_{-\infty}^{\infty}\delta\!\left(\omega-n\frac{2\pi}{\tau}\right)$	

TABLE V-1 Important Transforms (Continued)

Time operation	Frequency operation	Significance
Linear addition $af(t) + bg(t)$	Linear addition $aF(\omega) + bG(\omega)$	Linearity and superposition apply in both domains. The spectrum of a linear sum of functions is the same linear sum of their spectra (if spectra are complex, usual rules of addition of complex quantities apply). Further, any function may be regarded as a sum of component parts and the spectrum is the sum of the component spectra
Scale change $f(kt)$	Inverse scale change $\dfrac{1}{|k|} F\left(\dfrac{\omega}{k}\right)$	Time—bandwidth invariance. Compressing a time function expands its spectrum in frequency and reduces it in amplitude by the same factor. The amplitude reduces because less energy is spread over a greater bandwidth. For same energy pulse as for $k = 1$, multiply both functions by $\sqrt{|k|}$. The case where $k = -1$ reverses the function in time. This merely interchanges positive and negative frequencies; so for real time functions, reverses the phase
Even and odd partition $\frac{1}{2}[f(t) \pm f(-t)]$	Even and odd partition $\frac{1}{2}[F(\omega) \pm F(-\omega)]$	Any real function $f(t)$ may be separated into an even part $\frac{1}{2}[f(t) + f(-t)]$ and an odd part $\frac{1}{2}[f(t) - f(-t)]$. The transform of the even part is $\frac{1}{2}[F(\omega) + F(-\omega)]$, which is purely real and involves only even powers of ω. The transform of the odd part is $\frac{1}{2}[F(\omega) - F(-\omega)]$, which is purely imaginary and involves only odd powers of ω. Note: for $f(t)$ real, $F(-\omega) = \overline{F(\omega)}$.
Delay $f(t - t_0)$	Linear added phase $e^{-i\omega t_0} F(\omega)$	Delaying a function by a time t_0 multiplies its spectrum by $e^{-i\omega t_0}$, thus adding a linear phase $\theta = -\omega t_0$ to the original phase. Conversely a linear-phase filter produces a delay of $-d\theta/d\omega = t_0$
Complex modulation $e^{i\omega_0 t} f(t)$	Shift of spectrum $F(\omega - \omega_0)$	Multiplying a time function by $e^{i\omega_0 t}$ "delays" its spectrum, i.e., shifts it to center about ω_0 rather than zero frequency. Ordinary real modulation—by $\cos \omega_0 t$ say—produces the time function $\frac{1}{2}(e^{i\omega_0 t} + e^{-i\omega_0 t})f(t)$ with the spectrum $\frac{1}{2}[F(\omega - \omega_0) + F(\omega + \omega_0)]$

Operation	Transform property	Description
Convolution $\int_{-\infty}^{\infty} f(\tau)g(t-\tau)d\tau$	Multiplication (filtering) $F(\omega)G(\omega)$	The spectrum of the convolution of two time functions is the product of their spectra. In convolution one of the two functions to be convolved is reversed left to right and displaced. The integral of the product is then evaluated and is a new function of the displacement. Convolution occurs whenever a signal is obtained which is proportional to the integral of the product of two functions as they slide past each other—in other words, in any *scanning* operation such as in optical or magnetic recording or picture scanning in television. Transform theory states that such scanning is equivalent to filtering the signal with a filter whose transmission is the transform of the scanning function (reversed in time). Conversely, the effect of an electrical filter is equivalent to a convolution of the input with a time function which is the transform of filter characteristic. This function, the so-called "memory curve," of the filter, is identical with the filter impulse response, aside from dimensions. (NOTE: the convolution of a time function with a unit impulse gives the same function times the dimensions of the impulse)
Multiplication $f(t)g(t)$	Convolution $\dfrac{1}{2\pi}\displaystyle\int_{-\infty}^{\infty} F(s)G(\omega - s)ds$	The spectrum of the product of two time functions is the convolution of their spectra. This is the more general statement of the modulation property. For example, sampling a signal is equivalent to multiplying it by a regular train of unit area impulses. The spectrum of the sampled signal consists of the original signal spectrum repeated about each component of the (line) spectrum of the train of impulses (see pair 6S). For no overlap, highest frequency in signal to be sampled must be less than half sampling frequency. If this is true, the original signal spectrum (hence the signal) can be recovered by a low-pass filter (sampling theorem)
Differentiation $\dfrac{d^n f(t)}{dt^n}$	Multiplication by p $p^n F(p)$	The spectrum of the nth derivative of a function is $(i\omega)^n$ times the spectrum of the function. A "differentiating network" has (over the appropriate frequency range) a transmission $K(p/\omega_0)$ where K is dimensionless or has the dimensions of impedance or admittance. Thus the output wave is *proportional* to the derivative of the input
Integration $\underbrace{\displaystyle\int_{-\infty}^{t}\cdots\int_{-\infty}^{t}}_{n} f(\tau)(d\tau)^n$	Multiplication by $\dfrac{1}{p}$ $\dfrac{1}{p^n}F(p)$	The spectrum of the nth integral of a function is $(i\omega)^{-n}$ times the spectrum of the function. Thus, the response of any filter to a step function is the integral of its impulse response. An "integrating network" has (over the appropriate frequency range) a transmission $K(\omega_0/p)$, where K is dimensionless or has the dimensions of impedance or admittance. Thus the output is *proportional* to the integral of the past of the input

APPENDIX VI   MICROSTRIP DESIGN FORMULAS[41]

Boundary Conditions

A transverse cross section with coordinate representation of a typical pair of microstrip lines is shown in Fig. VI-1. The parameters used are consistent with those used in the literature, where H is the substrate thickness, W is the strip width, and S is the spacing between strips. The strip thickness is assumed to be negligible throughout the analysis.

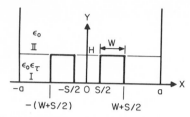

Fig. VI-1   Transverse cross section of microstrip lines.

Two TEM modes can propagate along such a structure: an even mode which corresponds to voltages and currents of equal and the same polarity on the conductors, and an odd mode which corresponds to voltages and currents of equal and opposite polarity on the conductor.

The boundary conditions for the two modes are slightly different and can be easily stated with the aid of Fig. VI-1.

a. Even mode

$$\nabla^2\phi = 0 \tag{1a}$$
$$\phi(x,0) = 0 \tag{2a}$$
$$\phi(x,\infty) = 0 \tag{3a}$$
$$\phi(\pm\infty,y) = 0 \tag{4a}$$
$$\frac{\partial\phi}{\partial x}(0,y) = 0 \tag{5a}$$
$$\phi(x,H) = V_e(x) \tag{6a}$$

b. Odd mode

$$\nabla^2\phi = 0 \tag{1b}$$
$$\phi(x,0) = 0 \tag{2b}$$
$$\phi(x,\infty) = 0 \tag{3b}$$
$$\phi(\pm\infty,y) = 0 \tag{4b}$$
$$\phi(o,y) = 0 \tag{5b}$$
$$\phi(x,H) = V_o(x) \tag{6b}$$

$V_e(x)$ and $V_o(x)$ are the potentials at the interface between the dielectric medium and vacuum above it for the even mode and odd mode, respectively. These two functions will be evaluated approximately later. Since the potential dies out rapidly in the x direction, boundary conditions (4a) and (4b) may be changed without any loss of accuracy to

$$\phi(\pm a,y) = 0 \qquad \text{where } a \gg \left(W + \frac{S}{2}\right) \tag{4}$$

This implies physically that two potential walls exist at $\pm a$.

Solution of the Boundary-Value Problem

a. Even mode   The potential anywhere may be found by solving Laplace's equation, subject to the boundary conditions stated previously. Since the transmission line is uniform in the Z direction, Laplace's equation reduces to

$$\frac{\partial^2\phi}{\partial x^2} + \frac{\partial^2\phi}{\partial y^2} = 0 \tag{7}$$

A product solution is assumed, or

$$\phi = \bar{X}(x)\,\bar{Y}(y) \tag{8}$$

Substituting (8) into (7) and rearranging, we get

$$\frac{\bar{X}''}{\bar{X}} = -\frac{\bar{Y}''}{\bar{Y}} = -\alpha^2 \tag{9}$$

Equation (9) is equated to a constant α, since its left-hand side is a function of x alone and the right-hand side is a function of y alone. Negative sign was chosen because a periodic field in the x direction is required. Solving (9), we obtain

$$\bar{X} = A\cos \alpha x + B\sin \alpha x$$
$$\bar{Y} = Ce^{\alpha y} + De^{-\alpha y}$$

and hence

$$\Phi = (A\cos \alpha x + B\sin \alpha x)(Ce^{\alpha y} + De^{-\alpha y}) \tag{10}$$

The constants may be found by imposing the boundary conditions. Thus, from (2), we obtain $C = -D$, and (3) gives $C = 0$ in region II only. Boundary conditions 5 and 4 give $B = 0$ and $\alpha_n = [(2n-1)/2](\pi/a)$, respectively. Thus, a solution for the fields in regions I and II yields

$$\Phi_I = A_n \cos \alpha_n x \sinh \alpha_n y$$
$$\Phi_{II} = B_n \cos \alpha_n x e^{-\alpha_n y}$$

Since the field has discontinuities at the interface, a series of such solutions is required, or

$$\Phi_I = \sum_{n=1}^{\infty} A_n \cos \alpha_n x \sinh \alpha_n y \tag{11}$$

$$\Phi_{II} = \sum_{n=1}^{\infty} B_n \cos \alpha_n x e^{-\alpha_u y} \tag{12}$$

$$\alpha_n = \frac{2n-1}{2}\frac{\pi}{a} \qquad n = 1, 2, 3, \ldots \tag{13}$$

A_n and B_n may be found using (6); thus

$$V_e(x) = \sum_{n=1}^{\infty} A_n \cos \alpha_n x \sinh \alpha_n H$$

Multiplying both sides of the above equation by $\cos \alpha_m x$ and integrating between the limits $\pm a$, we obtain

$$\int_{-a}^{a} V_e(x) \cos \alpha_n x\, dx = \int_{-a}^{a} \sum_{n=1}^{\infty} A_n \cos \alpha_n x \cos \alpha_n x \sinh \alpha_n H\, dx$$
$$= a A_n \sinh \alpha_n H$$

Hence

$$A_n = \frac{1}{a}\int_{-a}^{a} \frac{V_e(x) \cos \alpha_n x\, dx}{\sinh \alpha_n H} \tag{14}$$

Similarly

$$B_n = \frac{1}{a}\int_{-a}^{a} \frac{V_e(x) \cos \alpha_n x\, dx}{e^{-H\alpha_n}} \tag{15}$$

b. Odd mode   Similarly, the field solutions for the odd mode are

$$\Phi_I = \sum_{n=1}^{\infty} C_n \sin \gamma_n x \sinh \gamma_n y \tag{16}$$

$$\Phi_{II} = \sum_{n=1}^{\infty} D_n \sin \gamma_n x \, e^{-\gamma_n y} \tag{17}$$

$$\gamma_n = \frac{n\pi}{a} \qquad n = 1, 2, \ldots \tag{18}$$

where

$$C_n = \frac{1}{a} \int_{-a}^{a} \frac{V_o(x) \sin \gamma_n x \, dx}{\sinh \gamma_n H} \tag{19}$$

$$D_n = \frac{1}{a} \int_{-a}^{a} \frac{V_o(x) \sin \gamma_n x \, dx}{e^{-\gamma_n H}} \tag{20}$$

Approximate Expressions for the Interface Potentials $V_e(x)$ and $V_o(x)$

The potential along the surface of the substrate is known only at the conductors. For the even mode, both conductors are at a potential V_o, and for the odd mode, one of the conductors is at a potential V_o, while the other is at a potential $(-V_o)$.   To

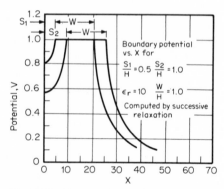

Fig. VI-2   Even-mode interface potential.

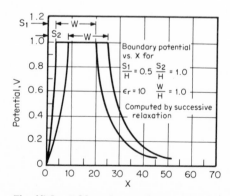

Fig. VI-3   Odd-mode interface potential.

find the surface potential elsewhere, a numerical analysis based on a finite difference or relaxation method is used.[8]

The results of the computer analysis are shown graphically. Figure VI-2 shows the interface potential for the even-mode case and Fig. VI-3 shows the interface potential for the odd-mode case. The curves were plotted for different S-to-H ratios. Using these plots, an approximate analytical expression for the interface potential may be found for both even- and odd-mode cases. The expression for the even-mode interface potential is given by

$$V_e(x) = V_o \left[A + \frac{8}{S^3}(1-A)x^3 \right] \qquad \frac{S}{2} > |X| > 0 \qquad (21)$$

$$= V_o \qquad\qquad\qquad W + \frac{S}{2} > |X| \, S/2$$

$$= V_o e^{-B[W+S/2]} \qquad\qquad |X| > W + S/2$$

where $A = \left[1 - \left(\frac{1 + \ln(K+5)}{10} \right) S/H \right]$

$B = \dfrac{1 + \ln \sqrt{K+2}}{W}$

K = relative dielectric

The expression for the odd-mode interface potential is given by

$$V_o(x) = \frac{4V_o x^2}{S^2} \qquad\qquad S/2 > |X| > 0 \qquad (22)$$

$$= V_o \qquad\qquad W + S/2 > |X|S/2$$

$$= V_o e^{-B[x-(W+S/2)]} \qquad |X| > W + S/2$$

Variational Solution to Capacitance of Microstrip Coupled Lines

The variational expression for the capacitance is given by [1]

$$2C = \frac{\epsilon}{V_o^2} \iint (\nabla\Phi_t \cdot \nabla\Phi_t) dA \qquad (23)$$

where dA is the differential cross-sectional surface and C is an upper-bound value for the transmission-line capacitance.

Equation (23) may be written as

$$2C = \frac{\epsilon}{V_o^2} \left\{ \int_0^h \int_{-a}^a \left[\left(\frac{\partial\Phi_I}{\partial x}\right)^2 + \left(\frac{\partial\Phi_I}{\partial y}\right)^2 \right] dx\, dy \right\}$$
$$+ \frac{\epsilon_0}{V_o^2} \left\{ \int_h^\infty \int_{-a}^a \left[\left(\frac{\partial\Phi_{II}}{\partial x}\right)^2 + \left(\frac{\partial\Phi_{II}}{\partial y}\right)^2 \right] dx\, dy \right\} \qquad (24)$$

Substituting Eq. (11) and (12) in (24) and simplifying, we obtain the even-mode capacitance C_e. Thus

$$2C_e = \frac{\epsilon}{V_o^2} \sum_{n=1}^\infty A_n^2 \frac{\alpha_n a}{2} \sinh 2\alpha_n y + \frac{\epsilon_0}{V_o^2} \sum_{n=1}^\infty B_n^2 \, \alpha a \, e^{-2\alpha_n H} \qquad (25)$$

A_n and B_n may be obtained from Eq. (14) and (15); thus

$$2C_e = \frac{\epsilon}{V_o^2} \sum_{n=1}^\infty \left[\int_{-a}^a \frac{\cos\alpha_n x \, V_e(x)\, dx}{a \sinh \alpha_n H} \right]^2 a \, \alpha_n \sinh \alpha_n H \cos \alpha_n H$$

$$+ \frac{\epsilon_0}{V_o^2} \sum_{n=1}^\infty \left[\int_{-a}^a \frac{\cos \alpha_n a \, V_e(x)\, dx}{a \, e^{-\alpha_n H}} \right]^2 a \, \alpha_n \, e^{-2\alpha_n H}$$

Simplifying, we obtain

$$2C_e = \sum_{n=1}^{\infty} \left[\frac{\int_{-a}^{a} \cos \alpha_n x \, V_e(x) \, dx}{V_o} \right]^2 \frac{\alpha_n}{a} (\epsilon \coth \alpha_n H + \epsilon_0) \quad (26)$$

The odd-mode capacitance C_o may be obtained in a similar manner; thus

$$2C_o = \sum_{n=1}^{\infty} \left[\frac{\int_{-a}^{a} \sin \gamma_n \, V_o(x) \, dx}{V_o} \right]^2 \frac{\gamma_n}{a} (\epsilon \coth \gamma_n H + \epsilon_0) \quad (27)$$

It is a simple matter now to substitute the expressions for $V_e(x)$ and $V_o(x)$ into Eq. (26) and (27) and carry out the integration. The results are

$$2C_e = \sum_{n=1}^{\infty} \frac{4}{\alpha a} \epsilon_0(\epsilon_r \coth \alpha H + 1) \left\{ -\frac{24(1-A)}{S^2\alpha^2} \sin \alpha S/2 \right.$$

$$+ \frac{24(1-A)}{S^3\alpha} \left(\frac{S^2}{4} - \frac{2}{\alpha^2} \right) \cos \alpha S/2 + \frac{48(1-A)}{S^3\alpha^3} + \frac{B^2}{\alpha^2 + B^2} \sin \alpha(W + S/2)$$

$$+ \frac{B\alpha}{\alpha^2 + B^2} \cos \alpha(W + S/2) + \frac{\alpha}{\alpha^2 + B^2} e^{-3B(W+S/2)} [\alpha \sin 4\alpha(W + S/2)$$

$$\left. - B \cos 4\alpha(W + S/2)] \right\}^2 \quad (28)$$

$$2C_o = \sum_{n=1}^{\infty} \frac{4}{\gamma a} \epsilon_0(\epsilon_r \coth \gamma H + 1) \left\{ \frac{4}{S\gamma} \sin \gamma \frac{S}{2} + \frac{8}{S^2\gamma^2} \left(\cos \gamma \frac{S}{2} - 1 \right) \right.$$

$$- \frac{B^2}{\gamma^2 + B^2} \cos \gamma(W + S/2) + \frac{B\gamma}{\gamma^2 + B^2} \sin \gamma(W + S/2)$$

$$\left. - \frac{\gamma}{\gamma^2 + B^2} e^{-3B(W+S/2)} [\gamma \cos 4\gamma(W + S/2) + B \sin 4\gamma(W + S/2)] \right\}^2 \quad (29)$$

a was taken as $a = 4 (S/2 + W)$. γ, α, B, and A were defined earlier.

$$Z_{oe} = \frac{1}{C \sqrt{C_{e_1} C_{e_2}}}$$

$$Z_{oo} = \frac{1}{C \sqrt{C_{o_1} C_{o_2}}}$$

$$V_e = \frac{C}{\sqrt{C_{e_2}/C_{e_1}}}$$

$$V_o = \frac{C}{\sqrt{C_{o_2}/C_{o_1}}}$$

The even- and odd-mode impedances and velocities may then be evaluated from the following set of equations, where

$$C_{e_1} = C_e|_{\epsilon_r=1}$$
$$C_{e_2} = C_e|_{\epsilon_r=\kappa}$$
$$C_{o_1} = C_o|_{\epsilon_r=1}$$
$$C_{o_2} = C_o|_{\epsilon_r=\kappa}$$

The expressions for the impedances were evaluated using the computer. The results are shown in Fig. VI-4. In order to verify the validity of these equations experimentally, two test circuits were constructed on alumina substrate. The first was a single-resonator, side-coupled filter and the second was an 8-dB directional coupler. The experimental results were in good agreement with the theoretical predictions, as can be seen from Fig. VI-5 for the case of the filter and Figs. VI-6 and VI-7 for the case of the 8-dB coupler.

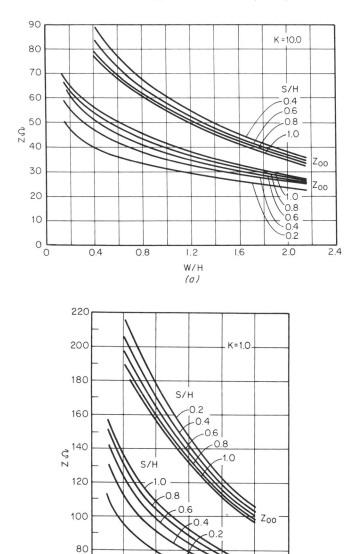

Fig. VI-4   Impedance of microstrip transmission line.

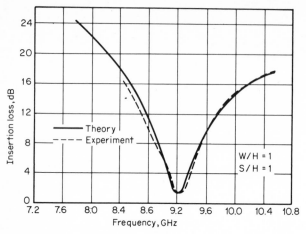

Fig. VI-5   Single resonator filter insertion loss (dB) vs. frequency.

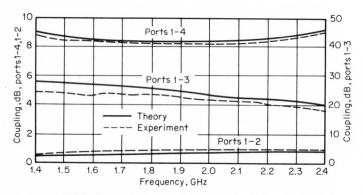

Fig. VI-6   Transmission vs. frequency for 8-dB coupler.

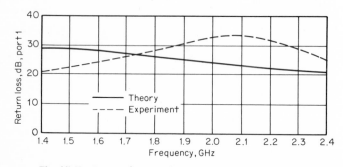

Fig. VI-7   Return loss vs. frequency for 8-dB coupler.

Chapter **7**

Resistors and Passive-Parts Standardization

EDWARD L. HIERHOLZER AND H. BENNETT DREXLER

Martin Marietta Aerospace, Orlando, Florida

WITH

JOHN H. POWERS

IBM Corporation, Hopewell Junction, New York

INTRODUCTION

Scope This chapter presents information regarding the principles of parts standardization and the selection and application of resistors for use in the design of electronic equipment. Capacitor selection and application is presented in Chap. 8.

Information Management. Techniques of standardization for resistors and capacitors have many common aspects and are approached first in this chapter. The remainder of the chapter treats fixed resistors and resistor networks with a brief treatment of trimming potentiometers used in electronic-circuit design. Chapters 7 and 8 are organized with a general part description and selection guide followed by paragraphs furnishing increased detail on current significant part classifications. The latter part of each section contains paragraphs on special application techniques and reliability considerations.

Cost Considerations. Part costs are treated in a relative manner, since markets for resistors and capacitors are competitive and are affected by many factors. The reader is cautioned that immediate circumstances of quantity, demand, material costs, special requirements, and available innovations or other aspects may change even the relative cost rankings given.

Suppliers. Numerous manufacturers and part types referenced in Chaps. 7 and 8 represent only a sampling of sources for products described. Several listing services provide adequate coverage of the industry and should be consulted to establish sources which can meet delivery, cost, and performance requirements. Some of these services are:

1. "Electronics Buyers Guide," McGraw-Hill, New York
2. "The Radio Electronic Master" and "Electronic Engineers Master," United Technical Publications Division, Cox Broadcasting Corporation, Garden City, N.Y.
3. "Thomas Register of American Manufacturers and Thomas Register Catalogue File," Thomas Publishing Company; New York
4. "Who's Who in Electronics," Electronic Periodicals, Inc., Cleveland, Ohio
5. "Electronic Design Gold Book," Hayden Publishing Co., Inc., Rochelle Park, N.J.

Trends in component design Developments in circuit-fabrication techniques, increasing complexity, and the demand for smaller package sizes have provided continued stimuli for innovation in resistor and capacitor design and production. The needs of semiconductor circuit designs for low-impedance networks, lower signal levels, and faster switching speeds have modified demands for functional value ranges and operating levels. Despite the more recent developments of smaller parts adaptable to hybrid circuits and other high-density assemblies, however, a healthy usage of many established designs has persisted. Many of the vintage designs, while appearing constant, have actually been made orders of magnitude more stable and more reliable through improved materials technology, better testing techniques, and vastly improved fabrication methods. Some, such as the carbon composition resistor are virtually unchanged but have remained popular because of their broad utility, practical cost, and satisfactory performance records. Product safety is of currently increasing concern; and as this is written, new self-fusing and nonflammable resistor designs are being offered by several producers for use in consumer products. Safety from shock hazards and insulation leakage are of concern; and guaranteed longevity for consumer, commercial, and military products made possible through solid-state active components have made ever greater demands on capacitor and resistor performance. Small chip-type capacitors and resistors are marketed for use in hybrid-circuit designs, and increasing varieties of resistor networks are available in packaged assemblies resembling some of the standard semiconductor-device packages. Larger values of capacitance in smaller volumes with improved stability and shelf life are needed constantly and are becoming increasingly available to fit the low-voltage, low-impedance requirements of semiconductor-circuit designs. Metal oxide films and other thick films for resistors are found to provide improved performance and increasingly competitive costs for many low-power fixed and variable types when compared with older types.

Standardization Considerations

Why standardize? The typical electronics assembly routinely uses proportionately large quantities of resistors and capacitors in comparison with other part types. This circumstance provides a fertile opportunity for large benefits from good standardization practices. An effective standardization program reduces the number of separate item types for procurement and stocking, and enhances ability to obtain pricing considerations for larger quantities of the selected standard item. The significant advantages that accrue from standardization justify a degree of compromise in part selection. For instance, a decision may be made to use two parallel-connected standard resistors at a small apparent disadvantage in cost and packaging to achieve a needed power rating, rather than to specify isolated usage of a higher wattage rating that may actually end up with a higher cost because of its small-quantity procurement and stocking. Likewise, two or more standard capacitors may be paralleled to obtain a larger nonstandard capacitance value. In some cases where resistance or capacitance tolerances are required to be tighter than standard values, the root-sum-square (rss) technique can be applied to obtain a combined tolerance of series or parallel elements. Cost of these standard parts can be traded off against the cost of a nonstandard tight-tolerance part.

Published standards Industry standards have been of great benefit to users and part manufacturers alike and can be used effectively as a starting basis for standardization. Beyond the basic plan, however, a detailed and diligent analysis of selection alternatives seldom fails to reveal new available economies. Industry standard specifications, in general, provide the technical base for part definition and performance to be followed by participating manufacturers. These standards seldom define a part with limited suppliers and as such provide good technical requirements for competitive procurement. In addition, standards provide guidelines for quality-assurance provisions (i.e., acceptance inspection, tests, and rejection criteria) and shipping requirements.

Parts having more than one manufacturer are the best basic choices, since they permit competitive procurement and frequently are proved articles with well-known performance. A part configuration manufactured by only a single supplier can sometimes be a good selection if good performance history has been established and pricing is competitive with alternates, but in general, an established standard with two or more sources is a superior choice.

Military standards and specifications are required to be observed by the Standardization Directives in the product assurance sections on most government contracts. These standards are available individually or by subscription through the single-point stocking center designated by the Department of Defense. Military standards contain adequate part descriptions, performance requirements, quality-assurance provisions, and packing and shipping requirements. Most of the military specifications have an associated Qualified Products List (QPL) which is revised as needed to list all the suppliers, their addresses, and the types and values for which the government has given qualification approval. The range of values and other conditions of approval are also stated. These lists must be consulted if valid military products are required in the application, since a specification listing alone does not guarantee availability of a particular part. A few military specifications do not have associated QPL's and thus assume the same standardization functions as do EIA standards. Others, MIL-R-15109 (ships), for example, which covers "Resistors and Rheostats, Naval Shipboard" have QPL's that list suppliers and their product lines that conform to the intent of the specification and which have at one time passed the qualification requirements. Even if a user does not require military qualified parts, the government specifications and standards provide a great deal of information regarding part-type capabilities, definitions of terms, standard test methods, and application data. The government also publishes alphanumeric indexes by both title and specification number for all current standards, specifications, and handbooks, showing latest revisions, amendments, and deletions. If government resistor and capacitor specifications are ordered, they should be supplemented with current copies of Military Standards MIL-STD-202, Test Methods for Electronic and Electrical Component Parts, and

MIL-STD-199, Resistors, Selection and Use of, or MIL-STD-198, Capacitors, Selection and Use of. Also, the "general specification" does not usually cover the details for each part. The specific items are depicted on separate satellite specifications called "detail sheets" or frequently "slash sheets," so named because they are identified by using the basic general specification number with an identifying number following a slash bar after the specification number. These must be ordered as separate specifications. In military and EIA specifications alike, letters following the identifying number signify revision level. Numbers in parentheses following military numbers are successive amendment identifiers that are interim to revisions. These specification activities are coordinated by the Defense Electronic Supply Center (DESC) in Dayton, Ohio. Military specifications and applicable standards contain much useful information and should not be eschewed as effective tools by commercial-equipment designers. In addition to the previously mentioned standards, MIL-STD-1470 (MI) provides a guided-missile preferred-items list which can be used as a general military-parts shopping list.

Standards organizations Organizations actively engaged in standardization are the American National Standards Institute ANSI),* International Electrotechnical Commission (IEC)†, Electronic Industries Association (EIA)‡, Aerospace Industries Association of America (AIA)§, and military agencies of the Department of Defense (DOD).¶ National Aerospace Standards (NAS) specifications and standards are published by the AIA. Publication references and designations of these organizations are used throughout this chapter. The membership of the industry associations includes users and suppliers, and the publications reflect agreed-upon test methods, packaging, quality assurance, and purchase specifications. They are useful aids in specifying the products they cover.

The EIA is a nonprofit organization representing manufacturers of electronic products, and confines its activities to the area of legitimate public-interest objectives under the policy direction of its board of governors. The EIA stimulates public awareness of the role of the electronic industry in national defense, space exploration, communications, education and entertainment, evolution of industrial technology, and improvement of living standards. The EIA advises the Department of Defense and armed services to effect the use of the most advanced and reliable products and scientific development from industry through the interchange of ideas and information.

The AIA is a nonprofit organization consisting of representatives from the aerospace industry and provides an interface with governmental agencies who are engaged in defining military standards and specifications.

Preferred values Fundamental standardization practices require the selection of preferred values within the ranges available. Values have been defined in Department of Defense publications, by industry standards, and in other trade publications, using a system proposed by Charles Renard in 1870 for use by the French army to reduce the ridiculous proliferation of cordage sizes he found to be specified for balloon moorings.[2]

Decade Progression. The system is based on preferred numbers generated by a geometric progression devised to repeat in succeeding decades. The general geometric progression is defined by

$$N = ar^{n-1}$$

where N is the nth term, a is the first term, and r is a chosen common ratio. If

* American National Standards Institute, Inc., 1430 Broadway, New York, N.Y. 10018.

† International Electrotechnical Commission Central Office, Case Postale 56, 1211 Geneva 20, Switzerland. Publications also available from ANSI.

‡ Electronic Industries Association, Engineering Department, Standard Sales, 2001 Eye Street, N.W., Washington, D.C. 20006.

§ Aerospace Industries Association of America, Inc., 1725 De Sales Street, N.W., Washington, D.C. 20036.

¶ DOD Single Stocking Point, Naval Publications and Forms Center, 5801 Tabor Ave., Philadelphia, Pa. 19120.

r is chosen to be the kth root of 10 and the first term is set at unity, then

$$N = 1 \times (\sqrt[k]{10})^{n-1} \qquad (1)$$

where k can be selected to provide a desired scale graduation. If, for instance, 3 values per decade are desired, k is 3 and the common ratio becomes 2.154. The three rounded-off values are 1.00, 2.15, and 4.64. Standard decades for resistors and capacitors have been chosen, having 3, 6, 12, 48, 96, and 192 terms, with common ratios being the appropriate roots of 10. The 192-value-per-decade system has use for high-precision capacitors and resistors, but the large number of values tends to defeat the standardization purpose. Table 1 shows preferred-number decade values with appropriate numbers of significant figures used to designate resistors, capacitors, and Zener diodes.

Tolerances. Service variability of a resistor or capacitor, frequently called "end-of-life tolerance," is an overall value tolerance composed of factors due to purchase tolerance, lapsed time, and stress, and short-term excursions due to local environment. Service variability constitutes a more useful and realistic factor in part selection for a given application than does the all too common use of the purchase tolerance alone. Decade-value progressions used in procurement of components should cover incremental values commensurate with part variability. Aggressive standardization practices base the decade common ratio r on expected service-life variability rather than on purchase tolerances, thereby decreasing the number of stock values. Thus, if a component purchased with a 5 percent tolerance is actually found to remain within a ±20 percent range in service, the service variability is 20 percent. Observing the values shown in the 10 percent decade having 12 values per decade, it is seen that steps are essentially 20 percent between values. Standard values for circuit-design selection, therefore, should be each alternating value in the 5 percent decade, or actually, the "10 percent" decade. A resistor or capacitor of a particular marked value is considered to be allowed to assume a value anywhere within the expected service tolerance band. Statistical distribution techniques can be applied to variation if more than one of the parts in question are significant to a particular circuit variation, but for each individual part the total variability must be considered possible. Use of a decade ratio graduated according to service variability will thus decrease stock varieties without significantly impairing ultimate selection utility, despite initial impressions to the contrary. This technique provides a major standardization benefit. One approach to estimation of service variability consists of performing a root-sum-square of the value change expected with purchase tolerance plus that for each environmental extreme expected for the part in its application. Except for initial tolerance, the most significant contributor is frequently that of long-term drift, which unfortunately is usually ignored by procurement specifications and must be assessed by experience. (See special application guidelines below.)

Part-type selection Application analysis is an important part of standardization, and standard-part selection listings for designers should be chosen to fit the needs of particular equipment function and use. Resistors, except for high-precision types, are chosen within general-purpose, medium-power, and high-power-dissipation categories to provide the desired service variability, functional adequacy, packaging utility, and cost. For capacitors, the categories include basically similar considerations, but specialized usage classifications dominate in determining the number of included types. In miniaturized semiconductor circuits, high capacitance—volume ratio becomes the limiting factor, and performance compromises are frequently necessary to achieve reduced size.

General-purpose usage parts should be selected to provide the fewest power and voltage ratings and the lowest acceptable precision that can be tolerated for the application. If, in a given equipment, a significant quantity of applications require higher precision, lower drift, lower noise, or better rf performance than a lower-precision norm, it may be possible to show that the small quantity of low-precision parts needed will cost more than just increasing the already large quantity of more expensive close-tolerance units, especially if procurement and stocking costs are considered.

Table 2 shows suggested type classifications for standardization purposes. A minimized number of selections within each category is suggested.

TABLE 1 Standard Decade Values (Industry and Military Standards), Preferred Values for Resistors, Capacitors, Zener Diodes

*	±1%	±2%	*	±1%	±2%	*	±1%	±2%	*	±1%	±2%	*	±1%	±2%	*	±1%	±2%	±5%	±10%	±20%
1.00	1.00	1.00	1.47	1.47	1.47	2.15	2.15	2.15	3.16	3.16	3.16	4.64	4.64	4.64	6.81	6.81	6.81	1.0	1.0	1.0
1.01			1.49			2.18			3.20			4.70			6.90			1.1		
1.02	1.02		1.50	1.50		2.21	2.21		3.24	3.24		4.75	4.75		6.98	6.98		1.2	1.2	
1.04			1.52			2.23			3.28			4.81			7.06			1.3		
1.05	1.05	1.05	1.54	1.54	1.54	2.26	2.26	2.26	3.32	3.32	3.32	4.87	4.87	4.87	7.15	7.15	7.15	1.5	1.5	1.5
1.06			1.56			2.29			3.36			4.93			7.23			1.6		
1.07	1.07		1.58	1.58		2.32	2.32		3.40	3.40		4.99	4.99		7.32	7.32		1.8	1.8	
1.09			1.60			2.34			3.44			5.05			7.41			2.0		
1.10	1.10	1.10	1.62	1.62	1.62	2.37	2.37	2.37	3.48	3.48	3.48	5.11	5.11	5.11	7.50	7.50	7.50	2.2	2.2	2.2
1.11			1.64			2.40			3.52			5.17			7.59			2.4		
1.13	1.13		1.65	1.65		2.43	2.43		3.57	3.57		5.23	5.23		7.68	7.68		2.7	2.7	
1.14			1.67			2.46			3.61			5.30			7.77			3.0		
1.15	1.15	1.15	1.69	1.69	1.69	2.49	2.49	2.49	3.65	3.65	3.65	5.36	5.36	5.36	7.87	7.87	7.87	3.3	3.3	3.3
1.17			1.72			2.52			3.70			5.42			7.96			3.6		
1.18	1.18		1.74	1.74		2.55	2.55		3.74	3.74		5.49	5.49		8.06	8.06		3.9	3.9	
1.20			1.76			2.58			3.79			5.56			8.16			4.3		
1.21	1.21	1.21	1.78	1.78	1.78	2.61	2.61	2.61	3.83	3.83	3.83	5.62	5.62	5.62	8.25	8.25	8.25	4.7	4.7	4.7
1.23			1.80			2.64			3.88			5.69			8.35			5.1		
1.24	1.24		1.82	1.82		2.67	2.67		3.92	3.92		5.76	5.76		8.45	8.45		5.6	5.6	
1.26			1.84			2.71			3.97			5.83			8.56			6.2		
1.27	1.27	1.27	1.87	1.87	1.87	2.74	2.74	2.74	4.02	4.02	4.02	5.90	5.90	5.90	8.66	8.66	8.66	6.8	6.8	6.8
1.29			1.89			2.77			4.07			5.97			8.76			7.5		
1.30	1.30		1.91	1.91		2.80	2.80		4.12	4.12		6.04	6.04		8.87	8.87		8.2	8.2	
1.32			1.93			2.84			4.17			6.12			8.98			9.1		
1.33	1.33	1.33	1.96	1.96	1.96	2.87	2.87	2.87	4.22	4.22	4.22	6.19	6.19	6.19	9.09	9.09	9.09			
1.35			1.98			2.91			4.27			6.26			9.20					
1.37	1.37		2.00	2.00		2.94	2.94		4.32	4.32		6.34	6.34		9.31	9.31				
1.38			2.03			2.98			4.37			6.42			9.42					
1.40	1.40	1.40	2.05	2.05	2.05	3.01	3.01	3.01	4.42	4.42	4.42	6.49	6.49	6.49	9.53	9.53	9.53			
1.42			2.08			3.05			4.48			6.57			9.65					
1.43	1.43		2.10	2.10		3.09	3.09		4.53	4.53		6.65	6.65		9.76	9.76				
1.45			2.13			3.12			4.59			6.73			9.88					
Values per decade:																				
192	96	48	192	96	48	192	96	48	192	96	48	192	96	48	192	96	48	24	12	6

* ±0.1, ±0.25, and ±0.5%.

TABLE 2   Standardization Classes of Fixed and Variable Resistors and
Capacitors

Fixed and variable resistors	Fixed capacitors
General-purpose, ≤ 2 W	General-purpose ceramic
Medium-power, >2 to 6 W	Stable and temperature-compensating ceramic
High-power, >6 to 210 W	Stable, low-loss, low-value glass, mica, porcelain
High-precision	General-purpose miniature electrolytic
Packaged networks	Power-filter electrolytic
Low-power controls	General-purpose medium-value paper and film
Medium-power controls	Stable, medium-value—plastic film
Power rheostats	AC power paper, electrolytic
Low-power trimmers	Energy-storage paper, electrolytic, mica, bentonite
Multiturn controls	Low-frequency RFI filter paper, paper-film
Function generators	High-frequency RFI filter ceramic, mica
	Packaged networks, chips

Identification marking   Standardized marking schemes have been devised for resistors and capacitors using color coding, shape conventions, and alphanumeric designations.   Agreement is, unfortunately, not complete but is nevertheless quite good for industry-wide standards.   In addition to value and tolerance marking, some methods also label part type, "failure rate level," lot control number, and date code along with the needed polarity and function markings for electrical terminals.

Color banding is particularly adaptable to cylindrical parts having axial leads, since it avoids manipulation of an installed part to expose typographical markings otherwise masked by orientation of the part body.   Significances of nonmetallic colors are well established in published standards with good agreement.   Military and industry standards now refer to EIA Standard RS-359 (ANSI C83.1-1969), reaffirmed by ANSI in 1973; for color hue, value, and chroma limits.   Table 3 shows general color—number significance and accepted standard abbreviations.   Colors retain their significance for terminal numbering except when convention provides an overriding functional significance for the color.   Capacitor color codes suffer from a lack of standardization in general but agree well in value and tolerance significance.

Resistor Color Codes.   Color coding for resistors is subject to damage from high dissipation temperatures.   Also, color bands do not adhere well to some encasing materials, although large improvements have been made in both respects by current materials technology.   Their use to designate value and tolerance therefore finds greatest utility in low-wattage general-purpose types, with alphanumeric methods being employed for others.   A few manufacturers use both.   Three-color value codes are easier to memorize to a degree of one-look recognition but are limited to two significant figures and cannot adequately identify decade values more finely graduated than 5 percent (see Table 1).   Four-band value codes are used in some military and EIA specifications and are widely specified for OEM in-house standards.   Figure 1 illustrates the standard resistor color banding in use by both military and EIA publications.

Variable resistor terminals are sometimes designated by standardized colors that do not completely follow the numerical significance of Table 3.   Figure 2 shows the MIL-STD-1285, EIA RS-345, and EIA RS-360 standard for lead-wire insulation color, terminal numerical designation, and required schematic labeling on lead-screw-actuated trimmers.   Colors shown may be used for pin or lug terminals but are mandatory for flexible insulated leads.   Both the EIA and military standards require schematic labeling for adjustment trimmer resistors.   Resistance value and power ratings of other variables as well as "taper" or function are covered by alphanumeric conventions (see information below).

Capacitor Polarities.   A number of methods are in use for designation of capacitor terminals (Fig. 3).

Capacitor Color Codes.   Standardization of capacitor color codes has been achieved only to a point of good agreement for capacitance-value designators.

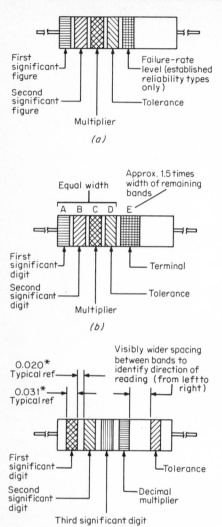

First significant figure

Second significant figure

Multiplier

Failure-rate level (established reliability types only)

Tolerance

(a)

Equal width

Approx. 1.5 times width of remaining bands

A B C D E

First significant digit

Second significant digit

Multiplier

Terminal

Tolerance

(b)

0.020* Typical ref

0.031* Typical ref

Visibly wider spacing between bands to identify direction of reading (from left to right)

First significant digit

Second significant digit

Third significant digit

Tolerance

Decimal multiplier

(c)

Fig. 1 Color coding of resistors. (*a*) MIL-R-39008 composition and certain wirewound types. MIL-STD-1285 specifies two significant figures for 5, 10, and 20 percent purchase tolerances. EIA RS-172 composition and RS-344 wirewound-type markings are identical to this except that the failure-rate band is omitted, the first band is wide for wirewound types, and the tolerance band is omitted for 20 percent tolerance. (*b*) Film type per MIL-R-22684. Gold D band is ±5 percent tolerance and red is ±2 percent. White E band denotes solderable leads. Although now superseded by non-color-coded MIL-R-39017 types, these parts may still be encountered. (*c*) Film type per EIA RS-196 with three significant digits. *NOTE: All band widths and spacing are for reference only and may be scaled up or down.

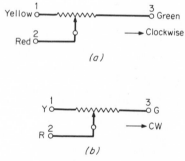

(a)

(b)

Fig. 2 Circuit diagram and terminal identification for variable resistors (rotation is operating shaft, viewed from operated end). (*a*) Required by MIL-STD-1285 and EIA RS-360. (*b*) Alternate allowed by MIL-STD-1285.

Colors are useful for quick recognition of different values where the basic generic type is obvious, but schemes to cover type designations, temperature characteristics, and other information have been largely unsuccessful because of their complexity and the profusion of capacitor types. While military specifications have virtually abandoned colors in favor of alphanumeric markings, EIA standards still define color codes, and color-coded MIL-C-20 ceramic capacitors or MIL-C-5 molded mica units may still be found. Internal company standard-value designator codes probably find widest usage. Tables 4 and 5 and Figs. 4 through 8 depict color codes for various capacitor types. The difficulties are easily seen.

Alphanumeric Labeling. Part-numbering systems incorporating functional value and other information, or "significant part numbers," are in almost universal use in both military and industry standards. In some instances, however, sequential nonsignificant dash numbers are used, and a value and type listing must be provided to all who have a need to identify parts. These lists are frequently keyed to a specification which must be referred to for value information. While sequentially

TABLE 3 General Color-Number Significance for Resistors and Capacitors and Approved Standard Abbreviations for Color Names

Color	Number	Multiplier, EIA-MIL		Resistor value[d] tolerances		Abbreviations			MIL-STD-1285 part-type identifier	MIL-STD-1285 failure rate[e]
		Resistor	Capacitor	MIL resistor, (±)%	EIA resistor, (±)%	MIL-STD-12	EIA 3-letter	EIA alternate		
Black	0	1	1	20		BLK	Blk	BK	Capacitor	L ( )
Brown	1	10	10	1	1	BRN	Brn	BR		M (10^4)
Red	2	10^2	10^2	2	2	RED	Red	R,RD		P (10^3)
Orange	3	10^3	10^3			ORN	Orn	O,OR		R (10^2)
Yellow	4	10^4	10^4c			YEL	Yel	Y		S (10)
Green	5	10^5			0.5	GRN	Grn	GN,G		
Blue	6	10^6			0.25	BLU	Blu	BL		
Violet	7	10^7			0.10	VIO	Vio	V		
Gray	8		$10^{-2}f$		0.05	GY	Gra	GY		
White	9		$10^{-1}f$			WHT	Wht	WH,W		
Gold[a]		10^{-1}	$10^{-1}g$	5	5	(a)	Gld			
Silver[b]		10^{-2}	$10^{-2}g$	10	10	SIL	Sil		Inductor	

[a] MIL-STD-12B uses the chemical symbol Au for gold.

[b] Metallic colors do not have EIA color standards.

[c] Not included in MIL-C-20.

[d] Capacitor value tolerance and stability characteristic designators vary widely and are covered in separate tables to avoid confusion.

[e] Failure rates are given in failures per 10^9 part-hours (FITS), symbol L has its value assigned by the individual part specification.

[f] EIA RS 198A ceramic TC and EIA RS 335-A composition capacitors.

[g] EIA RS 153B mica capacitors.

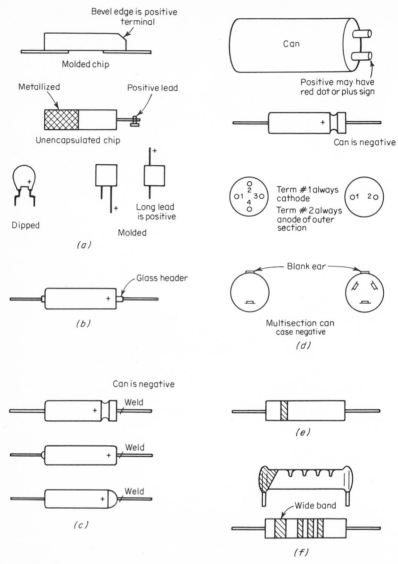

Fig. 3 Capacitor polarity conventions. (*a*) Tantalum, solid electrolyte (polarized). (*b*) Military tantalum (polarized). Hermetic (wet or solid) metal casing. (*c*) Tantalum, wet electrolyte (polarized). Red bands or dots are also sometimes used on electrolytic capacitors to denote anode. (*d*) Aluminum foil, electrolytic (polarized). (*e*) Paper, plastic film (nonpolarized). Band denotes connection for outside foil. (*f*) Ceramic (nonpolarized). Wide band and colored ends denote inner electrodes.

ordered dash-number part listings make for orderly arrays of categories in ascending value or size without disrupting the value progression with decimal multiplier numbers, the small clerical advantage is overwhelmed by the inefficiencies of part-label inscrutability. Part numbers should incorporate value designation for maximum utility.

Value designators are almost universally a three- or four-digit sequence denoting

TABLE 4   Color Coding of MIL-C-20 Temperature-Compensating Ceramic Capacitors

Color	Characteristic* ppm/°C	Characteristic* Letter designator	Nominal capacitance, pF First and second significant figures	Nominal capacitance, pF Multi-plier†	Capacitance tolerance For nominal capacitances greater than 10 pF, % (±)	Capacitance tolerance For nominal capacitances of 10 pF or smaller, pF (±)
Black	0	C–	0	1		2.0 (G)
Brown	–30	H–	1	10	1 (F)	
Red	–80	L–	2	100	2 (G)	0.25 (C)
Orange	–150	P–	3	1,000		
Yellow	–220	R–	4			
Green	–330	S–	5		5 (J)	0.5 (D)
Blue	–470	T–	6			
Purple (violet)	–750	U–	7			
Gray			8	0.01		
White			9	0.1	10 (K)	1.0 (F)
Gold	+100	A–				

* The characteristic is a two-letter symbol identifying the nominal temperature coefficient and the tolerance envelope for the temperature coefficient, respectively. However, the characteristic band or spot identifies only the nominal temperature coefficient.

† The multiplier is the factor by which the two significant figures are multiplied to yield the nominal capacitance. The lowest possible numerical multiplier is used to avoid alternate coding; for example 0.5 pF should be green, black, gray not black, green, white.

TABLE 5   Color Coding of EIA Standard RS-228 Tantalum Capacitors

Color	Significant figures	Multiplier	Capacitance tolerance, %
Black	0		
Brown	1		
Red*	2	100	
Orange	3	1,000	
Yellow	4	10,000	
Green	5	100,000	
Blue	6	1,000,000	
Violet	7		
Gray	8		
White	9		
No color			±20
Silver			±10
Gold			± 5

* Red may also be used to identify the positive lead.

respectively two or three significant figures followed by a decimal multiplier number. Resistor values are given in ohms, and capacitor values normally in picofarads. For resistor values with value order less than the number of significant figures, a letter R is sometimes inserted to depict the decimal-point position. Thus, 1R1 signifies 1.1 Ω. Initial value tolerance at time of purchase, or "purchase tolerance," is stated as part of the value code by means of well-standardized letter designators. Table 3 shows value-tolerance letters, and examples of value and tolerance marking. The complete part numbers of military and EIA standards on resistors and capacitors begin with three-letter groups to denote part variety and governing specification, followed by a two-number designation of subspecies. Various letter-number combinations are then added to give value and characteristics such as voltage rating, temperature coefficient, reliability rating, or other traits. Table 6 shows several examples of such labeling. Military designations have 10 to 13 characters, while complete EIA labels may have 15 or more.

Date coding is a useful practice, being of great aid in identifying blocks of articles likely to include a common variation. Convention follows the scheme required by military standards (see MIL-STD-1285). A four-digit year-week designator uses the last two digits of the calendar year followed by a two-digit week-of-the-year number. Preceding zeros are used for the first nine weeks of the year; thus 8204 designates 1982, the fourth week.

Military standard MIL-STD-1285 has a useful section covering priority for inclusion of labeling information in limited space, along with tabulations of spaces required for various letter type sizes, and space available on various part case sizes. Capital letters and arabic numerals in sans serif (Gothic or Futura) type, of color contrasting with the background are required. Paper labels attached to parts are prohibited by MIL-STD-1285.

Size, shape, and terminal standardization Resistor and capacitors of established design have achieved a degree of standardized shape and size provided largely by practical .considerations (see Fig. 3). Newer types or those where geometry is a matter of arbitrary choice are likely to be found in a profusion of shapes and sizes. Even standard types vary, however, and a not uncommon mistake is to develop a circuit layout using the dimensions of a sample part only to discover that when the standard item is ordered from an alternate supplier, the lead diameter, case length, or other dimensions are larger. This condition arises when standard dimensions are specified as "maximum" in specifications.

Standard etched-circuit center-to-center pin spacings are preferred in increments of 0.200 or 0.100 in. Leads and terminals are furnished in many

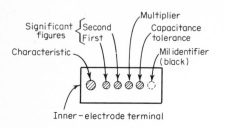

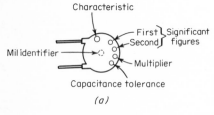

(a)

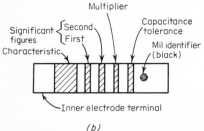

(b)

Fig. 4   Color-code marking of MIL-C-20 temperature-compensating capacitors. (*a*) Dot coding. (*b*) Band coding.

shapes, materials, and finishes for welding, soldering, plug-in, and other attachments. Each variation has advantages and limitations. Solder-filled hollow terminals are sometimes damaged by remelt during installation, or internal solder joints are melted by installation soldering to external terminals. Leads designed for weldability do not solder as easily as desired, and most readily solderable leads cannot be successfully resistance-welded. The most solderable leads after shelf storage appear to be those having a heat-fused tin-lead finish applied over a scrupulously cleaned copper surface, although glass-to-metal sealing needs sometimes preclude this choice. Tin-lead may be electroplate- and hot-oil-fused or hot-dipped. Nickel surfaces are difficult to solder unless activated fluxes and relatively high solder temperatures are used. Plated-on finishes may dissolve away in an initial soldering operation, leaving a poor surface for later resoldering if part removal and replacement is required. Tin-lead solders have a particular affinity for silver and gold, and too thick platings sometimes

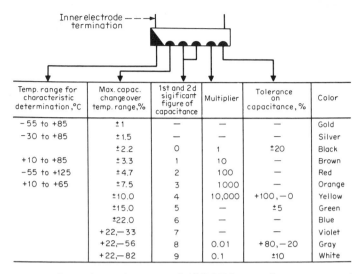

Inner electrode — — —→
termination

Temp. range for characteristic determination,°C	Max. capac. change over temp. range,%	1st and 2d sigificant figure of capacitance	Multiplier	Tolerance on capacitance, %	Color
−55 to +85	±1	—	—	—	Gold
−30 to +85	±1.5	—	—	—	Silver
	±2.2	0	1	±20	Black
+10 to +85	±3.3	1	10	—	Brown
−55 to +125	±4.7	2	100	—	Red
+10 to +65	±7.5	3	1000	—	Orange
	±10.0	4	10,000	+100,−0	Yellow
	±15.0	5	—	±5	Green
	±22.0	6	—	—	Blue
	+22,−33	7	—	—	Violet
	+22,−56	8	0.01	+80,−20	Gray
	+22,−82	9	0.1	±10	White

Fig. 5 Color-code marking of EIA standard RS198 general-purpose ceramic capacitors. (*Electronic Industries Association.*) 1. Use lowest decimal multiplier to avoid alternate coding; for example, 2.0 pF should be red, black, white, not black, red, black. 2. Listing of complete range of characteristics does not necessarily imply commercial availability of all values but is for the purpose of providing a standard identification code for future development.

degrade solder-joint strength by adulteration of the molten solder. Resistance welding requires good plating-thickness control, good lead-cross-section uniformity, smooth lead surfaces, and careful selection of platings. For welding, copper is generally to be avoided, as well as tin and tin-lead finishes, the former because of its high electrical conductivity compared with usual circuit interconnect materials and its malleability, the latter because of welding-electrode buildup and splatter. Alloys of nickel and iron, nickel, or iron-nickel cored wires are most used for welding. MIL-STD-1276 provides weldable and weldable-solderable lead-wire standards (see Table 7). Users are advised to minimize the number of sizes and types to reduce welding-machine pressure and current settings needed for producing a given assembly. Case-material mold flash and encapsulation menisci cause problems if they are present, or if removed by abrasive methods, lead surfaces may be roughened to a degree that will inhibit weldability. In general, circuit interconnection near resistor and capacitor body-to-lead transitions should be avoided because of the likelihood of poor finish and the proximity to internal joints, unless, of course, the part in question is of a design that avoids these problems.

Plastic coatings, cases, and encapsulants A large variety of organic materials and many fabrication methods are employed to encase and protect resistors and capacitors. Almost all plastics are eventually pervious to moisture and other fluids, though the degree of permeability varies widely. Good encapsulations can easily withstand short-term exposures to fluids without penetration and will provide protection from handling damage, as well as improve mechanical package strength. Vacuum wax-impregnated resins, once very popular, are giving way to plastic resin-impregnated coatings, and dipped, molded, and potted configurations. Circuit assemblies using resistors and capacitors are frequently encapsulated in foams, baking varnishes, or other systems which often impose package limitations. Curing temperatures of circuit assembly treatments as well as exothermic curing reactions must be carefully evaluated in conjunction with component-part package-temperature limitations. Assembly encapsulants can also exert surprisingly high stresses on component cases and terminal leads during thermal excursions and should also be considered when part encasements are evaluated. Conformal circuit assembly coatings of hard resins are notorious in this regard, as are injection or compression moldings using

TABLE 6 Examples of Labeling for Resistors and Capacitors

MIL-C-19978 capacitors*
Established reliability

CQR09	A	1	M	C	152	K	1	M
ER style	Terminal	Circuit	Characteristic	Voltage	Capacitance	Capacitance tolerance	Vibration grade	Failure-rate level

Standard

CQ09	A	1	M	C	152	K	1
Non-ER style	Terminal	Circuit	Characteristic	Voltage	Capacitance	Capacitance tolerance	Vibration grade

RS-401 capacitors

A	02	B	106	P	06	K	G
Case	Peak voltage	Characteristic	Capacitance	Finish	Terminal	Capacitance tolerance	Ground lug (optional)

RS-198 capacitors

CCXXXX	U2J	470	G	501
Style	Characteristic	Capacitance	Tolerance	Voltage

MIL-C-39006 capacitors*

Marking example:	M39006 01-3001	Part number
	JAN TM	JAN brand and trademark
	12345	Source code
	7015A	Date code and lot symbol
	15μF 15V	Capacitance and rated voltage

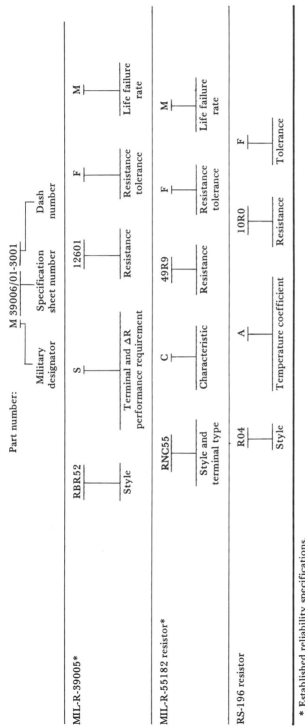

Part number: M 39006/01-3001

| | Military designator | Specification sheet number | Dash number | |

MIL-R-39005*

RBR52   S   12601   F   M

Style | Terminal and ΔR performance requirement | Resistance | Resistance tolerance | Life failure rate

MIL-R-55182 resistor*

RNC55   C   49R9   F   M

Style and terminal type | Characteristic | Resistance | Resistance tolerance | Life failure rate

RS-196 resistor

R04   A   10R0   F

Style | Temperature coefficient | Resistance | Tolerance

* Established reliability specifications.

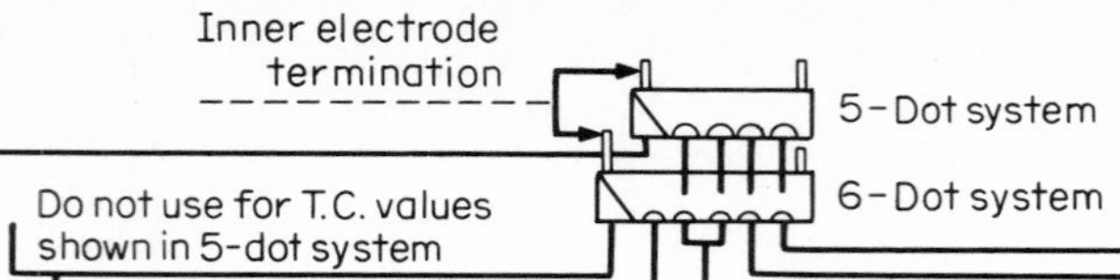

Temperature coefficient of capacitance (5-dot system) ppm/°C	Significant figure of temperature coefficient of capacitance (6-dot system)	Multiplier to apply to significant figure of temperature coefficient (6-dot system)	Color	1st and 2d significant figure of capacitance	Decimal multiplier of capacitance (use lowest possible numerical multiplier)	Tolerance of capacitance	
						Nominal 10 pF or less, pF	Nominal over 10 pF, %
0	0.0	-1	Black	0	1	±2.0	±20
-33	1.0	-10	Brown	1	10	±0.1	±1
-75	1.5	-100	Red	2	100		±2
-150	2.2	-1000	Orange	3	1000		±3
-220	3.3	-10000	Yellow	4	10000		
-330	4.7	+1	Green	5		±0.5	±5
-470	7.5	+10	Blue	6			
-750		+100	Violet	7			
General purpose (Note A)	General purpose (Note C)	+1000	Gray	8	0.01	±0.25	
General purpose (Note B)		+10000	White	9	0.1	±1.0	±10

Inner electrode termination

Do not use for T.C. values shown in 5-dot system

5-Dot system

6-Dot system

Fig. 6 Color-code marking of EIA standard RS198 temperature-compensating capacitors. (*Electronic Industries Association.*) Notes: (*A*) This is a general-purpose capacitor having any nominal temperature coefficient between +150 and −1500 ppm/°C; coefficient to be used at option of capacitor manufacturer. (*B*) This is a general-purpose capacitor having any nominal temperature coefficient between +100 and −750 ppm/°C; coefficient to be used at option of capacitor manufacturer. (*C*) This is a general-purpose capacitor having any nominal temperature coefficient between −1000 and −5200 ppm/°C; coefficient to be used at option of capacitor manufacturer. Use with multiplier color of black.

Fig. 7 Color-code marking of EIA standard RS335 composition capacitors with ratings as follows: Working potential, 500 V; Operating temperature, 85°C. (*Electronic Industries Association.*)

Color	First and second bands: first and second significant figure of capacitance	Third band decimal multiplier	Fourth band: capacitance tolerance, %	Capacitance-tolerance designation
Black	0	1		
Brown	1	10		
Red	2			
Orange	3			
Yellow	4			
Green	5			
Blue	6			
Violet	7			
Gray	8	0.01		
White	9	0.10		
Gold			± 5	J
Silver			± 10	K
No color band		Use lowest possible numerical multiplier	± 20	M

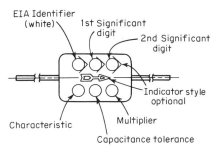

Fig. 8 Color-code marking of EIA standard RS153 mica capacitors. The multiplier is the factor by which the two significant figures are multiplied to yield the nominal capacitance.

Color		Characteristic		Capacitance		
	Letter designator	Temperature coefficient of capacitance, ppm/°C	Maximum capacitance drift	First and second significant figures	Multiplier	Capacitance tolerance, %
Black				0	1	± 20 (M)
Brown	B	Not specified	Not specified	1	10	± 1 (F)
Red	C	± 200	$\pm (0.5\% + 0.1$ pF)	2	100	± 2 (G)
Orange	D	± 100	$\pm (0.3\% + 0.1$ pF)	3	1,000	
Yellow	E	-20 to $+100$	$\pm (0.1\% + 0.1$ pF)	4	10,000	
Green	F	0 to $+70$	$\pm (0.05\% + 0.1$ pF)	5		± 5 (J)
Blue				6		
Purple				7		
Gray				8		
White				9		
Gold					0.1	$\pm \frac{1}{2}$ (E)
Silver					0.01	± 10 (K)

filled rigid thermoset resins. Standard specifications are frequently not specific about plastic materials and case fabrication, and this aspect should be determined if significant in the application. Thermoset resins are solidified at temperatures of 100°C and up, and composite assemblies, though stress-free at that temperature, become constantly stressed because of wide thermal-expansion-coefficient differences as they

TABLE 7   Weldable-Lead Standardized Types, MIL-STD-1276B
18 to 28 AWG wire and 0.004- by 0.012-in. ribbon

MIL—STD-1276 type	Trademark or name	Description (basic)	Plating*
D	Dumet	Cu-sheathed Fe-Ni alloy core	Ni + Au
K	Kovar	Fe-Ni-Co alloy	Au or Ni + Au
N-1	Nickel A	Ni	Bare
N-2	Nickel A	Ni	Au
N-3	Nickel A	Ni	Sn-Pb or Sn coating
C	Tinned copper	Copper (trace Ag)	Sn-Pb or Sn coating
F	Driver-Harris alloy 52	Ni-Fe alloy	Au or Ni + Au

* MIL-STD-1276 defines options on platings but controls limits on thickness and types. Limits on alloy constituents are also specified.

are cooled to room temperature. Soft-solder connections should never be placed in tension at any temperature the part will encounter, since even only a moderate force of 1 lb or so maintained overnight can result in creep-induced cracks, as can be easily demonstrated.

Ceramic and glass coatings and cases   Vitreous-enamel coatings used on some power resistors provide a hard, smooth, high-temperature encapsulation that is highly resistant to moisture penetration and solvents, is not flammable, and does not produce gases at ordinary elevated temperatures. The coatings are difficult to apply void-free, however, requiring layered applications and high-temperature firing. Enamels may contain active alkali salts in small concentrations. The major user problem is occasional susceptibility to cracking and crazing at terminal lead connections during installation and handling, and as a result of thermal shocks.

Glass encasement provides nearly inert truly hermetic high-temperature protection if well designed and executed. Glass-metal sealing is sometimes difficult, with material thermal-expansion differentials and glass-to-metal bonding the two major hurdles. Glass packages are naturally subject to the mechanical limitations of their materials, and glass cracking, chipping, and fracture are not uncommon handling and installation problems with some case designs.

Ceramic cases provide another means of achieving truly hermetic seals, usually through metal soldering of metallized areas. The metallization may be produced by painting on a metal-particle liquid suspension, then firing to blend the metal and ceramic interface and produce a solderable metal surface. Comparative mechanical characteristics of some common ceramics are presented in Table 8. Glass and ceramic packages tend to be large, heavy, and expensive in comparison with plastic packages. Though excellent for some applications, where their obviously superior protection is needed, standardization trends for resistors and capacitors do not depict increases in popularity of glass and ceramic-cased units.

Metal cases   Solder-sealed metal cases have been used for resistors, but present metal-cased varieties use cement or plastic-resin encapsulation within metal structures that provide a heat-dissipating function. Many capacitor types, by contrast, are commonly supplied in metal-cased designs. Automotive paper-foil units are enclosed in plated steel cases with usually a swaged-on plastic plug seal. Paper and paper-plastic radio-frequency filter units also use similar steel or brass cases, and military paper- and film-wound capacitors have, for years, used solder-sealed brass and steel cases having glass headers. Electrolytic capacitors present special encasement problems.

Aluminum electrolytic units use pure aluminum that cannot be hermetically sealed because of the necessity for maintaining an absolutely monometallic system, and the present nonavailability of a method of hard-sealing glass or ceramic terminals to aluminum to provide a durable hermetic seal. Aluminum units are therefore resin or plastic plug-sealed, usually with an elastomeric gasket. Tantalum electrolytic

TABLE 8 Nominal Comparison Characteristics of Common Ceramic and Glass Materials

Property	Units	Steatite, $MgO \cdot SiO_2$	Porcelain (electrical), $SiO_2 + 3Al_2O_3 \cdot 2SiO_2$	Alumina 96%, $Al_2O_3 + SiO_2$	Alumina 99.5% $Al_2O_3 + SiO_2$	Beryllia 99.5%, BeO	Glass (general)	Glass (quartz), 99.5% SiO_2
Coefficient of linear thermal expansion	$(10^{-6})\,°F^{-1}$ $(10^{-6})\,°C^{-1}$	3.9-5.6 (7.0-10)	2.2-3.3 (4.0-6.0)	3.6-4.4 (6.4-8.0)	3.3-4.4 (6.0-8.0)	2.3-5.2 (4.2-8.0)	0.4-0.7 (0.8-1.3)	0.22 (0.4)
Thermal conductivity at 25°C	$Btu\ in\ h^{-1}\ ft^{-2}\ °F^{-1}$ $J\ m\ s^{-1}\ m^{-2}\ °C^{-1}$	41 (5.9)	20 (2.88)	244 (35.19)	255 (36.84)	1,596 (230.17)	4.8-9.6 (0.7-1.4)	10.40 (1.50)
Thermal conductivity at 300°C	$Btu\ in\ h^{-1}\ ft^{-2}\ °F^{-1}$ $J\ m\ s^{-1}\ m^{-2}\ °C^{-1}$	29 (4.19)	18 (2.55)	119 (17.16)	128 (18.42)	813 (116.37)	4.8-10.3 (0.7-1.5)	10.82 (1.56)
Specific heat	$Btu\ lb^{-1}\ °F^{-1}\ cal\ g^{-1}\ °C^{-1}$	0.26	0.22	0.22	0.31	0.28	0.20	0.18
Heat capacity	$J\ g^{-1}\ °C^{-1}$	1.09	0.92	0.92	0.88	1.17	0.84	0.75
Melting temp (approx)	°F °C	2642 (1450)	3182 (1750)	3272 (1800)	3272 (1800)	4658 (2570)	1382 (750)	3110 (1710)
Operating temp (max practical)	°F °C	1832 (1000)	2192 (1200)	2822 (1500)	2912 (1600)	2732 (1500)	815 (435)	2012 (1100)
Density	$lb\ ft^{-3}$	168.2	149.5	230.5	237.2	180.7	155.7	137.1
Specific gravity	$g\ cm^{-3}$	(2.7)	(2.4)	(3.7)	(3.8)	(2.9)	(2.5)	(2.2)
Thermal-shock rating		Fair-poor	Good	Good	Good	Good	Fair	Excellent
Dielectric constant (at 25°C, 1 MHz)	K	5.9-6.1	5.0-6.5	9.1-9.4	9.6-9.9	5.8	3.5-15.0	3.5-4.0
Volume resistivity (at 100°C)	$\Omega\ cm^2\ cm^{-1}$	10^{16}	10^{14}	10^{16}	10^{16}	10^{16}	10^{12}	10^{14}-10^{18}
Compressive strength	$lb\ in^{-2}$ $kg\ cm^{-2}$	90,000 (6,330)	50,000 (3,500)	375,000 (26,360)	380,000 (26,720)	225,000 (15,800)	100,000 (7,030)	160,000 (11,250)
Tensile strength	$lb\ in^{-2}$ $kg\ cm^{-2}$	10,000 (700)	7,000 (490)	25,000 (1,760)	28,000 (1970)	15,000 (1,060)	4,000 (280)	7,000 (490)
Flexural strength	$lb\ in^{-2}$ $kg\ cm^{-2}$	20,000 (1,400)	10,000 (700)	48,000 (3,375)	48,000 (3375)	25,000 (1,760)	10,000 (700)	15,000 (1,060)
Young's modulus	$(10^6)\ lb\ in^{-2}$ $(10^6)\ kg\ cm^{-2}$	16 (1.12)	17 (1.20)	47 (3.5)	55 (4.1)	49 (3.45)	7-9.5 (0.49-0.67)	10 (0.70)
Hardness, Rockwell scales			15N-90	45N-78	45N-80	45N-65	N/A	N/A

types of the wet or semiwet variety have similar limitations and are glass-to-metal sealed only in units having special provisions (i.e., a titanium case or a secondary case with tandem seal). Platinum-plated silver, titanium, and a few other materials are used for the primary cases, with secondary cases being any convenient metal. Solid electrolyte tantalum units do not require metal encasement and can be obtained in a variety of plastic cases as well as in solder-sealed glass-terminal metal cases. Most types of metal-cased capacitors have one terminal connected to the case, and units are often sleeved with paper or plastic to prevent inadvertent short circuiting. Ceramic and mica radio-frequency suppression standoff and feedthrough types use hermetic or resin-filled metal cases having mounting means for optimized low-inductance connection to a conductive ground plane. These styles are discussed in more detail in Chap. 8.

FIXED RESISTORS

General Information

Resistance is the scalar property of an electric circuit which determines the rate at which electrical energy is converted into thermal energy while a given electric current is flowing. The property is analogous to viscous friction losses in mechanical systems. In its mathematical generalization resistance value is a function of the electric current, being equal to the dissipated thermal power divided by the square of the current. In practical usage, however, the resistance property is considered to be independent of the current flow. Fundamentally, a potential difference of 1 volt across a resistance of 1 absolute ohm is associated with a charge flow of 1 coulomb per second (1 ampere) and a thermal dissipation of 1 watt (1 joule per second). The international ohm, used prior to 1948, is 1.00049 absolute ohms, the absolute ohm being unity by definition, and now almost universally used. The unit is designated in honor of Dr. Georg Simon Ohm, who, in 1827, demonstrated the electromotive force–resistance current relationship.

Excluding, as a class, those devices designed specifically as heating elements, resistors are devices constructed for functional applications in electrical circuits. Basic design considerations are convenience of installation, stability of properties over a desired range of electrical and environmental exposures, and functional longevity. Further desired properties include small physical size, good mechanical strength, stability of resistance value, efficient thermal dissipation, ease and reliability of circuit connection, purity of resistive function over broad electrical frequency range, wide range of available resistance values, good dielectric strength, nonflammability, low noise production, good producibility, avoidance of catastrophic failure mechanisms, and low cost.

General construction Changes in resistance value are associated with physical, mechanical, and chemical changes in the functioning structure and materials of a resistor. Designs should therefore utilize stress-relieved mechanical systems, low operating hot-spot temperatures, and chemically and galvanically compatible stable materials in low-energy states. With few exceptions, the functional element of fixed and variable resistors is one of three types:

1. A core of electrically nonconductive material having resistance wire wound onto the core, with the core providing support for the wire and the terminals

2. A core of electrically nonconductive material having a film of resistance material bonded to the surface with the core also supporting the terminals

3. A solid pellet of resistance material formed around suitable terminations

Various terminal types and core materials are used with different resistive materials, cases, and exterior coatings. Cylindrical shapes are the most common, being volumetrically efficient as a heat dissipator for a two-lead device. Other shapes are used to obtain desired variation in performance or installation. Coaxial disks are made specially for radio-frequency applications, flat-chip types are provided for hybrid circuits, and even threaded studs on flat plates are provided for heat-dissipator mounting. Figure 9 shows an assortment of fixed-resistor construction methods. In general, the power rating of a given resistor is associated with its physical size and resulting ability to dissipate the heat resulting from its electrical-energy conversion at a given rate. Designed efficiency of the thermal system and varying high-temperature capabilities of constituent materials naturally modify this general circumstance.

Critical resistance value A maximum terminal-to-terminal voltage limitation is reached for a given case size and resistance-element design. This limit is expressed as a maximum voltage rating and is considered to be dc or rms at line frequency. Turn-to-turn and end cap–to–end cap dielectric properties under worst-case environmental conditions establish the part capability. This limitation leads to the definition of a "critical value of resistance" for each resistor size and power rating, being the highest-decade resistance value at which with voltage applied equal to rated voltage, power being dissipated at room ambient conditions does not exceed rated power. All parts of the same case size and power rating with resistance lower than the critical value are power-dissipation-limited, while those of higher resistance cannot dissipate rated power without exceeding the safe voltage that can be applied. Voltage ratings below the critical value of resistance are given by

$$\text{Voltage max } = \sqrt{\text{power rating} \times \text{resistance value}} = \sqrt{PR} \qquad (2)$$

If the ambient temperature in the area of a resistor under consideration is high enough to necessitate derating of the power dissipation, the voltage should be derated to a value determined by the square root of the ratio of the derated power to the full rated power. If operating environmental temperature range is precisely known and is always lower than the inflection point on a specified derating curve, it is possible to modify power ratings upward provided the voltage-rating maximum is observed and the hottest spot in the resistor is held to a temperature below that at which the resistor is rated to be capable of zero dissipation. Conservative application, of course, will stop well short of that temperature, particularly if low-resistance change in service is desired. Dissipation of one-half rated power is often used as a rule of thumb for conservative practice.

High-frequency performance Performance of conventional resistor shapes and types at high frequencies is of considerable interest to circuit designers. Unsuspected reactance of circuit elements sometimes causes elusive functional problems. Data derived from various sources, analytical studies as well as empirical data, show, as may be expected, that wirewound low-, medium-, and high-power designs made without inductance-cancellation winding techniques have rather spectacular impedance excursions, particularly in low-resistance values and in a range of about 10 to 200 mHz. Ayrton-Perry winding, which is probably the most effective, or other noninductive winding techniques, can be used to reduce this effect; but because of complexities of part configuration vs. resistance value, shape, size, frequency of interest, and their interactions, each part and its application must be separately considered. This applies to wirewound parts as well as to other types of resistors. The problems of resistance and high-speed, high-frequency performance are even more trying for precision wirewound low-power parts in high-resistance values. These parts require many turns of resistance wire and, despite the use of noninductive winding techniques, are limited to less than about 1 MHz for resistance of 1,000 Ω and up. Considerable difference exists between manufacturers.

Equivalent Circuit. A resistor can be considered to have an equivalent circuit as shown in Fig. 10. R_s is the series resistance of the element. In film resistors of value above 50 Ω, this property is essentially independent of applied frequency well into the gigahertz region. In very low resistance values, if not masked by other variations, skin effect in lead wires could be noticeable if lead wires are long and are of magnetic material. Skin effect also will theoretically cause variation of this basic quantity in wirewound types and in slug composition types, since cross-sectional current density becomes nonuniform. In practice, though, skin effect seems to be of minor importance for conventional resistor styles, and performance is dominated by other properties. It is convenient to define a series-inductive component L_s whose magnitude is important for wirewound types, as previously mentioned, but except for very low values of resistance is almost totally insignificant for all other types of resistors, even including helically spiraled film types. This is partially because of the very low value of inductance involved, but primarily because of the relatively more significant distributed and shunt capacitance, represented by C_p, and its increasing effect as frequency is increased.   Investigators report that lead length and dress seem to have more influence on inductive effect than variation in spiraling. This can be seen by calculating approximate inductance using a single-layer air-core coil-approximation formula

$$L \approx \frac{4\pi N^2 A}{B} \times 10^{-3} \qquad \text{microhenrys} \qquad (3)$$

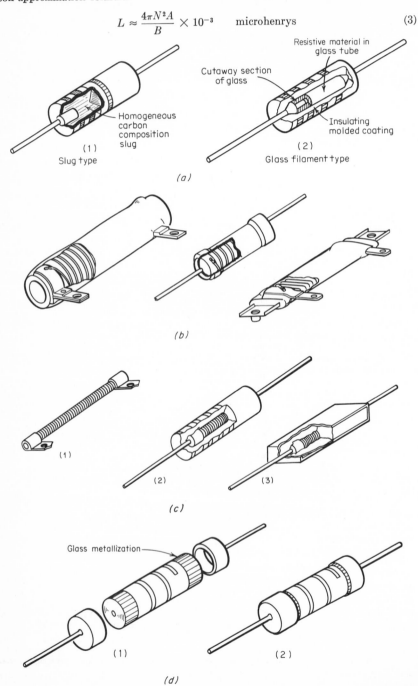

(a)
(1) Slug type — Homogeneous carbon composition slug

(2) Glass filament type — Resistive material in glass tube, Cutaway section of glass, Insulating molded coating

(b)

(c)
(1)   (2)   (3)

(d)
(1) Glass metallization   (2)

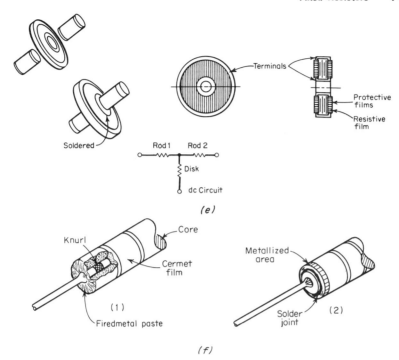

Fig. 9 Various resistor constructions. (*a*) Carbon composition resistor construction. (1) Molded composition pellet with phenolic outer layer. (2) Composition film on glass-tube filament with molded encasement, resistive material on glass tube. (*b*) Conformally coated wirewound power resistor construction. This basic shape runs from about 1¼ to 12 in. in length, with diameter and terminals scaled up in proportion, except that screw and nut terminals are furnished for 20 W and up. (*c*) Rod-type wirewound resistors, wound on resin-impregnated glass-fiber cores, chopped to length/value, compression-capped. (1) Automotive/appliance bare wire. (2) General-purpose molded case (note wide first color band). (3) Encapsulated in ceramic case with refractory cement. (*d*) End cap and lead to core assembly. Typical of most thin-film types, some thick films. (1) Core with spiraling ground in film, metallized termination areas. (2) End cap–lead assemblies pressed on, ready for coating, molding, or other encasement. (*e*) RF tee pad assembled from rods and disk. Pyrolytic-carbon deposition on ceramic with metallized terminal areas. (*KDI Pyrofilm.*) (*f*) Examples of film resistor lead attachments not using end caps. (1) Headed and knurled lead is pressed into alumina core center. Firing of metal paste bonds lead to core and forms conductive termination for cermet element. (2) Flat-headed lead is attached with high-temperature solder to copper plated on termination area at end of core and resistor element.

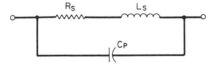

Fig. 10 Equivalent circuit of a fixed resistor at high frequency.

where *A* and *B* are the coil cross-sectional area and length, cm, respectively, and *N* is the number of turns. If a ⅛-in-diameter resistor having a ⅛-in spiraled length with 15 turns is assumed, the calculation yields

$$L \approx \frac{4\pi^2(0.062)^2(10)^{-3}(15)^2(2.54)}{0.125} = 0.7 \ \mu\text{H}$$

The reactance of this inductance at 100 MHz is $X_L = 2\pi f L = (2)(\pi)(100)(0.7) \approx$ 440 Ω, which is, for instance, 4.4 percent of a 10-kΩ nominal value. Lower resistances would have fewer spirals, and the shunt capacitance of about 0.1 to 0.8 pF will moderate and is likely to dominate the effect, so that inductance of ordinary film resistors appears only in the lowest values. At the same 100 MHz, for instance, 0.4 pF provides a shunt reactance of $X_C = (2\pi f c)^{-1} \approx 3,180$ Ω across our hypothetical $10,000 + j440$ Ω impedance.

Resistor Design Considerations. Special coaxial terminations which limit lead inductances, and the use of low resistances that are relatively insensitive to shunting capacitance and also do not require spiraling, can make film resistors useful far into the gigahertz range. For conventionally configured units above about 1,000 Ω, however, impedance starts to decrease with frequency above approximately 10 MHz. Nominal expectations for generic film types are shown in Fig. 11. These are the

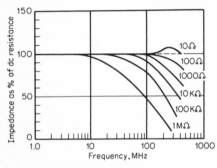

Fig. 11 High-frequency performance of thin-film resistors. (*MepCo—Electra.*)

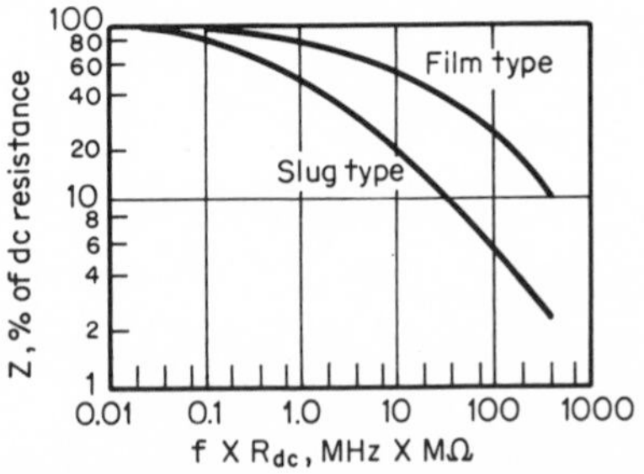

Fig. 12 Composite curve of frequency response for film- and slug-type fixed carbon composition resistors. (*Wellard.*)

usually encountered shapes of resistors. The best shape for axial-lead film resistors in high-frequency applications is obviously long and slender as opposed to short and thick. Carbon composition resistors, often used in rf applications, are also usable in low resistance values at high frequencies but drop off in resistance more rapidly than do film types as resistance value and frequency are increased. It is thought that capacitive-shunting effects occur in the dielectric particulates within the pellet structure. Composite curves are shown in Fig. 12. Glass-filament composition resistors are more nearly like film units and have somewhat higher-frequency resistance roll-off values than do pellet types.

Impedance. Specifying wirewound-resistor behavior in frequency bands where their reactances are significant is at best a complex problem. It is instructive to examine the analytical expressions involved in the interplay of frequency and impedance. The impedance of the circuit shown in Fig. 10 can be written as

$$Z = \frac{R_S X_C^2 + jX_C[X_L X_C - (R_S^2 + X_L^2)]}{R_S^2 + (X_L - X_C)^2} \quad (4)$$

where R_S is the true resistance, X_L is the reactance of the true series inductance, and X_c is the reactance of the equivalent parallel capacitance. Resistance may change slightly with frequency, as mentioned previously, owing to skin effect, and dielectric losses in the equivalent capacitance may introduce small variations, but neither effect is considered to be of major proportions and R_S is assumed to be constant. True inductance likewise is assumed constant. Capacitance will vary slightly as dielectric constants change with frequency and voltage, but again the effect is ignored here. Rewriting Eq. (4):

$$Z = \frac{R_S}{(R_S^2/X_C^2 + [(X_L - X_C)^2/X_C^2]} + j \frac{X_L - [(R_S^2 + X_L^2)/X_c]}{(R_S^2/X_C^2 + [(X_L - X_C)^2/X_C^2]} \quad (5)$$

allows inspection of the effects of frequency. At very low frequency, for instance, the X_c is very high, X_L is very low, the denominators of the real and reactive terms approach unity, and while the real part approaches the value of R_S, the imaginary part is nearly zero because of the extreme values of X_c and X_L. As frequency is increased, both real and imaginary portions are affected as the series and shunting effects interplay. Eventually the lowered capacitance reactance will dominate. The real part becomes reduced in value as its denominator increases. The imaginary portion is also reduced, because of signs and because of the arrangement of first-degree denominator terms above the line as opposed to second-degree denominator terms below the line. For resistors having low values of R_S, it is seen that X_L will play a more significant part in the impedance determination. After a "self-resonant" point is passed, the impedance always becomes capacitive and starts to decrease with frequency.

Noninductive Windings. The reactive effects can be reduced, if not eliminated, by the use of various "noninductive" winding techniques. Most users at first thought consider that if the inductance can be gotten rid of, the problem will be solved. The difficulties, of course, are not so simply eliminated, as we have seen, because not only is distributed capacitance also an important contributor, but leakage magnetic flux and unavoidable residual inductances remain. Single-layer close-wound or space-wound general-purpose and power resistors can easily use the Ayrton-Perry winding. This technique seems to be superior to others. Two wires are wound in opposite pitches and the two are connected in parallel. At the wire-crossing points, 180° apart on the core, the potentials are essentially equal. This low potential decreases the effect of distributed capacitance and allows the resistor, if desired, to be wound with uninsulated wire. Some suppliers use cores with flatted sides to allow precise location of the crossings and to allow the result with its double wire build at crossings to remain within a cylindrical envelope. A disadvantage of the Ayrton-Perry winding is that with the smallest usable resistance-wire diameter of the highest obtainable resistivity, the highest resistance value that can be obtained in a given finished-size part is halved because of the parallel connection.

A less effective noninductive winding, but one which does not require halving of the maximum value, is the Chapron winding. This technique simply reverses winding pitch in the center of the mandrel. Bifilar winding is one of the two most popular noninductive winding methods for precision wirewound multilayer resistors, but is seldom used for single-layer power units. For this method, the length of wire is bent in half and the resulting doubled wire is wound on the core. Each of the free ends then becomes a terminal, with one end being brought over the wound core to the end opposite that where the winding started. The result is two series-connected windings with the circuit current flowing in opposite directions in each with respect to the core. The potential difference between adjacent turns at the start ends is high, however, and the effects of distributed capacitance are correspondingly increased.

Another popular winding method for multilayer resistors, which is also not easily adaptable to power units, is the reverse-pi method. Here, the resistance winding is divided into an even number of approximately equal segments on a bobbin that is constructed with the needed segments separated by insulation barriers. One segment is wound, and the wire-running end is dressed into a slot to the next segment where the next segment is wound in a reversed direction. The process is repeated so that the end result is a series of oppositely wound segments. Leakage flux and distributed capacitance are not completely eliminated, however, and results are obtained about equivalent to that of the bifilar winding. With all currently used techniques the distributed capacitance eventually dominates as frequency is increased, resulting in a lowered impedance, with the roll-off effect occurring at highest frequencies for lower resistance. Inductive reactance and self-resonant peaks occur, causing impedance peaks for many wirewound values, particularly if they are not noninductively wound.

Determination of Frequency Effects. Manufacturers furnish data on product performance at radio frequencies, but the form of the information varies widely. Some specify inductance only—the parameter most suspected by circuit designers faced with the physical fact of a solenoidal coil. Others supply "resistance" as a percentage of dc resistance value. Some of these mean total impedance value, others mean the real component of the impedance, and most do not specify which is intended. The measured inductance is not the true inductance but is a composite of resistance and reactances and therefore has a frequency-sensitive value, so that measured inductance has little meaning unless a frequency is specified. One technique used is to wind a physical model of the resistor with copper wire and to measure the result. Examination of the equivalent circuit and its algebraic representation shows that this method has serious limitations. Calculated values using approximation formulas are also used, but such approximations have validity limited by freqency range and complexity of the coil configuration. For the user with a critical application, it will be prudent to obtain samples and conduct measurements or to construct worst-case in situ circuit tests. After the suitability of a given part is established, it can be assumed that resistors wound for the same value using wire of the same resistivity, of the same gage and on the same bobbins, using the same winding method, will closely repeat. It will probably be necessary to specify these variables. A measurement technique is described later in the chapter under Special Application Guidelines.

Temperature coefficient of resistance   Because resistors are heat-producing devices, their ability to remain stable in resistance value over a wide range of temperatures is important. The term coefficient is used to denote a variation that is small or essentially linear, while temperature characteristic is used to define a wide nonlinear variation, confined chiefly to carbon composition resistors. Manganin wire, for instance, has a parabolic resistance-temperature variation which is quite small and is still referred to as a coefficient. Coefficients given in parts per million per degree C (ppm/°C), percent per degree C (which is parts per hundred per degree C), or sometimes just a decimal coefficient number per degree C are valid only within the temperature range specified and are defined for the temperature average of the resistor body rather than the surrounding air temperature. Self-heating due to the power being dissipated thus must be included in determinations. Power-derating curves provide maximum hot-spot temperature information for free-air mountings but almost always will give a very conservative estimate. This is apparently because limitation on resistance changes in operating load life is a more compelling design constraint than hot-spot temperature limits.

For critical applications measurements are usually made in the laboratory. Carbon has a generally negative coefficient while metals tend to have positive coefficients. Some alloys, however, have been formulated to have zero or slightly negative values. Special-purpose devices are available to utilize linear positive temperature coefficients of special alloys, such as Balco,* which has an advertised plus 0.45 percent per degree C linear slope. In situations where temperature coefficient is especially critical, it may be necessary to consider retrace hysteresis and changes in temperature coefficient with part life. This variation is of secondary significance, however, and is not usually specified directly. Resistors are designed to operate at given maximum temperatures dependent upon the materials used. At full rated power, therefore, a high-temperature resistor will experience more resistance change for a given temperature coefficient of resistance than will a lower-temperature part or one that is operated at a fraction of its rated power. This consideration should be made when selecting resistor types. Resistance variation as a proportion of the total value tends to be greater with very low values of resistance, approximately 5 Ω and less. This is because lead-wire resistance, thermal elongations, and variables such as end-cap contact resistance that are insignificant at higher resistances play a greater part. Standards and specifications recognize this by allowing wider variations for low values.

Voltage coefficient of resistance   Changes in conductivity due to higher potential gradients across molecular interfaces cause resistivity to vary slightly with applied voltage. This is expressed as a coefficient of the nominal resistance (percent or parts per million) per volt. This quantity is specified to be independent of effect due to self-heating, and measurement is thus difficult. The effect varies from −700 ppm/V for the higher resistance values of carbon composition through about 5 to 30 ppm/V for carbon film and cermet, and from 10 to 0.05 ppm/V for metal film

* Trademark of Wilbur B. Driver Co., Newark, N.J.

and oxide films, although some thick-film types go as high as 400 ppm/V. Voltage coefficient is not usually of consequence in wirewound designs.

Resistor noise The noise output of a given resistor depends upon its thermal "white" noise plus a noise output due to the applied current. The latter portion depends upon the resistor design, while the former or Johnson noise depends upon the resistance and its temperature:

$$(Johnson)E_{rms} = 7.4 \sqrt{kT \,\Delta f} \times 10^{-12} \qquad (6)$$

where E is the noise voltage, R is resistance, T is absolute temperature in kelvins, and Δf is the frequency bandwidth of consideration.

A noise index is conventionally specified for a given resistor type. This is the ratio of rms noise voltage caused by a specified current flow through a resistor to the average dc voltage across the resistor, measured over one bandwidth decade at one specific hot-spot resistor temperature. The units are microvolts (μV) per volt or voltage-ratio decibels, where 0 dB is 1 μV:

$$\text{Noise index} = 20 \log_{10} \frac{\text{noise voltage, } \mu V}{\text{dc voltage}} \qquad (7)$$

The current-noise voltage from frequency f_1 to frequency f_2 is given by

$$(current)E_{rms} = V_{dc} \times 10^{\dfrac{\text{noise index, dB}}{20}} \times \sqrt{\log \frac{f_2}{f_1}} \qquad (8)$$

Where more than one frequency decade is of interest, the noise voltages add as the square root of the quantity of decades. 10 Hz to 100 KHz thus has an rms amplitude content twice that of 10 to 100 Hz (four decades have twice the noise of one decade). Except for possible noisy terminations and wire imperfections, wirewound resistors produce only the inevitable thermal noise. Metal-film resistors have about -30- to -40-dB indexes, with lower resistances having lower noise. Carbon film ranges about $+10$ down to -30 dB, with low values also having lower noise. Carbon composition ranges from about $+40$ dB for the highest values of $\frac{1}{10}$-W parts down to about 0 dB maximum for the lowest values of 2-W parts. Fired thick films range from 0 dB down to about -20 dB. For current-induced noise in the carbon types, the larger physical sizes have lower current densities for a given load, and are consequently less noisy. Figure 13 shows relative noise indexes for metal and carbon films. Total noise voltage is, of course, the sum of the thermal Johnson noise and the current noise.

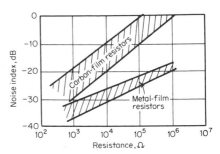

Fig. 13 Relative noise characteristics of carbon-film and metal-film resistors. (*MepCo-Electra.*)

Core materials Core materials are dictated by needs of the design. Filled thermoset organic resins such as epoxies are used as structural bases for low-power precision parts where power dissipation and high temperatures are not foremost considerations. Molded diallyl phthalate, phenolics, melamine, and similar materials also are used in variable resistors and in cost-efficient devices where power is low or where physical size can be conveniently made large enough to keep orperating temperatures within the capability of the materials. In comparison with metals, ceramics, and other inorganic materials, however, organic materials have large coefficients of thermal expansion, are relatively pervious to moisture, are more chemically active, and are generally less efficient thermal conductors. Ceramics and glasses are therefore extensively used in resistor designs. Nominal properties of various ceramics and glasses are given in Table 8. Comparison of thermal-expansion rates and thermal conductivities with those of nickel, chromium, iron, cobalt, brazing alloys, and other resistor materials will show a practical degree of structural compatibility. Excellent insulation properties, chemical inertness, and highly developed technologies account for their popularity. Automotive power wirewound resistors use woven-glass-fiber cores which combine mechanical flexibility with high-temperature capability and low cost. These types are often lug-mounted directly, which provides efficient heat transfer from winding to mounting. To miniaturize power resistors while controlling hot-spot temperatures, various techniques are used to improve the thermal system, one of which is to use core materials with higher thermal conductivity. Beryllia cores are widely used; also metallic aluminum-cored resistors can be purchased. Other types use thermally efficient jacketing to remove the heat, as is noted in the following paragraphs on power resistors.

Resistance wire Alloys used for resistance wire are mostly nickel-chromium, copper-nickel, and these metals alloyed with percentages of iron, aluminum, cobalt, manganese, aluminum, or other elements to obtain stability and tensile strength. Desired properties are high resistivity for large cross section, stability of resistance with temperature, chemical and physical inertness, high tensile strength in small diameters, workability, low thermoelectric effect when connected to other metals, and absence of deteriorative mechanisms associated with thermal cycling and high temperatures. Properties of some commonly used alloys are tabulated in Table 10. Manufacturer's type names are shown in Table 9. Melting temperatures of the alloy are seen to be generally quite high in comparison with temperatures at which electronic assemblies operate, which is a most welcome convenience in resistor-fabrication technology. In resistor service, however, considerations of stability and longevity require that operating limits be set at about 400°C for the most liberal situations involving the most durable alloys such as 80 percent nickel-chromium, while about 270°C is used for conservative ratings, with resistance standards and applications requiring close precision being even more limited. Surface oxidation and resulting cross-sectional changes, grain growth, embrittlement, insulation deterioration (if used), and similar effects generally progress faster as temperature is increased. One should not be puzzled by apparent inconsistency in use of those same alloys in heating-element designs operating at temperatures at 500°C and higher with good durability. These designs are usually more forgiving as to tolerable shifts in resistance, use relatively large-diameter wires, and avoid some of the embrittlement and grain-growth problems by their periodic excursions to the higher temperatures, which appear to have an annealing effect on many alloys. Resistivities and tensile strength in small cross section are dominating considerations in choices for providing higher resistance values in packages sizes that are thermally nominal for the dissipation rate needed. Wires are available to diameters as small as 0.0005 in and, in very special instances, even smaller. High-reliability military specifications have types qualified with 0.00075 diameter, but sizes smaller than 0.001 in diameter should be avoided if possible. For power resistors with adjustable taps requiring exposed wire, diameter should be no smaller than 0.005 in (AWG 36).

Resistance alloys are supplied in strip, ribbon, and wire of round and rectangular cross section. A wide range of insulations and finishes are available: bright finish, oxidized surface, oleoresinous enamel, wrapped silk and cotton fiber, glass fiber, ceramic, nylon coat, and the newer high-temperature resins. Polyvinyl acetal (Formvar*) and polyimides can be supplied. Good coatings are essential for use in designs such as high-value precision parts, where it is necessary to close-wind or random-wind the wire. Specifications and application guides for commercial magnet-wire coatings are useful in obtaining information on film coatings, since the materials are marketed under trandemarks well known in magnetic-apparatus design terminology. Temperature limitations and mechanical characteristics of these insulations are of primary importance in multilayer-resistor performances.

Resistance values Resistance-value selections should, if practicable, conform to established standard decades as covered at the beginning of this section. Most resistance bridges are capable of resolving values no closer than about 0.001 Ω. Thus if a 10-Ω value were specified, the closest percentage tolerance that could be reasonably measured would be 0.01 percent and a 0.1-Ω value should have a tolerance

* Trademark of Monsanto Chemical Company.

TABLE 9 Resistance Alloys: Typical Manufacturer's Types

ASTM class*	Type	Resistivity, Ω·cmil ft⁻¹	Wilbur B. Driver Co.	Driver Harris Co.	C. O. Jelliff Co.	Hoskins Mfg. Co.	Molecu-Wire Corp.	Kanthal Corp.	Nominal composition, %
3b	80-20 Ni-Cr	650–675	Tophet A‡	Nichrome V‡	Alloy A‡	Chromel A‡	Protoloy A‡	Nikrothal 8‡	80 Ni, 20 Cr
5a, b	Constantan†	294–300	Cupron†	Advance‡,§	Alloy 45‡	Copel‡	Neutroloy‡	Cuprothal 294‡	55 Cu, 45 Ni
6	Manganin	230–290	Manganin‡	Manganin‡	Manganin‡		Manganin‡		83–87 Cu, 10–13 Mn, 0–4 Ni
7	Alloy 180	180	Alloy 180	Midohm‡,§	Alloy 180	180 Alloy		Cuprothal 180‡	77 Cu, 23 Ni
9	Alloy 90	90	Alloy 90	Alloy 90§	Alloy 90		90 Alloy	Cuprothal 90‡	88–90 Cu, 10–12 Ni
10	Alloy 60	60	Alloy 60	Lohm‡	Alloy 60		60 Alloy	Cuprothal 60‡	94 Cu, 6 Ni
11	Alloy 30	30	Alloy 30	Alloy 30	Alloy 30		30 Alloy	Cuprothal 30‡	98 Cu, 2 Ni
8	Linear TC	120	Balco‡	Hytemco‡			Pelcoloy‡		70 Ni, 30 Fe
	Nickel A	60	A Nickel	Grade A Nickel‡	A Nickel	Alloy 651‡			99 Ni
	High purity Ni	58	Ballast‡ Nickel	99 Alloy‡	HP Nickel				99.8 Ni
	Iron	61.1							Fe
	Copper	10.73							Cu
1c	Ni base	800	Evanohm‡	Karma‡					75 Ni, 20 Cr, 2.5 Al, 2.5 Cu
1 (–)	Ni base	800							76 Ni, 20 Cr, Al, Fe
1 (–)	Ni base	800							61 Ni, 20 Cr, 17.5 Mn, 1.5 Mo
1 (–)	Ni base	800			Alloy 800‡	Chromel R‡	Moleculoy‡		74 Ni, 20 Cr
1 (–)	Ni base	800						Nikrothal LX‡	75 Ni, 20 Cr, 3 Al, 2 Co
1 (–)	Ni base	800							75 Ni, 20 Cr, Si, Mn
2§	Fe base	812							75 Fe, 20 Cr, 4 Al, 0.5 Co, +
2§	Fe base	825				Alloy 815‡	Mesoloy‡	Kanthal DR‡	72.25 Fe, 23 Cr, 4.75 Al
2§	Fe base	815							72.9 Fe, 22.5 Cr, 4.6 Al
2§	Fe base	815			Alloy K-20‡				73 Fe, 23.5 Cr, 4.5 Al
1 (--)	Ni base	825	Evanohm-S‡						75 Ni, 20 Cr, Al, Mn

* ASTM standard specification B267-68 (reapproved 1972) published by American Society for Testing and Materials.
† Constantan is often used as a term to denote this basic alloy processed especially for thermocouple use.
‡ Manufacturer type or trademark.
§ Similar.

TABLE 10 Characteristics of Resistance Wire

Alloy type[a]	Resistivity,[b] Ω cmil ft⁻¹ 20°C	Resistivity,[b] Ω cm² × 10⁶ cm⁻¹ 20°C	Resistance temp coefficient,[c] ppm °C⁻¹	Linear expansion thermal coefficient, cm/cm/°C × 10⁶, 20–100°C	Min tensile strength lb in⁻² 25°C[d]	Melting temp (approx), °C	Relative magnetic attraction	Density, g cm⁻³, 20°C	Heat capacity,[e] J g⁻¹°C⁻¹	Thermoelectric potential to copper[f] (approx), V °C⁻¹ × 10⁶
80-20 Ni-Cr	650-675	108-112	+60 to +90 ± 20	12-18	100,000	1400	None	8.41	0.435	+6.0
Constantan	294-300	49-50	0 ± 20	14.5	60,000	1350	None	8.90	0.393	-45
Manganin	230-290	38-48	0 ± 15[c]	18.7	40,000	1020	None	8.192-8.41	0.406	-3.0, +1
Alloy 180	180	29.9	+180 ± 30	15.7-17.5	50,000	1100	None	8.90	0.385	-37
Alloy 90	90	14.9	+450 ± 50	16-17.5	35,000	1100	None	8.90	0.385	-26
Alloy 60	60	9.97	+500 to +800 ± 200	16.2-16.3	50,000	1100	None	8.90	0.385	-22
Alloy 30	30	4.99	+1400 to +1500 ± 300	16.4-16.5	30,000	1100	None	8.91	0.385	-14
Linear TC[a]	120	19.9	+4500 ± 400	12-15	70,000	1100	Strong	8.46	0.523	-40
Nickel A	60	9.97	+4800	13	60,000	1450	Strong	8.90	0.544	-22
High purity Ni[a]	50	8.31	+6000	13.3-15	50,000	1400	Strong	8.90	0.544	-22
Iron	61.1	10.15	+5000 to +6200	11.7 (20°C)	50,000	1535	Strong	7.86	0.445	+12.2
Copper	10.37	1.72	+3900 to +4300	16.5 (20°C)	35,000	1083	None	8.90	0.385	0
Evanohm[g]	800	133	0 ± 5	12.6	100,000	1350	None	8.10	0.448	+3.0
Karma[h]	800	133	0 ± 20	13.3	180,000	1400	None	8.10	0.435	+3.0
Alloy 800[i]	800	133	0 ± 5	15	150,000	1260	None	7.95		+2.5
Chromel R[j]	800	133	0 ± 10	13.5	95,000	1398	None	8.1	0.448	~+1.0
Moleculoy[k]	800	133	0 ± 5	13.3	130,000	1395	None	8.12	0.435	+3.0
Nikrothal LX[l]	800	133	0 ± 5	12.6	150,000	1410	None	8.1	0.460	+2.0
Kanthal DR[l]	812	135	0 ± 20	11.9	100,000	1505	Strong	7.2	0.494	-3.5
Mesoloy[k]	825	137.2	0 ± 10	13.5	100,000	1500	Strong	7.15	0.481	-3.3
Alloy 815[j]	815	135.5	+82	15.9	~115,000	1520	Strong	7.25	0.460	~-3.7
Alloy K-20[i]	815	135.5	0 ± 20	13	100,000	1530	Strong	7.25	~0.460	-3.5
Evanohm S[g]	825	137	0 ± 5	13	100,000	1350	None	7.13	0.460	+0.2

a Refer to Table 9 for alloy composition and suppliers.

b Microhms per cm length of a 1-cm²-section was obtained by dividing the ohm-circular mil per foot value by 6.015.

c Various suppliers adjust minor constituents and processing to provide selected temperature coefficients or slightly different ranges. Values given cover generally a range of 0 to 100°C. Nonlinearity of the curve will cause slight deviations outside this range. Manganin, in particular, has a parabolic *TC* curve whose peak can be adjusted for the desired operating temperature. *TC* stated for 20 to 35°C for manganin. Values for 800 Ω cmil ft^{-1} and higher resistivity alloys cover a range of generally −55 to +150°C.

d Tensile strength varies considerably with sample shape and size. Values given are advertised minimum values and are stated primarily for comparison purposes.

e Heat capacity in joules per gram for a 1.0°C rise differs from "specific heat" referenced to calories per gram for a 1.0°C rise, or an identical numerical quantity referenced to Btu per pound for a 1.0° Fahrenheit rise. The conversion factor for any given specific heat to joules per gram for 1°C rise is therefore a multiplier of 4.186 J cal^{-1}.

f Thermoelectric-potential values given by different sources vary; also the alloy may be used with metals other than copper. Range of temperature for the values given is, in general, 0 to 100°C. (See "Metals Handbook," vol. I, 8th ed., American Society for Metals, 1961.)

g Registered trademark, Wilbur B. Driver Company.

h Registered trademark, Driver Harris Company.

i Registered trademark, C. O. Jelliff Corporation.

j Registered trademark, Hoskins Mfg. Company.

k Registered trademark, Molecu-wire Corporation.

l Registered trademark, Kanthal Corporation.

no closer than ±1 percent. Tolerances closer than 0.001 Ω can be measured using special techniques if necessary. Precision wirewound resistors for meter shunts and applications requiring comparable precision and exactness of value are usually procured to exact value and tolerance, rather than to a decade value. Mil-R-39005 and EIA RS-229A allow actual values to be specified in at least the tolerance values of 0.01, 0.02, 0.05, and 0.1 percent. Almost all other standards and specifications declare values not listed in the geometric decades to be "nonstandard."

Resistor Specifications and Standards

The resistors covered by this chapter are categorized by power dissipation and degree of stability and precision as shown in Table 2. Table 11 shows EIA and military resistor types current at this writing, and as can be seen, the power and usage categories are not necessarily separated by particular specifications. Military specifications are currently undergoing a substantial revision effort to standardize types better and generally to replace the older familiar specifications with more stringent versions that require manufacturers to establish reliability test levels as a condition of qualification.

National Aerospace Standards (NAS) documents have been issued only where they do not duplicate existing military specifications or standards. At this time, only three NAS resistor documents are current in AIA publications. NAS 710 is a comprehensive detailed specification for potentiometers, and NAS1484 and NAS1485 cover a family of precision wirewound resistors furnished by a single source. Further information on these publications is not included here, since military and EIA publications cover such resistors. Detail sheets, defined in the discussion of standardization, applicable to resistors are also shown in Table 12 for easy reference.

NEMA publications presently cover resistors only insofar as limits and standard test descriptions are provided for equipment using resistors. Publication ICS-1970, Standards for Industrial Motor Control Equipment, provides information on high-power resistors for some heavy-duty uses. Part ICS 2-213 covers resistors and rheostats, primarily for motor-starting, dynamic-braking, and speed-regulating services. Limits are established for temperature rise on resistor surfaces and air temperature near fully loaded units. Rating classes for various applications are enumerated and resistor tests are described. Intermittent-duty overload standards and altitude derating are stated. ICS-1970 supersedes IC-1.

The National Electrical Code, published by the National Board of Fire Underwriters, also currenly does not cover resistors specifically.

No established product safety standards are yet available for resistors, but considerable interest has been shown by both suppliers and consumer organizations in characterizing resistor overload performance to assure that coatings are flameproof and that the resistors open-circuit at safe temperatures should inadvertent overload occur.

Many suppliers and manufacturers publish extensive and highly informative product application data and categorized part listings; they are usually available upon request. Domestic and some overseas suppliers are listed in the industry directory publications named at the beginning of this chapter.

General-Purpose Resistors

As classified in Table 2, general-purpose resistors are considered to be those designs having rated power dissipation of 2 W and less, and resistance-value service stability from semiprecision downward. Included in this category are the carbon composition types and a multiplicity of thick- and thin-film types, as well as several wirewound designs. No single type provides the complete range of function and cost. Each type is discussed separately in the following paragraphs.

Carbon composition resistors Composition resistors are widely used and provide a reliable low-cost choice for many applications where their service-life variability can be tolerated. Their size is relatively small, and value range is the widest available in any resistor type.

Construction. Two basic types are furnished, one having carbon granules molded with an organic binder into the proper pellet shape around the terminal leads, and another employing a carbon-resin-paint deposition on a glass tube (called a filament) attached to the terminal leads by conductive cement, with a thick organic-resin case

TABLE 11 Fixed-Resistor Specifications, Types, and Power Ratings*

Use category	Resistor type	Power ratings	Specification	Letter type designation	Former specification coverage
General-purpose	Fixed, composition	1/8–2 W	EIA RS-172A	RRCXX. . .	MIL-R-11
General-purpose	Fixed, composition, insulated, established reliability	1/8–2 W	MIL-R-39008	RCRXX. . .	
General-purpose and semiprecision	Fixed, film	1/20–2 W	EIA RS-196A	RXX. . .	MIL-R-22684
General-purpose	Fixed, film, insulated, established reliability	1/8–2 W	MIL-R-39017	RLRXX. . .	
General-purpose, medium-power, high-power	Fixed, wirewound, insulated	1.25–15 W	EIA RS-344	CRUXX. . .	
General-purpose, semiprecision	Fixed, film, established reliability	1/20–1/2 W	MIL-R-55182	RNCXX. . .†	MIL-R-10509
Precision	Fixed, wirewound, accurate, established reliability	0.15–3/4 W	MIL-R-39005	RBRXX. . .	MIL-R-93
Precision	Fixed, wirewound, precision	0.1–0.5 W	EIA RS-229A	CRBXX. . .	MIL-R-26‡
General-purpose, medium-power, high-power	Fixed, wirewound (power type), established reliability	1–10 W	MIL-R-39007	RWRXX. . .	
General-purpose, medium-power, high-power	Fixed, wirewound, power	1–210 W	EIA RS-155B	CRWXX. . .	
High-power	Fixed, wirewound (power-type)	11–210 W	MIL-R-26	RWXX. . .	
Special-purpose	Fixed, meter multiplier, high-voltage, ferrule-terminal type	1 mA, 0.5 MΩ –20 MΩ	MIL-R-29	MFC. . .	
High-power	Resistors and rheostats, naval, shipboard	5–2500 W	MIL-R-15109 (Ships)	Not Applicable	
General-purpose, medium-power	Fixed, film (power-type)	1–4 W	MIL-R-11804	RDXX. . .	
Chip type	Fixed, film, chip, established reliability	10–225 mW	MIL-R-55342	RMXXXX. . .	
High-power	Fixed, wirewound (power-type, chassis mount)	75 W, 120 W	MIL-R-18546	REXX. . .	
Medium-power, high-power	Fixed, wirewound (power-type, chassis mount), established reliability	5–30 W	MIL-R-39009	RERXX. . .	MIL-R-18546‡

* Only currently released EIA specifications and coordinated military specifications are shown, except for MIL-R-15109 (Ships) which is not coordinated. QPL coverage is not considered. Service specifications exist for resistor decades (MIL-R-9991, USAF), resistor networks (MIL-R-83401, USAF), 1-W metal-film resistors, type RNC 75 (MIL-R-55182, Navy) and others, but are not shown.

† MIL-R-55182 types may have a third letter designate of R, N, or C, depending upon lead type. R: solderable only, N: weldable only (gold-plated nickel, type N-2 of MIL-STD-1286); C: solderable/weldable (tin-plated copper, type C of MIL-STD-1286).

‡ Supersession is only partial; former specification is still active.

TABLE 12 Current Military and EIA Fixed-Resistor Performance Ratings

Type	Specification	Type designation	Ohmic range[a]	Max power rating, W	Voltage max	Resistance tolerance, ± %	Temp coefficient, 0 ± ppm/°C	Max temp for full rated power, °C	Temp at fully derated zero power, °C	Body shape[b]	Terminal type	Terminal exit
Fixed, carbon composition	MIL-R-39008/4	RCR05	2.7–22 MΩ	⅛	150							
	RS-172	RRC05	Not specified									
	MIL-R-39008/1	RCR07	2.7–22 MΩ	¼	250							
	RS-172	RRC07	Not specified									
	MIL-R-39008/2	RCR20	2.7–22 MΩ	½	350	5, 10	See text; also Fig. 14	70	130	1	Wire	Axial
	RS-172	RRC20	Not specified									
	MIL-R-39008/3	RCR32	2.7–22 MΩ	1	500							
	RS-172	RRC32	Not specified									
	MIL-R-39008/5	RCR42	10–22 MΩ	2	500							
	RS-172	RRC42	Not specified									
Fixed, film, precision, and semi-precision	RS-196A-CLI	R05	4.7–150 kΩ	⅛	200							
	RS-196A-CLI	R07	10–301 kΩ	¼	250							
	RS-196A-CLI	R09	4.3–1 MΩ	½	350	1, 2, 5	A: 500, D: 200, K: 100	70	150			
	RS-196A-CLI	R11	10–2 MΩ	1	500							
	RS-196A-CLI	R15	10–1.5 MΩ	2	500					1	Wire	Axial
	RS-196A-CLII	R04	10–150 kΩ	¹⁄₂₀	200							
	RS-196A-CLII	R06	10–301 kΩ	¹⁄₁₀	200							
	RS-196A-CLII	R08	10–1 MΩ	⅛	250	0.1, 0.25, 0.5, 1.0	A: 500, C: 50, D: 200, E: 25, K: 100	125	175			
	RS-196A-CLII	R12	10–2 MΩ	¼	300							
	RS-196A-CLII	R14	24.9–2 MΩ	½	350							
	RS-196A-CLII	R16	49.9–2 MΩ	1	500							
Fixed film	MIL-R-39017/5	RLR05	4.7–300 kΩ	⅛	200							

Fixed, wire-wound low-power, also medium- and high-power	MIL-R-39017/1	RLR07	10-300 kΩ	1/4	250	1, 2	100	70	150	1	Wire	Axial
	MIL-R-39017/2	RLR20	4.3-1.0 MΩ	1/2	350							
	MIL-R-39017/3	RLR32	10-2.7 MΩ	1	500							
	MIL-R-39017/4	RLR42	10-1.5 MΩ	2	500							
	RS-344	CRU1A	0.1-1,000	1.25	$\sqrt{PR} = 35$				160	1c		
		CRU2A	0.1-2,000	2.5	$\sqrt{PR} = 70$							
		CRU2AB	0.18-2,400	2	$\sqrt{PR} = 70$							
		CRU3B	0.1-7,500	3	$\sqrt{PR} = 150$	5, 10	400 at >1 Ω / 800 at ≤1 Ω	25	275	3	Wire	Axial
		CRU5B	0.1-8,200	5	$\sqrt{PR} = 200$							
		CRU7B	0.1-18 kΩ	7	$\sqrt{PR} = 350$							
		CRU10B	0.1-30 kΩ	10	$\sqrt{PR} = 550$							
		CRU15B	0.18-30 kΩ	15	$\sqrt{PR} = 670$							
Fixed film, established reliability	MIL-R-55182/7	RNC50K[a]	10-604 kΩ	1/20	200	0.5, 1	110	125	175	1	Wire	Axial
	MIL-R-55182/1	RNC55K[a]	10-1.21 MΩ	1/10	200							
	MIL-R-55182/3	RNC60K[a]	24.9-3.01 MΩ	1/8	250							
	MIL-R-55182/5	RNC65K[a]	10-6.26 MΩ	1/4	300							
	MIL-R-55182/6	RNC70K[a]	10-15.0 MΩ	1/2	350							
	MIL-R-55182/7	RNC50H[a]	49.9-604 kΩ	1/20	200	0.1, 0.5, 1	50	125	175	1	Wire	Axial
	MIL-R-55182/1	RNC55H[a]	49.9-1.21 MΩ	1/10	200							
	MIL-R-55182/3	RNC60H[a]	49.9-3.01 MΩ	1/8	250							
	MIL-R-55182/5	RNC65H[a]	10-6.26 MΩ	1/4	300							
	MIL-R-55182/6	RNC70H[a]	10-15.0 MΩ	1/2	350							
	MIL-R-55182/7	RNC50J[a]	49.9-604 kΩ	1/20	200	0.1, 0.5, 1	25	125	175	1	Wire	Axial
	MIL-R-55182/1	RNC55J[a]	49.9-1.21 MΩ	1/10	200							
	MIL-R-55182/3	RNC60J[a]	49.9-3.01 MΩ	1/8	250							
	MIL-R-55132/5	RNC65J[a]	10-6.26 MΩ	1/4	300							

TABLE 12 Current Military and EIA Fixed-Resistor Performance Ratings (Continued)

Type	Specification	Type designation	Ohmic range[a]	Max power rating,[a] W	Voltage max	Resistance tolerance, ± %	Temp coefficient, 0 ± ppm/°C	Max temp for full rated power, °C	Temp at fully de-rated zero power, °C	Body shape[b]	Terminal type	Terminal exit
Fixed film, established reliability	MIL-R-55182/6	RNC70J[d]	10-15.0 MΩ	½	350							
	MIL-R-55182/10(Navy)	RNC75J[d,e]	10-20 MΩ	1	750							
	MIL-R-55182/1	RNC55C[d]	49.9-200 kΩ	1/10	200							
	MIL-R-55182/3	RNC60C[d]	49.9-499 kΩ	1/8	250	0.1, 0.5, 1	50	125	175	1[f]	Wire	Axial
	MIL-R-55182/5	RNC65C[d]	10-1.0 MΩ	¼	300							
	MIL-R-55182/6	RNC70C[d]	20-1.5 MΩ	½	350							
	MIL-R-55182/1	RNC55E[d]	49.9-200 kΩ	1/10	200							
	MIL-R-55182/3	RNC60E[d]	49.9-499 kΩ	1/8	250	0.1, 0.5, 1	25	125	175	1[f]	Wire	Axial
	MIL-R-55182/5	RNC65E[d]	10-1.0 MΩ	¼	300							
	MIL-R-55182/6	RNC70E[d]	20-1.5 MΩ	½	350							
	MIL-R-55182/10(Navy)	RNC75E[d,e]	20-2.0 MΩ	1	750							
Fixed film (high-stability)	MIL-R-10509/5	RN75B	10-20 MΩ	1	500	0.1, 1	25, 500	70, 125	150, 175	1	Wire	Axial
Fixed, wirewound, accurate, established reliability	MIL-R-39005/4	RBR55	0.1-332 kΩ[g]	0.15	200					1		Axial
	MIL-R-39005/5	RBR56	0.1-220 kΩ[g]	1/8	150	0.01, 0.02, 0.05, 0.1, 1, but no closer than ± 0.001Ω for low values	90 at $R < 1\Omega$; 30 at $1\Omega \le R < 10\Omega$; 15 at $10\Omega \le R < 100\Omega$; 10 at $R \ge 100\Omega$	125	145			
	MIL-R-39005/9	RBR75	0.1-316 kΩ[g]	1/8	150							
	MIL-R-39005/6	RBR71	0.1-150 kΩ[g]	1/8	150					2		Single end
	MIL-R-39005/3	RBR54	0.1-562 kΩ[g]	¼	300							
	MIL-R-39005/2	RBR53	0.1-1.10 MΩ[g]	⅓	300						Wire	
	MIL-R-39005/1	RBR52	0.1-1.21 MΩ[g]	½	600					1		Axial
	MIL-R-39005/7	RBR57	0.1-6.42 MΩ[g]	¾	600							
Fixed, wirewound precision		CRB63	1-26 kΩ	1/10	30							
		CRB62	1-100 kΩ	1/8	50							

Description	Spec	Type	Resistance range	Power, W	Max voltage	Tolerance, %	Temp. coeff., ppm/°C	°C	Max °C	No.	Terminal	Mounting
Fixed, wireound (power-type), established reliability	EIA RS-229A	CRB56	1–125 kΩ	1/8	150	0.05 for R ≥ 100Ω, 0.1 for R ≥ 50Ω, 0.25 for R ≥ 10Ω, 0.5 for R ≥ 1Ω, 1 for R ≥ 1Ω	20 at R ≥ 10Ω, 60 at 5Ω ≤ R < 10Ω, 100 at R ≤ 5Ω	125	145	1	Wire	Axial
		CRB55	1–225 kΩ	0.15	200							
		CRB54	1–330 kΩ	1/4	300							
		CRB53	1–750 kΩ	1/3	300							
		CRB52	1–1.0 MΩ	1/2	600							
	MIL-R-39007/9	RWR81[h]	0.1–464[h]	1	$\sqrt{PR} = 21.5$	0.1, 0.5, 1	90 at R < 1Ω, 50 at 1Ω ≤ R < 10Ω, 20 at R ≥ 10Ω	25	275	1	Wire	Axial
	MIL-R-39007/8	RWR80[h]	0.1–1.21 kΩ[h]	2	$\sqrt{PR} = 49$							
	MIL-R-39007/5	RWR71[h]	0.1–6.04 kΩ[h]	2	$\sqrt{PR} = 110$							
	MIL-R-39007/11	RWR89[h]	0.1–3.57 kΩ[h]	3	$\sqrt{PR} = 103$							
	MIL-R-39007/6	RWR74[h]	0.1–12.1 kΩ[h]	5	$\sqrt{PR} = 246$							
	MIL-R-39007/10	RWR84[h]	0.1–12.4 kΩ[h]	7	$\sqrt{PR} = 295$							
	MIL-R-39007/7	RWR78[h]	0.1–39.2 kΩ[h]	10	$\sqrt{PR} = 626$							
Fixed, wireound (power-type)	MIL-R-26/4	RW56	0.1–9.1 kΩ	14	$\sqrt{PR} = 357$	5, 10	400 at R < 20Ω, 260 at R ≥ 20Ω	25	350	4	Solder tab	Radial
	MIL-R-26/3	RW29	0.1–5.6 kΩ	11	$\sqrt{PR} = 248$							
	MIL-R-26/3	RW31	0.1–6.8 kΩ	14	$\sqrt{PR} = 309$							
	MIL-R-26/3	RW33	0.1–18 kΩ	26[i]	$\sqrt{PR} = 684$							
	MIL-R-26/3	RW35	0.1–43 kΩ	55[i]	$\sqrt{PR} = 1{,}538$							
	MIL-R-26/3	RW37	0.1–91 kΩ	113[i]	$\sqrt{PR} = 3{,}207$							
	MIL-R-26/3	RW38	0.1–150 kΩ	159[i]	$\sqrt{PR} = 4{,}884$							
	MIL-R-26/3	RW47	0.1–180 kΩ	210[i]	$\sqrt{PR} = 6{,}148$							
Fixed, wireound, power	EIA RS-155B	CRW71[j]	1–1.4 kΩ	1, 1.25	$\sqrt{PR} = 42$		Characteristic G: 30 for R ≥ 20Ω, 50 for 5Ω ≤ R < 20Ω, 100 for R ≤ 5Ω		275 (characteristic G)			
		CRW69[j]	1–3.57 kΩ	2.5, 3	$\sqrt{PR} = 103$							

TABLE 12 Current Military and EIA Fixed-Resistor Performance Ratings (Continued)

Type	Specification	Type designation	Ohmic range[a]	Max power rating, W	Voltage max	Resistance tolerance, ±%	Temp coefficient, 0 ± ppm/°C	Max temp for full rated power, °C	Temp at fully de-rated zero power, °C	Body shape[b]	Terminal type	Terminal exit
Fixed, wire-wound power	EIA RS-155B	CRW59[j,k]	1–3.57 kΩ		$\sqrt{PR}=103$	1, 5	Characteristic E: 30 for $R \geq 20\Omega$; 50 for $5\Omega \leq R < 20\Omega$; 100 for $R < 5\Omega$	25	350 (characteristics E, H, V)	1	Wire	Axial
		CRW67[j]	1–12.1 kΩ	5, 6.5	$\sqrt{PR}=280$							
		CRW57[j,k]	1–12.1 kΩ		$\sqrt{PR}=280$							
		CRW68[j]	1–39.2 kΩ	10, 11	$\sqrt{PR}=657$							
		CRW58[j,k]	1–39.2 kΩ		$\sqrt{PR}=657$							
		CRW17[j]	1–3.65 kΩ	10[l]	$\sqrt{PR}=189$	1, 5	Characteristic H (temp >25°C): 30 for $R \geq 20\Omega$; 50 for $5\Omega \geq R < 20\Omega$; 100 for $R < 5\Omega$. Characteristic H (temp <25°C): 60 for $R \geq 5\Omega$; 100 for $R < 5\Omega$. Characteristic V: 260 ppm/°C for 25Ω in^{-2} of winding area, and above, 400 ppm/°C below	25	350 (characteristics H, V)	7	Solder tab	Radial
		CRW18[j]	1–6.08 kΩ	15[l]	$\sqrt{PR}=302$							
		CRW19[j]	1–18.24 kΩ	20[l]	$\sqrt{PR}=604$							
		CRW20[j]	1–8.25 kΩ	21	$\sqrt{PR}=416$							
		CRW21[j]	1–16.2 kΩ	31	$\sqrt{PR}=709$							
		CRW22[j]	1–36.5 kΩ	53	$\sqrt{PR}=1{,}390$							
		CRW23[j]	1–51.1 kΩ	68	$\sqrt{PR}=1{,}864$							
		CRW24[j]	1–68.1 kΩ	91	$\sqrt{PR}=2{,}490$							
		CRW29[j]	1–5.62 kΩ	11	$\sqrt{PR}=249$	1, 5		25	350 (characteristics H, V)	4	Solder tab	Radial
		CRW31[j]	1–6.34 kΩ	14	$\sqrt{PR}=298$							
		CRW32[j]	1–10 kΩ	17	$\sqrt{PR}=412$							
		CRW33[j]	1–18.2 kΩ	26	$\sqrt{PR}=688$							
		CRW35[j]	1–40.2 kΩ	55	$\sqrt{PR}=1{,}487$							
		CRW36[j]	1–56.2 kΩ	78	$\sqrt{PR}=2{,}094$							

Category	MIL Spec	Type	Resistance range		√PR / rating	Tolerance (%)	Temp. coeff. / power				Terminal	Lead
		CRW37i	1–90.9 kΩ	113	$\sqrt{PR} = 3{,}205$							
		CRW38i	1–140 kΩ	159	$\sqrt{PR} = 4{,}718$							
		CRW47i	1–182 kΩ	210	$\sqrt{PR} = 6{,}182$							
Fixed, film (power type)	MIL-R-11804/2	RD60	10–487 kΩ	1	350	2, 5	500	25	235	1	Wire	Axial
	MIL-R-11804/2	RD65	10–1.54 MΩ	2	500							
	MIL-R-11804/2	RD70	20.5–4.22 MΩ	4	750							
	MIL-R-11804/1	RD31	10–196 kΩ	7	525							
	MIL-R-11804/1	RD33	19.6–196 kΩ	13	1,380							
	MIL-R-10804/1	RD35	19.6–487 kΩ	25	2,275	2, 5	500	25	235	4	Solder	Radial
	MIL-R-11804/1	RD37	19.6–487 kΩ	55	3,675							
	MIL-R-11804/1	RD39	30.1–1 MΩ	115	7,875							
Fixed, meter, multiplier, high-voltage, ferrule terminal	MIL-R-29	MFA	3.5–6 MΩ									
		MFB	1.0–3.5 MΩ									
		MFC	0.5–1.0 MΩ	1 mA current	Resistance in megohm units times 1kV	0.5	Zero + 200 ppm, −50 ppm/°C	25	110	5	Ferrule	Not applicable
		MFD	10 MΩ only									
		MFE	15 MΩ only									
		MFF	20 MΩ only									
Fixed, wire-wound chassis-mounted power type	MIL-R-18546/2	RE77	0.05–29.4 kΩ	75^m	$\sqrt{PR} = 1{,}485$	1	$R < 2\ \mathrm{k}\Omega$ 50	25	275	6	Solder lug	Axial
	MIL-R-18546/2	RE80	0.1–35.7 kΩ	120^m	$\sqrt{PR} = 2{,}070$	1	$R \geq 2\ \mathrm{k}\Omega$ 30	25	275	6	Threaded stud	Axial
Fixed, wire-wound chassis-mounted power-type, established reliability	MIL-R-39009/2	RER40n	1–1.65 kΩ	5^m	$\sqrt{PR} = 91$	1	$R < 1\Omega{:}100$ $1\Omega \leq R \leq 19.6\Omega{:}50$ $R \geq 20\Omega{:}30$	25	275		Solder lug	Axial
	MIL-R-39009/1	RER60n	1–3.32 kΩ		$\sqrt{PR} = 129$							
	MIL-R-39009/2	RER45n	1–2.80 kΩ	10^m	$\sqrt{PR} = 167$							
	MIL-R-39009/1	RER65n	0.1–3.32 kΩ		$\sqrt{PR} = 182$							
	MIL-R-39009/2	RER50n	1–6.04 kΩ		$\sqrt{PR} = 348$							

TABLE 12 Current Military and EIA Fixed-Resistor Performance Ratings (Continued)

Type	Specification	Type designation	Ohmic range[a]	Max power rating, W	Voltage max	Resistance tolerance, ± %	Temp coefficient, 0 ± ppm/°C	Max temp for full rated power, °C	Temp at fully derated zero power, °C	Body shape[b]	Terminal type	Terminal exit
	MIL-R-39009/1	RER70[n]	0.1–12.1 kΩ	20[m]	$\sqrt{PR} = 492$	1	Continued	25	275	6	Solder lug	Axial
	MIL-R-39009/2	RER55[n]	1–19.6 kΩ		$\sqrt{PR} = 626$							
	MIL-R-39009/1	RER75[n]	0.1–39.2 kΩ	30[m]	$\sqrt{PR} = 1{,}084$							
Fixed, film, chip, established reliability	MIL-R-55342/1	RM0502	5–100 kΩ	10 mW	40	$10(5, 1, 0.1)^{n}$	50	70	150	Chip	Various metallized areas and pads, solder, wire bonding	Part of body
				20 mW	40	$10(5, 1, 0.1)^{n}$	200					
	MIL-R-55342/2	RM0505	5–500 kΩ	25 mW	40	$10(5, 1, 0.1)^{n}$	50					
				50 mW	40	$10(5, 1, 0.1)^{n}$	200					
	MIL-R-55342/3	RM1005	5–1 MΩ[o]	50 mW	40	$10(5, 1, 0.1)^{n}$	50					
				100 mW	40	$10(5, 1, 0.1)^{n}$	200					
	MIL-R-55342/4	RM1505	5–5 MΩ[o]	100 mW	40	$10(5, 1, 0.1)^{n}$	50					
				150 mW	40	$10(5, 1, 0.1)^{n}$	200					
	MIL-R-55342/5	RM2208	5–15 MΩ[o]	200 mW	40	$10(5, 1, 0.1)^{n}$	50					
				225 mW	40	$10(5, 1, 0.1)^{n}$	200					
Fixed and variable, wire- and strip-wound (power)	MIL-R-15109/Ships	Form EW[p] Form IW Form G Form R Form P	0.0325–50 kΩ[p] 0.6–125 kΩ 0.126–8 0.041–6.3 0.5–10 kΩ		150, 300, or 600 V	5		50	Temp rise 300°C for embedded types, 375°C for exposed wire, ribbon strip	Varied—cylindrial, flat oval, grid	Varied—solder, bolts	Varied, primarily radial

a Specification ohms range does not guarantee availability or qualified status. Check QPL's and supplier information.

b Body shapes are keyed to illustrations shown in the key at the end of this table.

c RS-344 CRU1A, CRU2A have molded, color-coded bodies of appearance similar to carbon composition types.

d The third letter of MIL-R-55182 type designation is variable. R means solderable only, W means weldable only, and C means solderable and weldable (see Table 11).

e Type RNC75 is covered, at present, only by a noncoordinated detail specification.

f Hermetically sealed types. Type C leads may not be available.

g All RBR styles are noninductively wound, either bifilar or reverse-pi style. Ranges shown are for 0.0007 minimum diameter wire. Restriction to 0.0009 diameter wire reduces maximum values to RBR55: 226kΩ; RBR56: 100kΩ; RBR75: 71.5kΩ; RBR71: 100kΩ; RBR54: 255kΩ; RBR33: 500kΩ; RBR52: 806kΩ; RBR47: 1.37MΩ.

h First letter following style number for MIL-R-39007 types denotes lead type and winding type. S or W denote inductively wound, solderable or weldable, respectively. N denotes noninductive bilfiar or Ayrton-Perry winding and solderable leads. Maximum value must be reduced to one-half that shown for N designation.

i For center-tapped configuration power rating is 90% of that given (also derate voltage).

j First letter following style number designates characteristic. Styles 57 to 71 available in characteristics G, E, H, V; all others available in characteristic H, V only. Power rating for characteristic G is lower than for E, H, V (see power-rating column; lower rating is for characteristic G). Values of resistance and type designations are given for inductive windings. Noninductive-winding styles are designated NRW and have maximum resistance half the given value. Resistance wire is 0.001 in nominal diameter minimum for all styles.

k CRW 57 to 59 are uninsulated types.

l CRW 17,18, 19 require mounting on 10- by 0.040-in-thick steel chassis for rated wattage.

m Power ratings given as mounted on a metal chassis of size prescribed in applicable specification. (See text.)

n Styles RER 40, 45, 50, 55 are noninductively wound; styles RER 60, 65, 70, 75 are inductively wound.

o Values for chip types depend on resistance, tolerances; stated value range is for ±10% tolerance. All styles limited to 10 Ω min for ±5%. For ±1%, RM 0502 is 10 to 100 kΩ; RM 0505 is 10 to 300 kΩ; RM 1005 is 100 to 500 kΩ; RM 1505 is 10 to 1 MΩ; RM 2208 is 10 to 2 MΩ. For ±0.1%, RM 0502 is 100 to 100 kΩ; RM 0505 is 100 to 200 kΩ; RM 1005 is 100 to 300 kΩ; RM 0505 is 100 to 500 kΩ, and RM 2208 is 100-1 MΩ.

p EW = exposed wire, IW = embedded wire, G = grids, R = ribbon, P = plates. Value ranges are those listed in QPL 15109-21 and are not available in all power ratings or from all listed suppliers' Current QPL should be consulted for further information.

1. Molded color-coded

1. Molded or end-filled tubular case

1. Conformally coated

1. Hermetic, glass-to-metal

2. Molded or end-filled single-ended axial leads

3. Cement-embedded

4. Radial tab, cylinder

5. Ferrule-terminal cylinder

6. Chassis-mount power type (may have threaded stud terminals in some styles)

7. Flat-strip (flat oval) radial tab terminals

molded around the assembly (see construction-type illustrations in Fig. 9). All types employ organic exterior coatings and are supplied in standardized values, tolerances, and case sizes, usually with color-coded value and tolerance painted on (see previous paragraphs on color coding). Standard leads are solderable, but weldable versions can be supplied on special order, usually with consideration for some reduction in power rating due to the lower thermal conductivity of weldable alloys compared with the standard copper leads. Some suppliers furnish tin-lead-coated oxygen-free copper for "weldable" versions, which some users find difficult to resistance-weld reliably.

Resistance Variability. The chief disadvantage of carbon composition resistors is their variability with shelf and service life. Water absorption from exposure to humid atmospheres in nonoperating storage changes the resistance value dramatically; it is not unusual to observe out-of-tolerance resistance values on parts at incoming inspection. Moisture causes resistance generally to increase, with the change amounting to as much as several percent. The moisture can be safely removed and a partial value recovery effected by baking or operational-temperature rise, provided temperatures used are no higher than about 100°C. Permanent effects on the organic structure are accelerated at higher temperatures. The recovery is temporary unless subsequent moisture is scrupulously excluded, however. In service, normal relative humidities and irreversible operational aging can cause resistance value to vary well outside purchase tolerances, particularly for the 5 and 10 percent tolerance. Temperature characteristics are nonlinear and vary with resistance value, as shown in Fig. 14. Since the slope of the temperature-resistance curve is not constant,

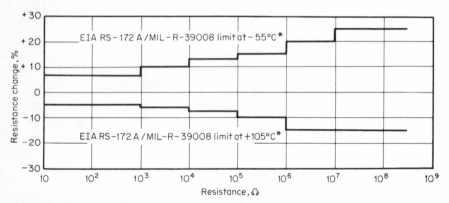

Fig. 14 Temperature-characteristics limits for carbon-composition resistors. °RS-172A and MIL-R-39008 actually allow a full ± excursion for both high- and low-temperature extremes, but carbon has an inherently negative temperature-resistance characteristic.

the property is called a characteristic rather than a coefficient. Noise is high compared with film and wirewound types because of the so-called carbon noise, which is proportional to load current and adds to the temperature-resistance-dependent Johnson noise (see resistor noise paragraph above). Availability of parts having significantly different constructional character under the same standardized part designations requires scrupulous evaluation in applications involving radio frequencies. Careful testing in the circuit and control of the source of supply are indicated for critical rf applications. Figure 12 shows nominal rf characteristics of composition resistors, which are generally not as good as those of film types.

Within their variability limitations, however, the reliability record of composition resistors is enviable, and designs furnished as domestic standards are covered by years of successful experience. A number of designs are produced in Japan that have proved to be equal to domestic parts. Other nondomestic types are marketed competitively as being interchangeable.

Performance. The functional requirements of EIA RS-172 and military specification MIL-C-39008 for composition resistors are nearly identical except for some differ-ences in available resistance values. Considerable consolidation of types and sizes has been accomplished in recent years. Although neither specification presently specifies shelf-aging drift rate, both contain qualification tests for installation soldering heat, lead strength, load life, temperature characteristic, voltage coefficient, dielectric strength, and humidity. The military humidity test is more severe, being a cycled-type test; also RS-192 bases standard compliance only on individual resistor perfor-mance within the test sample, while MIL-C-39008 also limits average resistance changes of the sample. The current MIL-C-39008 replaces the former MIL-R-11 and, except for more extensive data recording and better lot-quality controls, provides an essentially equivalent part. At this writing, four domestic producers have parts listed as qualified under MIL-C-39008. Four designs are supplied—three are molded-pellet types and one is an encapsulated carbon-coated glass type. Not all resistance values are qualified for each power rating and supplier; so the qualified products list (QPL) should be consulted if military requirements are involved in the intended applications. Table 13 shows current domestic military producers and

TABLE 13   Qualified Producers and Value Range for Carbon Composition Resistors TRW-IRC

MIL-R-39008 type	RS-172 type	Watt rat-ing	Volt rat-ing	MIL-R-39008 Qualified values of resistance			
				Allen Bradley	Airco Speer	Stack-pole	TRW-IRC
RCR05	RRC05	⅛	150	2.7 Ω– 22 MΩ			
RCR07	RRC07	¼	250	2.7 Ω– 22 MΩ	10 Ω– 1.0 MΩ	10 Ω– 1.0 MΩ	2.7 Ω– 22 MΩ
RCR20	RRC20	½	350	2.7 Ω– 22 MΩ	10 Ω– 22 MΩ	10 Ω– 22 MΩ	2.7 Ω– 22 MΩ
RCR32	RRC32	1	500	2.7 Ω– 22 MΩ	4.7 Ω– 5.6 MΩ	10 Ω– 22 MΩ	2.7 Ω– 22 MΩ
RCR42	RRC42	2	500	10 Ω– 22 MΩ		10 Ω– 22 MΩ	

Ref. MIL-R-39008, QPL.

qualified resistance-value ranges. Voltage ratings are the same in both specifications, as are the maximum voltages allowed for the short-time (5-s) overload test. Voltage maximums are listed in Table 12. Voltage rating is specified as rms, 100 Hz or less, or dc. Overload potentials require dc voltages for test.

Above the critical resistance value (i.e., that resistance value where the rated volt-age dissipates rated power for the design), the maximum rated voltage becomes also the power-dissipation limitation. Both specifications warn that nonsinusoidal voltages having peaks exceeding 1.4 times the rms rated voltage require careful evalu-ation, and that broad tolerance on resistance values (service variability as well as purchase tolerance) should be considered in order to prevent possible dissipative overloads as resistance changes in service.

High altitude or other causes of low atmospheric pressure decrease convection cooling and require evaluation. Although 50 to 75 percent of the heat dissipated by 2-W and lower-value resistors is normally through the leads, convection remains significant. Because of vulnerability to excessive soldering heat, it is recommended that lead length from resistor body to solder joint be no shorter than 0.25 in and that installation adjacent to other heat-producing parts be avoided. Allen Bradley states that permanent resistance shift due to soldering heat is lessened for their prod-uct if moisture is first baked out. Figure 15 shows specifications for power derating vs. ambient temperature at sea level for normally installed parts. Some of the sup-pliers advertise derating to zero power at 150°C, provided operating voltage is also derated by calculating the voltage associated with the derated power, or by derating maximum voltage by the square root of the power-derating factor associated with a given temperature. These limits are based upon an allowed change of 10 percent maximum in resistance value in 1,000 h operation. It is noted, however, that neither MIL-R-39008 nor EIA RS-172A requires testing of resistors at body temperatures of 130°C. Life testing is conducted at sea level, room temperature, full power. Resistor body temperatures under this condition can be well below the 130°C allowed maximum, and since performance tests such as the high-temperature exposure tests in many specifications are not conducted, it is prudent to limit dissipation to no more than half the rated values, particularly if stability of resistance is needed. Small constant dissipation levels dehumidify the parts, tending to reduce positive drift from moisture absorption, but the permanent long-term drift mechanisms, also essentially positive in direction, are accelerated by higher temperatures produced by constant dissipation of a major portion of the rated power, particularly if ambient environmental temperatures are high. Half-power derating is recommended by MIL-R-39008 and MIL-STD-199.

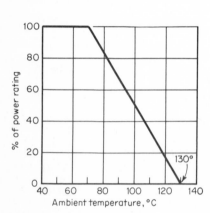

Fig. 15 Power derating of carbon-composition resistors per Mil-R-39008 and RS-172.

Fig. 16 Voltage coefficient of resistance, specification limits for carbon composition.

Voltage coefficient is of interest for some applications. Figure 16 shows the specified limits on voltage effects.

The Allen Bradley Company advertises good pulse performance capability of its line of carbon composition resistors, stating "insignificant effects" at the isolated single

TABLE 14   Pulse Performance Range for Allen Bradley Composition Resistors

Rated watts	Single pulse energy capability, W-s	Equivalent energy source
1/8	0.45	2 μF at 670 V
1/4	1.8	10 μF at 600 V
1/2	6.4	32 μF at 630 V
1	16	32 μF at 1,000 V
2	44	32 μF at 1,650 V

SOURCE: Allen Bradley Company.

pulse energies shown in Table 14. Caution is advised in applying this information to competitive designs and in extrapolation to other voltages and charge magnitudes.

RF performance of composition resistors is fairly good in the low-resistance range but becomes increasingly poor as resistance increases. The standard designs differ markedly in performance because of dependence upon design geometry and rf characteristics of materials. The effects are complex, being functions of applied frequency, resistance value, and installation configurations, as well as part design. While subject to resistance variation, composition resistors seldom fail catastrophically in service unless burned up by overload or damaged mechanically.

Film Resistors Film resistors fall naturally into two general classifications according to thickness of film employed. Those said to be thin films are 5 μm or less in thickness and include evaporated-metal films, pyrolytic carbon films, and tin oxide films. Normally thicker than 5 μm are palladium-silver cermets, tantalum oxide and other oxides, nitrides, carbides and metal combinations, generally deposited on a substrate as paint or paste suspensions, then fired to vitrify and fuse and reduce the material. The thickness line is not precise, however, and the achieved performance in temperature and resistance stability is often used instead to classify parts. Tin oxide surface diffusion into glass, for instance, produces what is by definition a thin film. Inherent temperature coefficient of resistance achievable in the past was more like vitrified cermet types than metal films, but comparatively recent development has improved performance and tin oxide types are now marketed as "metal film." Film resistors include types in virtually all performance classes, including high-power. The general-purpose category here covers those classes commonly denoted as precision, semiprecision, and general-usage.

Allowable operating temperature maximums for the materials involved establish the degree of miniaturization attainable. Physical size in these, as in other resistor types, largely determines the rate at which heat can be dissipated. This establishes the stabilized hot-spot temperature for a given design in a specified environment, while it dissipates a given power rate. Thus, use of high-temperature materials allows decreased physical size, provided that the resulting part design retains adequate mechanical strength and provides sufficient protection to the resistive element. Many types have hot-spot ratings well below their catastrophic-failure point but must be held at lower temperatures to avoid large resistance changes or temperature-coefficient changes. In high-resistance, high-voltage types, film resistors supplement wire-wound power types, which reach limitations of small wire diameter, long wire lengths, and degraded high-frequency performance. These are covered below under the special-purpose and high-power categories. In fractional-wattage ratings useful for low-level semiconductor electronic assemblies, film resistors offer excellent stability and value ranges in small physical sizes. While film resistors are in general higher-priced than carbon composition, carbon-film types are now commercially competitive, and both thick- and thin-film metals and oxides in their economical coated versions compete with some composition types. Deposited thin-film and some thick-film types are constructed by treatment of ceramic or glass cylindrical substrates to deposit the film and to apply metallized termination areas on the ends, to which lead assemblies can be subsequently attached. Value adjustment is accomplished on these types by grinding a spiral groove on the cylindrical surface to lengthen the film conduction path. A small fine trim is often done by light surface wiping of some soft films with a material such as paper. Although other methods are used, terminals are mostly pressed-on end-cap and lead assemblies which have interference fit with fired-on or evaporated metallized bands on the ends of the ceramic substrate (see general construction illustrations, Fig. 9). Termination quality and process controls are important considerations, and many functional difficulties arise in this area. Silver-loaded thermoset-resin cements enjoyed a brief popularity for termination use in the last decade, but reliability problems drove the method into relative disuse. Some thick-film types have their resistive material printed onto the substrate in a pattern near the final configuration, after which the substrate is fired to fuse the conductor onto the substrate. Some are not spiral-ground but may use various means of abrading away conductor cross section for fine trimming. Good-quality-film resistor spiraling is uniformly distributed over the longest practical length of core in order to avoid heat concentrations.

Metal thin films and carbon films are vulnerable to mechanical damage and to atmospheric moisture and must be protected immediately after desposition. Varnish coatings of the silicone variety are used directly over the film, being essentially inert and having sufficient high-temperature capability. Some films such as tin oxide are mechanically hard and also relatively inert chemically and do not need this treatment. One type of thin-film resistor uses a metal film deposited on the inside of a ceramic tube. Helical grooving for value adjustment is followed by application of terminal metallization and solder-sealed pressed-on end caps to produce a self-sealing package which is then epoxy-encapsulated. The resulting product is marketed as a hermetically sealed metal-film resistor.

Carbon Film. Pyrolytic carbon-film depositions on ceramic and glass substrates have been manufactured for several decades and, when used within their limitations, currently provide increased stability, lower noise, and much better high-frequency or high-speed performance in resistance values higher than about 100 Ω than competitively priced carbon composition. The more complex mechanical structure and higher functional current densities would seem to offer a greater opportunity for reliability—robbing failure mechanisms, but in practice, well-made products in commercial applications have given years of reliable service. The conformal coatings used by most competitively priced carbon films are sometimes less desirable for automated part-handling equipment than the uniform, rugged molded-case or composition types. Military specifications have essentially abandoned carbon-film designs, primarily because of increased availability of relatively low-priced replacements of other types with superior performance.

Pyrolytic carbon deposition requires considerable skill and control to produce consistent results. In general, the substrates are heated in a furnace and a carbon-bearing gas such as methane is metered into the enclosure, where the red heat pyrolyzes the gas. Substrates are rotated or agitated to obtain equal exposures. Film uniformity, thickness, hardness, and resulting characteristics all depend upon properties of the substrate, mechanical features of the deposition exposure, control of the atmosphere in the deposition furnace, time of exposure, and similar considerations. Resistivities of 10 to 10,000 Ω/square can be achieved with operating-temperature stability of about −250 ppm/°C at the low resistivity, progressively less stable to about −1,000 ppm/°C at 10,000 Ω/square. Custom variations in deposition processes allow improvement of this general range. Usually 100 to 2,000 Ω/square is used with a stability of about −400 ppm/°C. Parts are available in wide value ranges and tolerances—from 2 Ω to 125 MΩ in the largest physical-case sizes (2W) and 2 Ω to 400 kΩ in ¹⁄₁₀-W versions, with generally 1 Ω to 10 to 14 MΩ for intermediate sizes. Tolerances at purchase are 0.5, 1, or 2 percent.

Operational-temperature derating suggested in EIA specifications and in supplier catalog data is shown in Fig. 17. The total environment should be considered, such

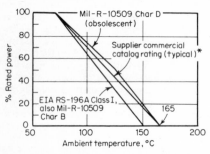

as proximity to other heat-producing components as well as operational environments of temperature and altitude. The flat portion of the curve obviously allows for more realistic extrapolation of dissipation capability if a single narrow range of conditions is known, and provided, of course, that voltage-rating limitations are observed. Again, however, as is the case with composition resistors, the specification load-life drift testing is done at room temperature, full load, without regard to achieved case temperatures, which can be well below the zero rated power temperatures. Such a test provides no value drift data for the higher ambient temperatures. It is still a good practice to keep

Fig. 17 Temperature-power derating for pyrolytic-carbon-film resistors. *Dale Electronics Type MC.

dissipated power to one-half the values allowed if ambient temperatures go much above 30 to 40°C. Voltage coefficient of resistance is 20 ppm/V or less in the center of the voltage rating. Working voltages are determined by the square root of the product of power rating and resistance value with maximums for each case size to determine the critical value of resistance (see General Information above). EIA RS-196 class I sizes rated for full power up to +70°C are covered by the advertised capabilities of some available carbon-film resistors, except that resistance range and voltage ratings may be more restricted than RS-196. (Other domestically available constructions such as coated metal film and thick-film types also are covered by the same RS-196 types and sizes.) Supplier's catalogs should be consulted for details on voltage ratings and power-derating curves. In general, domestic carbon-film types designed and tested against the now obsolescent MIL-R-10509 specification are physically larger for a given power rating than the RS-196 equivalents.

Typical carbon-film designs domestically developed are tabulated in Table 12. Table 15 shows nominal characteristics of domestic carbon-film resistors.

TABLE 15   Nominal Characteristics of Domestic Carbon-Film Resistors

Power rating, W	Ohms range* (1% tolerance)	Working voltage max	Length (nominal), in		Diam (nominal), in	
			Coated	Molded	Coated	Molded
1/10	1–400 kΩ	200	0.250	0.260	0.090	0.095
1/8	1–3 MΩ	300	0.343	0.406	0.109	0.135
1/4	1–5 MΩ	350	0.468	0.593	0.125	0.203
1/2	1–10 MΩ	500	0.562	0.730	0.137	0.250
1	1–15 MΩ	500	0.937	1.093	0.296	0.375
2	1–100 MΩ	750	2.063	2.188	0.296	0.375

* Below 400 kΩ all temperature coefficients of resistance are within –400 ppm/°C, below 100 MΩ within –600 ppm/°C.

Good carbon-film resistors can be expected to remain within about 2.5 to 3 percent of their initial values over a life of 5 years if applied conservatively. Derating curves are said to be based upon a resistance change of 1 percent over 1,000 h at rated loads. Additional allowances are necessary for installation, moisture, lead bend, and other likely stresses. Typical temperature-coefficient curves are shown in Fig. 18.

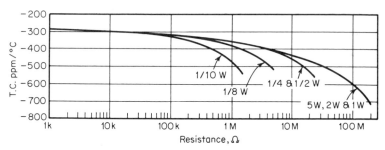

Fig. 18 Typical temperature coefficients of values and watt sizes for pyrolytic-carbon-film resistors. (*Dale Electronics.*)

High-frequency characteristics as with other film types are very good. Each resistance value and case size, as well as individual geometries, are naturally different. Predominant, however, is the roll-off in impedance at high frequencies for higher values of resistance due primarily to the shunt capacitance from end terminal to end terminal, which usually amounts to about 0.2 to 0.6 pF. Figure 11 shows nominal high-frequency characteristics applicable in general to most film-type resistors. The helixed conductor on these and other types of film resistors is of negligible effect except perhaps in the very lowest resistance values, the predominant effect being capacitive. Lead inductance is a major inductive consideration.

Coatings and encasements available provide varying degrees of mechanical strength and environmental protection. Moisture is damaging to carbon-film resistors and may cause catastrophic failure if liquid is allowed to accumulate on film surfaces. Small sizes need rigid coatings to enhance mechanical strength. Uniformly molded cases are generally stronger, easier to handle with automatic insertion equipment, and provide adequate moisture protection and dielectric strength. Changes in resistance value due to moisture are only partially recoverable, and amount to 1.0 to 1.5 percent in humid exposures. Cost for molded cases is of course higher than for roll coatings, lacquer coats, or other conformal coverings. True hermetic encasement is no longer available for pyrolytic-carbon general-purpose resistors, but soft-soldered ceramic-case designs are still procurable at this writing.

Coatings should be flame-retardant and should not produce toxic gases under extreme temperatures, in order to provide product safety. Coatings can be cracked and abraded in service, and should be considered when mounting resistors and adjacent conducting parts. Coatings should therefore be as good as practicable with regard to mechanical strength, moisture permeability, solvent compatibility, flammability, and general durability. Sealed construction, if used, must be extremely dependable, since the reliability of such construction depends absolutely upon the continued ability of the encasement to remain sealed. Small leaks in such cases tend to trap moisture and other detrimental materials in the very areas where they can do the most harm, whereas more pervious coatings can expel the damaging fluids as easily as they absorb them.

Tin Oxide. Depositions of tin oxide into glass-rod surfaces were introduced by Corning Glass Works in the middle 1940s. The process uses exposure of nearly

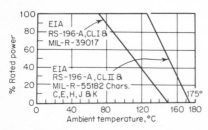

Fig. 19 Temperature-power derating for film resistors: premium metal films, general-purpose metal films, thick films, and certain oxide films.

molten glass rods being continuously extruded through an atmosphere containing the required gaseous compounds. Tin and antimony are said to be constituents of the gaseous medium. The result is a nominal tin oxide film diffused minutely into and onto the glass-rod surface. The film is under 5 μm in thickness, is smooth and glossy, and is on the order of tool-steel hardness. Because it is a relatively stable oxide in a low-energy state, chemical activity is minimal in most commonly encountered ionic combinations. The glass substrate is not as mechanically strong as some ceramic types. Also, the conchoidal fracture tendency of the glass substrate combined with the hardness of the film would appear to make spiral grinding somewhat more critical than for some other film types; it should be noted, however, that alumina is even harder than tin oxide. Initially, tin oxide was not produced at temperature coefficients of resistance closer than about zero ±200 ppm/°C, but process refinements have narrowed the range to ±50 ppm/°C and parts are now supplied under MIL-R-55182, with some value limitation at the upper end of the value range. Test experience has shown the glass–tin oxide resistor to have excellent drift characteristics and good high-temperature-life capability, along with the low-noise production and high-frequency performance common to metal-film types. Resistors supplied domestically are nearly immune to humidity-caused catastrophic failure and large-value drifts. Tests have shown that uncased resistors are capable of meeting most standard moisture tests. Imported tin oxide resistors using ceramic substrates are available and compete favorably in the general-usage and semiprecision ranges, as well as in medium-power. Tin oxide on glass is also used for high-frequency, high-power units. Semiprecision, precision, and general-usage tin oxide film resistors are marketed under obsolescent MIL-R-22684 and under MIL-R-55182 and MIL-R-39017 as well as various special requirements. Only 1/10- and 1/8-W sizes are currently qualified under MIL-R-55182. No hermetic versions are now available. For high-frequency performance, noise characteristics, temperature derating, see data on metal-film types and Fig. 19, which shows MIL-R-39017 power derating. In comparison tests for moisture resistance and overload stability, coated tin oxide resistors remain stable during exposures well beyond specification limits, while coated nickel-chromium metal-film resistors show better stability under nonsevere conditions but deteriorate rapidly as coatings are penetrated by severe humidity or overload conditions become severe. Fusible versions of both are commercially available to enhance product safety.

Current pricing of coated tin oxide resistors is competitive with semiprecision metal-film types in the temperature-coefficient range of 0 ± 50 ppm/°C and, in the broader temperature-coefficient and purchase-tolerance ranges, competes with thick-film glaze and film types. Corning Glass Works, Bradford, Pa., and Welwyn—Canada, marketed by Dale Electronics, Columbus, Neb., are the two of the current major suppliers of tin oxide resistors. (Welwyn does not use glass cores.)

Thick Film. Advances in technology of materials and increased understanding of solid-state physical properties have led to a promising development of low-cost resistors using thick films composed of stable combinations of metals and metal derivatives and other material combinations, usually in vitrified-glass-frit suspensions. Metals are usually gold, silver, palladium, ruthenium, iridium, and platinum for cermets, and tantalum, tungsten, or titanium for other types of film. After processing, these films are in stable states, with oxides, nitrides, and carbides being commonly formed, and are usually self-protecting by virtue of a passivating surface-coating conversion. Initial thick-film cermet resistors used palladium-silver on ceramic substrates. This cermet combination is somewhat susceptible to hydrogen exposure, and temperature coefficient of resistance for cermets is about ± 200 ppm/°C. Improvements in purity, better understanding of processes, and use of other materials have resulted in increased stabilities. In glaze-type thick-film construction, finely divided metal particles are mixed with glass powder and a volatile vehicle. The mixture is painted on, rolled on, screened on, dipped, or otherwise applied to a ceramic substrate. The resulting combination is fired at temperatures of about 1000°C, melting the glass powder, suspending the metal particles therein, and bonding the resulting film to the ceramic substrate. Atmospheric controls are provided to control oxidation and reduction during the firing while the glass is molten. Metallized bands are applied to ends for connection of terminal leads, which may be press-on end caps, soldered-on leads, or other methods of termination. Other thick-film types employ anodized, oxidized, or otherwise modified applied films. A good range of resistance value is possible within 0 to ± 200 ppm/°C temperature coefficient of resistance for most of the thick-film types, with some refined and controlled types being capable of 0 ± 50 ppm/°C, and selected batches of 0 ± 25 ppm/°C. Values from 10 Ω to 1.5 MΩ are made in standard sizes, with some limitation to about 500 kΩ in smaller sizes for good temperature coefficient. Special longer configurations and high-resistivity films extend the value range well into the high-megohm region for special-purposes uses, and special low-resistivity metal-doped films allow values down to 1Ω. These parts have performance comparable with that of metal-film types. Current-generated noise is generally a few decibels higher than for comparable-size metal films, but still well below carbon composition. Size, type of film, and resistance value strongly influence noise index. Stability varies with the film used and the care exercised in production. Load-life stability comparable with that of metal films is achievable with some thick films. Voltage coefficient is likely to be higher for some thick films than for thin-film units and, if important to applications, should be investigated. Applicable specifications for the general-purpose versions are MIL-R-39017 type RLR, replacing MIL-R-22684, type RL (now obsolescent); RS-196, types RO, R1; and the obsolescent MIL-R-10509. Temperature coefficients of resistance and the nonavailability of purchase tolerances closer than ± 1 percent keep currently available thick-film parts from equivalency with the best metal-film resistors, but the inherent low cost and basic stability of the films achieved makes this a most promising area for future development. Figure 19 shows the temperature-derating curve specified in MIL-R-39017 and RS-196A. MIL-R-39017 specifies testing at the zero rating temperature with limits on allowed resistance drift.

Thin Metal Film. Initial metal films were platinum-iridium or palladium-silver, but most are now nickel-chromium in the approximate ratio of 80 to 20 respectively. Partial reduction, oxidation, passivation, and other custom treatments modify characteristics to achieve desired traits. Most are produced by vacuum-vapor deposition at high temperatures onto ceramic substrates, but sputtered tantalum nitride in thin sections is also used, as are cobalt-chromium combinations, and a variety of other types.

Bulk temperature coefficients of resistance for pure metals are almost exclusively positive. In thin films, however, surface-conduction modes that are essentially independent of temperature become a large factor, and together with alloying and custom treatments, temperature coefficients very near zero are achieved. Resistivities depend upon film thickness primarily, and as in other film types, the best characteristics are obtained in the middle of the practical range. Low values are somewhat limited, and high resistances were in earliest designs much more limited as compared with carbon-film types. Nickel-chromium films of 2,000 to 3,000 Ω/square provide for good stability, but higher sheet resistivities need thicknesses of film of less than 20 Å, and drift stability begins to fall off rapidly. The addition of nonconductor materials to chromium in evaporation processes has allowed achievement of reasonably stable devices up to 20 kΩ/square. Chromium–silicon monoxide evaporated film, considered to be a "cermet" combination, has been successfully used in this regard. Resistors are finished by grinding grooves in the film to produce a helixed conduction path. This is done usually on a Wheatstone bridge against a predetermined value, with the grinding wheel automatically lifted as the bridge balance is approached. Mild surface abrasion is sometimes used as a final trim. Most good precision nickel-chromium film resistors, including hermetic versions, that are procured at 1 percent or better purchase tolerance and ±50 ppm/°C temperature coefficient or better can be expected to remain within about ±1.5 to 2.0 percent of their value as received, for a normal service life of 5 to 10 years of normal operation and storage. For a 1 percent procurement tolerance, therefore, 2.5 percent is a practical end-of-life tolerance. Changes tend to be greatest at installation and shortly thereafter but cannot be depended upon to track from part to part in either direction or magnitude, although general trend-change directions can be seen for some manufacturers and designs. These should not be depended upon unless empirically verified. Some manufacturers have responsibly promoted realistic design application-tolerancing schemes by which valid assessment can be made of future (long-term) variability. For further discussion of this technique, refer to the discussion below on resistor application techniques.

Metal films are able to remain stable at higher operating temperatures than carbon films and, along with other film types, have completely replaced carbon films in military standards. Various sealed designs have been developed. These are intended to exclude environmental conditions, and some have been very successful in use. One glass-metal sealed design is capable of direct transfer from 125°C oil to minus 180°C liquid without fracture or damage.

Nickel-chromium thin films cannot tolerate water in liquid form if ionizable salts are present. Electrolytic couples that otherwise would be of minor significance become formidable in terms of the thin cross sections of metal films. Hermetically sealed designs therefore must be extremely good to avoid entrapment of contaminants. Coated and molded encasements provide mechanical strength and moisture protection in proportion to their particular capabilities, and each must be measured in terms of its cost, character, and performance. High-frequency characteristics of metal films are extremely good compared with wirewound or composition types, particularly in resistance values greater than about 1,000 Ω. Helical spiraling has negligible inductive effect in lower resistances on the high-frequency characteristics. Figure 11 shows frequency characteristics applicable in general to metal films of ordinary geometrical dimensions. End cap–to–end cap and distributed capacitances dominate the performance in the higher values and frequencies, producing a shunting effect. Low resistance values remain very close to labeled value. Very high-frequency performance on low values can be improved by use of coaxial mounting to eliminate lead inductance. Nominal noise characteristics are shown in Fig. 13.

Drift characteristics are not well publicized. Military specifications, for instance, are written to define requirements to be met at the time of purchase and do not mention drift that can be expected as the resistors are stored in stockrooms. This tendency actually amounts to one of the most significant value changes that can be expected. Some users are dismayed to discover that resistance values outside labeled tolerances are encountered after brief storage periods, particularly on batches having value distributions skewed badly at the time of purchase. In critical equipment this circumstance causes uncertainties about where the manufactured equipment test limits should be drawn to allow for expected service-life drift. Obviously, the full expected value-drift limits should not be used, but neither can limits be based upon resistor purchase tolerance if parts have been stocked for any significant time. Unfortunately, drift is not always in the same direction, though it may show a tendency for a given batch of parts. Figure 20 shows empirical drift of a number

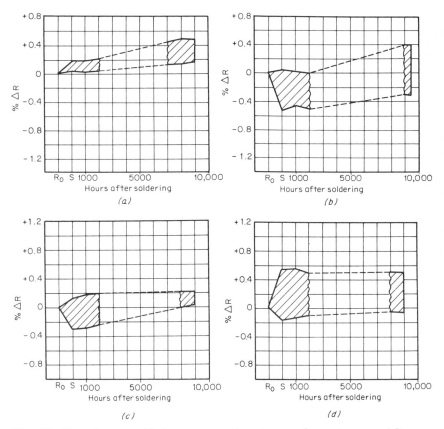

Fig. 20 Data for postsoldering, nonoperating storage of premium metal-film resistors (three standard deviations), normalized to observed value at time zero. (*a*) Supplier A, 13 kΩ. (*b*) Supplier B, 10 kΩ. (*c*) Supplier C, 147 kΩ. (*d*) Supplier B, 105 Ω. (*Martin Marietta Aerospace.*)

of resistors and the effect of soldering on their values. There seems to be a tendency for drift to occur much more rapidly while a resistor is new and to level off at a lower rate over the long term. Drift is a highly variable trait, and if circuit applications are too critical to make adequate allowance, the performance of the particular design or even particular batch of parts used should be determined experimentally. Drift mechanisms are not all identifiable, but structural-stress relaxations, molecular reordering of conductor structures, oxidation, and cross-section reductions through minute stress relaxations are known factors.

Initial tolerance, temperature coefficient of resistance, and encasement design control basic pricing. Least expensive and smallest physically are lacquer and roll coatings with the most expensive and largest physical size being hermetically sealed designs. The coating system used dictates the degree of resistance to humidity, and the cautions stated in the introductory paragraphs above apply. The selected material should, in addition to being resistant to moisture penetration, provide mechanical strength, solvent resistance, and mechanical uniformity. For product safety, the coating material should be nonflammable or flame-retardant and should not give off toxic gases at high temperatures.

The best nickel-chromium thin films will provide the stabilities stated above if conservatively applied. Extreme conditions or less refined production processes will, of course, result in degradation of stability or even vulnerability to catastrophic-failure mechanisms. Power-derating curves of MIL-R-55182 and RS-196 (class II) are shown by Fig. 19. MIL-R-55182 includes a high-temperature exposure test wherein the test sample is heated to the zero-power-rating point for 2,000 h without power applied. The allowed resistance change is controlled, thus providing a degree of assurance for use of the derating curve. Half-power further derating is still recommended, however. Chemical inertness due to low-energy states of the constituent materials enhances stability by reducing the rate of change in character, and systems employing such materials in stress-relieved systems with other materials of carefully matched compatibility enjoy the maximum stability and reliability. Electrolytic similarity and matched thermal-expansion rates are equally important considerations. Conductor cross sections are extremely minute, and small changes tend to produce relatively large effects.

Temperature coefficients of resistance of ± 25 ppm/$°C$ are routinely furnished and of ± 5 ppm/$°C$ are achievable with nickel-chromium films.

Reliability figures stated for military uses are based upon a predefined point of failure, usually some amount of change in resistance. These "failures" may not be catastrophic at all, and in most circuit applications may even be negligible. The worst troublemakers in applying well-made metal-film resistors tend to be the catastrophic-failure modes and mechanisms. Cracked cores, defective lead–to–end cap connection, defective coatings, electrolysis, solderability of leads, and actual breakage are factors that should be addressed by quality-assurance controls. Screening tests devised to locate film defects generally require overloading of the parts up to about five times rated power for a short time, the effect being the creation of localized hot spots at high-current-density points, resulting in either open circuiting or a large measurable change in resistance value.

Prices of metal-film resistors depend basically upon quality of materials and degree of control exercised in manufacture, and upon the encasement method used, but if stringent quality controls are required for elimination of marginal product, the cost of the assurance tests may equal or exceed the cost of the basic off-the-line product.

Metal-film resistors offer the highest temperature stability of any nonwirewound resistor. Their high-frequency and low-noise performances are excellent, and if they are well made and conservatively employed, they can be expected to provide at least 10 years of reliable service.

Current specifications covering various types of metal-film resistors are MIL-R-55182, MIL-R-39017, RS-196, and other chip and network assembly military specifications (refer to Table 12).

Low-power wirewound resistors Several companies market inexpensive wirewound resistors having molded or conformally coated cases. These resistors are of limited resistance range in the ratings of 2 W and below from about 0.1 Ω to 2 kΩ or even more limited depending upon manufacturer. They offer reduced size per wattage rating over comparable carbon composition styles, and temperature coefficients of 800 ppm/$°C$ in the low values down to about 100 ppm/$°C$ in higher values. These units operate at full power at high hot-spot temperatures, however, and estimates of change in resistance with applied power must consider this fact. EIA specification RS-344 covers this style of resistor in 1.25 and 2.5 W, styles CRU1 and CRU2, respectively, which are derated to zero at 275°C. A four-band color code is used, with the first band being noticeably wider than those remaining. Additional general-purpose and medium-power wirewound resistors are also covered by RS-344. For example, a CRU2A type, in a molded, potted, or case encasement having an essentially square body cross section, has a power rating of 2 W. Others of this configuration in RS-344 are included in the medium- and high-power categories in a range up to 15 W. Purchase tolerances of 5 and 10 percent are specified. Power-derating curves are shown in Fig. 21.

Medium-Power Resistors

Almost all resistors produced for power-dissipation ratings greater than 2 W are wirewound types which can operate at higher temperatures than metal film, carbon films, or composition with its organic binders. This capability allows physical sizes of wirewound resistors to be smaller in general than those of resistors of the film type with comparable power-dissipation ratings. With the predominant wirewound types, however, some power film resistors are available, and a few suppliers market film resistors with thermally massive integral heat dissipators to allow medium- or higher-power ratings.

Medium-power wirewound

Thermal Considerations. Application technques for wirewound resistors in the range of 2 to 6 W must allow for their higher operating temperatures over those of general-purpose types by arranging to take the produced heat away efficiently and by protecting adjacent parts from possible damage. Melting point for commonly used resistance-wire alloys is higher than 1000°C and ranges up to about 1500°C, but considerations of repeated thermal expansion, wire grain-structure changes, oxidation, and hot-spot occurrences combine to make a practical limit of 350°C or less external temperature for types where the wire is embedded in enamel or cement compounds. For even liberal practice, however, commercial and military standards alike use a 275°C maximum limit for styles having less than 10 W dissipation.

Figure 22 shows temperature derating recommended by specifications. Conserva-

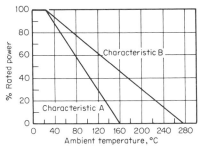

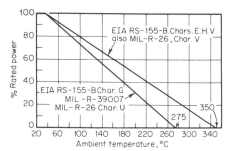

Fig. 21 Medium-power and general-purpose wirewound resistor power derating for high temperatures. EIA specification RS344.

Fig. 22 Medium- and high-power wirewound resistors power derating for high temperatures.

tive practices suggest use of lower limits, both from considerations of ultimate longevity of a given part population and to avoid resistance drift associated with physical changes. Exposed-wire types do not have the thermal resistance of an embedment to cope with and consequently can be used at higher temperatures on their surface. Smaller wire sizes naturally are susceptible to more reliability problems of a structural-mechanical type than are larger wire sizes which are therefore preferred for reliability enhancement.

Inductive Effects. Inductively wound resistors naturally present significant inductive and capacitive reactances over certain frequency ranges. Since number of turns depends upon the resistivity and gage of the wire used, and capacitance depends upon end-cap spacing, core material, turn-to-turn spacing, and dielectric constant of coatings, only general information is possible for presentation here. For resistances above 100 Ω, reactive effects start to become noticeable at 2 to 3 MHz, for these usually single-layer windings but do not become spectacular until about 20 MHz and beyond. Ayrton-Perry winding is the best method for inductance cancellation. Center-reverse or Chapron winding is less effective. Reverse-pi and bifilar windings cannot normally be used because of the high winding temperatures and the unsuitability of most wire insulations at high temperatures. Almost all designs are single-layer or Ayrton-Perry crossover types. If high-frequency requirements are severe, consideration should be given to use of power film resistors. Refer to the General Information discussion above.

Resistor Cores. Core material is a significant part of the thermal system, being in intimate contact with the resistance wire and usually well connected to the metal lead wires. Porcelains and pottery formulations are widely used. High-alumina ceramics are better thermal conductors and are much stronger than other ceramics. Beryllium oxide is an excellent thermal conductor and is used in special miniature power resistors, but its toxicity and high cost limit its use. Some small military types are qualified with BeO_2 cores. The effect of good core thermal conductivity is to lower the operating-temperature gradient between the center of the resistor body and its ends, thus reducing the hot-spot temperature. One design is available with metallic aluminum cores. Other materials are also used. Mineral-filled plastics, silicones, and impregnated glass fibers are used as cores and mounting frames, though not as extensively as in previous years. Glass-fiber cores do not, of course, provide an efficient thermal path. This construction with end caps crimped-on commonly has small diameter relative to length, and is used for low-cost wirewound units in low resistance values for exposed-winding automotive parts rated in watts per inch. Some encapsulated commercial parts, such as RS-344 molded parts, also use this basic element, with some designs providing an intimate and relatively massive encapsulation which increases the dissipation capability of the basic element. The attractiveness of the glass fiber core resistor lies in its capability for mass production of relatively reliable parts. The resistance wire is wound continuously on long lengths of the impregnated glass-cloth rod, then machine-chopped to length, capped, and crimped almost simultaneously. The end caps compress the wire and core sandwich, making good pressure contact with the winding, and the result is an easily produced part having about ±5 percent room-temperature tolerance. Some fabricators crimp on the end cap but make the electrical connection by resistance welding. The wirewound element is marketed as a bare unit for automotive and appliance applications, and is incorporated into the various embedded broad-tolerance commercial wirewound units.

Resistor Coatings. Considerable controversy exists among manufacturers and users alike with regard to respective virtues and disadvantages of vitreous-enamel and filled-silicone-resin coatings for outside encasements. Good vitreous enamels are dense, moistureproof, strong, hard, capable of high-temperature operation, and virtually solvent proof, and are good thermal-expansion matches with core and wire. Their glazed state holds the resistor element in compression. Silicone coatings, if properly impregnated and cured, meet most of these criteria. They are less dense, less moistureproof, can be made strong enough, not as hard, have generally adequate, but lower, rated temperature than vitreous enamels, and are vulnerable to some solvents. Both are prone to crack at lead menisci, with vitreous enamel being somewhat more prone to this problem. Silicones are much gentler in induced mechanical stress on resistor elements. Vitreous enamels may contain small residuals of alkalies, which also may cause functional variations, particularly at the high firing temperatures needed in resistor production. Close-wound wire also has a tendency to bunch up during the firing process, which can cause problems. In general, improved stability of resistance can be obtained with silicone coatings, but better environmental protection is obtained with good vitreous enamels. Cost is generally lower for the silicone. If silicones are used, particularly in automated solder operations, their solvent resistance should first be fully evaluated. Vitreous coatings still predominate in the higher power ratings, particularly where large wire can be used, resistance changes are less critical, and turn-to-turn spacing is not close. Such units are known to operate reliably for many years.

Resistance Wire. Resistance wires must be capable of withstanding the high firing temperatures if vitreous coatings are employed, and must withstand repeated thermal cycling and long exposures to elevated temperatures, while remaining stable and inert. For their stable products, most suppliers use one of the 800 Ω cmil ft^{-1} special alloys. The older nickel-chromium or copper-nickel alloys still are used when resistance value and stability allowances permit.

When temperature changes of resistance value are assessed in comparison with film types or precision wirewound parts, the increased temperature permitted for a given power rating should be noted. Refer to the construction material paragraph for further wire information.

Reliability Considerations. Good-quality resistors have windings evenly spaced on the core and will have well-made termination welds. Coatings should be capable of withstanding the specified high-potential tests when the resistor body is intimately wrapped in foil with voltage applied from foil to winding. Fillers for silicone and cement coatings are sometimes sources of ionic salts and, in the presence of environmental moisture, can cause galvanic corrosion. Resistance alloys and terminal hardware are in themselves usually noncorrosive, but galvanic dissimilarities can cause corrosion problems in the presence of ionizable salts. The resistance-wire joint at the termination is probably the point most vulnerable to corrosion damage. Most standardized resistors in the medium-power range are of the axial-wire-lead type, having conformally coated cylindrical bodies. Some designs have been encapsulated into metal tubes to allow efficient use of clip-type connection to chassis for better heat removal. The bare and molded automotive types mentioned previously are also popular. One available commercial design uses a hollow ceramic tube or lidless rectangular box with the automotive-type resistor element intimately encased inside in a ceramic-filled cement.

Heavy-bodied designs should be mounted by means other than their leads if shock and vibration environments are to be encountered, but care must be exercised to prevent breakage of ceramic and glazed parts by stress concentrations or tensile forces.

Power film Film resistors provide improved rf performance over wirewound types and, in addition, allow a wider choice of resistance values. Tin oxide on glass has been available in large physical sizes for many years, and varnish-coated carbon-composition-film and carbon-film types are used where sizes can be large enough to keep temperatures down and where environments permit. The newer thick films, fired-on oxide films, and modified metal–metal oxide thin films on ceramic cores are finding increasing uses where resistance value and frequency characteristics indicate a preference. A good selection is currently available with power ratings up to 15 W. High-power parts have tab-type band terminals or clip-mounting ferrule terminals. Lower-power types have end-cap axial lead wires. By improving the thermal dissipation, a resistor of relatively small physical size can be made to dissipate increased power without an increase in its hottest-spot temperature. Accordingly, Dale Electronics produces metal-film resistors molded into aluminum housings designed to mount on a chassis in an intimate mechanical fashion, thereby allowing improved thermal-dissipation rate. A good high-value range is provided (up to 2 to 2.6 MΩ) in power ratings up to 12 W in compact size and with the characteristics of metal film. Uses are primarily in military applications, although no current military specification coverage exists. As thick-film technology develops, use of medium-power film resistors should increase somewhat.

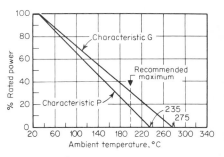

Fig. 23 Film resistors, power type per MIL-R-11804, power derating for high temperatures.

MIL-R-11804 covers power-film resistors in conformally coated types. Power-derating curves are shown in Fig. 23.

High-Power Resistors

High-power characterization is defined to include all resistors having rated power dissipation greater than 6 W, except for devices designed specifically to be electrical heating elements. Usages of high-power resistors are more frequent in heavy-duty power-equipment applications than in usual electronics apparatus, but applications in electronics and its peripheral and test equipment are frequent enough to warrant treatment here, particularly for the lower end of the power range.

Basic design considerations High-power-resistor designs are almost exclusively wire- or strip-wound, though there are some carbon-composition types, as well as tin oxide and thick vitreous-glaze film units. The high-temperature capability and stability of available resistance-wire alloys provide a large practical advantage for the wire- and strip-wound constructions unless fluid cooling is available, or unless high-frequency requirements make wound units less attractive. Reliable wound types of established design and ready availability can be procured from several domestic sources and can be separated into the following categories: (1) fully exposed winding, (2) partially exposed complete winding, (3) partially exposed winding segment for tap adjustment, (4) fully embedded or coated winding, and (5) wound element enclosed within an integral heat-dissipating metal case. Dealing with the large amounts of heat produced by these parts is the primary design consideration. Large resistor surface areas are needed to dissipate the energy and to keep hot-spot temperatures within safe and reliable ranges.

Terminals and Mountings. The large physical sizes and masses require considerable attention to mounting details, and the familiar self-supporting axial-lead-mounted cylinder is little used above 10 W rating. Radial tabs that are extensions of metal bands welded around flat oval or cylindrical core ends leave the hollow core centers open for various mounting methods and are the most widely used termination method. The resistance wire is welded or hard-soldered to the band, and the radial tab may be fitted with a variety of connection accommodations such as appropriately sized screw and washer terminals, long multistrand flexible or short solid-copper lead wires crimped and soldered to the tabs, or most frequently, a hole made in the tab to accommodate a wire for soft soldering. MIL-R-26, for example, requires that this hole be at least large enough to permit a fully affixed and crimped 14 AWG wire, which has a 0.065 in diameter. Lugs and terminal connections are sized larger proportional to the scale of the resistor and the current-carrying needs. Support of the resistor body by lead wires alone is increasingly less desirable up to about 20-W sizes, beyond which even the most benign mechanical-environmental considerations require body mounting. Some part designs require hard solid mounting to a metal chassis of a specified area to achieve their labeled power rating. Besides the radial-tab terminal, parts are supplied with fuse-clip ferrules, flat blades for pressure connections, long flexible insulated wires, and various standard Edison-type lamp screw bases. Lead wires are not as significant in the terminal dissipation process as in lower-power resistors. Figures 9 and 24 show some typical high-power-resistor designs.

Thermal Considerations. For conventional free-air-mounted 2-W and smaller resistors, lead wires account for 50 to 75 percent of the heat removal, but above 2 to 5 W this fraction diminishes rapidly and radiation quickly gains significance for free-air designs, particularly for the high temperatures of wound wire and strip units operating at rated capacity. Convection is also prominent, retaining about half the significance of radiation at maximum rated dissipation for wirewound styles. Resistor shape and surface-finish variations have large effects upon the convection rate, and the dissipation-mode proportions vary as the operating temperature achieved by a particular resistor changes. The relatively high operating temperature of wirewound units is a strong factor in the predominance of the radiation mode, since the rate of energy dissipation by radiation from a body is proportional to the fourth power of its absolute temperature, whereas thermal conduction and the related convection are essentially linear or second degree with temperature.

Material Types. Materials must remain stable in high-temperature service and should be well matched with respect to thermal-expansion characteristics to avoid physical damage or failure in repeated thermal cycling. The finished item should be capable of withstanding the effects of atmospheric humidity and condensed moisture along with moderate industrial atmospheric contaminants if it is intended for use in unprotected applications. Galvanic metal couples should be carefully avoided by use of compatible platings and materials, and critical circuit points should be well protected to impede corrosion. If wirewound, resistance wire of the largest possible cross-sectional size should be used, and the heat-producing winding should be evenly distributed over the physical body of the device to avoid hot spots.

Coatings. Protection coatings and encasements available are of three basic types: filled silicone resins, vitreous enamels, and ceramic cement using binders other than silicone resin. Silicone-coated parts are used where precision resistance stability is needed, since curing temperatures are low compared with those needed for vitrified materials, and wound resistance wire is less likely to be stressed and moved about during its curing process than is the case for the glossy vitreous enamels. Most suppliers limit silicone-coated units to about 275°C maximum operating temperature, but some are recommended by suppliers for operation up to 375°C maximum. EIA and military specifications warn about gaseous exudations from "some coatings at temperatures above 200°C." Presumably this refers to silicone types. RS-155 states further that this outgassing may be detrimental to open electrical contacts. Production cleaning solvents also may affect some silicone coatings.

Vitreous enamels are formulated to match thermal-expansion coefficients of wire, core, and terminals and, in general, are in a surface-stress condition after firing, so that the resistor wire and core are in a degree of compression. Surface crazing, particularly in the vicinity of windings having large wire, is sometimes difficult to avoid. Gas-free alloys are often necessary to avoid included bubbles and foaming effects in cured coatings, but since multiple layers of material are used and voids are generally isolated, high-voltage "V-block" or foil-wrapper tests can be met, as well as can humidity tests. Dye-penetrant pressurized soaks, followed by careful sectioning, may be effectively used to evaluate extent of element exposure. Vitreous enamel and silicone coatings provide handling and installation protection as well as improving turn-to-turn and element-to-mounting high-voltage capabilities. Protection from inadvertent short circuiting is provided by those types having completely embedded windings, and surface contaminants are prevented from accumulating across wire turns, as can happen in exposed-winding types. Although high surface temperatures for parts operating at near rated power are a factor in personnel safety, exposed conductors in accessible areas can be a much more serious hazard. Electrical-shock dangers are greatly reduced by use of fully coated windings. All coatings interpose a thermal impedance between resistance element and the medium surrounding the resistor, and under powered condition the resistance element will operate at a higher temperature than that measured on the outside surface of the coating. This condition is recognized in specifications by allowing exposed wire and strip windings to be rated at somewhat higher temperatures, measured on the wire itself, than is allowed for the coating surface on completely embedded types. General military specification MIL-R-26 does not allow exposed-wire designs having conductors less than 25.3 mils in diameter for round wire and even then requires that turns be fixed firmly in place by embedment between turns. Inadvertent voids in coatings are not permitted by MIL-R-26 to expose conductors smaller than 10 mils in diameter, even though V-block and humidity tests may be met. Film high-power resistors are sometimes furnished uncoated, or may have only a thin high-temperature varnish-type coating. Fired thick-film types in lower power ratings are usually heavily coated, but the carbon composition film and tin oxide film types are used primarily in protected installations and are lightly protected.

Cores and Cases. Cores for some film types are glass tubes, while most other high-power units use ceramic cores. The relative size of core needed to provide required surface dissipative area makes it practical to use hollow cores, which also allows for ease of mounting by several means. The tubular film units may be fitted with ferrule terminals or fired colloidal metallization in a ferrule shape, which allows design of fittings that can be used to circulate fluid coolants through the core. Ceramic core materials are a variety of pottery formulations, most being steatite or electrical porcelain (see information in chapter introduction). Cores are popularly cylindrical, flat ovals, cylindrical with flatted sides to allow precision in Ayrton-Perry noninductive windings, or helically grooved cylinders and flat ovals. Glass-fiber core units continuously wound in long thin rods, then chopped and terminated to size by length (automotive types) as described in the medium-power category above, are also produced in units having ratings greater than 6 W. They are furnished as bare windings for use in automotive and apppliance applications where they are used in protected enclosures and are rated in watts per inch. This type of element is also furnished as an encapsulated unit, frequently square or rectangular in cross section. Since the diameter-to-length ratio is small and the surface area is therefore limited, a thermal advantage is achieved when the unit is intimately embedded in a relatively thermally conductive ceramic cement. Allowed surface operating temperature of these units is naturally much less than that of the large-diameter windings with thin coatings, and the technique is used only in lower power ratings. These parts are inexpensive units suitable for many applications.

Some large, low-resistance-value parts are wound from heavy wire or strip and are designed to be supported by ribs or bars so that an open-wound unit is constructed. Those parts are usually not insulated or coated and are used in protective enclosures, often with forced-air cooling. Several designs as shown in Fig. 24 are available from established suppliers.

A variety of embedded or "tub-type" resistor units are available. These parts employ zig-zag noninductive windings on flat cards or ribs that are subsequently placed into a "tub" or depression in a fabricated base of ceramic or high-temperature molded material, there to be embedded in a refractory cement with terminals protruding from the cured embedment. These are convenient for mounting and require little height. One design shown in Fig. 24 is that of a flat disk having a mounting hole in its center with terminals brought out radially. Units of this configuration can be stacked on a long threaded rod, observing proper power derating, of course, or can be chassis-mounted using a screw-and-washer combination through the center hole.

Related to these types are the military chassis-mounted units that are cylindrical wirewound elements molded with thermoplastic into extruded aluminum cases. Chassis feedthrough, standoff, and horizontal mountings are available, and the technique allows quite large power ratings to be furnished in relatively small physical sizes, provided that enough chassis area is available for heat dissipation. These units are designed to remove as much heat as possible by conduction, and even this convection-radiation efficiency depends upon getting generated heat at the winding center to the outside case efficiently. Beryllium oxide solid cores are much used for that reason in these designs, being much better heat conductors than other ceramic materials; in fact most power resistors that are designed to provide the smallest possible sizes use beryllia cores. The effect is to reduce the thermal gradient in a given resistor size under a given thermal operating condition or to "spread out" the hot spot. Some power metal-film resistors are supplied in heat-dissipating-case designs, but currently availability is limited to about 12 W maximum. In addition to the above configurations most power-resistor manufacturers list in their catalogs a variety of perforated metal enclosures that have resistor-mounting provisions and outside wiring terminals. These cases provide a shield for the hot resistors and also protect bare wound resistors from accidental contact with personnel or with other circuit elements. Manufacturer's catalogs specify degree of derating for multiple mounting and for inclusion in enclosures. (Also refer to application notes later in this chapter.)

Resistance Alloys. For many years 80-20 nickel chromium was the predominant power-resistor alloy, and it is still used in many designs. Other alloys are extensively used today, and in the low-power region particularly, many suppliers use the improved stable high-resistivity alloys exclusively except for low resistance values where copper-nickel alloys are more practical. Noncorrosivity, good tensile strength, freedom from grain growth, weldability, and similar considerations are made in selection of a particular alloy by resistor fabricators. High-power resistors are not usually required to be precision devices, but purchase tolerances of ±1 percent can be held with minimal fabrication care for most wirewound types, and tolerances are not usually broader than ±5 percent. The inherent temperature and life stability of good resistance alloys is somewhat offset by the expected high operating temperatures, but well-built units will remain stable for many years with only 1 to 2 percent or less change. Resistance wire having a cross-sectional diameter less than 0.001 in, even in completely embedded designs, should be avoided, although some military parts are qualified with wire 0.00075 in in diameter. It is obvious that small wire-drawing defects, scratches, scrapes, corrosion, and other incidental imperfections are more significant for small wire than for large diameters. Reductions in cross section cause hot spots that can lead to accelerated localized deterioration and eventual early failure of the winding. Whether to set the limit at 0.001 in or at a smaller

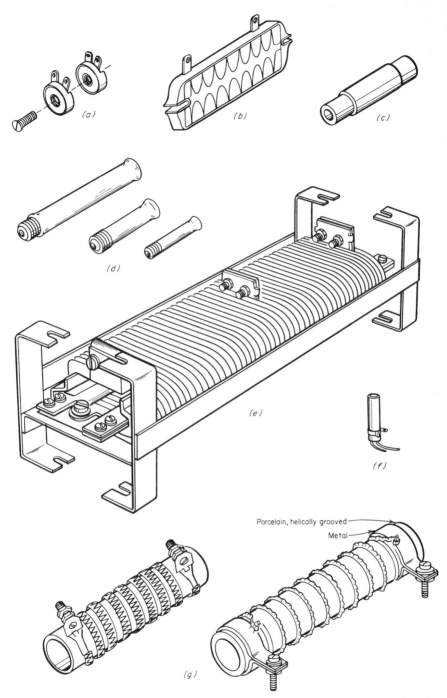

Fig. 24 High-power resistor designs. (*a*) Tub-type disk. (*Ward Leonard.*) (*b*) Tub-type resistor. (*Ward Leonard.*) (*c*) Ferrule fuse-clip vitreous-enamel-coated resistor. (*Ohmite.*) (*d*) Typical Edison-base resistors. (*e*) Edge-wound bare ribbon. (*Ward Leonard.*) (*f*) Appliance-motor control. (*g*) Corrugated edge-wound ribbon. This is one type. There is another type in which the space between the ribbon is filled with vitreous enamel, leaving just the ribbon edges exposed.

diameter is a matter of judgment. Exposed wire in small size should be approached even more carefully, especially if adjustable taps are furnished. A limit of 0.004 in and larger diameter is suggested for these parts. Exposed small wires and adjustable taps should be used only when necessary if maximum reliability is desired. Exposed-winding designs are limited by some specifications to larger diameters. (See previous paragraph on coatings.)

Ratings and performance There are two basic approaches to defining high-power-resistor power ratings. One is to specify a maximum temperature rise allowed, then increase size of a resistor design for a desired power dissipation until the temperature rise is not exceeded when the desired power is applied. NEMA, Underwriters' Laboratories, and MIL-R-15109 (Ships) ratings are based upon this principle. The second basic method, used in conjunction with temperature-rise limitations, is to limit the resistance change in life tests at full-power conditions. RS-155, MIL-R-26, MIL-R-39007, MIL-R-18546, and MIL-R-39009 all limit temperature rise but also have specifications of varying stringency on life-test-resistance drift limits.

Temperature Rise and Drift. Table 16 shows temperature-rise and life drift limits of current specifications. Military and EIA and most military specifications do not explicitly state temperature-rise limits but instead imply limits by specifying a recommended temperature-power derating curve. Since full-load power tests at room ambient do not necessarily result in full allowed rise, these tests do not confirm the ability of a given resistor to meet the implied performance at the high-temperature end of the derating curve. Military specifications have included, in some cases, high-temperature-exposure tests, in which the resistors are heated to the zero power rated temperature for 250 to 2,000 h. The magnitude of the allowed resistance changes in the specifications provides insight as to the effectiveness and need for such tests. See the special application guidelines paragraph below for further discussion.

Although specifications derating information shows linear derating from full power at room ambient to zero at maximum body temperature, the temperature rise is obviously not linear; if a given resistor is operated at one-half power in a 25°C ambient, the hot-spot temperature will not be one-half that for steady operation under full power but will likely be somewhat higher than half. The amount of expected drift with operating life is largely dependent upon the operating temperature, and if low drift and failure rates are desired, parts should be operated at lower than the rated power. One-half rated power is used as a rule of thumb for conservative practice and will result in an operating-temperature rise of about 60 to 65 percent of observed rise at full power, assuming free-air single-unit mounting under sea-level conditions.

Temperature-rise ratings are all based upon hot-spot measurements made on the highest-temperature portion of the resistor that is exposed. This can be logically deduced to be that portion of the assumedly evenly distributed heat-producing surface which has the highest thermal impedance to surrounding media. For a free-air-mounted tube supported at its extremities by a mounting of low thermal conductivity, this spot occurs in a band about the outer circumference of the cylinder at its thermal centroid, or about equidistant from the terminals if the unit is horizontally mounted. Empirically, on hollow steatite-cored designs of conventional porportions, the terminals with attached leads are found to be cooler than the ends of the ceramic core, which reaches a temperature about 60 to 65 percent of the maximum hot-spot temperature. If through bolts are used with heavy brackets for mounting, the thermal character is changed and the ends may be quite a bit cooler, while the center hot spot is not changed appreciably. Mounting the resistor vertically by small end-suspension clips shifts the hot spot vertically upward, and if a vertical bolt through the center of the hollow tube is used as a mounting to the chassis, the shift will be more pronounced and the spot may be noticeably lower in temperature depending upon the resistor size, the operating-temperature rise in the application, and the size of the bolt.

Each resistor size and design in a given application presents an individual set of conditions about which only qualitative data can be stated. See the paragraph on wirewound resistors below for further discussion of effects on power ratings by altitude, forced-air cooling, intermittent operation, and multiple stacking. Taps on resistors reduce the total power rating. Military specification MIL-R-26 resistors

TABLE 16 Resistor Power Ratings and Life-Test Drift Limits

Resistor type	Specification	Class/characteristic	Full-load life test — Test hours[a]	Full-load life test — Allowed resistance change	High-temperature exposure — Test hours	High-temperature exposure — °C	High-temperature exposure — Allowed resistance change	Temp stated for zero power dissipation rating, °C — Full power	Temp stated for zero power dissipation rating, °C — Zero power
Film	RS-196-A	Class I	1,000	2%				70	150
		Class II	1,000	0.2%				70	175
Film	MIL-R-55182	Characteristics C, H, F, J, K	2,000	0.5% + 0.01 Ω	2,000	175	0.5% + 0.01 Ω	125	175
			10,000[b]	2.0% + 0.01 Ω					
Film	MIL-R-39017		2,000	2.0%	2,000	150	2.0%	70	150
			10,000[b]	4.0%					
Film	MIL-R-10509	B, D	1,000	1.0% + 0.05 Ω				70	150
Carbon composition	RS-172-A		1,000	10%				70	130
Carbon composition	MIL-R-39008	G	1,000	10%				70	130
Wirewound low-power	RS-344	Characteristics A, B	1,000	1.0% + 0.05 Ω				25	A: 175, B: 275
Wirewound precision	MIL-R-39005	Characteristics L, U	2,000[c]	0.5% + 0.05 Ω	2,000	275	0.5% + 0.05 Ω	125	145
			10,000[b]	1.0% + 0.05 Ω					
		Characteristics W, S	2,000[c]	0.1% + 0.01 Ω	2,000	275	0.1% + 0.01 Ω	125	145
			10,000[b]	0.2% + 0.01 Ω					
Wirewound precision	RS-229A	Characteristic G	2,000[c]	0.5%				125	145
Film, power	MIL-R-11804	Characteristic P	1,000	2%	1,000	125[c]	2%	25	P: 235, G: 275
Wirewound, power	MIL-R-26	Characteristic V[d]	2,000	3% + 0.05 Ω	250	275	0.5% + 0.05 Ω	25	275
					250	350	2% + 0.05 Ω	25	350
Wirewound, power	MIL-R-39007		2,000	0.5% + 0.05 Ω	2,000	275	0.5% + 0.05 Ω	25	275
			10,000[b]	1.0% + 0.05 Ω					
Wirewound, power	RS-155-B	Characteristic E	2,000	3% + 0.05 Ω				25	350
		Characteristic G	2,000	1% + 0.05 Ω				25	275
		Characteristics H, V	2,000	5% + 0.05 Ω				25	350
Wirewound, power, chassis-mount	MIL-R-39009		2,000	1.0% + 0.05 Ω	2,000	275	1.0% + 0.05 Ω	25	275
			10,000[b]	2.0% + 0.05 Ω					
Wirewound, power, chassis-mount	MIL-R-18546	Characteristics G, N	1,000	1.0% + 0.05 Ω				25	275
Wirewound, high-power	MIL-R-15109(Ships)		400[e,f]	10%				Ambient	375, 300
Wirewound, high-power	NEMA ICS-1970		f					Ambient	375, 300

a Times given are for total test length. Most procedures require on-off cycling so that actual operating time is less than that stated.

b Established-reliability military specifications require extension of the operating-life qualification test, for the purpose of generating reliability data. The 10,000-h tests shown have failure thresholds for resistance change defined as listed.

c Test conducted at +125°C ambient. MIL-R-11804 parts have half power applied also.

d Allowed ΔR depends upon individual type and style and its detail specification. Those given are typical.

e Test is conducted at 125% rated power.

f NEMA standards and MIL-R-15109(Ships) specify temperature rise vs. ambient. 40°C is used as ambient by most suppliers of this type of resistor. Bare-wire styles are allowed to have 375°C rise, embedded-wire styles are limited to a rise of 300°C on the outside surface.

can be procured with single center taps, but power rating is cut by 10 percent. For movable taps and off-center taps the power rating of a segment is proportional to its fraction of the total power rating. It is expedient to calculate the maximum allowed current for full power, then use the value as a limit for any segment.

Voltage Ratings. The ability of power resistors to remain electrically isolated from their mounting or from adjacent conductors is evaluated in specifications by standard tests that apply potential between windings and a body wrapping of metallic foil or a metal V block, or just to the mounting hardware if it is integral. These tests are made at 500 V rms, 1,000 V rms, or higher voltages for special cases and for high-wattage resistors. The 30-W aluminum-housed MIL-R-39009 units must meet a 2,000-V test between winding and housing, and MIL-R-18546 requires 4,500 V for the 75- and 120-W chassis-mounted aluminum-case types. The coated tubular parts generally depend upon spacing from the conductive chassis and the addition of electrical insulation in the mounting for high-voltage isolation (refer to the paragraph below on mountings). Test voltages are applied for 1 min. For use purposes applied voltages should be held well below the high-potential test levels. A maximum of one-half the test voltage is suggested as a practice unless extra insulation is used. Resistor-element voltage ratings depend upon the power and resistance ($E_{max} = \sqrt{PR}$) up to the critical resistance value (refer to general information for resistors at the beginning of the chapter) and are limited primarily by turn-to-turn breakdown above the critical resistance value.

In accordance with most supplier recommendations and military specifications, voltage ratings for embedded designs follow a general rule of 500 V per inch of winding, extended to about 1,000 V per winding inch for units of 150 W and up. RS-155 does not cover voltage ratings, but the styles are similar to MIL-R-26 types and should be treated accordingly. Voltage ratings may become a limitation if steep leading edges of pulses are applied, since the inherent inductive reactance of the winding may cause high-voltage gradients. For bare round wires, such as those used on adjustable-tap units, voltage should not exceed 495 V per inch of winding. Arcing conditions cause very high energy densities, and the concentrated stress can rapidly deteriorate resistance windings. It is thus important to observe voltage ratings.

High-Frequency Performance. As in lower watt sizes, power film resistors have generally superior performance for high-frequency applications. Wirewound units can be procured in Ayrton-Perry noninductive configurations in either flat oval-cored units or round tubular-core units but have effectiveness limitations similar to those already discussed for medium-power wirewound parts. If inductance is a difficulty in the application, ferromagnetic mounting hardware should be avoided.

Large high-power resistors have higher inductances than smaller watt ratings owing to their geometrical dimensions. Refer to the general information paragraphs at the beginning of the chapter and to the application notes at the end of the chapter for further discussion.

Pulse Application. Resistors can withstand many times their rated power for short pulse durations. Application notes at the end of the chapter provide methods of estimating suitability of wirewound units for very short pulses. Power-resistor thermal masses are basically quite large and thermal time constants are long enough to allow intermittent overpower applications of relatively long durations without endangering part reliability, particularly for large strip-wound units. Film parts are not as good in this respect as wirewound units, and small wire diameters require more care than large wire. Manufacturers are able to supply information on intermittent duty for their parts, and ICS 1970 NEMA standards define standardized duty cycles for motor-starting and braking uses of high-power wire- and strip-wound resistors. Intermittent application of inputs on the order of ten times rated power are allowed there provided duty-cycle limitations are observed. Voltage ratings should, of course, never be exceeded.

Figure 25 illustrates the range of performance that can be expected on two size classes of vitreous-enamel-coated wirewound resistors produced by one manufacturer. Heavy strip-wound resistors limit, in general, at longer duty cycles than those shown. Specific data should be requested from particular manufacturers or laboratory tests should be made for other suppliers' parts, since this trait varies with resistor design.

It is worthy of note that very high current surges in helically wound solenoids can produce considerable magnetic force on resistance windings. The force is in directions that would increase the self-inductance if the wire were free to move; i.e., there is a radially outward force that places wire in tension, and also longitudinal forces that are directed axially from the ends of the winding toward the winding center. With well-constructed resistors, the effect should not be of major concern. In general, the force can be defined by

$$\mathbf{f}(\text{newtons}) = \tfrac{1}{2}i^2\,\frac{\partial L}{\partial S}\,\mathbf{s} \qquad (9)$$

where i is current in amperes, L inductance in henrys, and $\mathbf{s}$ a unit vector in the direction under consideration. A more detailed treatment is given in Refs. 6 and 7.

Wire- and strip-wound resistors Wire- and strip-wound products are described in general above. Specifications covering these types are listed in Table 12.

Air-mounted Types. Large suppliers of cylindrical high-power resistors list their product in categories of their standardized core-sizes. A listing of core inside diameters catalogs maximum available resistance values defined for the particular size. These cores are used alternatively to wind vitreous-enamel-coated wire units or for parts wound with corrugated rectangular-cross-section strip laid down on one of its narrow edges, so that the remaining narrow edge projects radially from the core. A generous coating of vitreous enamel fills in between the helices and in some units coats the exposed edges. Some, however, are left intentionally bare to take advantage of the NEMA allowance of increased temperature rise for bared resistor conductors and to permit movable bands for taps. These edge-wound units in high power ratings are rugged, reliable, and inexpensive.

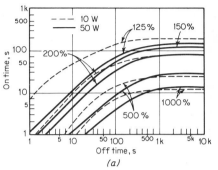

(a)

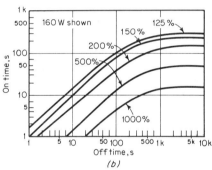

(b)

Fig. 25 Percent of continuous-duty rating for pulse operation of vitreous-enameled wire wound resistors. (*Ohmite Mfg. Co., Skokie, Ill.,* Bulletin 1100A, 1972. (*a*) Small to medium body size (10 to 50W). (*b*) Large body size (160W).

Supplier application data are extensive, and even though core diameters are not standardized, mountings are sufficiently standard to allow good "cross accommodation" of alternate supplier's parts. Specification coverage is afforded by MIL-R-15109 (Ships), MIL-R-26, and NEMA standards.

Some embedded-wire-type resistors are furnished with silicone-resin coatings, usually at reduced allowable temperature-rise limits but with better resistance stability. Most high-power resistors, however, utilize some form of vitreous-enamel coating. EIA specifications cover flat oval-core wirewound parts, and the flatted packages with their integral mounting brackets furnished with spacers for stacking provide a convenient mounting and inherent economy of space. Several units can be easily stacked provided that proper derating is observed.

Some styles in EIA specifications, called "miniature types," require chassis mounting of a given area to reach the rated power with which they are labeled. Orientation of stacked parts (see Fig. 26) should ideally be such that maximum air can flow across the flat surfaces without being first heated by neighboring units. Radiation from one unit to its neighbor is unfortunately higher than for comparably spaced round units with similar mounting and power-dissipation rates. Military specifica-

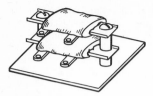

Fig. 26 Stacked flat-strip resistors.

tions no longer cover this flat configuration, which is unfortunate, since it has unique packaging advantages. Tapped units, Ayrton-Perry windings, and adjustable-tap flat units are furnished by suppliers, but these details are not addressed by RS 155. (See also the previous paragraphs on coatings and resistance wire.) Because of the dependence of electrical performance and reliability upon the operating temperature, it is important to consider application conditions that will affect the heat-dissipation process. Some suppliers have published good application information, though there is some disagreement among them. Examples of these data are illustrated in the application section under Power Ratings.

Chassis-mounted Types. Metal heat-dissipating cases with mounting provisions are covered by military established-reliability specification MIL-R-39009 for parts from 5 to 30 W, and by MIL-R-18546 for 75- and 120-W versions. These are all horizontally mounted parts, and all are available in noninductive windings. Efforts have been made to miniaturize the units while maintaining reliable function. Beryllia cores are used on the established-reliability parts, and the resistive element is carefully molded into the extruded aluminum housing. Power ratings in MIL-R-18546 are stated for resistors mounted on a chassis of aluminum 0.040 in thick and 4 by 6 by 2 in for units up to 10 W, 0.040 in thick and 5 by 7 by 2 in for 20- and 30-W units, and 0.125 in thick, 12- by 12-in panel for the 75- and 120-W parts. All are rated at +275°C case temperature maximum. The 75- and 120-W units are said by one supplier to use steatite or alumina cores, depending upon physical size. The same basic construction is used on parts designed for feedthrough or standoff mounting in a chassis or panel hole large enough to accommodate the body of the resistor. A large nut threaded onto the body mounts the part. Various terminal configurations are available. Chassis-mounted resistors shown in Figs. 27 and 28 are, as might be expected, more expensive than comparable power ratings

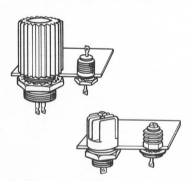

Fig. 27 Chassis feedthrough and standoff aluminum-cased power resistors—grounded and ungrounded types. (*Dale Electronics.*)

Fig. 28 Chassis-mount axial-lead power wirewound resistor. (*Dale Electronics.*)

in conventional types. Various other chassis-mounted resistor configurations are marketed as supplier specialties in the lower end of the power range, e.g., disks, plate types (see previous paragraph on cores and cases).

Mounting hardware and methods Flat oval-core resistors have mounting brackets furnished as integral parts of the unit. Some, mostly the miniaturized designs, have a metal strap completely through the core, with the rods bent to form mounting feet, while others have push-in brackets that are held in the ends of the core by dimples or toothed projections until installation provides rigidity. Figure 26 shows resistors of this type mounted in a stacked configuration. The horizontal position shown is less preferred than a vertical arrangement in which the flat sides are exposed to convected air that has not been preheated by its neighbor. For round tubular cores and those which have flatted sides for Ayrton-Perry windings, the mounting-hardware types can be divided into (1) those methods which require a long bolt through the hollow core center, (2) those which insert short spring clips into the hollow ends of the core, and (3) ferrule-type terminals designed for fuse clips. Military standard drawing MS 75009 covers threaded rod or long screws in 8-32 and 10-32 sizes and 3.375 to 12.750 in. in length, together with lock washer, centering washers of a cup type and an "eared" type, and right-angle brackets for single mounting of tubular resistors. Three sizes of brackets are depicted, ranging from 1.0- to 1.5-in center-hole height above the mounting plane. All threaded rod, screws, and brackets are zinc- or cadmium-Iridite-plated* steel. Centering washers may be brass or plated steel, and the split-type lock washers may be phosphor bronze or MS 35338 corrosion-resisting steel. Four sizes of mica end washers are also provided, all of 0.031- to 0.094-in thickness. These washers cushion the ceramic-steel junction at the ends of the core and also increase the high-potential surface creep distance to a grounded mounting bracket.

Figure 30 illustrates the use of the mica washer. Although it is not covered by MS specifications, resistor manufacturers list a variety of multiple mounting brackets that are basically MS-type mountings, but with provisions for mounting two or more units on the same bracket (Fig. 29). Perforated-metal protective housings for single

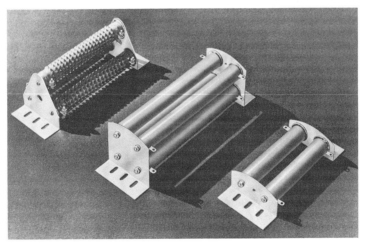

Fig. 29 Multiple mounting of large power resistors. (*Ward Leonard.*)

and multiple resistors also use the through-bolt basic mounting. Noninductively wound resistors should not have steel screws through their center because of possible enhancement of the leakage-flux effects. Also, if high voltages are present on the resistor with respect to the mounting, supplier catalogs list porcelain end bushings that can be bonded into the ends of the core, or just used as loose bushings on through bolts to serve a dual function as additional insulation and as a replacement for the normally used metal centering washer. When these bushings are used, the terminals are placed farther from the electrically conductive bracket. The through bolt must, of course, be lengthened accordingly. Figure 30 illustrates the porcelain-bushing use.

The second most-used mounting is the spring clip. This mounting method consists of inserting into the inside of the core end a clip that is configured to grip the inside of the tube by friction, brought to bear by spring pressure when the clip

---

* Iridite is a trademark of Allied Research Products, Inc.

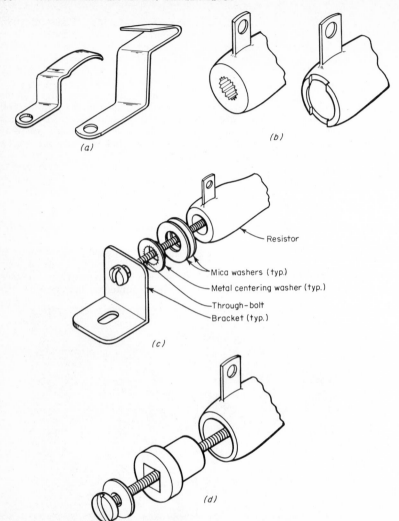

Fig. 30 Power-resistor mounting accessories. (*a*) Spring-clip types. (*b*) Nonturn features. (*c*) MS 75009-type mounting, opposite end similar but has a nut and a split lock washer. (*d*) Ceramic plug provides extra insulation to grounded mounting. (*Ohmite Mfg. Co.*)

is deformed at insertion.   These clips are available in brass, spring steel, and ordinary mild steel, with the brass parts being useful for mounting noninductively wound resistors.   Figure 30 shows examples of spring clips.

Resistors are available with fluted interior or notches on the core ends to provide nonturn features.   Figure 30 illustrates this feature.

Ferrules for standard fuse-clip mountings are also supplied, with resistor terminal leads brought out to the ferrules and attached by high-temperature soldering or welding.   The ferrules are usually brass.   Some types fit over an extension of the ceramic core, and others consist of cups attached to the ends by a through bolt. Fuse clips are of the standard electrical type.   One other mounting for tubular resistors that is worthy of note is the Edison lamp-base type.   This is available in several standard sizes.   Through bolts and fuse clips are obviously superior for application involving shock and vibration, while spring clips are less likely to place strain on the resistor body during thermal cycling. The lamp-base mount is, of course, not intended for high shock and vibration. Most other power-resistor designs of rib open-wound, tub-potted units, and others have integral mounting means or are configured to be mounted by the hardware and methods described above. High-power film units on glass cores with radial-tab leads can be mounted using the through-bolt method or by using spring clips. These and various high-power carbon-composition-film parts frequently use the ferrule fuse-clip mounting.

High-power resistors should not be solidly bonded to chassis or structure using resins or cements, and should generally not be encapsulated in organic materials. Bonding materials can cause extremely high stresses in the resistor structure, leading to core breakages. Encapsulation moves the radiating surface out to the surface of the encapsulant and depends upon the thermal conductivity of the encapsulant to transfer heat to that surface. Most encapsulants have difficulty in withstanding the high operating temperatures of power resistors.

Precision Resistors

Precision wirewound Resistors that are produced to provide the ultimate in accuracy and stability while still maintaining good installation and performance utility are called precision, or "accurate," resistors. They are differentiated from perhaps more accurate, but cumbersome and nonutilitarian parts fabricated to serve as laboratory standards, whose precision and stability they nevertheless approach. High-precision resistors are currently exclusively wirewound for the best achievable stability and are carefully produced to provide the lowest possible temperature coefficient of resistance and the minimum in-service resistance drift. Being wirewound, both noise and voltage coefficient of resistance are of negligible consideration. The manufacturer must, however, avoid wire junctions that produce significant thermocouple-type potentials, must provide near perfect joints from resistance wire to terminals, and must provide a design that avoids corrosion and disturbing stresses on the resistance wire.

Designs. The available designs are most commonly plastic-resin-encased elements having copper, Dumet, copper-nickel alloy, or nickel leads, with enamel- or plastic-film-coated resistance wire wound on a molded filled-resin bobbin. The design may have end caps or may just have leads embedded in the bobbin core. In all the best products, the resistance wire is joined to terminals by welding, but hard and soft soldering are sometimes used, with the latter being limited strictly to those types where operating temperatures are very low. Intermediate tabs are often used, with one end connected to lead wire and the other welded to the resistance wire. Various other different designs are marketed, frequently with very little environmental protection, which is of course acceptable for many applications. One frequently used design has a ceramic bobbin core on which the wire is wound, with radial-wire leads soldered around end grooves. The element is protected by paper or plastic tape. Precision resistors are larger in comparable power ratings than are corresponding general-purpose parts. In addition to the temperature limitations of the resistance-wire insulation films and served insulations, power ratings envision a minimum temperature rise to reduce temporary resistance changes due to temperature coefficient, as well as to reduce the probability of accelerating subtle drift mechanisms within the resistor that will result in permanent changes. Packages having internal cavities within cases intended to be sealed with epoxy or other resins avoid pressures on the resistance wire but should be carefully evaluated in the intended environment. Some end-sealing encapsulation methods may leave undesired water or other residues inside the unit, or may leak at the lead exit in service as the unit is thermally cycled. The dissimilar metals at element-to-terminal welds are very vulnerable to corrosive attack by electrolytic processes. Systems that provide mechanical cushioning of the winding and then provide an essentially void-free dense encapsulation of the protected winding would appear to better choices. Also, few materials are poorer thermal conductors than air, and elimination of interstitial spaces improves the thermal system. Avoidance of resistor-wire stress is very difficult in solid constructions, however. Figure 31 illustrates one available design. Corrosiveness of fluxes and of encapsulants and their fillers, and possible corrosive or aqueous by-products of the encapsulation process should be evaluated thoroughly. Although most resistance

Fig. 31   Precision-wirewound-resistor construction.   (*Shallcross Division of Cutler-Hammer, Inc.*)

alloys and terminal materials are in themselves noncorrosive, it is easily possible to develop corrosion through dissimilar-metal galvanic activity.

Care must be exercised in measurement of low-value resistors to exclude the effects of lead length and contact resistance.   Special techniques are usually necessary to resolve measurements closer than $\pm 0.001\Omega$.   Temperature coefficients and drift mechanisms also tend to have more effect on low-value parts when considered as a percentage of the resistance values, and most specifications recognize this circumstance by providing slightly increased allowances for low values.   For these low values, it should be remembered that copper circuit conductors have an approximate temperature coefficient of resistance of $+3,800$ ppm/°C.

Resistors above a few ohms involve multilayer windings, and the result is a large amount of inductance and distributed capacitance, frequently not desirable owing to the requirements of high-speed or high-frequency circuits, or perhaps because of use in sensitive high-gain circuits where unexpected resonances may cause oscillation.   Noninductive windings of the reverse-pi or bifilar type are most used.   Reverse-pi refers to the practice of separating the winding into segments by the use of a slotted bobbin as shown in Fig. 31, then reversing winding direction in each successive slot.   This technique is effective at low frequencies but is limited by the presence of leakage flux between sections, and the distributed capacitance, which effect is increased by the incidental adjacent turn potentials, which are higher than for some alternate methods.   (See the preceding resistor general information section for further discussion of reactive effects.)   Bifilar windings use two wires wound together, with the result being series-connected so that current flow is opposite in each winding.   Turn-to-turn potential is obviously the same for the two windings at their connected end, but high at the start ends, thus increasing the effects of distributed capacitance.   These winding methods are effective for low frequencies only, and when moderate resistance values are used above 20 kHz, reactive effects must be considered despite the noninductive windings.   All the practical "noninductance" methods fail at often disappointingly low frequencies.

Considerable technical development and skill are required to wind precision resistors, particularly in fine wire sizes.   Tensioning control is all-important because the resulting system must be as stress-free as possible to achieve in-service stability.   Just the act of bending wire around a mandrel (or core) produces stresses that affect stability, particularly for some alloys.

Wire sizes down to 0.0005-in diameter are used, but more conservative practice is to use wire no smaller than 0.001-in diameter.   Special temperature-stable alloys of about 800 Ω cmil^{-1} ft^{-1} are used almost exclusively for higher resistance values, while manganin or copper-nickel alloys are sometimes used for lower values.   (See general resistor information paragraphs for more discussion.)   Wires are insulated with film coatings like those used for transformer magnet wire and occasionally with glass fiber, silk, or other served fibers.   Polyvinyl acetal is rated for 105°C, and polyimide films such as Du Pont Pyre M.L. are rated as high as 220°C.

Stability.   After winding adjustment and encapsulation, the resistors are stabilized by temperature cycling, baking, and high-power run-in.   This step is particularly critical for manganin wire because of its tendency to change character significantly as winding strains are annealed out.   After stress relief, well-made resistors can be expected to maintain value within ± 0.1 percent and to retrace temperature coefficient of resistance within +10 to ±50 ppm/°C over the rated temperature range, depending upon resistance value. Fine trim adjustment by the manufacturer is sometimes accomplished by mechanically scraping a series service loop. The manufacturer must assure that adjustment gradients are small and wire cross section is not seriously reduced in this operation. The ready availability of stable precision wirewound resistors is due in large part to the development of special stable mechanically strong resistance alloys having repeatable properties and good handling and aging characteristics. The development of winding machinery capable of providing consistent controls has also been important, as well as the persistence of the relatively small number of producers specializing in this product, in providing higher-quality, better-controlled products.

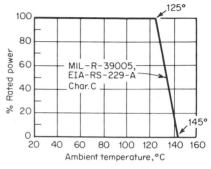

Fig. 32 Power derating for precision wirewound resistors at ambient temperatures above 125°C.

Power Ratings. Power dissipation is low for the physical sizes used. Figure 32 shows the RS-229A and MIL-R-39005 power-derating curves. For best stability, temperatures should be held as low as possible.

Voltage Rating. Voltage limitations are encountered because of the construction methods used. Turn-to-turn voltage must be limited, and consequently critical resistance values are reached where a resistor of a given power rating and size becomes dissipation-limited by voltage. (See resistor general information section.)

Specialized Resistor Types

A brief mention is made here of resistors for specialized uses (except for heating purposes). Because of the number of such uses, only a fraction of the total variety is covered.

High-frequency types For rf devices and stripline assemblies, resistor shapes and designs are fabricated to provide broad useful high-frequency traits and ease of circuit attachment. Thin-film resistive elements perform well at high frequencies but are difficult to protect from environmental exposure without conventional lead and case configuration. Waveguide terminations, rf attenuators, power dividers, tee pads, and coaxial terminations all make use of flat, shaped cards, rod elements, coaxial disk elements, and other specialized shapes having deposited films (Fig 9e). These may be thick or thin films deposited on mica, glass fiber–resin combinations, ceramic, glass, or other materials. Devices designed using these basic elements must allow for the differences in thermal-expansion rates with respect to interfacing structures, usually by some type of spring-pressure connection arrangement. Failure to consider this aspect of design will result in problems of cracked connection joints and damaged resistor elements. Various passivations are used. Thin quartz coatings have been available in the past, and high-quality silicone coatings are employed. Coaxial disks allow fabrication of film resistors having very good characteristics well into the microwave range. Stripline components are also produced, in long, flat strips that can be cut and fitted, as well as in pellet pillbox types that fit in stripline sandwich constructions between conductive pads. Chips are also produced for hybrid microcircuits. (See the paragraphs on thick- and thin- film networks.) These are mostly metal thin films or fired cermet screenings that have metallized pads for soldering or wire bonding by semiconductor bonding techniques. Military specification MIL-R-55342 covers both types in several terminal configurations, but at this writing, there are no qualified suppliers. This specification in its current coverage is listed earlier in the chapter in the tables of specifications.

Meter multipliers and shunts In addition to the precision wirewound resistors that find much use as meter multipliers and shunts, special resistor designs are manufactured for high-voltage meter multipliers. Also for high-current measurements, precision resistors of very low value are also available. MIL-R-29, listed in the specification discussion, covers large ferrule-terminal metal-film resistors of value from 0.5 to 20 MΩ. These resistors are sealed in glass tubes and are said to provide long-term stability. Their power rating is that which is dissipated when a 1.0-mA current is induced through the element. Extra-long-bodied wirewound parts are also supplied for meter multipliers. They are wound with very fine high-resistivity wire, and this, with the body length (to 10 in on a nominal 1½-in-diameter body), allows value up to 7.5 MΩ for a 1-mA rating. Low-resistance high-current meter shunts are made of shunt manganin or other resistance alloys in bar or heavy strip, mounted on a base having large, efficient screw-clamp terminals mounted directly onto the strip material. The strip element is calibrated by small grinding or filing adjustments to its cross section. Many sizes, shapes, and configurations of these high-current shunts are produced, many specially adapted to the mounting used in their application. Shunt manganium is manganin alloy that has its minor constituents and metallurgical conditioning adjusted to move the characteristic parabolic temperature-coefficient curve so that its low inflection point occurs at the operating temperature to be expected when the element carries heavy currents.

Power ceramic composition types   A type of resistor that is not widely produced, but which has unique capabilities, consists of a structure in the form of a thick-walled tube which is composed of a relatively strong homogeneous resistive material, and usually has ferrule-type terminals on the tube ends. The terminals are metallized areas on the composition structure and serve as both electrical termination and mechanical amounting. As furnished by The Carborundum Company, two basic materials are said to be carbon-ceramic and silicon-ceramic compositions. A ceramic and carbon or silicon putty is extruded under pressure, and the resulting body is fired to sinter the mixture. Resistivity is varied by adjusting the material proportions, and other mechanical and electrical properties are adjusted by materials and process controls. The result is a hard, rather durable structure. The ceramic carbons are said to be better for high-power pulse operation (see paragraph on pulse applications), while the silicon ceramics are superior for steady power application. Catalog listings show silicon-ceramic units from 22.5 W at 1.0 to 150 Ω up to 1,000 W at 1.0 to 500 Ω, with comparable sizes of carbon-ceramic resistors from 15 W, at 7.5 to 1,000 Ω over a range up to 150 W at 50 to 10,000 Ω. Purchase tolerances are 5, 10, and 20 percent. Full power is rated at 40°C, derated to zero at 230 and 350°C, respectively, for the carbon and silicon materials. Voltage coefficient of resistance is no greater than 1.0 percent per V per inch of resistive length for both, and temperature coefficient is said to be 750 ppm/°C maximum. These resistors have good high-frequency characterstics up to 50 MHZ, and excellent survivability under high-power short pulses. They are widely used as rf loads, antenna-termination resistors, and in high-power radar modulators. Epoxy coatings are applied to the carbon types if it is necessary to resist transformer oils.

High-resistance types   For high-voltage bleeders and dividers, and for other high-resistance uses, a class of resistors called "high-megohm" types is produced. These parts are almost all film types of various designs. One part that has been produced for a number of years, however, uses a stack of pressed and sintered carbon composition disks strung on a ceramic rod and connected in series. The film types are made long in body and use either finely helixed thin metal films, high-resistivity tin oxide films, oxide-doped metal films, or high-resistivity metal-frit fired enamels. The latter have durability advantages, being relatively inert, as also is tin oxide. Metal and pyrolytic-carbon films should be carefully protected, and because of the high resistance value, all the parts are vulnerable to surface leakage if moisture and contaminations are allowed to accumulate. Organic coatings must be sufficiently impervious to avoid volume resistivity decreases due to moisture absorption, at least during short-term exposures to condensing water. Temperature coefficient of resistance is large for those types which gain high value by increasing film resistivities to values that are beyond the ranges needed to maintain optimum temperature stability. Some of the recent developments of thick films appear to be quite good in temperature stability and environmental endurance for this use. Various glass-cased designs are well developed and are available for this use.

Fusible resistor   A wider use is being made of resistors having established fusing characteristics under overload conditions. This amounts only to characterization of established designs in some cases, but some are new designs by suppliers. Prior to the present growing concern with product safety and fire-hazard avoidance, resistor design goals were directed toward ability of resistors to withstand high-overload conditions while retaining stability and performance. The ability to fail quickly under moderate overloads is counter to design criteria used in the past (refer to the application paragraph on fusible resistors). Employment of this resistor type is likely to increase.

Positive-temperature-coefficient resistors and ballasts Ballast resistors are wound with wires having high positive temperature coefficients of resistance. They are used as voltage regulators and for related sensing functions. A resistor of this type in series with a power source to a load increases its resistance if the voltage increases and causes increased current flow and higher operating temperature, with the increase in resistance having a regulatory effect on the load voltage. Table 9 lists information on some ballast alloys.

Miscellaneous Thermistors and varistors are not treated here. Thermistors are more properly classed as semiconductors, and varistor materials are so specialized for their particular applications that generalizations are difficult. Laboratory standard resistors and accurate resistors for decade boxes are also not treated in detail, being usually of simple design and used in benign conditions. Resistance alloys of the most stable types are spool-wound. Wire often is insulated by served fibers of cotton, silk, or other types and may be enameled as well. With the state of development of resistance alloys, good accuracy and stability can be obtained with moderate attention to design and manufacturing controls.

Special Application Guidelines

Fixed-resistor trimming Trimming adjustment of resistor values is often desired in situations where subsequent retrimming is not desirable, or where environmental circumstances make the use of variable resistors undesirable. This can be accomplished as a production operation using selected values of fixed resistors. The incremental adjustment can be made arbitrarily small by using simple techniques. The total resistance of two parallel-connected resistors R_1 and R_2 is given by

$$R_T = \frac{R_1 R_2}{R_1 + R_2} \qquad (10)$$

By fixing R_1 at a predetermined "coarse" value and selecting R_2 to adjust the combination, the desired trimming is accomplished. Figure 33 shows R_T as a function of the ratios of R_1 and R_2. R_2 is denoted as a "star" value, its schematic value being conventionally denoted by an asterisk referring to a footnote. The technique em-

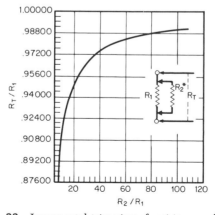

Fig. 33 Incremental trimming of resistance value.

$$R_T = \frac{R_T}{R_1} R_1 \qquad \frac{R_T}{R_1} = \frac{R_2}{R_1} \frac{1}{1 + R_2/R_1}$$

Examples: For $R_2/R_1 = 80$, $R_T = 0.98765 R_1$. For $R_2/R_1 = 8$, $R_T = 0.88889 R_1$.

ployed is to install R_1, then clip in a precision-switching-resistor decade standard for R_2, switch until the desired circuit condition is obtained, read the switch positions, and select the nearest standard marked value for installation as R_2. The procedure can be made relatively insensitive to moderate in-service variability of R_2, by judicious selection of ratios. Also, R_1 can be selected as a precision type while R_2 can be made less critical owing to the relative insensitivity to its variation. Radio-frequency circuits will of course require considerations of lead lengths, and resistor decade reactances for the bench adjustment.

The kit of values from which to select R_2 can be minimized by circuit analysis to determine the most probable needed values and calculating from an appropriate distribution curve. If the relative probability of needing a value step can be estimated, a Monte Carlo technique can be used to reduce the needed inventory. Range of adjustment can be increased and span of needed stock values for R_2 selection can be decreased by making a substitution selection of R_1 also. A number of pre-packaged networks using incremental adjustment to value are available from various suppliers. These are usually networks prepared on substrates by thick- or thin-film depositions, with terminal junctions brought out for connections and adjustment by terminal interconnection patterns to be accomplished after the package has been installed in its circuit location. Industry directories list manufacturers.

Short pulse-rating estimates In circuit applications, it is frequently necessary to determine the proper wattage size of resistors for uses involving pulses of short duration. It is evident that resistors should be able to withstand higher power than rated continuous power if the exposure is sufficiently brief. Film and composition resistors do not lend themselves to simple analysis in this regard, although data published for pellet-type composition resistors by one manufacturer are shown earlier in the paragraphs on carbon composition resistors. Films become an intimate part of the thermal mass of their substrate and overcoatings, but the current-carrying film is so thin that all the energy from a short pulse (less than 100 ms) can be considered to be concentrated in the very thin film, which limits the very short pulse-overload performance of film resistors. Thick-film types should be a little better than thin films, but core, coating materials, and design probably are as significant variables as film thickness. Because of the lack of empirical data and the difficulty of analysis, pulse overload operation of film resistors is not covered here. Wire-wound-resistor construction is comparatively easy to treat, and most designs are capable of safe operation under large overloads for short pulses. Basically, the ratings of voltage and power must be addressed, and if after the calculations are made, the resulting safety margins are small, then methods should be explored for obtaining empirical data. One suggested method is to test a number of sample parts at stepped pulse-power levels until a significant change in resistance value or catastrophic failure occurs; then use the observed levels of significant effect to plot a distribution and calculate a standard deviation for an assumed normal distribution. If the 3-sigma point of the damaging power-level distribution is outside the applied level, a case can be made for conservative adequacy.

For isolated pulses 100 ms or less in duration, it can be assumed that the pulse energy is absorbed completely by the resistance wire and that the wire temperature is raised to a level that is calculable if the mass and thermal capacity of the wire are known, the initial temperature is considered, and the total pulse energy is determined. A resistor can be disassembled to determine the diameter and length of the wire, or the manufacturer can be queried for the information. Heat capacities of most resistance alloys are given in Table 9. If the exact alloy is not known, the heat capacities are sufficiently similar to allow a good estimate as to whether the design application is marginal. First obtain the mass of the wire:

$$m = \pi \left(\frac{D}{2}\right)^2 L\rho(2.54)^3 \times (10^{-3})^2$$

where m = mass, g
D = diameter, mils
L = length, in
ρ = density, g cm^{-3}
2.54 = conversion factor, inches to centimeters

Then compute pulse power:

$$P = \frac{V^2}{R} t$$

where P = pulse energy, J (or W-s)
V = pulse voltage, average
R = resistance value, Ω
t = pulse-time duration, s

The wire temperature can now be computed as a function of the initial wire temperature plus the incremental heat generated:

$$T = T_0 + \frac{P}{\theta m} \tag{11}$$

where T and T_0 are total and initial temperatures, respectively, θ is the heat capacity of the wire in joules per gram -°C, and m and P are as defined above. Combining the above expressions using the defined quantities, the temperature of the wire is

$$T = T_0 + \frac{4V^2 t \times 10^6}{\pi \rho \theta R D^2 L (2.54)^3} \tag{12}$$

If the wire diameter is given in circular mils, it is proportional to cross-sectional area, being the diameter in mils squared; and it is necessary only to substitute D in circular mils for D^2 in the above expression. When the temperature is obtained, it should not exceed 275°C for conservative uses requiring resistance-value stability or 350°C absolute maximum. T_0 must be based upon the maximum ambient temperature for the resistor in its application. It should be noted that heat capacity in joules per gram -°C is not equivalent to the specific heat, which like specific gravity, is used to measure the material property against that of pure water at room temperature. Specific heat is given in calories per gram -°C or Btu per pound -°F. Both are the same number and can be converted to watt-seconds (or joules) per gram-°C by the use of a constant multiplier of 4.186 J-cal⁻¹. Specific gravity (more properly, density), when expressed in grams per cubic centimeter, provides convenient units for this purpose; the SI units for density are kilograms per cubic meter.

If inductance-resistance or resistance-capacitance discharge pulses are involved and meet the 100-ms criterion, the pulse energy is obtained by assuming that all the energy stored in the capacitor or inductor is dissipated. This is

$$E = \tfrac{1}{2}CV^2 \quad \text{for capacitors}$$

and

$$E = \tfrac{1}{2}LI^2 \quad \text{for inductors}$$

where E = energy, J
C = capacitance, F
V = voltage on the capacitor, V
L = inductance, H
I = inductor current, A

For a sinusoid, the average voltage of one-half cycle is 0.637 times the peak. Other pulse shapes are treated similarly, with the object being to obtain the total pulse energy. If trains of pulses are applied, it is necessary to determine the average hot-spot wire temperature for use as the initial temperature in Eq. 11. This can be done empirically using 30 AWG thermocouples on the center of the resistor body, or a worst-case estimate can be made by using the temperature-derating curves specified for the part in question. To determine worst-case temperature rise per watt, use the slope of the power-derating curve and express the result in degrees per watt. Using the calculated average power for the pulse train, multiply by this temperature rise per watt, then add to the expected maximum ambient environmental temperature to obtain the value of T_0 in Eqs. (11) and (12). Use the pulse-power-calculation technique for a single pulse, and add to T_0 to arrive at the total estimated temperature of the resistance wire. Temperature-rise estimation using the power-derating curve is not dependable for situations where safety margins are slightly exceeded when calculated as above, because the results are often unnecessarily pessimistic. In such cases measurements should be taken, or suppliers should be consulted for information on their particular product.

Voltage limitations for short pulses are not established, although it is known that good coated wirewound designs will withstand several times their steady-state voltage rating for millisecond-range pulses, with higher power ratings being somewhat better in this respect than smaller units. One manufacturer recommends a design figure of $\sqrt{10}$ times rated voltage as a maximum for 4-W and larger resistors if pulses are 100 ms to 5 s and $\sqrt{5}$ times rated voltage for the same pulse-width ranges on units smaller than 4 W.[13] For shorter pulses, it is stated that work has shown that 20-μs pulses of 20,000 V/inch of resistor can be applied if pulse energy–temperature rise guidelines are observed. Obviously, resistor designs having large resistance-wire thermal mass are better for short-pulse applications from the standpoint of power capability.

For pulses longer than 100 ms, the momentary-overload rating of the resistor can be utilized to assess the design margin. Military and EIA specifications for resistors have momentary-overload tests of five to ten times rated power for 5 or 3 s duration depending upon the resistor type and size. Resistance-value shifts of varying amounts are allowed in the test. Allowed changes for wirewound resistors are no more than 2 percent or ±0.05 Ω; some premium-performance styles allow only 0.2 percent. If changes of about 1 percent or more are observed in iterations of the momentary-overload test, the reliability of the part under those conditions is questionable. The specification test is an intentional stress on a sample of parts, and thus a derating factor of at least half the watt-second maximum test value is advised as a practice. For pulse width less than 3, 5, or 10 s, whichever test is specified, the power over-rating can be proportionally higher. A 1-s pulse would therefore be allowed to have an average power rate of five times that for a 5-s pulse. The longer pulses allow the resistor thermal system of core, leads, coating, etc., to begin functioning. From 100 ms to 1 s, the 1-s proportion should be used. For pulses longer than 5 s on parts of 20 W and higher, the intermittent service ratings for NEMA and UL requirements that are tabulated by some suppliers can be used as a guideline. The large parts possess longer thermal time lags. Short-pulse ratings on large parts, however, can be developed as outlined above. Additional insulation to ground may be needed on large power resistors in pulse applications.

Specialized resistors are available for high-power-pulse applications. Varieties that are, by experience, superior in ability to survive high-energy pulses, while also maintaining good high-frequency characteristics, are types of ceramic composition units made by the Carborundum Company. These parts, described under Specialized Resistor Types, probably represent the best overall compromise for high-power short pulses. Both carbon-ceramic and silicon-ceramic types are furnished. The carbon-ceramic type is rated higher for pulse energy but lower for steady-state dissipation than the silicon-ceramic type for comparable physical sizes. The 15- to 150-W and 40°C steady-power rated range in the carbon-ceramic type, for instance, is given a corresponding single-pulse energy range of 2,500 to 50,000 J and a standard peak-voltage rating range of 4 to 60 kV over the same range of sizes. Resistance values range from 7.5 to 10,000 Ω at ±5, ±10, and ±20 percent purchase tolerances.

Encapsulation of resistors   Many electronic packages are designed to embed fixed resistors in various encapsulating systems. As might be expected, temperature rise is affected by the encapsulation. The effect varies, depending upon the thermal conductivity of the embedment, the basic operating temperature of the resistor, and the location of the resistor with respect to adjacent thermal masses. Some empirical data are presented in Fig. 34. As could be predicted, rigid plastic foam caused temperatures to rise in all watt ratings, but only slightly so in fractional-watt sizes. The room-temperature-vulcanizing silicone rubber has a fairly good thermal conductivity and lowered operating temperatures in low watt ratings, but raised values in higher power ratings. Each situation is an individual case and should be evaluated experimentally. One further consideration that must be made is the ability of encapsulants to withstand the high body temperatures of resistors, especially the small wirewound units, which often surprise packaging designers by their high operating temperatures and ability to char circuit-board materials.

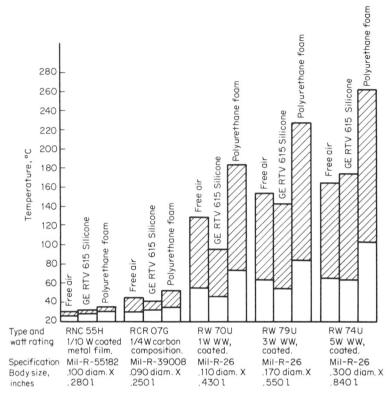

Fig. 34 Experimental data showing the effect of encapsulants on resistor body temperatures. Relevant data furnished by Martin Marietta Corporation, Component Engineering Laboratory:

- Foam density is 2 lb/ft³.
- Dimensions are in inches.
- Shaded portions show difference between half power (unshaded) and full power.
- Block of encapsulant is approximately 2.75 × 3.00 × 1.00 in. thick; parts are mounted on standoffs on 2.75 × 3.00 × 0.1-in. epoxy-glass laminate.
- Temperature sensor is a AWG 30 Cu-Constantan thermocouple.

Resistor service variability Specifications, standards, and some supplier application literature have lately taken responsible recognition of resistor in-service value changes. In the past, only occasional efforts were made to characterize service performance beyond the purchase inspection by supplier and user. With few exceptions allowances in specifications do not distinguish between possible test-result situations that could show all the individuals in a test sample to be barely within defined change limits, and results from the same test on a different set of samples that could show all individuals tightly distributed about the initial value. Both situations are equivalent in the scheme of attributes testing. It would, however, be unlikely for the first situation to occur unless some preselection process were to be employed either deliberately or inadvertently. Both military and EIA specifications require resistors to be subjected to series of tests, and limits of success must be defined by the specification for each test. When circuits are designed to use resistors, allowances must obviously be made for purchase tolerance and temperature-resistance coefficient. Less apparent, however, are the variational allowances that must be made for lead pull and twist, soldering, load life, humidity, temperature cycling, and other expected exposures. The circuit designer must recognize that any one individual resistor can have performance that falls anywhere within the defined limits and still be within specifications. Fortunately, however, most conditions of use are not as severe as the tests, nor is it usual for a particular resistor to vary the full allowed amount in the same direction for each test in a series.

Some empirical data on postsoldering resistance values of military-grade parts are shown in Fig. 20. The random variations in makeup and history of a resistor cause it to deviate from the ideal in a generally gaussian fashion, provided that a number of production batches are considered together. Certain conditional circumstances should be noted, however. For instance, a wirewound-resistor design may habitually change value in only one direction as winding stresses are relieved by operational temperatures and thermal cycling. The amount of value change for each individual part will be likely, however, to be normally distributed about an average value. Specifications do not usually recognize this type of trend, because to do so would sacrifice generality in coverage. Symmetrical tolerances are stated, and in this case, the supplier must hold his distribution about the posttest mean (arithmetic-average) value to tighter limits than would be necessary if his average change were zero, in order to remain within specification limits. Some suppliers, mostly of established-reliability military parts, have published performance information on their particular designs that is based upon large amounts of actual variables test data. Such information is extremely useful if the usages can be restricted to supplier's designs whose demonstrated performance is similar. Care should be taken that all significant causes of variation are included, however.

Other, more generalized prediction techniques can be used with good validity. If it is assumed that part traits are normally and symmetrically distributed in a population, and that each test result is independent of other test results and the part test history, then probability analysis may be used to predict service variability. The assumptions are not unrealistic and may provide conservative results if the average change is predictable as in the wirewound-resistor example above, and if the application is one like a two-resistor voltage divider where parallel "tracking" of changes cancels error. Under the assumption of randomness and independence of tests, each posttest distribution is assumed to be symmetrical about an average that is the same as the pretest average value, which is to say that the symmetrical tolerance allows the distribution to spread about the pretest mean but does not change the mean. In a perfectly normal distribution an equal quantity of resistors are above and below the average or, alternatively, if one resistor is subjected to a series of tests, it will change positively or negatively with equal probability. Under the above assumptions, each test in a series will have a normal distribution, and if the allowed maximum ± variation from the pretest value of resistance is taken as the ±3 standard-deviation points of the distribution (±3 σ points), then it can be said, with only a slight error due to the normalization of pretest resistances to the nominal resistance value, that the density distribution of resistance value will show 99.73 percent of all resistors within the defined limits after the test. The standard deviation σ is defined by

$$\sigma = \sqrt{\sum_{i=1}^{n} \frac{(X_i - \bar{X})^2}{n-1}} \qquad (13)$$

for n test articles having posttest value of $\bar{X}_i$, and arithmetic mean of $\bar{X}$. Deviation from posttest average value for each sample is seen to be

$$\text{Variation} = V_i = X_i - \bar{X} \qquad (14)$$

Bias of the mean is said to occur if the posttest average value is significantly different from the pretest average value. If the average pretest value does not coincide with the labeled resistance value, a small perturbation is introduced. Specifications define variation of individual resistors as an allowed percentage of their pretest value rather than a percentage of the labeled nominal value or of the posttest average value of the sample. This is seldom a serious effect, however. If a test, for instance, allows a 0.5 percent change and the resistor is already 1.0 percent high, the total allowed change could be $0.005 (R_0 + 0.01R_0) = 0.005R_0 + 0.00005R_0$, or a total error of 0.005 percent of R_0.

It can be shown that when the standard-deviation limits of a series of density distributions of gaussian form under the stated assumptions are added by a root-sum-square process, the deviation limits obtained by the process are identical to the deviation limits that would be obtained if all the variations were combined into a single distribution density. Thus a maximum limit to be expected on value variation for a series of tests that meet the normal distribution, independence, and zero-trend-bias criteria above can be determined by performing a root-sum-square of the limits of each of the tests in the series. Thus total variation V_T, defined usually as a percentage of the nominal resistance R_0 expressed as a total percent tolerance T_T, is

$$V_T \approx \frac{T_T}{100} R_0 \simeq \sqrt{\sum_{j=1}^{M} (V_j)^2} \simeq \sqrt{\sum_{j=1}^{M} \left(\frac{T_j R_0}{100}\right)^2} \tag{15}$$

for a series of M tests. It is valid to work with percentages in this case, since $R_0/100$ can be divided from both expressions involving percent tolerances. Using percentages should be approached with care, however, since calculations of averages and standard deviations must always use units of the parameter involved, rather than ratios. An example using the technique is shown in Fig. 35 as applied to general-purpose film-resistor specification MIL-R-39017B.

Test	% allowed T_j	$(T_j)^2$
Power conditioning	0.5	0.25
Temperature cycling	0.25	0.0625
Low-temp. storage	0.25	0.0625
Low-temp. operation	0.25	0.0625
Short-time overload	0.5	0.25
Terminal strength	0.25	0.0625
Resistance to soldering	0.25	0.0625
Moisture resistance	1.0	1.0
Shock and vibration	0.5	0.25
Life, 2000 h	2.0	4.0
High-temperature exposure*	2.0	4.0
TOTALS	$T_j = 7.75\%$	$(T_j)^2 = 10.0625\ (\%)^2$
Total tolerance		$\sqrt{10.0625\ (\%)^2} = \pm 3.16\%$

* 2000 h at 150°C, not powered. Probably a good accelerated shelf-aging test.

Fig. 35 Design-tolerance determination of MIL-R-39017 resistors, fixed-film (insulated), established-reliability, type RLRO5.

For MIL-R-39017, purchase tolerance (± 2 or ± 5 percent) and temperature coefficient of resistance (0 ± 200 ppm/°C) are needed to reach the total service-variability estimate. If the application is benign and avoids stresses of the types covered by the tests (high potential, for instance, was omitted in the example), it may be desired to omit that variation in the analysis, but shelf drift is not included, which is often significant. Also, in this case, life testing is limited to 2,000 h at rated load, while the 10,000-h continuation allows an additional ± 2 percent change. It should be noted that purchase tolerance distributions may not be normal, particularly if close-tolerance parts have been screened from the population.

Tolerance analysis is a useful technique and should be employed as a realistic evaluation of expected resistor performance. Military and EIA specifications provide test stress limits for most part types available domestically and should be used to evaluate service performance instead of the unrealistic practice of expecting resistors to maintain purchase tolerances in service life.

Measurement of high-frequency characteristics The determination and comparison of high-frequency performance of resistors is not a simple matter. Most treatments of the subject that provide empirical data or performance predictions for conventional fixed resistors at frequencies from 500 kHz through MHz provide information derived from measurements made on the Hewlett-Packard Company (formerly Boonton Radio Company) model 250A or 250B Rx meter. This instrument consists of a modified Schering-bridge* circuit combined with a 0.5- to 250-MHz signal generator and an intermediate-frequency-type null-balance indicator system. The measurement obtained is the value of the two-terminal equivalent of the *parallel*-connected resistance and reactance, which may be either inductive or capacitive, depending upon how the resistor appears at the particular frequency of measurement.

If the parallel reactance is inductive, a "negative" capacitance reading is obtained, which really amounts to a calibrated reading of the capacitance needed to parallel-resonate the unknown inductance. Data given graphically in many resistor suppliers' catalogs for resistor characteristics reflect this measurement technique. Most presentations show reactive components charted against a frequency scale but do not specify whether series or parallel components are meant. Values are often given in microhenrys or picofarads, which can be transformed into reactances and thence to total two-terminal series-equivalent impedances, if desired. The resistance reading is frequently not given and an assumption must be made that the dc true resistance is unchanged by frequency, which is basically correct except for carbon composition units. The series two-terminal impedance is calculated as shown in Fig. 36. It is thus seen that the real component of the *series*-equivalent impedance depends upon the ratio of the dc resistance ($\approx R_P$) to the reactive component. For further discussion of high-frequency effects, see the resistor general information section above. The measurements are essentially independent of signal amplitude within practical limits if no ferromagnetic materials are located in proximity, but if core or coatings

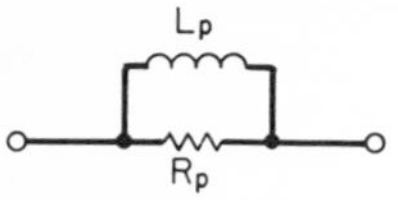

$$X_{Lp} = 2\pi f_1 L_p$$

$$Z_a(\text{series}) = \frac{R_p}{1 + \dfrac{R_p^2}{X_{Lp}^2}} + j\left(\frac{R_p^2}{X_{Lp} + \dfrac{R_p^2}{X_{Lp}}}\right)$$

(a)

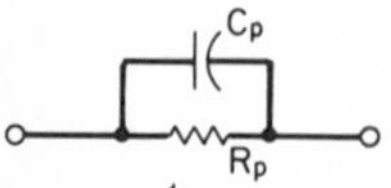

$$X_{Cp} = \frac{1}{2\pi f_2 C_p}$$

$$Z_b(\text{series}) = \frac{R_p}{1 + \dfrac{R_p^2}{X_{Cp}^2}} - j\left(\frac{R_p^2}{X_{Cp} + \dfrac{R_p^2}{X_{Cp}}}\right)$$

(b)

Fig. 36 Resistor high-frequency impedances. (*a*) Parallel equivalent circuit at frequency f_1. (*b*) Parallel equivalent circuit at frequency f_2.

are very lossy at high frequencies, the assumption of constant R_P may not be valid. Lead length and spacing to ground-plane areas affect values and should be considered when measurements are taken.

Power Ratings. Industry standards, military specifications, and supplier data furnish power-rating information on resistor designs. Except for the highest-power types, these ratings are based upon a specified change in resistance value with time at rated power dissipation. It is generally established that drift rates for resistors are increased for increased temperature, following the expectations for increased chemical and physical activity with temperature. Power rating, temperature rise, and physical size are thus intimately interrelated. For given types of materials and designs, practical compromises between temperature maximums and drift rates have been established to define power ratings. A maximum ambient temperature of 85°C will include almost all severe-environment applications, including most under-hood automotive levels and military equipment. Some of the most severe military and space environments or exceptionally severe industrial applications require a 125°C maximum temperature, but few are higher.

* See, for instance, "Reference Data for Radio Engineering," International Telephone & Telegraph Corp., New York.

Table 16 shows a number of commercial and military specification ratings. Recommended power-derating curves included in specifications are also included in the paragraphs applicable to the various resistor types. It is readily observed that almost all resistors classed as power types are rated for full labeled power dissipation only up to ambient temperatures of 25°C or sometimes 40°C. At higher ambient temperatures the dissipated power must be reduced. Some types also must be mounted on chassis of specified sizes in order to realize even the full 25°C rating. General-purpose types, by contrast, are rated for full power up to some reasonably high temperature, 70 to 125°C, then derated linearly to zero at a maximum temperature. Although these derating curves are included in power and general-purpose resistor specifications as recommended practice, not all test routines specified will determine the ability of a given resistor design to remain stable under high-temperature conditions. Only the more recent military specifications include tests that measure the resistance drift at temperatures approaching the zero-power-rated level. Most of the other specifications require life tests at full rated power and room temperature, and the hot-spot temperature reached by the test articles may be much less than the recommended maximum application temperatures. Care should therefore be exercised when using resistors at temperatures covered by the sloped portion of the derating curves. Specifications that require high-temperature-exposure tests make significant resistance-change allowances for the test, and it can be expected that similar drift rates will be experienced by those designs not so tested.

High-power resistors present special application-rating situations. It is frequently necessary to mount several units together in a bank, and power ratings must be reduced to allow for mutual heating. Forced-air cooling is sometimes used, and allows still-air power ratings to be exceeded without danger. At high altitudes air is less dense and convection cooling is less efficient. All these variables have complex effects, and the exact results for a given installation are only grossly predictable, barring a complete and detailed thermal modeling. General rules are published by suppliers, however, based upon experience and empirical data. A number of these guidelines are shown in Figs. 37 through 41. More detailed information is available, once application details and circumstances are known.

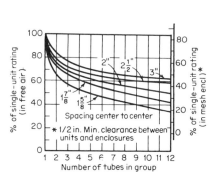

Fig. 37 Power ratings for multiple installation of tubular vitreous-enameled resistors, core diameters $\frac{7}{16}$ through $1\frac{7}{16}$ in. (*Ward Leonard.*)

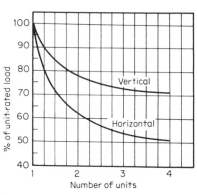

Fig. 38 Power ratings for typical stacked flat miniature wirewound strip resistors, vitreous-enamel-coated. "Vertical" means flat side vertical, parallel to panel. (*Ward Leonard.*)

Fusible Resistors. Resistors are furnished by several suppliers for applications in which it is desired that the unit open-circuit if overload conditions should occur. In general, resistors for this purpose are not cataloged as standard items because of the varied application circumstances that must be considered. Failure under moderate overloads is a trait usually to be avoided in resistor design, and deliberately designing for such a function involves a complete knowledge and consideration of the application situation. If the expected overload condition is precipitous and sub-

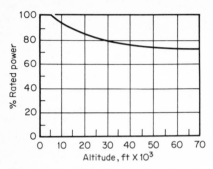

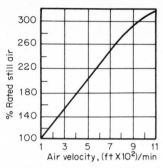

Fig. 39 Derating of power for altitude: high-power resistors. (*Ohmite Manufacturing Company*, Skokie, Ill., Bulletin 1100A, 1972.)

Fig. 40 Increase in dissipated power for forced-air cooling: large vitreous-enameled power resistors. (*Ohmite Manufacturing Company*, Skokie, Ill., Bulletin 1100A, 1972.

stantial, the problem is simplified, provided that a resistance value can be used that is high enough to allow small-diameter resistance wire to be used in a wirewound unit. Millisecond-range fusing is said to be available if the overload gradient is high enough. Reference to the tables on resistance alloys at the beginning of the chapter will show that these alloys melt at temperatures in the 1000 to 1500°C range, while good design temperatures for stable performance do not usually exceed 350°C. A reasonably well-cased wirewound unit will therefore withstand several times rated current for long periods without open circuiting. If an underrated unit is used, the continual high-temperature exposure will be likely to lower reliability in the normal function. Fifteen or twenty times rated power or even more should be provided for fusing if possible. Even then, it may be necessary to allow several seconds for action. The most common design furnished for fusible application is the small-diameter glass-fiber-core continuous-wound type of resistor. It is furnished as a molded general-purpose wirewound type by several suppliers and also is often cemented into ceramic cases and tubes to increase thermal mass and, thereby, power ratings. A high-current impulse into these units creates a high thermal drop to the case surface for the transient condition, but under long-term conditions the case-cement thermal mass and surface area dissipate well.

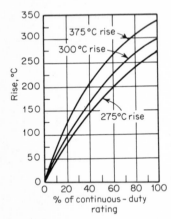

Fig. 41 Temperature rise vs. percent of continuous-duty rating of typical power resistors.

Under extremely high currents complex effects occur. If current is high enough, wire can even be made to detonate. Conditions are seldom this severe, however, but one of the concerns must be that of providing a mechanical package that will contain the fusing action and the resulting molten-metal debris. Estimates for short-pulse fusing can be made using the calculations described above for pulse applications if the melting temperatures of the alloys are substituted for the safe-temperature limits. Fusible designs other than those mentioned here are of course available. Film resistors have a better potential for this use than do wirewound units, and several film-type suppliers have characterized their products for fusing functions. Suppliers listed in this specific product category in industry directories (see chapter introduction) should be queried for current information on the subject if needed.

The drive for product safety is increasing the need for fusible resistors, and it is expected that developments will be made to broaden availability and performance.

THICK-FILM NETWORKS*

Introduction

Basic Technology. Thick-film networks are resistor and conductor materials comprising precious metals in a glass binding system which have been screened onto a ceramic substrate and fired at high temperatures. Originally developed in the late 1950s and early 1960s, thick films have established a broad range of applications and exhibited a substantial growth in demand. This was made possible by exploiting the inherent capabilities of the thick-film technology, which include high power, precision, high performance, packaging flexibility, and low cost. Thick-film networks can provide miniaturization at costs comparable with those of general-purpose discrete resistors and performance comparable with that of semiprecision types. They are inherently very reliable, since they typically exhibit predictable changes, have a rugged construction, and are usually not subject to catastrophic failure. Thick-film technology, by its nature, lends itself best to custom resistor networks where cost, performance, and size are optimized to provide advantages over equivalent discrete approaches.

A wide variety of thick-film resistor materials exist, but they are most generally comprised of similar basic types of constituents, including ceramic glasses which provide a binder and comprise most of the resistor volume (typically metal borosilicates such as lead); precious metals to provide the conductor (e.g., gold, silver, paladium, platinum, ruthenium); metal oxide semiconductors which are used to control electrical characteristics; ceramic flux to produce adhesion of metal and glass particles (e.g., bismuth trioxide); and opacifiers which are refractory-metal oxides to reduce contact resistance. The electrode materials employed are similar in formulation, but of course more conductive and usually solderable. Some of the common electrodes include silver, gold, palladium silver, and platinum gold.

Substrates are employed to provide a base or carrier for the thick-film network. Their selection is based on a number of important properties, including low electrical conductivity (insulator), high thermal conductivity (heat conductor), mechanical strength, ability to withstand high firing temperatures, and the absence of free alkalies (chemically inert). Since ceramics are well suited for most of these requirements, substrate materials such as alumina, steatite, barium titanate, and beryllia are most often employed. A completed network is usually encapsulated by means of molding, potting, or coating to provide moisture protection, mechanical protection, a marking surface, a modular package, and cosmetic appearance.

The ohmic value of thick-film resistors is a function of the sheet resistivity (i.e., ohms per square, which assumes a constant thickness) and the length-to-width ratio (i.e., $R = \rho L/W$). Resistor formulations are available with resistivities from as low as 1Ω/square to more than $1M\Omega$/square. The desired nominal resistor values are obtained by varying resistor geometry by way of the screen-printing process (i.e., changing the length-to-width ratio, which changes the number of effective "squares," as illustrated in Fig. 42). Resistor values after firing can be tailored to precise values or tolerances by means of an abrading or trimming process which increases the ohmic value by the removal of resistor material (typically sandblast or laser-trim techniques are employed). The trimming operation changes the length-to-width ratio, which results in an increase in the effective number of squares, as illustrated in Fig. 43.

A typical thick-film network manufacturing process therefore comprises screening, firing, and packaging steps such as those outlined in the flow chart shown in Fig. 44. Such a process can be assembled for relatively low cost to fabricate prototype networks or can be developed into a highly sophisticated mechanized line for high-volume production.

Scope of Components and Application. Thick-film technology lends itself to a wide variety of applications, including discrete resistors, resistor chips, hybrid sub-

---

* This section was written by John H. Powers, IBM Corporation.

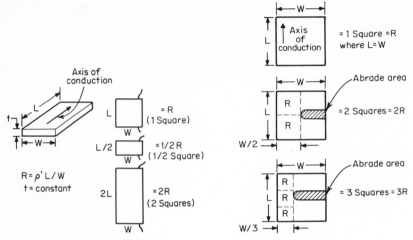

Fig. 42 Aspect ratio for thick-film resistors.

Fig. 43 Thick-film-resistor trimming geometry.

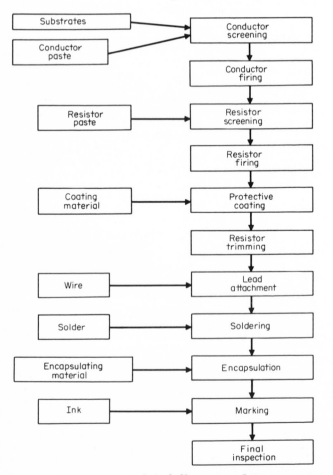

Fig. 44 Typical thick-film process flow.

strates, and *RC* networks as well as resistor networks, which will be the principal area covered here. Resistor networks are typically custom devices owing to the unique combination of values, configuration, and package employed in each design. Most applications can be categorized as either the general-purpose, precision, or power type, each of which has been developed in many forms over the years, as illustrated by some typical examples in Fig. 45. Note that most network devices

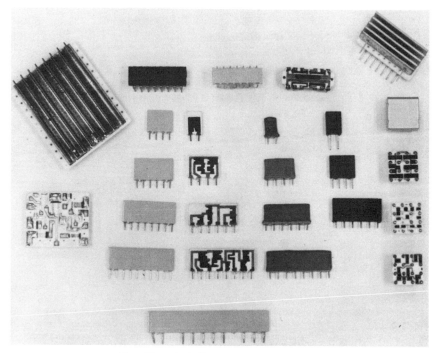

Fig. 45 Thick-film-network components.

take the form of stand-alone modular components employing a variety of substrate sizes, pin configurations, heat sinks, and package designs or encapsulations.

Resistor networks can be employed in any resistor circuit application in which the thick-film technology offers some advantage in cost, performance, or size over alternative approaches (e.g., discrete, thin-film, IC). This therefore includes such common resistor functions as line termination, bias, bleeder, voltage divider, current sense, and load. In most cases, economics dictate that as many resistors as possible be integrated into a network within the constraints of process yields, resistivity ranges, and package size. Production devices have been made which incorporate from 1 to 50 resistor elements, but 5 to 15 is a typical range. The capabilities of thick-film networks with respect to resistance values, tolerances, power ratings, and performance characteristics are a function of the specific materials and package designs employed, but capabilities indicative of the products generally available to industry today are discussed in the following sections.

General Terms and Definitions. Following is a summary of several terms (in alphabetical order) with brief definitions which are commonly used in association with thick-film networks.

Aspect Ratio. The geometrical relationship of the length and width (L/W) of a rectangular resistor element which is used in the layout of networks to establish "as-fired" resistance values (e.g., when $L = W$, the aspect ratio is 1:1, or 1 "square," and the resistor value is equal to the sheet resistivity).

DIP. Dual-in-line-package.

Formulation.   A specific mix of thick-film material where the conductor, glass, and other additive ingredients are formulated to provide certain properties such as a specific sheet resistivity or TCR.

Paste/Ink.   Screenable thick-film material comprising metals, oxides, and glasses in an organic vehicle which when fired produces a circuit element such as a resistor or conductor.

Power Density.   The power dissipation per unit area of a resistor or substrate (in watts per square inch) used to determine the optimum layout design of a network.

Screen.   The process of printing a network pattern of thick-film ink or paste onto a substrate by means of a squeegee applied to a photoetched wire-mesh "silk screen" or metal mask.

Sheet Resistivity.   The nominal resistance per unit area of a thick-film ink or paste which is usually expressed in ohms per square (assuming a constant thickness), where the design resistance value of the screened resistor is determined by $R = \rho L/W$.

Substrate.   The base or carrier for the thick-film network, which is usually a ceramic plate.

Thick-Film/Cermet/Metal Glaze.   Resistor and conductor materials comprising metals or metal oxides in a glass binding system which can be screened onto a substrate and fired to provide circuit elements or networks.

Tracking.   The inherent capability of resistors which have been made from the same formulation and screened onto the same substrate to exhibit similar performance characteristics (e.g., drift, TCR).

Trim/Abrade/Adjust.   The process of tailoring a thick-film-resistor element to a specific value or tolerance by the removal of resistor material (by means of sand-blasting or laser abrading), which increases the ohmic value.

Voltage Gradient/Field Strength.   The linear voltage stress applied across a resistor element (in volts per inch), used to determine the optimum geometry of a high-voltage resistor.

Design considerations

Application Requirements.   Thick-film technology is extremely versatile and can be adapted to many applications if its characteristics are understood and its inherent capabilities exploited.   If a successful thick-film product is to be developed, the component design must be optimized to match the requirements of its intended application.   Some of the principal application requirements which should be identified to permit effective component design include:

1. *Parametric requirements*—resistor values, ratios, network configuration, and power ratings
2. *Packaging constraints*—size, shape, assembly, and power-handling requirements
3. *Performance requirements*—initial tolerances, short-term stability (e.g., TCR, handling, overload), drift, tracking, useful life, failure rate
4. *Cost constraints*—vs. alternatives (e.g., discrete resistors, IC's), circuit cost, quantities, packaging costs (e.g., handling, assembly, space)
5. *Application environment*—ambient temperature, humidity, airflow
6. *Process compatibility*—handling, assembly, materials, temperatures
7. *Functional application characteristics*—frequency, noise, duty cycle, overload

Packaging Factors.   Once the application requirements are determined, a number of basic design factors must be addressed to establish a component package compatible with the needs identified, including:

1. *Substrate Material.*   The principal factors are thermal conductivity, thick-film compatibility, cost, mechanical strength, insulating properties, and physical configuration.   The most common substrates are ceramics such as alumina, beryllia, steatite, and barium titanate.

2. *Resistor Material.*   The principal factors are resistivity, TCR, drift characteristics, and cost.   Some of the most common systems are based on formulations of PdO, RuO_2, IrO_2, or AuO.

3. *Conductor and Termination Materials.*   The principal factors are conductivity, solderability, resistor compatibility, and cost.   Some of the most common conductors are based on formulations of Ag, Au, Pt-Au, Pd-Ag, and Pd-Au.

4. *Encapsulating Materials and/or Package.* The principal factors are environmental protection, physical configuration, mechanical integrity, process compatibility, cost, and cosmetic appearance. Some of the common packaging approaches are dipping, molding, potting, and roller coating.

5. *Interconnections or Terminals.* The principal factors are the selection of materials (e.g., copper, nickel, zirconium-copper), design (e.g., lead frame, pins, wires), and interconnection technique (e.g., staking, wire bonding, soldering).

6. *Joining Materials and Process.* The principal factors are process and material compatibility, mechanical strength, cost, and reliability. Solders such as 60/40 or 10/90 Sn/Pb are by far the most common approach.

7. *Element and Network Layout Geometry.* This is obviously influenced by resistor values, power dissipation, network configuration, substrate size, and the resistor materials available.

Typical Component Packages. Since such a wide variety of package styles, network configurations, material systems, and applications are feasible and available for thick-film resistor networks, it is obviously not possible to present and discuss all the potential combinations. Three basic styles of packages that have been employed widely for networks, however, can be used as a standard reference and in fact may satisfy most application requirements. The first one, illustrated in Fig. 46, is a single-in-line, stand-up-type package which is available in a number of sizes

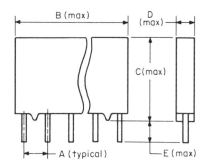

Fig. 46   Single-in-line package.

A	No. of leads	B	C	D	E
0.100	4	0.384	0.350	0.090	0.095
	6	0.584	0.350	0.090	0.095
	8	0.784	0.350	0.090	0.095
	10	0.984	0.350	0.090	0.095
0.125	2	0.234	0.350	0.110	0.095
	4	0.484	0.350	0.110	0.095
	6	0.734	0.350	0.110	0.095
	8	0.984	0.350	0.110	0.095
0.150	4	0.600	0.350	0.132	0.130
	6	0.900	0.350	0.132	0.130

(i.e., from 2 to 10 pins for 1 to 9 resistors typically) and lead spacings (i.e., 100, 125, and 150 mils). This has proved to be a popular approach to general-purpose applications requiring a network of only a few resistors in a small package at a low cost per element. The DIP package, illustrated in Fig. 47, has gained wide acceptance in recent years and is available in 8- to 18-pin sizes on 100-mil spacing. This approach is often employed when a larger network of resistors is required (typically 4 to 18 elements) and a standard DIP packaging and assembly scheme is adopted. As illustrated in Fig. 48, the third design is a square package (½-in, 16-pin, 125-mil in this case) which is used for both high-density resistor-network and hybrid-circuit applications in modular packaging systems.

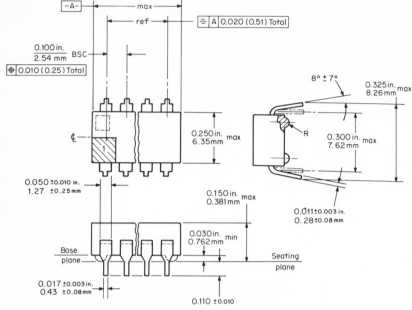

Fig. 47 Dual-in-line package.

No. of pins	A dimension	
	inch	millimeters
8	0.450	11.4
14	0.750	19.1
16	0.850	21.6
18	0.945	24.0

Parametric capabilities

Initial Resistance. The resistance values which can be achieved with thick-film technology are a function of resistor material (sheet resistivity), substrate area available, resistor geometry, trimming technique, and process reproducibility. Table 17

TABLE 17 Typical Parametric Capabilities of Thick-Film Resistor Networks

Parameter	Units	Range	Typical
Resistance value	Ω	1–100 M	10–1 M
Sheet resistivity	Ω/square	10–1 M	20–20 k
Initial resistor tolerance	± %	0.1–30	1–5
Initial resistance ratio	± %	0.1–20	1–5
TCR (absolute)	± ppm/°C	50–1,000	100–500
TCR (relative)	± ppm/°C	1–100	10–50
VCR	%/V		0.001–0.01
Current noise	dB	−50 to +10	−30 to −10

is a summary of the typical range of parametric capabilities available in production thick-film resistor networks. Sheet resistivities from 1 Ω/square to more than 1 MΩ/square are feasible, but the most stable and most commonly used resistor systems are typically between 10 Ω/square and 10 kΩ/square.   By using fractional and multiple-square resistor geometries as illustrated in Fig. 49, one can achieve a wide range of resistance values.   As-fired resistor values on reproducible production lots can generally maintain a ±20 to ±30 percent untrimmed initial tolerance.   Once trimming is employed, ±5 percent is readily achieved and is therefore often the maxi-

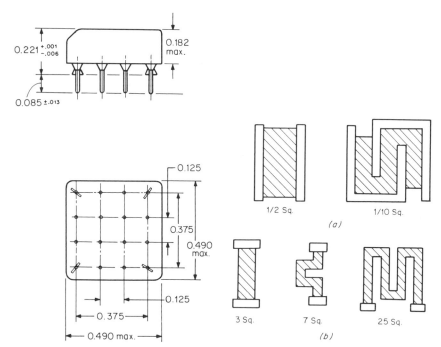

Fig. 48 Square package.

Fig.   49 Resistor-layout   geometries. (a) Fractional-square geometries. (b) Multiple-square geometries.

mum tolerance used for trimmed resistors.   Tighter tolerances are commonplace (±1 to ±5 percent) and true precision is feasible (i.e., less than ±1 percent), but process control and yields are obviously impacted which will increase cost.   Since a number of resistor elements can be fabricated simultaneously on a single substrate, it is possible and occasionally desirable to match them (on an absolute or ratio basis).   This is achieved at the trimming process where each resistor is trimmed in relation to the previous or reference resistor, but good process and geometry control are obviously required to match values within tenths of a percent.

Temperature Coefficient of Resistance (TCR).   The TCR characteristics of thick-film resistors are determined primarily by the resistor material and firing process. Midrange resistivities (e.g., 100 Ω/□ to 1 kΩ/□) generally offer the tightest TCR capabilities (i.e., ±100 ppm/°C typical) while the extremely low and high ranges tend to be more temperature-variant and difficult to control.   Resistor elements fabricated simultaneously on the same substrate, employing the same resistor material and geometry, exhibit similar TCR characteristics and can be specified to relative as well as absolute values (e.g., typically within 25 to 50 ppm/°C).   Although TCR is normally specified as a constant or linear parameter, the TCR of most resistors (i.e., the rate of change of resistance with respect to temperature) varies with both resistivity and temperature as illustrated in Fig. 50.

Voltage Coefficient of Resistance (VCR).   The VCR of a thick-film resistor is relatively negligible (i.e., typically 0.001 to 0.01 percent/V) and therefore usually of interest only in high-voltage applications where its effect on total resistance changes

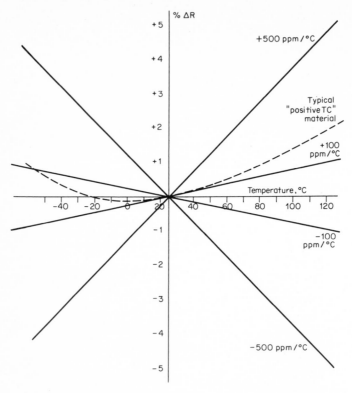

Fig. 50 Temperature coefficient of resistance characteristics.

may become significant. VCR is determined primarily by the resistor material or resistivity employed and usually varies with voltage. Owing to the heating effects of applied voltage, it is difficult to measure and separate the VCR and TCR characteristics of a resistor. Unless critical to the application, VCR is normally not a specified parameter.

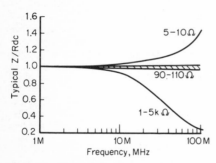

Fig. 51 Typical frequency characteristics of thick-film resistor networks.

Frequency. Although resistors are normally thought of as fixed dc elements, and thick-film networks by their nature are expected to have relatively little reactance, frequency-response characteristics do exist and can be important in some applications (e.g., line termination for a high-frequency switching load). The frequency characteristics of a thick-film resistor are influenced by its material composition and structure, physical geometry, terminations, conductors, and interconnections. Generalizations, therefore, may not apply well in all cases, and custom networks with frequency concerns should be characterized. Enough data exist, however, to indicate that most resistors can be envisioned as having an equivalent circuit comprising a series resistance (which varies from the dc value with frequency), a series inductance (generally fixed and in the range of 1 to 100 nH), and a parallel capacitance (major variable with each resistor type considered). An example of the impedance characteristics of some simple, uniform resistor networks is given in Fig. 51. In this case, it appears that the most ideal (i.e., non-reactive) characteristics can be achieved with resistor values in the vicinity of 100 Ω while lower values tend to be inductive and higher values predominantly capacitive. At very high frequencies, of course, the skin effect on signal transmission through the network will begin to influence the net response. In general, thick-film resistors can be assumed to be ideal or fixed with respect to frequency below approximately 10 MHz.

Current Noise. Current noise is a form of distortion generated within a resistor element when transmitting an electric current. In circuits sensitive to low signal voltages, such noise can affect performance, and the characteristics of the resistors to be used should therefore be understood. Current noise is usually expressed as the ratio of the ac "noise voltage" sensed with respect to the dc voltage applied in either μV/V or dB. Figure 52 provides a conversion between these two forms of noise expression. The noise characteristics of thick-film resistors is influenced by the magnitude of the dc voltage applied, the resistor geometry, material systems, resistor value, and trimming technique. In general, it has been found that current noise can be minimized by increasing the length-to-width ratio, using low or mid-range resistivities, and employing uniform (rectangular) resistor geometry with minimum trimming. Thick-film resistors have

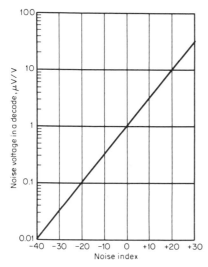

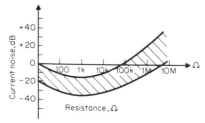

Fig. 52 Noise-conversion graph. (*Quan-Tech. Laboratories, Inc.*)

Fig. 53 Typical current-noise characteristics of thick-film resistor net-works.

typically been found to exhibit lower magnitudes of current noise than carbon-based resistor systems (i.e., discrete composition or film types), and although difficult to generalize, the range of noise index most types of thick-film resistors (including a variety of materials systems and geometries) is within −50 to +10 dB (depending on the range of resistance value) with −10 to −30 dB being typical, as illustrated in Fig. 53.

Performance characteristics

Package-Variant Characteristics. The performance characteristics of thick-film components are obviously a function of such things as the application, package design, and material systems employed. Therefore, generalizations about the performance of thick-film resistors can be made only in terms of their typical or relative stability, since the specific characteristics unique to each package must be determined. These include the short-term permanent changes in resistance due to handling and processing, and the effects of environment as well as the resistor operating temperature under load. Some typical short-term performance characteristics as they relate to common specification tests (such as MIL-STD-202) are presented in Fig. 54. In general, thick-film resistors exhibit stability under such conditions which is typically superior to carbon-composition- and film-type resistors and comparable with that of general-purpose wirewound and metal-film devices.

Since operating temperature is typically the major influence on resistor degradation, the temperature-rise characteristics of resistor networks under load conditions must

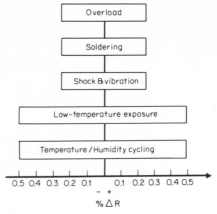

Fig. 54 Typical short-term test performance of thick-film networks.

be understood. Operating-temperature characteristics are obviously a function of package construction and materials and therefore can be related only to specific designs. Using the three standard network package styles described previously as examples (i.e., SIP, DIP, and square), Fig. 55 illustrates the average temperature rise for some standard package sizes and uniform network geometries in a natural-convection environment. Moving air ambients will provide a reduction in operating temperature, as illustrated in Fig. 56. These figures, because of their general "ideal

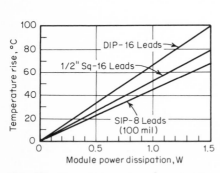

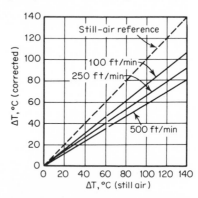

Fig. 55 Temperature-rise characteristics of typical resistor network modules (natural convection).

Fig. 56 Temperature-rise correction for resistor networks in moving air ambients.

design" nature, can be considered indicative of maximum power-handling capability. There are a number of other variables, however, even within these standard packages, which could have a significant effect on operating temperature and must therefore be compensated for, such as resistor size, resistor geometry, number of resistor elements, type of printed-circuit board, and application environment (particularly with respect to neighboring heat dissipators). These factors all tend to reduce power-handling capability from the ideal data presented and are unique to each specific network design and application.

Resistor Drift. Long-term resistor drift is comprised primarily of aging and degradation under stress. As shown by the aging data in Fig. 57, which is based on actual resistor performance during long-term room-ambient storage, thick-film resistors have excellent inherent stability (e.g., drifts typically less than a few tenths of a percent can be expected for up to 10 years). Typical resistor degradation rates

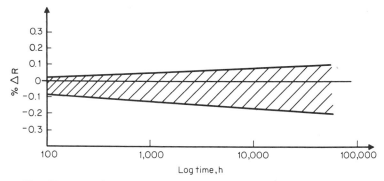

Fig. 57 Typical aging characteristics for thick-film resistor networks.

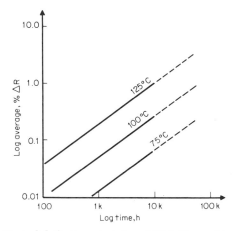

Fig. 58 Typical drift characteristics of thick-film resistor networks.

as a function of operating temperature are illustrated in Fig. 58, which is comprised of data representing the typical average drift exhibited by several different module designs when subjected to a variety of life-test conditions at low-voltage stress. From these data, it can be seen that average drifts of approximately 1 percent in 5 years can be expected at a maximum resistor operating temperature of 100°C.

TRIMMING POTENTIOMETERS*

Introduction

Scope of components and applications   Trimming potentiometers, which are also known as resistance trimmers and variable resistors, are adjustable three-terminal resistor elements commonly used as voltage dividers or rheostats. As illustrated by some typical examples in Fig. 59, trimming potentiometers are available in a wide variety of styles, sizes, and constructions.

Five basic resistor technologies are employed for potentiometers, with the wire-wound and cermet (thick-film) types being the most popular for a broad range of applications and the carbon-composition, carbon-film, and metal-film types used more for selective markets.   Potentiometers are available in both single and multiturn adjustment styles and are typically compatible with printed-circuit-board mounting.

Trimmers are used in a wide variety of applications where precise resistance values or voltage levels are required or where circuit functions must be set up or adjusted.

* This section was written by John H. Powers, IBM Corporation.

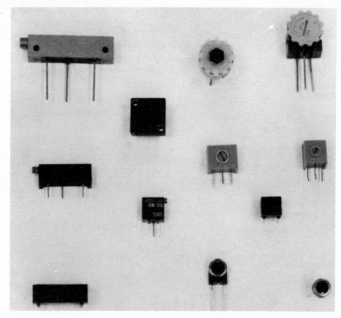

Fig. 59   Typical trimming potentiometers.

Some typical applications for potentiometers include setting biases for transistors, adjusting time constants for RC networks, setting reference voltages for control circuits, adjusting the gain of an amplifier, and varying or limiting current in a bleeder circuit.   The large number of styles, technologies, and constructions of trimming potentiometers available permit the user to select a device which optimizes cost, performance, and size for an application.   The capabilities and characteristics of potentiometers in general, as well as some of the most common styles, will be discussed in the following sections.

As an additional reference and for specified detail relating to standard specifications, the following documents are applicable to potentiometers:

Applicable military specifications
MIL-R-39015   Resistor, Variable, Wirewound (Lead Screw Actuated)
MIL-R-22097   General Specification for Resistors, Variable, Non-Wirewound (Adjustment Type)
Applicable EIA standards
RS-345   Resistors, Variable, Wirewound (Lead Screw Actuated)
RS-360   Resistors, Variable, Non-Wirewound (Lead Screw Actuated)
Applicable industry standards (Variable Resistive Components Institute)
VRCI-T-110   Terms and Definitions
VRCI-T-215   Inspection and Test Procedures

Basic technologies   Two basic types of resistor element are employed in potentiometers: one is incremental (i.e., wirewound), and the other is continuous (i.e., composition or film).   The resistor may be fabricated as either a linear or circular element depending upon the type of potentiometer desired.   Although many package styles exist, there are only two basic approaches to trimmer construction, i.e., single- or multiturn design.   Multiturn units typically employ lead-screw-actuated sliding contacts and may utilize either circular or rectilinear elements.   In the single-turn units, however, the wiper is usually driven directly by the screwdriver slot or shaft on a circular element.   Examples of several typical potentiometer constructions are illustrated in Figs. 60 to 62.   Following is a brief description of the five basic resistor technologies employed for potentiometers.

Wirewound.   The resistor element is fabricated by winding fine resistance wire

Fig. 60 Potentiometer construction: ½-in-square multiturn wirewound.

Fig. 61 Potentiometer construction, ¾-in rectilinear multiturn film.

Fig. 62 Potentiometer construction, ¼-in round single-turn cermet.

around an insulating bobbin or core which is circular in cross section and either rectilinear or circular in shape. The lead terminations are typically welded to the ends of the resistor element. Wirewound potentiometers are the most common type and usually exhibit low TCR and good stability but by the nature of the noncontinu-ous-element construction are limited in the resolution and noise performance they can provide.

Cermet. The resistor element is fabricated by screening a thick-film composition of precious metal and ceramic materials on a ceramic substrate. The element may be either rectangular or circular in shape and is typically terminated by soldering thick-film conductor pads to pins staked through the substrate. Cermet potentiome-ters provide a continuous element for high resolution as well as the capability for excellent performance.

Carbon Composition. The resistor element is comprised of carbon molded with an organic binder (similar to the discrete axial-leaded carbon-composition-resistor element) as an integral part of the potentiometer baseplate subassembly. This provides a relatively inexpensive potentiometer with a continuous element, but its performance is generally the least stable of the five technologies and is intended primarily for general-purpose application.

Carbon Film. The resistor element is fabricated by depositing a film of carbon (similar to the discrete axial-leaded carbon-film resistor) on a substrate or baseplate. This approach can provide an economical continuous-element potentiometer with improved performance over the composition type.

Metal Film. The resistor element is fabricated by depositing a thin metal film (similar to the discrete axial-leaded metal-film resistor) on a ceramic substrate with similar configurations and termination techniques to those employed for cermet types. This approach can provide a precision, high-performance potentiometer and perhaps offers the greatest potential to achieve ideal characteristics (in terms of resolution, linearity, noise, and stability), but it is generally more expensive than the conventional types.

General terms and definitions   Following are brief definitions of several general terms normally used in association with trimming potentiometers. Additional terms and more rigorous definitions can be found in the industry standard published by the Variable Resistive Components Institute (T-110, Terms and Definitions).

Total Resistance. The resistance measured between the two end terminals of a potentiometer.

End Resistance. The resistance measured between the wiper terminal and an end terminal, with the wiper element positioned at the corresponding end of its mechanical travel.

Travel. The clockwise or counterclockwise rotation of the wiper along the resistor element. Mechanical travel is the total rotation of the wiper between end-stop positions. Electrical travel is the total rotation between maximum and minimum resistance values. These are not identical, owing to discontinuities at the end positions.

Resolution or Settability. The ability of an operator to set the potentiometer to a predetermined ohmic value, voltage, or current. This is a measure of the sensitivity or degree of accuracy to which a potentiometer may be set. Wirewound potentiometers, having noncontinuous elements, typically are referred to in terms of a theoretical resolution which is the reciprocal of the number of turns of wire. Settability is affected by the material and the uniformity of the resistor and wiper elements, the length of the resistor element, and the design of the adjustment mechanism.

Noise. The effective contact resistance introduced in the wiper arm while rotating the wiper across the resistor element. Wirewound potentiometers are normally referred to in terms of equivalent noise resistance (ENR), which is caused primarily by variations in wiper contact during travel along the wire element. Although nonwirewound potentiometers have a continuous resistor element, there is usually a built-in dc offset due to a measurable contact resistance between the wiper and the resistor. Their noise is therefore normally referred to in terms of contact-resistance variation (CRV), which is caused primarily by the changes in contact resistance between the wiper and the resistor element during rotation.

Torque. The mechanical moment of force applied to a potentiometer shaft. Starting torque is the maximum torque required to initiate shaft rotation. Stop torque is the maximum torque which can be applied to the adjustment shaft at a mechanical end-stop position.

Setting Stability. The ability of a potentiometer to maintain its initial setting during mechanical and environmental stresses, normally expressed as a percentage change in output voltage with respect to the total applied voltage.

Rotational Life. The number of cycles of rotation which can be attained at certain operating conditions while remaining within specified allowable parametric criteria. A cycle comprises the travel of the wiper along the total resistor element in both directions.

Potentiometer Styles

Design considerations   When selecting a potentiometer, one should attempt to match and optimize the design factors which satisfy each of the basic requirements of the application. The principal application requirements which should be considered when specifying a potentiometer are:

Parametric requirements—total resistance, initial resistance tolerance, power rating

Performance requirements—TCR, resolution, noise, setting stability, drift, rotational life

Packaging constraints—size, configuration (e.g., top or side adjust), lead spacing

Cost constraints—cost of component, assembly, and adjustment

Application conditions—assembly process, operating environment, circuit characteristics.

The major design factors to be considered in the selection of a suitable potentiometer are:

Resistor technology—wirewound, composition, film

Adjustment style—single-turn (and mechanical rotation), multiturn (and number of turns)

Package design—shape (round, square, rectangular), size, top or side adjust, slot size, seal

Military designations Trimming potentiometers which have been qualified for military applications are classified by a standard type-designation system. The type number identifies the style, size, performance characteristics, terminals, resistance, and failure rate which are applicable to the designated potentiometer. Figure 63 presents a summary outline and explanation of some of the common type designations for military-grade trimmers.

Package designs Many potentiometer styles and sizes exist even beyond those identified in the military-type designation system. Only two basic kinds of package are available, however, i.e., the single-turn and multiturn, from which many variations have developed with respect to size, shape, adjustment position (i.e., top- or side-adjust), lead configuration, and resistor technology. Figure 64 illustrates several common single-turn potentiometer styles, including top- and side-adjust types, which are most often selected when low cost and small size are required and resolution is not critical. Some of the more popular styles of multiturn potentiometers are illustrated in Fig. 65, including rectilinear and circular-element types, which are usually employed when high resolution is required.

Performance Characteristics

A potentiometer's performance is comprised of many elements, the importance of which should be assessed for each application. The principal performance characteristics can be categorized into four groups which are discussed in the following sections: initial parameters, functional performance, package-variant characteristics, and drift.

Initial parameters There is only one basic initial electrical parameter of significance for a potentiometer, of course, and that is its total resistance. Each resistor technology has its own inherent capabilities with respect to resistance values, tolerances, and TCR characteristics which are summarized in Table 18. Total resistance is a function of the resistivity and length of the resistor element; therefore, the wirewound and metal-film technologies, which are restricted to low resistivities and simple geometry constraints, provide the most limited range of resistance values. The initial tolerance capability of a potentiometer is determined primarily by the nature and economics of the resistor-element fabrication process. Although the carbon and cermet types are generally the most economical to fabricate, their inherent process distribution is greater, which prohibits tight tolerances. The TCR characteristics of each technology are similar to those available with discrete resistors, so that the wirewound and metal-film types can normally be expected to be the most stable.

Functional performance A number of performance characteristics of potentiometers are functional in nature; that is, they relate directly to the adjustment of a potentiometer, and their relative importance varies according to the application. The principal functional-performance characteristics are noise, end resistance, setting stability, rotational life, and resolution. Typical examples of trimmer performance are presented in Table 19 and Figs. 66 and 67 and are discussed below.

Noise. The equivalent noise resistance of a wirewound potentiometer is influenced

Numerical Code Format

$$\underset{\text{Style}}{\text{XXX}}-\underset{\text{Size}}{\text{XX}}-\underset{\text{Char.}}{\text{X}}-\underset{\text{Term.}}{\text{X}}-\underset{\text{Res.}}{\text{XXX}}-\underset{\text{F/R}}{\text{X}}$$

Styles
RT—Wirewound
RJ—Non-wirewound
XXR—Established reliability

Sizes
10—1" rectilinear, multi-turn, 3/4W
11—1 1/4" rectilinear, multi-turn, 1W (0.280" wide)
12—1 1/4" rectilinear, multi-turn, 3/4W (0.190" wide)
22—1/2" square, multi-turn, 3/4W
24—3/8" square, multi-turn, 1/2W
26—1/4" square, multi-turn, 1/4W
50—1/4" round, single-turn, 1/4W

Characteristics

Designation	TCR (+ppm/°C)	Max. temp. (°C) at RP†	Max. temp. (°C) at NL‡
B	600	70	125
C	250	85	150
F	100	85	150
H	50	85	150
J	10	85	150

Terminals*
L—Flexible, insulated wire lead
P—Printed circuit pin (base mount)
W—Printed circuit pin (edge mount—top adjust)
X—Printed circuit pin (edge mount—side adjust)
Y—Printed circuit pin (staggered)

Resistance

Designation	Value, Ω
101	100
201	200
102	1,000
202	2,000
103	10,000
etc.	

Failure rate (F/R)

Designation	Failure rate (%/K hrs.)
M	1.0
P	0.1
R	0.01
S	0.001

* Color code, terminal identification, and schematic diagram conventions are shown in Fig. 7-2.
† Rated power
‡ No load

Fig. 63 Military-type designations for trimmer potentiometers.

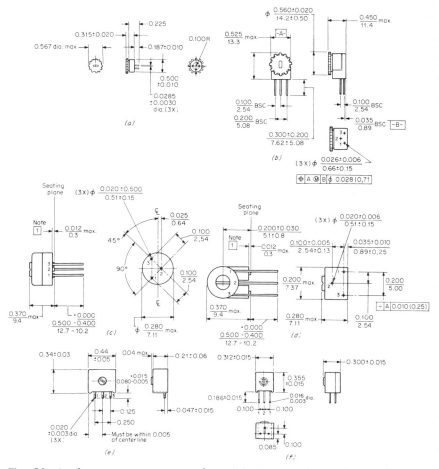

Fig. 64 Single-turn potentiometer styles. (*a*) ½-in round top-adjust. (*b*) ½-in round side-adjust. (*c*) ¼-in round top-adjust. (*d*) ¼-in round side-adjust. (*e*) Rectangular single-in-line. (*f*) Rectangular staggered lead.

TABLE 18 Initial Parameter Capabilities of Trimming Potentiometers

Parameter	Resistor technology			
	Wire	Carbon	Metal	Cermet
Total resistance:				
Typical range	100–10 kΩ	1 kΩ–1 MΩ	100–1 kΩ	100–100 kΩ
Max range	10–100 kΩ	100–5 MΩ	10–10 kΩ	10–1 MΩ
Resistance tolerance, ±%:				
Typical	10	20	10	20
Min	5	10	1	10
TCR, ± ppm/°C:				
Typical range	50–100	500–1,000	20–50	200–500
Min	20	500	10	100

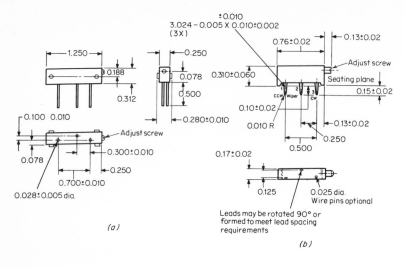

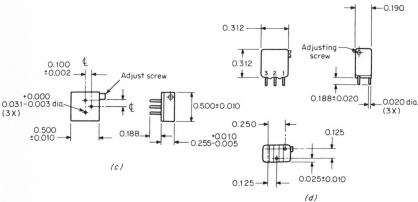

Fig. 65 Multiturn potentiometer styles. (*a*) 1¼-in rectilinear. (*b*) ¾-in rectilinear. (*c*) ½-in square. (*d*) ⁵⁄₁₆-in square.

TABLE 19 **Functional-Performance Characteristics of Trimming Potentiometers**

	Resistor technology			
Characteristic	Wire	Carbon	Metal	Cermet
ENR	10–100 Ω	2% or 10 Ω	2% or 10 Ω	3% or 20 Ω
End resistance	1% or 2 Ω	1% or 5 Ω	2.5% or 5 Ω	1% or 5 Ω
Setting stability, %	0.5–1.5	2	1	1
Rotational life, %	2	5	2	2

by the wire size and wiper design.   Nonwirewound potentiometers exhibit a contact resistance during rotation similar to that illustrated in Fig. 66, which is primarily a function of the surface characteristics of the resistor element and the wiper design.

End Resistance.   The end resistance is primarily a function of the termination technique and materials.   It is generally highest in the film-type potentiometers owing to the relatively high resistance of the termination metallurgy interface.

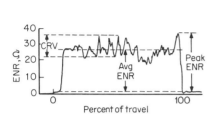

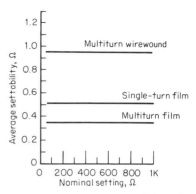

Fig. 66 Typical contact-resistance variation.

Fig. 67 Potentiometer settability data (1 kΩ).

Setting Stability. This is basically a function of potentiometer construction, and is influenced by the torque of the wiper and the geometry of the resistor element. In wirewound potentiometers setting stability generally improves with the number of windings and finer wire size (i.e., the higher resistance values are more stable).

Rotational Life. A common measure of rotational life performance for trimmers is a 100-cycle test. Total resistance and noise typically decrease during early rotation owing to the break-in of the wiper and resistor-element contact. Eventually, however, wearout and a loss of lubrication begin which will tend to reverse this process. Trimmers are generally not subjected to many rotations, and therefore wearout is not expected to occur.

Resolution. The settability of a potentiometer is determined primarily by the design of the adjustment mechanism and the nature of the resistor element. As summarized in Table 20, the theoretical resolution of wirewound potentiometers is a

TABLE 20 Theoretical Resolution of Wirewound Trimmers (Typical for ½-in-square, 20-Turn Style)

Total resistance	Max resolution, %
100	0.51
200	0.42
500	0.40
1,000	0.36
2,000	0.29
5,000	0.26
10,000	0.14
20,000	0.11

function of wire size and the number of windings and therefore improves dramatically for higher resistance values. Nonwirewound potentiometers have an infinite theoretical resolution since their resistor elements are continuous, but actual settability is limited by the mechanics of the adjustment technique and the contact-resistance variation. As illustrated by the example in Fig. 67, operational settability (i.e., actual setting trials) demonstrates that multiturn potentiometers and film-type elements provide a significant improvement in resolution capabilities. It should also be noted that resolution is generally limited by a fixed resistance, which means that higher resistor values should be expected to have a better resolution capability on a percentage basis.

Package-variant characteristics   Certain characteristics of the performance of potentiometers are a function solely of the package style and construction.   When these performance characteristics are critical to an application, therefore, the selection of a suitable potentiometer must include an assessment of the appropriate package-design factors. The major package-variant characteristics are power-handling capability, environmental performance, solvent resistance, and short-term effects each discussed briefly below.

Power-handling Capability.   The power-handling capability of a potentiometer, as represented by its power rating, is based upon an assumed level of performance (with respect to resistor drift) at an assumed maximum resistor operating temperature.   Each resistor technology exhibits its own inherent drift characteristics as a function of operating-temperature rise caused by power dissipation. Figure 68 presents a family of typical derating curves for the principal types of potentiome-

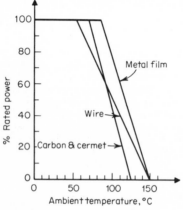

Fig. 68   Typical potentiometer power derating.

ters.   In most cases, trimmers are used at low enough power levels so that the power rating is not a significant concern.   When trimmers are used as rheostats (i.e., as a two-terminal variable resistor), it is recommended that the nominal power rating be reduced by 50 percent, in addition to other deratings, since only a portion of the resistance element will be powered and the full current will be drawn through the wiper.

Environmental Performance.   The principal concern in relation to the environmental performance of a potentiometer is the effects of moisture during shipping, storage, and operation.   This is obviously a function of the package materials and construction as well as the resistor technology selected.   As illustrated in Table 21, however, one may expect in general that under standard-humidity exposure-test conditions (i.e., 40°C/90% RH), wirewound devices will typically exhibit the most stable performance and carbon types the least.

Solvent Resistance.   A potentiometer's performance when exposed to solvents and cleaning processes is dictated solely by the package construction.   In general, the resistor and wiper elements cannot be expected to provide acceptable stability and noise characteristics if they are exposed to water, solvents, or fluxes.   Only sealed-type potentiometers, therefore, such as those employing O-rings as part of the lead-screw or slot assembly, are recommended for applications which require exposure to washing or cleaning processes.   Unsealed and open-construction units should therefore normally be considered nonwettable and should be assembled to printed-circuit boards manually.

Short-Term Effects.   As with any other component, potentiometers are subjected to a great deal of handling prior to actually performing in the circuit application. This includes shipping, storage, and assembly, which expose the potentiometer to environmental, mechanical, and thermal stresses.   A number of common tests are

TABLE 21   Standard Test Performance for Trimmers
(Maximum % Change in Total Resistance)

Test	Resistor technology			
	Wire	Carbon	Metal	Cermet
Humidity exposure	1	10	2	2
Temperature cycling	0.5–1.5	2	2	2
Soldering	1	2	1	1
Vibration	1	2	1	1
Load life	2	10	1	3

employed to evaluate these effects, such as those included in Table 21 (e.g., temperature cycling, resistance to soldering, vibration). Wirewound and film-type units generally exhibit similar performance characteristics under these conditions, which as would be expected, are typically superior to the carbon types.

Drift   Once a potentiometer is performing in its application, the only performance characteristic of major concern aside from the stability of the setting is the drift or aging of the resistor element. A small change in total resistance is generally not critical, but there should be an awareness of the expected drift characteristics, particularly under load. Potentiometer elements perform as discrete resistors which would utilize the same technology and are usually subjected to standard load-life tests to evaluate their long-term drift characteristics. As presented in Table 21, the metal-film devices are typically expected to be the most stable, followed closely by the wirewound and cermet types.

ACKNOWLEDGMENTS

The authors of this chapter express sincere appreciation for the technical support given by the many resistor manufacturing companies and technical agencies.

REFERENCES

1. Blackburn, John F. (ed.): "Components Handbook," McGraw-Hill, New York, 1949.
2. Dietz, R. E.: Preferred Numbers—A Standardization Tool, *IEEE Trans. Aerosp., Support Conf. Proc.,* August 1963.
3. Fink, Donald G., and John M. Carroll: "Standard Handbook for Electrical Engineers," McGraw-Hill, New York, 1969.
4. Henney, Keith, and Craig Walsh: "Electronic Components Handbook," McGraw-Hill, New York, 1957.
5. Littlejohn, H. F., Jr.: "Handbook of Power Resistors," Ward-Leonard Electric Co., Mt. Vernon, N.Y., 1959.
6. Neal, J. P.: "Electrical Engineering Fundamentals," McGraw-Hill, New York, 1960.
7. Parkinson, David H., and Brian E. Mulhall: "Generation of High Magnetic Fields," Plesham Press, New York, 1967.
8. Wellard, Charles L.: "Resistance and Resistors," McGraw-Hill, New York, 1950.
9. "Alloy Digest" (file Information), Engineering Alloys Digest, Inc., Upper Montclair, N.J.
10. "Industrial Controls and Systems," NEMA ICS-1970, 1970 National Electrical Manufacturers Association, New York.
11. "Insulation/Circuits, Directory/Encyclopedia" (annual), 1973, Lake Publishing Corporation, Libertyville, Ill.
12. "Metals Handbook," vol. I, 8th ed., American Society for Metals, 1961.
13. "Pulse Handling Capabilities of Wirewound Resistors," Dale Electronics Inc., Columbus, Neb.
14. Selection and Use of Resistors, Military Standard MIL-STD-199A, Notice 1 through 9, U.S. Department of Defense, May 1973.
15. Standard Specification for Wire for Use in Wirewound Resistors, Publication B-267, American Society for Testing and Materials.

Catalog and application data from the following companies:
Airco Speer Electronics, St. Marys, Pa.
Allen Bradley Co., Milwaukee, Wis.
American Lava Corp., Subs. 3M Co., Chattanooga, Tenn.
Caddock Electronics, Riverside, Calif.
Carborundum Co., Electronics Div., Niagara Falls, N.Y.
Coors Porcelain Company, Golden, Colo.
Corning Glass Works, Corning, N.Y.
CTS Corp., Berne, Ind.
Dale Electronics, Columbus, Neb., Subs. Lionel Corp.
Wilbur B. Driver Co., Newark, N.J.
Driver-Harris Co., Harrison, N.J.
Hoskins Mfg. Co., Detroit, Mich.
IRC—Boone Div., TRW, Inc., Boone, N.C.
IRC—Burlington Div., TRW, Inc., Burlington, Iowa
IRC—St. Petersburg Div., TRW, Inc., St. Petersburg, Fla.
C.O. Jelliff Mfg. Corp., Southport, Conn.
Kanthal Corp., Bethel, Conn.
KDI Pyrofilm Corporation, Whippany, N.J.
Kelvin Co., Van Nuys, Calif.
P.R. Mallory Co., Indianapolis, Ind.
Mepco-Electra, Inc., Morristown, N.J.
Molecu-Wire Corp., Scobyville, N.J.
Nytronics, Incorporated, Sage Electronics Div., Darlington, S.C.
Ohmite Mfg. Co., Nashua, N.H. (formerly Sprague Electric Co. Resistor Div.)
Ohmite Mfg. Co., Skokie, Ill.
Resistance Products Co., Harrisburg, Pa.
Shallcross Manufacturing Co., Subsidiary of Cutler-Hammer, Inc., Selma, N.C.
Stackpole Carbon Co., Kane, Pa.
Vamistor Corp., Cedar Knolls, N.J.
Ward-Leonard Electric Co., Subs. Riker Maxson, Mt. Vernon, N.Y.
Ward-Leonard Hagerstown Div., Angstrom Precision, Hagerstown, Md.

Chapter **8**

Capacitors

EDWARD L. HIERHOLZER
and
H. BENNETT DREXLER
Martin Marietta Aerospace, Orlando, Florida
with
HAROLD T. CATES, Ph.D.,
JOSEPH F. RHODES,
and
STEPHEN D. DAS
Martin Marietta Aerospace, Orlando, Florida

FIXED CAPACITORS

Definitions and General Characteristics

Capacitance It has been known since the time of early experimental discovery of electricity that weak electric currents can be stored for subsequent discharge in the form of intensive current pulses.   Experiments at the University of Leyden, about 1747, employed the use of electrical energy stored in a "Leyden jar" or "condenser" or "capacitor."   The later term "capacitor" is now officially used extensively to describe a device included in an electric circuit for the purpose of storing electrical charge.

Electrically, capacitance is present between any two adjacent conductors.   A capacitor consists basically of two parallel metal plates separated by a dielectric material.   When a voltage is applied across the plates, the capacitor will become charged, and the amount of charge will depend upon the source voltage, time of charge, and total energy available.   The capacitance of a capacitor is a factor of proportionality defined as the ratio of the charge acquired to the voltage applied, or

$$C = \frac{Q}{V}$$

where Q is the charge in coulombs (or ampere-seconds), V is the voltage in volts, and C is capacitance in farads.   A capacitor is said to possess 1 farad of capacitance if its potential is raised 1 volt when it receives a charge of 1 coulomb.   This unit of capacitance is inconveniently large and is submultiplied into "microfarad" (μf, one-millionth of a farad) and "micromicrofarad" ($\mu\mu$f) or picofarad (pf, one-million-millionth of a farad) for practical usage.

The physical unit of capacitance is defined as the capacitance of a sphere in free space with respect to a single point of reference.   Numerous sources such as Hayt[1] treat the expansion of Coulomb's law and Maxwell's field equations to provide the classical derivation of the formula for capacitance of varieties of geometries.   The fundamental formula for the capacitance of two parallel plates is

$$C = \frac{\epsilon A}{d} \qquad \text{farads (MKSA)}$$

where C is capacitance in farads, ϵ is the permittivity of the dielectric, A is the area of one plate in square meters, and d is the distance between the plates in meters.   Where the dielectric is isotropic, linear, homogeneous space, the permittivity is given as

$$\epsilon_0 = \frac{10^7}{4\pi c^2} \qquad \text{coulomb}^2/\text{newton-meter}^2$$

where c is the speed of light (2.9979×10^8 m/s).   The above formula for capacitance now becomes

$$C \doteq \frac{A}{36\pi d} \times 10^9 \qquad \text{farads (MKSA)}$$

or more accurately

$$C = \frac{8.854 A}{d} \qquad \text{picofarads (MKSA)}$$

If the dielectric is other than free space (with permittivity ϵ_0) a relative dielectric constant k is defined as being the ratio of the permittivity of the particular material ϵ to that of space, thus

$$k = \frac{\epsilon}{\epsilon_0}$$

and the fundamental formula for the capacitance of two parallel plates separated by an isotropic, linear, homogeneous material becomes

$$C = \frac{8.854 k A}{d} \qquad \text{picofarads (MKSA)}$$

or if A and d are measured in inches, the formula becomes

$$C = \frac{0.2249kA}{d} \quad \text{picofarads}$$

This formula is acceptable where the area A is very large with respect to distance d. For parallel plates where the area is small and the distance between them is approximately $10^{-3}A$ or greater, the effect of electric-field nonuniformity becomes more predominant. As shown in Fig. 1, the effective electric field extends beyond

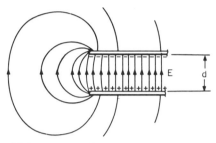

Fig. 1 The electric field near the edge of two parallel conducting plates showing fringe effects.

the edge of the plates and is known as fringe effect. The capacitance would be slightly higher, and the following corrections[2] should be added to the plate dimensions:

1. For straight edges, add $0.44d$ to the sides.
2. For circular edges, add $.011d$ to the radius.

Energy. The energy of the charge is stored as electrostatic energy in the dielectric and is equal to $\frac{1}{2}CV^2$. If this energy is absorbed at a uniform rate over a time t, the power required is

$$P_{av} = \frac{1}{2}\frac{CV^2}{t}$$

where P_{av} is the average power in watts, C is capacitance in farads, V is the voltage across the plates in volts, and t is charging time in seconds.

Under ac conditions the rate at which energy is supplied to and returned by the capacitor is

$$P_{av} = 2\pi CV^2 f$$

where f is the frequency in hertz and V is the rms voltage.

When a direct voltage is applied to a capacitor, the electric field within the dielectric is displaced and the bound electric charges are polarized or displaced from their normal position of equilibrium. Work is therefore done in charging the capacitor. The work done, in joules or watt-seconds, which is available as stored potential energy, is

$$J = \frac{1}{2}QV \text{ or } \frac{1}{2}CV^2 \text{ or } \frac{Q^2}{2C}$$

where J is work in joules or watt-seconds, C is capacitance in farads, V is voltage in volts, and Q is charge in coulombs or ampere-seconds.

Equivalent Circuit. A useful expression in the understanding of capacitance is that appearing in the differential equation defining the response of an electric circuit containing a constant voltage V_0, resistance R, inductance L, and capacitance C in series:

$$R\frac{dq}{dt} + \frac{Q}{C} + L\frac{d^2q}{dt} = V_0$$

or if the electric current is defined as $i = dq/dt$, then

$$iR + \frac{1}{C} \int i\, dt + L \frac{di}{dt} = V_0$$

so that C is seen to be analogous to the reciprocal of the spring constant in a dynamic mechanical system. Since practical capacitors embody inductance and resistance, as well as capacitance, these expressions are useful in evaluating capacitor-circuit behavior. The familiar expression for capacitive reactance to a sine-wave alternating current can be derived by inserting the sine function for the current and per-

TABLE 1   Common Expressions Used in the Application and Definition of Capacitors

Capacitance	$C = \dfrac{0.224\ kA\ 10^{-6}}{d}$	(1)
Total capacitance, capacitors in parallel	$C_T = C_1 + C_2 + C_3 + \cdots + C_n$	(2)
Total capacitance, capacitors in series	$\dfrac{1}{C_T} = \dfrac{1}{C_1} + \dfrac{1}{C_2} + \dfrac{1}{C_3} + \cdots + \dfrac{1}{C_n}$	(3)
Capacitive reactance	$X_C = \dfrac{10^6}{2\pi f C}$	(4)
Resonant frequency	$f_r = \dfrac{10^3}{2\pi \sqrt{LC}}$	(5)
Dissipation factor	$\mathrm{DF} = \dfrac{R}{X_C} = 2\pi f RC \times 10^{-6}$	(6)
Figure of merit	$Q_M = \dfrac{X_C}{R} = \dfrac{1}{\mathrm{DF}}$	(7)
Equivalent series resistance	$R = \dfrac{\mathrm{PF}}{2\pi f C} \times 10^6 = \dfrac{P_L}{I^2}$	(8)
Impedance	$Z = \sqrt{R^2 + (X_L - X_C)^2}$	(9)
Power factor	$\mathrm{PF} = \dfrac{R}{Z}$	(10)
Capacitance of bypass capacitor (nominal) f at lowest frequency to be bypassed	$C = \dfrac{10^7}{2\pi f R_B}$	(11)
Charge of a capacitor	$Q = CV \times 10^{-6}$ or $(I_{PK})\, t$ (short duration)	(12)
Stored energy of a dc capacitor	J or W-s $= \dfrac{CV^2 \times 10^{-6}}{2}$	(13)
Capacitor heating, total watts dissipated within the capacitor	$W_T = W_{\mathrm{dc}} + W_{ac}$ where $W_{\mathrm{dc}} = VI$ $W_{ac} = I_{ac}{}^2 R$	(14)
Voltage during charge	$V_c = V_0 (1 - e^{-t/RC10^6})$	(15)
Voltage during discharge	$V_c = V_0 e^{-t/RC10^6}$	(16)
Temperature coefficient of capacitance	$TC = \dfrac{C_1 - C_2}{(T_1 - T_2)C_1} \times 10^6$	(17)

TABLE 1   Common Expressions Used in the Application and Definition of Capacitors (Continued)

where A = area, in²
C = capacitance, μF
d = spacing between flat plates, in
DF = dissipation factor dimensionless (multiply by 100 for percent equivalent)
e = base natural logarithm
f = frequency, Hz
I = dc leakage current, A
I_{ac} = rms ripple current, A
I_{PK} = peak dc current, A
k = dielectric constant, dimensionless
L = inductance, H
PF = power factor, dimensionless (multiply by 100 for percent equivalent)

P_L = power lost, W
Q = capacitor charge, C or A-s
Q_M = figure of merit, dimensionless
R = equivalent series resistance of capacitor, Ω
R_B = resistance bypassed or shunted by capacitor, Ω
t = time, s
T = temperature, °C
TC = temperature coefficient, ppm/°C
V = dc volts across capacitor
V_c = voltage on capacitor during charge or discharge
V_0 = charging or charged potential
W = watts
X_C = capacitive reactance, Ω
X_L = inductive reactance, Ω
Z = impedance, Ω

forming the integration over a quarter cycle to obtain the reactance X_c in ohms:

$$X_c = \tfrac{1}{2}\pi f C$$

where f is the frequency in hertz of the sine function.

The general form of a capacitor equivalent circuit is shown in Fig. 2. The values of L, R_S, and R_P vary widely between capacitor types and will be defined for particular types as they are discussed. All these factors are important in the application of capacitors. The inductance and resistive losses control the utility of the capacitor at high frequency, and parallel resistance R_P normally expressed as insulation resistance affects performance in timing and coupling circuits. Series resistance R_S is a measure of the heat-dissipation loss in the capacitor and is expressed as power factor or dissipation factor.

Equations used to define capacitor characteristics discussed thus far along with other common expressions used in the application and definition of capacitors are listed for convenience in Table 1.

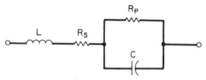

Fig. 2 Typical equivalent circuit of a capacitor.

C = capacitance
R_S = series resistance (leads, plates, electrical interfaces)
L = inductance (leads, plates, etc.)
R_P = parallel resistance (leakage current, dielectric absorption, insulation resistance)

Capacitor types Fixed capacitor types discussed in this handbook cover most common dielectric materials, case styles, and standard capacitance and voltage ranges. An overview of capacitor types by dielectric material is shown in Table 2 along with governing specifications. An understanding of the interrelationship of capacitor design and construction with their nominal characteristics is generally required for proper circuit application.

Dielectric Types. The characteristic which all dielectric materials have in common, whether they are solid, liquid, or gas, and whether or not they are crystalline in nature, is their ability to store electrical energy. This storage takes place by a shift in the relative positions of the internal positive and negative charges against the normal molecular and atomic forces. The ability of a dielectric material to retain stored charge in this manner was previously defined as the permittivity of the material, and the ratio with respect to free space is the dielectric constant k. Constants for common capacitor dielectrics are given in Table 3. Dielectric materials are chosen for a specific capacitor design because of their frequency, voltage, thermal,

TABLE 2   Cross Reference of Industry and Military Standard Capacitors

Capacitor type	EIA type	ANSI type	IEC type	Standard military specification	Established-reliability military specification
Tantalum:					
Type 1, foil	RS-228	C83.33			MIL-C-39006
Type 2, wet electrolyte porous anode	RS-228	C83.33	361	MIL-C-3965*	MIL-C-39006
Type 3, solid electrolyte porous anode	RS-228	C83.33	361	MIL-C-26655* MIL-C-55365	MIL-C-39003
Aluminum:					
Type 1, long life	RS-395	C83.22	103		
Type 2, general-purpose	RS-395	C83.22	103		
Extended life				MIL-C-62	
High performance				MIL-C-39018	
Ceramic:					
General purpose class 2, 3	RS-198	C83.4	108, 187, 324	MIL-C-11015*	MIL-C-39014 MIL-C-55681
Temperature-compensating class 1	RS-198	C83.4		MIL-C-20	MIL-C-20 MIL-C-55681
High-voltage, class 2, 10 to 30 kV	RS-171				
High-voltage, class 1 and 2, 1 to 7.5 kV	RS-165				
Variable				MIL-C-81	
Mica:					
General-purpose	RS-153	C83.85	116	MIL-C-5	MIL-C-39001
Potted	TR-109				
Button				MIL-C-10950	
Glass:					
General-purpose				MIL-C-11272*	MIL-C-23269
Variable, piston				MIL-C-14409	
Variable, ganged				MIL-C-28718	
Paper, paper plastic:					
AC, paper, RFI reduction				MIL-C-12889	
AC, paper	RS-392	C83.67	252†		
DC, general-purpose hermetic	RS-218A	C83.11	80	MIL-C-25* MIL-C-19978	MIL-C-14157* MIL-C-19978
Power semiconductor applications	RS-401	C83.97			
Feed through, RFI reduction	RS-361	C83.54	161	MIL-C-11693	MIL-C-39011* MIL-C-11693 MIL-C-83439 (USAF)
Metallized paper	RS-377‡	C83.62	166	MIL-C-18312*	MIL-C-39022
Paper-film nonhermetic H.V.	RS-164	C83.29	80		
AC power			70, 110, 143		
Plastic film:					
Polystyrene	RS-376‡	C83.61	275	MIL-C-19978	MIL-C-14157*
Polyethylene terephthalate (Mylar)¶	RS-376‡	C83.61	202	MIL-C-19978	MIL-C-19978
General-purpose, nonhermetic	RS-376‡	C83.61		MIL-C-27287*	MIL-C-55514
Polyester	RS-376‡	C83.61	202		
Polycarbonate	RS-376‡	C83.61			
Polytetrafluoroethylene (Teflon)¶	RS-376‡	C83.61			
Metallized film	RS-377‡	C83.62		MIL-C-18312*	MIL-C-39022
Metallized polycarbonate					MIL-C-83421 (USAF)§
Composition	RS-335	C83.89			
Air, variable			334	MIL-C-92	
Vacuum, fixed and variable				MIL-C-23183	

* These specifications are obsolete and not approved for new design; they are shown for reference only.
† This specification covers ac paper (run) capacitors as well as ac electrolytic (start) capacitors.
‡ Parts list supplements are covered in RS-376-1 and RS-377-1.
§ A high-performance type where no voltage breakdown or self-healing is permitted.
¶ Mylar and Teflon are registered trademarks of E. I. DuPont de Nemours & Co., Wilmington, Del.

TABLE 3   Relative Dielectric Constants of Common Capacitor Dielectric Materials

Material	Dielectric constant k	Organic	Inorganic
Vacuum	1 (by definition)		X
Air	1.0006		X
Ruby mica	6.5–8.7		X
Glass (flint)	10		X
Barium titanate (class I)	5–450		X
Barium titanate (class II)	200–12,000		X
Kraft paper	≈2.6	X	
Mineral oil	≈2.23	X	
Caster oil	≈4.7	X	
Halowax	≈5.2	X	
Chlorinated diphenyl	≈5.3	X	
Polyisobutylene	≈2.2	X	
Polytetrafluoroethylene	≈2.1	X	
Polyethylene terephthalate	≈3	X	
Polystyrene	≈2.6	X	
Polycarbonate	≈3.1	X	
Aluminum oxide	≈8.4		X
Tantalum pentoxide	≈28		X
Niobium oxide	≈40		X
Titanium dioxide	≈80		X

mechanical, and other desired characteristics. The majority of the dielectric constants are shown as approximate values, since they will vary with temperature, frequency, voltage, purity, chemical composition, and exact processing in manufacture. In addition to increasing the size and number of parallel plates in a capacitor, it can be seen from Eqs. (1) and (2) in Table 1 that the capacitance can be increased dramatically by the use of high-k dielectric materials. Also, aluminum oxide and tantalum oxide electrolytic capacitors have very thin oxide films developed between foil and electrolyte, thus providing extremely high capacitance values with a given area of capacitor because of the very small separation d between plates.

Capacitor characteristics The characteristics usually specified for a capacitor include capacitance, dissipation factor (DF), equivalent series resistance (ESR), insulation resistance (IR) or leakage current, and dielectric strength. The electrical and environmental parameters of most interest with respect to these basic measurements are temperature, voltage, and frequency. The reference temperature for most capacitors is 25°C, with a typical temperature range of −55 to 125°C. Voltage is dependent upon the rating applied by the manufacturer, and application frequency typically depends primarily upon the capacitor dielectric type. Recommended test conditions will be given later with data for each capacitor type.

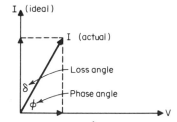

Fig. 3 Current and voltage relationship in a capacitor showing definition of loss and phase angles.

Losses in Capacitors. Dissipation factor (DF) and power factor (PF) are measures of loss in a capacitor, and either can be used depending upon the type of capacitor. Losses in large ac oil and paper capacitors are of interest to the user and should be specified as dissipation factor; however, losses in the majority of capacitors used in dc or low-level ac applications can be specified as power factor. Most manufacturers consider dissipation factor as a measure of process control and will specify loss in this manner for most cases. In the ideal capacitor, the alternating current will lead the voltage by 90°, as shown in Fig. 3.   In practice, the current leads the voltage by some lesser phase angle ϕ owing to the series resistance R, the complement of this angle is called the loss angle δ, and

Power factor $= \cos \phi$ or sine δ
Dissipation factor $= \tan \delta$

For small values of δ the tangent and sine are essentially equal, which has led to some common interchangeability of the two terms in the industry.   Dissipation factor is affected by frequency, capacitance, and resistance as shown by Eq. (6) in Table 1.   Resistance and capacitance are affected by temperature, which in turn affects the dielectric constant.   Data on dissipation factor are therefore given as a function of frequency, temperature, and dielectric material.   Typical DF variations with temperature characteristics of common dielectrics are shown in Figs. 4 to 6.   Loss variation with frequency will be shown in the treatment of each capacitor type.

The variation of dielectric permittivity with frequency is negligible so long as

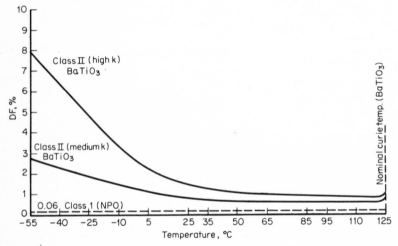

Fig. 4 Typical percent dissipation factor vs. temperature characteristic for ceramic dielectric capacitors.

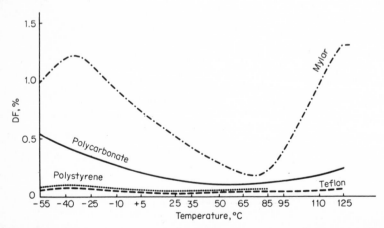

Fig. 5 Typical percent dissipation factor vs. temperature characteristic for dry section film dielectric capacitors. Mylar and Teflon are registered trademarks of E. I. Du Pont de Nemours & Co., Wilmington, Del. (Electron Products, Monrovia CA.)[3]

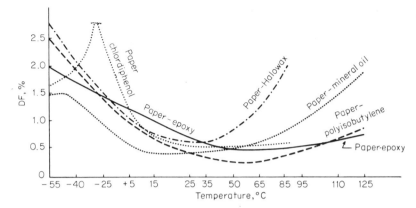

Fig. 6 Typical percent dissipation factor vs. temperature characteristic for impregnated-paper dielectric capacitors.   (Electron Products, Monrovia, CA.)[3]

the losses are low.   Increased losses occur when the process of alignment cannot be completed, owing to molecular collisions, and in these regions there is a loss in permittivity.   Molecular interference within the material limits the frequency at which full alignment can be carried out.   If the applied frequency is comparable with the limiting frequency, losses will become high.

Equivalent Series Resistance (ESR).   The ac resistance R of a capacitor reflecting both R_s and R_P shown in Fig. 2 expresses loss at a given frequency and is defined as ESR.   The ESR of electrolytics is normally specified in detail so that accurate loss and heating calculations can be made for a specific application.   It is common practice to equate ESR and PF by Eq. (8) in Table 1.   The values of PF can be calculated from manufacturer's data for ESR and capacitance at operating temperatures, and ripple frequency applied.

Insulation Resistance (IR).   All dielectric materials used in manufacture of capacitors will allow certain amounts of dc flow.   This current, known as leakage current, resulting from an applied voltage can be expressed as resistance in capacitors at a specified measurement voltage and temperature.   IR, therefore, is the resistance measured across the terminals of a capacitor and consists principally of the parallel resistance R_P shown in Fig. 2.   For small paper, film, mica, and ceramic capacitors the IR expressed in megohms and is usually in the order of 100,000 MΩ,   As capacitance values increase and hence the area of dielectric increases or thickness of dielectric decreases, the IR decreases proportionately, and hence the IR is often specified in ohm farads or more commonly megohm microfarads.   IR is inversely proportional to temperature and is usually specified over the capacitor design temperature range.

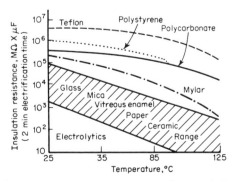

Fig. 7 Typical insulation resistance vs. temperature characteristic of common electrostatic capacitors.   Mylar and Teflon are registered trademarks of E. I. Du Pont de Nemours & Co., Wilmington, Del.   (Electron Products, Monrovia, CA.)[3]

Relative insulation resistances of various dielectrics vs. temperature are shown in Fig. 7. In the case of electrolytic capacitors where the R_P is low owing to the very high capacitance and thin dielectric, dc leakage current is normally specified rather than resistance. Typical leakage currents are shown in Fig. 8.

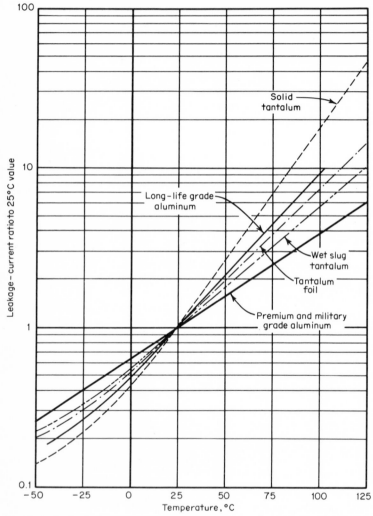

Fig. 8 Typical dc leakage current vs. temperature characteristic for electrolytic capacitors.

Dielectric Strength. The ability of a capacitor dielectric to withstand an applied dc voltage without breakdown is called dielectric strength and is normally specified for a given dielectric material as volts per mil at a specified temperature. Most electrostatic capacitor specifications list a dielectric withstanding voltage (DWV) which considers temperature effects and a design safety margin. The manufacturer's design safety margins are sufficiently above the DWV rating so as to assure good yield in manufacturing and reliable operation for the user. Dielectric voltage overstress is one of the key tests employed to screen out early capacitor failures and obtain improved reliability.

Dielectric Absorption. Charging current from a low-impedance undirectional current source continues to flow at a gradually decreasing rate into a capacitor of negligible series resistance for some time after the steady-state voltage has been reached. A steady value proportional to the capacitor parallel resistance is finally reached. This additional charge which is absorbed by the dielectric is called dielectric absorption. Conversely, a capacitor does not discharge instantaneously upon application of a short circuit but continues to drain gradually after the capacitor has been discharged. It is common practice to measure the dielectric absorption by determining the "reappearing voltage" which appears across a capacitor at some point in time after it has been fully discharged under short-circuit conditions. This reluctance of the dielectric to give up charge carriers is primarily due to the polarization effect that takes place in the dielectric whenever the capacitor is charged. These carriers become trapped in the dielectric during discharge, and when the shorting mechanism is removed these trapped carriers can freely move to the electrode. This results in the observed "recovery voltage" or "bounce back." Since dielectric absorption is a phenomenon of electron motion, voltage, time, and temperature are controlling factors. Dielectric absorption is an important consideration in timing and high-speed switching circuits, and in a nuclear-radiation environment. This characteristic is expressed as the ratio of the recovery voltage to the charging voltage and is given in percent.

Corona. The ionization of air or other gases and vapors which causes them to conduct current is called corona. It is especially prevalent in high-voltage capacitors but can occur with ac voltages as low as 250 V in air at reduced pressure as well as where high-voltage gradients occur. Voids in solids, sharp electrode points or edges, nonhomogeneous insulation, bubbles, etc., can promote the formation of corona. In air, corona produces ozone (O_3) and the several oxides of nitrogen. Chemically, ozone, as a powerful oxidizing agent, attacks most organic materials. The oxides of nitrogen in the presence of moisture lead to the formation of nitric and nitrous acids which will corrode metals and embrittle organic materials. Physically, corona discharges produce high-velocity electrons and ions which can erode organic materials, and even glass and ceramics. In the case of oils, corona leads to formation of wax which can cause voids and ultimate failure. In chlorinated aromatic dielectrics, corona causes erosion and eventually the formation of semicarbonized channels, with voltage breakdown and subsequent failure. Corona or its effects can be avoided by proper overall capacitor design and by the elimination of causes in both the capacitor construction and the installation. Detail data were obtained by Perkins and Brodhun in air by applying 0.25-in-diameter electrodes to various materials as shown in Fig. 9.[4] The voltage at which corona is initiated is called the corona

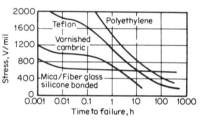

Fig. 9 Effect of corona discharges on dielectric strength of typical capacitor dielectric materials. Teflon is a registered trademark of E. I. Du Pont de Nemours & Co., Wilmington, Del. (Reprinted from "Dielectric Materials and Applications" by Arthur R. von Hippel by permission of the M.I.T. Press, Cambridge, Mass. Copyright, 1954 by the Massachusetts Institute of Technology.)[4]

start voltage, and the voltage at which corona extinguishes itself is the corona stop voltage. In high-reliability applications a capacitor design is such that the peak applied voltage including transients does not exceed the corona start voltage and the corona stop voltage is specified well above the working voltage of the capacitor.

Capacitor stresses The capacitor characteristics previously defined are all affected by the three primary capacitor stresses, voltage, temperature, and frequency.

Voltage. In addition to the voltage rating of a capacitor, there are certain types of ceramic and electrolytic capacitors in which applied voltage is of primary concern. The voltage sensitivity of these capacitors in ac and dc applications originates with the effect of the electric field on the minute polarized domains within the dielectric itself and will be treated in more detail later. Electrolytic capacitors, although exhibiting no dramatic parametric change, are sensitive to the effects of voltage because they are highly polarized devices. The correct polarity must be observed if failure is to be avoided. If operation with alternating current is desired, two units must be put back to back or built specially to make an ac or nonpolar type. The sensitivity to voltage reversal of most electrolytic capacitors even goes to the point of limiting ripple current in the case of many units such as wet-slug tantalum capacitors. Even in those cases where voltage application does not exceed the specified peak voltage or maximum voltage reversal of a capacitor, the voltage drop across the ESR of the capacitor will shorten the life expectancy of the capacitor through the acceleration effect of internal heating.

Temperature. As the ambient temperature changes, the dielectric constant and hence the capacitance of most capacitors change. In general, materials with lower dielectric constant tend to change capacitance less with temperature or with relatively predictable changes that are linear with temperature. Semistable or high-dielectric-constant cermics, general-purpose paper, and many films such as Mylar tend to have

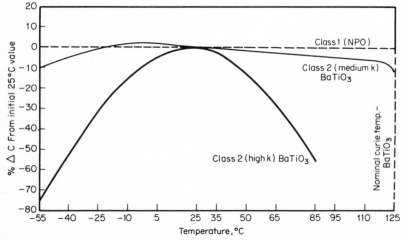

Fig. 10 Typical capacitance change vs. temperature characteristic for ceramic dielectric capacitors.

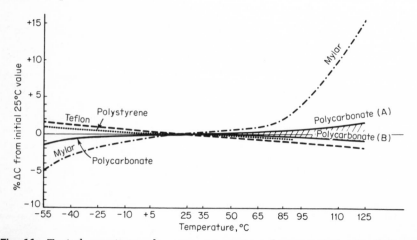

Fig. 11 Typical capacitance change vs. temperature characteristic for film dielectric capacitors. Mylar and Teflon are registered trademark of E. I. Du Pont de Nemours & Co., Wilmington, Del. (Electron Products, Monrovia, CA.)[3]

capacitance changes that are nonlinear and expressed as percent capacitance change over a temperature range.   These materials are available in general in capacitance changes ranging from the vicinity of ±3 percent over the temperature range of −55 to +85°C to changes as great as +22.82 percent over the same temperature range. Typical temperature/capacitance characteristics of ceramic, film, and impregnated-paper dielectric capacitors are shown in Figs. 10, 11, and 12, respectively.   Electrolytic capacitors change greatly in a nonlinear manner with a range of a few degrees.   Certain electrolytic capacitors are very sensitive to temperatures as low as −55°C.   Special application precautions and construction materials are required for these capacitors.   Increasing temperature usually reduces insulation resistance (as shown in Fig. 7), increases leakage current and power factor/dissipation factor, and reduces the voltage rating of the part.

Conversely, reducing temperature normally improves most characteristics; however, at cold-temperature extremes some impregnants and electrolytes may lose their effectiveness.   Study of detail specifications is recommended before applying capacitors at −55°C.   Typical temperature characteristics of electrolytic capacitors are shown in Fig. 13.   Mica, glass, and vitreous-enamel dielectrics provide predictable tempera-

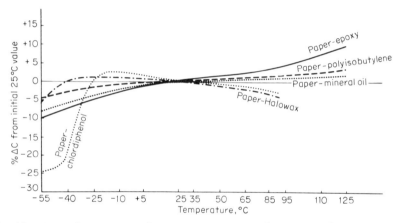

Fig. 12   Typical capacitance change vs. temperature characteristic for impregnated-paper dielectric capacitors.   (Electron Products, Monrovia, CA.)[3]

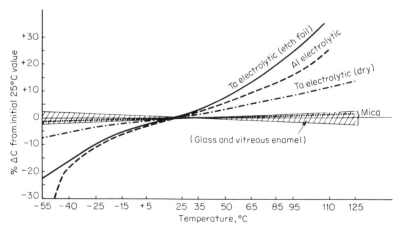

Fig. 13   Typical capacitance change vs. temperature characteristic for electrolytic and stable electrostatic capacitors.   (Electron Products, Monrovia, CA.)[3]

ture characteristics in that they are relatively linear over the entire usable temperature range.   The penalty for this feature, however, is large size.

Frequency.   This is a capacitor stress most often overlooked by the circuit designer.   As observed in Fig. 2, there are both inductance and capacitance in each capacitor.   Obviously, there is a resonant frequency, and depending upon the capacitor type, it may or may not fall in a range troublesome to a designer.   Effective series inductance is primarily caused by the lead configurations and the electrode geometry.   As the applied frequency increases, the capacitor eventually passes through self-resonance and becomes inductive with gradually increasing impedance. Even though a capacitor is beyond the self-resonant point, it blocks direct current and may have a low impedance, as shown in the frequency-vs.-impedance characteristic curves in Fig. 14.   In high-frequency applications, NPO ceramic, extended-foil

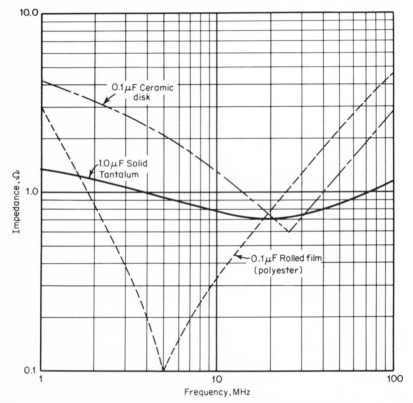

Fig. 14 Typical impedance vs. frequency characteristic for low-value capacitors showing relative effect of Q sharpness at resonant point.

film, and mica and glass capacitors are usually used.   Capacitors with essentially zero lead length called feedthrough or coaxial capacitors are generally superior in bypass applications to the axial- or radial-lead types.   However, they must be mounted on a bulkhead, as opposed to printed wiring board or point-to-point wiring, and are normally not used for energy-storage applications.   Differences in curve shapes are observed with the ceramic and film capacitors vs. tantalum capacitors.   The ceramic and film types, having lower equivalent series resistances (higher Q), exhibit a much sharper resonance.   The tantalum types, with their lower Q's, have flatter curves but also exhibit lower impedances over a very wide frequency range as a result of their relatively high capacitance value.

Fixed-Capacitor Application

Application categories   Capacitor dielectric types and mechanical designs exist over a wide range to satisfy a variety of application requirements for both electrical performance and physical size. Selection of a specific capacitor type should follow the standardization guidelines presented in Chap. 7.   Final capacitor selection will depend upon supplemental considerations of capacitance, value, tolerance, voltage rating, frequency characteristics, dissipation factor, insulation resistance, temperature, stability, and other basic characteristics related to the circuit application. As discussed earlier, the basic function of a capacitor is to store electrical energy. However, this function coupled with other circuit elements and a knowledge of the capacitor characteristics previously discussed will make it possible to use capacitors in numerous ways.   Principal characteristics of common capacitor types are shown in Table 4, which lists typical application categories along with characteristics and disadvantages for major dielectric families.   It should be noted that these are recommended applications and are by no means inclusive of the complete versatility of many capacitor types.   There are special applications where numerous capacitors can be used depending on cost, environmental requirements, and other circuit characteristics desired.

Capacitance and voltage ratings   Ranges of capacitance and voltage of commonly available capacitors for various dielectric families are shown in Figs. 15 and

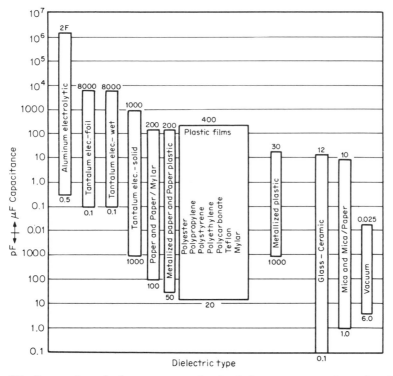

Fig. 15 Range of standard capacitance value by dielectric type.   Mylar and Teflon are registered trademarks of E. I. Du Pont de Nemours & Co., Wilmington, Del.

16, respectively.   Ranges shown are for standard configurations governed by industry and military standards.   They do not include special-purpose capacitors for uncommon applications for which specially designed capacitors are required. Initial tradeoffs can be made at this point to select the dielectrics most likely to satisfy the

TABLE 4 Capacitor Features by Dielectric Type

Type	Application	Advantages	Disadvantages
Paper	Blocking, buffering, bypass, coupling, and filtering at low frequencies Power-factor correction Contact protection Timing, photoflash, motor start and run	Readily available in a wide range of capacitance and voltage values Low cost per μF Reliable Medium stability Extensive test data	Medium capacitance-to-volume ratio High effective resistance at vhf
Film	Blocking, buffering, bypass, coupling, filtering to medium frequency Tuning and timing	Wide range of capacitance and voltage values High IR, low DF, good Q Stable Low TC High voltage	Medium cost per μF
Mica (dipped)	Filtering, coupling, and bypassing at high frequencies Resonant circuits, tuning High-voltage circuits Padding of larger capacitors	Low dielectric losses and good temperature, frequency, and aging characteristics Low ac loss (high Q), high frequency High IR Low cost Extensive test data, reliable	Low capacitance-to-volume ratio
Glass	Resonant circuits, tuning High-frequency bypass High-frequency coupling	High IR High operating temperatures Excellent temperature stability Low ac loss (high Q), high frequency Extensive test data Reliable	Low capacitance-to-volume ratio

Type	Applications	Advantages	Disadvantages
Ceramic (general-purpose)	Bypassing, coupling, and filtering to high frequency	High capacitance-to-volume ratio Chip style available Low cost	Poor temperature coefficients and time stability Susceptible to shock and vibration damage Poor reliability (particularly high-k commercial types)
Ceramic (temperature-compensating)	Compensation for temperature variations	Temperature characteristic can be closely controlled. Stable with time	Available only in low capacitance values Susceptible to shock and vibration damage Low capacitance-to-volume ratio
Variable ceramic and glass piston	Frequency tuning	Selectable TC (ceramic) High Q	Susceptible to humidity, shock, and vibration damage Low-capacitance-to-volume ratio Low capacitance values only High cost
Tantalum (solid dielectric)	Blocking, bypassing, coupling, and filtering in low-frequency circuits, timing, color-convergence circuits, squib firing, photo-flash firing	High capacitance-to-volume ratio Good temperature coefficients Extensive test data	Voltage limitation Leakage current Poor rf characteristics Medium cost
Tantalum (foil and wet-slug)	Bypassing and filtering in low-frequency circuits	High capacitance-to-volume ratio Highest voltage for tantalums Lowest leakage currents (slug) Extensive test data	Poor temperature characteristics, particularly at low temperature Susceptible to shock and vibration damage (particularly slug) Seal life Medium to high cost
Aluminum electrolytic	Blocking, bypassing, coupling, and low-frequency filtering Photoflash	Highest capacitance-to-volume ratio of electrolytics Highest voltage of electrolytics Highest capacitance Lowest cost per cv unit for commercial types High ripple capability	Affected by chlorinated hydrocarbons High leakage current Requires more reforming after periods of storage Medium to high cost for types with controlled reliability Susceptible to dynamic environments

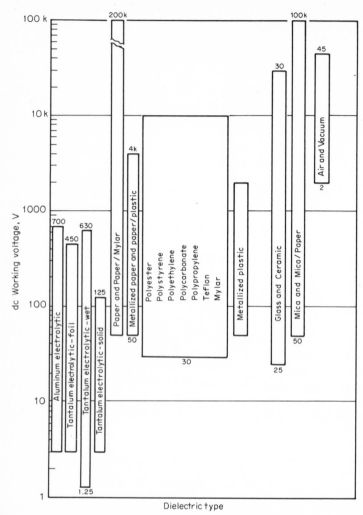

Fig. 16 Range of maximum dc-voltage rating by dielectric type.   Mylar and Teflon are registered trademarks of E. I. Du Pont de Nemours & Co., Wilmington, Del.

design need.   Dissipation factor by dielectric type is shown in Fig. 17 to provide an additional trade-off for losses.

Volumetric efficiency   Of concern to the design engineer is the volume required for a desired capacitance and the amount of space he must allow in the physical hardware.   This is one of the major trade-offs made in selecting a capacitor.   Figure 18 shows comparative volumetric efficiency for common capacitor types.   As discussed earlier, capacitor size is a function of the dielectric material used and the method of construction.   Electrolytic and general-purpose ceramic capacitors provide the greatest volumetric efficiency.   These capacitors are also low-voltage types by nature and are therefore more applicable in the majority of the new integrated-circuit electronic-equipment designs.   A point should also be made that because of case materials (for example, end seals, hermetically sealed cans, and lead-attachment areas) it is more practical to specify a larger value of capacitor than a small value, since the volume of seals and other mechanical functions will not increase proportionately.

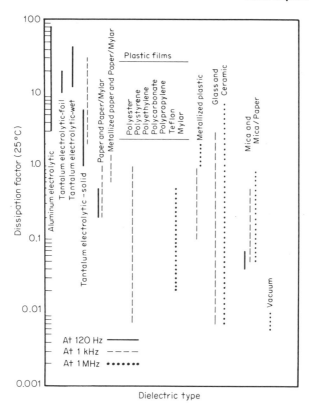

Fig. 17 Range of dissipation factor at various frequencies by capacitor dielectric type. Mylar and Teflon are registered trademarks of E. I. Du Pont de Nemours & Co., Wilmington, Del.

Environmental effects on capacitors The performance, storage life, and service life of all capacitors are highly dependent upon the environments to which they are exposed. Not only should the implications of each individual environmental factor be considered but also the effects resulting from various combinations of environments must be taken into account. A good capacitor can fail when designed into equipment that must see environmental conditions that exceed the design capabilities of the capacitor. The major environments affecting performance and life of a capacitor are ambient temperature, humidity or moisture, vibration, shock and acceleration, barometric pressure, and nuclear radiation. The individual effects of these environments will be discussed briefly.

Ambient Temperature. The maximum operating ambient temperature surrounding the capacitor in an application is of critical importance, since this is one of the determining factors affecting the dielectric. Service life of a capacitor will decrease with an increase in temperature. The magnitude of decrease in service life, however, depends upon the dielectric type and its temperature stability. One of the factors affecting wear-out of the capacitor is dielectric degradation resulting from chemical changes with time. This degradation can take the form of a drying out of a semiwet or wet dielectric, oxidation of paper and other materials, or chlorine activity due to improper cleansing of the material during manufacturing process or assembly into the electronic hardware. Capacitance will vary directly or indirectly with increasing temperature depending on the dielectric and construction. Temperature changes can cause two distinct actions to take place that will affect capacitance. Both dielectric constant of the material and the spacing between the electrodes can change,

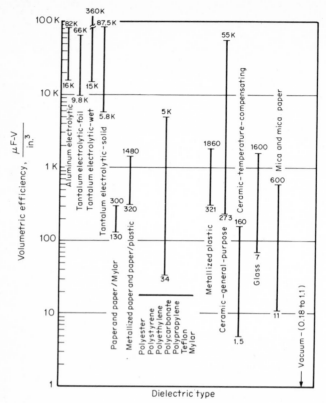

Fig. 18 Range of volumetric efficiency by capacitor dielectric type. Mylar and Teflon are registered trademarks of E. I. Du Pont de Nemours & Co., Wilmington, Del.

thus altering two of the factors in the equation for capacitance. Insulation resistance also will decrease with an increase in temperature due to the increased electron activity, which in turn can change dipole orientation and/or dipole response within the dielectric itself. As seen before, dissipation factor is a complex function with temperature and may vary up or down with increased temperature depending on the dielectric material. In some cases several variations will take place in the same style of capacitor over its listed usable temperature range. Dielectric strength behaves like insulation resistance. As the temperature increases, the peak applied voltage must be limited, and in no case should it exceed the manufacturer's rating. Where capacitor parameters are critical over the desired operating-temperature range, a thorough electrical characterization with temperature of the selected device type may be required. In general, however, curves depicting performance change with temperature are published and are adequate to support good design practice.

Humidity or Moisture. Sealing of the capacitor element within the case is an important consideration in the application of a capacitor. Where the capacitor is to be applied in an environment consisting of a high moisture content, it may be desirable to employ the use of hermetically sealed styles. In this case, when the temperature increases, there is an increase in the cell pressure due to the expansion of the dielectric material. If this pressure becomes great enough, the seal can be ruptured. Conversely, capacitor styles that are not hermetically sealed but employ an epoxy end seal, epoxy encapsulation, or an open lining which has been wax-impregnated or epoxy-dipped can fail owing to moisture absorption along the leads or through imperfections in the case or coating material. The effects of moisture on the capacitor can be detected from parametric changes, reduced service life, or catastrophic failure from gross moisture penetration. Although moisture will cause variations in capacitance-dissipation factor and dielectric-breakdown strength, the most noticeable parameter effect is the decrease in insulation resistance. Many non-hermetic capacitor designs use paper as part or all of the dielectric system and are much more vulnerable to moisture than those designs using film dielectrics. The reason is that the moisture can more readily penetrate into the paper and thus will be entrapped during the manufacture of the capacitor or penetrate the unit during service life or exposure to the moist environment. It is for this reason that most paper and paper-film dielectric capacitors are fabricated in hermetic cases. A number of tests are specified in Military Standard 202 (MIL-STD-202) and EIA Standard RS-186 which provide excellent measures of moisture resistance. These tests include immersion cycling for a period of anywhere from 15 to 60 min per cycle in fresh or salt water. Another test is humidity exposure, where the capacitors are subjected to exposure to 90 to 95 percent relative humidity at 40°C for a period of 96 h or more. A third test is moisture resistance. This is a combination of temperature cycling and humidity exposure and is used on a cycling basis for 10 cycles, each lasting 24 h. Subzero-temperature exposure and vibration are also included in the cycling phase. External corrosion of a hermetically sealed metal chassis under excessive moisture conditions can also lead to the development of current-leakage paths and shorting across the glass header. Care should be exercised when applying capacitors where prolonged dc voltages exist in a corrosive atmosphere. DC channeling across insulators can occur unless conductive surfaces are adequately protected by anodizing, coating, or other means.

Dynamic Environments. Vibration, shock, and acceleration are the principal dynamic environments which can mechanically destroy or damage a capacitor. Movement of the capacitor assembly inside a case can cause capacitance fluctuations, dielectric and insulation failures, and electrode-attachment failures. Most commercial capacitors are designed to withstand normal transportation, handling, and equipment-use levels of mechanical shock and periodic vibration. Capacitors designed to meet more severe stresses such as those experienced in military, aerospace, and automotive uses are usually more expensive and are the most advanced of a manufacturer's product line. Most high-reliability military standard capacitors have been qualified for use in electronic equipment subjected to several hundred g's acceleration, random vibration at levels of 1 g^2/Hz (power spectral density) up to 3,000 Hz, and 1,500 g's, $\frac{1}{2}$ ms, $\frac{1}{2}$ sine shock impulse. A good understanding of capacitor susceptibility to dynamic environments can be obtained by considering its physical construction. The larger the complex elements in the capacitor the lower will be the frequency of response for these elements. Conversely, high-frequency dynamic environment will excite smaller construction elements in a given capacitor. In most cases the solid molded capacitor or potted capacitor elements are relatively inert to dynamic environment except for piezoelectric effects exhibited in some ceramic types. Some tantalum capacitors with a free length of lead wire between the hermetic-seal eyelet and the tantalum slug and certain other hermetically sealed capacitors utilizing a winding with an internal ribbon lead (tab) attached to the case will resonate and fail under certain dynamic stresses. These should be considered and appropriate studies made before the capacitors are applied in severe mechanical environments. Large-volume capacitors where a significant mass exists will require special mounting features in the form of either brackets welded to the can or straps that will provide adequate support. All straps and mounting brackets should be large enough to prevent mechanical distortion in any one portion of the case, thereby causing a variation in the insulation resistance or dielectric strength of the capacitor. Where high dynamic environments are to be employed, most manufacturers can readily make special capacitor designs or alter their existing designs where necessary. The assembly of capacitors in modules, circuit boards, and special assembly methods must take into consideration mechanical strain on leads. Conformal coatings and plastic cases cannot be relied upon for mechanical strength.

Barometric Pressure. The altitude at which a hermetically sealed capacitor can be safely operated will depend largely on the design and integrity of end-seal case-wall strength, the voltage at which the capacitor is being operated, and the type of impregnant used in the dielectric material.   As altitude is increased, the dielectric strength across the given end seal will decrease.   This must be taken into consideration if such designs are to be utilized.   Likewise, as the altitude is increased with the barometric pressure reduced, the pressures inside the capacitor will increase the mechanical stress on the case and seal.   The end result could be a catastrophic failure of the case or seal.   Electrical failure will be in the form of a dielectric breakdown, and in addition the capacitance can be affected by the changing of internal dimensions due to affected pressure differentials.   This likewise can cause arc-overs within the capacitor.   Heat transfer in ac applications is also an important consideration.   Normally the heat is removed from the capacitor through the metal case (if it has one) or through its leads.   An increase in altitude will affect the thermal convection characteristics.

Electrolytic Capacitors

Many low-frequency filtering, long-term timing, coupling, decoupling, and certain bypass (self-biasing circuit) applications require high capacitance values.   These requirements for high capacitance and small volume are met with the use of electrolytic capacitors which have a unique capability of high capacitance times voltage (CV) to volume ratios and the lowest cost per microfarad of any capacitor type now in use.   In addition many electrolytic capacitors are used interchangeably with electrostatic capacitors in many special applications where severe volume limitations exist.   Electrolytic capacitors are characterized by very different construction and operating parameters.   Actual experience in the production of electrolytic capacitors has resulted in high-quality processing and acceptance testing and has now produced capacitors with the capability of up to 20 years of performance.   To apply electrolytic capacitors properly, therefore, it is essential that the nature and behavior of these thin-film devices be understood.

Basic theory   The dielectric material in an electrolytic capacitor consists of an anodically formed oxide of the anode material, which serves as the positive electrode of the capacitor.   The metals most employed are aluminum and tantalum, although titanium, niobium, zirconium, and others are anodized to form dielectric films of limited value.   Simplified electrolytic-capacitor diagrams are shown in Fig. 19.   When aluminum is used, the dielectric oxide is a combination of amorphous

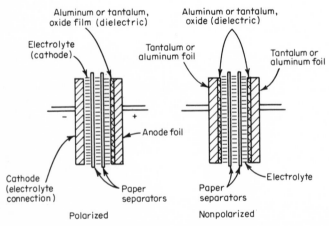

Fig. 19   General construction of typical aluminum or tantalum foil capacitors.

and crystalline (gamma) aluminum oxide with the basic formula AL_2O_3.   When tantalum is used, the oxide is Ta_2O_5.   These oxides can be "formed" by connecting the anode material as a positive electrode in an electrolytic cell containing a phosphate or borate solution.   The cathodic electrode used in the forming process normally is copper or nickel.   A voltage applied to the system oxidizes the aluminum or tantalum at its surface to form an adherent layer of the oxide to the parent metal. This oxide film, formed with a dc voltage, is highly polarized. The "forward" direction of an oxide film is characterized by a high insulation resistance. The "reverse" direction conversely is characterized by a low insulation resistance in which current will easily flow. The reverse current normally will not damage the capacitor unless it causes excessive heating or, as may occur with sintered-anode wet tantalums, causes a migration of silver molecules from cathode to anode. Reverse current can be in the form of a reverse dc voltage or ac voltages of low frequencies up to very high frequencies of sufficient magnitude to cause a voltage reversal. The higher the frequency of a reverse ac signal, the greater will be the heating effect. For proper application of electrolytic capacitors, therefore, it is important to consider ripple frequency, voltage, and current. The thickness of the oxide film depends on the voltage at which the film was formed. In the case of aluminum the forming voltage may be as high as 754 V dc. Other factors in the formation of the oxide film are temperature and time, and the type of formation that is used. The thickness of the oxide produced ranges between 11 and 15 Å/V applied. When tantalum is the parent metal, formation voltages rarely exceed 500 V dc. The oxide thickness ranges from 16 to 20 Å/V applied. The effective dielectric constant of pure aluminum oxide is 8.4, and that of tantalum oxide is approximately 28. In certain commercial-grade capacitors, the dielectric constant can be slightly lower for aluminum oxide and appreciably lower for tantalum oxide because of the impurities in the oxide. High-purity oxides and electrolytes are used in extended-life and high-performance grade capacitors.

Since the thickness of the film increases linearly with voltage, the capacitance per unit area of electrode decreases inversely according to Eq. (1) (Table 1).

Capacitor polarization Aluminum and tantalum electrolytic capacitors are produced in polarized, semipolarized, and nonpolarized designs.

Polarized Capacitors. This is the most common configuration and has an oxide film formed on only one foil or electrode called the anode or positive terminal. If a capacitor with a plain, unformed foil cathode and a formed anode is employed in a circuit with a large dc bias and small ripple voltage, an oxide film will also form on the cathode surface during the negative slope of the ripple voltage, with a resultant reduction in capacitance. This has the same effect as two capacitances in series, one between anode and the electrolyte and the second between the electrolyte and the cathode. The "cathode" capacitor, will reduce the total capacity of the device. Etching the cathode increases the cathode area, decreases the ripple-current density at the cathode, increases the cathode capacitance, and delays the decrease in total capacitance caused by the ripple voltage. Reversal of the polarity of the voltage applied to a polarized capacitor will cause either a dielectric breakdown or the formation of an oxide film on the cathode, with a resultant decrease in capacitance, depending on the magnitude and time of the reverse-polarity voltage. Polarized capacitors are used largely in dc circuits where the applied potential is unidirectional, with a specified maximum ac ripple.

Semipolarized Capacitors. These capacitors are similar to polarized capacitors in construction except that a thin oxide film has been formed on the cathode to minimize the effects of reverse voltage. These capacitors are for use in circuits where some specified dc potential, which is less than the rated voltage, may be applied in the reverse direction for extended periods of time.

Nonpolarized Capacitors. Capacitors of this design have equal thicknesses of oxide film formed on both the anode and cathode and are for use in dc circuits where a full range of voltage may be applied across the capacitor in either direction for extended periods or for use in ac circuits where the ac rated voltage is to be applied for a limited period of time, or where low ac voltages are applied for longer periods of time, the period being limited by the heating of the capacitor due to the high power factor and leakage current. The ac rating is always less than the dc rating with the exception of certain tantalum-foil types. AC nonpolarized capacitors are designed to minimize power factor, whereas dc nonpolarized capacitors are designed to minimize the leakage current.

Wet and dry electrolytics Electrolytic capacitors are produced in two basic styles. The sintered anode or pellet style with either wet or dry electrolyte is used only for tantalums. The foil style, which also includes aluminum electrolytics, employs a wet, dry, or paste electrolyte. Foils may be either etched or plain depending on the temperature characteristics and reliability desired. Plain foil types have a better capacitance-temperature characteristic but have significantly lower volumetric efficiency. Wet-slug tantalum capacitors and wet-foil tantalum and aluminum electrolytic capacitors are usually produced with higher capacitance per unit volume and are also characterized by a low equivalent series resistance (ESR). Most styles of these capacitors, however, cannot be easily hermetically sealed and are provided with a special seal or case wall which will incorporate a vent valve, elastomers or case-rupture zone to relieve excess pressure due to heating of the capacitor during high leakage current, or ripple-current stress periods (resulting in reduced service life). Dry and solid electrolytic capacitors, which are characterized by a slightly higher ESR, can be hermetically sealed and are normally used in military equipment where the danger of electrolyte leakage cannot be tolerated and longer life is required.

Voltage ratings The thickness of the oxide film which is formed both initially on the foil and during the reforming operation of the completed capacitor determines the maximum surge and dc voltages which may be applied. The relationship of these voltages to the forming voltage is shown in Fig. 20. Each capacitor has an

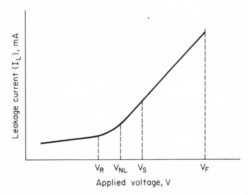

Fig. 20   Voltage relationships of electrolytic capacitors.

$$V_F = \text{formation voltage}$$
$$V_S = \text{surge voltage}$$
$$V_R = \text{rated voltage}$$
$$V_{NL} = \text{no-load voltage}$$

overvoltage (surge) rating which is greater than the rated dc voltage. The surge voltage is the maximum voltage to which the capacitor should be subjected under any conditions and should not be continuously applied. This includes transients and peak ripple at the highest line voltage. It may be based on a test specifying that the surge voltage be applied across the capacitor in series with a specified resistor which limits the peak current (not to exceed the allowable core-temperature rise) and that the voltage is applied for a maximum period of 30 s or other specified time, after which the capacitor is allowed to cool for a specified time before the voltage is again applied. Excessive voltage or a combination of voltage and excess temperature will cause the oxide film to break down. Under certain conditions, however, with potential applied to the capacitor, the electrolyte will form a new oxide film on the electrodes and cause the oxide film to be self-healing. This, however, is not to be depended upon, and in many cases the self-healing characteristics are not desirable, since it is difficult to locate intermittent faults in a system employing the capacitor.

Tantalum and aluminum foil capacitors which are polarized at high voltages will appear to be nonpolarized at dc voltages in the order of 5 V or less because of oxides on the cathode.   Therefore, operation of polarized capacitors with high voltage ratings at low voltage stress will cause a decrease in capacitance due to the thickness of the oxide film on the cathode.   The magnitude of the decreasing capacitance is dependent on the thickness of the cathode film compared with the thickness of the anode film.   Aluminum electrolytics operating in circuits considerably below the rated voltage may show a gradual decrease in capacitance due to the reformation of the oxide film.   For this reason, voltage deratings of electrolytics, with the exception of solid tantalums, will not provide a long-time voltage safety factor as in electrostatic capacitors.   Thus, a large temperature safety factor or voltage-derating factor from the manufacturer's best applied range may not be as beneficial as specifying a lower-voltage unit which has the optimum electrolyte for operation at the required voltage and temperature.   Tantalum capacitors have a more stable film which results in longer shelf life and less deterioration under the same low-voltage conditions.   The electrolyte in single capacitors is selected for the optimum voltage and temperature at which the capacitor will operate.   Dual-voltage capacitors, on the other hand, have both sections impregnated with the same electrolyte and cannot have an optimum electrolyte for each rating.   These units are not recommended where the best performance and high reliability are required.   A low-voltage section of a dual-voltage device will have a higher series resistance and power factor than a single-section capacitor of the same voltage and capacitance rating.   In general, higher-voltage electrolytics have a higher-resistance electrolyte to prevent voltage breakdown of the electrolyte.   Capacitors which are to operate at a higher temperature usually have a nonaqueous electrolyte, since aqueous electrolytes will gradually dry out and lead to a decrease in capacitance, increase in series resistance, and often dielectric failure.

Ripple   Factors affecting the safe ripple that can be superimposed under working dc voltage levels are voltage, current, frequency, and temperature.   Voltage limitation generally occurs in lower-capacitance devices.   Voltage limitation specified by the capacitor manufacturer can be calculated in several ways as follows:

1. The sum of the peak ac ripple voltage and the dc voltage applied may not exceed the maximum dc rated voltage limits specified.

2. The sum of the negative half-cycle of the peak ac voltage and the dc voltage applied shall not exceed the voltage-reversal limit specified.

3. The rms values, at 120 Hz, of the voltage sums from items 1 and 2 above shall be equal to or less than the rms voltage obtained from multiplying the specified maximum ripple current by the specified equivalent series resistance.

The maximum ripple current is limited by the ability of the capacitor to dissipate the heat generated in its ESR due to the dc leakage current and the ac rms current. In higher-capacitance devices the ripple current usually becomes the limiting factor. Ripple current is specified by most manufacturers at 120 Hz and at the maximum operating temperature.   Ripple-current limits can be adjusted for actual operating temperature and frequency.   Ripple-current adjustments for temperature and frequency are specified by the manufacturer.   Adjustment factors are not standard, since they are a function of case size, core mass, type of tabbing, and electrolyte used in the capacitor design.   All these ripple limitations must be considered for proper capacitor application and service life.   Typical derating factors for ambient temperature are shown in Fig. 21.

Current ratings   Electrolytic capacitors, in addition to a ripple-current rating, have a maximum allowable charge and discharge current rating.   Excessive current can burn off internal metal tabs or cause severe overheating and possible rupture.   In general, this current rating should be investigated when the peak charge and discharge current exceed 1 A.   Nonpolarized electrolytics for use in capacitor-start motors are usually designed for 20 starts per hour, with each start having a maximum duration of 3 s, or sixty 1-s starts per hour.   In applications where long life is desired, it is recommended that the case temperature not exceed 60°C.

Leakage current   Electrolytic capacitors pass appreciable currents known as leakage current, the magnitude of which is a function of the oxide materials used and their purity, the thickness of the oxide film, resistivity of the electrolyte, applied voltage, area of the foil, and operating temperature.   Most electrolytics have poor

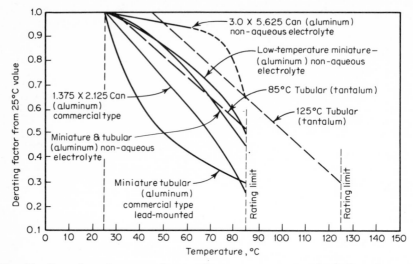

Fig. 21 Typical ripple-current derating factor vs. temperature for high-performance aluminum and tantalum foil capacitors.

resistance-temperature characteristics. An increase in leakage current occurs as the temperature is raised, with a resulting decrease in the breakdown voltage. The increase in dc leakage current with temperature further increases capacitor heating, which in turn further increases the dc leakage current. This may cause thermal runaway and eventual destruction of the capacitor. In some wet-electrolyte types there may be a compensating factor with time, in the drying out of the electrolyte resulting in increased resistance. This increase in resistance increases the power factor and decreases leakage current. However, an increase in leakage can again occur owing to the heating when ac or pulsating dc voltages are applied near the capacitor rating. The leakage current increases as the capacitance, voltage, and temperature of the unit increase. Capacitors are specified by the manufacturer for

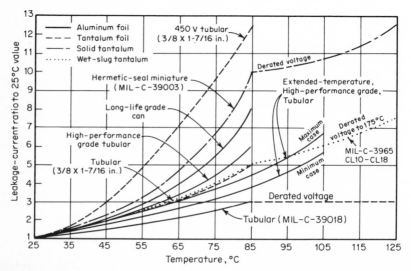

Fig. 22 Typical comparison curves of leakage-current variation with temperature for aluminum and tantalum electrolytic capacitors.

maximum leakage current at maximum operating temperature. Typical curves comparing leakage current as a function of temperature for aluminum and tantalum electrolytics are shown in Fig. 22.

Losses In electrolytic capacitors the equivalent series resistance (ESR) is usually specified as a definition of loss in the capacitor. The ESR is the total power in watts lost divided by the capacitor current squared (I^2) as given in Eq. (8) (Table 1). This includes dielectric losses in the oxide film, resistance of the electrolyte, contact resistance, and resistance in the foil and internal conductors (tabs). These losses vary with frequency and temperature. Power factor is defined as the ESR (R) divided by the impedance (Z) as given in Eq. (10) (Table 1). Power factor times the volt-amperes equals the watts dissipated in the capacitor. From this the temperature rise of the container can be determined by referring to Fig. 23. In applications involving an ac voltage on a dc bias and where the power factor is determined by an ac measurement only, the total watts dissipated in the capacitor W_T is the sum of the leakage current times the dc bias W_{dc} and the power factor times the ac volt-amperes as referenced by Eq. (14) (Table 1). Note that the application of a polarizing voltage to a capacitor may affect the power factor of the unit being measured. A figure of merit which is often used in evaluating losses in electrolytic capacitors is Q_M, which is the capacitive reactance (X_C) divided by the ESR (R) as given in Eq. (6) (Table 1). Measurements are made on a polarized capacitance bridge at a frequency of 120 Hz with bias applied. Without sufficient dc polarization voltage, the ac measurement potential must be small enough to prevent formation of oxides on the cathode of polar capacitors which will change the capacitance value.

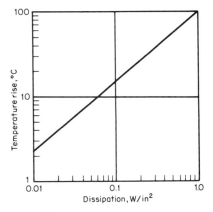

Fig. 23 Capacitor-case temperature rise as a function of watts dissipated per square inch of case surface area.

Temperature effects Characteristic curves for tantalum and aluminum electrolytic capacitors (Fig. 24) show that the capacitance drops off rapidly at low temperature. This characteristic can be controlled to some extent by the manufacturer, since it is a function of the chemical nature of the electrolyte. Most aluminum capacitors lose practically all their capacitance at −55°C, whereas tantalum lose up to about 20 percent of their capacitance at the same temperature. The power factor of the capacitor at these low temperatures is high. This condition may cause internal heating and increase the temperature. For this reason, equipment at low temperatures is powered up and time is given for the capacitance to rise before the equipment is fully operable. At higher frequencies the capacitance drops off more rapidly at low-temperature limits, and at 20 kHz the capacitance of electrolytics may only be 20 percent of the 25°C capacitance. Plain-foil aluminum electrolytics typically have lower losses and better stability for a wider range of temperature than etched-foil units. When operating aluminum capacitors at low temperature, careful consideration must be given to the capacitance-temperature characteristic. As the temperature is increased above 25°C, the capacitance normally increases and the power factor decreases. At still higher temperature, the power factor may again increase and may cause thermal instability with a resultant thermal-runaway condition. At the higher temperature the effect of frequency on capacitance is not as great as it is at low temperatures. The capacitance decreases and the power factor increases as the frequency increases. Aluminum electrolytic capacitors are available which are suitable for operation at reduced voltage up to 150°C for 2,000 h. For long life, however, the maximum temperature should not exceed 85°C, with most general-purpose styles not recommended above 65°C. Tantalum electrolytic capacitors are available for operation at reduced voltage to temperatures as high as 175°C,

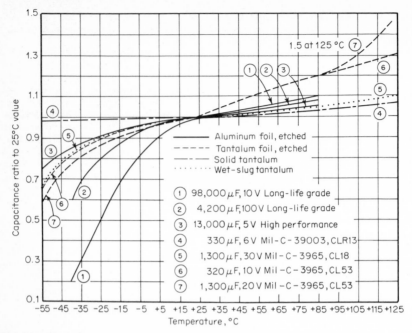

Fig. 24 Typical capacitance vs. temperature characteristic for aluminum and tantalum electrolytic capacitors.

although the low-temperature limit is −55°C. Series resistance increases sharply at low temperatures, and as a result effective capacitance decreases. These features do not exist after the equipment temperature rises, so that electrolytic capacitors are available which may be exposed to temperatures of −55°C.

Parallel Operation. To obtain a higher capacitance than can be obtained from a single capacitor, a number of units may be connected in parallel. However, the sum of the peak ripple and the applied dc voltage should not exceed the dc working voltage of the unit with the lowest voltage rate. The connecting leads of the parallel network should be large enough to carry the combined currents without reducing the effect of the capacitance due to series lead resistance.

Application precautions Several precautions are worth noting for proper application of electrolytic capacitors.

Shelf Life. Sudden application of rated voltage to an unformed capacitor after a period of shelf life may cause dielectric breakdown due to the thinness of the oxide film and the heat generated from excessive leakage current. It is recommended that electrolytics, particularly aluminum styles, having more than the maximum safe storage be checked and preconditioned* at room temperature for leakage current in accordance with the applicable ratings specified by the manufacturer.

Ripple Current. It should be emphasized that capacitors with plain unformed cathode foils receive an oxide coating during the negative slope of any voltage present. This film is similar to the anode film and is in series with it, causing a net reduction in capacitance. With etched-foil anode, the decrease in capacitance is much larger than with plain-foil anodes. This condition can be overcome by the use of an etched cathode foil or a semipolar or nonpolar capacitor. The decrease in capacitance with the application of negative-slope voltage to polar electrolytic capacitors may occur without any appreciable heating of the capacitor. To eliminate this reduction in capacitance due to cathode formation, the capacitors should always be operated within the ripple-current rating specified by the manufacturer.

---

* Preconditioning methods are specified in MIL-STD-1131.

Frequency Limitation. The high power factor of most electrolytic capacitors makes this type of capacitor unsatisfactory for most tuning circuits and also generally limits their use in ac circuits and in high-frequency bypass, coupling, and filter applications. The electrolytic capacitors designed for high-frequency operation implementing the low-loss electrolytes and improved mechanical design are now available for use in appications up to 10 MHz.

Mounting and Installation. Care should be taken to avoid mounting procedures which place undue strain on terminal elements or case. Potting procedures should be doubly checked to assure that excessive curing exotherms or shrinkage pressures are not present. Soldering techniques are important and should be designed to avoid excessive heat transfer to terminals which could possibly cause melting of the eyelet solder.

Venting. Electrolytic capacitors designed with wet electrolytes employ vented or elastomeric seals to accommodate thermal expansion and gases caused by excessive currents. Explosions can occur from both gas pressure and spark ignition of free oxygen and hydrogen liberated at the electrolyte. Adequate surge-current limitation and provision to minimize danger to personnel and equipment should be considered, especially when using aluminum electrolytics where these phenomena can exist.

Insulation and Grounding. Capacitors with uninsulated cases should have the case grounded or protected by means of a suitable insulated sleeve. Even if the case is not connected to any circuit element, a high potential may develop between a floating case and ground by leakage between capacitor elements and the case. Electrolytic capacitors have an inherently low indeterminate insulation resultance between the cathode terminal and the container. The container should therefore be considered at the same potential as the cathode terminal with respect to ground. Where a potential difference exists between the cathode terminal and the chassis or other metallic mounting surfaces, which are usually ground, the capacitor should be mounted in such a manner as to insulate the container from such mounting surfaces. In ac or pulsating dc applications, the capacitance between the metallic case and the mounting bracket, when the capacitor has an insulating sleeve, must be considered as well as the dc insulating properties of the insulating sleeve. When aluminum electrolytics with case vents are installed in mounting brackets, care must be taken to avoid clamping over the vent.

Cleaning Solvents. The aluminum electrolytic-capacitor internal assembly is affected by halogenated hydrocarbons in the presence of an electric field. A potential problem exists with these capacitors when immersed in certain cleaning-solvent baths commonly used in electronic assembly. A halogenated hydrocarbon by itself does not affect aluminum; however, if some of the material gets inside the capacitor and a voltage is applied, an electrochemical action takes place. The basic aluminum is attacked and can cause actual failure of the capacitor. This is a time-voltage effect and as such is an insidious problem in that it normally shows up as a failure in the field.

The effects of cleaning solvents on aluminum electrolytic capacitors must therefore be considered in selecting a capacitor package and end-seal type, or selecting the proper solvent and solvent-cleaning technique. Many users of these solvents specify the addition of an epoxy barrier over the seal to give added insurance against penetration. Table 5 shows typical cleaning solvents that have been determined to be safe or unsafe when solvent is allowed to progress past the elastomeric seal and to the inside section of the capacitor.

Life In addition to proper and judicious application, the ultimate useful life on an electrolytic capacitor depends on the use of high-purity materials, complete elimination of the slightest trace of contaminants, and maintenance of the electrolyte in proper chemical composition. Chloride contamination is of particular concern in the production of aluminum electrolytics, where greater than 1 ppm is considered dangerous for reliability. Except for obvious defects and short-time-test failures, the quality of the capacitor can be observed only by extensive life testing and the use of adequate controls by the manufacturer to assure reliable product. Capacitors that are poorly manufactured may have extensive corrosion, which can cause an open circuit by completely disintegrating sections of foils, and/or tabs. Aluminum-electrolytic-capacitor shelf life has been improved greatly by the use of high-purity

TABLE 5   Common Safe and Unsafe Cleaning Solvents For Aluminum Foil Capacitors

Safe	Unsafe
Xylene	Freon TF, TMC
Ethyl alcohol	Carbon tetrachloride
Butyl alcohol	Chloroform
Methyl alcohol	Trichloroethylene
Propyl alcohol	Trichloroethane
Calgonite (detergent)	ALL (detergent)
Naphtha	Methylene chloride
Water	Methylethyl ketone (MEK)
Toluene	
Methanol	
Methyl cellosolve	
Alkinox	

aluminum anode material and rigid process controls.   Typical life test for high-grade aluminum and tantalum electrolytics is 2,000 h at rated temperature and voltage, and 500 h for general-purpose-grade aluminums.   The application requirements of electrolytics and the reliability of the equipment will dictate the quality of capacitor that is required.   Most failures in electrolytic capacitors are from two causes, either the breakdown of the dielectric film because leakage resistance is too low, or the drying out of the electrolyte resulting in too high a leakage resistance.   Because of this second feature, the problem with wet electrolytics is to seal moisture in rather than seal it out as in most other electronic components.   When the electrolyte dries out, the capacitance decreases drastically.

The dielectric breakdown in electrolytic capacitors is entirely different from that occurring in electrostatic capacitors.   Normally, an electrolytic capacitor would be expected to break down at a voltage approximately 30 percent higher than the rated dc voltage or 15 percent higher than the specified surge voltage, which begins to approach the forming voltage of the oxide film (Fig. 20).   This breakdown may be permanent or may be healed, depending on the energy level present at the time of breakdown as well as the period of time involved.   Present military and industry specifications may permit 1 failure in 12 during rated voltage life test at maximum ambient temperature.   With the present state of the art this is a conservative rating, and manufacturers who are thoroughly experienced make capacitors with a much lower breakdown rate, perhaps 1 in 100.   Lowering the ambient temperature will improve this figure.   Lowering the test voltage will improve this figure only to a limited degree and at the risk of forming new cathode oxides or deforming the anode oxide.

Testing[5,6]   Standard tests and test methods have been specified in military and EIA specifications and provide a good degree of consistency among various manufacturers and users.   When specifying a capacitor, it is always advantageous to use accepted standard tests and test methods.   For aluminum and tantalum electrolytics the following tests are generally accepted by industry standards.

Visual Examination and Mechanical Inspection.   Where size permits, each capacitor is legibly marked with:

1. Polarity indicator per detail specification sheet (where applicable)
2. Manufacturer's name, source code symbol, or trademark
3. Rated capacitance in microfarads
4. Rated dc voltage
5. Terminal identification (where applicable)
6. Maximum rated temperature
7. Customer or manufacturer's part number (where applicable)
8. EIA code date
9. Tolerance on rated capacitance

Where size precludes this information, items 6, 7, 5, 8, and 9 may be omitted in that order or the information to be marked on the capacitor case may be revised, omitted, or abbreviated, as agreed between the manufacturer and the customer. Marking will remain legible through ordinary handling.

The capacitor dimensions should be checked to be within the limits of the detail-specification-drawing dimensions.

DC Leakage Current. Capacitors are initially preconditioned at standard atmospheric conditions by applying rated voltage to them through a suitable protective series resistor (usually 1,000 Ω) for a duration of 30 min at least 24 h and not more than 48 h for aluminums, and 5 min duration for tantalums before measurement of dc leakage current. This preconditioning is not to be repeated when subsequent leakage-current measurements are made such as are required in a schedule of qualification tests. After preconditioning, rated dc voltage is applied to the capacitor in series with the required current-limiting resistor and a dc milliammeter or microammeter. The buildup of voltage will occur within 60 s. The dc leakage is measured at the end of a 5-min period with a measurement accuracy of ±5 percent for aluminums and ±2 percent for tantalums. The leakage current is not to exceed the values listed in the manufacturer's data sheet.

Capacitance. The capacitance value must correspond with the rated capacitance, taking into consideration the tolerance, and is measured at 120 Hz by a bridge method or equivalent suitable for measuring series capacitance and dissipation factor. A typical bridge for measurement of capacitance and dissipation factor is shown in Fig. 25.

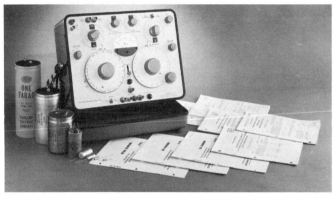

Fig. 25 Type 1617-A capacitance bridge measures electrolytic capacitors of 1 pF to 1.1 F, 20 Hz to 1 kHz with 1 percent accuracy. (*General Radio Company.*)

Dissipation Factor. The dissipation factor, when measured under the same test conditions as for the capacitance measurement, must not exceed the values specified in the manufacturer's data sheet. For multisection aluminum electrolytic capacitors with 175-V and up sections meshed with 75-V and lower sections, the DF values for 75-V (or lower) sections can be twice the values shown for single-voltage capacitors. For capacitors with a *CV* product greater than 100,000 the values of dissipation factor are generally negotiated between the manufacturer and the customer.

Impedance. The impedance measurement, when required, is performed at low temperature and other temperatures as needed. Both the minimum rated temperature and the values of the impedance limit are negotiated between the manufacturer and the customer. The impedance is measured at 120 Hz using any suitable method providing an accuracy of ±2½ percent. The ac-measuring voltage must be as small as practical and is applied for as short a time as practical in order that excessive voltage reversal does not occur or that it will not cause undue heating of the capacitor. The capacitor is brought to thermal stability at the applicable minimum rated temperature. Thermal stability will have been reached when no further change in impedance is observed between two successive measurements taken at 15-min intervals. Extreme care must be exercised in making low-impedance measurements by providing proper current limitations and by keeping safety vents unobstructed.

Solderability. Lead wires and solder terminals are subjected to the EIA Solderability Test Standard RS-178 or MIL-STD-202, Method 208, and will meet specified acceptance criteria for solder coating.

Terminal Strength. Strength of capacitor terminations is tested in accordance with the requirements of MIL-STD-202, Method 211, as applicable or of EIA Standard 186, Method 6 (Mechanical Robustness of Terminals), where the following details apply:

1. Types I, III, IV, and VI tests are applied where applicable.

2. The load and/or torque values to be applied are as specified in the individual detail-specification sheets.

3. The number of bends (type III test) are specified in the individual detail-specification sheets.

4. As a result of the above tests, no lead or terminal may fracture, break, loosen, or break away from egress of case. Threaded insert terminations cannot exhibit perceptive movement relative to their cover under the torque values specified in type VI test.

Pressure-Relief Test (Where Applicable). A vent to relieve excessive gas pressure is required in certain capacitor designs for safety reasons when abnormal conditions occur. The venting capabilities of an electrolytic capacitor depend on such factors as design, style, rating, size, and application of test voltages. The requirement for a vent and its test method should be a matter of agreement between the manufacturer and the customer. Four test methods are suggested to be used as guides in determining venting or pressure-relief capabilities of aluminum electrolytic capacitors.

1. The capacitor is subjected to an alternating current the values of which are shown in Table 6. These values are to be used only as a guide in selecting practical values for the test.

TABLE 6   AC Levels for Pressure-Relief Test of Aluminum Foil Capacitors

Nominal capacitance, μF	60-Hz test current, A (rms), as specified
Up to 3,000	1–100
3,000–20,000	85–150
Above 20,000	100–175

2. For capacitors rated 150 vdc and above, 110 to 125 V ac rms 60 Hz is applied through a 5-Ω ± 10 percent series-limiting resistor. When multisection capacitors are tested, the voltage is applied only to the input filter section.

3. The capacitor is subjected to a reversed-polarity dc voltage sufficient to allow a current of from 1 to 10 dc A to flow for a specified time.

4. This method is as established and agreed upon between the manufacturer and the customer.

Under the test conditions of the above four methods as applicable, the excessive internal pressure must be relieved without violent expulsion of the capacitor element or cover or the ignition of surrounding material. To illustrate the latter, the case is lightly wrapped with two layers of cheesecloth which must not ignite during the test. A short or open circuit does not constitute a failure. This is a hazardous test, and suitable precautions must be taken to prevent damage to equipment and personnel.

Vibration. Industry-standard commercial- and long-life-grade aluminum electrolytic capacitors are tested to low-frequency (10- to 55-Hz) sinusoidal vibration in accordance with Method 7, type III Test of EIA Standard RS-186. Test samples are mounted by their normal means. At some time during the last 15 min of the test, the capacitors are monitored 3 min by bridge, and there should be no evidence of intermittency. There should be no loosening of the element within the case as evidenced by shaking by hand following the test. Industry- and military-standard tantalum electrolytics and military-standard aluminum electrolytes are tested to various vibration levels (test conditions) per Method 204 of MIL-STD-202. Vibration levels in detail specifications are from 10 to 25g's, with a frequency range of 10 to 2,000 Hz. Capacitors are mounted by their bodies and subjected to sustained vibration in two mutually perpendicular directions, one being the axis through the leads, for specified lengths of time. A final portion of the test duration in each direction is monitored with detection-equipment sensitivity capable of detecting a 0.5-ms or greater interruption. Capacitors must show no sign of mechanical damage or leakage of electrolyte and should not exhibit a detectable interruption.

Case Insulation (Where Applicable). Aluminum electrolytic capacitors with outer insulating sleeves are tested in accordance with Method 13, Insulation Resistance Test, of EIA Standard RS-186. The standard V-block test, Method 302 of MIL-STD-202, may be used for tantalum capacitors. Test procedure is as follows:

1. Where Method 13 is used, the capacitor is wrapped tightly as with two turns of thin metal foil over the insulating sleeve. The foil must be no closer than ¼ in from either end of the capacitor. Three turns of 18 AWG bare copper wire are wound around the center of the foil.

2. Insulation resistance is measured between the capacitor terminations shorted together and the foil wrap or V-block at 500 V dc with an electrification time of 1 min.

Under the above conditions, the insulation resistance must not be less than 100 MΩ.

Within 5 min following the insulation-resistance test, the capacitor with foil wrap retained is tested for insulation-withstanding voltage. A voltage of 2,000 V dc is applied between the capacitor terminations shorted together and the foil wrap or V-block. This voltage is applied by increasing it from zero to 2,000 V dc at a rate not exceeding 500 V/s. Under these conditions, there must be no evidence of breakdown of the case insulation. This is determined after test by examining the case insulation for evidence of burning, charring, or arcing.

Coupling Impedance for Multisection Aluminum Electrolytic Capacitors (Where Applicable). The requirement for this test is a matter of agreement between the manufacturer and the customer. When performed as specified in RS-395, the coupling impedance must not exceed 0.015 Ω or other value as agreed between the manufacturer and the customer.

Temperature Cycle and Immersion. Industry-standard aluminum electrolytics must be capable of meeting the following container-seal test. Subject capacitors to two temperature cycles in circulated air in accordance with the following schedule:

1. Place capacitors in an oven maintained at the capacitor maximum rated temperature and hold for 15 min.

2. Remove from oven and allow capacitors to cool to room temperature.

3. Place capacitors in a cold chamber maintained at −20°C and hold for 15 min.

4. Remove from chamber and allow capacitors to attain room temperature. Capacitors are then immersed in hot water maintained at 90°C for a period of 3 min. During immersion, there shall be no chain of repetitive bubbling from any part of the capacitor.

Military-standard aluminum and tantalum electrolytics must be capable of meeting the temperature cycle, Method 102 of MIL-STD-202. Aluminum electrolytics in addition must meet the hot-water-immersion test of Method 104 of MIL-STD-202. Posttest inspections are similar to those of the moisture-resistance tests.

Characteristics at Minimum and Maximum Temperatures. The capacitor measurements shown in Table 7 are made in the order shown. Capacitors under measurement are brought to thermal stability at each step of testing. Thermal stability will have been reached when no further change in capacitance (or impedance) is observed between two successive measurements taken at 15-min intervals. The limits and/or parametric changes for the characteristic measurements shall be as specified in the applicable manufacturer's detail specifications or as negotiated between the manufacturer and the customer.

Humidity and Moisture Resistance. Industry-standard aluminum electrolytic capacitors are subjected to humidity tests without voltage. The test consists of five

TABLE 7   Typical Measurements at Temperature Extremes for Electrolytic Capacitors

Step	Temperature	Test
1	Standard atmospheric conditions	DC leakage, capacitance, DF
2	Minimum temperature −3, +0 °C as specified in the manufacturer's data sheet or as agreed upon with the manufacturer	Impedance (aluminums only); dc leakage, capacitance, DF (tantalums only)
3	Maximum rated temperature +3, −0 °C or as agreed between manufacturer and the customer. (65, 85, 125 °C steps to maximum rated temperature)	DC leakage, capacitance, DF
4	Standard atmospheric conditions	DC leakage, capacitance, DF

complete cycles of exposure to an atmosphere of 90 to 95 percent relative humidity without precipitation at +40°C for a period of 16 h; followed by 8 h exposure to standard atmospheric conditions, making a total of 120 h. Immediately after the completion of the fifth cycle, the dc leakage, capacitance change, and dissipation factor are measured and must meet the initial requirements.

Military-standard aluminum electrolytics and tantalum electrolytics are required to meet the moisture-resistance test of Method 106 of MIL-STD-202. This test is conducted with no voltage applied and the capacitors mounted by their normal mounting means. Test duration is ten 24-h cycles with a temperature excursion from room temperature to 56°C at a maximum of 98 percent relative humidity. Capacitors are required to meet dc leakage, capacitance change, dissipation factor, and insulation sleeve resistance as specified.

Reverse Voltage (Polarized Tantalum Foil Capacitors). Capacitors with a 6-V rating and above are subjected to a 3-V reverse potential for 125 h at 85°C, then to a rated forward voltage for another 125 h at the same temperature. Changes in room-temperature measurements of dc leakage must not exceed 125 percent of the value in the detail specification, and the capacitance must not exceed 110 percent of the initial measurement. Dissipation factor will conform to the requirement of the detail specification.

Surge Test. Capacitors are subjected to the applicable rated surge voltage shown in Table 8. Multiple-section aluminum electrolytic capacitors should have the surge voltage applied to the first filter section only. For aluminum electrolytics the voltage application should be at normal room temperature and through a series resistor of value of 1,000 ohms for capacitors below 2,500 μF and $R = 2.5 \times 10^6/C(\mu)$ ohms for capacitors of 2,500 μF and up. Surge-voltage application should be for a period of 30 s followed by 4½ min during which each capacitor is discharged through the charging series resistor (or separate resistor). The test is continued for 120 h. Tantalum electrolytic capacitors are tested in a similar manner but at the maximum rated temperature and using a surge resistor value specified in the manufacturer's detail specification. The discharge cycle is 5½ min with a total of 1,000 charge and discharge cycles.

Following the above test, the dc leakage current and DF should meet the initial requirements and the capacitance should not have changed more than 20 percent of the initial measurement for aluminums and 5 percent for tantalums. The capacitor should show no evidence of mechanical damage or leakage of electrolyte. Condensation on the terminals or cover of aluminums, without free-flowing (droplets) electrolyte, is permitted.

Life (Operating). Industry-standard aluminum and tantalum electrolytic capacitors are tested in accordance with the procedures of Method 12, Heat-Life Test, of EIA Standard 186. They must be mounted by their normal or supplemental means and be subjected to continuous rated dc voltage at maximum rated temperature. The length of the life test is specified for each capacitor type and grade. Upon completion of this test, the dc-leakage-current capacitance change and dissipation factor should not exceed the limits specified in the detail specification. In addition to meeting the electrical requirements, the capacitors should show no evi-

TABLE 8   Typical Surge Voltages Allowed for Aluminum and
Tantalum Capacitors at Maximum Rated Temperature

Aluminum foil		Tantalum foil and wet porous anode	
Rated dc volts	DC surge, V*	Rated dc volts	DC surge, V
3	4	3	3.4
5	7	6	6.9
6	8	8	9.2
7	10	10	11.5
10	13	15	17.2
15	20	25	28.8
20	25	30	34.5
25	30	50	57.5
30	40	60	69.0
35	45	75	86.2
40	50	100	115.0
50	65	125	144.0
60	75	150	172.0
75	95	200	230.0
100	125	250	287.0
150	175	300	345.0
175	200	375	431.0
200	250	450	518.0
250	300		
300	350		
350	400		
400	475		
450	525		

Tantalum, Solid Electrolyte

DC rated voltage		DC surge voltage	
85 °C	125 °C	85 °C	125 °C
2	1.3	2.6	1.7
3	2	4	2.7
4	2.7	5	3.5
6	4	8	5
10	7	13	9
15	10	20	12
20	13	26	16
25	17	32	21
30	20	39	26
35	23	46	28
50	33	65	40
75	50	98	64
100	67	130	86

* Surge-voltage limits are reduced for high-performance-grade aluminum foil capacitors.

dence of mechanical damage or leakage of electrolyte.   Condensation on the terminals or cover of aluminums without free-flowing (droplets) electrolyte is permitted.
   Shelf Life (Accelerated).   To demonstrate shelf-life capability of aluminum electrolytes, they are subjected to maximum rated temperature for 250 h with the
exception of general-purpose grades, which are subjected to 100 h, during which

time no voltage is applied.   Within 24 h after completion of the test, the capacitors are measured at standard atmospheric conditions for dc leakage, capacitance, and dissipation factor.   The dc leakage current must not exceed twice the value initially specified.   The capacitance should not change more than 10 percent from the initial measured value.   The dissipation factor should not be more than 1.5 times the value initially specified.   Capacitors should show no evidence of mechanical damage or leakage of electrolyte.   Condensation on the terminals or cover without free-flowing (droplets) electrolyte is permitted.

Aluminum electrolytic capacitors

Electrolyte.   The electrolyte commonly used in aluminum electrolytic capacitors is a combination of the electrolyte and the electrolyte separator and normally consists of an ionogen such as boric acid dissolved in and reacted with glycol to form a pastelike mass of medium resistivity.   This is normally supported in a carrier of a high-purity paper such as kraft or hemp.   In addition to the standard glycol borate electrolyte, low-resistivity nonaqueous electrolytes are used to obtain a lower ESR and increased performance.   The ESR of this material is approximately 20 to 30 percent less than the resistivity of the standard glycol borate electrolyte.   Ethylene glycol and dimethylene formamide are used as basic electrolyte systems to extend the operating-temperature range, improve ripple capability, and lower the characteristic impedance at high frequencies.   These electrolytes are treated with additives (driers) which vary widely between manufacturers.   The more complex electrolyte systems are used in high-performance stable capacitors but at a premium cost.

Foils.   The foils used in the production of aluminum electrolytic capacitors are either standard plain foil or a high-gain etched foil, depending on capacitor voltage rating desired.   It can be seen in Fig. 26 that a high-gain anode foil can be obtained

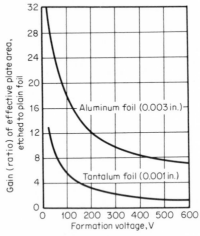

Fig. 26   Approximate gain of effective surface area of electrolytic-capacitor foils vs. formation voltage.

by forming a lower voltage on an etched surface, and thus it is possible to produce capacitors up to 1 F at 3 V.   This type of capacitor is popular in the design of low-voltage, high-current stable power supplies for computer and other IC equipment.   The anode material is always high-grade 99.99 percent pure aluminum.   The cathode material is normally of a purity comparable with or slightly less than that of the anode material.   Because aluminum anodizes so readily, it is desirable to form the cathode at a low voltage to prevent change of capacitance by a series capacitance effect.

Grading.   Aluminum-electrolytic-capacitor grading varies among manufacturers and is only partially standardized.   Aluminum electrolytics, however, can be conveniently graded into four groups as shown in Table 9.   Typical applications for

TABLE 9 Suggested Grading of Aluminum Foil Capacitors (Including EIA and Military Standards)

Grade and features	Standards		Life			Application
	EIA	Military	Shelf, years	Service, years	Test, h	
General-purpose: Low cost Multisection Mixed voltages Plug-in styles	RS 395 (polarized)	None	1	3–5	500–1,000	Consumer products Entertainment equipment Limited temperature range
Long-life: Moderate cost Plug-in styles High-gain foils	RS 395 (polarized)	None	2–3	Up to 10	1,000–2,000	Industrial and telecommunications equipment Avionics Business machines Control equipment
Extended-life: Welded construction High-gain foils Plug-in styles Low ESR		MIL-C-62D (polarized) CE11, CE13 CE34, CE35 CE36, CE44 CE45, CE56 CE57, CE58 CE71	3	>10	1,000–2,500	Wide temperature range Instrumentation Avionics Life stability Standard military RF communications Computer equipment Unattended equipment Automotive
High-performance: High cost Welded construction High-vibration-resistant construction Premium electrolytes Lowest ESR		MIL-C-39018 (polarized) CU13 CU16, CU17 CU71, CU81 CU74, CU15 (nonpolarized)	3–5*	>10	2,000–3,000	High-reliability military and industrial equipment High shock and vibration requirements Low- and high-temperature operation High-speed power switching Aerospace applications

*Refer to shelf-life criteria in MIL-STD-1131.

TABLE 10   Comparison of Aluminum Foil Capacitor Features among Grades

Feature	Grade				Grading factor*
	General-purpose	Long-life	Extended-life	High-performance	
Voltage limit, V	500	450	450	500	
Volume efficiency	1.8	1	1.5	1.2	Highest
High temp, °C	65/85	85	85/105	85/125	
Low temp, °C	−20/−30	−40	−40	−55	
Leakage current	2.0	1.8	1.2	1	Lowest
Equivalent series resistance	1.4	1.3	2	1	Lowest
Capacitance change with temperature	3	2	2	1	Least change
Capacitance change with time	3	2	2	1	Least change
Ripple	1.3	1.3	1.2	1	Best handling capability
Surge voltage	1.3	1.2	1.2	1	Best safety factor
Reliability	4	2	2	1	Highest
Normal life, years	3–5	5–10	10–20	10–20	
Cost	1	1.5	2	4	Lowest

* Lowest number is highest rating.

each grade are also shown but will vary because of equipment design intent, parts costs, and performance desired.   In most cases the general-purpose grade meets requirements for low cost in commercial and consumer equipment.   Higher capacitor grades are required as applications for broader ambient temperature, longer life, low impedance, and improved reliability are specified.   Table 10 shows a comparison

TABLE 11   Recommended Standard Capacitance Ratings and Tolerances for Aluminum Foil Capacitors*

Capacitance Value†

1	8	50	100	600	1,500
2	10	60	200	700	2,000
3	20	70	300	800	3,000
4	30	80	400	900	4,000
5	40	90	500	1,000	5,000

Capacitance Tolerance‡

Rated dc voltage, V	Single-section tolerance, %	Multisection tolerance, %
3 through 100	−10, +100	−10, +150
101 through 300	−10, +75	−10, +100
301 through 450	−10, +50	−10, +50

Values above 5,000 μf should be in 5,000 μf steps.

* EIA Standard RS-395.
† EIA has recommended that future standards be considered using the appropriate decade values shown in Table 1 in Chap. 7 (IEC Publication 63-E6).
‡ Tighter tolerances are usually available at increased cost.

of capacitor voltage range, temperature range, and performance features for each grade which can be used as a guide in performing trade-offs between cost and performance. Table 11 lists standard capacitance values and tolerances for aluminum electrolytics; however, most manufacturers also stock other values or will manufacture on special order. Some may not even produce the recommended standard values. Designers are urged to specify standard values to facilitate interchangeability and provide strength to the EIA and military standards. A guide to available capacitors within each grade is presented in Tables 12 to 15 using a sampling of the industry.

Case Styles. Aluminum electrolytic capacitors in each grade are produced in a few common case configurations. Standard cases are shown in Figs. 27 through 37 and are referred to in the capacitor selection guides for each grade.

Electrical Characteristics. Capacitor performance characteristics of particular interest to the designer are those affected by temperature and frequency. Performance curves vary among manufacturers and are controlled by electrolyte type, winding geometry, lead design, case size, and capacitance/voltage value. Typical curves of temperature effect on capacitance ratio, impedance ratio, and ESR for various capacitance values are shown in Figs. 38, 39, and 40, respectively. Typical curves of impedance variation with frequency for various capacitance values are shown in Fig. 41.

For applications where lowest available impedances are required to provide high-speed-switching regulator filtering, the four-terminal and stacked-foil types are used. Comparison curves[7] of stacked-foil, four-terminal, and standard-foil designs are shown in Figs. 42 and 43. Typical impedance variations with frequency for a range of capacitance values in stacked-foil designs are shown in Fig. 44.

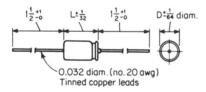

0.032 diam. (no. 20 awg)
Tinned copper leads

Fig. 27 Typical case style and dimensions of general-purpose-grade miniature axial-lead aluminum-foil capacitors. Dimensions listed are standard among many manufacturers.

Standard case dimensions (uninsulated case)		Case insulation thickness* (add to uninsulated case dimensions)		
Diam, in	Length, in	Insulation	Diam, in	Length, in
0.250	0.500	Polyester-film	0.010	0.036
0.250	0.687	Cardboard	0.052	0.063
0.312	0.687	Thermoplastic	0.062	0.062
0.312	0.812			
0.375	0.687			
0.375	0.812			
0.375	0.937			
0.375	1.125			
0.375	1.250			
0.375	1.500			
0.375	1.625			
0.375	2.187			
0.375	2.687			

* Insulation thickness will vary slightly among manufacturers.

TABLE 12 A Guide to General-Purpose-Grade Polarized Aluminum Electrolytic Capacitors[j]

Manufacturer's type[a]	Capacitance,[g] μF	Tolerance, %	Working-voltage range,[g] V dc	Max ripple current at 120 Hz at max temp,[g] A rms	Max ESR at 120 Hz at 25°C,[g] Ω	Max dc leakage current at 25°C,[g] mA	Temp range,[h] °C	Life test,[i] h	Max case size,[k] diam × length, in	Case and terminal type	Features
NLW[b]	1,500 / 35	-10, +75	3 / 150	1.100 / 0.270	0.31 / 5.70	0.015 / 0.053	-40/85	2,500	0.375 × 2.187		Miniature tubular axial-lead case, low cost
TT[c]	5,000 / 40	-10, +150 / -10, +100	3 / 150	0.950 / 0.200	0.16 / 5.00	0.10 / 0.055	-40/85	1,000	0.625 × 1.5 / 0.437 × 2.00	Fig. 27	
500D[d]	2,500 / 10	-10, +75 / -10, +50	3 / 450	1.25 / 0.075	0.18 / 20.0	0.020 / 0.062	-20/85	1,000	0.485 × 1.75		
MTA[c]	4,500 / 280	-10, +100 / -10, +50	3 / 100	0.920 / 0.425	0.19 / 0.48	0.41 / 0.84	-30/65	1,000	0.750 × 2.165	Molded plastic	
MTV[c]	1,400 / 75	-10, +100 / -10, +50	3 / 100	0.575 / 0.202	0.61 / 1.8	0.13 / 0.23	-30/65	1,000	0.5 × 1.5	Molded plastic	Miniature-size vertical mount. For printed-circuit boards. Low cost
VTT[c]	47 / 1.5	-10, +100	6.3 / 50		9.9 / 265.0	0.013 / 0.006	-20/85	500	0.236 × 0.433		
502D[d]	300 / 47	-10, +75 / -10, +50	3 / 50	2.74 / 1.0	1.50 / 5.32	0.009 / 0.004	-20/65	1,000	0.41 × 0.827		
503D[d]	3,300 / 330	-10, +75 / -10, +50	6.3 / 63	1.72 / 0.054	0.11 / 0.61	0.208 / 0.208	-20/85	1,000	0.630 × 1.614	Fig. 28	
PC[b]	1,000 / 6	-10, +150 / -10, +100	6 / 100	0.054 / 0.06	0.53 / 44.2	0.194 / 0.023	-25/65	500	0.5 × 1.0 / 0.375 × 0.5		

Type											Remarks
43D[d]	50,000	-10, +75	3	5.3	0.04	2.32	-20/85	500	1.375 × 4.125		Tubular aluminum case with axial leads. Wide capacitance and voltage range. Dual sections and various terminal options are available in most styles
	150	-10, +50	475	0.815	1.77	1.50			1.375 × 3.625	Fig. 29 (nonstandard sizes)	
WBR[b]	45,000	-10, +150	3	7.8	0.038	2.20	-40/85	2,000	1.375 × 4.125		
	225	-10, +50	450	0.82	0.97	1.90			1.375 × 3.625		
TCW[c]	30,000	-10, +100	3	2.60	0.04	1.8	-20/85	1,000	1.0 × 3.875		
	110	-10, +50	450	0.45	3.50	2.3					
TC[c]	2,000	-10, +100	50	1.200	0.15	4.0	-30/85	1,000	1.0 × 3.812		
	80	-10, +50	475	0.610	5.0	2.5			1.0 × 3.312		
052[e]	20,000	-10, +150	3	2.121	0.15	6.0	-40/85	500	1.0 × 3.562		
	100	-10, +50	450	0.250	2.50	4.3					
FP and PFP[c]	10,000	-10, +100	3	1.4	0.1	1.03	-30/85	1,000	1.0 × 2.5		Also available in multisections (up to 4) with various voltage and capacitance values as desired. Printed circuit and twist-prong mount
	300	-10, +50	450	1.3	0.9	2.20			1.375 × 5.0		
87F[f]	2,000	-10, +150	15	1.00	0.60	20.3	-20/85	500	1.00 × 2.5		
	80	-10, +50	450	0.20	3.12	3.5			1.00 × 3.5	Figs. 30 to 33	
60D[d]	4,000	-10, +100	15	1.50	0.30	1.5	-20/85	500	1.375 × 3.5		
	150	-10, +50	450	0.80	1.77	1.6			1.375 × 4.5		

[a] Manufacturers and capacitor types listed represent a sample of available types and are not to be construed as preferred items.

[b] Cornell Dubilier Electronics.

[c] Mallory Capacitor Company.

[d] Sprague Electric.

[e] Sangamo Electric Company.

[f] General Electric Company

[g] Parameters listed are applicable to the maximum case size for each manufacturer's type shown. Maximum capacitance values at working-voltage limits are listed.

[h] Maximum temperature shown applies to the listed dc working volts without derating.

[i] Standard manufacturer's life test. Capacitor measurements after test are within manufacturer's allowable limits, typically 20 percent maximum loss in capacitance, twice the initial ESR, and no increase in leakage-current limit.

[j] Nonpolarized capacitors are available in most styles on special order and at reduced maximum capacitance in each case size.

[k] Case dimensions are for bare can. Mylar, PVC, or cardboard sleeves are available for all electrolytic capacitors with a slight increase in dimensions.

(Mylar is a registered trademark of E. I. DuPont de Nemours & Co., Wilmington, Del.)

TABLE 13   A Guide to Long-Life-Grade Polarized Aluminum Electrolytic Capacitors[j]

Manufacturer's type[a]	Capacitance,[g] μF	Tolerance, %	Working-voltage range,[g] V dc	Max ripple current at 120 Hz at max temp,[g] A rms	Max ESR at 120 Hz at 25°C,[g] Ω	Max dc leakage current at 25°C,[g] μA	Temp range,[h] °C	Life test,[i] h	Max case size,[k] diam × length, in	Case and terminal type	Features
30D[b]	800 / 20	-10, +75 / -10, +50	3 / 150	0.40 / 0.064	0.56 / 10.0	2.8 / 22.5	-20/85	2,000	0.375 × 1.5		Miniature case size with good reliability at low cost
76F[c]	790 / 20	-10, +75 / -10, +50	3 / 150	0.75 / 0.12	0.51 / 10.0	3.0 / 22.5	-20/85	500	0.375 × 1.5	Fig. 27	
TTX[d]	4,500 / 40	-10, +150 / -10, +100	3 / 150	0.950 / 0.250	0.13 / 3.70	299.0 / 20.0	-55/85	2,000	0.625 × 1.5 / 0.437 × 2.0		
NLW[e]	1,500 / 35	-10, +75	3 / 150	1.10 / 0.220	0.31 / 5.7	15.0 / 53.0	-40/85 (105)	2,500	0.375 × 2.187		
39D[b]	18,000 / 75	-10, +75 / -10, +50	3 / 450	2.15 / 0.50	0.17 / 3.18	1,000 / 1,000	-20/85	500	1.0 × 3.625		Highest capacitance in tubular case. High ripple capability
84F[c]	18,000 / 100	-10, +75 / -10, +50	3 / 450	3.364 / 0.64	0.073 / 2.0	900 / 850	-20/85	1,000	1.0 × 3.625		
TCG[d]	22,000 / 100	-10, +75 / -10, +50	3 / 450	1.738 / 0.419	0.12 / 1.86	771 / 636	-40/85	500	1.0 × 3.625	Fig. 29	
WHB[e]	18,000 / 75	-10, +75 / -10, +50	3 / 450	3.65 / 0.465	0.060 / 2.70	464 / 368	-40/85	2,000	1.0 × 3.625		
066[f]	24,000 / 120	-10, +75 / -10, +50	3 / 450	2.58 / 0.64	0.063 / 2.08	1,000 / 1,000	-40/85	1,000	1.0 × 3.625		

Type									Remarks
36D[b]	270,000 1,200	−10, +75 −10, +50	3 450	20.6 6.3	0.020 0.21	6,000 6,000	−40/65 (85)	500	3.0 × 8.625	Highest capacitance range. Highest volumetric efficiency. High ripple capability
36DX[b]	650,000 3,100	−10, +75 −10, +50	3 450	30.7 10.2	0.012 0.11	6,000 6,000	−40/65 (85)	500	3.0 × 8.625	
86F[c]	370,000 1,900	−10, +100 −10, +50	5 450	28.0 7.6	0.011 0.11	4,000 4,000	−40/65 (85)	2,000	3.0 × 8.625 3.0 × 5.625	
86F500[c]	540,000 34,000	−10, +100 −10, +50	5 100	24.7 18.6	0.016 0.028	4,000 4,000	−40/65 (85)	2,000	3.0 × 8.625	
CGS[d]	530,000 1,700	−10, +75 −10, +50	3 450	25.0 4.6	0.005 0.141	6,000 5,250	−40/65	1,000	3.0 × 8.625 3.0 × 5.875	
FAH[e]	1,000,000 1,600	−10, +100 −10, +50	3 450	30.6 7.1	0.010 0.080	6,000 5,100	−40/85	2,000	3.0 × 8.625 3.0 × 5.625	Fig. 34
DCM[f]	540,000 13,000	−10, +75	5 150	26.4 18.6	0.009 0.018	6,000 6,000	−40/65 (85)	1,000	3.0 × 8.625	
DCM (ext)[f]	1,000,000 100,000	−10, +75	3 50	26.4 23.8	0.009 0.011	6,000 6,000	−40/65 (85)	1,000	3.0 × 8.625	

[a] Manufacturers and capacitor types listed represent a sample of available types and are not to be construed as preferred items.
[b] Sprague Electric Company.
[c] General Electric Company.
[d] Mallory Capacitor Company.
[e] Cornell Dubilier Electronics.
[f] Sangamo Electric Company.
[g] Parameters listed are applicable to the maximum case size for each manufacturer's type shown. Maximum capacitance values at working-voltage limits are listed.
[h] Maximum temperature shown applies to the list dc working volts without derating. Temperatures in parentheses are design limits but at reduced voltage.
[i] Standard manufacturer's life test. Capacitor measurements after test are within manufacturer's allowable limits, typically 20 percent maximum loss in capacitance, twice the initial ESR, and no increase in leakage-current limit.
[j] Nonpolarized capacitors are available in most styles on special order and at reduced maximum capacitance in each case size.
[k] Case dimensions are for bare can. Mylar, PVC, or cardboard sleeves are available for all electrolytic capacitors with a slight increase in dimensions. (Mylar is a registered trademark of E. I. DuPont de Nemours & Co, Wilmington, Del.)

TABLE 14 A Guide to Extended-Life-Grade Polarized Aluminum Electrolytic Capacitors[j]

Manufacturer's type[a]	Capacitance,[g] μF	Tolerance, %	Working-voltage range,[g] V dc	Max ripple current at 120 Hz at max temp,[g] A rms	Max ESR at 120 Hz at 25°C,[g] Ω	Max dc leakage current at 25°C,[g] μA	Temp range[h] min/max, °C	Life test,[i] h	Max case size,[k] diam × length, in	Case and terminal type	Features
40D[b]	550 15	−10, +75 −10, +50	3 150	0.47 0.077	0.45 13.33	2.0 12.0	−40/105	1,000	0.375 × 1.5		Medium-cost miniature with low impedance and good stability over 10 years
77F[c]	560 15	−10, +75 −10, +50	3 150	0.331 0.084	0.47 9.00	2.0 12.0	−40/105	2,000	0.375 × 1.5 0.375 × 1.25	Fig. 27	
78F[c]	790 20	−10, +75 −10, +50	2 100	0.324 0.084	0.34 6.75	2.0 15.0	−40/125	2,000	0.375 × 1.5		
TTQ[d]	300 10	−10, +100	3 150	0.15 0.07	1.0 3.0	1.7 15.5	−40/85	1,000	0.375 × 1.5		
TPG[d]	520 18	−10, +100 −10, +75	3 150	0.212 0.115	0.77 14.0	2.05 19.08	−40/85	1,000	0.375 × 1.625	Fig. 29	
556[e]	2,000 55	−10, +100	3 150	0.298 0.049	0.23 4.73	4.5 37.0	−40/105	1,000	0.5 × 2.125		

Type	Capacitance	Tolerance, %		ESR				Temp range, °C		Dimensions, in		
34D[b]	13,000 / 70	−10, +75 / −10, +50	2.5 / 450	3.750 / 0.520	0.056 / 3.0	360 / 355	−40/85	1,500	1.0 × 3.625	Fig. 29	High-reliability tubular style with good stability over life
056[e]	15,000 / 80	−10, +100 / −10, +50	3 / 450	2.449 / 0.268	0.030 / 2.75	424 / 380	−40/85	1,500	1.0 × 3.625		
32D[b]	200,000 / 900	−10, +75 / −10, +50	2.5 / 450	16.2 / 4.2	0.018 / 0.25	2,121 / 1,908	−40/85	1,500	3.0 × 5.625		Highest stability in 85°C capacitors
88F[c]	150,000 / 4,500	−10, +75 / −10, +50	5 / 150	15.8 / 9.7	0.015 / 0.039	4,000 / 4,000	−40/85	1,500	3.0 × 5.625	Fig. 34	
CG[d]	470,000 / 1,600	−10, +75 / −10, +50	3 / 450	15.0 / 2.8	0.006 / 0.15	3,562 / 2,545	−40/85	1,500	3.0 × 8.625 / 3.0 × 5.875		
500[e]	570,000 / 13,000	−10, +75	3 / 150	21.2 / 11.8	0.007 / 0.019	6,000 / 6,000	−40/85	1,500	3.0 × 8.625		

[a] Manufacturer's and capacitor types listed represent a sample of available types and are not to be construed as preferred items.
[b] Sprague Electric Company.
[c] General Electric Company.
[d] Mallory Capacitor Company.
[e] Sangamo Electric Company.
[f] Cornell Dubilier Electronics.
[g] Parameters listed are applicable to the maximum case size for each manufacturer's type shown. Maximum capacitance values at working-voltage limits are listed.
[h] Maximum temperature shown applies to the listed dc working volts without derating.
[i] Standard manufacturer's life test. Capacitor measurements after test are within manufacturer's allowable limits, typically 20 percent maximum loss in capacitance, twice the initial ESR, and no increase in leakage-current limit.
[j] Nonpolarized capacitors are available in most styles on special order and at reduced maximum capacitance in each case size.
[k] Case dimensions are for bare can. Mylar, PVC, or cardboard sleeves are available for all electrolytic capacitors with a slight increase in dimensions.

TABLE 15 A Guide to High-Performance-Grade Aluminum Electrolytic Capacitors[j]

Manufacturer's type[a]	Capacitance,[g] μF	Tolerance, %	Working-voltage range,[g] V dc	Max ripple current at 120 Hz at max temp,[g] A rms	Max ESR at 150 Hz at 25°C,[g] Ω	Max dc leakage current at 25°C,[g] μA	Temp range[h] min/max, °C	Life test,[i] h	Max case size,[k] diam × length, in	Case and terminal type	Features
600D[b]	1,000 27	−10, +75 −10, +50	5 200	3.333 0.968	0.265 5.0	10.0 15.0	−55/125	2,000	0.375 × 2.687		Meet requirements of MIL-C-39018 and wide-temperature-range applications. Replaces tantalum for cost saving
610D[b] nonpolarized	680 10	−10, +75 −10, +50	5 200	0.700 0.100	0.40 20.0	24.0 36.0	−55/125	2,000	0.375 × 2.750	Fig. 27	
HTA[c] (CU12)	1,000 12	−10, +75 −10, +50	5 275	1.356 0.148	0.265 16.575	10.0 21.0	−55/125	2,000	0.375 × 2.688		
UHL[d]	1,000 27	−10, +75	5 200	6.060 2.852	0.265 5.0	10.0 10.0	−55/125	2,000	0.375 × 2.687		
557[e]	1,000 39	−10, +75	5 150	6.060 3.090	0.265 3.333	10.0 10.0	−55/125	2,000	0.375 × 2.687		
HNLH[d]	600 15	−10, +75	3 150	0.365 0.068	0.45 9.0	2.1 12.0	−80/110	2,000	0.375 × 1.5		
UHT[d]	1,000 39	−10, +75	3 100	0.015 0.170	0.540 3.46	10.0 10.0	−55/150	2,000	0.375 × 2.687		
601D[b]	13,000 290	−10, +75 −10, +50	5 200	4.10 1.20	0.046 0.517	127.0 120.0	−55/85	2,000	1.0 × 3.625		High capacitance, stable types for wide temperature applications. Replaces tantalum for cost saving
HTA[c] (CU16)	12,000 120	−10, +75 −10, +50	7 350	7.92 0.079	0.044 1.66	127.0 112.0	−55/85	2,000	1.0 × 3.625	Fig. 29	
UHH[d] UHA[d]	13,000 290	−10, +75	5 200	2.967 0.774	0.040 0.48	120.0 120.0	−55/105	2,000	1.0 × 3.625		
057[e]	17,000 350	−10, +75 −10, +50	5 200	5.57 1.47	0.0294 0.428	146.0 132.0	−55/85	2,000	1.0 × 3.625		

Type										
UFT[d]	16,000 320	−10, +75	5 200	7.0[l] 2.5[l]	0.022 0.405	142.0 126.0	−55/105	2,000	1.0 × 3.625	Fig. 35	Extreme low inductance. Low loss, high capacitance, maximum reliability for medium- and high-power applications
602D[b]	220,000 6,700	−10, +75 −10, +50	5 100	28.3 6.3	0.010 0.030	1,573.0 1,228.0	−55/85	2,000	3.0 × 5.625	
92F[f]	180,000 6,500	−10, +50	5 150	22.2 15.7	0.011 0.022	1,700.0 1,267.0	−40/85	3,000	3.0 × 5.625	
UFH[d]	310,000 4,500	−10, +100 −10, +75	7(5)[m] 200(150)[m]	32.5(16.9)[m] 13.8(7.1)[m]	0.010 0.038	1,900.0 1,200.0	−55/85 (105)	2,000 at 105°C	3.0 × 8.625 3.0 × 5.625	
100[e]	250,000 36,000	−10, +75	3[m] 50[m]	31.0(10)[m] 29.0(9)[m]	0.0075 0.0087	2,000.0 2,000.0	−55/85 (105)	2,000 at 105°C	3.0 × 5.625	
432D[b] SFC[c]	100,000 10,000	−0, +100	5 50	52.0 18.0	0.0015 0.012	500.0 5,000.0	−40/85	2,000	3.0 × 5.625	Stacked-foil, ultra-low impedance. See Fig. 36
640D[b]	20,000 160	−20, +20	10(7)[n] 500(400)[n]	30.0(17.1)[n] 2.9(1.65)[n]	0.015 0.86	240.0 90.0	−55/85 (125)	2,000	Fig. 37	Hermetic-sealed rectangular case

[a] Manufacturer's and capacitor types listed represent a sample of available types and are not to be construed as preferred items.
[b] Sprague Electric Company.
[c] Mallory Capacitor Company.
[d] Cornell Dubilier Electronics.
[e] Sangamo Electric Company.
[f] General Electric Company.
[g] Parameters listed are applicable to the maximum case size for each manufacturer's type shown. Maximum capacitance values at working-voltage limits are listed.
[h] Maximum temperature shown applies to the listed dc working volts without derating. Temperatures in parentheses are designed limits but at reduced working voltage.
[i] Standard manufacturer's life test. Capacitor measurements after test are within manufacturer's allowable limits, typically 20 percent maximum loss in capacitance, twice the initial ESR, and no increase in leakage-current limit.
[j] Nonpolarized capacitors are available in most styles on special order and at reduced maximum capacitance in each case size.
[k] Case dimensions are for bare can. Mylar, PVC, or cardboard sleeves are available for all electrolytic capacitors with a slight increase in dimensions.
[l] Ripple current for a 10°C rise in case temperature.
[m] Rated values at 105°C.
[n] Rated values at 125°C.

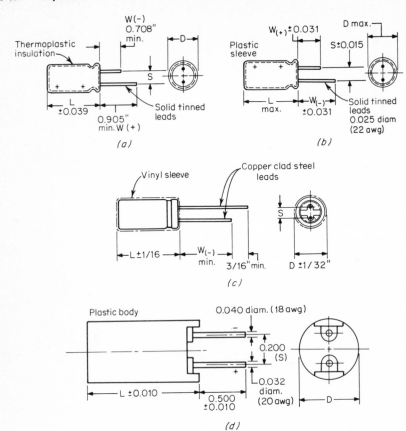

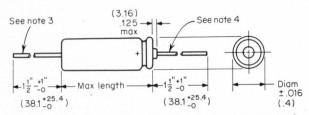

Metal - case axial lead tubular electrolytic capacitors (case negative)

Fig. 28 (Top, left) Typical single-ended general-purpose-grade aluminum foil capacitor designs showing variations and degree of nonstandardization. Close attention must be given in design use and replacement of this type of capacitor. Dimensions L, D, W, and S vary among manufacturers.

Fig. 29 (Bottom, left) Standard RS-395, style 30 tubular aluminum foil capacitor[2] configuration. (*Electronic Industries Association.*[5])

Standard Case Dimensions[1] (Basic Uninsulated Case)			
Max length		Diameter ±0.016 in (0.4)	
in	mm	in	mm
1.687	42.86	0.375	9.53
2.187	55.56	0.500	12.7
2.687	68.26	0.625	15.88
2.687	68.26	0.750	19.05
3.187	80.96	0.875	22.23
3.687	93.66	1.000	25.4
4.687	119.05	1.375	34.93

1. Metric equivalents tabulated and in parentheses (based on 1 in = 25.4 mm).
2. These capacitors are not intended to be mounted by their leads. They may be provided with a supplementary means of mounting such as a tangential bracket or wraparound bands.
3. Standard leads should be 0.032 in (0.79), 20 AWG for case diameters of 0.625 in (15.88) or less and 0.040 in (1.02), 18 AWG for case diameters greater than 0.625 in (15.88).
4. Case insulation (if required): (a) Plastic case insulation should lap over end or extend 0.014 in (0.4) minimum beyond each end. The increase in diameter should not exceed 0.032 in (0.79). (b) Waxed cardboard or acetate insulation should increase case length a maximum of 0.187 in (4.75). The increase in diameter should not exceed 0.062 in (1.59).

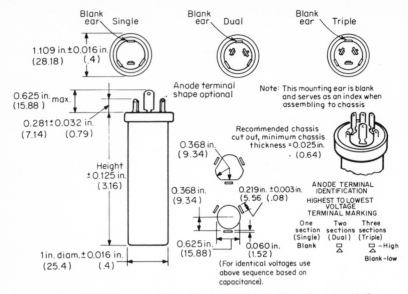

One-inch-diameter twist-mount aluminum electrolytic capacitors (case negative).

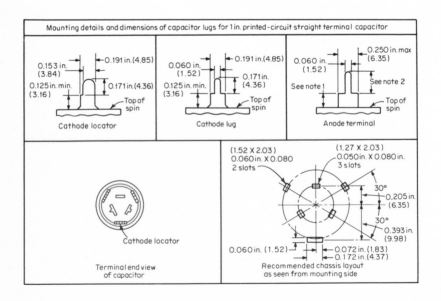

Fig. 30 (Top, left) EIA Standard RS-395, style 10 can-type aluminum foil capacitor configuration, general-purpose grade. (*Electronic Industries Association.*)[5]

Standard Case Heights

in	mm
1.5	38.1
2.0	50.8
2.5	63.5
3.0	76.2
3.5	88.9
4.0	101.6

1. Where no tolerances are given, manufacturer's normal tolerances shall apply.
2. Metric equivalents tabulated and in parentheses (based on 1 in = 25.4 mm).
3. Case diameter and heights are basic uninsulated dimensions.

Fig. 31 (Bottom, left) EIA Standard RS-395, style 11 one-inch-diameter printed-circuit mounting for electrolytic aluminum foil capacitors, general-purpose grade (case negative). (*Electronic Industries Association.*)[5]

1. Less than dimension from top of spin to cathode lug shoulder (manufacturer's standard).
2. Manufacturer's standard.
3. Case sizes and anode terminal markings are the same as style 10,1-in-diameter units.
4. Where no tolerances are given, manufacturer's normal tolerances should apply.
5. Metric equivalents in parentheses (based on 1 in = 25.4 mm).

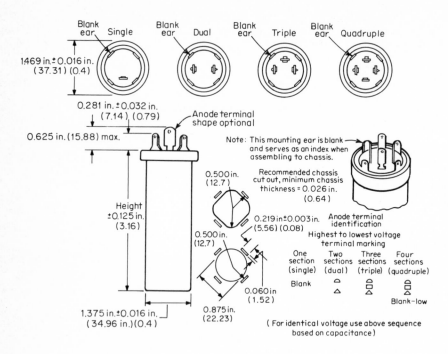

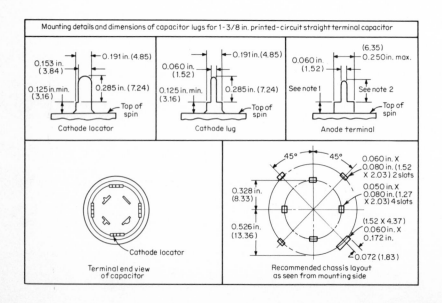

Fig. 32 (Top, left) EIA Standard RS-395, style 12 can-type aluminum foil capacitor configuration, general-purpose grade, maximum case size: 1⅜-in.-diameter twist-mount aluminum electrolytic capacitors (case negative). (*Electronic Industries Association.*)[5]

Standard Case Heights	
in	mm
2	50.8
2.5	63.5
3	76.2
3.5	88.9
4	101.6
4.5	114.3
5	127

1. Where no tolerances are given, manufacturer's normal tolerances shall apply.
2. Metric equivalents tabulated and in parentheses (based on 1 in = 25.4 mm).
3. Case diameter and heights are basic uninsulated dimensions.

Fig. 33 (Bottom, left) EIA Standard RS-395, style 13 printed-circuit mounting for maximum-case-size aluminum foil capacitors, general-purpose grade (case negative). (*Electronic Industries Association.*)[5]

1. Less than dimension from top of spin to cathode lug shoulder. (manufacturer's standard).
2. Manufacturer's standard.
3. Case sizes and anode terminal markings are the same as style 12, 1⅜-in-diameter units.
4. Where no tolerances are given, manufacturer's normal tolerances shall apply.
5. Metric equivalents in parentheses (based on 1 in = 25.4 mm).

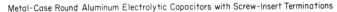

Metal-Case Round Aluminum Electrolytic Capacitors with Screw-Insert Terminations

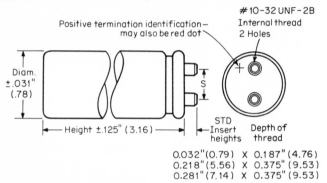

0.032"(0.79) X 0.187"(4.76)
0.218"(5.56) X 0.375"(9.53)
0.281"(7.14) X 0.375"(9.53)

Fig. 34 EIA Standard RS-395, style 40 can-type aluminum foil capacitor configuration. (*Electronic Industries Association.*)[5]

1. Where no tolerances are given, manufacturer's normal tolerances should apply.
2. Metric equivalents tabulated and in parentheses (based on 1 in = 25.4 mm).
3. Case diameters and heights are basic uninsulated dimensions.
4. Screws for inserts not specified.

Standard Case Dimensions

in	mm	in	mm	in	mm	in	mm	in	mm
				Case diameters					
1.375	34.93	1.75	44.45	2.0	50.8	2.5	63.5	3.0	76.2
				Insert spacing					
0.50	12.7	0.75	19.05	.875	22.23	1.125	28.58	1.25	31.75
				Heights					
2.125	53.98	2.125	53.98	2.125	53.98	3.125	79.38	3.625	92.08
2.625	66.68	2.625	66.68	2.625	66.68	3.625	92.08	4.125	104.78
3.125	79.38	3.125	79.38	3.125	79.38	4.125	104.78	4.625	117.48
3.625	92.08	3.625	92.08	3.625	92.08	4.625	117.48	5.125	130.18
4.125	104.78	4.125	104.78	4.125	104.78	5.125	130.18	5.625	142.88
4.625	117.48	4.625	117.48	4.625	117.48	5.625	142.88	6.625	168.29
5.125	130.18	5.125	130.18	5.125	130.18			8.625	219.09
5.625	142.88	5.625	142.88	5.625	142.88				

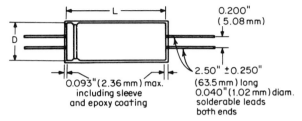

Fig. 35 Mechanical configuration of a four-terminal aluminum foil capacitor. (*Cornell-Dubilier Electronics.*)

D	L
$3\frac{3}{4}$	$1\frac{5}{8}$
$3\frac{3}{4}$	$2\frac{1}{8}$
$3\frac{3}{4}$	$2\frac{5}{8}$
$3\frac{3}{4}$	$3\frac{1}{8}$
$3\frac{3}{4}$	$3\frac{5}{8}$
$7\frac{7}{8}$	$1\frac{5}{8}$
$7\frac{7}{8}$	$2\frac{1}{8}$
$7\frac{7}{8}$	$2\frac{5}{8}$
$7\frac{7}{8}$	$3\frac{1}{8}$
$7\frac{7}{8}$	$3\frac{5}{8}$
1	$1\frac{5}{8}$
1	$2\frac{1}{8}$
1	$2\frac{5}{8}$
1	$3\frac{1}{8}$
1	$3\frac{5}{8}$

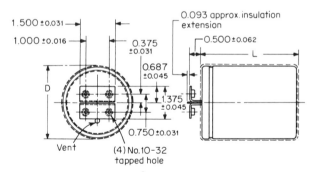

Fig. 36 Mechanical configuration of high-performance-grade stacked-foil aluminum electrolytic capacitor. (*Sprague Electric Company.*)

No outer sleeve		Outer insulation	
$D \pm 0.031$	$L \pm 0.062$	D max	L max
3.000	4.125	3.078	4.250
3.000	5.625	3.078	5.750

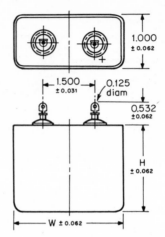

Fig. 37 Hermetically sealed rectangular-case aluminum foil capacitor type 640D. (*Sprague Electric Company.*)

Table of Dimensions, in

	Series B case				Series E case		
Case code	H	W	Weight, g	Case code	H	W	Weight, g
B1	1.500	1.593	83	E1	2.750	3.125	290
B2	1.875	1.593	105	E2	3.125	3.125	335
B3	2.250	1.593	124	E3	3.500	3.125	375
B4	2.265	1.593	140	E4	3.875	3.125	407
B6	3.375	1.593	178	E5	4.250	3.125	440
				E6	4.625	3.125	460
				E7	5.000	3.125	475

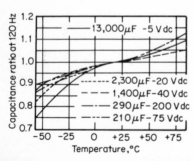

Fig. 38 Typical curves of capacitance as a function of temperature for high-performance-grade aluminum foil capacitors. (*Sprague Electric Company.*)

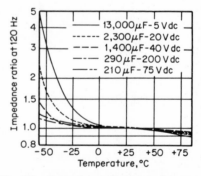

Fig. 39 Typical curves of impedance as a function of temperature for high-performance-grade aluminum foil capacitors. (*Sprague Electric Company.*)

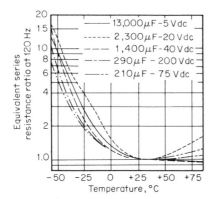

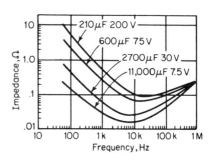

Fig. 40 Typical curves of equivalent series resistance as a function of temperature for high-performance-grade tubular aluminum foil capacitors. (*Sprague Electric Company.*)

Fig. 41 Typical impedance variation with frequency (at 25°C) for high-performance-grade tubular aluminum foil capacitors. (*Cornell-Dubilier Electronics.*)

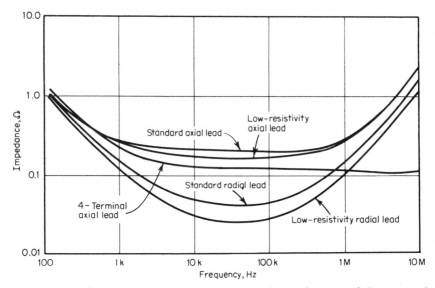

Fig. 42 Impedance characteristics of 1,000-μF, 7.5-V aluminum foil capacitor.[7]

Tantalum electrolytic capacitors This type of capacitor is smaller for equivalent ratings than the aluminum electrolytic type and is considerably more expensive. Tantalum electrolytics, however, possess superior characteristics for shelf life and load life, and they perform over a broad temperature range. The capacitance range

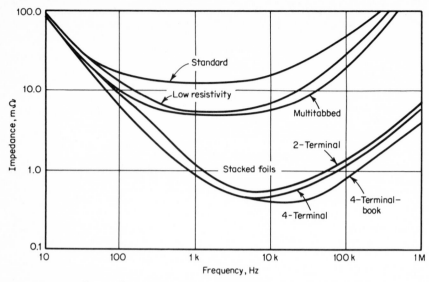

Fig. 43 Impedance characteristics of 150,000-μF, 5-V aluminum foil capacitor.[7]

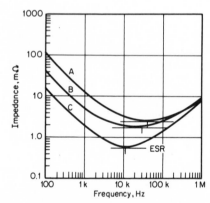

Fig. 44 Typical curves of impedance vs. frequency at 25°C for high-performance-grade stacked-foil aluminum electrolytic capacitor.

Curve	Capacitance, μF	Voltage rating, Vdc	Inductance, nH	ESR, MΩ
A	10,000	50	1.04	2.8
B	27,000	20	1.16	1.7
C	100,000	5	1.09	0.53

for tantalums is smaller than that of aluminum electrolytics. Tantalum electrolytic capacitors have been standardized to a good degree and governed by well-established industry and military specifications, as shown in Table 16. Unlike the aluminum electrolytics, tantalums need not be graded further.

1. The tantalum foil type, plain or etched, is manufactured like aluminum foil electrolytics and is made in polar and nonpolar configurations and in hermetic and nonhermetic cases.

TABLE 16   A Specification Guide for Tantalum Electrolytic Capacitors

| Type | EIA type | Military type | |
		Standard	High reliability
Foil	RS-228	MIL-C-3965^e	MIL-C-39006
Plain, polar	ECL-34,^{a} ECL-35^a	CL 30, CL 31	CLR 35
		CL 51	
Plain, nonpolar	ECL-36,^{a} ECL-37^a	CL 32, CL 33	CLR 37
		CL 52	
Etched, polar	ECL-24,^{a} ECL-25^a	CL 20, CL 21	CLR 25
		CL 53	CLR 53
		CL 70, CL 71	CLR 71
Etched, nonpolar	ECL-26,^{a} ECL-27^a	CL 22, CL 23	CLR 27
		CL 54	
		CL 72, CL 73	CLR 73
Sintered slug			
Wet, polar	RS-228	MIL-C-3965^e	MIL-C-39006
	ECL-64,^{a} ECL-65^a	CL 10, CL 13	CLR 10
		CL 14, CL 16	CLR 14
	ECL-66, ECL-67	CL 17, CL 18	CLR 17
		CL 55	CLR 65^b
		CL 66, CL 67	CLR 69
Dry, polar	RS-228	MIL-C-26655^e	MIL-C-39003
	ECS-12, ECS-13	CS 12, CS 13	CSR 13, CSR 09
	ECS-22, ECS-23		CSR 23
	ECS-15^b		MIL-C-55365
	ECS-16^b		CWR 01^c, CWR 02^c
	ECS-05^c		CWR 03^d, CWR 04^d
	ECS-09^d		CWR 07^f, CWR 08^b
Dry, nonpolar			CSR 91

^{a} Nonhermetic-seal type.
^{b} Plastic-encapsulated type.
^{c} Unencapsulated type.
^{d} Molded-chip type.
^{e} These specifications are not used for new design.
^{f} Cordwood design with molded case and radial leads.

2. The porous sintered-anode type is manufactured with either solid or wet electrolyte. The liquid-electrolyte type, commonly known as the wet-slug type, has the best performance characteristics and highest volumetric efficiency of any capacitor. The solid-electrolyte type possesses extremely long life characteristics and is produced in subminiature and chip styles.

3. The etched-wire type, similar to the porous-anode construction, is made in low capacitance and voltage values for use in hearing aids.

Tantalum electrolytic capacitors are more reliable than aluminum electrolytics because of the extreme chemical stability and corrosion resistance of the tantalum oxide film. Strong low-resistivity electrolytes may be employed with safety in tantalum capacitors.

Tantalum Foil Capacitors. This capacitor type is constructed of tantalum wires welded to each of two tantalum foils. The wires are then extended through the case seal and terminated in a weld to the external copper lead, as shown in Fig. 45. Inorganic salts dissolved in dimethylene formamide or neutral-type compounds such as lithium chloride or glycol borate are used as the basic electrolyte system. Where an electrolyte carrier or separator is required, paper or glass fiber may be used as barriers. In polarized devices the anode foil is formed and the cathode is usually plain tantalum foil.   In nonpolarized units both foils are formed alike. Tantalum foils, like aluminum foils, possess the feature of drying out with storage life or service life as a result of vapor transmission through elastomeric-case seals. These capacitors are particularly well suited to nonpolar requirements or any application where some reverse voltage is required.   These capacitors are manufactured in tubular styles as well as rectangular cases.   See Figs. 46–53.   Military styles and commercial styles available with capacitance range and other performance parameters are shown in Table 17.

Applications of tantalum foils compared with aluminum foils should be traded off with cost vs. lower leakage and longer shelf life characterized by tantalum foils. In addition to these features tantalum foils have a proved reliability history.   The foils used in this type of capacitor construction are usually 0.0005 in thick and may be either plain-etched or high-etched.   The approximate gain of etched tantalum foils shown in Fig. 26 indicates a gain in capacitance of 13 at 25 V and decreases to approximately 3 at 200 V.   The additional cost of manufacture in the etched- and high-etched-foil types sometimes makes the use of plain-foil construction in a larger case size more economical, especially where space is not at a premium.   Plain-foil types in general have superior temperature stability over etched foil in the same voltage rating.   The electrolyte systems employed throughout the voltage range may vary, however, and temperature-stability characteristics are dependent to a considerable degree on the system used.   When examining the construction of a tantalum foil capacitor, it should be noted that in the sealing of this type of capacitor aluminum or silver is crimped into a Teflon bushing with an elastomer end disk compressed to seal the package.   Teflon is used because it is inert to practically all other materials.   However, it has the property of cold flowing to relieve stresses; therefore, the crimping of a metal case of this type must be designed such that during temperature excursions no excess stress is built up.   Otherwise the shape will change and at lower temperatures the seal will become separated from the crimp wall and become a potential leak failure.   Some styles of tantalum foils are hermetically sealed by a multiple sealing system.   It should also be noted that tantalum wires are used

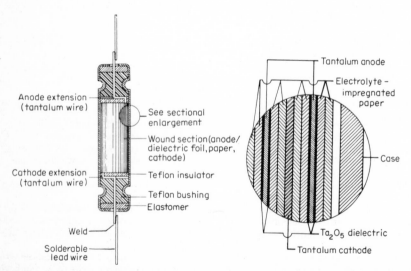

Fig. 45 Typical construction of a tantalum foil capacitor.   In nonpolar types the cathode foil is formed as an anode.   Teflon is a registered trademark of E. I. Du Pont de Nemours & Co., Wilmington, Del.

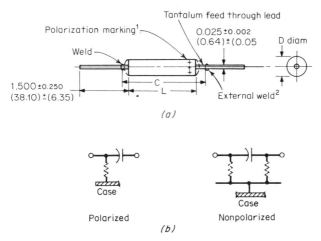

Fig. 46   (a) Standard tubular-case configuration for tantalum foil capacitors.   Case sizes include EIA and military standard sizes.   (b) Circuit diagrams.[6]

Case size			Dimensions[3]			
Refer-ence no.[b]	RS-228	MIL-C-3900b	L[4]	D (basic case) ± 0.016 (0.41)	D (insulated case) max	C max
1	C1	E1	0.688 (17.48)	0.188 (4.78)	0.219 (5.56)	1.188 (30.18)
2	C2	E2	0.906 (23.01)	0.281 (7.14)	0.312 (7.92)	1.406 (35.71)
3		G2	0.969 (24.61)	0.281 (7.14)	0.312 (7.92)	1.469 (37.31)
4	C3	E3 or G3	1.438 (36.53)	0.375 (9.53)	0.406 (10.31)	1.938 (49.23)
5	C4	E4 or G4	2.125 (53.98)	0.375 (9.53)	0.406 (10.31)	2.625 (66.68)
6	C5	E5 or G5	2.750 (69.85)	0.375 (9.53)	0.406 (10.31)	3.250 (82.55)
7			2.750 (69.85)	0.469 (11.91)	0.500 (12.70)	3.250 (82.55)
8			2.750 (69.85)	0.531 (13.50)	0.561 (14.28)	3.250 (82.55)

1. Polarization marking omitted on nonpolarized units.
2. Weld joint shall not be enclosed in the end seal.
3. Dimensions are in inches with millimeters shown in parentheses.
4. For insulated styles, nonshrinkable sleeving shall extend 0.016 in minimum, 0.062 in maximum, beyond each end of the case; shrinkable sleeving shall lap over the ends of the case.
5. There is an indeterminate resistance between the metal case and the negative terminal.
6. Case size reference numbers used for convenience in this handbook.

as foil tabs and are welded outside the package to a solderable lead.   Care must be taken in all cases to prevent the bending of the tantalum wire in installation in hardware, since bending of the wire again could exercise forces on the Teflon seal and induce leakage.   Leakage of this type would hasten the drying out of electrolytes in the capacitor.   Leads must not be clipped short of the weld if a reliable solderable joint is to be achieved.

Tantalum foil types are the only electrolytic capacitors capable of operating continuously to rated dc levels on unbiased ac voltages.   Ripple capability shown in Fig. 54 is applicable for unbiased ac voltages on nonpolar types and biased ac ripple voltages on polar types.   Peak ac voltages up to 150 V can be allowed if the dc voltage rating is not exceeded, the only limitation being the I^2R heating effect.   Etched-foil types have higher ESR and as such have only half the ac capabil-

TABLE 17 Selection Guide for Tantalum Electrolytic Capacitors

Construction type	Capacitance range, μF	Manufacturers type[a] (commercial)	EIA Standard RS-228 type[b]	Military standard type[b,c]	DC working-voltage range, V — At 85°C	At 125°C	DC leakage-current range, μA — At 25°C	At max temp	Max impedance range at −55°C, 120 Hz, Ω	AC ripple-capability range at 25°C, 60 Hz, mA	Case — Size reference code, length × diam, in	Style[b]	Seal
Plain foil, polarized	0.15–300	TFS,[d] TPF,[f] 29F,[g] TAND[h]	ECL34, ECL25	CLR35	3–450	2–300	1–100	2–1,800	9–21,400	24–1,670	1–5, Fig. 46	Metal, tubular, insulated or uninsulated	Hermetic and non-hermetic
Plain foil, nonpolarized	0.12–400	TFS,[d] TPF,[f] 29F,[g] TAND[h]	ECL36, ECL37	CLR37	3–300	2–250	1–100	2–525	6–17,500	16–1,940			
Etched foil, polarized	1.0–920	TFE,[d] TEF,[f] 29F,[g] TUF,[f] TAND[h]	ECL24, ECL25	CLR25, CLR71	15–150	10–100	2–70	4–750	≤3,385	35–2,540			
Etched foil, nonpolarized	0.5–590	TFE,[d] TEF,[f] 29F,[g] TUF,[f] TAND[h]	ECL26, ECL27	CLR27, CLR73	15–150	10–100	2–35	4–470	6.2–5,770	20–2,030			
Plain foil, polarized	40–960	RP,[f] 29F[g]		CL51	15–150	10–100	23–86	230–860	2.9–69.5	1,140–7,070	A, B, C (Fig. 47)	Metal, rectangular	Hermetic
Plain foil, nonpolarized	20–600	RP,[f] 29F[g]		CL52	15–150	10–100	15–65	150–650	4.64–139.0	806–5,590			
Etched foil, polarized	70–3,500	RE,[f] 29F,[g] RC-53[h]		CLR53	15–150	10–100	24–122	240–1,220	1.2–41.2	674–7,790	B	Insulated or uninsulated	
Etched foil, nonpolarized	35–2,500	RE,[f] 29F[g]		CL54	5–150	5–100	24–147	240–1,470	2.4–84.5	477–6,040	A, B, C		
Sintered anode, wet electrolyte, polarized	2–140	XTM-XTK,[d] 140D,[e] XNW[i]		CLR10[j]	8–360	7–310	5–12	40–96	30–1,000	108–213[m]	0.438 × 0.656 to 1.781 × 0.656	Metal, tubular, insulated or uninsulated	Hermetic
	3.5–200	XTL-XTH,[d] 140D,[e] XNW[i]		CLR14[l]	20–630	17.5–560	10–18	75–135	20–630	250–375[m]			
	12–1,300	XTV,[d] 141D,[e] XNW[i]		CLR17[l]	30–630	26–545	17–29	170–290	10–200	306–694[m]	0.600 × 1.125 to 2.812 × 1.125		
	1.7–1,200	TLW,[d] 138D,[e] ZNW,[i] 69F[v]		CLR65	6–125	4–85	1–10	2–40	20–1,250	50–750	0.453 × 0.188 to 1.065 × 0.375		Hermetic and nonhermetic
	6.8–2,200	138[g]		CLR69	6–125	4–85	2–12	9–48	13–300	50–750	T1–T3, Fig. 48		Hermetic
	1.7–560	TLW,[d] 137D,[e] WH,[j] 69F3000[v]	ECL66, ECL67		6–125	4–85	1–2	2–18	20–1,250	50–500			Hermetic
	1.7–560		ECL64, ECL65		6–125	4–85	1–2	2–18	20–1,250	50–500			
	22–1,800	109D/130D,[e] 69F2000[v]			6–75	4–50	2–9	9–36	10–100	50–750	0.453 × 0.188 to 1.062 × 0.375	Subminiature, fully insulated metal tubular	Nonhermetic
	2–330	TAH,[d] 145D,[e] 69F900[v]			6–60	4–60	1–7	4–28	30–850	40–250[m]	0.460 × 0.228 to 0.800 × 0.228		
	0.1–325	62F,[g] TE[j]			6–60		1–10	3–30	100–16,000		0.250 × 0.080 to 0.875 × 0.225		
	3.3–470	MTP,[d] 145D,[e] 69F900[v]			6–60		2–10	5–30	65–950		0.312 × 0.115 to 0.778 × 0.225		
	2–330	TAP,[d] TE,[j] 62F[v]			6–90		1–7	3–21	25–1,100		0.504 × 0.235 to 0.847 × 0.235		
	2–80	TNT,[d] 62F[v]			3–50		1–4.2	2–10.5	150–1,750		0.347 × 0.163 to 0.472 × 0.163	Subminiature	

Construction type	Capacitance range, μF	Manufacturers type[a] (commercial)	EIA Standard RS-228 type[b]	Military Standard type[b,c]	DC working-voltage range, V At 85°C	At 125°C	DC leakage-current range, μA At 25°C	At max temp	Max dissipation factor at 25°C, 120 Hz, %	AC ripple-capability range at 25°C, 60 Hz, mA	Size reference code, length × diam, in	Style[b]	Seal
Sintered anode, solid electrolyte, polarized (except where noted)	0.0047–330	TER,[d] 176D/376D,[e] T-212,[j] TR[k]	ECS12, ECS13	CSR13	6–100	4–67	0.5–20	6.3–250	3–8	Ripple voltage and currents for applications of this type of tantalum are considerably smaller than wet electrolytes and must be carefully computed at application temperatures and frequencies	A–D, Fig. 49	Metal tubular, insulated and uninsulated case	Hermetic
	1.2–1,000	178D,[e] XNS,[i] T-242[j]	ECS22, ECS23	CSR23	6–50	4–33	0.9–30	11–375	4–10				
	0.047–18	176D,[e] TNS,[i] T-222,[j] TR[k]		CSR09	6–75	4–50	0.6–2.5	8–51	3–6		0.250 × 0.090 and 0.390 × 0.138		
	0.0023–160 (non-polarized)	TER,[d] 177D,[e] RSP,[i] T-213,[j] TN[k]		CSR91	6–100	4–67	0.5–20	6.3–250	3–10		0.750 × 0.161 to 1.725 × 0.370		
	5.6–330	THH[d] (low Z)			6–50	4–33	4.5–20	54–250	0.5–1.7		Fig. 49, C and D		Nonhermetic and hermetic
	0.0047–330	TAE,[d] SNS,[i] T314,[j] CC[k]			6–50	4–33	0.1–10	1.2–120	3–10		0.250 × 0.090 to 0.810 × 0.360		
	0.0047–47	TNS,[j] T324,[j] CC[k]			6–50	4–33	0.5–2.8	6.3–55	6–10		0.250 × 0.090 to 0.410 × 0.178		
	0.0047–22	Z[j]			6–125	4–82	0.1–2.0	1.2–24	3–6		0.265 × 0.080 and 0.390 × 0.138		Hermetic
	0.10–330	158D,[e] T310[j]		CWR08	6–50		1.0–100	10–100	4–10		0.390 × 0.138 to 0.875 × 0.365	Epoxy molded cases in bullet-nose configuration	
	0.1–330	TDC,[d] DNS,[i] 368,[j] TK[k]	ECS15	CWR01 CWR02	6–50		1–20	10–200	5–10		A–D, Fig. 50	Molded body, radial leads	
	0.001–220	TC,[j] Modular,[k] MNS,[i] T370,[j] T372[j]	ECS16	CWR08	3–35	2–23	0.5–9.0	6.3–108	6–10		A–F, Fig. 51	Rectangular plastic-molded case, axial and radial leads	
	0.10–68	T411[j]	ECS05		4–50	2.7–33	0.5–3.0	8–45	4–6		A–E, Fig. 52	Tantalum chip, unencapsulated	
	0.10–47	193D[e]	ECS09	CWR03 CWR04	3–35	2–23	1.0–9.4	17–150	4–6		A–E, Fig. 53	Tantalum chip, molded-plastic encapsulation	

a Manufacturer's types listed are representative of types available and are not shown as preferred types or to be construed as the only sources available.
b Uninsulated cases are in styles with even type numbers, insulated cases are odd type numbers.
c Military type designators shown in Table 16.
d Mallory Capacitor Company.
e Sprague Electric Company.
f Transistor Electronics.
g General Electric Company.
h Cornell Dublier.
i National Components Industries.
j Union Carbide, Kemet Division.
k Corning Electronics.
l Commercial equivalents of these styles are available over a slightly broader capacitance and voltage range and derated to 175°C.
m Over temperature range of −55 to +125°C and at 120 Hz.

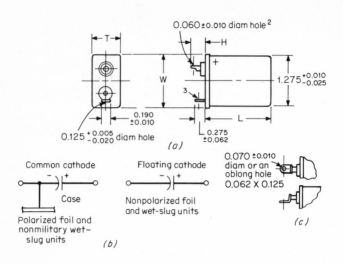

Common cathode

Case

Polarized foil and
nonmilitary wet-
slug units

Floating cathode

Nonpolarized foil
and wet-slug units

(b)

0.070 ±0.010
diam or an
oblong hole
0.062 X 0.125

(c)

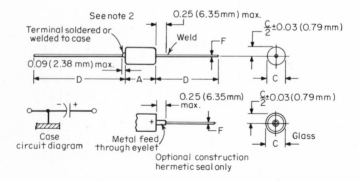

Case
circuit diagram

Metal feed
through eyelet

Optional construction
hermetic seal only

Fig. 47 (Top, left) (*a*) Standard case configuration for rectangular-style tantalum capacitors. (*b*) Circuit diagrams. (*c*) Optional alternate terminal constructions.

Capacitor type	Case size reference symbol	Case standard	Dimensions[1]			
			L ± 0.062	W max	T ± 0.062	H max
Foil	A[4]	Industry	1.375	1.378	0.783	0.426
	B[4]	MIL-C-39006	2.000	1.378	0.783	0.426
	C[4]	Industry	2.500	1.378	0.783	0.426
Sintered slug, wet electrolyte	A1[5]	Industry and	1.062	1.367	0.720	0.312
	A2[5]	MIL-C-3965	1.375	1.367	0.720	0.312
	A3[5]		1.625	1.367	0.720	0.312
	A4[5]		2.000	1.367	0.720	0.312
	A5[5]		2.500	1.367	0.720	0.312

1. All dimensions are in inches.
2. Orientation of terminal hole is not fixed.
3. In nonpolarized foil types and in wet-slug floating-case types this terminal is the same as the other.
4. Reference symbols are for convenience in this handbook.
5. Military standard case size designations.

Fig. 48 (Bottom, left) Standard case configuration for sintered-anode wet tantalum capacitors (EIA Standard RS-228, types 2.1, 2.2). (*Electronic Industries Association.*[6])

	Uninsulated dimensions[1]						Insulated dimensions[1,3]				
	in			mm			in			mm	
Code	Min	Nomi-nal	Max	Min	Max	Code	Min	Nomi-nal	Max	Min	Max
T1:											
A	0.437	0.453	0.484	11.10	12.29	A*			.546		13.87
C	0.172	0.188	0.204	4.37	5.28	C*			0.219		5.56
D	1.250	1.500	1.750	31.75	44.45	D	1.250	1.500	1.750	31.75	44.45
F	0.023	0.025	0.027	0.58	0.69	F	0.023	0.025	0.027	0.58	0.69
T2:											
A	0.625	0.641	0.672	15.88	17.07	A*			.734		18.64
C	0.265	0.281	0.297	6.73	7.54	C*			0.313		7.95
D	2.000	2.250	2.500	50.80	63.50	D	2.000	2.250	2.500	50.80	63.50
F	0.023	0.025	0.027	0.58	0.69	F	0.023	0.025	0.027	0.58	0.69
T3:											
A	0.750	0.766	0.797	19.05	20.24	A*			.859		21.82
C	0.359	0.375	0.391	9.19	9.93	C*			0.406		10.31
D	2.000	2.250	2.500	50.80	63.50	D	2.000	2.250	2.500	50.80	63.50
F	0.023	0.025	0.027	0.58	0.69	F	0.023	0.025	0.027	0.58	0.69

* Basic case dimensions for the insulated style should be the same as for the uninsulated.

1. Metric equivalents tabulated and in parentheses are for information only and are based upon 1 in = 25.4 mm.
2. Marking for polarity should be indicated as shown, near the positive terminal.
3. The case insulation should extend 0.015 in (0.38 mm) minimum beyond each end of the capacitor body. However, when a shrink-fitted insulation is used, it should lap over the ends of the capacitor body.
4. Leads solderable.

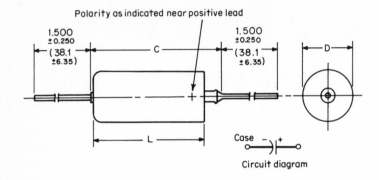

Polarity as indicated near positive lead

1.500 ±0.250 (38.1 ±6.35)

C

1.500 ±0.250 (38.1 ±6.35)

D

L

Case

Circuit diagram

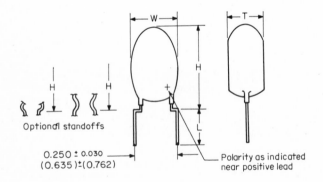

W

T

H

H

H

H

Optional standoffs

0.250 ± 0.030 (0.635)±(0.762)

L

Polarity as indicated near positive lead

Fig. 49 (Top, left) Standard case configuration for hermetic-sealed solid-electrolyte sintered-anode tantalum capacitors (EIA Standard RS-228, types 3.1, 3.2). (*Electronic Industries Association.*[6])

	Uninsulated						Insulated						
	C max		D +0.016 (0.41) −0.010 (0.25)		L ±0.031 (0.79)		C max		D +0.016 (0.41) −0.015 (0.38)		L ±0.031 (0.79)		Lead diam ±0.002 (0.05)
Case size	in	mm	in	mm	in	mm	in	mm	in	mm	in	mm	in	mm
A	0.422	10.71	0.125	3.17	0.250	6.35	0.422	10.71	0.135	3.43	0.286	7.26	0.020	0.51
B	0.610	15.49	0.175	4.44	0.438	11.12	0.610	15.49	0.185	4.70	0.474	12.04	0.020	0.51
C	0.822	20.87	0.279	7.08	0.650	16.51	0.822	20.87	0.289	734.	0.686	17.42	0.025	0.64
D	0.922	23.41	0.341	8.66	0.750	19.05	0.922	23.41	0.351	8.92	0.786	19.96	0.025	0.64

1. The case insulation shall extend 0.015 (.38 mm) minimum beyond each end of the capacitor body. However, when a shrink-fitted insulation is used, it shall lap over the ends of the capacitor body.

2. Metric equivalents tabulated and in parentheses are for information only and are based upon 1 in = 25.4 mm.

3. Leads solderable.

Fig. 50 (Bottom, left) Standard case configuration for dipped solid-electrolyte sintered-anode tantalum capacitors (EIA Standard RS-228, types 3.3, style ECS-15). (*Electronic Industries Association.*[6])

	H max		W max		T max		L		Optional[1] L-1 min		Lead diam	
Case size	in	mm	in	mm	in	mm	in ±0.032	mm ±0.79	in	mm	in	mm
A	0.450	11.43	0.400	10.16	0.225	5.71	0.187	4.76	0.900	22.86	0.020 or 0.025	0.508 or 0.63
B	0.550	13.97	0.450	11.43	0.250	6.35	0.187	4.76	0.900	22.86	0.020 or 0.025	0.508 or 0.63
C	0.750	19.05	0.600	15.24	0.350	8.89	0.187	4.76	0.900	22.86	0.025	0.63
D	0.920	23.37	0.600	15.24	0.400	10.16	0.187	4.76	0.900	22.86	0.025	0.63

1. If optional lead length L-1 is desired, specify.

2. It is necessary that parts must have a standoff. Inward hook, outward hook, or any other type of kink is acceptable as long as it serves the purpose of standoff. Regardless of configuration, the standoff must not pass through A 0.050 in (1.27) diam hole.

3. Metric equivalents tabulated and in parentheses are for information only and are based upon 1 in = 25.4 mm.

4. Leads solderable.

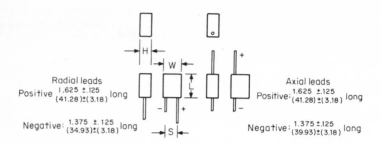

Radial leads
Positive $\frac{1.625 \ \pm.125}{(41.28)\pm(3.18)}$ long

Negative: $\frac{1.375 \ \pm.125}{(34.93)\pm(3.18)}$ long

Axial leads
Positive: $\frac{1.625 \ \pm.125}{(41.28)\pm(3.18)}$ long

Negative: $\frac{1.375 \ \pm.125}{(39.93)\pm(3.18)}$ long

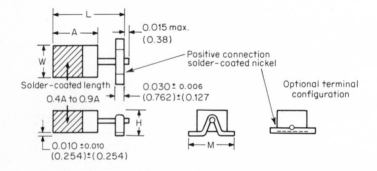

0.015 max.
(0.38)

Positive connection
solder-coated nickel

Solder-coated length
0.4A to 0.9A

0.030 ± 0.006
(0.762)±(0.127

Optional terminal
configuration

0.010 ±0.010
(0.254)±(0.254)

Fig. 51 (Top, left) Standard case configuration for molded solid-electrolyte sintered-anode tantalum capacitors (EIA Standard RS-228, type 3.4, style ECS-16). (*Electronic Industries Association.*[6])

Case size	L max in	L max mm	W max in	W max mm	H max in	H max mm	S in	S mm	Nominal lead diam in	Nominal lead diam mm
A	0.125	3.18	0.070	1.78	0.040	1.02	0.050 ±0.015	1.27 ±0.38	0.010	0.25
B	0.165	4.19	0.120	3.04	0.070	1.78	0.100 ±0.020	2.54 ±0.51	0.010	0.25
C	0.225	5.72	0.185	4.70	0.075	1.91	0.150 ±0.020	3.81 ±0.51	0.010	0.25
D	0.290	7.37	0.220	5.59	0.110	2.79	0.180 ±0.025	4.57 ±0.64	0.016	0.41
E	0.310	7.87	0.230	5.84	0.130	3.30	0.200 ±0.025	5.08 ±0.64	0.016	0.41
F	0.475	12.06	0.375	9.53	0.150	3.81	0.300 ±0.025	7.62 ±0.64	0.016	0.41

1. All dimensions in inches with metric equivalents in parentheses or tabulated (based on 1 in = 25.4 mm).
2. Leads solderable.

Fig. 52 (Bottom, left) Standard case configuration for unencapsulated chip-type solid-electrolyte sintered-anode tantalum capacitors (EIA Standard RS-228, type 3.5, style ECS-05). (*Electronic Industries Association.*[6])

Case size	W in ±0.015	W mm ±0.38	L in	L mm	H max in	H max mm	A in	A mm	M in ±0.010	M mm ±0.254
A	0.085	2.16	0.165 ±0.015	4.19 ±0.38	0.070	1.78	0.065 ±0.015	1.65 ±0.38	0.090	2.29
B	0.105	2.67	0.195 ±0.015	4.95 ±0.38	0.070	1.78	0.105 ±0.015	2.67 ±0.38	0.140	3.56
C	0.145	3.68	0.235 ±0.015	5.97 ±0.38	0.070	1.78	0.145 ±0.015	3.68 ±0.38	0.150	3.81
D	0.150	3.81	0.290 ±0.020	7.37 ±0.51	0.100	2.54	0.180 ±0.020	4.57 ±0.51	0.150	3.81
E	0.155	3.94	0.290 ±0.020	7.37 ±0.51	0.150	3.81	0.190 ±0.020	4.83 ±0.51	0.150	3.81

1. These capacitors are designed for mounting by reflow solder or conductive epoxy on circuit substrates.
2. These capacitors must be handled carefully to avoid damage.
3. Metric equivalents tabulated and in parentheses are for information only and are based upon 1 in = 25.4 mm.

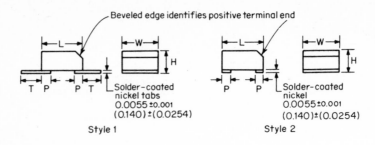

Beveled edge identifies positive terminal end

Solder-coated
nickel tabs
0.0055 ±0.001
(0.140) ±(0.0254)

Style 1

Solder-coated
nickel
0.0055 ±0.001
(0.140) ±(0.0254)

Style 2

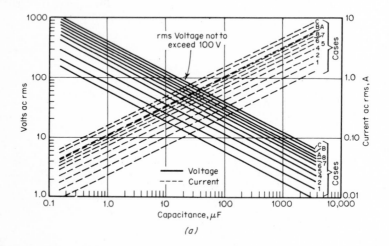

(a)

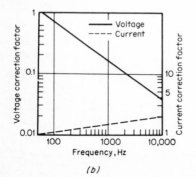

(b)

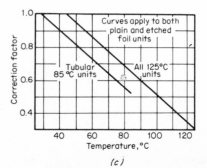

(c)

Fig. 53 (Top, left) Standard case configuration for molded-plastic chip-type solid-electrolyte sintered-anode tantalum capacitors (EIA Standard RS-228, type 3.5, style ECS-09). (*Electronic Industries Association.*[6])

Case size	Style no.	L		W		H		P		T	
		in ±0.005	mm ±0.127	in ±0.010	mm ±0.254	in ±0.005	mm ±0.127	in ±0.010	mm ±0.254	in ±0.010	mm ±0.254
A	1	0.180	4.57	0.100	2.54	0.070	1.78	0.030	0.762	0.150	3.81
A	2	0.180	4.57	0.100	2.54	0.070	1.78	0.030	0.762	No tabs	
B	1	0.180	4.57	0.100	2.54	0.100	2.54	0.030	0.762	0.150	3.81
B	2	0.180	4.57	0.100	2.54	0.100	2.54	0.030	0.762	No tabs	
C	1	0.320	8.13	0.180	4.57	0.070	1.78	0.050	1.27	0.110	2.80
C	2	0.320	8.13	0.180	4.57	0.070	1.78	0.050	1.27	No tabs	
D	1	0.320	8.13	0.180	4.57	0.100	2.54	0.050	1.27	0.110	2.80
D	2	0.320	8.13	0.180	4.57	0.100	2.54	0.050	1.27	No tabs	
E	1	0.320	8.13	0.180	4.57	0.195	4.95	0.050	1.27	0.110	2.80
E	2	0.320	8.13	0.180	4.57	0.195	4.95	0.050	1.27	No tabs	

1. These capacitors are designed for mounting by dip soldering, welding, reflow soldering, or other conventional means.

2. Metric equivalents tabulated and in parentheses are for information only and are based upon 1 in = 25.4 mm.

Fig. 54 (Bottom, left) AC ripple capability of typical tantalum foil capacitors at 60 Hz and 25°C. (*Basic curves, General Electric Company.*) (*a*) Permissible rms voltage or current for plain foil capacitors at 25°C and 60 Hz may be read directly from curve (*a*). For etched foil capacitors, use one-half this value. (*b*) To determine ac capability at some other frequency, multiply the current or voltage value obtained in step (*a*) by a correction factor from curve (*b*). (*c*) To determine ac capability at some other temperature, multiply the voltage or current value from step (*a*) or (*b*) by a correction factor from curve (*c*). NOTE: Case reference designations are from Figs. 46 and 47.

ity of plain-foil types. Typical impedance variation with frequency characteristic curves is shown in Fig. 55. Correction factors for impedance at various operating temperatures are shown in Fig. 56. Typical ESR characteristic for various standard case sizes, temperatures, frequencies, and voltages is shown in Fig. 57.

Sintered-Anode Wet Tantalum Capacitors. The sintered-anode wet-electrolyte and solid-electrolyte tantalum capacitors are similar in construction. The anode consists of high-grade tantalum powders sintered around a tantalum wire in the desired "slug" configuration and formed at the desired voltage. The sintered anode is then

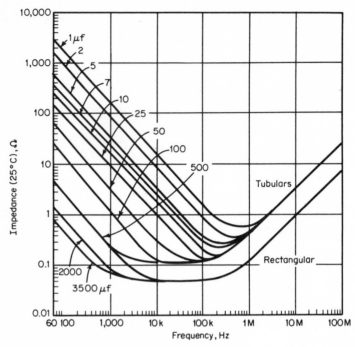

Fig. 55 Impedance variation with frequency for typical tantalum foil capacitors. (*General Electric Company.*)

impregnated with a sulfuric acid or lithium chloride liquid or a thixotropic electrolyte and encased in a solid-silver or silver-plated case with an elastomeric or hermetic seal. Newer sintered-anode wet tantalum capacitors are built with the slug mechanically held within the case for high vibration resistance. The tantalum wire extending from the slug through an elastomeric compression seal may also be bonded to a glass header, thereby allowing a true hermetic seal and eliminating any possibility of electrolyte leakage. Commercially available sintered-anode wet tantalum capacitors are sealed with only an elastomeric seal but may be backed up with an epoxy end cap. These capacitors are less expensive but are prone to leakage and consequent damage to the electronic equipment. Typical construction of a sintered-anode wet tantalum capacitor is shown in Fig. 58. It should be noted that tantalum-cased capacitors are available with increased ripple capability but at premium cost.

These capacitors are not suitable for applications involving any voltage reversal. They should not be operated on unbiased ac voltage or applied in nonpolar applications involving back-to-back connections. Refer to special treatment of back-to-back connections. Any ac ripple applied must be superimposed on sufficient dc bias voltage to prevent voltage reversal. Ripple current is limited to small values because progressive degradation of the unit will result if the cathode (silver case) becomes positive during the discharge cycle. See Fig. 59 for ripple limits. The common failure mode with this capacitor is peeling or migration of silver or silver compounds inside the case with subsequent shorting to the anode. This failure mode is greatly reduced with good manufacturing-process controls and proper application of the capacitor in circuits where ripple currents are experienced. Typical impedance variation with frequency characteristic curves is shown in Fig. 60. Correction factors

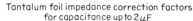

Tantalum foil impedance correction factors
for capacitance up to 2 μF

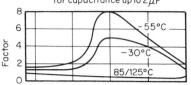

Tantalum foil impedance correction factors
for capacitance = 2–50 μF

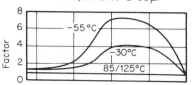

Tantalum foil impedance correction factors
for capacitance = 50 μF and over

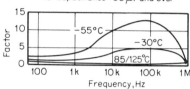

Fig. 56 Correction factors for impedance of typical tantalum foil capacitors at various operating temperatures. (*General Electric Company.*)

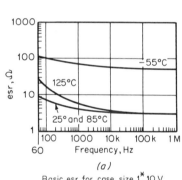

(a)

Basic esr for case size 1*, 10 V

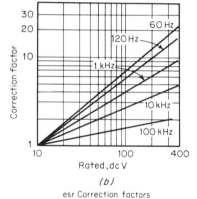

(b)

esr Correction factors

Fig. 57 Typical ESR for tantalum foil capacitors. (*General Electric Company.*)

ESR Correction Table

Case size*	Correction factor
1	1.0
2	0.255
4	0.0875
5	0.0386
6	0.0275
7	0.0155
8	0.0113
A	0.0124
B	0.010
C	0.009

* Case reference designations are from Figs. 46 and 47.

A. ESR in ohms for the smallest case size at 10 V may be read directly from curve *a*.

B. To obtain ESR for other voltage ratings, multiply the value from step A by a correction factor from curve *b*.

C. To obtain ESR for other case sizes, multiply the value from step A or B by a correction factor from the table.

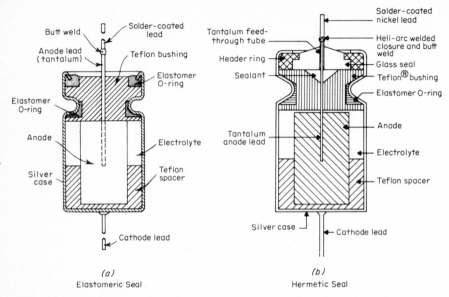

Fig. 58 Typical construction of a MIL-C-39006 sintered-anode wet tantalum capacitor. Teflon is a registered trademark of E. I. Du Pont de Nemours & Co., Wilmington, Del. (*General Electric Company.*)

for impedance at various operating temperatures are shown in Fig. 61. Typical ESR characteristic for various case sizes, temperatures, frequencies, and voltages is depicted in Fig. 62.

Sintered-Anode Solid Tantalum Capacitors. The sintered-anode solid-electrolyte tantalum capacitor is a miniature device offering the circuit designer an improved technological design and superior performance at highly competitive cost. This type of tantalum capacitor is inherently very reliable, and in fact significant evidence accumulated through millions of part hours of life testing indicates that the reliability improves with life, where applications comply with the basic ground rules of good design. Capacitance loss with age or corrosion problems associated with wet- or active-electrolyte systems are virtually eliminated. Solid-electrolyte tantalum-capacitor construction as shown in Fig. 63 employs a solid semiconductor material such as manganese dioxide or lead peroxide to conduct the current from the dielectric film (Ta_2O_5) to the cathode (case) of the capacitor. The sintered slug is formed at the desired voltage, thus creating Ta_2O_5 regions as shown in Fig. 64. The anode is normally coated with a carbon sheet and plated with silver or platinum silver to gain contact with the cathode structure. The cathode structure consists in most cases of a metal can in which the slug is soldered in place. Performance characteristics of solid-tantalum capacitors are well defined. Typical curves of dissipation factor as a function of frequency and temperature are shown in Figs. 65 and 66, respectively. Typical curves of capacitance change with frequency are shown in Fig. 67. Equivalent series-resistance variation with frequency is shown in Fig. 68. Factors for leakage-current variation with applied voltage and operating temperature are shown in Fig. 69. Dielectric-absorption characteristic of solid-tantalum capacitors is shown in Fig. 70. In the development of these curves capacitors were charged for 1 h at rated voltage and then discharged through a dead short for 1 min. Voltage recovery was measured with a high-impedance electrometer at intervals shown on the curve. Increase in the ambient temperature shifts the curve to the left and decreases the amplitude and does not affect the shape. Shortening charge time, lengthening discharge time, or decreas-

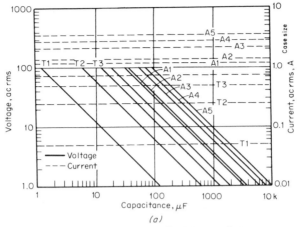

(a)

ac-Ripple Capability for Tantalum Wet-Slug Tubular and Packaged Units
at 60Hz, 25°C

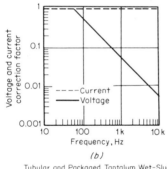

(b)

Tubular and Packaged Tantalum Wet-Slug
ac-Ripple Correction Factors

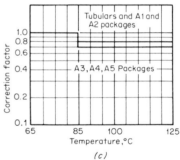

(c)

Tubular and Packaged Tantalum Wet-Slug
ac-Ripple Correction Factors

Fig. 59 AC ripple capability of sintered-anode wet tantalum capacitors. Case sizes for MIL-C-3965, MIL-C-39006, and EIA Standard RS-228 are as shown in Figs. 47 and 48. (*General Electric Company.*)

A. Permissible rms voltage or current at 25°C and 60 Hz may be read directly from curve *a*.

B. To determine ac capability at other frequencies, multiply the voltage or current value obtained in step A by a correction factor from curve *b*.

C. To determine ac capability at other temperatures, multiply the voltage or current value obtained in step A or B by a correction factor from curve *c*.

ing charging voltage results in the reduction of the peak amplitude of the curve but has little effect on its shape or relative position. AC ripple-voltage limitations of solid-tantalum capacitors in pure ac applications are generally 15 percent of the 25°C dc voltage rating of the capacitor. Ripple voltages are derated further by a factor of 0.7 at 50°C, 0.5 at 85°C, and 0.3 at 125°C. Derating for frequency is shown in Fig. 71. Voltage reversal is generally limited to 15 percent of the rated dc voltage at 25°C, 10 percent at 55°C, 5 percent at 85°C, and 1 percent at 125°C. Impedance change with temperature for solid tantalums varies from a factor of 1.1 at −55°C to 0.95 at 85°C. Typical variations with frequency are shown in Fig. 72.

Solid-Tantalum Voltage Derating. Unlike other types of electrolytic capacitors, the solid-tantalum capacitor may be operated continuously at any voltage ranging

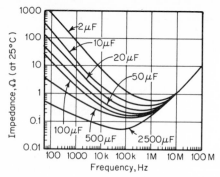

Fig. 60 Impedance variation with temperature curves for typical sintered-anode wet tantalum capacitors. (*General Electric Company.*)

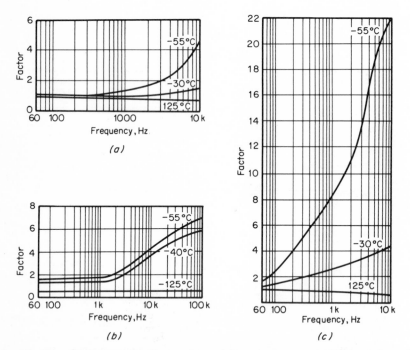

Fig. 61 Impedance correction factors for variations in temperature of sintered-anode wet tantalum capacitors. (*General Electric Company.*) (*a*) Capacitance up to 5 μF. (*b*) Capacitance 5 to 100 μF. (*c*) Capacitance 100 μF and above.

from zero up to its maximum rated voltage and at highest rated temperature. Operation at a voltage less than the rated value improves the reliability of the solid-tantalum capacitor. Subsequent performance at a higher voltage is not degraded by such low voltage use, since there is no reformation of the dielectric system or other adverse effects. The use of reduced operating voltage lowers the electrical stress placed on the dielectric, and thus decreases the probability of failure. As a general rule, operation of a solid-tantalum capacitor at one-half its rated voltage will improve the failure rate of the capacitor and service life by as much as three orders of magni-

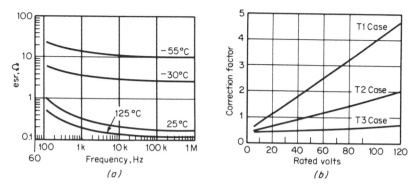

Fig. 62 Typical ESR for sintered-anode wet tantalum capacitors. Case sizes shown are EIA Standard RS-228 types. (*General Electric Company.*) (*a*) Basic ESR for tantalum wet-slug T1 case, 18 V. (*b*) ESR correction factors for other voltages and case sizes.

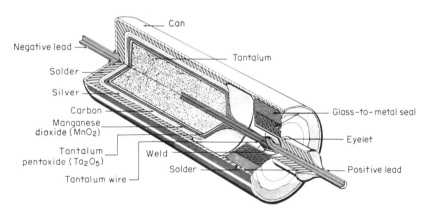

Fig. 63 Typical construction of a sintered-anode solid-tantalum capacitor. (*Kemet Div., Union Carbide, Inc.*)

tude. Expected reliability-improvement factors for other voltages and temperatures can be obtained from the nomograph shown in Fig. 73.

Solid-Tantalum High Current Transient. Some applications produce large current transient, and steady-state testing does not accurately predict performance of a capacitor under these conditions. The highest transients would naturally occur during abrupt switching between the voltage extremes defined by zero and maximum rating. The positive electrode of a solid-tantalum capacitor consists only of tantalum metal and does not present complicating factors. Conversely the negative electrode is a complex structure of heterogeneous materials. With the many opportunities therefore afforded it is reasonable to postulate large variations among the resistive paths of the terminals to the incremental capacitance areas. Heavy charge or discharge currents could produce localized hot spots, rapidly raising small sites to temperatures far above ambient. Normally amorphous Ta_2O_5 dielectric will crystallize at high temperatures, losing much of the dielectric strength in the process. A severely heated site in the dielectric could then fail catastrophically under an electric field below that produced by the rated voltage. In contrast to the extremely high rate

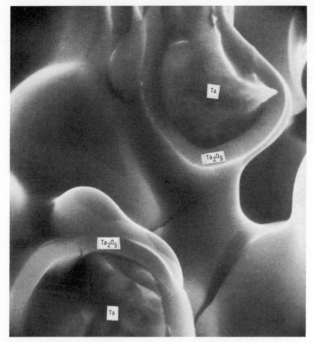

Fig. 64 Scanning-electron-microscope view of a sintered tantalum anode showing tantalum pentoxide formations. (*Martin Marietta Corp.*)

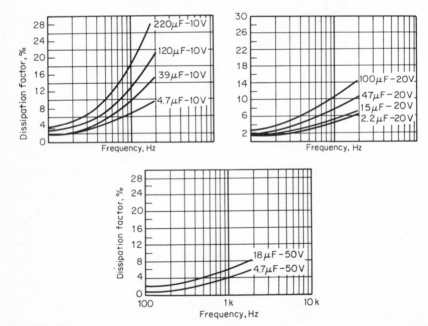

Fig. 65 Dissipation-factor variation with frequency for various capacitance and voltage ratings of typical sintered-anode solid-tantalum capacitors. (*Kemet Div., Union Carbide Corp.*)

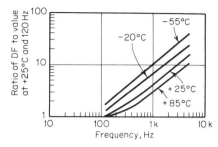

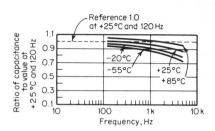

Fig. 66 Dissipation-factor variation with frequency and temperature for typical sintered-anode solid-tantalum capacitors. (*Sprague Electric Company.*)

Fig. 67 Capacitance change with frequency and temperature for typical sintered-anode solid-tantalum capacitors. (*Sprague Electric Company.*)

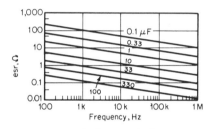

Fig. 68 Typical ESR variation with frequency for various capacitance values of sintered-anode solid-tantalum capacitors. (*Kemet Div., Union Carbide Corp.*)

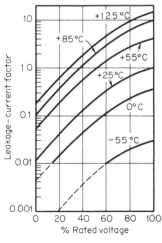

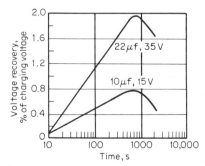

Fig. 69 Typical leakage-current factors with respect to percent of rated voltage and at various operating temperatures for sintered-anode solid-tantalum capacitors. (*Sprague Electric Company.*)

Fig. 70 Typical dielectric absorption of sintered-anode solid-tantalum capacitors at 25°C.

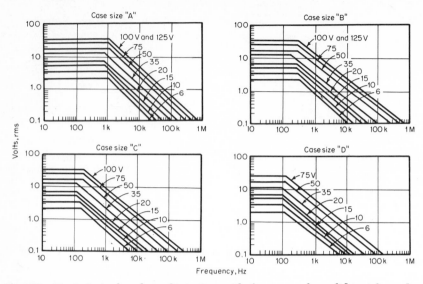

Fig. 71 Typical ripple-voltage limitation with frequency for solid-tantalum electrolytic capacitors. (*Sprague Electric Company.*)

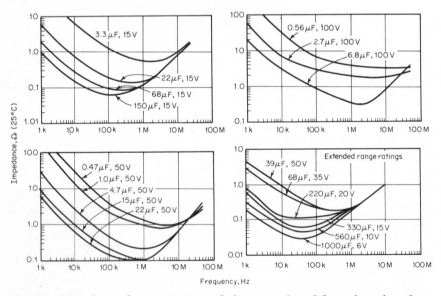

Fig. 72 Typical impedance variation with frequency for solid-tantalum electrolytic capacitors. (*Sprague Electric Company.*)

of heating possible during rapid charge or discharge, much slower heating is postulated at the time of incipient failure under steady-state dc conditions. In the latter case, reduction of the semiconducting manganese dioxide can provide a counteraction to isolate the fault and prevent catastrophic failure. The time difference is a function of the instantaneous circuit impedance, and it is further postulated that this difference explains the lack of correlation between failure rates under steady-state conditions and under high-current-transient conditions.

Reliability Alignment Chart

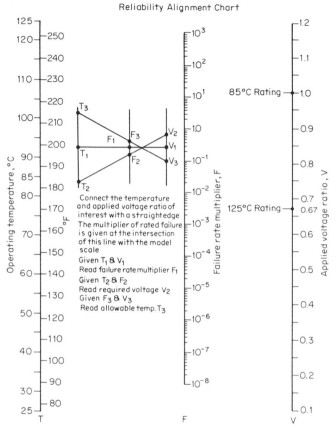

Fig. 73 A reliability-improvement chart for sintered-anode solid-tantalum capacitors. (*Kemet Div., Union Carbide Corp.*)

Back-to-Back Solid Tantalum Applications.[*] When designing circuits with long time constants, a common practice is to connect a pair of polarized electrolytic capacitors back to back (i.e., cathode to cathode) in order to obtain a nonpolar configuration. While this works quite well as a rule, the designer should be aware of certain facts concerning this method of operation. Although the authors' experience has been primarily with sintered-anode, solid-electrolyte tantalum capacitors, much of the material presented here would also apply to other types.

From a user's standpoint, the characteristic which distinguishes a polar from a nonpolar capacitor is the fact that in a polar capacitor, normal and reverse leakage currents are drastically different. In fact, it is this characteristic which makes back-to-back operation necessary in bipolar circuits.

When two perfect nonpolarized capacitors are connected in series, it is well known from linear-circuit theory that the effective capacity C_T is given by the equation

$$C_T = \frac{C_1 C_2}{C_1 + C_2}$$

[*] From a special study for this handbook by Joseph F. Rhodes.

The situation with back-to-back polarized capacitors is not quite so apparent or so well defined. Theoretically, the effective capacity can vary by as much as 2:1, depending on the time history of the applied signal and the characteristics of the particular units involved. In practice, the variation seldom approaches 2:1, except for applied voltages very near the maximum rating.

A polarized capacitor can be represented to a fair degree of accuracy by the model shown in Fig. 74. Limited test data indicate that typically $I_{01} \cong I_{02}$ and $\theta_2 \cong 10\ \theta_1$. For example, Fig. 75 shows V-I test data for a 47-mF, 20-V dc solid tantalum capacitor vs. calculated values using the empirically derived values of

$$I_{01} = I_{02} = 6 \times 10^{-9}\ \text{A}$$
$$\theta_1 = 0.21\ \text{V}^{-1}$$
$$\theta_2 = 2.1\ \text{V}^{-1}$$

Sintered-anode wet tantalums will, in general, have slightly lower forward (normal voltage) leakage and somewhat higher reverse voltage leakage for the same applied voltage. Also, in general, both forward and reverse leakage currents increase with increasing temperature.

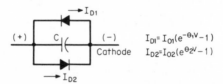

Fig. 74   Polarized-capacitor model.

To analyze this "diode effect" on circuit performance, we will first look at the case of $I_{01} = \theta_1 = 0$. I_{02} is finite but very small and $\theta_2 \simeq \infty$; i.e., the normal bias leakage is effectively zero and the reverse bias leakage is characterized by a perfect diode. Referring to Fig. 76, if C_1 and C_2 are initially discharged and a positive step voltage V_1 is applied to the circuit, the diode around C_2 acts as a short circuit and $V_2 = V_A$. In this case, the rise time of V_2 is obviously characterized by the time constant RC_1. Assuming V_1 is a zero-impedance voltage source, when V_1 drops to zero, C_1 discharges through R and back around into C_2, charging up C_2 in its normal direction (V_B minus). Since neither diode is conducting at this time, the discharge time constant is $R[C_1C_2/(C_1 + C_2)]$. Assuming $C_1 = C_2 = C$, after several time constants $V_A \doteq -V_B =$ one-half the maximum value to which C_1 was originally charged. If V_1 then goes to a minus value, C_1 will discharge and C_2 will charge, still with a time constant of $R[C_1C_2/(C_1 + C_2)]$ (or ½RC for the case of $C_1 = C_2 = C$), for as long as any finite charge remains on C_1. Thus in general, if an alternating signal V_1 is applied to the circuit, the circuit will respond with a time constant of $T = RC$ whenever one diode or the other is conducting, and $T = ½RC$ whenever the distribution of charges is such that neither diode conducts.

Unfortunately, from the standpoint of easy analysis, in real tantalum capacitors the reverse-conduction diode is never perfect and the normal-direction leakage is never zero. As a consequence, when a pair of typical solid tantalums are connected back to back in a circuit, the degree of deviation from what would be obtained with a pair of dielectric capacitors in series is a function of both the applied voltage and the recent time history of applied voltages.

Much of what has been written to date concerning polarized tantalum capacitors would lead one to believe that the diode seen when a reverse voltage is applied is near perfect and that whenever a significant voltage can be developed across the capacitor in the reverse direction, it is due to forming of the cathode, causing the capacitor actually to be semipolarized. The following tests were performed to examine this theory.

Several pairs of 15-μF, 20-V solid-tantalum capacitors were tested in an integrator circuit (Fig. 77) in which it was possible to short out each capacitor individually and the two as a pair. Since the gain of the integrator is $1/RC_T$

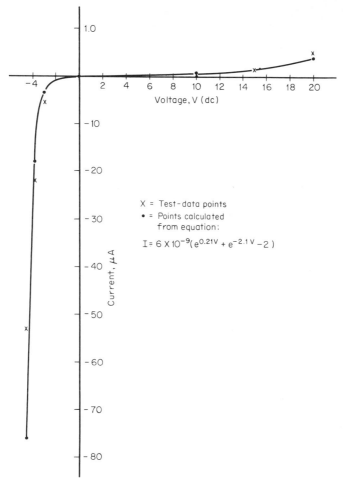

X = Test-data points
• = Points calculated
   from equation:

$$I = 6 \times 10^{-9}(e^{0.21V} + e^{-2.1V} - 2)$$

Fig. 75 Current vs. voltage of a 47-μF, 20-V dc solid-tantalum capacitor.

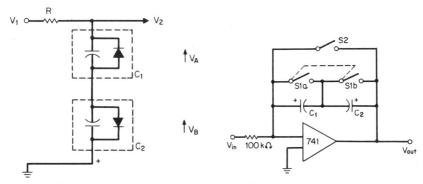

Fig. 76 Diode effect of electrolytic capacitors.

Fig. 77 Integrator test circuit for tantalum capacitors.

where C_T is the effective capacity, the output response to a step input should be a ramp which equals the input voltage when $t = 1/RC_T$. For this test, the input was a 750-ms pulse, so that the output would exactly equal the input for the case of $C_T = \frac{1}{2}C_1 = \frac{1}{2}C_2$. Three cases were examined with an input pulse of 8 V peak. Case 1 was also performed with an input of 1 V and 4 V peak.

1. Capacitors which had had no previous voltage applied for at least 12 h. In addition, S1 was closed for a few seconds just prior to the test.

2. The capacitors were cycled through plus and minus inputs to get some recent voltage history on them; then S1 was closed for a few seconds just prior to applying the test pulse.

3. The capacitors were again cycled several times and S2 closed for only a few seconds just prior to applying the test pulse to equalize the charge without dissipating it.

The results were photographed, and a typical plot of the results is shown in Fig. 78. The difference in the results obtained with different samples was quite small.

In addition, several tests were performed under identical conditions except that only one capacitor was used. Tests were performed with the capacitor being charged in both the normal and the reverse direction. Typical results of this test are shown in Fig. 79.

For example, the fact that case 1 and case 2 yielded essentially identical results indicates that these particular capacitors had little, if any, forming of the cathode, yet voltages greater than 10 percent of the rated voltage could readily be developed in the reverse direction with a reverse-charging current of only about $5\mu A\mu F^{-1}$. Thus the model shown in Fig. 74 is thought to be a good one, although the values of I_0 and θ would vary with capacitor type.

The results of the tests described plus considerable practical experience in using solid-tantalum capacitors in a back-to-back configuration for both coupling and signal shaping lead to the following guidelines:

1. For signals up to about 10 percent of the rated value, little difference will be seen between using back-to-back solid tantalums and using two dielectric capacitors in series. That is, for equal capacitors, the effective capacity is equal to one-half the value of either capacitor, regardless of time/voltage history.

2. As the applied signal level is increased, the effective capacity seen by the initial input increases nonlinearly to about 1.5 to 1.8 times the value with low voltages.

3. After the initializing process, the effective capacity drops back toward the

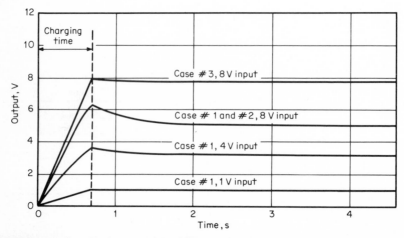

Fig. 78 Results of 15-μF, 20-V solid-tantalum capacitors tested in an integrator circuit.

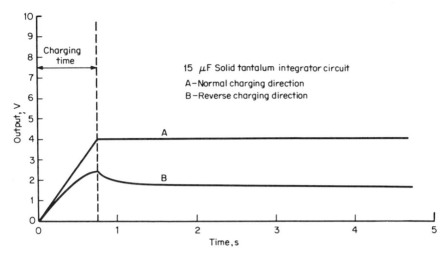

Fig. 79   Single solid-tantalum capacitor forward- and reverse-charging characteristic.

$\tfrac{1}{2}C$ value again.   In fact, for tests using steady-state sinusoidal inputs after a few cycles $C_T = \tfrac{1}{2}C$ applies almost perfectly even for very large signal inputs.

4. Apparent increases in effective capacity have also been observed for sinusoidal inputs of very low frequency (less than 1 Hz).   This effect appears to be essentially independent of input-signal level, however, and is probably related primarily to other factors such as dielectric absorption rather than polarization effects.

5. While tests on wet-slug tantalum capacitors have been performed with results similar to those obtained with solid (dry-slug) tantalum types except for a lower reverse breakdown potential, most manufacturers advise against using wet tantalums in a back-to-back configuration since reverse current through these devices can result in metallic deposits on the anode with a subsequent premature failure.   No such problem exists with solid- and foil-type tantalums, the only precautions required in this case being to limit reverse currents sufficiently to prevent excessive heating.

Ceramic Capacitors

General information   Ceramic dielectric capacitors are a unique family of capacitors with relatively high dielectric constant ranging from 6 to 10,000 and are easily manufactured to desired physical and electrical characteristics by applying the art of ceramic chemistry.   Ceramics used are based primarily on the mineral rutile (TiO_2) and on titanates (combinations of titanium dioxide and other oxides).   Some development has also been accomplished with other compositions from TiO_2.   By varying compositions, it is possible to produce a large variety of ceramic dielectrics with well-defined electrical properties to suit special requirements.   The variety of compositions and properties makes the field of ceramic-capacitor dielectrics quite broad but also complicated in that a great variety of capacitor types has been produced. Over the past 30 years the ceramic capacitor has been developed to permit higher capacitance per unit of volume ranging from 0.2 mF/in³ with large pressed disks in 1945 to 1,000 mF/in³ with high-k materials and monolithic construction in 1970.   Other improvements during this time include higher voltage ratings, lower leakage current, improved dissipation factor, better stability, and increased reliability.   The progress made in the high-dielectric-constant ceramic-chip capacitors, for example, has developed into one of the most effective solutions to the problem of obtaining capacitance in hybrid integrated circuitry.[8]   The ceramic-chip capacitor has become extremely popular because of its wide capacitance range, high volumetric efficiency, and relatively low cost.   Like any other capacitor, ceramic capacitors have their own set of characteristics, advantages, and disadvantages. Ceramic-capacitor dielectrics are divided into three classifications, as shown in Table 18.

Class I Ceramics. Capacitors of this dielectric type usually employ a basic titanium dioxide or calcium titanate system in which additives such as $Ba_2Ti_9O_{20}$, $MgTiO_3$, and $SrTiO_3$ may be used to achieve the desired characteristics. They are characterized by a nominal temperature coefficient defined over the range of 25 to 85°C. As shown in Table 18, the relative dielectric constant of these materials ranges from 6 to 500 and the power factor is 0.4 percent or less. Class I ceramic capacitors are suited for resonant circuits, high-frequency bypass and coupling, temperature compensation, and other applications where high Q and stability of dielectric constant are essential. The fact that ceramics in this classification are nonferroelectric is the prime factor responsible for the stability and temperature-compensating characteristics of the finished capacitor. In order to obtain the best dielectric properties, careful control of manufacturing processes is essential. Minute impurities may cause slight reduction of TiO_2 accompanied by a significant drop in Q value and insulation resistance. For the same reason, careful control of firing conditions such as time, temperature, and cleanliness is imperative.

Class II Ceramics. Capacitors of this dielectric type use barium titanate ($BaTiO_3$) as the basic dielectric material. A wide variety of compounds such as $SrTiO_3$, $CaTiO_3$, $CaZrO_3$, $Bi_2Sn_3O_9$, and Nb_2O_5 are reacted with $BaTiO_3$ to modify the dielectric properties as desired. Processing of these materials to achieve the desired electrical properties must be closely controlled as in class I ceramics. Barium titanate-based dielectrics are nonlinear (i.e., the charge accumulation is not a linear function of the applied voltage), primarily owing to the ferroelectric property of the unit cell.[9] The relative dielectric constant of class II materials is typically in the range of 250 to 10,000, as shown in Table 18, and decreases with age. The power factor ranges from 0.4 to 4 percent.

Class II ceramic capacitors are used where miniaturization is required for bypassing at radio frequencies, filtering, and interstage coupling where Q and capacitance stability can be compromised.

Class II ceramics are divided into two subgroups, stable and unstable (high-k), where the groups are defined in general by the temperature characteristics. Stable-k ceramic ranges in dielectric constant from 250 to approximately 2,400, has a nonlinear temperature characteristic defined over a temperature range of −60 to 125°C, and exhibits a maximum capacitance change of 15 percent from the 25°C value. Unstable or high-k ceramic ranges in dielectric constant from 3,000 to 10,000. These high k values are obtained by special formulations of titanates and additives which move the curie point* from approximately 125°C for stable-k ceramic to near room-temperature values for high-k ceramic. These ferroelectric ceramic compounds in finished capacitors exhibit severe changes in capacitance over a temperature range of −55 to 85°C or less (depending on formulation used) caused by k decreases of 30 to 80 percent. Bismuth additives are popular where improved capacitance stability with temperature is desired without causing a large decrease in dielectric constant. The use of bismuth, however, creates a compromise in ceramic technology. This additive, as a result of chemical instability, is dependent upon the use of platinum as an electrode for monolithic designs. The high cost of platinum leads to the use of other additives which will permit the use of less expensive noble metal electrodes.

Class III Ceramics. Capacitor dielectrics in this class are of the barrier-layer type. In this design an insulating ceramic disk is heat-treated in a reducing atmosphere so that the resistivity decreases to about 10 Ω cm. A thin layer of the surface of this body is then reconverted to the insulating state by firing in an oxidizing atmosphere. Silver electrodes are applied to the surface and are usually fired at the same time. A capacitor is thereby formed between the electrode and the semiconductor body. In an actual device electrodes are often applied to both sides of the disk and the finished capacitor is actually made up of two capacitors in series. Ceramic formulations are widely varied but are initially of the class II type of composition.

* The curie point for pure barium titanate is 120°C, above which the unit cell is cubic in form and nonferroelectric. Below the curie point the unit cell assumes a tetragonal form, is ferroelectric, and possesses a dipole moment sufficient to cause temperature, mechanical, and voltage effects on electrical properties. See Ref. 9.

TABLE 18   Classification of Ceramic Dielectrics

Attribute	Ceramic type			
	Class I	Class II		Class III
		Stable	Unstable	
Dielectric constant k at 25°C	6–500	250–2,400	3,000–10,000	900–5,100 (before reduction)
Operating temp range, °C	−55 to +125	−55 to +125	+10 to +85	−55 to +80
Temp coefficient of capacitance, ppm/°C*	P150 to N5,600			
Dissipation factor at 25°C, %	0.01–0.4	0.4–1.0	1.0–4.0	4–8
Max capacitance change, 25 to −55°C (ΔC in %)	Various (see Fig. 83)	+2, −8	+0, −90	−15 at 3 V −85 at 10 V −71 at 12 V −75 at 25 V −29 at 30 V
Max capacitance change, 25°C to max rated temp (ΔC in %) at °C	Various (see Fig. 83)	+2, −10 (125°C)	+0, −60 (90°C)	+13 at 3 V (80°C) +17, −60 at 10 V (80°C) +50 at 12 V (80°C) + 4, −30 at 25 V (80°C) +0.5 at 30 V (80°C)
Aging per decade time, %	Negligible	0.8–2.8	2.5–8.0	1–6
Usage	Tuned circuits, high frequency (microwave), high Q	Coupling, timing	Bypassing, filtering	Low-voltage coupling and bypassing in transistor circuits (3 to 50 V)
Major features	Ultrastable capacitance with respect to temperature, dc and ac voltages, frequency, and time. High Q. Widest temperature range	Small size. Wide range of characteristics. Low cost. Wide temp range	High volumetric efficiency	High capacitance. High volumetric efficiency. Low cost.
Major disadvantages	Large size, cost	Piezoelectric effect	Limited operating temp range (usually 0 to 65°C). Piezoelectric effect	Low leakage resistance. High voltage sensitivity

* Positive coefficient is prefixed with P; negative coefficient is prefixed with N.

Ceramic-capacitor construction   Ceramic capacitors with class I and class II dielectric materials are manufactured in numerous styles as shown in Figs. 80 to 82. Fabrication of the capacitor begins with cylinders or sheets of green ceramic which are formed to the desired shape and consist of a mixture of fine ceramic powders and suitable resin materials used as binders. The green ceramic materials are then silk-screened or painted with precious-metal paste consisting of platinum, palladium, or silver combinations to form electrodes. The electroded sheets are then stacked (for monolithic bodies) or processed as a single layer (for disk styles) and are subsequently cut into squares, disks, rectangles, or other desired shapes. In a monolithic design, several sheets are stacked together and compacted, with each cut section containing several electrodes which extend alternately to each cut end. Units are baked to remove organic materials and then sintered at about 1200 to 1450°C depending on the ceramic material used. The ends or edges of the ceramic section with exposed electrodes are then dipped into a precious-metal paste, generally silver or platinum silver, and are fired at about 750° to form solderable metal terminations. Capacitor sections at this point along with the class III types are further processed to completion in several configurations as shown in Table 19.

Chip Capacitors.   Ceramic sections of class I and II can be used as "chips" without the use of leads or body encasement of any kind. Chips with solderable terminations are eutectically bonded to substrates containing other circuit elements as well. These substrates are usually employed in microelectronic-circuit designs where the entire package is backfilled and sealed. Ceramic materials are nonhygroscopic and have practically no moisture absorption. Therefore, it is possible to produce unsealed, potted, or coated modules so long as other circuit elements or materials affected by moisture are not used or moisture resistance between terminations is not a problem. Chip capacitors, with the absence of lead inductance, have the highest frequency range.

Ceramic-chip capacitors are produced in several termination materials depending upon application. Fired-on silver is the simplest and least expensive termination. This type of termination is suitable when nondemanding techniques exist for attachment to the substrate metallization. If methods are not carefully controlled, loss of termination silver is possible during solder attachment. Also, during operating life the silver has a tendency to migrate (dendritic growth) and eventually cause electrical shorting. Silver alone should not be used where moisture is present.

Fired-on palladium-silver terminations are more costly than silver alone but are more forgiving during solder operations. Tests have shown that the silver is stabilized in this type of termination to a degree that arrests silver migration. This is one type of termination used in high-reliability microelectronic circuits.

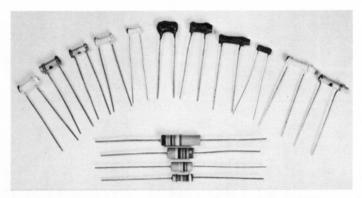

Fig. 80   Typical package styles for tubular ceramic capacitors. Top row—all dipped or painted; four dark units in center are dipped; all others are white enamel or clear-lacquer-coated. Bottom group—molded-axial-lead style, except that bottom unit is end-cap style with paint coating. (*ERIE Technological Products.*)

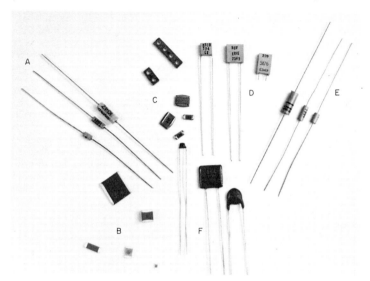

Fig. 81 Typical package styles for monolithic ceramic capacitors. (*ERIE Technological Products.*) (*A*) Glass-sealed axial style. (*B*) Standard chips. (*C*) Special chips and special heavy-solder chips. (*D*) Molded radial style. (*E*) Molded axial style. (*F*) Dipped style.

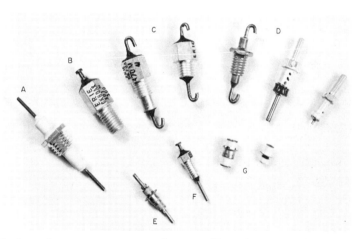

Fig. 82 Special ceramic capacitor styles. Feedthrough and standoff types shown above are a few of many styles available. (*ERIE Technological Products.*) (*A*) 1-kV threaded-bushing feedthrough. (*B*) Military-style threaded standoff. (*C*) Standard feedthrough styles. (*D*) Miniature threaded standoff. (*E*) Solder-eyelet feedthrough. (*F*) Miniature threaded-bushing feedthrough. (*G*) Solder-eyelet hollow feedthrough.

Presoldered silver or palladium-silver terminations are available wherein the manufacturer performs a "tinning operation" under carefully controlled conditions so that adequate solder coverage is obtained. This chip can then be more easily soldered on the substrate than the untinned chip.

Reflowed solder terminations are similar to the tinned termination. In this case sufficient solder is left on the chip to facilitate reflow during substrate attachment without the addition of more solder. This type is more expensive than the previous types discussed.

Plated terminations are identical to the silver and palladium-silver types except that a chemically or electrochemically deposited layer of nickel or copper is added. This plating serves as an inhibitor to prevent silver migration. Since nickel or copper plate is not reliably solderable, it is common to plate on a thin layer of gold also. This termination system is more costly. Fired-on gold, the most costly termination, is used where reliable thermocompression bonding is required. Gold is painted on and fired at a thickness of 0.001 in, nominal.

Molded Capacitors. Rectangular-chip capacitors, whether thick single layer or monolithic block, can be molded in radial or axial-lead rectangular packages or axial-lead cylindrical packages. Molded capacitors are fabricated by attaching leads by solder to the fired-metallic terminated ends of the capacitor chip. Resin undercoat is applied by dipping for the purpose of improving the bond of the molding material. This also provides improved resistance to moisture. The capacitors are then encapsulated in a modified epoxy resin by transfer-molding process. Parts are then marked with the appropriate information required by governing specifications and are individually checked for electrical parameters. Axial lead devices are then usually mounted on cardboard strips to facilitate automation of subsequent operations.

Glass-encased Capacitors. This type employs a multilayer ceramic chip or single-layer chip with axial leads attached sealed into a glass tube. In appearance the finished unit is similar to axial-molded capacitors. Fabrication is accomplished by assembling terminated ceramic chips, leads, and glass tubes, en masse, into machined carbon blocks with the aid of special fixtures. This assembly is then placed in a chamber where the air can be replaced by dried nitrogen. Passage of electric current through the carbon blocks causes localized heating, which in turn melts the glass sufficiently to seal the ends of the capacitors. By appropriate selection of materials· the lead can be caused to fuse the termination on the chip so that there is a metallic bond rather than just a pressure contact.

Potted Capacitors. In many ways, potted capacitors are interchangeable with molded capacitors. Some manufacturers prefer to pot instead of mold, and others reserve the potting approach to small-volume nonstandard sizes and physically large units. These capacitors are fabricated by the assembly of the chip to the leads in a manner similar to molded capacitors. However, instead of molding, the units are placed in a premolded epoxy case and filled with a liquid resin, which is then oven-cured.

Conformally Coated Capacitors. These capacitors, commonly known as dipped capacitors in both rectangular and disk styles, find great usage where precise geometry is not essential. These capacitors of all three classes are fabricated by the assembly of the ceramic chip to the lead in a manner similar to molded or potted capacitors. Conformal coating with resin can be achieved by dipping the ceramic chip in liquid resin or a fluidized powder-resin bed or an electrostatic fluidized bed. In the dip process, often more than one resin is used and usually some form of vacuum impregnation. In fluidized-bed coating, the parts are preheated before passing through an air-suspended cloud of resin powder. Successive passes through the resin provide the necessary coating thickness. These capacitors usually require a postcure. In the electrostatic fluidized bed, the units are usually not preheated. Powder is attracted to the surface of the ceramic which is maintained at ground potential with respect to a high-voltage electrode immersed in a fluidized powder. Heating of the units in an oven is needed to cure the powder coat. Several coats are usually required. Vacuum impregnation with wax or other suitable insulation material is common.

Dual-in-Line Packaging. Multilayer and single-layer ceramic capacitors can be provided in standard dual-in-line integrated-circuit-type packages. The finished capacitors are fabricated by soldering individual capacitor chips to lead frames and then molding or potting them. Where convenient, as when all capacitors are of the same value, the capacitors can be incorporated into a single block of ceramic.

Special Ceramic-Capacitor Designs. In addition to standard rectangular-chip capacitors previously discussed, a variety of special shapes are common. One variety,

TABLE 19 Ceramic Capacitor Selection Guide

Capacitance range, pF	Voltage rating, V dc	Military type[a]	EIA type RS-198[a]	Case Description	Case Figure ref.	No. of sizes
Class I		MIL-C-20				
1–10,000	200	CC(R) {75, 76, 77, 78, 79}		Molded tubular, axial leads	Fig. 80	5
82–39,000	100					
270–82,000	50					
1–680	200	CC(R) {81, 82, 83}		Conformal coated tubular, axial leads	Fig. 80	3
82–2,200	100					
270–5,600	50					
1–270	500	CC(R) {54, 55, 56, 57}		Conformal coated disk, radial leads	Fig. 81(F)	4
1–4,700	500		CC-030, 039, 044, 049, 059, 068, 076, 089, 110[e]			9
1–3,300	200	CC(R) {05, 06, 07}		Molded rectangular, radial leads	Fig. 81(D)	3
348–12,000	100					
1,870–100,000	50					
1–4,700	200	CC(R) {15, 16, 17, 18}		Conformal coated rectangular, radial leads	Fig. 81(F)	4
390–18,000	100					
2,200–68,000	50.					

TABLE 19 Ceramic Capacitor Selection Guide (Continued)

	Capacitance range, pF	Voltage rating, V dc	Military type[a]	EIA type RS-198[a]	Case Description	Case Figure ref.	Case No. of sizes
Class I	1–220	100	CC(R)13		Small-sized conformal coated rectangular, radial leads		2
	82–560	50	CC(R)14				
	200–5,100	50			Unencapsulated rectangular, radial leads	Similar to Fig. 81(F)	1
			MIL-C-55681				
	10–3,300[f]	100	CDR {01 02 03 04 05 06} BP[f]	CC {0805 1805 1808 2225} COG[f]	Uninsulated single or multilayer chip with solderable end terminations	Fig. 81(B)	4(EIA) 6(MIL)
	110–6,800[f]	50					
	240–10,000[f]	25					
Class II	47–330,000[f]	100	CDR {01 02 03 04 05 06} BX	CC {0805 1805 1808 2225} X7R or Z5U[f]	Uninsulated single or multilayer chip with solderable end terminations	Fig. 81(B)	4(EIA) 6(MIL)
	5,600–680,000[f]	50					
	12,000–1,000,000[f]	25					
			MIL-C-39014				
	10–10,000	200	CKR05, CKR06		Insulated (molded), rectangular, radial leads	Fig. 81(D)	2
	1,200–100,000	100					
	12,000–1,000,000	25					
	10–1,000,000	100	CKR11 through CKR 18		Insulated tubular, axial leads	Fig. 81(E)	8
	5,600–3,300,000	50					

Class	Capacitance (pF)	Voltage	Style	CC designators	Description	Figure	
Class II	510–3,900	1600	CKR64		Insulated (dipped), disk, radial leads	Fig. 81(F)	1
	47–47,000	500		CC-030, 039, 044, 049, 059, 068, 076, 089, 110[e]			9
	100–12,000	200					
	1,000–100,000	100					
	100–1,000	1500	CKR72		Feedthrough, mount, wire leads, 5/16–24 UNF–2A bushing	Fig. 81(A)	1
	10–1,500	500	CKR75		Same as above except 1/4–28 UNF–2A	Similar to Fig. 81(C)	1
	10–1,500	200	CKR76		Feedthrough, mount, turret and wire leads, 10–32 UNF–2A bushing	Fig. 81(F)	1
	10–1,500	500	CKR82		Standoff, mount, wire lead, 1/4–28 UNF–2A bushing	Similar to Fig. 81(B)	1
	10–1,500	200	CKR83		Same as above except turret terminal and 10–32 UNF–2A	Fig. 81(B)	1
	10–3,000	300			Feedthrough solder eyelet	Fig. 81(G)	b
	10–5,000	500				Fig. 81(E)	b
Class III	(Capacitance, µF) 0.02–0.47	12		5600[c]		Fig. 81(F)	d
	0.01–0.2	18		5700[c]			
	0.01–0.47	25		5800[c]			

[a] Specifications should be referred to for detail characteristics.
[b] Numerous size, styles, and assembly methods are available from ceramic-capacitor manufacturers.
[c] ERIE Technological Products, Inc., style designator. EIA part numbers are not available.
[d] Case sizes will vary with manufacturers.
[e] Last two digits of RS-198 part style number signify diameter in hundredths of an inch.
[f] Wider capacitance, voltage, and case-size ranges are available as nonstandard parts, but with limited sources.

the discoidal capacitor, is a disk with a hole through the center. Internal electrodes are brought out to the outside edge of the disk and to the edge of the hole. Metallizing with fired-on silver on these two edges completes the capacitor. This capacitor is commonly used in the shunt element of RFI filters. Multiple capacitors are also produced in many shapes and arrangements. One such arrangement is used in providing filtering of electrical connector pins. Here the connector pins passed through holes in a block of ceramic. Each hole is metallized and connected to internal electrodes within the ceramic block. Grounding is via metallizing around the periphery of the block. Other special designs include various feedthrough and standoff styles as shown in Fig. 82. These styles can be made from any of the ceramic dielectric classes as required.

Capacitance variations   Capacitance in ceramic capacitors varies with temperature, dc voltage, ac voltage, and frequency. The amount of variation caused by these stresses is dependent upon the particular dielectric thickness and formulation used and will vary considerably among manufacturers. The discussion of capacitance variation here is typical of those through the industry. Governing specifications and manufacturer's data sheets should be consulted where more detail is desired.

Temperature Coefficient (TC). In all ceramics the relationship between temperature and capacitance is, in general, not linear. It is important therefore to understand what is meant by the actual temperature coefficient of any given capacitor. Temperature coefficient is defined as the percent change in capacitance over a specified temperature interval and is given in parts per million per degree Celsius (ppm/°C). Temperature coefficient is calculated by measuring the capacitance change between 25 and 85°C and dividing by 60. Temperature coefficients themselves are nonlinear with respect to temperature. The TC calculation does not represent the change in capacitance to be expected for each degree of change in temperature. Therefore, the temperature coefficient is not exactly expressible by a single number. Class I dielectrics are identified by their nominal TC, and it is important to remember that the TC is determined from only two measurement points. These ceramics have a more negative TC as the temperature approaches −55°C. Should the temperature coefficient be critical over a range other than 25 to 85°C, the controlling specifications should be referred to.

There is a tolerance on nominal TC which varies from plus or minus 30 ppm for NPO ceramics to plus or minus 1,000 ppm for N5600 ceramics. Effectively an NPO ceramic is actually a zero plus or minus 30 ppm material with a temperature coefficient falling between plus 30 and minus 30 ppm limit. This limit includes the typical distribution of tolerance within a single lot of capacitors as well as a typical distribution from lot to lot. The ±30 ppm tolerance therefore becomes an envelope in which all capacitors from all manufacturers must fall in order to meet specification requirements. Temperature coefficients for common Class I ceramics are shown in Fig. 83. The range of temperature coefficients shown is obtained by use of ceramic formulations with varying percentages of high-k dielectric materials, such as titanium dioxide in a low-loss ceramic base. The temperature coefficient becomes increasingly more negative with the increase in dielectric constant. For example, titanium dioxide with a dielectric constant k of 85 has a temperature coefficient of −750 ppm/°C; low-loss ceramic with a dielectric constant k of 6 has a temperature coefficient of +100 ppm/°C. Therefore, for any given size of capacitor the relative capacitance will be high with a high negative temperature coefficient, and the converse will also be true. A high degree of reproducibility is obtained for all class I ceramics commonly used.

Temperature-characteristic curves for class II and III ceramics, both stable and unstable, are shown in Fig. 84; they vary considerably over a temperature range of −55 to +125°C. These capacitors should never be relied on for operation as a temperature-compensating component. EIA Standard RS-198 temperature-characteristic designations used throughout the industry are shown in Table 20.

Voltage Effects. Solid-state physics studies of ceramic dielectric properties as given by Kittel[9] and many others define the polarization properties of the unit cell in commonly used ceramic materials. Each of the unit cells which form crystals

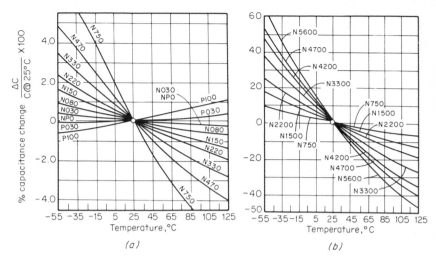

Fig. 83 Temperature coefficient of capacitance for Class I ceramic capacitors. Tolerances on the coefficients shown are given in governing specifications. (*a*) P100 to N750. (*b*) N750 to N5,600. (Note: Positive coefficient is prefixed with *P*; negative coefficient is prefixed with *N*.)

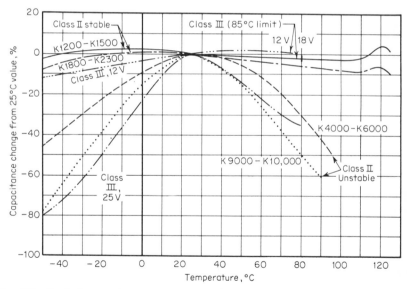

Fig. 84 Typical capacitance variation with temperature for ceramic dielectric capacitors.

for each ceramic compound, varying from the NPO ceramics through the high-*k* ceramics, has its own hysteresis curve, polarization voltage, and crystal form. As an example, the polarization voltage for K1200 ceramic is about 150 V/mil. It is obvious then that as ceramic dielectric thickness in monolithic types is reduced to 0.001 in to increase the capacitance in a given volume, the effects of impressed

TABLE 20   EIA Standard Temperature Characteristics for Class II Ceramic Capacitors as Coded in Significant Part Numbers (RS-198B)

Example part number: CCXX20 (Style, See RS-198) — Z 5 S (Temperature characteristics) — 22 1 M 15 1 (Detail part data)

Low-temp requirement °C	Letter symbol	High-temp requirement °C	Numerical symbol	Max capacitance change over temp range, %	Letter symbol	First and second significant figure of capacitance	Multiplier*	Numerical symbol	Tolerance on capacitance, %	Letter symbol	First and second significant figure (Voltage rating)	Multiplier (Voltage rating)	Numerical symbol (Voltage rating)
+10	Z	+45	2	±1.0	A		0	0				1	0
−30	Y	+65	4	±1.5	B		10	1	±5	J		10	1
−55	X	+85	5	±2.2	C		100	2	±10	K		100	2
		+105	6	±3.3	D		1,000	3	±20	M		1,000	
		+125	7	±4.7	E		10,000	4	+100 −0	P			
				±7.5	F		100,000	5	+80 −20	Z			
				±10.0	P								
				±15.0	R								
				±22.0	S		0.01	8					
				+22 −33	T		0.1	9					
				+22 −56	U								
				+22 −82	V								

Source: Electronic Industries Association.

* Use lowest decimal multiplier to avoid alternate coding; for example, 2.0 pF should be 209, not 020.

Listing of complete range of characteristics does not necessarily imply commercial availability of all values but is for the purpose of providing a standard identification code for future development.

X7R was formerly noted by industry as W5R.

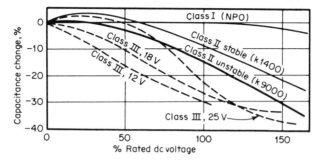

Fig. 85 Typical capacitance change with dc voltage for a standard ceramic-capacitor design. Capacitance change is referenced to 0 V dc and 1 V rms at 14-Hz signal.

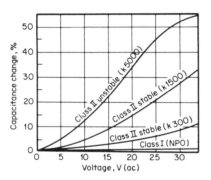

Fig. 86 Typical capacitance change with increasing ac voltage. These curves are based on standard designs and vary widely depending on ceramic formulation and dielectric thickness. Capacitance change is referenced to 1 V rms, 1 kHz measurement.

voltage will be more significant. Figure 85 shows typical effects of dc voltage on class I and III ceramics. In single-layer disk types, however, the dielectric thickness is typically 10 to 20 mils thick for mechanical strength and the voltage effect is not as predominant. Class I ceramics operating between −55 and +125°C are generally considered to have 0 effects from applied voltage. Class II and III ceramics, however, do have a significant voltage effect which must be considered when these capacitors are applied and when measurements are specified for them. In many cases it may be advisable to specify capacitance for high-k ceramics at the temperature and frequency of use. In contrast to a dc bias-voltage effect which produces a decrease in capacitance, an ac voltage will aid in the reordering of unit cells within polarization domains of the ceramic material, with the result of an increase in capacitance as ac voltage level is increased. Class II ceramics based on barium titanate material are designed with modifiers such as bismuth stannate or niobium or tantalum pentoxides. These dielectric compounds are ferroelectric and exhibit in varying degrees those characteristics for which the high-k ceramics are known. The affects of ac applied voltage on capacitance are shown in Fig. 86.

AC Voltage Rating. Typically, ceramic capacitors should be applied with the ac voltage limited such that the sum of the dc voltage and ac peak-to-peak voltage does not exceed the dc rating of the capacitor. Also, the heat generated must not cause the temperature rise of the capacitor body to exceed 35°C. As an example, capacitors such as CKR05 and CKR06 of MIL-C-39014 are considered ⅛-W devices. Maximum power rating for each capacitor size is calculated from voltage applied and specified dissipation factors.

Capacitance Change vs. Frequency. Effects of frequency vary considerably depending on capacitance values, body configuration, and lead length. Curves shown in Fig. 87 give an illustration of the range of frequency response for ceramic dielectric materials and also show typical differences between bismuth-additive and bismuth-

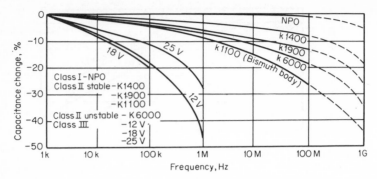

Fig. 87 Typical capacitance variation with frequency for ceramic capacitors.

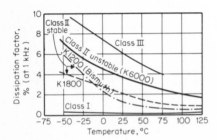

Fig. 88 Typical dissipation-factor variation with temperature for ceramic dielectric capacitors. These curves vary widely between specific ceramic formulations.

free ceramics with variations in frequency. For low-value ceramic chips in class I dielectric, the stability with increasing frequency is quite good, as illustrated. With increasing dielectric constant and higher capacitance values, capacitance will decrease. Typical curves are illustrated in the same figure for various dielectric constants.

Dissipation factor Dissipation factor is a parameter that will rarely affect circuit operation except in applications requiring high Q, where class I dielectrics are used. In class II ceramics the dissipation factor at room temperature will vary typically between 1 and 4 percent. Military specifications normally limit dissipation factor to 2.5 percent. Dissipation factor is used as a guide by the manufacturer to determine whether the ceramic formulation is consistent with previous lots and that the determinations and electrode construction are not defective. Dissipation-factor measurements can reveal contamination, cold-solder joints, lifted electrodes, delamination, or separated leads, etc. Figure 88 shows variations in dissipation factor with temperature. As may be noted, the dissipation factor at −55°C increases to as much as 10 percent depending upon ceramic formulation. The dissipation factor of the bismuth-free ceramic is significantly lower at −55°C. Measurement conditions for this parameter are similar to those of capacitance in that the value is affected by measurement voltage and frequency. Variations in dissipation factor with dc voltage and ac voltage are shown in Figs. 89 and 90, respectively. Variations in dissipation factor with frequency are shown in Fig. 91.

Ceramic capacitor Q The Q factor for ceramic dielectric capacitors will vary considerably according to the ceramic class, as shown in Fig. 92. Variations also exist from lot to lot and between ceramic formulations. These curves, however, will give the designer a reference to determine which ceramic fulfills the application. Measurement techniques used to measure Q will determine the validity of the measurement at higher frequencies. All leads and connections must be as short as possible, and the contact should be as near noninductive as possible to keep the inductive reactance from swamping the Q-factor measurement.

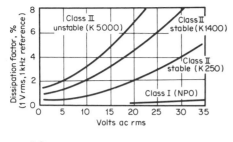

Fig. 89 Typical dissipation-factor change with ac voltage. These curves will vary widely depending upon dielectric thickness and formulation.

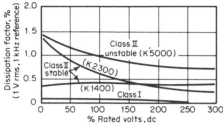

Fig. 90 Typical dissipation-factor change with dc voltage (applied during measurement).

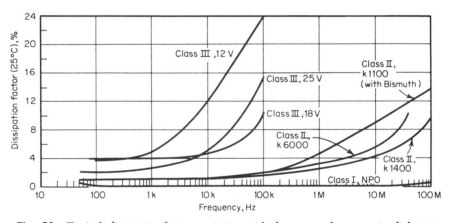

Fig. 91 Typical dissipation-factor variation with frequency for ceramic dielectric capacitors.

Insulation resistance Insulation resistance of ceramic capacitors consists of a combination of surface resistance (which is sensitive to surface contamination), moisture and gas absorption, and leakage resistance through the dielectric materials and between the electrodes. In general these resistances are lumped together and referred to as insulation resistance. It has become common practice to express insulation resistance of all electrostatic capacitors as the product of megohms times microfarads. The relationship between capacitance and leakage resistance is one of inverse proportionality. Since leakage resistance is primarily a function of leakage current through the dielectric, it is practical to assume that the leakage resistance is also a function of dielectric thickness. The more surface area in the capacitor the lower the leakage resistance. This correlates with the earlier statements concerning the common use of the term megohms-microfarads. Low capacitance values will have lower leakage resistance if they are made from many thin laminations of dielectrics. Leakage resistance varies greatly from one ceramic composition to another and varies as a function of manufacturing-process control. Generally the lower-k ceramic has a higher bulk resistance. Leakage resistance decreases with increasing temperature,

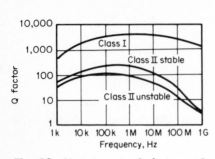

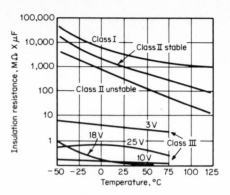

Fig. 92 Variations in Q factor with frequency for ceramic dielectric capacitors.

Fig. 93 Typical insulation-resistance variation with temperature for ceramic dielectric capacitors.

which is a result of increased ion activity. Insulation resistance is illustrated for all classes of ceramic dielectric in Fig. 93.

Aging and deaging effects in ceramics An understanding of aging in ceramic capacitors is necessary for proper application. This characteristic term is used to describe the negative logarithmic change in dielectric constant with time. Class I dielectrics, considered to be nonferroelectric, do not appear to exhibit an aging characteristic. Class II ceramics, as previously defined, exhibit a significant aging characteristic. This characteristic is due to several causes. Crystal-phase transformation in barium titanate ceramics is the primary cause of aging. As previously stated, barium titanates when maintained below 120°C have a ferroelectric tetragonal crystal structure. The dipole moment of the unit cell is such that when under voltage or mechanical stress the cells will, in time, orient themselves within the crystal-polarization domain. This orientation or movement of the unit cell will lower the dielectric constant, which in turn will lower the capacitance and dissipation factor. This effect is called aging. Each time the capacitor is heated above the curing point, all the negative-dielectric-constant change is recovered. In this condition the unit cell reverts back to the cubic form. This effect is called deaging. Most barium titanate ceramic capacitors will be completely deaged at a temperature of 150°C.

Another important parameter affecting capacitor aging is dc voltage. The application dc voltage near rating of the capacitor will cause a negative capacitance change, and when the voltage is removed the capacitor does not normally return to the original value. The causes of aging in ceramics under this condition are very similar to those in the crystal-phase transformation. In this case an electric field produces a coercive force that will tend to align the unit cells within polarization domains and when achieved will cause a reduction in capacitance and dissipation factor. The aging of barium-titanate-base ceramics follows the general equation of

$$k_t' = k_0' - m \log t$$

where k_t' is the dielectric constant at time t, k_0' is the dielectric constant at some initial time, set at 1, m is the rate of change from k_0' to k_t' as shown in Table 21. Owing to the logarithmic aging rate it may be noted that very little aging will take place after 1,000 h. Manufacturers sometimes allow for aging when measuring capacitance, and parts are sorted to be in tolerance for 3 months of storage. Longer shelf life can be guaranteed by preaging and by sorting capacitors to tighter guard bands. As previously mentioned, dissipation factor is affected by aging, with a gradual decrease occurring during storage. This change is favorable, is not of great magnitude, and is of only minor concern. Ceramic capacitors can be preaged by the application of dc voltage for a specified length of time. Preaging is beneficial

TABLE 21   Typical Aging Rates for Class II Ceramic Dielectrics

Dielectric constant* k	Temp characteristic (TC),* %	EIA TC code	Aging rate,† % per decade of time
250	±4.7	E	0.8
700	±6.0		1.5
1,200	±7.0		1.8
1,400	±7.5	F	2.0
2,000	±10.0	P	2.5
2,200	±15.0	R	2.8
2,500	±22.0	S	3.25
5,000	+22 −56	U	4.0
10,000	+22 −82	V	5−8

* Dielectric constant value can vary with formulation to obtain the same TC and aging effect.
† Rates based on 0 V dc applied.

where circuit designs cannot tolerate a capacitance change normally experienced during the first 1,000 h of capacitor shelf life. Another means of controlling aging effects is by the inclusion of certain chemical additives in the ceramic formulations. Research is continuing to obtain more information about the mechanism of aging with the object of obtaining lower aging rates. Variations in the aging rate of class II stable ceramic capacitors will vary as a function of applied voltage, as shown in Fig. 94.

Stable Capacitors

Dielectric types

Mica. Of several varieties of mica used in the construction of capacitors, muscovite mica is the most common form. It has good mechanical strength and can be used up to temperatures of 500°C. One form of muscovite, ruby mica, so called because of its color, has the best high-voltage and high-frequency dielectric properties. Another type, phlogopite mica, is softer than muscovite and does not have as good dielectric properties but can be used up to temperatures as high as 900°C. The dielectric constant of capacitor-grade mica varies from 6.5 to 8.5, as shown in Table 22. Chemically, natural micas are complex silicates of aluminum with potassium, magnesium, iron, sodium, lithium, and traces of many other elements.

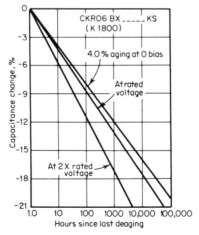

Fig. 94 Typical variation in aging rate with dc voltage applied for class II stable ceramic dielectric capacitors.

TABLE 22 General Characteristics of Stable-Capacitor Dielectric Materials

| Characteristic | Muscovite mica (ruby)* | Glass | | | Porcelain |
		Soft-lead-soda	Hard borosilicate	Quartz (fused)	
Loss (at 25°C):					
Limiting operating frequency, Hz, min	100	200	100	100	100
max	3×10^9	10×10^9	10×10^9	$>10 \times 10^9$	10×10^9
Dissipation factor at 10 MHz, %	0.01–0.05	0.1	0.1	0.02–0.04	0.1–0.7
Dielectric constant k, 1 kHz to 10 MHz	6.5–8.7	8.5	4.0	4.4	5.0–7.0
Dielectric strength (breakdown), V/mil	1,000, 2,000†	400	400	1,000	55–300
Operating temperature range,‡ °C, min	No limit	No limit	No limit	No limit	No limit
max	200	200	200	300	200
Capacitance drift (25°C) after temperature-excursion, %	0.05–3.0 (see Table 24)	0.1 or 0.1 pF, whichever is greater			0.1 or 0.1 pF, whichever is greater
Capacitance stability (after life test), % change	2 (molded stacked mica), 1 (molded metallized mica)	1	1		1.0 or 0.5 pF max
Capacitance stability (after climatic environments), % change	±2 or 1 pF, whichever is greater	0.5 or 0.5 pF, whichever is greater	1		1.0 or 0.5 pF, whichever is greater
Life test, h	2,000	2,000			2,000
Temp coefficient at 100 kHz, ppm/°C	From (±500) to (0, −50) (see Table 24)	+140 ±25			(+105, ±25) and (0, ±25)
Insulation resistance, MΩ, at 25°C	>100,000	>100,000			>500,000
at max temp	>5,000 (150°C)	>10,000 (125°C)			>100,000 (125°C)
Voltage coefficient	Negligible	None	None		None
Advantages	Various temperature coefficients; Low-inductance designs; High insulation resistance; Lowest dielectric absorption; High mechanical strength; Accommodates custom designs	Fixed temperature coefficient; Excellent retrace of temperature coefficient; High insulation resistance; Low dielectric absorption; Hermetic seal; Tolerates high humidity and temperature			Temperature stability; Low dissipation factor; High moisture resistance; High insulation resistance; Hermetic seal; Tolerates high humidity and temperature
Disadvantages	Silver-ion migration under dc stress at high temperature and humidity; Natural-resource depletion	Requires cost trade-off with mica and porcelain			Requires cost trade-off with mica and glass

* V-1 or V-2 grades in accordance with Ref. 10.
† Impregnated mica paper.
‡ Temperature range is usually limited by capacitor-encasement materials rather than the dielectric itself.

Muscovite can be termed a potassium aluminum silicate $KAl_3Si_3O_{10}(OH)_2$ and phlogopite a potassium magnesium aluminum silicate $KMg_3AlSi_3O_{10}(OH)_2$. Each type has varying amounts of iron, manganese, copper, and chromium. Mica crystals in the monoclinic system can be readily cleaved into sheets to a thinness of less than one-thousandth (0.001). Capacitors require flawless or near flawless mica of grade* V-1 or V-2 for capacitor standards, V-3 for transmitting capacitors, and V-4 for receiver and variable capacitors. The geographical origin of mica is an important factor in successful manufacture of precision stable mica capacitors. As a result, natural mica resources can become limited.

Reconstituted Mica. A paper made from shredded mica known as reconstituted mica has been developed over the past 15 years. Capacitor-grade mica paper is produced by processes similar to those used in the manufacture of cellulose papers. A slurry is prepared by thermal or high-velocity water treatment of readily available forms of mica. The slurry of small mica platelets suspended in water is fed into the bed of a modified Fourdrinier machine. As the sheet is formed on the screen, a series of dryers and calenders processes it into a continuous laminated structure consisting of minute inorganic mica splittings. The paper thus is made from purified mica with no chemical binders, and a rigid structure is maintained by the natural cohesive forces characteristic of pure mica itself. The dielectric strength of this mica paper is 800 V/mil, slightly less than that of ruby mica, but by the use of suitable impregnants such as silicon or polystyrene the dielectric strength of the mica paper can be increased to approximately 2,000 V/mil. The thermal capability of reconstituted mica is similar to that of muscovite natural mica. The impregnants used in conjunction with mica paper therefore limit the operating-temperature capability; that is, polyesters to 125°C, silicons to 315°C, and other inorganic impregnants to 400°C.

Glass. The use of glass as a capacitor dielectric is not new; in fact, glass is about the oldest capacitor dielectric, beginning with the Leyden jar. However, the development of glass ribbons down to 0.001 in thick permits the manufacture of capacitors comparable with mica. Finished glass capacitors have a temperature coefficient very close to 140 ppm/°C, as shown in Table 22, whereas the design and construction of mica capacitors affect their temperature coefficient considerably. Electrical-grade glasses, also considered optical-grade, exhibit high bulk resistivity. Conduction takes place by ion migration through the glass. Therefore, alkali-metal ions, which are quite mobile in many glasses, must be eliminated or demobilized through the "mixed alkalized effect," or the use of additives such as calcium or lead. Glass ribbon used in the manufacture of glass capacitors is a mixed alkali-lead silicate. The dielectric constant is slightly above average for glasses owing to the high lead content. The mixed alkali achieves a small interstitial volume and thereby produces a low-loss tangent (dissipation-factor) value.

Porcelain. Another type of dielectric used in the stable-capacitor group is one using a ceramic glaze or enamel as the dielectric. This type, developed by Du Pont under a Signal Corps contract, was produced in an effort to mass-produce capacitors when the demand for mica was high. Essentially, alternate layers of glaze and silver are sprayed and the entire unit is fired into a single block.

Stable-capacitor construction

Mica capacitors. Several significant packaging techniques have been developed to provide mica capacitors suitable for combinations of high frequency, high voltage, high-density packaging, and automatic-insertion applications; see Table 23.

Most mica-sheet capacitors employ electrodes of one or two types. For low-power applications electrodes are formed by screening a silver paste on mica slices gaged from 0.001 to 0.005 in depending on voltage and mechanical requirements. The silvered mica is then fired in an oxidizing atmosphere to obtain a permanent bond. By employing the screening process for application of silver, proper margins can be obtained. Fired mica wafers can then be stacked and clamped together to obtain the desired capacitance. For high-power rf and very-high-voltage capacitors, the

---

* Grades specified in Ref. 10, secs. B15 and B16.

TABLE 23 Standard Stable Capacitors

Dielectric type	Capacitance range, pF	DC working voltage, V	Standard capacitance tolerance, %	Standard temp coefficients	Military standard type	EIA standard type	Dissipation factor at 25°C, %	Operating temp range, °C	Life test, h	Operating frequency range, MHz	Case style
Ruby mica (sheet)	47–100,000	250–5 kV		B, C, D, E, G^a	MIL-C-5 CM65, CM70	TR-109 65, 70	0.35 at 1 MHz 0.15 at 1 kHz	−55 to +70	750 (EIA) 2,000 (MIL)	0.1–3	Fig. 95b
	47–100,000	1–30 kV	±2, ±5, ±10		CM75, CM80 CM85, CM90	75, 80 85, 90					High voltage Fig. 95a
	100–10,000	15–35 kV			CM95	95					
	1–20,000	100	±0.5, ±1	B, C, D, E, F^a	MIL-C-5 CM {15^c	RS-153 RCM {15	See Fig. 100	−55 to +150	250 (EIA) 2,000 (MIL)	Up to 100	Axial leads, molded case, Fig. 96
	1–16,000	300	±2, ±5		20	20					
	1–10,000	500	±10, ±20		30	30					
	1–5,700	1,000			35}	35}					
	1–10,000	100	±0.5, ±1	C, D, E, F^a	CM {04 and 09^d	RDM {10 and 10C^d					Radial leads, dipped case, Fig. 97
	1–68,000	300	±2, ±5		05 and 10	15 and 15C					
	1–51,000	500	±10, ±20		06 and 11	19 and 19C					
	1–32,000	1,000			07 and 12	20 and 20C					
					08 and 13}	30 and 30C					
						42 and 42C}					
	1–400	50	±0.5, ±1	C, E, F^a	MIL-C-39001 CMR {03		See Fig. 100	−55 to +150	2,000^e	Up to 100	Radial leads, dipped case, Fig. 97
	75,000–91,000	100	±2, ±5		04						
	56,000–68,000	300			05						
	1–51,000	500			06						
					07						
					08}						
	12,000–47,000	600	±2, ±5	D, E, F^a	MIL-C-5 CM {45		See Fig. 100	−55 to +150	2,000	Up to 80	Fixed terminal, Fig. 98
	4,700–33,000	1,200			50						
	47–16,000	2,500			55						
					60}						

Material	Capacitance range (pF)	Voltage	Capacitance tolerance, %	Temperature coefficient	MIL spec / Style	DF[b]	Temperature range, °C	Max voltage	Description
Ruby mica (sheet)	5–1,500								Button mica (see Fig. 99)
	5–2,400	500	±2; ±5, ±10	B, D[a]	MIL-C-10950, CB50	See Fig. 100	−55 to +85		Feedthrough threaded mounting, resin seal
					CB55, CB56, CB57				Feedthrough, solder mount, glass-sealed
	15–2,400				CB60, CB61, CB62		−55 to +150	2,000	Feedthrough, threaded mount, glass-sealed
	5–2,400				CB65, CB66, CB67			Up to 500	Standoff, solder mount, glass-sealed
Glass	220–10,000; 0.5–6,200	300; 500	±1, ±2, ±5	+140, ±25 ppm/°C	MIL-C-23269, CYR { 10, 15, 20, 30	0.1			Axial-lead hermetic-sealed glass case
	1–2,400	300	±0.25 pF or ±2, ±5 %		CYR { 51, 52, 53	0.1–0.2	−55 to +125	2,000[e] Up to 100	Radial lead, epoxy case
Porcelain	220–5,600; 0.5–3,300	300; 500		105, ±25 ppm/°C	CYR { 13, 17, 22, 32	0.7 (up to 4.7 pF) 0.3 (from 5.6 to 27 pF) 0.1 (from 30 pF up)	−55 to +125	2,000[e]	Axial-radial lead, vitreous-enamel case
	270–1,200; 0.8–680; 68–390; 0.5–56	50; 100; 300; 500	±1, ±2, ±5	0, ±25 ppm/°C	CYR { 41, 42, 43	0.1		Up to 100	Radial lead, vitreous-enamel case

[a] See Table 24.
[b] DF is measured at 1 MHz for capacitance values up to 1,000 pF and 1 kHz for capacitance values above 1,000 pF.
[c] Capacitance and voltage ranges are reduced from EIA ranges.
[d] Style numbers in second column have crimped leads.
[e] Continuing life tests are performed to establish failure rate.

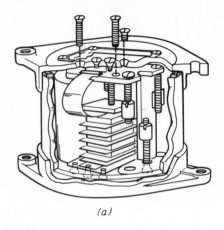

(a)

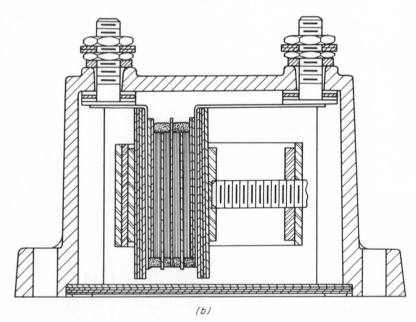

(b)

Fig. 95 Standard high-voltage transmitting mica capacitors.   Refer to applicable specifications for dimensions.   (*Pictures from Sangamo Electric Company.*)

MIL-C-5 type	RTMA standard TR-109 type
(a)	
CM 75	75
CM 80	80
CM 85	85
CM 90	90
CM 95	95
(b)	
CM 65	65
CM 70	70

mica leaves are gaged for thickness and then stacked with interleaving tin-lead foils to obtain the desired capacitance.

Transmitting-type mica capacitors are enclosed in a molded phenolic case or in low-loss ceramic or porcelain cases. Mica sheets are assembled into sections with interleaving foils. These sections are then connected in series and/or parallel into stacks, depending on the design requirements. All foils are metallurgically bonded to form permanent connections. The stacks are arranged so as to provide low-inductance and low-resistance connections. Mica stacks are vacuum-impregnated in a low-loss wax. This process assures the removal of all air and adds protection against failure of the capacitor due to corona. The internal assembly is then clamped in nonmagnetic clamps for mechanical strength and electrical stability. The assembly is then encased in a low-loss wax or resin, which provides protection against moisture absorption and transmission and protection from shock and vibration. These capacitors can be supplied oil-filled where severe duty, such as pulsing, would result in increased risk of corona and subsequent breakdown. Transmitting mica capacitors, shown in Fig. 95 have screw-type terminals which are soldered to the foil electrodes. The entire assembly of mica-foil sections, terminals, and encasement is mounted in a molded phenolic case which is filled with an epoxy/polyester encapsulating compound.

Ceramic-case high-voltage transmitting mica capacitors, shown in Fig. 95, have end caps made of cast aluminum providing very low contact resistance and convenient mounting surface. These capacitors are stacked for a variety of series or series-parallel capacitor banks. A major advantage is the long distance between the terminals, which gives an excellent creepage distance and reduces the possibility of external corona.

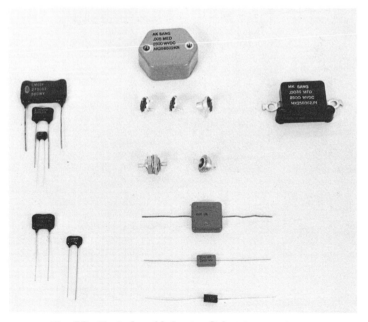

Fig. 96 Typical molded mica-dielectric capacitors.

Molded mica axial-lead capacitors, as shown in Fig. 96, consist of silvered-mica plates stacked into sections in a precisely determined pattern to maintain miniaturization. Wire leads are staked to pressure clips, and the assembly is solder-coated. These assemblies are then attached to the completed mica electrode section. Wire leads are made of tinned copper for soldering properties.   The molded case of these capacitors is made of an impregnated phenolic or polyester material with good insulation properties.   Resistance to moisture absorption and transmission is achieved with these materials.   The molded case also imparts rigidity to the capacitor in the event the capacitor is subjected to vibration or shock.

Dipped mica radial-lead capacitors, as shown in Fig. 97, are constructed of silvered mica sheets clipped into stacks similar to molded mica capacitors.   The standard

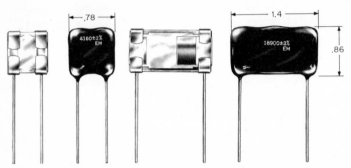

Fig. 97 Typical dipped mica-dielectric capacitors.   Left view of each size is epoxy-coated and shows clamping of mica stack and lead attachment.   Blackbody is multiple-coated configuration.   (*Electromotive Mfg. Co., Inc.*)

dipped case consists of multiple coatings of electrical-grade phenolic and epoxy resins which exhibit high insulating properties and high resistance to moisture.   This case easily resists the high temperatures of soldering.   The coatings are applied automatically to avoid contamination due to handling.   When case size is a critical factor in the selection of a dipped mica capacitor, but moisture resistance must be maintained at moderate levels, a single dip case can be used.   This capacitor case consists of a single coating each of phenolic and epoxy resins.   In applications such as filters, networks, and delay lines, even smaller size is required.   For such applications, a capacitor with a single coating of epoxy resin can be obtained.   This capacitor should be used only in circuitry which will later be hermetically sealed or receive some other environmental protection, since their greatly reduced size has been achieved by sacrificing the moisture-resistance characteristics of the standard dipped mica capacitor.

Fig. 98 Typical fixed-terminal molded-body mica-dielectric capacitors.   (*Sangamo Electric Company.*)

Fixed-terminal mica capacitors, as shown in Fig. 98, are constructed of accurately gaged sheets of mica interleaved with foil.   The accuracy with which both the mica and foil sheets may be dimensioned permits very close adherence to the capacitive rating.   The low dissipation factor and the stacked-capacitor construction contribute to a low internal inductance and account for the suitability of these capacitors for high-frequency applications.   Certain capacitive and voltage ratings require the use of silvered mica sheets, rather than mica and foil, in order to maintain the specified electrical characteristics. Terminals are attached to the silvered or foil stack by means of a pressure clip. A tapped brass terminal insert is soldered onto this clip, or a copper clip is extended through the molded case, providing an electrical connection with a minimum of contact resistance. Capacitors are molded in low-loss, electrical-grade phenolic molding materials.

Button mica capacitors are composed of a stack of silvered-mica disks connected in parallel. This assembly is encased in a metal case with a high-potential terminal connected through the center of the stack. The other terminal is formed by the metal case connected at all points around the outer edge of the electrodes. This design permits the current to fan out in a 360° pattern from the center terminal, providing the shortest rf current path between the center terminal and chassis. The internal inductance is thus kept small. The use of relatively heavy and short terminals results in minimum external inductance associated permanently with the capacitor. The units are then welded and hermetically sealed or resin-sealed. Typical styles are shown in Fig. 99.

Fig. 99   Typical button mica capacitors.   (*ERIE Technological Products.*)

Reconstituted-Mica Capacitors.   The flexible and highly workable nature of mica paper enables it to be wound on conventional capacitor-winding machines. Highly purified aluminum foil is used as the electrode. Completed windings after impregnation are compressed while the impregnant is still in an unpolymerized state. Pressure is maintained until polymerization is complete. This results in a totally solid capacitor section ready for packaging. The capacitor sections can be packaged singly or in multiples using encasements of Mylar wrap with epoxy end fill, transfer-molded epoxy, or cast epoxy tubes of either rectangular or cylindrical form. Thermal and environmental capability is predicated on the type of encasement used. The totally solid construction enables this type of capacitor to meet rigorous corona, vibration, and shock requirements without the loss of liquid impregnants commonly associated with paper and polyester dielectrics. They are well adapted to special packaging techniques because the form factor is variable, allowing more license in the design where compactness is essential.

Glass Capacitors.   A typical construction of these capacitors consists of alternate layers of alkali-lead glass ribbon and high-purity aluminum foil stacked to meet the capacitance required. This stack is then fired into a monolithic block. Copper-clad nickel-iron (Dumet) leads are welded to electrode and terminations, and the assembly is then enclosed between two larger blocks of the same glass composition as the dielectric material. The assembly is then fired to obtain a hermetic seal along the leads as well as around the periphery of the capacitor body. Refer to case configurations and parameter ranges shown in Table 23.

Porcelain Capacitors. These capacitors consist of a monolithic block of vitreous-enamel dielectric and silver electrode material encased in a heavy porcelain body. External leads are brazed onto the electrodes along the edge of the monolithic block. Small capacitor styles shown in Table 23 which are designed to meet the military-standard requirements have a slotted case at lead egress so that the lead can be bent in either axial or radial configuration. Commercial configurations are built to voltage ratings as high as 2,500 V dc with a test voltage of 6,250 V dc for 5 min. Porcelain capacitors are also available in chip form and are constructed in a monolithic block similar to ceramic monolithic chip capacitors.

Electrical characteristics

Temperature coefficient and capacitance drift. Measurements of capacitance for computation of temperature coefficient and capacitance drift are performed at 1 MHz for capacitance values of 1,000 pF or less and at 1 kHz for capacitance values above 1,000 pF. As an alternate, a frequency of 100 kHz may be used. Measurements for TC are performed, in order, at room temperature (25°C), −55°C, then in selected steps (including 25°C) up to the maximum rated temperature and back to room temperature. The steps are selected depending upon the points at which TC is to be computed [see Eq. (17) in Table 1]. Capacitance drift is computed by dividing the greatest single difference between any two of the three capacitance values measured at 25°C by the second value recorded at 25°C. Typical temperature coefficients and capacitance drift are shown in Table 24.

TABLE 24   Standard Temperature Characteristics for Mica Capacitors

Specification					Charac-teristic code	Temp coefficient (max), ppm/°C	Capacitance drift (max)
MIL-C-5	MIL-C-10950	MIL-C-39001	TR-109	RS-153			
X	X		X	X	B	±500*	±(3.0% + 1.0 pF)*
X		X	X	X	C	±200	±(0.5% + 0.5 pF)
X	X		X	X	D	±100	±(0.3% + 0.1 pF)
X	X	X	X	X	E	±100, −20	±(0.1% + 0.1 pF)
X		X		X	F	+70, 0	±(0.05% + 0.1 pF)
			X		G	0, −50*	±(0.1% + 0.1 pF)*

* This value specified only in TR-109 for transmitting mica capacitors.

Dissipation Factor (DF). Measurement of this characteristic is made in a manner similar to that of capacitance and usually at the same time using a suitable capacitance bridge. Standard DF limits for mica, glass, and porcelain capacitors are shown in Fig. 100. Typical DF change with temperature and frequency is shown in Figs. 101 and 102, respectively. In general the loss characteristics of glass and porcelain are superior to those of mica. However, the availability of wider capacitance range, and lower cost for micas provide a trade-off for improved loss and moisture-resistance characteristics in glass. A comparison in Q factor between glass and mica is shown in Fig. 103.

Insulation Resistance (IR). In most stable capacitors the insulation resistance is measured at 100 V dc but never in excess of the rated voltage. High-voltage micas are measured at 500 V dc. Electrification time is nominally 60 s for 100 V dc measurements and 120 s for 500 V dc measurements. Insulation resistance is a function of both ambient temperature and nominal capacitance. Higher capacitive ratings and temperatures necessarily entail a reduction in insulation resistance. The minimum IR limit for mica capacitors over the capacitance range is shown in Fig. 104. A comparison of typical insulation resistance among mica, glass, and porcelain capacitors is shown in Fig. 105. These curves will vary widely among actual capacitors depending upon construction style, impregnation used, and the manufacturer's own limits.

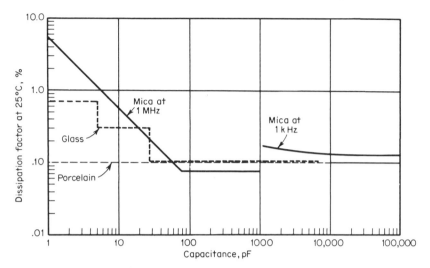

Fig. 100 Dissipation-factor limits over capacitance range for stable capacitors.

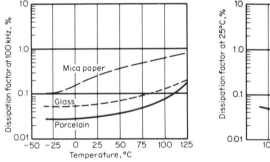

Fig. 101 Typical dissipation-factor change with temperature for stable capacitors.

Fig. 102 Typical dissipation-factor change with frequency for stable capacitors.

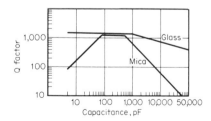

Fig. 103 A comparison in Q factor between glass and mica capacitors as a function of capacitance.

Current Rating. Transmitting-type fixed-terminal mica capacitors are designed to operate with a 20°C case temperature rise (at 30°C ambient). RF current ratings for various capacitance values in each case size are shown in Fig. 106 and are based on free air on all sides not used as mounting surfaces. Forced-air cooling should be used where these capacitors are used in high ambient temperatures and with a duty cycle in excess of 10 percent.

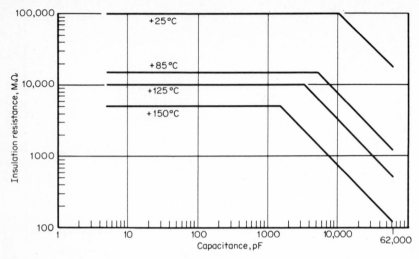

Fig. 104 Insulation resistance (minimum limits) at various temperatures and capatance values for mica capacitors.

Paper and Film Capacitors

Dielectric materials

Kraft Paper.[10]   Practically all paper capacitors produced in the United States are made from kraft paper (tissue), paperboard, with the addition of suitable impregnants necessary to meet performance requirements.   Kraft (from the German word meaning strength) papers are composed of cellulose fibers obtained from wood (largely coniferous) by an alkaline pulping process.   Wood chips (⅛ to ¾ in long) are cooked in a mixture of caustic soda and sodium sulfide to remove lignin (wood's natural binder) fats, gums, minerals, etc., and leave only pure cellulose.   Such pulps are distinguished by the length and strength of their fibers, the attrition to which they may be subjected, their pliability, and the electrical life of products made from them.

Capacitor-grade kraft paper is an especially high grade of paper that is made to rigid specifications to ensure that proper fiber length, chemical composition, and

TABLE 25   Analysis of a Typical High-Quality Paper[14]

Thickness	0.30 mil (average)
Apparent density	1.010 (dry)
Moisture	5.6%
Base weight	4.73 lb/ream 24 × 36 (500) BD
Coverage	91,147 in²/lb
Porosity	0.7 cm³/15 s (average)
Conducting paths	1.0/ft² (average)
Ash	0.271%
pH	7.0
Chlorides	4–6 ppm
Sulfates	No detectable amount
Dielectric strength	1,227 V/mil, dry
Resistivity of water extract	283,000 Ω-cm, at 25°C
Power factor at 60 Hz	30°C, 0.150%
	100°C, 0.214%
Dielectric constant	2.23

Reprinted from "Dielectric Materials and Applications" by Arthur R. von Hippel by permission of the M.I.T. Press, Cambridge, Massachusetts; copyright © 1954 by The Massachusetts Institute of Technology.

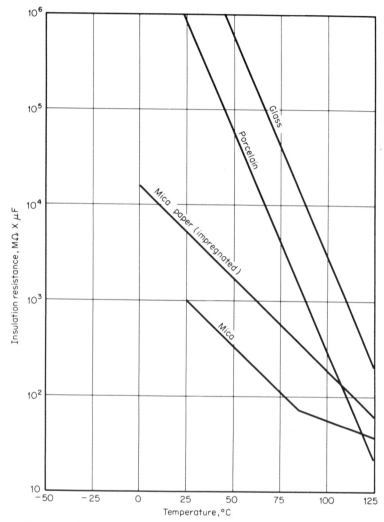

Fig. 105 Typical insulation-resistance variation with temperature for stable capacitors.

purity characterize each lot of pulp. Water used in making this grade of paper has to be exceptionally clean and is generally deionized. Careful control is necessary throughout the entire papermaking process to ensure a pinhole-free matrix and close dimensional tolerances. Contaminants, such as a small amount of lubricating oil, can have disastrous effects on the electrical properties of the final product. Power factor, uniformity, freedom from conductive particles and foreign matter, and insulation resistance are other factors which may be of importance in selecting this paper. Paper is supplied in thicknesses from 0.00017 to 0.005 in, while heavier materials used as wrapping and spacers range from 0.020 to 0.250 in. Typical high-quality kraft paper is defined by the analysis shown in Table 25. Paper is produced in sheet form and slit into specified widths, then dried and rolled for easy handling on capacitor-winding machines.

Film Dielectrics.[10] A great variety of flexible plastic films are gaining prominence in the capacitor industry and can now replace mica and paper types in many applica-

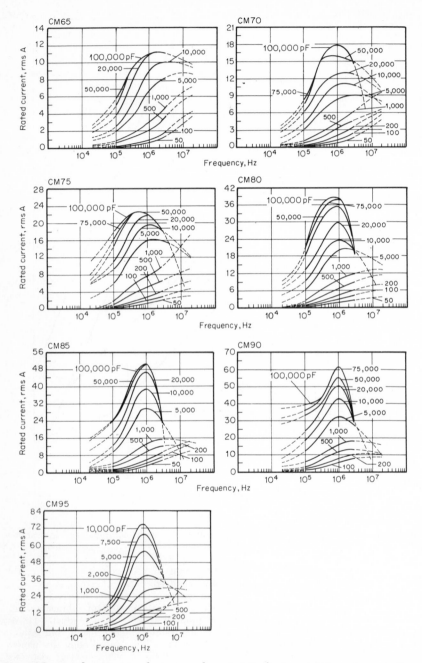

Fig. 106 Rated current vs. frequency characteristic for transmitting mica capacitors. (*Sangamo Electric Co.*)

tions. Plastic films are characterized generally by their low porosity, low moisture content, high dielectric strength, and predictable temperature characteristics. Some of the films used in capacitor construction are described below and are listed in Table 26 with their general characteristics.

   1. Fluorinated ethylene propylene (FEP). This is a transparent flexible film, a copolymer of tetrafluoroethylene and hexafluoropropylene, and is produced in rolls of film thickness down to 0.0005 in and in widths suitable for capacitor winding. FEP fluorocarbon film has a melting point of 260 to 280°C and can be easily metallized. It is relatively free of pinholes and does not absorb moisture. FEP fluorocarbon film is inert to virtually all known chemicals and solvents, although alkali metals and certain complex halogenated compounds at elevated temperatures can affect it. In addition, this film is characterized by a relatively high dielectric strength, a low and essentially constant dissipation factor over the temperature and frequency ranges, and a high surface and volume resistivity.

   2. Irradiated polyethylene. Bombarding polyethylene film with high-energy electrons produces a film which is crosslinked and nonmelting. The irradiated film will possess all the mechanical and electrical properties of polyethylene including toughness, flexibility, low specific gravity, chemical inertness, negligible water absorption, high dielectric strength, and very low loss at all frequencies. Irradiated polyethylene is completely resistant to environmental cross cracking, and though it is swollen by oils and solvents, it is not soluble and not destroyed by solvents, as is normal unirradiated polyethylene.

   At the crystalline melting point (about 110°C for low-density polyethylene and about 135°C for high-density polyethylene), irradiated polyethylene undergoes a transition from the normal plastic state and becomes elastomeric. General-purpose irradiated low-density and high-density polyethylene is available as tape, film, or sheet in thickness from 0.002 to 0.010 in. The high dielectric strength and the very low loss of irradiated polyethylene make it suitable for capacitors.

   3. Polycarbonate. Films made of polycarbonate resin offer the combination of dimensional stability, stable electrical properties over a broad range of environments, heat resistance, toughness, flexibility, and clarity. In addition, polycarbonate film has a high tensile strength which decreases very slowly with rising temperature. While these properties may vary somewhat, they are essentially the same for both extruded and solvent cast film.

   An oriented and crystallized film is designed for capacitor use. It has a maximum useful temperature approximately 25°F higher than the noncrystallized forms. Its chemical resistance is also higher.

   Electrical properties include a dielectric constant which is virtually unaffected by temperature or frequency, a low power factor at 60 Hz that is nearly independent of temperature, high dielectric strength, and high volume resistivity. Polycarbonate film is produced in thickness of less than 0.001 in and can be either metallized or used with foil electrodes.

   4. Polyethylene Terephthalate. This is a polyester film made from the condensation product of ethylene glycol and terephthalic acid. It is produced in the United States as a biaxially oriented film under the trade names Mylar, Scotchpar, and Celanar polyester film. It is transparent and flexible and is available in rolls or sheets in gauges from 0.15 to 14 mils and in widths from $\frac{1}{4}$ to 120 in. The electrical, physical, chemical, and thermal properties in the standard film make it useful in a wide variety of electrical-insulation applications. It is characterized by high dielectric strength, relatively low dissipation factor, and high surface and volume resistivity, high tensile strength (25,000 to 30,000 lb in^{-2}), great flexibility, general chemical inertness, and good thermal endurance. For the most part, the film is selected for application in direct competition with materials that are cheaper on a per pound basis. Since less thickness of film is required for equivalent performance, the usual result is actual material cost savings, either directly or through overall size reduction, and with upgraded performance. This film is used as the dielectric in a wide variety of capacitor applications. Its long-term reliability and stability under voltage at temperatures up to 125°C, plus its high insulation resistance, have proved useful in this application. The development of a 0.00015-in-thick (15-gage) type has added emphasis to the fact that the film can be used in thinner total sections with

equivalent or superior performance and compete economically with other dielectric materials.

5. Polyethylene.   Polyethylene is a pale, translucent, waxlike thermoplastic resin made by polymerizing ethylene gas.   It is the configuration of the long-chain polymer of these ethylene molecules that gives polyethylene its wide density range and versatility in its properties.   Polyethylene is usually classified as low-density (branched or high-pressure), medium-density, or high-density polyethylene (i.e., linear, low-pressure, Ziegler- or Phillips-type-process polyethylene).   Most of the polyethylene film used as electrical insulation is the low-density type, although there is a trend toward higher-density films because of increased tensile strength and service temperatures.   Polyethylene film is used to a considerable extent as a capacitor dielectric where high-frequency stability is important.   It is produced in thickness from 1 to 10 mils and slit to widths suitable for capacitor winding.

6. Polyimide.   This film is the product of the condensation of pyromellitic dianhydride with an aromatic diamine.   It is among the most flame-resistant of organic materials and exhibits very high radiation resistance.   The film has maintained useful properties after an exposure of 10^{10} rd in the Brookhaven pile.   High-temperature polyimide film provides excellent physical and electrical properties over a wide temperature range.

7. Polypropylene.   This film is similar to polyethylene in electrical performance except that it has a much higher melting point.   Capacitor films are usually a heat-set biaxially oriented polypropylene film with low shrinkage and elongation, high melting point, low power factor, high resistance, excellent moisture sensitivity, and good oil resistance.

8. Polystyrene.   This film is made from essentially pure polystyrene of 190°F minimum heat-distortion temperature (ASTM D648) and biaxially oriented to develop optimum physical characteristics for applications requiring stable dielectric-loss properties.   It has a maximum useful temperature of about 85°C and is resistant to acids, alkalies, alcohol, and oils.   Available in thicknesses from 0.25 to 5 mils, polystyrene film is used in precision capacitors for instruments and standards.

9. Polytetrafluoroethylene (TFE).   This is a chemically inert, low-loss, temperature-resistant film made from polytetrafluoroethylene (PTFE) resin.   It was developed and produced in pilot-plant quantities during the latter part of World War II to serve as a gasketing and packing material for the atomic-energy program and as insulation for some of the associated electrical equipment.   Skived tape was the earliest available type, and it is still used.   Literally, it is shaved or skived continuously from the surface of a cylindrical block of molded resin.   Cast films, free of voids and physical pinholes, are now formed in a multiple-dip-casting process.   They are available in thicknesses ranging from 0.25 to 4 mils.   TFE is relatively soft but for most electrical applications has a sufficiently low flow or creep to make it useful.

Dielectric constant is relatively unchanged with time, frequency, and temperature and remains in the 2.1 range, dependent upon the density of the tape.   Losses are constant and typically less than 0.0002.   Dielectric strength is in the same range as for other plastic dielectric materials.   In high-voltage applications, where corona occurs, TFE resin deteriorates, as do other organic materials.

10. Polyvinyl Fluoride (PVF).   This film combines suitable electrical properties with high mechanical strength, hydrolytic and thermal stability, and outstanding inertness.   As a member of the fluorocarbon family, PVF inherits the stability of the fluorocarbons but has the ease of fabrication associated with a vinyl.   Film is currently supplied in thicknesses ranging from 0.0005 to 0.002 in and in unoriented and biaxially oriented forms.   PVF films have a high dielectric strength coupled with an unusually high dielectric constant.   The volume resistivity and dissipation factor are adequate for special dc and ac capacitors.

11. Parylene.   Unlike other plastic films discussed, this film is deposited directly on aluminum-foil electrode material.   In a process developed by Union Carbide, poly-para-xylylene is deposited under vacuum on a moving belt of aluminum foil.   A film is then formed by an evaporating process in which heat is applied to form free radicals in the gas phase; polymerization occurs as the gas condenses again on both sides of the aluminum foil.   The film is relatively uniform in thickness and free of pinholes when compared with other film types.   Parylene films conform to

TABLE 26 Typical Properties of Plastic-Film–Capacitor Dielectrics*

Property	Fluorinated ethylene propylene (FEP)	Irradiated polyethylene	Polycarbonate film	Polyesters	Polyethylene	Polyimide	Polypropylene	Polystyrene	Polytetrafluoroethylene	Polyvinylfluoride	Parylene
Chemical resistance to other materials	Good for practically all except fluorine above 200°C	Good	Poor	Good	Good	Good for all except bases which will degrade film	Good	Fair	Good	Good	Good
Melting point, °C†	280			250–265	110	None			327	180	280–405
Moisture absorption, %	<0.01	0.01	0.35	0.5	<0.02	1.3 (50% RH)	0.03	0.1			
Specific gravity	2.15	1.1	1.2	1.39	0.916–0.960	1.43–1.47	0.90–0.91	1.05	2.2	1.76	1.103–1.289
Coefficient of thermal expansion at room temp, in/in/°C	8.3×10^{-5}	20×10^{-5}	3.9×10^{-5}	5.3×10^{-5}	9.5×10^{-5} to 7.0×10^{-5}	2.5×10^{-5}	11×10^{-5}	3.3×10^{-5} to 4.8×10^{-5}	5.5×10^{-5}	8.5×10^{-6}	2.5×10^{-6}
Useful continuous temp range, °C**	−255 to +200	−55 to 100 (low density) −55 to 135 (high density)	−65 to 270	−60 to 150	−45 to 95	Up to 400	Up to 125	−55 to +85	Up to +250	−55 to 135	−55 to 200
Dielectric strength at 25°C, V/mil (1 to 2 mil thickness)	5,000	2,500	2,250 (cast) 1,500 (extruded)	7,500	4,700	7,000	5,000	5,000	1,500 (8 mils)	4,000	7,000
Dielectric constant k at 25°C	2.0 ± 0.1	2.3	2.93–2.99	3.25	2.2	3.5	2.1	2.5	2.1	8.5	2.65
Over frequency range, Hz	100–1,000 M	60–10,000 M	1 k to 1 M	60–1k	60–1 M	1 k	1 M	1 M	60–10 M	1 k	1 k
Dissipation factor at 25°C, %	0.02–0.15	0.05	0.13–1.1	0.5	0.03	0.3	0.05/0.03	0.03	0.05–0.02	1.6	0.02
Over frequency range, Hz	100–1,000 M	60–10,000 M	1 k to 1 M	1 k	60–1 M	1 k	1 k/1 M	1 M	60/10 M	1 k	1 k
Volume resistivity at 25°C, Ω-cm	$>10^{18}$	$>8 \times 10^{15}$	4.7×10^{16} (cast) 15×10^{16} (extruded)	10^{18}	10^{17}	10^{18}	5×10^{16}	1.6×10^{15}	$>10^{12}$	$>10^{13}$	8.8×10^{10} to 1.4×10^{17}
Temperature coefficient, ppm/°C‡			Up to ±350	+1,150				−120 ± 30			−100

* Data in this table were compiled from Refs. 10 to 13. ** Capacitor core temperature at full rated conditions.

† Melting point.

‡ Temperature coefficients for finished capacitors are altered as desired by selection of impregnants.

the electrode material without the presence of winding stresses, thus promoting stability in the finished capacitor.

Impregnants. Even highly calendered kraft capacitor papers are somewhat porous and, additionally, contain 5 to 10 percent water. The water is removed by vacuum drying the rolled capacitor element at an elevated temperature, which increases the porosity. The same applies to film dielectrics but to a much lesser degree. To replace the air-filled spaces with a dielectric, vacuum impregnation immediately follows the vacuum drying. The interstices of the dielectric are filled with a liquid or solid dielectric to increase the overall electrical capacitance and the dielectric strength of the combined separator.

Many liquid-oil impregnants are used, most common of which are mineral oil, chlorinated diphenyl isomers, and polyisobutylene. The chlorinated diphenyl isomer impregnant is polar in nature and has a high dielectric constant. The mineral oil and polyisobutylene are hydrocarbons, possessing a low dielectric constant with superior temperature-coefficient and power-factor characteristics. When selecting an impregnant, the characteristics listed in Table 27 must be evaluated for manufacture

TABLE 27   Characteristics for Consideration in Selection of Capacitor Impregnation

Capacitance stability	Specific gravity
Dielectric constant	Chemical stability
Power factor	Decomposition products in an electric arc
Insulation resistance	Toxicity (personnel safety)
Viscosity	Process economics
Pour points	Dynamic environments
Volume coefficients of expansion	Nuclear radiation
Material compatibility	

and performance of the finished capacitor as well as for the liquid or compound itself.

Electrical properties of common liquid and solid impregnants and the capacitor characteristics to be realized when these impregnants are used in the construction of paper capacitors are shown in Table 28. A comparison between capacitors im-

TABLE 28   Impregnant and Capacitor Characteristics for Paper Capacitors[14]

Material	Impregnant*			Capacitor		
	k	Power factor, %	ρ, Ω-cm	μF/mil	Power factor, %	IR, $M\Omega/\mu F$
Mineral oil	2.23	0.03	10^{14}	600	0.20	20,000
Mineral wax	2.2	0.05	10^{15}	600	0.35	10,000
Petroleum jelly	2.2	0.05	10^{12}	600	0.35	10,000
Polyisobutylene	2.2	0.03	10^{15}	600	0.20	25,000
Silicone liquids	2.6	0.05	10^{16}	560	0.35	22,000
Castor oil	4.7	0.08	10^{12}	455	0.5	2,000
Chlorinated diphenyl	4.9	0.05	10^{13}	430	0.2	15,000
Chlorinated naphthalene	5.2	0.2	10^{11}	390	0.4	5,000

* At 100°C, except chlorinated diphenyls at 65°C and chlorinated naphthalene at 125°C.

Reprinted from "Dielectric Materials and Applications" by Arthur R. von Hippel by permission of the M.I.T. Press, Cambridge, Massachusetts; copyright © 1954 by The Massachusetts Institute of Technology.

pregnated with mineral oil and with chlorinated diphenyl is shown in Table 29. It can be seen that when space is not a prime factor the mineral-oil impregnant is preferred. Table 30 lists typical characteristics of liquid impregnants used in paper and film capacitors. The combinations of selected impregnants, paper, film, or

TABLE 29   Comparison of Two Types of Impregnated Capacitors[14]

	Mineral oil	Chlorinated diphenyl
Volume	1	0.667
Weight	1	1
Insulation resistance at 25°C, MΩ/μF	4,000–22,000	8,000–12,000
Insulation resistance at 85°C	125–400	75–250
Power factor, 60 Hz, 25°C, %	0.20	0.20
Power factor, 1,000 Hz, 25°C, %	0.3	0.35
Capacitance at −55°C compared with 25°C, %	94	73 at 60 Hz, 55 at 1 MHz

Reprinted from "Dielectric Materials and Applications" by Arthur R. von Hippel by permission of the M.I.T. Press, Cambridge, Massachusetts; copyright © 1954 by The Massachusetts Institute of Technology.

paper/film will produce the desired electrical-performance characteristics in the finished capacitor.

Paper and film capacitor construction   A variety of package styles, electrode materials, terminations, and dielectric systems are used depending upon the intended application.   Several of the more common design features will be discussed here. Paper and film dielectric materials may be used with either interwoven-foil electrodes or a thin layer of metallization applied directly to the dielectric.

Foil and Dielectric Capacitors.   The trend toward miniaturization, closer electrical tolerances, and higher operating temperatures is being met by the use of thin-plastic-film dielectrics in the construction of capacitors.   The greatest advantage of plastic-film dielectrics over natural dielectrics (such as paper and mica) is that the plastic film is a synthetic that can be made to meet specific requirements (such as thickness of dielectric and high heat resistance).   Many plastic-film capacitors are not impregnated but are wound and encased "dry."   Plastic-dielectric capacitors have insulation-resistance values far in excess of those for paper capacitors, and since they are nonabsorbent, their moisture resistance is superior to that of mica.

Some manufacturers use only one sheet of plastic film for those with low-voltage ratings, whereas at least two sheets of paper are used in conventional paper types. The principal advantage of polyethylene terephthalate dielectric capacitors is the high order of insulation-resistance values available over the dielectric's temperature range of −55 to +125°C; however, for some military styles, the high-temperature limit is +85°C.   Polyethylene terephthalate dielectric capacitors have an insulation resistance that is normally about 100,000 MΩ μF^{-1} at room temperature and about 25,000 MΩ μF^{-1} at +85°C.   These insulation-resistance values decrease considerably when polyethylene terephthalate dielectric capacitors are impregnated. However, a higher volt per mil rating is possible by impregnation and the possibility of corona and catastrophic failures due to pinholes in the dielectric is minimized.

Paper and/or film strips of selected widths are interleaved with aluminum foil and wound to the desired capacitance value.   Lead attachment is made by one of two methods:

1. Inserted-tab construction, which consists of metal-strip terminals which are inserted between the windings of the capacitor roll and contact the appropriate electrodes.

2. Extended-foil construction in which the electrodes are extended beyond the dielectric material during winding.   One of the electrodes extends out of one end of the capacitor roll and the other electrode extends out the opposite roll end.   These two ends are then pressed to form a smooth surface, and terminals are added by soldering or welding.

After lead attachment, the capacitor roll is metal-encased, potted, dipped, or molded as desired for the finished product.

TABLE 30   Typical Characteristics of Capacitor Liquid Impregnants[10]

Property	Polybutenes Paper impregnant	Polybutenes Capacitor liquid	Askarels				
Viscosity, SUS:							
25°C	8,000[a]	300,000[a]					
37.8°C			40–42	44–51	82–92	185–240	1,800–2,500
99–100°C	176[a]	2,200[a]	30–31	31–32	34–35	36–37	44–48
Viscosity, CST:							
25°C							
37.8°C			4.6	6.9	17.2	45.3	46.4
99–100°C			<1.8	<1.8	2.5	3.2	6.14
Flash point, open cup, °C	160[b]	252[b]	146.1[y]	153.3[y]	182.2[y]	192.8[y]	None
Acidity, mg KOH/g	0.01[i]	0.01[i]	0.010 max	0.010 max	0.010 max	0.010 max	0.010 max
Pour point, °C	–23[c]	1.7[c]	1.1	–35.5	–19.0	–7.0	10.0
Specific gravity:							
15.6°C	0.870[o]	0.905[o]					
25°C			1.18	1.26	1.38	1.45	1.54
Coefficient of expansion, cm^3/cm^3/°C	0.00076		0.00071	0.00073	0.00068	0.00070	0.00066
Thermal conductivity: (g-cal/s)(cm^2)(°C/cm) (Btu/h)(ft^2)(°F)(ft)	0.062[l]	0.062[l]	0.067	0.063	0.058	0.057	0.054
Boiling point at 760 mm, °C			275.0	290.0	325.0	340.0	365.0
Volatility, weight loss							
Dielectric strength, kV/0.1 in (0.254 cm)	>35[e,r]	>35[e,r]	>35	>35	>35	>35	35
Dielectric constant:							
60 Hz							
10^3 Hz			4.5	5.7	5.8	5.6	5.0
10^6 Hz	2.16[f,q]	2.22[f,q]					
Dissipation factor:							
60 Hz	0.0005[f,s]	0.0005[f,s]					
10^3 Hz			0.001	0.001	0.001	0.001	0.001
10^6 Hz							
Volume resistivity, Ω-cm	>1 × 10^{14p,s}	>1 × 10^{14p,s}	>5 × 10^{12}	>5 × 10^{12}	>5 × 10^{12}	>5 × 10^{12}	>5 × 10^{12}

[a] ASTM D446.
[b] ASTM D92.
[c] ASTM D97.
[d] ASTM D1250.
[e] ASTM D877. Suitability of D877 for high-viscosity oils such as polybutenes listed in table has not been determined.
[f] ASTM D24.
[g] ASTM D974.
[h] Weight loss on a 15-g sample in a 50-ml beaker at 150° for 24 hr.
[i] ASTM D664.
[j] ASTM 1169.
[k] Weight loss after 48 h at 200°C.
[l] 100°F.

Silicones				Organic ester (Dielectric grade castor oil)	Mineral oil (Capacitor oil)	Ethylene glycol	Silicate ester-base fluid	Cyanoethyl sucrose
	360	720	180		103[a]			$\approx 5 \times 10^5$
	137.9				38[a]			
10.5	100	200	500					
	80	160	400		21		12.2	
	29	58	148	98	3.5		3.95	
	>30	>315	>325	291	154.4[b]	116	187.8	>190
					0[g]		0.15 max	
-90	-55	-52.7	-50.0	-23	-45.6[c]		<-59.5	29[c]
					0.907[d]			
0.940	0.970	0.971	0.973	0.959		1.1154	0.887	1.20 at 20°C
0.00095	0.00097	0.00097	0.00097	0.00066		0.00062	0.000863	0.000605
0.00034		0.00037						
	0.086	0.087	0.088	0.103	0.00031[m]	0.00063	0.000327	
					0.076[n]		0.080	
						196-206	>371.1	>300
31%[k]	1%[h]	<2%[k]						
35[v]	35	35	35		>30[e]		27[e]	
2.7[u,x]				3.74[s,t]				38.0
	2.75	2.75	2.75				2.65	35.2
2.7[u,x]	2.75	2.75	2.75			41	2.65	25.0
0.00015[m,x]				0.06[u]	0.001[f,s]			0.010
	<0.0001	<0.0001	<0.0001				0.0336	0.080
0.00001[w,x]		0.00005					0.0042	0.108
1×10^{14}[j]	$>1 \times 10^{14}$[j]	$>1 \times 10^{14}$	$>1 \times 10^{14}$	3×10^{10}[s]			9×10^{10}	5×10^{11}

[m] 0°C.
[n] -8°C.
[o] ASTM D287.
[p] ASTM D257.
[q] 23°C.
[r] 80°C.
[s] 100°C.
[t] Hz unknown.
[u] 100 Hz.
[v] ASTM D149.
[w] 10^5 Hz.
[x] ASTM D150.
[y] Askarels have true flash point.

Metallized Dielectric Capacitors. Metallization of paper or film is performed under vacuum and consists of vapor-deposited or sprayed-on layers (typically 1,000 Å) of metal, usually zinc or aluminum. Metallized paper or film is then rolled in an extended-foil configuration. The rolled ends are sprayed with a finely divided molten metal to which the capacitor lead wires are soldered.

This construction type offers the advantages of high volumetric efficiency and a self-healing capability. However, volumetric efficiency has limitations. Working-voltage limitations are given by volume factors of the conventional-foil design. For example, a 200-V metallized capacitor has a volume factor of 0.75; at 600 V the factor increases to 0.8, and above 600 V the metallized capacitor usually offers no size advantage. Capacitance limitations also exist. For small capacitance ratings (under approximately 0.1 μF) the metallized capacitor offers little size advantage. This is evident since the extended-foil end terminations, casings, and end seals form a disproportionate part of the volume.

Self-healing is a property wherein a momentary short (through pinholes or due to high-dielectric-voltage stress) is eliminated by the burning away of the electrode at the short. Heat generated by the short is sufficient to vaporize the metallized surface. This action, also known as self-clearing, occurs at an energy level of approximately 10 μWs for most capacitors. Where the energy for this clearing action comes from the capacitor charge itself, the capacitance/charging-voltage relation for safe clearing is shown in Fig. 107. This curve will vary with circuit series resistance and dielectric material used but is given as typical for most styles. Most of the self-healing occurs during burn-in, dielectric-strength, and other voltage-stress tests. Clearing action during service life is reduced to an insignificant amount, especially in dc applications, when voltage and temperature deratings are used. Another form of self-healing appears as a low insulation resistance at a spot on the film upon initial voltage application. Breakdown does not occur, and eventually the insulation resistance increases to normal value. These capacitors are not recommended in ac applications where high currents exist. This current limitation is caused by the lower power-handling capability of the metallized film as compared with that of the foil designs. AC ratings are given in detail specifications or may be obtained from the manufacturer.

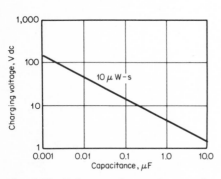

Fig. 107 Voltage and capacitance relationship for capacitor energy.

Feedthrough Capacitors. The feedthrough type of construction is essentially a three-terminal network in which the case constitutes a common or ground terminal. This construction, which is analogous to a coaxial cable, requires a center conductor to carry rated line current, and a capacitor section disposed concentrically around the center conductor so that the path of rf currents flowing between the center conductor and the case is symmetrical. The capacitor illustrated schematically in Fig. 108 represents a circuit arrangement which attains the ultimate in reduction of lead length and, therefore, lead inductance. By carrying the line to be bypassed through the center of the capacitor, connecting leads are virtually eliminated.

Certain defects in feedthrough-capacitor construction may cause the insertion-loss characteristic to depart from the ideal. For example, resistance in the foils, or in the solder connections at the edges of the foils, may slow the rate of insertion-loss rise below that for the ideal capacitor and level out the curve in the high-frequency region. A downward slope in the insertion-loss curve at high frequency, beyond 100 MHz, is usually indicative of incomplete or asymmetrical foil-to-ground connections.

Since a feedthrough capacitor must carry all the line current through its center conductor, it is rated not only in terms of the capacitance and the voltage it has to withstand but also in terms of the current it can safely carry to the load.

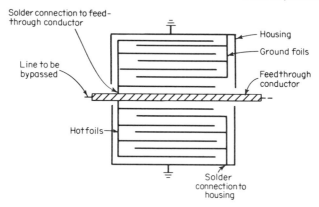

Fig. 108   Schematic diagram of a feedthrough capacitor.[15]

These capacitors are enclosed in hermetically sealed cases which will prevent leakage of the impregnant or filling and, in addition, will protect the capacitor element from moisture and mechanical damage.

Case Construction.   Most rolled capacitor sections with either metallized or plain foils can be wound to practically any shape desired. The most practical shape for low-cost high-volume production is the cylindrical section wound on a mandrel or a rectangular section (for high-density packaging) wound in a similar manner. Where metallic cases are used, the capacitor section can be encased with both leads insulated (floating) from the case or with the outside foil grounded to the case. These two circuit configurations are annotated in the alphanumeric part number commonly used in specifying capacitors (see Chap. 7). Insulation sleeving is also optional for most tubular metal cases and is also annotated in the alphanumeric part number. Large capacitor values and those with liquid impregnants are encased in fabricated cases, bathtub cases, or deep-drawn cases. Where high corona resistance is desired, a well-insulated, hermetically sealed, vacuum-filled design is required. Most paper and film capacitors are either liquid- or solid-impregnated, but a few dry sections are produced in low-cost designs and where only direct current is applied.

Nonmetallic case designs are becoming more popular. The oldest design is a wax-dipped and molded encasement similar to those used in ceramic capacitors. Non-dipped cardboard sleeves have given way to more modern dipped and molded encasements similar to those used in ceramic capacitors. Capacitors with nonmetallic cases are produced in axial-lead, radial-lead and radial-crimped-lead configuration. The two radial-lead types lend themselves to high-density packaging on printed-circuit cards and are formed in circular, rectangular, or oval shapes. Some film capacitors are also produced in stacks and assembled in a manner similar to the monolithic ceramic capacitor. This design, though desired for high-density packing, should be used with caution. Internal termination and encasement techniques should be evaluated for proper performance in thermal, moisture, and dynamic environments. As a general rule the nonmetallic capacitor case should always be a "protective cover" and not be required for mechnaical integrity of the winding and lead assembly.

Radio-frequency interference (RFI) capacitors   Special capacitor designs made for suppressing unwanted noise from electronic circuits minimize its passing from one stage to another. This noise, generated by commutating, pulsing, and other abrupt current-waveform distortions, is known as RFI and is ideally shunted to ground at a convenient point in the electronic assembly. Feedthrough- and bypass-capacitor constructions are commonly used. Standard RFI-suppression capacitors are listed in Table 31.

Feedthrough-Capacitor Application.   These capacitors are intended for the bulk-head or through-panel type of mounting. Only when the capacitor is mounted through and makes complete circumferential contact with the grounded bulkhead or chassis will maximum suppression effectiveness, particularly at high frequency,

TABLE 31   Selection Guide for Standard Paper, Paper/Film, and Film Capacitors

EIA type	Dielectric	Dissipation factor at 25°C, %	Current rating, A	Principal application	Case style
RS-361					
Z23, Z24	P, P/F, F	1.5 (1.0)	10 dc	RFI-suppression feed-through type	Tubular hermetic-sealed can
Z33	Metallized paper	1.5	50 dc		Bushing mount, solder terminal
Z32	Metallized paper	1.5	300 dc		Flange mount, screw terminal
	Paper	1.5	100 dc		Flange mount, stud terminal
	Paper	1.5	20 dc		Bracket mount, screw terminal
Z50	Paper	1.5	5 dc		Flange mount, solder terminal
	Paper			RFI-suppression bypass type	Tubular can, clamp mount with screw terminal
	Paper	1.5			
	Paper				Rectangular case, clamp mount, solder terminal
RS-376					
	Polysulfone	0.15			Tubular hermetic-sealed can
	Polycarbonate	0.15		AC and dc application	Wire leads
	Polyethylene terephthalate	1.0		High reliability	Wire leads
	Polyethylene terephthalate	1.0			Clamp mount, wire leads
	Polyethylene terephthalate	1.0	See Fig. 116	High insulation resistance	Bushing mount, wire leads
	Polyethylene terephthalate	0.6			Wire leads
	Polyethylene terephthalate	0.6			Clamp mount, wire leads
	Polyethylene terephthalate	0.6			Bushing mount, wire leads
	Polyethylene terephthalate	1.0			Ceramic case, axial lead
C08, C09	Polystyrene	0.1		AC and dc industrial applications	Hermetic seal, axial lead
C08, C09	Polyethylene terephthalate	0.6			Hermetic seal, axial lead
	Metallized film	0.15		AC and dc voltages with specified ac and where no dielectric clearing is permitted High-reliability requirements	Hermetic seal, axial lead
	Metallized film	0.15			
	Metallized film	0.15	f		
	Metallized film	0.15			
	Metallized film	0.15			
RS-377					
	Polyethylene terephthalate	0.5		DC voltages where clearing can be tolerated	Axial lead, nonmetallic case
	Polycarbonate	0.3			
	Polyethylene terephthalate	0.5			Radial lead, nonmetallic oval case
	Polyethylene terephthalate	0.5		Military and industrial equipment. Small size	Axial or radial leads in molded plastic case
	Polycarbonate	0.3	f		
	Poly-paraxylylene	0.25			
	Metallized polycarbonate	0.3			Axial lead, nonmetallic case
	Metallized polyethylene terephthalate	1.0			Axial or radial leads in non-metallic case
C26	Metallized polyethylene	1.0			Axial lead, oval wrapped case
RS-164					
164C20M	Paper and paper polyester film	1.0		General-purpose dc, commercial applications (also limited ac)	Axial lead, ceramic case, plastic end seal
164C20C	Paper and paper polyester film	1.0	See Fig. 117		Axial lead, plastic case
164C30D	Paper and paper polyester film	1.5			Axial lead, molded case
164C40B	Paper and paper polyester film	1.0			Dipped or plastic case, radial leads
164C50B	Paper and paper polyester film	1.0			
RS-218					
P04 through P13	Paper	1.0		General-purpose dc applications, high voltages, high capacitance (also limited ac)	Axial lead, miniature tubular
P25 through P29	Paper	1.0			Axial lead, small tubular
P40, P41	Paper	1.0	See Fig. 118		Solder lug or screw terminals, large tubular
P53 through P35	Paper	1.0			Solder lug terminal, small bathtub cases
P61 through P69, and P91	Paper	1.0		Multisection	Solder lug terminal, small rectangular cases
P70, P72	Paper	1.0			Solder lug and pillar terminals, large rectangular cases
P73	Paper	1.0			Solder lug terminals, oval case
RS-401					
Various (see RS-401)	Paper, paper/film, or film oil-impregnated	0.4	25 and 50 ac	SCR commutating, snubbers, harmonic filters, EMI/RFI suppression, static power supplies, frequency changers	Threaded pillar terminals, deep-deep-drawn oval or rectangular cases

d Also available as CZR types with established reliability.
e Style codes 01–05 are high reliability, high IR; style codes 06–00 are standard military grade.
f Only where specified by the manufacturer or detail specification sheet.

Capacitance range, µF	Standard capacitance tolerances, %	DC working-voltage range, V	Operating temp, °C		Max capacitance change from 25°C value, %		Military standard type
			Min	Max	Low	High	
							MIL-C-11693
0.047-2.0	±10	100-600	-55	+85 (125)	±10	+20, -10, (±10)	CZ23,[d] CZ24[d]
2.0	±20	100	-55	+85	±10	±10	CZ23[d]
1.8	-10, +20	100	-55	+85	±10	±10	CZ32
0.1	±10	250 ac, 400 Hz	-55	+85	±10	±10	CZ42[d]
0.32	±10	400	-55	+85	±10	±10	CZ41
0.001-0.1	-10, +20	1,200	-55	+85			CZ50
							MIL-C-12889
0.01-0.5	+50, -10	100	-55	+85			CA32, CA33, CA34
0.01-0.25	-10, +50	500 ac and dc	-55	+85	+10, -30	±10	CA36, CA37, CA38
0.01	-30, +0	250 ac and dc	-55	+85			CA47
							MIL-C-19978
0.001-1.0	±5, ±10	50-600	-55	+125	±2.5	+1, -4	CQR01
0.001-1.0	±5, ±10	50-600	-55	+125	±2.5	+1, -4	CQR07
0.001-1.0	±2, ±5, ±10	200-1,000	-65	+125	-10	+10	CQR09
0.001-1.0	±2, ±5, +10	200-1,000	-65	+125	-10	+10	CQR12
0.001-1.0	±2, ±5, ±10	200-1,000	-65	+125	-10	+10	CQR13
0.001-10.0	±2, ±5, ±10	30-1,000	-65	+85	-7	+5	CQR29
0.001-10.0	±2, ±5, ±10	30-1,000	-65	+85	-7	+5	CQR32
0.001-10.0	±2, ±5, ±10	30-1,000	-65	+85	-7	+5	CQR33
0.001-1.0	±10	1,000-15,000	-65	+125	-10	+10	CQ20
0.001-0.47	±1, ±2, ±5, ±10	200-600	-55	+85	+0.5 to +1.4	-0.3 to -1.0	
0.001-10.0	±2, ±5, ±10, ±20	30-600	-55	+125	-10	+20	
							MIL-C-83421[e]
0.001-22.0		30[a]	-55	+100[a]	-2	±2	CRH01, CRH06
0.001-10.0	±0.25, ±0.5	50[a]	-55	+100	-2	±2	CRH02, CRH07
0.001-10.0	±1, ±2, ±5	100[a]	-55	+100	-2	±2	CRH03, CRH08
0.001-4.0	±10	200[a]	-55	+100	-2	±2	CRH04, CRH09
0.001-2.0		400[a]	-55	+100	-2	±2	CRH05, CRH00
							MIL-C-55514
0.001-1.0	±2, ±5, ±10	100-600	-55	+85	-10	+10	CFR02
0.001-1.0	±1, ±2, ±5, ±10	100-200[b]	-55	+85[b]	-2.5	±1	CFR02
0.0047-2.2	±2, ±5, ±10	100-400	-55	+85	-10	+10	CFR03
0.001-0.1	±2, ±5, ±10	200-400	-55	+85	-10	+10	CFR04
0.001-0.1	±1, ±2, ±5, ±10	100-400[b]	-55	+85[b]	-2.5	±1	CFR04
0.001-1.0	±1, ±2, ±5, ±10	50[a]	-55	+85[b]	+1.8	-1.8	CFR04
0.001-10.0	±1, ±2, ±5, ±10	200-400	-55	+85[b]	-2.5	±1	CFR05
0.01-10.0	±2, ±5, ±10	200-400	-55	+85	-10	+10	CFR06
0.01-10.0	±5, ±10, ±20	100-600[a]	-55	+85 (+125)[a]	-10	+10 (+20)	
0.001-1.0	±10, ±20	200-1,600	-40	+85	-10	+5	
0.001-1.0	±10, ±20	200-1,600	-40	+85	-10	±5	
0.001-1.0	±10, ±20	200-1,600	-55	+100	-10	+10	
0.001-1.0	±10, ±20	200-1,600	-55	+125	-10	+15	
0.001-0.47	±5, ±10, ±20	200-1,600	-55	+85 (+125)[a]	-10	+10 (15)	
0.001-1.0	±10, ±20	100-1,000	-55	+125	±10	±10	
0.003-0.1	±10, ±20	1,500	-55	+85	±10	±5	
0.25-4.0	±10, ±20	600-1,500	-55	+85	±10	±5	
0.05-4.0	±10, ±20	600-100	-55	+85	±10	±5	
0.01-1.0	±10, ±20	400-1,000	-55	+85	±10	±5	
0.1-75	±10, ±20	600-12,500[c]	-55	+85	±10	±5	
0.1-35	±10, ±20	600-2,000	-55	+85	+10, -30	±10	
0.25-50	±6, ±10	200-2,000	-40 and -30	+85 and +85	±3 and ±5		

[a] Derated to 50 percent rated voltage at 125°C.
[b] Derated to 65 percent rated voltage at 125°C.
[c] Capacitors 10 kV and up must be installed with terminals on top (vertical position).

be realized. The square or oval flange-type bracket and the threaded-housing or threaded-neck type of mounting effectively provide this required degree of contact. The wraparound bracket which permits mounting only on flat surfaces should not be used for feedthrough capacitors, since the bracket introduces inductance in series with the capacitor exactly as does a lead; the capacitor then exhibits all the characteristics of a lead-type capacitor, including resonance and poor high-frequency effectiveness. Also, when the capacitor is mounted on a flat surface with little or no shielding between input or output, appreciable coupling around the capacitor can exist.

The ideal location for the feedthrough capacitor in any suppression system is at the point of exit of the wiring from a chassis or housing surrounding the source of interference. The capacitor may thus at the same time serve as an insulating bushing. This preferred mounting location is in conformance with a basic principle of suppression which requires shielding around a potential interference source and all the immediately associated circuitry, using the existing chassis, housing, or other metallic enclosure wherever possible, thus filtering or bypassing all wires emerging from the enclosure at their point of exit.

Selection of the optimum feedthrough capacitance for any given application usually necessitates a compromise between capacitance sufficiently large to provide good insertion loss at the lowest radio frequency to be suppressed, and capacitance not so large as to cause appreciable loading at 60 Hz or other power frequencies. Generally, for the most practical suppression purposes, the range of capacitance values is from 0.01 to 2.0 μF, inclusive.

Insertion Loss. Figure 109 illustrates the insertion-loss characteristic for typical feedthrough capacitors compared with that for the ideal capacitor. The curve showing the insertion loss of feedthrough capacitors does not strictly follow the ideal; it contains a characteristic dip below the ideal, followed by a recovery during which the insertion loss rises with frequency faster than does the ideal. This dip is termed a "transmission-line resonance" and occurs at the frequency at which the transmission line, formed by the total series of concentric foil and paper cylinders making up the capacitor section, becomes a half wavelength. Subsequent dips, not shown, may also occur at multiples of the half-wavelength frequency. Transmission-line resonance is inherent in the extended-foil type of rolled capacitor section.

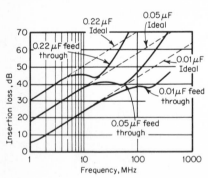
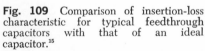
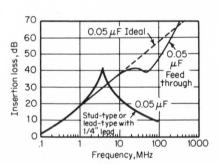

Fig. 109 Comparison of insertion-loss characteristic for typical feedthrough capacitors with that of an ideal capacitor.[15]

Fig. 110 Comparison of insertion-loss characteristic for typical feedthrough capacitor with that of a lead-type capacitor.[15]

The overall superiority of the feedthrough capacitor is well demonstrated by the curves in Fig. 110, which compare the insertion loss for a feedthrough capacitor with that for a lead-type capacitor of similar value. These curves illustrate graphically the prime reason for preferring and employing feedthrough capacitors. For example, at 10 MHz the feedthrough capacitor shown is approximately 12 dB better than the lead type, and at 100 MHz, about 40 dB or 100 times more effective; moreover, the spread would be even greater for larger capacitance values.

Impedance Limitations. The suppression effectiveness of a capacitor in any installation is a function of the ratio of circuit impedance to that of the capacitor. Hence any combination which includes a capacitor with high impedance (which inevitably occurs at low frequencies) or low circuit impedance (which conceivably may occur at any frequency) will result in poor suppression effectiveness. For example, a capacitor located at a voltage null point on an electrically long, open-circuited, or short-circuited line would produce no effect on the line voltages beyond that point. Complete nulls seldom occur in practical installations, but standing waves occur on its position on the line relative to the voltage minimums and maximums, respectively. Also, placing a capacitor across an inductive-source impedance may actually increase the voltage on the line beyond the capacitor at that frequency for which the capacitor and source inductance are resonant. With the values of inductance normally encountered in practical circuits and the capacitance normally used for suppression purposes, the resonant frequency is quite low, well below 150 kHz and therefore is of little concern in connection with radio-interference suppression.

A properly-constructed feedthrough capacitor comprises a design for which suppression effectiveness approaches that of an ideal capacitor, wherein insertion loss rises indefinitely with frequency, and the capacitor thereby becomes an extremely broadband device. Impedance vs. frequency for ideal capacitors of values commonly used for radio-interference bypassing is shown in Fig. 111.

A decision is frequently required when to use a feedthrough capacitor or a filter. The question is difficult to answer except in general terms. If, in any new application, an opportunity for trial and experimentation exists, the feedthrough capacitor should be tried first. The common low-pass, *LC* type of radio-interference-suppression filter does have an advantage over a capacitor alone in that, generally, it provides substantially higher suppression effectiveness in the low-frequency region (below 5 MHz), although the filter attains its superiority in this region at the expense of introducing a line drop into the circuit. Above 5 MHz, the feedthrough capacitor becomes and remains very much superior to the *LC* filter, unless the filter also incorporates feedthrough capacitors; however, many commercial filters do not. In general, therefore, unless the circuit to be suppressed contains high levels of radio-interference voltages at low frequencies, a feedthrough capacitor alone may be depended upon to provide the necessary bypassing effectiveness.

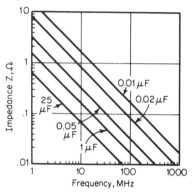

Fig. 111   Typical impedance vs. frequency characteristic for RFI capacitors.[15]

$$Z = 1/WC = \tfrac{1}{2}\,\pi\,fc$$

where Z = absolute magnitude of impedance, Ω
f = frequency, Hz
c = capacitance, F

Paper and film capacitor performance

Electrical Parameters. Standard measurements of capacitance, capacitance change with temperature, dielectric strength, dissipation factor, and insulation resistance are defined in governing industry (EIA) and military specifications. Capacitance and dissipation factor are usually measured at 1,000 or 60 Hz as specified in governing test procedures and using a suitable ac bridge. Typical curves of capacitance change with temperature for dry-section film and impregnated-paper capacitors are shown in Figs. 11 and 12, respectively. In addition, capacitance change with temperature varies slightly with capacitance value and frequency, as shown, for example, in Figs. 112 and 113, respectively, for Mylar capacitors. Maximum capacitance variation at temperature limits for standard capacitors is given in Table 31. Dissipation-factor variation for typical film and impregnated-paper capacitors is shown in Figs. 5 and 6, respectively. Capacitance and frequency variation of dissipation factor over the

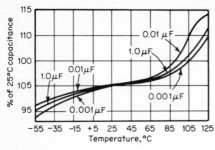

Fig. 112 Capacitance change vs. temperature for various capacitance values in typical Mylar-film capacitors. (*General Electric Co.*[16])

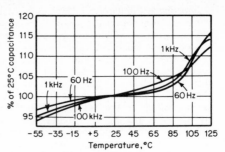

Fig. 113 Capacitance change vs. temperature for a typical 0.01-μF Mylar-film capacitor measured at various frequencies. (*General Electric Co.*[16])

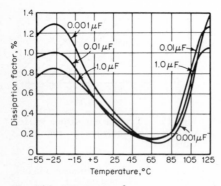

Fig. 114 Dissipation factor vs. temperature for various capacitance values in typical Mylar-film capacitors. (*General Electric Co.*[16])

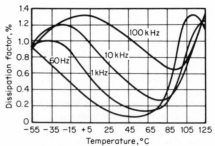

Fig. 115 Dissipation factor vs. temperature for a typical 0.01-μF Mylar-film capacitor measured at various frequencies. (*General Electric Co.*[16])

operating-temperature range is illustrated for a typical Mylar capacitor design in Figs. 114 and 115, respectively. Maximum dissipation-factor values at 25°C for standard capacitors are given in Table 31. Insulation resistance of typical film capacitors is shown in Fig. 7.

Capacitor parameters and characteristic curves vary among manufacturers depending upon their own proprietary processes, which in some cases can result in a competitive edge. Therefore, standard specifications set limits broad enough to include major manufacturers and provide a limit within which the design engineer can base a circuit design.

AC Rating. Application of paper and film capacitors in ac circuits or where ripple or pulse conditions exist follows. rules similar to those for electrolytic capacitors. Capacitor phenomena which must be considered to achieve reliable ac performance safely are:

1. Dielectric stress
2. Corona
3. Internal heating
4. Current-carrying capability

Dielectric stresses result from the dc voltage applied, rate of change of the ac voltage or pulses, magnitude of the ac voltage or pulse, variations in dielectric thickness, and pinholes in the dielectric. Capacitor voltage ratings are normally specified for a given dielectric system such that the sum of the dc voltage and peak ac voltage can be used to select the proper capacitor voltage rating. In addition, ac voltage (ripple) limitations with frequency are specified by most manufacturers and stan-

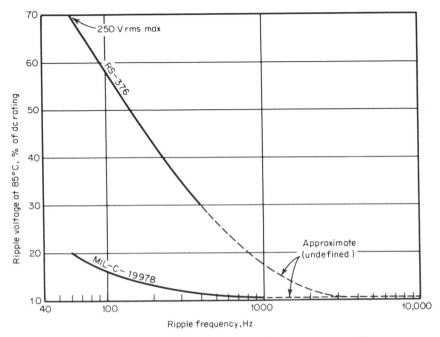

Fig. 116 Safe ripple voltage vs. frequency for typical metal-encased film capacitors.

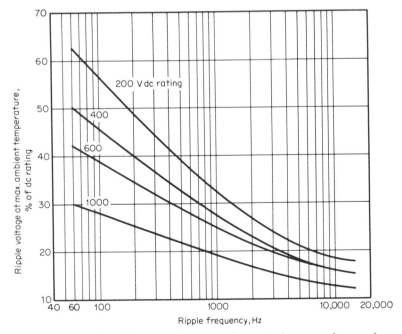

Fig. 117 Safe ripple voltage vs. frequency for typical nonmetal-encased paper, paper/film, and film capacitors.

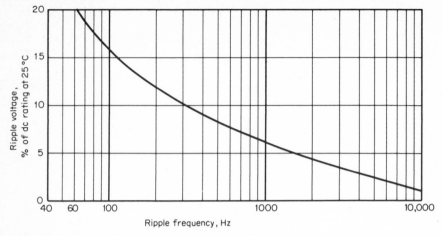

Fig. 118 Safe ripple voltage vs. frequency for typical dc metal-encased paper capacitors.

dards, some of which are referenced in Table 31. See also Figs. 116–118. In pulse and energy-discharge applications such as for flash tubes, it is common to find a voltage reversal on the capacitor. This is analogous to ringing in pulse terminology. The peak voltage reversal is treated as an ac voltage and is added to the voltage summation to determine the voltage rating needed. Energy-discharge capacitors are designed with dielectric systems capable of repeated discharges at very fast fall times with currents in the thousands of amperes. These voltage stresses on the dielectric with the accompanying rapid change in stress magnitude will, over a period of time, cause weak spots in the film to break down and short the capacitor. This failure mode is reduced in energy-discharge capacitors by multileaved extended-foil designs; and where small size is required, a metallized dielectric system can be employed which will "clear" points of high dielectric stress.

Corona is an important consideration in determining the ac rating of a capacitor. Corona caused by ionization within the capacitor is controlled to the greatest extent possible by selection of high-quality dielectric materials and the proper impregnant. Vacuum impregnation is always recommended to minimize voids with gases or trapped air which can become ionized. High voltage gradients also can cause corona; therefore, frequency and pulse limitations are provided by the manufacturer. Corona is damaging to the capacitor, and once started it will result in a catastrophic failure unless voltage is removed first. In the latter case, serious degradation in the dielectric and insulation system often results.

Internal heat generated in the capacitor through its ESR and dielectric losses is also considered in the ac rating. This is defined by the power factor and is one means of comparing ac capability. The physical construction and mounting provisions of a capacitor are such that heat can be conducted or radiated away from the dielectric itself. Therefore, ac ratings will vary among case designs. Extended-foil designs are usually used for high-current or high-frequency designs (1 kHz and up) to minimize internal lead resistance and inductance.

Current-carrying capability is closely related to internal heating in establishing ac ratings for capacitors. Factors which must be considered are size and material of leads, type of lead to electrode connection, method of connection (metallurgical, weld, pressure), electrode-material properties, frequency, maximum operating temperature, and capacitance. Where a high percentage of alternating current is present, a thorough knowledge of the capacitor construction and adherence to manufacturer's ratings will minimize failures caused by this stress.

AC capacitors designed for use in power-factor correction and motor-starting applications are usually specified for 60-Hz line voltages with a specified case-temperature rise. As frequency increases, say for 400-Hz applications, the internally generated heat will be excessive. Ruptured cases can result from excessive heat. A sizable increase in capacitor volume and surface area is required to dissipate this excessive heat adequately. A higher voltage rating can be selected, which will provide adequate safety factor. Manufacturers generally will suggest the proper capacitor size when consulted. A safety factor of 2 has been successfully employed for 60-Hz ac paper capacitors in 400-Hz applications. However, it is recommended that temperature rise be measured; it should not exceed the manufacturer's specified value.

Reliability

Reliability in capacitors, similar to that in many other components, is a function of the part construction, the care in manufacture, the type of testing it receives, the manner in which it is applied in the circuit, the environments in which the part is to operate, and the expected life of the equipment in which it is used. A decision to use a high-reliability part, a standard military part, or a commercial part is one of the primary decisions that must be made before a parts selection can be completed. Considerations for the use of a high-reliability part should include cost, equipment usage, the repairability of the equipment, the life and duty cycle expected for the equipment, and the combined environments.

Design considerations The reliability of a capacitor is, in general, the degree of success which has been achieved in housing the capacitor element in a mechanically and environmentally secure enclosure. Capacitor elements including internal lead construction must be mechanically and electrically sound before the encasement is applied. Only in the case of hermetically sealed metal-enclosed capacitors can case and lead configuration be trusted to provide mechanical and climatic protection. In no case should an encapsulated, dipped, or molded capacitor be expected to withstand dynamic environments such as high levels of shock and vibration commonly found in high-performance equipment unless additional mechanical support is provided. Likewise, in the construction of the capacitor element, nonmetallic encasement material should never be relied upon to provide mechanical holding power for lead attachment or for holding capacitor sections together. Metallurgical bonds and reinforcing materials such as straps and lead wires should be provided for mechanical integrity.

Temperature has a drastic effect on complex material systems used in capacitor construction. The temperature coefficient of expansion of various materials used in capacitor construction in many cases can provide stresses sufficient to cause capacitor failure. Failures have been found where capacitor elements have split open, the dielectric material becomes separated, or internal leads become separated from the capacitor element.

These types of failures must be considered in the selection of a component for high-reliability applications, in particular those where repair is inaccessible by virtue of either hardware inaccessibility or the fact that the component is buried so deeply within the equipment that troubleshooting and repair would be too costly.

In the selection of components for high-reliability applications, the design engineer must select parts that are manufactured on well-established production lines and by reputable manufacturers who have both the research and the manufacturing capability to maintain a quality production line continuously. Normally, by adhering to the industry or military standards defined in this chapter, the designer can find suppliers for capacitors that meet these specification requirements who will also provide parts with adequate quality built in that will preclude both early and latent failures.

Reliability screening In addition to the proper manufacture of a part, it is obvious that no matter how carefully a part is built, impurities and imperfections in construction will occur. Therefore, it is imperative that proper screening tests be performed on 100 percent of the product for high-reliability use. These screening tests as defined in standard specifications usually consist of a voltage-temperature burn-in test, limited life test, temperature-cycling test, and temperature-immersion or moisture-resistance test. These tests, when performed on standard products, will yield parts that can be expected to meet reliability predictions. Many producers of high-reliability equipment such as computers, calculators, aerospace-equipment missile systems, and man-rated equipment will generate their own in-house specifications to effect the degree of quality control necessary to match the reliability requirements for their products. These quality-control specifications usually follow the guidelines specified in military and industry standards and specify test methods and electrical-performance limits that have been historically demonstrated and can be duplicated by the many qualified testing laboratories.

Application for reliability   In addition to proper manufacture and screening of parts for high-reliability applications, the proper application of the components is the third criterion. In the application of capacitors defined in this chapter much has been said about proper ratings and designing to perform within these ratings. If one keeps in mind the three most damaging stresses that can be imposed upon a capacitor—temperature, frequency, and voltage stresses—it can be seen that reliable application of capacitors must include voltage, frequency, and temperature derating. Temperature derating can be achieved by selecting a dielectric material capable of performance in excess of the desired temperature limit as well as selection of encasement materials and a case size large enough to dissipate the internal heat generated. Heat generated entirely through the capacitor internal series resistance and through frequency effects can be readily calculated. Voltage deratings on capacitors, except for some commercial-grade wet tantalums and aluminum electrolytics, are an effective means of improving part reliability. Frequency derating is also required to prevent failure from oveheating.

Failure rates   Reliability definitions, theory, and calculations are well defined in MIL-HDBK-217[17], and treat most capacitor styles discussed in this chapter. Failure-rate curves have been generated from life tests at maximum rated voltage and temperature and are continually being performed, with billions of part hours now accumulated. Failure-rate-improvement factors with derated voltage and tempera-

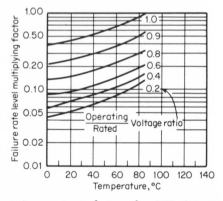

Fig. 119   Failure-rate-improvement factors for MIL-C-39006 nonsolid-electrolyte tantalum capacitors (60 percent confidence level). To use this chart, multiply observed failure-rate level at 85°C by the multiplying factor for failure rate level at desired operating condition.[15]

ture are also available for most capacitor types. Figure 119 shows failure-rate-improvement factors for nonsolid-electrolyte tantalum capacitors at a 60 percent confidence level. Figure 120 shows failure-rate-improvement factors for solid-electrolyte tantalum capacitors at a 60 percent confidence level. As discussed in the treatment of solid-tantalum capacitors, the reliability of this type of capacitor can be improved by limiting the surge currents available to the device. Additional reliability improvement is obtained by increasing the series resistance up to a minimum of 3 Ω/V. Multiplying factors are shown in Fig. 120 for circuit-impedance values up to 3 Ω/V. Failure-rate-improvement factors for ceramic capacitors at 90 percent confidence level are shown in Fig. 121. Mica-capacitor failure-rate-improvement factors at a 90 percent confidence level are shown in Fig. 122. Failure-rate-improvement factors for glass and porcelain capacitors at a 90 percent confidence level are shown in Fig. 123. RFI feedthrough-capacitor failure-rate-improvement factors at a 90 percent

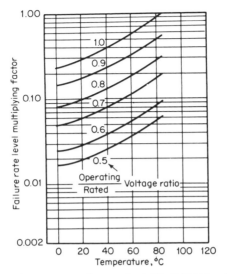

Fig. 120 Failure-rate-improvement factors for MIL-C-39003 solid-electrolyte tantalum capacitors (60 percent confidence level). To use this chart, multiply the observed failure-rate level at 85°C by multiplying factor for failure-rate level at the desired operating condition. Multiply the value read from the curves by the applicable multiplying factor below.[15]

Circuit impedance, Ω/V	Multiplying factor	Circuit impedance, Ω/V	Multiplying factor
3 or greater	1.0	0.5	4.5
2	1.4	0.4	5.1
1	2.8	0.3	6.1
0.9	2.9	0.2	7.5
0.8	3.2	0.15	9.0
0.6	4.0	0.10	12.0

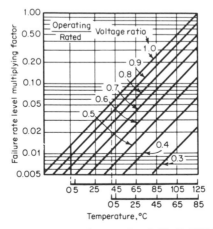

Fig. 121 Failure-rate-improvement factors for MIL-C-39014 ceramic capacitors (90 percent confidence level. To use this chart, multiply the observed failure-rate level at high ambient temperature by the multiplying factor for the failure-rate level at the desired operating condition.[15]

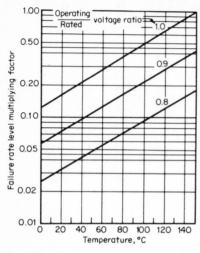

Fig. 122 Failure-rate-improvement factors for MIL-C-39001 mica capacitors (90 percent confidence level). To use this chart, multiply the observed failure-rate level at 150°C by the multiplying factor for the failure-rate level at the desired operating condition.[15]

Fig. 123 Failure-rate-improvement factors for MIL-C-23269 glass and porcelain capacitors (90 percent confidence level). To use this chart, multiply observed failure-rate level at 125°C by multiplying factor for failure-rate level at desired operating condition.[15]

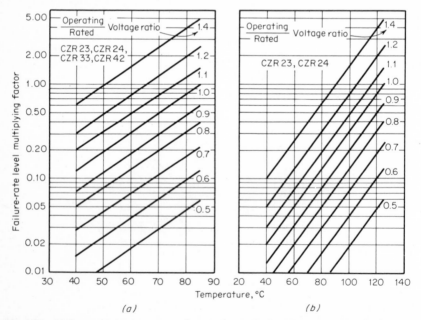

Fig. 124 Failure-rate-improvement factors for MIL-C-11693 type RFI feedthrough capacitors (90 percent confidence level).[15] (a) 85°C type (see Table 31). To use this chart, multiply observed failure-rate level at 85°C by the multiplying factor for the failure-rate level at the desired operating condition. (b) 125°C type (see Table 31). To use this chart, multiply the observed failure-rate level at 125°C by the multiplying factor for the failure-rate level at the desired operating condition.

confidence level are shown in Fig. 124.   Both the 85 and 125°C characteristics are
shown.   Figure 125 shows failure-rate-improvement factors for film and paper-film

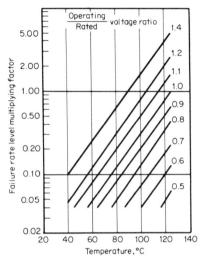

Fig. 125   Failure-rate-improvement factors for MIL-C-19978 film and paper-film
capacitors (CQR09, CQR12, CQR13 as shown in Table 31, 90 percent confidence
level).[15]   To use this chart, multiply the observed failure rate level at 125°C by the
multiplying factor for the failure-rate level at the desired operating condition.   Curves
for 100 percent of rated voltage and higher are based on the fifth-power rule, and
curves for 90 percent of rated voltage and below are based on the fourth-power rule.

capacitors at a 90 percent confidence level.   All these failure rates represent the
useful life of a capacitor, as shown in a typical failure-rate bathtub curve in Fig.
126.   It is evident that adequate screening and burn-in tests can remove essentially

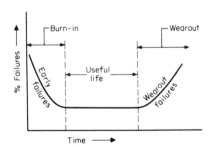

Fig. 126   Typical failure-rate bathtub curve.

all infant mortality, or early failures, from the capacitor lot population, leaving only
the lowest-failure-rate parts for usable life.   The failure-rate values given for each
capacitor type represent the length of time before wearout can be expected.

Failure modes in capacitors

Electrolytic Capacitors.   Failure modes in electrolytic capacitors were discussed
in detail earlier; however, some are worth mentioning again.   Mechanical-failure
modes in wet-electrolytic capacitors consist of electrolyte leakage, which is caused
primarily by defective vent diaphragms, insufficient compression seal, leakage at the
weld on the bottom of the can in axial-lead devices, and leakage around the aluminum or tantalum terminals in molded plastic headers or seals.   Most of these defects
can be eliminated by screening after temperature cycling.   Open circuits caused by

poor mechanical connection or embrittled foil are usually eliminated by 100 percent electrical tests. Poor welds or pressure connections in these capacitors are subject to becoming open-circuited after a short shelf life or operating life. This type of failure is usually eliminated by vibration or shock testing, where they are subsequently measured for capacitance change. Short circuits in foil electrolytics, caused by mechanical defects, consist of defects in the paper or paper alignment, burrs on edges of foils or at tab connections, wrinkles in the foil, and conductive particles which cut through the oxide film. Short-circuit defects are removed by temperature burn-in test and final electrical inspection. Electrochemical defects in electrolytic capacitors are caused by improper chemical composition of material used in their manufacture. The addition of contaminants such as chlorides is also a predominant cause of capacitor failure. This can be eliminated only by proper processes and in-process handling of capacitor elements and finally by a thorough burn-in test. Some known instances of failures for electrolytic capacitors are described as follows:

1. A shorted wet tantalum foil capacitor was found to have silver bridging across the Teflon seal from the case to the tantalum slug. The most probable cause of shorting was the application of high ripple currents or low-level reverse voltage.

2. In a solid-tantalum hermetically sealed capacitor a short was found to be due to reflow of internal solder caused by excessive heat applied during retinning operations or installation of the capacitor in the equipment.

3. Also, in a solid-tantalum capacitor high internal resistance between cathode and case was found to be caused by the absence of sufficient solder between the solid-tantalum slug and the capacitor case. This defect is normally detected by radiographic inspection.

4. In an aluminum foil electrolytic capacitor short circuits or low resistance between electrodes was found after burn-in tests and after short operational life. Shorts were found to be copper particles embedded on the oxide surface of the aluminum foil. This type of failure is extremely difficult to detect, especially when it survives a very severe burn-in test. Failures of this type can be eliminated only by in-process inspections and proper corrective action with the manufacturer.

5. In tantalum foil capacitors with wet electrolytes, severe drying out of electrolyte was found after approximately 10 years of operation, thus causing high internal impedance. Equipment failure resulted and capacitors were replaced. Failure modes of this type in equipment with long service life can be eliminated by either periodically replacing capacitors or using hermetically sealed capacitors.

6. Etched tantalum foil capacitors were found open-circuit or noisy in completed electronic assemblies. Cause of failure was cracked or completely separated foil near the internal tantalum lead. The problem was attributed to an embrittled foil, probably caused by overetching combined with in-process stresses applied to the lead. This type of failure is extremely difficult to detect if parts pass the electrical test screen. Nonstandard tests must be devised to screen for this problem.

Ceramic Capacitors. Monolithic ceramic-capacitor elements are themselves extremely reliable and have a very low failure rate. Most failures in this type of capacitor are due to encapsulating or encasement materials used to protect the capacitor and lead assembly from external environments. Two common causes of electrical degradation are thermal expansion of encapsulants and moisture between the coating and capacitor section. Poor soldering techniques and terminal design often result in loose or detached leads which cause intermittent or open failures. Some known failures in ceramic capacitors are:

1. Barium titanate changes from a dielectric to a resistor and is caused by elevated temperature and voltage stress. This results from poor manufacturing processes or improper ceramic formulation.

2. Glass in silver electrodes used in single-plate capacitors will cause degradation of insulation resistance when excessive elevated temperature causes it to become fluid and flow into microscopic cracks in the dielectric. The resulting failure is a short.

3. Silver migration is also a failure mechanism in single-plate capacitors. Under dc voltage and free water silver migrates as a chloride or hydroxide, resulting in decreased insulation resistance and early short.

4. Melting of solder used to attach the leads to the ceramic block. Melting

is caused by excessive heat from the external soldering process. This is prevented by the use of high-temperature solder in capacitor construction.

Paper and Film Capacitors. These capacitors, owing to the great amount of manufacturing experience and the high degree of refinement in the processing of films and paper, are some of the most reliable capacitors made. This is evidenced by the 90 percent confidence level shown in the failure-rate curves. Failure modes in this type of capacitor consist of the majority of failure modes for electrolytic capacitors with the exception of the drying out of the electrolyte. Seal leakage, however, is prevalent in poorly manufactured oil-impregnated capacitors. Seal integrity is vitally important where all-liquid impregnants are used. In general hermetically sealed cases provide an adequate degree of reliability to prevent impregnant leakage. Mechanical failures in these capacitors often are caused by fracture of the electrode tab at the point of attachment to the electrode or to the external lead. It is important that the capacitor and lead assembly be mechanically contained rigidly in the capacitor case to prevent movement and subsequent failure of the internal leads. Rough edges on foil electrodes also cause corona and premature shorting. These failures can be screened out with proper voltage burn-in tests.

RADIATION EFFECTS ON CAPACITORS*

Radiation environments The major types of radiation environments that electronic components must be designed to operate or survive in are (1) radiation from nuclear reactors, (2) radiation from nuclear warheads detonated by offensive or defensive weapon systems such as the atomic bomb, (3) space radiation including the Van Allen belt, and (4) radiation-simulation machines such as the cyclotron, linear accelerator, flash x-ray, or neutron generator.

When electronic components are exposed to these radiation environments, changes are observed in their rated electrical parameters. The magnitude of the changes is a function of such things as the radiation type, the intensity and energy spectrum of the radiation, the radiation time history, and a host of factors relating to the surrounding environment of the electronic components, the components materials, the packaging, and the application.

The major forms of radiation (excluding effects from thermal, air pressure, shock, electromagnetic pulse, etc.) which can affect an electronic component or circuit/system performance significantly are either particles (protons, neutrons, electrons) or electromagnetic radiation (photons). Generally, the remaining radiation products, such as radioactive isotopes and charged particles, will have little or no effect on the electronic components.

Protons. The proton is an elementary particle in the nucleus having a mass approximately 1,837 times greater than an electron and a positive charge with magnitude equal to that of the negative charge of the electron. In essence, the proton is the positive nucleus of the hydrogen atom. Protons are encountered in interplanetary space as a result of solar winds and solar flares (i.e., protons are released by the sun and travel through space with energies ranging from 1 to 100 MeV) and are found in the form of belts around planets (i.e., the magnetic fields of the planets trap charged particles).

Neutrons. The neutron (n) has no charge and has approximately the same mass as the proton. Since neutrons are uncharged particles, they can penetrate the electron cloud of an atom and interact with the nucleus. Neutrons are found in a nuclear-weapon explosion and nuclear reactor. Needless to say, they can be found in a space mission as a result of an on-board nuclear reactor and in electrical-power-generation plants (for example, the SNAP—systems for nuclear auxiliary power—or the RTG—radioisotope thermoelectric generator).

Electrons. Electrons are nuclear particles that have a rest mass of 9.107×10^{-28} g and a negative charge of 1.6×10^{-19} C. These particles are much more penetrating than protons. In space, the energetic electrons are found in the form of belts around planets. Energetic electrons are also produced as secondaries from the interaction of electromagnetic radiation (photon) with matter, as a result of being emitted from radioactive-fission products, and in various radiation-simulation machines.

---

* This section was written by Harold T. Cates.

Photons. Photons have neither mass nor charge and consist of electromagnetic energy that travels at the speed of light. They are described as either low-energy x-rays or high-energy gamma rays. Gamma rays are usually referred to as electromagnetic radiation emitted from a nuclear reaction but of much higher energy than ordinary x-rays. X-rays are electromagnetic radiation of wavelengths less than 100 Å which can be produced by bombarding a dense target with high-energy electrons or protons. The basic kinds of x-rays are bremsstrahlung and characteristic x-rays. Characteristic x-rays are produced as the result of the transition of electrons from outer atomic orbits or high-energy states to inner atomic orbits or low-energy states. Conservation of energy is maintained when the inner atomic orbit is made available as a result of the bombarding electron's giving enough energy to the low-energy electron to free it from the atom. Bremsstrahlung (a German word meaning "braking radiation") x-rays are produced when electrons bombarding a target lose kinetic energy in passing through the strong electric fields surrounding the target nuclei, thereby producing a continuous x-ray spectrum as the linear accelerator or flash x-ray.

Other Radiation Phenomena. A nuclear-weapon burst results in the generation of several other phenomena such as intense heat waves, air overpressure, and shock wave. The gamma-ray flux produced can create high-intensity electromagnetic fields through production of electronic currents by Compton interaction with environment molecules, creating an ionized sphere. This effect is generally called the EMP (electromagnetic pulse) effect. Its associated fields pose a potential threat to any system that must operate during the exposure and/or after the fields have degenerated. The EMP may be coupled into a system of electronic components by antennas, long cables, and missile shells. For example, in the case of missiles, a complete missile system forms an antenna which responds as a whole to the EMP. It is also possible for the electronic components to interact directly with the EMP field. Such a direct interaction is relatively simple to consider and is usually of minor significance in determining system vulnerability. The most difficult type of EMP interaction to analyze and eliminate is the indirect effect of induction of currents/voltages entering one part of the system and causing a disruption in another part of the system. The time-varying electric and magnetic fields induce circulating currents in conducting loops found in a compact subsystem. Associated with these circulating currents are voltages determined by characteristic impedance of the loops. These voltage differences appear to systems as signals and may cause severe disruption in system operation or if of sufficient magnitude may damage the components. In order to minimize the EMP effect, the designer must harden the system by proper circuit and component design, filtering, and shielding.

A factor which must be considered is the interaction of the effect of radiation and the standard environments (dynamic, climatic, electrical, etc.). Those materials which become brittle during nuclear radiation are more likely to fail in the subsequent electrical, vibration, or shock environments. For example, an insulator can lose its dielectric strength with radiation exposure, and subsequent electrical failure will become more prevalent. Also many parts are grossly affected by the internal temperature rise associated with the nuclear environment. Therefore, the selection of component parts must be based on overall ruggedness as well as nuclear hardness.

Radiation Units. Radiation is usually measured and expressed in terms of absorbed energy per second (dose rate), absorbed energy (dose), particles per square centimeter-second (flux), or particles per square centimeter (fluence). For ionization effects, the most commonly used unit is absorbed dose (dose rate) expressed as rad (rad/s) where the rad is defined as 100 ergs (ergs/s) absorbed from the radiation field per gram of irradiated material. In defining an absorbed dose, the material being irradiated must be defined. For example, 234 rads deposited in silicon would be expressed as 234 rads (silicon). For displacement effects the radiation of the particles (neutrons, protons, and electrons) is generally measured in fluence. Neutrons are sometimes expressed in NVT, where N is neutrons per unit volume, V is the velocity, and T is time, with the velocity expressing the energy spectra. For this discussion, the neutron will be given in fluence, with the energy spectrum understood to be a fission spectrum ($E > 10$ keV).

Nuclear-reactor and space environments (Van Allen belts, solar wind, solar flares, etc.) are considered to be steady-state environments as opposed to transient-radiation environments (less than ~1 s duration) produced by a nuclear-weapon burst, fast-burst reactor, and pulsed-radiation-simulation machines. The word TREES, an acronym for *transient-radiation effects on electronic systems*, is commonly used. The word transient in this case connotes the time history of the radiation environment and not the resultant duration of the effect, which may be either permanent or transient.

Radiation effects on components   Past experimental evidence and theory have shown that semiconductor devices (unipolar and bipolar transistors, silicon-controlled rectifiers, diodes, integrated circuits, unijunction transistors, etc.) are the components most sensitive to nuclear radiation and are responsible for current and voltage signals that perturb an electronic system. However, capacitors either discharge or, if they are uncharged, acquire charge from the radiation burst, which can also give false signals. In addition, semiconductors usually suffer significant permanent degradation before other components do, which can lead to system failure. The four common types of component radiation damage caused by a nuclear blast are surface effects (usually due to charge collection and migration in the surface passivation or packaging of the device), transient effects, permanent effects, and catastrophic failures. Therefore, a nuclear-hardened system requires reduction of the radiation sensitivity and nuclear hardness and qualification of electronic components.

The result of interaction of nuclear radiation with electronic components generally can be grouped in three categories: (1) displacement effects, (2) ionization effects, and (3) chemical effects.

Displacement Effects. Radiation particles behave like high-speed projectiles having random collisions with the atoms of the irradiated component material, which causes physical displacement of the atoms (physical properties change). As a result, some atoms (nuclei) of the structure can be moved from their normal lattice position (creates a vacancy) to an intermediate position (creates an interstitial), which produces permanent damage and is referred to as the displacement effect. The vacancy-interstitial pair, commonly referred to as a Frenkel defect, creates new allowable energy levels in the forbidden bandgap. This results in changes in such things as carrier mobility, electrical conductivity, minority-carrier lifetime, and thermal conductivity. In semiconductors, minority-carrier lifetime and electrical-conductivity changes are the most rapid and are the predominant mechanisms responsible for changes in the electrical characteristics of the device. The minority-carrier lifetime decreases owing to introduction of a new recombination center caused by the vacancy-interstitial pair. The electrical conductivity decreases owing to a process called carrier removal, which removes (traps) majority carriers from the conduction process. Thus the displacement of atoms is important in devices that depend on orderly lattice structure for proper operation.

The displacement of an atom from its normal lattice position requires that the bombarding radiation particle impart sufficient energy to the struck atom to overcome the displacement threshold value of the lattice. Generally the heavier radiation particles are more efficient in producing displacement than the lighter particles because the energy transfer is a function of their respective masses. The displacement capability of a given particle is also a function of the particle energy.

Energetic photons can cause displacement effects if the photon-generated electrons receive sufficient energy to displace atoms along their path.

Ionization Effects. Nuclear radiation can interact with the atomic electrons of the material being irradiated and imparts sufficient energy to free electrons from their atomic orbits. This phenomenon is known as ionization. Both particles and electromagnetic radiation are effective in producing ionization effects which raise the conductivity of the irradiated material through the creation of electron-ion (hole) pairs. For example, if an electronic circuit is exposed to a pulse of ionizing radiation, the condition of the circuit will be modified by passing from its electrical condition before the radiation pulse to a state of excitation that will depend upon the intensity and duration of the pulse. When the radiation pulse has ceased, an analog circuit will generally relax to its previous electrical condition within some characteristic time depending on the circuit components. However, in some digital circuits (i.e., flip-flops, counters, etc.), the output-voltage level may change states from the preirradiation state if the ionization level gets high enough. This transient effect is due to the generation of excess electron-ion pairs in the circuit components by the ionizing radiation. This ionization effect causes unwanted currents (photocurrents) to flow in the electronic parts and surrounding environment (for example, the gas or air around the part), which make semiconductor junctions and insulators behave more like conductors and appear like electrical shorts during and immediately after the radiation. The electronic components will usually regain their preirradiation electrical characteristics shortly after the radiation ceases. However, if the photocurrents are allowed to reach high enough current levels, catastrophic failures (for example, semiconductor metallization-interconnect burnout) can occur, leaving the devices inoperative. In addition, the ionization near the surface or in the passivating/isolation layer of components can cause semipermanent changes in the electrical properties of devices. These changes in electrical properties result from charge migration and collection in the surface, passivation, or isolation layer which may persist for periods from hours to years after the radiation exposure.

X-rays deposit their energy as ionizing radiation through x-ray absorption whose absorption cross section varies as the fourth power of the atomic number (Z^4). If the radiation environment contains enough x-rays, the subsequent deposited thermal energy in component material can lead to thermomechanical destruct mechanisms such as melt and shock-wave-induced fractures. The effect can render devices inoperative through solder melt, package fracture, lead liftoff, die liftoff, etc. The low-energy characteristic of the x-rays can be rapidly absorbed by the air in a relatively short distance. The absence or presence of an x-ray environment in an equipment specification is dependent upon whether the equipment is employed on the ground, in air, or in the exoatmosphere.

Chemical Effects. Changes in the molecular composition through rearrangements of the chemical bonds are called chemical effects. Generally changes in electrical properties due to chemical changes are too small to be of practical significance as opposed to displacement or ionization effects.

Capacitor performance in a radiation environment As previously pointed out, radiation particles degrade the electrical performance of semiconductor devices by the introduction of displacement defects (vacancy-interstitial pairs). The capability of various radiation particles to produce displacement effect in silicon is a function of the particle type and energy. As a rule of thumb, a fission neutron (n) produces displacement in a silicon transistor equivalent to 0.031 proton (15 MeV) or 28 electrons (1.3 MeV) or 2,800 cobalt 60 gamma rays. Only fission-spectrum displacement damage will be discussed here. Additional particle-radiation degradation can be estimated with Refs. 18 and 19.

Many current semiconductor devices can be hardened to a fast-neutron environment between 10^{13} to 10^{14} n/cm^2 and a few selected devices to 10^{15} n/cm^2. Almost all other electronic components are relatively insensitive at these neutron fluence levels to displacement defects which cause concentration and minority-carrier lifetime changes and are considerably less sensitive to radiation damage than semiconductors. Semipermanent effects associated with the ionization portion of the radiation environment can often be very important in these components, with such effects occurring predominantly in insulating materials. At the particle fluences of concern for hardening most electronic systems, semiconductor-device displacement effects are quite severe; they can be considered to be minor in other materials and components.

The principal radiation effects in insulating materials (both organic and inorganic) are related to the ionization dose produced by the particle or photon. The major changes in the macroscopic properties (thermal, mechanical, electrical, and optical) of insulating materials resulting from ionizing radiation are: (1) increase of ionic conductivity and dielectric loss, resulting in decrease of the dielectric Q; (2) changes in dimensions; (3) modification of tensile strength, yield point, and plastic and elastic properties; (4) gas evolution; (5) small changes in the dielectric constant and dielectric strength; and (6) increased optical absorption. Generally the dominant radiation degradation of inorganic materials results from increased conductivity of the materials; whereas for organic materials, mechanical changes are usually the dominant effects. Mechanical changes in organic material occur because of modification of the organic polymer structure caused by radical interaction and formation.

The neutron-radiation sensitivity of various types of components is shown in Fig. 127. Analysis of this figure shows that permanent change in capacitance value, dissipation factor, and leakage current is not considered severe at a fission neutron-fluence spectrum less than about 10^{14} n/cm². For most capacitor applications this limit is about 10^{17} n/cm², with the exception of paper and paper-plastic capacitors.

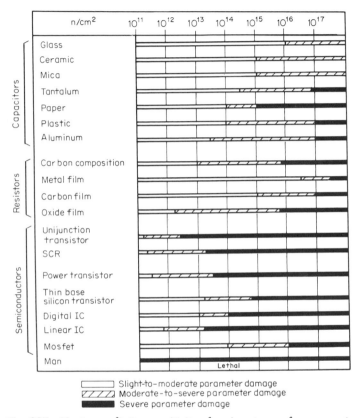

Fig. 127   Neutron-radiation sensitivity of various types of components.

The principal cause of radiation-induced capacitance changes is dimensional changes in the interelectrode spacing due to gas evolution and swelling. These changes are more pronounced in organic-dielectric capacitor construction. Gamma heating and changes in dielectric constants of capacitor dielectrics have been rare and can be considered a second-order effect especially for inorganic materials. Generally, capacitors using organic materials such as polystyrene, polyethylene terephthalate (Mylar), and polyethylene are less satisfactory in a radiation environment by about a factor of 10 than those using inorganic dielectrics. However, it should be noted that even for these types, experimental data indicate no significant permanent changes occurred for exposures up to about 10^{15} n/cm². On the other hand, usage of tantalum and aluminum electrolytic capacitors indicates that both types show capability of surviving extended radiation exposure, with the tantalum being more radiation-resistant. Both decreases and increases in capacitance value have been observed. For example, changes from -10 to $+25$ percent for tantalum and from -6 to $+65$ percent for aluminum types have been observed for exposure up to 10^{17} n/cm². These changes were observed during radiation exposure, with some recovery in the electrical characteristics noted in several days after the end of the radiation exposure. This recovery in some cases is rather slow, and in many instances complete recovery was never attained.

Generally, wet-electrolyte electrolytic capacitors are not permitted in high-reliability equipment. Permanent changes in electrical characteristics of these capacitors began at about 5×10^{13} n/cm². The principal mechanism is gas evolution caused by the interaction of the ionizing radiation with the electrolyte, which tends to rupture the capacitor.

Capacitor Conductivity in an Ionizing-Radiation Environment. The capacitor dielectric can exhibit significant increases in its conductivity in an ionizing-radiation environment. This phenomenon can result in a charged capacitor discharging, which may seriously affect the operation of the electronic circuit. The electron-ion pairs that are produced in the capacitor dielectric by the ionizing radiation drift in the presence of an applied electric field or diffuse in a concentration gradient to create a dielectric current. This induced dielectric current disappears very rapidly after cessation of the radiation because of the small recombination time of the free carriers. However, a portion of this radiation-induced conductivity continues after the ionizing radiation pulse with a long time delay. The general postirradiation behavior of the dielectric has been characterized by a set of exponential functions which are assumed to be associated with the various trap-release processes. A reasonably good empirical relationship relating the shunt conductivity σ to the gamma dose rate $\dot{\gamma}$ is obtained from the following relationship:[20]

$$\frac{\sigma(t) - \sigma_0}{\epsilon\epsilon_0} = \left(\kappa_p \dot{\gamma}^n + \sum_{i=1}^{2} \kappa_{di} \int_{-\infty}^{t} \exp\left(-\frac{t-\tau}{\tau_{di}}\right) \dot{\gamma}(\tau)^n \, d\tau \right) \tag{18}$$

where σ_0 = dielectric conductivity in the absence of radiation, mho-cm⁻¹ or (Ω-cm)⁻¹
ϵ = relative dielectric constant
ϵ_0 = permittivity of free space, 8.85×10^{-14} F/cm
κ_p = prompt conductivity constant, mhos/F [rad (material)/s]⁻ⁿ
$\dot{\gamma}(\tau)$ = dose rate, rad (material)/s
n = empirical radiation exponent that varies between 0.5 and 1.1 depending on the dielectric material
κ_{di} = empirical derived coefficient for the delayed components of radiation-induced conductivity effect, s⁻² [rad (material)/s]⁻ⁿ
τ_{di} = ith time constant for the free carrier in the dielectric associated with the κ_{di} coefficient, s

The conductance of the capacitor may be obtained by multiplying the above equation (whose units are mhos/farad) by the capacitance.

Equation (18) includes the summation of two faltung (convolution) integrals, and it may not be possible to evaluate the expression in closed form. Therefore, it is advantageous to select a geometric approximation of $\dot{\gamma}(\tau)$ to simplify the integration and employ some approximation that will give sufficiently accurate results. One of the most frequently used environments assumes a square pulse of ionizing radiation of amplitude $\dot{\gamma}_p$ and duration t_p, with the result that the conductivity equations become

$$\frac{\sigma(t) - \sigma_0}{\epsilon\epsilon_0} = \kappa_p + \sum_{i=1}^{2} \kappa_{di}\tau_{di}\left[1 - \exp\left(\frac{-t}{\tau_{di}}\right)\right]\dot{\gamma}_p^n \qquad t \leq t_p \tag{19}$$

$$\frac{\sigma(t) - \sigma_0}{\epsilon\epsilon_0} = \sum_{i=1}^{2} \kappa_{di}\tau_{di} \exp\left(\frac{-t}{\tau_{di}}\right)\left[\exp\left(\frac{t_p}{\tau_{di}}\right) - 1\right]\dot{\gamma}_p^n \qquad t > t_p \tag{20}$$

A useful approximation is to assume t_p approaches infinity and $t \gg \tau_{di}$ so that Eq. (19) can be written as

$$\frac{\sigma_{ss} - \sigma_0}{\epsilon\epsilon_0} = \left(\kappa_p + \sum_{i=1}^{2} \kappa_{di}\tau_{di}\right)\dot{\gamma}_p^n \tag{21}$$

where σ_{ss} is the steady-state ionization-induced conductivity. In general $R = d/\sigma A$ and $C = \epsilon\epsilon_0 A/d$; therefore, Eq. (21) can be expressed as

$$\frac{(\sigma_{ss} - \sigma_0)C}{\epsilon\epsilon_0} = \frac{1}{R_{ss}} - \frac{1}{R_p} = C\left(\kappa_p + \sum_{i=1} \kappa_{di}\tau_{di}\right)\dot{\gamma}_p{}^n \tag{22}$$

or

$$\frac{1}{R_{ss}} = \frac{1}{R_p} + \frac{1}{R_{\dot{\gamma}}} \tag{23}$$

where $1/R_{\dot{\gamma}} = C\left(\kappa_p + \sum_{i=1}^{2} \kappa_{di}\tau_{di}\right)\dot{\gamma}_p{}^n$, $1/R_{ss} = \sigma_{ss}C/\epsilon\epsilon_0$, and $1/R_p = \sigma_0 C/\epsilon\epsilon_0$.

Generally in a high-intensity $\dot{\gamma}$ environment, $R_p \gg R_{\dot{\gamma}}$ and Eq. (22) can be approximated by

$$\frac{1}{R_{ss}} = \frac{1}{R_{\dot{\gamma}}} + \frac{1}{R_p} \simeq \frac{1}{R_{\dot{\gamma}}} = C\left(\kappa_p + \sum_{i=1}^{2} \kappa_{di}\tau_{di}\right)\dot{\gamma}_p{}^n \tag{24}$$

It should be noted here that the term κ_p is referred to as a prompt component and that it applies only when the radiation is bombarding the capacitor. κ_p then drops out after the radiation environment ceases and the delayed terms (κ_{d1} and κ_{d2}) are responsible for the time it takes a capacitor to recover and in turn for the circuit to recover. The total effect, however, is the sum of these two. Using the assumption that only the conductivity effects need to be considered to predict the behavior of capacitors under an ionizing-radiation exposure, the equivalent circuit as shown in Fig. 128 is applicable. From this equivalent circuit it becomes apparent that the

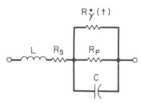

Fig. 128 Equivalent circuit of a capacitor in an ionizing-radiation environment.

C = capacitance
R_S = series resistance (leads, plates, electrical interfaces)
L = inductance (leads, plates, etc.)
R_P = parallel resistance (leakage current, dielectric absorption, insulation resistance)
$R_{\dot{\gamma}}$ = ionizing-radiation-induced leakage resistance

gamma-leakage resistance ($R_{\dot{\gamma}}$) will severely limit the ability of a capacitor to function in a radiation environment.

The variations of the capacitor radiation parameters given in Eq. (18) are shown in Table 32. Equations (18) through (24) along with Table 32 can be used to compare the shunt-leakage resistances of various dielectric materials used in capacitors. For example, if it is required for the circuit delayed response to be minimized and not controlled by the radiation-induced delayed response, capacitors with small values of κ_{d1} and κ_{d2} should be chosen (for example, mica). Another example would be for the minimum capacitor shunt-leakage resistance to be maintained as high as possible during the radiation exposure. If the choice was between a 0.1-μF ceramic or oil-impregnated paper, application of Table 32 and Eq. (24) in a $\dot{\gamma}_p$ environment of 10^9 rads/s would reveal that the worst-case shunt-leakage resistance was 1,381Ω and 72Ω for ceramic and oil-impregnated paper, respectively. Thus, a ceramic capacitor would be better than the oil-impregnated paper for this application.

Capacitor Voltage Discharge in an Ionizing-Radiation Environment. As previously discussed, a capacitor shows a transient change in the conductivity of the dielectric as the result of ionizing radiation, which can be represented as a dose-rate- and time-dependent resistance across the capacitor as shown in Fig. 128. In most hand

TABLE 32   Radiation-induced Capacitor Parameters

Dielectric material	n	κ_p, 10^{-5} s^{n-1}/ rad^n	κ_{d1}, s^{n-2}/rad^n	τ_{d1}, 10^{-6} s	κ_{d2}, 10^{-4} s^{n-1}/ rad^n	τ_{d2}, s	γ_c,* 10^5 rad
Mica	0.98	0.9	ND†	ND	0.03	1.3	1.11
Ceramic (barium titanate)	1.0	0.03	0.06	24	0.05	1.1	5.75
Tantalum pentoxide	1.0	≤ 0.05	4	0.9	360	2×10^{-4}	2.4
Paper (dry)	0.7	100	0.07	200	0.01	0.5	0.0099
Paper (oil-impregnated)	1.0	5	0.4	400	0.01	1.0	0.048
Polystyrene	1.0	0.2	0.07	10	470	$\geq 10^{-3}$	3.7
Teflon §	1.0	90	NR‡				

* $\gamma_c = (\kappa_p + \kappa_{d1}\tau_{d1})^{-1}$
† ND = not detected.
‡ NR = not reported.
§ Teflon is a trademark of E. I. DuPont de Nemours & Co., Wilmington, Del.

analysis, the effects of R_s and L can be neglected and they are set equal to zero. During high exposure rates $R_{\dot\gamma}(t)$ in parallel with R_p is equal to $R_{\dot\gamma}(t)$ [i.e., $R_{\dot\gamma}(t) \ll R_p$] and σ_0 can be neglected in Eq. (22).

In many applications of capacitors it is the decrease of capacitor voltage (or loss of charge) during the radiation environment that controls the design of the circuit. If it is assumed that the external charging voltage source $e_{\text{ext}'}$ and its associated charging resistor R_c shown in Fig. 129 cannot replenish the charge loss during the

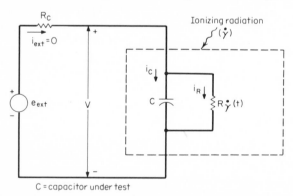

Fig. 129 Simplified model of capacitor under ionizing radiation with no external supplied current.

ionizing radiation (i.e., high-impedence external circuit and radiation pulse short compared with the circuit time constant), the following analysis develops a single expression for computing the approximate voltage loss. Analyzing Fig. 129 and using Eq. (18) with $n = 1$ and σ_0 neglected, the following expressions can be written:

$$\frac{dV}{dt} = \frac{i_c}{C} = \frac{-i_R}{C} = \frac{-V/R_{\dot\gamma}}{C} = \frac{-V\sigma}{\epsilon\epsilon_0}$$

$$\frac{dV}{dt} = -V\left[\kappa_p\dot\gamma + \sum_{i=1}^{2} \kappa_{di} \int_{-\infty}^{t} \exp\left(-\frac{t-\tau}{\tau_{di}}\right) \dot\gamma(\tau)\, d\tau \right] \quad (25)$$

which suggests that the rate of voltage loss is independent of the capacitor value. τ_{d2} is generally large and can be comparable with the total circuit time constant. Therefore, its effect will be neglected (i.e., let $\kappa_{d2} = 0$ and $\tau_{d2} = 0$), since in this time frame the external circuit will start to replenish the charge loss to the capacitor (i.e., $i_{ext} \neq 0$). Separating variables in Eq. (25) and integrating from $t = 0$ to ∞ gives the result

$$V_f = V_i \exp\left(\frac{-\gamma}{\gamma_c}\right) \qquad (26)$$

where V_f = limit $t \to \infty$ $V(t)$, final voltage immediately following radiation exposure, V

V_i = initial voltage prior to radiation exposure, V

$\gamma = \int_0^\infty \dot{\gamma}\, dt$, total dose received by capacitor, rads (material)

$\gamma_c = (\kappa_p + \kappa_{dT}d)^{-1}$ rads (material)

The variation of V_f/V_i vs γ for various types of capacitors is shown in Fig. 130.

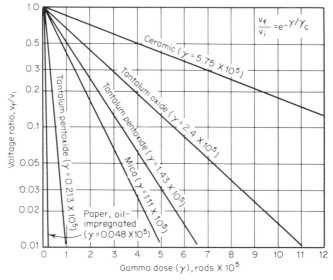

Fig. 130   Voltage ratio V_f/V_i vs. gamma dose γ for various types of capacitors.

It should be noted that Eq. (26) should be used only as a rule of thumb to obtain preliminary design information, as there are other variables which will modify this result. For example, Fig. 131 shows the variation of $1/\gamma_c$ and γ_c vs ambient temperature and working voltage for solid-electrolyte tantalum pentoxide capacitors. Examination of Fig. 131 reveals that as the working voltage increases (dielectric-thickness increase) or the ambient temperature increases, the value of γ_c decreases or conversely the percentage of voltage discharge increases. Thus in order to minimize the radiation response of a tantalum capacitor, the ambient temperature and working voltage of the capacitor should be minimized consistent with the circuit application requirements. In addition, when an unbiased solid-electrolyte tantalum (also aluminum) capacitor is exposed to an ionizing radiation, it has been observed that a voltage builds up across the capacitor, with the tantalum terminal (anode) found to be charged negative. A photovoltaic model is used to explain the voltage buildup with the assumption that the energy band of the tantalum pentoxide dielectric is bent near the tantalum slug–tantalum pentoxide interface owing to the formation of a Schottky junction. The band bending is caused by the electro-chemical-potential difference (contact potential) between the tantalum and manganese dioxide contacts, and the band bending produces an electric field in the oxide. A typical ionized-radiation response of an unbiased tantalum capacitor is shown in Fig. 132 with peak voltage/1,000 rads [V_p/krad(Ta)] varying between 10 and 30 mV/krad(Ta) for a 10-kΩ load resistor and $\gamma \ll \gamma_c$. The peak signal is weakly dependent on the load resistor and varies as $R_L^{1/6}$ to $R_L^{1/7}$. A reasonable empirical

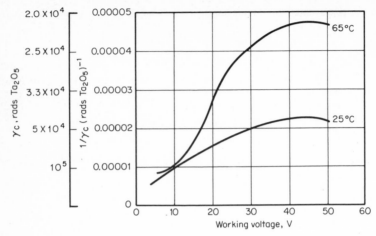

Fig. 131 Radiation-induced voltage-discharge variation of sintered-anode solid-tantalum capacitor vs. working voltage and ambient temperature.

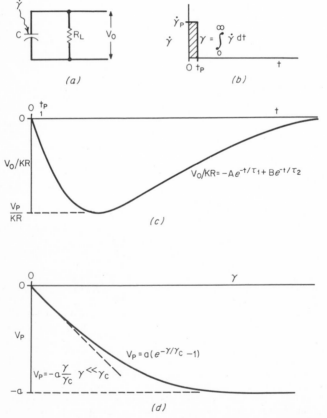

Fig. 132 Unbiased sintered-anode solid-electrolyte tantalum capacitor voltage response vs. time and total dose. (*a*) Experimental circuit. (*b*) $\dot{\gamma}$ vs. time. (*c*) $V_o/Krad(Ta)$ vs. time. (*d*) V_p vs. γ.

fit to the time history of the voltage response is given by[21]

$$V_p = -A \exp\left(\frac{-t}{\tau_1}\right) + B \exp\left(\frac{-t}{\tau_2}\right) \qquad (27)$$

where A and B are empirical constants and τ_1 and τ_2 are time constants related to the shallow and deep traps in the oxide and the RC time constant. τ_1 is approximately equal to $0.1RC$ and τ_2 varies between RC and $2RC$.

A reasonable approximation to the unbiased tantalum response to the ionizing dose with a narrow pulse width ($t_p \lesssim 5 \ \mu s$) is given by

$$V_p = a\left[\exp\left(\frac{-\gamma}{\gamma_c}\right) - 1 \right] \qquad (28)$$

where a is the built-in contact potential and is equal to $1.1 \pm 0.5V$ for a tantalum capacitor.

As long as the ionizing-dose pulse t_p is short compared with the emission time for a carrier trapped in the oxide and $\gamma \ll \gamma_c$, the peak signal is linear with ionizing dose and not dose rate. That is, $e^{-x} \approx 1 - x$, or Eq. (28) becomes

$$V_p \approx a\left(1 - \frac{\gamma}{\gamma_c} + 1\right) = \frac{a}{\gamma_c}\gamma \qquad \text{for } \gamma \ll \gamma_c \qquad (29)$$

Combining the result given in Eq. (28) with the previous biased capacitor [Eq. (26)], Eq. (26) becomes

$$V_f = (V_i + a) \exp\left(\frac{-\gamma}{\gamma_c}\right) - a \qquad (30)$$

and is shown in Fig. 133.

Radiation-hardening techniques In selecting capacitors for a radiation-environment application, a survey of available piece-parts radiation data should be performed in order to determine whether radiation data exist on that particular part or a similar part. If no data exist, a radiation analysis should be made of the materials that make up the capacitor to try to narrow down the number of different candidates. A radiation exposure is then performed on a few samples of the remaining candidates to reduce their population to one or two. These remaining candidates then receive extensive investigation in terms of radiation characterization. However, after a capacitor becomes qualified to a nuclear environment, there is always the problem of maintaining this nuclear qualification in any future procurement (i.e., lot-to-lot radiation-quality assurance). In reality, manufacturing processes do change and the vendor does not always inform the procuring agency. Depending on the criticality of the application, some type of screening for usage in radiation environment may have to be performed, which can vary from validating no change in manufacturing techniques (baselining) to electrical screens, to lot-to-lot sample radiation or to 100 percent radiation screening.

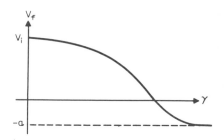

Fig. 133 Biased sintered-anode solid-electrolyte tantalum capacitor voltage response vs. total dose.

Hardening for Fast Neutrons. For neutron fluence less than about 10^{14} n/cm² (generally the maximum level most semiconductors devices can survive), changes in capacitance value, dissipation factor, and leakage resistance are considered to be minimum if at all detectable. For some capacitors, a fast-neutron fluence of about 10^{17} n/cm² is necessary before the radiation damage becomes severe. Glass, ceramic, mica, and tantalum are quite radiation-resistant, as shown in Fig. 127, and they are preferred in that order. Generally, the radiation deterioration of organic dielectric materials is more severe than that of the inorganic.

Hardening for Ionizing Radiation. [*] Capacitors compatible with the circuit requirement made with dielectric materials that have low prompt conductivity ($\kappa_p \dot{\gamma}_p{}^n$) and
minimum delayed conductivity at the maximum $\dot{\gamma}_p$ to be encountered should be
used by the designer to minimize conductivity changes. However, if the time it
takes for the radiation-induced leakage current to get below a specified value is
the important effect, then κ_{d1}, κ_{d2}, τ_{d1}, and τ_{d2} need to be considered. There
is no simple capacitor-selection criterion that can be given independent of radiation
pulse shape; but the solution given in Eq. (24) will generally determine the proper
capacitor selection. For example, ceramic, glass, or tantalum dielectric material improves the capacitor radiation resistance over paper-dielectric capacitors. In order
to minimize the radiation capacitor voltage change, the dc voltage and capacitor
working voltage should be kept low. Where possible, Zener diodes should be substituted for capacitors in voltage-blocking applications, and utilization of transistor constant-current generators to supply emitter bias rather than emitter bypass capacitors
is recommended.

In low-voltage tantalum-capacitor applications (i.e., when the maximum applied
voltage is much less than the contact potential), it is possible to cancel out a large
portion of ionizing-radiation-induced volt-
age seen by the external circuit by using
two tantalum capacitors in series or parallel. These two techniques are shown
in Fig. 134. In the parallel configuration
the net radiation-induced charge is the
difference between the radiation-induced
charges of each capacitor; whereas in
the series or back-to-back configuration,
the external circuit sees only the difference between the radiation-induced voltage of each capacitor, since the induced
voltages are of opposite polarity. However, it should be noted that in either
configuration the cancellation effectiveness depends on how well the capacitors are matched and how equally
they are irradiated, The cancellation
will not apply when the applied voltage is large compared with the contact
potential.

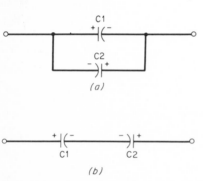

Fig. 134 Electrical configuration of
sintered-anode solid-electrolyte tantalum
capacitor for ionizing-radiation hardening.
(*a*) Parallel configuration. (*b*) Series
or back-to-back configurations.

Tantalum capacitors are frequently used in the back-to-back configuration shown
in Fig. 134, where large capacitor values in a nonpolar type of application are required. The problems and circuit considerations associated with this type of operation in a nonradiation environment were discussed in the paragraphs on tantalum
capacitors. Although long-term exposure to neutron and/or gamma radiation can
eventually cause permanent degradation of component parameters, the design engineer
should be aware of the effects of short pulses of high-level gamma (or x-ray)
radiation.

As shown in Figs. 132 and 133, the voltage across a biased or unbiased tantalum
capacitor tends to approach a value of approximately -1.1 V when the capacitor
is exposed to a pulse of high-level gamma radiation. The actual voltage change
for a given capacitor is a function of both gamma level and initial charge, and of
course varies from unit to unit for a given set of gamma and initial-charge conditions.
These facts are not changed by connecting capacitors back to back, but the results
seen by the circuit are affected. The greatest benefit from back-to-back operation
is obtained for the unbiased state, i.e., when there is a net charge of zero across
the capacitor pair. As noted previously, this does not necessarily imply zero voltage
across each capacitor, but rather that their voltages are equal and opposite. Whenever this condition exists, the net result of a gamma burst is a voltage across the
pair equal to the difference in the change in voltage across each capacitor. This
net resulting voltage may be of either polarity or may even be initially of one polarity
and then change to the opposite polarity.

[*] Includes results from a special study by Joseph F. Rhodes.

The preceding theoretical considerations backed by considerable data yield the following guidelines for analysis and design of circuits using back-to-back tantalum pairs:

1. Owing to cancellation effects, induced voltage across a pair of zero-biased capacitors is roughly an order of magnitude less in amplitude than for a single tantalum capacitor, but the decay time is of the same order of magnitude.

2. Since the induced voltage seen is the difference between two induced voltages, it may be of either polarity.

3. The lowest $\dot{\gamma}$ response is generally obtained by using capacitors of the lowest possible voltage rating consistent with the circuit and reliability requirements.

4. The voltage induced in large capacitors is no greater than that induced in small ones. In circuits involving operational amplifiers, therefore, it generally pays to use large capacitors and small resistors rather than vice versa to obtain a given time constant, since this tends to minimize the effect of operational-amplifier (op-amp) offset current. This consideration can become quite important in circuits exposed to neutron radiation, since high neutron dosage causes op-amp offset currents to increase rather drastically. For example, the zero-bias 3-sigma response for a group of back-to-back 22-μF, 20-V tantalum capacitors was 4.0 mV/krad, while the 3-sigma response for a group of 100-μF, 10-V capacitors from the same manufacturer was only 2.7 mV/krad. The time constant of the circuit was the same in both cases.

5. Back-to-back tantalum capacitors with a net charge (bias) across the pair will generally lose charge at a rate of approximately 1.0 to 1.5 percent per krad (25°C). Of course, for very small biases (a few millivolts), the voltage loss may be masked by the random voltage which is normally induced on zero-bias capacitors. If it is imperative that the initial charge be retained in spite of high gamma dosage, tantalum capacitors should not be used. Tests on a limited number of high-k ceramic capacitors indicate that these devices yield about the same level of induced charge as a pair of back-to-back tantalum capacitors, while losing an order of magnitude less initial charge. One pays a considerable penalty in both size and cost by using ceramics in lieu of tantalum; however, solid-tantalum capacitors are also generally more stable with age and temperature.

6. Where the gamma radiation to which the capacitor pair is exposed is of short duration (a few microseconds or less), the response is almost purely gamma-dose- rather than gamma-dose-rate-dependent. That is, the response to a pulse of 1×10^9 rads/s for 10μs would have essentially the same effect as a 100-ns pulse of 1×10^{11} rads/s amplitude.

7. Virgin (previously unradiated) parts usually respond more than previously irradiated parts. This seems to be true of ceramic as well as tantalum capacitors and occurs even when several hours or even days pass between exposures, although it is much more pronounced for the case where the second burst follows the first by only a few seconds.

TRIMMER CAPACITORS*

Introduction   Trimmer capacitors, also referred to as variable capacitors, are invaluable in the design of electronic equipment. They are commonly used in applications where exact capacitance values cannot be computed using normal design procedures. In this case a trimmer capacitor is employed to provide the needed capacitance range.

Trimmer capacitors come in a variety of sizes and shapes. The variety is due to the method used to vary the capacitance. If we examine the basic formula for capacitance [Table 1, Eq. (1)], it becomes evident that there are only three ways to vary capacitance. Since it is usually impractical to change dielectric material, the common means left are to change the effective area of the plates or the distance between them.

Some of the most widely used trimmers in the industry today are ceramic, glass and quartz, air, plastic (including Mylar), and mica. It is evident that trimmer types are classed by dielectric. It therefore follows that the various trimmers enjoy

* This section was written by Stephen D. Das.

the same benefits (and suffer the same problems) as their fixed counterparts discussed elsewhere.

Basic construction

Air. The air variable capacitor is the forerunner of all trimmer capacitors, first finding use in radio as the tuning element. It is simply a larger version of the air trimmer and as such is constructed in basically the same way. These units contain two basic parts: the rotor and the stator. The rotor is the rotating part and

Fig. 135 Typical air trimmer capacitors.

is usually constructed of several semicircular disk or plates affixed to a shaft. The stator is the fixed portion and consists of similar-shaped plates mounted such that the rotor plates can be intermeshed with the stator plates by rotating the rotor shaft. The dielectric is of course air, and because of its dielectric constant of 1.00+ (Table 3), these units are relatively large compared with other types. They exhibit one of the best capacitance vs. temperature curves owing to the relative ability of air to maintain its dielectric constant over a wide temperature range. Electrical contact to the rotor is made through brushes or bearings of various quality commensurate with the quality of the overall part. Figure 135 shows typical open-plate air trimmers.

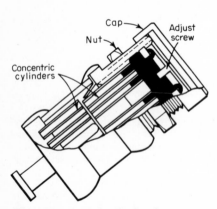

Fig. 136 Typical cylindrical air trimmer capacitor.

A miniaturization approach to the design of air variable capacitors is one in which the plates take the form of cylinders of various size. The stator cylinders are fixed in one end of an insulating tube (usually glass or ceramic). The rotor cylinders are mounted on a movable screw which is fed in from the opposite end of the assembly. Contact to the movable rotor screw is usually made through multiple low-resistance spring fingers. Capacitance is varied by changing effective plate area by intermeshing rotor and stator cylinders. These units are generally smaller than the open type of air trimmer having the same capacitance range. Figure 136 shows various configurations of a sealed air trimmer capacitor popular for high-reliability RF applications.

Ceramic. The ceramic trimmer contains three basic parts: the rotor, dielectric, and stator. The rotor carries a semicircular pattern. This pattern consists of metallization deposited, painted, or screened on the rotor to provide an electrode. The rotor pattern is usually semicircular to utilize maximum area and maximum resolution capability during tuning. The stator is essentially composed of a metallized pattern deposited, painted, or screened onto a low-loss class I ceramic base. The dielectric, particularly the titanate ceramics, is placed between the rotor and stator electrodes. To assure good electrical stability, the interfaces of all the bearing surfaces must be in intimate contact. This is accomplished by lapping. All parts involved are held together by spring pressure, and rotation of the rotor varies the capacitance. Figure 137 shows typical ceramic trimmers.

Fig. 137 Typical ceramic trimmer capacitors.

Glass. As with the air and ceramic trimmers mentioned above, the glass trimmer also varies the capacitance by changing effective plate area. The general construction is as follows: A glass (or sometimes quartz) tube is coated on the outside with a conducting material. This metallization is deposited, painted, or screened from one end of the tube toward the other end and forms the element analogous to the stator in the ceramic and air devices. The element analogous to the rotor is a piston, usually on a threaded rod, and inserted from the other (nonmetallized) end of the tube. This piston, when moved in and out of the metallized section of the tube, varies the capacitance by changing the effective plate area of the capacitor. Connection to the device is made by means of straps or clamps embedded in the metallization on the outside of the tube and by brushes in intimate contact with the threaded rod on the piston. These units are usually very low in capacitance and limited in range and therefore are used where very precise trimming is necessary. Figure 138 shows typical glass trimmers.

Fig. 138 Typical glass trimmer capacitors.

TABLE 33 Selection Guide to Trimmer Capacitors

Type	Capacitance range, pF	Min Q (typical) at 20 MHz	Voltage	Temp range, °C	Temp coefficient	Applicable* MIL specification	Applicable styles	Configuration
Air open-plate	1.3–6.0 to 9.0–143.0	1,500	700	−55 to +85	45 ± 15 ppm/°C	MIL-C-92B	All	Fig. 135
piston (or cylinder)	0.8–10.0	2,000	250	−55 to +125	±50 ppm/°C	MIL-C-14409	PC 25, PC 26	Fig. 136
Ceramic	1.0–3.0 to 7.0–40.0	500	100	−55 to +125	At −55°C: At +125°C: −4.5 −14.0 +14.0% +3.4%	MIL-C-81	All	Fig. 137
Glass	0.6–1.8 to 1.0–120.0	250–1,500	1,500	−55 to +150	+50 to ±100 ppm/°C −0	MIL-C-14409	PC 38, 39, 42, 43, 48, 52	Fig. 138
Plastic	1.0–5.0 to 5.0–150.0	1,500	1,000	−55 to +85	45 ± 15 ppm/°C	None†	None	Similar to Fig. 135
Mica	1.0–15.0 to 1,400–3,055	150	500	−30 to +85	+(2.5% + 0.5 pF) −(2.0% + 0.5 pF)	None†	None	Fig. 139

* At this time, there are no EIA specifications covering variable capacitors.
† These parts are generally commercial parts. Lists of suppliers will be found in the listing services outlined in Chap. 7.

Plastic. Film-dielectric trimmers are generally of the same construction as the air trimmers. That is, they use semicircular plates separated by plastic materials (including polyethylene terephthalate). The use of a dielectric material other than air allows the plates to be spaced closer together than in the air units, with a resultant increase in capacitance-to-volume ratio and usually an increase in voltage rating.

Mica. Mica trimmers are classified as compression trimmers. These capacitors are made up of stacked-mica-dielectric capacitance units, with a basic capacitor consisting of a thin film of mica between two spring nonferrous-metal conducting plates; the stacked units are mounted within a ceramic container or base. By alternating mica film, metal plate, mica film, metal plate, and paralleling these units, one can obtain any desired capacitance within the physical limitations of the ceramic container. A captive adjusting screw inserted through the center of all plates and mica films provides variable compression on the formed metal plates, thereby varying plate separation and hence capacitance, which is inversely proportional to plate separation. Figure 139 shows a typical compression mica trimmer.

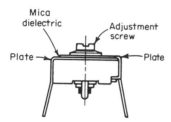

Fig. 139   Typical mica trimmer capacitor.

Trimmer-capacitor application   The selection guide in Table 33 lists the parameters most useful in selecting a variable capacitor. The usable frequency range depends highly upon the mechanical application; however, in general the upper limit is comparable with fixed capacitors using similar dielectric materials. The principal features of trimmer capacitors are summarized as follows:

Mica
  Low cost
  Good stability
  Good temperature coefficient
  Smooth adjustment over several turns
  Good for all transmitter bands
  Low inductance
  Low power dissipation
  Large rf power ability
  Good in moderate shock and vibration environments
  Good C/V ratio
Ceramic
  Low cost
  Good C/V ratio
  Fairly high Q
  Predictable temperature coefficient
  Low inductance (high-frequency capability)
  Brittle—poor in high shock and vibration environments
  Limited to 180° rotation maximum
  Overvoltage can cause permanent damage
Glass
  High-voltage capability
  Can be environmentally sealed (capped)
  Can be adjusted up to 10,000 times
  Can be fairly brittle, depending on mount
  Relatively large
  Smooth adjustment over several turns

Little if any backlash
Nearly linear capacitance vs. rotation curve
Air
Excellent stability
Low temperature coefficient
Good shock and vibration resistance
High-torque units are available
Excellent for use where frequent adjustments are required
180° maximum tuning (open type)
Multiturn tuning (cylinder type)
Good voltage rating
Good Q
Low inductance
Easily environmentally sealed (capped)

ACKNOWLEDGMENTS

The authors of this chapter express sincere appreciation for the technical support given by the many capacitor manufacturing companies and technical agencies. Of note are the contributions by Mr. John Scienicki and Mr. Arthur Hoffman of Erie Technological Products, Inc., and the technical support from the engineering staffs of the General Electric Company, Capacitor Division, the Mallory Capacitor Company, Cornell Dubilier Electronics, Sangamo Electric Company, and Sprague Electric Company. Special appreciation is also given to Mr. James Burkhart of the Defense Electronic Supply Center, and Colonel Charles Flint and his staff at the Electronic Industries Association Engineering Department for their assistance.

REFERENCES

1. Hayt, William H., Jr.: "Engineering Electromagnetics," chaps. 2, 5, McGraw-Hill, New York, 1958.
2. Frugel, Frank B. A.: "High Speed Pulse Technology," vol. 1, Academic, New York, 1965.
3. Johnson, F. L.: "Which Capacitor?" Marshall Industries, Capacitor Division (now Electron Products, Monrovia, CA), 1965.
4. Perkins, J. R., and C. G. Brodhun: "Teflon-Polytetrafluoroethylene Resin as an Electrical Insulator," Annual Report of The National Research Council Conference on Electrical Insulation, p.31, 1949.
5. Polarized Aluminum Electrolytic Capacitors, EIA Standard RS 395, Electronic Industries Association, Washington, D.C., 1972.
6. Fixed Electrolytic Tantalum Capacitors, EIA Standard RS 228, Electronic Industries Association, Washington, D.C., 1972.
7. Powers, John H.: Low-Impedance Capacitors for Power Distribution Applications, *Proc. 22nd Electron. Conf.*, 1972.
8. "Ceramic Chip Capacitor Handbook," San Fernando Electric Manufacturing Co., San Fernando, Calif.
9. Kittel, Charles: "Elementary Solid State Physics," chap. 4, Wiley, New York, 1962.
10. "Insulation/Circuits, Dictionary/Encyclopedia," vol. 18, no. 7, Lake Publishing Corporation, Libertyville, Ill., June/July 1972.
11. "Military Standardization Handbook; Plastics," MIL-HDBK-700 (MR), 1965.
12. "Machine Design, Plastics Reference Issue," Penton, Cleveland, Ohio, 1973.
13. "Materials Engineering, Materials Selector," Reinhold, Stamford, Conn., 1974.
14. Von Hippel, Arthur R.: "Dielectric Materials and Applications," sec. IV B 1, 3, Wiley, New York, 1954.
15. Selection and Use of Capacitors, MIL-STD-198.
16. Johnson, F. L.: "Characteristic Properties of Lectrofilm-B Capacitors," General Electric Company, Bull. GET 3394, 1963.
17. "Reliability Stress and Failure Rate Data for Electronic Equipment," MIL-HDBK-217A.

18. Horne, W. E., et al.; "Literature Search and Radiation Study on Electronic Parts," Boeing Document D2-126203-3, May 1970.
19. Hank, C. F., and J. D. Hamman: The Effect of Nuclear Radiation on Capacitors, *REIC Rept.* 44, December 1966.
20. Jones, D. C., et al.: "Transient Radiation Effects on Electronics Handbook," DASA 1420, Feb. 26, 1965.
21. Flanagan, T. M., and R. E. Leadon: "Effects of Ionizing Radiation on Tantalum Capacitors," GULF-RT-10609, Apr. 19, 1971.

General References

22. Tinnell, J. T., Jr., and F. W. Karpowich: Skipping the Hard Part of Radiation Hardening, *Electronics,* Mar. 4, 1968, p.122–127.
23. Tinnell, J. T., Jr., et al.: Equivalent Circuit Estimate Damage from Nuclear Radiation, *Electronics,* Oct. 30, 1967, pp.73–82.
24. Ricketts, L. W.: "Fundamentals of Nuclear Hardening of Electronic Equipment," Wiley, New York, 1972.
25. Olesen, H. L.: "Radiation Effects on Electronic Systems," Plenum, New York, 1966.
26. "Technical Information for Mica Compression Trimmers and Single Padders," The Electro Motive Mfg. Co. Inc., Willimantic, Conn., Nov. 19, 1970.

Chapter **9**

Transformers and Inductive Devices

E. N. HENRY

Systems Development Division
Westinghouse Electric Corporation

and

L. E. WILSON

Systems Development Division
Westinghouse Electric Corporation

INTRODUCTION

In elementary terms a transformer consists of two or more wirewound coils which are coupled inductively. When an alternating voltage is applied to one winding (usually called the primary), a corresponding alternating voltage is induced in the other winding (usually called the secondary). The magnitude of this voltage is determined by the number of turns in each of the coils; and if the coils are adequately coupled magnetically, it is expressed as

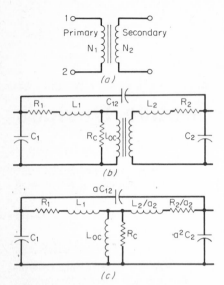

$$\frac{V_1}{V_2} = \frac{N_1}{N_2} \qquad (1)$$

or the winding voltages are directly proportional to the winding turns. This principle applies to all transformers regardless of the application or the operating frequency. Figure 1a, b, and c shows a simple two-winding, single-phase transformer and its equivalent circuit.

Basic relationship If the voltage drops and various losses are considered to be negligible (a perfect transformer) the power received by the transformer will equal the power delivered to the load.

$$V_1 I_1 = V_2 I_2 \qquad (2)$$

or

$$\frac{V_1}{V_2} = \frac{I_2}{I_1} \qquad (3)$$

and from Eq. (1),

$$\frac{I_2}{I_1} = \frac{V_1}{V_2} = \frac{N_1}{N_2} \qquad (4)$$

or the winding currents are inversely proportional to the winding turns.

Replacing the primary of the transformer with an equivalent impedance Z_1 and the secondary load resistance R_1 with impedance Z_2, the primary and secondary currents may be expressed as

$$I_1 = \frac{V_1}{Z_1} \qquad (5)$$

and

$$I_2 = \frac{V_2}{Z_2} \qquad (6)$$

substituting these expressions in Eq. (3),

$$\frac{Z_1}{Z_2} = \left(\frac{V_1}{V_2}\right)^2 \qquad (7)$$

and from Eq. (1),

$$\frac{Z_1}{Z_2} = \left(\frac{N_1}{N_2}\right)^2 \qquad (8)$$

Fig. 1 Transformer equivalent circuit. (a) Schematic diagram. (b) Equivalent circuit. (c) Equivalent circuit referred to primary.

N_1 = primary turns
N_2 = secondary turns
a^2 = N_2/N_1
C_1 = primary capacitance
C_2 = secondary capacitance
C_{12} = primary-to-secondary capacitance
L_1 = primary leakage inductance
L_2 = secondary leakage inductance
L_{oc} = primary open-circuit inductance
R_1 = primary resistance
R_2 = secondary resistance
R_c = equivalent core-loss resistance (shunt)

Transformer parameters The values of the various inductances, capacitances, and resistances will vary with the size of the transformer, the materials used in its construction, the type of transformer, and its application. It can be seen that as the operating frequency and the magnitude of the various parameters change, the performance of the transformer as a circuit element will change. Normally the various secondary parameters are reflected to the primary by the ratio of the square of the turns of the two windings. Figure 1c reflects this. This ratio is reasonably accurate for most power, audio, and pulse transformers; however, in if and rf transformers, where the magnetic path is largely through air, the equivalent circuit depends on the coupling of the windings. Table 1 shows the effect of each of these parameters on the operation of power audio and pulse transformers for steady-state conditions, i.e., the condition of normal operation. The operation of a transformer or inductor during turn-on, turn-off, switching modes, circuit faults, or other transient-inducing modes of operation is discussed elsewhere.

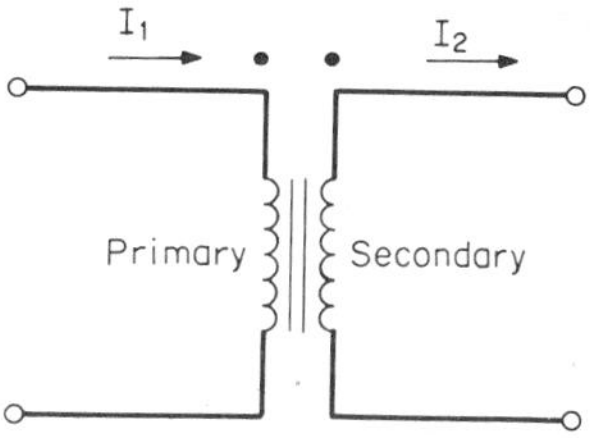

Fig. 2 Transformer polarity.

Transformer operating characteristics
Polarity. Figure 2 shows schematically the convention used to indicate transformer polarity. The dots establish the terminals of the same polarity. The + and − signs indicate instantaneous polarities; i.e., V_1 and V_2 reach positive maximums at the same time. The dots also show that current flows into the primary terminal and out of the secondary terminal of the same polarity.
Regulation. Transformer regulation is the ratio of the difference in secondary voltage between no load and full load to the full-load voltage; it is usually expressed as a percentage:

$$\text{Percent regulation} = \frac{100(V_{\text{NL}} - V_{\text{FL}})}{V_{\text{FL}}} \tag{9}$$

Efficiency. Efficiency is the ratio of the power out of the transformer to the power into the transformer and is expressed:

$$n = \frac{\text{output power}}{\text{input power}} \tag{10}$$

The input power is equal to the output power plus the transformer losses.
Power Factor. The power factor is the ratio of the transformer input power to the input volt-amperes and is expressed

$$\text{Power factor} = \frac{\text{input power}}{\text{input volt-amperes}} \tag{11}$$

ELECTRONIC TRANSFORMERS

The use of alternating current for the generation, transmission, and distribution of electrical energy is largely due to the reliability, efficiency, and convenience of the static transformer. In the application of electronic transformers it is found that they often perform many functions besides the basic changing of voltage and/or current values. The development by design of the various characteristics will not be discussed except to bring out the application and to show the effect on size and complexity of manufacture.

Power conversion Perhaps the best-known use of transformers is for the conversion of ac power from one voltage to another, generally at relatively low frequencies. The generation frequency has been chosen primarily for transmission considerations and to a lesser extent for reduced losses in the transformers and equipment both in the magnetic material and in the windings. Commercial power in the United States is almost entirely 60 Hz, while 50 Hz is used extensively in Europe. Where

TABLE 1   Effect of Transformer Parameters on Transformer Characteristics

Parameter	Transformer characteristic affected				
	Power transformer	Inverter transformer	Wideband transformer	Pulse transformer	Inductor
R_1	Regulation, efficiency, temperature rise	Regulation, efficiency, temperature rise	lf response, hf response, efficiency, temperature rise	Efficiency, temperature rise	Temperature rise, voltage drop, Q factor
L_1	Regulation, commutation	Efficiency, high-voltage spikes, hf response, primary balance, loss of inverter switches	hf response	Front-edge response	
C_1	Turn-off-pulse suppression	hf response	hf response	Front-edge response, trail-edge response	hf impedance
C_{12}	EMI	EMI	EMI	Front-edge response	
R_c	Temperature rise, exciting current efficiency	Efficiency, temperature rise, exciting current	Efficiency, temperature rise	Efficiency, temperature rise	Temperature rise, Q factor
L_{oc}	Exciting current	Exciting current	lf response	Top response, pulse droop, trail-edge response	Inductance
R_2	Regulation, efficiency, temperature rise	Regulation, efficiency, temperature rise	lf response, hf response, efficiency, temperature rise	Efficiency, temperature rise	
L_2	Regulation, commutation	Efficiency, high-voltage spikes, hf response, secondary balance, loss of inverter switches	hf response	Front-edge response	
C_2	Turn-of-pulse suppression	hf response	hf response	Front-edge response, trail-edge response	

transmission over great distances is not required, as on ships and in aircraft, 400 Hz and even higher frequencies permit a reduction in the size of devices which use magnetic materials such as generators, motors, and transformers.

Improved magnetic materials, semiconductors, and coil construction have enabled power transformers to operate at considerably higher frequencies when the power is converted to direct current close to the transformer. Modern practice using inverters transforms power in kilovolt-ampere blocks at frequencies up to 100 kHz for airborne equipment. As more power is required in airborne or space applications, the size and weight considerations will encourage the development of techniques to use these higher frequencies at higher power levels.

The conversion of power from one voltage to another has many applications in operation of ac equipment, but in electronics application conversion to direct current at various voltages is common. In general, older equipment using hard vacuum tubes requires higher voltages than does more modern solid-state circuitry. This use of lower voltages naturally means higher currents for equivalent power.

Power sources Commercial power is standardized into insulation classes each of which is assigned a basic impulse level (BIL). The BIL is primarily the ability to withstand lightning surges, which are a transmission hazard. It follows that the BIL will apply to the primary or ac winding of transformers supplied directly from commercial power lines. Transformers supplied for local generators or from in-plant distribution sources may not require BIL ratings.

When local generation is used, the frequency may be subject to choice. Rotary generators usually operate at frequencies below 2,000 Hz, but oscillators or inverters can generate frequencies limited only by the switching speeds obtainable.

The number of phases will also require consideration. Many applications use polyphase power to reduce the size of equipment for equivalent power. This may be especially important for rectifier operation where ripple frequency and magnitude are directly related to number of phases, rectifier circuit used, and primary frequency. Overall performance including filtering and regulation needs to be considered.

Alternating current at high frequencies obtained from inverters introduces transmission limitations. The waveshape, especially when combined with pulse-width regulators, may introduce transformer problems. Increased losses in magnetic materials and in windings from eddy currents and skin effect may result. If transformed and rectified at the inverter, this method of producing dc power at various voltages can be desirable, especially when size and weight are primary considerations.

Tolerances. Two types of tolerances must be provided. Short-time voltage and/or current transient usually presents a hazard rather than out-of-specification performance, although it may affect performance too. Voltage surges may breakdown insulation in the transformer as well as in other places in the circuit. Frequently protective devices are used to limit this voltage.

Inrush current at turn-on can reach several times normal load current and may operate overload devices such as fuses or circuit breakers. This hazard can be eliminated by gradual turn-on instead of switching on at full voltage. Another method is to place a current-limiting impedance in series with the primary, which is removed from the circuit after steady-state conditions have been reached.

Improper performance may result from unusual voltage variations, source impedance, transformer regulation, auxiliary equipment impedance, etc. In rectifier circuits at light loads filter capacitors tend to charge to the peak voltage. Thus for zero load a single-phase full-wave rectifier with filter capacitor will reach 1.57 times average voltage. The effect is less with polyphase rectifiers and two-way rectifier circuits, but it is present to some extent in all rectifier circuits. This can be eliminated by using a bleeder which prevents the load from dropping into the critical region. Transformer regulation may be limited by design. Other voltage changes may be provided for with transformer taps.

Environment and life expectancy

Industrial Transformers. Industrial transformers supplied from public utilities fall into two classifications. Outdoor transformers, usually liquid-insulated, are expected to operate unprotected from weather over the ambient-temperature range from -40 to $+40°C$. Indoor transformers are frequently classified as dry-type and are protected from weather including lightning surges. Both types must have a life expectancy of 20 years or more.

Communication Transformers. Communication transformers including radio- and television-broadcast transformers may be either of the above, or they may be even more protected in air-conditioned environments. Computer transformers may operate in carefully controlled environments where both temperature and humidity are controlled within a narrow range. This permits very limited regulation and may make life expectancy independent of thermal aging.

Military Transformers. Military transformers usually must operate at ambients from −55 to +95°C and sometimes even higher. They also may be required to withstand many special environments such as salt spray, high humidity, fungus, contamination, and high altitude. Fortunately the required life expectancy of these transformers is much less than that of industrial transformers, but offsetting this shorter life is an increased reliability requirement. This results in more carefully specified transformers for military usage.

Special Requirements. Minimum size and weight are always desirable, and there are several ways to accomplish this requirement. Frequency is the most effective way, since the magnetic-core size is inversely proportional to frequency. A 400-Hz transformer may be only about 25 percent the size of a 60-Hz transformer of equal power rating; however, there are limitations to this reduction, since transformers are not exclusively core. High-temperature insulation is another method, but small transformers may find regulation does not permit full realization of available temperature. Special cooling such as forced air or gas or circulated liquid may also aid in size reduction. Some of the more exotic methods will add appreciably to the cost.

MIL Standard Transformers. MIL standard transformers, especially airborne military transformers, may have structural requirements such as shock and vibration and environmental protection. Many of these transformers are designed for the particular application and resist attempts to standardize parts and construction. In general standardization does not permit close tolerances or precise requirements. In precise applications the use of standard or off-the-shelf transformers involves relaxation of requirements and usually some loss of efficiency.

Partial discharge Partial discharge (corona) in transformers is undesirable for two reasons. The historic objection is the effect on insulation. Corona damage to insulation is well known and treated extensively in literature. Many inorganic insulations such as mica, glass, and ceramics are relatively tolerant of corona. There is the effect of corona on radio and television broadcasting and on communication in general which has led to limits of RIV (radio-influence voltage). More recently the computer has forced the suppression of the EMI (electromagnetic influence) of corona to extremely low levels.

Transformer construction In most instances the circuit designer who uses a transformer is not particularly concerned with what materials are in the transformer. The primary concern is that the device meet a set of requirements, with the ultimate selection of a part often being made on the basis of its cost.

However, the circuit designer should have some knowledge of the materials used in the construction of transformers and of the various configurations of transformers

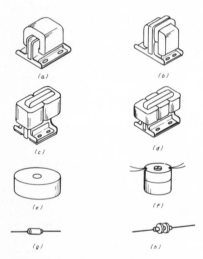

Fig. 3 Transformer configurations. (*a*) Simple (one-phase). (*b*) Shell (one-phase). (*c*) Core (one-phase). (*d*) Three-phase. (*e*) Toroidal, (*f*) Cup core. (*g*) Single-layer solenoid. (*h*) Pie.

with the advantages and disadvantages of each. Usually no specific configuration is indigenous to a particular type of magnetic device. Figure 3 shows various configurations of transformer and inductors, and Table 2 lists the usual applications of each type together with advantages and disadvantages of each.

TABLE 2   Application of Transformer Types

Type	Uses	Comment
Simple	Power transformers, inductors, pulse transformers	Only one coil to wind, low leakage inductance, large mean turn
Core	Power transformers, inductors, wide band transformers, pulse transformers, inverter transformers	Permits winding balance in push-pull application, minimizes external magnetic field, permits lower winding capacitance in high-voltage transformer
Shell	Power transformers, inductors, wide band transformers	Better winding efficiency than core type, smaller mean turn than simple type
Three-phase	Power transformers	E core not affected by unbalanced currents in primary
Toroid	Power transformers, pulse transformers, inductors, wide band transformers, inverter transformers	Minimum leakage inductance, more difficult to wind, usually degraded by direct currents in windings, may be difficult to cool, can function over wide frequency ranges
Cup core	Pulse transformers, inductors, wide band transformers, pulse transformers, inverter transformers	Available only in ferrite-type core materials. Low external magnetic field. Relatively fragile. Can adjust to given inductance. Difficult to wind with larger magnet wire
Single-layer solenoid	if and rf transformers, if and rf inductors	Can be wound on magnetic or nonmagnetic form
Pi winding	if and rf inductors	Lowest interwinding capacitance

An electronic transformer can be considered to consist of several physical systems, each of which has an effect on the performance of the device.

Magnetic Circuit. The magnetic circuit consists of the core, which may be any type of magnetic material; and the magnetic path, which may also include air spaces, gaps, and leakage paths.   In some applications, usually of high frequencies, the magnetic path may include little or no magnetic material.

Transformer Cores. The component most responsible for the shape, size, and weight of transformers and inductors, except air-core transformers, is the magnetic core. Transformer cores are basically iron, sometimes with nickel or cobalt in various proportions and usually with small amounts of other elements. Power, audio, and large pulse transformers are usually fabricated from thin sheet or strip core laminations. The laminations must be insulated from each other to prevent eddy currents which represent losses, and their thickness is a function of the operating frequency of the transformer.

Laminated cores are fabricated in many shapes, the simplest being a toroid. The ratio of core to winding may be varied over a limited range, and the cross-sectional shape of the iron circuit may also be varied a limited amount. Constrictions on transformer shape usually affect the core configuration and result in a reduction of efficiency and/or increase in transformer cost.

In addition to laminations, cores may be made of powdered metal bonded with a resin or of oxides in a ceramic structure called a ferrite. Powdered-metal cores are usually toroids, and ferrite cores may be of many shapes, including toroids and pot cores which completely surround the windings. Powdered iron and ferrites are used for the higher-frequency applications.

Figure 4 lists the types of core configuration often used. Table 3 lists many of the types of core material used.

Electric Circuit. The electric circuit consists of successive turns of wire, usually single strands of insulated copper or aluminum wire; the wire may be round or for higher currents square or rectangular. As the operating frequency of the transformer goes up, skin effect may affect losses and regulation. To reduce these effects litz wire may be used. Litz wire is made of multiple strands of individually insulated wires. The need for litz wire is determined by the current being conducted and the operating frequency.

Insulation System. The insulation system consists of the insulation on the magnet

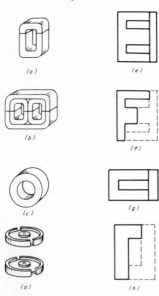

wire; sheet insulation which may be used between layers of a winding and between windings themselves; various plastic impregnants used to provide mechanical support, heat transfer, and/or environmental protection; various gaseous or liquid dielectrics; and miscellaneous tapes and adhesives. These materials may be organic or inorganic and are generally selected for one or more of the following: their compatibility with each other, the expected life of the transformer, the allowable hot-spot temperature, their compatibility with the anticipated environment, the magnitude and type of voltage stress, their effect on size and weight, and their cost. It is also important to recognize that many factors affect both the temperature and voltage at which an insulation system may be used.

Voltage Considerations. In a transformer the voltage to which a combination of insulation may be stressed is not the summation of voltages at which given thickness of each material break down. Rather it is usually a substantially reduced voltage determined by the dielectric constants of the materials, the geometry of the windings, the geometry of associated ground planes, the atmospheric pressure, and other factors. It is also affected by the

Fig. 4 Transformer cores. (*a*) Tape-wound cut, C core (one-phase applications). (*b*) Tape-wound cut, E core (three-phase applications). (*c*) Tape-wound. (*d*) Cup core (used in pairs). (*e*) E-I lamination. (*f*) F lamination (used in pairs). (*g*) U-I lamination. (*h*) L lamination (used in pairs).

nature of the voltage stress: alternating current, direct current, impulse, or combinations of these.

Thermal Considerations. In a transformer the temperature at which a combination of insulations may be used is determined by their compatibility and the anticipated life of the transformer. Transformer life is essentially temperature-related, being determined by the rate of thermal decomposition of the insulation (within reasonable bounds). The Arrhenius equation is often used to relate the life of a transformer manufactured with organic insulation to temperature:

$$\log (\text{rate}) = \log A - \frac{E}{RT} \tag{12}$$

where A, E, and R are constants based on the materials and T is the temperature in degrees Kelvin

The corresponding graph of log life vs. $1/T$ (Kelvin) results in a straight line. The slope of this line leads to the often-used 10° rule, i.e., that the life of an insulation system is reduced 50 percent for each 10°C increase in temperature. Modern insulation systems often have life plots with slopes providing 8 to 14° rules. The graph in Fig. 5 depicts anticipated life of an transformer insulation system which consists of organic materials and which is impregnated with an organic resin.

Thermal Circuit. The thermal circuit consist of (1) all elements which generate heat in the transformer: winding losses, core losses, and possibly dielectric losses; (2) all elements which conduct heat out of and away from the transformer: the windings, the core, the mounting structure, the various insulating materials; (3) all elements which create thermal barriers: the insulation materials and the transformer geometry; and (4) the environment.

Mechanical System. The mechanical system consists of all elements which support the transformer and maintain its physical integrity. It can include parts of the insula-

TABLE 3   Core Materials

Material	B_{max} (T)	Suggested frequency range	Uses	Forms
Silicon-iron alloy:				
Oriented, Hipersil	17.6	To 20 kHz and pulse	A,B,C,D,E,H,J	1,2,3
Nonoriented, AISI M-19	15.0	To 400 Hz	A,B,D,I	3
Nickel-iron alloys:				
50% Ni, square loop	16.0	To 50 kHz and pulse	B,C,E,H,J	1,2,3
48% Ni, round loop	15.0	To 20 kHz	A,D,F	3
79% Ni, square loop	8.0	To 50 kHz and pulse	B,C,E,H,J	1,2,3
79% Ni, round loop	8.0	To 20 kHz	A,D,F	2,3,7
Supermalloy	7.0	To 50 kHz and pulse	A,C,E,F	2
Cobalt-iron alloy,				
Supermendur	21.0	To 5 kHz	A,B,D	1,2
Powdered moly permalloy	2.0	To 100 kHz	A,D	4,7
Ferrites	5.0	To 300 MHz	A,B,C,D,F,G,H	4,5,6
Iron powders	5.0	2 kHz to 300 MHz	B,G	7,8,9,10

Uses:

A. Audio
B. Power
C. Pulse
D. Inductor
E. Saturable reactors

F. Wideband
G. rf and if
H. Inverter
I. Communications
J. Instrument transformer

Forms:

1. Tape-wound cut cores, C and E
2. Tape-wound cores, toroidal and rectangular
3. Laminations
4. Toroidal

5. Cup cores
6. Cast or machined shapes
7. Bars
8. Rods
9. Threaded rods
10. Sleeves

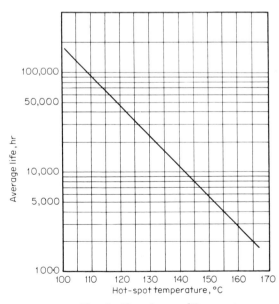

Fig. 5   Transformer life.

tion system, the windings, and the core as well as additional hardware used for support or protection.

Power Transformers

Power transformers   Power transformers are used primarily to change the voltage and current of electrical energy.   Additional functions may include isolation of circuits for safety or insulation purposes, change in number of phases, or change of phase angle including polarity reversal.   Special-purpose transformers which are current-limiting, constant-current, constant-voltage, regulating, or power-factor-correcting also may be classified as power transformers.   By far the most common use of electronic power transformers is in rectifier circuits.

Power transformers may be single- or polyphase with single- and three-phase the most common.   Systems beyond three phases are multiples of three.   The principal use of these 6-, 12-, and 24-phase arrangements is in rectifier circuits to reduce current in rectifiers and/or to reduce filter size and harmonic voltages.

Transformer Connections.   Three-Phase.   Figure 6 shows several different configurations of polyphase transformations.   There are many more combinations of 3-, 6-, 12-, and even 24-phase transformer configurations which are usually dependent upon the rectifier circuit being used.   Figure 6 contains only delta primary windings;

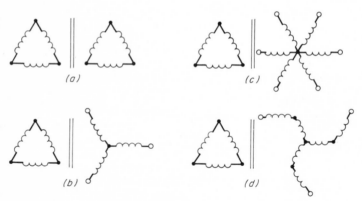

Fig. 6   Three-phase transformer connections.   (*a*) Delta-delta.   (*b*) Delta-wye. (*c*) Delta-double wye.   (*d*) Delta-wye, zigzag.

however, each can be replaced with a wye winding.   These transformations, while offering some advantages, are not generally recommended for electronic circuits, primarily because if a neutral ground is not provided they do not provide a path for the third-harmonic currents generated by the nonlinearity of the transformer core; this may also cause unbalanced voltages.   If neutral is grounded, the third-harmonic current flowing in the ground may create an EMI problem.

The main advantages of a delta winding are that it provides a circulating path for triple harmonics and it requires a smaller conductor for given line current than does a wye winding.   The main disadvantages of a delta winding are that it must hold off the full rms line-to-line voltage and it is more difficult to change taps in a tapped delta than in a tapped wye.

The main advantages of a wye winding are that it must sustain only $1/\sqrt{3}$ of line-to-line voltage, the neutral may have graded insulation, and it is easier to change taps in a tapped wye than in a tapped delta.   The main disadvantages are that it must carry full line current and, unless the neutral is grounded, it does not permit a path for triple harmonics, thereby creating a possible line-voltage unbalance.

Three-Phase/Two-Phase.   The main transformation configurations for converting from three-phase to two-phase or vice versa are the Scott and LeBlanc connections (see Fig. 7). In general the LeBlanc connection will result in a smaller transformer. However, when the LeBlanc connection is used to convert from two-phase to three-phase, it performs essentially as a three-phase wye-wye connection and may therefore require a delta tertiary to eliminate possible line unbalance.

Single-Phase. When either a single-phase or three-phase source of power is available, single-phase transformers are normally used only to supply relatively low power loads, usually less than 200 W. Large single-phase loads will unbalance the phases of a three-phase power system. Virtually all transformers used in home-entertainment equipment and general low-power commercial applications are single-phase.

Tap Changing. The adjustment of voltages in a transformer is accomplished by arranging windings in parallel or in series or by taps which add or eliminate portions of windings. Changing taps or windings is usually done with power off; however, tap changing under load is possible with proper equipment. The most common example of this is the continuously variable autotransformer such as the Variac.*

Autotransformer. An autotransformer may be either single-phase or polyphase, the principle of operation being identical. The connection of a single-phase autotransformer is shown in Fig. 8.

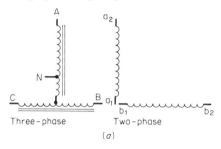

Three-phase   Two-phase

(*a*)

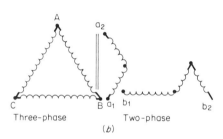

Three-phase   Two-phase

(*b*)

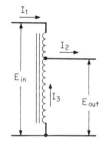

Fig. 7 Three-phase–two-phase connection. (*a*) Scott T. (*b*) LeBlanc.

Fig. 8 Single-phase autotransformer.

The transformer has a single winding tapped in such a way that a fraction of the primary voltage is across the secondary load; Fig. 8 shows a step-down transformer; a step-up transformer would have reversed connections.

The volt-ampere rating of autotransformers is dependent on the ratio of the primary and secondary voltages. In Fig. 8 the ratio of $100 \times E_{out}/E_{in}$ equals the percent tap. If p = percent tap divided by 100, then $I_2 = I/p$ and $I_3 = (1/p - 1)I_1$. Then

$$\text{VA above the tap} = (1 - p)E_{in}I_1 \qquad (13)$$

$$\text{VA below the tap} = PE_1I_3 = (1 - p)E_{in}I_1$$

When ratio p is close to 1, the VA rating is small, resulting in a small transformer. When the ratio p is small, there is not much advantage in size as compared with a two-winding transformer. In practice the autotransformer offers little physical advantage over a two-winding transformer when $100 \times E_{out}/E_{in}$ is less than 50 percent.

The main advantage of the autotransformer is possible reduction in size with reduction in regulation and leakage inductances when compared with similarly rated two-

* Trademark of General Radio Co.

winding transformers. A major disadvantage of the autotransformer is that there is no isolation between primary and secondary.

Power-Transformer Theoretical Relationship. If all transformer proportions are kept the same, the following theoretical relationships may be applied:

Dimensions vary as $VA^{1/4}$
Weight varies as $VA^{3/4}$
Cost varies as $VA^{3/4}$
Losses vary as $VA^{3/4}$

If transformer voltages or the method of cooling are changed appreciably, the foregoing factors may be considerably affected.

Power-Transformer Specification. The circuit designer must, when selecting a power transformer for a given application, consider many factors, all of which will have an effect on the rating, size, and availability of the transformer.

The block diagram in Fig. 9 shows the elements which may be included in power-transformer applications. The simplest application is that in which a transformer is directly connected to an ac load.

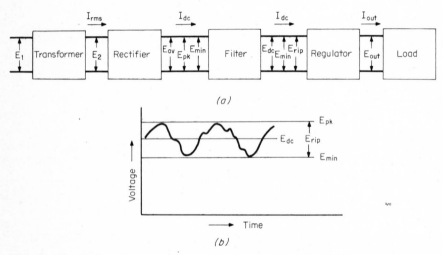

(a)

(b)

Fig. 9 Electronic power-transformer application. (*a*) Block diagram. (*b*) Output-voltage waveshape (symbolic), where transformer may be single-phase or three-phase rectifier may be single-phase or three-phase one-way or two-way; filter may be any type; regulator may be any type.

E_1 = rms input voltage; may be single-phase or three-phase
E_2 = rms output voltage; may be single-phase or three-phase
E_{av} = average rectified voltage
E_{pk} = peak of rectified voltage
E_{min} = minimum of rectified voltage (also called voltage)
E_{dc} = rectified voltage with filtering
E_{rip} = ripple voltage, specified as a percentage of E_{dc} in either rms or peak-to-peak
E_{out} = voltage into the load
I_{rms} = rms current in the output of the transformer
I_{dc} = average current in the output of the rectifier or filter
I_{out} = average load current

The requirements for power transformers may be established in two ways. First the transformer output may be specified in rms volts and amperes for a given input voltage. Second, in applications in which the load is direct current, the circuit designer may specify the dc voltage and current required at the load or input to a regulator; this permits the transformer designer to establish the transformer voltage and filter parameters.

Practical Limitations. The following are practical limitations in the design and operation of power transformers which must be considered when specifying power-transformer requirements.

1. The normal tolerance on the absolute voltage of a single-output, 'low-voltage transfomer is ±3 percent.

2. The normal tolerance on the absolute voltage of a multiple-output low-voltage transformer is ±5 percent.

3. Normal regulation of a single-output, low-voltage transformer, no load to full load, at nominal input voltage is 5 percent or less.

4. Normal regulation of a multiple-voltage, low-voltage transformer, no load to full load, at nominal input voltage is a function of the percent volt-amperes, the secondary winding being considered is of the total volt-amperes. A heavily loaded winding, >>30 percent total volt-amperes, would be about 5 percent. More lightly loaded windings would have less regulation.

5. Compensation for line-voltage drop and rectifying-diode drop must be designed into the transformer.

6. The maximum differential between the minimum of the rectified voltage at full load ($E_{\min}$) and the peak of the rectified voltage at minimum load (E_{pk}) may be 15 percent of E_{dc}. The effect of input-voltage variation must also be added to this variation to determine the maximum possible variation in output voltage.

7. Volt-ampere safety factors are not normally designed into a transformer; the current specified is the current used to establish core and wire sizes.

Transformer Requirements. In order to approach the optimum transformer, the circuit designer must specify:

Power source. Voltage with limits of deviation including surges, number of phases and phase unbalance, frequency including tolerances, line impedance, turn-on provisions, and protective devices

Load. Load type, power factor, rectifier-circuit filtering, regulation limits, duty cycle, load application and removal taps

Loads may be continuous, intermittent, or variable. Intermittent and variable loads permit duty-cycle rating so that more volt-amperes can be realized than are available for continuous operation. The power rating of transformers with duty less than continuous is

$$P = \sqrt{\frac{H_1{}^2 S_1 + \frac{1}{3}[(H_2{}^2 - H_3{}^3)/(H_2 - H_3)]S_2 + \text{etc.}}{S}} \qquad (15)$$

where P = equivalent rating in current or volt-amperes
H_1, H_2, H_3, etc. = various current, etc., levels
during
S_1, S_2, etc. = time
S = total time of one cycle

In this formula H_2 and H_3 represent the maximum and minimum of power level, which changes at essentially a constant rate during time S_2.

Environment. Ambient-temperature limits including cooling available, operating-temperature limits, protection from weather and/or contamination, and altitude limits

Other: Size and weight limitations, special mounting, shock and/or vibration, interface requirements, auxiliary equipment, corona limits, and shielding required.

Rectified Loads and Circuits. A discussion of electronic power transformers must also include a discussion of rectifier circuits and their impact on the design and operation of the transformer in the circuit. Although some applications rectify line energy, and the simple two-way circuits can operate from ordinary single-phase or three-phase transformers, the majority of rectifier applications must have the transformer designed to match the circuit for the desired dc output.

The single-phase half-wave rectifier is the simplest possible rectifier circuit, consisting of a transformer, a rectifying element, and a load. The transformer in this arrangement has a utility factor (UF) of only 0.287, meaning the transformer is large in proportion to the power output. Utility factor is defined as the ratio of the dc power output to the volt-ampere rating or size of the transformer. The ripple factor, defined as the ratio of the amplitude of the first harmonic of the ripple to the dc output voltage, of this circuit is 1.57.   In contrast, the full-wave two-way or bridge rectifier circuit which required four rectifying elements has a UF of 0.90 and a ripple factor of 0.67.   This means a much smaller transformer and less filtering.

The single-way rectifier circuit, as the name implies, causes current to flow in one direction in the secondary of the transformer.   This means that current flows only during the part of the cycle when the voltage causes a rectifier to conduct, and thereby creates a duty cycle less than unity.   It also means that the current when it is flowing must be greater than it would be with uninterrupted current flow, the increase being equal to the reciprocal of the duty cycle.   The transformer-secondary size increase is proportional to the current increase multiplied by the square root of the duty cycle:

$$P = \sqrt{I^2 \frac{t}{T}} = I \sqrt{\frac{t}{T}} \tag{16}$$

where P = transformer volt-ampere rating
$\quad I$ = current in winding during time t
$\quad t/T$ = fraction of cycle current flow = duty cycle

The transformer primary current and voltage cannot contain a dc component because it is connected to an ac line.   Therefore, a magnetizing current flows in the primary when no-load current is flowing which has an average value equal and opposite to the load current.   The primary of the transformer must be designed to handle the load current plus the magnetizing current.

ANSI Standard C57-18 recognizes more than 60 different rectifier circuits.   Table 4 contains the more frequently used of these circuits together with some of the more important constants which may influence selection of a suitable circuit.

Power-Transformer Application and Protection.   Most power transformers and inductors are operated from low-impedance sources.   These sources will continue to supply power to the transformes until the circuit is opened by either the opening of a switch or the operation of some protective device.

Usually a transformer fails because it is subjected to localized or general overheating.   The conditions which usually cause the device to overheat are component failures or other conditions which place a short circuit or overload on the transformer or inductor.   There may be instances where dielectric failure occurs within the transformer or inductor, but these usually happen in a device which is improperly designed for the application or which is improperly applied.

The most common approach to protecting transformers is to prevent the overheating of the insulation.   This can be done either by removing the source of power to the fault or by isolating the fault itself.   A fault can be any condition of operation which imposes a current, voltage, or temperature in excess of the safe limit for the parts or circuitry affected by the fault and which, if not removed or prevented, will cause the failure of a part or circuit.   A fault can be a short circuit, an overload, a transient condition, or an open circuit.

Table 5 and Fig. 10 show a variety of fault conditions which may be experienced, with their probable effects on the transformer primary current.   If a fault occurs within the load, the increase in load current is reflected by an increase in the transformer primary current.   The power-interrupting device must be capable of isolating the transformer and its load from the ac source.   A fuse or circuit breaker must respond to the current for which it is rated before the temperature within the transformer exceeds acceptable limits.   This critical overload current is that which causes the temperature of the winding to reach a point where projected life of the transformer is reduced to one-fiftieth of the design life under normal operating conditions.   The reduction in life is based on the Arrhenius relationship (10° rule).

Guides to Transformer Protection.   These guides should be considered by the circuit designer in the application of transformers:

1. Do not assume that any transformer or inductor can be operated under conditions other than those for which they were designed.   Safety factors are not normally designed into transformers used in electronic equipment.

TABLE 4   Rectifier Circuits with Transformer

Name (Single way)	Single-phase half-wave	Single-phase full-wave	Three-phase delta-wye	Three-phase delta-zigzag	Three phase delta-diametric
Number phases primary	1	1	3	3	3
Number rectifiers	1	2	3	3	6
Circuit					
E_{dc}	$0.45\,E_s$	$0.90\,E_s$	$1.17\,E_s$	$1.17\,E_s$	$1.35\,E_s$
$I_{rectifier}$ AV	I_{dc}	$0.5\,I_{dc}$	$0.33\,I_{dc}$	$0.33\,I_{dc}$	$0.167\,I_{dc}$
I_s RMS	$1.57\,I_{dc}$	$0.707\,I_{dc}$	$0.577\,I_{dc}$	$0.557\,I_{dc}$	$0.408\,I_{dc}$
Secondary U.F	0.287	0.636	0.675	0.585	0.552
I_p RMS	$1.21\,I_{dc}$	I_{dc}	$0.471\,I_{dc}$	$0.471\,I_{dc}$	$0.578\,I_{dc}$
Primary U.F.	0.373	0.90	0.827	0.95	0.78
Ripple factor	1.57	0.667	0.25	0.25	0.057
Ripple frequency	f	2f	3f	3f	6f

Name (double way)	Single-phase full-wave(bridge)	Three-phase delta-wye	Three-phase wye-delta	Three-phase delta-zigzag
Number phases primary	1	3	3	3
Number rectifiers	4	6	6	6
Circuit				
E_{dc}	$0.90\,E_s$	$2.34\,E_s$	$2.34\,E_s$	$2.34\,E_s$
$I_{rectifier}$ AV	$0.5\,I_{dc}$	$0.33\,I_{dc}$	$0.33\,I_{dc}$	$0.33\,I_{dc}$
I_s RMS	I_{dc}	$0.816\,I_{dc}$	$0.472\,I_{dc}$	$0.816\,I_{dc}$
Secondary U.F.	0.90	0.95	0.95	0.95
I_p RMS	I_{dc}	$0.816\,I_{dc}$	$1.41\,I_{dc}$	$0.816\,I_{dc}$
Primary U.F.	0.90	0.95	0.95	0.95
Ripple factor	0.667	0.057	0.057	0.057
Ripple frequency	2f	6f	6f	6f

TABLE 5   Effects of Faults on Transformers

Fault	Fig. 10	Loads	Voltage	Effect on primary fuse
Short in Z	a			Probably open
Short in Z_1	b	$P_{Z1} \lll P_{Z2}$		May not open
Short in Z_1	b	$P_{Z1} = P_{Z2}$		Probably open
Short in Z_1	b	$P_{Z1} \ggg P_{Z2}$		Will open
Short in Z_1	c	$P_{Z1} = P_{Z2}$	$V_1 = V_2$	Probably open
Short in Z_1	c	$P_{Z1} = P_{Z2}$	$V_1 \ggg V_2$	Will open
Short in Z_1	c	$P_{Z1} = P_{Z2}$	$V_1 \lll V_2$	Probably open
Short in Z_1	c	$P_{Z1} \ggg P_{Z2}$	$V_1 \ggg V_2$	Will open
Short in Z_1	c	$P_{Z1} \ggg P_{Z2}$	$V_1 \lll V_2$	Probably open
Short in Z_1	c	$P_{Z1} \lll P_{Z2}$	$V_1 \ggg V_2$	May not open
Short in Z_1	c	$P_{Z1} \lll P_{Z2}$	$V_1 \lll V_2$	May not open
Short line to ground	d			May not open
Short line to line	d			Will open
Line open	d			Probably will not open
Short primary to ground	e			Will open faulted line

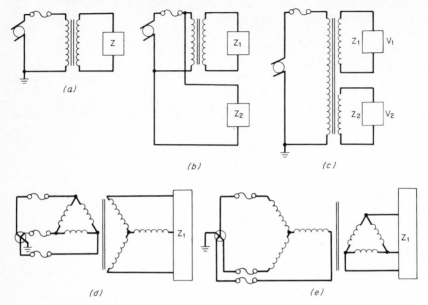

Fig. 10 Transformer faults (see Table 5 for description).

2. Do not assume that any transformer or inductor can operate in any environment other than that for which it was designed. A transformer designed to operate in air should not be subsequently embedded in a plastic resin even though the transformer losses are in the milliwatt range.

3. Do not assume that a fuse or circuit breaker in the primary of a single-phase transformer will operate for all types of faults on the transformer secondary or within the transformer. The impedances of the transformer windings and the related circuitry may limit currents to a level which may permit the transformer to overheat but not cause a fuse to open or circuit breaker to function.

4. Do not assume that fuses in all three input lines to a three-phase transformer will isolate the transformer if a fault occurs in the load or within the transformer. It is likely that a three-phase transformer will operate in a single-phase mode for some time if only one input line is opened.

5. Do not assume that a "short-circuit"-proof circuit will keep a transformer from overheating. The overload current may be below the cutoff threshold of the protective circuit.

6. Do consider all plastic materials as potentially hazardous even though they pass requisite military, ASTM, or other specification with respect to flammability, dielectric strength, and arc resistance.

7. Do perform a careful electrical-stress analysis on all insulation systems used. Remember that the distribution of stress within a system which utilizes several insulations with different dielectric constants will vary with the dielectric constants and the thickness of the materials. It is possible that while the average stress between two points of different electrical potential is well within maximum limits, the actual stress on the lower-dielectric-strength material is much greater than is safe.

8. Do perform a careful thermal analysis of all embedded assemblies, subassemblies, or parts, particularly those which contain heat-dissipative elements. Most materials which are good electrical insulators are also good thermal insulators.

Inverter Transformers

Inverter transformers   Inverter transformers may be thought of as rectifier transformers operated in reverse. Inverter operation in its simplest form can be explained by considering a single-way, single-phase, full-wave rectifier transformer as is shown in Table 4, with dc voltage applied to the center top of the secondary and the end terminals alternately switched to the opposite dc polarity. Current will then flow alternately in each half of the winding and in opposite directions. This establishes the requirements for inducing alternating voltage in the windings with the center-topped secondary now acting as the primary. The switching frequency controls the ac voltage frequency and is limited only by the characteristic of the switch used.

The use made of the alternating current thus generated is unlimited, and inverters have been made in ratings from milliwatts to thousands of kilowatts. Both single-phase and multiphase designs are used, and they may be self-excited or separately excited (switched). Frequently inverter transformers also have rectifier-circuit secondaries. Power may be received from one dc circuit, transformed as alternating current, and delivered from a secondary through rectifiers at a new dc voltage. As in other power transformers, multiple secondaries are possible so that dc power may be distributed at several dc voltages.

Inverters often operate at a frequency much higher than is ordinarily considered power frequency. When the rectification takes place close to the transformer so as to limit transmission losses, modern practice uses from 5 to 100 kHz. The advantage thus gained is reduction in transformer and filter size. Limits on this reduction are switching, recovery time, core-material losses, and conductor fabrication since high frequency may introduce skin effect.

One popular arrangement of inverter circuits is to control the switching by feedback from the secondary, thus regulating the output dc voltage. In this arrangement, the alternating current may approach a square wave or may be alternating-polarity pulses of various widths. This produces a duty cycle in the winding and excitation conditions which the transformer designer must take into consideration.

Inductors

Inductors are circuit elements of lumped inductance which are normally designed so that their impedance at a specified frequency or over a specified frequency range is predominantly inductively reactive. They are considered with transformers because the same theory is used in design and the same general construction methods and materials are used for both. Some electronic transformers have inductance requirements and some inductors actually perform voltage, current, or impedance transformations.

Inductor applications   Inductors have many uses in electronic circuits. The most common are in wave filters, for current limiting and in tuned circuits. They are also used for energy storage, for differentiating and integrating, and for waveshaping. One of the more common uses at power frequencies is in the LC filter which reduces the ripple on the dc output of rectifier circuits.

Current-limiting inductors may be either air-core or magnetic-core devices depending upon the linearity requirements. In either case the winding must be adequate to carry the current, it must be insulated for the voltage induced across the winding, and it must be capable of withstanding mechanical forces created by high currents.

Inductor requirements   Table 6 lists the electrical, environmental, and other requirements which should be specified for various inductor applications.

Inductor Q.   The equivalent circuit of an inductor can be derived from that of the transformer (Fig. 1). The parameters which apply

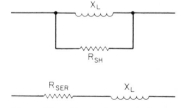

Fig. 11   Equivalent inductor core-loss resistance.

to the inductor are R_1, L_N, and R_N. In this circuit R_N depicts a resistance equivalent to core losses and is in shunt with the winding. This equivalent resistance can also be shown in series with the winding as shown in Fig. 11 and is expressed

$$R_{\text{ser}} + jX_L = \frac{jX_L R_{SH}}{R_{SH} + jX_L} \tag{17}$$

TABLE 6   Inductor Requirements

Type	Specify
1. Linear inductor for ripple reduction	Minimum inductance Type of rectifier circuit Fundamental frequency Rectified voltage Direct current Maximum working voltage Transient voltages Maximum dc resistance Thermal and mechanical environment
2. Swinging inductor for ripple reduction	Same as type 1 except maximum and minimum inductance must be specified together with range of direct currents which correspond to inductances
3. Charging inductor	Inductance with tolerance Type of charging circuit DC charging voltage Average current Peak current Pulse-repetition rate DC resistance Inductor Q Thermal and mechanical environment
4. Current-limiting inductor	Inductance and tolerances Maximum alternating current Nominal alternating current Maximum voltage drop Nominal voltage drop Frequency Thermal and mechanical environment

The Q of a coil is the ratio of coil reactance to ac resistance for large values of Q_1 greater than 5:

$$R_{SH} \approx \frac{X_L}{R_{\text{ser}}} \tag{18}$$

Since winding resistances are not negligible, the expression for Q becomes, with R_1 equal to winding resistance,

$$Q = \frac{X_L}{R_1 + R_{\text{ser}}} \tag{19}$$

If the value of R_{ser}, obtained from Eq. (18), is substituted, the expression becomes approximately

$$Q = \frac{1}{R_1/X_L + X_L/R_{SH}} \tag{20}$$

This equation relates Q to winding resistance, open-circuit inductance, core losses, and operating frequency.

Saturable inductors Saturable inductors are iron-core inductors in which the impedance is varied over a wide range by varying the flux density in the magnetic core with a bias current in an auxiliary winding.   The bias current is usually direct current and the bias winding contains many turns of small wire, making control with a small direct current possible.

Most iron-core transformers are designed to operate at a flux density in the magnetic circuit which does not reach saturation.   Under this condition, the open-circuit impedance is virtually inductive reactance with a very small resistance component.

If operating conditions are changed so that the flux density is increased and the core saturates for part of each cycle, the inductive reactance decreases. If the core is completely saturated, the inductance becomes virtually the same as if no iron core were present, and the impedance becomes the resistance of the winding.

A saturable inductor is similar to a transformer with an ac winding adequate to carry the full-line current and connected in series with the line and a dc winding with which the impedance of the ac winding can be controlled over a wide range by means of dc-flux saturation of the magnetic core. An application of saturable inductors familiar to all is theater-light dimmers.

Magnetic Amplifiers. Magnetic amplifiers are saturable reactors arranged to control large amounts of power with small power inputs. Power gains in excess of 10^6 can be achieved. Feedback is often used to obtain automatic control. Magnetic amplifiers have wide application in industrial control especially when large power handling is involved. Their use in electronic circuits is gradually giving way to solid-state amplifiers, which are smaller and lighter.

Wideband Transformers

Wideband transformers Included in this classification of transformers are the types listed in Table 7. The primary difference between electronic power trans-

TABLE 7   Wideband Transformer Types

1. Audio-frequency transformers which operate from vacuum-tube voltage source. Full-power frequency response normally ranges from 20 Hz to 20 kHz
2. Audio-frequency transformers which operate from solid-state current source. Full-power frequency response normally ranges from 20 Hz to 20 kHz
3. Modulation transformers for amplitude-modulated transmitters. This type of transformer is usually designed for specific application and not for general use
4. Driver transformers used to supply power to the grids of class AB and class B amplifier tubes
5. Line-matching transformers which receive power from one impedance and deliver it to another impedance
6. Load-matching autotransformers which provide impedance matching with taps
7. Control-system transformers used in open or closed system feedback. These transformers operate on a carrier frequency modulated by an error signal. Either amplitude or phase modulation may be used
8. Transducer service transformer which are usually step-down ratio and in which the load impedance may vary with frequency
9. Random-noise transformers used in vibration-machine applications. These operate on random frequencies of varying amplitude and duration
10. Ultrasonic transformers used in various communication and industrial applications where the lowest frequency is higher than 10 kHz
11. Carrier transformers which operate at a specific frequency and transmit intelligence by amplitude, frequency, or phase modulation
12. Video transformers which are exceptionally wideband and may operate over part or all of the range from 10 Hz to 10 MHz
13. Baluns are used to provide balanced to unbalanced impedance transformations, usually in the ratios 1:1 and 4:1. Typically they are used to drive balanced antennas and other balanced loads

formers and wideband transformers is that the latter operate over a range of frequencies where leakage inductance and winding capacitances affect performance while these parameters usually have little effect in properly designed electronic power transformers. Pulse transformers are a special case of wideband transformers which require the passage of a broad spectrum of harmonics. Additional considerations are necessary in pulse transformers to produce the correct pulse shape; these must be provided in the design.

Wideband transformers operate over a frequency range of less than 1 Hz to greater than 100 MHz. The ratio of the highest frequency to the lowest frequency in the more common audio-frequency transformers is about 2,000:1. Most rf transformers have relatively narrow bandwidth.

In a wideband transformer for a given source impedance and core material the product of the bandwidth and the turns ratio is an approximation of size.

By adapting the transformer equivalent circuit in Fig. 1c to wideband transformers, the two circuits in Fig. 12 for low- and high-frequency response are developed.

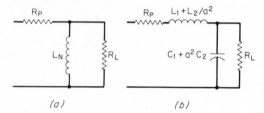

(a)   *(b)*

Fig. 12 Wideband transformer equivalent circuits. (*a*) Low-frequency equivalent circuit. (*b*) High-frequency equivalent circuit.

$$R_P = \text{source impedance plus } R_1$$
$$R_L = \text{load impedance plus } R_2$$

To obtain proper operation of wideband transformers, factors external to the transformer which affect performance must be considered; these include (1) frequency, (2) source and load impedances, and (3) the linearity of these impedances.

In addition to these external factors it can be determined from Table 1 and from an examination of the equivalent circuits in Fig. 12 which transformer parameters also affect the response.

Impedance Matching. Since wideband transformer operation usually requires the matching of the source and load impedance, the ratio of the input and output voltages is usually expressed in decibels (dB) according to the expression

$$\text{dB} = 20 \log_{10} \frac{E_1}{E_2} \tag{21}$$

However, to be meaningful, voltage and power must be related to a reference level. The standard reference level is 1 mW and is expressed as zero dBm. The voltage for zero dBm across 50 Ω is $\sqrt{50 \times 0.001} = 0.224$ V.

Power level. The power level of a wideband transformer is expressed in dB as

$$\text{dB} = 20 \log_{10} \sqrt{\frac{P_1}{P_2}} \tag{22}$$

A transformer whose maximum power level is expressed as 200 dBm can deliver 100 mW.

Pulse Transformers

Pulse transformers Pulse transformers generally perform a coupling function or are used as a part of a circuit which generates pulses.

Pulse-coupling Transformers. In these applications the transformer must pass a square wave or pulse which approaches a square wave. The square wave differs from the sine wave in that the rise and fall of the pulse is very steep. Applications involving pulses which are not square are not discussed because a transformer which can pass a square wave can also pass pulses with a sloping rise or fall and with nonflat tops. The standard pulse waveform is shown in Fig. 13.

From Fig. 13 it can be seen that the pulse consists of a front edge, a top, and a trailing edge. By adapting the transformer equivalent circuit (Fig. 1c) to pulse transformers, the three equivalent circuits in Fig. 14 for front edge, top, and trailing edge are developed.

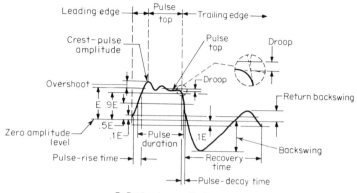

E = Peak pulse amplitude

Fig. 13 Pulse waveform.

As in wideband transformers the pulse transformer is affected by factors external to the transformer. These include (1) the source and load impedance, (2) the linearity of these impedances, (3) the pulse width, (4) the repetition rate, and (5) the magnitude of the pulse voltage.

In addition to these external factors, it can be determined from Table 1 and from an examination of the equivalent circuits in Fig. 14 which transformer characteristics affect the pulse response.

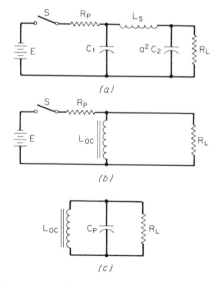

Fig. 14 Pulse transformer equivalent circuits. (*a*) Front edge of pulse. (*b*) Top of pulse. (*c*) Trailing edge of pulse.

E = voltage
S = switch
$R_P = R_1$ + source impedance
$R_L = R_2$ + load impedance
L_{oc} = primary open-circuit inductance
L_S = leakage inductance referred to primary
C_1 = primary capacitance
C_2 = secondary capacitance
C_p = primary equivalent capacitance

Pulse Formation. Transformers and inductors are often used in circuits which generate pulses. These include blocking oscillators, pulse modulators, and sweep generators. The transformers and inductors in these circuits perform pulse-forming, pulse-coupling, charging, and switching functions.

Air-Core Transformers

Air-core transformers In radio-frequency applications, transformers without iron cores are widely used to couple circuits. In transformers with magnetic cores, the exciting current required for inducing the secondary voltage is a small percentage of the total primary current. In air-core transformers, all the current is exciting current and induces a secondary voltage proportional to the mutual inductance.

A maximum power transfer between primary and secondary occurs when the reactance due to the mutual inductance equals the geometric mean of the resistances of the primary and secondary circuits. The primary and secondary may be tuned to appear pure resistance.

$$X_m = \sqrt{R_p R_s} \tag{23}$$

The coupling coefficient k is the ratio of the mutual inductance to the geometric mean of the self-inductance of the primary and secondary coils.

$$k = \frac{L_m}{\sqrt{L_p L_s}} \tag{24}$$

The value of k is never greater than unity and may be as small as 0.01 or lower at high frequencies.

PASSIVE NETWORKS

Passive Network

The frequency-dependent characteristics of inductive and capacitive reactances are often used in electronic circuits to attenuate, delay, and form selected frequencies or pulses. These include pulse-forming networks, electromagnetic delay lines, and electric-wave filters.

Pulse-forming networks In a pulse-generating circuit the pulse-forming network (PFN) performs two functions, storing energy and discharging the energy into a load in a specified pulse shape. The network usually consists of inductances and capacitances and may be designed to store the energy in either.

Those which store energy in the capacitances are called voltage-fed networks; those which store the energy in the inductances are called current-fed networks. Voltage-fed PFN's are almost exclusively used because the gaseous-discharge switches normally used in radar-pulse generators cannot be used with the current-fed networks.

Electromagnetic delay lines The delay line delays the transmission of electric energy while maintaining the fidelity of the transmitted waveshape. Electromagnetic delay lines include the lumped-constant and distributed-constant types. The lumped-constant line consists of a network of series inductances and parallel capacitances. The time delay is determined by the time required to charge the capacitances through the inductances. The distributed-constant line functions in principle like the lumped-constant line. The manner in which this line is manufactured permits a distribution of inductance and capacitance along the line.

Delay times of 5 ns to 5 ms can readily be achieved.

Electric-wave filters The electric-wave filter is normally a network consisting of inductances and capacitances which attenuates certain frequencies and passes others. In general they are of three types, low-pass, high-pass, and bandpass, as shown in Fig. 15. The characteristics normally specified are the passband, the cutoff frequency or frequencies, the insertion loss, and the ripple in the passband. Insertion loss is the ratio of the power delivered to the load without the filter to the power delivered to the load with the filter in the circuit at a reference frequency and with

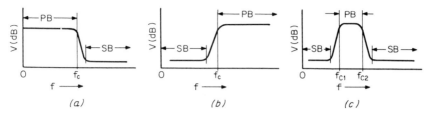

Fig. 15   Electric-wave filters.   (a) Low-pass.   (b) High-pass.   (c) Bandpass.

$$f = \text{frequency}$$
$$f_c = \text{cutoff frequency}$$
$$PB = \text{passband}$$
$$SB = \text{stop band}$$
$$V = \text{voltage in dB}$$

a constant source voltage.   It is calculated with the following formula:

$$IL(\text{dB}) = 20 \log \frac{E_1}{E_2} \tag{25}$$

where E_1 = load voltage without the filter
E_2 = load voltage with the filter

GENERAL CONSIDERATIONS

High Voltages

High voltage is a relative term and often is considered to be any voltage which is a safety hazard.   From the standpoint of dielectric materials it can be considered any voltage at which ionization can occur, about 250 V in 1 atm of air.   In an insulation system high voltages are considered from the standpoint of corona, point-to-point flashover, surface creepage, and dielectric failure.

Corona, partial discharge   The effect of corona or partial discharge in transformers may be twofold; it may be destructive to the insulation system and it may create undesirable electromagnetic interference.   Partial discharge is not a phenomenon indigenous to transformers.   Any device, circuit element, wire, cable, etc., may be a source and may be damaged by its existence.

The destructive effects of partial discharge under ac stress are accelerated as the frequency of the voltage is increased.   While under dc stress the occurrence of discharge is intermittent, the resistivity of the insulation which dissipates the charge developed by the discharge being the limiting factor.

Flashover and creepage   Flashover is arcing in a gas between points of differing voltage.   Creepage is flashover across the surface of insulation between points of differing voltage.   Figure 16 provides design limits for flashover and creepage in air.   Table 8 provides a comparison of flashover and creepage in air with those in sulfur hexafluoride, SF_6.

Dielectric strength   The dielectric strength of typical insulating materials is given in Table 9.   This table also provides recommended maximum stresses for these insulations as they apply to transformers.

Solids typically have much higher dielectric strength than liquids or gases.   The utilization of their inherent strength is often difficult owing to conditions existing adjacent to the solid insulation where gas may form corona.   The inherent dielectric strength of solids can be realized only by achieving a uniform field and avoiding series gas gaps.

The dielectric strength of liquids and solids also decreases with increasing thickness.   The presence of gas bubbles in liquids reduces the dielectirc strength drastically, and liquids saturated with gas will break down at lower stresses when subjected to lower pressures.

High-voltage insulation systems   Some considerations in selecting an insulation system for a high-voltage transformer and in applying a high-voltage transformer to an electronic circuit are:

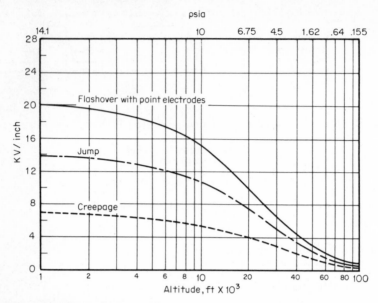

Fig. 16 Crest-voltage design guide. Jump clearances and creepage over clean, dry surfaces; maximum temperature, 125°C. Data from literature sources.

TABLE 8   Flashover and Creepage in Air and SF$_6$

1. Creepage is a function of the dielectric constant of the insulation; the higher the dielectric constant the lower the voltage to flashover. Creepage has the least effect (as compared with flashover) when the dielectric constant is low and when the surface is parallel to the electric field
2. Small series of air gaps (cracks or surface irregularities) reduce the ac flashover in air but have little effect in SF$_6$ below 2 atm
3. The ratio of impulse (unidirectional pulse) creep over glass in SF$_6$ to air is approximately 2.00
4. The ratio of ac creep over glass in SF$_6$ to air at 1 atm ≈ 2.25; at 2 atm ≈ 2.75
5. Impulse ratio (unidirectional pulses to ac crest) in air and in SF$_6$ (approximately—varies with spacing):

	Air		SF$_6$	
	1 atm	2 atm	1 atm	2 atm
Flashover	1.19–1.45	1.20	1.2–1.4	1.1
Creep (glass)	1.21–1.65	1.16–1.57	1.0–1.28	0.99

1. The shape of elements which are at high potential with respect to ground and other elements on the breakdown voltage of the insulation between these points. The Paschen's curve in Fig. 17 is for uniform field. A utilization factor of 0.33 is sometimes used for point electrodes, but as indicated in Fig. 18 at small spacings the utilization factor approaches unity. The effect of electrode shape is most pronounced in gases, although it is obvious that stress concentrations will occur at point or sharp-edge electrodes in liquids and solids. In this connection it is well to remember that a small wire is a sharp edge.
2. When more than one insulation material is used in series between circuit elements of ac potential difference, the stress in each is inversely proportional to the dielectric constant ϵ of each material. Thus the insulation with the lowest ϵ

TABLE 9   Recommended Voltage Stress and Temperature Limits for Various Insulations

Type	Material	Dielectric constant, 100 Hz, ϵ	Dielectric strength V(rms)/0.001	Max stress V(rms)/0.001	Hot-spot max, °C
Gases	Air	1.0	See Fig. 17	See Fig. 18	
	Sulfur hexafluoride (SF$_6$)	1.0	See Fig. 17	See Fig. 18	180
Liquids	FC75 or FC77*	1.9	350	150	105
	Coolanol 20⁺	2.5	350	150	105
	Askarel	5.5	350	150	105
Solids	Epoxy resin	4.2	>500	200	130
	Polyester resin	4.0	550	150	130
	RTV	3.0–3.5	>300	150	200
	Laminate, epoxy glass	5.2	500	150	130
	Diallyl phthalate (DAP)	4.2	>300	150	130
Films	Mylar† (0.001-in. thick)	3.3	>3,000	200	130
	Kapton† (0.001-in. thick)	3.5	>5,000	250	200
	Teflon (0.002-in. thick)	4.2	>900	200	180
Sheet	Kraft paper, unimpregnated	2.0	>250	75	105
	Kraft paper, resin-impregnated	3.5	>500	200	130
	Nomex,† unimpregnated	2.7	150	75	180
	Nomex,† liquid-impregnated	2.8	>850	200	180
	Nomex,† resin-impregnated	2.9	>350	150	155
	Pressboard	2.0	>200	100	130
Other	Mica	5.8	3,000	250	200
	Wire enamel	3.2	>800	200	105

* 3M Co. trademarks for fluorocarbon liquids.
⁺ Monsanto Co. trademark for silicate-ester-base fluid.
† Trademark of E.I. du Pont de Nemours & Co., Inc.

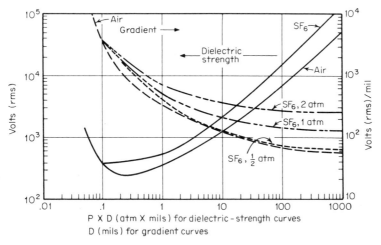

Fig. 17   Paschen's-law curve for air and SF₆; SF₆ at 25°C.

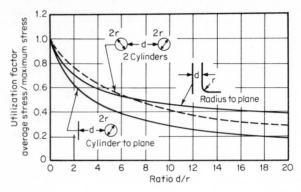

Fig. 18 Utilization factor as a function of spacing-to-radius ratio.

has the highest stress. Often this is air or gas, which also usually has the lowest dielectric strength.

When the voltage is dc, the voltage drop is simply IR. Therefore, the material with the highest resistance will have the highest stress. One should not forget that the total resistance may be composed of both volume and surface resistances.

When the stress is unidirectional pulses, there is an ac-voltage component which will be the source of ac stresses and may be the source of corona. Even a large ripple on dc voltages may produce ac stresses sufficient to require investigation and may be a corona source.

With unidirectional pulses above the corona-inception level, charges are deposited on the surface. These will increase the next pulse-breakdown level. With ac voltage the effect is reversed; i.e., the corona-inception voltage will be lowered.

3. Great care must be taken to ensure that liquid and gaseous dielectrics are kept free of contaminants.

4. In processing rigid or flexible plastic dielectric materials, great care must be taken to ensure that no voids, cracks, or physical interfaces exist within the plastic casting.

5. When possible, stresses within a transformer, between layers and between turns, should be kept below the Paschen's minimum for air.

Voltage stress on insulations in series The ac voltage stress on each insulation element in series may be calculated by

$$G_x = \frac{V}{\epsilon_x(T_1/\epsilon_1) + T_2/\epsilon_2 + , \cdot \cdot \cdot + T_n/\epsilon_n} \tag{26}$$

where V = total volts across all insulation
$\quad\quad\quad\quad G$ = stress, V/mil
$\quad\quad\quad\quad T$ = thickness, mils
$\quad\quad\quad\quad \epsilon$ = dielectric constant
$\quad\quad$ Subscript x = 1, 2, . . . , n

Shielding

Shielding in transformers can be of two types, electrostatic and electromagnetic.

Electrostatic shielding Electrostatic shielding prevents or reduces the transfer of voltage through interwinding capacitances. It is usually required to isolate the power-input circuit to the transformer from transient voltages or high-frequency noise which may appear on the transformer secondary winding. An electrostatic shield is usually placed between the primary and secondary windings, with the shield being suitably grounded.

Electromagnetic shielding Electromagnetic shielding is required to attenuate the magnetic field formed as a part of a power-transformer magnetic field. These fields may induce small voltages in amplifier input transformers which are then in turn amplified by high-gain circuitry. Since the main purpose of shielding is to reduce the effect of stray magnetic fields, several techniques may be employed.

1. Separate the input to the high-gain circuit from the power transformer as much as possible. This is most effective, since the field varies as the reciprocal of the distance cubed.

2. Use the core-type construction in the power transformer. This is most effective in single-phase transformers because the leakage fluxes generated in the two legs effectively cancel at a distance from the transformer. The three-phase transformer wound on an E-type core generates a third-harmonic field because they cannot circulate in the core.

3. Orient the power transformer and/or the input circuitry for minimum pickup. The axis of an input-transformer coil should be 90° to the stray field.

4. Provide magnetic shielding around the circuitry being affected by the magnetic field.

Placing the shielding around the power transformer is generally less effective, because a large amount of the lines of flux which originate at the transformer would be essentially perpendicular to the shield and would pass through it.

Shielding usually consists of ferromagnetic material or layers of thin high-permeability material often interleaved with sheets of nonmagnetic material like copper.

Cooling

All transformers have power losses which raise the temperature of the various parts until reaching the point of thermal equilibrium (where the losses generated equal the power dissipated).

The normal transformer is manufactured from many different materials, most of which have different coefficients of thermal conductivity. Generaly the materials used with the best thermal conductivity also generate the bulk of the losses, while the electrical insulators are usually thermal insulators. Transformer cooling is usually accomplished by providing a combination of solid insulation (varnishes or solventless-resin systems) with a mechanical structure which will permit the most efficient conduction of the winding and core losses to the surface of the transformer if it is to be cooled by convection or to a mounting surface if it is to be cooled by conduction.

There are approximations which permit reasonably accurate determination of the temperature rise of transformers cooled by convection. Figure 19, which is based on these approximations, relates temperature rise to transformer weight and loss per unit weight. Table 9 lists the hottest-spot temperature limitation for various insulation materials.

The determination of temperature rise in transformers cooled by conduction only is more complex and should be determined by more rigorous thermal analysis.

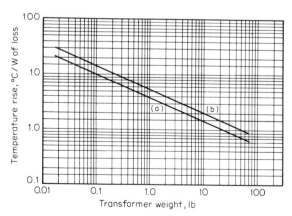

Fig. 19 Temperature rise of transformers with convection cooling. (a) 70°C rise at 95°C ambient. (b) 50°C rise at 20°C ambient.

Color Code for Electronic Transformer

The color code for electronic-transformer winding identification is as shown in Table 10.

TABLE 10   Color Code for Transformer

Color	Power	Audio	if
Black	Primary	Grid return	Grid return
Brown	Filament No. 2	Plate lead when primary is center-tapped	
Brown/yellow stripe	Center tap, filament No. 2		
Red	HV plate	B^+	B^+
Red/yellow stripe	Center tap, HV plate		
Yellow	Rectifier filament winding	Grid lead when center-tapped	
Yellow/blue stripe	Center tap, rectifier filament winding		
Green	Filament No. 1	Grid lead to secondary	Grid lead
Green/yellow stripe	Center tap, filament No. 1		
Gray	Filament No. 3	Plate lead of primary finish	Plate lead
Gray/yellow stripe	Center tap, filament No. 3		

Chapter **10**

Relays and Switches

VIRGIL E. JAMES
Professional Engineer; Member, IEEE, WSE, and ISA

INTRODUCTION

An electrical *switch* is simply defined as a device for making, breaking, or changing the connections in an electric circuit. A *relay* is an electrically operated switch. This definition is nonrestrictive enough to embrace both solid-state (semiconductor) relays and electromagnetic or electromechanical and hybrid types.

RELAYS

Definitions of a Relay

The National Association of Relay Manufacturers (NARM) defines a relay as an electrically controlled device that opens or closes electrical contacts to effect the operation of other devices in the same or another electric circuit. The reference to contacts obviously restricts this definition to only the most common form, the electromechanical relay. Webster's definition, "an electromagnetic device . . . actuated by a variation in conditions of an electric circuit and that operates in turn other devices (as switches, circuit breakers) in the same or a different circuit," can be made to suit today's needs just by replacing the word electromagnetic with electric.

A solid-state relay (SSR) is a device without any moving parts which performs a relaying or electrical-switching function. It employs semiconductors in both input and output electric circuits to perform essentially the switching function normal to the simple electromagnetic relay. A hybrid relay is an electrical device or unit having a solid-state input and electromagnetic output, or vice versa, to perform an electrical-switching function.

Solid-state relays and hybrid relays will be treated here as special forms.

The electromagnetic/electromechanical relay is still the form most often encountered, the simplest and most readily understood, and it demonstrates best the basic switching principles and problems. Unless otherwise noted, it is the type of relay discussed here. Thermal relays, solid-state relays, hybrid relays, and such will be treated separately later on and clearly identified.

Figures 1 through 4 illustrate some of the most common relay types.

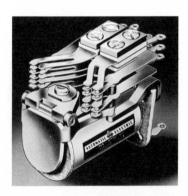

Fig. 1 A common type of relay having a clapper armature and heavy-duty contacts. (*Magnecraft Electric Company.*)

Fig. 2 A long-frame telephone-type relay with medium-duty contacts. *GTE Automatic Electric Company, Inc.*)

Fig. 3 A dry-reed contact type of relay. (*C. P. Clare & Co.*)

Fig. 4 Examples of a solid-state relay. (*Hamlin, Inc.*)

Relays must be considered by the engineer as a basic component and not an accessory. The kind, overall shape, method of mounting, and aesthetics are somewhat a matter of choice, but usually the coil characteristics, kind and amount of power required, overall size, nature of the contacts required to handle the known load, etc., are largely dictated by the limitations of the job to be done. For example, it is unrealistic to expect a microminiature relay to handle a heavy-duty load requiring 25-A contacts.

Relay terms As with many other electrical devices, custom as much as usage has resulted in the terms commonly applied to relays. Many are confusing and some are contradictory as commonly used. To some extent relay manufacturers have created and used terms peculiar to their own product, sometimes at variance or in conflict with those used by another manufacturer. The terminology common to one field of application may bear little resemblance to that of another field. In an attempt to improve this situation, The National Association of Relay Manufacturers (NARM) and American National Standards Institute (ANSI) agreed upon some standardized definitions. They are given in the Glossary. Use of these terms will be found helpful in discussion of relays and relay problems with vendors and with other engineers.

Relay classifications Before a meaningful discussion of relays can be carried on with the vendor, broad classification is necessary. Good system design using relays necessitates a consideration of relay characteristics from the output-requirements end forward. That is to say, to ensure that the relay is large enough, sensitive enough, rugged enough, etc., the nature and requirements of the job the relay has to do must be first established and a choice made, working from the output or load-handling end forward to the input requirements. All too often the tendency is to practically finish the design, with relays hung on almost as an afterthought. This can nullify an otherwise good design and lead to unjust criticism of relays in service.

Relays classified by output

1. Low-wattage dc load
2. Medium-wattage dc load
3. High-wattage ("power") dc load
4. Low-wattage ac load
5. Medium-wattage ac load
6. High-wattage ("power") ac load
7. Specialized loads (e.g., coaxial switching of high-frequency power)
8. Specialized contacts (e.g., sealed contacts or solid-state switching)

Relays classified by input

1. Direct current: neutral, polarized, or thermal
2. Alternating current: commercial power or other low-frequency sources, frequency-sensitive (e.g., tuned resonant reed), or thermal

Relays classified by duty rating

Contact Performance. Because much of a relay's size, shape, method of terminating, method of mounting, coil power requirements, protection from ambients, etc., depends upon the nature of the contact load, it is important that this matter receive early attention. Large electrical-contact loads usually require a larger relay of a particular type than small loads. This can be a factor in method of mounting as well as terminating. It is therefore essential that the vendor be made aware of all pertinent data regarding contact service requirements initially.

Service Life. Another early factor classifying the relay is service life. If the need is for longevity with a lot of operations, a certain relay type may be precluded from favorable consideration in favor of a type capable of providing the required life. On the other hand, if other factors are ruling, such as the need for small size or resistance to shock and vibration as in airborne service, for example, then this needs to be recognized early so that a proper choice will be made.

Relays classified by custom and usage Relays are sometimes most importantly classified by one or more of the following:

Commercial
Industrial

Military
Communications
Railway

Customer or user acceptance, training and experience of service personnel, field operating extreme conditions, or safety may dictate that relays of a certain type, as specifically associated with one of the above classifications, be chosen.   An industrial application, for example, may dictate the use of a relay type meeting National Electrical Manufacturers Association (NEMA) standards, while such a relay might be totally unsuited to a military or railway application.

Relays classified by performance Relays are classified by what they can do and are required to do.   Three broad classifications are:

1. General-purpose
2. Special-purpose
3. Definite-purpose

Specific performance classifications are these:

1. Marginal (with respect to pickup and/or dropout current)
2. Timing (either or both allowable operate and/or release times, operate and/or release delays)
3. Sensitivity (capability to recognize narrow pickup and/or dropout currents, or ability to operate on much less than normal current)
4. Latching
5. Sequencing
6. Frequency-sensitive
7. Thermal-response
8. Stepping

Relay Performance

Electromechanisms do not switch instantaneously.   Relay performance presents a series of sequential events on both energization and deenergization.   These events are identified as to sequence and defined in Figs. 5 and 6.   A look at these figures, when associated with a study of the most common terms applied to the relay from Table 1, provides an understanding of basic relay performance.   The relay terms of Table 1 are defined in detail in the Glossary.   Figure 7 is helpful in placing the events of Figs. 5 and 6 in their proper time frame.

TABLE 1   Most Commonly Used Terms Relating to Relay Performance[1]

Preferred	Not preferred
Hold, measured	Nondropout, measured
	Nonrelease, measured
Hold, specified	Maximum dropout
	Nondropout, specified
	Nonrelease, specified
Nonpickup, measured	Nonoperate, measured
Nonpickup, specified	Minimum pickup
Pickup, measured	Operate, measured
	Pull-in (or pull-on) value, measured
	Operate value, just
Pickup, specified	Operate, specified
	Pull-in (or pull-on) value, specified
	Operate value, must
	Maximum pickup
Operate time	Pickup (or pull-in) time
Dropout, measured	Release, measured
Dropout, specified	Release, specified
	Minimum dropout
Release time	Dropout (or drop away) time
Transfer time	

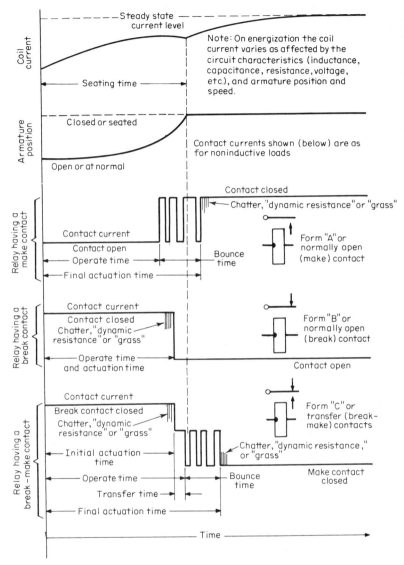

Fig. 5 Time traces typical of relay pickup.[1]

Relay Construction

All electromechanical relays consist of at least three basic elements: the actuating coil, linkage to transform coil energization into output, and a change of output conditions due to coil energization (switching).

It is customary to picture relay coil and contact details by symbols. Basic contact symbols are presented in Figs. 8 and 9 and coil symbols in Fig. 10.

In Fig. 8, the heavy arrow indicates the direction of operation. The armature contact spring (indicated by the long spring in each example) moves downward. In forms D and I some electrical discontinuity may be caused by contact chatter. The symbols are taken from ANSI C83.16-1959, Y32.2-1962, and Y32.2a-1964.

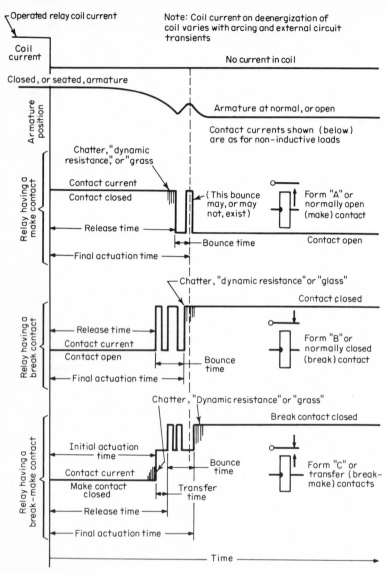

Fig. 6   Time traces typical of relay dropout.[1]

When abbreviations in Fig. 8 are used to designate a contact assembly, the following order is used: (1) poles, (2) throws, (3) normal position, and (4) double make or break (if applicable).   Example: SPST NO DM refers to single-pole single-throw, normally open, double-make contacts.

It is significant in the symbolic presentation of contact combinations in Fig. 8 that although Form A comes before Form B alphabetically, in a normal relay contact assembly the closed contacts are closer to the armature than the open contacts (see Fig. 11b). This prevents any armature spring tension from going to waste by keeping the back contacts closed with as much pressure as possible.   Thus an order calling for a relay having 1A, 2B, 1C contact combinations will usually be arranged in the order of 2B, 1C, 1A, unless, otherwise specified by the purchaser, and for a

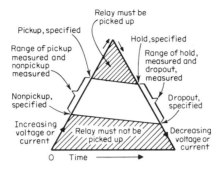

Fig. 7   Relationship of relay performance to definitions.[1]

good reason. If an "early make" is required, it must be specified (as in Fig. 11c, where the X associated with a make combination indicates that the circuit requires one A combination to be preliminary). In Fig. 11c the contacts are drawn in vertical alignment with the coil symbol. Movable, armature, or lever contacts are drawn as if attracted to the coil on energization. The contacts are numbered in sequence from the mounting surface outward; No. 1 is closest to the mounting and No. 10 the farthest away. X contacts are preliminary and operate first. Y contacts are break contacts and operate last. As make contacts normally operate last, they are not so identified. Battery and ground symbols are shown connected to the coil (negative battery to inside terminal, positive battery grounded) to minimize electrolysis in permanent installations. Switching ground is telephone practice at 48 V dc, nominal. This is not permitted above 50 V dc by National Electrical Code.

It is notable that the 2B, 1C, 1A contact-assembly designation referred to above appears to be simpler and less likely to be misunderstood than the equivalent "one DPSTNC, one SPDT, and one SPSTNO."

The association of coil and contact symbols in a simple relay as frequently seen is covered in Fig. 11a. There is ordinarily no picturing of the linkage between actuating coil and contacts, but the assumption is made that the movable contact is attracted toward the coil on energization, returning to the pictured position on deenergization unless otherwise stated.

Joint Industry Conference and National Machine Tool Builders Association have adopted relay coil and contact symbols as seen in Fig. 11d.

Symbols for thermal relays, which will be discussed in full in a section devoted solely to them, are as pictured in Fig. 11e.

Symbols used in motor-control circuits (JIC/NMTBA) are as pictured in Fig. 11f. An explanation of the method of operation for this slightly more complicated diagram follows:

1. Two-wire control is generally thought of in relation to a pilot device such as a thermostat or pressure switch, or to a simple maintained SPDT toggle or push-button switch. As the term implies, these devices require the use of only two wires between the control unit and the starter. The device is connected in series with the main contactor coil of the starter, and the opening or closing of the pilot device directly controls the deenergizing or energizing of the starter. The major feature of a two-wire control system is low-voltage release. The starter drops in the event of a power failure but operates automatically when the power is restored.

2. In three-wire control systems the main contactor coil of the starter is wired in series with its own NO auxiliary contacts. The start-stop push-button station, which requires the use of three wires between the control and the starter, is connected in parallel with the coil. In the event of a power failure, the starter will drop out and remain deenergized until the start button is depressed. Since the starter drops out when there is a power failure and will not operate again until the start button is depressed, this control system provides low-voltage protection.

Two of the most common forms of the relay are demonstrated by Figs. 12 and 13. The linkage that converts coil energization to contact switching is obvious. The

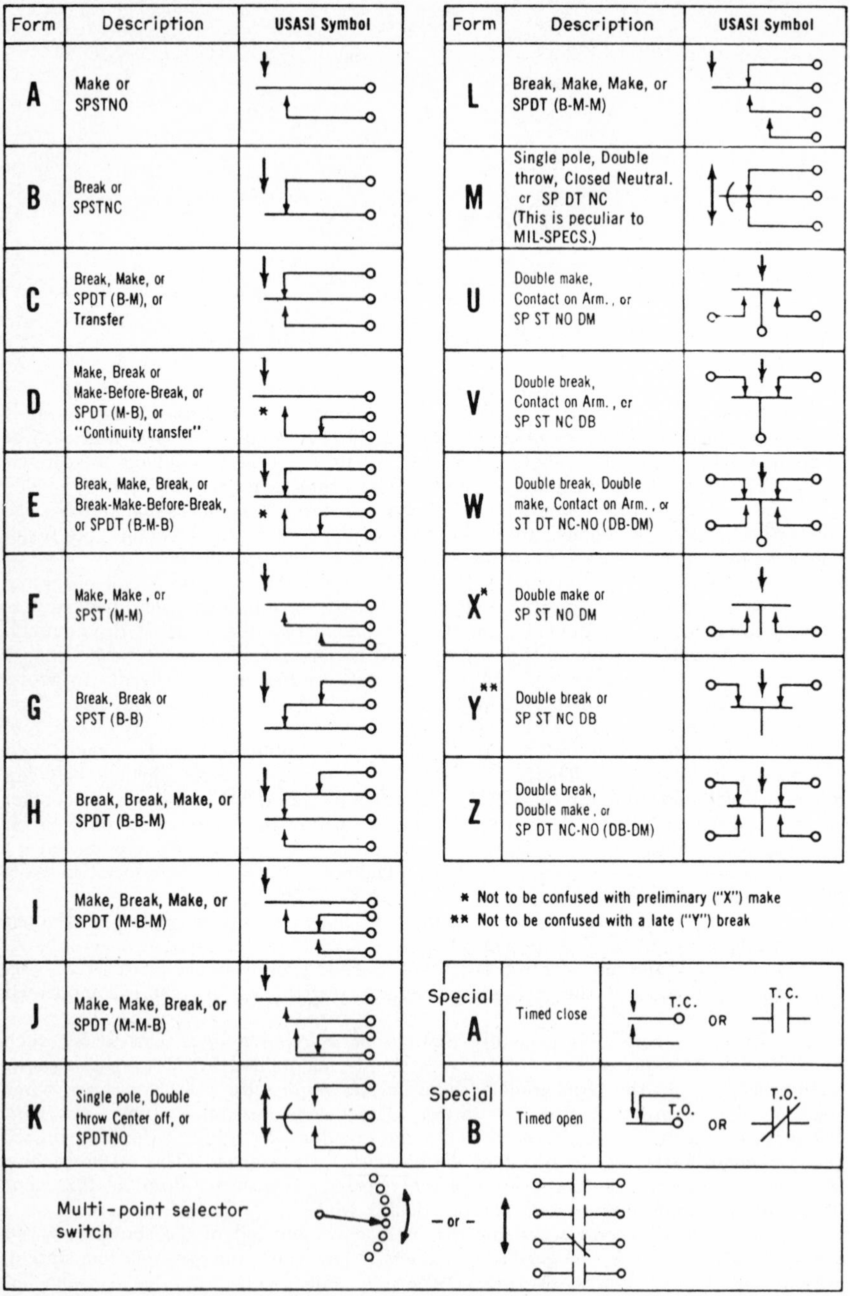

Form	Description	USASI Symbol
A	Make or SPSTNO	
B	Break or SPSTNC	
C	Break, Make, or SPDT (B-M), or Transfer	
D	Make, Break or Make-Before-Break, or SPDT (M-B), or "Continuity transfer"	
E	Break, Make, Break, or Break-Make-Before-Break, or SPDT (B-M-B)	
F	Make, Make, or SPST (M-M)	
G	Break, Break or SPST (B-B)	
H	Break, Break, Make, or SPDT (B-B-M)	
I	Make, Break, Make, or SPDT (M-B-M)	
J	Make, Make, Break, or SPDT (M-M-B)	
K	Single pole, Double throw Center off, or SPDTNO	

Form	Description	USASI Symbol
L	Break, Make, Make, or SPDT (B-M-M)	
M	Single pole, Double throw, Closed Neutral. or SP DT NC (This is peculiar to MIL-SPECS.)	
U	Double make, Contact on Arm., or SP ST NO DM	
V	Double break, Contact on Arm., or SP ST NC DB	
W	Double break, Double make, Contact on Arm., or ST DT NC-NO (DB-DM)	
X*	Double make or SP ST NO DM	
Y**	Double break or SP ST NC DB	
Z	Double break, Double make, or SP DT NC-NO (DB-DM)	

* Not to be confused with preliminary ("X") make
** Not to be confused with a late ("Y") break

Special A — Timed close

Special B — Timed open

Multi-point selector switch — or —

Fig. 8 Symbols for relay-contact combinations established by American National Standards Institute (ANSI).[1]

Form	Decription	IEC, JIC and NMTBA symbols	Other IEC symbols	Mod. tel. symbols
A	Make or SPSTNO		or	
B	Break or SPSTNC		or	
C	Break, make or SPDT (B-M), or transfer		or	2 —— R, 1, 3
D	Make, break or make-before-brake or SPDT (M-B) or "continuity transfer"	CT		3 —— R*, 2, 1
E	Break, make, break or break-make-before break, or SPDT (B-M-B)			R**, 1, 2, 3, 4
F	Make, make, or SPST (M-M)		(Time sequential closing)	

Fig. 9 Alternative symbols for relay-contact combinations.[1] Sources of symbols: IEC—International Electrotechnical Commission; JIC—Joint Industry Conference, Electrical Standards for Industrial Equipment; NMTBA—National Machine Tool Builders Association, Electrical Standards; Mod. Tel.—modern telephone practice. CT indicates continuity transfer, one asterisk denotes make before break, and two asterisks denote break make before break.

force acting upon the armature when the coil is energized is directly applied to cause contact switching, and its effectiveness is somewhat proportional to the degree of coil energization.

A design rapidly increasing in use is that called "permissive make," as demonstrated in Fig. 14. In this type of relay, contact switching takes place when the energized coil provides sufficient force to overpower a pretensioned spring that held the contacts in a biased (unoperated or normal) position. When the biasing force is overcome by sufficient armature pull owing to coil energization, switching of the contacts takes place. When the coil is deenergized, the contact springs return to their unoperated position because the biasing force of the restoring spring is now unopposed. Advantages of this design are many, but principally they are relative freedom from need of readjustment, long life, predictably adequate and uniform contact pressures, a capacity for handling a large number of contacts with a minimum expenditure of consumed power, and independence from effects of gravity due to mounting positions.

An increasingly popular relay design is called "dry reed." No mechanical linkage is employed. The contacts are switched as a result of being placed directly into the magnetic field of the coil. Figure 15 shows in simplified form how this is accomplished. Much more complicated relays of this type are possible, wherein one or more coil windings control a battery of reed-type contacts. These relays too are explained in more detail later.

Specialized forms of the reed-contact relay use a mercury coating of the reeds to increase contact life and rating. These differ little from the dry reeds in size

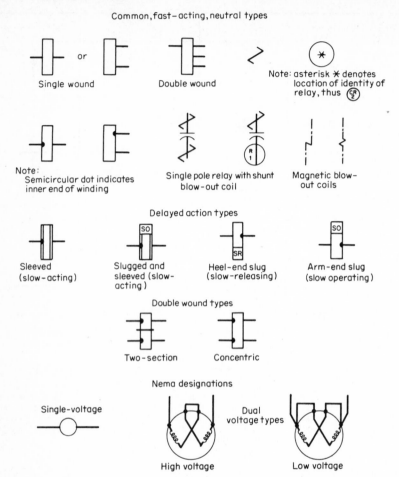

Fig. 10   Commonly used symbols for relay coils.[1]

and method of operation but are usually position-sensitive. They are covered in more detail later.

Not to be confused with the above is the so-called mercury-contact relay. It employs a reed as one of the contacting elements, but the size and design are considerably different. The cutaway view in Fig. 16 shows much of the design detail. This relay too will be examined in greater detail later.

The use of mercury as a wetting element of the contacting surfaces in both the above cases reduces and stabilizes contact resistance and increases load-carrying capabilities.

Classes of Service

Some relay characteristics must be considered early owing to the nature of the service.

Aircraft, commercial The environmental requirements for relays in this category are now rather well understood, and features are designed into the relay to make it compatible with the service needs. Danger exists when the vendor is not made aware of all the facts. For example, if a vendor was aware that the relay was to mount in communications equipment but was unaware that said equipment was to be airborne, some deficiency might exist.

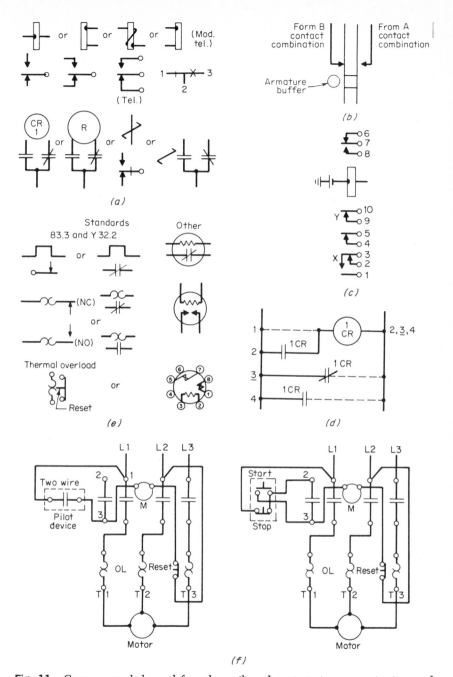

Fig. 11 Common symbols used for relay coils and contacts in communications and general systems.[1] (*a*) Common symbols used to show a relay with a form C contact combination. (*b*) The preferred order of contact arrangement in a simple relay pileup. (*c*) A relay with a large contact-spring pileup showing preferred order of arrangement. (*d*) Location of contact symbols with respect to coil on JIC/NMTBA simplified diagrams. Sequence of numbers at right of coil symbol locate associated contacts in lines numbered in left column. An underscored number signifies a normally closed contact. (*e*) Symbols for thermal relays. (*f*) Symbols used in motor-control relay circuits (JIC/NMTBA); 1, two-wire control; 2, three-wire control.

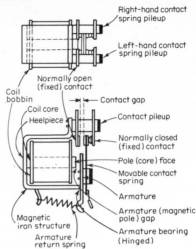

Fig. 12   A relay commonly referred to as general-purpose.[1]

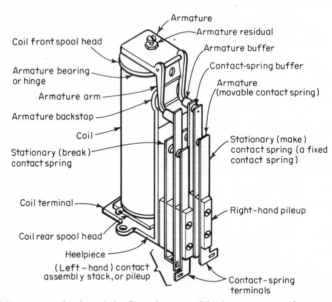

Fig. 13   A typical relay of the long-frame and leaf-spring type of construction.[1]

Aircraft, military   In general the relays required for this kind of application are of special design, required to meet a particular military specification referred to in the government contract, which must be made known to the vendor at time of ordering.

Air conditioning (and heating)   Standards that apply for air conditioning and heating are under the cognizance of the Underwriters' Laboratories (UL) or other national equivalent such as Canadian Standards Association. This must be made known to the vendor initially.

Appliances, household electrical   Frequently Underwriters' Laboratories (UL) approval of relays will be required with respect to not only flame prevention but also safety for the operator of such equipment. One approach is to find out, quite early in the design, what the UL ruling will be. Submission of the entire device so that the relay gets approved as a portion of the whole is a desirable approach

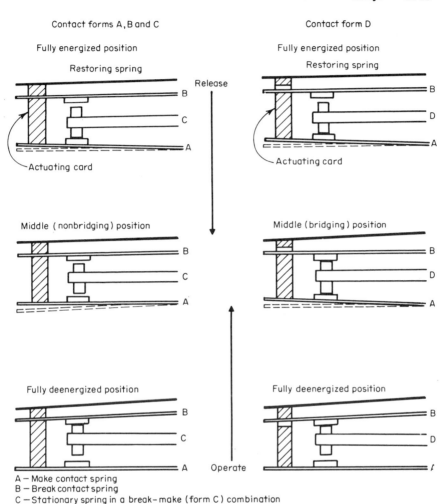

Contact forms A, B and C

Fully energized position

Restoring spring

Release

B

C

Actuating card

A

Middle (nonbridging) position

B

C

A

Fully deenergized position

B

C

A

Operate

Contact form D

Fully energized position

Restoring spring

B

D

Actuating card

A

Middle (bridging) position

B

D

A

Fully deenergized position

B

D

A

A — Make contact spring
B — Break contact spring
C — Stationary spring in a break-make (form C) combination
D — Stationary spring in a make-before break (form D) combination

Fig. 14 Permissive make contacting. Contact forms A, B, and C: When the coil is energized, the armature operates the actuating card, moves spring 1 into an open (break) position, and then permits spring 2 to close (make) with the stationary contact 3. On deenergization, the contacting sequence is in reverse. Contact form D: When the coil is energized, the armature operates the actuating card, which is shaped to cause spring 2 to close (make) with stationary contact 4 before spring 1 opens (breaks) its contact with stationary contact 4. In fully operated position spring 1 does break contact with 4. On deenergization, the contacting sequence is reversed.

where a UL listed relay of the kind needed is not readily available. Use of an unlisted relay will always require adherence to recognized UL standards of insulation, creepage path, and flameproof materials before eventual approval can be obtained; so it is well to start in this direction in order to save time and frustration. Contact life is usually not required to be of a high order, but its magnitude should be expressed.

Automobiles and trucks Environmental requirements are usually quite severe. The worst of these in respect to shock, vibration, dust, temperature, humidity, etc., should be made known to the vendor at once. Contact life is not usually of a high order, but it needs to be carefully established.

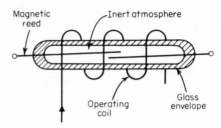

Fig. 15 Basic parts of a dry-reed contact relay. (*C. P. Clare & Co.*) Basically, the switch is composed of only three parts, a glass envelope and two reeds. The reeds are flat and made of a magnetically soft material. The glass seals at the ends of the envelope support the reeds as cantilevers so that their free ends overlap and are separated by a small gap. If the capsule is now enclosed in an operating coil and coil current is applied, a magnetic flux will be present in the reed gap. This flux causes an attractive magnetic force to act on the reeds and pull them together.

Business machines Relay requirements for such an application are for fast-to-operate, long-lived relays, free from the need of too frequent maintenance. Contact life expectancy, usually without the need for readjustment or routine maintenance, ordinarily exceeds 100 million operations. The relay is usually operated in a favorable environment, and if such is not the case the vendor should be made aware of it.

Coin-operated machines Frequently, low cost is of prime consideration, but the degree of sophistication and reliability required is such a variable across the design spectrum that there is definite need for good communications between designer and vendor. Also, environment is a large variable that needs full exploration in each case. In view of the cost factors, low first cost vs. too frequent maintenance must be carefully evaluated.

Communications equipment Long life, reliability, freedom from too frequent maintenance, and favorable environment are usually the requirements of prime interest in this category. Telephone-type relays fortunately are available in just about any design form needed and are ideal for this kind of application.

Computer input-output devices Owing to the heavy-duty requirements, a maximum life expectancy with utmost reliability is needed. Quick disconnect mounting, so that a unit in need of attention can be instantly replaced by a standby, is frequently a requirement. Low coil power consumption is usually specified, but if of a high order electrical-interference suppression must be provided and carefully chosen. Environment is usually not a problem.

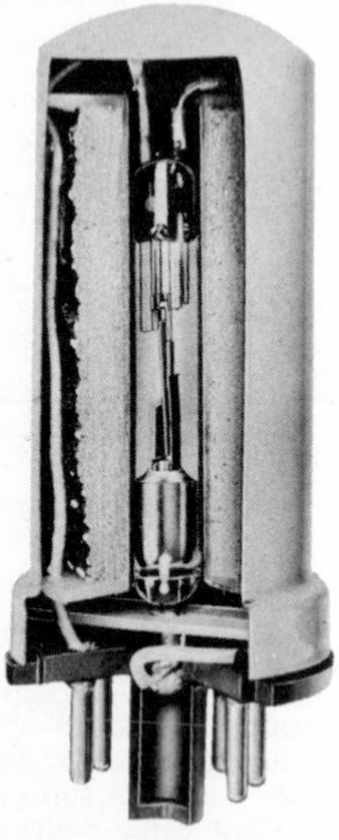

Fig. 16 Cutaway view of mercury-wetted contact relay. *GTE Automatic Electric Company, Inc.*)

Electric-power control In this field, relays are used principally in supervisory equipment. Long life, reliability, and freedom from too frequent maintenance are important. The environment is usually favorable to relays.

Electronic data processing Process control is a frequent user of relays in this rather broad field. Gathering of data from sensors and the reduction of such to usable form frequently employs relays. The most troublesome area is usually contact selection to accommodate a high variation in contact load. For example, very low level signals as generated by thermocouples, strain gages, and the like make necessary the use of especially reliable contacts (usually gold-plated or gold alloy), but the other functions scanned by even the same relay may be of a destructive type of load. Great care must therefore be taken in describing all contact-load functions to the vendor. Environment may be a great variable in this field, varying from an air-conditioned room to a highly contaminating manufacturing-plant area. It should be fully discussed with the vendor.

Laboratory test instruments   Maximum reliability with good life and freedom from too frequent servicing is the ruling requirement for relays in this field. Environment is normally of no consequence.

Machine-tool control   In general, the relays used in this field are of a particular design in accordance with NMTBA recommendations and to meet NEMA and UL standards. Some of the associated equipment is exempt from these requirements. It is therefore necessary to have a clear understanding with the vendor on this matter. Environment is frequently such that maximum protection from the ambient extremes is necessary.

Production test equipment   Cable testers, circuit checkers, automatic continuity, voltage circuit testers, etc., make rather extensive use of relays. The required properties are that the relays not introduce any faults of their own. This means a high order of insulation resistance and dielectric withstanding voltage with low contact resistance.

Military applications   This is a highly varied category. The requirements are ordinarily so specific to the job that only by taking the vendor entirely into the designer's confidence can an adequate choice of relay be made. Because the military requirements are so often quite stringent and a ruling military performance specification for the whole job frequently and too often is extended to the relay maker, the best relay for the job is not always obtained. The question must be asked early whether the relay is to function or only survive the worst of the specification. In meeting a *performance* requirement of a specification, a relay is sometimes not as reliable, in general, as would have been the case had it only been required to *survive* maximum temperature, shock, vibration, etc.

The four main military and space applications in which relays find usage are these: military aircraft, land-based equipment, missiles and aerospace, and naval shipboard. The logic on which specification writing is based is quite different for each, although some overlap is frequent.

Factors in Relay Selection

Power input, ac or dc   As was pointed out above, relays are not something to be added on after all design work is finished. A suitable power supply should be a part of the basic design. A decision on choice of operation from alternating or direct current is required early in the design. Commercial alternating current usually offers economic advantages, but direct current is most often employed, for reasons discussed below.

If the only requirement is that the relay simply shall operate when a switch to it is closed and release when that switch is opened, it probably matters little with respect to performance in the circuit whether it is powered by alternating or direct current. But many relay applications are so complex that dc power in some form is required. The bulk of the discussion in this book therefore relates almost entirely to dc operated relays. In addition to the usual inductance and induced-noise problems that a supply of ac power is likely to produce, there can be a number of specific reasons for going the dc route.

DC relays usually have longer life. The contacts of ac relays flatten prematurely as a result of wear due to noticeable ac vibrations during their closing and opening, as always occurs to some degree just before the armature has sealed in. There is also some perceptible light chatter of the armature for relays operated on alternating current even while fully operated. Adding to longer life of dc relays is the reduced bearing wear resulting from absence of ac vibration at this point.

DC relays usually have greater sensitivity. Since there is no tendency to chatter, lighter energizing forces may be employed than is the case for alternating current, where early saturation of the shading coil is required and where poor power factor may cause the coil current to be large. In other words, the power that is adequate on direct current to start armature movement is usually more than adequate to maintain a securely operated position.

The heat loss of dc coils is usually noticeably lower. There are both fewer iron losses (no hysteresis on direct current) and fewer copper losses (because usually the required holding power is less).

On a grams per contact operated-pressure basis, dc relays for the same contact load, can be kept smaller than ac relays. This reduced size may be quite important at times, especially when mounting on a printed-circuit (PC) board.

Costs should favor dc relays. DC coils are less expensive to make than ac (a solid coil core is cheaper than laminated, and no shading coil is required for direct current).

DC relays, especially if heavily loaded, can accommodate to a wider voltage range than ac relays.

Desired timing variations are almost impossible of achievement when operating conventional relays on alternating current.

Reliability is usually in the dc relay's favor because no compromises in contact pressure, number of contact springs, or adjustment refinement need be made in the interest of quieting the relay.

If the battery type of dc operation is economically feasible, the relay equipment is practically independent of commercial power failure. This kind of reliability, as would exist for battery-powered relays with the battery and rectifier floated across the alternating current, may be what is needed in many cases.

If alternating current is chosen, there is usually no concern as to the adequacy of the power source to maintain the required operating voltage and wattage. If direct current is to be provided, however, there is frequently need to determine carefully in advance that there is voltage stability, especially if the power-supply source is a rectifier. Most dc relays will function well on rather poorly filtered rectified power supply, but this must be determined well in advance of the final design.

Rectification of Alternating Current to Provide DC-Relay Operation. Some simple arrangements to permit operating dc relays from an ac power supply are shown in Fig. 17. The following paragraphs present various schemes for providing relays with a variety of rectification circuits for permitting the use of the relay directly on an ac line.

Figure 17*a* shows a full-wave bridge rectifier circuit using semiconductor diodes

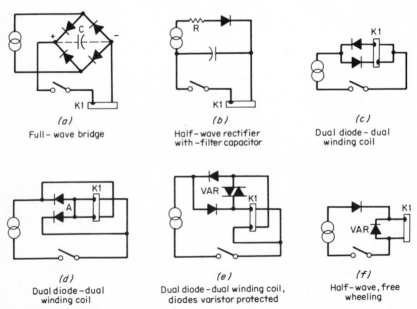

(a)
Full-wave bridge

(b)
Half-wave rectifier
with –filter capacitor

(c)
Dual diode-dual
winding coil

(d)
Dual diode-dual
winding coil

(e)
Dual diode-dual winding coil,
diodes varistor protected

(f)
Half-wave, free
wheeling

Fig. 17   Rectification of alternating current to provide dc-relay operation.[2]

as the rectifying elements. To improve the quality of the output, a suitable filter is sometimes placed at position **C**.

Figure 17*b* represents a more common arrangement for producing satisfactory dc power for relays. Capacitor **C** is a large electrolytic (at 115 V ac applied power, a commonly employed capacitor is 40 μF, 230 V test with a protective resistor in series with the diode of approximately 30 to 40 Ω). The economy of this arrangement is apparent.

Figure 17*c* shows a circuit requiring a relay with two equal windings, one of which functions on one half-cycle, the other on the other half-cycle. The diodes alternately block and conduct with respect to each other in order to maintain unidirectional flux.

A modification of this circuit is shown in Fig. 17*d*. Here connection *A* ties a diode directly across each winding, providing a low-impedance path for circulating current.

Figure 17*e* is essentially the same as Fig. 17*c* except that a varistor has been employed to protect the diodes from line surges.

Figure 17*f* is a half-wave, inexpensive circuit found satisfactory for relays inherently not too fast to release. Flux established during the current *on* half-cycle remains in force owing to the effect of the counter emf during the current *off* half-cycle. The current freewheels through VAR and Kl in series during the half-cycle when no external voltage is being applied, hence the name that has been given to this technique. Relays that have a tendency to release fast may not perform satisfactorily because of a tendency to chatter.

Timing Because of the exponential rise in current through the relay coil on circuit closure, there is considerable elapsed time between initial relay-coil energization and operation of the contacts as seen in Fig. 5. There is a similar delay in release time following circuit opening (Fig. 6). The inherent operate and release-time delay may be of little or great importance. Usually operate and release timing as short as possible is desired, but sometimes considerable delay on either pickup or dropout (or both) is needed. The operate and release times of relays are controllable to some degree, depending upon the particular relay design and the constants of the circuit in which the relay is applied. If the relay coil is of sufficient size, delay factors can be built into the coil. However, where space is at a premium as in very small relays, series-connected or shunting capacitors, resistors, and diodes may be used to modify the relay operate and release times.

The following discussion is presented to demonstrate how the operate and release times of medium- to large-sized relays are affected by coil design.

Slow-operating Relays. Slow-operating relays as demonstrated by Fig. 18 use coils with a large copper collar or slug at the armature end of the core. The copper collar, acting as a short-circuited secondary winding, retards the building up of the magnetic field. These coils are most effective when used with a large armature stroke and a heavy spring load. The large armature stroke reduces magnetic pull on the armature, and the large spring load prevents pickup until the magnetic field has been built up to, or near, its full value, thus providing the maximum operate-time delay. Relays with armature-end slugs are also somewhat slow to release.

Slow-releasing Relays. Slow-releasing relays as in Fig. 19 have a copper collar or slug around the heel end of the coil core, a small residual gap, and a light contact load. The copper collar retards the collapse of the magnetic field once it has been established. This, together with the small residual gap and the light contact load, permits the armature to remain in its operated position until the magnetic field has died down to a very low value. Where the greatest possible release delay is desired, or where permanence of adjustment is very important, a short-lever armature is helpful.

Slow-acting Relays. Slow-acting relays use coils with a full-length copper sleeve about the coil core to provide release-time delay. Since the sleeve extends full core length, it also causes some delay in operate time as well as the sought-after release-time delay. The main reason sleeve coils are used is to save copper and reduce weight. Maximum release-time delay equal to that achieved by a full-sized slug is accomplished with about one-half the copper volume by using a sleeve.

Where small relay size makes the use of slugs and sleeves impractical, circuit ele-

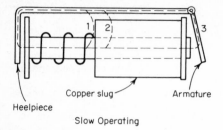

Slow Operating

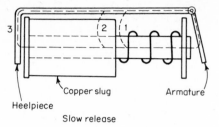

Slow release

Fig. 18 Effect of copper armature-end slug on relay operate time.   1. Magnetic-flux path immediately after coil-circuit closure.   2. Magnetic-flux path a moment later.   3. Final flux links armature to cause pickup.

Fig. 19 Effect of copper heel-end slug on relay release time.   1. Magnetic-flux path immediately after coil-circuit closure.   2. Magnetic-flux path a moment later.   3. Final flux links slug.   Slug can now delay dropout on coil-circuit opening.

ments such as diodes, capacitors, resistors, and thermistors are employed to alter the relay's inherent timing.   Examples of some of these techniques follow, together with the explanation of how the timing is affected by associated circuit components. If a more extensive discussion is desired, it may be found in Ref. 2.

Slow Release by Noninductive Shunt (Fig. 20).   This circuit provides for a slight increase in the normal release time of the relay by means of a noninductive resistor R connected across the relay coil.   The shunt path provides means for the back emf (created by the decay of flux in the core when any relay coil is deenergized) to circulate and somewhat prolong the operated condition of the relay.

Slow Release with Shunt Diode.   If the resistor of Fig. 20 is replaced by a properly poled diode, a longer release time will be achieved than for the use of the resistor. Normally, for a dc electromagnetic relay, dropout will occur following a delay of several milliseconds after the coil is shunted with a diode, such as might be used for transient voltage suppression in transistor circuits or to protect electrical contacts from voltage surges from an inductive load.   The delay can be lengthened to 20 to 30 ms or longer.   The delay is obtained by circulating the exponentially decaying current through the diode shunt after the coil circuit is opened.

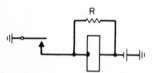

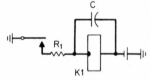

Fig. 20 Slow release by use of noninductive shunt.[2]

Fig. 21 Slow operate and slow release by series resistor and shunt condenser.[2]

Slow Action Using Series Resistor and Shunt Capacitor.   A simple application of an RC circuit to provide a brief time delay in the operation of a relay is illustrated in the circuit of Fig. 21.   When the circuit is closed, capacitor C, uncharged, provides a direct shunt around the relay coil.   As the voltage across C builds up and the current through it decays at the same rate, a design point is reached where the current diverted through the relay provides sufficient ampere-turns to operate it. Resistor R limits the current available for charging the capacitor and operating the relay, but it also makes the timing voltage-sensitive.   Since the capacitor is charged when the relay circuit is opened, the capacitor discharges slowly through the relay coil, giving a release time delay.

Slow Operate by Switched RC Shunt.   The circuit of Fig. 22 provides the same operate delay as for that shown in Fig. 21, but once the relay is operated, the RC network is discharged locally so that no release delay is introduced.   Operate delay will be a function of R, C, the applied voltage, and the operating characteristics of the relay. Because of the marginal arrangement of the relay coil and the resistor, this circuit is voltage-sensitive. Timing will change somewhat as the voltage changes.

Slow Operate and Slow Release by Coil-Winding Interaction. Short delays in operate or release times are possible for multiwinding relay coils by making use of the circuit conditions of Fig. 23.

Slow Release by Shorted Heel-End Winding.

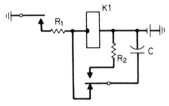

Fig. 22 Slow operate by switched condenser shunt and series resistor.[2]

The relay shown in Fig. 23a has a two-section quick-acting coil which is energized and deenergized on its No. 1 (armature-end) winding. When the relay is energized, a pair of auxiliary contacts will short the No. 2 heel winding, forming somewhat of a delay slug. This shorted winding is not nearly as efficient as a solid-copper slow-operate slug but will provide some increase in release time. The connection between the in terminals of the two coil windings is provided so that both windings will have a negative potential standing on them when the relay is idle. This eliminates any possibility of electrolytic corrosion on the windings and associated wiring.

Slow Operate by Shorted Armature-End Winding. The circuit of Fig. 23b is the converse of Fig. 23a. In this case the relay is energized and deenergized on its

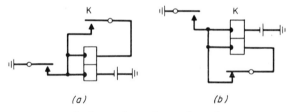

(a)   *(b)*

Fig. 23 Slow operate and slow release by coil-winding interaction.[2] (a) Slow release by shorted heel-end winding. (b) Slow operate by shorted armature-end winding.

No. 2 heel-end winding, and prior to operation, the No. 1 armature-end winding is shorted through a pair of normally closed contacts on the relay. This arrangement is not as efficient as a solid-copper slow-operate slug but will introduce some operate delay. The connection shown between the in terminals of the two coil windings is provided as before so that any possibility of electrolytic corrosion on the windings and associated wiring is eliminated.

Circuit fundamentals The requirements that a relay take the operated state when its operating-coil winding has power applied and restored when deenergized is of course basic and in need of no further recognition. A few other requirements are almost as basic but do require recognition before a relay selection can be made. The most elementary of these circuit fundamentals are described below. More sophisticated relay circuits can be found in many places such as Refs. 1 and 2.

Locking through Own Contact(s). Figure 24a shows a basic locking circuit in which the relay is energized by the closure of S1 and is held energized, or is locked up electrically, through one of its own make contacts and S2. It is now held operated independent of S1 and will release when S2 is opened after S1 is closed.

A variation of the above is shown in Fig. 24b. Here closures of S1 and S2 are required to operate the relay, which is then held through one of its own make contacts and S2. The relay restores to normal when S2 is opened.

Another form of lockup utilizes a continuity (make-before-break, or form D) contact combination. In this circuit, illustrated in Fig. 24c, the relay is operated through

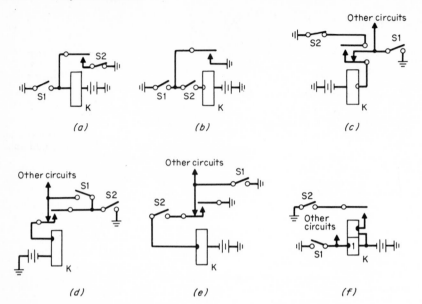

Fig. 24 Six methods of electrically locking up a relay through one of its own contacts.[2]

S1 and the normally closed portion of the form *D* contact. After operation, the relay is locked up through the normally open portion of the form *D* contact and S2. After the relay is operated, S1 is free to control other circuits without being affected by the ground potential to the winding of relay *K*. A make-before-break contact is used in this case because a break-make would fail to secure the holding circuit and the relay would buzz in a partially operated position.

Variations in continuity lockup circuits that require closure of both S1 and S2 before the relay will operate are illustrated in Fig. 24*d* and *e*.

In Fig. 24*f* a method is illustrated for isolating an operating and lockup circuit through use of two electrically isolated relay windings with a common source of power. Operation of S1 will energize the relay, but it can be locked in only if S2 is closed before S1 is released. Both S2 and S1 must be open to release the relay.

Use of a Preliminary Make Contact. A more desirable way of arranging the contacts to provide for a locking circuit is by use of form *A* make-type spring combination. However, if the operating signal is of extremely short duration, such that there is some doubt as to whether or not the relay will have time to operate fully, the locking contact can be provided as a preliminary make. Once these locking contacts have closed, complete operation of the relay is assured. If the operating signal is still present, the armature will be driven by both the locking winding and the operating winding.

To provide this preliminary make in a telephone-type relay, it is necessary that the relay be equipped with a heavy normally closed back contact. The balance of the spring pileup above the preliminary make is then supported in a position such that the preliminary make is free to move when the relay is energized. This heavy, normally closed, contact must be provided even though the controlled circuit does not require a normally closed contact.

Avoiding Malfunction from Feedback. If a relay locks itself in with the simple circuit of Fig. 24*a* as a function of some remote signal, it is possible that the locking potential could be fed back over the operating lead and cause a malfunction in some other part of the circuit. These sneak or feedback circuits can be eliminated by the addition of a blocking diode in series with the operating lead.

A simpler, more direct means of eliminating this feedback problem is to use a dual-coil relay with windings aiding and to use an independent locking contact. In this circuit (Fig. 25) the relay is operated on its No. 1 winding and locks itself in the operated condition on its No. 2 winding.

Forced Release of Relays. The holding circuit of Fig. 26a uses dual coils but differs from that of Fig. 24f in that the two coil windings are connected in opposition and two Form A contacts are employed. It provides lockup indefinitely until the release key is closed to energize the opposition winding. The only precaution essential to successful use of this dual-opposed winding relay is that the release circuit be connected to the most powerful winding (usually the armature-end section). This precaution is taken to assure that the flux is driven through zero, so that the relay will release positively.

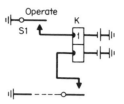

Fig. 25 Locking up a relay on an auxiliary winding.[2]

A somewhat different circuit for forced release of a dual-opposed coil relay is shown in Fig. 26b. This mode of operation assumes that the operating circuit was opened prior to the closure of the release circuit.

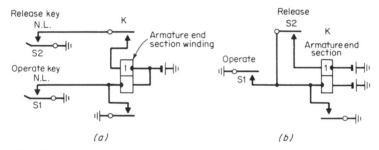

Fig. 26 Two methods of forced release of a relay by use of an opposed winding.[2]

In either case, such forced release requires considerably more time to effect restoration of the relay to normal than would have been required had the relay been restored by merely opening the holding circuit. This increased release time is the sum of the time for the counter magnetic flux to build up to the point where the relay is caused to release plus the armature-movement time necessary to restore the relay.

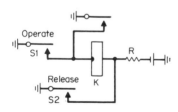

Fig. 27 Release of a relay by shunting it down.[2]

Shunt Control of Relays. The shunt-release relay shown in Fig. 27 is always operated in series with a resistor and is provided with a local locking circuit. This locking circuit is not necessary if the remote operating circuit is maintained. To release the relay, closure of the remote-release (RLS) contact effectively shorts out the relay coil and leaves the resistor in series with the power supply. If the RLS contact is maintained, this resistor must have sufficient wattage to operate satisfactorily under this condition.

When the relay is shorted out to effect its release, the low-resistance short across the relay coil provides means for the back emf to circulate for the complete exponential decay time of the flux in the magnetic circuit of the relay. The relay remains in the operated condition until the flux density has decayed to the point where the spring load in combination with the residual gap used can restore the relay. Normally, larger residual gaps are required for this type of service than would ordinarily be expected for a particular relay.

Generally speaking, shunt circuits of this type operate more satisfactorily on higher than on lower voltages. When the voltages are lower, the current which must be handled by the operate and locking contacts as well as the release contacts must be higher in order to achieve the required relay wattage.

Lockdown Operation by Coil Shunt. A variation in shunt release of a relay is found in the lockdown circuit, in which a relay prevents itself from being operated through one circuit path until another circuit path has been opened. Two forms of relay lockdown circuits involving a shunt-resistor path around the relay coil are shown in Fig. 28. In both circuits, the relay cannot be operated through switch

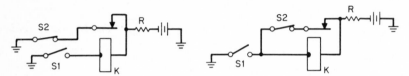

Fig. 28   Two forms of lockdown circuit using a coil shunting resistor.[2]

S1 until S2 has been opened. However, after the relay has been operated through S1, opening and closing of S2 will not affect the operated state of the relay. Since the resistor remains in series with the relay winding during operation, it can have an appreciable effect on operate time. These circuits have the disadvantage of dissipating power in the shunt resistor during the lockdown state.

Contact load

Contact Choices. Since the prime purpose of a relay is to establish the flow of electricity through a mating pair of contacts in the controlled circuit(s), the choice of materials assuring that this is accomplished becomes a first priority. The contact must be large enough so that there is no deterioration due to destructive melting or welding, yet not so large that the current density is below a critical value. The material must be highly conductive yet hard enough to meet the required number of closures without excessive wear (flattening).

For each kind of load condition there is a best choice of material, shape, size, and pressure when in contact. To determine what is best for the specific job at hand the vendor must be fully informed by having the purchaser include such information as ambient temperature, humidity, and degree of dust and dirt likely to reach the contacts. The required service life is also a factor in the choice to be made, also frequency of operation. Voltage and current to be handled are naturally of prime concern. The nature of the load with respect to whether it is resistive only, capacitive, inductive, or combinations of these must be stated. If resistive, is it fixed, lamp (surge), or subject to overload currents? If inductive or capacitive, what is the likelihood of surges? What is the chance of a flashover? What is the duty cycle? How many sets of contacts are required, and in what sequential order?

One common error made by relay users of little experience is to assume that if a contact is large enough, or rated above the known load, it will perform perfectly. This frequently leads to real trouble. Each contact material has a critical minimum or threshold voltage and current for its composition, size, and shape. The best contacting results when there is sufficient electrical pressure (voltage) and current, along with sufficient mechanical pressure on the contacts, to cause some fusing of contact surfaces on each operation. This fusion is of a low order, but it means that a very small amount of welding occurs on each operation, with a rupture of that weld each time the contacts are separated. If the critical or threshold voltage is not reached or exceeded, if the current density is inadequate to melt and weld even one tiny spot, if the contacting surfaces are too large and flat to give the critical current density necessary to get surface softening, or if the mechanical pressure is inadequate to push any insulating film aside that may be covering the surfaces, then contact failure results. This minimum electrical requirement is known as minimum (reliable) current.

Table 2 shows some critical values applying to the most commonly used contact

TABLE 2   Typical Softening, Melting, and Boiling Voltages of Commonly Used Contact Materials (Holm)[1]

Material	Softening	Melting	Boiling
Silver	0.09	0.37	0.67
Gold	0.08	0.43	0.90
Palladium		0.57	1.30

materials (which presumes that there is sufficient mechanical pressure to make conducting spots on the contact faces touch through any insulating film that may exist). It is difficult to generalize on contact materials as related to load current and voltage handling capabilities. This is why in the final analysis the vendor's judgment has to be relied upon. As an aid in understanding how a decision is reached, Table 3 indicates some generally accepted practices. From this it might be assumed that

TABLE 3   Considerations Affecting Choice of Material for Relay Contacts

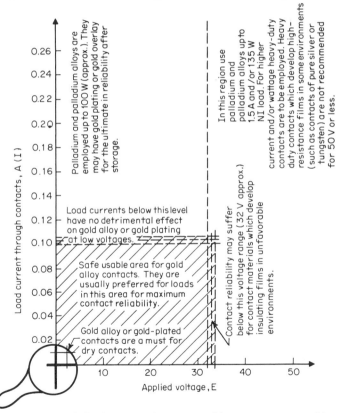

gold contacts are not used often enough. Pure gold is not an acceptable material because it cold-welds without even the presence of current. Impurities, to prevent cold welding and to add some mechanical hardness for improvement of life, are deliberately added to so-called gold contacts to make them acceptable for use in relay circuits. Even so, their mechanical life is not very great, and for that reason other noble metals of greater hardness are more often used. If conditions permit, an alloy of palladium and silver is probably the best practical contact.

Cost factors favor the use of silver and silver alloys wherever service requirements are favorable. Silver and some of its alloys tarnish readily in industrial atmospheres, and the tarnish is of very high resistance, high enough in many cases to look like an insulator. In general silver is not satisfactory for use much below 50 V, or even above 50 V if the current flow is light owing to large circuit resistance.

Understandably, most concern is normally shown by the user regarding maximum-allowable-current figures. Typical values are shown by Table 4. Because of maximum concern with what can be seen rather than what is invisible, it must again be emphasized that failure to reach or exceed minimum (reliable) current has caused as many, or more, contact failures than overload. That is why the choice of kind of material, size, and shape of contact is important.

Contact Characteristics. While the final choice of contact is a matter best left to the vendor in the end, the user must give it some consideration before even a tentative choice of relay type and vendor can be made. In other words, kind of relay, relay size, and relay availability depend to a considerable degree on the nature of the contact load to be handled and the contact life required. The need to use a contact small enough so that the minimum current is adequate for keeping the contacting faces in a conducting state has already been discussed. The opposite situation is now to be examined. If relays such as dry reeds, mercury-wetted contacts, miniature size, or other kinds having specialized contacts are the choice, there is little latitude for the user to explore. But if the more common general-purpose relay, telephone-type relay, or power-type relay is being investigated, the kinds, sizes, and shapes of contacts available vary extensively. As an example of what is available in the way of contacts on a large- to medium-sized telephone-type relay, one manufacturer offers as standard the contact variety of Table 4. All manufacturers have similar data applicable to the relays of their manufacture.

In interpreting Table 4, it should be recognized that advertised wattage ratings of relay contacts are based on fixed-resistance resistor (noninductive) loads, and care must be exercised to avoid introducing inductance or capacitive reactance accidentally, and to recognize its effect when intentional. When setting up ratings for inductive circuits, it is common practice to derate the contacts to one-half (or less) of the wattage for noninductive circuits. Adequate arc and spark suppression must also be provided with inductive loads; otherwise the contacts should be derated even further.

Contamination by small amounts of airborne dust or fumes is another common cause of reduced contact life. Even accidental fingerprints on the contacts can accelerate their rate of deterioration.

More detailed discussion of contact capability will be given for each of the relay types examined below in the discussion of individual relay characteristics. Table 3, for example, shows a relationship between contact load and kind of contact material that is applicable to many kinds of relays other than large- and medium-frame-size telephone types such as many kinds of general-purpose relays and some specialized relay types.

Contact Protection. A relay operating near sensitive circuits may cause trouble in electronic equipment from arcs generated as the contacts function. Also the coil transients represent both radiated rf interference and conducted radio frequency back over the power leads. Some form of suppression must be applied as electrical protection of the controlling contacts. Ideally it can serve the purposes of both contact protection and interference suppression. To determine that both objects are achieved, some considerable experimental testing may be involved. This subject is much too complex for complete analysis here, but from a practical standpoint it may be rather easily solved by adjusting the contact protection for maximum effectiveness (see Refs. 2 and 4).

Closing Contacts, DC and AC Loads. Contacts can be damaged on both closure and opening. Contact damage at closure is frequently due to current surges because contact forces at this instant are light, permitting contact sliding and bouncing to take place. This is particularly bad, because the load current is often many times the steady-state value at this instant. A microscopic weld or "bridge" will frequently form at the point of contact closure. In dc circuits this bridge usually ruptures asymmetrically at the next contact opening, resulting in metal transfer. In ac circuits,

TABLE 4 Typical Data for User Guidance in Choosing Telephone-Type Relay Contacts

Code	Material	Break or make, load	Will carry load	Remarks	Diam, in	Height, in	Shape
0-20	Palladium-silver	50 W (max 1 A N.I.)* 135 W (max ½ A) inductive	135 W (max 2 A)	See Code 0-18	0.055	0.020	
3-18	Gold alloy			For low-level circuits			
0-18	Palladium-silver	135 W (max 3 A N.I.)	150 W (max 3 A)	Resistant to tarnish, and nonmicrophonic	0.067	0.020	
9-18	Platinum-ruthenium	Same as Code 0-18	Same as Code 0-18	More resistant to wear than Code 0 contacts			
4-18	Palladium	135 W (max 3 A N.I.)	150 W (max 3 A)	See Code 0-18			
4-14	Palladium	150 W (max 3 A N.I.)	150 W (max 3 A)	See Code 0-18	0.084	0.031	
9-14	Platinum-ruthenium	Same as Code 4-14	Same as Code 4-14	See Code 9-18			
0-14 Flat-or domed	Palladium-silver	150 W (max 3 A N.I.)	150 W (max 3 A)	See Code 0-18			
Tungsten	Tungsten	450 W (max 2 A N.I.)	450 W (max 2 A)	Use in highly inductive, low-current circuits	0.125	0.063	
Laminated silver	Laminated silver	450 W (max 4 A N.I.)	450 W (max 4 A)		$3/16$	$3/64$	
Tungsten silver	Tungsten silver alloy	575 W (max 5 A)	1,150 W (max 10 A)		¼	$1/16$	
Silver cadmium oxide	Silver cadmium oxide	Meets NARM Grade B life—100,000 operations N.I. load—20 A, 28 vdc; 10 A, 48 vdc; 1 A, 110 vdc; 5 A, 115 vac	Will handle 300-W lamp loads (allow for inrush currents) or the current drain of an average ¼ hp or a well-designed ½-hp motor	Especially applicable to circuits where high current density causes welding of other material	$3/16$	0.040	
Snap-action switch, ac		AC N.I. Loads Break 10 A, 115 V Make 6 A, 115 V Break 5 A, 230 V Make 3 A, 230 V		Also available for loads up to 30 A, 220 V, ac on special order			
Snap-action switch, dc		On special order, magnetic blowout switches up to 15 A dc					

SOURCE: Industrial Products Division, GTE Automatic Electric Company, Inc.
* N.I. = noninductive.

there is usually a net loss of contact material. The metal vapor that condenses in the vicinity of the actual contact area is normally black and is frequently mistaken for carbon.

Loads that produce transients at contact closure are as follows:

1. Tungsten lamps whose cold resistance is 7 to 10 percent of their hot resistance.

2. Transformers and ballasts that may cause transients five to twenty times their normal currents when switched in their inputs.

3. AC solenoids and some kinds of motors.

4. Capacitors placed across contacts or loads with inadequate series-connected current-limiting resistance.

To meet these circuit conditions, the relay designer may elect to employ:

1. Heavy-duty contacts and a high contact force to minimize contact bounce on closure.

2. Contact materials with the highest possible electrical and thermal conduction usually silver or silver alloys.

3. Contact-material additives to inhibit welding, such as cadmium or cadmium oxide.

The circuit designer can usually add small values of series resistance to the circuit to reduce current surges.

Opening Contacts, DC Load.   There are almost certain to be transients on contact opening. When the circuit to an inductive dc load is opened, much of the energy stored in the load must be dissipated as arcing at the contacts unless some alternative means of energy absorption is provided. Some of the load energy is dissipated as heat in the load resistance, in eddy-current losses in its magnetic circuit, and in the disbributed capacitance of the coil winding. For dc circuits, a number of simple solutions are available to lessen or inhibit contact arcing:

1. A semiconductor diode may be connected across the inductive load (see Fig. 29), so that it blocks the applied voltage at contact closure but allows the stored

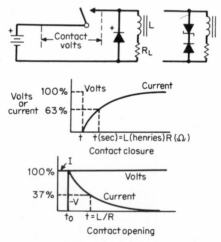

Fig. 29 Method of using semiconductor diode to suppress voltage surge from inductive load at contact opening.[2]

energy in the load to recirculate through it at contact opening. The time for the load current to decay to 37 percent of its steady-state value equals L/R. The blocking diode will prevent any inductive transients from appearing across the contacts during the switching operation. For loads below the minimum arcing current, the time required for load deenergization can be materially reduced by adding a Zener diode, resistor, or varistor in series with the blocking diode, thus increasing R. The rate of energy dissipation after the contacts are opened is thereby increased as the load current circulates back through this additional voltage drop. The instantaneous voltage plus the source voltage should not exceed 320 V.

2. Either the load or the contacts may be shunted with a resistor-capacitor combination (see Fig. 30). For load currents in the stable-arc range, the resistor R_C can be selected to match the load resistance, or it may be ½ or 1 Ω V⁻¹ of the power source. For smaller load currents than can use a stable arc, the resistor can be higher in value.

A reasonable value is one resulting in a voltage transient of less than 300 V for the sum of the source voltage and the instantaneous voltage generated by the load current in the resistor, calculated thus:

$$R = \frac{300 - E_{\text{source}}}{I_{\text{load}}}$$

The resistor is essential and must be large enough to limit the current transient from the capacitor discharge (or charge) on contact closure to prevent contact welding. The capacitor should be large enough to accept the stored energy of the load without permitting an electric breakdown of the contact gap, normally at greater than 320 V. An oscillograph is the best way to determine when these transients are adequately suppressed. In Fig. 30 the capacitor can be connected at either

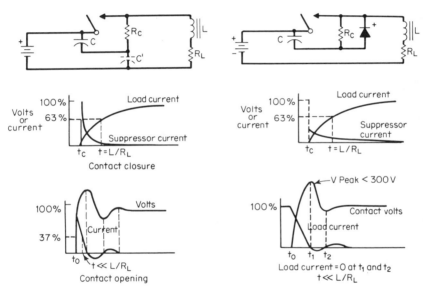

Fig. 30 Use of capacitor-resistor combination to suppress surge from inductive load.[2] Capacitor may be at either *C* or *C'*.

Fig. 31 Use of capacitor-resistor-diode combination for arc suppression with a highly inductive load.[2]

C or *C'*. Connection at *C* is preferred, since it protects against source and line, as well as load, inductance.

3. A varistor (voltage-sensitive resistor) or thyristor may be used to shunt the load. If such a device carries 10 percent as much current as the load, the maximum switching transient will be about twice the source voltage. This method is also suitable for ac circuits.

4. For extremely inductive loads, for the longest possible life, or for load power and contact-gap length above the minimums for a stable arc, the circuit of Fig. 31 may be used. In this circuit, the capacitor is charged through the diode but can discharge only through the resistor. This arrangement gives essentially zero contact-voltage drop at the instant of contact opening. The capacitor value should

be such that when the energy transfer from the load is complete, the peak voltage to which it charges will not cause a breakdown of the diode, the contact gap, or itself. Usually the peak voltage should not exceed 200 to 350. For dc inductive loads for which the conditions for a stable arc may be satisfied by the partially opened contacts, the circuit of Fig. 31 permits the contact gap to be established without drawing an arc, and the stored-energy transfer is accomplished more quickly than would have been the case if the contacts had been allowed to arc. The reason for this is that the integrated inverse voltage to which the capacitor charges is greater than the voltage drop in an arc, were arcing permitted.

5. Where the inductive load of a relay coil presents a hazard to transistor drive circuits, coils with dual windings (called bifilar coils) may be used with one winding shorted. This arrangement provides a pronounced damping effect on the rate of change of magnetic flux in the iron and hence provides a significant moderating effect on the induced voltage. It is, however, wasteful of coil winding space and increases relay cost.

Opening Contacts, AC Load. AC loads are most commonly treated in a different manner from dc loads because of the fact that a stable arc will normally be terminated when the current passes through zero and reverses at the end of the first half-cycle following contact separation. Fairly common practice is to use arc-resistant contact material, preferably in a relay in which the contacts separate slowly, and let arcing be terminated by the reversal of the current rather than by the continuing separation of the contacts. When load currents get too heavy for safe interruption by small relays (greater than 10 to 25 A), the current-reversal effect can be supplemented by magnetic or air blowout, multiple break contacts, or arc-gap cooling labyrinths, or by evacuating the contact chamber.

Under moderate arcing conditions, contact life may be greatly increased by shunting the load with a resistor-capacitor-diode combination whose time constant is equal to that of the load:

$$R_cC = \frac{L}{R_1}$$

or assuming R_c equals the load resistance R_v,

$$C = \frac{L}{R_2}$$

This network makes the load characteristics essentially resistive. When the maximum possible contact life is required, either of the capacitor-diode combinations shown in Fig. 32 may be justified. For 115-V ac service, the diodes should have a peak inverse voltage rating of 400, the capacitor should have a dc working voltage of 200 V dc, and there should be a 100-kΩ resistor, which will dissipate nearly 1 W. The capacitor discharge time after a switch closure may be as long as 1 s.

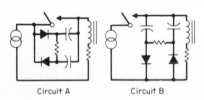

Circuit A   Circuit B

Fig. 32 Use of capacitor-resistor-diode combination for suppressing contact arc on ac inductive load.[2]

Cautions. The transient voltages developed when the contacts open the load circuit may exceed the dielectric withstanding voltage between contacts and another part of the relay. In some circuits, these voltages may be high enough to cause breakdown of another circuit component. These transients often cause interference in adjacent or associated circuits. Usually a resistor-capacitor network, as recommended by the relay vendor's literature, will reduce the voltage to a level that suitably protects the contacts and avoids dielectric breakdown. However, it is sometimes necessary to use diodes to eliminate radio interference from arcing. For the latter cases, no general rules can be formulated because the interference is closely associated with the particular circuits.

In general, careful attention to contact protection can increase life expectancy as much as three orders of magnitude. System reliability may be greatly improved by elimination of high-voltage transients, and the speed of response and its consistency are often substantially improved.

Cost Good engineering always necessitates much consideration of cost. If only one relay or a few are involved, this consideration might be slight. Where the quantities per device are great or the production heavy, this matter may justify a great deal of thought. Several items must be balanced against first cost before a decision to reduce relay purchase cost is made. These are cost of maintenance, cost of field replacement, customer satisfaction, and value of product enhancement accomplished by using the most reputable relay.

Mountings and enclosures Sometimes the method of mounting becomes of prime importance. For example, if all the other equipment is PC-board-mounted, the relays are probably most acceptable if mounted that way too. In some cases, mounting preferences will limit the kind of relay types that can be used.

The least expensive relay with respect to original purchase price is usually without extraneous materials, using only two or more screws into tapped holes and without dust covers. Simple angle-iron mounting brackets are usually the next least costly. Mounting on strips, with or without dust covers, usually represents modest added cost.

Enclosures are used mostly for protection from ambient dust, dirt, oil, metallic chips, etc. Enclosures may also be employed with intent to keep out prying fingers. Enclosures run the wide gamut from simple drawn metallic covers and molded-plastic covers to elaborately tailored enclosures and hermetically sealed enclosures. Hermetic sealing protects the contacts thoroughly from the environment but has several disadvantages that make its use less than ideal in all cases. For example, contact troubles develop in the confined area of the contacts that do not occur in air, and hermetic sealing is expensive and prevents inspection of the working elements of the relay when in service.

Relays which plug in exist in many forms. A common variety that is very popular is demonstrated by Fig. 33. The increase in the use of relays directly on PC boards

Fig. 33 A covered general-purpose relay having an octal plug-in base. (*Potter & Brumfield, Div. of AMF, Inc.*)

Fig. 34 A relay in a DIP enclosure for mounting on a PC board. (*Magnecraft Electric Co.*)

has resulted in the development of the DIP method of mounting, as demonstrated by one form of this relay in Fig. 34 (see also Fig. 3).

Before design is made final, it is well to examine the matter of accessibility. The ideal arrangement in this regard is to have relays so mounted that they are accessible for service but discourage tinkering. Use of a relay having a cover firmly in place and practically impossible to remove, as on the relay of Fig. 35, is one solution. This relay mounts on a PC card, and the entire assembly is replaced when being serviced. It is discouraging as well as expensive for an authorized service person to have to dismantle a complicated device partially in order to see if the contacts on a relay are in physical contact or in need of cleaning, for example.

Terminals and connections A variety of wire terminations as used for connecting the relay into a circuit are available for many types and kinds of relays. Some have only one kind, possibly by decree of the governing industry association. Machine-

Fig. 35 A small telephone-type relay designed for insertion into and soldering directly to a PC board.   (*GTE Automatic Electric Company, Inc.*)

tool people, for example, prefer screw terminals.   Connections to relays can be made as follows: lead wires (not usually a good idea), terminals shaped for insertion into a PC board and then to be soldered, solder terminals of the tab, tang, eyelet, or other types, screw terminals, quick connectors (AN and similar types), taper tabs, taper pins, solderless (gastight) wrap, Termi-point,° plugs and sockets, and DIP. Some relays accommodate to several of the above, some to only one or a few.   The manufacturer's catalog is usually sufficient for finding out what is available in a particular relay type, but sometimes it becomes necessary to ask about specific needs.

   Environment   For some kinds of relays any environment can be accommodated. Other designs may do a good job in only one kind of environment.   Usually, however, there is one type of relay that functions best in any specific environment.   It is not good engineering, therefore, to try to fit a personally preferred type of relay to an environment to which it is not ideally suited.   For example, it is not good practice to use military-type relays of the hermetically sealed and shock-resistant variety in a stable communications setup, or conversely to use a telephone-type relay in a military application where severe shock, vibration, excessive humidity, etc., are to be encountered.

   The relay chosen should be just as accommodating to the environment as possible but not overengineered.   Some environmental extremes occasionally encountered for which relays may have to be specially engineered are the extremes of shock, vibration, humidity, radiation, temperature, etc., especially as encountered in airborne service and space applications.

   Circuit requirements   Some circuit requirements can dictate the relay type or size.   For example, operate and release times of some circuits may require the use of copper collars, slugs, and sleeves in order to provide the needed time delays, as discussed elsewhere in this section.   Such relays have to be as large as a certain minimum size for best economy because the timing that can be generated by slugs or collars and sleeves of copper requires a certain mass and volume in order to give the desired results.   This would preclude the use of a very small relay.   But if size and weight are of prime importance, as on some airborne or space applications, then the required timing delays in the performance of the relay will have to be accomplished by other means, such as capacitors, diodes, or thermistors. Also, larger conventional relays are normally slower than smaller conventional relays, or relays designed for fast operation, such as reed contact relays, mercury-wetted contact relays, and polarized relays.   Timing may therefore require a great deal of consideration during the early design activities.

   Sequence of relay operation, exclusive of the operate or release time of any individual relay, may dictate that a particular kind of circuit performance be provided. Not all kinds of relay designs permit just any kind of circuit feature; so this too must be taken into account during the relay-selection stages.

   Vendor capability   In the end, the vendor is probably the best able to determine whether relays can be provided that will do a required job.   Some original search

° Trademark of American Pamcor Corp., Paoli, Pa.

TABLE 5   Checklist of Possible Relay Requirements for an Airborne Application[1]

I. Relay function
II. Description of equipment in which relay
   is used
III. Class of application
   A. Military
   B. Commercial
   C. Industrial
   D. Electronic
   E. Communications
   F. Commercial airborne
   G. Other
IV. Applicable documents
   A. Military specifications
      1. MIL-R-5757E
      2. MIL-R-6106F
      3. MIL-STD-202C
      4. Other
   B. Underwriters' Laboratories (UL)
   C. Canadian Standards Association
      (CSA)
   D. National Electrical Manufacturers
      Association (NEMA)
   E. Electronic Industries Association
      (EIA/RETMA)
   F. Quality assurance specifications
   G. Reliability specifications
V. Environmental tests
   A. Nonoperative
      1. Thermal shock
      2. Sealing
      3. Salt spray
      4. Humidity
   B. Operational
      1. RF noise
      2. Vibration
      3. Altitude
      4. Shock
      5. Temperature range
         (a) −55 to +85°C
         (b) −65 to +125°C
         (c) Other
      6. High- and low-temperature
         operation
      7. Temperature cycling
      8. Acceleration
      9. Random drop
VI. Contact specifications
   A. Form designation
   B. Loads (specify each pole separately
      if loads are different)
      1. Current
      2. Voltage
      3. AC or dc
      4. Frequency
      5. Resistive

   6. Inductive
      (a) Power factor
      (b) L/R ratio
   7. Motor
      (a) Starting-current transient
      (b) Locked-rotor current
   8. Lamp
      (a) Inrush current
      (b) Time to reach steady-state
         current
   C. Transient conditions (provide cali-
      brated CRO photograph)
   D. Circuit diagram
   E. Rate of operation
   F. Overload
VII. Coil specifications
   A. Resistance
   B. Impedance
   C. AC or dc
   D. Frequency
   E. Voltage
      1. Nominal
      2. Minimum
      3. Maximum
   F. Current
      1. Nominal
      2. Minimum
      3. Maximum
   G. Duty cycle
      1. On-off ratio
      2. Magnitude of on time
         (a) Minimum
         (b) Maximum
   H. Repetition rate
   I. Circuit diagram
VIII. Electrical characteristics specifications
   A. Contact resistance
   B. Insulation resistance
   C. Dielectric strength
      1. Sea level
      2. High altitude
IX. Operational specifications
   A. Pickup values
   B. Dropout values
   C. Operate time
   D. Release time
   E. Contact bounce
   F. Contact chatter
   G. Instrumentation
   H. Temperature
X. Enclosures
   A. Open
   B. Dust cover
   C. Hermetically sealed
   D. Size limitations

TABLE 5   Checklist of Possible Relay Requirements for an Airborne Application[1] (Continued)

XI. Mounting methods
XII. Termination
  A. Terminal type
    1. Solder
    2. Screw
    3. Wedge
    4. Solderless wrap
    5. Pin type (printed-circuit or plug-in)
  B. Method of connection
    1. Welding
    2. Soldering
  C. Terminal strength
XIII. Marking
  A. Type designation
  B. Part number

  C. Date code
  D. Manufacturer's code
XIV. Life expectancy
  A. Mechanical
  B. Electrical
XV. Failure criteria
  A. Minor
  B. Major
  C. Catastrophic
XVI. Qualification tests
XVII. Acceptance tests
XVIII. Procurement factors
  A. Quantity required
  B. Delivery schedule
  C. Cost limitations

of catalogs, literature, and engineering aids will be necessary on the part of the design engineer, however, in order to determine if the kind of relays the job needs seem to be available and what and how many manufacturers can provide them. Also vigilance is required to make certain the vendor does not overlook an important requirement.

If a vendor is to be reasonably certain to have what is needed and will satisfy the job requirements, all information pertinent to the performance of the required relay must be available. Specifically regulatory codes, specifications, and practices are pertinent, and/or any prescribed performance abnormalities must be known. The designer should not overspecify. This not only can forestall the use of the ideal relay but may increase costs needlessly and cripple the performance of even an ideal relay. On the other hand, the vendor must know all the pertinent facts such as voltage extremes, circuit peculiarities, shock and vibration extremes, humidity and temperature extremes, military specifications, sensitivity needs, power-supply limitations, contact performance, life, marginal operating conditions, timing required and/or delays in operation and release that cannot be tolerated, contact insulation withstanding voltage, and permissible insulation leakage.

Some commonly overlooked pitfalls are shunt circuits that were not noticed, lamps in series or parallel with the relay coil, other relay coils in series or parallel with the operating coil of the relay, capacitors that charge or are discharged in series or in shunt with the relay coil, and large inductances that affect the performance of the relay either by induction, due to proximity, or through an electrical feedback into the relay coil.

The National Association of Relay Manufacturers has prepared several specification checklists oriented to specific industries. A checklist for airborne applications, one of the broadest, is reproduced in Table 5. The type of purchase specification that might result is given in Table 6.

A too common mistake is that of a relay purchaser who attempts to apply a specification, such as one of the military specifications (see IV of Table 5), to a job for which it does not fit, is not needed, or is a handicap to obtaining the best-performing or most economical relay. In order to aid both vendor and buyer in this matter, NARM offers the use of a specification as shown by Table 7.

Incoming tests and inspections It is advisable to make adequate tests and inspections of representative samples of the purchased relays as soon as incoming shipments are being received. All too often just simple visual inspections are made of the early relay shipments by the purchaser, who may merely count and then store the relays. It is too serious a hardship to both user and vendor to discover only as the relays are being checked out later in completely manufactured units that something is wrong or missing. Losses in both time and money could have been prevented or lessened had adequate attention been paid to the relay shipments as received. If an actual functioning unit in which the relays will be used is not available for checkout purposes at the time the relays start arriving from the vendor,

TABLE 6 Example of a Detailed Relay Specification Resulting from Use of Table 5[1]

Numerical identification after each item in this sample specification refers to the corresponding section of Table 5. In actual use, of course, these references would be omitted.

Item	Checklist reference
Relay is required to switch audio-frequency circuitry in radio receiver	I
Model 9999 airborne equipment	II
This equipment will be used on commercial airlines, and the following documents are applicable: MIL-R-5757E, MIL-STD-202C, and MIL-R-6106F. Only those paragraphs specifically mentioned in this detail specification apply. In case of any discrepancy, the detail specification shall govern	IIIF and IVA
The following environmental specifications apply:	
1. Thermal shock per Test Condition B of Method 107 of MIL-STD-202C	VA 1
2. Sealing Test II per Paragraph 4.8.4.2 of MIL-R-5757E	VA 2
3. 100-h salt spray test per Paragraph 4.8.13 of MIL-R-5757E	VA 3
4. Humidity per Moisture Resistance Test Method 106A of MIL-STD-202C, except eliminate Paragraph 2.4.2	VA 4
5. Vibration Test I of Paragraph 4.8.11.1 of MIL-R-5757E	VB 2
6. Shock Test of 30 *g* per MIL-R-5757E Shock Type 4, Paragraph 4.8.16.1	VB 4
7. High-altitude performance at 70,000 ft.	VB 3
8. Relay shall operate over ambient temperature range of −65 to +125°C	VB 5b
9. High- and low-temperatue test per MIL-R-5757E, Paragraph 4.8.9 shall apply	VB 6
This relay shall have a contact form C (SPDT)	VIA
Both A and B portions of the pole shall handle similar loads of audio-frequency levels of 30 mA min to 1 A max at voltages of 100 mV to 8 V. Load will be basically resistive with power factor exceeding 0.8	VIB
Normal rate of operation in equipment will be 4 c/min with equal on and off times	VIE
The relay coil resistance shall be 250 Ω min at 25°C	VIIA
The nominal dc coil voltage shall be 28 V dc with a range of 24–32 V dc	VIIC and VIIE 1, 2, 3
Coil shall be capable of continuous duty over temperature range of −65 to +125°C	VIIG
Contact resistance shall not exceed 0.02 Ω initially when checked by voltmeter-ammeter method with an open-circuit voltage of 1 V dc and a closed-circuit current of 100 mA	VIIIA
Relay contacts shall be closed before applying test-circuit voltage	
Dielectric strength at sea level shall be required by Paragraph 4.8.5.1 of MIL-R-5757E	VIIIC 1
Dielectric strength at 70,000 ft shall be in accordance with Paragraph 4.8.5.2. of MIL-R-5757E	VIIIC 2
Insulation resistance of 1 mΩ determined per Paragraph 4.8.6 of MIL-R-5757E	VIIIB
Relay shall pick up at 20 V dc max over the temperature range of −65 to +125°C, and shall drop out at 1 to 10 V dc over the temperature range of −65 to +125°C	IXA and IXB
Relay shall be hermetically sealed	XC
Size, mounting, and solder terminals to be per drawing 9999-1 (to be included as part of the specification)	XD, XI, and XIIA 1
Relay shall be marked per Paragraph 3.39 of MIL-R-5757E, items b, e, f, and g only	XIII

TABLE 6 Example of a Detailed Relay Specification Resulting from Use of Table 5[1] (Continued)

Numerical identification after each item in this sample specification refers to the corresponding section of Table 5. In actual use, of course, these references would be omitted.

Item	Checklist reference
Electrically loaded life expectancy shall be per Paragraph 4.8.34 of MIL-R-5757E	XIVB
Failure criteria are categorized as follows:	XV
1. Minor	
(a) Marking dimensions in error	
(b) 0.1-V deviation beyond allowable limits of pickup and dropout	
2. Major	
(a) Contact resistance exceeds 0.02 Ω but is less than 0.5 Ω	
3. Catastrophic	
(a) Failure of normally open contacts to make contact when coil is energized at 28 V	
(b) Open coil circuit	
5,000 relays required, with delivery to begin 60 days after receipt of order at a rate of 100 relays per week	XVIII

it is well worthwhile for the purchaser to simulate an operating unit accurately for purposes of relay testing. Relay failure found at this stage can be due to inability of the furnished product to meet the specification, or it may have resulted from an inadvertent omission in the specification or even a failure on the part of the purchaser to recognize a necessary requirement.

Relay characteristics When it comes time to make at least a tentative choice, or choices, of relay(s) for a particular job, and the circuit and contact needs have been established, it is the individual relay type characteristics that dictate what kind of relay is best for an application. So that individual relay types can be properly evaluated, they will be examined below and their particular strengths and weaknesses assessed. There is no particular significance to the order in which they have been chosen for purposes of discussion. It is mostly chronological, with the oldest being examined first.

The General-Purpose Relay. A generalized kind of relay has become designated by its manufacturers and most users as general-purpose. This is not in accordance with the NARM definition, which says that *any* relay not a special-purpose or a definite-purpose relay is a general-purpose relay. Thus, by NARM's definition almost any type of relay could qualify as general-purpose, regardless of shape, size, or construction. This is logical, but usage and habit decree ortherwise. The commonly recognized general-purpose (GP) relay usually has a clapper-type armature, leaf springs, button contacts, and an L- or U-shaped heelpiece, with the coil pulling directly on the clapper-type armature, and the movable contacts attached to the armature. General-purpose relays come in roughly three (or more) duty ranges, light (up to 2 A), medium (2 to 5 and 10 A, as in Fig. 36), and heavy, or power-type general-purpose (contacts rated 15 A or more, as in Fig. 37). It needs to be restated that a contact that is too large for the job will often fail because the actual current is insufficient to break down surface insulation buildup. Therefore, a relay of this type is not the solution for all problems even though the contact loads to be handled do not exceed the ratings.

Some advantages for the general-purpose relay are relatively low first cost and good availability. It is usually a shelf item from stocking distributors. Disadvantages are that a general-purpose relay may not fit any particular job well in that the coil is a generalization and not readily tailored for marginal current and/or specific timing, it is frequently position-sensitive, the life is less than for many other designs, and its very size, shape, and arrangement invite tampering. It is usually not shock- or vibration-resistant, and it frequently presents mounting problems.

TABLE 7 NARM Standard Specification: Electromechanical Relays for Industrial and/or Commercial Application[1]

11.1 Scope

    11.1.1 This standard covers general-purpose electromechanical relays for industrial and/or commercial applications

    11.1.2 It is not intended to cover contactors, specific-purpose types of relays, circuit breakers, choppers, timers, or smaller allied devices

11.2 Reference Documents

    11.2.1 "Engineers' Relay Handbook" sponsored by NARM, dated 1969

11.3 Environmental Section

    11.3.1 Altitude

    11.3.1.1 Altitude will not exceed 10,000 ft above sea level

    11.3.2 Ambient

    11.3.2.1 Operating-temperature range—enclosed relays: -20 to $+40°$C (refer to Par. 12.4.1.1)

Open relays: -20 to $+40°$C

    11.3.2.2 Storage-temperature range—enclosed and open relays: -55 to $+70°$C

    11.3.3 Humidity

    11.3.3.1 Relative humidity up to 50% will be considered standard

    11.3.4 Shock and Vibration

    11.3.4.1 Only normal shock and vibration conditions encountered in handling and shipping are considered applicable

    11.3.5 Unusual Service Conditions

    11.3.5.1 The use of relays at altitudes, ambients, humidity, shock or vibration other than that specified in Par. 12.3.1, 12.3.2, 12.3.3, and 12.3.4 shall be considered as a special application. Other unusual service conditions where they exist will be called out to the manufacturer, such as:

(a) Excessive dust

(b) Excessive fumes

(c) Excessive sprays

(d) Excessive corrosion

(e) Excessive oil and oil vapor

(f) Excessive dampness

11.4 Mechanical and Physical Requirements Section

    11.4.1 Physical

    11.4.1.1 Enclosure: The relay enclosure refers to a protective enclosure which is fastened to the relay as an integral part at the place of manufacture and not an enclosure into which the complete relay is mounted with or without other components. The following enclosures shall be considered standard:

(a) Dust cover

(b) Gasket sealed

(c) Hermetically sealed

    11.4.1.2 Terminals: The following terminals shall be considered standard:

(a) Screw type

(b) Threaded stud with nut and hardware

(c) Solder lug

(d) Plug-in for socket mounting

(e) Printed-circuit board mounting

(f) Quick connect (disconnect)

(g) Clamp- or crimp-type terminals

(h) Solderless wrap

    11.4.1.3 Coils: Construction: Coil winding may be untreated, molded, vacuum-impregnated, varnish-dipped or brushed

11.5 Electrical Requirements

    11.5.1 Voltages: The following nominal voltage ratings shall be considered standard:

    11.5.1.1 AC voltage: 6 V; 12 V; 24 V; 48 V; 120 V; 208 V; 240 V; 480 V; 600 V

    11.5.1.2 DC voltage: 6 V; 12 V; 24 V; 48 V; 120 V; 240 V

TABLE 7   NARM Standard Specification: Electromechanical Relays for Industrial and/or Commercial Application[1] (Continued)

11.5 Electrical Requirements (Cont.)

11.5.2 Coils:

11.5.2.1 Range of operation: Relays are to operate satisfactorily over a range of voltage from 85 to 110% of rated nominal voltage on ac coils, and 80% of rated nominal voltage to 110% of rated voltage on dc coils. Relays will be required to pickup and seal at the minimum voltage with the coil at ultimate operating temperature due to the nominal coil voltage. The coil shall be able to withstand 10% above rated nominal voltage without injury. The above tests to be conducted at 25°C

11.5.2.2 Duty cycle: Continuous-duty coils will be considered standard except pulse-operated coils as used in latching, stepping, etc., may be considered standard when operated within their specified duty cycle

11.5.2.3 Temperature rise: The temperature rise of the coil or coils shall be limited to the allowable rise for the insulation used. An optional method to determine coil-temperature rise is that specified in UL Standard 508

11.5.2.4 Winding Tolerance: If relay coil resistance is rated or specified, the standard winding tolerance at 25°C shall be ±10%

11.5.3 Contacts:

11.5.3.1 Unless specified otherwise, contacts are rated on basis that:

(1) Each pole is capable of controlling the rated load

(2) All circuits controlled by a given pole are of the same polarity

11.5.3.2 Contacts may be rated in these terms for the following types of loads as specified by the user:

(1) Resistive—in terms of continuous current and nominal voltage

(2) Motor load—

    (a) Horsepower—in terms of horsepower and voltage

    (b) Continuous current and in-rush current at nominal voltage

(3) Lamp—in terms of type, watts, and volts

(4) Inductive—

    (a) DC—DC inductive rating shall call out the amount of inductance by specifying either maximum number of henrys or a maximum L/R ratio

    (b) AC—AC inductive rating is specified by indicating minimum power factor

11.5.3.3 Contact must be able to control rated load (service rating). An optional method to determine performance is that specified in UL Standard 508

11.5.3.4 Life—Ratings must be determined by application requirement. Relay life varies with the application and is not directly related to ratings. When life is specified, the following levels are preferrred or should be used as a guideline for actual application requirements:

No. of operations (electrical)

(1)   10,000
(2)   25,000
(3)   50,000
(4)   100,000
(5)   250,000
(6)   500,000
(7)   1,000,000
(8)   3,000,000
(9)   5,000,000
(10) 10,000,000
(11) 100,000,000 and over

11.6 Test Section

The performance for relays meeting this standard shall be tests performed on units in accordance with the requirements listed below. All tests shall be performed at room temperature with the relay in its normal mounting position. Performance tests shall include:

11.6.1 Visual inspection to ensure compliance with outline drawings and standards of good workmanship

TABLE 7   NARM Standard Specification: Electromechanical Relays for Industrial and/or Commercial Application[1] (Continued)

11.6 Test Section (Cont.)

  11.6.2 Pickup voltage—All relays are to pick up at 85% of rated nominal ac voltage and at 80% of rated nominal dc voltage.

  11.6.3 Contact and coil continuity—Check coil continuity and/or resistance. Check contact continuity with the relay energized and deenergized

  11.6.4 Dielectric strength—Test dielectric strength to values as specified by the manufacturer

  11.6.5 Life testing—All relays shall meet manufacturer's standard and/or special ratings when applicable and tested as specified by the manufacturer

11.7 Marking

  11.7.1 Relays manufactured to satisfy this standard will include as visible minimum marking:

  (a) Manufacturer's name or trademark

  (b) Manufacturer's part number

  (c) Coil rating (voltage and frequency when applicable)

  11.7.1.1 The following would be considered optional marking requirements:

  (a) Contact rating (voltage and current)

  (b) Customer's designation

  (c) Circuit-connection diagram

The contact rating for relays of this type is usually stated in the vendor's catalog as 2, 5, or 10 A at 120 V ac, 0.8 power factor, or 28 V dc. If it is capable of handling motor and such kinds of loads, the horsepower rating is given. Underwriters' recognition will be indicated if applicable, as is the case for the relay of Fig. 36.

Fig. 36 A typical general-purpose relay. (*Potter & Brumfield, Div. of AMF, Inc.*)

Fig. 37 A Typical Underwriters' listed power relay. (*Potter & Brumfield, Div. of AMF, Inc.*)

Contact resistance is the Ohm's-law resistance measured at closed contacts and is usually in the order of 50 to 100 mΩ for new, clean contacts.

Insulation resistance is typically 1,000 MΩ, minimum, at 500 V dc.

Life expectancy for the contacts may be specified at some value such as 100,000 operations at a specifically indicated electrical load. The mechanical life of the mechanism is usually indicated as 10 million operations. For lighter than rated loads the contacts could presumably last as long as the mechanism.

General-purpose relays are employed principally in the fields of air-conditioning and heating equipment, household electrical appliances, coin-operated machines, control of low-wattage motors, some lighting controls, and some kinds of elevator controls. A special design of this type of relay is used on automobiles and trucks.

The Power-Type Relay. The appearance of most power-type relays is much like the general-purpose relay, only larger or more rugged (see Fig. 37). Usually the insulation is thicker or of superior material, the terminals larger, usually screw type, and in general favorably looked upon by the underwriters; hence it may be UL listed.   The contacts are adequate for quite heavy current and highly inductive loads, with large armature strokes and contact gaps.   Thus the sensitivity is not great. Contact-current rating is usually 20 to 25 A or more.   They are frequently position-sensitive and usually not resistant to shock or vibration.   There is not much latitude in mounting method or location.

The advantages of relays of this kind are they can best handle heavy contact loads, heavy fixture wire is easily attached to the coil and contacts, repairs can be made by relatively inexpert maintenance personnel, without sophisticated tools, and visual inspection of the contacts to determine probable remaining life is relatively easy.   Also, in power-handling and switching situations the rugged appearance builds confidence in the user's mind that the best relay for the job has been employed.

Power-type relays usually are employed in these fields: commercial aircraft (when of a specially designed type), air-conditioning and heating equipment, household electrical appliances, electric power control, machine-tool control, and some military applications (when of a specially designed type).   The specific job is usually electric-motor control.

The Telephone-Type Relay.   Telephone-type relays were developed and perfected during decades of application to the switching and signaling needs of wired communications, where the contacts of the same relay are required to successfully carry wide ranges of power extending from "dry" voltage voice circuits to the medium power levels employed in the actual switching of relays and other electromagnets.   This versatility with respect to handling a variety of power levels made relays of this type valuable commercially in other fields with similar requirements.

The original telephone-type relays were approximately 4 in long, 1½ in high, and varied in width from approximately 1¾ to 2¼ in, depending upon the number of contact springs in the pileup(s).   Figure 13 is a generalized concept of such a relay.

Size and weight reduction were problems that eventually faced the designers of telephone-type relays, resulting in an intermediate-sized relay, as demonstrated by Fig. 2.   This relay has roughly one-half the volume of the full-sized telephone-type relay, with the reduction being principally one of length.

Still later, designs incorporating the permissive make technique (described earlier), and demonstrated by Figs. 14 and 35, evolved.   These relays sacrifice something in the way of controlled timing, contact versatility and flexibility but lend themselves particularly well to the modern mounting methods and circuit-design concepts.

A comparison of timing delays obtainable from full-sized telephone-type relays and reduced-size but similar relays is shown by Table 8.   The miniature, permissive

TABLE 8   Timing Limits of Telephone-Type Relays

Slug or sleeve position	Large size		Reduced size	
	Operate time, ms	Release time, ms	Operate time, ms	Release time, ms
Armature end	100–150 max, 25–40 min	300–750 max, 75–200 min	25–40 max, 10–15 min	40–120 max, 15–25 min
Heel end	4–25	300–750 max, 75–200 min	7–10	40–120 max, 15–25 min
Sleeve (full length of coil core)	25–75	300–750 max, 75–200 min	10–25	40–120 max, 15–25 min

SOURCE: Industrial Products Division, GTE Automatic Electric Company, Inc.

make type of relays employ coils too small for the addition of sleeves or slugs; hence any timing desired beyond what is inherent requires the addition of capacitors, diodes, or other extraneous circuit devices to accomplish what is wanted.

The advantages for telephone-type relays in general are high mounting density, a practically unlimited kind and quantity of contact forms and materials, good sensitivity, high order of contact reliability, good control of and practical variation in timing, fairly insensitive to mounting position, capable of withstanding moderate shock and vibration both while operating or at rest, and moderately light and small. Telephone-type relays are not readily adaptable to heavy-duty contact loads, usually are not capable of operating under heavy shock or vibration conditions, are not readily adaptable to underwriters' requirements for insulation resistance and voltage breakdown, and are difficult for inexperienced personnel to service on the job (contact springs are small and closely spaced). In an attempt to alleviate the latter difficulty, a great variety of enclosures and mountings are available.

Telephone-type relays have appeared in practially every field of application, sometimes by dint of heavy modification. In general, they are readily applicable to business machines, coin-operated machines, telephone and all other wired communications systems, radio and microwave systems, computer input-output devices, electronic data processing, laboratory test instruments, lighting controls, machine-tool control logic, production test equipment, street-traffic control, and military ground-defense systems. When properly modified and/or protected from environment shock and vibration, telephone-type relays have been used on commerical aircraft and military aircraft, in aerospace, and on naval shipboard.

The Dry-Reed (Contact) Relay. Another telephone-industry design innovation has captured a large share of the relay market. In this case the relay electromagnet (coil) generates a flux that acts directly on the contacts without employing any linkage in the form of an armature. Two normally separated, electrically conducting, and magnetic-flux-conducting elements, in a sealed glass envelope, provide a portion of the main flux path of the coil, so that when the coil is energized these elements are attracted to each other to form a closed contact. They may also be permanent magnetic-flux-biased so that they are normally closed but are open when the coil is energized sufficiently to neutralize the permanent-magnet flux. This kind of a relay in its simplest form is pictured in Fig. 15.

Up to this point in this discussion we have been concerned with relays having armatures to actuate the contacts. Such relays have long been referred to as electromagnetic, because the armature was attracted by the flux from an electromagnet, as opposed to thermal relays, operating from heat, electrostatic relays, operating from opposite electrical charges, etc. Many people would now like to reserve the designation of electromechanical for relays with armatures, and electromagnetic for the reed relays. Currently such a practice is more confusing than clarifying, but perhaps in the future this kind of a distinction might be embraced. If or until then this class of relays will be most often recognized by the term reed relay with further identification as to type, such as dry reed, mercury-wetted reed, and Fereed* (principally used in telephony, and not ordinarily available as a commercial item). These all differ from the resonant-reed relay, which is a frequency-sensitive mechanically resonant device, to be discussed later.

The reed switches used in reed relays come in a wide variety of forms and sizes. It was mentioned above that the basic switch is a normally open, or Form A. When biased with a permanent magnet, it becomes a Form B, or normally closed. Originally all Form C relays were a pair of these wired together. Now there are "true Form C" contacts in which the movable contact member is caused to move from its normally closed (biased) position to the other position when the operating coil is energized. Sizes of contacts are difficult to categorize because there is little standardization among manufacturers. Roughly they can be described as (1) the "regular" Bell Laboratories original design, which was approximately $3\frac{1}{4}$ in in length (2 in glass length) and $\frac{7}{32}$ in in diameter; (2) the so-called minature, which is about $1\frac{5}{8}$ in long with a diameter of approximately $\frac{1}{10}$ in; and (3) microminiature, having a length of $1\frac{3}{8}$ in or less and a diameter of less than $\frac{1}{10}$ in. Other appelations are micromicro, pico, etc. In order to determine what is meant by these, a detailed examination of the manufacturer's literature is required, as no standardized terms yet exist regarding size.

There is a tendency to size similarity that makes some comparisons possible. Most

* Trademark, Western Electric Company.

relays are grouped roughly into the above three categories for contact size, and the contact ratings and capabilities result in something similar to Table 6 for dry reeds with gold-plated or rhodium (special) plated contacts. Other ratings are achieved by wetting the contacts with mercury, differently shaping the glass envelopes, potting, encasing in epoxy, etc.

There is no clearly defined limit to the number of reed capsules that may be put into a single relay for operation from a common coil. Many manufacturers offer custom packaging, if the job size warrants it. All vendors have multicapsule standard offerings with all makes (Form A), combinations of makes and breaks (Forms A and B), or transfers. As an example of the variety offered, see Fig. 38.

Fig. 38 An example of possible variations in number of dry-reed switches available in a relay. (*C. P. Clare & Co.*)

Dry-reed relay structures can be categorized as open assemblies, enclosed assemblies, potted or molded assemblies, and hermetically sealed assemblies.

Among printed-board relays, the open assembly is frequently used. The basic structure, typically molded of an electrical grade plastic, functions as a coil bobbin and provides a means of supporting the reed switches and relay terminals. In many cases, the switch terminals are formed so that they will insert directly into the printed board.

This same structure is often placed in a metal or plastic box, which is then filled with a potting material. In some instances, the box provides terminal support, further simplifying the internal structure or permitting use of a self-supporting coil. These assemblies may offer improved resistance to environmental or handling stresses.

Molded relays are similar to the potted relays, but differ in materials used. In these assemblies, the molding material provides the primary mechanical support for the switches and coil and also produces the finished external surfaces.

Assemblies similar to the potted relays can be hermetically sealed, using a metal cover and a base with terminals mounted in glass-to-metal seals.

Plug-in and wire-in relay assemblies follow the general patterns described for printed-board relays. Plug-in relays usually use a potting material to support the coil-switch structure within the enclosure carrying the plug. Wire-in relays follow both the open and potted patterns. The DIP has become very much in demand. See Fig. 3.

In general, dry-reed devices can be characterized as quite susceptible to the influences of external magnetic fields. For this reason, and to improve magnetic coupling of the coil to the switches, many relays incorporate some form of magnetic shielding. Metal cases serve this function, as do internal wraps or plates affixed to the coil. In some instances, the magnetic shield is connected to a terminal which may be grounded to provide electrostatic shielding, but in most cases the electrostatic shield is nonferrous and separate from the magnetic shield.

Special Dry-Reed Contact Relays. High Voltage. Typical dry-reed switches are rather limited in their ability to withstand high voltages across their open contacts, with the standard switch rated at 500 V rms. For special applications, switches which can withstand voltages as high as 10,000 V can be incorporated in assemblies similar to those described earlier. Terminal spacing is modified as required to withstand the voltages.

While modifications for higher voltage typically incorporate special versions of the standard dry-reed capsule, similar versions of the miniature switches are also available. They are limited to about 2,000 V.

Power. Reed switches capable of handling power in excess of the typical 15-VA rating of the standard switch capsule are available. Relays incorporating these switches, rated at 50 to 350 VA, are defined as power relays.

High Insulation Resistance. Reed switches manufactured under carefully controlled processes provide an insulation resistance between contacts of in excess of 10^{12} Ω. While most standard relay assemblies provide shunting paths which appreciably lower insulation resistance, special structures using appropriate materials and processes can preserve the basic high-insulation-resistance capability of the switch.

Low Thermal Voltage. Relays typically produce a voltage between contact terminals as a result of differing temperatures between the junctions of materials in the assembly. Changing ambient temperatures or heat produced by the relay coil cause temperature gradients within the relay. Dry-reed relays incorporating materials and assembly techniques which minimize these effects are available.

Low Noise. The cantilever reed members in a switch continue to move for a few milliseconds following contact closure. This motion can produce a variation in contact resistance, and it does cause a voltage to be generated between switch terminals. Relays using reeds and structural techniques which minimize the latter effect are called low-noise relays.

Low Capacitance. Since the contact-overlap area of most reed switches is small, capacitance between contacts is small. When the switch is installed in a coil, this capacitance is paralleled by the comparatively large capacitance of individual reed blades to the coil. The resulting increased capacitance across contacts and the capacitive coupling from coil to reeds can be objectionable in some applications.

By interposing an electrostatic shield between the reed switch and the coil, the paralleling capacitances can be greatly reduced, with capacitance across contacts approaching basic switch capacitance. In multipole relays, the electrostatic shield can be interposed between the switch group and the coil, or can also be interposed between individual switches.

Cross Point. Relays used in matrix applications are called cross-point relays. Reed relays adapt readily to the various schemes of "no response to one input—response to two inputs" and have been used extensively in matrices.

Logic Devices. Reed relays readily adapt to the performance of logic functions through the addition or subtraction of magnetic fluxes produced by multiple coils.

Electrical Characteristics. Contact ratings are shown on a generalized basis in Table 9. Specific figures from individual manufacturers may vary somewhat from these values. Other data, as presented here for such things as timing, sensitivity,

TABLE 9 Dry-Reed-Relay Contact Capabilities[1]

Reed size	Ratings
Standard	Load: 15 VA, 1 A max, 250 V ac max, 3.0 A (carry) Withstanding voltage: 500 rms, 60 Hz Insulation resistance: 10^{11} Ω min Initial contact resistance: 40–100 mΩ Life: 20 million operations at rated load
Miniature	Load: 10 VA, 0.75 A max, 200 V dc max, 1.0 A (carry) Withstanding voltage: 250 rms, 60 Hz Insulation resistance: 10^{10} Ω min Initial contact resistance: 100–200 mΩ (regular), 100 mΩ (with special plating)
Microminiature	Load: 10 VA, 0.50 A max, 100 V dc max Withstanding voltage: 200 rms, 60 Hz Insulation resistance: 10^{10} Ω min Initial contact resistance: 100–250 mΩ (regular), 100 mΩ (with special plating)

contact bounce, and capacitance, may also be at variance with the claims of any specific manufacturer or product.

Sensitivity   Power input required to operate dry-reed relays is determined by the sensitivity of the particular reed switch used, by the number of reeds operated by the coil, by the permanent-magnet biasing used, and by the efficiency of the coil and the effectiveness of its coupling to the reeds.   Minimum input required to effect closure ranges from the very low milliwatt level for a single-capsule "sensitive" unit to several watts for a standard multipole relay.

Operate Time.   Coil time constant, overdrive, and the characteristics of the reed switch determine operate time.   With maximum overdrive, standard reed switches will operate in just under 1 ms; miniature reeds in 500 μs, and microminiature reeds in less than 200 μs.   At normal drive levels, operate times will be two to three times these values.

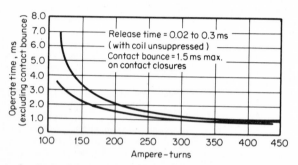

Fig. 39 Typical relationship of operate time to operating coil power for a standard-size switch in a dry-reed relay. (*C. P. Clare & Co.*)

The other end of the operate time spectrum is less definable, since coil time constant and drive level are the primary determinants.   However, with the low inductance typical of reed relay coils, operate times of even the standard reeds rarely exceed 10 ms.   Figure 39 shows the operate time of one manufacturer's standard reed relay with a single Form A switch.

Release Time.   With the relay coil unsuppressed, dry-reed switch contacts release in a fraction of a millisecond.   Miniature and microminiature Form A contacts open in as little as 10 to 20 μs.   Standard switches open in 100 μs.   Magnetically biased Form B contacts and normally closed contacts of Form C switches reclose in from 100 μs to 1 ms.

If the relay coil is suppressed, release times are increased.   Resistor-capacitor suppression usually has the least effect.   Zener-diode suppression stretches release time somewhat more. Diode suppression can delay release for several milliseconds, depending on coil characteristics, drive level, and reed-release characteristics.

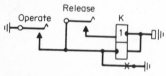

Fig. 40 Forced release of a dry-reed relay.

Figure 39 also shows a value for expected release time with an unsuppressed coil.   Faster and more positive release may be obtained by using two exactly equal but opposed windings, as in Fig. 40.

Bounce.   As with other hard contact switches, dry-reed contacts bounce on closure.   The duration of bounce is typically quite short, and is in part dependent on drive level.   In some of the faster devices, the sum of operate time and bounce is relatively constant as drive is increased, operate time decreasing, and bounce increasing.

While normally closed contacts—those which are mechanically biased—bounce more than normally open contacts, magnetically biased Form B contacts exhibit essentially the same bounce as Form A's.

Typical bounce times (mechanically biased normally closed contacts) in milliseconds are:

Standard
    Normally open   0.50
    Normally closed   2.5
Miniature
    Normally open   0.25
    Normally closed   2.0
Microminiature
    Normally open   0.25

Capitance. Reed capsules typically have low terminal-to-terminal capacitance. However, in the usual relay structure where the switch is surrounded by a coil, capacitances from each reed to the coil act to increase basic capacitance many times. If the increased capacitance is objectionable, it can be reduced by placing a grounded electrostatic shield between the switch and coil.

Typical capacitance values, in picofarads, for relays using Form A switches are:

Relay type	Unshielded		Shielded	
	Across contacts	Closed switch to coil	Across contacts	Closed switch to shield
Standard	1.0	4.0	0.2	5.0
Miniature	0.7	3.0	0.1	3.5
Microminiature	0.4	2.0	0.08	2.5

Where the capacitance from contact to shield is objectionable, greater spacing or unique methods of coupling the coil to the contacts may be employed.

Thermal EMF. Since thermally generated voltages result from thermal gradients within the relay assembly, relays built to minimize this effect often use sensitive switches to reduce required coil power and thermally conductive materials to reduce temperature gradients. Latching relays, which may be operated by short-duration pulses, are often used if the operational rate is such that the potential benefit of reduced duty cycle can be realized.

Measurements of thermal emf are specified in a number of ways, each suited to a particular application. One of the more standard, and the one documented in MIL-R-5757E, is measurement at maximum ambient with continuous coil input for a time sufficient to ensure temperature stability. Measured in this manner, relays can be supplied to specifications with limits as low as 10 μV.

Noise. In reed relays, noise is defined as a voltage appearing between terminals of a switch for a few milliseconds following closure. It occurs because the reeds are moving in a magnetic field and because voltages are produced within them by magnetostrictive effects. From an application standpoint, noise is important if the signal switched by the reed is to be used in the few milliseconds immediately following closure, if the level of noise comparses unfavorably with the signal level, or if the frequencies constituting the noise cannot be filtered conveniently.

When noise is critical in an application, a peak-to-peak limit is established, with measurement made a specified number of milliseconds following application of coil power. Measurement techniques, including filters which are to be used, are also specified. MIL-R-5757E, for example, sets a peak-to-peak limit of 50 μV at 10 ms with frequencies below 600 Hz and above 100 Hz attenuated.

Vibration. Except at resonant points, reed switches do well when subjected to vibration. With vibratory inputs reasonably separated from the resonant frequency, the relay will withstand relatively high inputs, 20 g's or more. At resonance of the reeds, the typical device will fail at very low level inputs.

Typical resonant frequencies in hertz are

Standard   800
Miniature   2,500
Microminiature   5,000

Shock. Dry-reed relays withstand relatively high levels of shock. Form A contacts are usually rated as able to pass 30 to 50 *g*'s, 11 ms, half-sine-wave shock, without false operation of contacts. Switches exposed to a magnetic field tending to close them, such as in the biased latching form, demonstrate somewhat lower resistance to shock. Normally closed contacts of mechanically biased Form C switches may also fail at somewhat lower levels.

Radiation. The basic reed switch is quite resistant to radiation and has been used to perform control functions in hot environments. Coils and supporting structures utilizing appropriate materials permit the construction of reed relays resistant to radiation.

Dry-reed relays are quite adaptable to electronic applications and hence are found in electronic equipment in many fields such as logic circuits. Some fields to which they are especially suited are business machines, telephone and other wired communications systems, radio and microwave systems, computer input-output devices, electronic data processing, laboratory test instruments, machine-tool control (logic circuits, particularly), production test equipment, and military ground-defense systems.

Advantages existing for dry-reed relays are that they are small, fast-acting, provide good isolation of input to output, lend themselves well to use with solid-state circuitry, mount easily on PC cards, and require relatively small amounts of power. Their chief weakness is a tendency toward contact sticking and unwanted welding. At times it is difficult to locate the source for this kind of a problem. Such an unsuspected thing as accumulated cable capacitance may be the cause of this kind of failure. Lamps are also a frequent source of such trouble.

Resonant Reed Relays. This type of relay is designed to respond to a given freqency of coil input current. It consists of an electromagnetic coil that, when energized, drives a vibrating reed with a contact at its end. When the coil input frequency corresponds to the resonant frequency of the reed, the reed will vibrate and cause its contact to touch a stationary contact and thereby close a circuit once each electrical cycle. At other frequencies, the reed does not respond. Sometimes the reed is surrounded for a portion of its length with a permanent-magnet field to provide a constant magnetic bias. Since the vibrating reed closes its contact only for a portion of each cycle, it is often necessary to provide an output circuit that will store these pulses long enough to operate a conventional relay for control purposes.

Resonant-reed relays can be built with a number of reeds having frequency responses in discrete steps, thus providing a device that will give signals on either side of a desired frequency for control purposes. Resonant-reed relays are also used for a variety of applications where response to frequency only is desired, such as communications, selective signaling, data transmission, and telemetry.

The following material is from Ref. 3.

The physical construction of a typical resonant-reed relay is shown in Fig. 41. Notice how closely it resembles a conventional relay, having a coil, a relay armature

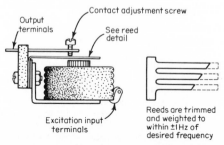

Fig. 41   The physical construction of a typical resonant-reed relay.[3]

(resonant reed) and a fixed contact. In this case, the fixed contact is adjustable to obtain maximum dwell time from the reed. When the reed is gold plated, it resists environemntal contaminants and provides an excellent conductive path. The upper portion of Fig. [41] shows how the reeds are "tuned."

These reeds respond to almost any frequency in the audio range. For practical operation, the range of 100 Hz to 10 kHz can be considered useful for exciting resonant-reed relays. Harmonic rejection is usually quite good within this range, and simultaneous tone sensing is possible under most circumstances. Thus, if a resonant-red relay has eight or ten tuned reeds, it is quite possible that two or three could be excited at the same time by a like number of input tones. For practical purposes, the designer would probably select reed frequencies that nullify harmonic interaction if concurrent excitation is required.

The circuit configuration of these devices is shown in Fig. [42]. Obviously, the resonant-reed relay is a very elementary device whose theory of operation is within

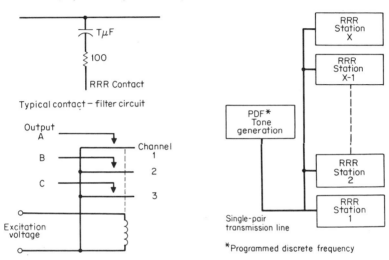

Fig. 42 Circuit configuration and typical contact-filter circuit of a resonant-reed relay.[3] ° Programmed discrete frequency. Although only three channels are shown, as many as ten to twelve resonant reeds may be excited by a single coil.

Fig. 43 Simplified diagram of data-link hookup between tone generator and resonant-reed relay.[3]

the understanding of almost anyone. However, its simplicity can be misleading to those who do not understand its versatility and usefulness. A resistance/capacitance filter, such as the one shown in Fig. 42, may be used to reduce the dc ripple in the output. However, buffer amplifiers working from these devices must be biased below the ripple, because it is not practical to eliminate all of it. The common connection for the reeds and excitation coil is optional and may be undesirable in some cases.

The functional aspects of resonant-reed relays are their biggest selling point. Regardless of the number of relays used, the input consists of a tone generator and a single-pair transmission line. Theoretically, the tone generator will produce programmed discrete frequencies. The programming may consist of presetting potentiometers, or more elaborate means for frequency control.

Figure [43] is a simplified diagram illustrating the data-link hookup between the tone generator and the resonant-reed relay network. Noise spikes that could be dangerous to semiconductor circuitry will have little effect on these devices, and therefore, elaborate filtering precautions are not necessary. The resonant frequencies of the reeds may be intermixed or repeated in almost any fashion in the relay stations. There is little or no interaction, even when a reed frequency is repeated on the same relay.

The power output requirement for the programmed discrete-frequency tone generator is naturally dependent on the number of relays it is driving. A wide range of relay coil impedances may be used, but an impedance of approximately 5 kΩ should be used for rule-of-thumb calculations. In practice, impedances may vary from near 100Ω to more than 10 kΩ. The relays are tolerant of considerable tone-signal distortion, although it is helpful to keep the signal as pure as practical.

If the resonant-reed relay is to drive a conventional slave relay, as shown in Fig. [44], nothing more than a simple filter circuit is needed for the interface. The pulsating dc output signal can be adjusted to provide approximately a 50 percent duty cycle. With the additional filtering and the response characteristics of the slave relay, this direct connection works quite well.

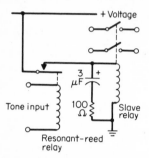

Fig. 44 Method of driving a slave relay from contacts of resonant-reed relay.[3]

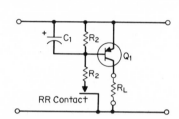

Fig. 45 Method of using a buffer amplifier between a resonant-reed relay and the load.[3]

Restrictive output criteria may call for buffer amplifiers or other considerations. One such buffer amplifier is shown in Fig. [45]. The number of possible buffer-amplifier configurations is no doubt apparent to the designer.

Since resonant-reed relays are, in effect, gates for tone-modulated signals, they may also be used as logic elements in some applications. To be sure, these are unusual logic elements, but it is not unusual to find sequential tone coding which could be considered as a form of logic. With this as with other applications, the designer would do well to consider all of the tradeoff factors before ruling out the use of resonant-reed relays.

Resonant-reed relays have same disadvantages. The contacts do not close with a firm positive closure; hence many kinds of equipment cannot be easily controlled from the contacts. These relays sometimes demonstrate undesired frequency drift, due to temperature extremes, tampering, shock, vibration, or other environmental extremes. Advantages, besides those covered in detail above, are simplicity and low cost.

Because of the nature of the contacting action, loads other than those of the type described above are not advisable.

Mercury-wetted Contact Relays. The switching element in mercury-wetted contact relays is considerably more sophisticated than the deceptively simple-appearing dry-reed switch. There are essentially two design sizes and types of mercury-wetted capsules in common usage. The original type of design, of Bell Laboratories parentage, was a maximum 5-A load-current-bridging (Form D) capsule. See Fig. 46. High-pressure gas, sealed in the capsule, provides a maximum voltage rating of 500 and a load capability of 250 VA, but a resistor-capacitor protection of the contacts is essential.

A smaller capsule that may be either bridging or nonbridging and rated at approximately 2A or less depending upon required switching speed is demonstrated by Fig. 47. The maximum load capacity for the smaller contact is 100 VA, 500 V, and resistor-capacitor contact protection is required.

In both contact types the electrical contacting is mercury to mercury, with the contacting faces renewed by capillary action drawing a film of mercury over the surfaces of the contact switching members as the movable contact member is moved from one transfer position to the other. The mercury film is drawn up from a pool at the bottom of the capsule, between the stationary members to provide bridging, or Form D, contacting (Fig. 48). Contact bounce and chatter are eliminated by

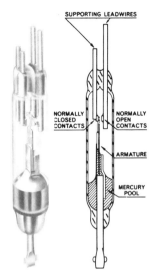

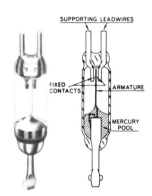

Fig. 46 The basic design of mercury-wetted contact relay. (Contact Form D). (*C. P. Clare & Co.*)

Fig. 47 A smaller and faster mercury-wetted contact relay (can be either Form C or Form D contact). (*C. P. Clare & Co.*)

the dampening effect of the mercury films. No solid metal to solid metal contacting takes place; so the contacts are actually renewed on each operation. As a result, wide ranges of signal and power levels can be reliably switched without having the natue of the load affect either contact life or performance.

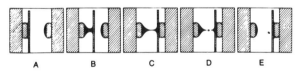

Fig. 48 Form D action (SPDT make before break) in a mercury-wetted contact. (*C. P. Clare & Co.*)

The action of the mercury is similar for Form C (nonbridging) contacting, except that spacing is such that a droplet of mercury falls out before bridging can occur (see Fig. 49).

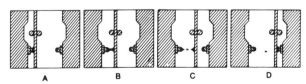

Fig. 49 Form C action (break before make) in a mercury-wetted contact. (*C. P. Clare & Co.*)

For one form of relay from one to four capsules are operated by a single coil and housed in an octal-base electron-tube type of metal can. Examples of this kind of relay are provided by Fig. 50. More compact forms, suitable for mounting directly on printed-circuit boards, are also available. With regard to mounting, it must be

Fig. 50   Octal-base types of mercury-wetted contact relays.   (*C. P. Clare & Co.*)

remembered that these relays are quite position-sensitive and must be used with the capsule right side up and with its axis tilted less than 20 to 30° from the vertical, depending upon type and manufacturer.   After having been inverted, the contacts are flooded from the mercury pool and may not perform properly for some time   Another problem sometimes encountered is failure at low temperature due to the fact that mercury becomes solid at $-38.8°C$ ($-37.8°F$).

Operating speeds for relays using the large capsule are of the order of 6 ms to transfer and 4 ms to restore.   For the smaller capsule these figures are about 1 ms each.   An example of the small relay is Fig. 51. One manufacturer's performance data for the various kinds of this relay are shown in Table 10.

Contact protection is a matter to be fully explored with the vendor, whose recommendations should be followed exactly if good contact performance is to be achieved.   A formula in common usage for a choice of capacitor-resistor for this purpose is

$$C = \frac{I_2}{10 \ \mu\text{F}}$$

$$C_{\min} = 0.001 \ \mu\text{F}$$

$$R = \frac{E}{10I^{(1+50/E)}} \ \Omega$$

$$R_{\min} = 0.5 \ \Omega$$

in which I = current immediately prior to contact opening
E = source voltage

Fig. 51   A small mercury-wetted contact relay for PC-board mounting.   (*C. P. Clare & Co.*)

A list of the advantages for the mercury-wetted contact relay would include these items: good operating sensitivity, fast contact switching, long contact life, great contact reliability, low and stable contact resistance, bounce-free and chatterless contacting freedom from detrimental effect of environment such as dust, humidity, and changes in atmospheric pressure, good contacting capability in the low-power region, protection from igniting hazardous atmospheres, and excellent isolation between coil input and contact output.   Shortcomings for this relay may well be that it is position-sensitive, cannot work at low temperatures, generates too much thermal emf under some conditions where it would otherwise be ideal, and is costly.   It is ideal for pulsing highly inductive electromagnets, such as rotary stepping switches.

Fields of application for this kind of relay can be air conditioning and heating, business machines, wired   communications, computer input-ouptut devices, electric power control, electronic data processing, laboratory test instruments, lighting controls, machine-tool controls, production test equipment, and street-traffic control.

Mercury-wetted Reed Relays.   These relays are basically of the same type as the dry-reed relays, except that mercury has been added to the reed capsule during

TABLE 10 Typical Mercury-wetted Contact-Relay Capsule Characteristics at Optimum Drive Conditions[9]

Capsule	Timing (speed) Operate, ms*	Timing (speed) Release, ms†	Timing (speed) Frequency, Hz	Load (max) VA	Load (max) V	Load (max) I	Form C transfer open time, µs Min	Form C transfer open time, µs Max	Form D bridging time, µs Min	Form D bridging time, µs Max	Sensitivity, mW	Min on and off time, ms	Contact noise at closure, 1-kHz bandwidth, µV
High-speed	1.0	1.0	400	50	500	1		‡		‡	100	1.25	30
Intermediate-speed	1.25	1.25	250	100	500	2	50	350	50	800	100	2.5	60
Standard	2.0	2.0	125	100	500	2	50	500	50	800	100	4.0	75
Heavy-duty	4.0	2.5	80	250	500	5			150	900	500	5.0	50

* Measured at twice minimum operating voltage.
† Measured with nominal coil suppression.
‡ Form T contact; see switch action discussion.

manufacture so that contacting takes place from mercury film to mercury film.   There is little standardization, and the ratings and characteristics vary so from manufacturer to manufacturer that if this type of relay seems desirable for a job, the first order of business is to consult with the vendor.

The advantages for this kind of relay are obviously many of those which exist for the dry-reed relay, plus the advantages for mercury-to-mercury film contacting. Disadvantages are essentially the same as those listed above for mercury-wetted contact relays.   Some vendors claim to have solved the position-sensitivity and tilt problem.

The main purpose in adding the mercury to the capsule was to reduce and stabilize contact resistance, and to increase contact load-handling capability and life.   The maximum allowable load varies so much from one reed capsule design to another that the vendor's catalog must be relied upon entirely in determining what is available and whether or not it will do the job.   One word of caution to be observed is that use of the vendor's prescribed contact protection is a must.

Application areas for these relays are essentially the same as listed above for mercury-wetted contact relays, with obviously the fields of aviation, automobiles and trucks, and military being largely incompatible.

Heavy-Duty Power-Type Mercury Contact Relays.   Since contact erosion, with resulting need for maintenance, is the largest single problem for relays handling heavy power levels on their contacts, it was inevitable that the continuous contact-renewal capabilities of mercury would lead to early consideration of this metal for switching power.   Rather than the making of contacts through a thin mercury film, as was discussed above, the conduction in power-type mercury contacts is through a pool of mercury.   There are two principal ways of accomplishing this: the tilted mercury tube, which causes the terminals to be bridged when the tube containing the mercury is in one position and non-bridged or open in the other position, and the mercury-displacement technique where a plunger is pulled down into the mercury of the pool so that a bridge of conducting mercury extends from one terminal to the other, thus closing a circuit over a dam that otherwise isolates one terminal from the other.   In the latter method, when the coil is deenergized, the plunger floats back up, the mercury returns to refill the pool, and the circuit is opened.

Fig.   52 A heavy-duty, power-type mercury contact relay.   (*Magnecraft Electric Co.*)

Contact ratings usually vary from 10 to 35 A for 115 V ac and approximately half as much current at 230 V ac.   Some types are rated up to 100 A and are UL and CSA listed.   DC ratings are considerably reduced from the above because of the sustained arc possibilities.   One example of a relay of this kind is shown in Fig. 52.

Advantages for the heavy-duty mercury contact relays are these: the contacts give some protection against igniting an explosion in hazardous atmospheres (they are sealed); they are protected from corrosive, dirty, or moist ambient conditions; they are not noisy when operating; there is no appreciable maintenance; and usually there is some saving in weight and size.   Disadvantages are of course those associated with a liquid such as mercury, which are position sensitivity, inability to function at low temperatures, inability to function where there is severe shock or vibration, and a tendency to delay functioning longer than is the case for the more common power-type relays.

Crystal-Can Relays.   The size and appearance of relays changed greatly in World War II from what was in common usage previously.   Detrimental environmental service conditions dictated that relays be hermetically sealed, shock- and vibration-resistant, and be made as small and lightweight as possible.   In the absence of service-tested new designs aimed specifically at the above, much of the relay use in

World War II had to be satisfied by existing designs modified as required to meet the service requirements. In many cases this resulted in compromises with regard to space, weight, shock, and vibration that caused extensive design effort in the direction of producing relays specifically aimed at the military's needs. Today's design is usually called a crystal-can relay or some fraction thereof, such as half crystal can or one-sixth-size crystal can. The reference is to the size and shape of the relay housing, originally used to house a frequency-control type of quartz crystal. The dimensions of a full-sized crystal can are usually assumed to be 0.4 by 0.8 by 0.97 in. A picture of one kind of such relay with three different kinds of terminals is shown in Fig. 53. Various methods of mounting, as well as kinds of terminals, are used. This is a specific kind, size, and shape of relay, but with many variations on its exterior.

The distinguishing design features of this relay are dynamically balanced armature, designed-in resistance to shock and vibration, and hermetic sealing. A majority of them have their terminals spaced on a 0.2-in grid, which makes them readily inserted into printed-circuit boards. Also, most relays of this type offer only two Form C contact combinations. Because the relays are small and lightweight, only relatively small contacts with fairly light pressures can be operated; hence the contact ratings must be restricted to relatively light loads.

Fig. 53 Typical crystal-can relays. (*C. P. Clare & Co.*)

Relays of this type are furnished both as conventional energized-on and deenergized-off relays, and as bistable magnetically latched relays. Although developed for aerospace applications, their small size and convenience of mounting have made them a popular choice for many other uses. Later modifications in design in the interest of still smaller size and lighter weight have resulted in the half and one-sixth crystal-can sizes. A still newer and smaller design is the TO-5, a small relay in the transistor case indicated by its title. This item will be discussed separately later, since it is of a different size and shape.

Advantages for crystal-can relays have already been enumerated as small, light, vibration- and shock-resistant, adaptable to printed-circuit boards and solid-state circuitry, hermetically sealed, plentiful, and moderately inexpensive considering the complexity of design and manufacture. Disadvantages are that the inside mechanisms are inaccessible during use for inspection of remaining life, or even initial correctness and adequacy of adjustment. Because they are rather expensive relays to build, price is sometimes a disadvantage if a vendor does not have overrun or other kinds of residual from a military-contract job.

TO-5 Relays. These relay designs were mentioned above but will be briefly discussed separately because they differ from crystal-can relays in size, shape, and some intended applications. Many are essentially hybrid relays, which will be discussed later. They were aimed directly at logic circuits and relay sensing of low-level signals, originally in aerospace and military applications but now in such fields as computers and computer-allied equipment.

Besides the one Form C basic relay, there are these options: with internal diode for coil transient suppression, with internal diodes for coil transient suppression and polarity-reversal protection, and with internal transistor driver and diode coil suppression.

In addition to the two Form C basic relay, options are these: with internal diode for coil transient suppression and with internal diodes for coil transient suppression and polarity-reversal protection. There is also a magnetic-latching version. Four poles, single-throw, may also be provided.

Relays of the TO-5 type have the advantages of being quite compatible with logic circuits and low-level signals; so they are found in computers and in control equipment. Their resistance to shock and vibration suggests their use in military, airborne, and aerospace applications.

Time-Delay Relay Units (TDR's). An earlier discussion indicated how individual relays had their inherent operate and release times modified by coil modifications and series and shunt circuits connected to the coil windings. In the case of sleeves and slugs, the time-delay features are built into the coil and are barely perceptible externally. The series and shunt circuits are generally introduced into the wiring external to the relay and hence are not really a unit assembly with the relay proper. Recently, as solid-state devices became common and it became relatively easy to package the delay elements on or in the same enclosure as required for the relay, delay units consisting of hybridized circuits plus the relay were offered commercially. It has been noted that several models of the TO-5 are hybridized, but such hybrids are not especially aimed at producing time-delay features. The most popular TDR's use a more or less conventional relay plus the required hybrid circuitry (usually on a printed-circuit card or hung on the frame of the relay), plus an enclosure employed to combine all these elements into a unit. Adjustable timing, when required, is accomplished by altering the pot settings by means of a knob that can be turned externally, or a slotted shaft for screwdriver setting may be used. Figure 54 shows such a unit. A popular form of these is octal-plug mounted, although relays of this kind appear in all kinds of enclosures, with or without the adjustable knob accessible from the outside. The adjustable knob may or may not be calibrated for identifiable time intervals, depending upon need and allowable cost.

Fig. 54 A hybridized plug-in relay with knob-adjustable time delay. (*Potter & Brumfield, Div. of AMF, Inc.*)

Since there is no industry-wide standardization for relays of this type, as to either design parameters or operating characteristics, a search of possible vendor's literature is usually necessary in order to locate a satisfactory source. All the kinds of timing functions can be handled, such as operate time delay, release time delay, generation of a delay interval with reset, momentary actuation, sequence timing with repetition, pulse generation, and interval timing. Repeat *accuracy* is a problem.

Because devices of this type may employ any kind of relay, the only applications limitations for these timed relays is that which has already been noted for that particular kind of relay with regard to shock, vibration, temperature, mounting position, etc.

The above discussion has ignored timing relays of the thermal type, since they will be treated as a specific type later. Also it is not within the scope of a handbook to examine so diverse a lineup of products as the various kinds and makes of dashpot relays (both pneumatic and oil-retarded), or motor-driven timers with associated relay or relay-contact type of output. A good technical discussion of this subject is given in Ref. 5. As is pointed out there, some vendor's literature does a good job of stating what can and cannot be expected from their products in the way of accuracy at voltage extremes, repeatable accuracy with various values of elapsed time, and effect of temperature extremes on initial timing accuracy.

Thermal Relays. Not all thermal relays are employed to create intentional delay, but there is an unavoidable time lag between current application and contact movement for any relay operating on this principle. Therefore, this relay is suitable for generating time-delay functions. The initial operate time can be well established, but there are two kinds of reoperate times: the operate time when the heater is starting from ambient and the operate time when there is residual heat from a prior operation. Release time is also affected by the length of time the heater was on. For a relay that had reached maximum temperature on operation, called saturation, the release time is sometimes several times longer than the operate time. Thermal relays are a heat-integrating device, hence voltage-sensitive. Thus it will be seen that if only an operate-time delay of no specific value is required, with adequate time for cooling, the thermal relay can be a satisfactory solution. One obvious use is to take advantage of the thermal time delay to prevent giving an alarm when an infrequently occurring unstandard condition of unlikely frequent repeatability has to be accommodated. The telephone industry, for example, has many such situations. Sometimes the voltage sensitivity referred to above is made use of in supply-voltage sensing and control. Thermal relays operate equally well on alternating or direct current and are free of the too frequent tendency of other relays to contact chatter on alternating current. Thermal relays usually are not position-sensitive or affected by stray magnetic fields, and these facts are sometimes used advantageously. See Fig. 55.

Contact capabilities vary from design to design and manufacturer to manufacturer; so the vendors' literature is the best guide in determining whether or not the load at hand can be handled.

It is difficult to identify fields of application in which relays of this type are used as against fields in which they are not applicable, because their usefulness is oriented more to function than to class of service. If they seem applicable to a particular job for the reasons given above, it is time to consult with the vendor to make certain of a complete fit to the job at hand.

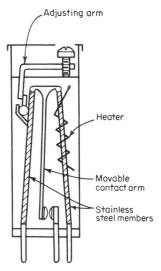

Fig. 55 A schematic diagram of one kind of thermal relay.[1]

Hybrid Relays. Defined as a combination of mechanical switch and solid-state circuitry, it is important to note that the mechanical switch may be on either end of the hybrid relay. For example, mechanical contacts are on the output end with time delay but on the input end when gating a triac. As so often happens, usage of a term seems at variance with definition. For example, it has become a custom in some industries, and even by some relay manufacturers, to limit the term hybrid relay to relays composed of a reed switch and a solid-state device in combination, for example, a reed relay input and a triac output, or a transistor input to a reed relay. Surely these are hybrid relays, but it was never intended to so limit the term. Any relay and solid-state combination is a hybrid as originally defined. By definition, then, some forms of the relays discussed above can be considered hybrids, especially some of the TDR's. However, the combination of relay and one or more solid-state component(s) into a unit deserves considerable attention in a work of this kind; so the matter will be treated in detail. The danger in doing this is that the solid-state art is advancing so rapidly that the sought-after solutions of today may become too cumbersome or uneconomical for tomorrow. An example is the effect of LED's on solid-state relays. Impractical and almost unknown as a means of isolating input from output of such relays at the time much of what follows was first written, they are in use for exactly that purpose today.

In 1967, a NARM committee was commissioned to study hybrid combinations of relays and solid-state devices. They pointed out that since there were advantages for relays over solid-state units such as transistors and diodes in some situations and the reverse was true for other cases, it appeared logical that a hybrid could in many cases outdo either alone and must certainly be the best engineering for those situations.

The general characteristics of relays vs. solid-state units were examined side by side, and where conclusions could be drawn, they were listed in tabular-comparison form. These tabular details are as shown in Table 11 and should provide the designer with valuable guidance in making a choice of whether to use conventional relays, hybrids, or solid-state only for any specific case.

Circuits making use of a marriage between electromechanical/electromagnetic relays and solid-state components to form a hybrid unit will be found in Figs. 56

TABLE 11 Characteristics of Relays and Solid-State Components as Switching Devices[1]

Characteristic	Device	
	Relay	**Solid state**
	LIFE	
Cyclic	The normal mechanical life of relays (expressed in number of operations) may vary from less than one million operations to hundreds of millions, with some (such as mercury-wetted contact relays) capable of many billions of operations, depending on type and design. The electrical life rating is normally a function of the particular electrical loads being switched	Theoretically, correctly applied semiconductors do not have known wearout modes, since they are essentially mechanically static devices. Cyclic limitations may be encountered which are dependent upon the design, fabrication, construction or application of the semiconductor. (Life is normally expressed in number of hours rather than operations. A good transistor switch can make one million or more operations in 1 s without impairing its useful life.)
Static	The static life of electrically functioning relays may be limited by physical or chemical deterioration of their components. The nature of the conventional relay is such that its contact forms can function in a prescribed manner (blocking or conducting) with or without coil power, depending on type and design. In their blocking condition, relay contacts are inherently immune to transients. Relay contacts dissipate very little power while conducting, because of their normal condition of low contact resistance. In some cases, excessive contact resistance may be encountered owing to contamination, corrosion, or oxidation. Other possible limitations may be coil deterioration (seldom, if ever, encountered in the absence of electrical stresses) and galvanic action between certain dissimilar metals. The design, materials, and manufacturing processes of the relay–along with the application and environmental surroundings–are the ultimate factors that determine static life, and are normally chosen to minimize or effectively eliminate these problems	The static life for electrically functioning semiconductors may be limited by chemical or physical changes affecting the intended function of their junctions. Semiconductors usually require a continuous external driving source, except for latching types, which are internally driven by their output. The maximum junction temperature for semiconductors limits the power dissipated. This internally dissipated power is caused by the forward voltage drop across the device and by the requirements of the device drive. Above-rated voltage transients can destroy or cause a device to go into an unwanted condition. The environmental surroundings, application, design, and fabrication of the semiconductor are the ultimate factors that determine static life

Shelf	With proper storage, shelf life is normally not a problem. Hermetically sealed units have a potentially longer shelf life than open units	Generally nonapplicable

Failure modes	Failure modes may be contacts sticking, transferring, or welding; high contact resistance; mechanical failure; coil opening or shorting. Contact sticking and high contact resistance may be intermittent and regarded as misses instead of failures for some applications. Coil failures are usually attributable to excessive voltage, electrolysis or other chemical reactions, or harsh environments. Excessive temperature, especially if prolonged, may deteriorate the insulation, causing the coil to become defective. Most relay failures are fairly easily detected because of visual evidence of failure	Failure modes may be permanent shorts (although opens do occur), inability to block voltage, or leakage current reaching failure proportions.
		
		  General failure factors related to semiconductors are: exceeding of maximum voltage ratings, e.g., transients; thermomechanical fatigue caused by cyclic temperature surges; chemical reactions, such as channeling; physical changes, such as crystallization of materials; and other associated packaging problems which generally cause greater than intended power dissipation within the device.
		
		  Most failures are hastened with prolonged temperature increases.
		
		  Specific failures for semiconductor devices are secondary breakdown found in bipolar transistors, and di/dt and dv/dt found in thyristors. If the commutating dv/dt of a thyristor is exceeded, it will not turn off; and if the static dv/dt of the device is exceeded, it may go into unwanted conduction. Semiconductor-failure detection can become quite involved depending on the knowledge, experience, and equipment required. In many instances there is no visual evidence of failure unless it is heat discoloration

ENVIRONMENT

General	Commercial atmospheres are reasonably well tolerated by most relays in either an unenclosed (open) condition or, if conditions warrant it, in an enclosure. Extreme problems of atmosphere, moisture, particles, etc., may require hermetic sealing. Relays may be designed with radiation-hardened materials	The types of packaging and small mass of semiconductors make them inherently immune to most environments, particularly shock and vibration. For radiation applications, shielding must be provided

TABLE 11 Characteristics of Relays and Solid-State Components as Switching Devices[1] (Continued)

Characteristic	Device	
	Relay	Solid state
	ENVIRONMENT (Cont.)	
Temperature	Generally, the ability to withstand heat is ultimately limited by the type of insulating materials employed. Maximum or temperatures above maximum rating, if sustained, will produce a faster deterioration and decomposition of most insulating materials. Above-rated, elevated-ambient temperatures for reasonably short durations can usually be tolerated by most relay designs without causing irreversible changes to the unit. Relay designs are available that can operate in maximum ambients of 125°C, with specials good to 200°C. In general, the contact rating applies over its specified operating-temperature range without derating. Coil resistance varies directly with its temperature, according to the temperature coefficient of the coil-wire material (copper is used almost exclusively)	Essentially, the ability to withstand heat (internal losses plus ambient temperature) is ultimately limited by junction temperature considerations. Above-rated, elevated-ambient temperature surges usually have sufficient inertia to cause irreversible changes in the semiconductor if it is functioning near its maximum capacity. Generally, semiconductors can withstand junction temperature overshoots caused by current surges of several milliseconds. The cumulative effects of the internally dissipated power, combined with the ambient temperature, must not exceed the maximum permissible junction temperature. As the internal power dissipation increases, the maximum allowable ambient temperature decreases. The type of heat sink employed substantially affects semiconductor performance. The proper design for this heat sink may become quite involved, depending upon application and if electrical isolation is required. Prolonged heat exposure hastens chemical and other types of failure. Depending on type, many semiconductors can operate in ambients of 125°C or above if properly applied. Gate sensitivity and gain usually fall off with low temperatures, particularly below −20°C
Contamination	Contamination is of most concern with contacts. Where contact contamination is encountered, the result may vary from slightly increased contact resistance to an electrically open condition. Relay coils, depending on insulation, may be susceptible to certain contaminants which will chemically deteriorate the coil and may result in electrical breakdown and shorting	Contamination is mostly of concern when it is an internal type on a semiconductor element. Where semiconductor pellet contamination is encountered, a decrease in blocking voltage and an increase in leakage current normally results

External contamination	Unenclosed relays may be affected by undesirable gases and other contaminants, which may require either hermetic or nonhermetic enclosures. Contaminants, such as oxides, on connecting terminals may present difficulties	Semiconductors are essentially immune to external contamination except when encapsulation flaws permit atmospheric impurities to reach the sealed semiconductor junction. Contamination on leads may offer connecting problems
Internal contamination	Internal contamination is generally a result of outgassing of various insulation materials (e.g., in hermetically sealed units). Electrical switching of contacts, at levels producing sparking or arcing, in the presence of outgassing from various organic volatiles, may form contaminants and may promote contact erosion. Internal outgassing and contamination are normally controlled by proper choice of materials, design, and manufacturing methods	Internal contamination consists of entrapment or inclusion of ionizable material inside the sealed package, which may lead to failures. Manufacturing techniques and processes have been developed that provide a high degree of freedom from contamination within the sealed package

RELIABILITY

Rating method	Failure rate is generally expressed in percent per 10,000 operations	Failure rate is usually expressed in percent per 1,000 h
Degree	Relays have demonstrated high component reliability using the above rating method; however, this depends greatly on relay type and use	Under reasonably ideal conditions, extreme reliability can be obtained from semiconductors with the above rating method
Failure rate	The failure rate tends to follow a "bathtub" distribution curve; i.e., it decreases after each consecutive, successful operation, levels off, and does not appreciably increase until mechanical wearout begins	It is generally assumed that the failure rate ranges from constant to slightly decreasing with time, once the infant-mortality period is passed. There is no upturn in failure rate with life
Run-in or burn-in	In some instances where added reliability or stability is desired, relays are given a number of prelife operations (run-in) under predetermined conditions, related to intended use. This tends to minimize early failures. However, extensive run-in will only use up a portion of the useful life	The initial burn-in for semiconductors tends to eliminate devices which would normally fail during the first few hours of operation. It is effective for semiconductor devices having a high initial failure rate followed by a decreasing failure rate. The extent of such testing is usually limited by economic considerations

TABLE 11 Characteristics of Relays and Solid-State Components as Switching Devices[1] (Continued)

Characteristic	Device	
	Relay	Solid state
RELIABILITY (Cont.)		
System	System reliability is reduced according to the cumulative failures of all the components used. Where the choice of relay or solid-state system is considered, the complexity required for each may be a greater factor in system reliability than is the reliability of the individual component	
Hybrid	The greatest reliability is achieved when the strong points of a certain component offset the weak points of another, as is done in many hybrid devices using both relays and semiconductors	
ELECTRICAL ISOLATION		
Output/input	Relays have inherently high isolation between output circuits, between output and control (input) circuits, and between control circuits. (Insulation resistance of $\geq$1,000 MΩ and dielectric withstanding voltage of the order of 500 to 1,000 vac are typical.)	Generally, a high degree of electrical isolation between control and output circuits cannot be achieved with junction-type semiconductors. In a limited area of application, a high degree of isolation can be achieved with FET's
High voltage	Isolation is very little affected by voltages which are relatively high compared with nominal system voltages. The loss of dielectric due to momentary exposure to excessively high voltage is usually temporary. The degree of recovery depends on the type of insulating material used	If maximum rated voltage values as given by the manufacturer (usually at 25° C) are exceeded even momentarily, many semiconductor devices will be permanently damaged
Variation	Relay contact isolation normally does not vary substantially with time, temperature, radiation, voltage, etc, unless there is a complete failure under extreme conditions	Semiconductor leakage current is a variable of temperature, time, radiation, and voltage. If device limitations are not exceeded, the variation is reversible, except where due to time (aging)
Electrical noise and magnetic fields	Relays, because of the power and time required for operation, are essentially insensitive to electrical noise. Sensitive relays can be subject to false operation in high magnetic fields unless	In many applications, shielding and signal conditioning are required to prevent false operation caused by electrical noise and electromagnetically induced currents. They do not nor-

shielded. RFI, produced by relay contacts during opening and closing but not while carrying current, is difficult to control. EMI, produced by coils and magnetic circuits, may be suppressed to some extent

mally show sensitivity to static magnetic fields (except Hall devices). Semiconductors generate RFI during turn-on and turn-off while switching ac if they are not turned on at zero current. Thyristors using phase-control techniques may be a serious RFI source because of turning on during each half-cycle (or alternate half-cycles). Various techniques for RFI suppression are possible for some applications

OFF/ON CHARACTERISTICS

Characteristic		
Off/On impedance ratio	The off/on impedance ratio is extremely high	The off/on impedance ratio is moderately high. (FET's may provide significantly higher ratios.)
Power loss	Relays have the ability to handle power with extremely low loss because of low resistance of their closed contact circuitry. The coil power must also be considered to obtain the complete power-consumption picture	Semiconductor ability to handle power is limited by inherent on voltage drop, dependency on heat sinks, and ambient temperature. The base or gate drive power is small compared with the output. However, for some transistors, as output current increases, input current must be disproportionately increased (beta decreases) to obtain desired output saturation
On voltage	The voltage drop across a closed relay contact is the IR product (generally less than 100 mV)	The semiconductor usually has a forward voltage drop from 0.3 to 2.5 V. The voltage drop per junction is approximately 0.3 V for germanium and 0.6 V for silicon, in addition to the voltage drop due to bulk resistivity
On resistance	The on (closed contact) resistance of the relays may vary slightly from cycle to cycle and with life in terms of operations and load. Under adverse conditions, misses may occur because contacts do not close electrically (particles, film, welds on opposite sets of contacts, etc.) or resistance exceeds some predetermined value. The on resistance remains essentially constant with clean contacts. It can increase slightly because of heating of the contact circuitry, and increases with current	Forward voltage drop across a semiconductor is consistent from cycle to cycle. This drop varies with junction temperature. The on resistance generally decreases as current increases, and for some devices such as transistors can be varied with base drive

TABLE 11 Characteristics of Relays and Solid-State Components as Switching Devices[1] (Continued)

	Device	
Characteristic	Relay	Solid state
	OFF/ON CHARACTERISTICS (Cont.)	
Off resistance	The off resistance (open contacts) is affected very little by temperature, voltage, etc.	The off resistance varies with time, temperature, voltage, and radiation. Leakage current increases exponentially with temperature (e.g., may double with every 8 to 10°C increase)
	INPUT CONSIDERATIONS	
Operating power	Relays are available with operating power from milliwatts to watts, and specials operate on microwatts. Duration of power pulse required for latching relays may vary from less than 1 ms to several milliseconds. Relays having ferrite magnetic circuits can be made to operate from pulses of less than 5 μs duration. Latching relays are normally reset with a power pulse equal to (in some cases less than) the value required to latch. Proper polarity coil voltage must be observed for devices designed to latch magnetically. In special cases, manual-reset features may be provided. Relays do not normally require regulated power supplies	Different semiconductor devices are available with operating power from microwatts to milliwatts. Latching semiconductors, such as thyristors, generally require a 2-μs pulse or greater for latching (conduction turn-on). Latch is lost when conducting current is reduced below holding value. Semiconductors require well-regulated, transient-free, dc power supplies
Operating voltage	Relays operate in response to a wide range of ac or dc coil input voltages as determined by design. Relay coils are generally designed to operate within a ± tolerance of a specified nominal voltage. While insufficient voltage will not permit the relay to operate properly, if at all, greater than maximum specified coil voltage may cause coil deterioration depending on duration and magnitude. Conventional, nonlatching, relays will drop out (return to unenergized condition) when coil voltage is removed or reduced to a value which may be varied widely by design and/or adjustment. Latching relays	Semiconductors easily adapt to a wide range of input voltages via appropriate circuitry. The absolute maximum voltage ratings for the input (and output) of semiconductors were well specified by the manufacturer and are not to be exceeded without risk of permanent damage. The range of operating voltages is predetermined by design

	require an input voltage of a specified magnitude and duration for latch and reset. (Proper polarity must also be observed for magnetically latched relays.)	Semiconductors are current-operated devices. Transistors are essentially current amplifiers; i.e., a given input current determines a given output current for a set of conditions. Thyristors require minimum gate-current drive to ensure latching, if load conditions permit. Generally, gate drive can be removed after thyristor turn-on. Semiconductors usually turn off within a matter of microseconds following removal of drive current, whether supplied externally or internally
Operating current	Relays operate over a wide range of predetermined current levels. Even when voltage levels are specified, the electromagnetic relay is essentially a current-operating device whose operation is accomplished when its inherent ampere-turn requirements are met. Again, those comments made under Operating Voltage apply to current-operated relays	
Transients	Relay coils are generally insensitive to transient voltages	High-voltage, short-duration transients can be particularly damaging
Duty cycle	Generally nonapplicable. In some cases, coils may be rated for intermittent duty because of temperature considerations	As the percent duty (conduction time) decreases, the drive-current rating usually increases as long as maximum junction temperature is not exceeded. Large drive currents may be required in power transistors to obtain saturation, and in thyristors to minimize di/dt stresses

OUTPUT CONSIDERATIONS

Multiple switching	Relays are available with various types and numbers of contact forms which can be operated by a single input. For most, the choice of circuits switched by each contact form may be of a widely different voltage and frequency from that switched by the others; this choice requires little if any predetermination in the design of the relay	Solid-state systems feasibly can perform any switching function. However, switching more than one circuit simultaneously using discrete semiconductor components requires proper combination into a workable assembly. The complication, cost, and size of this assembly will increase substantially with the number of poles, magnitude of current, and degree of electrical isolation. (Where the level of switching permits integrated circuits to be used, considerable economic and size advantages may be realized.)

TABLE 11 Characteristics of Relays and Solid-State Components as Switching Devices[1] (Continued)

Characteristic	Device	
	Relay	Solid state
	OUTPUT CONSIDERATIONS (Cont.)	
Switching range	Contacts are generally adaptable to a wide current, voltage, and frequency range, since they simply physically connect electrical conductors. However, relay families are usually designed for dry, low, intermediate, or high-level switching. The upper frequency range which they are capable of switching also varies greatly with design	Semiconductors and their assemblies are usually designed for a specific voltage, current, and frequency range. The operating current ratio (output to input) for many semiconductors is extremely high. The frequency range depends greatly on the design of the semiconductor
Voltage	Relay contacts can generally tolerate an exceptionally wide operating range of load voltages without design changes, and they usually recover from short breakdowns or excessive overvoltages. Special designs may be required for high voltages or extreme environmental conditions	Maximum output-voltage ratings are particularly well specified, and it is essential that semiconductors be operated within these ratings to avoid permanent damage
Current	Relays can generally tolerate overload currents of various degrees. As overload current increases, contact sticking or contact welding during break and make will increase. An empirical determination is usually necessary to find out if the relay will function reliably under given overload-current conditions	While most switching-type semiconductors can handle surge currents many times their steady-state ratings, overload currents of relatively sustained duration cannot be tolerated. The junction temperature of the semiconductor is ordinarily the limiting factor in determining the magnitude and duration of its surge-current rating
Transients	The inherent design of relays is such that they are essentially immune to transients	High-voltage, short-duration transients can be particularly damaging to semiconductors. Special protective devices and means are frequently required
Duty cycle	Generally output is independent of conduction duty; however, the maximum permissible cyclic rate of switching is limited by the magnitude of the load being switched	As percent duty (conduction time) decreases, current rating usually increases as long as the maximum junction temperature is not exceeded

Contact bounce	Contact bounce is usually present to some degree during contact make and, in a few designs, during break. (Duration may be from fraction of a millisecond to several milliseconds, and may consist of several closures and openings per contact operation.) Mercury-wetted relays are an exception, since their contacts have essentially no bounce due to the masking effect of the mercury	No contact bounce exists in semiconductors
Amplification	Amplification is possible only in that small signals may control large ones. Extremely high amplification factors are possible using a single relay	Semiconductors amplify the input signal. Some devices are used as switches in that small input signals control large output signals

LOGIC

Systems	The logic of simple control systems can be used most economically with control relays, particularly since special power supplies and noise-suppression techniques are not required	Solid-state devices lead the field where extensive and complex logic systems are involved or very high-speed operation is required
Speed	Pickup times range from 0.5 to 5.0 ms for reed and other relatively fast operating types to 5.0 to 50 ms for conventional control relays. Dropout times for relays are generally somewhat faster with some reeds having dropout times of 20 μs. Use of overdriving coil voltages to achieve a lower pickup time may increase the severity of contact bounce during make	Operating times for transistors range from nanoseconds for computer types to microseconds for power types. Typical turn-on time for thyristors varies from 0.5 to 5.0 μs while commutation requires from 10 to 50 μs. If thyristor turn-off is achieved when conducted ac goes through zero, turn-off time can be approximately as long as a half-cycle. The above values vary widely with type of semiconductor, circuit, and application
Memory	Memory functions can be easily achieved by latching or stepping relays. In such cases, memory can generally be retained in spite of power loss	Memory is normally accomplished with cores, flip-flops, and integrated circuits. Power loss generally means memory loss, except for core logic or where special auxiliary memory circuits are used
Electrical noise	Electrical noise is normally not a problem because of relay operating speed and power requirements	Signal conditioning and filtering for noise, overshoot, and transients are often required

TABLE 11   Characteristics of Relays and Solid-State Components as Switching Devices[1] (Continued)

Characteristic	Device	
	Relay	Solid state
	LOGIC (Cont.)	
Fan-out	Fan-out logic functions per input for conventional relays is generally not a problem. One relay contact can drive many relay coils. Fan-out speeds for relays are usually in milliseconds	Fan-out limits the number of gates which can be driven by one logic element. One gate can fan out to about 10 identical gates. Power gates are available with a fan-out of about 20. Switching times, generally in nanoseconds (both rise and fall time), usually increase with fan-out
Fan-in	Fan-in is the number of inputs to a logic element. Relays are often considered single-input, multiple-output devices. Multiple inputs may be obtained by using diode drivers, and to a limited extent, by separate windings	IC's are often considered to be multiple-input, single-output devices. Rise time is fairly independent of fan-in (fall time increases with fan-in)
Interfacing	Relays can be driven directly by solid-state circuitry. Relays requiring low operating power, such as reed relays, are ideal interfacing devices between solid-state circuits and relays or motor starters	Discrete solid-state power gates can interface directly with relays, solenoids, stepping switches, etc. Integrated microcircuits usually use a transistor to interface with relays and other electromagnetic devices
	MAINTENANCE	
Installation	While generally not critical in hookup, reasonable care should be exercised in handling of unenclosed relays	Care should be exercised so that the maximum permissible temperature for the semiconductor is not exceeded during solder hookup or potting. Handling of devices is not generally critical except where damage to termination must be considered
Troubleshooting	Technicians are normally able to diagnose failures with reasonable success. (Unenclosed or transparently enclosed relays give visual evidence of operation.)	Specially trained technicians and special test equipment are often needed to analyze problems. Assembly packages consisting of discrete components require module evaluation rather than simple component evaluation

SIZE

Intermediate to high-level switching

In most cases where power or multipole switching are required, the relay is generally smaller and far less complex than the equivalent solid-state unit. However, the life of the equivalent solid-state unit may be many times greater than that of the corresponding relay. The relay does easily adapt to various contact forms and combinations as the number of pole requirements increases. Although the size of the contact and motor assemblies for the relay are usually larger than the required semiconductor devices, this is often offset by the size of the peripheral components (including heat sink) required for the semiconductor devices. The relatively small power dissipated in the relay is usually accommodated by its own radiating surfaces. By contrast, the power dissipated in the semiconductor device, because of its forward voltage drop, generally imposes special heat-sink requirements which will often substantially increase size

Low to intermediate-level switching

Low-level switching applications that permit extreme reductions in heat-sink requirements, especially where extensive logic is performed, favor semiconductors because of savings in weight and size. This is particularly true with microelectronics. For very low-level switching applications, however, difficulties may be encountered with semiconductors where direct device operation is necessary, especially if voltage and current requirements are below that of device operation. Special relays often adapt directly to fulfill this application. While semiconductors can also work in this region, appropriate supporting circuitry, power supplies, and peripheral considerations may become increasingly complex and expensive as the switching level becomes extremely low. How critical these considerations are cannot be sharply defined, but applications in the microamperes—millivolts region should be examined to see which method, or combination of methods, of switching is best suited

COST

Switching levels

Relays are generally economical devices where heavy-load, high-voltage, dry-circuit, or multipole switching is desired

Semiconductors are desirable and economical where low-level, multiple-input switching is used or for applications requiring unusually long life or high speed. Where extensive complex logic switching is required, IC systems are the only practical choice

Application

Worst-case condition studies are only moderately difficult and often done by the relay manufacturer

In general, when solid-state switching circuits are to be mass-produced, particularly from discrete components, comprehensive and thorough worst-case studies usually become quite involved

to 78. These are obviously generalizations of circuit possibilities, and many relay vendors, as well as users, have their own versions of circuits to do these and other tasks, which for the most part they are willing to share. Hybrid circuits seem destined to continue to grow in number and scope as the components change and improve.

Typical Hybrid-Relay Applications. In the following paragraphs relays are shown in hybrid combinations with solid-state components. Sensors, light-sensitive devices, thermistors, and other devices are illustrated as circuit elements functioning with relays and other solid-state devices.

Time Delay (*Pull-in*). Figure 56 is a circuit designed to delay a relay on pull-in. Upon application of power, capacitor $C1$ begins charging through $R1$ until it reaches the firing point of unijunction transistor $Q1$. The unijunction transistor triggers the SCR, which energizes the relay $K1$. Resistor $R2$ and Zener diode $D1$ form a voltage regulator to allow for a wide range of voltage inputs.

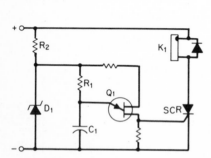

Fig. 56 Hybrid relay to provide time delay on pull-in.[1]

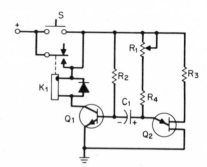

Fig. 57 Hybrid relay to provide time delay on dropout.[1]

Time Delay. The circuit of Fig. 57 simply and accurately delays the dropout of a relay after it is energized. In the quiescent state, no power is applied to the operating components. When S is momentarily closed, transistor $Q1$ turns on and relay $K1$ pulls in. Voltage to the circuit is then maintained through the relay contact so that the relay remains energized when S is opened. After a time interval determined by the values of $R1$, $R4$, and $C1$, unijunction transistor $Q2$ triggers and the discharge of $C1$ turns off $Q1$, allowing the relay to drop out. If S is open, voltage to the circuit is removed and the circuit reverts to its quiescent state.

An output voltage can be obtained from the relay contacts, as shown, or extra contacts on the relay may be used.

Depending on the supply voltage and the relay used, $R2$ provides sufficient base current to $Q1$ to allow the relay to pull in. The size of the capacitor provides sufficient off time for $Q1$ to allow the relay to drop out. $R1$ provides the time delay required and the maximum peak-point current of the unijunction transistor. $R3$ provides the required overall temperature compensation.

Time Delay (*Dropout*). The circuit of Fig. 58 provides a time delay on dropout using an auxiliary voltage. A control voltage is applied to $Q1$, causing it to turn on and essentially short out $C1$. The auxiliary voltage causes $Q3$ to turn on and allows relay $K1$ to pull in. When the control voltage is removed, $Q1$ turns off, allowing $C1$ to charge through $R3$ and $R5$. When the unijunction transistor $Q2$ fires, the SCR is triggered, turning off $Q3$ and deenergizing the relay. $R3$ and Zener diode $D1$ allow for a wide range of voltage inputs.

Time Delay—No Recovery Time. The circuit of Fig. 59 provides a dropout time delay without requiring recovery time. When a positive pulse is applied to the input terminal, the silicon-controlled rectifier (SCR) is turned on and current flows through it to turn on $Q1$ and charge capacitor $C1$. When $C1$ is sufficiently charged, current through the SCR drops below its required holding-current level. The SCR then turns off and $C1$ begins discharging through $R1$ into the base of $Q1$. Meanwhile, $Q1$, having been turned on, has pulled in relay $K1$, providing the output connection. As $C1$ discharges, current in the collector of $Q1$ and in the relay decreases until it drops below the relay dropout level. Then the relay opens, breaking the output connection. The circuit may be retriggered at any time after SCR shutoff, provided the trigger amplitude is sufficiently greater than the charge on $C1$ at that time.

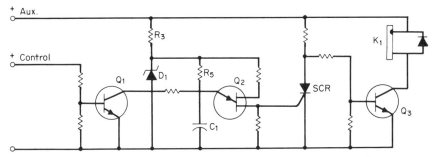

Fig. 58 Hybrid relay with delay on dropout using an auxiliary voltage.[1]

Time Delay (Pull-in) for Two-Coil Relay. A circuit for the time delay of a two-coil relay is shown in Fig. 60. The two coils of relay $K1$ are so wound that when power is applied, the resultant magnetomotive force is approximately zero. This does not allow the relay to pull in.

Capacitor $C2$ is charged through the relay coils and $R2$. When unijunction transistor $Q1$ fires, it triggers the SCR into conduction. This essentially shorts out one coil of the relay, allowing a magnetomotive force to develop and pull in the relay.

Repeat-Cycle Timer. The circuit of Fig. 61 allows the repeat cycling of a relay at a predetermined frequency. Upon application of power, capacitor $C1$ charges

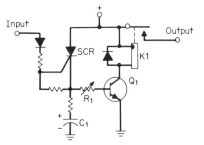

Fig. 59 Hybrid-relay circuit to provide release-time delay without recovery time.[1]

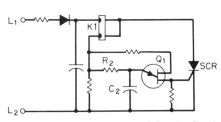

Fig. 60 Hybrid-relay time delay (pull-in) using a dual coil.[1]

through $R3$. $R4$ provides base current to transistor $Q2$, turning it on and energizing relay $K1$. The on time of the relay is determined by $R3C1$. When the firing point of unijunction transistor $Q1$ is reached, $C1$ discharges, causing a negative shift in the voltage on the base of $Q2$, turning it off and deenergizing the relay. $C1$ continues to discharge through $R4$ until the base voltage of $Q2$ is positive enough to turn on $Q2$, thus reenergizing the relay. This cycle repeats itself.

Amplifier Relay. In the circuit of Fig. 62 power required of the controlling source can be greatly reduced by combining a relay with an amplifier-driver. This allows use of a low-level input while retaining the inherent ability of a relay to control power circuits of various voltages and currents.

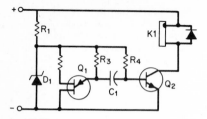

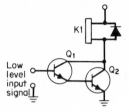

Fig. 61 Hybrid relay as a repeat-cycle timer.[1]

Fig. 62 Hybrid relay functioning from a low-input signal.[1]

A Darlington amplifier-driver is shown in Fig. 62. Application of a positive input to the base of Q1 produces an amplified drive to Q2, causing relay K1 to operate.

Triac Relay Driver. The triac relay driver (Fig. 63) is ideally suited to control ac-operated relays with small input signals. The presence of an input signal causes triac T1 to conduct and in turn operate relay K1. Absence of an input signal causes the triac to assume a blocking state, and the relay deenergizes.

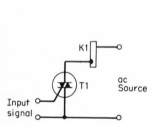

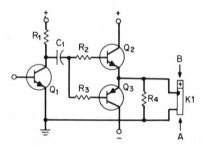

Fig. 63 Hybrid relay functioning from a triac.[1]

Fig. 64 Bistable polar relay as a hybrid functioning from an amplifier driver.[1]

Relay Driver. In the circuit of Fig. 64 a bistable polar relay is controlled by a one-polarity low-level input to the amplifier-driver circuit.

Application of a positive input to Q1 causes it to conduct, lowering the voltage at the R1C1 junction. C1 discharges through Q1, the baseemitter junction of Q3, and R3, turning Q3 on for a period determined by the time constant of the circuit. The short-duration pulse in direction A through K1 causes the relay to assume one of its stable positions. At the termination of the positive input to Q1, Q1 ceases to conduct, the voltage at the R1C1 junction rises, and C1 charges through R1, R2, and the base-emitter junction of Q2. Q2 conducts for a period determined by the circuit time constant, producing a pulse of current in direction B in K1, causing K1 to assume the other of its stable positions.

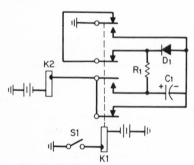

Fig. 65 Hybrid relay for fast operation.[1]

Fast Operation. Figure 65 shows a circuit that decreases the operating time of relays. When S1 is closed and the contacts of K1 transfer, K2 is operated by the sum of source voltage and the charge on capacitor C1. Diode D and resistor R provide a holding circuit which conducts when the voltage across C1 assumes a polarity opposite to that shown and a magnitude adequate to cause diode turn-on. Momentarily the voltage is doubled.

Diode $D1$ does not conduct until the voltage across capacitor $C1$ falls to the voltage equal to the drop across resistor $R1$ resulting from the holding current for $K2$.

Voltage Sensor. Figure 66 shows a basic dc voltage sensor. It may be used to sense ac voltage or current. The alternating current is rectified and filtered to provide a dc level. For ac sensing, a current transformer is used, with the output rectified and filtered. The Zener diode $D1$ establishes a reference voltage at the emitter of transistor $Q1$. The voltage divider $R1R2$ is set so that when the desired input voltage is reached, $Q1$ turns on. This allows transistor $Q2$ to conduct and causes relay $K1$ to pull in.

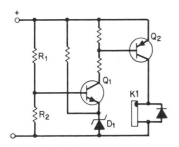

Fig. 66 Hybrid relay as a voltage sensor.[1]

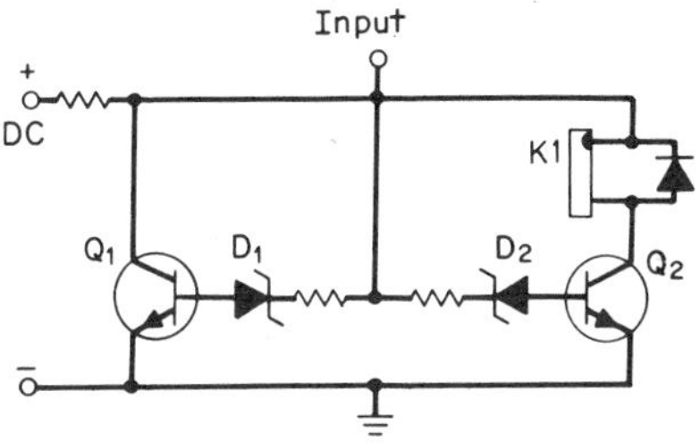

Fig. 67 Hybrid relay as a voltage-limits sensor.[1]

Voltage-Limits Sensor. The circuit of Fig. 67 can be in one of two states: if the input voltage is within the specified limits, the contacts are closed; if it is either higher or lower, the contacts are open. The desired voltage limits are determined by the Zener diodes $D1$ and $D2$.

When the voltage at the input is within the specified limits, $Q2$ conducts and relay $K1$ contacts are closed. If the voltage rises, $Q1$ also begins to conduct, effectively shorting out the relay coil. Consequently, the contacts open. If the voltage goes down, both transistors cut off and the contacts again open.

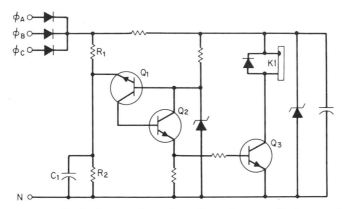

Fig. 68 Hybrid relay as a three-phase, four-wire overvoltage sensor.[1]

Three-Phase, Four-Wire Overvoltage Sensor. The circuit of Fig. 68 senses the highest voltage of the three-phase input. The time constant of $R1$ and $C1$ is such that a phase voltage after rectification that is higher than the predetermined level as set up by $R1$ and $R2$ causes regenerative transistor switch $Q1$ and $Q2$ to turn on. This in turn allows transistor $Q3$ to conduct and to pull in relay $K1$. When all three phases are below this predetermined point, $K1$ drops out.

Four-Wire Undervoltage Sensor. The detection of voltage below a reference is provided by the circuit of Fig. 69. Zener diode *D4* establishes a voltage reference level to the emitter of transistor *Q1*. As long as the average dc level of *L1*, *L2*, and *L3* through diodes *D1*, *D2*, and *D3* and voltage divider *R1* and *R3* remains higher than this reference voltage, *Q1* remains off. This in turn allows transistor

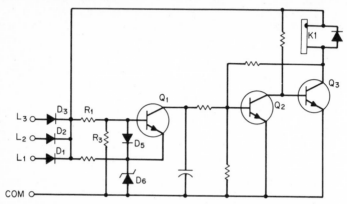

Fig. 69   Hybrid relay as a four-wire undervoltage sensor.[1]

Q3 to turn on and pull in relay *K1*. When the average level of the line voltage drops below the predetermined level, *Q1* conducts and turns on *Q2*, causing *Q3* to turn off since the base and emitter now are essentially at the same voltage level. This causes *K1* to drop out.

Current Sensor. The circuit of Fig. 70 monitors the level of an alternating current. The output is a relay, and SPDT contacts may be used to determine overcurrent or undercurrent.

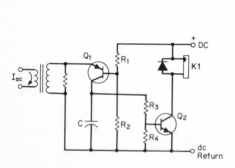

Fig. 70   Hybrid relay as a current sensor.[1]

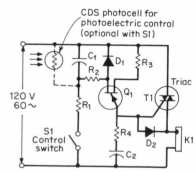

Fig. 71   Hybrid relay as a photoelectric switch.[1]

When direct current is applied in the absence of any alternating current, transistor *Q1* is biased off by *R1* and *R2*. The relative values of *R1* and *R2* determine this back bias, which is the set point. *Q2* is also off, and relay *K1* remains deenergized.

As the ac level increases, the half-cycle peaks reach a value to forward-bias *Q1* and cause conduction. These pulses charge *C1* and cause current flow in *R3* and *R4*. When this current becomes great enough, *Q2* is turned on, causing *K1* to pull in.

Synchronous Photoelectric Switch. Synchronous switching is turning on only at the instant the ac supply voltage passes through zero and turning off only when current passes through zero. The circuit of Fig. 71 theoretically provides this function in response to either a mechanical switch or a variable resistance, such as a cadmium sulfide photocell. It should be used with caution, since erratic behavior may be caused by too long an operate time of the relay used, and coil inductance may cause a current lag, producing too long a time on turn-off.

Current Sensor. The circuit of Fig. 72 can detect current level. When dc voltage is applied, relay $K1$ pulls in through $Q2$. $Q1$ is in the cutoff mode. As long as the current remains below a predetermined value, tunnel diode $D1$ maintains a very small differential voltage between base and emitter of $Q1$. When the current increases above the desired value, the tunnel diode acts to increase this differential voltage, causing $Q1$ to conduct. This in turn effectively shorts out the relay coil, causing it to drop out.

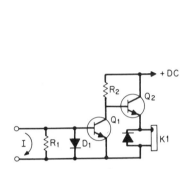

Fig. 72 Hybrid relay as a current sensor.[1]

Fig. 73 Hybrid relay as low-current detector.[1]

Low-Current Detection. The circuit of Fig. 73 is a discriminator that detects $1\text{-}\mu A$ currents with a 1 percent accuracy. This is made possible by driving field-effect transistor $Q1$ with the output from backward diode $D1$. When the input signal exceeds a preset threshold, the sum of the currents through the backward diode switches the diode to its highest voltage stage. This voltage is then amplified by the FET output stage.

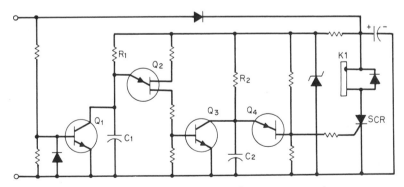

Fig. 74 Hybrid relay as an overfrequency sensor.[1]

The detection threshold may be varied over a broad range by adjusting resistor $R1$.

Overfrequency Sensor. The circuit of Fig. 74 detects frequencies in excess of a predetermined frequency. This circuit will cause relay $K1$ to be operated if input frequency exceeds a preset value determined by the time constant of $R1C1$.

On each half-cycle, $Q1$ is turned on and discharges $C1$. During alternate half-cycles, $C1$ charges, and at frequencies below the preset limit, it will attain the firing voltage of $Q2$.   Firing of $Q2$ turns on $Q3$, which discharges $C2$.

$C2$ and $R2$ establish a time constant long enough so that a period exceeding one cycle is required for the charge on $C2$ to reach the firing voltage of $Q4$. If the frequency exceeds the preset limit, $Q2$ and $Q3$ will not be turned on. $C2$ can then charge to the firing point of $Q4$, which will turn on the SCR and $K1$.

Underfrequency Sensor.   The circuit of Fig. 75 detects frequencies below a predetermined frequency.   This circuit will cause $K1$ to be energized if the input frequency falls below a preset value determined by the time constant $R1C1$.

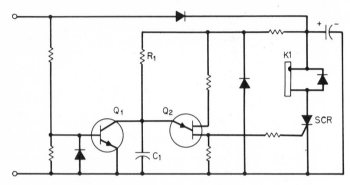

Fig. 75   Hybrid relay as an underfrequency sensor.[1]

When the line frequency is above the set trip point, $Q1$ discharges $C1$ on each positive half-cycle of the line voltage before the charge voltage on $C1$ reaches the intrinsic level of the unijunction transistor (UJT).   Thus the UJT never fires when line frequency is above the sense point.

If the line frequency should drop below the set trip point, $C1$ is allowed to charge to the firing level of $Q2$.   This pulses the SCR, and fault detecting relay $K1$ is energized. $R1$ can be variable to allow for an adjustable trip point.

Thermistors.   Thermistors may be used with a relay to provide a time delay or present a constant impedance to a source supply.   For many designs the preferred thermistor type provides a resistance that decreases over a wide temperature range (NTC).   Simplified circuits are shown in Fig. 76 and normally are not used in ac relay applications.

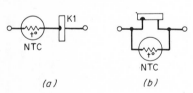

(a)   *(b)*

Fig. 76 Hybrid relay used with thermistor.[1]

Variable-Rate Driver for Stepping Switches.   Spring-driven stepping switches can be driven at controlled rates of less than one step per minute to 20 steps per second by the circuit of Fig. 77.   Coil on time is the minimum necessary, keeping power dissipation at a minimum.   The circuit may also be used to drive a relay to produce pulses at a controlled rate.

With voltage applied to the circuit, $C1$ charges at a rate determined by its value and the setting of $R1$.   When its voltage reaches the level at which $Q1$ conducts, $C1$ discharges through $Q1$ and the gate circuit of the SCR.   The SCR turns on and energizes the switch drive coil.   As the switch cocks, the interrupter contact opens, turning the SCR off.   The switch steps, recloses its interrupter contact, and the solid-state circuit repeats its cycle.

Reset Control For Direct-Drive Stepping Switch.   Two-coil direct-drive (minor) switches may be reliably and economically reset by solid-state control of the rotary magnet (reset) coil.   See Fig. 78.   Closing S1 applies voltage to the rotary (step) coil, advancing the switch wiper arm one step, and also applies a positive bias to $Q1$, holding it off.   When the switch is advanced to position N, positive voltage

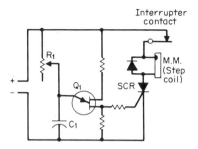

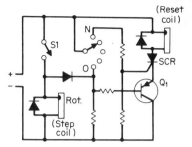

Fig. 77 A variable-rate driver for use with stepping switches.[1]

Fig. 78 Reset control for a direct-drive stepping switch.[1]

gates the SCR, but Q1 remains off until S1 is opened. With S1 opened, Q1 and the SCR conduct, energizing the release magnet (reset) coil. As the wiper arm reaches O position, positive bias is applied to Q1, stopping conduction through the release magnet coil.

Solid-State Relays (SSR). As the name implies, the solid-state relay (see Fig. 4) has no moving parts. But it does have the equivalent of a coil (the control) and contacts (the controlled output). In its simplest form the SSR's output is functionally either a make or a break. More complicated designs offer multiple contacts, but practical limitations of contact forms usually end with three (called three-phase controller). Thus the circuit engineer is limited in the circuit logic that can be practically accomplished by the use of solid-state relays. For most present fields of application this is not much of a handicap in that the logic is obtained in the circuits themselves, to which the relays are attached for implementation.

The solid-state relay has the immediate advantage of appearing to represent the latest in the switching art and thus appealing to both the design engineer and the user. Its use may in some cases contribute to the salability of a product because of this appearance of modernity, reliability, and freedom from maintenance. Also, many design engineers feel an empathy for this kind of relay that may be lacking for conventional relays.

A solid-state relay (SSR) has been identified by one authority as a semiconductor switching device with input terminals isolated from the output switch path. The output switch may be an FET for low-level switching, a bipolar device for medium and power switching, or a triac for ac power switching.

There is as yet little standardization in SSR types other than that which develops from usage, ratings, and packaging, but there are two commonly recognized broad classifications: ac power switching and ac/dc low-power switching. For a majority of the SSR's in use today the load is ac, with the Triac employed as the controlled element. Since a Triac always breaks at zero current, its many advantages become immediately apparent in handling heavy current loads and eliminating RFI.

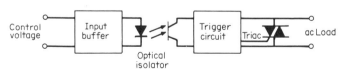

Fig. 79 Schematic of a solid-state relay using an optical isolator. (*Monsanto Commercial Products Co.*)

It is frequently important that the input signals be isolated electrically from the output (a feature the conventional relay comes by naturally). To accomplish this in SSR's it has been a common practice in the past to use a pulse transformer (magnetic coupling). In a more recent technique, shown in Fig. 79, a gallium arsenide (GaAs) infrared (IR) light-emitting diode (LED) is used to gate the triac. The LED and a photodetector are mounted in close proximity and housed in a six or eight lead DIP.   When the LED carries current, its emitted light excites the photo-detector and couples the input signal to cause triggering of the triac.   External interference is eliminated by an opaque covering.   No feedback occurs because the coupling is strictly optical.

Advantages for this kind of relay are the following:

1. Compatibility with DTL, TTL, and MOS logic
2. High sensitivity (typically 8 mW at 4 V)
3. High noise immunity and freedom from false actuation
4. Elimination of some components in earlier designs such as coil suppression, diodes, and transistor buffers, thus providing some cost reduction

For SSR's in general there are so many variations in what various manufacturers' relays can do that it is difficult to list any meaningful application data.   One manufacturer's data appear as Table 12.   AC power-switching relays handle contact loads that vary from a fraction of an ampere up to 40 A, at voltages of 120 and 240, 47 to 63 Hz.   The input signal can be either ac or dc, but most often it is a low-level dc signal in the 3- to 32-V range.   Thus SSR's readily accept logic-type input to handle an ac line load at the output.   In general, loads that can be handled by SSR's of various types extend from microvolts into high impedance, to 440 V into low impedance.   Loads may be resistive, inductive, capacitive, and lamp.   Allowable junction temperatures must not be exceeded.   To make certain of this, the vendor will provide the maximum allowable case and ambient temperatures, since these are all that the user can measure.   Rated ambient temperatures vary from relay to relay but are usually in the range of $-30/-20°C$ to $+80/+100°C$, with allowable percent of rated load current dictating allowable maximum temperature.   The allowable range of storage temperature is somewhat wider.   A typical curve sheet for load vs. temperature is shown in Fig. 80.   One surprising aspect of this

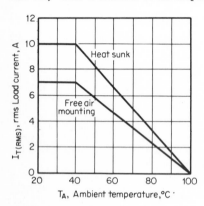

Fig. 80   Typical load current vs. temperature curves for one kind of 10-A triac output, solid-state relay. (*Monsanto Commercial Products Co.*)

heat-dissipation problem is that for very heavy load relays the required heat sinking usually means that the SSR is larger than the conventional power-type relay equivalent.

Early in the decision-making process the design engineer should find out from the prospective vendor the specific identities of the solid-state components in the prospective SSR relay, such as kind of transistor, SCR, triac, and FET, so that causes of failures due to limitations in one or more of these can be avoided and a determination can be made as to what degree of satisfaction and reliability can be expected to result from its use in a particular application.

In general the advantages for the SSR, besides those noted above, are practically unlimited life (barring catastrophic failure due to abnormal circumstances), high resistance to shock and vibration, ready mounting on PC boards (for many kinds), extreme operating speed, insignificant operating power, and in general a high degree of compatibility with electronic circuitry.

For those whose chief interest is in the SSR power-type relay application. Ref. 6 will be helpful.

Fields in which SSR's are already being used in considerable quantity are computer and computer-associated equipment, medical electronics, and industrial control. Their advantage in a hazardous atmosphere is readily apparent.   Their vulnerability to noise, temperature extremes, and power surges, and their somewhat higher cost are disadvantages that make conventional relays more attractive in some areas.   Also, large numbers of varied contact combinations are hard to provide on SSR's, leaving the many fields where this need is extensive to conventional relays, at least for the foreseeable future.

TABLE 12 Typical Solid-State-Relay Specifications as Listed for One Manufacturer's Products

CONTACT-CLOSURE CONTROL

	AC control of ac load, part A	DC control of dc load, part B
Load circuit:		
Load voltage	50–135 vac, 45/60 Hz	6–48 vdc
Load current	6 A with panel mount	
Switch on resistance	0.2 Ω typical	
Switch off resistance	20 kΩ typical	1,600 MΩ typical
Control circuit:		
Control voltage	115 vac typical	Same as load voltage
Control current	0.2 mA typical	100 mA typical
Dielectric strength, device to case	1,500 vac	
Insulation resistance, device to case	3 MΩ at 200 vdc	3 MΩ at 50 vdc
Other characteristics:		
Operating temperature	−55 to +85°C max, switching rated load, continuous duty	−55 to +100°C max, switching rated load, continuous duty
Thermal shock	−65 to 100°C	
Vibration	20g's, 10–2,000 Hz	
Mechanical shock	100g's, 6 ms sawtooth, 1,500g's, 5 ms half-sine	

DC CONTROL OF AC LOAD

	Transformer isolated			LED isolated	
	Part C	Part D	Part E	Part F	Part G
Load circuit:					
Output current	0.25 to 6 A		0.1 to 6 A	6 A	
Load voltage	50 to 140 vac			35 to 140 vac	
Frequency range	45 to 70 Hz				
Transient protection (does not turn on) 0.1 μs rise time	400 V min			300 V min	
Turn-on time (60 Hz)	0.05–0.2 ms			9 ms max	
Turn-off time (60 Hz)	9 ms max				
Off-state leakage	4 ma dc max				
Surge current (1 cycle at 60 Hz)	100 A				
Switch-on voltage drop	0.8–1.6 vac				
Switch-off resistance	10³ MΩ min				
Control circuit:					
Control voltage	10–30 vdc	4.5–10 vdc	3.5–30 vdc	2.0 vdc min	
Control current	7.0–22 mA	1.0–8.0 mA	1.0–8.0 mA	Approx 10–40 mA	
Turn-off voltage	4.5–6.2 vdc	2.8–3.3 vdc	2.8–3.2 vdc	1.1 vdc min	
Transient input voltage (does not turn on)	14 V peak min (100 μs duration)			Min excitation voltage dependent on turn-on voltage	
Isolation:					
Contacts (output) to case	10⁸–10¹³ Ω			10⁸–10¹³ Ω	
Coil (input) to case (output)	10⁷–10¹⁰ Ω			10⁹ Ω min	
Coil (input) to case	10⁷–10¹⁰ Ω			10⁹ Ω min	
Dielectric strength:					
Contacts (output) to case	1,000 vac (rms) min				
Coil (input) to contacts (output)	1,000 vac (rms) min				

SOURCE: Grayhill, Inc.

TABLE 13 A Selection Guide and Summary of Electromechanical/Electromagnetic Relay Operational Characteristics by Types As averaged from the literature of several manufacturers

Kind of relay	Required operating power	Operate time	Release time	Max contact-load current	Max contact arrangement	Size and weight	Remarks
General-purpose (GP) (no particular design shape or arrangement exists that is recognizable as common to all manufacturers)	200 MW to 6 or 8 W, depending upon size of contact load and volume of operating coil. (Sensitive versions of this general relay type are able to operate on approximately 100 mW per pole)	10–25 ms depending upon contact spring load and coil L/R. Operate-time delay up to 50 ms available on some models	5–20 ms depending upon contact spring load and degree of saturation. Release-time delay up to 100 ms available on some models	1, 2, 3, 5, or 10 A, depending upon nature of load and size, shape, and kind of contact. In some cases rating given in horsepower and voltage; e.g., $1/3$ hp at 120 V ac	One to eight or nine poles	Less than 1 to 8 in³, depending upon type and number of contacts, ½–6 oz	This class of relays is so highly varied that it is difficult to establish limits. A thorough search of vendors' literature is necessary to determine the suitability of any specific relay for a particular problem. Advertised life varies from 100,000 operations electrical to 10 million operations mechanical depending upon manufacturer and type of relay
	Unless otherwise specified, the insulation-voltage breakdown test is 500 V rms 60 Hz, or more, between all insulated parts. Contact resistance is usually 50 to 100 mΩ, new. Typical insulation resistance, specified, is 1,000 MΩ.						
Power type	2–8 W dc, 3–10 VA, ac	In excess of 10 ms	In excess of 10 ms	10–50 A	One to three poles, single- or double-throw	5–20 in³, 2–12 oz	This class of relays is highly varied as to size, shape, and style, making it difficult to establish limits. A thorough search of vendors' literature is necessary to determine a relay's suitability for any specific application.
	Contact-load ratings are frequently given in specific load capability, such as 1 or 2 hp. Also the maximum current figure is sometimes indicated as for a noninductive load. It is significant that maximum current rating is often the same for 115/120 V ac and 24/28 V dc. Insulation-voltage breakdown test is usually 500 V rms 60 Hz between open contacts, 1,000 V rms elsewhere						

| Telephone type | 200 MW to 12 W, depending upon size of contact load and the copper volume of the coil. In general large relays use 2–8 W, medium 1.5–5 W, and miniature 1–3 W | 2 to 15 ms depending upon contact spring load and coil L/R. Operate-time delay of 50 to 100 ms obtainable on larger relays | 5 to 20 ms depending upon contact spring load. Release-time delay of 50–100 ms relatively easy to obtain. Release-time delay up to 500 ms available on largest relays with short-lever-ratio armatures | Approximately 1 A for standard contacts. Special contacts are available on some models extending the maximum allowable current up into the light- to medium-duty power-relay ranges | Depending upon the size and design of relay, this is rather easily varied from single-pole single-throw to as many as 13 or more contact springs, single- or double-throw, of two or more pileups | 1–20 in^3, depending upon type of delay and number of required contacts | Advertised life varies from 50,000 to 100,000 operations electrical and 3 to 10 million operations mechanical, depending upon manufacturer and type of relay | In general this is a somewhat clearly defined design shape, but the size is quite varied. The contacts are usually either leaf or wire types, stacked in a pileup of considerable height, and fastened to an L-shaped frame. The relay is usually arranged for wiring in a plane on the opposite surface to which the relay mounts |

The features which make telephone-type relays attractive are usually the highly varied kinds of contacts and the allowable timing variations. Unless otherwise stated, the insulation-voltage breakdown test is 500 V rms 60 Hz, between all insulated parts and open contacts. Maximum contact resistance is usually less than 50 to 100 mΩ, new.

TABLE 13 A Selection Guide and Summary of Electromechanical/Electromagnetic Relay Operational Characteristics by Types (Continued)

Kind of relay	Required operating power	Operate time	Release time	Max contact-load current	Max contact arrangement	Size and weight	Remarks
Dry-reed and mercury-wetted reed relays	80 MW for lightest load to 6 W for maximum size load. Wattage that coils can dissipate usually varies from two times just-operate wattage to five or more times that value	From 1 ms (max) for single capsule, exclusive of bounce, to 6 ms (max) for large multicapsule units, including bounce	½ ms (max) for single capsule exclusive of bounce to 6 ms (max) for large multicapsule units, including bounce	Varies from 0.10 to 1.0 A depending upon size of capsule and VA of load. The large, or standard, size reed is rated at 1 A switching, 15 VA, 250 V maximum, and 5 A carry	From single-pole single-throw to 10 or more poles, with up to four double-throw combinations	Size varies from a fraction of a cubic inch (single-pole, microminiature) to 9 or more cubic inches for the largest multicapsule assemblies. Weight varies from 5 g to 2 oz	Life varies from 5 to 25 million operations at rated load, fully protected, depending upon capsule size and kind. Within allowable and specified conditions the mercury-wetted reeds have longer life and more constant contact resistance than the nonwetted
	Insulation tests vary for various types and sizes. Vendor's information will have to be referred to in each case. Reed relays are attractive because they are compatible with the design concept of discrete components having a low profile when attached to a PC board, and their low coil self-inductance makes them suitable for direct drive from a transistor						
Mercury-wetted contact relay (Although there are only two sizes and shapes of capsules, these may be housed in electron-tube octal-plug housings of from one to four capsules per relay or in	100 MW for small-capsule relays, 500 MW for large-capsule relays	1–2 ms for small capsule and 4 ms for large capsule	1–2 ms for small capsule and 2.5 ms for large capsule	1–2 A for small capsule, 5 A for large capsule	1 Form C or 1 Form D, only, for the small capsule, and 1 Form D only for the large capsule	The size varies from about 4 in³, cylindrical, with a 1.1 in diameter to 0.75 in³ in the form of a rectangular module. The weight varies from about 3 oz for small single-capsule	The popularity of these relays is restricted somewhat by their obvious limitations of weight (mercury is heavy), mounting position (the relay cannot be tilted much), low temperature (mercury becomes a solid in the vicinity of −40°), shock and
	These relays have unlimited life; the contact resistance is nonfluctuating and predictable. They are fast and handle heavy loads quietly, efficiently, and safely						

Case/type	Power	Operate time	Current rating	Contacts	Weight	Application notes
flat-sided rectangular-module cases with PC pin terminals, a popular arrangement)	100 MW, or less, to 750 MW	2 to 5 ms not including bounce. Bounce does not exceed 1.5 to 2 ms for some models, does not exceed 0.250–1.0 ms for others			relay to 10 oz for large 4-capsule relay	vibration (in normal temperature mercury is a liquid), and cost (mercury is expensive)
Crystal-can relays (both the *bathtub* conventional shape and the *top hat*, TO-5, shape)		1.5–4 ms	Up to 5 A, 28 V dc resistive for full-sized can. About 2 A maximum for half size can. Still lower maximum current allowed for the smallest can	2 Form C	⅙ crystal-can size to full crystal-can size, and TO-5 size. 0.5 oz or less, up to 0.8 oz for crystal can. TO-5 weighs 0.08 oz or less up to 0.15 oz	Relays of this type were designed for and originally used in military and space-age applications. It was inevitable that because of size and weight reduction other uses would be found, particularly in the sensing of low-level signals. In most industrial applications the inherent resistance to shock and vibration is of little consequence. One of the TO-5's called an op-amp universal relay contains in one TO-5 case, besides the relay, an op-amp, driver, and surge suppressor

Life is usually not required to exceed 100,000 operations at rated load, for these relays

Special-purpose relays such as TDR's, latching, coaxial, and thermal are omitted because their operating characteristics are peculiar to their design types and the need they fill. The text discussion for these special relays should be adequate for determining early circuit design and application interest in them. They are not, in most cases, directly competitive with the basic types. For example, thermal relays cannot be compared directly with the basic types because their actuation has a nonspecific inherent time factor and there is no standardization in this class of relay to make such comparison meaningful.

Special-Purpose Relays. Relays designed for a special purpose, such as the coaxial relay, and relays that are identifiable as a general class but have special features added such as magnetic-latching crystal-can relays are referred to as special-purpose relays. The coaxial relay is not discussed in detail elsewhere, but the other special relays have already been examined in detail.

Coaxial Relay. Figure 81 shows one make of this kind of relay. It has at least one set of contacts for switching radio frequency but may have additional conventional contacts as well. Since the aim here is to switch high-frequency current with minimum loss, the contact mechanisms are enclosed in a metal chamber with dimensions such that it forms a cavity whose characteristic impedance matches that of the coaxial cable to be attached to it, as nearly as possible. A measure of the effectiveness of the design is its voltage standing-wave ratio (vswr). If the characteristic impedance of the switching mechanism of the relay exactly matches the cable impedance, the ratio is 1:1. This is never realized, but a good ratio that is frequently attained is 1.02:1.

Fig. 81 A typical coaxial relay. (*Magnecraft Electric Co.*)

A relay of the coaxial type is used in microwave switching, as in switching from transmitters to receivers, and also in switching from antenna to antenna and in multiple-antenna systems. Many television installations employ coaxial relays for switching purposes.

Besides the chambered contact type of relay in Fig. 81 there are also dry-reed contact coaxial relays.

Stepping Relays. See the discussion of Stepping Switches for the distinction between stepping relays and stepping switches.

Latching Relays. There are basically two kinds of latching relays, mechanical and magnetic latching. In the first kind some variety of mechanical catch or toggle action causes the relay to remain in the operated position after the actuating pulse ceases. A second pulse through the same or a different winding restores the relay to its original position. Because most mechanical latches are additions to relays of a type discussed earlier, they will not be examined in detail again. Mechanical latching is subject to early and rapid wear, maladjustment (usually the result of unauthorized tinkering), and failure from shock or vibration. It has the advantage of simplicity and is easy to understand.

Magnetic latching uses either a permanent magnet to hold the armature in the operated position after the operating pulse ceases or a coil with a permanent-magnet material where its core takes a magnetic set to hold the armature operated after the actuation takes place. In either case the armature is restored by having the holding power of the permanent magnetism overcome by a reverse pulse through the same or a different winding. Both these techniques are usually employed on relays of a general type, as already examined above. Most common magnetic-latching relays are of the general-purpose, telephone, or crystal-can (military) types. Relays of this kind are used for memory, overload response, or to aid in resistance to vibration and shock.

Definite-Purpose Relays. In almost all manufacturers' catalogs one or more kinds of relays designed to handle a definite task are shown. Some of these indicate or control specific voltages, sequence of operations, or specific machine operations. Since these are not likely to be of general interest, they will not be examined further here.

The Relay Choice. The decision as to what kind of relay to choose can be based upon an analysis of the foregoing information and a search of possible vendors' literature. Tables 21 and 22 at the end of this chapter contain lists of relay vendors and agencies.

In deciding which way to go, the circuit designer has to weigh many things. Tables 13 to 15 may be of assistance. The EMR is a device with the possibility of a large number of outputs for one or more inputs. The dry-reed relay has the possibility of up to 10 to 12 outputs for a single (or double) input. The SSR is one solid-state element per switching function. The EMR, because of its mechanical coupling, has a definite wear-out disadvantage. Sometimes its exposed contacts are considered a disadvantage. The dry-reed relay, by eliminating mechanical coupling, has some advantage from the reduced-wear standpoint and provides sealed contacts. The SSR has all the advantages listed above for such a relay plus some others shown in Tables 14 and 15, but it also presents many problems, most of which are instantly recognized by electronics engineers. The question of what relay

TABLE 14   A General Comparison of Electromechanical/Electromagnetic Relays and Solid-State Relays[10]

Parameter	EMR	Dry-reed relay	SSR
Life	Dependent upon load and number of operations	Dependent upon load and number of operations	Barring accident, dependent only upon time
Response time	2–5 to 25–50 ms	0.5–2 to 6–20 ms	Fractional microseconds
Power gain per $	Excellent	Good	Poor
Contact functions	Multiple	One or more	One
Size-weight	Poor	Good	Excellent

TABLE 15   A Comparison of Reed, Solid-State, and Hybrid Relay Characteristics[7]

	Reed	All SS	Hybrid
Input specifications:			
Coil voltage	0.6–96 V	3–280 V	2.4–48 V
Coil current	0.66–241 mA	0.5–20 mA	1–33 mA
Coil sensitivity	66–700 mW	1.5–7 mW	60 μW–290 mW
Must pickup voltage (typical 75% of coil V)	0.34–46 V	3.5–90 V	1–36 V
Must dropout voltage (typical 10–20% of coil V)	0.07–15 V	0.4–20 V	0.5–2 V
Pull-in speed	0.5–20 ms	5 μs–1½ cycles	1–25 ms
Dropout speed	60 μs–17 ms	5 μs–1½ cycles	First 0 crossing–½ cycle
Output specifications:			
Max output voltage	500 V	280 V	240 V
Max output current	5 A	40 A	15 A
Max output power	360 VA	9.6 kVA	840 VA
Contact resistance	15–200 mΩ	0.1–2 Ω	0.1 Ω
Contact offset voltage	Virtually 0 V	1 mV–1.6 V	1.0–1.5 V
Off-state leakage current	None	1–5 mA at 120 V	1–5 mA at 120 V
Min on-state current	None	20–100 mA	100–200 mA
Max contact operating frequency	1 kHz	100 kHz	1 kHz
Max lifetime cycles	10^8–10^{10}	$>10^{10}$	5×10^8
Contact bounce	Hg, no; dry, yes	No	Yes and no
Min dV/dt	N/A	20 V/μs	1 V/ms

is to be selected therefore becomes one of weighing advantages against disadvantages and recognizing the necessities of the situation. For example, ask the questions: How necessary is speed of response? Does it outweigh the inherent advantages of the other alternatives? Usually the economic factor will be the deciding one.

Cautions: Coil-winding Polarity. Unless conventional types of relays as designed for low-voltage circuits are to be used in relatively short-lived equipment, well insulated from ground, it is considered best practice to connect negative potential permanently to the outer coil terminal(s) and control the relay by switching grounded positive potential to the inside terminals of the relay winding(s). This reduces electrolysis to a minimum and, particularly for coil windings of relatively fine wire, adds years of life to relay coils that otherwise might deteriorate from electrolysis in a very short time. The low voltage referred to above is 50 V or less as encountered, for example, in wired communications equipment. A notable exception to this rule exists for 24/28-V equipment intended for the military, where switching of positive ground is frowned upon as uncomfortable for maintenance personnel to work around, although telephone maintenance personnel accept it as a necessary situation and have done so for many years.

Contact-Spring Polarity. The same potential, whichever it is, should be connected to all movable springs. This is an aid to maintenance personnel in that they do not have to consult a blueprint continually for that information, which makes servicing faster and simpler. Also it lessens the chance of accidental short circuiting, which can destroy relay contacts in an unguarded instant, as may occur when the maintenance man has no recognized standard arrangement to rely upon.

Interaction of Windings. Relays should not be operated with windings in parallel if at all possible. Occasionally when relay windings are connected to the same bus on one side and are operated from a common contact closure on the other, they affect each other adversely when their operating circuits are opened, owing to interaction from the back emf generated by their collapsing magnetic fields. This problem may exist for (1) two or more nonpolarized relays with parallel coils, (2) a single relay with more than one winding (bifilar or polarized), (3) dc relays using diodes for operation from an ac power source, and (4) polarized magnetically latching relays.

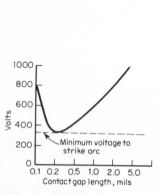

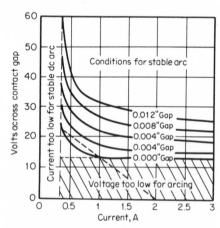

Fig. 82 Breakdown voltages for various air gaps at 1 atm (Paschen effect).[14]

Fig. 83 Voltage–current gap–length relationship for stable and unstable dc arcs.[14]

Arcing Problems.[14] A number of means exist to accelerate the arc extinction when interrupting large voltages and currents. These include:

    Large contact gaps (see Figs. 82 and 83)
    Double-break switch structures
    Magnetic or air-blast arc blowouts for heavy-duty relays
    Arc-suppression circuits

If arcing generates enough plasma to reach from the live circuit to some other circuit or ground that bypasses any current-limiting resistance, an explosively catastrophic flashover occurs. This hazard particularly affects polyphase power and polarity-reversal circuits. Mandatory for other considerations, these standard design and construction practices often accentuate problems:

Relays must switch *power* circuits between the power source and the load, not on the ground side of the load.

Relays used for switching 115 V ac or higher must have a grounded frame (and can, if enclosed) so that electrical leakage or short circuit in either load or relay does not constitute an operator hazard.

Fault Circuits.[14] These several examples illustrate fault circuits in common control circuits:

In three-phase Y-connected (or four-wire) circuits (Fig. 84) a high-voltage (up to 280 V on 120-V systems) a low-resistance circuit exists between legs during the arcing just before the instant of current reversal in any leg. If the conductive plasma cloud reaches from one leg to another, only the capability of line and source limit the current.

If contacts are physically near a grounded structure (i.e., a relay frame or can), relay contact chattering can cause a flashover. Flashover can occur if the ac coil power changes slowly, if the relay is improperly adjusted, or as a result of switching or lightning-induced line transients.

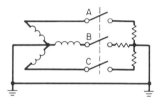

Fig. 84 Conditions for three-phase flashovers in 120-V, Y-connected, four-wire switching circuits.[14] Three-phase flashovers in 120-V Y-connected, four-wire switching circuits can occur with appropriate phase polarities and one contact opening before the other two. In the presence of a conductive plasma an arc can flash if *B* is negative and either *A* or *C* is positive.

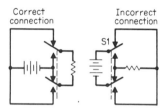

Fig. 85 Problem facing a relay with DPDT contacts when reversing a load.[14] (*a*) Connected as shown at right, power source can short if one contact sticks (dashed S1 contact). (*b*) Opposed polarities across fixed switch contacts (especially "correctly connected" left circuit) enhances chance of plasma-arc flashovers.

The familiar double-pole reversing switch can prove troublesome even when connected correctly. Unfortunately this trouble is seldom recognized for what it is. Of course, the switch must be connected so that progressive contact engagement momentarily shorts the load, not the power source (Fig. 85). Unfortunately, rapidly repeated switch operation (or circuit-induced contact chatter) may generate a power-source-shorting plasma cloud. If transient current and voltages exceed the stable arc minimums, the plasma may permit a flashover.

Ironically, safety codes force us to live with potentially catastrophic fault circuits. The problem would not exist if the load-circuit ground leg rather than its hot leg could be switched. On the other hand, this may explain why ungrounded relays constitute personnel hazards (short-circuit fault currents > 2 to 5 mA) even though the fault current cannot exceed the load current and a catastrophic fault could never become established. Such a hazard to personnel exists with ungrounded relay cans whenever load voltages and currents exceed the minimum arcing conditions and the design permits the plasma clouds to reach the can.

The causes of these faults and the circuit conditions which contribute to vulnerable situations must be recognized. Generally, the most economical remedies become self-evident:

Relays must be rated for the voltages they will experience.

Interleg spacings near contacts in Y-connected 120-V three-phase circuits should be rated at 208 V. There should be adequate space and barriers between contact sets.

A three-pole relay switching on and off a three-phase 115-V ac load contains about 170 V peak across the contact gap or between any contact and ground. If the same relay is used to switch a load between two unsynchronized three-phase power sources, as much as 340 V peak can appear across a contact gap.

Provision for grounded frame and case is mandatory in small hermetically sealed relays in 120-V ac single-phase applications. If the load current and voltage generate enough plasma to reach from the contacts to a grounded part, internal flashover prevention is mandatory.

When a 28-V dc inductive load is reversed, a catastrophic fault can occur only during the very brief contact transfer. With stable arc conditions sustained for the duration of contact transfer, the fault path simply jumps whichever fixed contact gap opens first.

These problems can be solved by rearranging the circuit to eliminate uncontrolled fault current, using a relay which tolerates the arc, or employing arc suppression to reduce the arcing to an insignificant amount. Above all, a suitable relay should be used for power-supply changeover or, in fault-protection service, a heavy-duty power contactor or circuit breaker.

SWITCHES

The term switch was defined at the beginning of this chapter so that it could be used in the simplest relay definition. In the rest of this chapter, the word switch may have a far more comprehensive meaning. It will be applied to all kinds of switching devices, from a simple pair of mating contacts, manually actuated, to electrically operated multipole, multiposition devices such as stepping switches. Such a range of identities makes this subject difficult to deal with in the space allowable. A recent listing of switch manufacturers consisted of over 160 names, covering a wide range of types of products (see Table 23). Consequently the simpler, well-known, and recognized switches will be dealt with sparingly, and the newer and/or more complicated switching devices will be examined in more detail.

Matters such as contact load-handling capabilities, contact protection, and arc and noise suppression are so similar to those of relay contacts of the equivalent type, material, and size that no repetition of ways and means of dealing with such problems will be repeated here. One caution on the subject of contact problems is in order, however. For manually operated switches, not of the toggle or snap-action type, it is possible for an individual to so delay the opening or closing of a pair of contacts or to so reduce the contact pressure by deliberately slow motion that the load-handling capability is seriously reduced. It is not too unusual to see a failed contact set on a manually operated switch that proved inadequate to a load situation where relay contacts of the same size, shape, material, and mechanical separation and pressure did the job well. For that reason manually operated switches are frequently used to close the circuit to a relay, which in turn handles the actual load. There may well be a wire-size advantage also in doing it that way.

Kinds and Types of Manually Operated Switches

A listing of kinds of switches to be manually operated is quite extensive, and includes the following:

    Push-button switches (also called push keys)
        Illuminated and nonilluminated
        Single and multiple, or ganged
        Snap action, wiping action, or butt action
        Locking (push-pull) and nonlocking
        Alternate (on-off)

Slide switches
   Single and ganged
Lever switches (also called lever keys)
   Locking and nonlocking
   Toggle
Turn-button switches (also called turnkeys)
Thumbwheel switches
Leverwheel switches
Rocker switches
Keyboards
Rotary selector switches
Matrix-type selector switches

This is obviously a somewhat generalized listing, and many specials may not fit any of these catagories. One vendor even has a listing of his "weird and wonderful switches." Types of mountings, method of attaching circuit connections (such as through the printed-circuit edge of card connectors), open or enclosed bodies (frames), and customized variations of all kinds could lead to further categorization.

Push-button switches Push-button switches are available with the contacts remaining operated after the button has been depressed (locking or latching) and with nonmaintained operation after finger removal (nonlocking or nonlatching). There is also a form with a button that is pulled out to latch and pushed back in to unlatch. In most cases visual observation is depended upon to determine whether or not latching push buttons are in the operated state. An indication light may be either separate from the button or self-contained. Lighted push buttons of two types are shown in Fig. 86. The lighting may be transmitted through translucent colored buttons or may be projected as a colored light. Split-screen displays are common. Contact ratings and life vary to such an extent from one type of switch to another and from manufacturer to manufacturer that it is impossible to present meaningful general data. For example, one vendor will promise a life of 25,000 operations electrical, 50,000 mechanical, at rated load, indicating only limited life regardless of contact load, but another vendor will quote a life of 10,000,000 operations for a light contact load on a similar switch. Vendor's claims in these matters are to be relied upon. This same problem of inability to establish generally meaningful load and life rating will apply to all the manually operated switches which follow, also.

Multiple-station push-button switches have a common frame with interaction between switches so that they perform interrelated functions that a single switch cannot perform. Also it is possible to get sequential protection from such a grouping that is impossible to achieve mechanically any other way. Solenoid-operated multistation switches can be actuated from a remote or local position. In one form this provides an important security feature to prevent aimless or unauthorized pushing of buttons.

Table 16 is a list of prominent manufacturers of lighted push-button switches.

Slide switches As the name suggests, slide switches take either one of two positions or one of three positions, by means of lateral displacement of the button. They come in single-switch and ganged arrangements. The three-position switches present somewhat of a hazard in correctly identifying the midposition. Contacts may be either toggle-actuated or straight deflection-actuated. The contact closure may be momentary, locking, or a combination of both. A slide-switch example is shown in Fig. 87a.

Lever switches The telephone-industry relative of the lever switch is not available with toggle action and is called a lever key. Commercial and industrial versions of these devices may be either toggle-actuated or straight deflection-actuated and either locking or nonlocking. Both two- and three-position switches are available. The three-position switches can be locking to one side and nonlocking to the other side, locking on both sides, or nonlocking on both sides. The telephone-type switch, or "key," because of the nature of its original purpose, has quite a long life (around 10 million operations) when used within the limits of its contact rating.

Turn-button switches This is primarily an item made by telephone manufacturers and offered for sale as an industrial item. It operates its contacts when a button

(a)   (b)

Fig. 86 Lighted push buttons. (*a*) Dual lamp, exploded view. (*Switchcraft, Inc.*) (*b*) Round or square button for panel or subpanel mounting. (*Grayhill, Inc.*)

is rotated approximately a quarter turn. One of its advantages is that it is easily made either locking or nonlocking and gives ready visual indication of whether it is operated or not operated.

Thumbwheel and lever switches   Rotating a serrated wheel that protrudes above the front surface or operating a lever that sticks out in front activates these essentially similar switches having 8, 10, 12, or 16 discrete dial positions. Modular thumbwheel switches require minimal panel space in setting up multidigit inputs. Integral encoder assemblies convert the thumbwheel position directly to output code. The switch normally contains the circuitry and requires no added wiring (see Fig. 88). Some typical truth tables for the more common codings are shown in Table 17. Built-in illumination is available in some models. Many mounting variations and custom assemblies are also offered.

Push-button switches, coding   Push-button switches which provide essentially the same functions as thumbwheel switches in that they have the same kind of numer-

TABLE 16   A Partial List of Manufacturers of Lighted Push Buttons[13]

Airpax Electronics 1836 Floradale Ave. El Monte, CA 91733	International Electro Exchange 8081 Wallace Rd. Eden Prarie, MN 55343
Alco Electronic Products 1551 Osgood Street N. Andover, MA 01845	Ledex 123 Webster St. Dayton, OH 45401
Arcoelectric Switch PO Box 348 N. Hollywood, CA 91603	Licon Div., Illinois Tool Works 6616 W. Irving Park Rd. Chicago, IL 60634
Arrow-Hart 103 Hawthorn St. Hartford, CT 06106	Master Specialties 1640 Monrovia Costa Mesa, CA 92627
Burgess Switch Co. 777 Warden Ave. Scarborough, Ont. Canada	Maxi-Switch 3121 Washington Ave., N. Minneapolis, MN 55411
Carling Electric 505 New Park Ave. W. Hartford, CT 06110	Micro-Switch, Div. of Honeywell 11 Spring Street Freeport, IL 61032
Chicago Switch 2035 W. Wabansia Ave. Chicago, IL 60647	Molex 2222 Wellington Ct. Lisle, IL 60532
Clare-Pendar PO Box 785 Post Falls, ID 3854	Oak Industries, Switch Div. Crystal Lake Ave. Crystal Lake, IL 60014
Compu-Lite 17795 "C" Sky Park Cir. Irvine, CA 92707	Seacor 598 Broadway Norwood, NJ 07648
Dialight 60 Stewart Ave. Brooklyn, NY 11237	Staco-Switch 1139 Baker St. Costa Mesa, CA 92626
Drake Manufacturing 4626 North Olcott Ave. Harwood Heights, IL 60656	Switchcraft 555 N. Elston Ave. Chicago, IL 60630
Furnas Electric 1000 McKee Batavia, IL 60510	Symbolic Displays 1762 McGaw Irvine, CA 92705
Grayhill 561 Hillgrove Ave. La Grange, IL 60525	UID Electronics 4105 Pembroke Rd. Holywood, FL 33021
Industrial Devices Edgewater, NJ 07020	Unimax Switch Ives Road Wallingford, CT 06492
Industrial Electronic Engineers 7720 Lemona Ave. Van Nuys, CA 91405	Waldom Electronics 4625 West 53rd St. Chicago, IL 60632

TABLE 17   Typical Truth Tables for Thumbwheel Switches

Table a Code: BCD (binary-coded decimal)

Readout symbol	Common C connected to terminals = ●			
	1	2	4	8
0				
1	●			
2		●		
3	●	●		
4			●	
5	●		●	
6		●	●	
7	●	●	●	
8				●
9	●			●

Table b Code: 10 position decimal

Readout symbol	Common C connected to terminals = ●									
	0	1	2	3	4	5	6	7	8	9
0	●									
1		●								
2			●							
3				●						
4					●					
5						●				
6							●			
7								●		
8									●	
9										●

Table c Code: binary-coded octal

Readout symbol	Common C connected to terminals = ●		
	1	2	4
0			
1	●		
2		●	
3	●	●	
4			●
5	●		●
6		●	●
7	●	●	●

Table d Code: BCD complement

Readout symbol	Common C connected to terminals = ●							
	1	2	4	8	1	2	4	8
0					●	●	●	●
1	●					●	●	●
2		●			●		●	●
3	●	●					●	●
4			●		●	●		●
5	●		●			●		●
6		●	●		●			●
7	●	●	●					●
8				●	●	●	●	
9	●			●		●	●	

Table e Code: BCO complement only

Readout symbol	Com. connected to complement of binary bit output No. = ●		
	1	2	4
0	●	●	●
1		●	●
2	●		●
3			●
4	●	●	
5		●	
6	●		
7			

Table f Code: Hexadecimal plus complement plus 2 commons

Common X connected to terminals = ●
Common Y connected to terminals = ○

Readout symbol	Interconnection from common X to binary bit output numbers				Interconnection from common Y to complement of binary bits output number			
	1	2	4	8	$\bar{1}$	$\bar{2}$	$\bar{4}$	$\bar{8}$
0					○	○	○	○
1	●					○	○	○
2		●			○		○	○
3	●	●					○	○
4			●		○	○		○
5	●		●			○		○
6		●	●		○			○
7	●	●	●					○
8				●	○	○	○	
9	●			●		○	○	
10		●		●	○		○	
11	●	●		●			○	
12			●	●	○	○		
13	●		●	●		○		
14		●	●	●	○			
15	●	●	●	●				

SOURCE: Cherry Electrical Products Corporation, Waukegan, Ill.

ical display, both digital and binary codes, come sealed or not sealed, lighted or unlighted, and are available in bidirectional or unidirectional types.   Some fit in the same dimension of panel openings as thumbwheel switches.   An example of these devices is seen in Fig. 88a.

Rocker switches   Rocker switches get their name from appearance and feel.   The actuating button is pivoted so that it rocks to either of two operated positions and usually has a neutral or off midposition. The contact switching elements are usually toggle-activated. The appeal is largely aesthetic because of its good appearance, although its ease of actuation, toggle-switch action, and ease of position identification make it attractive from the good-engineering stand- point. Illumination can be provided by edge lighting of the panel in such a way that light is conducted into the button when tilted. One form is Fig. 87*b*.

Keyboards A wide variety of keysets, keyboards, touch pads, and similarly named components provide a convenient and inexpensive way of providing manual input to electrical and electronic devices. Some are identical in appearance and similar or identical in con- struction to the touch calling keysets of automatic telephony, while others were obviously designed with noncommunicaiton applications in mind. The key arrangement varies from the 12-button (arranged in 4 rows, 3 buttons per row) with button caps labeled 1 through 0 plus ✱ and #, and telephone-type alpha- numeric, to elaborate key arrangements of 20 or more keys following adding-machine and/or typewriter but- ton-cap identity sequence and arrangement. Since it would be impossible to cover all these in this work, only the simplest will be examined. Such is probably the telephone-type keyset, without tone generators, al- though that form too is readily available for industrial use if proper frequency-decoding equipment is at hand for receiving, identifying, and making use of the re- ceived tones.

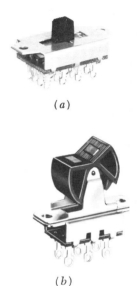

(*a*)

(*b*)

Fig. 87 Typical slide switch (*a*) and rocker switch (*b*). (*Switchcraft, Inc.*)

The keyset in Fig. 89 has its 12 push buttons ar- ranged in the familiar pattern that is standard for all touch-calling telephone instruments. However, it is not equipped with a tone generator. The keyset will find application where serial information is to be en- coded in a system and where the cost of tone-decoding equipment cannot be justified.

Signaling with the keyset in Fig. 89 requires a multiple number of leads (the num- ber required depends on the code chosen). Unless considerable distance is involved, dc signaling on a multiple number of leads is quite economical.

Each of the push buttons on the keyset controls two normally open contacts of the permissive make design so that contact pressure does not depend on contact overtravel. In addition to the contacts operated by each push button, a common switch is actuated whenever any push button is depressed. The contacts on the common switch operate after the push-button contacts close and open before the push-button contacts reopen. By conneting the common contacts in series with all push-button contacts, loads of up to 1 A, 50 VA, may be switched. Figure 90*a, b,* and *c* shows typical applications.

There are seven actuator bars per keyset. Each actuator bar operates one contact form. When any of the 12 buttons is pushed, it moves a unique combination of two actuators to produce a two-out-of-seven contact-code output. This code format can readily be adapted to most control requirements.

Because they are standard telephone keysets, these keysets have a familiar "feel" which is conducive to fast, accurate operation. Figure 91 shows how the two-out- of-seven contact code is derived.

The common switch may be connected in series with the push-button contacts to ensure simultaneous switching of those contacts. Used in this manner, when a button is pushed, the two sets of contact springs prepare isolated paths for the two outputs. Immediately thereafter, the common switch closes the circuit through the push-button contacts. Upon release, the common switch opens first to break the circuit, and then the push-button contacts open.

Two methods are shown for converting the two-out-of-seven code generated by the keyset into decimal code in Fig. 92*a* and *b*.

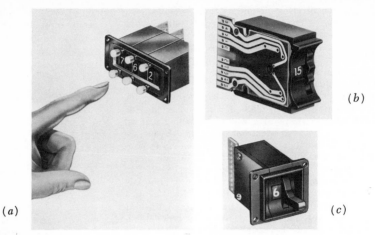

Fig. 88   Typical push button (a) (*Chicago Dynamic Industries, Inc.*), thumbwheel (b), and lever (c) coded switches.   (*Cherry Electrical Products Corporation.*)

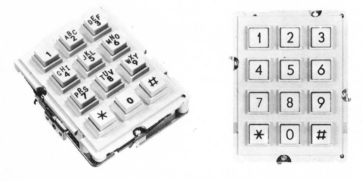

Fig. 89   A typical telephone-type keyset.   (*GTE Automatic Electric Co., Inc.*)

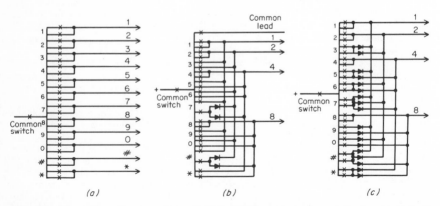

Fig. 90   Examples of connections to the telephone-type keyset for some typical coding.   (a) Straight decimal coding.   (b) Binary coding.   (c) Binary with parallel contacts.

Rotary selector switches One of the oldest forms of manually operated switches is the rotary switch, so common and so simple it needs little discussion other than to point out what range is available. As will be seen from Fig. 93, multideck switches are common, almost any number of contacts in each deck are offered, the decks may be enclosed or open, the contacting may be momentary or maintained, and the shaft rotation may be continuous, in either direction, or limited to just 360° of rotation, after which the knob must be reversed. One of the advantages inherent in a rotary switch is that contacting must be in a predetermined sequence. This can, of course, also be a disadvantage preventing its use. The latter problem is sometimes circumvented by using a special form of the rotary switch where it is necessary to push the handle or knob in to actuate the connection. Selection of contact layer is achieved in this way on some models. Another special form of the rotary switch is the familiar tap switch, usually associated with the switching of heavy electrical loads.

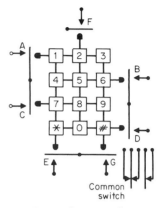

Fig. 91 Schematic drawing showing two-out-of-seven. Code arrangement of telephone-type keyset.

Grades of manual switches Manual switches are like relays in that they may be graded by class of service. For example, four common classifications are commercial (as used in office-type equipment), communications (as used in telephone central-office and subscriber equipment), appliance (as used in low-cost devices on vending machines, home appliances, television sets, and automobile dashboards), and industrial (rugged devices for use on machine tools, industrial materials handling, and such equipment) where the device has to withstand operator abuse, oils, coolants, chemicals, and dust and dirt. The latter needs are recognized by NEMA in Standard ICS 1-110, with these designations: Type 1—General Purpose, Indoor; Type 3—Dusttight, Raintight, and Sleet (Ice) Resistant, Outdoor; Type 4 and 4X—Watertight and Dusttight, Indoor; Type 7—Hazardous Locations, Class I (explosive gases and vapors); Type 9—Hazardous Locations, Class II (explosive dusts); and Type 13—Oiltight and Dusttight, Indoor, for protective housing of pilot devices such as push-buttons and selector switches.

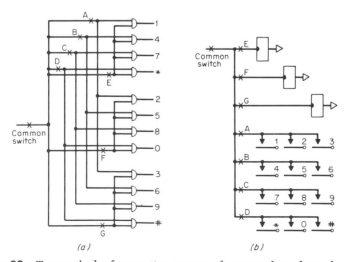

(a)   *(b)*

Fig. 92 Two methods of converting two-out-of-seven code to decimal code.

If equipment is being designed for any of the above types of location, the ruling NEMA standard should be obtained and the applicable portions included in the purchase specifications.

Sensing Switches

Switches that are not operated manually or electrically are usually those operated from a mechanical stimulus, such as by a float or cam-operated lever. Switches of this type recognized by NEMA are cam-operated, control cutout, drum, float, isolating, limit (both control and power circuit), master, pressure, proximity, temperature, and selector. No matter what the input, a pair or more of contacts open, close, or transfer to control, indicate, or both. The contacts are usually snap or toggle switches, so that small increments of changes will still result in very positive contact action.

Snap-acting switches A snap switch gets its action from a specially formed and prestressed main spring or blade (see Fig. 94). The slotted bipositional blade is prestressed (usually by heat treating) so that the center section is compressed but the two outside sections are in tension, causing it to reside in an unoperated or normal position as shown in Fig. 94. Depressing the center section by means of a plunger disturbs the forces within the blade to cause it to assume its other (operated) position with a rapid over-center action. This fast transfer of contacts, or snap action, not only aids in extinguishing arcs because of the speedy contact transfer but also results in good contact pressure, which permits handling relatively heavy loads. In manufacture, the force required for operation and the distance required to deflect the plunger to get snap-over are rather easily controlled within strict tolerances. This permits using the snap switch for relatively precise applications. The repeatability is also rather closely limited because there is only one moving part. Life is good because there are no localized extensively deflected areas.

Fig. 93 A typical multideck rotary switch (manually operated). (*Grayhill, Inc.*)

Obviously the snap switch is adaptable to almost any kind of switching device whether manually, mechanically, or electrically actuated. As a result it may be found in manually controlled switches, on relays, or mechanically actuated as a limit switch in the control of motion (as on a machine tool), temperature (as in a thermostat), time (as on a motor-driven timer), etc. The big advantage, of course, is that the snap switch can directly control motors, heaters, household appliances, and such from small stimuli without the use of an interposing relay. Movement differentials as small as 0.0005 in permit a snap switch to be rated 25 A and 2 hp at 250 V ac.

There are several classifications or grades of the snap switch. Precision grades, as the name implies, are those whose actuating forces and distance of plunger travel are the most closely controlled. The appliance grade does not require very close tolerance, relatively speaking. There is a size classification of miniature, subminiature, and miniature-subminiature. Some are enclosed in plastic cases, some are open, and some are housed for special environments. The latter may take the form of a metal housing for attachment to a conduit system; it may be sealed against dust, moisture, or corrosive atmosphere; or it may even be immersionproofed and explosionproofed.

The same precautions apply to the need to pick the correct size and kind of contact to match the load here as were discussed with respect to relay contacts. Ratings given in manufacturers' catalogs principally apply to alternating current (rapid snap-action contact movement and the zero current on each half-cycle aid these devices in attaining a relatively high load rating for their contact size and distance of separa-

TABLE 18   Typical DC Loads for a 20-A, 250-V, AC General-Purpose Precision Snap-acting Switch[15]

Contact separation, in	Noninductive			Inductive	
	Direct current, V	Heater load, A	Lamp load, A	At sea level, A	At 50,000-ft altitude, A
0.010	6–8	20.0	3.0	8.0	7.0
	12–14	20.0	3.0	5.0	5.0
	24–30	2.0	2.0	1.0	1.0
	110–115	0.4	0.4	0.03	0.02
	220–230	0.2	0.2	0.02	0.01
0.020	6–8	20.0	3.0	20.0	15.0
	12–14	20.0	3.0	10.0	8.0
	24–30	6.0	3.0	5.0	2.0
	110–115	0.4	0.4	0.05	0.03
	220–230	0.2	0.2	0.03	0.02
0.040	6–8	20.0	3.0	20.0	15.0
	12–14	20.0	3.0	20.0	15.0
	24–30	10.0	3.0	10.0	5.0
	110–115	0.6	0.6	0.1	0.05
	220–230	0.3	0.3	0.05	0.03
0.070	6–8	20.0	3.0	20.0	15.0
	12–14	20.0	3.0	20.0	15.0
	24–30	20.0	3.0	10.0	7.5
	110–115	0.75	0.75	0.4	0.2
	220–230	0.3	0.3	0.2	0.1

tion), but dc ratings which can apply are those of Table 18. Life expectancy at various ac loads is typically that of Fig. 95.

Proximity switches   The name indicates that switches of this type function without the switch being physically contacted by the stimulus. There are many kinds of proximity switches, some consisting of so many elements and of much complexity that they are really systems. Examples of this are those which operate from a sensing head that recognizes radio frequency, magnetic bridge, inductive imbalance, photoelectricity (light), ultrasonic beams, or ferromagnetism. This discussion will be confined to the simple proximity switch, principally the dry reed, or the mercury-wetted reed. Anything that can be arranged to move a permanent-magnet past, or close enough, to a reed switch can be the stimulus for operating a reed-type proximity switch. This principle is applied in some designs of keyboards, for example. Others are elevators and escalators, conveyors, machine-tool carriage tranverse, etc. Most of the reed-switch manufacturers have literature suggesting how these switches can be operated from a permanent magnet in proximity.

Electrically Operated Switches

Electrically operated switches come in many forms. As was pointed out, the relay is one example. However, it has been treated separately because of its extremely large number of variations and versatility. The following will deal with some of those remaining, and some, such as motor-driven switches, are so specialized in performance and application that they will not be examined in detail at all.

Stepping switches   There has long been confusion in the use of the terms stepping switch and stepping relay. The line of demarcation is not easily recognized. Since the operating mechanisms may even be identical, or else quite similar, a distinction between stepping relay and stepping switch is hard to make, except by precise definition. Somewhat arbitrarily, therefore, NARM has made a distinction which seems to be working rather well. NARM defines a stepping relay as a relay having many rotary positions, ratchet-actuated, moving from one step to the next in successive operations, and usually operating its contacts by means of cams. There are two forms: (1) directly driven and (2) indirectly driven, where a spring produces the forward motion on pulse cessation, as in Fig. 96.

The shorter of two definitions NARM gives for a stepping switch will be found in the Glossary. The longer one is as follows:

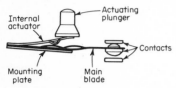

Fig. 94 Schematic diagram of the arrangement of a snap-action switch. Snap-action switches derive their snap action from a springlike main blade. This slotted bipositional piece is pre-stressed by heat treating so that the center member is compressed and the two outside members are under tension. Depressing the actuating plunger forces the internal actuator against the two outside members of the main blade. The relationship between the members is thereby changed, and the blade snaps from the normal position to a second positive position. When the actuating plunger is released, a bias in the blade returns it to the normal position. Advantages: (1) High contact pressures result in excellent resistance to vibration. (2) Fast transfer time limits arcing and increases load capacity. (3) Variety of operating forces obtainable by varying thickness of blade. (4) Good repeatability, due to only one moving part. (5) One-piece tumbled blade has no wear points and provides long life.

A class of electromagnetically operated, multiposition switching devices. Wipers, or groups of wipers, are mounted on a rotatable shaft, which is rotated in steps so that contact is successively made between the wiper tips and contacts that are separated electrically from each other and mounted in a circular arc called a bank. The wiper positioning is done electromechanically on successive pulses to the actuating coil. There are two general kinds in common usage, rotary stepping switches and the Strowger two-motion switch: (1) direct-acting (two-coil) rotary stepping switch—a directly driven rotary stepping switch is a two-coil switch in which one electromagnet (called a rotary magnet, motor magnet, or step coil) and its associated armature provide forward stepping immediately on energization, by ratchet action, advancing the wipers; one step for each pulse received, to the desired contacting position. It remains in this position without further coil energization. The rotor is spring-restored to normal, or home, position, returning in reverse to the route over which it advanced under control of a second electromagnet (called a release magnet or release coil) from a single pulse. (2) Indirect drive, or spring-driven, (one-coil) rotary stepping switch—The indirectly driven rotary stepping switch advances the wipers on the return action, or release of the armature, following each pulse to the motor magnet (coil). The rotation is unidirectional, one step for each pulse, on pulse cessation. The switch is returned to the normal, or home position, by being stepped forward to the home position either from externally produced pulses or by being self-interrupted. (3) Bidirectional (two-coil) rotary stepping switch—A rotary stepping switch of the ratcheting direct acting type, having two coils and associated stepping mechanisms and capable of rotation either clockwise or counterclockwise. (4)

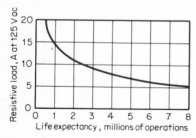

Fig. 95 Life-expectancy curve for a general-purpose precision snap-acting switch.

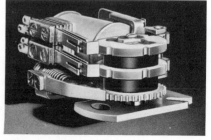

Fig. 96 A typical stepping relay of the indirectly (spring driven type. (*GTE Automatic Electric Co., Inc.*)

Strowger two motion switch—a large capacity switch having 100 discrete positions and used principally in telephone switching.

In short, the stepping relay differs mainly from the switch in that its contact springs take one of two or three positions as the stepping-relay rotor moves around for a complete revolution. It may do so many times in one complete rotation (see Fig. 96), while the stepping-switch wipers sequentially make mechanical contact with *different* output circuits on each step for a full sweep across the bank (see Fig. 98). In other words, if the movable springs of the contacting elements go only up and down, it is a relay; if they sweep around like the hands of a clock mechanically contacting different output contacts each time, it is a stepping switch.

The contacts of a stepping relay are usually as large as on any other kind of relay and capable of handling quite heavy electrical loads. The contacts of a stepping switch are usually limited to relatively light contact loads, since they wipe only across their mating contacts rather than push solidly against them. One bit of added confusion sometimes arises from the fact that stepping switches usually have relatively heavy-duty contacts, too, called interrupter contacts, which operate each time the armature operates, plus some other rather heavy-duty contacts called off-normal contacts, which operate after each passage over or sweep of the bank contacts.

Stepping switches are considerably more versatile than relays from the standpoint of the ways in which they can be used for switching. All stepping-switch manufacturers have extensive and informative literature on the ways to use their switches. A good place to start in designing them into a job is to get this helpful information from the manufacturer. Although a stepping switch looks large to an electronics engineer, appears to be cumbersome to mount and wire in to other equipment, and is relatively expensive, one switch can do so much and replace so many other items that its use shrinks all the above objections to insignificance in many cases.

The rotary stepping switch[16] A stepping switch is basically an electromechanical device used to connect one or more input circuits rapidly to an output circuit chosen from a sizable group of such circuits. The switch responds to current pulses supplied by an external source, or operates by interruption of its circuit through interrupter springs on the switch. Stepping switches are widely used to count, sequence, program, select, and control. They are often applied in machine-tool controls, conveyor systems, test equipment, and communication switching.

Rotary stepping switches are available in a variety of sizes and shapes, primarily dependent on the number of contact points in the bank assembly (Fig. 97). Two basic types are compact switches, which are approximately 4 by 3 by 2 in, and larger switches, which measure approximately 7 by 6 by 3 in.

Switch Construction. As shown in Fig. 97, stepping switches consist of three basic parts: (1) driving mechanism, (2) bank assembly, and (3) wiper assembly.

Fig. 97 The three basic components of a rotary stepping switch.

Driving Mechanism. Rotary stepping switches are stepped by a pawl-and-ratchet mechanism, making one step for each current pulse applied to the switch coil. There are two types of driving mechanisms—direct and indirect. With direct drive, the armature advances the wiper to the next position when the switch magnet is energized. Thus, the driving force varies with the power supplied to the coil.

In the indirect-drive, the wipers are advanced with coil deenergization and resulting release of the armature. The driving force does not vary, since it is provided by a spring in which potential mechanical energy is stored when the armature is attracted to the coil.

Indirect-drive switches are most widely used because they offer higher speeds, greater efficiency, and longer life than can ordinarily be expected of direct-drive switches.

Bank Assembly. The bank assembly, containing individual contact levels built up as a unit, is of a semicircular form in which the bank contacts are firmly held. The individual levels are composed of contacts insulated from the next level by a phenolic insulation or by molding the contacts in a plastic which has the necessary electrical and physical properties. These individual levels are stacked on top of each other, and are assembled together under pressure to assure complete tightness. Compact switches generally have 10, 11, or 12 points per level, and from 1 to 12 levels—for a minimum or 10 points to a maximum of 144.

By special wiper arrangements a larger number of positions per cycle can be provided. For instance, a 10-position switch has its 10 contacts in an arc of 120° and the wiper is triple-ended. As one wiper tip leaves contact number 10, another wiper tip touches contact number 1. The same ratchet mechanism can provide a 30-point switch. Single-ended wipers on 3 successive levels are staggered 120° apart. Thus contacts 1 to 10 are on the first physical level, 11 to 20 on the second physical level, and 21 to 30 on the third physical level.

If fewer positions are needed in a particular application, the excess positions can easily be skipped automatically by wiring the switch to self-interrupt past certain contacts.

On larger switches, between 25 and 624 points are available per full bank capacity. These switches are also available for 50- or 52-point operation by special arrangement of the wiper.

Wiper Assembly. The wiper assembly is the portion of a rotary stepping switch which rotates, making electrical contact with the stationary bank contacts. The wiper assembly consists of a shaft-and-hub assembly; the wiper blades, which do the actual contacting; the ratchet-wheel indicating disk, which shows the wiper positioning on the bank; and a cam, which operates the off-normal contact assembly.

Each bank level has its own corresponding wiper level with which it makes contact. Each wiper is made up of two separate phosphor-bronze blades assembled to and properly spaced on the shaft of the wiper assembly. Both ends of each wiper blade are formed into a wiping tip, and the two blades engage both sides of the bank contacts. When one end is in contact with the bank terminals, the other is off the bank. That is, when one end of the wiper is leaving the last terminal, the other end is approaching the next terminal. Thus, the wipers are engaged with the intended associated bank contacts at all times.

Within a wiper assembly there may be basically two types of wipers—bridging and nonbridging. A nonbridging wiper leaves one bank contact before engaging the next. Bridging wipers have long flat tips, which permit the wiper to engage the next bank contact before leaving the preceding one. Bridging wipers are used when the circuit through them must be continuous and unbroken—as in self-interrupted stepping through a bank level. Typical examples of this are absence-of-ground searching cricuits and homing circuits.

A comparatively new adaptation is the use of normally closed (NC) contacts. Two physical levels are used to make one electrical level of NC contacts by tensioning together the mating contact points. These are opened one at a time by rotation of a finger on the associated wiper assembly.

In addition to the contacts on the bank-and-wiper assemblies, auxiliary spring assemblies of the off-normal and interrupter type are available. The off-normal spring assembly is a set of contacts actuated by a cam on the wiper assembly. It is used to control an auxiliary circuit or to home the switch—that is, return the wipers to the start position. The interrupter combination is used mainly for self-cycling operation, and sometimes for controlling auxiliary circuits.

Stepping Relays. Stepping-switch mechanisms have been adapted for use as cam switches or stepping relays as shown in Fig. 98 (lower right) and Fig. 96. Cams are cut to provide operation or restoration of the associated contact assemblies per the customer's specification.

Fig. 98 Some typical rotary stepping switches and a stepping relay (lower right). (*GTE Automatic Electric Co., Inc.*)

Cam switches with 30, 32, or 36 steps per rotation and with up to 8 cams are available. Cam switches have the advantage that the program is determined by the way the cam is cut, so that wiring is usually simplified. Also, cam contacts can switch larger loads (up to 5 A) than can standard stepping-switch contacts.

Electrical Characteristics. Operate and Release Times. The operate and release time ranges for rotary switches depend on mechanical and electrical characteristics associated with the construction of a particular switch. Generally, the cocking time of an indirectly driven switch is approximately 20 to 25 ms, and the armature release time 8 to 12 ms. Performance timing within these ranges varies between families of switches, and also between switches of the same type.

Operate and Release Voltages. Rotary stepping switches are designed to operate on a particular nominal voltage, over a limited voltage range. This range must be held to an allowable ± 5 V dc at a nominal value of 48 V dc (or slightly more depending upon ambient conditions and other influencing factors). Any voltage rating under 48 V dc should have a corresponding smaller voltage range, such as $\pm 2\frac{1}{2}$ V at a nominal value of 24 V dc. Limiting factors in the allowable voltage variation include temperature, shock and vibration, series resistance in control leads, and pulse rate.

Coil Resistance. Rotary-stepping-switch coils are designed to be self-protecting for an indefinite amount of time, while being pulsed at the voltage for which they are designed.

If the coil must be energized for long periods, it should be protected by a current-limiting resistor. A10-W resistor of approximately twice coil resistance is placed in series with the coil. A normally closed interrupter contact in parallel with the resistor shorts the resistor until the switch is almost fully cocked. When the interrupter contacts open, coil current is reduced to a holding value which can be tolerated continuously. Typical coil resistances are shown in Table 19.

Contact Bounce. Contact bounce is usually not a problem, because nonbridging wipers are usually switched without any power on them—or with only minute current and voltages. Bridging wipers do not bounce appreciably, but if the switch is subjected to extreme vibration, a wiper might bounce beyond its insulated limits and touch the adjacent wiper level. Because of the possibility of the wipers contacting under unfavorable conditions, it is not advisable to place opposite potentials on adjacent wipers.

TABLE 19 Typical Rotary-Stepping-Switch Coil-Resistance Values[15]

	Coil resistance, Ω	
Voltage, V dc	Regular switches	Oversize switches
24	30	28
48	120	100
120	650	480
6	1.9	1.5

Power consumption varies from 18 to 30 W, depending on resistance and voltage ratings.

To prevent contact bounce in the wiper assembly, snubbing washers or barrier insulators between individual wiper levels can be provided in specially engineered assemblies. The snubbing washers tend to damp vibration and the barrier insulator mechanically insulates adjacent levels. Interrupter and off-normal spring combinations are not as likely to respond violently to vibration or shock, but anticipated shock and vibration should be emphasized when ordering a stepping switch.

Intercontact Capacitance. Typical values of intercontact capacitance and Q losses are shown in Table 20 for two switch families.

TABLE 20 Typical Capacitance and Q-Loss Values for Rotary Stepping Switches[15]

	Compact switches		Large switches	
	Capacitance,* pF	Q loss,† %	Capacitance,* pF	Q loss,† %
Between adjacent bank contacts (same level)	1.0	12	1.7	14
Between adjacent bank contacts (same level, with bank-mounting screw between the two)	1.2	13	1.8	14
Between bank contact and frame	1.2	13	1.3	11
Between bank contact and frame (adjacent to bank-mounting screw)	1.5	16	1.5	12
Between same contact (adjacent levels)	1.2	13	1.7	14
Between wipers (adjacent levels, with wipers sitting on contact)	13.0	48	13.8	46
Between wipers (adjacent levels, with wipers floating between contacts)	11.7	40	11.9	38

* Approximate values at 10 MHz.
† Percentage change from standard Q value of 120 at 10 MHz.

Dielectric Strength. Minimum dielectric or voltage-breakdown value which can be expected from a standard rotary stepping switch is 500 V ac at 60 Hz between all mutually insulated points for a period of 1 s. This applies generally to coils and interrupter spring assemblies. Wiper and bank insulation is generally good for 1,250 V ac, 60 Hz for a 1-s period.

Contact Rating. Standard rotary-switch bank and wiper contacts are commonly rated at 3-A carrying capacity. It is usually good practice to design the circuit so that the wiper is disconnected as it advances from one step to the next. In this way the contacts are merely carrying the load and are not making or breaking it. However, if no more than 100 mA is interrupted, the full mechanical life of a switch should be reached without detrimental contact effects.

Standard wipers and bank contacts are usually made of phosphor bronze, a material generally corrosion-resistant except in marine atmospheres. For use in corrosive environments, it is recommended that the switch be hermetically sealed. Phosphor bronze has high electrical conductivity which is more than adequate when switching circuits at normal contact power levels. Where accuracy and reliability of readings taken through the banks and wipers of a rotary switch are necessary, gold-plated banks and wiping contacts should be used. Gold plating ensures the ability of the switch to provide a constant and low-level resistance path for measurement, indication, and monitoring purposes.

Large switches usually have interrupter contacts composed of tungsten, rated at 450 W at 2 A maximum. Compact switches are rated at 150 W at 3 A maximum. Large switches usually have off-normal spring assemblies with palladium-alloy contacts, rated as 150 W at 3 A noninductive. Compact switch, off-normal, spring assemblies are composed of various materials such as palladium or platinum alloys, and are rated at 150 W at 3 A.

The primary function of interrupter and off-normal spring-assembly contacts is to make and break the coil circuit, or the coil circuit of an associated switch. This can be accomplished if the combinations are kept in adjustment, and if adequate contact-protection devices are employed. This contact protection can be provided by using a resistor-capacitor network, or a varistor or thyrector across the coil.

The varistor is a nonlinear, resistance-varying silicon-carbon device which draws little current when voltage is applied to the switch coil. However, when the coil circuit is opened, the varistor offers a low-resistance path to short-circuit the high-voltage surge caused by the collapse of the magnetic field.

Physical Characteristics. Life Expectancy. While rotary stepping switches have been known to operate for long periods of time without attention—even under quite adverse operating conditions—switch life can be extended by proper maintenance. This consists of periodic cleaning and lubrication with the correct lubricant. By exerting proper care, life can be expected to be between 200 and 250 million operations when run self-interruptedly, and about one-half of this when pulsed from an external source.

Temperature Limits. Most standard rotary stepping switches will function satisfactorily over a temperature range of -10 to $65°C$. Switches treated with special lubricants and employing special insulating materials may operate over a temperature range of -55 to $85°C$. A maximum of $125°C$ can be tolerated in special cases.

Shock and Vibration Resistance. Normally, rotary switches need not be specially designed for extremes of shock or vibration. Standard rotary stepping switches in an operative condition will withstand 5 g vibration at 5 to 55 Hz, and a 20 g shock load.

Mounting Methods. Stepping switches can be mounted directly on a panel, or on rubber cushions to damp sound and limit resonant vibrations. Also, they can be shelf- or base-mounted. Special brackets may be used to meet certain specific mounting requirements. If a switch is to be subjected to extreme shock or vibration, special high-shock mounting frames are available.

Terminal Types. Standard solder-lug terminals are most commonly used for stepping switches. These terminals are always pretinned for soldering convenience. Terminals are also available which mate with taper-tab connectors. These connectors slide over the taper-tab terminal and are crimped into place with a small hand tool.

Enclosure Types. It is often necessary to seal a rotary stepping switch hermetically. When sealed, the switch cannot be tampered with, corrosive elements cannot reach it, dust and dirt cannot adversely affect it, and the switch is rendered harmless in an explosive atmosphere.

Rotary-switch enclosures vary, primarily depending on the size of the switch and the terminal connections desired. The headers or connectors available are many, but the most common are the solder hook, the plug-in connector, the MS (Military Standard) connector, and the gold-plated ribbon connector.

Things to Avoid When Using Stepping Switches

Do not put opposite potentials on adjacent wipers or bank levels, and preferably not on adjacent bank contacts.

Do not switch live circuits exceeding the 100-mA noninductive circuit-load rating (or its inductive equivalent) on the wiper-to-bank contacts of nonbridging wipers, unless willing to accept some life reduction.

Do not load the bank of a rotary stepping switch to the point where arcing or burning at the wiper tips and bank contacts occurs.

Do not overpower the driving mechanisms with excessively high voltage.

Do not locate the power supply for a rotary stepping switch too far away from the switch. Good performance requires full rated voltage applied to the coil.

Do not locate the switch too close physically or electrically to sensitive electronic circuitry. The operating coil of the switch is quite inductive and generates all kinds of noise and interference from which sensitive electronic circuits must be protected.

Solid-State Drivers.   Something that may appeal to electronics engineers who have need of it is programmed control of rotary stepping switches using solid-state components for the control.   There are many solutions to this problem, but using solid-state circuitry wired directly to the switch itself is of particular interest, as follows:

Solid-State Control of Stepping Rate of the Rotary Stepping Switch (Drivers).[17] Solid-state drivers are successfully employed, and can be either designed by the user or purchased commercially.   One bank of the stepping switch can be employed to vary the circuit constants and thus control self-stepping speed from one step to the next, as desired.   This permits the creation of an automated-process control, with widely variable preset timing between individual steps, when using only a stepper and appropriate solid-state components.

An example of what can be done in this area is shown by Fig. 99, which illustrates a somewhat typical but uncomplicated circuit arrangement, providing the perfor-

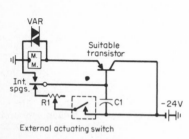

Fig. 99   A simple schematic circuit for a transistor-driven slow-stepping and/or variable-stepping-speed rotary stepping switch.[17]

Fig. 100   Typical speed curves for a rotary stepping switch operating in the circuit of Fig. 99.[17]

mance indicated in Fig. 100 with respect to stepping speed of a specific switch. There will be variations in speed from switch to switch.   The $R1$ and $C1$ portions of the circuit can be wired through individual bank contacts to give the desired variation in R and C values, and thus control the contact dwell in each wiper position.

Even slower and more varied stepping-time intervals have been achieved by the use of more elegant circuitry, employing additional transistors, SCR's, etc.   A more elegant circuit is that of Fig. 77.

Figure 101 shows the circuit of a commercially available driver for which less than ±5 percent drift is claimed, with a stepping speed controllable by means of rheostat

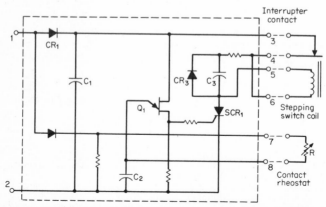

Fig. 101   Basic schematic circuit of a solid-state driver intended for use with a rotary stepping switch.   (*Electro Seal Corp., Div. of C. P. Clare & Co.*)

setting from one step in 5 s to 15 steps per second.   By using one set of contacts on the rotary stepping switch to select external timing resistors, the stepping can be programmed to dwell on any particular position for a predetermined time.   This permits a program of sequential control that gives the desired dwell on any or all individual contacts.

Exclusive of the control rheostat, the 100 to 130 V ac unit provides the source of dc required by the stepping switch and its control circuitry, in addition to providing stepping-switch interrupter-contact protection.   Similar units are available for operation on 24, 48, and 110 V dc.

Theory of Operation.   The basic schematic circuit is shown in Fig. 101.   When 115 V ac is applied to input terminals 1 and 2, capacitor C1 is charged through

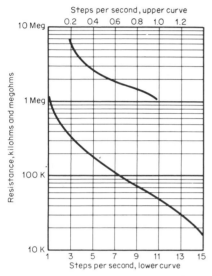

Fig. 102 Typical stepping rate for a stepping switch operated from the driver circuit of Fig. 101.   (*Electro Seal Corp., Div. of C. P. Clare & Co.*)

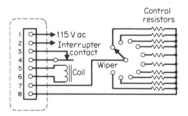

Fig. 103 A circuit illustrating use of a solid-state driver to program a rotary stepping switch for varied and controlled dwell times. (*Electro Seal Corp., Div. of C. P. Clare & Co.*)

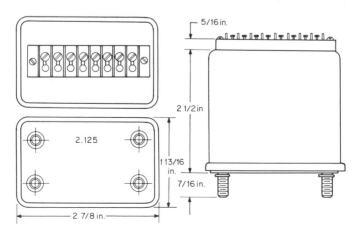

Fig. 104 Dimensional drawing of a typical solid-state driver for operating a stepping switch.   (*Electro Seal Corp., Div. of C. P. Clare & Co.*)

TABLE 21   A Partial List of Manufacturers of Relays[15]

AMP, Inc.
ASEA, Inc.
Action Electronics Co.
The Adams & Westlake Co.*
Agastat Div., Amerace Esna Corp.
Aircraft Appliances & Equipment, Ltd.
Airpax Electronics, Inc.
Alco Electronic Products, Inc.
Allen-Bradley Co.
Allied Control Co., Inc.*
Allis Chalmers
American Design Components
American Solenoid Co., Inc.
Amtron, Inc.
Arrow-Hart, Inc.
Artisan Electronics Corp.
Auto-Matic Products Co.
Babcock Electronics Corp.*
B/W Controls, Inc.
Bach-Simpson, Ltd.
Beckwith Electric Co., Inc.
California Electronic Mfg. Co., Inc.
C. P. Clare & Co.*
Clark Control Div., A. O. Smith Corp.
Cole-Hersee Co.
Compac Engineering, Inc.
Computer Components, Inc.
Controls Div., Ingraham Industries, Div. of
  McGraw-Edison Co.
Cook Electric Co.
Cornell-Dubilier Electronics*
S. H. Couch Div., ESB, Inc.
Cox & Co., Inc.
Cutler-Hammer, Inc.*
Davis Electric Co.
Delaval, Gems Sensors Div.
Deltrol Controls/Div. of Deltrol Corp.*
Deutsch Relays, Inc.*
Diversified Electronics, Inc.
Durakool, Inc.*
Dynage, Inc.
E-T-A Products Co. of America
Eagle Signal, A Gulf & Western Systems Co.*
Edison Electronics, Div. of McGraw-Edison
  Co.
The Electric Tachometer Corp.
Elec-Trol, Inc.*
Electronic Applications Co.*
Electronic Specialty Div., Daytron System,
  Inc.*
Elmwood Sensors, Inc.
Essex International, Inc.
F & B Mfg. Co., Omega Electronics Div.*
Federal Pacific Electric Co.
Fifth Dimension, Inc.
Flight Systems, Inc.
Foster & Allen, Inc.

Frost Controls Corp.
Furnas Electric Co.
GTE Automatic Electric
Gemco Electric Co.
General Automatic Corp.
General Devices, Inc.
General Electric Co.
General Electric Co., General Purpose Control Products Dept.
General Electric Co., Semiconductor
  Products, Electronic Components Div.
Gordos Corp.*
Grayhill, Inc.*
Greentron, Inc.
Grigsby Barton, Inc.*
GTE Automatic Electric Co., Inc.*
Guardian Electric Mfg. Co.*
H-B Instrument Co.
Hamlin, Inc.*
Hartman Electrical Mfg., Div. of A-T-O, Inc.
Heinemann Electric Co.
Hi-G, Inc.*
Hoagland Instruments
Harvey Hubbell, Inc., Industrial Controls
  Div.
ICS, Inc.
I-T-E Imperial Corp.
ITT Jennings Industrial Products*
Imtra Corp.
Ingraham Industries—Special Products Div.
Instrument Components Co., Inc.
International Rectifier, Semiconductor Div.
Jaidinger Mfg. Co.
Jettron Products, Inc.
Jewell Electrical Instruments, Inc.
Kilovac Corp.
Kilovac Corp., Dow-Key Div.
King Seeley Div., King Seeley Thermos Co.
Klockner-Moeller Corp.
Kratos
LaMarche Mfg. Co.
Larson Instrument Co., Inc.
Ledex, Inc.
Line Electric Co., A Unit of Esterline Corp.
Logitek, Inc.
MKC Electronics Corp.
Mack Electric Devices, Inc.
Madison Electric Products, Inc.
Madison Laboratories, Inc.
Magnecraft Electric Co.*
Master Electronic Controls
McGraw-Edison Co., Edison Instrument
  Div.
MEKontrol, Inc.
The Mercoid Corp.
Midtex, Inc., Aemco Div.*
Micro Switch, A Div. of Honeywell, Inc.

TABLE 21   A Partial List of Manufacturers of Relays[15] (Continued)

Monsanto Co., Electronic Special Products Div.*	Smiths Industries, Inc.
N. P. E./Wabash	Solid State Electronics Corp.
North American Philips Controls Corp.*	Sprague Electric Co.
North Electric Co., Electronics Div.	Square D Co.*
Oak Industries, Inc., Switch Div.	Sterer Engineering & Mfg. Co., Logic Systems Div.
Omnetics, Inc.	Struthers–Dunn, Inc.*
Payne Engineering Co.	Sunshine Scientific
Peco Corp.	Syracuse Electronics Corp.
Potter & Brumfield, Div. of AMF, Inc.*	Systems Matrix, Inc.
Power Control Corp.	T-Bar, Inc., Switching Components Div.*
Prestolite Co., Div. Eltra Corp.,	Tech Laboratories, Inc.
Princo Instruments, Inc.	Teledyne Crystalonics
Raytheon Co., Industrial Components Operation	Teledyne Relays
Regents Controls, Inc.	Tempo Instrument, Inc., Industrial Products Div.
Relay & Control, Div. of A. W. Sperry Instruments, Inc.	Tenor Co.
Relay Specialties, Inc.	Thermosen, Inc.
Relays, Inc.	O. Thompson, Inc.
Renfrew Electric Co., Ltd.	Tokyo Electric Co., Ltd.
Ril Electronics, Inc.	Universal Relay Corp.
Rohde & Schwarz Sales Co.	Vanguard Relay Corp.
Ronk Electrical Industries, Inc., System Analyzer Div.	Vapor Corporation*
Ross Engineering Corp.	Vectrol, Inc.
Rowan Controller, Inc., Subs. ITE-Imperial Corp.	Wabash Relay and Electronics, Inc.
F. A. Scherma Mfg. Co., Inc.	Wabco Aerospace Department
Schrack Electrical Sales Corp.	Wapco Mfg. Co.
Selco Electronics, Inc.	Ward Leonard Electric Co., Inc.*
Shared Technical Services, Inc., An I. E. C. Affiliated Co.	Warner Electric Brake & Clutch Co.
Shigoto Industries, Ltd.	Warren GV Communications
Siemens Corp.	Watlow Electric Mfg. Co.
Sigma Instruments (Canada), Ltd.	Western Electric Co., Inc.*
Sigma Instruments, Inc.*	Westinghouse, Control Products Div.
Simpson Electric Co.	Westinghouse Electric Corp.
Slocum Industries, Electronics Div.	Westinghouse Electric Corp., Semiconductor Div.
	Weston Instruments, Inc.
	Wilmar Electronics, Inc.
	Zenith Controls, Inc.

* Members of National Association of Relay Manufacturers (NARM).

diode *CR*1. This voltage appears across the stepping-switch coil and controlled rectifier SCR 1 through the normally closed interrupter contacts of the stepping switch. Since SCR 1 is off at this time, the coil is not energized.

Simultaneously, capacitor *C*2 charges through diode *CR*2 at a rate determined by the (external) control rheostat. When *C*2 charges to the emitter peak point voltage of unijunction transistor *Q*1, the latter conducts, gating on *SCR* 1. Current flows through the switch coil, cocking the switch. This opens the interrupter contact, disconnecting the coil and turning off *SCR* 1. The switch then steps, and the action repeats. The control rheostat changes the time required to charge *C*2 to the firing point of *Q*1 and changes the stepping rate accordingly. *CR*3 and *C*3 provide arc suppression for the interrupter contacts. Typical stepping rate is shown by Fig. 102.

Application. The stepping-switch driver is designed to operate stepping devices at fixed rates sufficiently slow to permit the bank or wafer contacts to pull in and/or drop out relays and contactors. This may require 10 to 100 ms of contact dwell before stepping to the next position.

The driver offers the further advantage of programming such a device to stay in each position a different length of time. This circuit is illustrated in Fig. 103. Terminal 7 is connected to the wiper of an extra bank of contacts on the stepping

TABLE 22   Nongovernmental Organizations Having Publications Relating to Relays[1]

AAR	Association of American Railroads 1920 L Street, N.W. Washington, DC 20036
AIA	Aerospace Industries Association 1725 De Sales Street, N.W. Washington, DC 20036
ANSI	American National Standards Institute 1430 Broadway New York, NY 10018
EIA	Electronic Industries Association 2001 I Street, N.W. Washington, DC 20006
IEEE	Institute of Electrical and Electronics Engineers 345 East 47th Street New York, NY 10017
JIC	Joint Industrial Council 2139 Wisconsin Avenue Washington, DC 20007
NARM	National Association of Relay Manufacturers P. O. Box 1649 Scottsdale, AR 85252
NEMA	National Electric Manufacturers Association 155 East 44th Street New York, NY 10017
SAE	Society of Automotive Engineers, Inc. 2 Pennsylvania Plaza New York, NY 10001

Governmental specifications relating to relays and relay equipment may be obtained from:

Commanding Officer
U.S. Naval Supply Depot
5801 Tabor Avenue
Philadelphia, PA 19120

switch.   As the switch steps, the value of resistance that will give the desired delay for the next step is automatically connected to terminal 8.   Delay of as long as 5 s between steps may be obtained.   An approximation of the value of resistance required for a given delay may be obtained from the curves in Fig. 102.   Resistors under 26,000 Ω must be capable of carrying 2 W.   Higher values may be rated 1 W up to 0.25 MΩ and ½ W for delay time greater than ¼ s (4 steps per second).   A dimensional drawing of the driver in its usual form is shown in Fig. 104.

Other stepping switches

Rotary Solenoid Ratcheting.   One form of nontelephone-type rotary stepping switch consists of a rotary switch, having one or more decks, quite similar in appearance to the manually actuated rotary switch of Fig. 93 except that it is driven by a ratcheting mechanism powered from an electromagnet.   The electromagnet is usually of the rotary solenoid type.   Unidirectional stepping switches of this type employ only one rotary solenoid, functioning at one end of the rotary shaft.   Bidirectional types are available, having a ratcheting solenoid at each end of the shaft, one turning the wipers clockwise and the other turning them counterclockwise.   This is particularly useful where an add and subtract requirement has to be met.

100/200 Point Selection.   A telephone-type switch that can select any one of 100 or 200 bank contacts by first moving the wipers to any one of a 10 bank-contact levels and then rotating them to any one of 10 bank positions is provided by the so-called step-by-step, or Strowger (telephone-type), switch.   Connection by means of relays that can choose which of two sets of contacts in each rotary position are to be contacted permits a choice of one set of contacts out of a possible 200 sets.

TABLE 23   A Partial List of Switch Manufacturers[13]

Agastat Div., Amerace Esna Corp.
Airflyte Electronics
Airpax Electronics, Inc.
Alco Electronic Products, Inc.
Allen-Bradley Co.
American Zettler, Inc.
Analog Devices, Inc., Pastoriza Div.
Ansley Div., Thomas & Betts Corp.
Arrow-Hart & Hegeman
Automatic Electric Co.
Automatic Metal Products
Automatic Switch Co.
Beckman Instruments, Inc.
Bristol Instrument Div., American
  Chain & Cable
C & K Components, Inc.
CTS Corp.
Candy Manufacturing Co.
Capitol Machine & Switch
Carling Electric, Inc.
Carter Mfg. Corp.
Centralab/Elexs Div., Globe-Union, Inc.
Cherry Electrical Products
Chicago Dynamics Industries
Chicago Switch, Div. F & F Enterprises
C. P. Clare & Co.
Clare-Pendar Co.
Cole Instrument
Collectron Corp.
Computer Products Div., Wyle Labs
Consolidated Controls Corp.
Control Products, Inc.
Controlotron Corp.
Controls Co. of America, Control Switch Div.
Corning Glass Works
Cunningham Corp.
Custom Component Switches
Cutler-Hammer, Inc.
Daburn Electronics & Cable
Daven Div., McGraw Edison
R. B. Denison, Inc.
Dialight Corp.
Digitran Co.
Disc Instruments, Inc.
Double A Products Co.
Dresser Industrial Valve & Instrument Div.
Dwyer Instruments, Inc.
Electric Regulator Corp.
Electro-Mec Instruments
Electro-Miniatures Corp.
Electro-Products Labs.
Electro Switch Corp.
Electronic Components for Industry
Electronic Controls, Inc.
Electronic Engineering Co. Cal.
Electronic Resources, Inc.
Elmwood Sensors, Inc.
Euclid Electric & Mfg.

Farmer Electric Products Co., Inc.
Fifth Dimension, Inc.
Film Microelectronics, Inc.
Furnas Electric Co.
Gemco Electric Co.
The Gems Co., Inc.
General Devices, Inc.
General Electric
General Equipment & Mfg. Co.
General Reed Co.
Gordos Corp.
Gorn Corp.
Grayhill, Inc.
Hamlin, Inc.
Haydon, A. W. Co.
Haydon Switch & Instrument
High Vacuum Electronics, Inc.
Hi-Tek Corp., Switch Div.
Honeywell Industrial Div.
Humphrey, Inc.
IBM Corp, Industrial Products Div.
ITT Jennings
Imtra Corp.
Instrumentation & Control Systems
International Electro Exchange of
  Minneapolis
Interswitch
Janco Corp.
Jay-El Products, Inc.
Jordan Controls, Inc.
Langevin Electromechanical
Ledex, Inc.
Leecraft Mfg. Co., Inc.
Leeds & Northrup Co.
Leviton Mfg. Co.
Licon-Ill. Tool Works
Linemaster Switch Corp.
Liquid Level Lectronics
Litton Industries, Clifton Div.
Litton Industries, USECO Div.
Mack Electric Devices
Magnetrol, Inc.
Marco-Oak Industries, Div. of Electro/Netics
  Corp.
Mason Electric Co.
Master Specialties Co.
McDonnell & Miller, Inc.
McGill Mfg. Co., Inc.
Mead Fluid Dynamics
Mechanical Enterprises, Inc.
Metrix Instrument Co.
Micro-Lectric, Inc.
Micro Switch, Div. of Honeywell
Milliswitch Corp.
MINELCO (Miniature Electronic
  Components Corp.)
Miniature Electronic Components Corp.
  (MINELCO)

TABLE 23 A Partial List of Switch Manufacturers[13] (Continued)

Molex Products Co.	Shallcross Mfg.
Donald P. Mossman, Inc.	Shelly Associates, Inc.
Nanasi Co.	Sparton Southwest
Nelson Electric Div.	Spectrol Electronics Corp.
New England Instrument Co.	Stackpole Components Co.
New Product Engineering, Inc.	Staco, Inc.
N. M. Ney Co.	Standard Controls, Inc.
Northern Precision Labs.	Standard Instrument Corp. Div., Automatic
Oak Mfg. Co.	Timing and Controls, Inc.
Ohmite Mfg. Co.	Subminiature Instrument Corp.
Otto Controls	Switchcraft, Inc.
Philadelphia Scientific Controls, Inc.	Synchro Start Products, Inc.
Pollak, J. Corp.	Tann Controls Co.
Potter & Brumfield	Tapeswitch Corp. of America
Precision Mechanisms Corp.	Tech Labs, Inc., Bergen & Edsall
Precision Sensors, Inc.	Tele-Dynamics, Div. of AMBAC Industries,
Pressure Controls, Inc.	Inc.
Qualitrol Corp.	Texas Instruments, Controls Products Div.
RBM Controls	Therm-O-Disc
RCL Electronics, Inc.	Torq Engineered Products
Reed Switch Development Co., Inc.	Truco, Inc.
Remvac Components, Inc.	UMC Electronics Co.
Robertshaw Controls, Acro Div.	Unimax Switch, Div. Maxson Electronics
Ross, Milton Co.	Corp.
Rotary Controls Div., Ledex Inc.	United Control Corp., Subs Sundstrand
Rowan Controller Co.	Corp.
Sage Laboratories, Inc.	Wabash Magnetics, Inc.
Scaico Controls Div.	Waneteck Data Communications
Schrack Electrical Sales Corp.	Wheelock Signals, Inc.
Seacor, Inc.	Zenith Controls, Inc.
Sealectro Corp.	

A telephone-type stepping switch that is quite different in appearance with similar bank-choice capabilities is the X-Y switch.

Another of the large telephone-type switches capable of 100-point selection, but operating on a quite different principle, more like a gigantic relay, is the crossbar switch.

A typical 10- by 10-point crossbar switch consists of 100 sets of contacts, arranged in a 10 by 10 grid. At each grid intersection (called a crosspoint) mechanical linkage causes operation of the contact set at that point when a particular pair of electromagnets is actuated. Twenty electromagnets per switch are involved. Ten are arranged along one side of the switch, to control the selection of a horizontal row, and a second set of 10 electromagnets are arranged across the top of the switch, to control the vertical columns. When a horizontal row and vertical column are selected, the contact set at the intersection of the selected row and column is operated. Obviously this is not a small device. The telephone-type version of a crossbar switch has long been in production by certain telephone-equipment manufacturers, but a commercial version by a nontelephone manufacturer is also available. This commercial version comes in other matrix sizes than 10 by 10, such as 16 by 10 and 21 by 10.

Summary The above by no means exhausts the subject of kinds, types, and applications of switches, both manually and electrically operated. Vendors' catalogs are a source of additional information, and a rather complete list of switch vendors appears in Table 23. Manufacturers whose names appear in this listing collectively manufacture every type of switch that has been offered commercially. A careful review of their product literature will reveal something for almost any possible need.

Continuing developments Technology in this field continues to expand. One special area which should be mentioned is the continuing progress on standardization of solid-state relays with respect to size, shape, configuration, and operating characteristics. Up-to-date information is available at any given time from the National Association of Relay Manufacturers.

Another recent development which typified continuing developments is an optically encoded keyboard for which all the encoding is retained in the key switch, a single piece of stamped steel, free from eroding wear and contact bounce. It is claimed that improved efficiency and simplicity with lower cost result in a keyboard immune to environmental disturbances.

REFERENCES

1. National Association of Relay Manufacturers, "Engineers' Relay Handbook," Hayden Book Co., New York, 1969.
2. Oliver, Frank J.: "Practical Relay Circuits," Hayden Book Co., New York, 1971.
3. Kear, Fred W.: Take Another Look at Resonant-Reed Relays, *EDN*, vol. 18, no. 10, June 20, 1973.
4. Kaetsch, Philip W.: Suppressing Relay Transients, *Electro-Technology*, December 1968.
5. Lippke, James A.: Time-Delay Relays, *EEE*, September 1968.
6. Andriev, N.: Power Relays—Solid State vs. Electromechanical, *Control Eng.*, January 1973.
7. Thompson, Stephen A.: Relays: Form versus Function, *Electron. Eng.*, October 1971.
8. "Correed Handbook and Application Manual," GTE Automatic Electric Co., Inc., Northlake, Ill.
9. "Technical Applications Reference for Mercury-wetted Contact Relays, Dry Reed Relays and Mercury-wetted Reed Relays," C. P. Clare & Co., Chicago, Ill.
10. Deeg, W. L., and R. H. Marks: "Clareed Control Concept and Its Systems Application." C. P. Clare & Co., Chicago, Ill.
11. "Designers' Handbook and Catalog of Reed and Mercury-wetted Contact Relays," Magnecraft Electric Co., Chicago, Ill.
12. "Designers' Handbook and Catalog of Time Delay Relays," Magnecraft Electric Co., Chicago, Ill.
13. 1972/73 Systems Designers' Handbook, *Electromech. Des.*, 1972–1973.
14. 1973/1974 Systems Designers' Handbook, *Electromech. Des.*, August 1973.
15. 1973–1974 Electric Controls Reference Issue, *Mach. Des.*, Apr. 26, 1973.
16. Ashby, J. D.: Stepping Switches, *Mach. Des.*, Apr. 26, 1973.
17. "How to Use Rotary Stepping Switches," GTE Automatic Electric Company, Inc., Northlake, Ill.

Chapter **11**

Connectors and Connective Devices

BERNARD R. SCHWARTZ

RCA Corporation, Missile and Surface Radar Division, Government and Commercial Systems, Moorestown, New Jersey

INTRODUCTION

Historically, an electrical connector was generally considered only a hardware item and, as such, was often the last item considered in the design and packaging of a piece of electrical or electronic equipment. Requirements were unsophisticated and were met by a number of standard, off-the-shelf connectors. These were reasonably satisfactory for most early applications of electronic equipment, in which signal voltage and current levels were relatively high. Thus, the selection and application of connectors posed no real problems.

Power connectors and cable connectors for signals between associated pieces of equipment were the first requirements for easily removable connections. In many cases, soldered connections and screw-type terminal boards were used to interconnect various subassemblies.

With the rapid growth in the field of electronics, equipment became more complex, and more and more connectors were needed to interconnect electronic functions in a practical, modular form that was both manufacturable and maintainable. In today's packaging concepts, connectors have become a very vital link in forming or making up a complete electronic system, and as such, they are in fact a very important component part of that system, rather than just being merely another item of necessary hardware. Thus, as with any component part, connector requirements must be evaluated, and connectors must be selected just as carefully as are the other components before a package design is frozen, rather than after the fact when volumetric considerations can dictate a compromise connector selection based almost entirely on size, which could be detrimental to reliability—or on the resultant need for a "special" that could be overly costly.

The design engineer, approaching the problem of selecting a practical connector system, is confronted with an extraordinary variety of available items and associated techniques. In addition, the expanding role of connectors and interconnecting systems in the accelerating pace of the electronics arts is continuing to provide even greater variety.

It could be expected that this wide availability would lead to confusion. Consider such connector-system requirements as electrical and mechanical characteristics, cost, available accessories, installation and tooling, reliability, maintenance and spares, and supplier production and engineering support. Add to this connector-system engineering task-time allotments and it becomes apparent that the conditions are such that, in desperation, poor choices are made.

This chapter attempts to provide a rational connector-system selection technique that can be taken by design engineers and which, in many respects, uses a decision-tree approach. The material will define packaging levels in which connector selections are required, describe the variety of designs available, and discuss the applicable advantages and disadvantages of each. In this manner the designer is literally led through a series of decisions that must be resolved, and then is given sequencing to other levels to provide the best possible connector-system selection.

Consideration must be given to all electrical, mechanical, and environmental stresses to which the connector will be likely to be subjected in use, as well as to the compatibility of physical form and dimension with the intended packaging concept of the equipment in which it will be used.

Not only will consideration of all available application information, followed up by careful connector selection, help to prevent misapplication; it should, in addition, result in functionally satisfactory and reliable interconnections.

It would be impossible in a single chapter to give all the available technical data and design specifications on all the multitude of types, sizes, and varieties of connectors that are presently available. Also, connector technology and requirements are both advancing so rapidly that whole new connector concepts are constantly being developed by connector manufacturers to meet the constantly advancing needs of the user; therefore, the content of this chapter will be limited to an introduction to and a discussion of the various connector families, broad engineering parameters, and design considerations that should provide the design engineer with adequate guidelines for the intelligent selection of connectors and interconnection devices. It cannot be stressed too heavily that for specific applications, manufacturers' literature, technical data, and engineering services should be consulted by the user.

BASIC SELECTION OF CONNECTOR SYSTEMS

Systems Interconnection Levels

The first consideration is to categorize the various connector requirements in typical electronic and electrical equipment systems. This structure will provide the major

TABLE 1   Connection Levels in Modern Electronic Systems

Level number	Connection type	
0	Connection inside component case	
1	Connection from component to printed-circuit board or wire	
2	Connection from printed-circuit board to wire or another printed-circuit board on a chassis (Usually internal)	
3	Connection from internal chassis to another internal chassis in the same package	
4	Connection from one piece of equipment to another (Usually external)	

Fig. 1   Typical interconnections within integrated-circuit package (level 0).

initial step in the overview required for the selection and definition task. Table 1 provides a breakdown into five distinct levels for which connectors and interconnecting systems are likely to be required. These five levels were arbitrarily selected,[1] but it has been found that they satisfactorily embrace the variety of types necessary to support modern systems. All these levels may not exist in any particular system, but conversely systems problems can generally be made to fit into one of the five levels shown.

The problems associated with level 0 will not be discussed because they are not of the type usually associated with connector-system selection. Figure 1 shows interconnections within an integrated-circuit package, which is typical of this level. The reader is referred to Chap. 1 for details associated with this interconnection (not a connector) level.

Level 1 connectors are primarily associated with part replacement, e.g., an integrated-circuit-element socket. This level is characterized by termination problems such as that illustrated in Fig. 2, which shows the use of a socket about to receive

Fig. 2   Typical part-mounting connection (level 1).   (*Barnes Corp.*)

an in-line integrated-circuit package. Thus levels 0 and 1 are basically shown for identification and to avoid confusion.

Levels 2 through 4 will be treated in detail and will include a discussion on the following points:

> Item to be connected
> Generic connector types
> Termination method
> Contact spacing
> Typical connector

A brief discussion of each of these points will provide clarification prior to entering the detail itself.

Item to be connected   In typical systems considerations, we find that the electrical conductors which must be provided an engineered break come in a wide variety of configurations: conventional wire, printed-circuit-board conductors, coaxial conductors, flat cable, etc. The choice of conductors arises from a wide variety of factors and is usually not constrained by connector selection.

In addition, there are the variations in weight, cost, flexibility, reliability, and temperature requirements demanded by undersea, airborne, ground-computer, and manned-spacecraft systems. Another major consideration might be the cost of terminating the conductors. Automated and semiautomatic solderless wrapping techniques may demand a solid conductor, while the desirable use of mass flow-soldering machines will require the use of the printed-circuit-board type of flat conductor. This then is the first major consideration in connector selection. Once the choices of conductor have been made, generic types of connectors can be considered, as discussed below.

Generic connector type   This choice is heavily influenced by the specific application. Just as these requirements affected the choice of conductors, so they now demand consideration in selection of the generic connector type. To these (environmental and cost) are added a wide variety of problems that are associated with frequency of maintenance and quality of service personnel. These factors will be discussed for each of the connector types which are found at each of the levels.

Termination method   This choice is intimately tied to factory operation. Experience has shown that a significantly high proportion of the total costs involved with connectors is associated with terminating and installing conductors. It is best first to determine the availability and costs of different types of terminating facilities. Also, since specific skills are required to operate and maintain these tools, suitable personnel must be obtained or trained. Field maintenance of the required facility may present the significant factors in the ulimate selection of the termination method. Quality is more readily controllable on some termination techniques at the installation site than on others. These inherent factors may represent the crucial decision point when system performance-reliability achievement is emphasized. These then are some of the considerations involved in choosing a termination method. Once the termination method is established, consideration can be given to contact spacing.

Contact spacing   In this area the designer is confronted with an interesting set of trade-offs. On one hand there is the natural desire to obtain the smallest possible connector (through the closest possible contact spacing); on the other is an impressive list of penalties, with varying degrees of intensity for each characteristic. A brief list of these is:

Higher part cost
Higher installation cost
Greater part fragility
Potential reliability risks (due to more critical design)

High-density connectors are required in specific applications where the benefits exceed the potential disadvantages. There is no need to specify them indiscriminately.

The designer is now ready for the selection of a specific connector. The remainder of this chapter will provide pertinent data and information regarding connector accessories, assembly tools, industrial and military specifications, reliability, and materials considerations in connector selection.

Illustration of a typical connector   A picture of a typical connector will be shown for each category shown in each table. This will serve to aid the reader in visualizing and identifying at least one example of such a connector. Obviously there are many more that cannot be shown because of space limitations. There are many variations in designs, materials, finishes, costs, etc. The ultimate selection will consider all these and the technical merits of the specific application.

CONNECTION LEVEL 2

Packaging Concepts

Table 2 is the "road map" for level 2. It shows the various packaging concepts used at the "part to first interface" level. This table shows that a printed-circuit board is involved in each case. Experience has indicated that this is true for the solution to a high percentage of electronic packaging problems. This printed-circuit board is then connected to solid or stranded insulated copper wire, or to coaxial, shielded, or twisted-pair cabling, or to another printed-circuit board. The table shows these three variations. As previously mentioned, the conductor choice here is not normally dependent upon connector considerations, and the designer can therefore continue with a detailed discussion of the types of generic connectors involved.

TABLE 2   Connection Level 2

Level	Item to be connected	Generic connector type	Termination method	Contact spacing	Illustration of typical connector
Level 2	Printed-circuit board to standed or solid wire	One-piece edge-on	Crimp, snap-in	0.156 0.100 0.078	
			Solderless wrap	0.200 0.100	
		Printed-circuit board is one half of connector	Solder or weld	0.200 0.156 0.100 0.050	
		Two-piece plug and receptacle	Crimp, snap-in	0.156;0.150 0.100	
			Solderless wrap	0.200;0.150 0.100	
			Solder or weld	0.200;0.150 0.100 0.050	
		Side-entry custom built	Solderless wrap	0.125	
	Printed-circuit board to printed circuit board	One-piece edge-on	Printed-circuit board is 1/2 of connector Solder or weld	0.156 0.100 0.050	
		Two-piece plug and receptacle	Solder or weld	0.156 0.060 0.050	
	Printed-circuit board to coaxial cable	Two-piece plug and receptacle	Solder plug Crimp receptacle contacts	0.270	

Generic Connector Types

One- and two-piece connectors   Two basic types of connectors are used with printed-circuit boards: one-piece, *edge*, and two-piece, *plug and receptacle*. The major difference between the two is that in the one-piece construction, the printed-circuit board itself is used to provide one-half of the connector system. In the two-piece concept, the printed-circuit board is usually soldered to the pins of the plug half of the two-piece type; the printed-circuit board then does not act as a connector element.

One-Piece Connectors.   Modern electronic production practice related to printed-circuit boards has developed proved techniques to permit the plated copper tabs to operate reliably.[2] The attendant gains are in lower connector cost, less weight, and more efficient use of space. However, the necessary controls must be placed on the preparation of the printed-circuit board. If one is unable or unwilling to exercise the requisite attention to assure that the proper levels and types of board cleaning and plating are maintained, then the two-piece connector is probably the better choice:

Two-Piece Connectors.   This design permits the control of all elements of materials, finishes, and geometry to remain with the connector manufacturer. They are then fully responsible for the connector, whereas in the one-piece design this responsibility is shared between the printed-circuit-board and connector manufacturers, Connector applications engineers generally view this design as more reliable than the one-piece version[3] for several reasons:

1. The break in the electrical conductor is better engineered; selection of materials and platings is more easily solved.

2. Traditional male/female contact designs are used, usually pin and socket or blade and tuning fork.

3. Materials dimensions and finishes are more easily controlled.

4. Cost of the plug is at least partially offset by additional costs of board-preparation steps such as buffing, cleaning, skiving, and plating.

Side-Entry-Type Connectors.   This type of printed-circuit-board connector was designed to overcome a major problem, pin limiting, associated with this packaging concept. This connector type makes at least three of the four sides of the printed-circuit board available for connections. The obvious result is to permit more active component parts (and peripheral passive parts) on the printed-circuit board. At the same time it also permits the board to become larger (more pins are available). While these connectors usually have contact designs normally associated with the one-piece connector type, it is possible to use a combination of both connector types where low-level electronic signals (less than 50 mV) must be handled. Only a limited number of designs have been made using the side-entry connector type, but it should be mentioned, as it will probably see increased service in the packaging trend associated with medium- and large-scale integrated circuits.[4]

Termination Methods

In the Termination Method column of Table 2 are shown the prevalent techniques for terminating conductors: crimp, solderless wrap, and solder or weld. The latter two are shown as a single category, since in both cases a permanent fusion of metals is achieved. A great deal of literature has been written on these subjects, and the reader who may be unfamiliar with any of these methods should obtain sufficient background elsewhere. At this point, only the highlights of each will be reviewed.

Crimp contacts, snap-in   This connection technique has now been firmly established for terminating virtually all types of wire, solid or standard, coaxial, shielded, and twisted-pair. From the reliability viewpoint crimping offers the ability to control quality by nondestructive-inspection techniques and by using calibrated tools. Because almost all crimping tools and semi- and fully automatic machines have full-cycle features, operator judgment is not a consideration. This use of low-skill personnel makes crimping attractive for the obvious reasons related to cost.

The high-pressure crimp termination is gastight when properly made. As a result the crimp termination is capable of withstanding extreme stresses—mechanical, environmental, and chemical—which might exist in a severe application. Low installed costs are possible using high-speed automatic terminating machines and by the use of preharnessing, which is practical because the connectors are designed to accept snap-in pins. This same feature provides a significant flexibility feature; change and repair are made possible by the easy removal and reuse capability of the snap-in contacts.

Solderless wrap   This technique is limited to the use of solid wire. Reliability of performance is obtained from the wrapping-wire intercepts at the post corners to obtain gastight joints. Extensive use in telephone switching centers and military ground and shipborne electronic installations has verified ultrastable performance over an extended period. This method can yield the lowest installed cost of terminated wiring by the use of preprogrammed, automated wrapping machines. Hand-held tools require prior training of operators but are fast and light, with no heat or fumes generated. Quality control of materials, tools, and operators is obtained through well-established procedures. Solderless-wrap posts must be given allowance for extra wraps to obtain needed flexibility for repairs and changes.

Crimp clip post   This technique is generally available for applications similar to those where solderless-wrap methods are used. This method permits the use of stranded wire, gages 32 through 22. Termi-Point* is a proprietary termination whereby a spring clip is slid onto a tang while holding the stranded wire against the tang face under high normal pressure. Semi- and fully automatic attachment tools are available, and their use results in reduced installed cost, although the cost of the individual clip must be included in such analysis. Long-term experience is being generated. Available data indicate gastight joints are obtained.

Solder or weld   Both these techniques, by creating fusion of metals, give excellent connections when properly made. They have a serious disadvantage, however, in the high degree of operator skill required. There is also the difficulty of controlling quality because of the destructive-sampling techniques used. Visual observation to determine quality is considered to be unreliable. Choice of materials and finishes is also very important to avoid brittle joints due to the creation of intermetallic compounds in the joint. Although the cost of solder is quite low, the labor cost is generally not competitive with the other techniques. The major exception to this is the use of mass-soldering techniques, such as dip, wave, flow, infrared, or jet

* Trademark of AMP, Inc.

soldering. The general practice for terminating many printed-circuit-board conductors simultaneously is by one of these methods. The lowest cost of any technique is thus achieved. However, flexibility for repairs and changes is severely restricted. The mass-soldering technique is limited to printed-circuit-board applications.

Contact Spacing

Table 2 shows that the highest density of contact spacing is obtained in connectors with the solder or weld type of termination. The advantage of higher density for any given application should be considered against the drawbacks before this decision is reached. These penalties may include less ruggedness (barrier walls are thinner), increased cost (tighter tolerances), and a potential reliability hazard (more critical parameters to control).

Another factor to be remembered in contact-spacing selection is the overall diameter of the wires to be used with the contacts. That dimension will definitely limit the contact spacing that can be achieved.

CONNECTION LEVEL 3

Packaging Concepts

At this level the necessity for making connections to circuit conductors on printed-circuit boards is found much less frequently than at level 2. When it occurs, the problem can essentially be treated the same as for a level 2 problem. However, it is much more likely that, for level 3, connections for wire to wire of various types will be needed. Where there are major distinctions, they will exist in the type of wire that is involved. Thus, it makes a significant difference whether a single conductor, coaxial, shielded, or twisted pairs is used. In the last two cases, another distinction will be made for the cases where the shield conductor is taken directly

TABLE 3   Connection Level 3

Level	Item to be connected	Generic connector type	Termination method	Contact spacing	Illustration of typical connector
Level 3	Printed-circuit board to printed-circuit board... ...or wire	When printed-circuit board circuits are involved as conductors in level 3 then all of the details shown for level 2 apply.		0.320; 0.150 0.160 0.150 0.200; 0.150	
	Wire to wire solid or stranded	Rack & panel drawer type	Crimp,snap-in	0.250 0.200	
			Solder	0.150 0.150;0.200 0.150;0.200	
		Plug & receptacle	Crimp,snap-in		
			Solder	0.200	
	Coax or twisted pair to coax or twisted pair	Rack & panel drawer type	Crimp,snap-in	0.350;0.320 0.325 0.150	
		Plug & receptacle	Crimp,snap-in	0.250;0.150 0.325 0.150	
	Coax or twisted pair to same	Rack & panel drawer type	Crimp,snap-in	0.350;0.320 0.150	
		Plug & receptacle	Crimp,snap-in	0.150	

to system ground at the connector. These variations are covered in Table 3 as four different categories under Item to Be Connected.

Generic Connector Types

For the latter three categories, the two different types of connector generic types that are used are the rack-and-panel and the plug-and-receptacle. The connectors for these two types may appear similar and in some cases may actually be the same connectors with different accessories. Nevertheless, there are some important distinctions between them.

Rack-and-Panel Connectors. Rack-and-panel connectors are those designed specifically for use in applications where "blind mating" occurs. Often, one-half of the connector is connected to a removable portion of the equipment, such as a box or drawer. The second half is usually attached to the fixed portion of the equipment, such as a panel or cabinet. The connector mating takes place during the act of inserting the removable part into the fixed part of the equipment. The relatively large size of a typical drawer, for example, compared with a connector, makes it virtually impossible for the operator to see and "feel" the connector halves going together. Indeed, when mechanical devices such as toggles and jackscrews are used with the drawer, they completely mask any possible feel relative to the connector. Figure 3 shows the method of floating and guiding rack-and-panel connectors.

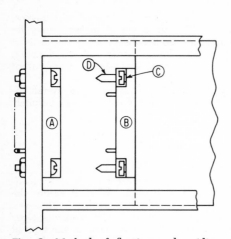

Fig. 3 Method of floating and guiding connectors; rack-and-panel application.

Rack-and-Panel Connectors—Special Features. Figure 4 shows the parts and construction of these types of rack-and-panel connectors, which need special features to overcome the problems just discussed. These features are briefly mentioned below.

1. Extra long contacts, or spring mounting—to compensate for variations in the final axial position of the drawer relative to the cabinet

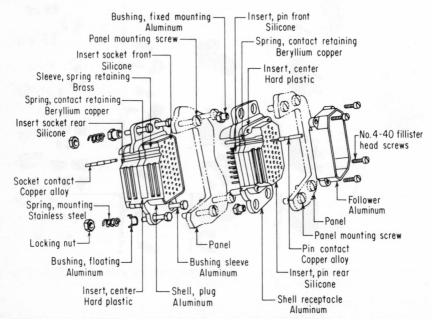

Fig. 4 Parts of a typical rack-and-panel connector.

2. Floating mountings—to compensate for tolerances
3. Durable, rugged lead-in to pick up large variations in position without damage
4. Indexing technique—to prevent accidental insertion of a drawer in a wrong cabinet position

Plug-and-receptacle connectors   When the application permits easy access by the operator (typically, where the receptacle is fixed on the front panel, or where two cables are to be hand-mated), the plug-and-receptacle connector needs none of the special features just described for rack-and-panel connectors. In this case, the ability to see, feel, and guide the connectors going together makes most of those features unnecessary. One of the distinguishing features of plug-and-receptacle connectors is that they will often require engaging and/or locking devices built right into the connectors (rack-and-panel connectors usually depend on the cabinets and/or drawers they are mounted to for this). Jackscrews and locking clips often serve this function very well. Another distinction is that they usually require hoods and/or cable clamps for the connector plug, whereas rack-and-panel connectors rarely have wiring exposed to mechanical and environmental stresses. In level 3 applications there is also another significant difference in connectors. This occurs when shielded coaxial or twisted-pair conductors are used and it is necessary to make all the ground (earth) circuits common and either carry them to ground or carry them through the equipment. This problem of terminating and jumping the ground conductors can be accomplished in several ways.

1. Crimp splices—results in bulky assembly.
2. Heat-shrinkable solder-sleeve splice—results in smaller assembly but is more costly and somewhat more difficult to inspect.
3. Use of a conductive coating on the connector body and the outer body of the coaxial contact—results in no additional termination to accomplish the desired function. Note that *all* shields are then made common and that *the specific application must permit this.*

Termination Methods

The termination methods used in level 3 connectors have the same features as those described for level 2, except that the options available to the designer have been markedly reduced. Solderless wrap is almost never considered, since there is little reason for its use at this level. Soldering has been eliminated for the shielded and coaxial types of conductors because of the problems which have been encountered in installation and reliability. This has been especially true for the miniature type of coaxial cable, where soldering has been difficult to accomplish on a production basis. For this reason only crimp, snap-in has been shown for these types of conductors.

Contact Spacing

The comments made for level 2 relative to contact spacing also hold true for level 3, except that they may be reemphasized even more because ruggedness, as dictated by the applications normally found in level 3, is even more important. Experience has also shown that level 3 connectors experience a far greater level of disconnect and remating than level 2 applications.

CONNECTION LEVEL 4

Just as there is an overlap between levels 2 and 3, so a comparable overlap exists between levels 3 and 4. In fact, many users find it convenient and desirable to use rack-and-panel connectors in level 4 applications. When this is so, the connectors are usually modified by the use of accessories, such as hoods, cable clamps, and mechanical engaging and locking devices. In a small number of cases the rack-and-panel connectors are further modified with sealing grommets to provide environmental and/or radio-frequency-interference protection. In these ways they are made more suitable for level 4 applications. As Table 4 for level 4 shows, most of the details shown for rack-and-panel connectors for level 3 will also apply. The conductors found in level 4 are virtually the same as those for level 3 except that now printed-circuit boards are not used at all.

TABLE 4   Connection Level 4

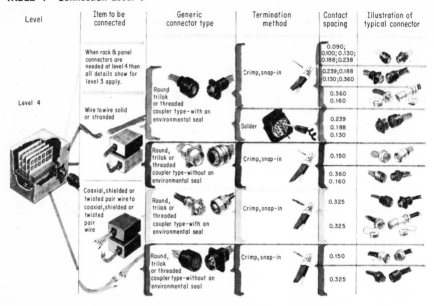

Level	Item to be connected	Generic connector type	Termination method	Contact spacing	Illustration of typical connector
Level 4	When rack & panel connectors are needed at level 4 then all details show for level 3 apply.	Round trilok or threaded coupler type–with an environmental seal	Crimp, snap-in	0.090; 0.100; 0.130; 0.188; 0.238	
				0.239; 0.188 0.130; 0.360	
				0.360 0.160	
	Wire to wire solid or stranded		Solder	0.239 0.188 0.130	
		Round, trilok or threaded coupler type–without an environmental seal	Crimp, snap-in	0.150	
				0.360 0.160	
	Coaxial, shielded or twisted pair wire to coaxial, shielded or twisted pair wire	Round, trilok or threaded coupler type–with an environmental seal	Crimp, snap-in	0.325	
				0.325	
		Round, trilok or threaded coupler type–without an environmental seal	Crimp, snap-in	0.150	
				0.325	

Generic Connector Types

The generic connector types most often used at this level are of the cylindrical, multicontact, coupled types. The couplers are either threaded, bayonet, or push-pull quick-coupling types. Threaded couplers are considered better suited for applications where vibration and shock considerations are important (e.g., aircraft and shipboard applications). Bayonet couplers are the most popular type for most applications. Figure 5 shows the parts of a typical cylindrical, multicontact connector. A major consideration for level 4 is the need for sealing against extreme environmental

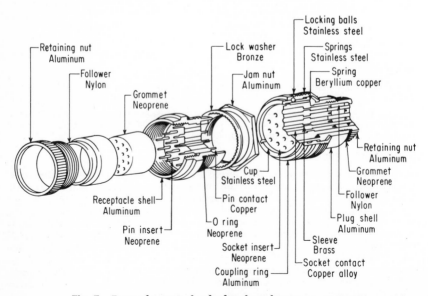

Fig. 5 Parts of a typical cylindrical, multicontact connector.

conditions. Many uses, such as military or high-humidity areas, leave no doubt as to the need for sealing. There are also many places where such performance requirements are unrealistic and result in higher cost, weight, and other technical disadvantages. For most of the crimp, snap-in contacts, the designs tend toward retaining clips in the connector body rather than on the contact itself. This is done to avoid tearing of the sealing grommets. Another approach has been the use of a plastic contact-retention disk to replace the retaining clip.

The use of elastomers at the various interfaces and at the wire grommet seal has been found necessary to permit connectors to withstand chemical attack and environmental extremes, including wide temperature-exposure ranges. On the other hand, for nonsealed connectors a wide variety of plastics are already available which have demonstrated ability to meet the less stringent requirements.

As mentioned in the level 3 discussion, some connectors of the unsealed type are also available with the capability to accept single-conductor or coaxial contacts interchangeably in any position. This feature may have a significant effect on factory costs as well as subsequent costs of maintenance. The key to the decision might well be the combined cost of manufacturing and maintaining a system throughout its expected life.

Termination Methods

The termination methods shown for level 4 further minimize the use of solder and promote the use of the crimp, snap-in approach. Solderless-wrap terminations remain out of contention at this level, for the same reasons as for level 3.

The need for ruggedness continues to increase as the application is characterized by level 4. The external-use aspects of this level subject it to greater abuse. In addition, still more disconnects and reconnects are generally required for levels 2 and 3. Where the specific application includes these requirements, greater emphasis should be made to avoid unnecessary miniaturization in the connectors.

SUMMARY OF SELECTION APPROACH

The material presented in this portion of the chapter has, of necessity, attempted to cover a very broad topic in a general way. Of course, there will be valid exceptions to the given guidelines for reasons dictated by a specific application, economics, or any of a dozen other factors. However, the approach has been found to follow logically the selection techniques used by experienced connector-application engineers. It is expected that the packaging engineer who is venturing into this area for the first time will find the task somewhat simplified and that connector choices will be made which will fit well within system needs.

Table 5 displays an assortment of common applications of many types of connectors.

DETAIL OF CONNECTOR-SELECTION FACTORS

The designer who has followed the connector-system-selection technique described in the first part of this chapter may now wish to consider specific details. In the remainder of this chapter various factors are discussed in greater detail with accompanying figures and tables. The factors included are electrical, mechanical, materials, contacts, environmental, and reliability. Because these are interrelated differently for different specific applications, a selection guide is not offered. Reference material on military and industrial specifications is included.

ELECTRICAL FACTORS

Contact resistance, constriction resistance   In most types of military connectors, the contact resistance is determined by measuring the millivolt drop from tail to tail for the mated set of contacts with a specified current flowing. Thus, the resistance of the contact material as well as of the actual point or points of contact is investigated. Generally, the contact resistance is 0.001 Ω or less, although the resistance including the contact lengths may exceed that value.

TABLE 5   Common Applications

Application	Remarks	
Cabling and harness	Readily available Low cost Solder and crimp contacts	
Rack and panel	Readily available Low cost Solder contacts Coaxial contacts	
Aircraft cabling and harness	Bayonet Quick-disconnect Crimp contacts For 200°C crimp contacts	
Missile cabling and harness	Lightweight, short shell, crimp contact, E seal Solder contacts, non-environmental For 200°C crimp contacts	
Replaceable assemblies	To 321 size 16 and 20 and coaxial crimp contacts	
Modular circuits	Custom design	

TABLE 5   Common Applications (Continued)

Application	Remarks	
Chassis plug-in units	To 50 size 20 contacts, chamfered socket lead-in	
Chassis connectors	To 104 size 22 crimp contacts, jackscrew	
Printed-circuit standard size	0.156-in. centers, for 0.062-in. board	
Miniature	0.100-in. centers	
Coaxial	Quick-disconnect, shield, crimp	
	Quick-disconnect, for RG-188/U, to 70,000 ft	
Triaxial	For high-voltage application	

The performance of an electrical contact can best be understood and, to some degree, predicted by the design engineer if (1) he is aware of the fundamental processes taking place at the contact surface and (2) he knows how they are influenced by the surface and bulk properties of the materials.

The development of subminiature connectors for use in low-level circuits places greater importance on the performance of electrical contacts. Low contact resistance is considered necessary in many present-day low-level applications, i.e., millivolt-microampere range of power.

Low electrical levels, or "dry-circuit" area of contact loading, is that area in which only mechanical forces (impact, pressure, wipe, abrasion, strain hardening) can change the condition of the contact surfaces; there can be no thermal effects (softening, melting) or electrical effects (arc transfer, bridge transfer, etc.). Thus the absolute-dry-circuit area is defined as that in which the open-circuit voltage does not, for an appreciable time, exceed the softening voltage for the contact material used. Dry circuit cannot be exactly defined, but it should apply to any low-voltage low-force circuit, since the exact condition might be different in each case.

The degree of electrical resistance or, conversely, conduction between two surfaces designed to make or break an electric circuit is limited by the mechanical qualities and resistivities of the materials. Basically, in a contact system, there are three controlling factors or conditions:

1. Bare metallic contacts where the continuity of current flow is assured by the nature of the material

2. Contacts between surfaces covered with an adsorbed layer in the magnitude of a few molecular diameters, where electrical flow may be slightly limited because of the impurity adsorbed

3. Coherent foreign films or individual large particles, both consisting of either insulators (oil, plastics, fibers) or semiconductors (oxides, sulfides) which act as limiting factors in conduction

The resistance of a pair of clean contacts depends on the bulk resistance of the material (which is so small that it can be neglected) and the resistance due to the squeezing together of the current lines through the contact spot, which is relatively small in diameter. Where there is plastic deformation of the contact, the constriction resistance conforms as follows:

$$R_C = 0.9p \sqrt{\frac{H}{P}} \qquad (1)$$

where R_C = constriction resistance, Ω
p = resistivity, Ω-cm
H = hardness, kg/cm^2 (approximately equal to the Brinell hardness)
P = force, kgf

For elastic deformation on two contacts with equal spherical radii, constriction resistance may be considered as

$$R_C = 0.6p \sqrt{\frac{E}{3\,Pr}} \qquad (2)$$

where R_C = constriction resistance, Ω
p = resistivity of the contact material, Ω-cm
E = modulus of elasticity, kg/cm^2
P = force, kgf
r = radius of the contacts

The factor 0.6 contains the Poisson numbers of the contact materials—gold, silver, and palladium being about equal. Both equations are derived using some simplifying assumptions, but they are accurate enough for practical purposes and show which material parameters influence the constriction resistance. Therefore, where a low constriction resistance is desired, a material with low resistivity, low modulus of elasticity, and low hardness should be chosen.

The amount of contact resistance added to the constriction resistance of clean contacts because of the presence of a thin adsorbed film can usually be neglected for practical purposes.   The resistance of a thin film is a function of its thickness and the contact material; the film offers little resistance to the electrons, which can travel through as a result of the tunnel effect.   The work function is another material parameter, representing the energy needed for an electron to leave the metal lattice.   Tunneling is facilitated to a greater degree as the work function is lowered.

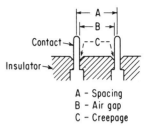

A - Spacing
B - Air gap
C - Creepage

Fig. 6 Spacing, air gap, and creepage.

The resistance of a thicker adsorption film (above 50 Å) is a function of the applied voltage.   If a low voltage is applied, resistance is high (in the magnitude of 10^6 Ω).   However, the resistance decreases slowly if the voltage is gradually increased, up to a point where it drops sharply.   In this procedure, a metallic path is established by a mechanism which is not yet fully understood.   The voltage necessary for this so-called fritting is a nonlinear function of the film thickness.   The field required for breakdown is in the magnitude of 10^6 V/cm.

Breakdown voltage   The breakdown voltage of a connector is dependent upon its contact spacing and geometry, shell spacing and geometry, and insert and seal materials and geometry.   The terms spacing, air gap, and creepage are defined by Fig. 6.

Contact spacings and test, working, and flashover (breakdown) voltages for three commonly used types of connectors are given in Tables 6 to 8.

TABLE 6   Voltage Rating, Standard Size Rectangular, MIL-C-8384, with Molded Insert*

Spacing, in.	Creepage, in.	Voltage at sea level			Voltage at 50,000 ft			Voltage at 70,000 ft		
		Test, V rms	Working		Test, V rms	Working		Test, V rms	Working	
			dc, V	ac, V rms		dc, V	ac, V rms		dc, V	ac, V rms
1/32	1/16	1,000	490	350	375	190	125	300	125	90
1/16	7/64	1,800	840	600	675	315	225	450	210	150
1/8	3/16	3,300	1,550	1,100	1,065	490	350	675	310	225
3/16	1/4	4,500	2,100	1,500	1,350	630	450	825	375	275
1/4	5/16	5,400	2,500	1,800	1,500	700	500	975	455	325
5/16	3/8	6,300	2,900	2,100	1,725	810	575	1,065	500	355

* These values vary with various insert materials and methods of sealing connectors.

Guidelines for determining the test voltage and working voltage for a given connector are

$$V_{\text{working}} = \tfrac{1}{3} V_{\text{test}} \tag{3}$$

$$V_{\text{test}} = \tfrac{3}{4} V_{\text{flashover}} \tag{4}$$

Insulation resistance   Insulation resistance is a measure of the inability of a material to conduct electricity.   The measurement of insulation resistance is, in fact, a measurement of the leakage current that flows not only through the material but along its surface.

The degree of homogeneity achieved during fabrication of the material directly influences the path of the leakage current.   With more current flowing though certain areas, a greater heat rise occurs.   As the temperature rises, the insulation resistance decreases, causing still higher operating temperatures.

TABLE 7   Voltage Rating, Standard Size Circular, MIL-C-5015, with Resilient Insert*

Nominal distance		Standard-sea-level conditions		Pressure altitude 50,000 ft		Pressure altitude 70,000 ft	
Air space	Creepage	Min flashover voltage, vac rms	Test voltage, vac rms	Min flashover voltage, vac rms	Test voltage, vac rms	Min flashover voltage, vac rms	Test voltage, vac rms
1/32	1/16	1,400	1,000	550	400	325	260
1/16	1/8	2,800	2,000	800	600	450	360
1/8	3/16	3,600	2,800	900	675	500	400
3/16	1/4	4,500	3,500	1,000	750	550	440
1/4	5/16	5,700	4,500	1,100	825	600	480
5/16	1	8,500	7,000	1,300	975	700	560

* These values vary with various insert materials and methods of sealing connectors.

TABLE 8   Voltage Rating, Miniature Circular

Specification	Creepage, in.	Spacing, in.	Flashover (min)		
			Sea level	At 70,000 ft	At 100,000 ft
MIL-C-26482C:					
Service I......	0.093	0.046	2,000 vac	500 vac	280 vac
Rating II.....	0.125	0.078	2,000 vac	650 vac	260 vac
MIL-C-26500B	.....	.....	2,000 vac	1,300 vac	1,300 vac

The proportion and conductivity of contaminants in the material also can greatly reduce the insulation resistance. One contaminant that should not be overlooked is the ink used for marking and identification. Some inks are conductive and can lower the insulation resistance of high-density connectors if letters or numbers are closely spaced.

The percentage of moisture absorbed by the material also degrades insulation resistance. Moisture can bridge conductive elements such as certain types of salts, which may be part of the material, surface deposits, etc.

Military specifications give three ways to measure insulation resistance. MIL-C-8384 and MIL-C-21097 provide for measuring between adjacent positions and each position adjacent to the shell. MIL-C-26500 and MIL-C-26518 suggest measurement between any three positions in close proximity and any three contacts adjacent to the shell.

Commoning, or "ganging," tends to yield lower insulation-resistance values, since many current paths are created. When the insulation resistance is measured between two positions, fewer current paths are created.

When less dense connectors are tested, the method of measurement can be significant, depending upon the proximity of the contacts in relation to each other. In higher-density configurations the commoning of additional positions beyond a certain minimum does not reduce the insulation-resistance reading significantly.

Insulation-resistance measurements can be used to determine if a connector is functioning properly. If the input to a circuit is not adequate, for example, current may be leaking through the connector. Erratic voltages may mean a faulty connector due to a metallic chip, moisture, etc. Problems such as those described above can be quickly located with an insulation-resistance measurement. For problems such as intermittent operation or loss of signal, causes other than low insulation resistance should be investigated.

Contact-current rating   Connectors are built to accommodate a large range of wire sizes and types. Crimp- and solder-type contacts are available in sizes 0, 4, 8, 12, 16, 20, and 22 and will accommodate wire from 0 to 22 AWG. Although some crimp and solder contacts may physically accommodate smaller wire or more than one wire, it sometimes is impractical to do so because of the low wire-retention force (crimp contacts) and the necessity for environmental seals. Since copper is an excellent conductor of heat, wires of larger sizes conduct more heat away from the connector, thus lowering its operating temperature. The thermal effects of various wire sizes are shown in Fig. 7. Figure 8 shows current-carrying capacity for various contact sizes.

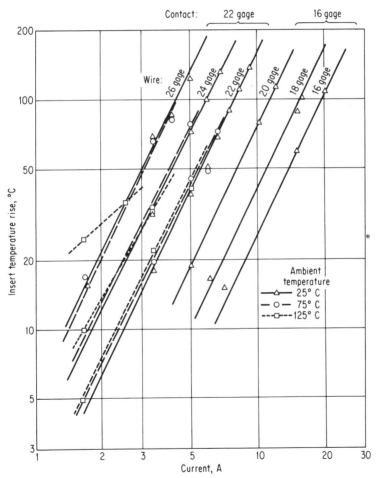

Fig. 7   Thermal effect of wire size.   (*Microdot, Inc.*)

Effect of clustering of active contacts   Distribution of active contacts which results in clusters provides a concentration of heat and a localized hot spot. In actual practice, heavy-duty contacts tend to be clustered. The same is true when power and signal contacts are separated into individual clusters. The end effect is to reduce the maximum permissible current per contact that other considerations would permit.

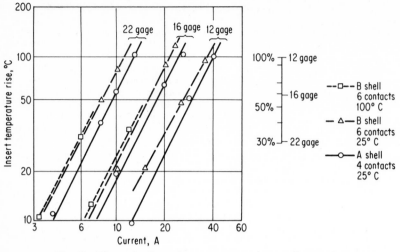

Fig. 8 Thermal effect of contact size. (*Microdot, Inc.*)

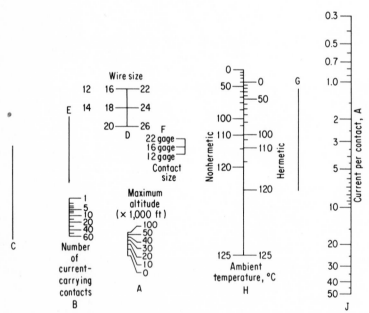

Fig. 9 Current rating of multicontact connectors.[5] (*Microdot, Inc.*)

Nomograph for current rating of multipin connectors Many of the factors discussed here have been combined to form the nomograph given in Fig. 9. Care has been taken that each derating factor is independent of the others.

From altitude *A* lay straightedge through Number of Contacts *B* and determine point on pivot line *C*. Lay straightedge from *C* point to Wire Size *D* (use left column for 12-gage wire, center for 16-gage, and right for 22-gage) and determine point on pivot line *E*. From this point lay straightedge through Contact Size *F* and determine point on pivot line *G*. Lay straightedge from Ambient Temperature *H* through point *G* and read the maximum allowable Current per Contact *J*.

MECHANICAL FACTORS

Mating methods Mating consists of two operations: polarizing and locking. Polarizing is necessary in order to position the entry of the pin contacts into the corresponding socket contacts. The locking mechanism then mates the pin contacts with the socket contacts and securely holds the two halves of the connector together. The most widely accepted coupling is the three-pin bayonet owing to its ability to remain coupled under severe vibration and shock environments. Threaded couplings can be satisfactory in extreme environment if fine mating threads are used. In most instances, spring action of some type is required in order to maintain tension between the plug and receptacle. For rectangular connectors the shape of the inserts and shells is varied, by a bevel or rounding a corner so that plug and receptable cannot be mated in the wrong position. Axial forces required to engage and to disengage a connector are specified. For a bayonet coupling the torques required to engage and disengage are also specified.

A certain amount of float is necessary in the mating of contacts. The ability of the contacts to move slightly (float) helps in alignment and reduces the mating force. In late designs of miniature connectors the probability of mating is improved by floating the pin contact in a resilient insert material. The socket contact is mounted in a rigid insert material. The entry holes in the rigid insert are chamfered. By this arrangement a connector with a No. 20 pin contact (diameter 0.040 in) misaligned as much as 0.042 in can be successfully mated.

Effect of shell size In connectors having enough contacts so that the distribution of heat is essentially uniform, the difference in temperature rise between the connectors of differing shell size, but with the same number of contacts, is identical within the limits of measurement error. Evidently most of the heat is carried along the connecting wires with little conducted out through the dielectric. The temperature rise depends more upon the density of active contacts than upon the shell size.

Effect of mounting on a metal panel A series of tests indicated that a connector mounted on a $\frac{1}{16}$-in panel requires about 10 percent more current to reach a given temperature than one in quiet air. Increasing the thickness of the panel does not produce a measurable reduction in the connector temperature. This parameter is not included in the charts and nomographs. This effect will provide a safety factor for panel-mounted connectors.

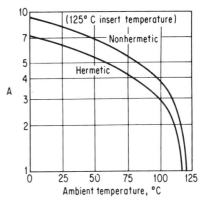

Fig. 10 Derating for hermetically sealed connectors.

Effect of hermetic seal Hermetically sealed connectors have higher contact resistances than other types. For a particular ambient temperature contact currents cannot be as high. Figure 10 is a derating curve comparing the current capacity for hermetic and other types of connectors.

MATERIALS

Contact Materials The choice of contact material has a direct bearing on all other connector design considerations and an important influence on the electrical characteristics of the connector. The contact is the heart of any connector, and thus it not only must function as an electrical conductor but must be adequately able to withstand all the projected mechanical and environmental conditions to which the connector will be exposed in service. Some commonly used contact materials will now be discussed.[6]

Beryllium Copper. Beryllium copper is an especially suitable spring contact material for connector applications because of its combination of good mechanical properties, electrical conductivity, thermal conductivity, and resistance to wear and corrosion. It is the best electrical conductor of any spring alloy of comparable hardness, and it is stronger and more resistant to fatigue than the other copper-base alloys. Practically all internal stresses caused by rolling, drawing, or forming are relieved during the heat treatment. The material also exhibits low mechanical hysteresis.

Phosphor Bronze. Phosphor bronze, specifically alloy grades A and C, is a widely used copper alloy because it has good corrosion resistance and fair conductivity, and it is easily formed. Alloy C is usually preferred where moderately high mechanical properties are required. Alloy A is used because of its lower cost where strength is not an important design factor. It is a good, general-purpose spring material for limited applications (105°C maximum).

Spring Brass. Spring brass is especially useful in low-cost electrical applications where high temperature or repeated flexing at high stresses is not a consideration. Although it has relatively low spring properties, it is often used in conjunction with beryllium copper spring members which provide the electrical contact pressure. It is readily crimped, electroplated, welded, brazed, and soldered.

Low-leaded Brass. This material is used primarily as rod stock for male (pin) contacts. Its machinability characteristics make it suitable for high-speed, high-production screw-machine work. It is therefore, an ideal material for high-volume rack-and-panel pin contacts. It has good electrical properties and good resistance to general corrosion and stress-corrosion cracking. It has relatively low spring properties, but it is often used in conjunction with beryllium copper spring members, which provide the electrical contact pressure.

Contact plating Plating of the basis metal of connector contacts is normally required to prevent deterioration of the mating surfaces of the contacts, mechanically or chemically. Such deterioration eventually results in the inability of the mated contacts to perform properly.

The two main problems to be considered are wear and chemical environment. The plating should be adequate to cover and protect the basis contact material in the "worst-case" conditions of both these factors.

Hard-gold plating on socket contacts and soft gold on pin contacts is used where numerous insertions and withdrawals are anticipated. This combination of hard and soft plating results in a burnishing action that improves wear resistance and also actually improves the contact resistance factor.

An overplating of gold with an underplating of a less precious metal is often used for specific environments as an added protective factor in the event that the gold overplating wears through.

Table 9 lists some commonly used platings for contacts.

Selective Plating.[8] The concept of selective plating of connector contacts is gaining greater interest among connector manufacturers as newer and less costly techniques are developed for plating only in those areas of the contact where it is actually necessary. The concept is not new. However, the cost of preparing parts initially for selective plating can exceed the value of the precious metal saved. Several manufacturers have developed methods for selective plating of connector contacts which not only reduce the amount of precious metal used but also permit rapid automated means for application. Selective plating may be the answer to some of the problems the designer is faced with in developing high-density highly reliable connectors. In most connector designs the pressure between mating surfaces has to be of sufficient magnitude to wipe away any contaminants and maintain good contact under severe environmental conditions. In these designs the plating (usually gold) is removed after several matings and unmatings, resulting in increased resistance caused by insulating contamination (corrosion) on the contact surface. A heavy or thick gold plate will reduce the effects of contaminants, especially in a corrosive environment, which will reduce the amount of pressure required for a reliable connection.

Consideration is also being given to the surface condition of the contact (pin) before plating. Specification MIL-C-39029 specifies the surface roughness to be 32 μ in or less for certain pin contacts. Much can be said for this requirement, since an ideal contact surface would enhance plating and control the amount of friction between mating contacts.

Lubricants. During the last 10 years many connector manufacturers have been

TABLE 9   Platings for Contacts[7]

Material and thickness, in.	Remarks
0.0003 hard or soft gold over 0.0002 (min) silver	Suitable for most crimped contact applications. Provides low contact resistance for signal circuits and good wear resistance. Corrosion resistance is good unless excessive wear exposes the silver underplating, which is then sensitive to sulfide atmosphere. Porous gold or thickness less than the specified 0.0003 will also result in sulfide contamination
0.00005 hard or soft gold over 0.0002 (min) silver	Same as above, but much more resistant to wear, with subsequent higher resistance to chemical deterioration
0.00005 gold over 0.0002 nickel....	Excellent for wear resistance and hostile environments
0.00005 gold over copper flash....	Excellent for low-level circuit applications with low to medium insertion and withdrawal requirements, as in many data-processing and computer applications
0.0005 to 0.001 silver............	Suitable for power contacts with relatively high contact forces
0.0003 electrotin...............	For use in low-cost applications where few disconnects are anticipated

developing and applying lubricants to their contacts to reduce wear and improve reliability. Lubricated gold-plated contacts can result in a substantial reduction in the thickness of gold required to ensure a reliable connection.

Several basic facts have been established with respect to improved contact reliability:

1. Failure of a connector through increased resistance can be caused by (a) insulating contamination on the contact surface or (b) excessive wear resulting from friction of the mating parts.

2. An increase in the connector loading force intended to overcome an increase in resistance caused by insulating contamination may lead also to excessively high connector insertion and withdrawal forces. This may cause interference with operation of the connector.

3. Since the power available at low signal levels is too small to break through the insulation barriers between contacts, contamination at such signal levels constitutes a greater problem than it does at power levels used for switching operations. Thus, insulating deposits that are normally innocuous can cause contact failures.

The most common materials used in present sliding-type miniature connectors consist of a copper-base alloy with either gold or silver plating, or with a gold-over-silver finish. Excessive wear, which removes the gold or silver plate, thus exposing base metal, contributes greatly to both organic and inorganic contamination. The amount of wear depends on many parameters, including number of connector-mating cycles, mechanical loading force, coefficient of friction, relative hardness of the mating parts (which is in turn related to metal-grain size), and the affinity of the two metals (the greater the tendency for pairs of metals to form solid solutions, the greater the wear).

The best indicated approach has been to apply surface films capable of reducing friction and wear, yet preserving contact resistance in the milliohm range. Ideally, it is desired that an electrical metallic junction have a low electrical contact resistance, even if the metallic terminals have not been used for several years, and remain stable during steady or repeated contact life, subject to a variety of temperature, humidity, and other atmospheric conditions. In addition, this performance should be obtained with minimal wear and low sliding frictional forces for long life. This may seem to be asking too much from a contact, but contacts are important and sometimes critical elements in modern electronic circuitry, and are occurring in ever-increasing numbers because of the complexity of new equipment.

Several years ago, a study[9] of the various types of contact designs commonly used in connectors revealed that the frictional forces increased rapidly, often exceeding their maximum specification requirements early in their life cycle. Figure 11 displays the force characteristics during mating for several typical contacts. The cycling duration was between 100 and 150 cycles. On most contacts the peak forces occurred within 50 cycles and were maintained for the duration of the cycling. These data display the characteristic "stick-slip" process of sliding. Metallic transfer between the surfaces was apparent from photographic examination. The pronounced wear exposes the base metal, which becomes corroded upon exposure to the various atmospheric conditions and, in turn, leads to erratic and sometimes high electrical-contact resistance.

In spite of the various contact designs, the actual area of contact A (the ratio of the normal load W to the yield pressure of the softer material P_m, i.e., $A = W/P_m$) was about the same for all designs since the contact-pin material was brass in all cases and the normal forces were similar. Since frictional force F is equal to the coefficient of friction μ times the normal load W, it follows from these relations that $F = \mu A P_m$. Accordingly, to reduce the frictional force, the coefficient of friction (or the condition between sliding surfaces) would have to be modified for a given material and load. The surface-contact area was therefore lubricated by coating the contact pin with graphite, resulting in a significant change in the friction and wear characteristics (Fig. 11). Since a graphite coating is considered impractical

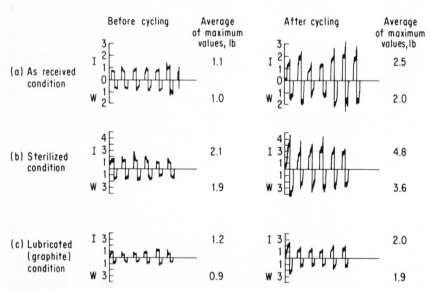

Fig. 11 Insertion—and withdrawal—force characteristics of typical contact, 16 AWG, with and without lubrication.[9]

for general use in electrical connectors, a program was initiated to develop a suitable lubricant for this application.

The main effort was directed toward boundary lubrication. Monomolecular layers of octadecylamine hydrochloride (ODA-HCl) applied to gold-plated surfaces exhibited the best results of all the materials tested. The friction characteristics at several conditions of loading are shown in Fig. 12. The contact resistance was essentially metal-to-metal contact (i.e., 0.001 Ω). Based on the results obtained, an ODA-HCl–lubricated gold-plated contact of 50 μin can provide adequate protection from wear as well as corrosion without adverse effect on electrical performance. Lubricated contacts have not been included in any military specification, mainly because adequate performance and/or control requirements have not been established.

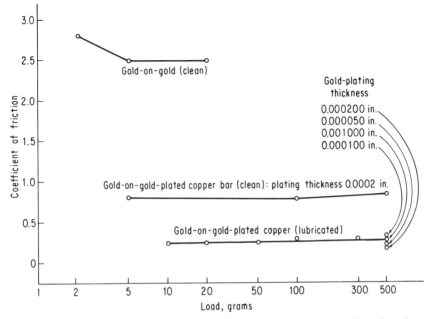

Fig. 12 Coefficient of friction vs. load of gold-on-gold surfaces with and without lubrication.[9]

ODA-HCl [$CH_3(CH_2)_{16}CH_2NH_2HCl$] was prepared from octadecylamine (Armeen 18D) obtained from Armour and Company.

Plastic molding materials[10] The choice of the plastic insulator material to be used in a particular connector application is governed not only by the related design and size of the part (i.e., wall-thickness variations, type of inserts, size, etc.) but also by the electrical, mechanical, thermal, and chemical-resistance requirements of the application. The chemical-resistance is a requirement that is many times overlooked but is most important where chemical cleaning of solder joints on terminals or printed-circuit boards is involved.

From the cost standpoint, the chosen material should offer the lowest material cost per cubic inch of component part, and also the shortest molding cycle while still satisfying all the application requirements.

Molding materials, whether for use in commercial or in military-grade connectors, are generally designated by the connector manufacturer according to type, as classified in Specification MIL-M-14, Molded Plastics and Molded Plastic Parts, Thermosetting.

It should be mentioned that much tailoring of materials can be done by suppliers and that better than average compounds exist for nearly any given design feature. Further, material improvements are constantly being made. Hence, consultation with suppliers is always advised in cases where any doubt exists.

As shown in Table 10, the principal thermosetting materials are the

Alkyds	Phenolics
Aminos	Polyesters
Diallyl phthalates	Silicones
Epoxies	

Alkyds. Alkyds are widely used for molded electrical parts, since they are easy to mold and economical to use. Molding dimensional tolerances can be held to within ±0.001 in/in. Postmolding shrinkage is small. Their greatest limitation is in extremes of temperature (above 350°F) and in extremes of humidity. In these respects silicones and diallyl phthalates are superior—silicones especially with respect

TABLE 10   Typical Properties of Materials Used for Connector Designs[10]

Property	Diallyl phthalate		Silicone		Epoxy	
	Mineral-filled	Glass-filled	Mineral-filled	Glass-filled	Mineral-filled	Glass-filled
Mold shrinkage, in./in.	0.004–0.006	0.002	0.006–0.007	0–0.005	0.001–0.008	0.001–0.002
Specific gravity	1.50–1.60	1.55–1.70	1.81–2.82	1.68–2.0	1.6–2.06	1.8–2.0
Tensile strength, psi	3,000–8,000	5,000–10,000	3,000–3,500	4,000–5,000	5,000–7,000	....
Compressive strength, psi	18,000–25,000	20,000–30,000	15,000–18,000	10,000–15,000	18,000–25,000	....
Flexural strength, psi	6,000–10,000	10,000–20,000	7,000–8,000	10,000–14,000	10,000–15,000	....
Impact strength (ft-lb/in. of notch) ($\frac{1}{2}$ by $\frac{1}{2}$-in. notched-bar Izod test)	0.30–0.50	1.0–10.0	0.26–0.35	3–15	0.25–0.40	....
Water absorption (24 hr, $\frac{1}{8}$-in. thickness), %	0.2–0.3	0.1–0.3	0.13	0.1–0.2	0.1	0.5–0.095
Thermal conductivity, 10^{-4} cal/(s)(cm^2)(°C)(cm)	7.0–10.0	7.0–10.0	....	7.51–7.54	7–18	7–10
Thermal expansion, 10^{-5} per °C	2.0–3.0	2.0–3.0	....	0.8	2.5–5.0	1.1–3
Resistance to heat, °F (continuous)	350–450	350–500	600	600	300–500	330–500
Volume resistivity, 50% RH, (Ω-cm)	10^{13} +	10^{13}–10^{16}	10^{14}	10^{10}–10^{14}	10^{14}	10^{14}
Dielectric strength, V/mil:						
Short time, $\frac{1}{8}$-in	395–420	395–450	200–400	200–400	300–400	300–400
Step-by-step, $\frac{1}{8}$-in	390–400	395–400	380	125–300	300–400	300–400
Dielectric constant:						
60 Hz	5.2	4.3	3.5–3.6	3.3–6.2	3.5–5.0	3.5–5.0
1 kHz	4.8–5.3	4.1–4.4		3.2–5.0	3.5–5.0	3.5–5.0
1 MHz	3.9–4.0	3.4–4.5	3.4–6.3	3.2–4.7	3.5–5.0	3.5–5.0
Dissipation factor:						
60 Hz	0.03–0.06	0.01–0.05	0.004–0.005	0.004–0.03	0.01	0.01
1 kHz	0.03–0.10	0.004–0.009	....	0.0035–0.02	0.01	0.01
1 MHz	0.02–0.04	0.009–0.014	0.002–0.005	0.002–0.02	0.01	0.01
Arc resistance, s	140–190	125–180	250–420	150–250	150–190	120–180

to temperature, and diallyl phthalates with respect to humidity. The electrical-insulation resistance of alkyds decreases considerably in high-continuous-humidity conditions.

Alkyds are chemically somewhat similar to polyester resins. Normally, the term alkyds is used to denote liquid resins, or low-pressure molding compounds. Alkyd molding compounds are commonly available in putty, granular, glass-fiber-reinforced, and rope form.

Aminos. Amino molding compounds are economical to mold, and they result in parts that are hard, rigid, and abrasion-resistant, and have high resistance to deformation under load. They can be exposed to subzero temperatures without embrittlement, and have excellent electrical-insulation characteristics. In addition, they are unaffected by common organic solvents, greases and oils, and weak acids and alkalies.

Of the two major types of aminos, namely, melamines and ureas, melamines are superior to ureas in resistance to acids, alkalies, heat, and boiling water, and are preferred for applications involving cycling between wet and dry conditions or rough handling. Both provide excellent heat insulation, and temperatures up to the destruction point will not cause parts to lose their shape.

When amino moldings are subjected to prolonged elevated temperatures, a loss of certain strength characteristics occurs. Some electrical characteristics are also adversely affected. Arc resistance of some industrial types, though, remains unaffected after exposure at 500°F.

Ureas are unsuitable for outdoor exposure, while melamines experience little degradation in electrical or physical properties after outdoor exposure, although color changes may occur.

Diallyl Phthalates. Diallyl phthalates are among the best of the thermosetting plastics with respect to high insulation resistance and low electrical losses. They

Alkyd and polyester		Phenolic		Melamine		Polyethylene oxide, unfilled	Polysulfone	Polycarbonate, unfilled
Mineral-filled	Glass-filled	Flock-filled	Glass-filled	Flock-filled	Glass-filled			
0.004-0.010	0.002-0.006	0.004-0.009	0.006-0.009	0.006-0.007	0.001-0.004	0.006	0.007	0.005-0.007
1.60-2.30	1.8-2.30	1.32-1.45	1.75-1.95	1.50-1.55	1.8-2.0	1.06-1.10	1.24-1.25	1.2
3,000-8,000	4,000-10,000	6,500-9,000	5,000-10,000	7,000-9,000	5,000-10,000	7,800-9,600	10,200 (at yield)	8,000-9,500
18,000-25,000	20,000-30,000	22,000-36,000	17,000-26,000	30,000-35,000	20,000-35,000	16,000-16,400	13,900 (at yield)	10,300-10,800
6,000-10,000	12,000-20,000	8,500-12,000	10,000-16,000	13,000	15,000-23,000	12,800-13,500	15,400 (at yield)	12,200-12,700
0.30-0.50	1.5-16.0	0.24-0.60	10-50	0.4-0.45	4.0-6.0	1.7-1.8	1.3 (¼-in. bar), 1.2 (⅛-in. bar)	12.0-18.0
0.5-0.05	0.06-0.28	0.3-1.0	0.1-1.2	0.16-0.3	0.09-0.21	0.06	0.22	0.15
7.0-15.0	7.0-10.0	4-7	....	....	11.5	....	1.8 Btu/(hr)(ft²)(°F)(in.)	4.6
3.5-5.0	2.5-3.3	3.0-4.5	1.6	....	1.5	5.2	3.1×10^{-5} in./(in.)(°F)	6.7-7
300-350	300-350	360-500	350-500	250	300-400	250	300	250
10^{13}-10^{15}	10^{15}	10^9-10^{13}	7×10^{12}	....	2.0×10^{11}	10^{16}-10^{17}	5.0×10^{16}	2.1×10^{16}
350-450	350	200-400	140-400	350-400	170-300	400-550	425	400
300-350	300	100-375	120-270	250-350	170-240	400-550	400	364
5.1-7.5	5.7	5.0-13.0	7.1	6.2-7.6	9.7-11.1	2.64	3.14	2.97-3.14
5.0-6.2	5.4	4.4-9.0	6.9	6.0-7.5	....	2.64	3.13	3.02
4.6-5.5	5.2	4.0-6.0	4.6-6.6	4.7-7.0	6.6-7.5	2.64	3.10	2.92-2.98
0.009-0.06	0.010	0 05-0.30	0.05	0.019-0.035	0.14-0.23	0.0004	0.0003	0.0009
0.007-0.03	0.007	0.04-0.20	0.02	0.013-0.034	....	0.0004	0.0010	0.0021
0.006-0.04	0.008	0.03-0.07	0.012-0.026	0.032-0.060	0.013-0.015	0.0009	0.0034	0.010
75-190	180	Tracks	T-190	95-135	180	75	75-122	10-120

maintain these favorable properties up to 400°F or higher, and in the presence of high-humidity environments. This makes them an excellent choice for use in environmental-type connectors. Diallyl phthalate resins are also easily molded and fabricated.

There are several chemical variations of diallyl phthalate resins, but the two most commonly used are basic diallyl phthalate (DAP) and diallyl isophthalate (DAIP). The primary application difference is that DAIP will withstand somewhat higher temperatures than will DAP.

The excellent dimensional stability of diallyl phthalates is demonstrated in Fig. 13, which compares them with other materials at various temperatures.

Epoxies. For electronic applications, epoxies are among the most versatile and widely used plastics. This is primarily because of the wide variety of formulations possible, and the ease with which they can be used. Formulations range from flexible to rigid in the cured state, and from thin liquids to thick pastes in the uncured state. Conversion from the uncured to the cured state is made by the use of hardeners and/or heat.

The main uses of epoxies in the electronics field are for embedding applications (potting, casting, encapsulating, and impregnating) and in laminated constructions, such as metal-clad laminates for printed circuits and unclad laminates for various types of insulating and terminal boards.

Epoxies are available for the most part as liquid or solid resins, and as powdered molding compounds. The molding compounds are widely used to mold cases for electronic modules, and even more widely to mold modules directly by the transfer-molding technique. They are also used for transfer and compression molding of may other types of electrical parts. The liquid resins find their broadest use in embedding applications and in the fabrication of laminate boards. Advanced multi-layer circuit boards are almost exclusively based on epoxies.

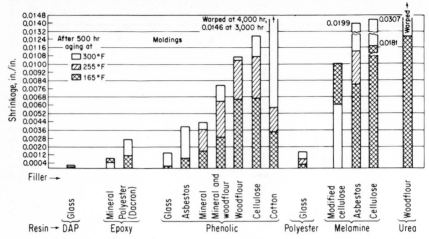

Fig. 13   Shrinkage of various thermosetting molding materials as a result of heat aging.[10]

In addition to their versatility and good electrical properties, epoxies are also outstanding in their low shrinkage, their dimensional stability, and their adhesive properties. Their shrinkage is often less than 1 percent, and the "as-molded" dimensions of an epoxy part change little with time or environmental conditions, other than excessive heat. Because of the low shrinkage and good strength properties of epoxies, cured epoxy parts resist cracking—both upon curing and in thermal shock—better than most other rigid thermosetting materials.

Based on the excellent bonds that can be obtained to most substrates with epoxy resins, epoxy formulations are broadly used as adhesives in electronic-equipment applications. Even when not specifically used as adhesives, epoxies have bonding properties that often provide a better seal around inserts, terminals, and other interfaces than do most other plastic materials.

Phenolics. Phenolics are among the oldest and best-known general-purpose molding materials. They are also among the lowest in cost and the easiest to mold. An extremely large number of phenolic materials are available, based on the many resin and filler combinations. Though phenolics are broadly used for general-purpose insulating uses, it should be noted that many special-purpose electrical grades are now available, including low-loss types.

While it is possible to get various grades of phenolics for different applications, they are, generally speaking, not equivalent to diallyl phthalates and epoxies in resistance to humidity, shrinkage, dimensional stability, and retention of electrical properties in extreme environments. Phenolics are, however, quite adequate for many applications where the electrical requirements are not severe. Further, the glass-filled heat-resistant grades are outstanding in thermal stability up to 400°F and higher.

Polyesters. Polyesters are versatile resins that, from a chemical standpoint, are similar to the alkyds and from a handling and application standpoint are much like the epoxies. They are available in forms ranging from low-viscosity liquids to thick pastes or putties. The liquids are used for embedding applications and laminated products, much like the epoxies, and the pastes are used for molding applications.

The major advantages of polyesters over epoxies are lower cost and lower electrical losses for the best electrical-grade polyesters. Some important disadvantages of polyesters, as compared with epoxies, are lower adhesion to most substrates, higher polymerization shrinkage, a greater tendency to crack during cure or in thermal shock, and greater change of electrical properties in a humid environment.

Silicones. Silicones are thermosetting polymers that can be classified as elastomers, when the cured product is rubberlike; classified as embedding materials, when the basic plastic form is a castable liquid (either rubberlike or rigid); or classified as molding compounds, especially low-pressure transfer-molding compounds for electronic-module embedment. They find wide usage, not only in electronic packaging but for laminated products, molded parts, and adhesives.

The most important properties of silicones for electronic applications are excellent electrical properties, which do not change drastically with temperature or frequency over the safe-operating-temperature range of the material. In addition, silicones are among the best of all plastic materials in resistance to temperature-aging effects. Usable temperatures of 500 to 700°F are possible with available silicone materials. Hence, silicones are well suited for high-temperature electronic applications, especially those applications requiring low electrical losses.

The mechanical properties of silicone molding compounds are not as stable with increasing temperature as are the electrical properties. Most mechanical properties, however, are stable in other extreme environments, such as humidity and vacuum. This has led to the wide use of silicones in both military and space applications.

The following thermoplastic materials, because of their excellent combination of properties, are presently being used to a limited extent in connector designs. These materials are susceptible to stress cracking due to either heat-moisture cycling or exposure to certain common chemical solvents.

Polycarbonates. Polycarbonates have an excellent combination of properties for use in electronic applications. Their basic electrical properties are very good (Table 10) and remain reasonably stable up to about 150°C, as well as in high-humidity environments. The dielectric-constant value changes little up to nearly 10^5 cycles, whereas the power factor or loss factor does increase somewhat in this frequency range.

Mechanically, polycarbonates excel in impact strength, heat resistance under load, dimensional stability, creep resistance, outdoor weatherability, and low-temperature strength. The Izod notched impact strength is approximately four times better than that of nylons or acetals. The useful upper temperature limit of polycarbonates, from the standpoint of their mechanical properties, is 250°F or higher, slightly exceeding that of acetals and exceeding that of nylons by a larger margin. Polycarbonates are also available in glass-reinforced combinations.

Polyphenylene Oxides. The polyphenylene oxides represent another relatively new and important class of plastics that are very useful for electronic applications. They are products of General Electric Company and are known as PPO and Noryl. Polyphenylene oxides have good all-round mechanical and electrical properties, with some of the PPO materials being especially good in their combination of mechanical properties and high-frequency electrical properties. For electronic-application purposes, polyphenylene oxides would compete with acetals, polycarbonates, polysulfones, and nylons (polyamides).

The electrical properties of polyphenylene oxides remain relatively constant with frequency and temperature over the rated temperature range of the material. The electrical properties are also relatively unaffected by humidity. The effects of frequency and temperature on dissipation factor for polyphenylene oxide and other materials are shown in Table 10.

The heat-deflection temperature of the polyphenylene oxides ranks high among the more rigid thermoplastics. Polyphenylene oxides also rate well in tensile strength vs. temperature, tensile modulus, and tensile creep. Water absorption is relatively low, as is the attendant change in dimensions and weight associated with water absorption. Impact strength, however, is not as good as that of polycarbonates at elevated temperatures.

Polysulfones. Polysulfones are relatively new and very useful thermoplastics for electronic-design applications. They have excellent strength vs. temperature properties, good electrical properties (though not outstanding for high frequency), and outstanding strength retention over long periods of aging up to 300°F or over. The electrical properties of polysulfones are maintained to approximately 90 percent of their initial values after exposure at 300°F one year or more. The properties are generally stable up to about 350°F, and under exposure to water or high humidity.

The dissipation factor vs. frequency for polysulfones is shown in Table 10. The dielectric constant of the material is approximately 3.1 up to 10^6 Hz, and decreases slightly at 10^7 Hz.

The heat-deflection temperature of polysulfones rates high among engineering thermoplastics, as does the flexural modulus vs. temperature. The tensile-strength properties of polysulfones are generally similar to those of polycarbonates and polyphenylene oxides, as are the dimensional changes due to absorbed moisture. The retention of strength properties after prolonged heat aging is perhaps the most outstanding feature of polysulfones.

Reversion of connector potting compounds   In 1960, certain users began to experience hydrolytic instability of urethane potting compounds. This instability ultimately resulted in severe reduction in hardness and, in some cases, reversion to a liquid state. Formulation improvements were made by suppliers, but as late as 1967 potted wired assemblies in aircraft stationed in Southeast Asia experienced the same difficulty. The U.S. Naval Avionics Facility, Indianapolis, Ind., developed a new accelerated tropical service life test and published results in 1968.[11] They concluded that the hardness test at 100°C and 95 percent relative humidity is a convenient, reliable, and inexpensive means of screening polyurethanes and polyacrylates for hydrolytic stability. They also indicated that the humidity degradation is permanent and cumulative and thus it is possible that the high-temperature test results may be useful in predicting the service life of these materials in humid environments. A major supplier (Baker Castor Oil Company) reported one year later that they had reversion-resistant polyurethane products which could now meet the navy hydrolytic-stability test.[12] Polysulfides are also widely used for connector potting.

Miscellaneous materials   Connector shells and hoods are mainly made from formed sheet aluminum and die-cast aluminum alloys, although in a few cases, cold-rolled steel and stainless steel are used. Locking devices and cable clamps are made from cold-rolled steel, spring steel, and stainless steel. Guide pins and miscellaneous screws and nuts are made from brass, cold-rolled steel, and stainless steel. Jackscrews are made from cold-rolled and stainless steel.

TABLE 11   Contact Types

Contact type	Applications	Advantages	Disadvantages
Pin and socket:			
Open entry	Low-cost connectors	Low cost	Probe damage
	Tube sockets, test points	Easily available	High insertion force, not suitable with crimp
	Coaxial center contact		
Closed entry	Military/space applications—cylindrical, rack/panel connectors	Best suited for crimp High density	Expensive
	Printed-circuit-board connectors	Lower insertion force	
	Sockets, test points	Reliable	
Blade (metal) and tuning fork	Printed-circuit-board connectors	Low cost Rugged	Limited Applications
	Plug-in modules	Reliable	Nonenvironmental Single readout
Bellows	Printed-circuit-board connectors	Low cost Low insertion force Available in single and double readout Reliable	Probe damage Humidity accumulation in connector cavity
Formed wire	Plug-in module	Extreme contact density	Expensive
	Strip cable connectors		Delicate
	Small cable connectors	Light weight	Unproved reliability
Taper pin	Terminal blocks	Low cost	Limited military/
	Laminated bus bars	Fast, easy to mate/	space application
	Commercial appliances	unmate	

Protective coatings used are anodizing, chromate, tin and cadmium plating, and paint. If a bright silverlike finish is desired for appearance, a clear chromate finish over cadmium is used.

CONTACTS

Contact types Contacts are available in many shapes and sizes. Machining or stamping processes are used to form the various shapes of contacts. Table 11 lists the widely specified contact types. The most versatile and widely used contact is the round pin and socket. It has a history of proved reliability and is readily available in all practical sizes. The closed-entry design of the pin and socket is preferred for use in removable-contact connectors, since it is most resistant to damage. The bellows form of contact is preferred for printed-circuit-board connectors.

Contact-termination methods The four basic methods of terminating connector contacts to wire leads are:
1. Solder
2. Crimp
3. Weld
4. Solderless wrap

Most of these are then subdivided into specialized techniques. Table 12 provides a comparison of standard and newly developed contact-termination techniques. Figure 11 shows common types of contact terminals to accommodate these techniques.

TABLE 12   Comparison of Contact-Termination Techniques

Method	Principal advantage	Features of connection	Skill level	Requirements
Resistance soldering	Even heating of joint; low heat radiation; power source can also be used for stripping	Large contact area	High	Conductors must be cleaned and fluxed; equipment moderately sophisticated
Heat-conduction soldering	Simplicity of equipment; applicable to most interconnections	Large contact area	High	Conductors must be cleaned and fluxed
Resistance welding	Low heat radiation; speed in making connections	Small size; high tensile strength	High	Sophisticated equipment
Solderless wrapping	No flux or cleaning required; connections readily unwrapped	Large contact area	Minimum	Tool space needed around connection
Crimping........	No heat or flux; no subsequent cleaning of connection	High mechanical strength	Minimum	Tool space needed around connection
Electron-beam welding	Negligible heat radiation; microminiature applications	Small size; high tensile strength; high-density connection—unit area	Very high	Complex equipment; vacuum or gas atmosphere
Flow soldering....	Simultaneous connection of many joints	Large contact area	High	Critical heat control
Percussive arc welding	Joining of very thin materials; little thermal distortion	Very high tensile strength	Very high	Sophisticated equipment

Contact-retention methods[13]

Basic Retention Systems. Although there are many different trade names for crimp-type insertable/removable contacts, there are only a very few basic design concepts for the contact-retention systems.

For printed-circuit edge-connector and some blade-type contacts, there is a single basic method of retaining the contact in the connector molding. The body of the contacts contains a simple protruding tine that is compressed as the contact is inserted into the molding and then extends out again and is retained by a shoulder inside the molding. The contacts are inserted by simply pushing them into the molding by hand pressure or with the aid of a simple push-type tool.

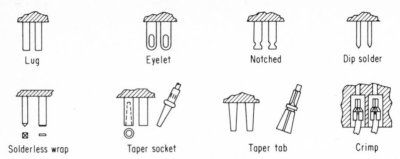

Fig. 14   Common contact-terminal types.

Removal is accomplished by the use of a blade-type tool that is pushed into the contact cavity to depress the tine, thus allowing the contact to be pushed or pulled from the molding (Fig. 15a).

For pin-and-socket contacts there are two basic retention methods:

1. A cyclindrical collar or clip containing one or more protruding tines is fabricated to the basic contact, or in some cases, the tines are fabricated as part of the contact, especially in the case of formed contacts.

The contacts are inserted into a restrictive cavity hole in the molding, and when they are in position, the tines expand over a retaining shoulder which locks the contact into position.

Insertion is accomplished either by hand pressure or with the aid of an insertion tool that applies pressure against the contact, thus forcing it into position in the molding.

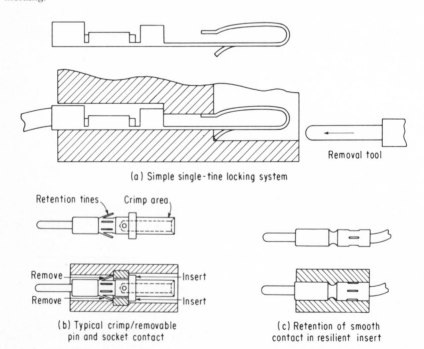

Fig. 15   Various contact designs.   Note especially (b) showing contact inserted and locked in connector body.   Arrows on the right indicate where pressure is applied by insertion tool.   Arrows on the left indicate where hollow cylindrical removal tool is inserted to compress retention tines, thus allowing contact to be removed.

Removal is accomplished by the use of a hollow cylindrical tool that fits over the outside diameter of the contact and, with pressure, compresses the tines, thus allowing the contact to be removed from the cavity in the molding. Some types of tools simply compress the tines, thus allowing the contact to be pulled from the molding. Other types have a plunger, either manually operated or spring-loaded, that exerts force on the contact after the tines have been compressed, thus expelling the contact from the molding (Fig. 15b).

NOTE: Some connectors merely have molded retaining shoulders in the connector inserts or moldings, while others have a molded-in insert, usually of stainless steel. The molded-in insert is more durable than the plastic molded shoulder and allows for many more insertions and removals without degradation of the retaining shoulder.

2. The contact has a circumferential groove that retains the contact when it is pushed into a resilient insert. Simple push-type insertion and removal tools are required to insert and remove this type of contact (Fig. 15c).

Other Contact-Retention Systems. A unique retention system is that of the patented REMI° contact. In this concept, retention sleeves in the connector molding are employed to receive the contacts, which are snapped into the sleeve rather than being retained by the shape of the molded contact cavity. During insertion

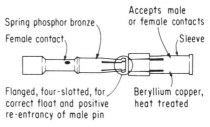

Fig. 16   REMI contact concept.[13]

or removal, all mechanical stresses are confined between metallic elements rather than between metal and plastic (Fig. 16). The contacts snap in with finger pressure and are removed with a tool provided by the manufacturer. An additional feature of this concept is that either male or female contacts can be inserted into the same sleeve. The retention sleeve is so designed that it provides the closed-entry feature.

Another unique contact-retention system is one in which a single-piece molded wafer of resilient material (Lexan,† nylon, or polyarylsulfone) contains integrally molded, cone-shaped contact retainers which lock behind the rear shoulder of the contact. The front of the shoulder butts against a barrel insulator, thus locking the contact into the connector body (Fig. 17). The contacts are inserted or removed

° Trademark of U.S. Components, Inc.
† Trademark of General Electric Company.

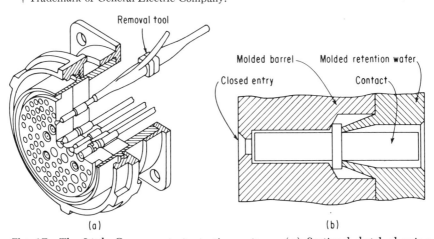

Fig. 17   The Little Caesar contact-retention system.   (a) Sectional sketch showing rear-release technique.   (b) Detail of barrel, retention-wafer cone, and female contact.   (*ITT Cannon Electric, Div. of ITT Corp.*)

with a simple, inexpensive plastic tool provided by the manufacturer. Contacts are inserted from the rear of the connector and held by cone-shaped retainers. Removal is accomplished by use of a simple plastic removal tool that fits around the terminal end of the contact, expanding the retaining cone in the insert and thus allowing the contact to be removed.

ENVIRONMENTAL FACTORS

The environmental conditions which most adversely affect the reliability of connectors are moisture, altitude, and temperature. Sand, dust, oil, and hydraulic-fluid contamination can also contribute to performance deterioration. For internally mounted connectors the two major environmental conditions are temperature and altitude.

Temperature Temperature failures constitute a principal source of environmental failures. At high temperatures metals lose strength, insulation weakens, dimensions and fits change, finishes are destroyed, and seals crack. At low temperatures the effects are similar but the deterioration due to ice must be included. The maximum current rating of a multicontact connector should be understood by design engineers, as the current flow in a connector directly affects the useful life of the connector. When an optimum connector selection must be made, the following should be considered:

1. Does the maximum current pass through all the contacts at once or only through some of them?

2. If the maximum current is not passing through all the contacts, how much can the rating be increased when only one or two of the contacts are fully loaded?

3. How much allowance must be made for temperature extremes and altitude?

4. What is the temperature effect when the wire size is varied?

The manner in which the operating temperature within the connector affects the operating life is shown in Fig. 18. For example, at room temperature with a current

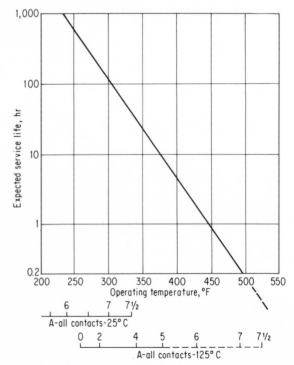

Fig. 18 Expected service life vs. operating temperature (from MIL-C-26482).

of 7 A through all contacts, the operating temperature is about 300°F and the connector will last just over 100 h.   At 125°C with a current of 5 A the connector will last about 9 h.   In rating a connector for current-carrying capacity, the parameter that must be controlled is the temperature of the insert.   If a horizontal line is drawn across Fig. 19 at any desired insert temperature, a new curve can be plotted

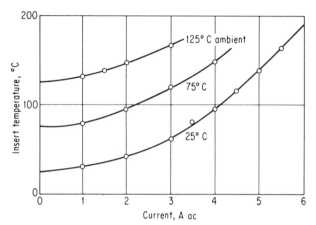

Fig. 19 Insert temperature vs. current for various ambient temperatures.   (*Microdot, Inc.*)

relating the allowable ambient temperature to the current for the chosen insert temperature.   Such a plot is shown in Fig. 20*a* for derating for ambient temperatures.

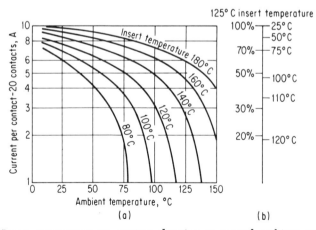

Fig. 20 Insert temperature vs. contact derating at several ambient temperatures. (*Microdot, Inc.*)

Figure 20*b* shows the relationship between insert temperature and the permissible percent of maximum current.

Effects of altitude
   Cooling. At lower altitudes much of the cooling of the connector is done by convection currents, both inside and outside the connector.   As the density of cooling air is decreased through an increase in altitude, the effectiveness of the cooling dimin-

ishes. Consequently, it is necessary to reduce the current rating of the connector as the operating altitude increases.

Figure 21 shows how the current required to produce a given insert temperature decreases at various altitudes. The curves indicate that the percentage derating for a given altitude is independent of the exact insert temperature used.

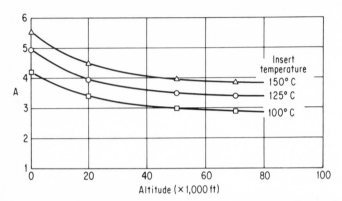

Fig. 21   Effect of altitude on insert temperature.   (*Microdot, Inc.*)

Cooling by convection at 100,000 ft and other high altitudes is negligible. The 100,000-ft level can be used for all greater altitudes. At the altitude at which the mean free path of the air molecules begins to be an appreciable fraction of the dimensions of the air space in the connector, a different mechanism of heat transfer occurs. The performance of the connector at high altitudes is similar to its performance when it is in a confined space or partially surrounded with insulating material that prevents the natural movement of cooling air. Connectors used under such circumstances can be rated as if they were to operate at 100,000-ft altitude.

Corona.   Corona is a leakage discharge distinct from flashover. It occurs at low atmospheric density. If the connector is environment-resistant and potted, this effect moves out to the cable or wire bundle. Figure 22 shows corona starting volt-

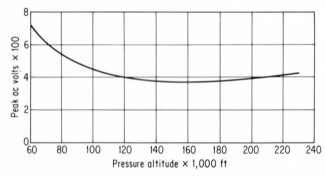

Fig. 22   Corona starting voltage vs. altitude.[14]   (*Bendix Corp., Electrical Components Division.*)

age vs. altitude for a miniature connector. Consideration should be given to additional derating that may be necessary because of transients or surges in specific applications.

Breakdown Voltage.   Figure 23 shows breakdown voltage vs. altitude for two-pin spacings commonly used in miniature connectors.

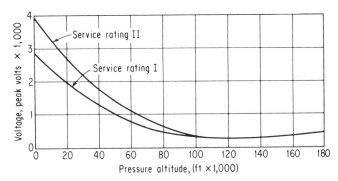

Fig. 23 Flashover voltage vs. altitude for miniature connectors.[14] (*Bendix Corp.,*
Electrical Components Division.)

RECTANGULAR CONNECTORS

Rectangular connectors generally fall into two generic types: rack and panel, and
plug and receptacle. A brief discussion of each was provided earlier in this chapter
(see connection level 3).

Available types of rectangular connectors[13] An almost endless variety of
types, sizes, and variations of rack-and-panel connectors is available today. A simple
list of them all would require a whole book, rather than just a chapter, and would
be of little help to the designer. What was considered a miniature connector yester-
day has been largely replaced by the subminiature connector of today—which, in
all probability, will be replaced by the microminiature connector of tomorrow. Contact density, contact numbers, and overall connector size are all relative to the state
of the art from year to year, and in a handbook such as this it would be almost
meaningless to attempt to catalog all the presently available connectors in all the
multitude of types, styles, sizes, pin numbers, and configurations, etc., because new
specific designs and constantly changing equipment requirements would rapidly make
many of today's connectors obsolete in favor of tomorrow's required concepts.

Many of the various miniature rack-and-panel types are available already as subminiature connectors. If size and weight are significant factors, these subminiatures
can be very useful. However, it is well to keep in mind that they are more fragile
and cannot stand rough handling. Installation, wiring, and maintenance all require
careful training of personnel, as with other subminiature component parts.

As a typical comparison of the size reduction that can be achieved in these subminiature designs, Fig. 24a shows the dimensional outline and pin spacing on a miniature
50-pin connector and Fig. 24b shows the comparable dimensions on a subminiature
with a like number of pins.

Typically, the breakdown voltage of the connector in Fig. 24a would be about
2,800 V ac rms with a contact rating of 7.5 A. The connector in Fig. 24b would
have a breakdown voltage of about 1,800 ac rms with a contact rating of only
3 A.

Examples of typical rectangular connectors A wide variety of standard, miniature, and subminiature rack-and-panel connectors are available—from rugged heavy-
duty types to very-high-density, light-duty types. Associated hardware, such as
hoods, cable clamps, locking devices, and various mounting methods, are available
for most connectors. All of them can be used as true rack-and-panel connectors,
or as cable-to-receptacle or cable-to-cable connectors. Contact ratings are dependent
on contact size (see Table 13).

Heavy-Duty Rectangular (*Fig. 25*). Illustrated are several examples of typical
rugged, heavy-duty connectors especially suitable for heavy sliding-drawer applications. Terminals are available for solder and for taper pins. Crimp/removable contacts are also available as standard parts.

Miniature Rectangular (*Fig. 26*). This family is one of the most widely used of
the rack-and-panel connectors. It is readily available as a plain rack-and-panel con-

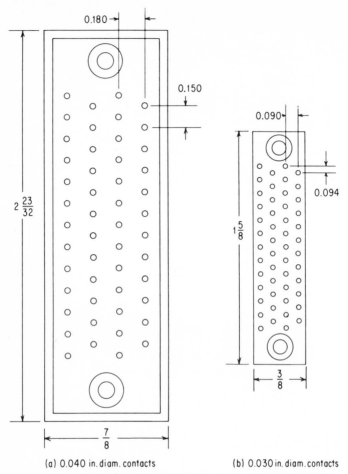

(a) 0.040 in. diam. contacts   (b) 0.030 in. diam. contacts

Fig. 24 Dimensional comparison of miniature and subminiature rectangular connectors.

nector with polarizing guide pins (Fig. 26a); as a cable-to-receptacle connector with hood and cable clamp, either utilizing plain guide pins or with jackscrews for ease of mating and locking (Fig. 26b); and with the male half as a hermetic-seal connector for bulkhead mounting (Fig. 26c).

Solder and taper terminals as well as crimp/removable contacts are all standard.

Miniature Rectangular with Center Jackscrew (Figs. 27 and 28). Various shapes, from square to rectangular to long and narrow, are available, with a great variety of contact sizes and numbers.

Environment-resistant (Fig. 29). Several shapes, sizes, and contact arrangements are available. Sealing is accomplished by interfacial contact of resilient insert moldings in one or both connector halves; by a pressurized seal around each contact and a grommet seal around the wires at the termination end; and by a lip-seal barrier between the mated shells. Crimp/removable contacts with a closed-entry feature are available.

Rack-and-Panel Coaxial (Fig. 30). Various of the rack-and-panel shell types are available with contacts for standard and miniature coaxial and shielded cable. The individual coaxial connectors can be isolated by the use of an insulating insert, or the shields can all be brought to a common potential by the use of metallic inserts.

TABLE 13 Mechanical and Electrical Characteristics of Contacts

Connector specification	Applicable contact specification	Contact plating	Contact size	Accept wire gage (AWG)	Contact-terminal type	Contact rating, A†
MIL-3-5015		Gold over silver	16		Solder*	22
			12			41
			8			73
			4			135
MIL-C-22992		Gold	0			245
MIL-C-26482		Gold over silver	20		Solder	7.5
			16			13.0
			12			
	MIL-C-23216	Gold over silver	20	24	Crimp	3.0
				22		
				20		7.5
			16	20		7.5
				18		
				16		13.0
			12	14		17.0
				12		23.0
MIL-C-26500	MIL-C-26636	Rhodium over silver or nickel	20	24	Crimp	3.0
				22		5.0
				20		7.5
			16	20		7.5
				18		16.0
				16		22.0
			12	14		32.0
				12		41.0
NAS 1599	NAS 1600	Gold or silver or nickel	20	24	Crimp	3.0
				22		5.0
				20		7.5
			16	20		7.5
				18		15.0
				16		20.0
			12	14		25.0
				12		35.0
MIL-C-25955		Gold over silver	20	30	Crimp	
				24		
				22		
				20		7.5
MIL-C-27599		Gold over silver	16		Solder	13.0
			20			7.5‡
			22			5.0
			22M			3.0
MIL-C-38999		Gold over silver	16	16	Crimp	13.0
				18		10.0
				20		7.5
			20	20		7.5
				22		5.0
				24		3.0
			22	22		5.0
				24		3.0
				26		2.0
			22M	24		3.0
				26		2.0
				28		1.5

* Available with crimp/removable contacts under manufacturer's part numbers.
† Maximum contact rating for individual contact.
‡ Hermetic contacts rated at 5.0 and 10.0 A, respectively.

Fig. 25 Typical heavy-duty connectors. (*a*) Polarization accomplished by cut-corner shape of metal shells. (*b*) Polarization accomplished by contact arrangement. (*c*) Two inserts in single housing. Center screw for ease of mating and positive locking.

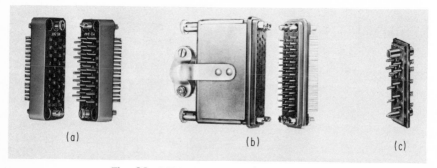

Fig. 26 Miniature rectangular connectors.

General-Purpose Rectangular (*Fig. 31*). This is one of the oldest of the rack-and-panel connectors, but it is still generally useful. It is available with several combinations of No. 12, No. 16, and No. 20 contacts, which makes it suitable for both power and signal circuits in a single connector block.

Contacts are available molded in, with solder terminations, or removable, with crimp termination. Hoods and cable clamps are provided.

Fig. 27 Miniature rectangular connector with center jackscrew.

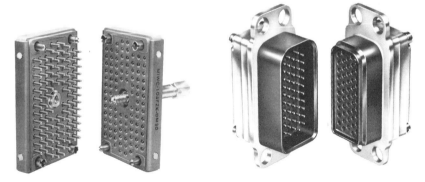

Fig. 28 Conventional rectangular connector with center jackscrew.

Fig. 29 Typical environment-resistant rack-and-panel connector.

Fig. 30 Rack-and-panel coaxial connectors.

Fig. 31 General-purpose rectangular connectors.

Jones Connectors (*Fig. 32a*). This family of connectors has been an industry standard for over a generation. They are designed for medium- or heavy-duty power applications, or for any application where a very rugged connector is required.

The male contacts are blade-type and the female contacts are "knife-switch" type that provides a solid contacting area on both sides of the male contact. They are available in from 2 to 33 contacts with solder teminations, and are related at from 4.5 to 15 A, depending on the series.

Fig. 32 (*a*) Jones connector. (*b*) Mini-Jones connector.

Mini-Jones Connectors (*Fig. 32b*). "Mini-Jones" connectors have the same contact types as the standard Jones connectors but are somewhat smaller and feature crimp-on, snap-in contacts that have a rating of 4.0 A. They are available in from 19 to 68 contacts.

In both the Jones and Mini-Jones families, male and female contacts are available in either the plugs or the receptacles. Associated hardware includes hoods, straight-locking devices, and surface and flush-mounting receptacles.

D Subminiature-Series Connectors (*Fig. 33*). These are available in arrangements of from 9 to 50 contacts in size 20, and also with various combinations of coaxial and shielded connector contacts. The associated hardware include floating mountings, locking devices, hoods, and cable clamps.

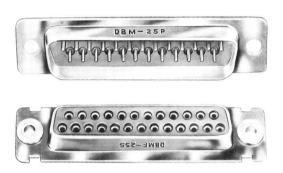

Fig. 33 D subminiature-series connector.

Blue Ribbon* Connectors (*Fig. 34*). This family of connectors is an extremely rugged, nonenvironmental, general-purpose connector featuring easy insertion and extraction. Contacts are ribbon-type with generous contacting surfaces that are self-winding and self-cleaning.

A variety of configurations and styles are available, including the barrier-polarization type without shells (Fig. 34*a*); the pin-polarization type without shells; the

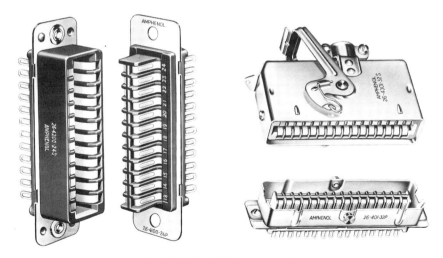

Fig. 34 Blue Ribbon connectors.

barrier-polarization type with latching-type keyed shells (Fig. 34*b*); and the barrier-polarization type with plain keyed shells. Straight-through and right-angle hoods with cable clamps are available, as are floating mountings for panel-mounted receptacles. The hardware is nonmagnetic, and they are available with 8, 16, 24, and 32 five-ampere contacts.

* Trademark of Amphenol-Borg Electronics Corporation.

Micro-Ribbon* Connector (Fig. 35). This family of connectors is nonenvironmental. Contacts are ribbon-type with solder terminations. The trapezoidal shape of the metal shells polarizes the connector halves. It features smooth, easy insertion and withdrawal, with self-wiping, self-cleaning contacts. It is used extensively in

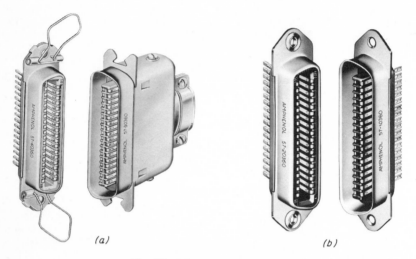

(a)   (b)

Fig. 35 Micro-Ribbon connectors.

telephone communications equipment. It is available in cable-to-chassis style (Fig. 35a), cable-to-cable style, and as a plain rack-and-panel connector (Fig. 35b). Hoods, latching devices, and floating mountains are available. The hardware is nonmagnetic, and it is available with 14, 24, 36, and 50 five-ampere contacts.

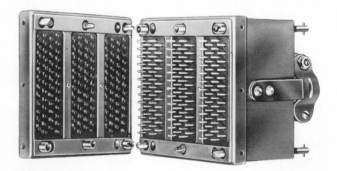

Fig. 36 Miniature rectangular connector with floating molded inserts.

Miniature Rectangular with Floating Molded Inserts (Fig. 36). This family features rugged die-cast frames in which the molded inserts containing the contacts are mounted in such a manner as to allow relative float. Mechanical stresses of mounting, engaging, and disengaging are carried by the frame rather than by the molded inserts. Guide pins align the contacts, and jackscrews provide force for engaging, disengaging, and locking.

* Trademark of Amphenol-Borg Electronics Corporation.

High-Contact-Density Miniature Rectangular (Fig. 37). The connector illustrated is typical of a whole new family of miniature rectangular connectors now available. Contact spacing is 0.100 in or less. The contacts are available in crimp/removable and solder-termination types, and also with dip-solder tails either straight out or bent at right angles. The bodies are molded of high-impact-strength plastics.

Other Molded Rack-and-Panel Configurations (Fig. 38). A variety of connector types and sizes are now being completely molded from high-impact-strength plastic materials. They contain no metallic parts, except for the contacts. Mounting ears, guide pins, polarizing devices, and shells are all a part of the basic insert molding.

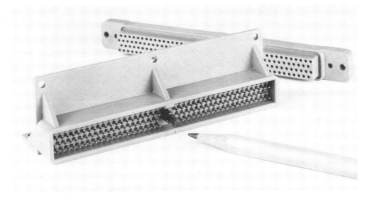

Fig. 37   High-contact-density miniature rectangular connector.

Functionally, similar connectors manufactured by different manufacturers are usually quite similar, but there are some design differences in hoods, shells, clamps, etc. Although it is possible to mate connector halves of similar types that have been produced by two different manufacturers, especially those manufactured to military specifications, it is generally not considered good practice because of sometimes incompatible manufacturing tolerances among manufacturers, and in the advent of problem areas there is no single source of responsibility.

Fig. 38   Typical examples of completely molded rectangular and cylindrical connectors.

With the exception of the ribbon-contact connectors shown, all the other types and families illustrated are available with either factory-assembled contacts or crimp/removable contacts. The terminations available vary from manufacturer to manufacturer, but generally speaking, the terminal styles available include solder, wire-wrap, taper-pin, Termi-point, and dip-solder.

One of the interesting features of many of today's rectangular rack-and-panel connectors employing removable contacts is the ability to intermix instantly various sizes of pin-and-socket contacts as well as miniature and subminiature coaxial contacts within the same connector block (Fig. 39). This is a practical and economical way of using a single connector to accommodate voltage, logic, and signal circuits.

Fig. 39 Mix or match contacts.   (a) Stranded wire crimped to machined contacts. (b) Stranded wire crimped to sheet-metal formed contacts.   (c) Shielded wire in miniature coaxial contacts.   (d) Twisted-pair in miniature coaxial contact.

Another advantage is that in development and prototype work it is extremely helpful to have the ability to change from single-wire leads to twisted pairs or coaxial cable if noise is a problem, as well as to accommodate circuit changes or additions conveniently.

CYLINDRICAL CONNECTORS

General description of AN-MS connectors[13]   Each complete AN-MS connector consists of two parts, a plug assembly and a receptacle assembly coupled by means of a coupling device which is part of the plug assembly.   Standard AN-type connectors are coupled with a threaded coupling ring except for MS 3107, which has a friction coupling. Miniature MS connectors, a smaller lightweight version of the AN-type connector, are coupled by means of a threaded ring, bayonet lock, or push-pull coupling (see Fig. 40).

The receptacle is usually the "fixed" part of the connector, and is attached to a wall, bulkhead, or equipment case. The plug is the removable part of the connector and includes the coupling ring. When the two parts are joined by the coupling device, the electric circuit is made by pin-and-socket contacts inside the connector. The "live" or "hot" side of the circuit usually has socket (female) contacts. Either the plug or the receptacle may contain the live parts of the circuit. The contacts are held in place and insulated from each other and from the shell by a dielectric insert.   Inserts and contacts are housed in a metal shell.

Cylindrical-connector types per MIL Specification MIL-C-5015   AN-MS connectors are separated into types and classes, with manufacturer's variations in each type and class. These variations are in the method of meeting specification requirements, and in appearance. The variations are minor and do not affect the ability to mate plugs and receptacles made by different manufacturers. There are six AN types of standard MS connectors as listed in Table 14. The connectors are further separated into seven classes.

AN connector types   The six types of standard AN-type connectors are described as follows:

MS 3100.   A receptacle with flange for mounting to wall or bulkhead. Contains front shell, insert-retaining ring, insert, contacts, and black shell. Connectors with resilient inserts omit the insert-retaining ring.

MS 3101.   A plug used at the end of a wire or wire bundle where mounting is not necessary. Similar to MS 3100 except that it has no mounting flange.

MS 3102.   A receptacle with flange for mounting to a junction box or equipment case. Similar to MS 3100, except that it has no back shell.

MS 3106.   A straight plug, used at the end of a wire or bundle. Consists of front shell, coupling nut, insert-retaining ring, insert, contacts, and back shell. Connectors with resilient inserts omit the retaining ring.

WALL MOUNTING RECEPTACLES

MS 3100 E
Environmental
Resistant

MS 3100 A
Solid Shell

MS 3100 B
Split Shell

MS 3100 C
Pressurized

CABLE RECEPTACLES

MS 3101 E
Environmental
Resistant

MS 3101 A
Solid Shell

97 3101 B
Split Shell

BOX MOUNTING RECEPTACLES

MS 3102 E
Environmental
Resistant

MS 3102 A
Solid Shell

MS 3102 C
Pressurized

QUICK DISCONNECT PLUGS

MS 3107 A
Solid Shell

MS 3107 B
Split Shell

STRAIGHT PLUGS

MS 3106 E
Environmental
Resistant

MS 3106 A
Solid Shell

MS 3106 B
Split Shell

ANGLE PLUGS

MS 3108 A
Solid Shell

MS 3108 B
Split Shell

Fig. 40 Typical cylindrical-connector shell types.

MS 3107. A "quick-disconnect" plug, used where fast-pull disconnection from the receptacle is necessary. It is similar to MS 3106 except that it is coupled to an MS receptacle by means of a friction ring instead of a coupling nut.

MS 3108. A right-angle plug, used where wiring must make an abrupt change in direction as it leaves the plug.

AN-MS connector classes There are seven classes of AN-MS connectors. All have aluminum-alloy shells except class K, which has a steel shell to achieve fire resistance.

Class A. General-purpose connector with solid one-piece back shell.

Class B. Connector with back shell split in two, lengthwise, used where it is important to be able to get at soldered connections easily. The two halves of the

TABLE 14   Classes of AN-Type Connectors—MIL-C-5015

MS class	Application	Shell	Availability					
			3100	3101	3102	3106	3107	3108
A	General-purpose	Solid aluminum alloy	Yes	Yes	Yes	Yes	Yes	Yes
B	General-purpose	Split aluminum alloy	Yes	Yes	No	Yes	Yes	Yes
C	Pressurized-receptacle	Solid aluminum alloy	Yes	No	Yes	No	No	No
E	Environment-resistant	Solid aluminum alloy with strain-relief clamp	Yes	No	Yes	Yes	No	Yes
K	Fire- and flame-resistant	Solid steel	Yes	Yes	Yes	Yes	No	Yes
R	Environment-resistant	Solid aluminum alloy						

back shell are held together by clamping ring or by screws.   See Fig. 41 for exploded views of typical split-shell MS connectors.

Class C.   Pressurized connector, used on walls and bulkheads of pressurized equipment.   Externally it looks the same as a class A receptacle, but the inside sealing arrangement is different.   Inserts of class C connectors are not removable.   Mating the class C receptacle to the other class plugs does not affect the sealing qualities of the class C receptacle.

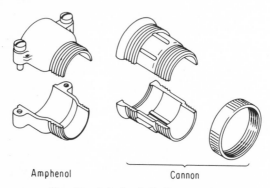

Amphenol   Cannon

Fig. 41   Split-back-shell connector.

Class E.   Environment- (moisture- and vibration-) resisting connector, used in areas where changes in temperature may cause condensation, or where there is likely to be vibration.   Class E connectors have a sealing grommet in the back shell.   The wires pass through tight-fitting holes in the grommet and are thereby sealed against moisture.   The contacts are supported in a resilient insert.   For proper performance class E receptacles should be mated to class E plugs.

Class K.   Fireproof connector, used where it is vital that current continue to flow even though the connector may be exposed to continuous open flame.   Class K connectors are longer in overall length than other classes, and have a shell made of steel instead of aluminum alloy.   Inserts of class K connectors are of special fire-resistant material, and have crimp-type contacts instead of solder-type.

Class R.   Lightweight environment-resisting connector similar to the class E connector, and intended to replace it where shorter length and lighter weight are required.   An O-ring is provided in the MS 3106 and MS 3108 plugs for additional sealing.

Class RC. Environment-resisting; identical to class R except connector is supplied with strain-relief clamp.

Contacts See Table 13. Thermocouple contacts are available in various material combinations. The number of contacts ranges from 1 to 104, and they can be had in all the same size or in combinations of sizes. Typical examples are shown in Fig. 42.

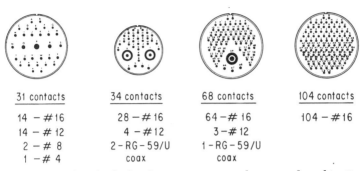

31 contacts	34 contacts	68 contacts	104 contacts
14 – #16	28 – #16	64 – #16	104 – #16
14 – #12	4 – #12	3 – #12	
2 – #8	2 – RG-59/U	1 – RG-59/U	
1 – #4	coax	coax	

Fig. 42 Examples of cylindrical-connector contact layouts and combinations.

Materials The shells and coupling rings are aluminum alloys. The finish is cadmium plate, dull olive drab in color, and is electrically conductive. Inserts are hard plastic or resilient materials or by the use of O-rings when hard inserts are used. Terminations are sealed by potting or by grommet seals.

Shell styles Many plug-and-receptacle shell types are available to suit a variety of physical and environmental application requirements. Figure 40 illustrates the shell styles that are available for this series connector, that are typical of the shell styles presently available in many of the various families of cylindrical connectors, and that will probably be available eventually for some of the newer MIL-specification types.

Miscellaneous hardware Figure 43 illustrates the miscellaneous hardware available for many cylindrical-connector families, dust caps (metal and plastic), conduit fittings and ferrules, reducing adapters, cable clamps, and telescoping bushings (to adapt wires to cable clamps). Table 15 indicates how to select MIL-C-5015 connectors.

MIL-C-26482 and MIL-C-26500 miniature MS connectors Two military specifications cover the miniature MS connectors most commonly used in aircraft: MIL-C-26482 and MIL-C-26500. Connectors manufactured to the requirements of MIL-C-26482 may have contacts of either the conventional solder type or crimp type; MIL-C-26500 connectors have crimp-type contacts only. Connectors to both specifications have contacts in sizes 20, 16, and 12 only. The types and classes of miniature MS connectors with solder-type contacts are listed in Table 16. Miniature MS connectors with crimp-type contacts are available in the types listed in Table 17, and are of the environment-resisting classes.

Miniature MS connector types—solder contacts The following types of miniature MS connectors with solder contacts are used in aircraft and ground equipment.

MS 3100. A receptacle with flange for mounting to a wall or bulkhead; is coupled by means of a bayonet lock.

MS 3112. A receptacle for mounting to junction box or equipment case similar to MS 3110 except that it has no back shell; has bayonet-lock coupling.

MS 3114. A rear-mounting receptacle, with jam nut instead of flange; has bayonet-lock coupling.

MS 3116. A straight plug for use at end of wire or wire bundle; has bayonet-lock coupling.

MS 3119. A through-bulkhead mounting receptacle; has bayonet-lock coupling.

MS 3130. A receptacle similar to MS 3110, except that it has a push-pull (ball-lock) coupling.

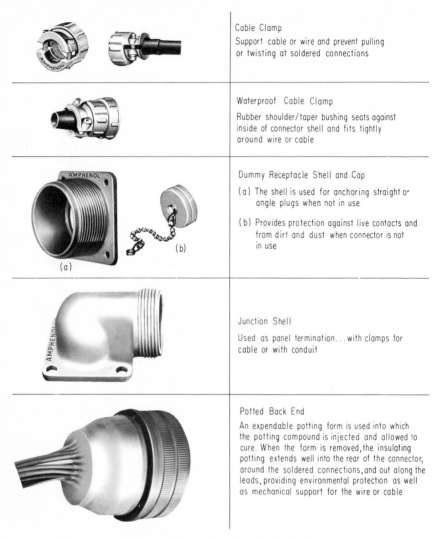

Cable Clamp

Support cable or wire and prevent pulling or twisting at soldered connections

Waterproof Cable Clamp

Rubber shoulder/taper bushing seats against inside of connector shell and fits tightly around wire or cable

Dummy Receptacle Shell and Cap

(a) The shell is used for anchoring straight or angle plugs when not in use

(b) Provides protection against live contacts and from dirt and dust when connector is not in use

Junction Shell

Used as panel termination...with clamps for cable or with conduit

Potted Back End

An expendable potting form is used into which the potting compound is injected and allowed to cure. When the form is removed, the insulating potting extends well into the rear of the connector, around the soldered connections, and out along the leads, providing environmental protection as well as mechanical support for the wire or cable

Fig. 43 Typical miscellaneous hardware for cylindrical connectors.

MS 3134. A single-hole mounting receptacle, similar to MS 3114, except that it has push-pull coupling.

MS 3137. A straight plug similar to MS 3116 except that it has push-pull coupling.

MS 3138. A plug with lanyard; has push-pull coupling.

MS 3139. A through-bulkhead mounting receptacle, similar to MS 3119 except that it has push-pull coupling.

Miniature MS connector types—crimp contacts These connectors are similar to the miniature solder-type connectors but have removable contacts to which wires are crimped with a standard crimping tool, instead of soldered. Connectors with crimp-type contacts are available in the following types:

MS 3120. A receptacle with flange for mounting to a wall or bulkhead; is coupled by means of a bayonet lock.

TABLE 15   How to Select MIL-C-5015 Connectors (According to Cataloged Information)

```
97   3101  A  20—27  S      (432)
CP   3102  E  36—10  P   X
CA   3108  B  28—21  S      (115)
MS   3106  E  28—21  P   W
```

Prefix ──┘   │   │   │   └──── Modification code
Shell style ─────────┘   │   └──────── Polarization
Service class ──────────────┘   └──────── Contact arrangement
Shell size ──────────────────┘   └──────── Insert number

	Prefixes
AN	Replaced by MS
CA	Cannon
CT	Cannon E series
FC	Flight connector
GP	Bendix gold contacts
MS	Military Standard
SC	Bendix A type
SF	Bendix modification to E type
SG	Bendix modification to E type
SB	Bendix modification to E type
97	Amphenol

Shell style

3100	Wall receptacle
3101	Cable receptacle
3102	Box receptacle
3103	Wall receptacle for potting
3106	Straight plug
3107	Quick disconnect
3108	90° angle plug
25183	Straight plug for potting
25183A	Straight plug for potting; with ground plug

Steps to determine part number
1. Select number of contacts
2. Select contact type and sizes
3. Select shell type and size
4. Select alternate position for inserts (polarization)
5. Select service class
6. Add proper prefix

Service class

A	Solid shell
AF	E type with threaded end bell and special coupling nut
B	Split shell
C	Pressurized
E	Environmental
ER	Cannon potted type
ES	Cannon potted type
F	E type with threaded end bell
K	Firewall
M	Replaced by E
P	Bendix potted type
PR	Replaced by C
R	Environmental with O-ring under coupling nut

Shell size
Outside diameter of mating portion of receptacle in $\frac{1}{16}$-in. increments

Insert number
Obtained from illustrated layouts in manufacturer's catalogs

Contact arrangement
P for pin (male); S for socket (female)

Alternate positions
Rotational position of insert in shell in relation to key and keyway; used to obtain polarization

Modification code
Used for changes in plating of shells or contacts, different insulation material, different types of contacts, etc.

TABLE 16   Types and Classes of MIL-C-26482 Miniature MS Connectors with Solder Contacts

MS type	Nomenclature	Availability				
		Class E	Class F	Class P	Class H	Class J
Bayonet coupling:						
MS 3110........	Wall-mounting receptacle	Yes	Yes	Yes	No	Yes
MS 3112........	Box-mounting receptacle	Yes	No	Yes	Yes	No
MS 3114........	Rear-mounting jam-nut receptacle	Yes	No	Yes	Yes	No
MS 3116........	Straight plug	Yes	Yes	Yes	No	Yes
MS 3119........	Through-bulkhead mounting receptacle	Yes	No	Yes	No	No
Push-pull coupling:						
MS 3130........	Wall-mounting receptacle	Yes	No	Yes	No	Yes
MS 3132........	Box-mounting receptacle	Yes	No	No	Yes	No
MS 3134........	Single-hole mounting receptacle	Yes	No	Yes	Yes	Yes
MS 3137........	Short plug	Yes	No	Yes	No	Yes
MS 3138........	Lanyard plug	Yes	No	Yes	No	Yes
MS 3139........	Through-bulkhead mounting receptacle	Yes	No	No	No	No

TABLE 17   Types of Miniature MS Connectors with Crimp Contacts

MS type	Nomenclature
1. MIL-C-26482	
Bayonet coupling:	
MS 3120..........	Wall-mounting receptacle
MS 3122..........	Box-mounting receptacle
MS 3124..........	Rear-mounting jam-nut receptacle
MS 3126..........	Straight plug
Push-pull coupling:	
MS 3140..........	Wall-mounting receptacle
MS 3144..........	Single-hole mounting receptacle
MS 3147..........	Plug
MS 3148..........	Lanyard plug
2. MIL-C-26500	
MS 24264.........	Flange-mounting receptacle
MS 24265.........	Single-hole mounting receptacle
MS 24266..........	Straight plug

MS 3122.   A receptacle for mounting to junction box or equipment case similar to MS 3120 except that it has no back shell; bayonet-lock coupling.

MS 3124.   A rear-mounting receptacle, with jam nut instead of flange; bayonet-lock coupling.

MS 3126.   A straight plug for use at end of wire or wire bundle; bayonet-lock coupling.

MS 3140.   A flange-mounting receptacle, similar to MS 3120 except that it has push-pull coupling.

MS 3144.   A single-hole-mounting receptacle, similar to MS 3124 except that it has push-pull coupling.

MS 3147.   A plug for use at end of wire or wire bundle; push-pull coupling.

MS 3148.   A plug with lanyard; push-pull coupling.

MS 24264.   A receptacle with flange for mounting to a wall or bulkhead.

MS 24265.   A receptacle with jam nut for panel mounting.

MS 24266. A straight plug used at the end of a wire or a wire bundle.

MS 27034. Connector receptacle, electrical pin insert, cylindrical miniature, hermetic, solder mount.

Miniature MS-connector classes There are five classes of miniature MS connectors with solder-type contacts. These are:

Class E. An environment- (moisture- and vibration-) resisting connector, moisture-proofed by means of a multihole wire grommet seal and clamping nut.

Class F. An environment-resisting connector, similar to class E, with addition of a strain-relief clamp (MIL-C-26500; class F is fluid-resisting).

Class H. Hermetic-sealed receptacle which has a glass insert fused to the contacts and the shell where leakage rate in the order of 0.1 μm/(ft²) (h) is required.

Class J. A connector incorporating a gland seal for sealing a jacketed cable.

Class P. Connectors supplied with a plastic potting mold, so that the connectors may be sealed by the application of a potting compound.

MS-connector marking Each MS connector is marked on the shell or coupling ring with a code of letters and numbers giving all the information necessary to identify the connector (see Fig. 44). A typical code is as follows: MS 3114E12-10 PW.

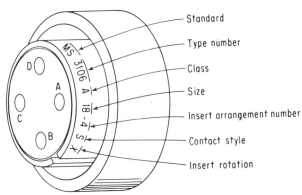

Fig. 44 MS-connector marking.

MS Number. The letters MS indicate that the connector has been made according to government standards.

Shell type. Numbers such as 3114 indicate type of shell, and whether plug or receptacle; see paragraph above.

Shell Design. Class letter indicates design of shell, and for what purpose connector is normally used; see paragraph above.

Shell Size. Numbers following class letter indicate shell size by outside diameter of mating part of receptacle in $\frac{1}{16}$-in increments, or by the diameter of the coupling thread in sixteenths of an inch. For example, size 12 has an outside diameter or a coupling thread of ¾ in.

ML-C-26500 connectors have an additional letter to indicate type of coupling between the shell-size and insert-arrangement code numbers. These letters are T for thread coupling, B for bayonet coupling, and Q for push-pull coupling, for example, MS 24264R18B30P6, where B indicates type of coupling.

Insert Arrangement. Numbers following hyphen indicate insert arrangement. This number does not indicate the number of contacts. Military-standard drawings cover contact arrangements approved for service use.

Style. First letter following number indicates style of contact.

Insert Position. Second letter indicates alternate insert position. Insert-position letters W, X, Y, or Z indicate that the connector insert has been rotated with respect to the shell a specified number of degrees from the normal position. Alternate positions are specified to prevent mismating when connectors of identical size and contact arrangement are installed adjacent to each other. These alternate positions are shown on governing military-standard drawings. If no letter appears, the insert is in the normal position. On connectors with multiple keyways the degree of rotation is measured from the widest keyway. See Fig. 45 for typical alternate-position arrangements.

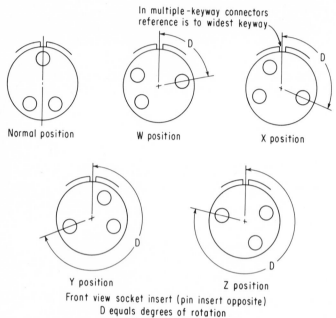

Fig. 45   Alternate position of connector inserts.

Alternate insert positions on MIL-C-26500 connectors are indicated by numbers 6, 7, 8, 9, and 10 instead of by letters.

AN-MS connectors MS 3100 through MS 3108 with socket contacts have the letter C stamped on the connector after the code-identification marking, indicating that the required prod-damage test has been met.

MS connector contacts   Two kinds of contacts are found in MS connectors: solder type and crimp type. Crimp-type contacts are removable. Contact sizes are related to AN wire sizes, but not all wire sizes have corresponding contacts.

Contacts should be used only with the connector for which they were designed. When they are replaced, the replacement contact should be identical with the contact being replaced.

MS connector-cable clamps   Connector-cable clamps are used at the back end of MS connectors, except potted connectors, to support wiring and to prevent twisting or pulling on soldered connections.

MS 3057.   Consists of a clamp body, two washers, and a clamp saddle held on the clamp body by two screws and lock washers.

MS 3057A.   Consists of a clamp body and two saddles held on by screws and lock washers. Used with AN telescoping bushings.

MS 3057B.   One-piece clamp with no separate cap or saddles. Used with AN 3420A bushings.

Manufacturers' variations in MS connectors   Standard AN-MS plugs and receptacles made to the requirements of a military specification may show differences in appearance between one manufacturer and another. Also minor changes in disassembly and installation instructions may be required. The text and illustrations to follow will show differences in detail.

MS Potting Connectors. These connectors are used only where potting is required. They are similar to other standard types, except that they have a shorter body shell and include a potting boot. MS potting connectors are available in the following types:

1. MS 3103—a receptacle with flange for mounting to a wall or bulkhead
2. MS 25183—a straight plug used at the end of a wire or wire bundle
3. MS 25183A—similar to MS 25183, with the addition of a grounding screw

MIL-C-26482, Connectors, Electric, Miniature, Quick-disconnect (Fig. 46). This specification covers environment-resisting, quick-disconnect miniature connectors, with

Fig. 46 MIL-C-26482 plug and receptacle.

either solder-type or crimp/removable-type contacts and with a maximum operating temperature of 125°C. They are available in five classes as indicated above.

Contacts (Table 13). The number of contacts available is from 2 to 61 in various combinations. Coaxial contacts are not covered in the specification but are available under manufacturer's part number. The shells and coupling rings are made of aluminum alloys, and the hermetic shells are made of material suitable for soft soldering.

MIL-C-26500, Connectors, General-purpose, Electrical, Miniature, Circular, Environment-resisting, 200°C Ambient Temperature (Fig. 47). This specification covers

Fig. 47 MIL-C-26500 connector.

an environment-resisting family of miniature circular connectors designed essentially voidless and to meet the higher-altitude and higher-temperature requirements of missiles and space vehicles. There are four basic classes:

Class R. For use in environment-resisting applications.

Class H. Hermetic; for receptacles where leakage rates are in the order of 0.01 μm/(ft^3) (h).

Class G. Grounding, environment-resisting; for use where grounding to mounting structure is required, the anodized coating is removed to provide a conducting path.

Class F. Fluid-resisting.

These connectors are available with threaded couplings (type T), bayonet coupling (type B), and push-pull coupling (type Q). The shells are aluminum alloy, stainless steel, or a material suitable for soldering or brazing for class H. Sealing is accomplished by interfacial interference of the inserts as well as an O-ring seal which provides shell sealing before mating is accomplished. The terminals are grommet-sealed.

Shell Styles
    Square-flange receptacle
    Single-hole mounting receptacle
    Straight plug
    Solder-flangd receptacle (hermetic only)

Contacts (see Table 13)   Coaxial contacts are available.

MIL-C-38300. This specification covers an upgraded version of the MIL-C-26500 design and was established primarily to cover closed-entry-design contacts.

NAS 1599, Connectors, General-purpose, Electrical, Miniature, Circular, Environment-resisting, 200°C Maximum Temperature. These connectors are much the same as the MIL-C-26500. The contacts are designed for rear insertion and removal. The sockets are closed-entry. See Table 13.

The connectors are capable of continuous operation between the temperature limits of −55 and +200°C. Hermetic receptacles are available to perform to the same temperature requirements as the environment-resisting construction. They are available in two basic classes:

Class R.   Environment-resisting.

Class H. Hermetic; receptacles intended for use in applications where pressures must be contained by the connector across the walls or panels on which they are mounted. Good to 0.01 μm/(ft³) (h) leakage.

The design requires that the combination ambient temperature and contact-current flow not allow the temperature of the connector assembly to exceed 200°C.

MIL-C-25955, Connectors, Electrical, Environment-resisting, Miniature, with Snap-in Contacts. This specification covers miniature connectors with threaded coupling provided with holes for safety wiring. They are available in two classes:

Class E.   Environment-resisting.

Class H.   Hermetic receptacle

Contacts   Specification covers size 20 contacts (see Table 13). Size 16 and 12 contact arrangements are available under manufacturer's part numbers.

MIL-C-27599, Connectors, Electrical, Miniature, Continuously Shielded, Quick-disconnect. This specification covers the requirements for two types of bayonet-locking miniature connectors. One series is designed to provide shell-to-shell mechanical orientation and electrical continuity of shells prior to mating of contacts. The pin-and-socket contacts are so located that they cannot be damaged during mating or unmating operation. The design is such that plug contacts cannot make contact with receptacle contacts, regardless of the angle of entry, until proper orientation is achieved. Spring fingers in the shell provide electrical continuity to the receptacle during mating and unmating. The connectors are available in two classes. The other series provide a low silhouette for minimum size and weight. This specification employs fixed pin-and-socket contacts intended for solder terminations.

Class T.   General-duty

Class H.   Hermetically sealed receptacles

For contact information see Fig. 45.

MIL-C-38999. This specification covers two series of miniature, quick-disconnect, bayonet-lock connectors. The two series are not interchangeable or intermatable. Both series have removable pin-and-socket contacts and are intended for crimp terminations. Both series are designed to assure proper orientation of the mating halves prior to electric-circuit closure. In addition one series (LJT) is designed so that electrical continuity is maintained between mating shells prior to mating of pin-and-socket contacts and in reverse order during disconnect, thereby providing protection against EMR, so as to protect the contacts against handling or field abuse and inadvertent electrical contact. The other series (JT) is designed to provide a low silhouette for minimum size and weight and includes connectors with positive shell-to-shell grounding when mated.

The various connector configurations are intended for use as follows:

1. *Class RE.* Environment-resistant; it has provisions for sealing around wires on rear end.

2. *Class RP.* Intended for use in applications where sealing around wires is accomplished by potting.

3. *Class RT.* General-duty, without rear accessories.

4. *Class Y.* Intended for use in applications where pressure must be maintained by the connectors across panels or walls.

MIL-C-81511. This is a miniature, environmentally resistant, very-high-contact-density cylindrical connector family. It features a monoblock internal construction that eliminates air voids between contacts, damageproof mating with pins recessed beyond the reach of shells, closed-entry hard inserts for the socket contacts, and prodproof socket contacts. There is a grommet seal on the terminal end that will accommodate a range of wire diameters from 0.020 to 0.054 in.

Environmental sealing is done internally by means of interfacial and shell O-ring seals. The mounting of the receptacle is single-hole and may be done on the front or rear panel. This type is available in six shell sizes with from 4 to 85 contacts. There are grounding springs in the receptacle that mate prior to electrical engagement of contacts. Shielding is provided for electromagnetic- and radio-frequency-interference protection (Fig. 48).

Fig. 48 High-density environmental connectors.

MIL-C-10544 and MIL-C-55116, Audio Connectors. The former is a 10-contact connector and the latter is a 5-contact audio connector. Both are widely used in Signal Corps ground communication equipment. They are both bayonet-locking and are available with either rigid contacts or nonrigid spring-loaded contacts (Fig. 49).

MIL-C-55181 and MIL-C-12520. These are power connectors. They are polarized and waterproof, with self-sealing cable clamps on the plugs. The cable plugs feature a center-locking screw for positive engagement. Both plugs and receptacles may be specified with pin or socket contacts.

There are five different arrangements in the MIL-C-12520: 4, 9, 14, 19, and 30 contacts (see Fig. 50).

The MIL-C-55181 connector is available with both removable and fixed contacts and with three contact arrangements, 4, 9, and 18 contacts.

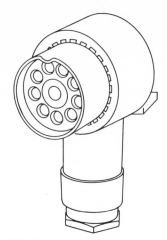

Fig. 49 Audio conector.

Fig. 50 Size 16 underwater power connector per MIL-C-12520.

MIL-C-22992, Connectors, Electrical, Waterproof, Quick-disconnect, Heavy-duty Type. This specification covers heavy-duty, multicontact, waterproof electrical connectors, with either solder-type or crimp/removable-type contacts and with a maxi-launching equipment, aircraft ground equipment, and ground radar. The contacts and inserts are basic MIL-C-5015 design. They are available in three basic classes:

Class R.   Environmental
Class R.   Environmental
Class C.   Pressurized

PRINTED-WIRING CONNECTORS

Virtually all the interconnection methods and devices available for rigid printed-wiring boards are independent of the inherent design of the board. Thus to take advantage of the many well-designed and -tooled products available, it is usually desirable to bring all termination points of conductors (including internal layers) to either external surface and locate them on standard centers to fit the available connectors. Table 18, in conjunction with Fig. 51, gives a listing of the various styles of readily available rigid-printed-wiring connectors and some description of their design capabilities.

There are basically five interconnect methods common to rigid boards: solder eyelets, individual contacts, direct reflow of flexible cabling, edge-board (or card-insert) connectors, and two-piece (or card-mounted) connectors.

Solder eyelets Perhaps the simplest method to interconnect two boards is to attach discrete wires to terminals or eyelets which have been inserted as tie points for other circuitry. This method has been common in the radio and television industry for many years—first by point-to-point soldering of wires and more recently by the use of wire-wrap connections. Although point-to-point wiring can be quite inexpensive, it does not lend itself to removal, replacement, testing, or high-density packaging.

Flexible cabling This product is usually a single layer of parallel conductors sandwiched between two insulating films, although ground planes can be laminated on to effect better circuit isolation and specific impedance characteristics. Conductors can have either rectangular or round cross sections. Aside from the obvious attraction of having well-defined electrical characteristics (such as crosstalk and impedance), this interconnect method can achieve extremely high-density packages. For example, 50-Ω lines can be fabricated with conductors on 0.025-in centers, and

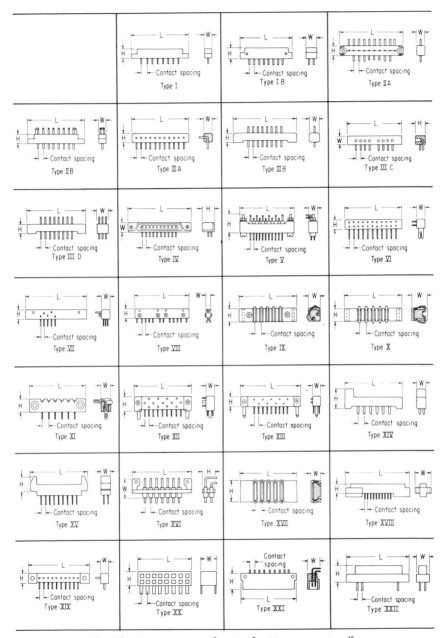

Fig. 51 Various types of printed-wiring connectors.[15]

these conductors can then be reflow-soldered directly to etched pads on a rigid board. Alternately, connectors are available which can interface ribbon cable to rigid boards utilizing an insulation-piercing technique (see Fig. 52). This approach offers the advantage of using stranded conductors and low-temperature insulations. An example of such a connector is shown in Fig. 53.

TABLE 18   Printed-wiring Connector Types[15]

Type no. (see Fig. 19)	Type of board accommodated (max no. of layers)	Type of contact design	Typical contact spacing, in.	Typical current rating, A
IA	Single	Tuning fork	0.156	5
IB	Double	Ribbon	0.100	5
IIA	Header	Blade	0.156	5
IIB	Header	Blade	0.156	5
IIIA	Multilayer	Pin	0.200	5
IIIB	Multilayer	Pin	0.200	5
IIIC	Multilayer	Socket	0.200	5
IIID	Multilayer	Pin	0.100	3
IV	Double	Pin	0.156	3
V	Double	Pin	0.200	5
VI	Double	Fork	0.200	5
VII	Multilayer	Pin	0.100	3
VIII	Double	Ribbon	0.156	3
IX	Single	90° pressure	0.050	1
X	Single	180° pressure	0.050	1
XI	Single	Fork	0.156	3
XII	Multilayer	Pin	0.200	5
XIII	Multilayer	Pin	0.200	5
XIV	Single	Ribbon	0.156	5
XV	Double	Ribbon	0.125	3
XVI	Multilayer	Blade	0.150	3
XVII	Single	180° pressure	0.050	1
XVIII	Double	Ribbon	0.050	1
XIX	Multilayer	Pin	0.200	3
XX	Double	Socket	0.150	3
XXI	Multilayer	Pin	0.150	3
XXII	Double	Ribbon	0.100	3

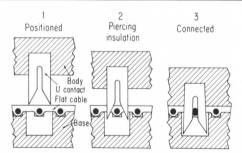

Fig. 52   Insulation-piercing termination for flat cable.   (*3M Company*).

Fig. 53   Typical insulation-piercing connector.   (*3M Company.*)

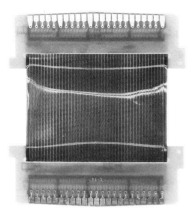 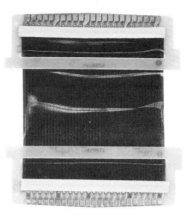

Fig. 54 Flexible jumper cable. (*Burroughs Corp.*)

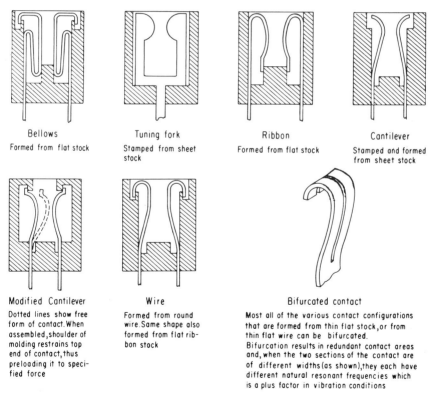

Bellows	**Tuning fork**
Formed from flat stock	Stamped from sheet stock
Ribbon	**Cantilever**
Formed from flat stock	Stamped and formed from sheet stock

Modified Cantilever
Dotted lines show free form of contact. When assembled, shoulder of molding restrains top end of contact, thus preloading it to specified force

Wire
Formed from round wire. Same shape also formed from flat ribbon stock

Bifurcated contact
Most all of the various contact configurations that are formed from thin flat stock, or from thin flat wire can be bifurcated.
Bifurcation results in redundant contact areas and, when the two sections of the contact are of different widths (as shown), they each have different natural resonant frequencies which is a plus factor in vibration conditions

Fig. 55 Basic edge-receptacle contact configurations.[13]

One popular method for interconnecting between groups of printed-wiring boards is to use flexible jumper cables (see Fig. 54).

Individual contacts These devices usually take the form of test points and seldom represent the primary interconnect system, although individual coaxial connectors have been employed in boards requiring good high-frequency response.

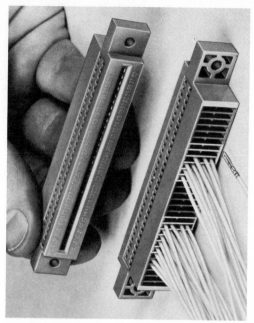

Fig. 56 Typical crimp/removable edge connector. (*Burndy Corp.*)

Edge-Board Connectors

The card-insert type of connector was the original and still is the most popular rigid-board connector type. Being a so-called one-piece connector, it is designed to accept a board having plated conductor tabs as the connecting medium between the board circuit and the connector contacts. A wide range of performance capabilities can be achieved by proper selection of the contact design (see Figs. 55 and 56).

Polarization is accomplished via a keyway in the board and a key in the receptacle or by integral molding of card guides of different lengths on each end. Improvements in board and connector materials have enabled these connectors to meet most performance and design requirements without an attendant cost increase. The only significant restriction on the use of edge-board connectors is contact density. Figure 57 shows the relative improvement which can be achieved when board size or surface

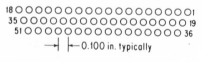

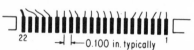

Fig. 57 Comparative contact density of two-piece vs. one-piece design.

area is at a premium. In addition, at 0.050-in centers, board tolerances are less rigorous for two-piece connectors.

Two-Piece Connectors

This type usually consists of one part (the plug) soldered to the board and the mating part (the receptacle) mounted on a chassis or another board. The most common contact design is pin-and-socket, but this is mainly historical and the type is generally versatile enough to accommodate any contact design which may show to advantage (see Fig. 58).

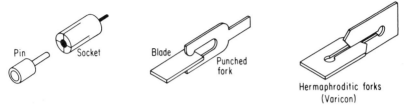

Fig. 58   Two-piece contact design.

Aside from their higher initial cost (typically three times that of edge-board types) the two-piece connectors usually add to the board cost (because of extra holes—normally plated through) and the assembly cost in production, since a good design uses auxiliary fastening devices to secure the plug to the board.   This relieves the dip-soldered contact connection of providing both electrical continuity and mechanical support against torsional-shear forces during mating.

Polarization is accomplished by the use of guide pins of greater length than the contacts or by offsetting the rows of contacts to preclude layout symmetery.   The former method is preferred, since it provides alignment assistance prior to contact mating.   The most common failure mode of this type of connector is pin misalignment due to tolerance accumulation, excess mounting float, or bent pins.   Some advantages in the use of two-piece connectors rather than edge-board types are first, their improved density capability (see Figs. 57 and 59) and second, the freedom

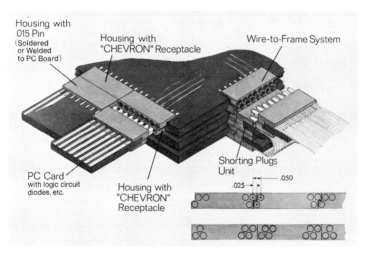

Fig. 59   Typical two-piece connector type.   (*AMP, Inc.*)

to select overall board thickness and tolerance.   The usual limit of board thickness for edge-board types is 0.125 in, and the required thickness tolerance is about ±0.007 in.   There may also be a problem if electrical requirements such as impedance between signal and ground conductors dictate greater vertical separation between layers than can be accommodated by a particular edge-board-connector design.   It is also easier to add contacts to a two-piece interconnect design than to an edge-board design when all available positions on the edge of a board have already been used.

Coaxial Board Connectors

In order to achieve maximum signal isolation between boards or controlled-impedance connections, it is sometimes necessary to use coaxial cables for board interconnecting.   One method of accomplishing this is to solder belted coaxial cable directly
onto an interface board.   This board is then plugged into an array of connectors.
Another method is to use a crimp, snap-in edge-board connector like the one shown
in Fig. 60.   Aside from the improved production-termination technique, a pluggable
interface is accomplished without using excessive board area.

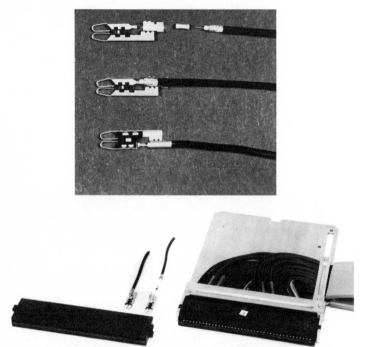

Fig. 60   Crimp coaxial edge-board connector.

If the impedance mismatch between cable and board cannot be tolerated, matched-
impedance edge-board connectors utilizing microstrip principles are available.   However, these usually require solder terminations (see Fig. 61).

Zero-Mating-Force Connectors

Two factors may lead a designer into the need for zero-insertion-force connectors;
one is the use of more than one edge for interconnecting wires and the other is
an overhwhelming mating force resulting from hundreds of contacts.   A low-inser-
tion-force connector as pictured in Fig. 62 is intended to provide extremely high
contact forces during the operating mode, and extremely low contact forces during
inserting or removing of the printed-wiring board.   These characteristics are achieved
by cam devices which open and close a split-receptacle body, which in turn transmits
high forces to the contacts when the connector is closed or locked and releases these
forces when the connector is unlocked.

COAXIAL CONNECTORS

In recent years a significant change has occurred in the coaxial-connector industry;
specifications for the procurement of coaxial connectors have changed from detailed
dimensional design requirements to performance requirements with the minimal di-

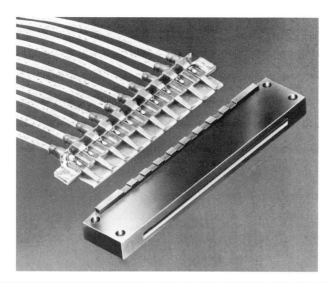

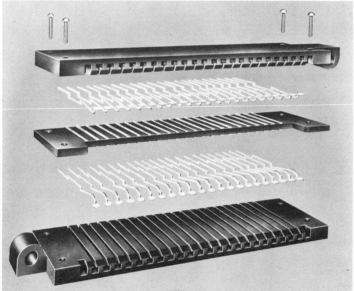

Fig. 61 Matched-impedance edge-board connector. (*Texas Instruments.*)

mensional requirements to assure intermatability. This change has had profound effects on both the user and the manufacturer of the connectors; and although the performance of cable assemblies has improved, it has introduced greater variety in the methods of assembly of connectors to cables. It is therefore essential that potential users be adequately aware of all the differences to assure that the optimum design is obtained for their transmission line in regard to performance, assembly and maintenance procedures, and cost.

Connector Design Parameters

There are three basic areas in a coaxial-connector design which are critical to achieving a stable, well-matched cable assembly over the desired frequency range,

Fig. 62 Typical zero-insertion-force connector. (*Cinch Mfg. Co.*)

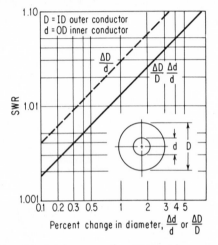

Fig. 63 Effect of diametrical tolerances of a 50-Ω cable on swr. (*Amphenol RF Division.*)

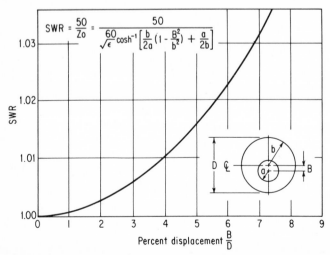

Fig. 64 Variation of swr with displacement of center conductor from center of a 50-Ω cable. (*Amphenol RF Division.*)

and environmental operating conditions. These areas are the dielectric structure, the coupling mechanism, and the cable clamp or assembly procedure. Each of these areas will be discussed in relation to the basic function of a coaxial connector, which is the efficient transmission of rf energy between cable sections, or between cable and terminal, with convenient means for rapidly and reliably coupling or uncoupling the connectors.

Impedance matching The major cause of variation of impedance in a coaxial transmission line is the variation in dimensions and position of the conductors. The effect on the standing-wave ratio can be readily seen in Figs. 63 and 64. When a cable is terminated in a connector, the transmission line experineces abrupt changes in conductor diameters. To maintain the characteristic impedance of the transmission line, the change of conductor diameters must be made to provide a constant imped-ance along the transmission line; however, at the point of discontinuity, a section of high impedance of sufficient width must be designed into the dielectric structure to compensate for the discontinuity capacitances over the desired frequency range. In any practical connector there are several points of discontinuity or impedance variation from the characteristic impedance.

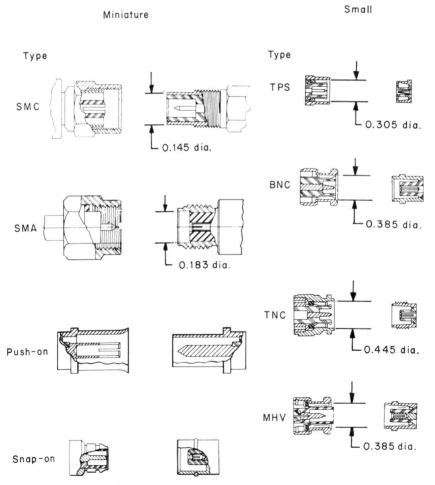

Fig. 65 Typical couplings and interfaces of miniature and small coaxial connectors (TPS, three-point bayonet; BNC and MHV, two-point bayonet; TNC, SMA, and SMC, thread).

In practice, it is extremely difficult to establish the value for s precisely to achieve optimum impedance compensation for low standing-wave ratio. In the past, once the appropriate dimensions were established for a given matched connector design,

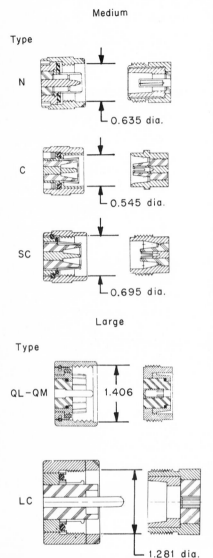

Fig. 66 Typical couplings and interfaces of medium and large coaxial connectors (C, two-point bayonet; N, SC, QM, QL, and LC, thread).

detailed drawings with very precise tolerances were prepared as standards. Because of the tedious and difficult control and measurement procedures of all the essential dimensions to assure good vswr performance, it was not always possible to verify the performance of the connector. Until recently vswr measurements were very time-consuming because of the discrete-frequency step-by-step technique of measurement, and very high peak values within the frequency range of interest could not always be reliably detected. New techniques of sweep frequency and time-domain reflectometry now provide the tools to permit performance testing of vswr with relative ease and high speed. The designer and manufacturer are presently relieved of the detailed dimensional requirements, and have greater freedom and flexibility to incorporate improvements in the connector.

In most cases, coaxial connectors have been designed for a 50-Ω characteristic impedance which has become the standard for radio-frequency and microwave applications. Fifty-ohm connectors can be readily used with cables of the same OD but with higher impedances of either 75 or 95 Ω, since the center conductor of the higher-impedance cables is smaller than that of the 50-Ω cable. The applications of the 50-Ω connectors to 75- to 95-Ω cables are generally limited to frequencies whose wavelengths are at least twenty times the electrical length of the connector; accordingly, the resultant impedance mismatch is not significant. Such general usage of 50-Ω connectors with various flexible cables has the great advantage of standardization; however, the growing importance of a matched 75-Ω transmission line for very-fast-rise-time pulses (a few nanoseconds) will probably broaden the use of existing 75-Ω connectors and lead to the design and development of additional 75-Ω types.

Coupling designs The most popular methods of coupling connectors are screw-thread and bayonet devices. Several other couplings are also available, such as push or snap-on, and hermaphrodite types. Most of these couplings and the cross sections of their interfaces are illustrated in Figs. 65 and 66. The coupling device not only provides a convenient and relatively quick method to connect or disconnect a coaxial transmission line but is critical to stable electrical performance and environmental protection. For use at very high frequencies, connectors must be positively and firmly coupled. Nontwist-type couplings have been designed for use in applications without sufficient access space to turn a coupling nut.   In addition, they are useful for multiple coaxial connection in rack-and-panel connectors or multicontact circular connectors.   Other considerations involve ruggedness to resist damage, ability to stay coupled during shock and vibration, and their ability to operate in a variety of environments, such as mud, sand, and ice.

The double-lead coarse-thread design (used on the QM and QL connectors) combines the best features of bayonet and thread couplings, namely, quick connect-disconnect, rugged, nonfouling, vibration-resistant, and electrically stable.   This coupling thread has been experimentally applied to smaller connectors (Fig. 67) and is being considered for future-generation equipment.

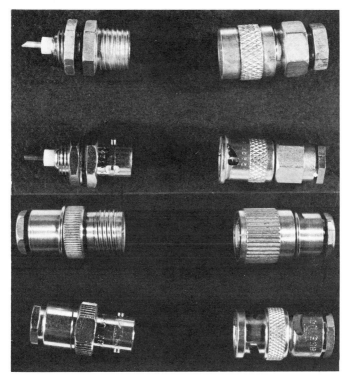

Fig. 67   Comparison of double-coarse-thread coupling with bayonet coupling on small connectors.

The coupling designs have generally been based on flexible-cable applications, and because of intermatability requirements with equipment, special coaxial devices, and test equipment such as couplers, attenuators, and slotted lines, the same basic couplings have also been used for semirigid cables, even though they may not be optimumly designed for them.

Cable clamp (or assembly procedures)   Until recently, when detailed dimensional requirements governed the procurement of connectors, standard cable clamps involving careful combing of the braid wires were in common use.   With the advent of performance-test requirements and the development of crimping techniques, crimp braid clamps are now commonly used (Fig. 68).   Several types of improved clamping without the use of special tools have also become available.   Typical examples of these common methods of assembly are illustrated in Fig. 69.   The development of all these techniques has generally simplified the procedure for assembly and has improved both mechanical and electrical performance; however, they have also introduced greater variety and more complex overall maintenance and have eliminated

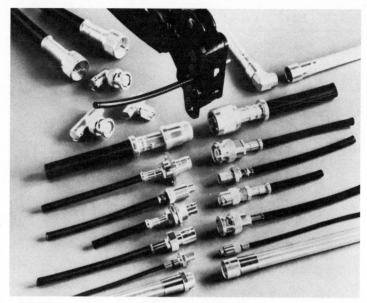

Fig. 68 A few of the many coaxial connectors designed specifically for crimp.[16] (*AMP, Inc.*)

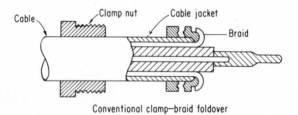

Conventional clamp—braid foldover

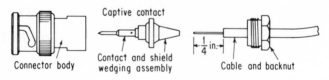

Ferrule braid clamp—no foldover

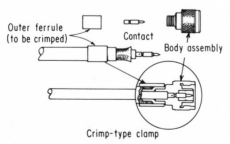

Crimp-type clamp

Fig. 69 Several common methods for clamping connectors to cable.

a common standard for cable preparation. It should be noted that there could be a significant variation in vswr performance from one type of crimp to another because of the different deformations of their conductors (Fig. 70).

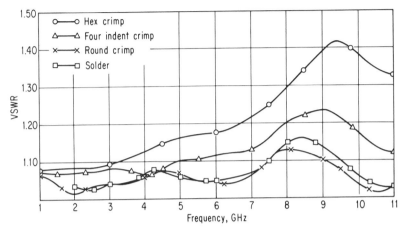

Fig. 70 vswr vs. frequency comparison of N-series plugs having soldered and crimped center contacts.[16] (*AMP, Inc.*)

In selecting the type of cable clamp for flexible cable, the user must take into account the following requirements: electrical stability, flex life, tensile pull, environmental resistance, and field repair. The choice of connectors for semirigid cable is generally limited by the choice of semirigid cable because of their proprietary nature and unique dimensions. Semirigid cables are not furnished to standard dimensions nor do they necessarily have the same type of dielectric; accordingly, each cable manufacturer has designed his own type of connector with somewhat different clamping methods (see Fig. 71). Most manufacturers use the standard coupling devices in order to satisfy intermatability requirements. Figure 72 illustrates a connector splice for joining two sections of semirigid cables.

Connector Types for Flexible and Semirigid Cable

A wide variety of fittings is available within a family of connectors (i.e., all have the same type of coupling), and Fig. 73 illustrates typical connective devices available within a family. A family of connectors is generally identified by a letter or combination of letters which are usually related to the size and type of coupling. In addition to connectors within a family, adapters between families are available to provide junctions between different types and sizes of coaxial cable. The following paragraphs describe briefly the pertinent technical characteristics of the popular types of coaxial connectors. The 50-Ω connectors are generally used with higher-impedance cables and many special types of cable, since matching is not critical at the frequency of application of these cables. For information on the numerous connectors and adapters available, Ref. 18 and manufacturers' catalogs should be consulted for specific applications (see Table 19).

Miniature connectors These connectors (Fig. 65) are intended primarily for use on relatively short lengths (usually not more than a few feet) of flexible and semirigid coaxial cables, including the plated grooved cable, where size and weight must be kept to a minimum. In general, they are used to interconnect rf or microwave devices and assemblies within a protected environment, and at peak voltages between 100 and 500 V. Depending upon the size and impedance match of the cable and connectors, they may be used at frequencies between 10 and 30 GHz. Only the SMA type of connector when assembled to semirigid cable (0.141 in OD)

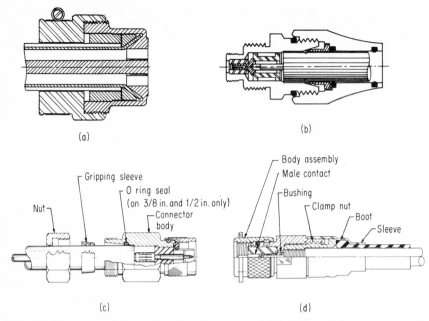

(a)

(b)

(c)

(d)

Fig. 71 Various techniques for clamping to semirigid cables. (*a*) Aluminum tube is flared and clamped. (*b*) and (*c*) Clamps with teeth or sharp edges are used to grip surface of aluminum tube. (*d*) Clamp is threaded on the plated grooved outer conductor of this miniature type.

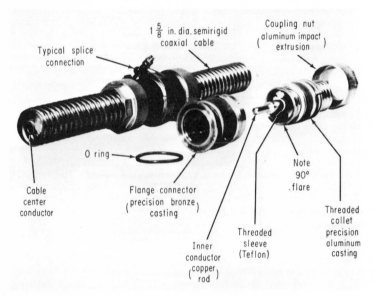

Fig. 72 Connector splice for joining two semirigid cables together.[17] (*Courtesy of B. Waters, Raytheon Co.*)

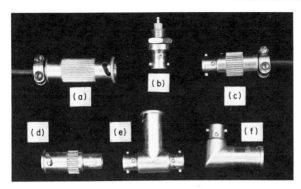

Fig. 73 Examples of the types of connectors which constitute a family (based on common coupling interface). (*a*) Cable plug (male). (*b*) Panel-mounted receptacle. (*c*) Cable jack (female). (*d*) Male-female adaptor. (*e*) Tee adapter (F-M-F). (*f*) Right-angle adapter (M-F). Not shown are panel-mounted jack and adapters, and such adapters are F-F, M-F-M.

exhibits a maximum vswr of 1.05:1 up to 12 GHz, whereas when assembled to plated grooved cable of similar size, it exhibits a maximum vswr of 1.2:1 up to 18 GHz. These connectors are not as well matched or electrically stable when assembled to the flexible cables. The cables most commonly used with the miniature connectors are the 0.141- and 0.085-in -OD semirigid cables and RG-174, -180, -316, and -371/U types of flexible cable. The push-on and snap-on are used primarily for plug-in packages or multiple, coaxial, rectangular, and circular connectors at relatively low radio frequencies.

Small connectors   These connectors are types used primarily with such flexible cables as RG-58, -59, -62, -122, and -223/U in both protected and exposed environments in small equipments. Except for the MHV types, which are rated for 5 kV, these connectors are limited to 500 V peak. The MHV type look similar to the BNC because of their similar bayonet coupling, but they do not intermate with the BNC because of their longer, overlapping dielectric needed to achieve their higher voltage rating. For quick-disconnect applications, the BNC and TPS connectors provide reasonably good electrical performance (see Fig. 74). BNC connectors

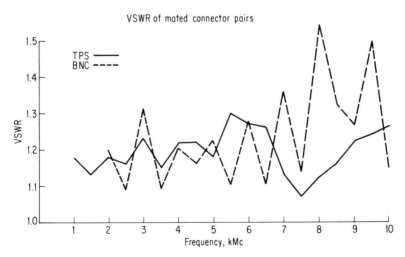

Fig. 74   vswr comparison for type BNC and TPS connectors.

TABLE 19   Coaxial Connectors for Standard Cables[a-f]

Cable RG- /U	Series	Applicable cable connectors UG- /U (MIL-C-39012 specification sheet no.)		
		Plug—male	Plug—female (jack)	Receptacle— jam-nut mounting (panel jack)
6	C	626(6-1)A	633(7-2)A	630(11-1)A
	N	18(1-1)A	20(2-1)A	159(3-1)A
		91[g]	92[g]	93[g]
	SC	(35-1)A	(36-2)A	(38-2)A (square flange)
11	BNC	959		
	C	573(6-2)A	572(7-1)A	570(11-2)A
		1752(6-11)C	1759(7-8)C	1777(11-8)C
	N	21(1-2)A	23(2-2)A	160(3-2)A
		(1-23)C	(2-26)C	(3-20)C
		94[g]	95[g]	96[g]
	SC	(35-2)A	(36-1)A	
		(35-13)C	(36-10)C	(40-21)C
22	Twin	421	. . . .	423
				422/R
23[h]	Twin			
25	Pulse	180	264/R	181
26	Pulse	180	264/R	181
27	Pulse	36	. . . .	158 or 1141
				37 and 38/R
28	Pulse	174	222/(adapter)	166
34	UHF	357	. . . .	358/R
58	BNC	88(16-1)A	89(17-1)A	909(19-1)A
		(16-13)C	1794(17-13)C	1804(19-13)C
	TNC	(26-1)A	(27-1)A	(28-1)A
		(26-10)C	(27-10)C	(28-10)C
	C	709(15-2)A	. . . .	704(9-2)A
		1779(15-3)C	. . . .	(9-3)C
	N	536	. . . .	556
	SC	. . . .	. . . .	(37-2)A
				(37-5)C
	TPS	1366	1415	1364
	SMA	(55-4)A	(57-4)A	(59-4)A
		(55-29)C	(57-29)C	(59-29)C
59	BNC	260(16-2)A	261(17-2)A	910(19-2)A
		(16-15)C	1798(17-15)C	1808(19-15)C
	TNC	(26-2)A	(27-2)A	(28-2)A
		(26-12)C	(27-12)C	(28-12)C
	C	627(15-1)A	. . . .	631(9-1)A
	N	603	602	593
	SC	. . . .	. . . .	(37-1)A
62	BNC	260(16-2)A	261(17-2)A	910(19-2)A
		(16-15)C	1799(17-15)C	1809(19-15)C
	TNC	(26-2)A	(27-2)A	(28-2)A
		(26-12)C	(27-12)C	(28-12)C
	C	627(15-1)A	. . . .	631(9-1)A
	N	603	602	593
	SC	. . . .	. . . .	(37-1)A
63	N	1003		
64	Pulse	180	. . . .	181
				264/R
65	C	1032		

TABLE 19   Coaxial Connectors for Standard Cables (Continued)

Cable RG- /U	Series	Applicable cable connectors UG- /U (MIL-C-39012 specification sheet no.)		
		Plug—male	Plug—female (jack)	Receptacle— jam-nut mounting (panel jack)
71	BNC	260(16-2)A (16-17)C	261(17-2)A (17-17)C	910(19-2)A (19-17)C
	TNC	(26-2)A (26-14)C	(27-2)A (27-14)C	(28-2)A (28-14)C
	C	627(15-1)A	. . . .	631(9-1)A
	N	603	602	593
	SC	. . . .	. . . .	(37-1)A
108[h]	Twin			
114	N	1003		
115	N	1185	1186	1187
119	LN	530	. . . .	531
122	BNC	1033(16-3)A (16-16)C	1056(17-3)A 1800(17-16)C	1055(19-10)A 1810(19-16)C
	TNC	(26-3)A (26-13)C	(27-3)A (27-13)C	(28-3)A (28-13)C
	C	709(15-2)A	. . . .	704(9-2)A
	N	536	. . . .	556
	SMA	(55-3)A (55-27)C	(57-3)A (57-27)C	(59-3)A (59-27)C
125[h]	Low capacitance			
130	Twin	1060	. . . .	1057/R
133	BNC	959		
	C	573(6-2)A	572(7-1)A	570(11-2)A
	N	21(1-2)A	23(2-2)A	160(3-2)A
142	BNC	88(16-1)A (16-14)C	89(17-1)A 1797(17-11)C	909(19-1)A 1807(19-11)C
	TNC	(26-1)A (26-11)C	(27-1)A (27-11)C	(28-1)A (28-11)C
	C	709(15-2)A	. . . .	704(9-2)A
	N	536	. . . .	556
	SMA	(55-4)A (55-28)C	(57-4)A (57-28)C	(59-4)A (59-28)C
144	BNC	959		
	C	573(6-2)A 1752(6-11)C	572(7-1)A 1759(7-8)C	570(11-2)A 1777(11-8)C
	N	21(1-2)A (1-23)C	23(2-2)A (2-26)C	160(3-2)A (3-20)C
	SC	(35-2)A (35-13)C	(36-1)A (36-10)C	(40-21)C
156	Pulse	1291	. . . .	1292/R
157	Pulse	1295	. . . .	1296/R
158[h]	Pulse			
164	C	708(6-5)A		
	N	167(1-4)A		
	LC	1258	. . . .	352/R
165	BNC	959		
	C	573(6-2)A 1753(6-12)C	572(7-1)A 1760(7-9)C	570(11-2)A 1788(11-9)C
	N	21(1-2)A (1-21)C	23(2-2)A (2-24)C	160(3-2)A (3-18)C
	SC	(35-2)A (35-11)C	(36-1)A (36-7)C	(40-19)C
174	SMA	(55-2)A (55-26)C	(57-2)A (57-26)C	(59-2)A (59-26)C

TABLE 19   Coaxial Connectors for Standard Cables (Continued)

Cable RG- /U	Series	Applicable cable connectors UG- /U (MIL-C-39012 specification sheet no.)		
		Plug—male	Plug—female (jack)	Receptacle— jam-nut mounting (panel jack)
174	SMB	Under consideration for MIL-C-39012.		
	SMC	See Specification MIL-C-22557		
176	C	1032		
177	C	708(6-5)A		
	N	167(1-4)A		
	LC	1258	. . . .	352/R
178	SMA	(55-1)A	(57-1)A	(59-1)A
		(55-25)C	(57-25)C	(59-25)C
	SMC	1460	1462	1463
179[h]	SMB	Under consideration for MIL-C-39012.		
	SMC	See Specification MIL-C-22557		
180[h]				
181	Twin	1253	. . . .	422/R
185[h]				
186	C	1032		
189	LC	1189	. . . .	352/R
	QL	1372	. . . .	1533(47-2001)/R
210	BNC	260(16-2)A	261(17-2)A	910(19-2)A
		(16-15)C	1799(17-15)C	1809(19-15)C
	TNC	(26-2)A	(27-2)A	(28-2)A
		(26-12)C	(27-12)C	(28-12)C
	C	627(15-1)A	. . . .	631(9-1)A
	N	603	602	593
	SC	. . . .	. . . .	(37-1)A
211	C	711(6-4)A	. . . .	569(14-1)/R
	SC	(35-4)A	. . . .	(42-1)/R
	LT	1305	. . . .	1314
212	C	626(6-1)A	633(7-2)A	630(11-1)A
		1748(6-7)C	1758(7-7)C	1773(11-4)C
	N	18(1-1)A	20(2-1)A	159(3-1)A
		(1-16)C	(2-19)C	(3-13)C
	SC	(35-1)A	(36-2)A	(40-1)A
		(35-10)C	(36-9)C	(40-14)C
213	BNC	959		
	C	573(6-2)A	572(7-1)A	570(11-2)A
		1749(6-8)C	1754(7-3)C	1774(11-5)C
	N	21(1-2)A	23(2-2)A	160(3-2)A
		(1-17)C	(2-20)C	(3-14)C
	SC	(35-2)A	(36-1)A	(40-2)A
		(35-11)C	(36-7)C	(40-15)C
214	BNC	959		
	C	573(6-2)A	572(7-1)A	570(11-2)A
		1750(6-9)C	1755(7-4)C	1775(11-6)C
	N	21(1-2)A	23(2-2)A	160(3-2)A
		(1-18)C	(2-21)C	(3-15)C
	SC	(35-2)A	(36-1)A	(40-2)A
		(35-12)C	(36-8)C	(40-16)C
216	BNC	959		
	C	573(6-2)A	572(7-1)A	570(11-2)A
	N	21(1-2)A	23(2-2)A	160(3-2)A
		(1-24)C	(2-30)C	(3-21)C
	SC	(35-2)A	(36-1)A	(40-17)C

TABLE 19   Coaxial Connectors for Standard Cables (Continued)

Cable RG- /U	Series	Plug—male	Plug—female (jack)	Receptacle— jam-nut mounting (panel jack)
		Applicable cable connectors UG- /U (MIL-C-39012 specification sheet no.)		
217	C	707(6-3)A		
	N	204(1-3)A		
		(1-19)C	(2-22)C	(3-16)C
	SC	(35-3)A	. . . .	(40-4)A and (40-18)C
	QM	1394(48-2001)	. . . .	1399(49-2001)
218	C	708(6-5)A		
	N	167(1-4)A		
	SC	(35-5)A	. . . .	(40-5)A
	LC	1258	. . . .	352/R
	QL	1392(44-2001)	. . . .	1397(45-2001)
220	LC	156	. . . .	1370/R
	QL	1393(44-2002)	. . . .	1398(44-2002)
222	C	626(6-1)A	633(7-2)A	630(11-1)A
	N	18(1-1)A	20(2-1)A	159(3-1)A
223	BNC	88(16-1)A	89(17-1)A	909(19-1)A
		(16-14)C	1795(17-14)C	1805(19-14)C
	TNC	(26-1)A	(27-1)A	(28-1)A
		(26-11)C	(27-11)C	(28-11)C
	C	709(15-2)A	. . . .	704(9-2)A
		1780(15-4)C	. . . .	1768(9-4)C
	N	536	. . . .	556
	SC	. . . .	. . . .	(37-2)A
				(37-6)C
	TPS	1412	1416	1413
	SMA	(55-4)A	(57-4)A	(59-4)A
		(55-28)C	(57-28)C	(59-28)C
225	BNC	959		
	C	573(6-2)A	572(7-1)A	570(11-2)A
		1751(6-10)C	1756(7-5)C	1776(11-7)C
	N	1185(1-5)A	1186(2-3)A	160(3-2)A
		(1-22)C	(2-25)C	(3-19)C
	SC	(35-2)A	(36-1)A	(40-2)A
		(35-12)C	(36-8)C	(40-16)C
226[h]				
280[h]				
281[h]				
301[h]				
302	BNC	(16-18)A	(17-18)A	(19-18)A
		(16-20)C	(17-20)C	(19-20)C
	TNC	(26-2)A	(27-2)A	(28-2)A
		(26-12)C	(27-12)C	(28-12)C
	C	627(15-1)A	. . . .	631(9-1)A
	N	603	602	593
	SC	. . . .		(37-1)A
303	BNC	(16-11)A	(17-11)A	(19-11)A
		(16-13)C	(17-13)C	(19-13)C
	TNC	(26-4)A	(27-4)A	(28-4)A
		(26-10)C	(27-10)C	(28-10)C
	C	709(15-2)A	. . . .	704(9-2)A
		1779(15-3)C	. . . .	1767(9-3)C
	N	536	. . . .	556
	SC	. . . .	. . . .	(37-2)A
				(37-5)C
	TPS	1366	1415	1364
	SMA	(55-5)A	(57-5)A	(59-5)A
		(55-29)C	(57-29)C	(59-29)C

TABLE 19   Coaxial Connectors for Standard Cables (Continued)

Cable RG- /U	Series	Applicable cable connectors UG- /U (MIL-C-39012 specification sheet no.)		
		Plug—male	Plug—female (jack)	Receptacle— jam-nut mounting (panel jack)
304	C	626(6-1)A	633(7-2)A	630(11-1)A
		1746(6-6)C	1757(7-6)C	1772(11-3)C
	N	18(1-1)A	20(2-1)A	159(3-1)A
		(1-10)C	(2-11)C	(3-7)C
	SC	(35-1)A	(36-2)A	(40-1)A
		(35-10)C	(36-9)C	(40-14)C
307[h]				
316	SMA	(55-2)A	(57-2)A	(59-2)A
		(55-26)C	(57-26)C	(59-26)C
326	QL	1832	1876	1533(47-2001)/R
327[h]				
328[h]				
329[h]				
371		Miniature types under consideration for MIL-C-39012		
389	QL	1372	. . . .	1533(47-2001)/R
391[h]				

/R   noncable receptacle.

[a] See MIL-HDBK-216 and specifications of Table 20 for additional information on cable connectors, receptacles, and adapters for the various families.

[b] Numbers not in parentheses are UG numbers; numbers in parentheses are abbreviated MIL-C-39012 part numbers [e.g., (6-1) is (06-0001)]. See para. 1.2.3 of specification MIL-C-39012 for detailed explanation of part number.

[c] Connectors in this list are Class 2 types which are intended to provide mechanical connection within an rf circuit providing specified rf performance.

[d] Crimp connectors of MIL-C-39012 as designated by the letter C are category C (field replaceable) connectors, which require only standard military crimping tools per MIL-T-55619 and standard cable-stripping dimensions to assemble.

[e] The noncrimp connectors of MIL-C-39012 as designated by letter A are category A (field serviceable) types which do not require special tools to assemble.

[f] The following is a cross reference of NATO type numbers with U.S. Military Specifications for coaxial connectors:

Size, DOD	Series	Military specification sheet	NATO NEPR no. 70 type numbers
Miniature (0.034–0.060 in., 0.86–1.52 mm)	SMA	39012/55–62	NUG 313–324
	SMB	Proposed for 39012	NUG 308–312
	SMC	22557/10–19	NUG 301–307
Small (0.116–0.146 in., 2.95–3.71 mm)	BNC	39012/16–24	NUG 101–107
	TNC	39012/26–34	
	TPS	55235	
Medium (0.285 in., 7.24 mm)	N	39012/1–5	NUG 205–210
	C	39012/6–15	NUG 201–204
	SC	39012/35–41	
Large (0.680 in., 17.27 mm)	QL	39012/44–47	NUG 403, 406, 408

[g] 70-Ω connectors.

[h] Standard connectors not available.

exhibit a maximum vswr of 1.3:1 but should be limited to a maximum frequency of 5 GHz because their two-point bayonet coupling does not provide adequate electrical stability above this frequency. The TPS connector has a three-point bayonet which provides vswr characteristics similar to those of the BNC connector but is more stable up to 10 GHz. It is also somewhat smaller than the BNC while providing superior clamping and comparable coupling features. The TNC connector is identical to the BNC except that it has a thread coupling in place of .the bayonet coupling. It offers all the same performance as the BNC but with greater stability up to 10 GHz. All these connectors can be obtained with crimp cable-clamp features. They provide a gasket seal to the cable and a gasket seal between mated pairs. Although used in exposed environments, they are not especially rugged and should therefore be used with care. In addition to their use with the flexible cables, BNC and TNC connectors have also been designed for small semirigid cables. Because of the mechanical stresses which can be encountered, the thread coupling should be preferred in these applications. Special designs are also available for use with medium-sized cable such as RG-213/U.

Medium connectors   The types illustrated in Fig. 66 are the more commonly used medium connectors for flexible cables such as RG-212, -213, and -214/U. In addition, special designs are available for both small- and large-sized flexible cables such as RG-58/U and RG-216 and -218/U as well as for medium-sized semirigid cables. These are probably the most popularly used connectors for general-purpose rf applications for interconnection between an antenna and receiver or transmitter. They have a peak voltage rating of 1,500 V and a maximum vswr of 1.3:1 up to 10 GHz. Because of the mechanical instability of the two-point bayonet, the type C connector should be limited to a maximum frequency of 5 GHz. The C connector does have the advantage of a quick coupling, whereas the N and SC types use a fine-thread design which is more susceptible to mechanical damage and cross threading. The C and SC offer an interlocking dielectric structure, whereas the N has an air and dielectric bead structure and is available in 70-Ω as well as 50-Ω impedances. Sealing is achieved by gaskets between the mated pair, and between the connector and cable. These connectors are also available with crimp clamping. The connectors are not exceptionally rugged but perform reasonably well in applications where mechanical hazards are not extreme.

Large connectors   A greater variety of large connectors is available than those illustrated in Fig. 66, but most of them are considered special types. The LT type (not illustrated) is similar to the LC except that it is designed for a specific Teflon cable. The LC connectors are limited to applications up to 1 GHz, whereas the QM and QL (Fig. 75) may be used up to 5 GHz. These connectors have voltage rating in the range of 5 to 10 kV depending upon the specific cable and connector. The maximum vswr is 1.3:1. The QM and QL connectors, although capable of use with flexible solid-dielectric cable, have been used primarily with flexible air dielectric, low-loss types such as RC-189, -326, and -389/U. They feature a dielectric structure, with a butt interface to eliminate air and attain good corona voltage characteristics, and a tapered ferrule for good clamping. For a solid or spline cable dielectric a small internal thread on the clamp is used to grip the dielectric for outstanding cable retention. The QM and QL were also designed to be very rugged and be able to withstand exceptional mechanical hazards. They utilize a course double-lead-thread coupling for quick connect-disconnect and resistance to damage.

Many more connector types are available, but the above represent a good cross section of probably the most widely used rf connectors for rf applications. Such connectors as the BN, HN, LN, QDC, SKL, SM, and UHF are probably in the inventory of both connector and equipment manufacturers, but are either very special or decreasing in use. Data on these connectors can be obtained from Ref. 18.

Precision connectors   Of special interest are the recently developed 14- and 7-mm precision connectors for use in precise high-frequency measurement techniques and/or critical transmission applications. The cross section and vswr capability of these connectors are shown in Fig. 76. It should be noted that the bead structure and inner contact have been standardized for use in connectors with different coupling mechanisms, and thereby allow the coupling mechanism to be replaced without disturbing the internal electrically matched structure. Both sexless (hermaphrodite)

Fig. 75 Type QL connector for RG-326/U cable, and QL to LT adapter. A cross section of the mating surface of the QL connector, shown in upper left-hand corner, indicates the solid-dielectric construction.

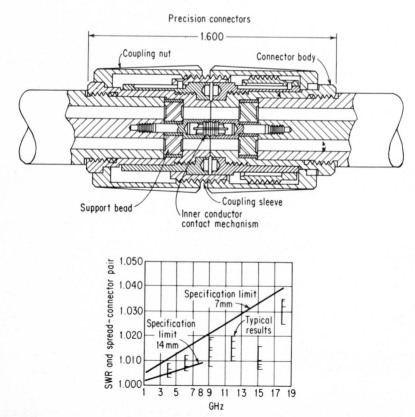

Fig. 76 Cross section and swr data for precision 7- and 14-nm coaxial connectors.

and conventional male-female couplings are available. In addition, a precision-type N connector has also been established. Comparative data on these precision connectors are given in Table 20.

TABLE 20   Comparative Connector Performance

	14 mm	7 mm	Precision Type N	Standard Type N
Upper frequency limit, GHz....	8	18	18	10
VSWR max.................	1.01	1.039	1.08	1.25
Characteristic impedance.......	50 ± 0.1 %	50 ± 0.2 %	50 ± 0.4 %	50
RF leakage, dB below signal....	120 min	120 min	90 min	NA
Insertion loss, dB per pair......	0.005 at 3 GHz	0.028 at 16 GHz	NA	0.02 at 3 GHz
Electrical length per pair.......	35.00 mm ±0.05 mm	18.36 mm ±0.03 mm	NA	NA
Contact resistance, mΩ max:				
Inner.....................	0.05	1.0	1.0	1.0
Outer.....................	0.07	0.1	1.1	0.15

Cable-Assembly Requirements

General   The electrical efficiency and stability of a coaxial transmission line consisting of a cable terminated at each end with an impedance-matched connector are especially dependent upon the quality of matching. The matching of the connector and cable is not merely a function of their individual impedance values but requires exact procedures in assembling the connector to the cable. In applications at the upper frequency range of 1 to 10 GHz or at voltages of 5 kV and above, it cannot be assumed that the assembly will automatically be satisfactory if the individual cable and connector meet their respective requirements. An interdependent relationship of electrical characteristics exists between the cable and connector which can greatly alter the net electrical characteristics of the coaxial transmission line.

With radio-frequency cable assemblies for microwave use, an improperly prepared assembly will introduce discontinuities with resultant reflections of energy back to the source. In low-power applications, loss of signal could result. In high-power applications, the magnitude of the standing waves which are set up could exceed the power rating of the assembly. In the case of high-power pulse-cable assemblies, care must be taken that no air voids are present in the electric field between the inner and outer conductors. The voltages which are normally applied in these applications could be sufficiently high to initiate corona in these voids, with resultant deterioration of the dielectric structure and eventual breakdown.

Specifications

MIL-HDBK-216[18] provides assembly instructions for the various types of connections used with the different coaxial cables. In addition, assembly instructions are provided by the manufacturer in the catalog and packaging information. Table 19 lists the connectors suggested for use with the standard types of coaxial cables. Table 21 provides a list of military and industrial specifications and standards currently in use to procure both cable and connectors which meet the·established minimum standards for performance. Because these documents are frequently revised, only the basic identification number is provided; however, the latest issue or revision of these documents should be used at the time of procurement.

FLAT-CABLE CONNECTORS

Suitable connectors and termination techniques have been evolving to exploit fully the advantages of flat-cable systems that have been apparent for several years.

Initially, the interconnections of flat cable were accomplished using existing connectors, both printed-circuit types and round-wire types, including cylindrical as well as rectangular. Many applications today still use such connectors requiring moderate to extensive tailoring of the flat cable to fit the connector. Increasingly, however, connectors specifically designed for flat cable are making their appearance.

Typical Configurations

In terms of flat-cable applications, connectors can basically be classified into two categories—those specifically designed for flat cable and those designed for round-wire cables or printed-wiring board which are used for flat-cable interconnections. All connectors are readily classified by shape, i.e., cylindrical, rectangular, etc. Further classification is by the method of termination, i.e., discrete solder, discrete crimp, reflow solder, simultaneous (wave) solder, simultaneous crimp, welding and finally

TABLE 21   Specifications and Standards for Coaxial-Cable Connectors

Cable Specifications and Standards

MIL-C-17—Cables, Radio Frequency, Coaxial, Dual Coaxial, Twin Conductors and Twin Lead
MIL-C-22931—Cables, Radio Frequency, Semi-rigid Coaxial, Semi-air-dielectric
MIL-C-23020—Cables, Coaxial (for Submarine Use)
MIL-C-23806—Cables, Radio Frequency, Coaxial, Semi-rigid, Foam Dielectric
MIL-L-3890—Lines, Radio Frequency Transmission (Coaxial, Air Dielectric)
NEPR No. 3—NATO Electronic Parts Recommendation for Radio Frequency Coaxial Cables
EIA Standard, RS-199—Solid Dielectric Transmission Lines
EIA Standard, RS-225—Rigid Coaxial Transmission Lines, 50 Ohms
EIA Standard, RS-258—Semi-flexible Air Dielectric Coaxial Cables and Connectors, 50 Ohms
EIA Standard, RS-259—Rigid Coaxial Transmission Lines and Connectors, 75 Ohms
IEC Publication No. 78—Characteristic Impedances and Dimensions of Radio Frequency Coaxial Cables
IEC Publication No. 96—Radio Frequency Cable

Connector Specifications and Standards

MIL-A-27434—Adapters, Connector, Coaxial, Radio Frequency, between Series
MIL-C-3607—Connector, Coaxial, Radio Frequency, Series Pulse
MIL-C-3643—Connector, Coaxial, Radio Frequency, Series HN
MIL-C-3650—Connector, Coaxial, Radio Frequency, Series LC
MIL-C-3655—Connector, Coaxial, Radio Frequency, Series Twin
MIL-C-22557—Connectors, Coaxial, Radio Frequency, Miniature (Screw-on)
MIL-C-25516—Connector, Electric Miniature, Coaxial Environmental Resistant Type
MIL-C-26637—Connector, Coaxial, Radio Frequency, Series LT
MIL-C-39012—Connector, Coaxial, Radio Frequency (Series BNC, C, N, QM, QL, SC, TNC)
MIL-C-55235—Connector, Coaxial, Radio Frequency, Series TPS
MIL-F-24044—Flanges, Coaxial Lines, Rigid Air Dielectric
MIL-STD-1327—Flange, Coaxial and Waveguide and Coupling Assemblies, Selection of
MIL-T-55619—Tools, Crimping, Category C, Radio Frequency Coaxial Connectors
NEPR No. 70—NATO Electronic Parts Recommendation for Radio Frequency Coaxial Connectors
IEC Publication No. 159—Dimensions of the Mating Parts of Radio Frequency Connectors
IEC Publication No. 169—Radio Frequency Connectors

MIL Specifications are coordinated by Defense Electronic Supply Center, Dayton.
NEPRs are coordinated by AC/67(SWG/10) Special Working Group on Radio Frequency Transmission Lines of the North Atlantic Treaty Organizations.
EIA Standards are published by the Engineering Department of Electronic Industries Association.
IEC Publications are prepared by the International Electrotechnical Commission, an affiliate of the International Standardization Organization.

piercing, self-stripping, or otherwise displacing or removing insulation to expose the conductor and make contact. Connectors which are designed specifically for flat cable have several basic construction forms.

Flat conductors used as contacts   These form a reasonably well developed concept. The flat conductors are exposed on one side by stripping, then are plated. The cable is then mechanically secured or molded into a suitable male-plug housing or into a female-receptacle housing utilizing a backup spring shape to maintain contact. This type of construction is limited to a single layer of cable per connector.

Wafer concept   In the most recent design each flat-cable layer terminates to a separate "wafer." Each flat cable is terminated to a row of contacts in a suitable insulated carrier, the contact-carrier assembly being designated a wafer. In this concept, each flat cable can thus be inserted or removed and replaced in a connector without changing or relocating other flat cables in the same connector. It also permits termination outside the connector without changing or relocating other flat cables in the same connector and termination outside the connector shell for production efficiency. Several sizes are available with conductor center spacings ranging from 0.200 down to 0.050 in. The mating connector half consists of equivalent wafers with flat cable or an insulator block with round-wire contacts.

Pierce-crimp   The desire for low-cost high-reliability termination of flat cable to connector contacts has promoted a connector concept which does not require a discrete cable-stripping operation. The method provides for sequentially crimping a suitable pin or socket contact through the insulation on either side of the conductor and then displacing the insulation on both the top and bottom conductor surfaces to achieve a gastight crimp. A fully automatic machine is available which takes contacts off a reel and with a step cam automatically terminates the cable. A hand tool is also available. Figure 77 shows the termination.

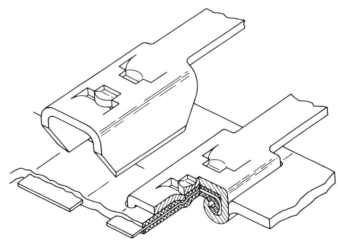

Fig. 77   Insulation-piercing crimp termination. (*AMP, Inc., "Unyt" system.*)

Pierce-pressure contact   Ribbon cable can be terminated by a connector wherein the contacts are spaced to match the conductor spacings and each contact pierces the insulation on either side of the round wire. Each conductor is simultaneously forced between two cantilever spring members which form a wedge. Examples of this type of connector are shown in Fig. 53.

Strip-crimp   The advantages of being able to terminate all conductors of one flat cable simultaneously are obvious. A military version of a connector has been developed whereby the cable is stripped and the flat conductors are then simultaneously folded in a suitable fixture, inserted into crimp pots, and crimped. See Fig. 78.

Conventional In a vast number of applications the flat cable is still terminated to conventional connectors, i.e., those designed for round wires or printed-circuit edgeboards.   While the rectangular shapes generally have a more suitable form factor, in many applications the flat cable is terminated to cylindrical connectors. Where the shape or contact spacing of the connector does not readily match that of the flat cable, several techniques are employed.   The flat cable can be slit and folded to achieve proper position.   Jumper conductors may also be employed.

Fig. 78 Ganged crimp junction.

Ribbon cables are more easily terminated to conventional connectors.   The individual wires are slit out, and then stripped and terminated as with standard hookup wire.   Thus ribbon cables provide the designer with a unique compromise where flat cables are desirable, but conventional hardware must be used.

Clearly, flat cables are being built with existing connectors of every conceivable configuration, chosen for their availability, cost, electrical and mechanical performance, military qualification, suitability to the packaging installations, or any of the various other factors used in choosing connectors.

Specifications and Standards

Of those connectors specifically designed for flat cable, there are those intended for military/aerospace environments of extended temperature, vibration, shock, humidity, moisture, immersion, and pressure sealing.   Connectors meeting most or all of these stringent requirements are described in the Flat Conductor Cable Connector Specification MIL-C-55544 in a series of slash sheets.   Examples are shown in Figs. 79 and 80.

Connectors designed for commercial/industrial applications are described in the Institute for Printed Circuits Specification IPC-FC-240.   In addition, several connector manufacturers have designed and tooled connectors not presently included in either of these two specifications. Attention is also called to Ref. 24.

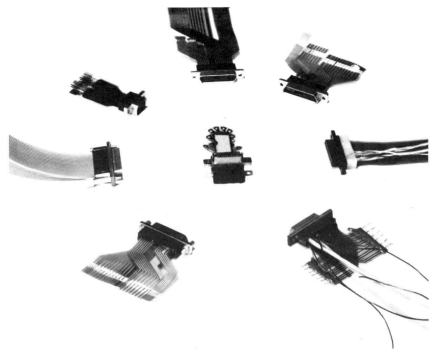

Fig. 79   Metal-shell flat-conductor-cable connector.   (*Microdot, Inc.*)

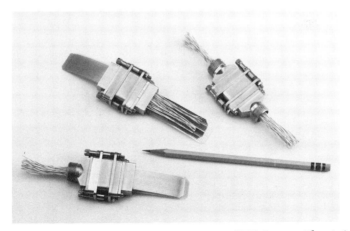

Fig. 80   Typically MIL-C-55544 connectors.   (*ITT Cannon Electric.*)

HERMETICALLY SEALED CONNECTORS

Definition   A hermetically sealed connector is one that offers a gas- or airtight interconnecting junction through a wall or bulkhead of a "black box" electronic package, or wherever it is necessary to conduct electrical energy or signals between ambient air pressure and either a pressurized or a vacuum chamber or container.

A practical definition: A hermetic seal must be gastight and be able to conduct an electric current into a sealed container with minimum disturbance to the circuit.

The performance must be maintained for extremes of pressure, temperature, humidity, thermal and physical shock, vibration, and corrosive atmospheres.

Under these conditions, the connectors must withstand rated current and voltage and must maintain high insulation resistance. They should produce no corona and should not have excessive shunt capacitance or dielectric losses.

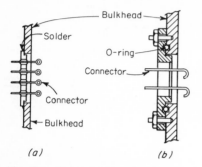

Description Simply described, a hermetically sealed connector is one in which individual contacts are mounted in a metal connector body and insulated from it either with separate glass beads surrounding each individual contact, or with all the contacts sealed in a larger piece of glass. The metal body is then attached to the container or chamber bulkhead by brazing or soldering, or in some instances of vacuum applications, it is held in place by bolts and sealed to the bulkhead with an O-ring type of gasket (Fig. 81).

Fig. 81 Cross section showing two methods of mounting hermetically sealed connectors. (*a*) Individual bead seal. Applicable to either round or rectangular connectors. (*b*) Single glass seal. Primarily suited only for round connectors.

The hermetic seal in a connector depends on a bond between the glass insulator and the individual contacts and between the glass and the metal body of the connector. This bond, or glass-to-metal seal, is produced in three ways: a soft-glass frit, matched glass, and compression glass. Although each type depends on a bond between a metal and the glass, the method of obtaining the bond differs.[19]

The soft-glass type uses an enamel-like glass bonded to mild steel. It makes an economical seal that is useful over a limited temperature range.

The matched-glass type uses low-expansion metal alloys, such as Kovar, Rodar, or Therlo, and glasses with matched expansion coefficients. This type is used for withstanding thermal shock and wide temperature ranges.

The compression-glass type has a steel outer shell shrunken into the glass insulator so that the glass is under compression throughout the operating-termperature range. The thermal coefficients of the glass and of the metal are deliberately mismatched, which aids in maintaining a seal over a wide temperature range. It is especially useful in miniaturized and multiple-contact connectors.

A routine temperature-range requirement for hermetically sealed connectors is from −65 to +400°F, although some connectors are available that will operate in temperatures from −200 to +600°F.

A variety of special glasses is used for hermetic seals, and along with the mild steel and special alloys already mentioned, several of the austenitic stainless steels are also used where mechanical stresses are great.

A variety of round, flat-pierced, formed, and hollow terminal types are available for soldering and welding connecting wires. Wire-wrap terminals are also available, but crimp-type are not.

The two most common types of hermetically sealed connectors are the rectangular rack-and-panel type, where the panel-mounting male half is the sealed

Fig. 82 Miniature rectangular hermetically sealed connector (male). (*U.S. Components, Inc.*)

half (Fig. 82) and the multipin cylindrical type with the seal is the male bulkhead-mounting half. The mating female connector halves are the conventional types.

As in all connector selection, complete application requirements should be supplied to the manufacturer who is to supply or design a hermetically sealed connector, because of the normally critical specifications of most such applications.

PLATE CONNECTORS

Description

The plate connector is so named because it consists basically of a metal baseplate on which contact assemblies and/or connectors are precisely mounted in very accurately positioned holes located in a predetermined grid pattern.

The principal advantage of this concept is that, without any special tooling, a plate connector can be designed to accommodate almost any conceivable combination of plug-in modules, cable connectors, printed-circuit boards, and patch cords. These can be keyed in a number of ways, including variations in group patterns, variations in terminal orientation, and the use of stand-off keys and washers. In addition, visual keying may be achieved by means of color variation in the plastic insulators used.

Rather than a simple connector, the plate connector is, in fact, a packaging system with great flexibility. A major constraint upon this flexibility is that the contacts must necessarily be confined to one of the several grid systems for which tooling exists, both for the plate and for the contact assemblies. However, the fact that all the contacts are on a predetermined grid pattern makes possible the use of automatic machinery to wire the contact terminals as a back plane. Wire wrap and Termi-point° are two of the most common and economical methods of wiring, although solder, weld, and crimp-type terminals are available.

Four basic connector contact types are available: the conventional pin-and-socket type, the blade-and-fork type, the printed-circuit edge-receptacle type, and the hermaphroditic Elco Varicon (Fig. 83). Bus strip contacts and terminal studs are also available.

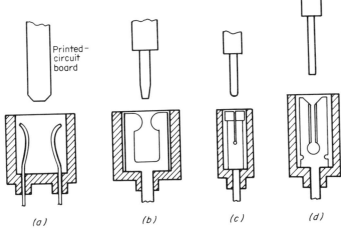

Fig. 83 Basic plate-connector contact types. (a) Printed-circuit edge-connector type, single- and double-sided. (b) Blade-and-fork type. (c) Pin-and-socket type. (d) Hermaphroditic type. (*Elco Corp.*)

The contacts are available individually, loose, and assembled in insulators; in modules containing two or four contacts assembled in insulators; or as complete connectors designed to mate with specific printed-circuit connector boards. The individual contacts can also be arranged to accommodate module headers or conventional mating connectors (Fig. 84).

At assembly, the insulators are pressed firmly and securely into the holes in the metal baseplate, with the terminations extending out the opposite side. Where con-

° Trademark of AMP, Incorporated.

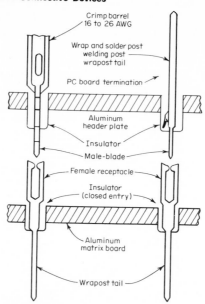

Fig. 84 Basic plate-connector concept illustrated by Malco's WASP construction. (*Malco, Div. of Microdot, Inc.*)

tacts are to be grounded to the metal plate, metal bushings are substituted for the insulating bushings. These metal bushings make intimate contact between the individual contact and the metal plate, thus providing a solid ground connection right at the contact.

Application

The size and shape of the basic plate are dependent on its final usage. The most commonly used plate material is an aluminum alloy with a chromate finish; however, other materials and finishes are also used in some instances.

The baseplate must be a rigid, self-supporting structural member that can withstand, without deformation, the forces required to insert and withdraw mating connecting assemblies, whether they be printed-circuit boards, module packages, or mating connectors.

Contact patterns are usually arranged in rows and groups, allowing for unpunched areas in between, which act as strengtheners and, on very large plates, permit the use of additional structural support to the plate.

The holes in the plate must be distributed in a pattern that is compatible with the automatic wiring equipment to be used. Deviation from true location of any terminal, which is a combination of hole location and the perpendicular attitude of the terminal post in relation to the plate, is limited by the tolerance specification set by the manufacturer of the terminating machinery and must be considered in the plate layout and overall design.

The basic plate design is, of course, dependent on the overall packaging concept, and is the responsibility of the user. All the manufacturers of plate connectors will supply engineering assistance in the design of the plate, especially in the specific details associated with the component contact parts and assemblies. Although the same functional designs are available from several manufacturers, the component parts and design details are not quite the same and are not interchangeable.

Manufacturer's data sheets should be used as references in working out the basic layout and design of the plate, after which it can be submitted to one or several manufacturers for their proposals. At present, all the manufacturers prefer to fabricate the plates and assemble the contacts to it. They have the necessary tools and equipment to do the job, and an important advantage to the user is that there is a single responsibility for the finished assembly. All associated hardware, such as studs and terminals, should be assembled at the same source and the finished plate delivered ready to be wired and installed. Many manufacturers are also equipped to do the wiring, with automatic programmed machinery where applicable.

After the manufacturer has been selected, details of component parts, dimensions, and tolerances can be worked out as a joint effort. Before proceeding with the fabrication job, the manufacturer will supply the user with engineering drawings of the details and assembly of the plate, based on the user's layout for final checking and approval.

Engineering Data

Plate Size. Back-plane plates have been made as large as 24 by 48 in; the only restrictions are that the size must be one that is practical to handle and must not exceed the capacity of the available fabrication tooling. Plates of just a few inches in area can be made for individual module headers and for mating connector bases.

Plate Thickness. At present, contacts and insulators are available for plate thicknesses of 0.080 and 0.125 in. Future developments will undoubtedly allow for other plate thicknesses.

Hole Grid Patterns. Presently available grid spacings, compatible with automatic wiring machinery, are shown in Fig. 85. A hole-tolerance location of 0.003 in nonaccumulative can be held.

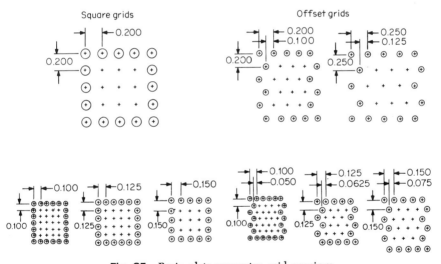

Fig. 85   Basic plate-connector grid spacings.

Contact Materials and Finishes. Phosphor-bronze contacts with various platings are readily available. The manufacturer should be consulted for particular applications.

Insulator Materials. Nylon and polycarbonate are available in many colors for identification and programming.

Assuming a flat back-plane application, the plate-connector concept is very versatile and readily adaptable to high-density packaging.

CONNECTOR ACCESSORIES

Accessories are devices which supplement the basic connector to adapt it to each unique application. Some of the more commonly used accessories are described on the following pages.

Hoods One of the most important accessories to be chosen with a connector is the hood. It is best described as an enclosure attached at the back of a connector to contain and protect wires and cables attached to the terminals of a connector. The hood gives both protection and strength in the form of providing support and strain relief for the cable.

Hoods are available in various shapes and sizes for cylindrical and rectangular connectors. Many provide access plates for servicing the wires. Almost all connector manufacturers supply their own hoods. If special designs are required, the connector manufacturer is usually best suited to design and fabricate these parts.

Figures 86 through 89 illustrate various designs of hoods that are readily available. In most cases the hood may be applied to either the plug or receptacle connector. Figure 86 shows a typical hood available for rectangular, miniature connectors and indicates the availability of a pin protector and 90 and 180° shields. Figure 87

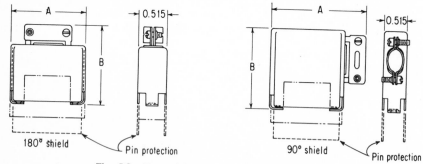

Fig. 86 Typical hood for rectangular connectors.

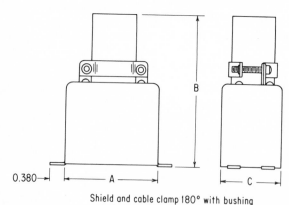

Shield and cable clamp 180° with bushing

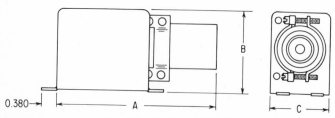

Shield and cable clamp 90° with bushing

Fig. 87 Hoods with bushings.

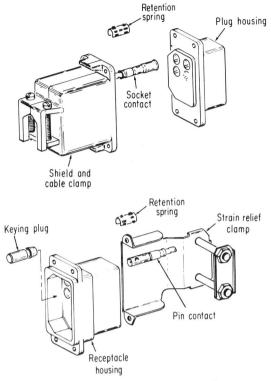

Fig. 88 Hood with strain-relief clamp.

shows hoods with cable bushings in place. These can be used singly or in multiples to reduce a small cable from a larger one. Figure 88 shows a special molded hood on the plug housing, while the receptacle housing uses a strain-relief clamp. Figure 89 also shows various designs of hoods with different cable openings. A special vibration-lock hood design is also shown. Note that this hood has no screw locks and must depend upon the vibration lock for locking purposes.

Boots Boots come in two categories. In the first, they are used as a potting cup. In this application, the boot is used as a mold in potting the region between the rear of the insert and the breakout of the cable or wire bundle. In some models the potting-cup boot is discarded after potting; in others it becomes a part of the shell. In the second application, the boot is used to prevent the entry of moisture into the rear of a connector. Figure 90 shows a series of heat-shrinkable boots used with cylindrical and subminiature rectangular connectors.

Protective shells The protective shell is an outside case, usually metallic, into which the insert and contacts are assembled. Shells of mating halves usually provide for proper alignment and polarization, as well as for protection of the projecting contacts. Figure 91 shows an example of shells with this polarizing slot.

Radio-frequency-interference/electromagnetic interference-filters Several manufacturers provide in their product line connectors with low-pass pi-type ferrite filters designed into their contacts. One type has 240 pF minimum capacitance, 50,000 pF maximum capacitance. Attenuation of size 20 contacts measures 60 dB minimum over 100 MHz through 10 GHz range. The contacts can accept either 20, 22 or 24 AWG size wire. By using this technique, external filters are eliminated for this frequency range. This also eliminates many extra solder joints for improved reliability performance. Filter-pin contacts have also been designed to be used in shell sizes called out in Military Specifications MIL-C-26482 and MIL-C-26500.

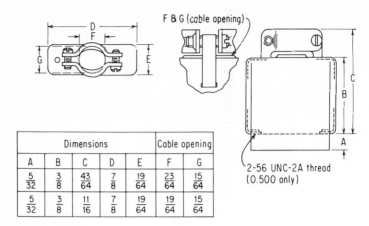

Dimensions					Cable opening	
A	B	C	D	E	F	G
$\frac{5}{32}$	$\frac{3}{8}$	$\frac{43}{64}$	$\frac{7}{8}$	$\frac{19}{64}$	$\frac{23}{64}$	$\frac{15}{64}$
$\frac{5}{32}$	$\frac{3}{8}$	$\frac{11}{16}$	$\frac{7}{8}$	$\frac{19}{64}$	$\frac{19}{64}$	$\frac{15}{64}$

2-56 UNC-2A thread (0.500 only)

(a) Top cable opening hoods

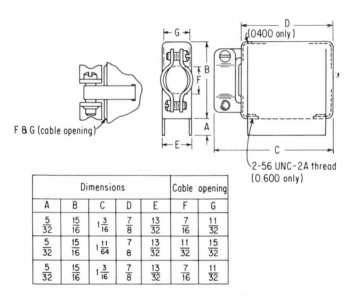

Dimensions					Cable opening	
A	B	C	D	E	F	G
$\frac{5}{32}$	$\frac{15}{16}$	$1\frac{3}{16}$	$\frac{7}{8}$	$\frac{13}{32}$	$\frac{7}{16}$	$\frac{11}{32}$
$\frac{5}{32}$	$\frac{15}{16}$	$1\frac{11}{64}$	$\frac{7}{8}$	$\frac{13}{32}$	$\frac{11}{32}$	$\frac{15}{32}$
$\frac{5}{32}$	$\frac{15}{16}$	$1\frac{3}{16}$	$\frac{7}{8}$	$\frac{13}{32}$	$\frac{7}{16}$	$\frac{11}{32}$

2-56 UNC-2A thread (0.600 only)

(b) Side cable opening hoods

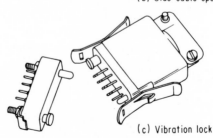

(c) Vibration lock

Fig. 89 Various hood designs: special vibration lock. (*Winchester Electronics Div., Litton Industries.*)

Fig. 90   Various connector boots.   (*ITT Cannon Electric.*)

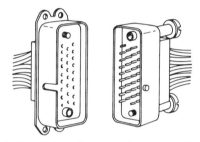

Fig. 91 Protective shells with polarizing slot. (*Winchester Electronics Div.,
Litton Industries.*)

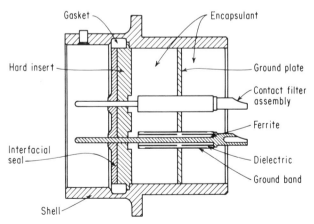

Fig. 92   Cross section of filter-pin contact connector.

Figure 92 shows a typical installation of a filter-pin contact. Figure 93 shows an
actual connector with the filter pins removed. Figure 94 demonstrates a method
for improving the RFI shielding of a connector. In this application the rectangular
connector was modified by sealing the backshell with an aluminum screen, neoprene,
and Monel° RFI gaskets and splitting the back assembly to facilitate servicing of

° Trademark of International Nickel Co.

Fig. 93 Filter-pin connector with pins removed. (*Amphenol Connector Div., Bunker-Ramo Corp.*)

Fig. 94 Shielded-connector assembly.

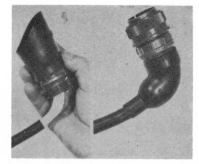

Fig. 95 Heat-shrinkable boots for RFI control. (*Emerson and Cuming, Inc.*)

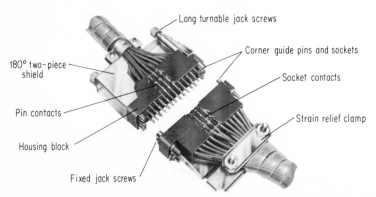

Long turnable jack screws

Corner guide pins and sockets

180° two-piece shield

Socket contacts

Pin contacts

Strain relief clamp

Housing block

Fixed jack screws

Fig. 96 Connector assembly showing jackscrews, clamps, shield, and guide pins and sockets.

wiring and contacts. Figure 95 shows another type of RFI shielding using a conductive, heat-shrinkable plastic boot over the shell and wires.

Mechanical Mating Assists

Connecting and disconnecting of mating halves of connector assemblies can be achieved in various manners; the most common involving mechanical assistance is the jackscrew. See Fig. 96. With dual jackscrews, care must be taken to rotate the jackscrews alternately to prevent damage from cocking. Rectangular or square multipin connectors are available in both single- (center) and double-jackscrew models.

Still another type is the quick-disconnect latch. Figure 97 shows an example on this mechanical assist. It consists basically of a hooked spring on the plug and a retainer on the receptacle. When the plug and receptacle are mated, the spring

Fig. 97 Quick-disconnect latch.

snaps onto the retainer and locks the plug and receptacle together. The two halves are separated by simply depressing the latch springs on both sides.

Other Accessories

Bail An attachment that provides a point of application of the separating force, used with a multipin connector to prevent damaging forces from being applied to wires and contacts. In some designs, it also serves as a latching device. A lanyard is a cable or chain which serves as a point of application of the separation force.

Cap An accessory that covers the face of an unmated connector to protect it from dust, moisture, or physical damage.

Clamp A device that clamps the wire bundle or cable and attaches it to the shell to prevent bending of wires too close to the contact and to apply the separation force to the shell rather than to the contacts.

Grip A woven-wire accessory placed about the cable and attached to the shell by an adapter. The grip prevents sharp cable bends near the connector and provides a suitable point of application of separation force.

Connector Assembly Tools

There are four major, common methods of attaching conductors to connector contacts: soldering, crimping, welding, and solderless wire wrapping. There are also variations on some of these such as infrared or hot-air soldering, taper tabs, and reflow soldering. Since these have been touched upon earlier in this chapter, they will not be discussed here. Instead, the method of crimping, which is the dominant termination method, will be discussed in greater detail.

Many advantages may be gained by crimping. Primarily, reliability is achieved in the strength and accuracy control of the crimping tool, not the operator. The human element is almost eliminated.

Automation levels

Hand. A basic hand tool is normally used for prototype production and experimental work. Normally, these hand tools are designed incorporating a ratchet. This ratchet keeps the tool pressure-locked until the jaws are brought together with the controlled pressure needed to form a perfect crimp. Figures 98 and 99 illustrate several different manufacturers' versions of hand tools.

The crimp tool actually bonds the wire and contact together to form an integral joint which is both mechanically and electrically sound.

Much of the reliability attributed to crimp-type terminations is attributable to the crimp tool itself.

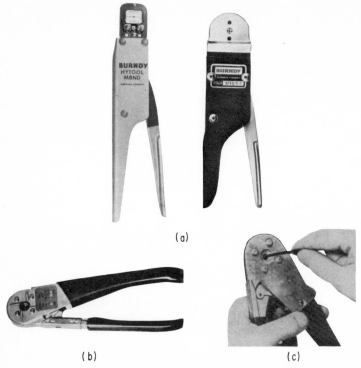

Fig. 98 Various hand-crimping tools. [(*a*) *Burndy Corp.* (*b*) *Buchanan Electrical Products.*]

For a crimp tool to function properly, the following parameters should be checked:
1. Full cycle, ratchet-controlled
2. Indenter configuration
3. Control of crimp depth
4. Performance of crimped joints
5. Gaging capability

As previously stated, crimping now is the dominant method for the following reasons:

1. *Contact Strength.* The crimped joints can be stronger than the wire itself.
2. *Reliability.* The human element is virtually eliminated, since the overall reliability is controlled by the crimping tool, not by the operator.
3. *Speed.* Crimping is fast and easy. Tools are uncomplicated and inexpensive.
4. *Ease of Repair and Modifications.* Terminations can be repaired or modified in the field exactly as in the factory using the same tools and techniques.
5. *Visual Inspection.* Viewing the wire through an inspection hole in the contact makes visual inspection quick, easy, and sure, by both the operator and the inspector.

A word of caution: Care must be taken to see that the correct crimping tool is used for the type and size of contacts being crimped. Also, the pin or socket must be inserted into the tool properly. Then the wire must be stripped of insulation and fully inserted into the contact. The crimping process itself relies on the tool, not on the operator—and is therefore practically foolproof.

Semi automatic. Not unlike some hand tools, semiautomatic tools can be portable hand tools or hand- and/or foot-operated. The latter two can be bench-mounted. Contacts can be supplied in disposable plastic carry strips as shown in Fig. 100*a*.

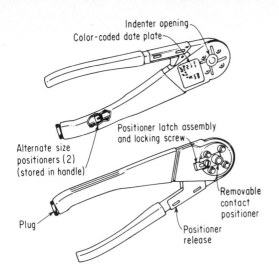

(a) MS No. MS 3191-3
    Adjustable crimping tool
    for pins and sockets with
    MS 3190 crimp barrels.
    Includes MS 3191-4 pliers
    and MS 3191-3T turret
    head for following sizes

Contact size	Color code	Wire size
No. 20	Red	20-22-24
No. 16	Blue	16-18-20
No. 12	Yellow	12-14

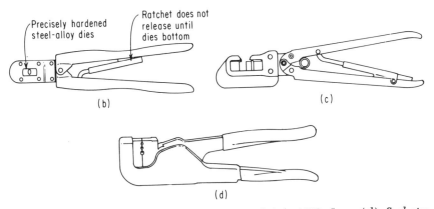

Fig. 99 Various hand-crimping tools. [(b) and (c) *AMP, Inc.* (d) *Sealectro Corp.*]

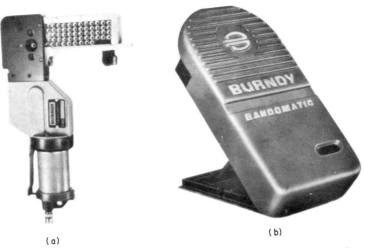

Fig. 100 Various semiautomatic crimping tools. (*Burndy Corp.*)

The carry strips are loaded directly into a magazine from a box so that within a few seconds the magazine is filled with approximately 70 terminals. These tools operate on 85 to 105 lb in^{-2} air-line pressure and will crimp-install contacts as fast as the operator can insert the wire and press the trigger.

These penumatic-powered tools provide the flexibility and portability of a manually operated hand tool but with increased production. Figure 101 illustrates several

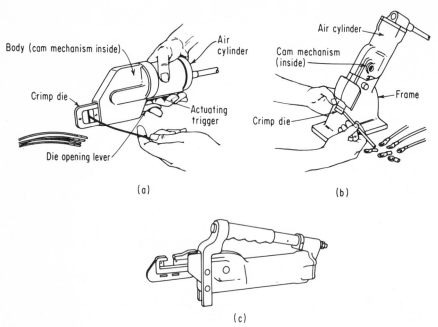

(a)

(b)

(c)

Fig. 101   Various semiautomatic crimping tools.   [(*a*) and (*b*) *Burndy Corp.* (*c*) *AMP, Inc.*]

models where loose piece contacts can also be crimped. Removable dies can be used. A built-in pneumatic cycling control system ensures completion of a stroke before the dies can be reopened.

Automatic. These machines are but one step away from fully automated crimp contact installation. The only requirement is that the operator insert prestripped wire into the crimp area. These machines have numerous features such as:

1. Feeding from a reel which can contain up to 5,000 contacts on a continuous belt.

2. Full cycling controls which can achieve rates to 4,000 installed contacts per hour.

3. Foot pedal on electric trip, which requires the operator to touch the prestripped wire to the trip area to acutate the machine.

4. The machines are constructed of heavy-duty material for continuous usage on high-volume production.

5. Loading on a new reel of terminals is simple and may be accomplished in less than a minute.

6. The machine crimps the terminal on the wire, ejects the crimped terminal from the crimp area, cuts off or feeds out the belt on which the terminals are carried, and feeds a fresh terminal into place for the next crimp.

As shown in Fig. 102*a*, high-speed automatic machines do not take up a lot of room and can be bench-mounted in a line on the production floor. Figure 102*b* shows a typical operator feeding wire into the same machine. Figure 102*c* illustrates

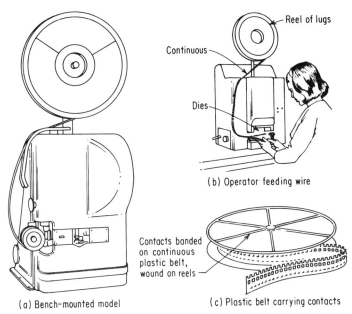

Reel of lugs

Continuous

Dies

(b) Operator feeding wire

Contacts banded
on continuous
plastic belt,
wound on reels

(a) Bench-mounted model

(c) Plastic belt carrying contacts

Fig. 102 Automatic crimping machine. (*a*) Bench-mounted model. (*b*) Operator feeding wire. (*c*) Plastic belt carrying contacts. (*Burndy Corp.*)

the plastic belt that houses the contacts. In these machines, die changes can be made quickly and easily, even by an inexperienced operator.

In addition, some manufacturers produce an automatic stripper/crimper machine which strips the wire and applies both pins and sockets. Figure 103 illustrates this type.

Fig. 103 Stripper/crimper automatic machine. (*AMP, Inc.*)

Crimping is a fast 1-2-3 process. The present crimping tool automatically controls the crimp depth and eliminates the possibility of overcrimping or undercrimping.

Once the crimping tool is inspected for proper setting, it consistently provides identical production terminations. Visual inspection through the control inspection hole immediately verifies proper wire insertion.

The four- indent crimp provides such excellent wire confinement that high pullout forces exceed the minimum wire-breaking strength before they pull the wire out of the contact. Figure 104 illustrates a machine for penumatically operating four-indenter dies for crimping military standard pin-and-socket contacts.

Fig. 104 Pneumatically operated, bench-mounted, automatic-feed crimping tool with four-indenter dies. (*Amphenol Connector Division, Bunker-Ramo Corp.*)

A big advantage to using crimp-type connections is the easy termination accessibility. Problems with closely spaced contacts and close quarters are minimized because contacts can be crimped to leads, then inserted into connectors.

Explosive atmospheres and lack of electric power—problems with soldering—are no longer problems because hand tools need no power and create no hazard. In addition, crimp terminations can be used in high-termperature applications where solder might soften or melt.

Contact Installation

Front insertion Many modern connectors have removable contacts, held in place by a spring clip that snaps into a retaining groove. Such contacts can be removed for wire termination, repair, or modification, even in the field with very simple hand tools.

Insertion and/or removal may be from the front of the connector, or from the rear; each system has advantages and disadvantages. (The wires enter a connector from the "rear"; the interface between plug and receptacle is the "front.")

Contacts that are inserted from the front usually are removed from the front also—but released from the rear. To insert, the wire is threaded through the grommet and then joined to the contact by crimping. Then the contact is worked back into place in the insert where it is held by the spring clip. This method offers the advantage that, during assembly, only the smooth surface of the wire insulation comes in contact with the insert, thereby avoiding the possibility of tool damage to the insert surface. However, the important advantage of front insertion is that it permits a positive stop to rearward motion; that is, the solid portion of the contact can butt against a solid stop in the insert and effectively eliminate any failure due to contact push-backs.

Figure 105 shows a step-by-step procedure for contact insertion from the front.

Contact Termination

The wires to be terminated are first inserted through the rear of the connector insert until they extend far enough beyond the front interface of the connector to provide adequate working space on the wire bundle.

With a hot wire stripper adjusted to the minimum heat setting necessary to strip the wire being terminated, melt through the insulation of the wire 1/8 to 3/16 in. from the wire end.

NOTE: IT IS IMPORTANT THAT THESE STRIPPING DIMENSIONS BE CLOSELY HELD.

Contact Insertion

Draw the terminated wire back — through the insert until the contact is halfway into the contact cavity.

If pin contacts are being inserted, use the end of the insertion tool marked "pin". This is a sleeve that is located over the engaging portion of the contact. Push straight into the contact cavity until a positive stop has been reached.

If socket contacts are being inserted, use the end of the insertion tool marked "soc". This is a probe that is placed in the flared portion of the socket contact front. Push straight into the contact cavity until a positive stop has been reached.

NOTE: An audible "click" will indicate when you have reached a positive stop and the contact is fully seated in the insert retention device.

Fig. 105   Procedure for front insertion of contacts.

Rear insertion   Contacts that are inserted from the rear are removed from the rear, and they may also be released from either the front or the rear.

Rear assembly is usually easier.   The wire is crimped to the contact, and then the contact is inserted into the connector insert.   Figure 106 illustrates a three-step procedure using a rear-entry tool.

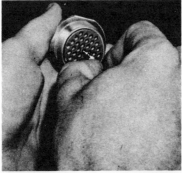

Hold the rear of the connector toward you in one hand, and with the other, push the contact and wire into the contact cavity, being certain not to push it all the way in.

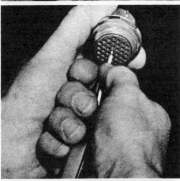

Place the insertion tool against the back shoulder of the contact and push the tool straight into the connector cavity until the contact "snaps" into its retained position in the center insert.
See Note Below.

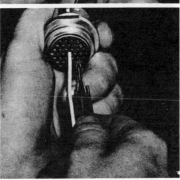

As soon as the retention spring seats itself the contact will be in a positive position of alignment. The operator can sense this and the tool should then be pulled straight back, out of the connector.

Fig. 106   Procedure for rear entry of contacts.

Contact Removal

Front removal   Figure 107 illustrates a typical example where a contact-wire assembly is released from the rear but removed from the front.   Care must be exercised that removal tools are always held and pushed straight against the contact.

In this case, the entire contact-wire assembly can be completely removed for repaired by adding a new contact and reinserting, all from the front of the connection.

Contact Removal

Hold the connector with the rear insert facing you. Fit the removal tool probe around the wire and into the cavity of the contact to be removed. Push *straight* against the contact until it is displaced. Pull the tool straight out of the rear of the cavity.

Contact Removal

Pull the displaced contact and wire from the front of the connector until sufficient working room exists. Cut the contact from the wire, terminate another contact and insert it according to the instructions given in this section.

If the contact cavity is not to be reused, withdraw the wire through the rear of of the insert.

Fig. 107   Procedure for front removal of contacts.

Rear removal   Ease of repair or modification in the field depends upon whether the contact is released from front or rear.   If release is from the front, the same situation holds as for front-removal contacts released from the rear.   If the connector is on a bulkhead, the technician must release the contact from one side of the bulkhead, then go around to remove the contact.

In addition to overcoming these disadvantages of release and removal from opposite sides of the connector, a specific rear-release system has the advantage that only one simple, fail-safe tool is required for insertion, release, and removal, and can be used with pins and sockets both.   The tool is made to break before any damage can be done to the contact or to the insert by improper use, either during assembly or in the field.   Figure 108 illustrates a step-by-step procedure for rear removal. Figure 109 illustrates another type of connector where the removal is from the rear but the release is from the front.

Insertion and removal tools are normally not interchangeable between one manufacturer's contact and another, and damage to contacts and/or moldings can easily result if the proper tools are not used.

SPECIFICATIONS AND STANDARDS

Military Connectors, Low-Frequency and Coaxial

Tables 22 and 23 provide a list of military procurement specifications covering a wide variety of styles, terminations, contact types, coupling, design temperature range, etc.   Some represent a single application.   Many are broad in scope.   These are further documented into individual Military Standard items (MS's), which are far too numerous to present in a single tabulation.   The designer should be aware of this proliferation of documents, if for no other reason than to be prepared.

With the connector rear toward you, snap the *white* end of the appropriate size double-ended plastic tool over the wire of the contact to be removed.

Slowly slide the tool along the wire into the insert cavity until it engages the contact rear and a positive resistance is felt. At this time, the contact retaining clip is in the unlock position.

Press the wire of the contact to be removed against the serrations of the plastic tool and pull both the tool and the contact-wire assembly out of the connector.

Fig. 108 Procedure for rear removal of contacts.

Fig. 109 Example of rear removal, front release of contacts.   Insert bit (into mating end of female contacts, over mating end of male contacts), and push the contact out.

Table 24 represents an attempt to select the most commonly used connector specifications for military applications and to present the various characteristics.   Many designers have used this as a convenient initial selection chart.

Industry Connectors, Low-Frequency

Industry standardization groups have traditionally been active in supporting the various military activities.   The Electronics Industries Association P5.1 Working Group on Connectors and the Society of Automotive Engineers A-2c Connector Committee are the main such groups.   The Aerospace Industries Association, National Aerospace Standards Committee, representing aerospace users only, has from time to time felt the need for special activity in this area.   Table 25 shows the National Aerospace Standards promulgated by that committee.   The NAS 1599 document and its contact specification NAS 1600 have designer support for some critical airborne applications and remain an inventory item with qualified suppliers.   Also shown in Table 25 is an audio connector standard that has been issued and maintained by EIA P5.1.

Specifications for Crimp Contacts and Tools

Table 26 shows the most popular specifications issued for crimp contacts and tools.

CONNECTOR RELIABILITY

General discussion and definition   The reliability of connectors has been the subject of discussion of many meetings and seminars.   Some of these have attempted to treat the subject on a theoretical basis, that is, determine the basic mathematical model.   However, these attempts have conspicuously failed to gain acceptance by coworkers.   On the other hand, when connector reliability is approached from a practical, applied basis, there seems to be great unanimity of opinion regarding the underlying rules, dos and don'ts, etc., which will result in a reliable application.

Within this chapter both aspects will be presented.   To be sure, the mathematical treatment will not represent a treatise on the subject of reliability as such.   Failure-rate-calculation techniques have a basic premise that a connector which is properly defined, well-engineered, and correctly used will be basically a reliable connector.

Reliability has been defined in many different ways by various segments of the electronics industry.   A simple definition that should be easily understood when

TABLE 22 Military Connectors, Low-Frequency

Specification	Date of revision	Title	Description	Preparing activity
W-C-596 Int. (3).......	A, 1-6-66	Connector, Plug, Electrical, Connector, Receptacle, Electrical	Connector, plug, electrical, connector, receptacle, electrical	Navy
W-C-00596...........	B, 10-22-67	Connector, Plug, Electrical, Connector, Receptacle, Electrical	Connector, plug, electrical, connector, receptacle, electrical	Army
AN-3113...........	6, 6-28-57	Connector, Receptacle—External Power, 115 Volt D.C.	Connector, receptacle—external power, 115 V dc	Air Force
MIL-C-3767, Supp. 1 NTC. 3	B, 10-13-66	Connectors, Electrical Power, Bladed Type, General Specification for	Connectors, electrical power, bladed type, general specification for	Navy
MIL-C-005015, Supp. 1	E, 11-7-69	Connectors, Electrical, AN Type	Connectors, electrical, AN type, standard size, cylindrical, environmental and nonenvironmental, solder and removable crimp contacts	Navy
MIL-C-7192...........	B, 10-27-58	Connectors, Cable, Plug and Receptacle, Molded, and Connection Pins	Connectors, cable, plug and receptacle, molded, and connection pins	Navy
MIL-C-7974, Supp. 1..	A, 4-27-60	Cable Assemblies, Plugs and Receptacles, External Power	Connectors used to connect ground power units to aircraft. Covers external power harness assemblies (cables and plugs) and mating receptacles	Navy
MIL-C-8384, Supp. 1..	6-20-63	Connectors, Plug and Receptacle, Electrical (Molded Body) and Accessories	Nonenvironmental connector used in equipment protected from elements. Has all molded plugs and receptacles in rectangular and circular shapes	Navy
MIL-C-10544...........	C, 8-31-64	Connector, Plug and Receptacle, Electrical, Audio, Waterproof, Ten Contact, Polarized	Connector, plug and receptacle, electrical, audio waterproof, ten contact, polarized	Army
MIL-C-12520, Supp. 1A	C, 11-5-65	Connectors, Plug and Receptacle, Electrical, Waterproof, and Accessories, General Specifications for	Ground-support-type cylindrical connectors used in weatherproof applications	Army
MIL-C-18148...........	B, 8-10-65	Connector, Plug, Electrical, Quick Disconnect, Battery	Connector, plug, electrical, quick-disconnect, battery	Navy
MIL-C-18832...........	9-26-61	Connector, Breakaway, Electro-mechanical	Connector, breakaway, electromechanical	Navy
MIL-C-21097, Supp. 1A	B, 11-15-66	Connectors, Electrical Printed Wiring Board, General Purpose, General Specification for	Miniature, multicontact connectors for $\frac{1}{16}$-, $\frac{3}{32}$-, and $\frac{1}{8}$-in. printed-wiring boards	Navy
MIL-C-21617...........	3-5-59	Connectors, Plug and Receptacle Electrical, Rectangular, Polarized, Shell, Miniature Type	Multicontact, rectangular, polarized shell, miniature-type electrical back and panel connectors	Navy

MIL-C-22249.........	4-19-68	Connector Sets, Electrical, Hermetically Sealed, 175-138, 72, and 44 Pin, Submarine	Covers watertight, cylindrical connectors for specialized Polaris connector application on submarine missile tubes	Navy
MIL-C-22857, Supp. 1	C, 2-15-68	Connectors, Electrical, Rectangular Crimp Type, Removable Contact, for Rack and Panel and Associated Applications	Nonenvironmental rack and panel connectors with crimp removable contacts and center and slide screw-lock devices	Navy
MIL-C-22992, Supp. 1C	C, 6-1-66	Connectors, Electrical, Waterproof, Quick Disconnect, Heavy Duty Type	Heavy-duty, waterproof cylindrical connector used with jacketed cables	Navy
MIL-C-23353.........	A, 8-28-64	Connector, High Reliability, Electrical Plug and Receptacle, Polarized Shell, for Printed Wiring and Related Applications	Printed-circuit wiring connectors, used in environment-protected applications	Navy
MIL-C-25955, Supp. 1 NTC. 1	2-5-60	Connectors, Electrical, Environment Resisting Miniature, with Snap-In Contacts	Miniature quick-disconnect cylindrical environmental connector with removable crimp contacts	Air Force
MIL-C-26290.........	B, 12-18-67	Connectors, Lead Assemblies, and Adapters, Electrical, Low Temperature, 300-, 600-, and 5,000-volt Airfield Lighting	Connectors, lead assemblies, and adapters, electrical, low temperature, 300-, 600-, and 5,000-V airfield lighting	Air Force
MIL-C-26482, Supp. 1 NTC. 3 Int. (3)	D, 5-10-66	Connectors, Electrical, Circular Miniature, Quick Disconnect	Miniature circular connectors of the bayonet, screw, and push-pull disconnect type. Cover crimp and solder-type contacts and are environmentally sealed. Used primarily in airborne and light-duty ground-support applications	Navy
MIL-C-26500, Supp. 1	C, 6-26-67	Connectors, General Purpose, Electrical, Miniature, Circular, Environmental resisting, 200°C Ambient Temperature	Miniature circular connectors of the threaded, or bayonet quick-disconnect type. Uses crimp removable contacts. These environmental resisting connectors are used primarily in airborne applications and where high temperatures are prevalent. Established reliability	Air Force
MIL-C-26518, Supp. 1	B, 1-27-61	Connectors, Electrical, Miniature Rectangular, Rack and Panel, Environment Resisting, 200°C Ambient Temperature	This is a rack and panel design using crimp removable contacts, MIL-C-26636	Air Force
MIL-C-27599, Supp. 1	A, 2-26-68	Connector, Electrical, Miniature, Quick Disconnect (for Weapons Systems), Established Reliability	Connector, electrical, miniature, quick-disconnect (for weapons systems), established reliability, solder type, contacts	Air Force
MIL-C-27699, Supp. 1	10-2-61	Connector, Plug, Electrical Connector, Receptacle, Electrical, Heavy Duty, waterproof	Connector, plug, electrical connector, receptacle, electrical, heavy-duty, waterproof	Army

TABLE 22 Military Connectors, Low-Frequency (Continued)

Specification	Date of revision	Title	Description	Preparing activity
MIL-C-38300, Supp. 1A	A, 5-5-61	Connector, Electrical, Circular, Multi-contact, High Environment, Quantitative Reliability, General Requirements for	Miniature cylindrical threaded and bayonet locking types environmental (200°C), crimp removable contacts, with closed-entry hardfront socket insulator option	Air Force
MIL-C-38999, Supp. 1A (1)	A, 1-10-69	Connectors, Electrical, Miniature, Quick Disconnect, Removable Crimp-type Contacts, Established Reliability	Connectors, electrical, miniature, quick-disconnect removable crimp-type contacts, for special weapons-systems circuitry, established reliability	Air Force
MIL-C-55181	A, 4-29-68	Connectors, Plug and Receptacle, Intermediate (Electrical, Waterproof), General Specification for	Connectors, plug and receptacle, intermediate (electrical, waterproof), general specification for	Army
MIL-C-55243	10-25-63	Connectors, Plug and Receptacle, Electrical, Quick Disconnect, 12 Contacts, Mechanical Power	Connectors, plug and receptacle, electrical, quick-disconnect, 12 contacts, mechanical power	Army
MIL-C-55302	A, 8-15-69	Connectors, Printed Circuit Subassembly and Accessories	Connectors, plug and receptacle, electrical, for use with single-sided and multilayer printed-wiring boards, or both	Army
MIL-C-55544	12-26-68	Connectors, Electrical, Environment Resistant, for Use with Flexible Flat Conductor Cable and Round Wire	Connectors, plug and receptacle, electrical, for use with flexible flat conductor cable (MIL-C-555543)	Navy
MIL-C-81511	....	Connector, Electric, Circular, High Density, Quick Disconnect, Environment	High-density, quick-disconnect, environment-resisting 150°C crimp and solder-type connectors	Navy
MIL-STD-1130	11-25-65	Connectors, Electrical, Solderless Wrapped	Connectors, electrical, solderless wrapped	Navy
MIL-STD-1344 NTC 2	5-5-70	Test Methods for Electrical Connectors	Test methods for electrical connectors	Navy

applied to parts within an electronic system is:[20] Reliability is the probability that a given device will perform without failure to a given set of requirements for a specified length of time.

Mathematical Calculations

Military Handbook MIL-HDBK-217B (Proposed), July 1973 This material represents an update of connector failure-rate calculations documented in Rome Air Development Center Reliability Notebook, vol. II, September 1967.

Introduction to failure-rate calculations Failure-rate calculations for connectors are based upon a two-part mathematical model, as follows:

$$\lambda = \lambda_b(\pi_E \times \pi_p) + N\Sigma_{\text{cyc}} \qquad \text{failures}/10^6 \text{ h} \qquad (5)$$

where N = number of active pins

λ_b = base failure rate

π_E = environmental failure-rate modifier

π_p = failure-rate modifier for number of active contacts

Σ_{cyc} = failure-rate modifier for mating/unmating cycling rate

(NOTE: The term containing Σ_{cyc} may be ignored for connectors not experiencing high cycling rates.)

The major factors that affect connector reliability are:

1. Type of insert material
2. Current load in contacts
3. Size of contacts
4. Insert-temperature rise
5. Ambient temperature
6. Maximum operating temperature
7. Number of active contacts
8. Application and environmental service
9. Mechanical or insertional factors

The following material explains how these factors are used to obtain the calculated failure rate.

Type of Insert Material. Four basic types of insert materials are considered in connection with reliability predictions. These are shown in Table 27. Ultimately several additional categories or basic types may be established. At present there is insufficient evidence to justify more than four distinct types. As an aid in determining which of these basic types to use, the connectors associated with the common military specifications have been listed in Table 28. Users of commercial-grade connectors should select an insert-material-type code letter from Table 27, using the operating-temperature range and notes as guides. In using these tables, the reader is cautioned to exercise engineering judgment in assigning a connector to a given specific type.

Contact Current vs. Temperature Rise. The temperature rise of the insert depends on the power dissipated by the current in the active contacts. Figure 110 relates the contact current and the size of the contacts to the insert-temperature rise. This chart can be used in a variety of ways depending on the design and application. The following instructions apply to different circumstances:

All contacts are the same gage and carry the same current: If all the contacts carry the same current, determine the insert-temperature rise by entering the bottom of Fig. 110 at this current value.

Wide variety of currents (all contacts the same gage): If there is a wide variety of currents in the different active contacts and the more heavily loaded contacts are uniformly dispersed in the contact pattern, use an average contact-current value to enter Fig. 110.

Unbalanced load symmetry (all contacts the same gage): If the more heavily loaded contacts are concentrated in a group, calculate the host-spot rise by averaging the current in the loaded group, and use this value to enter Fig. 110.

Unusual load configuration: For unusual situations not covered above, use a weighted average of load currents. Base the weighting factor on engineering judgment considering the configuration, the size of contacts, the size of the connector,

TABLE 23 Military Connectors, Coaxial

Specification	Date of revision	Title	Description	Preparing activity
MIL-C-3607	A, 8-30-61	Connectors, Coaxial, Radio Frequency, Series Pulse, General Specification for	Weatherproof, high voltage, series pulse, radio-frequency coaxial connector operating at a range of 0–100 MHz	Navy
MIL-C-3608, Supp. 1A	A, 9-3-63	Connectors, Coaxial, Radio Frequency, Series BNC, and Associated Fittings	Covers series BNC, having a nominal impedance of 50 Ω, rated at 500 V peak and operating within a frequency range of 0 to 10,000 MHz. Used with "small-sized" coaxial cables	Army
MIL-C-3643, Supp. 1A	A, 10-8-65	Connectors, Coaxial, Radio Frequency, Series HN, and Associated Fittings, General Specification for	Covers series HN, having a nominal impedance of 50 Ω and an operating voltage of 1,500 rms, and an operating frequency of 0–10,000 MHz	Army
MIL-C-3650, Supp. 1A	A, 1-16-64	Connector, Coaxial, Radio Frequency, Series LC	Covers series LC, having a nominal impedance of 50 Ω in a radio-frequency range of 0–1,000 MHz. Used with "large-sized" coaxial connectors	Navy
MIL-C-3655, Supp. 1A	A, 5-5-60	Connectors, Coaxial, Radio Frequency, Series LT, General Specification for	Connectors, coaxial, radio-frequency, series LT, type UG-421B-U, general specification for	Army
MIL-C-3989, Supp. 1A	A, 5-25-64	Connectors, Coaxial, Radio Frequency, Series C, and Associated Fittings	Covers series C, having a nominal impedance of 50 Ω in a radio-frequency range of 0–10,000 MHz. They are designed for "medium-sized" cables and are similar to type N except for quick-disconnect coupling feature	Army
MIL-C-18867	A, 9-21-61	Connectors, ODS, for Radio Frequency Cables	Connectors, ODS, for radio-frequency cables	Navy
MIL-C-21367	C, 11-1-63	Connectors and Associated Fittings for Flexible, Solid Dielectric, Radio Frequency Cables	Connectors and associated fittings for flexible, solid dielectric, radio-frequency cables	Navy
MIL-C-21474	A, 2-24-65	Connector Plug, Telephone Cable, Marine	Connector plug, telephone cable, marine	Navy
MIL-C-22557, Supp. 1	A, 1-26-67	Connectors, Miniature, Radio Frequency, Screw on, General Specification for	Covers miniature connectors with a nominal impedance of 50 Ω and rated at 500 V	Navy
MIL-C-25516, Supp. 1	C, 4-3-68	Connectors, Electrical, Miniature, Shielded or Unshielded, Coaxial, Environmental Resisting Type, General Specification for	Connectors, electrical, miniature, shielded or unshielded, coaxial, environmental-resisting type, general specification for	Air Force
MIL-C-26637	12-7-63	Connectors, Coaxial, Radio Frequency, Series IT, General Specification for	Connectors, coaxial, radio-frequency, series IT, general specification for	Air Force

MIL-C-39012, Supp. 1	A, 5-23-66	Connectors, Coaxial, Radio Frequency, General Specification for	Specification for general requirements on all rf coaxial connectors. Types are covered by slash sheets. Class I covers rf connectors to 10 GHz; class II covers general-purpose types	Army
MIL-C-55074..........	C, 5-15-67	Connector, Plug, Electrical—U-125/G Connector, Receptacle, Electrical—U-186/G Connector, Receptacle, Electrical—U-187/G Contact Assembly	Connector, plug, electrical—U-125/G connector, receptacle, electrical—U-186/G connector, receptacle, electrical—U-187/G contact assembly	Army
MIL-C-55081..........	B, 5-22-67	Connector, Plug, Electrical—U-176/G Connector, Plug, Electrical—U-319/G Connectors, Receptacle, Electrical—U-121/G, Connector, Receptacle, Electrical	Connector, plug, electrical—U-176/G connector, plug, electrical—U-319/G connectors, receptacle, electrical—U-121/G, connector, receptacle, electrical	Army
MIL-C-55116..........	A, 11-15-65	Connectors, Miniature, Audio—Five Pin	Waterproof, polarized, five-contact, electrical connectors (plugs and receptacles) for use in audio-frequency circuits at 60 V maximum potential and 0.5 A maximum potential and 0.5 A maximum current	Army
MIL-C-55169..........	4-13-62	Connectors, Electric, Type Q, Multi-contact.	Quick-coupling, general-purpose connectors	Army
MIL-C-55235..........	7-2-63	Connectors, Coaxial, Radio Frequency, Series TPS	Connectors, coaxial, radio-frequency, series TPS	Army
MIL-C-55459..........	A, 5-22-67	Connector, Plug—Electrical—U-226/G	Connectors, plug—electrical—U-226/G	Army

TABLE 24 Characteristics of Common Military and Industry Connectors

MIL-C-	Style: Cylindrical	Edgeboard	Rack and panel	Flat flexible cable	Termination: Solder	Crimp	Weld	Pierced	Taper pin	Contact size: 0	4	8	12	16	20	22	24	Temp. range °C: −55 to +75	−55 to +125	−55 to +150	−55 to +177	−55 to +200	−55 to +260	−65 to +85	−65 to +125	−65 to +200	Coupling: Thread	Bayonet	Push-pull
5015	X		X		X	X				X	X	X	X	X	X	X			X	X							X		
8384	X				X				X					X	X													X	
10544	X	X	X		X	X					X	X	X	X	X				X						X		X		
12520	X				X						X		X	X	X				X										
21097	X		X		X	X			X			X	X	X	X												X		
21617	X				X	X								X	X				X								X		
22249	X				X	X								X	X				X								X		
22539	X	X			X									X	X														
22857	X		X		X					X	X	X	X	X	X				X						X		X		
22992	X				X	X								X	X				X								X		
23353	X				X	X								X	X	X			X								X		
24217	X		X		X	X							X		X												X		
24231					X	X									X			X											
24308	X				X										X	X													
25955	X				X										X														
26482	X				X	X							X	X	X	X						X				X			
26500	X				X	X							X	X	X	X					X				X		X	X	
26518	X				X								X	X	X													X	
27599	X				X									X	X														
27699	X				X									X	X													X	
38300	X				X	X				X		X	X	X	X											X		X	X
38999	X				X	X			X				X	X	X											X		X	
55116	X				X	X							X		X	X												X	
55181	X	X		X	X	X	X	X					X		X	X	X		X									X	
55243					X											X												X	
55302					X										X										X				
55544					X																		X						
81511	X				X								X	X	X	X			X	X		X						X	
81582	X				X								X	X	X	X						X						X	
83723	X				X								X	X	X	X						X					X	X	
NAS 1599	X				X	X				X	X	X	X	X	X							X					X	X	

TABLE 25 Industry Connectors, Low-Frequency

Specification	Date of revision	Title	Description	Preparing activity
NAS-713	7-58	Connector, Electrical, for Printed Wiring	Connector, electrical, for printed wiring	NAS*
NAS-714	7-58	Connector, Electrical, Female—for Printed Wiring	Connector, electrical, female—for printed wiring	NAS
NAS-715	7-58	Connector, Electrical, Male—for Printed Wiring	Connector, electrical, male—for printed wiring	NAS
NAS-1599	11-30-66	Connectors, General Purpose, Electrical, Miniature Circular, Environment Resistant, 200°C. Maximum Temperature	Miniature cylindrical environmental-resisting connectors of the bayonet and threaded-coupling types. Uses crimp removable rear release contacts. Used primarily in airborne and ground-support applications. Threaded version based on MIL-C-26500 hardware and insert arrangements. Bayonet version based on MIL-C-26482 hardware and insert arrangements. Both use same contacts, back shells, accessories, and crimping tool	NAS
RS-297	A, 6-70	Cable Connectors for Audio Facilities for Radio Broadcasting	Connector, electrical; 3-contact mating and locking connectors used in audio circuits; covers male/female for cable and male/female for wall or panel mounting	EIA†

* National Aerospace Standards Committee, Aerospace Industries Association.
† Engineering Department, Electronic Industries Association.

TABLE 26 Military and Industrial Crimp Contacts and Tools

Specification	Date of revision	Title	Description	Preparing activity
MIL-C-23216, Supp. 1B	B, 3-1-68	Contacts, Crimp Type, Electrical Connector, General Specification	Crimp type, removable contacts for use in electrical connectors (MIL-C-26482, MIL-C-005015, and MIL-C-81511). Covers contact sizes 24, 23, 22, 20, 16, 12, 8, 4, and 0	Navy
MIL-C-26636 (2)	A, 8-15-63	Contacts, Crimp Type, for Electrical Connectors	Covers crimp removable contacts for use in electrical connectors (MIL-C-26500). Covers contact sizes 20, 16, and 12	Air Force
MIL-C-39029	11-8-67	Contacts, Electrical, General Specification for	The proposed specification is intended to be the tri-service general specification covering all contacts used in military connectors	Navy
NAS-1600	12-30-65	Contacts, Crimp Type, for Electrical Connectors	Contacts used in NAS-1599 connectors. Covers 12, 16, and 20 size contacts	NAS
MIL-C-22520	C, 6-6-67	Tool, Crimp Type, for Contact of Electrical Connectors	Crimp tool for termination of conductors in crimp-type contacts of connectors using crimp removable contacts such as NAS 1599, MIL-C-26482, MIL-C-26500, MIL-C-38300	Navy

TABLE 27 Temperature Ranges of Insert Materials

Type	Common insert materials	Operating temp range, °C
A*	Diallyl phthalate, vitreous glass, alumina ceramic	−40 to +250
B	Melamine, neoprene, silicone rubber	−40 to +200
C	Polytetrafluoroethylene (Teflon†)	−55 to +125
	Chlorotrifluoroethylene (Kel-F‡)	
D§	Polyamide (nylon), polychloroprene, polyethylene	−20 to +100

 * Units of extreme design. (Extreme design in this usage could mean multiple current-loaded contacts spaced closely, high filler content in the insert material, or application under conditions of unusual stress or stress cycling.)
 † Trademark of E. I. du Pont de Nemours & Co., Inc.
 ‡ Trademark of 3M Company.
 § Not necessarily accurate for all makes of connectors and all types of installation. (For example, a resilient connector with large contacts in polychloroprene inserts for both male and female shells and used at moderate temperatures may achieve as good a failure rate as rigid diallyl phthalate under the same conditions.)

TABLE 28 Configuration, Applicable Specification, and Insert Material for Connectors

Configuration	Specification	Insert material (see Table 20)			
		A	B	C	D
Rack and panel	MIL-C-8384	X			
	MIL-C-26518		X		
	MIL-C-26517		X		X
	MIL-C-22857	X		X	X
	MIL-C-26482	X			
	SCL-6020	X			
Printed-wiring board	MIL-C-21097			X	
	MIL-C-55544	X			
Printed-wiring board (ER)	MIL-C-23353			X	
Cable, circular	MIL-C-5015		X		X
	MIL-C-26482	X			X
	MIL-C-25955				X
	MIL-C-26500	X	X		
	MIL-C-38300		X		
	MIL-C-38999	X			
	MIL-C-81511	X			
	NAS-1599		X		
Power	MIL-C-3767				X
Coaxial, rf	MIL-C-22557			X	
	MIL-C-3608			X	
	MIL-C-3989			X	
	MIL-C-71			X	
	MIL-C-39012			X	

and the amount of current in relation to the maximum allowable current for the contact size. For a mix of contact sizes, use the current and combination which indicate the highest temperature rise.

 Signal-voltage load only: When no appreciable current is carried by the active contacts, dispense with Fig. 110 and assume no insert-temperature rise above ambient.

 Dry circuits: When most of the active contacts are carrying very minute currents and low signal voltages (dry circuits), derate the connector to a lower material class for reliability calaculation.

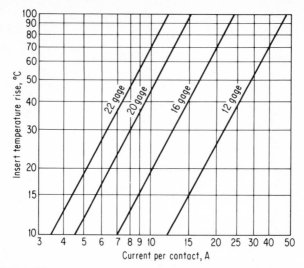

Fig. 110 Relationship of contact current and contact size to insert-temperature rise.

Sometimes the size and style of the lead wire are also important considerations. Large wire and heavy shields can constitute an effective heat sink. Heavy lead wires for small currents are sometimes justifiable on this basis alone.

Operating Temperature. The operating temperature of the connector is usually assumed to be the sum of the ambient temperature surrounding the connector plus the temperature rise generated in the insert. If the connector is mounted in a suitable heat sink (hot or cold plate), the temperature of this sink is usually taken as the ambient. For circuit-design conditions which generate an insert hot spot, the hot-spot temperature is added to the ambient to obtain the operating temperature for reliability-calculation purposes. Reliability calculations are valid only if the maximum temperature for the material is not exceeded. The exception to this rule is if a known high-temperature material is derated to a lower material class because of circuit or loading considerations other than maximum temperature.

Base failure rate Figure 111 shows the relationship between the operating temperature (as obtained above), the code letters related to insert material, and the

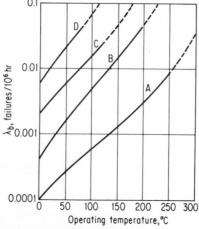

Fig. 111 Relationship of operating temperature to base failure rate λ_b. (NOTE: Letters *A*, *B*, *C*, *D* refer to material in Tables 27 and 28.)

base failure rate λ_b. The base-failure-rate value is then inserted into the reliability-model formula with the suitable additional factors (to be developed next) for the final calculation.

Base Failure-Rate Model λ_b. The mathematical model for the base failure rate is given below

$$\lambda_b = A \exp\left[\left(\frac{T + 273}{N_T}\right)^G + \left(\frac{T + 273}{T_0}\right)^P\right] \tag{6}$$

where T = ambient operating temperature, °C (see Fig. 111)

Constants	Insert material (see Tables 27 and 28)			
	A	B	C	D
A	0.0324	0.69	0.18	1.23
T_0	473	423	373	358
N_T	−1,592	−2,073.6	−1,298	1,528.8
G	−1	−1	−1	−1
P	5.36	4.66	4.25	4.72

For the hot-spot temperature curve, see Fig. 110.

Number of active contacts vs. insert type Connector reliability is affected by the number of contacts electrically active in circuits through the connector. Although the spacing and mechanical configuration of contacts are also important, the simplified approach used here assumes that these other factors are either insignificant or are taken care of in other ways. For example, if it is known that these other factors are significant for a type A insert-material connector, the reliability prediction can be made using one of the lower-temperature types, B, C, or D.

Table 29 provides values of the failure-rate modifier, based upon the number of active contacts. As indicated in Note 2, calculations were based on an exponential formula which provides for sharply increasing values of the modifier between 1.00 and 100.00 for number of active contacts between 1 and 200.

Application and environmental service Table 30 provides values of the failure-rate modifier for application and environmental-service conditions. These factors are applied to the calculation of failure rate to account for the relative severity which the intended application places upon the connector. Although these values are somewhat subjective in nature, extensive studies of filed failure data from military applications have indicated that these values are reasonable in the indication of relative severity. Until field reporting systems are significantly improved to the point where incoming data are screened by engineering experts for failure analysis, the status of this type of data will not be materially changed.

Mechanical and insertional factors Mechanical and insertional factors do not contribute on the same order as the previously mentioned areas until the mating/unmating rate becomes large. The reliability impact of this factor is directly related to the number of pins. The mathematical model for the Σ_{cyc} factor is

$$\Sigma_{\text{cyc}} = A e^{f/f_0} \tag{7}$$

where A = 0.001
f_0 = 100
f = mating/unmating cycle frequency per 1,000 h

Figure 112 displays this mathematical model, in terms of failures/10^6 h as a relationship to the number of cycles per 1,000 h.

TABLE 29 Values of Failure-Rate Modifier π_p for Number of Active Contacts (Pins) in a Connector (See Notes 1 and 2)

N (No. of active contacts)	π_p	N (No. of active contacts)	π_p
1	1.00	65	13.20
2	1.36	70	14.60
3	1.55	75	16.10
4	1.72	80	17.69
5	1.87	85	19.39
6	2.02	90	21.19
7	2.16	95	23.10
8	2.30	100	25.13
9	2.44	105	27.28
10	2.58	110	29.56
11	2.72	115	31.98
12	2.86	120	34.53
13	3.00	125	37.22
14	3.14	130	40.07
15	3.28	135	43.08
16	3.42	140	46.25
17	3.57	145	49.60
18	3.71	150	53.12
19	3.86	155	56.83
20	4.00	160	60.74
25	4.78	165	64.85
30	5.60	170	69.17
35	6.46	175	73.70
40	7.42	180	78.47
45	8.42	185	83.47
50	9.50	190	88.72
55	10.65	195	94.23
60	11.89	200	100.00

NOTE 1: For coaxial and triaxial connectors, etc., the shield contact is counted as an active pin.

NOTE 2: π_p is a function of the number of active pins, as follows:

$$\pi_p = e \left(\frac{N-1}{N_0}\right)^q$$

where $N_0 = 10$
$\quad q = 0.51064$
$\quad N =$ number of active pins

TABLE 30 πE Based on Application and Environmental-Service Condition

Environment	πE	
	UQG*	LQG*
Laboratory...............	1.0	10.0
Space flight...............	2.0	10.0
Ground, fixed...............	3.0	12.0
Naval, sheltered...............	4.0	15.0
Ground, mobile...............	5.0	16.0
Naval, unsheltered...............	8.0	16.0
Airborne, inhabited......... ..	9.0	19.0
Airborne, uninhabited...........	10.0	20.0
Satellite, launch...............	15.0	30.0
Missile, launch...............	15.0	30.0

* UQG and LQG represent upper quality grade and lower quality grade, respectively. Use LQG value until specific vendor data are available to justify UQG.

Examples of connector failure-rate calculations

GIVEN: A connector, pin size 22 gage, type B insert material with 10 active pins The connector will be installed in equipment operating at an ambient temperature of 25°C and an expected load current of 5 A in a ground fixed environment.

FIND: The catastrophic failure rate of the connector.

STEP 1. In determining the base failure rate, the hot-spot temperature of a connector must first be calculated. The conductor heat rise is first determined from Fig. 110 as 19°C, knowing the conductor current and conductor size. Then the approximation formula is applied:

Hot-spot temperature = ambient + heat rise

Hot-spot temperature = 25°C + 19°C = 44°C

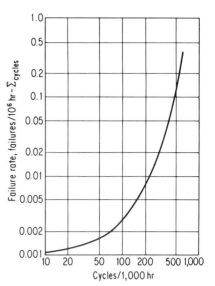

Fig. 112 Relationship between cycles/ 1,000 h and Σ_{cyc} failure rate.

STEP 2. By using Tables 27 and 28, typical insert-material types can be determined for a particular type of connector, if no information is otherwise available in the specifications or manufacturing literature. The insert material given in this example is type B.

STEP 3. Knowing the type of insert material and the hot-spot temperature, the base failure rate λ_b is determined from Fig. 111 to be 0.0013 failure/10^6 h.

STEP 4. As shown in Table 30, the value of π_E for a ground fixed environment is 12.0.

STEP 5. The failure-rate modifier π_p is determined from Table 29 to be 2.58 for 10 active pins.

STEP 6. For connectors not experiencing high cycling rates, the value of Σ_{cyc} is not applicable.

STEP 7. The failure rate of the connector is computed by substituting the values of λ_b and the π terms in the part failure-rate model:

$$\lambda = \lambda_b(\pi_E \times \pi_p) \tag{8}$$
$$= 0.0013(12 \times 2.58)$$
$$= 0.0402 \text{ failure}/10^6 \text{ h}$$

Connector Experiencing High Cycling Rates

GIVEN: A connector, pin size 20 gage, type C insert material with seven active pins. The connector will be installed in equipment operating at an ambient temperature of 25°C and an expected load current of 5 A in a ground mobile environment, with a replug cycling rate of 200 cycles/1,000 h.

FIND: The catastrophic failure rate of the connector.

STEP 1. Knowing that the conductor current is 5 A and the pin size is 20 gage, the heat rise is determined from Fig. 110 to be 12°C. Then, computing the hot-spot temperature by the approximation method:

Hot-spot temperature = ambient + heat rise

Hot-spot temperature = 25°C + 12°C = 37°C

STEP 2. The insert material is given as type C and referenced in Tables 27 and 28.

STEP 3. Knowing the type of insert material, type C, and the hot-spot temperature of 37°C, the base failure rate λ_b is determined from Fig. 111 to be 0.0042 failure/10^6 h.

STEP 4. The value of π_E for ground mobile environment is 16.0, as shown in Table 30.

STEP 5. The failure-rate modifier π_p is determined from Table 29 to be 2.16 for seven active pins.

STEP 6. Σ_{cyc} for 200 cycles/1,000 h is shown in Fig. 112 to be 0.0083 failure/10^6 h.

STEP 7. Substituting the values determined above, and $N = 7$ (for 7 active pins), in the part failure = rate model:

$$\lambda = \lambda_b(\pi_E \times \pi_P) + N\,\Sigma_{\text{cyc}} \tag{9}$$
$$= 0.0042(16.0 \times 2.16) + 7(0.0083)$$
$$= 0.203 \text{ failure}/10^6 \text{ h}$$

Military Handbook MIL-HDBK-217A, December 1, 1965[21]   This A version of the Military Handbook presented the first part-reliability failure-rate data for predictions in a triservices document.   The material is presented here because many designers may still find that this material is contractually invoked and that the publication date of MIL-HDBK-217B is still uncertain.   Material not pertinent to the failure-rate calculation has been omitted from this reprint of information contained in Paragraph 7.9 of MIL-HDBK-217A.

Connector failure-rate determination   For purposes of failure-rate determination, connectors were grouped into circular environment-resistant, rack-and-panel environment-resistant, non-environment-resistant, radio-frequency, and established-reliability types.   Failure rates for the following MIL types are shown and include both male and female portions of the connectors.

Circular, Environment-Resistant
   MIL-C-26500: Connector, General Purpose, Electrical Miniature, Circular, Environment Resisting, 200°C ambient temperature
   MIL-C-26482: Connector, Electric, Circular, Miniature, Quick Disconnect
Rack-and-Panel, Environment-Resistant
   MIL-C-26518: Connectors, Electrical, Miniature, Rack and Panel, Environment Resisting, 200°C ambient type
Non-Environment-Resistant
   MIL-C-5015: Connector, Electrical, AN type
   MIL-C-8384: Connector, Plug and Receptacle, Electrical, Molded Body
   MIL-C-21097: Connector, Electrical, Printed Wiring Board, General Purpose
   MIL-C-21617: Connector, Plug and Receptacle, Electrical, Rectangular, Polarized Shell, Miniature Type
   MIL-C-22857: Connector, Electrical, Rectangular Crimp Type, Removable Contact, for Rack and Panel
Radio-Frequency
   MIL-C-71: Connector, Coaxial, Radio Frequency, Series N
   MIL-C-3607: Connector, Coaxial, Radio Frequency, Series Pulse
   MIL-C-3608: Connector, Coaxial, Radio Frequency, Series BNC
   MIL-C-3655: Connector, Coaxial, Radio Frequency, Series Twin
   MIL-C-3989: Connector, Coaxial, Radio Frequency, Series C
   MIL-C-22557: Connector, Coaxial, Radio Frequency, Miniature Screw On
Established-Reliability
   MIL-C-38300: Connectors, Electrical, Circular, High Environment, Quantitative Reliability
   MIL-C-23353: Connectors, Plug and Receptacle, Electrical, Polarized shell, Established Reliability
The failure rate for a given connector should then be determined as follows:
Non-Environment-Resistant, Environment-Resistant Rack-and-Panel, and Environment-Resistant Circular Connectors
   STEP 1.   Classify the connector as to specific type, applicable military specification.   For example: MIL-C-8384, Non-environment-resistant.
   STEP 2.   Determine the number of active pins.   For example: assume a 24-pin connector wired as follows:
   a. 15 pins wired to active parts of the circuit
   b. 2 pins wired to test jacks
   c. 1 pin wired to ground
   d. 2 pins wired to meters
   e. 4 pins unused
In the description above, *a* and *c* are active pins, and *b* and *e* are not counted when failure rate is determined.   The pins wired to the meters are considered active pins only if a failure in a meter circuit would cause loss of an essential equipment function.

If one "active" pin is assumed in the meter circuit, the total number of active pins will be 17.

STEP 3.   Determine the static failure rate (i.e., in failures per 10^6 h) for the connector under controlled conditions by looking in the appropriate figure for the particular connector type (Fig. 113 to 115).

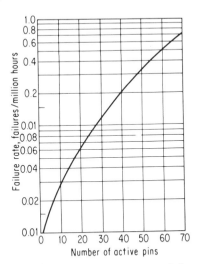

Fig. 113 Failure rates (λ_e in failures per 10^6 h) for MIL-C-26482 and MIL-C-26500, circular, environment-resistant connectors.

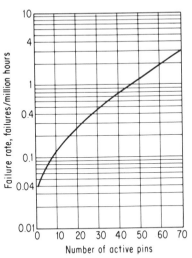

Fig. 114 Failure rates (λ_e in failures per 10^6 h) for MIL-C-26518, rack-and-panel environment-resistant connectors.

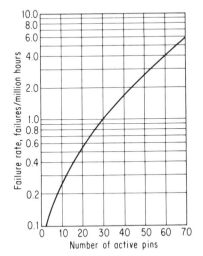

Fig. 115 Failure rates (λ_c in failures per 10^6 h) for MIL-C-5015, MIL-C-8384, MIL-C-21097, MIL-C-21617, or MIL-C-22857, non-environment-resistant connectors.

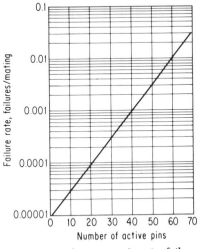

Fig. 116 Failure rates (λ_m in failures per mating) for MIL-C-26482 and MIL-C-26500, circular environment-resistant connectors.

STEP 4. Estimate the number of matings N_m the connector will see in 10^6 h. For example, suppose it is estimated that the connector will be mated once every

500 h.   Converting units as follows:

$$\frac{1 \text{ mating}}{500 \text{ h}} = \frac{N_m \text{ matings}}{10^6 \text{ h}} \tag{10}$$

then N_m = 2,000 matings per 10^6 h

STEP 5. Determine the mating failure rate (λ_m in failures per mating) under controlled conditions by looking in the appropriate chart (Figs. 116 to 118).

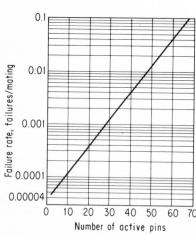

Fig. 117 Failure rates (λ_m in failures per mating) for MIL-C-26518, rack-and-panel environment-resistant connectors.

Fig. 118 Failure rates (λ_m in failures per mating) for MIL-C-5015, MIL-C-8384, MIL-C-21097, MIL-C-21617, or MIL-C-22857, non-environment-resistant connectors.

STEP 6. Determine the total failure rate for the connector under controlled conditions using the formula

$$\text{F. R.} = \lambda_e + N_m \lambda_m \tag{11}$$

where λ_e = static failure rate determined in step 3
   N_m = estimated number of matings in 10^6 h determined in step 4
   λ_m = mating failure rate determined in step 5

STEP 7. Multiply the failure rate determined in step 5 by the K factor corresponding to the part type and anticipated end use (see Table 31).

TABLE 31   Application K Factors

Part type	Ground	Vehicle mounted ground	Shipboard	Airborne	Missile
Connectors, general	1.1	6	0.58	6	60
RF connectors	1.1	4	0.58	4	40

RF Connectors
   STEP 1. Use a failure rate of 0.04 failure/10^6 h for λ_e and 0.00002 failure per mating for λ_m under normal conditions.
   STEPS 2 THROUGH 5. Same as steps 4 through 7 for the preceding type.

High-Reliability Connectors (MIL-C-38300 and MIL-C-23353)
STEP 1.   Use the following values:

Reliability-Level Symbol	Failure Rate (Failures per 10^6 h)
M	10.0
P	1.0
R	0.1
S	0.01

Reliability Considerations

The most generally overlooked contribution to connector reliability—or unreliability—is the handling of the connectors during installation and maintenance. Mishandling physical damage, particularly to contact pins, dirt, improper or careless attachment of wire or cable, improper mounting, etc., can result in reliability problems with an otherwise normally good, reliable connector.

Reliability rules   The following list of rules and suggestions should be of considerable help to the design engineer concerned with reliability in the selection and application of connectors.[22]

1. Unlike other components, a connector does not contribute to the function or characteristics of an electric or electronic circuit. Its function is a mechanical one, to provide a means of engaging and disengaging electrical circuitry that provides energy or routes intelligence from one point to another. Therefore, of prime consideration is that it be of sound mechanical design and construction, and that all materials used—basis contact material, platings, insulating, and structural materials—be adequate and compatible with the physical and environmental stresses to which it will be subjected.

2. Be sure that connectors are properly mounted. If mounted in a panel or frame, there must be sufficient rigidity so that any mechanical stresses on the mounting will not be transmitted to the connector. Proper float relationship between connector halves must be maintained. Make sure that wiring or cabling is properly dressed and supported so that it does not put undue strain on contacts (especially crimp/removable types), and also so that it does not inhibit connector float. Consider the mounting of connectors in relationship to their availability for maintenance and servicing.

3. Connectors must be installed and serviced by trained, capable personnel. Mishandling, from installation and wiring through maintenance, is probably the greatest single cause of connector failure. Be sure that all terminations are properly made using proper tools and processes. Be sure that miscellaneous hardware items are properly applied.

4. Make sure that the supplier or manufacturer has complete specifications for the application. The competitive situation is such that a manufacturer must naturally try to keep costs down and still supply what he feels will do a satisfactory job. It is possible that one manufacturer's price for a particular connector might be significantly lower than all others; if this is so, insist that he prove comparable quality and performance. Remember: you can always buy an item cheaper—and get less.

5. Know the qualifications of the connector manufacturer, and buy from a connector manufacturer, not a hardware manufacturer or assembler. Proper connector engineering is a most important initial step, and adequate quality control during manufacturing is of prime importance. Quality and reliability are synonymous.

6. Make sure that reliability claims made by a manufacturer are valid and applicable to the actual use requirements. Do not assume that a connector that has been qualified to a specification in the laboratory will necessarily perform as well under normal service conditions. A qualification test only demonstrates the capability of the connector to meet specified requirements under closely controlled laboratory conditions. Rough handling, for instance, is normally not a part of testing. Be sure that any anticipated physical stresses and the actual environmental conditions are incorporated into test and evaluation procedures. For example, a large-scale computer is normally installed in an air-conditioned room on a solid foundation, which

would be an ideal environmental atmosphere. However, keep in mind that it must be transported from the manufacturer to the customer, during which time it can be subjected to appreciable shock and vibration conditions that are not applicable to the end-use environment. It must be capable of withstanding these conditions if it is to arrive at its permanent location in good working condition.

7. When measurements are made during testing, especially in low-level work, they must be extremely accurate if they are to be meaningful, either pro or con. Also, it is important that personnel involved in test work be trained in the handling and terminating of connectors.

8. Keep in mind that indicated current ratings do not necessarily mean that all contacts in a connector can be operated simultaneously at full rated current. Check the connector specifications carefully, with the manufacturer if necessary.

9. Determine whether the stated connector thermal limit is under operating or nonoperating conditions. Some connectors, rated as high-temperature units, would burn up if operated at the stated ambient temperature.

10. Avoid a radical and unproved design unless the manufacturer can prove that this design will best meet your particular requirements fully.

11. Another significant cause of connector failure is misapplication. A broad rule of thumb is that a connector should be physically compatible with package size, structure, and handling requirements. For example, a delicate microminiature connector used in an otherwise rugged external cable application would most certainly be the weakest link in the system and, as such, would be subject to physical damage and eventual failure. A rugged application dictates a rugged connector, whereas in a light, protected application, a microminiature connector can well be utilized.

Reliability summary   A connector is a vital component part of a system. It is a mechanical component, and if proper consideration is given to all the many mechanical facets involved—design, construction, quality control during manufacturing, mounting, and installation termination, and compatibility with the mechanical and environmental application—a high level of confidence can be expected initially which, when backed up by a minimum of reliability testing, should result in solidly reliable connections.

OTHER CONNECTIVE DEVICES

In the past, the selection of terminals has been more a matter of individual preference, manufacturing technique, types of terminals in stock, and mechanical design than it has been a question of electrical or electronic design. For years design engineers considered the selection of terminals and terminations not to be their problem. With the stress put on equipment reliability, however, design engineers are giving more attention to the selection of terminals than they have in the past. The following discussion will describe the various terminals in common use and give their advantages and disadvantages. Terminals are made in many basic designs and several variations of each design.

Terminal Lugs

Terminal lugs are divided into two major classes: the solder type and the solderless type, which is also called the pressure or crimp type. The solder type has a cup in which the wire is permanently held by solder, whereas the solderless type is connected to the wire by special tools that deform the barrel of the terminal and exert pressure on the wire to form a strong mechanical bond and electrical connection. Solderless-type terminal lugs are raidly increasing in number and are replacing solder types in commercial, industrial, and military equipment. These terminal lugs come in a variety of designs. Some of the more common recommended terminal lugs are the ring-tongue, round-end, and flag types (Figs. 119 and 120).

Insulation-supporting-sleeve terminal lugs   One of the major sources of trouble when a terminal lug is connected to a wire has always been the breakage of the wire near its junction with the terminal. Wire failures have been decreased by adding a sleeve to the basic terminal. The inside diameter of the sleeve is slightly larger than the inside diameter of the barrel. In the crimping operation, when the barrel is fastened to the end of the wire, the insulation-supporting sleeve is fastened around

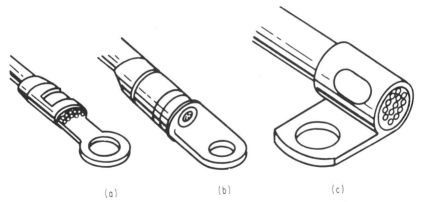

Fig. 119 Terminal lugs, classified according to tongue shape.

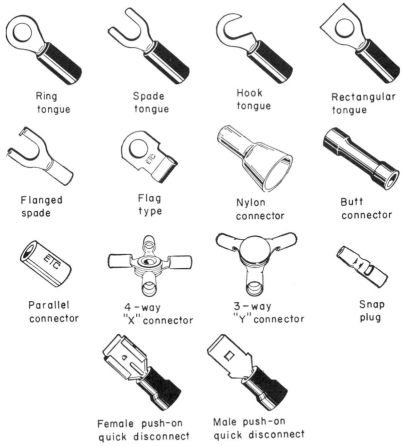

Ring tongue

Spade tongue

Hook tongue

Rectangular tongue

Flanged spade

Flag type

Nylon connector

Butt connector

Parallel connector

4-way "X" connector

3-way "Y" connector

Snap plug

Female push-on quick disconnect

Male push-on quick disconnect

Fig. 120 Typical terminal lug and connector types.[23]

the insulation. This additional support prevents excessive bending of the wire at the point where it enters the barrel of the terminal, and also prevents fraying of the insulation or braid that is over the wire. This type of terminal lug is illustrated in Fig. 121.

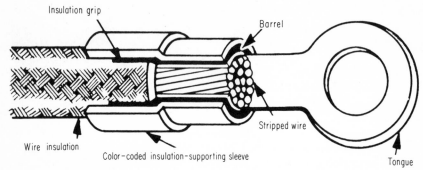

Fig. 121   Insulation-supporting-sleeve terminal lug.

Specifications *MIL-T-7928, Terminals; Lug and Splice, Crimp Style, Copper.* This specification is a combination of two older specifications, MIL-T-5042 and MIL-T-007928B(ASG). It covers terminals for use on wires conforming to Specifications MIL-W-5086, MIL-C-7078, and MIL-W-8777 in locations where the temperature of the conductor is limited to 175°C for type I terminals and 105°C for type II terminals. Because of this temperature limitation, use of type II terminals on MIL-W-8777 wire is not recommended.

The lug and splice terminals included in this specification are divided into types and classes as follows:

Type I: Lugs and splices, uninsulated

Type II: Lugs and splices, insulated

Type III: Water seal

Class 1: Lugs and splices that meet the requirements of this specification when installed with the applicable crimping tool or crimping dies

Type I: Crimping tools conforming to drawings AN3427 and AN3428

Type II: Crimping tools conforming to Standard MS 25037 for sizes AN-22 through AN-10 wire, and by tools approved by the activity responsible for qualification for sizes AN-26 and AN-24, and AN-8 through AN-0000 wire

Class 2: Lugs and splices that are replaceable by class 1 terminals and that, when installed with a tool recommended by the terminal manufacturer, meet the performance requirements of this specification.

The terminals should be made of soft copper in accordance with Specifications QQ-C-502 and QQ-C-576, soft temper, or of copper tubing in accordance with Specification WW-T-799. Commercial bronze, composed of 90 percent copper and 10 percent zinc, is an optional barrel material. The conducting parts of the terminals must be tin-plated. Outline drawings of the three styles of terminals included in this specification are shown in Fig. 122.

For type II lugs the insulating sleeve is a colored material as shown in Table 32. For type II splices the insulation need not be colored, but an identifying color in accordance with Table 32 must be on some portion of the splice. The insulating material has to be noncorrosive, noncombustible, and resistant to abrasion and fungus.

When lugs or splices are connected to the appropriately sized wires, the initial voltage drop should not exceed the value shown in Table 33 when the current is as shown.*

---

* The voltage drop of lugs is measured from the intersection of the tongue and barrel to a point on the wire at least ¼ in from the open end of the barrel. When lugs have insulation supports, this point is at least ⅛ in from the end of the support. On insulated lugs and splices, these points are at least ⅛ in from the ends of the insulation.

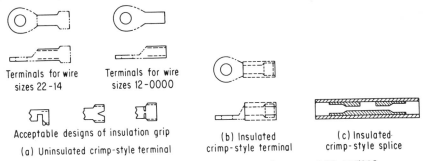

Fig. 122 Outlines of terminal lugs and splices per MIL-T-7928.

TABLE 32 Identifying Colors of Insulated Crimp-Style Terminals (MS25036) and Splices (MS25181) per MIL-T-7928

Wire size	Insulation color	
	Terminals	Splices
26–24	Black	
22–18	Red	Red
16–14	Blue	Blue
12–10	Yellow	Yellow
8–0000	Clear to white	

TABLE 33 Requirements for Terminals per MIL-T-7099 and MIL-T-7928

Copper wire and terminals		Aluminum wire and terminals		Voltage drop, max, mV		Tensile strength, lb min	
Wire size MIL-W-5086 designation	Test current, A dc	Wire size MIL-W-7072 designation	Test current, A dc	Copper	Aluminum	Copper	Aluminum
AN-26	3	....	...	8	...	7	
AN-24	4.5	....	...	8	...	10	
AN-22	9	....	...	8	...	15	
AN-20	11	....	...	7	...	19	
AN-18	16	....	...	7	...	38	
AN-16	22	....	...	7	...	50	
AN-14	32	....	...	6	...	70	
AN-12	41	....	...	5	...	110	
AN-10	55	....	...	5	...	180	
AN-8	73	AL-8	55	5	9	225	120
AN-6	101	AL-6	73	5	7	300	150
AN-4	135	AL-4	101	5	7	400	225
AN-2	181	AL-2	135	5	7	550	300
AN-1	211	AL-1	155	5	7	650	450
AN-0	245	AL-0	181	5	7	700	600
AN-00	283	AL-00	211	5	7	750	750
AN-000	328	AL-000	245	5	7	825	900
AN-0000	380	AL-0000	283	5	7	875	1050

MIL-T-7099, Terminals; Lugs, Crimp Style, for Aluminum Aircraft Cable. This specification covers straight and flag-type, crimp-style terminals for aluminum aircraft cable, sizes AL-8 to AL-0000, conforming to MIL-W-7072. The terminals must be tinned over their entire surface. The voltage drop from the intersection of the tongue and barrel to a point on the cable ¼ in from the open end of the barrel (or, when terminals have insulation supports, ⅛ in from the end of the support), measured before and after vibration, current-overload, and corrosion-resistance tests, with the test currents shown in Table 33, should not exceed the values shown in that table. Tensile-strength requirements are also shown in Table 33. All terminals have to withstand temperature-rise, current-overload, and corrosion-resistance tests as covered in the latest issue of the specification.

Other solderless-terminal specifications The above specifications were discussed in some detail because they are applicable to solderless terminals that are used with electronic equipment. Other specifications that are concerned with solderless terminals for other uses are given here for reference only.

MIL-T-55156, Terminals, Lugs; Splices, Conductor; Screw Type. This specification covers the general requirements for screw-type lug terminals and conductor splices for solid and stranded copper conductors. The style is identified by the two-letter symbol LP followed by a two-digit number. The letters identify screw-type lug terminals and conductor splices, and the number denotes the slash number of the detail specification. The metal used in the body and other current-carrying parts is copper or a copper alloy containing not more than 7 percent zinc.

MIL-T-13513(Ord), Terminals; Lug, Electrical, Solderless Splices; Conductor, Electrical, Solderless (for Automotive Use). This specification covers crimp-type solderless terminals and splices, which are commonly known as solderless cable lugs and are used with ordnance electrical cable for automotive vehicles.

MIL-E-16366(Ships), Electrical Clamps and Lug Terminals: Pressure Grip This specification covers electrical clamps and lug terminals used for shipboard electrical-cable connections. They are intended for (1) connecting feeder cables to distribution-panel busbars, (2) permanently splicing two cables, (3) emergency means for quick splicing two cables (4) connecting branch lines to feeder mains, and (5) those applications where the operating temperatures are too high to use solder-lug types.

MIL-T-15659, Terminals; Lug, Solder Type, Copper

  Class I: Stamped copper sheet
  Class II: Punched copper tubing
  Class III: Stamped phosphor-bronze or tin-brass sheet

The terminals are fabricated of soft copper. When they are dipped in a noncorrosive flux and then in a tin-lead solder (composition SN40 of Specification QQ-S-571) at a temperature between 550 and 660°F, the solder should adhere smoothly and firmly to the finished terminal so that it cannot be lifted with a sharp-edged instrument. Stamped-sheet terminals should be able to be formed around the conductors for which they are designed without showing any evidence of cracking of the base metal or flaking or peeling of the finish.

The type of tongue to use on the required terminal depends upon the part to which the terminated wire is to be connected, space limitations, whether there is any need for ready removal, and the required degree of security of the connection. When space is at a premium, the ring shape gives the smallest outside dimension for a given current-carrying capacity. When terminals must be held under screw heads for easy removal, the slotted type is sometimes considered. Its major disadvantage is that the terminal may become disconnected if the screw becomes loose, and for this reason its use is not recommended in military applications.

Stud Terminals

Description Stud terminals are constructed as either insulated or uninsulated types. The latter type is generally a single metal brass piece processed to provide for the connection of wires at one or both ends. Assembly to a mounting surface is afforded by a stud design, with male or female threads. If the stud is swaged, as a mounting means, it is knurled to prevent rotation. In some designs, the knurled construction is press-fitted into the mounting surface—either metal or insulating material—as the only means of retention, providing a reliable assembly with respect to torque, tension, and bending.

In the insulated constructions, a suitable insulating material is molded as a barrier electrically isolating the mounting means from the terminal section intended to serve as a junction point for one or more wires. Materials in common usage are glass-filled melamine, diallyl phthalate, ceramics, and Teflon (TFE). Mounting means for the uninsulated type are applicable, with the added assembly for the Teflon design. In this case the Teflon is press-fitted into the mounting hole of diameter smaller than the outside diameter of the Teflon stud. The memory characteristic of Teflon constrains to clamp the terminal to the mounting piece.

To ensure adequate solderability, the terminal is copper-plated followed by electro-tin plating or hot-tin dip. Other parts of the terminal, not in the area of termination, may be nickel-plated.

Terminals of the crimp design are essentially a formed copper or brass barrel terminating in a lug for assembly to threaded studs or by other mechanical means. The throat of the barrel receives the bared conductor, and a crimping process done with hand or automatic power tool tightly and permanently squeezes the bard conductor and terminal to form a reliable integral unit. Both insulated and uninsulated designs are used, the latter adding to outer insulating sleeve.

General use   Terminals of the basic stud design are primarily intended for use as miniature terminations for wiring and/or components. Their use in ac or dc circuits does not involve a problem, although consideration must be given to operating voltage, current, and frequency. Choice of insulating material, terminal size, and spacings between the terminal connector and mounting section of the terminal must be compatible with these parameters for reliable operation.

Terminals of the crimp design are intended for use as a termination for one or more wires, and its subsequent assembly on a threaded stud with a nut. Here also, the construction may be either the insulated or uninsulated designs, depending on the application.

Characteristics   The basic stud types provide for single- or double-turret designs. The choice of insulating material is a compromise of all the various properties desired in the final assembly. Diallyl phthalate is rated high with respect to dielectric and mechanical properties. However, its dielectric constant is high compared with Teflon, and in this event, the shunt capacitance of the terminal is correspondingly high. This may be a problem in high-frequency applications.

The ceramic-insulated types excel in low-loss application at high frequencies. However, their mechanical deficiencies on impact and bending suggest that care must be taken if this type is chosen. In the event of high-voltage break-over at the surface, ceramic types are not susceptible to tracking, as is the case with most organic insulations.

The Teflon-insulated types, while possibly necessary for specific applications noted above, are not generally desirable unless care is taken during assembly. Teflon is resilient and will "heel" upon application of pressure normal to the terminal axis, a condition which may occur during assembly. When spacings between the terminal head and other nearby parts of opposite polarity are critical, this factor could be catastrophic.

The crimp-type terminal, when properly crimped to one or more wires, provides a reliable means of termination. However, due consideration must be given to conductor size and insulation wall thickness, load current, and operating voltage.

Feedthrough Terminals

The function of a feedthrough terminal (sometimes called feedthrough insulators) is to provide an electrical connection through a sealed panel or surface. Electrically, the terminal should interfere with the function of the circuit as little as possible. The terminal is usually composed of a metal lead surrounded by insulating material, around which is a mounting flange or a bushing for sealing to metallic containers. A few of the many types of feedthrough terminals are shown in Fig. 123. In selecting feedthrough terminals, several factors must be considered: for example, corona level, flashover and breakdown voltages, and insulation resistance for the temperature and humidity conditions that will have to be met by the equipment; current-carrying

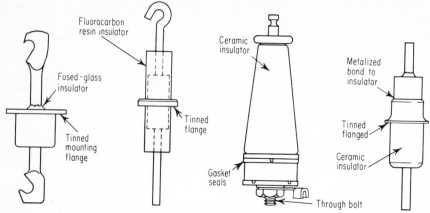

Fig. 123   Types of feedthrough terminals.

capacity; leakage rate; capacitance; loss factor; the temperature range through which the terminal must operate and the thermal shock that the terminal must withstand; the mechanical shock and vibration that the terminal will be subjected to; and space limitations in the equipment.

Most of the feedthrough terminals specified in MIL-T-55155 use either glass, ceramic, or a fluorocarbon resin as the insulating material. The metal-to-glass terminals may be subject to electrodeposition of metal on the glass surface when they are exposed to repeated temperature and humidity cycling while a dc polarizing voltage is applied between the conducting lead and the flange or bushing. This condition will lead to a decrease in insulation resistance and a lowering of the dielectric strength of the terminal, with eventual breakdown between the conductor and the bushing.

When voltages above 3,000 V dc (operating) or 6,000 V dc (test are encountered, ceramic terminals are generally used. This type of terminal falls into two broad classes: the compression type and the solder-sealed type.

The compression-type terminal is composed of an insulator body consisting of two sections and a central feedthrough terminal that is threaded on one end to provide axial compression. The sealing is provided by gaskets that are made of natural or synthetic rubber or a resin. The insulator body may be either hollow (to permit the use of insulating oil after fabrication is completed) or solid. Although the oil-filled terminal may have somewhat better electrical characteristics than the solid type, the solid type is generally preferred in military equipment, to avoid the hazard of flammability of most insulating oils.

The solder-sealed-type terminal is composed of a one-piece ceramic body and a central feedthrough conductor. A mounting flange is attached to the ceramic body. This terminal forms a hermetic seal when it is soft-soldered or brazed to the can that contains the circuit element with which the terminal is associated.

Test-Point Connectors

Application of connectors to provide isolated test ponts on printed-wiring boards has greatly expanded the variety of types available. Test-point connectors were formerly either an additional solder lag, specified terminals on a barrier terminal board, or a tip jack. Military Specification MIL-C-39024 has now been revised to include, besides tip jacks, a wide variety of printed-wiring-type test-point connectors. The latter types consist of: (1) single-point type, two- and three-leg mounting, low-voltage; (2) single-point panel type, threaded mounting; (3) 4, 6, or 8 test points, end mounted, contacts on 0.156-in centers; (4) 16 test points end-mounted, contacts on 0.200-in centers; (5) panel types, single-point, push-on mounting; (6) a panel type, single-point, subminiature with a snap-in contact; and (7) a single-point, high-voltage contact for panel mount. Figure 124 shows commercial equivalents of the single-test-point, low-voltage printed-wiring-board type.

Fig. 124   Commercial printed-wiring-board test-point connectors.   (*AMP, Inc.*)

Headers

Headers are used when a number of terminals are required in a limited area in connecting to a circuit a component that is to be hermetically sealed.   The terminals are of the feedthrough type.   Headers are usually composed of hard glass and a special alloy, Kovar,° which has a thermal expansion that matches that of glass over the entire operating range of components.

Headers come in two main types.   In one, the Kovar collar surrounds a glass plate through which pass the required number of terminals; in the other, a perforated metal plate has individual bead terminals passing through it.   Both types are illustrated in Fig. 125.

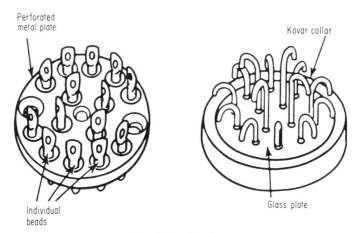

Fig. 125   Headers.

For a given number of terminals in the same area, the individual bead type is usually considered stronger mechanically.   The all-glass type is considered electrically superior, however, because the greater distances between the conducting metallic parts give them higher corona starting, flashover, and breakdown-voltage ratings.   As in the selection of practically all other components, the choice of which type of header to use is a compromise and is dependent upon which condition is the most critical.

° Trademark of Westinghouse Electric Corp.

Stand-off Terminals

Stand-off terminals (sometimes called stand-off insulators) are similar in appearance to feedthrough terminals except that the conducting portion does not pass through the insulating portion. Stand-off terminals are used to mount components a given distance away from a metal base, panel, or chassis.

Binding Posts

The BP-1 (hole-in-head) binding post per MIL-P-55149, Binding Post type PB01, should be used for terminating WD-1/TT or 16 AWG and smaller wire in applications requiring insulated caps, seal-to-panel, and quick disconnect. A special feature of this binding post is the self-tightening wire grip. It can be mounted on ½-in centers for miniature applications. Other types covered by this specification are insulated (type PB01) and noninsulated (types PB02, PB03, PB05, and PB06). Waterproof, insulated binding posts are also available per types PB07 and PB08. A side-slot, spring-cap binding post is covered by MIL-P-55419/8. It can be mounted on ¾-in centers (Fig. 126).

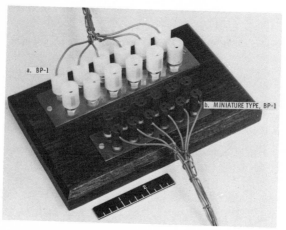

Fig. 126 Binding posts.

Terminal Boards

Terminal boards have two basic designs: the plain style (MS 27212) and the barrier style (MIL-T-55164). Either kind can have almost any type of terminal, such as binder-head screw, solder type, solderless or crimp type, turret type, or any special type. With the plain-type terminal board, shown in Fig. 127, no insulating material separates the individual terminals. The barrier type has ridges between terminals, as illustrated in Fig. 128.

The barriers between terminals increase the creepage distance between terminals and at least decrease, if they do not eliminate, the possibility of short circuits caused by wires extending too far from the terminals to which they are connected.

Clip-type terminals, in which fuses and other ferrule-terminated components are mounted, may be of the screw or soldered type. In the screw type, the binding screw is generally on the same side of the board as the clip. In the soldered type, the soldered tab may be on either side of the board.

Specifications *MIL-T-55164, Terminal Boards.* This specification covers terminal boards that are used in making connections in electric and electronic circuits. Terminal boards are designated in the following manner:

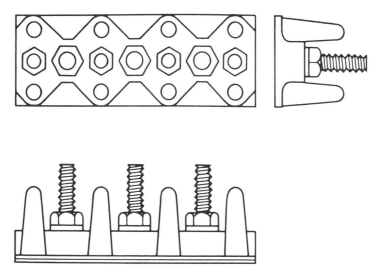

Fig. 127 Typical plain terminal board per MS 27212.

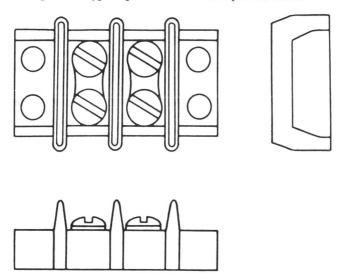

Fig. 128 Typical barrier-type terminal board per MIL-T-55164.

The class indicates a terminal board of a given construction and application, and the number of terminals indicates the number of insulated connection points on the terminal board.

The molded insulation material for all boards should be in accordance with type MAI-60 of MIL-M-14 or type GDI-30F of MIL-P-19833. For high-temperature terminal boards it should be type MSI-30 of MIL-M-14. The material for terminal nuts and stud connectors should be brass, in accordance with QQ-B-626. The material for terminal studs should be, in order of preference, (1) corrosion-resisting steel, class 304, condition E, of QQ-S-763; (2) manganese bronze, hard or half hard, in accordance with class B of QQ-B-728; and (3) phosphor bronze, in accordance with composition A of QQ-B-750. The bronze material should have a minimum yield strength of 60,000 lb in^{-2} and a minimum tensile strength of 80,000 lb in^{-2}.

When terminals are wired in a specified manner and carry rated current, the temperature rise should not exceed 50°C in an ambient temperature of 25°C. The dielectric strength between adjacent terminal studs of each terminal board should be 1,500 V rms for boards with 300-V ratings, 2,200 V rms for boards with 1,000-V ratings. In addition, front-connection terminal boards should withstand these voltages between terminal studs and ground. The insulation resistance should be not less than 10 MΩ when 600 V dc is applied for 1 min in the same manner as in the dielectric-strength test (see Table 34).

TABLE 34   Terminal-Board Types per MIL-T-55164

Class	Description	Max voltage rating	Max application set*	Max no. of wires per terminal	Wire-terminal specification MIL-E-16366
3TB	Single row, front connection	600	C	4	L-80
4TB	Double row, front connection	600	C	4	L-80
5TB	Single row, through connection	600	C	4	L-80
6TB	Double row, front connection	600	B	4	L-81
7TB	Single row, through connection	600	B	4	L-81
8TB†	Double row, linked, front connection	300	B	2	L-83
			A	3	
9TB	Single row, front connection	300	B	4	L-81
10TB	Double row, front connection	600	B	4	L-84
11TB	Single row, through connection	600	B	4	L-84
15TB	Double row, front connection	600	B	4	L-82
16TB	Double row, front connection	1,000	C	4	L-80
17TB†	Double row, linked, front connection	600	C	3	L-80
			B	4	
18TB	Single row, front connection	600	C	4	L-80
25TB	Single row, front connection	300	A	3	L-85
26TB†	Double row, linked, front connection	300	A	2	L-86
				3	
27TB	Single row, through connection	300	A	3	L-85

* Application sets are as follows:

Set A: These terminal boards are intended for use in equipment where the effect of a short circuit is limited to the terminal board and where normal operation powers up to 50 W are involved.

Set B: These terminal boards are intended for use where secondary short-circuit protection in the form of fuses, circuit breakers, and other parts is provided in the circuit, and where normal operating power does not exceed 2,000 W per terminal.

Set C: These terminal boards are for power applications in excess of 2,000 W (per terminal) but still protected by secondary devices in the circuit which can safely interrupt resultant short-circuit currents.

† The stud connector for these terminal boards is considered part of the stud, and is required, unless otherwise specified. With the stud connector removed, the terminal board will provide twice the number of connection points and allow for an additional wire per terminal; however, the application set will be reduced as shown.

Sockets (Tubes, Relays, Crystals)

Sockets for military applications should, in general, conform to the requirements of MIL-S-12883, General Specifications for Sockets and Accessories for Plug-in Electronic Components. Included are items for electron tubes, crystals, relays, and plug-in components for individual and printed-wiring-board mounting. The seven-pin miniature and nine-pin (noval) sockets should use the E-Z mate polarized contour to ensure quick and easy mating of tube and socket in blind locations. Printed-wiring-board tube sockets should be so designed that they are attached and held to the board by means other than the contacts. Contacts should not transmit the insertion and withdrawal forces to the solder joint.

Application   The application of a socket in high-voltage circuits is limited basically by the assigned voltage rating. However, it is not sufficient to design equipment to provide a tube-socket operating voltage at or below this rating. A number of factors may prevail which mitigate against satisfactory operation, especially from a high-reliability standpoint:

1. The commonest degrading influence is the accumulation of foreign material, particularly on the surface between adjacent socket clips. The resultant decrease

in dielectric capabilities, attendant upon voltage breakdown and/or loss of insulation resistance, may result in a circuit malfunction.

2. The operating voltage of the circuit containing the tube socket should be investigated not only from the standpoint of applied dc or rms, but also for waveform. Peak voltage pulses with steep wavefront superimposed on the socket terminal voltage may result in excessive corona and contribute to RFI and ultimate failure of the socket. Operation at air pressures below sea level aggravates this mode of failure.

3. A frequently overlooked detail is the casacding effect of operation of any two adjacent contacts at voltages to ground within the maximum rated limit. Inspection of the circuit may reveal a series configuration, e.g., transformer winding with center tap grounded; and in this event the voltage between socket clips connected to the undergrounded legs of this winding will represent the sum of voltage at each leg.

4. Another item meriting consideration is the effect of loss of load of other components in a parallel circuit. Owing to inherent circuit regulation, the applied voltage at the tube socket may increase with resultant degradation or failure of the socket.

Sockets (Integrated-Circuit Packages)

Leaded integrated-circuit sockets The major utilization of these sockets is to eliminate the delicate and time-consuming task of unsoldering integrated circuits (IC's) from expensive, densely packed printed-wiring boards. Sockets are used to simplify maintenance tasks: troubleshooting, both in-house and in the field, is accomplished by unplugging the suspected device and replacing it quickly and safely with one that is known to be good. Versions are available with the industry-wide standards for dual-in-line IC packages: 14, 16, 24, 28, and 40 contact layouts. These sockets are generally directly soldered to the printed-wiring board. Solderless-wrap connection posts are also available. Dimensions for a typical 16-contact solder type are shown in Fig. 129. Note that the individual terminal tabs are spaced on 0.100-in

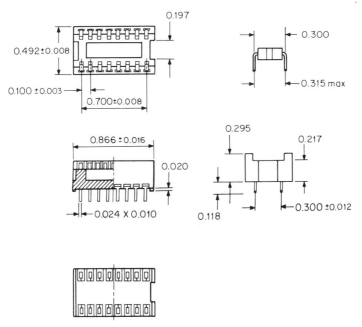

Fig. 129 Typical 16-contact, solder-type integrated-circuit dual-in-line package socket.

centers. The body material is available in type 66 nylon, phenolic (glass-filled, type MFH, per MIL-M-14), diallyl phthalate (glass-filled, type SDG, per MIL-M-14), on polyphenylene oxide. Contacts are generally made from brass (half hard, composition 2 per QQ-613) or beryllium copper (alloy 172, quarter hard per QQ-C-533). Contact finishes are generally gold-over-nickel (30 μin per MIL-G-45204 over 100 μin nickel). Tin plate is also available, per MIL-T-10727.

Leadless integrated-circuit sockets Recent developments in leadless ceramic substrates, which carry hybrid or monolithic medium-scale-integrated (MSI) and large-scale-integrated (LSI) circuits (generally more than 150 semiconductor devices), have eliminated the lead-frame contribution of the package to yield losses and packaging costs. Despite the absence of leads, the packages are made pluggable

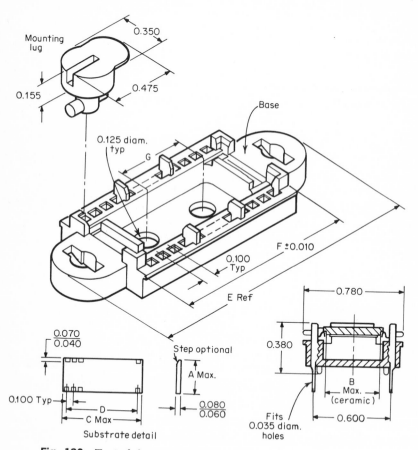

Fig. 130 Typical face-mount, face-contact socket. (*AMP, Inc.*)

No. of dual positions	Substrate dimensions				Connector dimensions		
	A	B	C	D	E	F	G
24	0.510	0.420	1.210	1.100	2.060	1.350	0.500
28	0.510	0.420	1.410	1.300	2.260	1.550	0.600
40	0.510	0.420	2.020	1.900	2.860	2.150	1.300
	0.520	0.480	2.040	1.900	2.860	2.150	1.300

by a variety of packaging capabilities for ceramic substrates having functional circuitry as well as displays.

Face-mount, Face-Contact Connectors. These connectors are used for face-metallized, leadless substrates. Substrates with top contact pads are simply inverted and installed in the connector in the same manner as those with bottom metallization. These zero-entry force connectors feature stored-energy spring contacts and are available in 40-, 28-, and 24-poistion, dual-in-line configurations. Gold- or tin-plated contacts are housed in a glass-filled thermoplastic base on 0.100- by 0.600-in centers. Positive retention of the substrate package is provided by cam-locking holddowns. See Fig. 130 for a typical connector of this type.

Face-mount, Side-Contact Connectors. These connectors are used for edge-metallized, leadless substrates and are available to accept the same configurations as the face-contact connectors. See Fig. 131 for a typical connector of this type.

Edge-mount Connectors. These connectors provide low-cost, high-density packaging for edge-mount-type leadless LSI substrate packages, and permit easy insertion and extraction. Three versions are available, including a 40-position connector that accepts single-layer substrates, a 40-position connector that accepts multilayer substrates, and an 80-position 15° angle connector which also accepts multilayer substrates. These connectors are furnished with contacts in a dual staggered-grid pattern which produces a close center-to-center spacing of 0.050 in. Figure 132 shows the connector/substrate detail for the 40-position (multilayer-substrate) connector.

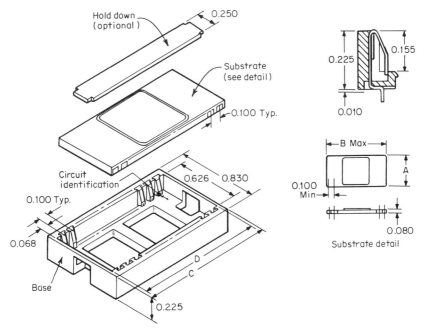

Fig. 131 Typical face-mount, side-contact socket. (*AMP, Inc.*)

No. of dual positions	Substrate dimensions		Connector dimensions	
	A	*B*	*C*	*D*
24	0.565/0.620	1.220	1.315	1.235
28	0.565/0.620	1.420	1.515	1.435
40	0.565/0.610	2.020	2.115	2.035

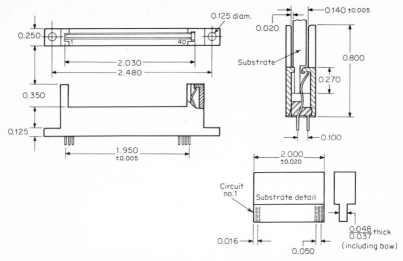

Fig. 132   Typical edge-mount 40-position socket.   (*AMP, Inc.*)

REFERENCES

1. Lazar, M. D.: Engineering Department Report BEDR-20, Burndy Corporation, Norwalk, Conn.
2. Stelling, R.: Failure Mode Analysis of New Printed Circuit Board Type Connector Systems, 1966 NEPCON.
3. Lane, W. L.: An Analysis of the Two-Part Printed Circuit Connector vs. the One-Part Printed Circuit Connector for Military Applications, *1970 Electron. Components Symp. Proc.*, pp. 185–189.
4. Eisenberg, R., M. D. Lazar, and A. J. Munn: New Concepts in Connectors and Packaging for High Speed Computer, 1968 INTER-NEPCON.
5. Goodman, F. R.: Optimum Specification of Current Carrying Capacity of Multipin Connectors under Widely Different Environments, Second Symposium on Connectors, Los Angeles, Calif., Feb. 26–27, 1964.
6. Burndy Corp. Catalog RP, p. 5, 1963.
7. Harper, C. A. (ed.): "Handbook of Electronic Packaging," p. 6–64, McGraw-Hill, New York, 1969.
8. Witte, R.: Selective Precious Metal Plating of Connector Contacts, *Electron. Packag. Prod.*, April 1967.
9. Spergel, J., E. F. Godwin, and G. Steinberg: A Thin Film Lubricant for Connector Contacts, *Proc. Electron. Components Conf.*, 1965.
10. Harper, C. A.: Insulating Materials, an Electronic Design Special Report, *Electron. Design*, vol. 17, no. 12, June 7, 1969.
11. Gahimer, F. H., and F. W. Nieske: Navy Investigates Reversion Phenomena of Two Elastomers, *Insulation*, August 1968, pp. 39–44.
12. Advertisement, *Insulation*, February 1969, p. 49.
13. Saunders, R.: Connectors and Interconnection Devices, in C. A. Harper (ed.), "Handbook of Electronic Packaging," chap. 6, McGraw-Hill, New York, 1969.
14. Coats, A. L.: High Altitude Problems with Electrical Connectors, Bendix Corporation *Tech. Rept.* TR 602, Sidney, N.Y.
15. Printed Circuit Connectors, *Electromech. Des.*, August 1970, pp. 16–23.
16. Forney, E.: *Proc. 1969 Ann. Connector Symp.*, Electronic Connector Study Group, Cherry Hill, N.J.
17. Waters, B.: *Proc. 1969 Ann. Connector Symp.*, Electronic Connector Study Group, Cherry Hill, N.J.
18. "RF Transmission Lines and Fittings," MIL-HDBK-216, Standardization Division, Defense Supply Agency.
19. Stasch, A.: The Hermetically Sealed Connector and Its Capabilities, *Electron. Ind.*, May 1964, p. 72.
20. Reliability Statistics for Electromechanical Devices, Amphenol-Borg Electronics Corp.

21. "Reliability Stress and Failure Rate Data for Electronic Equipment," Military Standardization Handbook, MIL-HDBK-217A, Dec. 1, 1965.
22. Ganzert, A. E.: Preventing Multiple-Contact Connector Problems, *Assem. Eng.*, January 1967, p. 22.
23. ETC, Inc.: "Crimp-Type Connectors," Product Data Brochure.
24. Van Ness, Robert F.: Termination Procedures and Connectors for Flat Conductor Cable and Flexible Printed Circuits, *Tech. Rept.* 3707, Picatinny Arsenal, Dover, N.J., February 1968.

Glossary of terms

ABC: Automatic brightness control.

Accessories: Mechanical devices such as cable clamps, added to connector shells and other such hardware which is attachable to connectors to make up the total connector configuration.

Accordion: A type of connector contact where a flat spring is given a Z shape to permit high deflection without overstress.

Accumulative Lock: See All-Lock definition.

Active Device: A device displaying transistance, that is, gain or control. Examples are transistors, diodes, and vacuum tubes.

Actual Interlace: The number of times a gapped array or electron beam is actually stepped per frame.

Adder: A circuit for performing the addition of two binary numbers with the generation and inclusion of carry bits.

Aging: In capacitors, the loss of dielectric constant K by dielectric relaxation. Expressed as a percent change per decade of time.

All-Lock: Mechanical function whereby actuated switch station stays in the "in" position when operated and releases only when the associated "release station" push button is depressed or a "release" solenoid is operated. Also known as "accumulative lock."

All-Lock/Nonlock Combination: Split function having certain prearranged switch stations operating with all-lock mechanical function and other switch stations operating with nonlock mechanical function.

Alternate Action: Mechanical function whereby a push button is depressed once to actuate switch circuits. Push button returns to normal (full out) position when pressure is removed from button (switch circuits stay actuated). Push button is again depressed and pressure is released to return switching circuits and push button to normal (full out) position.

Alternating-Current (AC) Relay: A relay designed for operation from an alternating-current source.

Ambient: The temperature of the environment immediately adjacent to the component part under test.

Ambient Temperature: The temperature of the medium (usually air) surrounding the part and into which the heat from electrical and related losses in the part is dissipated.

Ampere-Turns: The product of the number of turns in an electromagnetic coil and the current in amperes passing through the coil.

Analog Computer: A computer that processes data which are continuously variable in nature.

AND *Gate:* The logic circuit such that the output can only be in the one state if all the inputs are in the one state.

1

Anisotropic: In magnetics, capable of being magnetized more readily in one direction than in a transverse direction.

Anode: The electrode from which the forward current flows within the device.

Antenna Switching Relay: A relay designed to switch radio-frequency antenna circuits with a minimum of losses.

Arc, Contact: The electrical (current) discharge that occurs between mating contacts when the circuit is being disestablished.

Arm, Lockout Coupler: Arm which transfers the lockout function between rows of a gang switch assembly. Locking a station in one row moves the arm to effect the lockout in the next row.

Arm, Rocker: Transfers interlock and lockout mechanical functions between rows of a ganged switch assembly.

Armature: The moving magnetic member of an electromagnetic-relay structure that converts electrical energy into mechanical work.

Armature backstop: See Backstop, Armature.

Armature, Balanced: A relay armature that rotates about its center of mass and is therefore approximately in equilibrium with respect to both gravitational (static) and acceleration (dynamic) forces.

Armature Lever Ratio: The distance through which the armature buffer moves divided by the armature travel (see Armature Travel). Also, the ratio of the distance from the armature bearing pin (or fulcrum) to the armature buffer in relation to the distance from the bearing pin (or fulcrum) to the point on the armature that strikes the coil core.

Armature, Long-Lever: An armature with its contact-actuating arm longer than the distance from the armature hinge, bearing, or fulcrum to the portion of the armature opposite the pole face.

Armature, Short-Lever: An armature with its contact-actuating arm shorter than the distance from the armature hinge, bearing, or fulcrum to the portion of the armature opposite the pole face.

Armature Spring: The movable contact spring of a combination. This member is also sometimes referred to as the swinger spring.

Armature Travel: The distance that the armature moves in going from its unoperated to its operated position (usually measured at the center of the associated pole piece) or from one specified operated position to another.

Aspect Ratio: The ratio of the length to the width of a two-dimensional resistor, such as a thick- or thin-film resistor.

Auxiliary Contacts: Contacts used to operate a visual or audible signal to indicate the position of the main contacts, establish interlocking circuits, or hold a relay operated when the original operating circuit is opened.

Avalanche Photodiode (APD): A photodiode designed to take advantage of avalanche multiplication of photocurrent. As the reverse-bias voltage approached the breakdown voltage, hole-electron pairs created by absorbed photons acquire sufficient energy to create additional hole-electron pairs when they collide with substrate atoms; thus a multiplication effect is achieved. *Note:* APDs are especially suited for low-noise and/or high-speed applications.

Average Noise Figure or Average Noise Factor: The ratio of (1) the total output noise power within a designated output frequency band when the noise temperature of the input termination(s) is at the reference noise temperature T_0 at all frequencies to (2) that part of (1) caused by the noise temperature of the designated signal-input termination within a designated signal-input frequency band.

Back Contacts: Sometimes used for Contacts, Normally Closed.

Back Mounted: When a connector is mounted from the inside of a panel or box with its mounting flanges inside the equipment.

Back Panel: A relatively large piece of insulating material on one side of which components, modules, or other subassemblies are mounted and on the other side of which interconnections between them are made; also sometimes referred to as mother board.

Backplane Panels: An interconnection panel into which PC cards or other panels can be plugged. These panels come in a variety of designs ranging from a PC mother board to individual connectors mounted in a metal frame. Panels lend themselves to automated wiring.

Backstop, Armature: That part of the relay which limits the movement of the armature away from the pole face or core. In some relays a normally closed contact may serve as a backstop.

Bail: Loop of wire used to prevent permanent separation of two or more parts assembled together. Example: the bail holding dust caps on round connectors.

Ball Bonding: Process of wire bonding which uses gold wire whose end is shaped into a ball by a small flame. In this process the ball is attached to the bond pad using thermal compression.

Bank: One or more contact levels of a stepping switch.

Barrel: Cylindrical portion or portions of a terminal, splice, or contact accommodating the conductor or conductors.

Base Metal: Metal from which the connector, contact, or other metal accessory is made and on which one or more metals or coatings may be deposited.

Bayonet Coupling: A quick-coupling device for plug-and-receptacle connectors, accomplished by rotation of a cam-operating device designed to bring the connector halves together.

Beam Divergence in a Given Plane: The half angle of divergence of the laser emission at which the intensity of radiation is one-half the peak intensity.

Beam Lead: Lead formed on surface of a semiconductor, with tabs or lead protruding from edges. Bonding can be face down or face up in a recessed substrate. Thermocompression bonding is typical.

Beam-Lead Phototransistor: A phototransistor chip with thick-film leads formed on the chip which project cantilever-style beyond the chip periphery for attachment to a separate substrate.

Bell Mouth: Flared or widened entrance to a connector barrel, permitting easier insertion of the conductor.

Bialkali: A photocathode material having a spectral response similar to S11 but having lower thermionic emission.

Bias Driver: The circuit used to provide the reference level for the differential amplifier in an emitter-coupled logic circuit.

Bias, Electrical: An electrically produced force tending to move the relay armature toward a given position.

Bias, Magnetic: A steady magnetic field applied to magnetic circuit of a relay.

Bias, Mechanical: A mechanical force tending to move the armature toward a given position.

Bidirectional Relay or Add-and-Subtract Relay: A stepping relay in which the rotating wiper contacts may move in either direction.

Bifurcate: Describes lengthwise slotting of a flat spring contact, as used in a printed-circuit connector, to provide additional independently operating points of contact. Example: Bifurcated contact.

Bimetallic Element: An actuating element consisting of two strips of metal with different coefficients of thermal expansion bound together in such a way that the internal strains caused by temperature changes deflect the compound strip.

Bipolar: Circuitry using either npn or pnp transistors.

Bipolar Transistor: A transistor that utilizes charge carriers of both polarities.

Bit: Binary digit.

Blade Contact: Used in multiple-contact connectors, a flat male contact designed to mate with a tuning fork or a flat formed female contact.

Blades: Sometimes used for Springs, Contact.

Blocking: A state of a semiconductor device or junction which essentially prevents the flow of current.

Blowout Coil: An electromagnetic device that establishes a magnetic field in the space where an electric circuit is broken and helps to extinguish the arc by displacing it.

Blowout Magnet: A permanent magnet or an electromagnet located in such a position as to place a magnetic field in the space whee dc electric circuit is to be broken. This field causes the arc to be displaced, thus lengthening it and helping to extinguish it more rapidly.

Bobbin Lugs: Mounted in plastic or paper bobbins, lugs serve to connect coil wires to external lead wires.

Body: Main, or largest, portion of a connector to which other portions are attached.

Bond Pad: A region of metallization on the integrated-circuit chip to which the bonding wire will be attached.

Bonded Assembly: A connector assembly in which the components are bonded together using an electrically appropriate adhesive in a sandwich-like structure to provide sealing against moisture and other environments which weaken electrical insulating properties.

Boot: A form placed around wire terminations of a multiple-contact connector to contain the liquid potting compound before it hardens. Also, a protective housing usually made from a resilient material to prevent entry of moisture into a connector.

Break: The opening of closed contacts to interrupt an electric circuit.

Break-before-Make Contacts: Contacts that interrupt one circuit before establishing another.

Break Contacts: See Contacts, Normally Closed.

Breakdown: A phenomenon occurring in a reverse-biased semiconductor junction, the initiation of which is observed as a transition from a region of high small-signal resistance to a region of substantially lower small-signal resistance for increasing magnitude of reverse current.

Breakdown Region: A region of the volt-ampere characteristic beyond the initiation of breakdown for an increasing magnitude of reverse current.

Breakdown Voltage: The voltage measured at a specified current in a breakdown region.

Bridging: (1) Normal Bridging: The normal make-before-break action of a make-break or D contact combination. In a stepping switch or relay, the coming together momentarily of two adjacent contacts, by a wiper shaped for that purpose, in the process of moving from one contact to the next. (2) Abnormal Bridging: The undesired closing of open contacts resulting from bounce or caused by a metallic bridge or protrusion developed by arcing.

Bridging Contact: A contact combination designed to provide bridging.

Brightness: See Luminance.

Brush Spring: The spring in a rotary stepping switch bank that contacts the associated wiper.

Buffer, Armature, or Bushing, Armature, or Lifter, Armature, or Plug, Armature, or Pusher, or Stud, Armature: An insulating member attached to the armature that transmits the motion of the armature to an adjacent contact member.

Buffer, Spring, or Buffer, Contact Spring, or Bushing, Spring, or Stud, Spring: An insulating member that transmits the motion of the armature from one movable contact spring to another in the same pileup.

Burn-in: The process of exposing a component part to the elevated temperatures with a voltage stress applied to the part, for the purpose of screening out marginal parts.

Bus Driver: A specialized circuit used to impress signals on a data bus.

Butt: Placing two conductors together end to end (but not overlapping) with their axes in line.

Cable Clamp: A device used to give mechanical support to the wire bundle or cable at the rear of a plug or receptacle.

Cable Clamp Adapter: A mechanical adapter that attaches to the rear of a plug or receptacle to allow the attachment of a cable clamp.

Capacitance: Property of a capacitor which determines its ability to store electrical energy when a given voltage is applied, measured in farads, microfarads, or picofarads.

Capacitance Tolerance: The part manufacturer's guaranteed maximum deviation (expressed in percent) from the specified nominal value at standard (or stated) environmental conditions.

Capacitive Reactance: Opposition offered to the flow of an alternating or pulsating current by capacitance, measured in ohms.

Capacitor, Liquid-filled: A capacitor in which a liquid impregnant occupies substantially all the case volume not required by the capacitor element and its connections. (Space may be allowed for the expansion of the liquid under temperature variations.)

Capacitor, Liquid-impregnated: A capacitor in which a liquid impregnant is dominantly contained within the foil-and-paper winding but does not occupy substantially all the case volume.

Capacitor, Temperature-compensating: A capacitor whose capacitance varies with temperature in a known and predictable manner.

Card, Armature: An insulating member used to link the movable springs to the armature.

Case Temperature: The temperature measured at a specified location on the case of a device.

Cathode: Negative electrode of a capacitor.

Cerdip: A dual in-line package composed of a ceramic header and lid, a stamped-metal lead frame, and frit glass which is used to secure the structure together.

Cermet, Frit, Ink, Paste: Interchangeable terms used to designate the raw material for screened and fired components.

Cerpack: A flatpack composed of a ceramic base and lid, a stamped-metal lead frame, and frit glass which is used to secure the structure together.

CGI: Commercial government inspection.

Chamfer: Angle on the inside edge of the barrel entrance of a connector which permits easier insertion of the cable into the barrel.

Charging Inductor: A coupling inductor used to charge a pulse-forming network (PFN) to approximately twice the dc voltage.

Chatter, Armature: The undesired vibration of the armature due to inadequate ac performance or external shock and vibration.

Chatter, Contact: The undesired vibration of mating contacts during which there may or may not be actual physical-contact opening. If there is no actual opening but only a change in resistance, it is referred to as dynamic resistance and appears on the screen of an oscilloscope having adequate sensitivity and resolution. Chatter may result from contact impingement during normal relay operation and release, uncompensated ac operation, or external shock and vibration.

Chemical Milling: The process in which metal is formed to intricate shapes by masking certain portions and then etching away the unwanted material.

Chip Component: An uncased component, as semiconductor dice or chips.

Chopper: A special form of pulsing relay having contacts arranged to interrupt rapidly or reverse alternately the dc polarity input to an associated circuit.

Circuit: The interconnection of a number of components in one or more closed paths to perform a desired electrical or electronic function.

Circuit Breaker: A device designed to open and close a circuit by nonautomatic means, and to open the circuit automatically on a predetermined overload of current, without injury to itself, when properly applied within its rating.

Clapper: An armature that is hinged or pivoted. (See Armature.)

Clapper-Type Relay: A large and varied family of relays using clapper-type armatures.

Clock Driver: A circuit optimized for driving the clock lines of either a number of bipolar circuits in parallel or the highly capacitive clock line of a MOS circuit.

Close-Differential Relay: A relay having its dropout value specified close to its pickup value.

CML: Current-mode logic. Basically equivalent to ECL except that ECL is usually operated in the saturated mode. CML has greater speed because it operates in the nonsaturated mode.

CMOS: Complementary MOS. An MOS or IC device involving both *p*-channel and *n*-channel MOSFET's. This technique increases logic speed but requires additional processing steps that reduce circuit density and raise cost per function.

Coaxial Relay: A relay that opens and closes an electrical contact switching high-frequency current as required to maintain minimum losses.

Cofired: A term denoting the firing of two or more thick-film materials in one operation, such as simultaneous firing of resistors and conductors.

Coherent Radiation: Radiation in which the phase difference between any two points in the radiation field is constant throughout the duration of the radiation.

Coil: An assembly consisting of one or more magnet-wire windings, usually wound over an insulated iron core on a bobbin or spool, or self-supporting, with terminals, and any other required parts such as a sleeve or slug.

Coil, Concentrically Wound: A coil with two or more insulated windings, wound one over the other.

Coil-Form Terminals: Used on small transformer coils to connect coil wires. Terminals are attached to coil base or collar.

Coil, Parallel-Wound: A coil having two windings wound simultaneously with the turns of each winding being contiguous.

Coil, Sandwich-wound: A coil usually used for voice-current battery feed over wired communications lines. It is normally wound in three sections of two windings of equal total turns and resistance. This averages out the coil-core leakage flux and gives better line balance.

Coil, Tandem-wound: A coil having two or more windings, one behind the other along the longitudinal axis. Also referred to as a two-, three-, or four-section coil.

Coil Covering: A protective layer of insulating material over the outermost (surface) turns of wire.

Coil Terminal: A device, such as a solder lug, tab, binding post, or similar fitting, on which the coil winding lead is terminated and to which the coil power supply is connected.

Coil Thermal Equilibrium: The condition of an energized coil when the heat generated is equal to that dissipated and there is no additional temperature rise.

Coil Winding: An electrically continuous length of insulated wire wound on a bobbin, spool, or form.

Coil (Winding) Final Mean Temperature: The average temperature of an energized coil winding at thermal equilibrium, as determined from a measurement of dc resistance.

Coil (Winding) Mean Temperature: The average temperature of an energized coil winding as determined by a measurement of its dc resistance.

Coil (Winding) Power Dissipation: The electrical power (watts) consumed by the energized winding or windings of a coil. For most practical purposes this is calculated from I^2R or $E_1 \cos \theta$ of the specific case.

Coil (Winding) Resistance: The total terminal-to-terminal resistance of a coil at a specified temperature. A tolerance of measured value from a nominal specified resistance is usually allowed.

Color Temperature: The temperature of a blackbody whose radiation has the same visible color as that of a given nonblackbody radiator. Typical unit: K (kelvin, formerly °K).

Component: An individual functional element in a physically independent body which cannot be further reduced or divided without destroying its stated function, e.g., resistor, capacitor, diode, transistor.

Component Density: The number of components per unit volume (cubic inch or cubic foot).

Configuration: Specific configuration and arrangement of contacts in a multiple-contact connector.

Connector: Used generally to describe all devices used to provide rapid connect/disconnect service for electrical cable and wire terminations.

Contact Alignment: Defines the overall side play which contacts shall have within the insert cavity so as to permit self-alignment of mated contacts. Sometimes referred to as amount of contact float.

Contact Area: Area in contact between two conductors or a conductor and a connector permitting flow of electric current.

Contact Arrangement: The number and types of contact combinations of a relay.

Contact Bounce: The intermittent and undesired opening of closed contacts, or closing of open contacts, of a relay, due to the following causes: (1) Internally caused contact chatter: (*a*) from impingement of mating contacts; (*b*) from impact of the armature against the coil core on pickup, or against the backstop on dropout; (*c*) from momentary hesitation, or reversal, of the armature motion during the pickup or dropout stroke. (2) Externally caused contact chatter: (*a*) from shock impact experienced by the relay or the apparatus of which it is a part; (*b*) from vibration or shock outside the relay but transmitted to it through its mounting.

Contact, Bridging: See Bridging Contact.

Contact Bunching: The undesired simultaneous closure of make-and-break contacts during vibration, shock, or acceleration tests.

Contact Cavity: A defined hole in the connector insert into which the contacts must fit.

Contact Chatter: See Chatter, Contact.

Contact Combination, or Contact Set, or Contact Form: A single-pole or basic contact assembly.

Contact Compliance: The reciprocal of contact stiffness.

Contact Engaging and Separating Force: Force needed to either engage or separate pins and socket contacts when they are in and out of connector inserts. Values are generally established for maximum and minimum forces. Performance acceptance levels vary by specification and/or customer requirements. Sometimes contact engaging and separating force is measured not only initially but also after a specified number of engagements and separations.

Contact Follow: For compliant contacts this is the distance two contacts travel together after just touching. Also called contact overtravel. In the case of noncompliant contacts this is the distance the relay armature travels after the contacts touch one another.

Contact Force: The pressure exerted by a movable contact against a fixed contact when the contacts are closed. Also referred to as contact pressure.

Contact Gap: The distance between a pair of mating relay contacts when the contacts are open; same as contact separation.

Contact Inspection Hole: A hole in the cylindrical rear portion of contact used to check the depth to which a wire has been inserted. Crimp-type contacts usually have inspection holes; solder-type seldom do, except larger sizes in which the function of the hole is to allow solder and air to bleed out during soldering.

Contact Length: Length of travel made by one contact in contact with another during assembly or disassembly of connector.

Contact Load: The electrical-power demands encountered by a contact set in any particular application.

Contact Materials: (*a*) Brass. Low-cost, excellent electrical conductor. Reaches its yield point at lowest deflection force, hence deforms easily and fatigues quickly.

(*b*) Phosphor bronze. Medium-priced, harder than brass. Retains its resiliency longer, better fatique resistance. (*c*) Beryllium copper. Higher-priced material. Properties superior to those of brass and phosphor bronze. Recommended for contact applications requiring repeated extraction and reinsertion. Other materials may be used to meet special requirements.

Contact Miss: Failure of a contact mating pair to establish the intended circuit, electrically. This may be a circuit resistance in excess of a specified maximum value.

Contact Plating: Plated-on metal applied to the basic contact metal to provide the required contact resistance and/or wear resistance.

Contact Pressure: See Contact Force.

Contact Rating: The electrical power-handling capability of relay contacts under specified environmental conditions and for a prescribed number of operations as defined by the manufacturer. The Aerospace Industries Association of America, Inc., defines contact rating as "The maximum current for a given type of load (i.e., voltage, frequency and nature of impedance) which the relay (contacts) will make, carry, and break (unless otherwise specified) for its rated life."

Contact Resistance: The electrical resistance of operated contacts as measured at their associated contact-spring terminals.

Contact Resistance: Maximum permitted electrical resistance of pin-and-socket contacts when assembled in a connector under typical service use. Electrical resistance of each pair of mated pin-and-socket contacts in the connector assembly is determined by measuring pin to the extreme terminal end of the socket (excluding both crimps) when carrying a specified test current. Overall contact resistance includes wire-to-wire measurement.

Contact Retention: Defines minimum axial load in either direction which a contact must withstand while remaining firmly fixed in its normal position within an insert.

Contact Separation: See Contact Gap.

Contact(ing) Sequence: The order in which contacts open and close in relation to other contacts and armature motion.

Contact Shoulder: The flanged portion of the contact which limits its travel into the insert.

Contact Size: Defines the largest size of wire which can be used with the specific contact. By specification dimensioning it also defines the diameter of the engagement end of the pin.

Contact Spacing: The distance between the centers of contacts within an insert.

Contact Spring: A current-carrying spring to which the contacts are fastened.

Contact Spring: The spring placed inside the socket-type contact to force the pin into a position of positive intimate contact. Depending on the application, various types are used, including leaf cantilever, napkin-ring, squirrel-cage, hyperbolic and "chinese-finger" springs. All perform the function of wiping and establishing good contact. Various metal alloys are used. For example, beryllium copper is used where high conductivity and long life are required. Stainless steel, while its conductivity is only about 2 percent, is used in high-temperature applications.

Contact Stack: All the contact springs in one assembly. A contact pileup.

Contact (Spring) Tension: The contact pressure developed, usually resulting from the specified adjustment of movable contacts against mating stationary springs, when the relay is unenergized.

Contact Transfer Time: The interval between opening of the closed contact and closing of the open contact of a contact combination.

Contact Weld: (1) The point of attachment of a contact to its support when accomplished by resistance welding. (2) A contacting failure due to fusing of contacting surfaces under load conditions to the extent that the contacts fail to separate when expected to do so.

Contact Wipe: The scrubbing action between mating contacts resulting from contact overtravel or follow.

Contactor: A term for a power-type relay with heavy-duty contacts. (See also Power-Type Relay.)

Contacts: The surfaces of current-carrying members at which electric circuits are opened or closed.

Contacts: The conducting members of a connecting device which are designed to provide a separable through connection in a cable-to-cable, cable-to-box, or box-to-box situation.

Bellows Contact: A contact in which a flat spring is folded to provide a more uniform spring rate over the full tolerance range of the mating unit.

Bifurcated Contact: A contact (usually flat-spring) which is slotted lengthwise to provide additional independently operating points of contact.

Butt Contact: A mating-contact configuration in which the mating surfaces engage end to end without overlap and with their axes in line. This engagement is usually under spring pressure with the ends designed to provide optimum surface contact.

Buttonhook Contact: A contact with a curved, hooklike termination often located at the rear of hermetic headers to facilitate soldering or desoldering of leads.

Closed-Entry Contact: A female contact designed to prevent the entry of a pin or probing device having a cross-sectional dimension (diameter) greater than the mating pin.

Crimp Contact: A contact whose back portion is a hollow cylinder to allow it to accept a wire. After a bared wire is inserted, a swaging tool is applied to crimp the contact metal firmly against the wire. A crimp contact often is referred to as a solderless contact.

Dressed Contact: A contact with a permanently attached locking spring member.

Fixed Contact: A contact which is permanently included in the insert material. It is permanently locked, cemented, or embedded in the insert during molding.

Nude Contact: A contact with a locking member that remains in the insert at all times.

Removable Contact: A contact that can be mechanically joined to or removed from an insert. Usually, special tools are required to lock the contact in place or remove it for repair or replacement.

Open-Entry Contact: A female-opening contact unprotected from possible damage or distortion from a test probe or other wedging device.

Sheet-Metal Contacts: Contacts made by stamping and bending sheet metal rather than by the machining of metal stock. They are available in a wide variety of configurations and are usually less expensive than machined contacts.

Solder Contact: A contact having a cup, hollow cylinder, eyelet, or hook to accept a wire for a conventional soldered termination.

Socket Contact: A female-type contact (usually completely surrounded by insert material).

Spade Contact: A contact with fork-shaped female members designed to dovetail with spade-shaped male members. Alignment in this type of connection is critical if good conductivity is to be achieved.

Two-Piece Contact: A contact made of two or more separate parts joined by swaging, brazing, or other means of fastening to form a single contact. This type provides the mechanical advantages of two metals but also has the inherent electrical disadvantage of difference in conductivity.

Contacts, Armature: (1) A contact mounted directly on the armature. (2) Sometimes used for a movable contact.

Contacts, Back: See Contacts, Normally Closed.

Contacts, Bifurcated: A forked, or branched, contacting member so formed or arranged as to provide some degree of independent dual contacting.

Contacts, Break: See Contacts, Normally Closed.

Contacts, Break-Make: A contact combination in which one contact opens its connection to another contact and then closes its connection to a third contact. Same as transfer contacts.

Contacts, Continuity Transfer: Sometimes used for bridging contacts. Note: Although there is mechanical continuity, electrical discontinuity may occur as a result of bounce.

Contacts, Double-Break: A contact combination in which contacts on a single spring simultaneously open electric circuits connected to the contacts of two independent springs. In the case of power-type relays having noncomplaint contacts, double break refers to stationary contacts shorted by a bridging bar.

Contacts, Double-Make: A contact combination in which contacts on a single spring simultaneously close electric circuits connected to the contact of two independent springs. In the case of power-type relays having noncomplaint contacts, double make refers to stationary contacts shorted by a bridging bar.

Contacts, Double Break-Make: A contact combination employing the shorting-bar principle, similar to the double-make and double-break contacts defined above but combining both into a double-make and a double-break.

Contacts, Double-Throw: A contact combination having two positions, as in break-make and the like.

Contacts, Dry-Circuit: (1) Contacts that neither break nor make current. (2) Erroneously used for low-level contacts.

Contacts, Early: A contact combination that is adjusted to open or close before other contact combinations when the relay operates. A preliminary contact combination is so arranged and adjusted.

Contacts, Interrupter: On a stepping relay or switch, a set of contacts, operated directly by the armature, that opens and closes the winding circuit, permitting the device to self-step.

Contacts, Late: A contact combination that is adjusted to open or close after other contact combinations when the relay picks up.

Contacts, Low-Level: Contacts that control only the flow of relatively small currents in relatively low-voltage circuits, for example, alternating currents and voltages encountered in voice or tone circuits, direct currents in the order of microamperes, and voltages below the softening voltages of record for various contact materials (that is, 0.080 V for gold and 0.25 V for platinum, and the like). Also defined as the range of contact electrical loading where there can be no electrical (arc transfer) or thermal effects and where only mechanical forces can change the conditions of the contact interface.

Contacts, Make: See Contacts, Normally Open.

Contacts, Make-Break: See Bridging Contact.

Contacts, Movable: The member of a contact pair that is moved directly by the actuating system. This member is also referred to as the armature (contact) spring, or swinger spring. In reed-type contacts motion is caused magnetically.

Contacts, Nonbridging: A contact arrangement in which the opening contact opens before the closing contact closes.

Contacts, Normally Closed, or Contacts, Back: A contact pair which is closed when the armature is in its unoperated position.

Contacts, Normally Open, or Contacts, Front: A contact pair that is open when the armature is in its unoperated position.

Contacts, Off-Normal: Contacts on a relay or switch that are in one condition when the relay or switch is in its normal position and in the reverse condition for any other position of the relay or switch.

Contacts, Preliminary: A contact combination that opens, closes, or transfers in advance of other contact combinations when the relay picks up. (See also Auxiliary Contacts and Contacts, Early.)

Contacts, Snap-Action: A contact assembly having two or more equilibrium positions, in one of which the contacts remain with substantially constant contact pressure during the initial motion of the actuating member until a condition is reached at which stored energy overcomes restraint and contacts rapidly assume a new position.

Contacts, Transfer: See Contacts, Break-Make.

Continuity-Transfer Contacts: See Contacts, Continuity Transfer.

Continuous-Duty Relay: A relay that may be energized with maximum rated power indefinitely without exceeding specified temperature limitations.

Conventional Component: Discrete component, as opposed to chip component. This type of component is readily available from a number of suppliers, and data are available on its physical size and electrical characteristics. The upper limit of sophistication is the transistor.

Conversion Efficiency (of a Photoemissive Device): The ratio of maximum available luminous- or radiant-flux output to total input power.

Cordwood Module: The arrangement of components within a module so that the component bodies are parallel and in close proximity to one another. Often referred to as three-dimensional packaging.

Core, Coil: The portion of the magnetic structure of a relay about which the coil is usually wound.

Coupling Ring: A device used on cylindrical connectors to lock plug and receptacle together. It may or may not give mechanical advantage to the operator during the mating operation.

Crimp Termination: Connection in which a metal sleeve is secured to a conductor by mechanically crimping the sleeve with pliers, presses, or automated crimping machines. Splices, terminals, and multicontact connectors are typical terminating devices attached by crimping. Suitable for all wire types.

Crosstalk: Signals erroneously transmitted from one data line to another by inductive pickup.

Crystal-Can Relay: A term used to identify a microminiature relay housed in a hermetically sealed enclosure that was originally used to enclose a frequency-control type of quartz crystal. The most common size of crystal-can relay housing is approximately 0.4 by 0.8 by 0.97 in.

Curie Point: In ceramic capacitors, the temperature at which the crystalline structure of the ceramic changes form radically, causing a change in dielectric constant.

Dark Current (I_D): The current that flows through a photosensitive device in the dark condition. Note: The dark condition is attained when the electrical parameter under consideration approaches a value which cannot be altered by further irradiation shielding.

Darlington-connected Phototransistor: A phototransistor the collector and emitter of which are connected to the collector and base, respectively, of a second transistor. The emitter current of the input is amplified by the second transistor and the device.

Dashpot: A device that employs either a gas or liquid to absorb energy and retard the movement of the moving parts of a circuit breaker or other electric device to produce an operate-time delay.

Dashpot Relay: A relay employing the dashpot principle to develop a time delay.

DC Fanout: A term characterizing the driving capability of an output; specifically, the number of inputs of a specified type that can be driven simultaneously by the output.

DC Transfer Ratio (of an Optically Coupled Isolator): The ratio of the dc output current to the dc input current.

Deenergization: The removal of power from a relay coil or heater. Also commonly used to indicate a change in coil or heater applied power adequate to produce dropout.

Definite-Purpose Relay: A readily available relay having some electrical or mechanical feature that distinguishes it from a general-purpose relay. Types of definite-purpose relays are interlock, selector, stepping-sequence, latch-in, and time-delay.

Delay Line: An *LC* network which delays a signal, from input to output, by a time interval usually specified in microseconds and determined by the length of the network. Special design may also be used to shape the signal.

Delay Relay: A relay having an assured time interval between energization and pickup or between deenergization and dropout, or both. Also called time-delay relay.

Delay Time (t_d): The time interval from the point at which the leading edge of the input pulse has reached 10 percent of its maximum amplitude to the point at which the leading edge of the output pulse has reached 10 percent of its maximum amplitude.

Dielectric Absorption: Property of an imperfect dielectric whereby all electric charges within the body of the material caused by an electric field are not returned to the field.

Dielectric Constant: Property of a dielectric material that determines how much electrostatic energy can be stored per unit volume when unit voltage is applied (i.e., the ratio of the capacitance of a capacitor filled with a given dielectric to that of the same capacitor having a vacuum dielectric).

Dielectric Strength: Maximum voltage that a dielectric material can withstand without rupturing. (The value obtained for the dielectric strength will depend on the thickness of the material and on the method and conditions of test.)

Direct-Current (DC) relay: A relay designed for operation from a direct-current source.

Dissipation Factor (DF): The ratio of resistance to reactance, measured in percent.

Dropout, Measured, or Release, Measured: The maximum current or voltage at which the relay restores to its unoperated position.

Dropout, Specified, or Release, Specified: The specified maximum current or voltage at which the relay must restore to its unoperated position. Sometimes referred to as minimum dropout.

Dry Circuit: A mechanically closed circuit with no appreciable applied voltage during contacting.

Dry-Reed Relay: A reed relay with dry (non-mercury-wetted) contacts.

Dual-Coil Relays: Relays designed to take advantage of the dual-coil operating principle.

Duty Cycle: A statement of energized and deenergized time in repetitious operation, for example, 2 s on, 6 s off.

Duty Factor (du): The product of the pulse duration and the pulse-repetition frequency of a wave composed of pulses that occur at regular intervals.

Dynamic Contact Resistance: A change in contact electrical resistance due to a variation in contact pressure on contacts mechanically closed.

EBCDIC: A code that relates alphanumeric characters to binary numbers.

Edge Speed: The rate of rise or fall on the leading or trailing edge of a digital signal pulse.

Electrode (Semiconductors): An electrical and mechanical contact to a region of a semiconductor device.

Electrode: In capacitors, the conductive metal film which separates the layers of dielectric material.

Electroluminesence: The direct conversion of electrical energy into light.

Electrolyte: Current-conducting solution (liquid or solid) between two electrodes or plates of a capacitor at least one of which is covered by a dielectric film.

Electromagnetic Relay: A relay whose operation depends upon the electromagnetic effects of current following in an energizing winding.

Electromagnetic Wave Filter: An *LC* network designed to attenuate selected frequencies in an electrical signal and to provide relatively unattenuated transmission of desired frequencies.

Electronic Transformer: Any transformer intended for use in a circuit or system utilizing electronic or solid-state devices.

Electrostatic Relay: A relay in which operation depends upon motion of two or more insulated conductors caused by electrostatic effects.

Electrostatic Shield: A metallic shield or foil, usually grounded, used between reed switches or other types of contacts, between a reed switch or other types of contacts and coil, or between adjacent relays, to minimize crosstalk effects.

Electrostatic Spring Shield(s): Grounded metallic shield(s) between two relay springs to minimize crosstalk.

Electrostrictive Relay: A relay in which operation depends upon the dimensional changes of an electrostrictive dielectric.

Electrothermal Expansion Element: An actuating element in the form of a wire strip or other shape having a high coefficient of thermal expansion.

Emission-Beam Angle between Half-Power Points (HP): The angle centered on the optical axis of a light-emitting diode within which the relative radiant-power output or photon intensity is not less than half of the maximum output or intensity.

Enclosed Relay: A relay having the contacts or coil, or both, contained in an unsealed cover as protection from the surrounding medium.

Energization: The application of power to a coil or heater winding of a relay. With respect to an operating coil winding, or heater. Use of the word commonly assumes enough power to operate the relay fully, unless otherwise stated. (Examples of the latter are partially energized and half energized.)

Equivalent Series Resistance (ESR): All internal series resistances concentrated or "lumped" at one point in the circuit and treated as one resistance.

Etching: The process wherein unwanted metallic substances bonded to the base are removed by chemical milling or electrolytic processes. The chemical roughing of a smooth surface to increase the effective area.

Exciting Current: That component of primary current which produces the magnetic flux necessary to induce voltage in the windings. Exciting current lags the load current by approximately 90°.

Face-Down Bonding: Any process by which a chip is inverted in order to mount it into a package or hybrid construction.

Fall Time (t_f): The time duration during which the trailing edge of a pulse is decreasing from 90 to 10 percent of its maximum amplitude.

Fanout: The number of similar inputs that the output of the given circuit can drive.

Fast-Operate, Fast-Release Relay: A high-speed relay specifically designed for both short operate and short release time.

Fast-Operate Relay: A high-speed relay specifically designed for short operate time but not necessarily short release time.

Fast-Operate, Slow-Release Relay: A relay specifically designed for long release time but not necessarily short operate time.

Ferroelectric: A crystalline material that exhibits spontaneous electrical polarization, hysteresis, and piezoelectric properties.

Film, Ceramic: The ceramic tape that has been cast and dried and is free from effluents.

Film, Plastic: All hydrocarbon or similar plastic materials formed in thin sheets and used for capacitor dielectrics.

Filter: A network for separating waves on the basis of their frequency. Filters for rectifier circuits are *LC* or *RC* combinations intended to reduce the ac components in the rectified direct current.

Fixed-Resistor Terms

Aqueous Extract Conductivity of Core: Electrical conductivity of solution made from distilled water and powdered core material. Test for ionizable impurities in core material.

Core: The mechanical-electrical structure, typically cylindrical or oval cross section in shape, upon which a resistance film or winding is deployed.

Effective Wire Coverage: Axial length of a helical winding measured between points of radical departure from the normal pitch.

End Cap: A thin-walled metal cup, usually a press fit over the end of a resistor case, sometimes clinched on, to which terminal leads and resistance-element leads are electrically connected.

Hermetic Seal: An enclosure that provides a high degree of atmospheric isolation between enclosed elements and the ambient. Degree of hermeticity is specified by leak rates in atmosphere cubic centimeters per second. In military equipment, only fused glass-to-metal, welded metal, or solder sealing in some instances, are considered to be true hermetic seals.

Hot-Spot Temperature: The measured maximum temperature at the external surface of a resistor operating at a given steady power input, under specified ambient conditions, at the point on the body where the temperature is highest.

Substrate: The base, either a core or other surface upon which a resistive film or metallization is deposited.

Winding Pitch: The axial distance traversed by one complete turn of resistance wire in a helical winding.

Flasher Relay: A self-interrupting relay, usually of the thermal type.

Flash Point: For capacitor impregnating fluids and solids, the temperature to which the material must be heated to give off sufficient vapor to form a flammable mixture.

Flip-flop: A circuit or relay with two stable states, as used for storing one binary bit of information.

Forward Bias: The bias which tends to produce current flow in the forward direction.

Forward Direction: The direction of current flow which results when the p-type semiconductor region is at a positive potential relative to the n-type region.

Forward Voltage (V_F): The voltage across a semiconductor diode associated with the flow of forward current. The p region is at a positive potential with respect to the n region.

Frequency-sensitive Relay: A relay that operates when energized electrically at a particular frequency or within specific bands; a resonant-reed relay.

Frit: Finely powdered glass, used in metal inks to bond the metal particles to a substrate when heated to the melting point of the glass.

Gaging, Relay-Contact: The setting of relay-contact spacing by means of the use of thickness gages inserted between coil core and armature residual so as to determine the point in the armature's stroke at which specified contacts are to function.

Gain-Bandwidth Product (*of an Avalanche Photodiode*): The gain times the frequency of measurement when device is biased for maximum obtainable gain.

Gasket-sealed Relay: A relay contained within an enclosure sealed with a gasket.

Gating Electrode: Electrode provided to allow swinging of electrode voltage from cutoff value to focus value for high-speed gating; as short as 5 ns in certain types.

Glitch: An unwanted pulse or spike of short distance in the output state of a circuit.

Green Ceramic: Unsintered ceramic, usually a ceramic-loaded plastic system.

Half Adder: An adder without provision for the use of generation of carry bits.

Header: The surface in an integrated-circuit package on which the chip is mounted.

Hermetic: Seal by fusion of glass, ceramic, or metal so as to maintain a contaminant-free environment.

Hermetically Sealed Relay: A relay contained within an enclosure that is sealed by fusion or other comparable means to ensure a low rate of gas leakage over a long period of time. This generally refers to metal-to-metal or metal-to-glass sealing.

Hertz: A unit of frequency replacing or used as an alternative to cycles per second; hertz is abbreviated Hz.

Hexadecimal Display: A solid-state display capable of exhibiting numbers 0 through 9 and alpha characters A through F. *Note:* The TIL311 is a hexadecimal display with an integral TTL circuit which will accept, store, and display 4-bit binary data.°

High-Voltage Relay: (1) A relay adjusted to sense and function in a circuit or system at a specific maximum voltage. (2) A relay designed to handle elevated voltages on its contacts, coil, or both.

Homing Relay: A stepping relay that returns to a specified starting position prior to each operating cycle.

° TI stands for Texas Instrument, Inc.

Hot-Wire Relay: A relay in which the operating current flows directly through a tension member whose thermal expansion actuates the relay. A form of thermal relay.

Hum: The sound caused by mechanical (lamination) vibration resulting from alternating current flowing in the coil, or in some cases by unfiltered rectified current.

Hybrid Relay (HBR): A relay composed of a combination of electronic and electromechanical components.

Hybrid Circuit: A heterogeneous circuit, combining elements of several types of circuit technology.

Hysteresis: The lag in dielectric response which results from viscous damping of a variable force impressed upon the dielectric.

Illumination (E_v): The luminous flux density incident on a surface; the ratio of flux to area of illuminated surface. Typical units: $lm\ m^{-2} = 10.764\ lx$.

Image Tube: A camera tube which utilizes a photosensitive surface to produce an electron image and a video signal. *Note:* TI manufactures solid-state image tubes which have a silicon-diode matrix consisting of 620,000 diodes per square centimeter. These VID-series image tubes are superior to conventional image tubes because they have higher sensitivity, extremely broadband sensitivity, and are not damaged when subjected to high light levels up to $6 \times 10^5\ lm\ ft^{-2}$.

Impedance (Z): In capacitors, the total opposition offered to the flow of an alternating or pulsating current, measured in ohms. (Impedance is the vector sum of the resistance and the capacitive reactance, i.e., the complex ratio of voltage to current.)

Impregnant: A substance, usually liquid, used to saturate a paper- and plastic-film dielectric and to replace the air between its fibers and in pinholes. (Impregnation usually increases the dielectric strength and the dielectric constant of the assembled capacitor.)

Impregnated Coils: Coils that have been permeated with a phenolic varnish or other protective material to protect them from mechanical vibration, handling, fungus, and moisture.

Impulse Relay: (1) A relay that follows and repeats current pulses, as from a telephone dial. (2) A relay that operates on stored energy of a short pulse after the pulse ends. (3) A relay that discriminates between length and strength of pulses, operating on long or strong pulses and not operating on short or weak ones. (4) A relay that alternately assumes one of two positions as pulsed. (5) Erroneously used to describe an integrating relay.

Inductance: The property of an electric circuit whereby it resists any change of current during the building up or decaying of a self-induced magnetic field, and hence introduces a delay in current change with resulting operational delay. For convenience and standardization, rather than any technical significance, winding inductances are measured at a stated frequency, usually 1,000 Hz, with the armature held in its operated position unless otherwise specified. True inductance at any instant is very much affected by the degree of magnetic saturation, the presence of any steady current component, the armature position at instant of consideration, and the like.

Inductive Winding: A coil having an inductance, as contrasted with a noninductive winding. A coil in which all turns are wound in the same direction, or in which the turns wound in one direction are more effective than those wound in the opposite direction so that there is a net inductance.

Inductor: A device consisting of one or more associated windings, with or without a magnetic core, for introducing inductance into an electric circuit.

Inertia Relay: A relay with added weights or other modifications that increase the moment of inertia of its moving parts in order either to slow its operation or to cause it to continue in motion ofter the energizing force ends.

Infrared Emission: Radiant energy which is characterized by wavelengths longer than visible red, viz., 0.78 to 100 μm.

Infrared Light-emitting Diode: An optoelectronic device containing a semiconductor pn junction which emits radiant energy in the 0.78- to 100μm wavelength region when forward-biased.

Injection Laser: A solid-state semiconductor device consisting of at least one pn junction capable of emitting coherent or stimulated radiation under specified conditions. The device will incorporate a resonant optical cavity.

Input Reflection Coefficient: The ratio of the voltage reflected from the input port to the voltage incident on the input port with the output port terminated equal to the impedance of the source of the incident voltage.

Instrument Relay: A sensitive relay in which the principle of operation is similar to that of instruments such as the electrodynamometer, iron vane, D'Arsonval galvanometer, and moving magnet. This type of relay has a high ratio of nonpickup to pickup current or dropout to pickup current and therefore responds to small increases or decreases in the energizing source.

Insulation Resistance (of a Device): Resistance of insulation measured (in ohms) at a specified dc voltage and under ambient conditions, after current becomes constant. The resistance to leakage current of an intended insulator.

Integrated Circuit: A functional structure fabricated on a single crystal by a batch process.

Integrated-Circuit Chip: The active semiconductor element which is the functional part of an integrated circuit.

Interlock Relay: A relay with two or more armatures, each with an associated contact-spring pileup, or pileups, the positioning of which is according to the following: (1) Mechanical interlock relay—an interlock relay having a mechanical linkage whereby the position of one armature permits, prevents, or causes motion of another armature. (2) Electrical interlock relay—an interlock relay having an electrical interconnection such that the position of one armature permits, prevents, or causes operation of the other armature. (3) Combination interlock relay—an interlock relay having both mechanical linkage and electrical interconnection such that the position of one armature permits, prevents, or causes operation of the other armature. *Note:* If the position of one armature prevents operation of another, the relay is sometimes called a lockout relay.

Intermittent-Duty Relay: A relay which must be deenergized at intervals to avoid excessible temperature, or a relay that is energized at regular or irregular intervals, as in pulsing.

Irradiance (H or E_e): The radiant flux density incident on a surface; the ratio of flux to area of irradiated surface. Typical units: W ft^{-2}, W m^{-2} $\times$ 1 W ft^{-2} = 10.764 W m^{-2}.

JK Flip-flop: A binary with asynchronous set and reset terminals.

Lamp Driver: A circuit designed to deliver sufficient current and voltage to illuminate a lamp.

Lasing Condition (or State): The condition of an injection laser corresponding to the emission of predominantly coherent or stimulated radiation.

Latch-in Relay: A relay that maintains its contacts in the last position assumed without the need of maintaining coil energization.

Leakage, DC (DCL): In capacitors, the stray direct current of relatively small value which flows through the capacitor when voltage is applied across the terminals.

Leakage Flux: That portion of the magnetic flux that does not cross the armature-to-pole face gap.

Leakage Inductance: Inductance due to flux which fails to link both the primary and secondary windings. Leakage inductance affects output voltage and rectifier regulation. In wideband transformers response at the lower frequencies is dependent on limiting the leakage inductance.

Light Current (I_L): The current that flows through a photosensitive device, such as a phototransistor or a photodiode, when it is exposed to illumination or irradiance.

Light-emitting Diode (LED): See Infrared Light-emitting Diode and/or Visible Light-emitting Diode.

Line Driver: A special circuit designed to drive a specified class of transmission line over a long distance.

Low-Voltage Relay: A relay adjusted to sense and function in a circuit or system at a specific minimum voltage. Also a term used to designate a relay whose coil or contacts are designed to handle only ordinarily encountered voltages.

Luminance (L) (Photometric Brightness): The luminous intensity of a surface in a given direction per unit of projected area of the surface as viewed from that direction. Typical units: fL, cd ft^{-2}, cd m^{-2}. 1 fL = (1/n) cd ft^{-2} = 3.4263 cd m^{-2}.

Luminous Flux (ϕ_v): The time rate of flow of light. Typical unit: lm. *Note:* Luminous flux is related to radiant flux by the eye-response curve of the International Commission of Illumination (CIE). At the peak response (= 555 nm). 1 W = 680 lm.

Luminous Intensity (I_v): Luminous flux per unit solid angle in a given direction.

Magnetic Core: The magnetic circuit of ferromagnetic material which carries the magnetic flux.

Magnetic Latching Relay: (1) A relay that remains operated from remanent magnetism until reset electrically. (2) A bistable polarized (magnetically latched) relay.

Magnetomotive Force: The force that establishes the magnetic flux in the magnetic circuit. Mathematically, it is the magnetic flux multiplied by the reluctance. It is directly proportional to ampere-turns. The measurement unit is the gilbert.

Make-before-Break Contacts: Double-throw contacts so arranged that the moving contact establishes a new circuit before disrupting the old one.

Margined Relay: A relay that functions in response to predetermined changes in the value of coil current or voltage. Margining is frequently employed during relay adjustment to establish the desired amount of normally closed contact pressure when the relay is in the deenergized position.

Mask Set: A set of photolithographic artwork which is used to transfer the pattern of the integrated circuit onto the silicon slice.

Mechanical Latching Relay: A relay in which the armature or contacts are held in the operated or unoperated position until reset manually or electrically.

Mercury Relay: A relay in which the movement of mercury opens and closes contacts.

Metallization: The process of applying a conductive metal film to the surface of the integrated-circuit chip. This metallization is used both for interconnection on the chip and to define a place for the attachment of bond wires.

Mismatch, Connector-Impedance: Terminal or connector having a different impedance from that for which the circuit or cable is designed.

MOS (Metal-Oxide Semiconductors): Integrated circuits using field-effect transistors which are formed on the chip.

Multiposition Relay: A relay that has more than one operated and/or nonoperated position, for example, a stepping relay.

n Channel: MOS structures having a negative region as the channel.

NAND *Gate:* The logical inversion of the AND gate.

Neutral Relay: A relay whose operation is primarily independent of the direction of the coil current, in contrast to a polarized relay.

Noise Current, Equivalent Input: The noise current of an ideal current source (having a source impedance equal to infinity) in parallel with the input terminals of the device that, together with the equivalent input-noise voltage, represents the noise of the device.

Noise-Equivalent Power (P_n, NEP) (of an Avalanche Photodiode): The broadband output noise voltage divided by the responsivity in volts per watt (or the broadband output-noise current divided by the responsivity in amperes per watt) and the square root of the noise bandwidth. Noise bandwidth is $\pi/2$ times the 3-db bandwidth of the system.

Noise Temperature: The uniform physical absolute temperature (kelvin) at which a network (and all its sources, if a multiport) would have to be maintained if it (and its sources) were passive in order to make available (or deliver) the same random-noise power per unit bandwidth (spectral density) at a given frequency as is actually available (or delivered) from the network.

Noise Voltage, Equivalent Input: The noise voltage of an ideal voltage source (having a source impedance equal to zero) in series with the input terminals of the device that, together with the equivalent input-noise current, represents the noise of the device.

NOR *Gate:* The logical inversion of an OR gate.

NPO (Negative Positive Zero): Refers to capacitors with temperature coefficients having a nominal slope of zero. Highly stable capacitors.

Off-State Collector Current (of an Optically Coupled Isolator) [I_c (off)]: The output current when the input current is zero.

On-State Collector Current (of an Optically Coupled Isolator) [I_c (on)]: The output current when the input current is above a threshold level. *Note:* An increase in the input current will usually result in a corresponding increase in the on-state current.

Open Circuit: A circuit in which halving the magnitude of the terminating impedance does not produce a change in the parameter being measured greater than the required accuracy of the measurement.

Optical Axis: A line about which the radiant-energy pattern is centered; usually perpendicular to the active area.

Optically Coupled Isolator (Optical Coupler, Transoptor): An optoelectronic device consisting of a photoemissive device and a photosensitive device integrated into a single entity and intended for the transfer of a signal from the input to the output.

Optoelectronic Device: A device which detects and/or is responsive to electromagnetic radiation (light) in the visible, infrared, and/or ultraviolet spectra regions; emits or modifies noncoherent or coherent electromagnetic radiation in these same regions; or utilizes such electromagnetic radiation for its internal operation.

OR *Gate:* The logical circuit such that the output is in the one state when either of the inputs are in the one state.

Output Reflection Coefficient: The ratio of the voltage reflected from the output port to the voltage incident on the output port with the input port terminated in a purely resistive reference impedance equal to the impedance of the source of the incident voltage.

Overload Relay: An alarm or protective relay that is specifically designed to operate when its coil current reaches a predetermined or unsafe value above normal.   Time delay may be introduced as a requirement of overload.

Parallel Excitation: The process of simulating operation by applying power pulses to the inputs of integrated circuits during life test.

Parasitic Capacitance: Capacitance whose source is inherent in the structures used for interconnection and fabrication but which is not included by design.

Permissive Make Contact: A term applied to a contact combination in which the movable contact spring is pretensioned so that it will move from its normal or unoperated position as a result of its own force when the relay is energized.   In the deenergized position, it is restrained by a member acted upon by the armature return spring.

Photocurrent: The difference between light current I_L and dark current I_D in a photodetector.

Photocurrent Gain (of an Avalanche Photodiode): The ratio of photocurrent at high bias voltage to that at low bias voltage.   (See also Avalanche Photodiode.)

Photodiode: A diode which is used to detect light.   A diode which has been optimized for light sensitivity.   *Note:* The photodiode is characterized by linearity between the input radiation and the output current.   It has faster switching speeds than a phototransistor.

Photometric Axis: The direction from the source of radiant energy in which the measurement of photometric parameters is performed.

Photometric Brightness: See Luminance.

Photon: A quantum (the smallest possible unit) of radiant energy, a photon carries a quantity of energy equal to Planck's constant times the frequency.

Phototransistor: A transistor (bipolar or field-effect) which is used to detect light.   A transistor which has been optimized for light sensitivity.   *Note:* The base region or gate may or may not be connected to an external terminal.

Piezoelectric: Electricity or electric polarization caused by pressure.

Pin Contact: A male-type contact, usually designed to mate with a socket or female contact.   It is normally connected to the "dead" side of a circuit.

Polarization: A mechanical arrangement of inserts and/or shell configuration (referred to as clocking in some instances) which prohibits the mating of mismatched plugs and receptacles.   This is to allow connectors of the same size to be lined up side by side with no danger of making the wrong connection.   Coded arrangements of contact, keys, keyways, and insert position are used.   In rectangular connectors, the shells are so designed that mating usually is possible in only one way.

Polarized Relay: A relay the operation of which is primarily dependent upon the direction (polarity) of the energizing current(s) and the resultant magnetic flux.   The opposite of neutral relay.   Also called polar relay.

Power Factor (PF): The ratio of resistance to impedance, measured in percent.

Power-Type Relay: A term used for a relay designed to have heavy-duty contacts usually rated 15 A or higher.   Sometimes called a contactor.

Prelasing Condition (or State): The condition of an injection laser corresponding to the emission of predominantly incoherent or spontaneous radiation.

Pullout: Force needed to separate a cable from a connector by pulling them apart.

Pulse-forming Network: An *LC* network whose parameters are selected to give a specified shape to a pulse signal.

Pulse Time: The time duration from the point of the leading edge which is 90 percent of the maximum amplitude to the point on the trailing edge which is 90 percent of the maximum amplitude.

Pulse Transformer: A transformer with which pulses (usually steep-front and flat-top) are transmitted from one circuit to another while maintaining pulse shape.   The magnitude of the voltage is often changed.   In some cases the transformer may form the pulse, e.g., a blocking oscillator.

Quality Factor (*Q*): The ratio of reactance to resistance, for capacitors.

Quantum Efficiency (*QE*): The quantum efficiency of a source of radiant flux is the ratio of the number of quanta of radiant energy (photons) emitted per second to the number of electrons flowing per second, e.g., photons per electron.

Quick Disconnect: A type of connector shell which permits rapid locking and unlocking of two connector halves.

Rack-and-Panel Connectors: A rack-and-panel connector is one which connects the inside back end of the cabinet (rack) with the drawer containing the equipment when it is fully inserted. The drawer permits convenient removal of portions of the equipment for repair or examination. Special design and rugged construction of the connector allows for variations in rack-to-panel alignment.

Radiant Efficiency of a Source of Radiant Flux (*η*): The radiant efficiency of a source of radiant flux is the ratio of the total radiant flux to the forward power dissipation.

Radiant Flux (*Radiant Power*) (*φ*): Radiant flux is the time rate of flow of radiant energy. It is expressed preferably in watts, or ergs per seconds.

Radiant Intensity (*I*): The radiant intensity of a source is the radiant flux proceeding from the source per unit solid angle in the direction considered, e.g., watts per steradian.

Radiant-Pulse Fall Time (*t_f*): The time required for a photometric quantity to change from 90 to 10 percent of its peak value for a step change in electrical input.

Radiant-Pulse Rise Time (*t_r*): The time required for a photometric quantity to change from 10 to 90 percent of its peak value for a step change in electrical input.

Radiation Pattern: The representation of the intensity of emission as a function of direction, in a given plane. The axes are to be specified with respect to the junction plane and the cavity face.

Radio Interference: Undesired conducted or radiated electrical disturbances, including transients, which may interfere with the operation of electrical or electronic communications equipment or other electronic equipment.

Rated Coil Current: The steady-state coil current on which the relay is intended to operate for the prescribed duty cycle.

Rated Coil Voltage: The coil voltage on which the relay is intended to operate for the prescribed duty cycle.

Receptacle: Usually the fixed or stationary half of a two-piece multiple-contact connector. Also the connector half, usually mounted on a panel and containing socket contacts.

Rectifier Transformer: A transformer intended for use with a specific rectifier circuit. These may be single- or multiphase and single-way (half-wave) or two-way (full-wave). Most rectifier circuits require the transformer to be designed for the circuit when maximum efficiency and minimum size are desired.

Rectifying Junction: A junction in a semiconductor device which exhibits asymmetrical conductance.

Reed Relay: A relay using glass-enclosed, magnetically closed reeds as the contact members. Some forms are mercury-wetted.

Reference-Noise Temperature: A specified absolute temperature (kelvin) to be assumed as a noise temperature at the input ports of a network when calculating certain noise parameters, and for normalizing purposes. When the reference noise temperature is 290 K, it is considered to be the standard reference-noise temperature.

Relay: Most simply defined as an electrically controlled device that opens and closes electrical contacts to effect the operation of other devices in the same or another electric circuit.

Responsivity (N, $\bar{R}_m$): The quotient of the rms value of the fundamental component of the electrical output of the detector to the rms value of the fundamental component of the input-radiation power density when the radiation is incident normally on the detector surface. Typical units: $V\ W^{-1}$, $A\ W^{-1}$.

Reverse Bias: The bias which tends to produce current flow in the reverse direction.

Reverse Current, DC: The direct current that flows through a semiconductor junction in the reverse direction.

Reverse Direction: The direction of current flow which results when the *n*-type semiconductor region is at a positive potential relative to the *p*-type region.

Reverse-Transmission Coefficient: The ratio of the voltage at the input port to the voltage incident on the output port with the input port terminated in a purely resistive reference impedance equal to the impedance of the source of the incident voltage.

Reverse Voltage, DC: The dc voltage applied to a semiconductor junction which causes the current to flow in the reverse direction.

RF Switching Relay: A relay designed to switch frequencies that are higher than commercial power frequencies with low loss.

Ring-Counter Configuration: A technique to simulate operating life by using the outputs of circuits to drive the inputs of adjacent circuits.

Ripple Voltage (or Current): The ac component of a unidirectional voltage or current (the ac component is small in comparison with the dc component).

Rise Time: The time duration in which the leading edge of a pulse is increasing from 10 to 90 percent of its maximum amplitude.

Rotary Relay: (1) A relay whose armature is rotated to close the gap between two or more pole faces (usually having a balanced armature). (2) Incorrectly used for stepping relay.

Rotary Solenoid Relay: A relay in which the linear component of motion of the plunger or armature is converted mechanically into rotary motion.

RS Flip-flop—Set Reset flip-flop: A binary with the ability to be both set and reset.

Sealed Relay: A relay that has both coil and contacts enclosed in an airtight cover. Not to be confused with hermetically sealed relay.

Semiconductor Device: A device whose essential characteristics are governed by the flow of charge carriers within a semiconductor.

Semiconductor Diode: A semiconductor device having two terminals and exhibiting a nonlinear voltage-current characteristic; in more restricted usage, a semiconductor device which has the asymmetrical voltage-current characteristic exemplified by a single pn junction.

Semiconductor Junction (Commonly Referred to as Junction): A region of transition between semiconductor regions of different electrical properties (e.g., nn^+, pn, pp^+ semiconductors) or between a metal and a semiconductor.

Sensing Relay: A relay responding to a condition of overcurrent, overvoltage, under current, undervoltage, and the like. See also Thermal Relay, Overload Relay, and the like.

Sequential Relay: A relay that controls two or more contact combinations in a pre-determined sequence.

Shielding: The metal sleeving surrounding one or more of the conductors, in a wire circuit to prevent interference, interaction, or current leakage. Usually grounded, the shielding is carried through the connector shell or through a special internal shell in the case of individual coaxial contacts.

Short Circuit: A circuit in which doubling the magnitude of the terminating impedance does not produce a change in the parameter being measured that is greater than the required accuracy of the measurement.

Sinter: To heat, without melting, to cause a refractory dielectric material to become a rigid body free of binders, contaminants, etc.

Slices: Sections of ingots of single-crystal material which have been sawed from the ingot.

Small Signal: A signal which when doubled in magnitude does not produce a change in the parameter being measured that is greater than the required accuracy of the measurement.

Solid-State Relay (SSR): A relay which switches electric circuits by use of electronic components without moving parts or conventional contacts.

Spectral Bandwidth (between Half-Power Points): The wavelength interval in which a photometric or radiometric spectral quantity is not less than half its maximum value.

Spectral Output (of a Light-emitting Diode): A description of the radiant-energy of light-emission characteristic vs. wavelength. *Note:* This information is usually given by stating the wavelength at peak emission and the bandwidth between half-power points or by means of a curve.

Spectral Radiant Flux (ϕ_λ): Spectral radiant flux is the radiant flux per unit wave-length interval at wavelength λ, e.g., watts per nanometer.

Spectral Response (of a Photosensitive Device): A description of the electrical-output characteristic vs. wavelength of radiant energy incident upon the device. *Note:* This information is usually given by means of a curve.

Spinning: Process of coating a semiconductor slice with a photosensitive emulsion by placing the slice on a rotating chuck and then dropping the emulsion on the surface. The combination of the centrifugal acceleration and the adhesion of the liquid forms a uniform film of the emulsion on the surface.

Spot-Noise Figure or Spot-Noise Factor: The ratio of (1) the total output noise power per unit bandwidth (spectral density) at a designated output frequency when the noise temperature of the input termination(s) is at the reference-noise temperature T_0 at all frequencies to (2) that part of (1) caused by the noise temperature of the designated signal-input termination at a designated signal-input frequency.

Spring, Contact: (1) A current-carrying spring, usually a flat leaf type or wire member, to which the contacts are fastened. Contact-spring bearing contacts are assembled between insulators in the contact assembly, or pileup, to form contact combinations. (2) A non-current-carrying spring that positions or tensions a contact-carrying member.

Static Value: A nonvarying value or quantity measured at a specified fixed point, or the slope of the line from the origin to the operating point on the appropriate characteristic curve.

Stepping Relay: A relay having many rotary positions, ratchet-actuated, moving from one step to the next in successive operations, and usually operating its contacts by means of cams. There are two forms: (1) "Directly driven," where the forward motion occurs on energization. (2) "Indirectly (spring) driven," where a spring produces the forward motion on pulse cessation. The term is also incorrectly used for stepping switch.

Steradian (sr): A unit of solid angular measurement equal to the solid angle at the center of a sphere subtended by a portion of the surface area equal to the square of the radius; there are 4π steradians in a complete sphere.

Storage Temperature: The temperature at which the device, without any power applied, is stored.

Storage Time (t_s): The time interval from a point at which the trailing edge of the input pulse has dropped to 90 percent of its maximum amplitude to a point at which the trailing edge of the output pulse has dropped to 90 percent of its maximum amplitude.

Substrate: Any of the materials used in hybrid circuits as a support for other elements, e.g., alumina, phenolic, glass, quartz, sapphire.

Surge Voltage (or Current): Transient variation in the voltage or current at a point in the circuit; a voltage or current of large magnitude and short duration caused by a discontinuity in the circuit.

Telephone-Type Relay: A term most often applied to an armature relay with an end-mounted coil, an L-shaped heelpiece, and contact springs mounted parallel to the long axis of the relay coil. Originally used mainly in telephone systems.

Temperature Coefficient (TC): In capacitors, the change in capacitance per degree change in temperature. It may be positive, negative, or zero and is usually expressed in parts per million per degree Celsius (ppm/°C).

Terminal: An externally available point of connection to one or more electrodes.

Termination: The metal contact that connects the internal electrodes of a capacitor to external circuit elements.

Thermal-Compression Bonding: The process of bonding wire to the semiconductor chip by the combination of pressure and temperature in order to cause solid flow.

Thermal Relay: A relay actuated by the heating effects of an electric current.

Thermal Resistance, Case-to-Ambient: The thermal resistance (steady-state) from the device case to the ambient.

Thermal Resistance, Junction-to-Ambient: The thermal resistance (steady-state) from the semiconductor junction(s) to the ambient.

Thermal Resistance, Junction-to-case: The thermal resistance (steady-state) from the semiconductor junction(s) to a stated location on the case.

Thermal Resistance (Steady-State): The temperature difference between two specified points or regions divided by the power dissipation under conditions of thermal equilibrium.

Transformer Inrush Current: A transient phenomenon consisting of high magnetizing current when power is switched on. The magnitude varies according to the magnetic state of the core and the polarity of the voltage at the instant of turn-on and may reach several times load current for several cycles.

Transient Thermal Impedance: The change of temperature difference between two specified points or regions at the end of a time interval divided by the step-function change in power dissipation at the beginning of the same time interval causing the change of temperature difference.

Transistor: An active semiconductor device capable of providing power amplification and having three or more terminals.

Ultrasonic Bonding: The process of using ultrasonic energy to cause a solid-flow weld of a wire bond to the bond pad.

Undervoltage Relay: An alarm or protective relay specifically designed to function when its energizing voltage falls below a predetermined safe value.

V_{BB}: The bias-voltage power supply.
V_{CC}: The positive-collector power supply.
V_{DD}: The drain power supply.
V_{EE}: The negative-emitter power supply.
V_{SS}: The source power supply.

Variable Resistor: A variable resistor is a mechanical electrical transducer dependent upon the relative position of a moving-brush contact (wiper) and a resistance element for its operation. It delivers a voltage output that is some specific function of applied voltage and shaft position.

Variable-Resistor Electrical and Mechanical Input/Output:

1. Total Applied Voltage: The total applied voltage is the voltage applied between input terminals.
2. Variable Output: The variable output is the difference between the maximum and minimum output.
3. Actual Output: The actual output is the electrical output between the wiper terminal and one input terminal.
4. Output Ratio: The output ratio is the ratio of the actual output voltage to the total applied voltage.
5. Loading Error: The loading error is the difference between the actual output with an infinite-resistance brush load and the actual output with the specified-resistance brush load, at a specific position.
6. Contact-Resistance Variation: The apparent resistance seen between the wiper and the resistance element when the wiper is energized with a specified current and moved over the adjustment travel in either direction at a constant speed. The output variations are measured over a specified-frequency bandwidth, exclusive of the effects due to roll-on or roll-off of the terminations and are expressed in ohms or percent of total element resistance.
7. Setting Stability: The amount of change in the output voltage, without readjustment, expressed as a percentage of the total applied voltage.
8. Total Resistance: The total resistance is the resistance between the fixed or end terminals with the shaft positioned so as to give a maximum resistance value.
9. Minimum Resistance: The minimum resistance is the lowest resistance between the brush terminal and a fixed terminal as the shaft is rotated.
10. End Resistance: The end resistance is the resistance between the brush terminal and an end terminal with the brush in the extreme position allowed by the adjacent mechanical stop.
11. Total Variable Resistance: The total variable resistance is the resistance encompassed by the actual effective travel.
12. Equivalent-Noise Resistance: (Wirewound only.) Any spurious variation in the electrical output not present in the input, defined quantitatively in terms of an equivalent parasitic, transient resistance in ohms, appearing between the contact and the resistive element when the shaft is rotated or translated. The equivalent-noise resistance is defined independently of the resolution, functional characteristics, and total travel. The magnitude of the equivalent-noise resistance is the maximum departure from a specific reference line. The wiper of the potentiometer is required to be excited by a specific current and moved at a specific speed.
13. Direction of Rotation: The direction of rotation is the shaft rotation defined as clockwise (cw) or counterclockwise (ccw) when viewing the specified mounting end of the resistor.
14. Total Mechanical Travel: The total mechanical travel is the total travel of the shaft between integral stops under specified torque. In resistors without stops, the mechanical travel is continuous.
15. Mechanical Overtravel: The mechanical overtravel is the shaft travel between each end point and its adjacent mechanical stop. Mechanical overtravel is applicable only to units having stops.
16. Electrical-Continuity Travel: The electrical-continuity travel is the total travel of the shaft over which electrical continuity is maintained.
17. Actual Electrical Travel: The actual electrical travel is the total travel of the shaft between the end points.
18. Electrical Overtravel: The electrical overtravel is the shaft travel over which there is continuity between the wiper terminal and the resistance element beyond each end of the actual electrical travel. (In cases where absolute linearity or

absolute conformity is specified, "theoretical electrical travel" is substituted for "actual electrical travel" in this definition.)

19. Theoretical Electrical Travel: The theoretical electrical travel is the shaft travel over which the theoretical function characteristic extends.

20. Flat Zone: The flat zone is any portion of the actual electrical travel throughout which there is no change in the output voltage.

21. Taper: Used in nonprecision potentiometers to describe the change in resistance at the brush terminal with respect to mechanical shaft movement. Linear taper means a constant change, while nonlinear means a lack of constancy in change; A taper is linear.

22. End Point: The end point is the shaft position immediately before the first measurable change of output ratio is observed as the shaft moves the brush in a specified direction from the overtravel region onto the region of actual electrical travel. The other end point is the shaft position at which the final (maximum or minimum) output ratio first occurs while the shaft is still moving in the same direction.

23. End Voltage: The end voltage is the voltage between the brush terminal and an end terminal when the shaft is positioned at the end point, and is expressed as a percentage of the total applied voltage.

24. Jump-off Voltage: The jump-off voltage is the first measurable voltage change as the shaft moves the brush from the overtravel region onto the actual electrical-travel region.

25. Tap: A tap is the fixed electrical connection made to the resistance element.

26. Tap Location: The tap location is the position of a tap from some reference point. (This is commonly expressed in terms of resistance, voltage ratio, or shaft position. When a shaft position is specified, the tap position is measured at the center of the effective tap width.)

27. Effective Tap Width: The effective tap width is the travel of the shaft during which the voltage at the brush terminal and the tap terminal are essentially the same as the brush is moved past the tap in one direction.

28. Absolute Minimum Resistance: The absolute minimum resistance is the resistance between the brush terminal and end terminal with the shaft positioned so as to give a minimum value.

29. Minimum Resistance, at Stop: The minimum resistance at a stop is the resistance at a stop between the brush terminal and an end terminal when the shaft is positioned at the corresponding stop.

30. Index Point: The index point is a point reference fixing the relationship between a specified shaft position and the output ratio, used to establish a shaft-position reference.

31. Phasing: Phasing is the relative alignment of the sections of a gang with respect to the position of the brushes on their respective electrical elements.

32. Phasing Point: Phasing point is a point of reference fixing the relationship between the shaft position and the output ratio, or between the shaft position and a tap location, for each electrical element in a multisection potentiometer.

33. Simultaneous Conformity Phasing: Simultaneous conformity phasing is the alignment of the electrical elements of a multisection potentiometer so that the output ratios fall within their respective conformity limits over the theoretical electrical travel, using a common shaft-position reference. This definitition applies only to potentiometers with absolute-conformity or absolute-linearity function specifications.

34. Voltage-Tracking Error: The voltage-tracking error is the difference at any shaft position between the output ratios of any two commonly actuated similar electrical elements expressed as a percentage of the single total voltage applied to them.

35. Resistance-Tracking Error: The resistance-tracking error is the difference at any point between the actual resistances, from the brush to the other terminal or far end of integral resistor, of commonly actuated electrical elements of similar function expressed in ohms or percentage of nominal total resistance, or in other mutually satisfactory units.

36. Adjustability: Defines the precision with which the output of a device can be set to the desired value, expressed in terms of resistance or output-voltage ratio.

Variable-Resistor Parts:

1. Cup: A cup is a single mechanical section of a variable resistor which may contain one or more electrical elements. Usually refers to a precision variable resistor or control.

2. Gang: A gang is an assembly of two or more cups on a common operating shaft.

3. Shaft: The shaft is the mechanical input element of a variable resistor.
4. Shaft Position: The shaft position is an indication of the relative position of the wiper and some reference point.
5. Terminal: The terminal is an external device used to provide electrical access to the variable-resistor resistance element or wiper contact.
6. Integral Resistor: An integral resistor is an internal or external resistor preconnected to the electrical element and forming an integral part of a cup assembly to provide a desired electrical characteristic.
7. Resistance Element: The fabricated fixed resistor, usually connected between two electrical terminals, and having access surface for the wiper contact.
8. Panel Bushing: A threaded bushing, usually concentric with the operating shaft, intended to provide a through-panel mounting for a variable resistor.
9. Lead Screw: A captive threaded screw used to move a circular gear or threaded follower, thereby transmitting rotating-screw motion to circular or translatory wiper motion.
10. Test Point: The test point is an additional terminal used only to facilitate measurements.
11. Override Clutch: Feature for mechanical disengagement of the operating shaft from the wiper as an end stop is reached, thus avoiding damage to stop or mechanism.

Variable-Resistor Resolution and Conformity:
1. Resolution: Resolution is the measure of sensitivity to which the output ratio of the potentiometer may be set. Most wirewound units vary in incremental steps, while film and composition units are said to have "infinite" resolution.
2. Theoretical Resolution: Theoretical resolution is used in wirewound linear potentiometers only; it is the reciprocal of the number of turns of the resistance winding in the actual electrical travel, and is expressed as a percentage.
3. Travel Resolution: Travel resolution is the maximum value of shaft travel in one direction per incremental voltage step in any specified portion of the resistance element.
4. Voltage Resolution: Voltage resolution is the maximum incremental change in the output ratio with the shaft travel in one direction in any specified portion of the resistance element.
5. Theoretical Function Characteristic: The theoretical function characteristic is the relationship between the output ratio and the shaft position.
6. Actual Function Characteristic: The actual function characteristic is the relationship between output ratio and the shaft position.
7. Slope: The slope is the rate of change of the output ratio with shaft travel.
8. Average Slope: The average slope is the slope of the straight line joining the end points of the specified function, or the extremities of any specified portion thereof.
9. Deviation: Deviation is the plus or minus differences between the actual function characteristic and the theoretical function characteristic for a given shaft position.
10. Conformity: Conformity is the relationship between the actual function characteristic and the theoretical function characteristic.
11. Absolute Conformity: Absolute conformity is the maximum deviation expressed as a percent of the total applied voltage, of the actual function characteristic from a theoretical function characteristic extending between the specified output ratios which are separated by the theoretical travel. An "index point" on the actual output is required.
12. Linearity: Linearity is a specific type of conformity where the theoretical function characteristic is a straight line.
13. Absolute Linearity: Absolute linearity is the maximum deviation expressed as a percent of the total applied voltage, of the actual function characteristic from a straight reference line drawn through the specified minimum and maximum output ratios which are separated by the theoretical electrical travel. Unless otherwise specified, minimum and maximum are, respectively, 0 and 100 percent total applied voltage. An "index point" on the actual output is required.
14. Terminal-based, Linearity: Terminal-based linearity is the maximum deviation, expressed as a percent of the total applied voltage, of the actual function characteristic from a straight line drawn through the specific minimum and maximum-output-voltage ratios which are separated by the actual electrical travel. Unless otherwise specified, minimum and maximum output ratios are, respectively, 0 and 100 percent of total applied voltage.
15. Zero-based Linearity: Zero-based linearity is the maximum deviation, expressed as a percent of the total applied voltage of the actual function characteristic

from a straight reference line drawn through the specific minimum output ratio, extended over the actual electrical travel, and rotated to minimize the maximum deviations. Any specified end-voltage requirement limits the rotation of the reference line. Unless otherwise specified, the specified minimum output ratio will be zero.

16. Independent Linearity (Best Straight Line): Independent linearity is the maximum deviation, expressed as a percent of the total applied voltage, of the actual function characteristic from a straight reference line with its slope and position chosen to minimize the maximum deviations over the actual electrical travel, or any specified portion thereof.

Vibrating-Reed Relay: A relay in which the application of an alternating or a self-interrupted voltage to the driving coil produces an alternating or pulsating magnetic field that causes a reed to vibrate and operate contacts.

Visible Emission (Visible Light): Radiation which is characterized by wavelengths of about 0.38 to 0.78 μm.

Visible Light-emitting Diode (VLED): An optoelectronic device containing a semiconductor junction which emits visible light when forward-biased. *Note:* Material is usually gallium phosphide (GaP) or gallium arsenide phosphide (GaAsP).

Voltage Regulation: The drop-in voltage due to load current, usually expressed in percentage.

Voltage (Sensing) Relay: (1) A term correctly used to designate a special-purpose voltage-rated relay that is adjusted by means of a voltmeter across its terminals in order to secure pickup at a specified critical voltage without regard to coil or heater resistance and resulting energizing current at that voltage. (2) A term erroneously used to describe a general-purpose relay for which operational requirements are expressed in voltage.

Voltage Standing-Wave Ratio (VSWR): In a relay, the contacts of which handle radio frequency (rf), the power loss due to the mismatch introduced into the line by the coaxial relay contacts, expressed as a ratio of the highest voltage to the lowest voltage found in the rf line.

Wafer Probe: The process of making electrical contact and measurement to the semiconductor devices while they are still part of the slice.

Wavelength at Peak Emission (λ_p): The wavelength at which the power output from a light-emitting diode is maximum. Typical units: Å, μm, nm. 1 Å = 10^{-4} μm = 0.1 nm.

Wavelength of Peak Radiant Intensity: The wavelength at which the spectral distribution of radiant intensity is a maximum.

Wideband Transformer: A transformer which transforms a range of frequencies. Many special types are covered by this classification, including audio-frequency, modulation, impedance-matching, ultrasonic, carrier, and video transformers.

Winding, Bias: An auxiliary winding used to produce an electrical bias.

Wipe, Contact: The sliding or tangential motion between two mating contact surfaces when they are coming to rest. This is developed from pivoting about noncommon centers.

Wiper: The moving contact on a rotary stepping switch or relay.

Wire, Magnet: Any coated conductor used to wind an electromagnetic coil in order to develop and maintain a magnetic field under prescribed conditions.

Wire-Spring Relay: A relay design in which the contacts are attached to round wire springs instead of the conventional flat, or leaf, spring.

Index

A

Accelerated testing of transistors, **4**-50
Aluminum foil capacitors (*see* Electrolytic capacitors, aluminum foil)
Amplifiers (*see specific part*)
Analog-to-digital converters, **3**-9
Arcs (*see* Light sources; *and specific component*)
Artwork, integrated circuits, **1**-3
Attenuators (*see* Microwave components, attenuators)
Auger processes, light sources, **5**-49
Automotive circuits, **3**-36

B

Basic logic circuits (*see* Digital integrated circuits).
Beam-lead sealed-junction devices, **4**-42
Biasing (*see specific part*)
Bipolar (*see specific part*)
Blackbodies (*see* Electrooptical components, blackbodies)
Boltzmann's constant, **5**-3, **5**-20
Bonding, wire **1**-6, **1**-7

C

Calculator circuits, **3**-25
Cameras, **5**-105
(*see also under* Image devices)

Capacitors:
capacitance, **8**-1
of dielectrics, **8**-15 to **8**-18
capacitance ratings, **8**-15
ceramic (*see* Ceramic capacitors)
characteristics of, **8**-1, **8**-4, **8**-7 to **8**-11
comparative features of, **8**-16, **8**-17
corona, **8**-11
current-voltage relationship of, **8**-7
definitions of, **8**-1, **8**-4
dielectric absorption, **8**-11
dielectric materials, **8**-7, **8**-15 to **8**-18
dielectric strength, **8**-10, **8**-11
dielectric types of, **8**-5 to **8**-7, **8**-15 to **8**-18
electric field, **8**-3
electrolytic (*see* Electrolytic capacitors)
energy, **8**-3
environmental effects on, **8**-19
environmental stresses, **8**-19
equations, **8**-4, **8**-5
equivalent circuit, **8**-3, **8**-5
equivalent series resistance, **8**-9
failure modes, **8**-136 to **8**-137
failure-rate curve, **8**-135
failure-rate improvement, **8**-132 to **8**-135
failure rates, **8**-132
film (*see* Film and paper capacitors)
fixed, **8**-15 to **8**-22
(*See also specific type*)
application of, **8**-15 to **8**-22
frequency stresses, **8**-14
fringe effects of, **8**-3
glass (*see* Stable capacitors, glass)

1

D

G

H

I

Z

£29-60